Carolyn J. Miller
304 Lawndale NE
Grand Rapids, MI
49503

616-454-2460

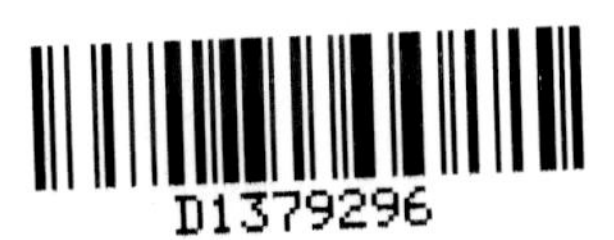

CELLS
PRINCIPLES OF MOLECULAR STRUCTURE AND FUNCTION

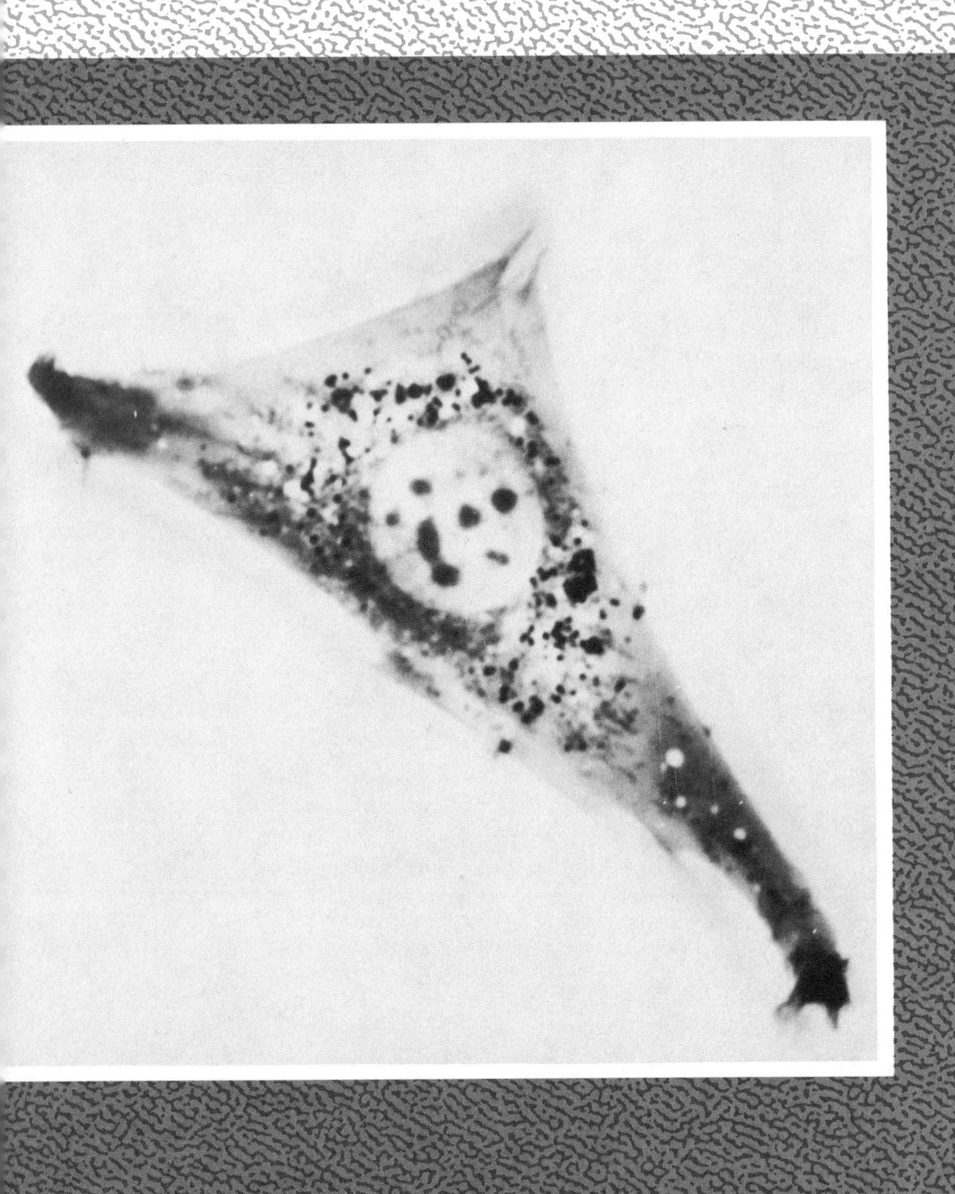

PRINCIPLES OF MOLECULAR STRUCTURE AND FUNCTION

DAVID M. PRESCOTT

Distinguished Professor

Department of Molecular, Cellular and Developmental Biology
University of Colorado, Boulder

Past President of the American Society for Cell Biology

JONES AND BARTLETT PUBLISHERS
BOSTON • PORTOLA VALLEY

Editorial, Sales, and Customer Service Offices
Jones and Bartlett Publishers
20 Park Plaza
Boston, MA 02116

Printed in the United States of America
10 9 8 7 6 5 4 3 2

Library of Congress Cataloging-in-Publication Data

Prescott, David M., 1926–
Cells: principles of molecular structure and function / David M. Prescott.
p. cm.
Includes bibliographies and index.
ISBN 0-86720-092-8
1. Cytology. I. Title.
QH581.2.P73 1988
574.87—dc19 87-31096
CIP

ISBN: 0-86720-092-8

Manuscript editor	Eta Fox Wayne
Text design	Piñeiro Design Associates
Illustrations	Art Ciccone; Bob Gallison, Textbook Art Associates; Christy Krames
Cover design	Rafael Millán
Typesetting	Bi-Comp, Inc.
Cover illustrations	Front cover: A living rat kangaroo cell in mitosis stained with fluorescent dyes. Back cover: Living rat kangaroo cells stained with fluorescent dyes. Courtesy of Mark Ladinsky and J. Richard McIntosh.

To Ryan, Jason, Lavonne, and Gayle

THE JONES AND BARTLETT SERIES IN BIOLOGY

Basic Genetics
Daniel L. Hartl, Washington University School of Medicine; David Freifelder, University of California, San Diego; Leon A. Snyder, University of Minnesota, St. Paul

General Genetics
Leon A. Snyder, University of Minnesota, St. Paul; David Freifelder, University of California, San Diego; Daniel L. Hartl, Washington University School of Medicine

Genetics
John R. S. Fincham
University of Edinburgh

Genetics of Populations
Philip W. Hedrick
University of Kansas

Genetic Principles: Human and Social Consequences (nonmajors text)
Gordon Edlin
University of California, Davis

Microbial Genetics
David Freifelder
University of California, San Diego

Cells: Principles of Molecular Structure and Function
David M. Prescott
University of Colorado, Boulder

Essentials of Molecular Biology
David Freifelder
University of California, San Diego

Introduction to Biology: A Human Perspective
Donald J. Farish
California State University at Sonoma

Introduction to Human Immunology
Teresa L. Huffer, Shady Grove Adventist Hospital, Gaithersburg, Maryland, and Frederick Community College, Frederick, Maryland; Dorothy J. Kanapa, National Cancer Institute, Frederick, Maryland; George W. Stevenson, Northwestern University Medical Center, Chicago, Illinois

Molecular Biology, Second Edition
David Freifelder
University of California, San Diego

The Molecular Biology of Bacterial Growth (a symposium volume)
M. Schaechter, Tufts University Medical School; F. Neidhardt, University of Michigan; J. Ingraham, University of California, Davis; N. O. Kjeldgaard, University of Aarhus, Denmark, editors

Molecular Evolution: An Annotated Reader
Eric Terzaghi, Adam S. Wilkins, and David Penny, all of Massey University, New Zealand

Population Biology
Philip W. Hedrick
University of Kansas

Virus Structure and Assembly
Sherwood Casjens
University of Utah College of Medicine

Cancer: A Biological and Clinical Introduction, Second Edition
Steven B. Oppenheimer
California State University, Northridge

Handbook of Protoctista
Lynn Margulis, John O. Corliss, Michael Melkonian, and David I. Chapman, editors

Living Images
Gene Shih and Richard Kessel

Early Life
Lynn Margulis
Boston University

Functional Diversity of Plants in the Sea and on Land
A. R. O. Chapman
Dalhousie University

Plant Mineral Nutrition: An Introduction to Current Concepts
A. D. M. Glass
University of British Columbia

Molecular Cloning Manual
Alfred Bothwell, Yale University School of Medicine; Fred Alt, Columbia University; Hans Lehrach, Imperial Cancer Research Fund

CONTENTS

FOREWORD

Our understanding of the molecular basis of cell structure and function has increased dramatically in recent years. New methods of looking at cells and their constituent molecules show incredibly complex and beautiful structural patterns in one organelle after another. The electron microscope, greatly improved optical microscopes, and x-ray crystallography, all backed with the analytical power of computers, reveal details of macromolecular organization only hinted at before. A whole host of biochemical techniques use radioactivity or enzymatic reactions to give information about single molecules or small numbers of molecules. And perhaps most significantly, biological processes themselves are being used to amplify the signals from single molecules, as shown most dramatically in gene cloning and the production of monoclonal antibodies.

The flood of new information makes cell biology a lively and fascinating subject, but it also forces those who teach it at the university level to assimilate an enormous amount of material. As a result, the typical cell biology course is likely to involve team teaching. For much the same reason we are seeing more multi-authored textbooks of cell and molecular biology. Nevertheless, the usefulness of a single-authored text is great—the single author is in a better position than an editor to impress his perspective on the subject as a whole, and he can tailor information to the needs of a particular kind of course, such as the typical one-semester university course in cell biology. Prescott is well equipped for this task. He is a distinguished cell biologist, a former President of the American Society for Cell Biology, whose own research focuses on problems of the cell cycle and on the unusual chromosome organization of ciliated protozoa. For many years he has taught the kind of course for which this book is intended, and he is aware of the needs of both undergraduate and graduate students in biology. By the judicious choice of examples he covers a wide range of topics without overwhelming the reader with technical detail. And he writes well, conveying a sense of excitement to each topic. This volume is not meant as a comprehensive treatise on the cell. Instead it is an up-to-date textbook, manageable in size, and well suited for a first course in cell biology at the college level. It can also serve as an introduction for anyone else who wants to gain an overview of the exciting field of cell biology.

Joseph G. Gall
Carnegie Institution
Baltimore, MD

PREFACE

This textbook was derived from long experience in teaching of cell biology. I have taught for many years the first semester of the two semester introductory course in Molecular, Cellular, and Developmental Biology in the Department of MCD Biology at the University of Colorado. The first semester of the course covers cell biology and the second semester is mostly about developmental biology. Molecular biology is heavily woven into the texture of both semesters. The course has undergone continuous reshaping, refining, and adjusting to try to make it an efficient and interesting learning experience for students. The 16 chapters in this book are the outcome of the evolution of this course. The chapters present principles of cell function and structure that are meant to give the student a good grasp of the cellular world and a foundation on which to build in more advanced courses in biology.

Since the 1960s the study of cells has become one of the dominating activities in all of **Science.** Many disciplines—biochemistry, molecular biology, genetics, biophysics, cytology, physiology, etc.—now contribute to the continuing growth in understanding about how cells function and how they are constructed. This is an exciting enterprise, which, as an intellectual achievement, contributes greatly to our understanding of the phenomenon of life. On a practical level research on cells provides an expanding foundation for advances in medical science and for new industrial and agricultural technologies.

Writing a textbook about cells is difficult because cells are complex creatures that have evolved in many directions. Writing a short book on cell biology, which I have tried to do here, is even more difficult. Research has produced a great wealth of information about cells—tens of thousands of research papers having to do with cells are published in scientific journals throughout the world every year. Hundreds of hard decisions had to be made to leave out this or that interesting experimental finding. Too much experimental detail, however fascinating, makes a long book that can overwhelm students and obscure the path to learning the principles of cell function and structure. This book is as short as I could make it without feeling that it had become cryptic. The first draft of the manuscript was twice as long as the final product. One reviewer summed up my purpose better than I could myself when he wrote that the book was **not** intended "as an imperial text intent on intellectual conquest" but rather as a text of manageable scope "that fits the rhythm of the academic year" and still maintains a "current, rigorous and disciplined approach to cell biology."

Research in cell biology is advancing rapidly, and it would be easy to miss some of the most recent discoveries. To try to minimize oversights and in an effort to present the most up-to-date accounts possible about cell functions and structures, I relied heavily on reviews published in the *Annual Reviews of Cell Biology, Biochemistry,* and *Genetics* as well as in many monographs and symposium volumes. I consulted over 6000 research papers published in first rank journals. Many of these papers were published in 1987.

It was important to include a few experimental details in order to give students a sense of how research on the cell is done. For this purpose some techniques are described. These I have integrated into the text. Understandably, students find separate chapters or appendices on techniques boring. Techniques are most effectively described in the context of research questions about the cell to which they apply, and I have tried to do that.

The first chapter presents an overview of cell function and structure. I have not included a drawing of a fictional composite cell, but instead used descriptions of five different cells, a bacterium, a yeast cell, a protozoan, a plant cell, and an animal cell to introduce the main features of cells. Millions of cell types live on this planet, but only a few thousand have been studied in detail. From this experience we can assert that a single set of principles underlies the operation of all cells; exposition of these principles begins in **Chapter 1** and is continued throughout the entire book. An understanding of cells requires knowledge of their molecular makeup. **Chapter 2** presents a summary of cell molecules from water through lipids, carbohydrates, proteins, and nucleic acids. **Chapter 3** discusses how cells use enzymes to manage their chemical affairs. **Chapter 4** deals with capture and manipulation of energy to drive the synthesis of molecules and other cellular operations. **Chapter 5** is about the relationship of cells to their environments, particularly the role of the plasma membrane in the traffic of substances in and out of the cell. **Chapter 6** is an overview of some of the structural organization in cells, focusing mostly on the fibrous elements that make up the cytoskeleton. **Chapters 5 through 11** are about the genetic operation of cells. These chapters cover the structure, function, expression, and replication of genes and chromosomes. **Chapter 12** is a discussion of how cells reproduce, one of the most important functions of cells. Cancer, which in part results from failure to regulate cell reproduction, is considered in **Chapter 13.** This includes brief discussions of the causes of cancer, the nature of the disease, the properties of cancer cells, and the study of oncogenes. Cells have evolved several mechanochemical means of motility, and these are analyzed in **Chapter 14**. Much of this chapter deals with the specialized function of muscle, and it is followed in **Chapter 15** with discussions of several kinds of cells differentiated to perform highly specialized tasks in animal tissues. **Chapter 16** is about the origin and evolution of cells. This topic is placed last because it cannot be thought about and analyzed intelligently without a solid understanding of cell function and structure.

Each chapter is followed by a set of questions. These are meant to test comprehension of major points. Answers to all the questions are contained in the chapters and easily found. If a student does not know the answer to a question, it is better that he/she find it in context within the chapter before consulting the answer in the back of the book.

Further readings are given at the end of the book. This includes references to reviews and general articles. These are the most effective ways to follow up some topic in depth without having to work through highly specialized research papers. Finally, a glossary is included as a quick reminder of the meaning of specialized terms.

DAVID M. PRESCOTT

ACKNOWLEDGMENTS

I express here my thanks to the many people who gave me help of many kinds. Particularly, I thank my lovely wife, Gayle Prescott, who typed the manuscript, and with great patience and ingenuity attended to endless details of preparation of the work. I am grateful to Arthur C. Bartlett for his unwaivering, enthusiastic encouragement, and to Maureen Cunningham-Neumann and Rafael Millán for their skillful work in the production of the book.

I am especially grateful to Jerry Feldman, University of California, Santa Cruz, who wrote the section on circadian rhythms in Chapter 12 and reviewed the entire manuscript.

David Freifelder, University of California, San Diego, author of several successful science textbooks, helped me enormously as a scientific editor in all phases of manuscript preparation. David died in 1987 from lung cancer, and I miss his friendship and his generous, intelligent counsel.

I thank the following colleagues who reviewed parts or all of the manuscript: Cedric I. Davern, University of Utah; Mark Dubin, University of Colorado; Abraham S. Flexer, Boulder, Colorado; Joseph G. Gall, Carnegie Institution of Washington; Elissa Guralnick, University of Colorado; Michael W. Klymkowsky, University of Colorado; Peter L. Kuempel, University of Colorado; J. Richard McIntosh, University of Colorado; Sheldon Penman, Massachusetts Institute of Technology; Jean Paul Revel, California Institute of Technology; and L. Andrew Staehelin, University of Colorado.

And finally, I am grateful to many colleagues who generously contributed micrographs from their research for many of the illustrations:

Guenter Albrecht-Buehler, Northwestern University
Bruce N. Ames, University of California, Berkeley
Clara F. Armstrong, University of Pennsylvania
Dorothy F. Bainton, University of California School of Medicine, San Francisco
A. Benichou-Ryter, Institut Pasteur
Kurt Benirschke, University of California Medical Center, San Diego
Howard Berg, Harvard University
Lester I. Binder, Yale University
Hans R. Bode, University of California, Irvine
Jose J. Bonner, Indiana University
B. R. Brinkley, University of Alabama

Breck Byers, University of Washington
Ivan L. Cameron, The University of Texas Health Science Center at San Antonio
A. Kent Christensen, University of Michigan Medical School
David A. Clayton, Stanford University School of Medicine
Roger Craig, University of Massachusetts Medical School
Kathleen J. Danna, University of Colorado
Etienne de Harven, University of Toronto
Ellen R. Dirksen, University of California School of Medicine, Los Angeles
René Dohmen, University of Utrecht
Karen A. Dyer, University of Washington
Don W. Fawcett, Harvard University Medical School
Charles J. Flickinger, University of Virginia School of Medicine
Sidney W. Fox, University of Miami
David Freifelder, University of California, San Diego
Daniel S. Friend, University of California, San Francisco
Joseph G. Gall, Carnegie Institution of Washington
Robert C. Gallo, National Cancer Institute
Stanley M. Gartler, University of Washington
Ian R. Gibbons, University of Hawaii
Thomas H. Giddings, Jr., University of Colorado
Robert D. Goldman, Northwestern University Medical and Dental Schools
Kathleen Green, Northwestern University Medical School
Anne Hales, Imperial Cancer Research Fund
Barbara A. Hamkalo, University of California, Irvine
Henry Harris, University of Oxford
Christine J. Harrison, Paterson Institute for Cancer Research
Volker Herzog, University of Munich
J. A. Hobot, University of Wales College of Medicine
Bessie Huang, Research Institute of Scripps Clinic
Ann L. Hubbard, The Johns Hopkins University School of Medicine
Joel A. Huberman, Roswell Park Memorial Institute
H. E. Huxley, Brandeis University
Richard O. Hynes, Massachusetts Institute of Technology
Susumu Ito, Harvard University Medical School
James D. Jamieson, Yale University School of Medicine
Lincoln V. Johnson, University of Southern California School of Medicine
R. T. Johnson, University of Cambridge
Arthur Kelman, University of Wisconsin-Madison
James Kezer, Eugene, Oregon
Helen Kim, The University of Alabama at Birmingham
John C. Kinnamon, University of Colorado
Marc W. Kirschner, University of California, San Francisco
Michael W. Klymkowsky, University of Colorado
S. Knutton, St. George's Hospital Medical School
Richard Kolodner, Dana-Farber Cancer Institute
Irwin R. Konigsberg, University of Virginia
Mark Ladinsky, University of Colorado
Charles P. Leblond, McGill University
Edwin R. Lewis, University of California, Berkeley
Donald R. Lowe, Louisiana State University
Anthony P. Mahowald, Case Western Reserve University
Nadia Malouf, The University of North Carolina at Chapel Hill
Vincent T. Marchesi, Yale University School of Medicine
Mark P. Mattson, Colorado State University
Joyce McCann, University of California, Berkeley
Maclyn McCarty, Rockefeller University
J. Richard McIntosh, University of Colorado
Uel J. McMahan II, Stanford University School of Medicine
Mark McNiven, The University of Maryland
O. L. Miller, Jr., University of Virginia
Mark S. Mooseker, Yale University
Enrico Mugnaini, University of Connecticut
A. M. Mullinger, University of Cambridge
K. G. Murti, St. Jude Children's Research Hospital
Nanne Nanninga, University of Amsterdam
Eldon H. Newcomb, University of Wisconsin-Madison

Ada L. Olins, University of Tennessee, Oak Ridge Graduate School of Biomedical Sciences, Oak Ridge National Laboratory, Biology Division
Donald E. Olins, University of Tennessee, Oak Ridge Graduate School of Biomedical Sciences, Oak Ridge National Laboratory, Biology Division
Mary Osborn, Max-Planck-Institute for Biophysical Chemistry
Mary Lou Pardue, Massachusetts Institute of Technology
David E. Pettijohn, University of Colorado Health Sciences Center
David M. Phillips, Rockefeller University
Jeremy Pickett-Heaps, University of Colorado
G. Pontecorvo, Imperial Cancer Research Fund
Keith R. Porter, The University of Maryland
Ryan M. Prescott, Boulder, Colorado
Potu N. Rao, University of Texas, MD Anderson Hospital
Shmuel Razin, The Hebrew University-Hadassah Medical School
Michael K. Reedy, Duke University Medical Center
Mary B. Rheuben, Michigan State University
Peter N. Riddle, Imperial Cancer Research Fund
Hans Ris, University of Wisconsin
J. David Robertson, Duke University Medical Center
William A. Robinson, University of Colorado Health Sciences Center
Joel L. Rosenbaum, Yale University
John A. Rupley, University of Arizona
S. Z. Salahuddin, National Cancer Institute
Robert E. Scott, Mayo Clinic
Jerry W. Shay, University of Texas Health Science Center at Dallas
Michael B. Shimkin, University of California School of Medicine, San Diego
Lee Simon, Waksman Institute, Rutgers University, New Brunswick, New Jersey
Tim Spurck, University of Colorado
L. Andrew Staehelin, University of Colorado
Herbert Stern, University of California, San Diego
Barbara J. Stevens, Centre National De La Recherche Scientifique
A. T. Sumner, MRC Clinical and Population Cytogenetics Unit
Krishna K. Tewari, University of California, Irvine
F. R. Turner, Indiana University
Victor D. Vacquier, Scripps Institution of Oceanography
D. von Wettstein, Carlsberg Laboratory
Maud M. Walsh, Louisiana State University
Marta Walters, Santa Barbara Botanical Garden
Fred D. Warner, Syracuse University
W. P. Wergin, US Department of Agriculture
Fred D. Williams, Iowa State University
Gary E. Wise, Texas College of Osteopathic Medicine
Jong Sik Yoon, Bowling Green State University
Jorge J. Yunis, University of Minnesota Medical School
Dorothea Zucker-Franklin, New York University Medical Center

CELLS

PRINCIPLES OF MOLECULAR STRUCTURE AND FUNCTION

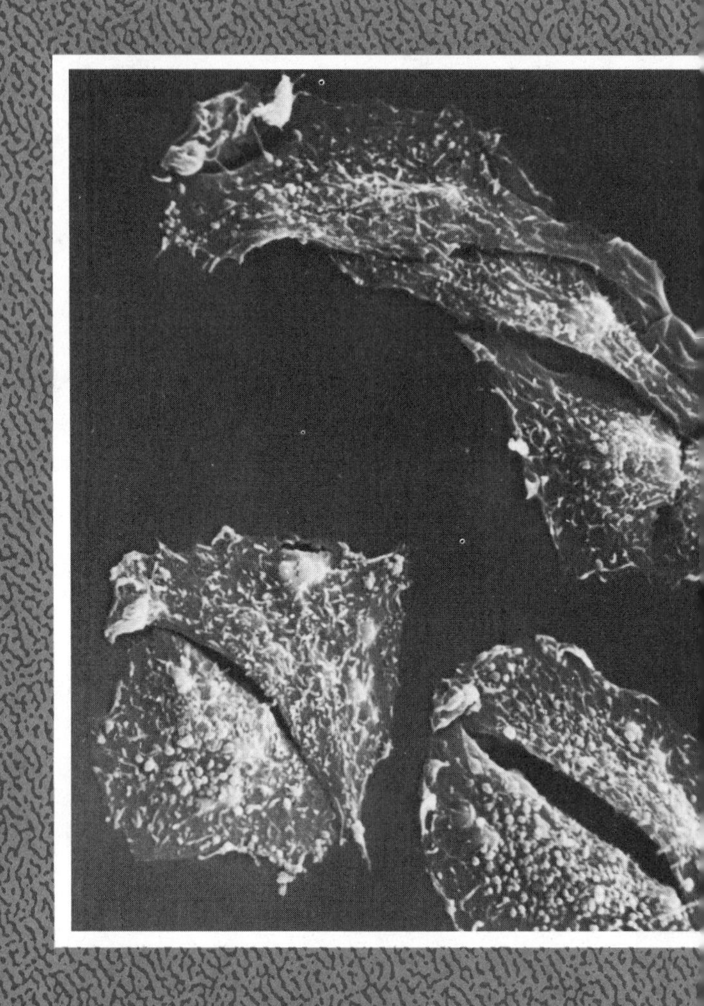

Introduction for the Student

A variety of reasons impels us to study cells. The most general is a natural curiosity about the structure and operation of all the organisms, from bacteria to humans, that populate the Earth—all are composed of cells. Cells enter into our thinking about a variety of issues. For example, the theory of evolution in its modern form can only be fully understood and evaluated with a detailed knowledge of genetic and other mechanisms that underlie the operation of cells. Essentially all medical science is centered on the function and structure of cells. Infectious diseases, degenerative diseases, developmental problems like birth defects, aging, death, and other medical problems arise through injury to cells or failure of cells to perform adequately. How antibiotics kill bacterial cells without damaging human cells or why cancer chemotherapeutic drugs are so toxic for a cancer patient depend on knowledge of cell properties and functions. How radiation effects us, how wounds heal, why we require vitamins and other substances in our diet are all matters that are determined by the properties and activities of the cells of which we are made.

Indeed, all of the properties of any living organism derive from the function and structure of the cells that make up that organism. Many organisms, like bacteria, yeast, and protozoa, consist of a single cell, and a microscope is needed to see them. In looking at such cells with a microscope it is easy to recognize that one cell equals one organism. Many other organisms, such as plants and animals, are each composed of enormous numbers of cells, but virtually all of them begin their existence as a single cell, a fertilized egg. A human egg divides into two cells within 24 hours after fertilization. By three days the embryo consists of 72 cells. The newborn human nine months later is made up of about 10 trillion (10^{13}) cells.

Within multicellular organisms groups of cells have developed specializations in structures and biochemical properties that give them particular functional capacities, for example, contraction by muscle cells, conduction of signals by nerve cells, transportation of oxygen by red blood cells, and absorption of nutrients from the soil by plant root cells. Understanding how an organism functions must be based on knowledge of how its many kinds of cells operate and cooperate. However, the cellular nature of such organisms is not a fact that is part of our general awareness. When we see a tree, a dog, or another human being, it does not strike us that we are looking at large collections of cells. It doesn't register in our minds that everything about the shape, size, behavior, state of health, and so forth, of an organism like an adult human is the sum of the cooperative workings of many trillions (10^{12}) of cells. An eye reads the words on this page because millions of light-sensitive cells in the retina are stimulated by light from the page, and the stimuli are carried as signals deep into the brain by nerve cells. There, groups of many millions of nerve cells interpret the signals as a particular kind of visual information and integrate it into the activities of many millions of other nerve cells. The interactions of nerve cells produce consciousness and higher functions of the brain like thought and memory, although as yet we understand rather little about how nerve cells achieve these astonishing feats.

Naturally, learning about cells and their operations is crucial to anyone entering fields in the life sciences and medicine. However, learning about cells is also essential for anyone who strives to be educated about the functioning of his/her own body or about a wide variety of issues dealing with health, drugs, illness, medical treatment, reproduction, environmental questions, evolution, and life in general.

In many ways our knowledge of cells is incomplete. At various places in this book appears the notation that one or another aspect of cell function or structure is incompletely understood. These gaps in our knowledge continue to be filled in by research. Today, research on thousands of kinds of cells is the full time occupation of tens of thousands of researchers in thou sands of laboratories throughout the world. These ef forts, continuously producing new discoveries, are major part of the human endeavor to understand th nature of life.

THE DIMENSIONS OF CELLS

Almost all cells are so small that one needs a micro scope to see them. A few, like the amoeba shown o the opposite page, are large enough to see as a tin speck without the aid of magnification, but little o nothing about the structure and function of an amoeb can be discerned without a microscope. Within th world seen with a light microscope the sizes of cell cover a vast range, as illustrated by the eight light mi crographs of living cells on the facing page all photo graphed at the same magnification. The longest dimen sion for each cell type is given in the list of names. Fo most purposes it isn't necessary to know the exact di mensions of cells, but a comparative sense of cell size is certainly useful. The illustrations here are an intro duction to relative sizes of some kinds of cells. Othe examples appear throughout the text.

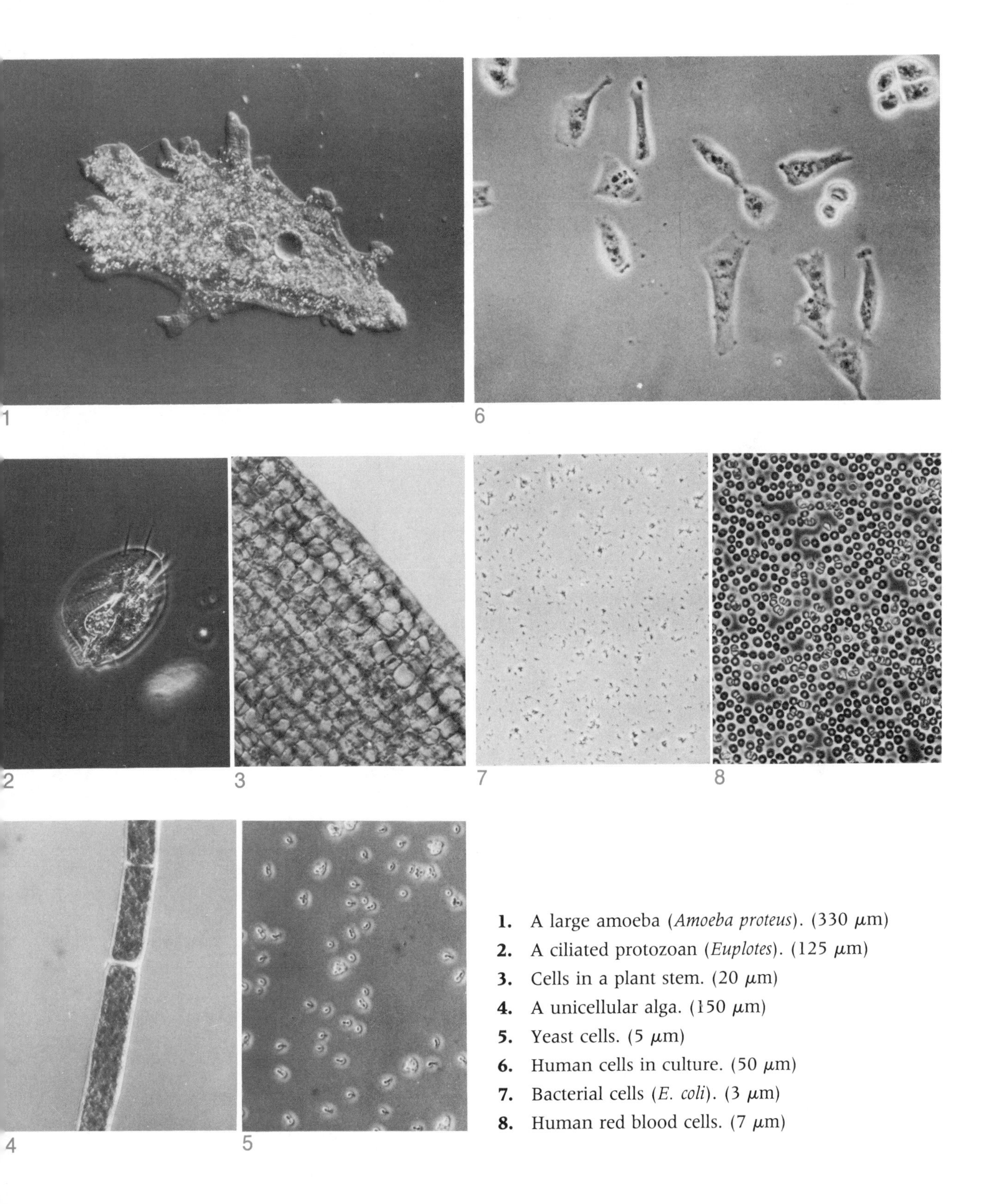

1. A large amoeba (*Amoeba proteus*). (330 μm)
2. A ciliated protozoan (*Euplotes*). (125 μm)
3. Cells in a plant stem. (20 μm)
4. A unicellular alga. (150 μm)
5. Yeast cells. (5 μm)
6. Human cells in culture. (50 μm)
7. Bacterial cells (*E. coli*). (3 μm)
8. Human red blood cells. (7 μm)

CHAPTER 1

Cells

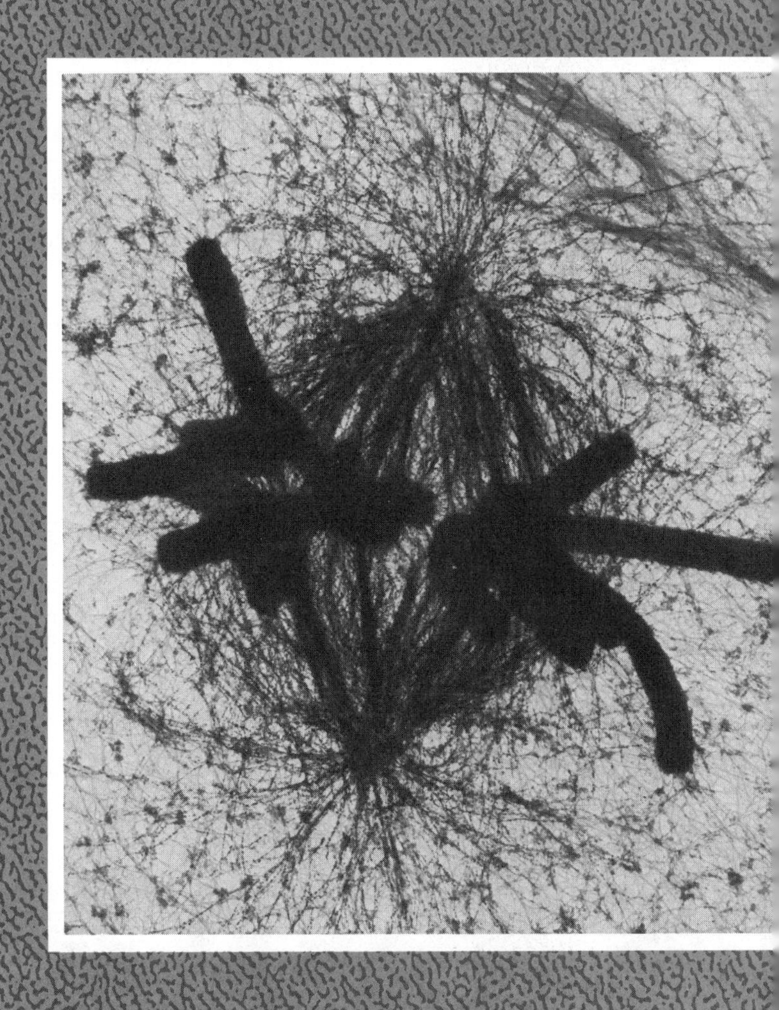

See Figure 1-29.

All organisms from the simplest to the most complex plant or animal are composed of cells. Some organisms, like a bacterium or an amoeba, consist of a single cell. A complex organism like an adult human is made up of more than one hundred trillion (10^{14}) cells. The appearance, behavior, activities, and functions of all organisms represent the sum total of the properties of the cell or cells of which they are made. Therefore, the study of the cell is overwhelmingly important for understanding life in all its forms.

The contemporary study of the cell is an enormous enterprise that combines many scientific disciplines and methods, including microscopy, genetics, biochemistry, biophysics, and physiology. New information has accumulated rapidly, particularly in the last 30 years, and a great deal about cells is now understood—how they function, how they are constructed, and how they work together within a multicellular organism. However, much remains to be learned.

The beginnings of cell biology go back more than three centuries to discoveries by the first pioneering microscopists. In 1665 the Englishman Robert Hook observed the cell walls of cork with a primitive micro scope and also discovered the cellular nature of a livin plant leaf. A leaf was easily studied with Hooke's m croscope because a leaf is thin enough for light to pa through, allowing its large cells to be seen (Figure 1-1 Hooke applied the term *cell* to describe the microscop units he discovered. A contemporary of Hooke, Antc nie van Leeuwenhoek, discovered sperm cells, re blood cells, and many kinds of microorganisms using microscope consisting of a single lens. These early ob servations set the stage for almost two centuries of m croscopic studies. As microscopes gradually improve the cell nucleus was discovered, and the cellular natu of all plants and animals came to be recognized. By th 1830s enough observations had been accumulated t allow Schleiden, a botanist, and Schwann, a zoologis to propose that *all organisms are constructed of cells,* statement now known as the **cell doctrine.** This pro nouncement constitutes the first major tenet upo which the contemporary science of cell biology

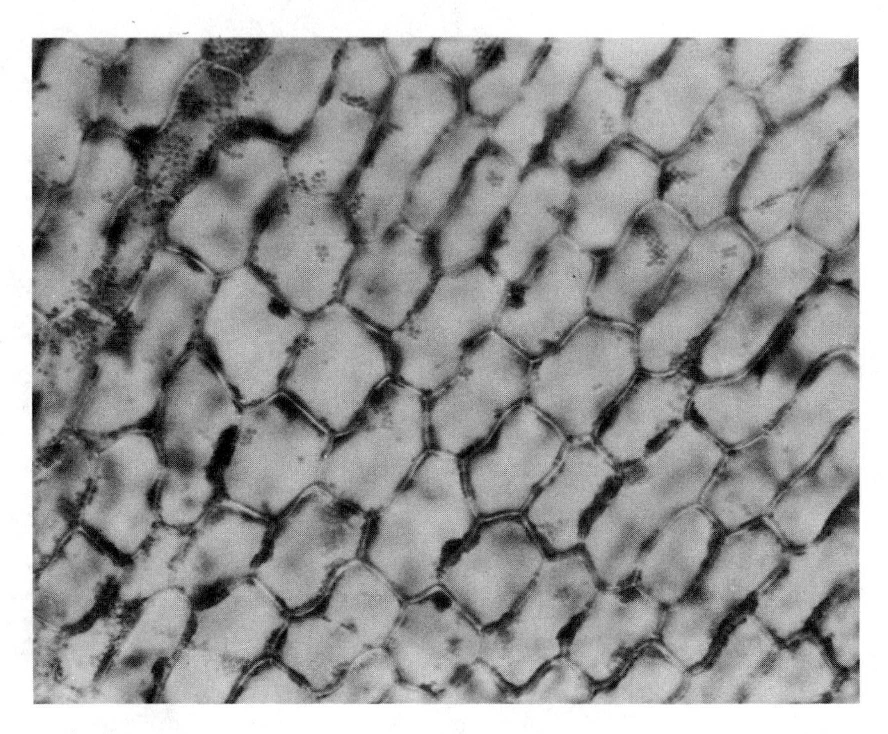

FIGURE 1-1
Light micrograph of living cells in a water fern. The walls of plant cells are prominent and were readily seen by microscopists using a single-lens microscope. In this micrograph most of the cell volume is occupied by the large central vacuole, obscuring visibility of all internal structures except a few chloroplasts (small spheres).

ɔunded. Following the formulation of the cell doc-
ine, biologists established that new cells do not arise
e novo but come only by **cell division,** that is, division
f a preexisting cell into two daughter cells. By the end
f the nineteenth century, cell biologists had discovered
hromosomes and described **mitosis**—the distribution
t cell division of chromosomes to daughter cells. Stud-
es that followed soon led cell biologists to conclude
at chromosomes are the hereditary material, and that
uring cell division, the process of mitosis distributes
is material equally to the two daughter cells.

Although microscopic studies from 1665 to 1900 onvinced cell biologists that the cell is the building lock of all multicellular organisms, a major question emained: Were microorganisms such as bacteria and rotozoa single-celled organisms or were they instead oncellular or acellular? As biologists came to undertand the genetic, physiological, and biochemical properties of cells, it became clear that microorganisms are ingle, free-living cells. All modern research recognizes at in both unicellular and multicellular organisms *the ell is the fundamental structural unit,* housing the genetic aterial and the biochemical organization that account or the existence of life.

This book describes the main elements of what has een learned so far about cells: how they are contructed and how they carry out such activities as selfaintenance, growth, cell division, differentiation, ovement, sensory perception, and cell–cell communcation. The discussions of cell activities and structures re based on integration of various kinds of information erived both from the more traditional tools (microsopy, physiology, and conventional genetics) and from e more recent approaches of biochemistry, macromoecular chemistry, and molecular genetics. Integration f information derived from these diverse disciplines hould eventually provide a complete understanding of ll the functional and structural properties of cells.

ALL CELLS POSSESS THE SAME BASIC PROPERTIES

We speak of *the* cell, yet one must be aware of the xistence of millions of different species of cells that show tremendous diversity in structure and metabolic capabilities. A bacterium, a yeast, an amoeba, a plant cell, and a liver cell would appear to be so different in structure and life-style that they might seem to have little in common, yet the similarities among these diverse cell types are more profound than the differences. The main properties in common are the following:

1. All cells store information in genes made of DNA.
2. The genetic code used in the genes of all cells is, with minor exceptions, the same.
3. All cell types decode the genes in their DNA by an RNA system that translates genetic information into proteins.
4. All cells synthesize proteins using a structure called the ribosome.
5. Proteins govern function and structure in all cells.
6. All cells need energy to maintain their internal environment and to drive the synthesis of their complex constituents. All cells use the molecule ATP as the currency for transfer of energy from energy sources to energy needs.
7. All cells are enclosed by a plasma membrane composed of proteins and a double layer of lipid molecules.

Because of these similarities, it is possible to generalize about the structures and functions of cells. One can begin constructing a set of principles of function and structure that has been present throughout the evolution of the great diversity of cell types that now exists.

CLASSIFICATION OF CELLS—PROKARYOTES AND EUKARYOTES

On the basis of certain major properties, all cell types currently found on this planet fall into two major groups: the **prokaryotes** and the **eukaryotes.** *Karyote* is a word root meaning nucleus. The word *prokaryote* (before a nucleus) therefore designates cells that do not have a structurally delineated unit containing the ge-

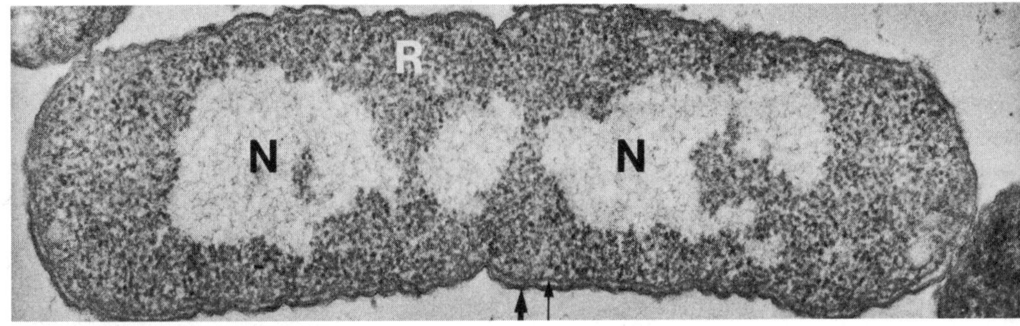

FIGURE 1-2
A section through a bacterial cell (*E. coli*) observed with an electron microscope. The cell is enclosed by a wall (thick arrow). The plasma membrane (thin arrow) is just inside the cell wall. The granular appearance is due to ribosomes (R). The lighter regions are portions of the nucleoid (N), which is separating into two parts. The cell is about to divide, as indicated by the slight indentation of the cell wall in the middle of the cell. [Courtesy of Nanne Nanninga. From *Molecular Cytology of Escherichia coli* (1985) N. Nanninga, ed. Academic Press, London.]

netic material (a nucleus). *Eu* means true; hence, the word *eukaryote* (true nucleus) designates cells that have a well defined nucleus whose chromosomal material is separated from the remainder of the cell contents (the cytoplasm) by a nuclear envelope. Prokaryotes do not have a nuclear envelope. The nucleus of a prokaryote is less well delineated and is usually called a **nucleoid** (see Figure 1-2). The prokaryotic cells are the bacteria, and all are unicellular. These include the blue-green bacteria, or *cyanobacteria,* which were until recently called blue-green algae. All other cell types are eukaryotes. These include many unicellular organisms, for example, protozoa, fungi, and some algae, as well as all multicellular plants and animals.

Originally, prokaryotes were distinguished from eukaryotes on the basis of a single characteristic, the absence of a nuclear envelope. Subsequently, other equally distinct differences have been discovered. The main differences are listed in Table 1-1. Some of the characteristics listed in Table 1-1 need explanations, for example introns in genes; the explanations are given in later chapters. Prokaryotic cells are both smaller and structurally less complex than eukaryotic cells. In addition, eukaryotes contain far more DNA than prokaryotes. Among the hundreds of eukaryotes examined, those with the least amount of DNA (yeast cells) stil have several times more than the prokaryote thought tc possess the most DNA (the bacterium *Serratia marces cens*). This difference in DNA content reflects the fac that eukaryotes are genetically more complex than pro karyotes.

It is thought that prokaryote-like cells precede eukaryotic cells in evolution and that the first euka ryote evolved from some type of prokaryote, probably a billion or more years ago. We have no clear idea which difference between prokaryotes and eukaryotes was the first to arise; hence, we do not know which of the differences was fundamental in the origination of euka ryotes. Presumably, a prokaryotic cell acquired one o another of the key characteristics for eukaryotes liste in Table 1-1, and this created the potential for subse quent evolution of the other distinguishing characteris tics of eukaryotes. For instance, the acquisition through evolution of an unusually large amount of DNA coulc have been the innovative step that first separated ar emerging eukaryote from prokaryotes. What coulc have brought about an increase in DNA content, how ever, is not known. Whatever the first step in the origi nation of eukaryotic cells, that step must have create the potential for great evolutionary diversification. The range in function and structure among contemporary cells is many times greater among eukaryotes than it i among prokaryotes.

Throughout this book comparisons between pro karyotes and eukaryotes enter into discussions of cel ınction and structure. Such comparisons often yield ısight into cell organization, operation, and evolution. he following section briefly characterizes prokaryotic ells. The general properties of eukaryotic cells are hen outlined through descriptions of a yeast cell, an moeba, a plant cell, and several mammalian cells.

PROKARYOTIC CELLS

Prokaryotes are divided into two major groups, the **ubacteria** and the **archaebacteria.** Both groups hare many properties—those listed in Table 1-1—that learly separate them from eukaryotes. Most species of acteria are eubacteria. Eubacteria can be divided into vo major groups: the **nonphotosynthetic** and **pho-osynthetic bacteria.** These two groups are discussed ı the next two sections. The bacterium *Escherichia coli* used as an example of nonphotosynthetic eubacteria.

Nonphotosynthetic Eubacteria

Escherichia coli is a bacterium that commonly inhabits the intestinal tract of humans and other animals. It is a cylindrical cell about 2 μm long and 1 μm in diameter (Figures 1-2, 1-3), with a volume of about 1.6 μm^3. On its surface are a number of filamentous appendages called **flagella,** usually six, by which it rapidly propels itself. One cm^3 (about one gram) of packed *E. coli* contains about 50×10^9 cells. An individual cell grows by increasing its length while maintaining a constant diameter. The cell divides into two daughters by the forming of a partition through the middle of the cylinder (Figure 1-4).

The genetic constitution of *E. coli* allows the organism to grow and divide in a medium containing only a few kinds of inorganic ions and a source of organic carbon, for example, the sugar glucose. Thus, the DNA of this bacterial cell contains genes for all of the en-

TABLE 1-1

Major Differences Between Prokaryotic and Eukaryotic Cells

Prokaryotes	Eukaryotes
1. No nuclear envelope	Nuclear envelope present
2. No nucleolus	One or more nucleoli
3. No histone proteins associated with DNA	Histones bound to DNA
4. DNA content ranges from 750,000 base pairs to 5×10^6 base pairs	DNA content ranges from about 1.5×10^7 to 1.5×10^{11} base pairs
5. Genes generally lack introns	Most genes contain introns
6. One chromosome	Two or more chromosomes
7. Small cell size; usually no more than several μm^3 in volume	Large cell size; usually from several μm^3 to several mm^3 in volume
8. Lack well developed intracellular membranous organelles except for photosynthetic membranes in some bacteria	Contain extensive membrane systems and membranous organelles such as mitochondria, chloroplasts, Golgi complex, and endoplasmic reticulum
9. Lack microtubules, microfilaments, and intermediate filaments	Contain microtubules, microfilaments, and intermediate filaments
10. Sterols usually not present in the plasma membrane	Sterols present in the plasma membrane

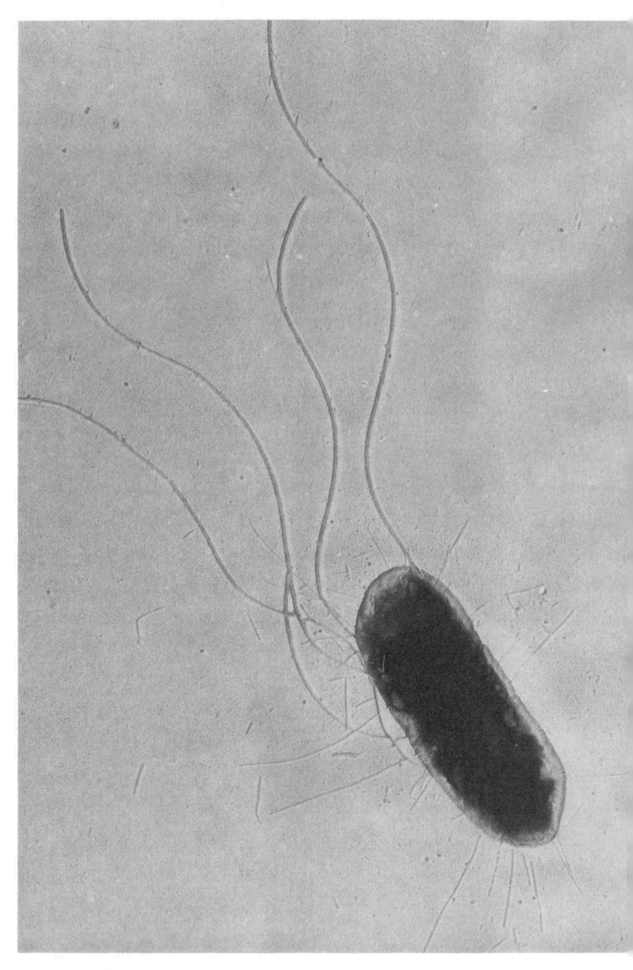

FIGURE 1-3
Electron micrograph of an *E. coli* bacterium with several flagella. [Courtesy of Howard Berg.]

zymes (Chapter 3) needed for the synthesis of all the amino acids, nucleosides, fatty acids, and other components needed to make macromolecules, using only a simple organic molecule, such as glucose, and inorganic salts as the starting materials.

In glucose-containing medium, *E. coli* doubles in size and divides every 40 minutes at the optimum temperature of 37°C. If amino acids, nucleosides, and other nutritionally useful organic molecules are added to the minimal culture medium, the rate of cell growth and division increases. This increase occurs because provision of useful nutrients relieves the cell of the need to synthesize those components. By addition of a rich variety of nutrients to the medium, an *E. coli* cell approaches the upper limit in its reproductive rate, doubling all of its contents and dividing every 20 minutes. The upper limit to growth rate is probably set by the maximum rate at which *E. coli* can synthesize macromolecules. With a generation time of only 20 minutes,

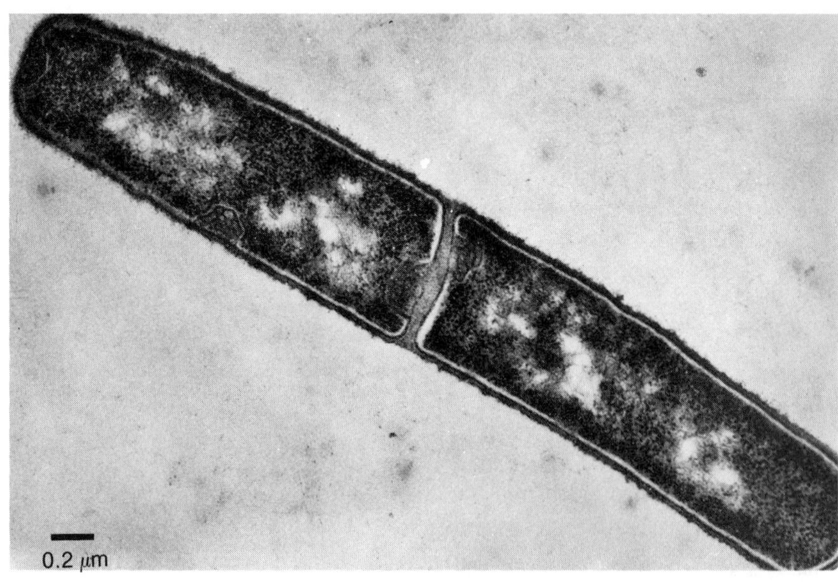

FIGURE 1-4
An electron micrograph of a section through a dividing bacterium. A wall has been laid down across the center of the cell. The wall later splits, producing two separate daughter cells. The light areas are occupied by DNA, and the dark areas are filled with ribosomes. [Courtesy of A. Benichou-Ryter.]

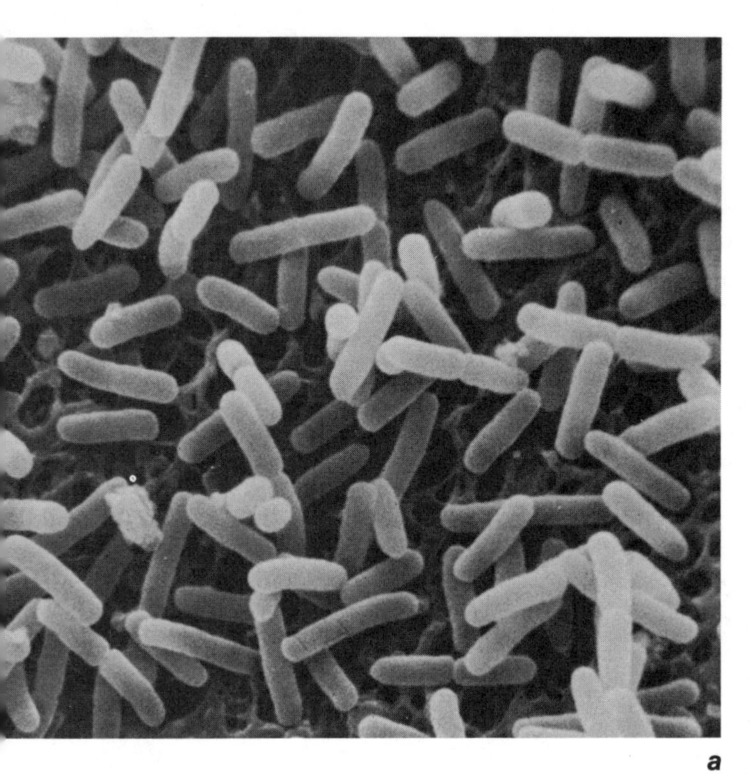

a

FIGURE 1-5
Scanning electron micrographs of three forms of whole bacteria. (a) Bacilli—rod-shaped bacteria. (b) Cocci—ovoid-shaped bacteria that sometimes form chains. (c) Spirillae—spiral-shaped bacteria. [(a) Courtesy of Nanne Nanninga, (b and c) are from G. Shih and R. Kessel. 1982. *Living Images.* Jones and Bartlett.]

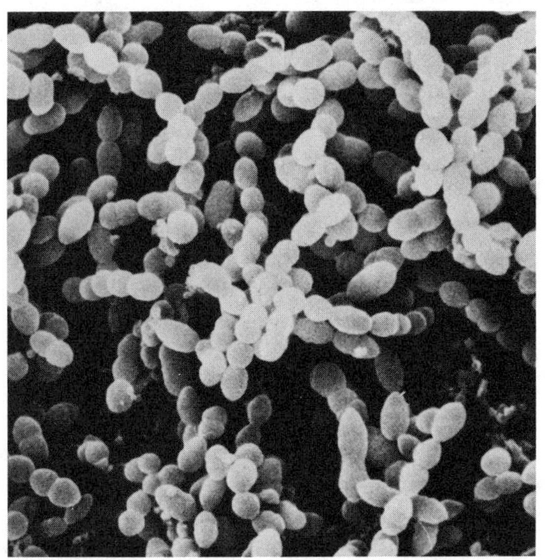

b

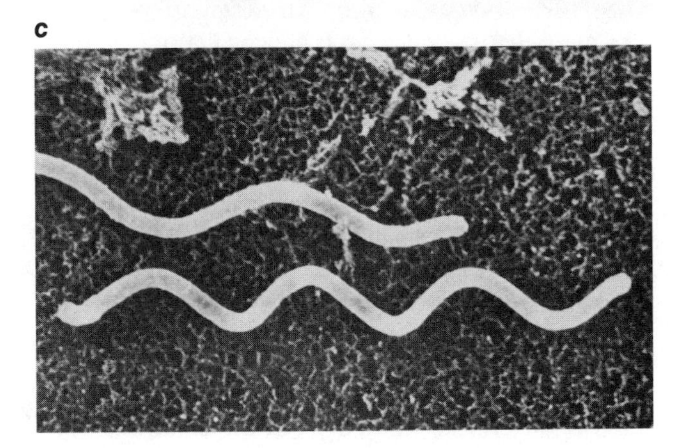

c

single cell could give rise to 2^{36} cells (more than 3 ×)^{11} cells) in just 12 hours.

Prokaryotic cells usually grow and reproduce much ore rapidly than eukaryotes. For instance, for a mam-alian cell, seven hours is about the shortest **genera-on time** or **cell cycle time**—the time a cell needs to from one cell division to the next. Some protozoa, hich are unicellular eukaryotes, have generation nes as short as two hours in a nutritionally rich me-um, and one kind of yeast cell can divide every 75 inutes; but none of the eukaryotes approaches the pid proliferation rates common among prokaryotes.

Prokaryotes are structurally simple cells. Most pes have a rigid **cell wall** made of polysaccharides, ptides, and lipids laid down outside the cell. A few pes, for example, the small bacteria known as *Myco-asmas,* lack an extracellular wall. The rigid wall of d-shaped bacteria maintains the cylindrical shape of e cell (Figure 1-5(a)). Other species of bacteria have alls that produce a spherical (Figure 1-5(b)) or a spiral form (Figure 1-5(c)). The cell wall provides mechanical protection, particularly against osmotic pressure (Chapter 5).

Immediately inside the cell wall of *E. coli* is the **plasma membrane** (Figure 1-2), which completely encloses the cell. The plasma membrane consists of a double layer of lipid molecules (Chapter 5) with many associated protein molecules. In contrast to the cell wall, which is porous and therefore penetrable by molecules and ions, the plasma membrane severely restricts the diffusion of molecules and ions in and out of the cell. Thus, the membrane serves the critical role of retaining desired substances inside the cell, although it also limits diffusion into the cell of environmental substances necessary to sustain cell metabolism. Certain specialized proteins bound to the lipid bilayer of the plasma membrane greatly enhance the inward passage of inorganic ions, sugars, amino acids, nucleosides, and other dissolved materials that are useful to the cell. Other proteins bound in the plasma membrane of a bacterial cell catalyze the process by which the energy contained in organic molecules is converted into a chemically usable form.

The intracellular contents of prokaryotes such as *E. coli* are present in two major structural parts, a nucleoid and the cytoplasm (Figure 1-2). The nucleoid consists of a single DNA molecule (the chromosome) condensed into an irregularly shaped, fibrous network, which occupies a few percent of the total cell volume. It is thought that the nucleoid is attached at one point to the plasma membrane. This attachment of the chromosome to the membrane may help both in the control of chromosome replication and in the separation of daughter chromosomes during cell division.

The cytoplasm of *E. coli* contains approximately 25,000 tiny particles called **ribosomes,** floating in a solution called the **cytosol** (Figure 1-2). Each ribosome is a machine for synthesizing proteins. The cytosol, which contains a large variety of ions, small organic molecules, and enzymes, is where the cell carries out most of its metabolic activities.

Much has been learned about the molecular biology of the cell from the study of prokaryotes, in particular from the study of *E. coli*. In part, bacteria were chosen as research materials because they are functionally and structurally far less complex than any of the eukaryotic cells. In addition, the fast growth rate and low

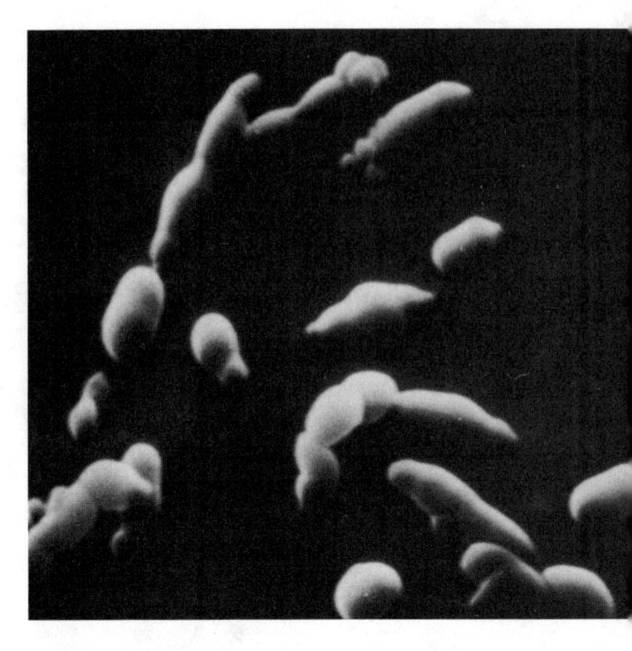

FIGURE 1-6
Scanning electron micrograph of the smallest known bacteria--mycoplasma. These bacteria lack cell walls and have irregular shapes. The cells are less than 0.5 μm in width. [Courtesy of Shmuel Razin.]

number of nutritional requirements of bacteria such a *E. coli* constitute a great practical advantage for researc because large numbers of cells can be obtained in a fe hours with a simple, inexpensive culture medium.

E. coli is not the smallest type of cell known. Som bacteria, the mycoplasmas (Figure 1-6) have volum as small as 0.02 μm^3, compared to a minimum volum for *E. coli* of 1.6 μm^3. Mycoplasmas lack cell walls, an their chromosomes can be as small as one-fifth of th chromosome in *E. coli*. These are the smallest chromo somes known among bacteria. The mycoplasmas we identified about 1900 as the cause of respiratory diseas in animals and gained attention during World War II a the causative agents of pneumonia among U.S. Arm recruits. Mycoplasmas are sometimes referred to a **PPLO,** which stands for pleuropneumonia-like orga isms. Given their small size and small amount of DN the mycoplasmas are no doubt genetically and fun

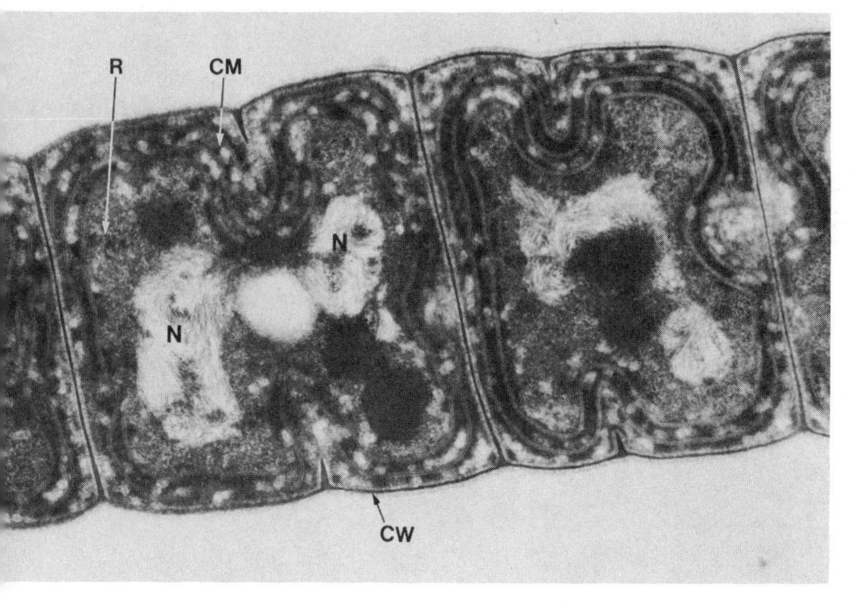

FIGURE 1-7
Electron micrograph of photosynthetic (blue-green) bacteria. The cells of this species remain attached to one another, forming chains. Main structural features are the nucleoid (N), ribosomes (R), cell wall (CW), and cytoplasmic membranes (CM), in which photosynthesis is carried out. [Courtesy of Thomas H. Giddings, Jr.]

ionally less complex than *E. coli.* However, they require a nutritionally complicated medium for growth nd grow slowly and hence are less convenient to use n research. Nevertheless, the study of mycoplasmas as intensified during recent years, and these simplest f known cells may well provide unique insight into rinciples of cell organization and operation. Mycolasmas are sometimes referred to as the *minimum* cell ecause they approach the minimum genetic and moecular complexity necessary to sustain the life and reroduction of a cell.

Photosynthetic Bacteria

The photosynthetic bacteria probably arose from onphotosynthetic bacteria very early in the course of volution, perhaps as early as 3.1 billion years ago. Most photosynthetic bacteria are **obligate photoautotrophs.** Photoautotroph means requiring only ght, water, inorganic ions, and CO_2; obligate means hat for growth light is necessary because these bacteria annot use organic compounds like sugars as an alterative source of energy. Photosynthetic bacteria are widely distributed in fresh and salt water and in soil. The enormous mass of photosynthetic bacteria growing in the oceans generates much of the oxygen in the Earth's atmosphere.

Figure 1-7 shows an electron micrograph of a section through a photosynthetic bacterium. The cell is enclosed by a rigid wall and, immediately inside the wall, by a plasma membrane. As in other bacteria, the cytoplasm is rich in ribosomes, and a nucleoid is present. In contrast to other kinds of bacteria, however, photosynthetic bacteria often have extensive internal membranes that contain light-absorbing pigments and the machinery for photosynthesis. Photosynthesis is the process by which the energy of light is captured and used to synthesize sugar, starting with carbon dioxide and water (Chapter 4).

EUKARYOTIC CELLS

All eukaryotic cells share certain basic properties, but nonetheless the eukaryotes are an extremely diverse group. Four cell types—a yeast cell, an amoeba (both unicellular organisms), a plant cell, and an animal cell—are described here to present an overview of the general features of eukaryotes. However these cells

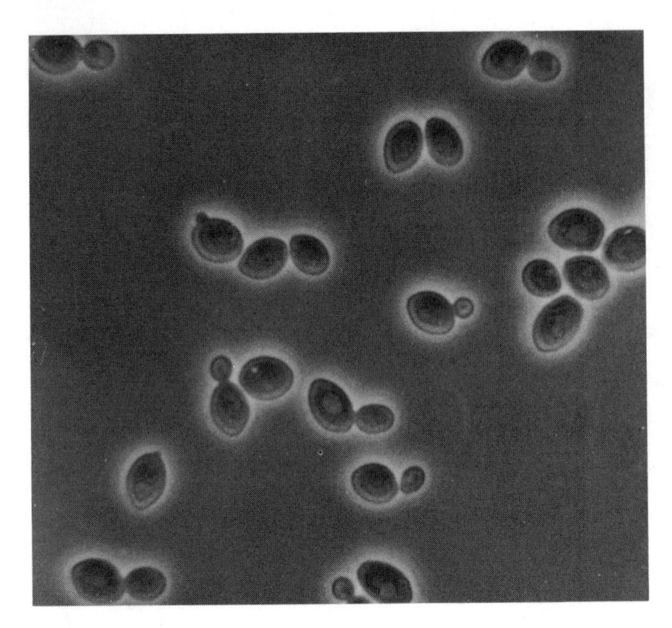

FIGURE 1-8
A light micrograph of the budding yeast *Saccharomyces cerevisiae.* Several cells have buds of various sizes. [Courtesy of Breck Byers.]

only begin to indicate the full range of diversity found in the group. Many kinds of eukaryotic cells are used in research, because different cells have special features that make them suitable for experiments on particular aspects of cell function and structure. The amoeba, for example, is well suited for experiments requiring transplantation of a nucleus from one cell to another. Yeast cells are particularly favorable for study of the genetic basis of cell operations. Mammalian cells in culture are favorable for studies of cell growth and division. Plant cells are important for the analysis of photosynthesis. Experiments with these and many other cell types will eventually lead to a detailed understanding of the cell in general.

The Yeast Cell

Yeast is a unicellular eukaryote classified as a fungus. In cell size, structural complexity, and DNA content the various species of yeasts are among the simplest known eukaryotic cells. The yeast, *Saccharomyces*

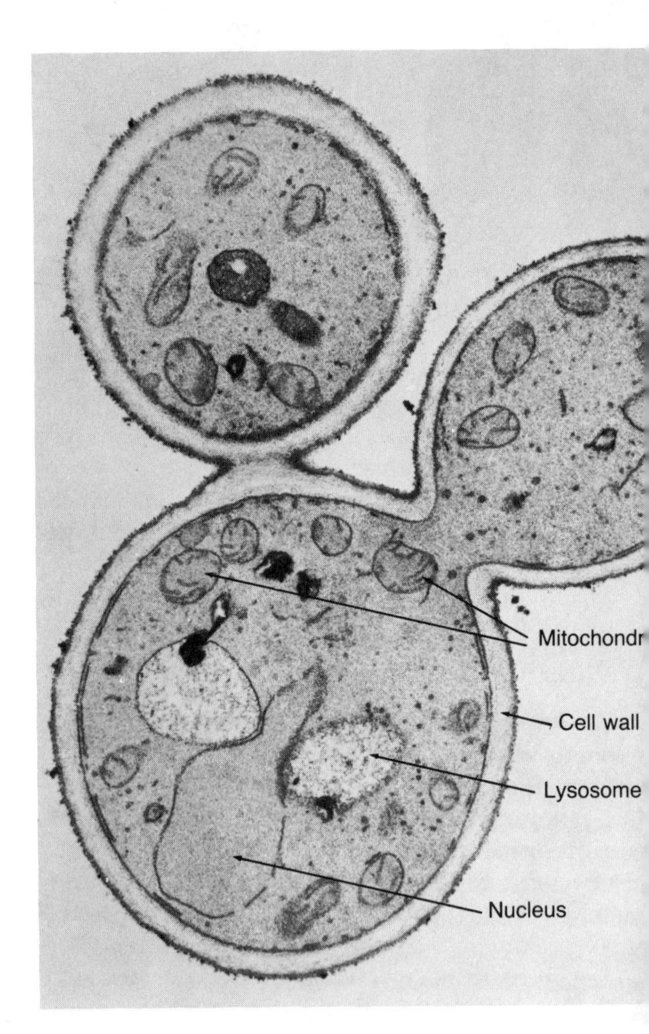

FIGURE 1-9
Electron micrographs of sections of the budding yeast *Saccharomyces cerevisiae.* (a) A cell with a completed bud, which is still attached (upper left), and a new bud in an advanced stage of formation. The completed bud appears smaller because the section is not through the center of the cell. (b) A yeast cell preserved with a different chemical fixative than in (a). The cell wall is not visible. The nucleolus is the more darkly staining material in the nucleus. The granular appearance of the cytoplasm in both pictures is due to ribosomes. [Courtesy of Barbara Stevens.]

erevisiae, is an oval-shaped cell measuring about 6 μm n its maximum diameter (Figure 1-8). Its name comes rom *saccharo,* which means sugar—a food that the rowing organism consumes in large quantities—from *nyces,* the Greek word for fungus—and from *cerevisia,* he Latin word for beer.

Yeast cells have a rigid cell wall that governs the hape of the cell and gives it mechanical protection. The vall of the yeast cell is much thicker than the wall of a acterium and consists primarily of a material (polynerized sugars) that is chemically related to the cellulose of plant cell walls (Figure 1-9). The wall is outside the plasma membrane. The plasma membrane regulates the passage of ions and molecules in and out of the cell. Unlike bacteria, eukaryotes do not have in their plasma membranes the enzymatic machinery for converting energy to a usable form; in eukaryotes the comparable function is carried out in cytoplasmic organelles called **mitochondria.**

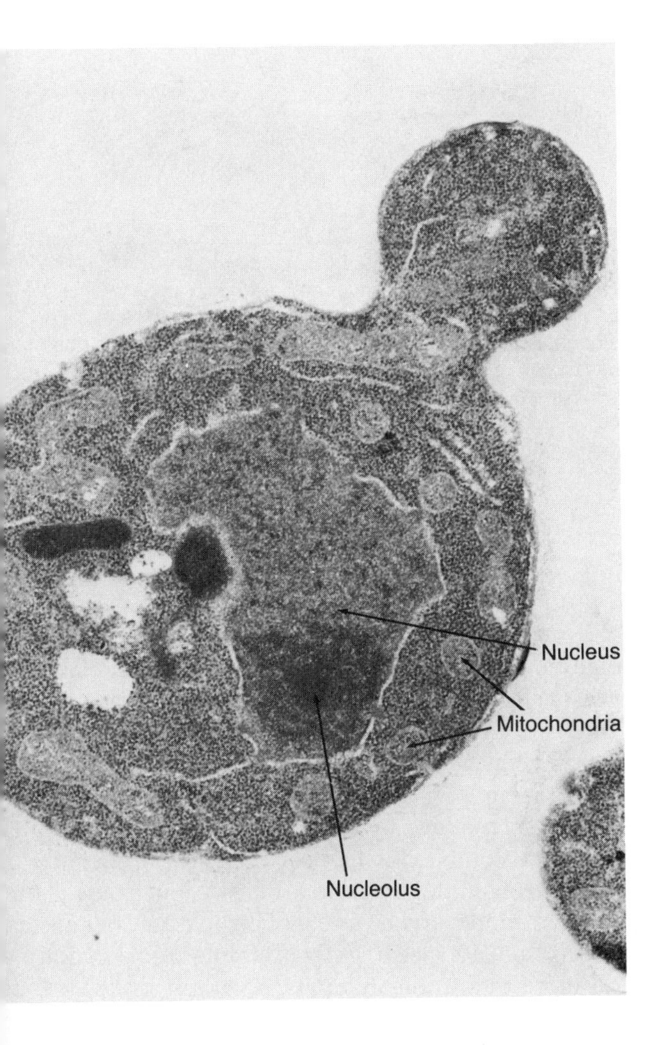

The yeast cell has two major compartments, the nucleus and the cytoplasm. The nucleus contains the **chromosomes,** which appear in the electron microscope as a tangled array of finely dispersed fibers (Figure 1-9), and which are separated from the **cytoplasm** by the nuclear envelope. The 17 chromosomes in a yeast cell have an aggregate DNA content of 1.5×10^7 base pairs. Hence, this simple eukaryote contains about four times more DNA than does *E. coli.*

The nuclear envelope of the yeast cell and all other eukaryotes is a **double membrane** (Figure 1-9), whose function is not well understood. It almost certainly helps to regulate the movement of materials between the nucleus and the cytoplasm. The presence of **pores** in the envelope is a general feature of nuclear envelopes (but not visible in Figure 1-9). These pores allow the entry and exit of such materials as proteins and RNA, which are too large to penetrate the membranes themselves. The nucleus also contains a roughly spherical body, the **nucleolus** (Figure 1-9), within which the RNA component of ribosomes is synthesized. From these RNA molecules, and from proteins that are produced in the cytoplasm, ribosomes are assembled in the nucleolus and then move through the pores of the nuclear envelope to the cytoplasm, which contains hundreds of thousands of ribosomes working to synthesize the proteins of the cell.

The cytoplasm of a yeast cell contains a number of mitochondria (Figure 1-9). Mitochondria are cytoplasmic organelles (an organelle is a structural component in a cell) consisting of an outer delimiting membrane and an array of inner membranes organized into folds called **cristae** (Chapter 4). The outer membrane is permeable to ions and molecules moving between the mitochondrion and the surrounding cytosol. The system of inner membranes, the cristae, contains the enzymes that oxidize organic molecules to provide the chemical energy that fuels the many metabolic activities of the cell. The inner membranes of mitochondria, in other words, perform the same energy-transforming function as the plasma membrane in bacteria. The mitochondria of yeasts and other eukaryotes are the organelles that convert the chemical energy in food into a form that is readily used by the cells. In nonphotosynthetic cells, they form 95 percent of the fuel for cellular activities. The remaining 5 percent is produced by the partial breakdown of sugars in the cytosol. In photosynthetic cells (algae and plants), the situation is a little different, since chemical energy is produced in chloroplasts as well as in the cytosol and mitochondria.

A mitochondrion also contains ribosomes, which are used in the synthesis of a few of the many proteins needed to carry out mitochondrial activities. In addition, each mitochondrion possesses a small DNA molecule (containing about 75,000 base pairs in yeast) that contains genes for synthesizing some of the RNA and protein components needed for mitochondrial function and structure. However, most of the proteins of the mitochondrion are encoded by genes in the cell nucleus. The presence of DNA in the mitochondrion provides this organelle with a limited amount of autonomy, although growth, multiplication, and metabolism of the mitochondrion are closely coordinated with the metabolism of the rest of the cell through mitochondrial-cytoplasmic-nuclear interactions.

The cytoplasm of yeast cells contains several other important functional components in addition to ribosomes and mitochondria. Most of these other components are discussed later in this chapter in connection with other kinds of cells in which they are more readily identified by electron microscopy.

In yeast cells that are no longer growing, one or two large **lysosomes** are prominent (Figure 1-9). Lysosomes are vesicles containing high concentrations of enzymes that break down proteins, polysaccharides, lipids, and nucleic acids. As long as these enzymes remain packaged in lysosomal vacuoles, the cell is protected from the digestive action of its own digestive enzymes. If the cell dies, the lysosomal enzymes are released into the cytoplasm, where they cause rapid destruction of cellular components. This process of self-digestion is known as **autolysis.** The cytoplasm of yeast cells contains **lipid droplets,** which are storage depots of lipid molecules that can be broken down to yield energy.

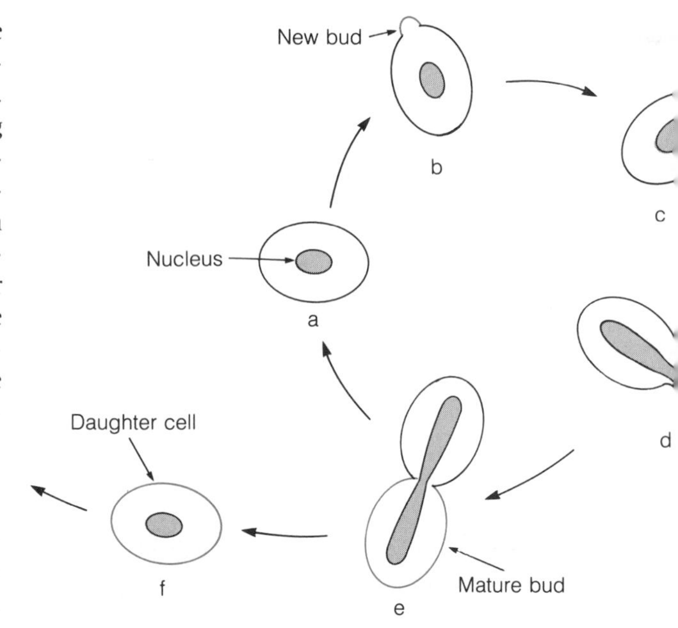

FIGURE 1-10
The yeast *Saccharomyces cerevisiae* reproduces by budding. The bud begins (b) as a small bleb (red) on (a) the mother cell and grows (c,d,e) into a full-sized daughter cell. The nucleus elongates into the bridge (d,e) between mother and the bud and undergoes (e) mitosis. Unlike most eukaryotes the nuclear envelope does not break down during mitosis. After mitosis is completed the two cells separate, and each (a,f) enters the next cell cycle.

In summary, the most simply structured of eukaryotes, exemplified by the yeast cell, has much greater structural complexity than bacterial cells in the form of more DNA, multiple chromosomes, a nucleolus, a nuclear envelope, and a variety of cytoplasmic organelles.

Most yeast species grow in simple media, requiring no more than inorganic ions, a carbohydrate such as lucose, and a few vitamins. In such a medium yeast ells double in number every few hours. Some species eproduce by **fission,** a process whereby a cell splits into wo. Budding yeasts like *Saccharomyces* reproduce by orming a bud, which is an unusual mode of cell reprouction among eukaryotes (most divide by fission). The rocess begins with the formation of a small bud at the ell surface (Figure 1-10). The bud grows into an incipent daughter cell, and subsequently the nucleus in the arent cell divides by mitosis; the daughter cell receives ne of the daughter nuclei while the parent cell retains he other. After mitosis the bridge connecting the nother and daughter cells is closed off and the cells eparate from each other.

It might seem more logical for cell biologists to ave begun the study of the eukaryotic cell by a thorugh analysis of the properties and behavior of one of he simple cell types, such as yeast. This was not done artly because the relative simplicity and experimental dvantages of yeasts were not widely recognized until ecent years. Furthermore, the importance of studying he simplest available eukaryote was perhaps not adequately appreciated. Also, in other kinds of eukaryotic ells some of the cytoplasmic organelles are more abundant and therefore more immediately amenable to experimental manipulation. Now that molecular biology as begun to reveal a uniformity in the basic molecular rganization and operation of all eukaryotes, the logic of intense study on yeasts has become more compelling. Research on this cell type, particularly on the genetic basis of cell properties, is rapidly expanding.

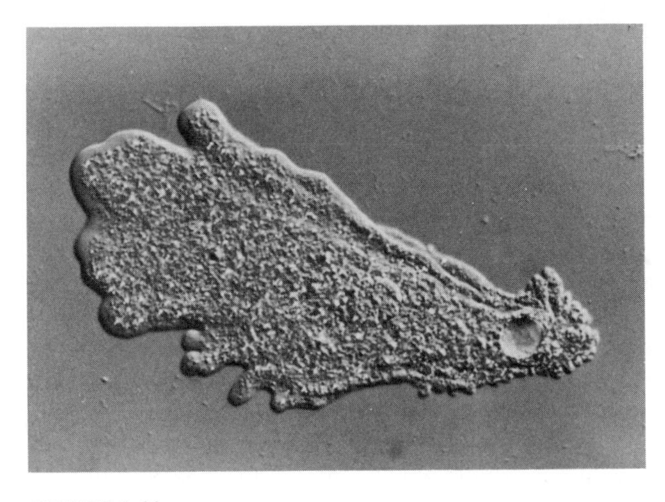

FIGURE 1-11
Light micrograph of a living amoeba (*Amoeba proteus*). The amoeba is moving from right to left. The clear sphere in the tail is the contractile vacuole.

Amoeba

The large, freshwater amoebae present a sharp contrast to the yeast cell. Amoebae are not only among the largest of eukaryotic cells, they are also among the most complicated. As the following brief description indicates, the common idea that the amoeba is a simple primitive cell is incorrect. The amoeba most studied by cell biologists is *Amoeba proteus.* (*Proteus* comes from the Greek word *proteios* meaning first rank; the word *protein* has the same origin.) This cell has a length of

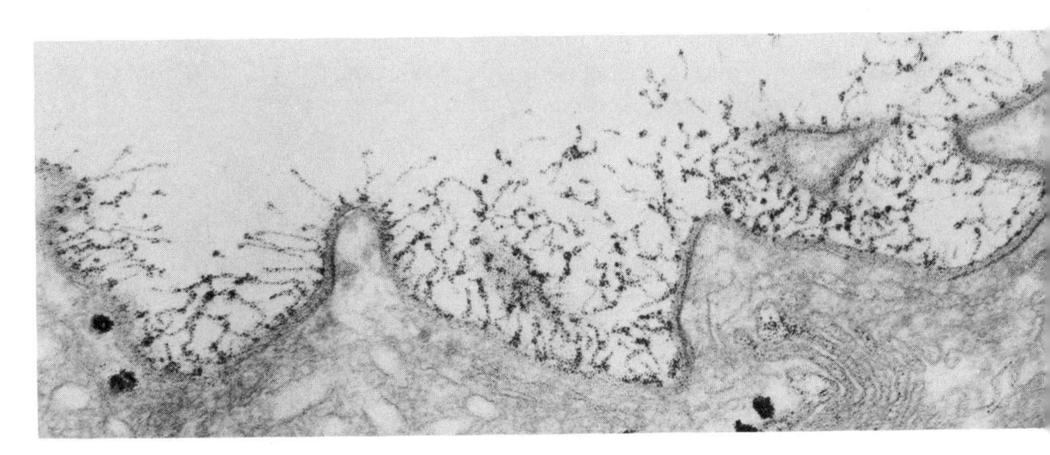

FIGURE 1-12
Electron micrograph of the surface coat on the plasma membrane of *Amoeba proteus.* [Courtesy of Charles J. Flickinger and The Company of Biologists Limited.]

one millimeter or more when stretched out into its streaming form (Figure 1-11), and has a volume 5000 times greater than the volume of a single yeast cell and 300,000 times that of *E. coli.*

The amoeba has no rigid cell wall, but its plasma membrane possesses a surface coat. Most if not all cells that lack a cell wall have a surface coat consisting of carbohydrate and protein molecules attached to the outside surface of the plasma membrane. Because of its thinness, the surface coat can be seen only with the electron microscope (Figure 1-12). The function of surface coats in eukaryotes is only partially understood. It probably contributes mechanical strength without counteracting the flexibility that the cell surface must have for amoeboid movement. In amoebae, it may also have a role in recognizing food organisms that are ingested by phagocytosis. In other kinds of eukaryotes, molecules of the surface coat serve as receptors for a wide variety of chemical signals from outside the cell.

The nucleus of the amoeba is extraordinarily large (an ellipsoid 30 μm in diameter) and contains dozens of nucleoli arranged in a layer immediately inside the nuclear envelope (Figure 1-13). A large number of nucleoli is not uncommon among protozoa, but in plant and animal cells there are usually only two to several nucleoli. The nucleus of the amoeba contains several hundred small chromosomes. The nuclear envelope is more elaborate than in yeast. On the inside of the envelope is a well developed honeycomb structure whose function is not known (Figure 1-13). This structure is called the **nuclear lamina,** a structure that is present in a less extensive and conspicuous form in other eukaryotic cell types. The lamina is a framework of proteins that stabilizes the nuclear envelope and serves for attachment of chromosomes. At the base of each honey-

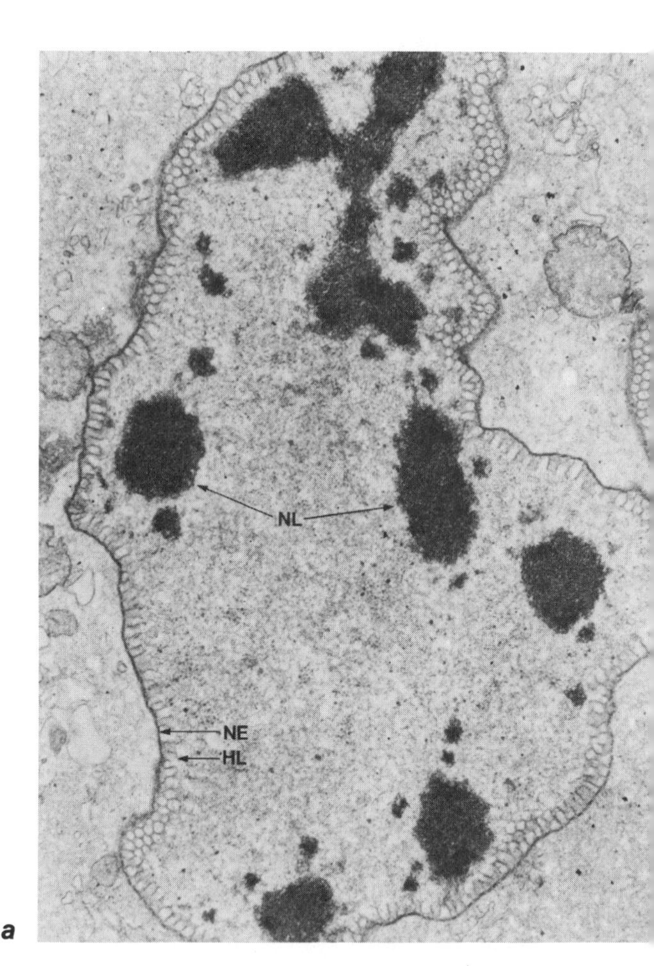

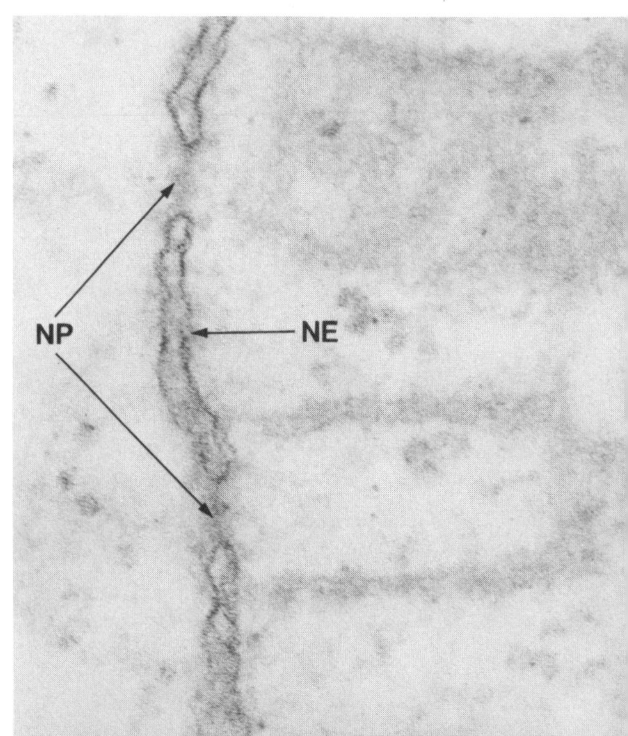

FIGURE 1-13
(a) Electron micrograph of *Amoeba proteus* showing the nucleus and part of the cytoplasm. The inner side of the nuclear envelope (NE) is covered by a honeycomb layer (HL) (nuclear lamina). The dark bodies in the nucleus are nucleoli (NL). (b) A part of the nuclear envelope at a higher magnification than in (a), showing the two membranes of the nuclear envelope (NE) and nuclear pores (NP). [Courtesy of Charles J. Flickinger. (a) From *J. Ultrastructure Res.*, 23:260–271 (1968).]

:omb cavity of the amoeba nucleus is a pore in the ıuclear envelope (Figure 1-13). As in the yeast cell, hese pores are a route for the movement of materials ɔetween nucleus and cytoplasm. For example, subunits ɔf ribosomal particles, which are produced in the nucleoli, move to the cytoplasm by passing through the pores in the envelope.

The cytoplasm of amoebae appears structurally more complex than that of the yeast cell. In addition to ribosomes and mitochondria a large number of membrane-bound vesicles collectively make up the **endoplasmic reticulum**, or ER (Figure 1-14). The endoplasmic reticulum is also present in the yeast cytoplasm but is far less conspicuous. The function of the endoplasmic reticulum will be discussed more extensively

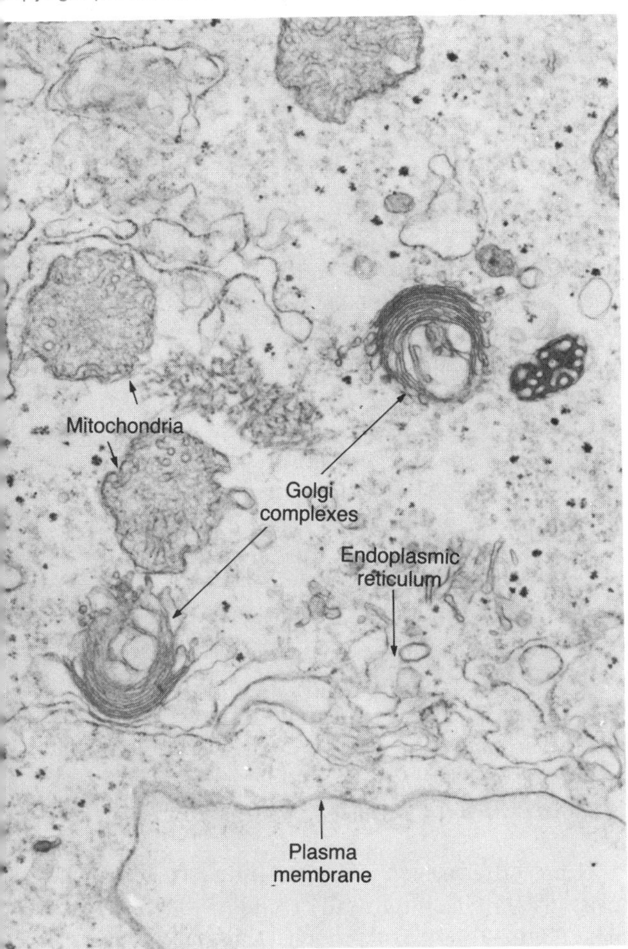

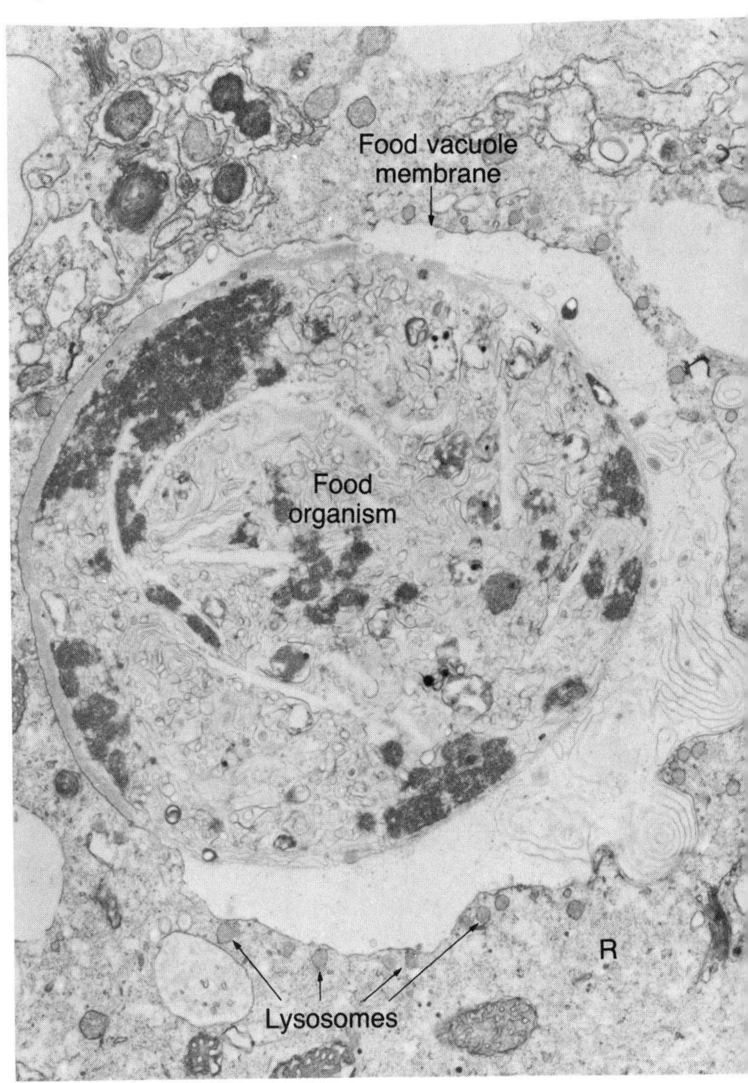

ꟳIGURE 1-14
:lectron micrographs of the cytoplasm of *Amoeba proteus*. a) Amoeba cytoplasm with mitochondria, Golgi :omplexes, endoplasmic reticulum, and plasma membrane. (b) Amoeba cytoplasm with lysosomes and a arge spherical food vacuole containing an unidentified ood organism. The granular material in the region labeled vith an R is ribosomes. [Courtesy of Charles J. Flickinger. *The Journal of Cell Biology*, 43:250–262 (1969) by :opyright permission of the Rockefeller University Press.]

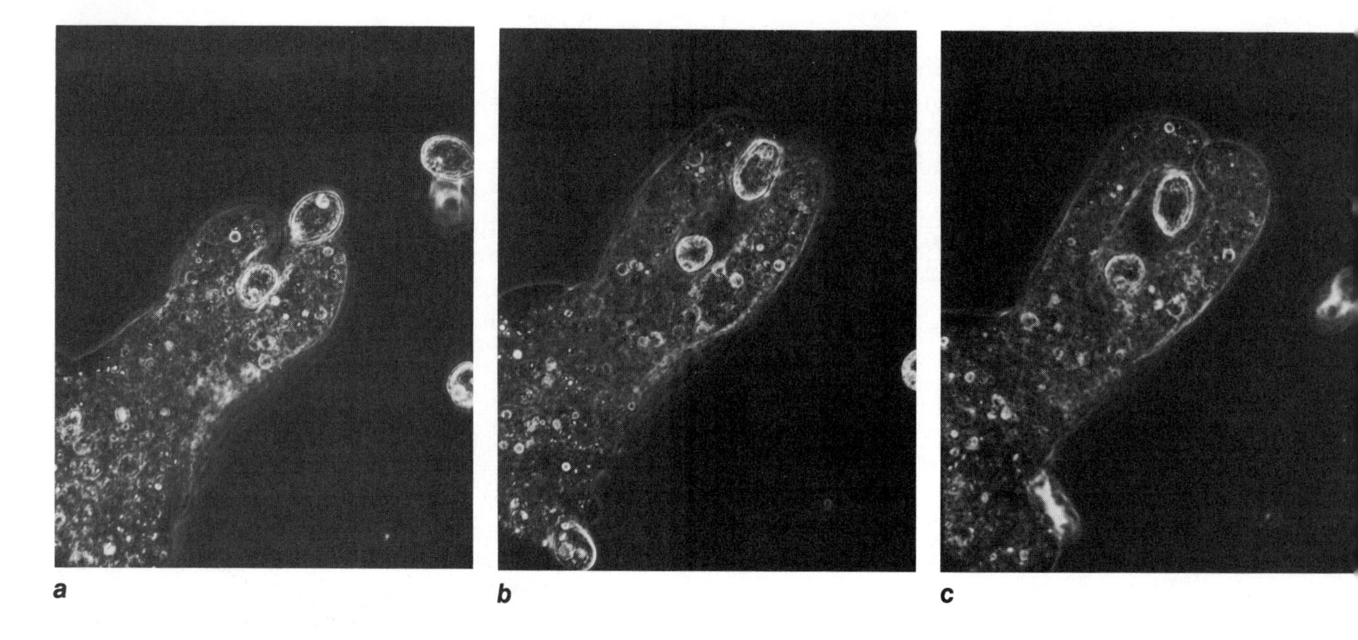

FIGURE 1-15
Amoeba proteus capturing two ciliated protozoans (*Tetrahymena*) by phagocytosis. (a) Formation of a phagocytic invagination containing one ciliate. (b) Extension of the pseudopod to engulf a second ciliate. (c) Closing of the invagination to make a food vacuole containing both ciliates. Fusion of the membrane will close off the vacuole. [Courtesy of O. L. Miller, Jr.]

when animal cells are considered later in this chapter. The endoplasmic reticulum is involved in the synthesis of lipids and proteins, some of which are destined for transport to the cell surface, where they become part of the cell's surface coat.

The cytoplasm of amoebae contains several **Golgi complexes.** The Golgi complex, which is an aggregate of membranes (Figure 1-14), functions in processing, concentrating, and packaging into vesicles the macromolecules synthesized in the endoplasmic reticulum. These macromolecules will become part of the surface coat or be exported from the cell. In the case of amoebae, the surface coat must be continually renewed because of turnover of the surface that occurs during amoeboid movement and **phagocytosis.** Phagocytosis (derived from the Greek words *phagein* meaning to eat and *cytosis* meaning cell) is the process by which cells engulf particulate material (Figure 1-15). Amoebae acquire nutrients by ingesting organisms such as bacteria, algae, and other protozoa, which they then digest. In phagocytosis a portion of the plasma membrane surrounds the food organism to form a **food vacuole** within the cytoplasm (Figure 1-16).

Like the cytoplasm of most eukaryotic cells, the cytoplasm of amoeba contains numerous lysosomes. As in other cells, the lysosomal enzymes are released if an amoeba dies, and self-digestion, or autolysis, results. The lysosomes have another important function in amoebae and in other cells that ingest particulate matter. After ingestion of organisms by phagocytosis, lysosomes fuse with the newly formed food vacuole (Figure 1-16). In effect, they empty their enzymes into the vacuole, and the ingested organisms are digested. The products of digestion (amino acids, nucleosides, sugars, fatty acids, etc.) pass through the membrane of the food vacuole into the cytoplasm and are then used by the amoeba.

Since the amoeba obtains nutrients by phagocytosis, its cytoplasm naturally contains many food vacuoles (Figure 1-14). The amoeba is apparently extremely

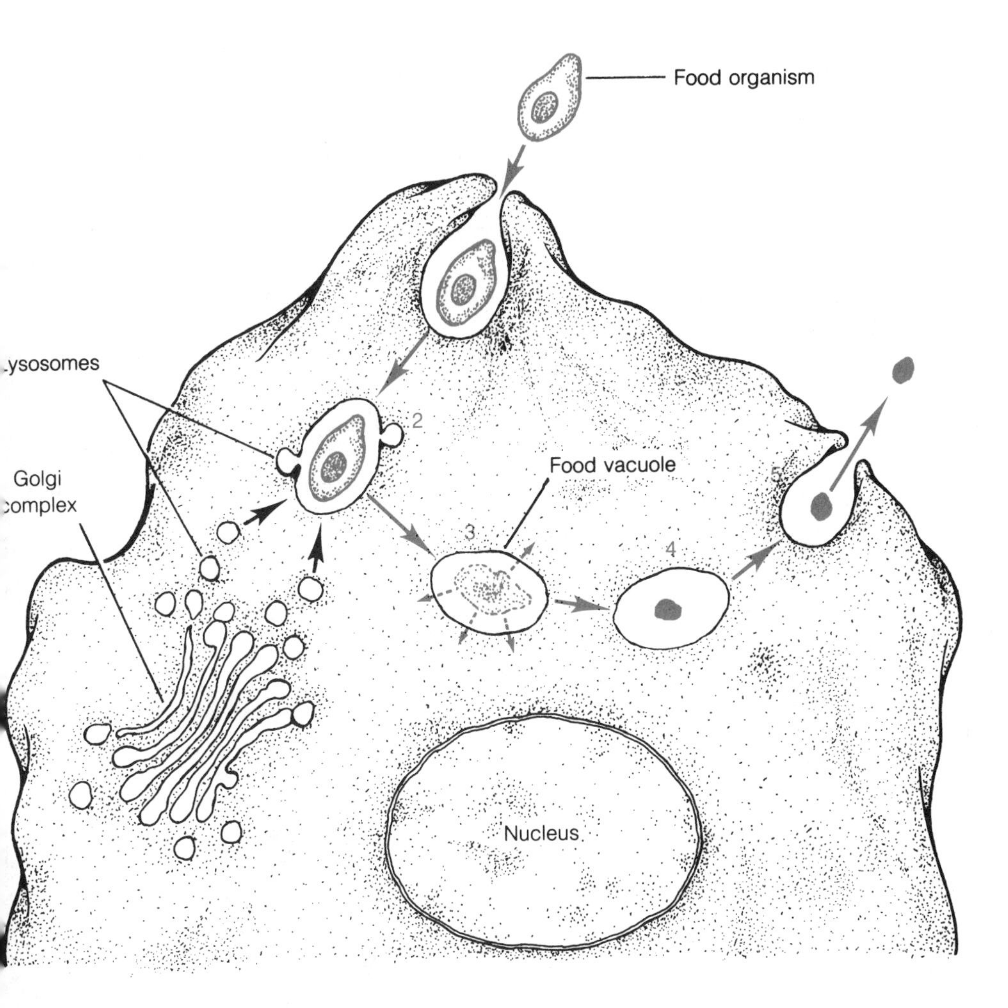

FIGURE 1-16 Phagocytosis of a ciliated protozoan by an amoeba. (1) Capture of a food organism and formation of a food vacuole; (2) Fusion of lysosomes with the food vacuole; (3) Digestion of the food organism and passage of digestion products into the surrounding cytoplasm; (4) Food vacuole with an undigestable remnant of the food organism; (5) Fusion of the food vacuole with the plasma membrane and expulsion of the remnant of the food organism.

fastidious in its nutritional requirements since the cell can survive only by ingestion of other *living* organisms. Attempts to grow amoebae in complex nutrient solutions or even on dead organisms have not been successful.

A cytoplasmic organelle in amoebae that has been relatively little studied is the **contractile vacuole** (Figure 1-17). A primary function of this organelle is the excretion of water. Contractile vacuoles are found primarily among protozoa living in fresh water. In an amoeba, the vacuole begins as a collection of tiny vesicles that arise in a group. These vesicles grow and coalesce to form a single, large, membrane-bound vacuole. The membrane of the vacuole then fuses with the plasma membrane, creating an opening to the outside through which the contents of the vacuole are discharged. The vacuole collapses and disappears, and a new vacuole begins to form amidst the remnants of the old vacuole. The contractile vacuole forms and empties every few minutes, pumping out excess water that has entered the cell across the plasma membrane. How water is pumped into the vacuole during its expansion is not known, but mitochondria are believed to participate because they aggregate on the surface of the vacuole, perhaps providing the necessary chemical energy for pumping water into the vacuole.

Like other eukaryotes, the cytoplasm of amoebae contains fibrous elements known as microtubules and

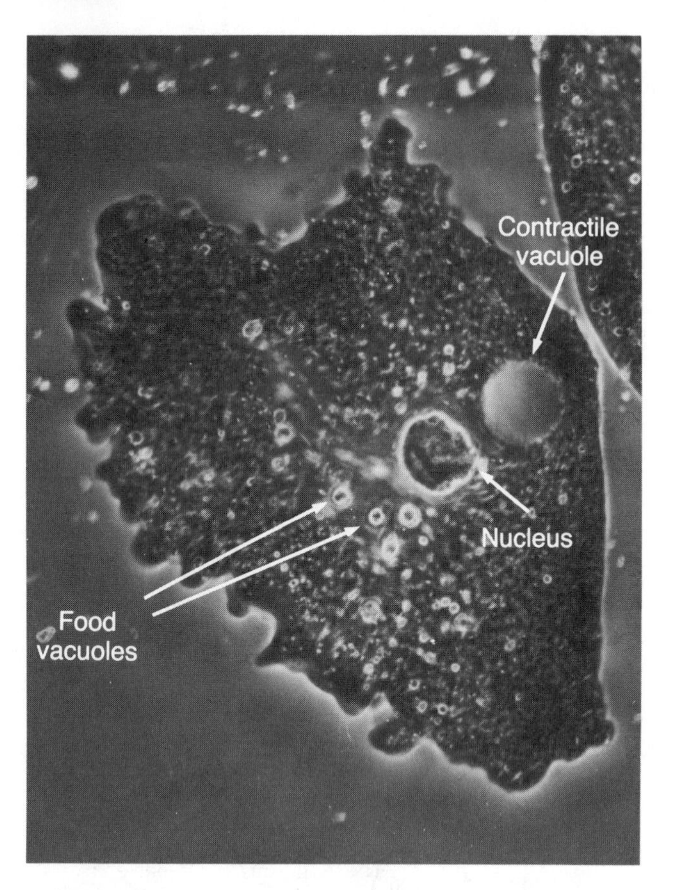

FIGURE 1-17
A light microscope photograph of *Amoeba proteus* showing the contractile vacuole. Ten seconds after the picture was taken the contractile vacuole moved to the cell surface and emptied its contents outside the cell.

microfilaments. These are not easily observed in amoeba and will be discussed later in this chapter in animal cells, where they are more conspicuous.

The amoeba has been useful in studies of the interaction between the nucleus and the cytoplasm because the cell is large enough for microsurgical operations. The amoeba was one of the first cell types to be cut into two parts with a fine needle, one part containing the nucleus and the other containing only cytoplasm (Figure 1-18). The nucleated half-cell regains its lost cytoplasm by growth and subsequently resumes proliferation. The enucleated half-cell lives for two to three weeks, during which time its metabolic activities decline steadily. The enucleated half-cell is unable to maintain its structural integrity and eventually disintegrates. By comparing the activities of nucleated and enucleated half-cells, cell biologists first learned about the roles of the nucleus and cytoplasm in various cell functions.

The amoeba is one of the few types of cells in which a nucleus can be transplanted from one cell to another. Essentially, the operation consists of inserting an extremely fine glass needle into an amoeba and pushing the nucleus through the plasma membrane into an adjacent cell (Figure 1-19). The needle must be moved by an instrument called a **micromanipulator** because even the surest hand is by itself too unsteady. The micromanipulator transmits movements of the hand to a tool like a microneedle but reduces the magnitude of the hand movement. For example, movement of one centimeter of the hand in a given direction is

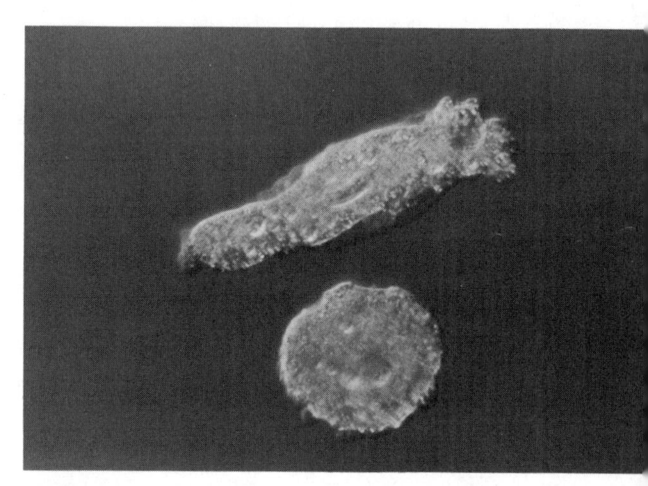

FIGURE 1-18
Light micrograph of living nucleated and enucleated halves of an amoeba (*Amoeba proteus*). The cell has just been cut with a microneedle. The nucleated half-cell has elongated and begun to move. The ability of the enucleated half-cell to move is impaired, and it has become spherical.

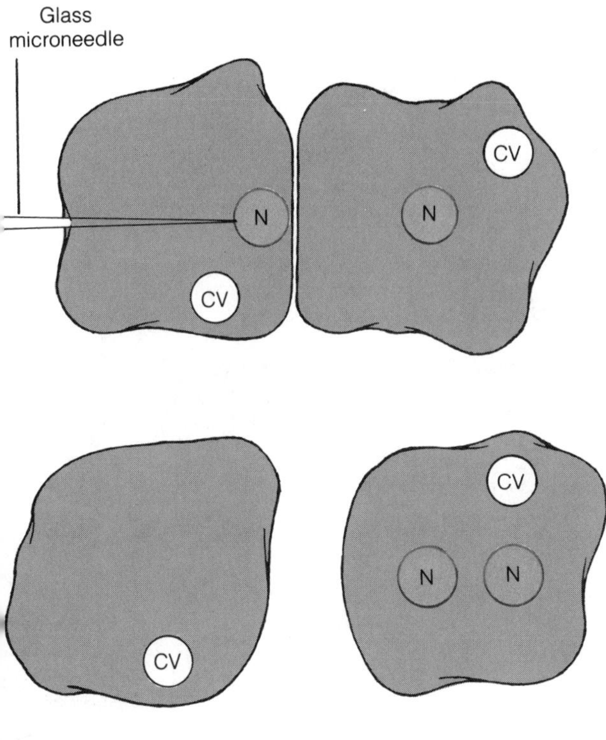

IGURE 1-19
uclear transplantation in amoebae. A glass microneedle ontrolled by a micromanipulator is inserted into an amoeba and sed to push the nucleus through the plasma membranes into an nmediately adjacent cell. Exposure of the nucleus to the urrounding medium for even a few seconds causes irreversible ıjury to the nucleus. The products of this particular experiment re an enucleated cell and a binucleated cell. CV = contractile acuole.

educed to movement of a few micrometers (10^{-4}cm) f the microneedle. Nuclear transplantation experi- nents in amoebae provided the first evidence that cer- ain kinds of proteins constantly migrate back and forth etween the nucleus and cytoplasm.

Plant Cells

Most plant cells are enveloped by a thick, rigid wall hat gives the cell its shape and protects it both from nechanical injury and from osmotic swelling (Figure -20). The wall is made up of cellulose fibers embedded in a matrix of other polysaccharides and proteins. Materials that make up the cell wall are synthesized inside the cell, then packaged just inside the plasma membrane, released from the cell, and deposited in the cell wall.

Plant cells, like all eukaryotic cells, contain a nucleus enclosed by a nuclear envelope. The cytoplasm contains ribosomes, mitochondria, a Golgi complex, membranes of the endoplasmic reticulum, microfilaments, microtubules, and several other structures.

Two additional structures present in plant cells are the **chloroplasts** and the **central vacuole** (Figure 1-20). The chloroplast is a membranous organelle that, like the mitochondrion, has a surrounding outer membrane and a closely apposed inner membrane. The space enclosed by the inner membrane contains a complex solution of enzymes and stacks or layers of additional membranes that largely fill the organelle (Figure 1-20). Associated with these layers of internal membranes are the molecules of chlorophyll, which absorb energy in the form of light. Also contained in the membranes is the enzymatic machinery for converting the energy of light absorbed by chlorophyll into chemical energy. A favorite plant for the study of photosynthesis is spinach, whose leaves are particularly rich in chloroplasts. Typically there are 20 to 30 chloroplasts per cell in a plant leaf, although in unicellular algae the number varies from one to several thousand depending on the species examined.

Like mitochondria, chloroplasts contain DNA and ribosomes. The DNA in chloroplasts contains information for the synthesis of ribosomal RNA and some of the proteins needed for chloroplast function and structure. The many other proteins present in the chloroplast are coded for by DNA in the cell nucleus. Through interactions with the cytoplasm and nucleus that are not yet well understood, the function, growth, and reproduction of chloroplasts are closely coordinated with the activities and growth of the cell.

The central vacuole of the cytoplasm in a plant cell (Figure 1-20) contains water, sugars, organic acids, proteins, inorganic ions, and pigments. Many of these are waste and storage products that accumulate during the life of the cell.

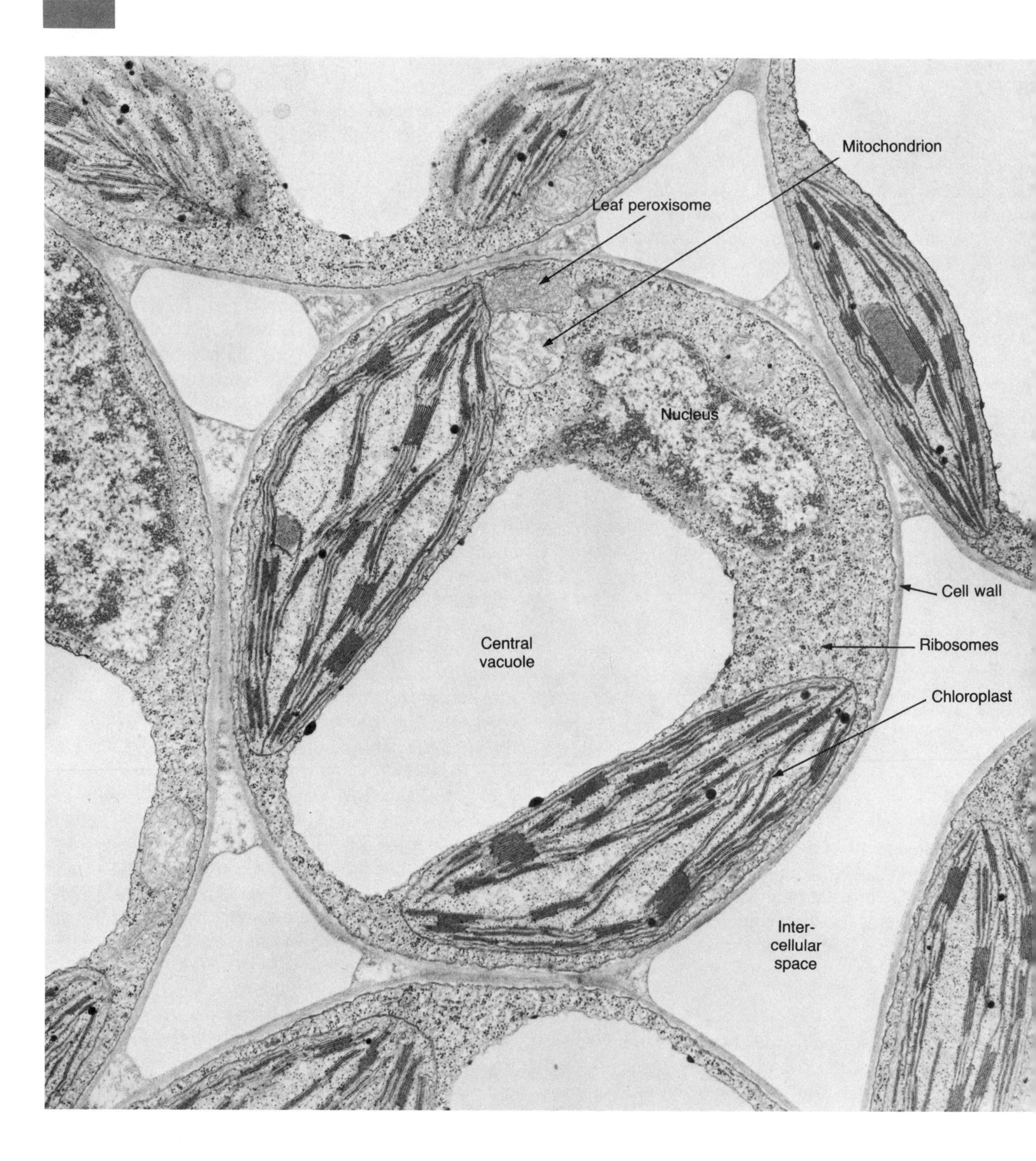

Mitochondrion
Leaf peroxisome
Nucleus
Cell wall
Central vacuole
Ribosomes
Chloroplast
Inter-cellular space

The nuclei of plant cells (Figure 1-20) closely resemble the nuclei of other kinds of eukaryotes. Nuclei contain the chromosomes and one or more nucleoli and are enclosed by an envelope penetrated by numerous pores. As in other cells, the nucleus produces all the RNA present in the cytoplasm, except for the small amount of RNA produced in mitochondria and chloroplasts. These cytoplasmic organelles account for less than one percent of the total RNA synthesis of the cell.

Although individual cells from a multicellular plant can now be cultured and grown in a nutrient medium of known constitution, they have been used less often than animal cells in research (except in the study of photosynthesis). Research on plant cells has recently become intense, particularly because of the potential for genetic engineering to introduce major new properties into plants of enormous economic importance. For molecular analyses of photosynthesis and other activities, plant cells are usually broken open by mechanical means, such as grinding or agitation in a kitchen blender; the chloroplasts or other organelles are then separated from other cell constituents by sedimentation in a centrifuge, described later in this chapter.

Plant and algal cells (and the photosynthetic bacteria) are uniquely important, because almost all the energy needed to sustain life on Earth is derived from the photosynthetic capture of radiant energy from the sun and because these photosynthetic organisms produce much of the oxygen in the atmosphere.

Animal Cells

Complex animals such as mammals are composed of about two hundred different types of cells. Each type has become specialized during embryological development to perform a particular function in one of the many tissues of the animal. These specialized differentiated cells include blood cells, liver cells, kidney cells, skin cells, bone cells, germ cells, and many others. Some of these cells can grow and divide in culture vessels if they are provided with a complex nutrient medium containing amino acids, inorganic ions, sugars, vitamins, and many other substances (Figure 1-21).

As in amoebae, the surface of the plasma membrane of cultured animal cells is coated with a thin layer of **glycoproteins**—proteins with various kinds and numbers of sugar molecules attached to them. One function of the surface is to provide the means by which cells can adhere to one another in tissues. As in other cells, the plasma membrane consists of a lipid bilayer with attached proteins (Chapter 5). Many of the membrane-associated proteins function in the absorption of nutrients from the medium, and others serve as receptors for signals such as hormones.

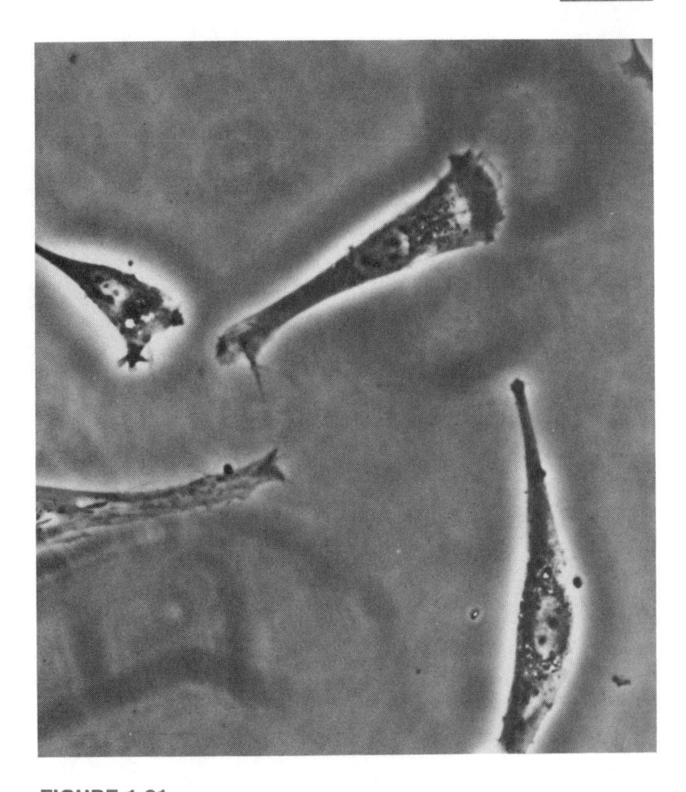

FIGURE 1-21
Light microscope photograph of hamster cells in culture. The dark spots in the nuclei are nucleoli.

FIGURE 1-20
Electron micrograph of a cell in a leaf of *Coleus blumei*. The principal structures visible are the cell wall, chloroplasts, central vacuole, nucleus, mitochondrion, a peroxisome, and ribosomes. [Micrograph by W. P. Wergin, courtesy of Eldon H. Newcomb.]

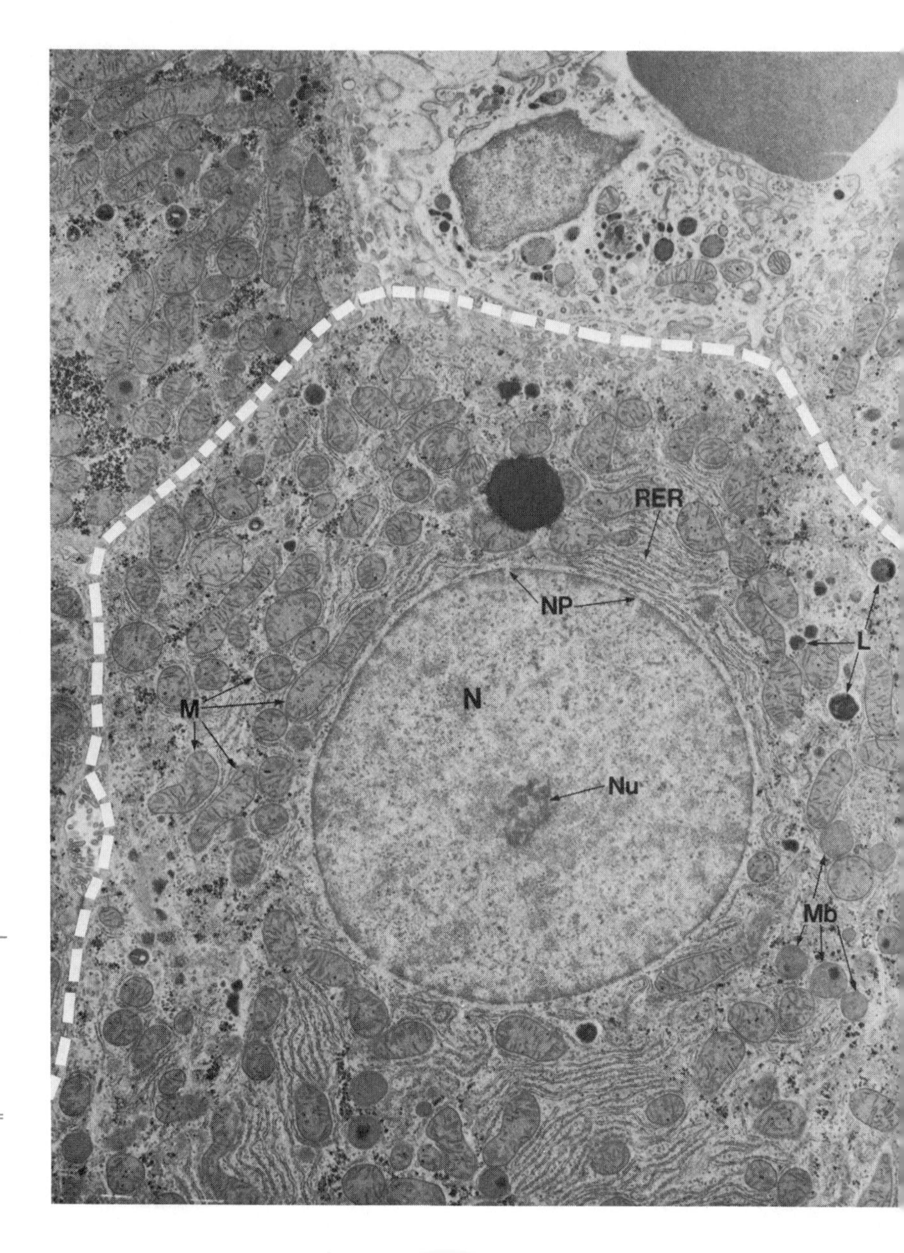

FIGURE 1-22
Electron micrograph of a rat liver cell showing some of the structures generally present in eukaryotic cells. The dashed line indicates the general outline of the cell. N = nucleus, Nu = nucleolus, NP = nuclear pores in the nuclear envelope, RER = rough endoplasmic reticulum, M = mitochondria, Mb = microbodies, and L = lysosomes. [Courtesy of Keith R. Porter.]

The cytoplasm of animal cells (Figure 1-22) contains ribosomes, endoplasmic reticulum, Golgi complexes, mitochondria, lysosomes, lipid droplets, and five structures generally present in eukaryotic cells but not yet described: microtubules, microfilaments, intermediate filaments, microbodies, and centrioles.

The two fiber structures, **microtubules** and **microfilaments,** that generally occur in eukaryotes are particularly prominent in some kinds of protozoa and in animal cells. Microtubules are long, hollow structures with a diameter of 30 nm and a wall thickness of about 8 nm (Figure 1-23). Microtubules are assembled

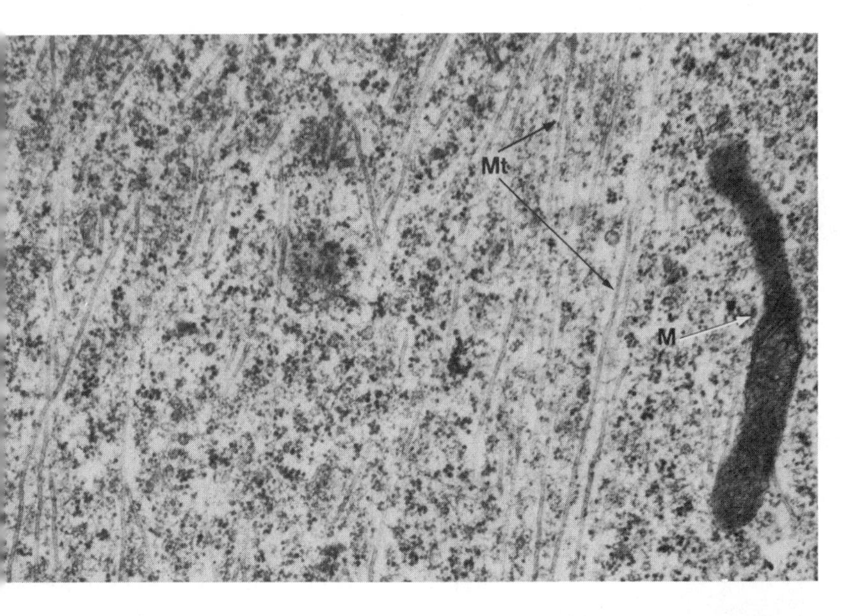

FIGURE 1-23
Electron micrograph of part of an animal cell showing microtubules (Mt). M = mitochondrion. [Courtesy of J. Richard McIntosh.]

rom subunit protein molecules called **tubulin.** Micro-ubules occur in bundles or singly and impart stiffness nd form to a cell. They thus provide a structural ramework or **cytoskeleton.** For example, long exten-ions of cytoplasm, such as the **axons** of nerve cells, ontain prominent longitudinal bundles of microtu-ules (Figure 1-24). Microtubules are also the main tructural element in cilia and flagella. Except for those n cilia and flagella, microtubules are labile structures hat disassemble into tubulin subunits and reassemble nto new microtubules during changes in cell shape. A articularly noteworthy example of this process of as-embly and disassembly occurs during cell division. The **mitotic apparatus,** which serves to distribute the hromosomes to the daughter cells, consists in large art of microtubules that assemble from tubulin sub-nits and disassemble once mitosis is complete.

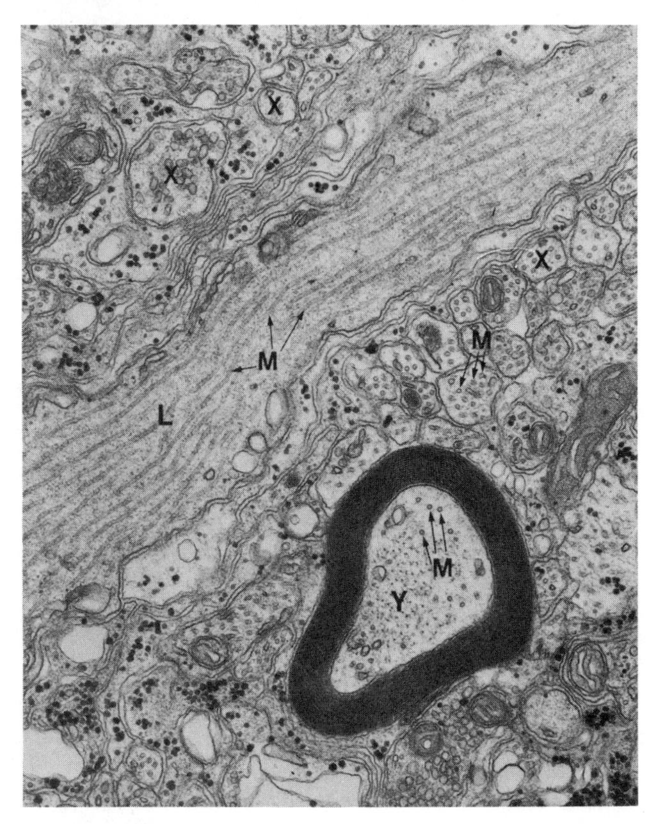

FIGURE 1-24
Electron micrograph of frog brain. Microtubules (M) cut in cross section are visible in the many small nerve processes (X) and in a large nerve process (Y). The nerve process cut in longitudinal section (L) contains many parallel microtubules. [Courtesy of Keith R. Porter.]

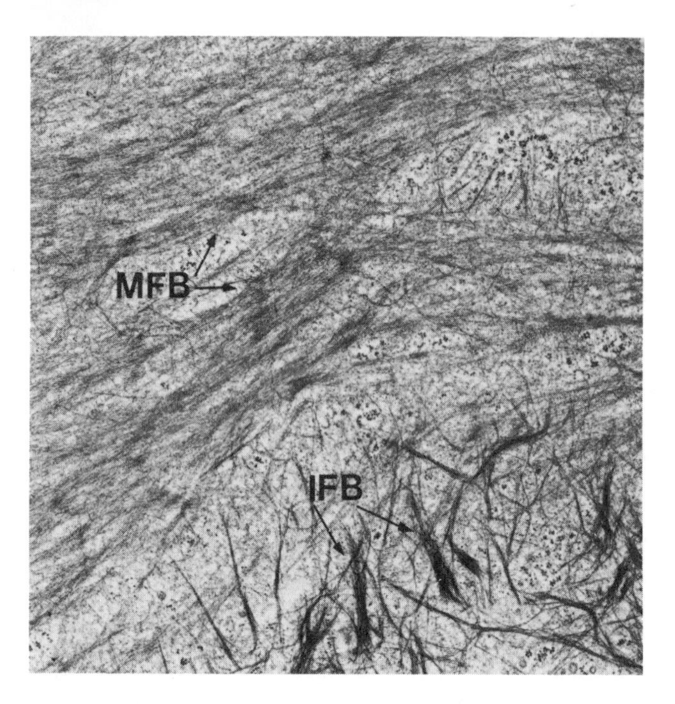

FIGURE 1-25
Electron micrograph of a portion of a mouse skin cell in culture with many bundles of microfilaments (MFB) and intermediate filaments (IFB). [Courtesy of Kathleen Green and Robert D. Goldman.]

Another fibrous element found in all eukaryoti cells is the microfilament. These are long, thin fibers 9.5 nm in diameter, made of proteins. Microfilament are often conspicuously present in bundles in anima cells in culture (Figure 1-25). Individual microfila ments are sometimes anchored to the plasma mem brane, either at their ends or along their lengths. The are made largely of **actin,** one of the major proteins i the contractile apparatus of muscle cells. Actin is one o the most abundant proteins in eukaryotic cells. In non muscle cells actin microfilaments are important i amoeboid movement and other kinds of cell move ment.

A third type of fiber structure is the **intermediat filament.** These are present in most if not all euka ryotes but are particularly numerous in certain kinds o animal cells (Figure 1-26). A variety of kinds of inter mediate filaments has been found, each composed o a different kind of protein subunit, and often severa kinds of proteins in a single filament. For example, ski cells contain intermediate filaments composed of pro teins called keratins; their presence in large number gives skin cells their special mechanical toughness Nerve cells are rich in intermediate filaments calle

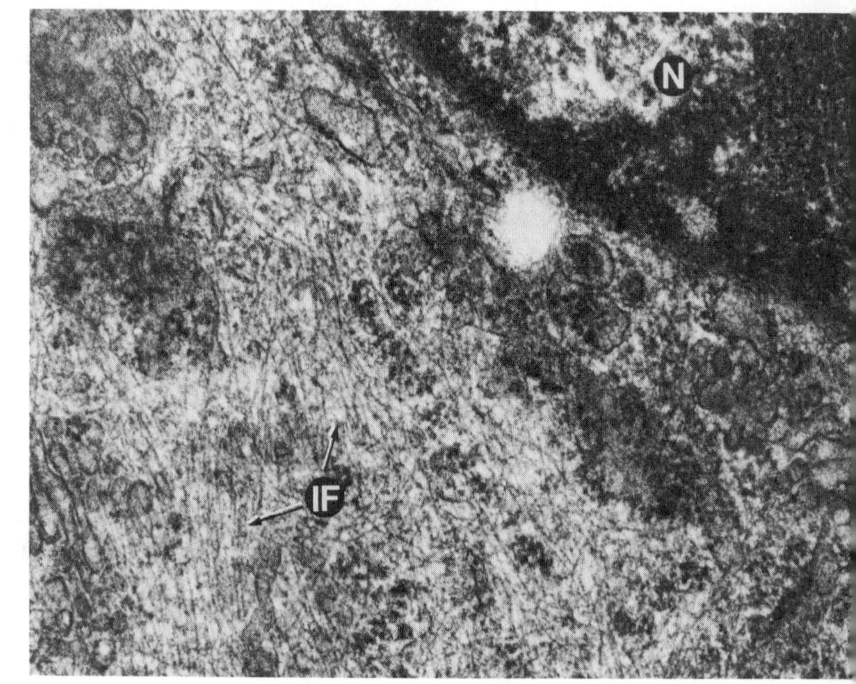

FIGURE 1-26
Electron micrograph of a portion of a mouse skin cell with many intermediate filaments (IF). N = nucleus. [Courtesy of Kathleen Green and Robert D. Goldman.]

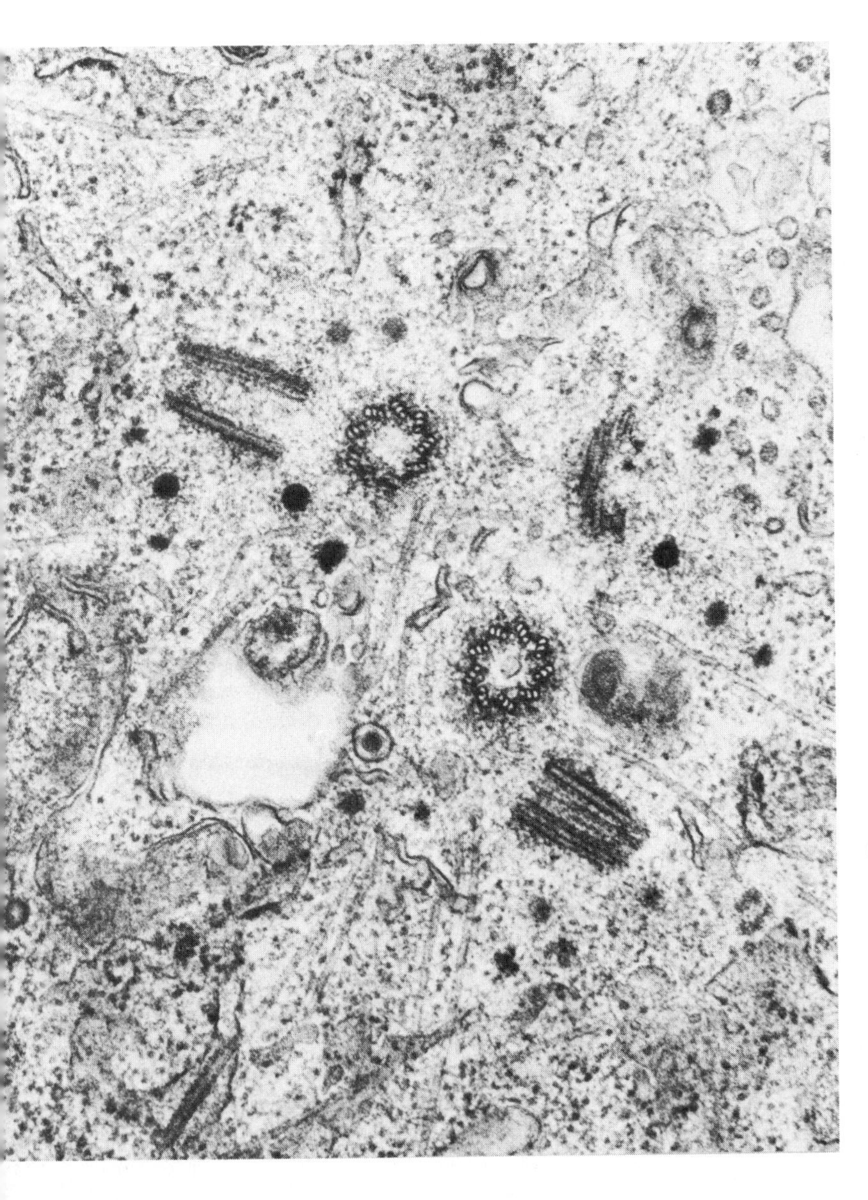

FIGURE 1-27
Electron micrograph of two pairs of centrioles in a cultured hamster cell. Two pairs are present because the centriole has duplicated during preparations for mitosis. One centriole in each pair is cut in cross section and shows the nine triplets of microtubules. The other centriole is cut in longitudinal section. [Courtesy of B. R. Brinkley. From M. McGill, D. P. Highfield, T. M. Monahan, and B. R. Brinkley. 1976. *J. Ultrastructural Research*, 57: 43.]

neurofilaments composed of several other kinds of proteins. Muscle cells contain their own kind of intermediate filaments. In every case, however, the role of intermediate filaments seems to be a passive, structural one; they form mechanical connections between different parts of the cytoplasm.

Microtubules, microfilaments, and intermediate filaments are the major elements making up a fibrous network in the cytoplasm of eukaryotic cells known as the cytoskeleton. The cytoskeleton, which has the major role in giving a cell its particular shape, is discussed in Chapter 6. In Chapter 14 the roles of microtubules and microfilaments in cell motility are discussed.

Centrioles are present in animal cells and in many kinds of unicellular eukaryotes (Figure 1-27). A centriole is a cylindrical structure about 0.15 μm in diameter and 0.3 to 0.5 μm long. It is made up of nine triplets of microtubules in a pinwheel-like arrangement that runs the length of the centriole (Figure 1-27). A matrix of unidentified, amorphous material surrounds the trip-

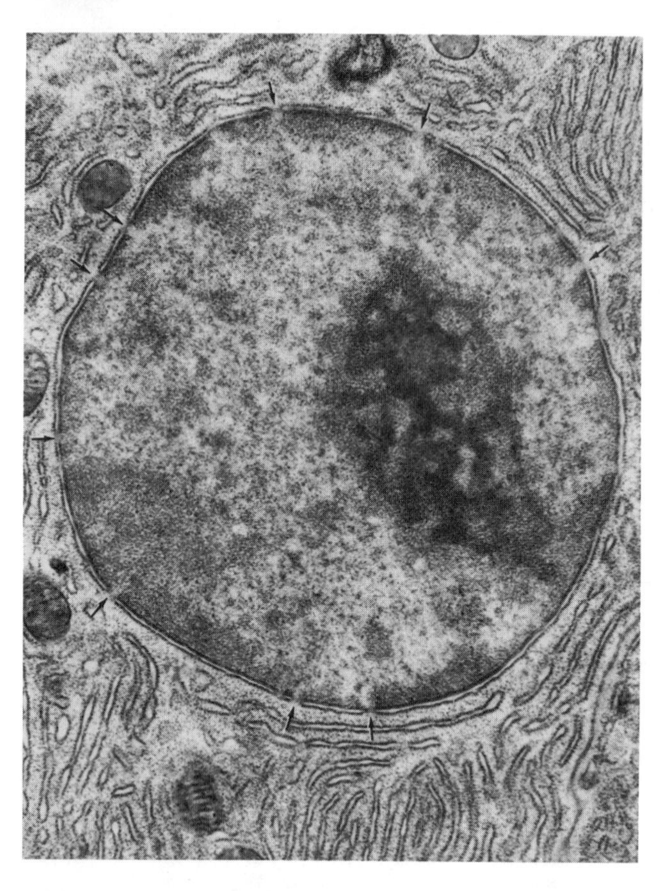

FIGURE 1-28
Electron micrograph of part of a pancreas cell. Nuclear pores are indicated by arrows. The nuclear envelope consists of two membranes. The dark mass in the nucleus is the nucleolus. The lighter areas of the nucleus contain chromosomal fibers. The cytoplasm is largely filled with endoplasmic reticulum with attached ribosomes. [Courtesy of Don W. Fawcett. From D. W. Fawcett (1981) *The Cell*, 2nd Ed. W. B. Saunders and Co.]

lets of microtubules. Centrioles almost always occur in pairs. The two centrioles in a pair lie with their long axes at right angles to each other (Figure 1-27). Centrioles have traditionally been believed to play a role in mitosis although their importance is uncertain, since many cells have no centrioles. For example, plant cells, amoebae, and many algae lack centrioles but nevertheless divide mitotically. In cells that are preparing to divide, two pairs of centrioles are present, and during cell division each daughter cell receives one pair.

Microbodies are small vesicles (usually 0.5 to 2 μm in diameter) composed of a granular matrix enclosed by a single membrane. Many different enzymes are present in microbodies. Most of these enzymes are involved in the oxidation of organic molecules, for example, amino acids and fats by molecular oxygen. The makeup of enzymes in microbodies varies somewhat from one type of cell to another, which has at times raised doubts about classifying them as a single entity. However, one common feature of microbodies is the production of hydrogen peroxide (H_2O_2). Because of the formation of hydrogen peroxide as a result of oxidative processes in microbodies, they came to be known as **peroxisomes.** Hydrogen peroxide is a strong oxidizing agent and potentially harmful to the cell, but it is rapidly destroyed by an enzyme **catalase** within the peroxisome to yield water and oxygen:

$$2H_2O_2 \rightarrow 2H_2O + O_2$$

In cells of plant seedlings rich in stored fat, for example peanut and cucumber seedlings, vesicles similar to microbodies and known as **glyoxysomes** convert fats to simpler molecules that can then be used in other parts of cell metabolism, including the generation of usable chemical energy.

The nucleus of an animal cell, like that in other eukaryotes, has a nuclear envelope with numerous pores, one or more nucleoli, and a network of chromosomal fibers (Figure 1-28). During the period between cell divisions, it is impossible to discern complete chromosomes in the nucleus because each chromosome is an extremely long, twisted thread of DNA and protein jumbled into a network with threads of other chromosomes. However, during mitosis the sinuous thread that comprises each chromosome condenses into a short, rod-shaped object, the mitotic chromosome (Figure 1-29).

Cells within a tissue of an animal usually retain a definite shape that results in part from structural properties acquired during differentiation and in part from physical constraints imposed by adjacent cells. Because such influences are greatly diminished when cells are

IGURE 1-29
icrograph of a portion of a whole, cultured cell of the rat angaroo taken with a high voltage electron microscope. The cell in mitosis (metaphase), and the chromosomes (dark, d-shaped structures) are attached to the center of the mitotic pindle. Dense arrays of microtubules, some in bundles, extend om each centrosome toward the center of the spindle. dditional microtubules radiate away from the spindle. [Courtesy J. Richard McIntosh.]

emoved from a tissue and grown in culture vessels, ultured cells are constantly moving and changing hape (Figure 1-30), though usually at rates too low to

IGURE 1-30
ight micrographs showing the movement of a mouse cell in ulture. Beginning at zero time in the micrograph in the upper eft, the time at which each successive micrograph was taken is narked on the pictures. The cell changes shape as it extends nd retracts ruffles and long filaments of cytoplasm. [Courtesy of Guenter Albrecht-Buehler.]

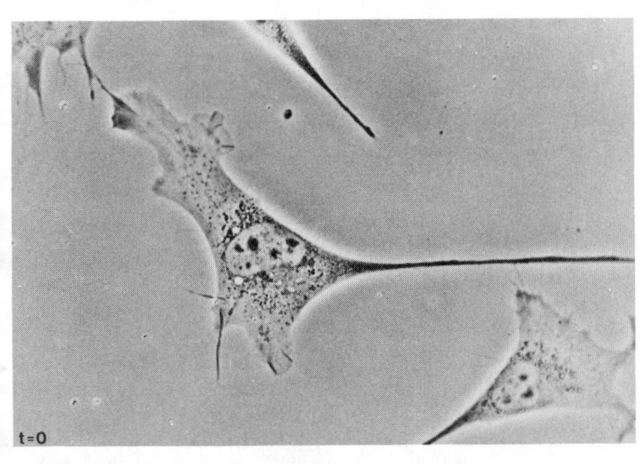

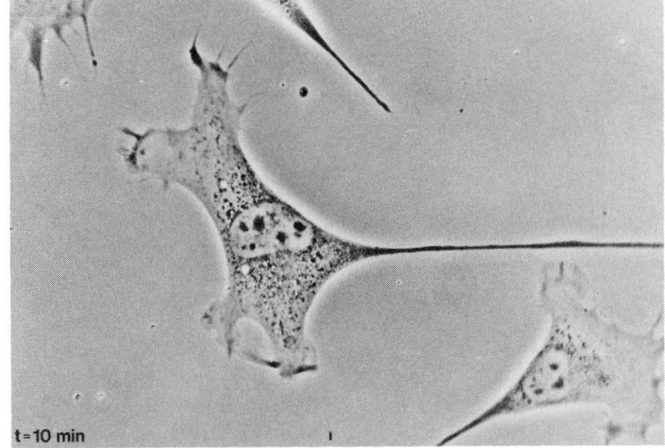

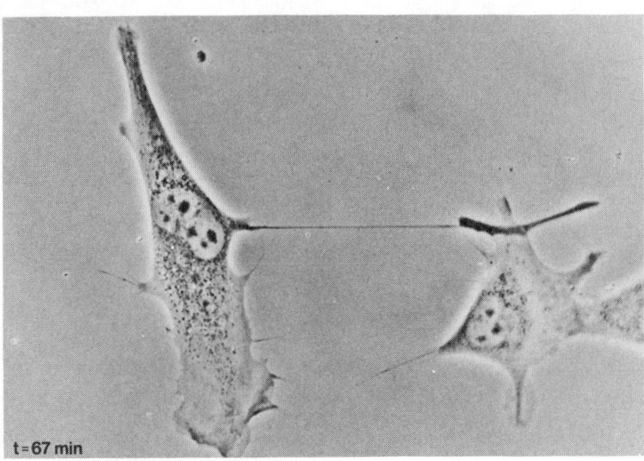

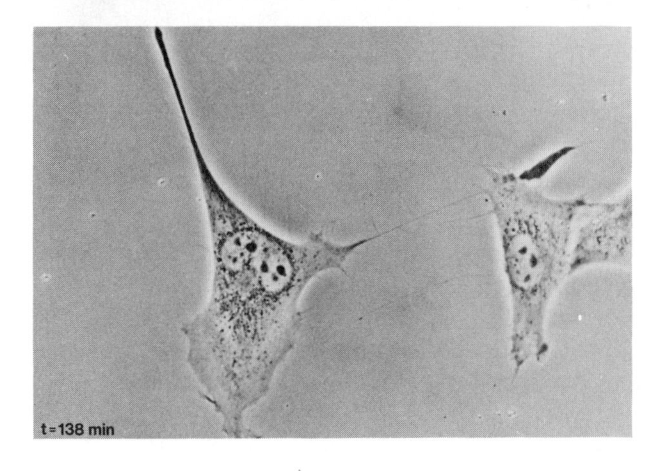

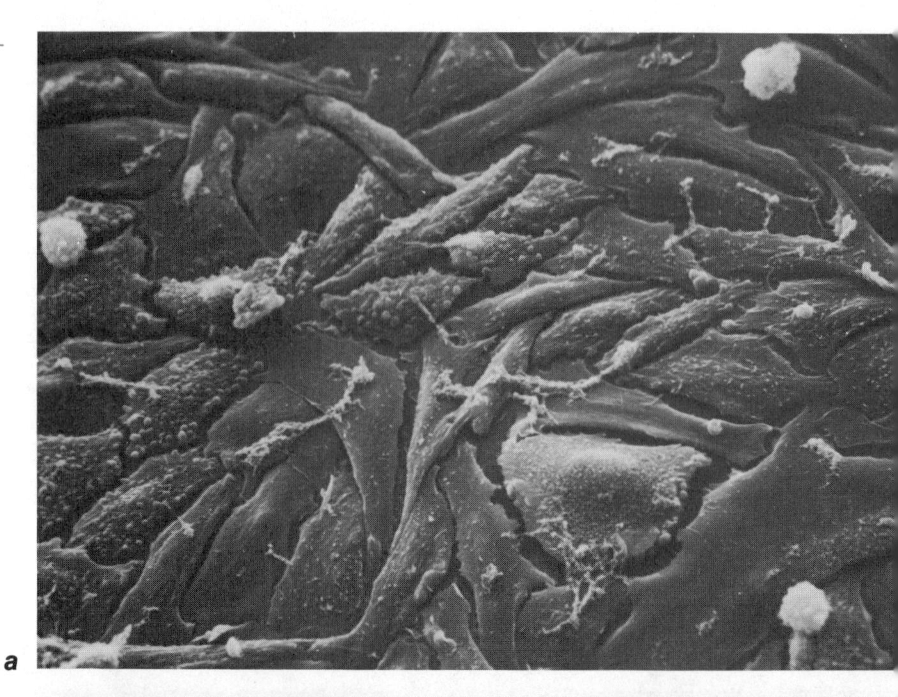

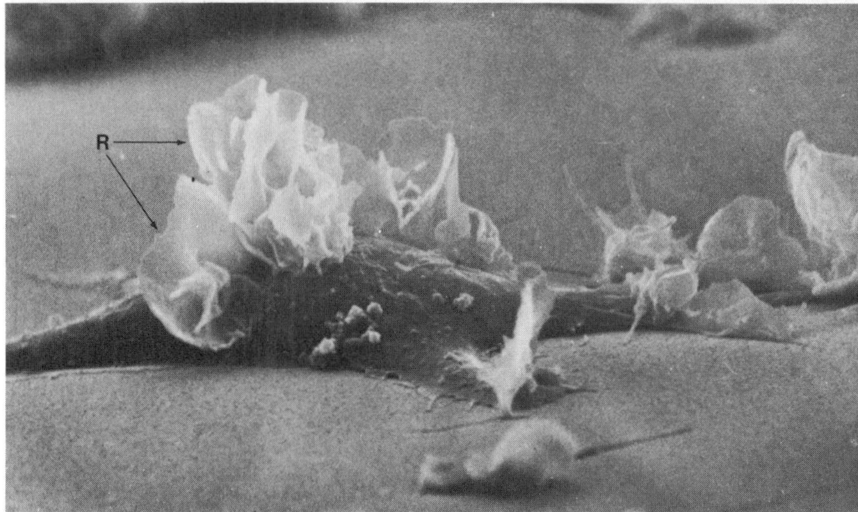

FIGURE 1-31
Scanning electron micrographs of animal cells. (a) Cells cultured from a hamster. The surfaces of some cells are covered with blebs and short filamentous projections. Some cells have smooth surfaces. Strands of proteins have precipitated from the medium onto the surface of a few cells. (b) A scavenger cell (macrophage) from a mouse. Large folds of cytoplasm (ruffles, R) extend from the cell surface. [Courtesy of Keith R. Porter.]

be detected by eye in a microscope. Movement is normally detected by time-lapse photography.

The electron microscope reveals that the overall form of an animal cell is considerably more complex than the light microscope would indicate. The view is particularly dramatic in a scanning electron microscope, which provides a high-resolution image of the surface of an object rather than showing its internal structure. A scanning electron microscope photograph of a cell in culture (Figure 1-31) reveals that the cell

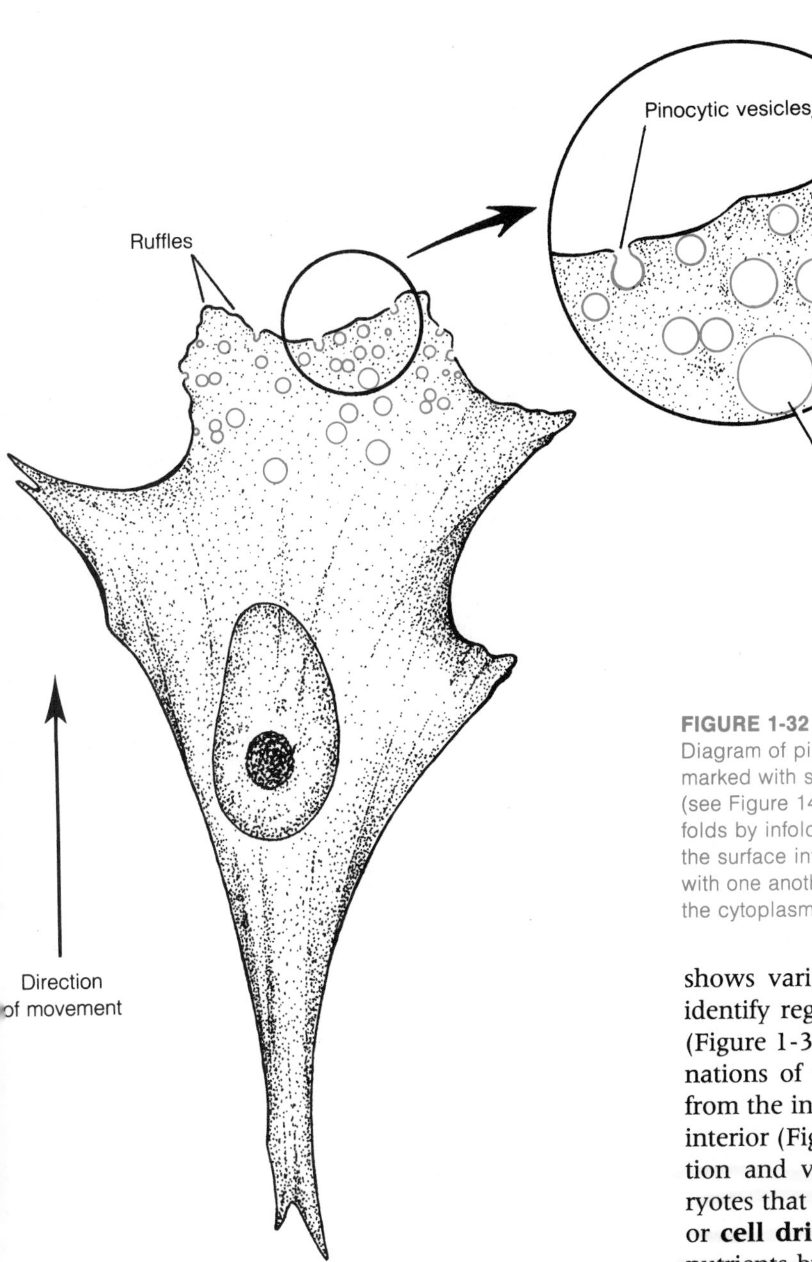

FIGURE 1-32
Diagram of pinocytosis. The advancing edge of an animal cell is marked with sheetlike projections called *lamellipodia* or ruffles (see Figure 14-19). Pinocytic vesicles form rapidly within these folds by infolding of the plasma membrane and pinching off from the surface into the cytoplasm. The tiny pinocytic vesicles fuse with one another (fused pinocytic vesicles) and drift deeper into the cytoplasm, where they fuse with lysosomes.

surface is covered with a variety of hairlike projections and blebs of different sizes. Although the significance of these structures is not well understood, the projections give the cell a much greater total surface area than had previously been assumed. The periphery of the cell also shows variable form. Ruffles at the edge of the cell identify regions of active movement of the cytoplasm (Figure 1-31). Along with this movement, tiny invaginations of the cell surface form vacuoles that detach from the inner surface of the cell and drift into the cell interior (Figure 1-32). This process of surface invagination and vacuole formation, which occurs in eukaryotes that lack a rigid cell wall, is called **pinocytosis** or **cell drinking.** Cells can take in large amounts of nutrients by pinocytosis. Some viruses also gain entry into cells by inclusion in pinocytic vacuoles. Overall pinocytosis is similar to phagocytosis. Pinocytosis is the ingestion of fluid and phagocytosis is the ingestion of particulates such as bacteria. Pinocytic vesicles are generally much smaller than phagocytic vesicles.

Specialization of Cells for Particular Functions

The cells that make up the many tissues of an animal are, for the most part, differentiated to perform various specialized functions. To carry out a specialized function a cell may develop specialized structures in the cytoplasm: for example, contractile filaments develop in muscle cells (Figure 1-33(a)); axons and dendrites in nerve cells (Figure 1-33(b)); flagella in sperm cells (Figure 1-33(c)); and cilia on some kinds of epithelial cells (Figure 1-33(d)). Other cells acquire the ability to synthesize specialized chemical products such as digestive enzymes, which are produced in cells of salivary glands and the pancreas. Glandular cells produce hormones. Cells in bone, cartilage, and connective tissue synthesize and release collagen, a protein that gives these tissues their mechanical strength.

The specialized biochemical activities of particular cells are frequently reflected in the structural organization of the cytoplasm and nucleus. A liver cell, or **hepatocyte,** has a high concentration of ribosomes in the cytoplasm (Figure 1-34(a)) and a nucleolus that is especially active in ribosome production. These structures are present because liver cells synthesize large amounts of protein (e.g., the blood protein serum albumin) that is released from the cell to become a major constituent of the blood. Most of the ribosomes in a liver cell are bound to membranes of the endoplasmic reticulum (ER), which is the characteristic location of ribosomes working to synthesize proteins destined for export from the cell via the Golgi complex. In contrast, proteins retained by the cell in the cytosol are synthesized mostly on ribosomes that are not attached to the ER. That portion of the ER to which ribosomes are attached is called the **rough ER** because of its studded appearance in electron micrographs. The cytoplasm of pancreas cells (Figure 1-34(b)) is dominated by the rough ER, because these cells synthesize digestive enzymes (proteins) that are released from the cell and then carried by the pancreatic duct to be emptied into the intestine.

In all cell types, a part of the ER lacks attached ribosomes and is called the **smooth ER.** The membranes of the smooth ER contain, among other things,

FIGURE 1-33
Some specialized structures of the cytoplasm. (a) An electron micrograph of part of a skeletal muscle cell. The fine fibers are composed of contractile proteins. (b) Light micrograph of a neuron isolated from a rat brain. The nucleus (N) is visible in the cell body. Several dendritic processes (D) are present, and the axon (A) bifurcates into two main branches from which small processes extend. (c) Scanning electron micrograph of a sea urchin sperm cell, specialized for propulsion by the long flagellum. (d) Electron micrograph of several epithelial cells of the mouse oviduct. The surfaces facing the lumen of the oviduct are covered with cilia (C), whose beating propels material along the lumen. [(a) Courtesy of Keith R. Porter. (b) Courtesy of Mark P. Mattson. (c) From G. Shih and R. Kessel. 1982. *Living Images*. Jones and Bartlett. (d) Courtesy of Ellen R. Dirksen.]

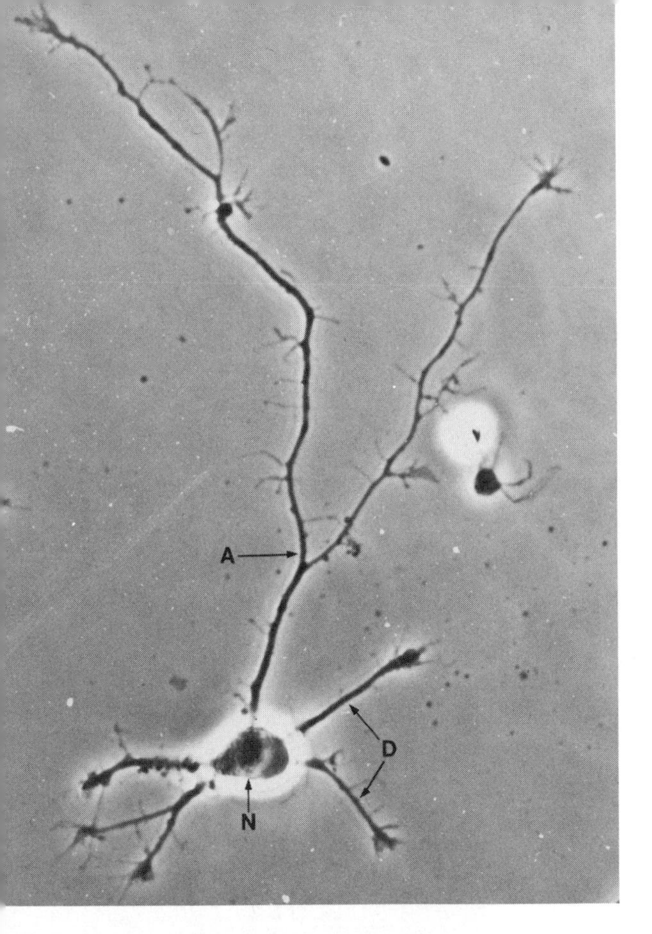

enzymes that catalyze the synthesis of steroids. Therefore, the smooth ER is prominent in cells that synthesize and secrete steroid hormones, for example, some cells in the testis (Figure 1-35) and those in the cortex of the adrenal gland. In skeletal muscle cells the molecular composition of the smooth ER is modified to play a major role in controlling muscle contractions. Membranes of both the smooth and rough ER synthesize lipids; they also join lipids to proteins to form lipoproteins, some of which are destined to become part of the plasma membrane. Built into the membranes of the smooth and rough ER of hepatocytes are enzymes that detoxify drugs such as barbiturates, tranquilizers, and ethyl alcohol, as well as environmental pollutants such as insecticides, herbicides, dyes, and some food preservatives.

A SURVEY OF CELL FUNCTIONS

Essentially all cell activities are devoted to one or more overall functions: cell self-maintenance, cell motility, cell growth and reproduction, and the specialized

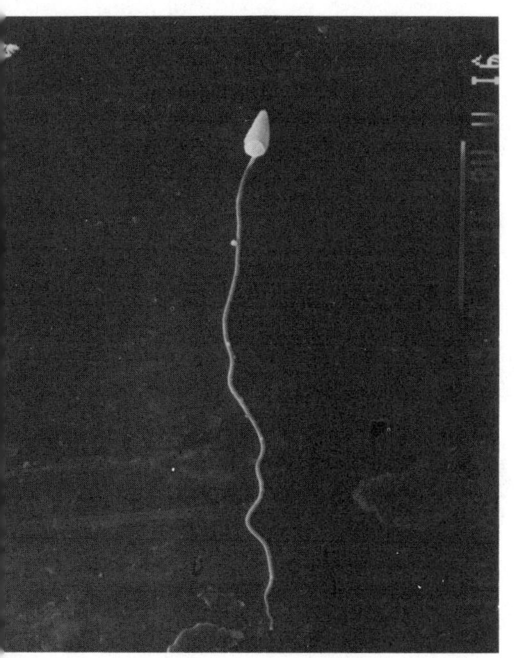

d

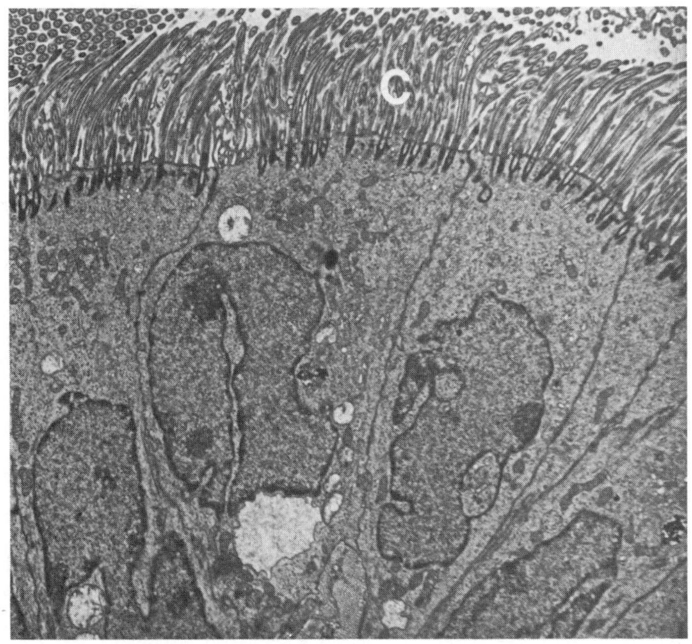

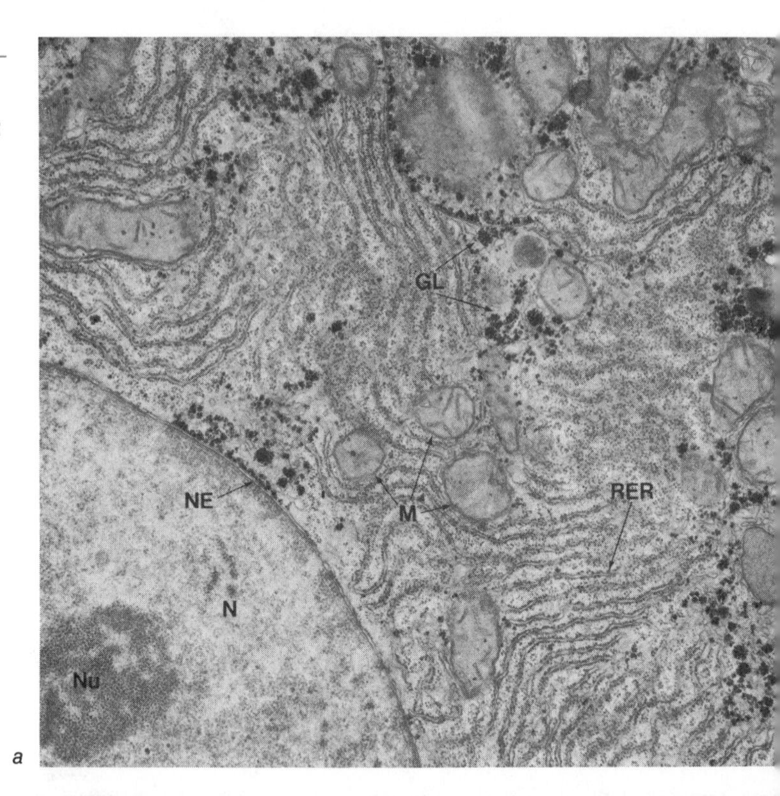

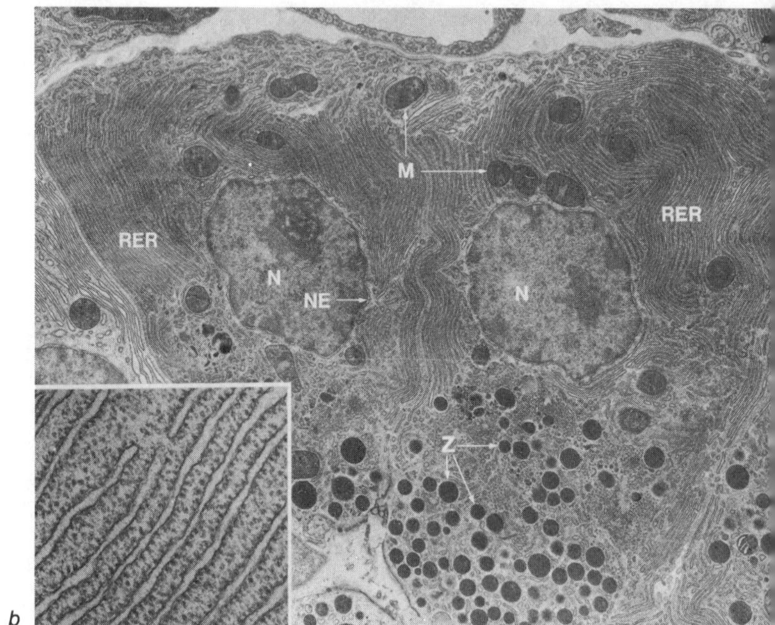

FIGURE 1-34

Electron micrographs of cells with high concentrations of rough endoplasmic reticulum (RER). (a) Portion of a liver cell. (b) Two pancreas cells. The inset shows RER at high magnification. N = nucleus, NE = nuclear envelope, Nu = nucleolus, M = mitochondria, GL = glycogen, and Z = zymogen. [Courtesy of Keith R. Porter.]

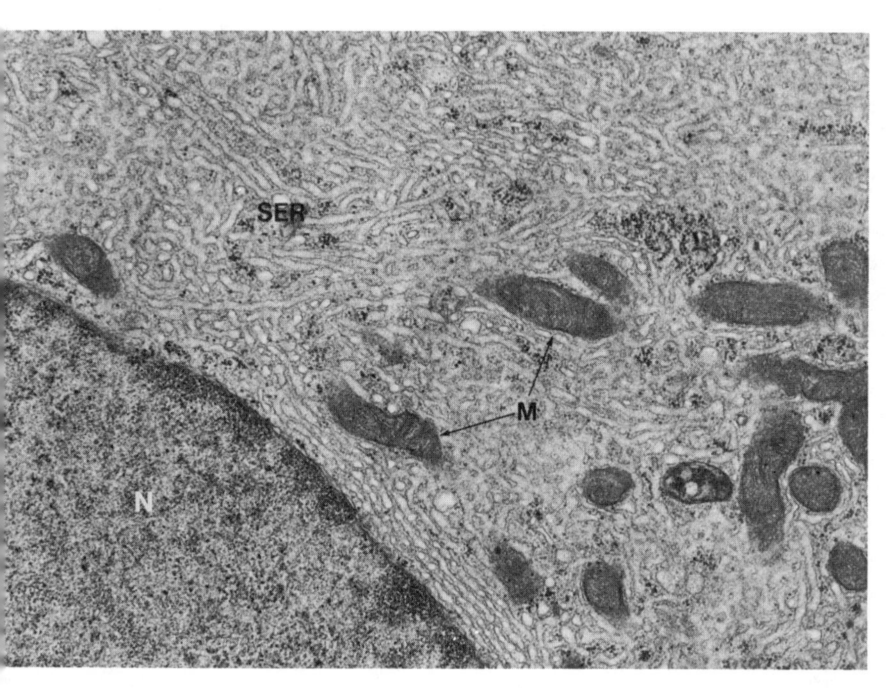

FIGURE 1-35 Electron micrograph of a part of a Leydig cell in guinea pig testis. A portion of the nucleus (N) is shown in the lower left. The cytoplasm is filled with smooth endoplasmic reticulum (SER), which is characteristic of cells that synthesize steroid hormones. Leydig cells synthesize the male hormone testosterone. A few mitochondria (M) are visible. [Courtesy of A. Kent Christensen. *The Journal of Cell Biology*, 26:911–935 (1965) by copyright permission of the Rockefeller University Press.]

work of differentiated cells. In effect, these functions encompass everything cells do.

Self-maintenance consists of the continuous synthesis of new molecules to replace those that become damaged. Within a cell, enzymes, the components of ribosomes, membranes, and other parts are constantly subject to damage from the highly energetic motions of ions and molecules. The tendency of cytoplasmic components to become damaged is particularly evident when the nucleus of a cell is removed. Without a nucleus, the cell is deprived of the genetic information required for protein synthesis; hence, replacement of its cytoplasmic parts cannot continue. In the first hours after enucleation, the cell continues to synthesize proteins and lipids, mobilizes energy, takes in nutrients, retains motility, and constantly changes shape. By 12 hours after enucleation, the rate of protein synthesis in the cell has declined to a fraction of its original rate; by 18 hours later, protein synthesis is no longer detectable. Experimentally enucleated animal cells live as long as several days, steadily decreasing in size, until self-maintenance fails completely and the cell disintegrates.

Self-maintenance also includes exchanges between the cell and its environment, particularly the taking in of nutrients that supply the cell with sources of chemical energy and raw materials to use in synthesizing replacement macromolecules. Self-maintenance is a housekeeping operation consisting of activities necessary for the functional and structural health of a cell. The cost of such housekeeping is met by the continual intake of sources of chemical energy and of nutrients along with the concomitant excretion of metabolic end products.

Cell motility takes several forms: cytoplasmic streaming, movement of organelles and particles inside cells, active changes in cell shape, amoeboid movement, propulsion by cilia and flagella, and muscle contraction. All forms of motility appear to be based on a particular set of proteins whose interactions with one another inside the cell produce movement. All forms of movement require chemical energy.

Cell growth and **reproduction** consist of those activities by which a cell increases the amount of its metabolic machinery and structural components, which are eventually distributed to two daughter cells at division. One of the most remarkable aspects of cell growth is the integration of the synthesis of thousands of different proteins and other molecules. A large and complex set of interactions, about which we still know very little, coordinates all these activities. Thus in a growing eukaryotic cell, chromosomes, ribosomes, mitochondria, endoplasmic reticulum, plasma membrane, and other structures, as well as thousands of proteins and other macromolecules, all increase in a precisely controlled and coordinated fashion. Growth prepares the cell for reproduction, which is the separation of the cell into two daughters by cell division. Cell division consists of two closely coordinated events: one is nuclear division, called **mitosis** or **karyokinesis,** which accomplishes the precise distribution of chromosomes to the daughter cells; the other is **cytokinesis,** the dividing of the cytoplasm to form two cells.

Cell self-maintenance is necessary for the survival of every species. It is perhaps less immediately evident that cell growth and reproduction are also stringent requirements for species survival. Without cell proliferation, no species of organism, either unicellular or multicellular, could reproduce to replace those individuals of its kind that die.

Unicellular organisms have special elaborate structures to carry out specialized activities, for example cilia, which are organelles of motility. However, this kind of differentiation is limited and does not approach the diversity, flexibility, and scope of the differentiation that occurs in multicellular plants and animals. The evolutionary success of multicellular organisms resides in their ability to produce differentiated cells to carry out specialized functions. The nature of differentiation is apparent in the development of an organism such as a human. Development begins with the formation of a single cell, the fertilized egg cell or zygote. Undergoing a rapid series of cell divisions, the human zygote consists of 64 cells by the end of 3 days. Most of these are destined to form the placenta, and a few to form the embryo. Only 21 days after fertilization the embryo consists of hundreds of thousands of cells that have differentiated into dozens of types. By this early point in development the heart has formed and begun to contract rhythmically, pumping newly formed blood cells through a rapidly developing system of arteries and veins. At birth the fetus is comprised of 10^{13} cells consisting of about two hundred or more distinctly differentiated types.

Cell growth and reproduction, as well as differentiation, continue to take place in the adult organism to replace cells that wear out. For example, each adult human has about 2.5×10^{13} red blood cells. The average red blood cell lives about 120 days, or 10^7 seconds. Therefore, the body must produce 2.5×10^{13} new red blood cells every 10^7 seconds, or 2.5×10^6 new red blood cells every second. In order to maintain this rate of production among those cells in the bone marrow that give rise to red blood cells, 2.5×10^6 cells must divide each second. Although few tissues in an organism produce cells at this rate, the total number of cells produced and differentiated in all of the many tissues of the human body probably exceeds 10^7 per second.

A SURVEY OF CELL NUTRITION

All of the activities carried out by a cell are fueled by an environmental supply of energy and depend on an environmental source of carbon, hydrogen, oxygen, nitrogen, phosphate, sulfate, and other components from which the molecules of the cell can be built. How these requirements are fulfilled varies greatly from one kind of cell to another. However, despite the variability, cells can be adequately characterized with only a few groups depending, first, on the chemical form of carbon they require, and second, on the form of the energy they obtain from the environment.

Those cells that can build all their organic molecules from only carbon dioxide (CO_2) are called **autotrophs** (derived from the Greek *autos* meaning self and *trophe,* nurturing). Photosynthetic cells and some bacteria make up the autotrophs. Cells that must fulfill their carbon requirement by taking in chemically reduced carbon in the form of organic molecules (for example, sugars) are called **heterotrophs** (Greek *hetero,* other). Most nonphotosynthetic bacteria, fungi, protozoa, and animal cells are heterotrophs. The degree of heterotrophy varies over a wide range; some cells, such as *E. coli*, can utilize a single organic compound, glucose, for building all of the complex molecules within the cell. Other heterotrophs require a broad spectrum of organic molecules; for growth the human cell requires ten preformed amino acids, sugar, numerous vitamins, and several fatty acids.

Cells are classified as **phototrophs** and **chemotrophs** depending upon how they obtain energy from the environment. Phototrophs are those cells that use the energy of light in order to synthesize sugar by photosynthesis. Chemotrophs are those that must derive energy through chemical reactions in which substances, usually carbohydrates, are broken down into simpler substances.

Almost all phototrophs are autotrophs; that is, they obtain energy by photosynthesis and use CO_2 as a source of carbon from which to build all their organic molecules. A few phototrophs require one or another organic molecule, and hence are called **photoheterotrophs.** For example, the alga *Euglena* requires the vitamins thiamine and B_{12}. Almost all chemotrophs are heterotrophs, requiring organic molecules as building blocks and digesting organic molecules to obtain chemical energy. Ultimately, chemotrophs-heterotrophs, such as animals, must ingest phototrophs-autotrophs (plants) to obtain a supply of organic molecules from which to obtain both building blocks and energy.

This brief review of cell functions and structures will be followed, in later chapters, by an expanded view of cell properties. Explanations of how cells are built and how they work, however, presuppose some knowledge of cell chemistry. The discussion of cells therefore continues in the next chapter with a review of the chemical components of and chemical interactions that take place in cells.

A BRIEF COMMENT ON METHODS IN CELL RESEARCH

This chapter shows that many methods have been devised for the study of cell function and structure. The oldest tool in cell research is the microscope. Microscopy has had a long history of development and improvement and continues to be invaluable in the study of cell structure. **Microsurgery** has also been mentioned as a means of analyzing cell functions, for example, the transplantation of a nucleus from one cell to another, as in amoebae. Other research tools and techniques are described, as appropriate, in succeeding chapters.

Modern cell research relies heavily on hundreds of techniques of biochemical analysis to determine the molecular nature of cell functions and structure. Biochemical analysis often requires large amounts of a purified cell component and therefore depends on the ability of the cell biologist to lyse cells and fractionate them into their structural parts. Cell fractionation continues to be central to much of cell research and is briefly described below. Indeed, identification of the biochemical composition and function of many cell parts and organelles was made possible by cell fractionation.

Cell Fractionation

Cell fractionation is the breaking open of cells and separation of the parts into pure fractions. The technique generally requires a large number of cells. Convenient sources are cells of plant and animal tissues and animal cells grown in culture, although bacteria, protozoa, algae, and fungi are frequently used. The breaking open of cells is called **lysis** or **homogenization** and is usually accomplished by mechanical disruption of the plasma membrane with an homogenizer (Figure 1-36). Sometimes the plasma membrane is dissolved with a detergent solution. A homogenizer consists of a tube and a close fitting pestle. The cells are placed in the tube in an appropriate solution of inorganic ions and low-molecular-weight organic molecules (e.g., sucrose) that will maintain as far as possible the functional and structural properties of the cell parts once the cells are broken open. The pestle is inserted into the tube and rotated as it is drawn in and out of the tube. The motion of the pestle creates a shearing action that breaks open the cells. Some cell structures such as mitochondria, chloroplasts, lysosomes, microbodies, nuclei, and ribosomes remain largely intact during homogenization while the endoplasmic reticulum, Golgi complex, and plasma membrane are usually fragmented. (However, special methods have been devised to isolate the Golgi complex and plasma membrane intact.) The endoplasmic reticulum is broken into small fragments that spontaneously form small sealed vesicles called **microsomes.** The major structural components can then be separated into pure fractions by centrifugation. Separation is possible because the velocity with which a particular cell structure sediments in a centrifugal field depends on its size, density, and shape.

Although there are various ways to fractionate a cell homogenate (lysate), the following method, known as **differential centrifugation,** illustrates the principle. A cell homogenate is placed in a centrifuge tube, which is then centrifuged (Figure 1-37). Nuclei are usually the largest intact structures in the homogenate (and are denser than water) and hence will sediment most rapidly and will be the first to accumulate at the bottom of the tube, forming a pellet of nuclei (Figure 1-38). At this point, the centrifugation is stopped and the cell homogenate (the supernatant) removed. The nuclear pellet is left behind, while the cell homogenate is placed in a second tube and centrifuged again, this time at higher centrifugal speed. This second centrifugation causes the next largest intact structures (usually the mitochondria) to sediment into a pellet. The process is repeated several times, with increasingly stronger centrifugal forces to obtain successively smaller structures. The final structures to sediment into a pellet are the ribosomes. The supernatant that remains after sedimentation of the ribosomes is called the **soluble fraction.** It contains many enzymes as well as all of the low-molecular-weight components (sugar, amino acids, etc.) of the cytosol.

Once organelles have been obtained as pure fractions they can be analyzed for biochemical makeup and function. Some organelles can be subfractionated; purified mitochondria can be broken and separated into the inner and outer membrane parts, and nucleoli can be isolated from purified nuclei. These separations enable further refinement to be made in the analysis of function and structure.

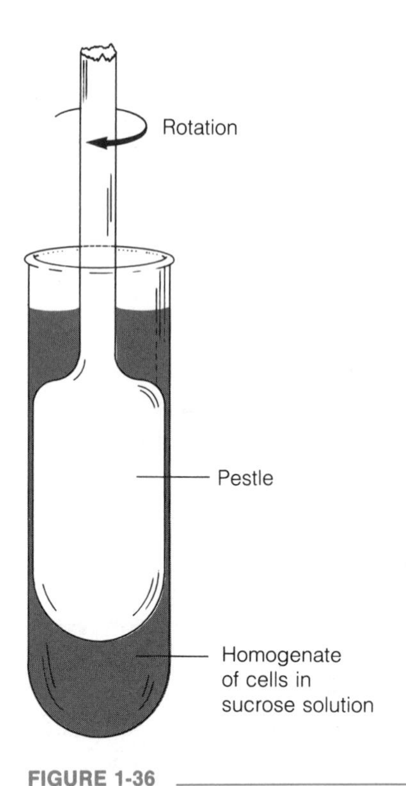

FIGURE 1-36
A glass homogenizer. Cells are broken open to release nuclei, mitochondria, and other structures by the shearing force between the pestle and the inner wall of the tube. The force is created by rotating the pestle as it is slowly moved up and down in the tube. The result is a cell homogenate from which cell parts may be purified.

Radioactive Isotopes in Cell Biology

Biochemical analysis of the functions and structures of cells is aided enormously by the use of radioactive isotopes as tracers of cell chemistry. They permit precise and detailed measurements that would otherwise be impossible. An isotope is a form of an element in which an extra neutron is present or a neutron is missing from the atomic nucleus. An abnormal number of neutrons often makes the atomic nucleus unstable, and they disintegrate, emitting radiation. For example, hydrogen contains one proton and no neutrons in its

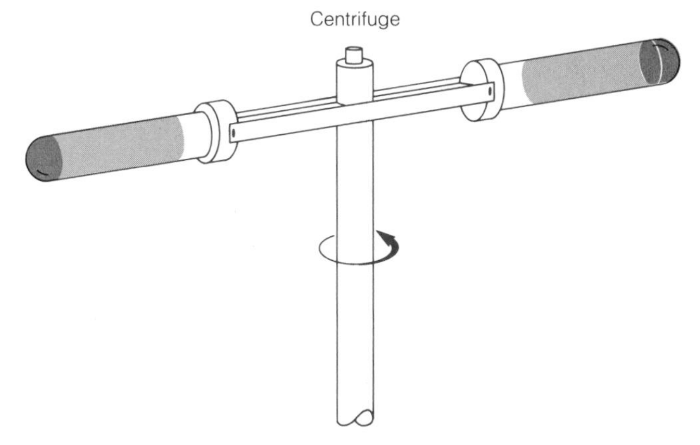

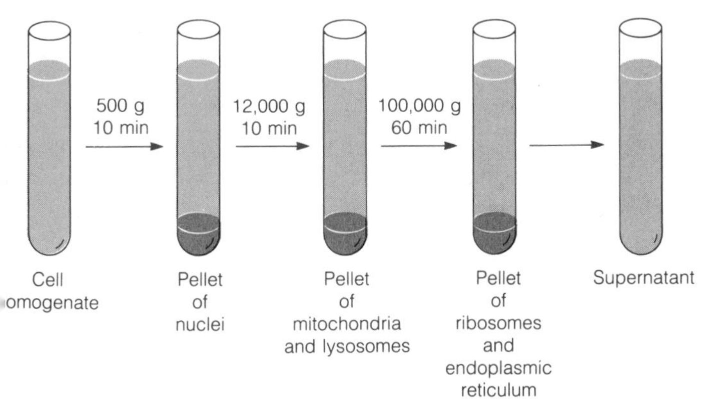

FIGURE 1-37

Different structural components present in a cell homogenate can be separated by centrifugation. A brief centrifugation at a low force (500 × gravity) for 10 minutes results in sedimentation of nuclei into a pellet. Centrifugation of the remaining supernatant at a higher force (12,000 × gravity) for 10 minutes causes sedimentation of mitochondria and lysosomes. Centrifugation of the supernatant once again but at a higher force (100,000 × gravity) for a longer time (60 min) causes sedimentation of the smallest structures, particularly ribosomes and fragments of the endoplasmic reticulum. The remaining supernatant contains the soluble components of the cell.

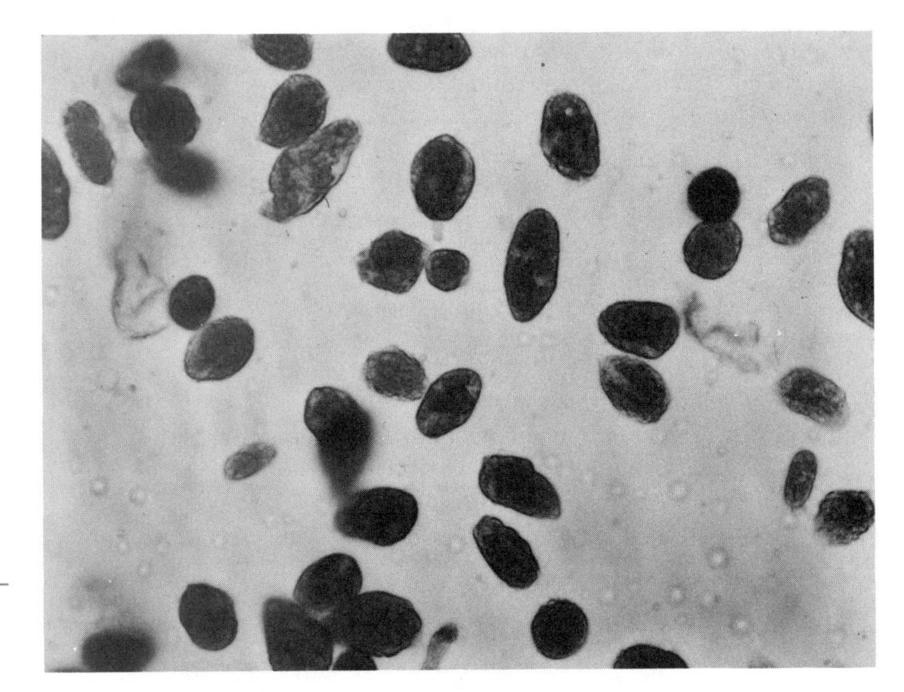

IGURE 1-38

light micrograph of nuclei purified from an omogenate of ciliated protozoa by entrifugation.

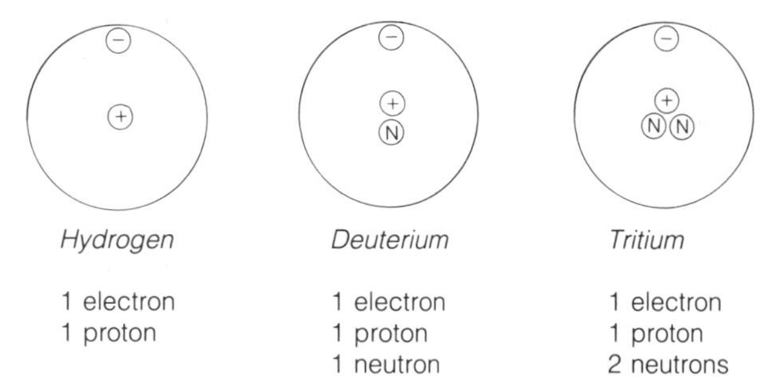

FIGURE 1-39
Hydrogen exists in three isotopes, differing in the number of neutrons in their atomic nuclei. *Ordinary hydrogen* has one proton and one electron. It accounts for 99.98 percent of all hydrogen found on the Earth. **Deuterium,** or heavy hydrogen, has one proton and one neutron in its atomic nucleus and one electron. It makes up 0.02 percent of the Earth's hydrogen. **Tritium** has one proton and two neutrons in its atomic nucleus and one electron. Tritium is an unstable isotope (a radioactive isotope) and also does not occur naturally on Earth.

nucleus (Figure 1-39). Deuterium is an isotope of hydrogen containing one proton and one neutron. Tritium has one proton and two neutrons. Deuterium is stable but tritium is unstable; one of the neutrons splits into a proton and an electron (Figure 1-40). The proton remains in the atomic nucleus but the electron, called a ***β* particle,** is ejected from the tritium atom and can be detected with a radioactivity counter or with photographic film.

Four radioactive isotopes widely used to investigate cell chemistry are tritium (^{3}H), radioactive carbon (^{14}C), radioactive phosphorus (^{32}P), and radioactive sulfur (^{35}S). Properties of these isotopes and kinds of molecules into which they become incorporated are listed in Table 1-2. Illustrations of how they are used to study cell functions and structures will appear frequently in the chapters that follow.

Each kind of radioactive isotopes decay at a characteristic rate called its **half-life** (Table 1-2), the time required for half of the radioactive atoms present to disintegrate. The half-life of ^{32}P is 14.3 days, which means that starting with a given amount of the isotope, 50 percent of the atoms will have disintegrated (decayed) in 14.3 days; half of the remaining will decay in the next 14.3 days, and so on. By 42.9 days (equal to half-lives) 87.5 percent of the atoms will have decayed.

Because they can be substituted for normal atom in organic molecules, ^{3}H, ^{14}C, ^{32}P, and ^{35}S are particu larly useful in cell research. For example, it is possibl to produce glucose molecules in which one or mor carbon atoms is replaced by ^{14}C. The radioactivel labeled glucose can be given to cells and the metaboli fate of glucose determined by observing which mole cules in the cell become radioactive when the glucose i broken down. Most of the glucose is ultimately de graded to carbon dioxide (CO_2) and water, so most c the ^{14}C atoms leave the cell as $^{14}CO_2$. Amino acids ca be labeled by substituting ^{14}C for one of the norma carbons, ^{3}H for a hydrogen, or ^{35}S in one of the tw sulfur-containing amino acids. When labeled amin acids are given to a cell, they are used for protein syn thesis. The rate at which cells make proteins can the be determined by measuring the rate at which radioac tivity appears in cellular proteins. The elements ^{3}H, ^{14}C or ^{32}P can be used in a similar manner to follow nuclei acid synthesis and breakdown.

The amount of radioactivity in a cell in the form c labeled molecules can be measured electronically o photographically. Electronic measurements are typi

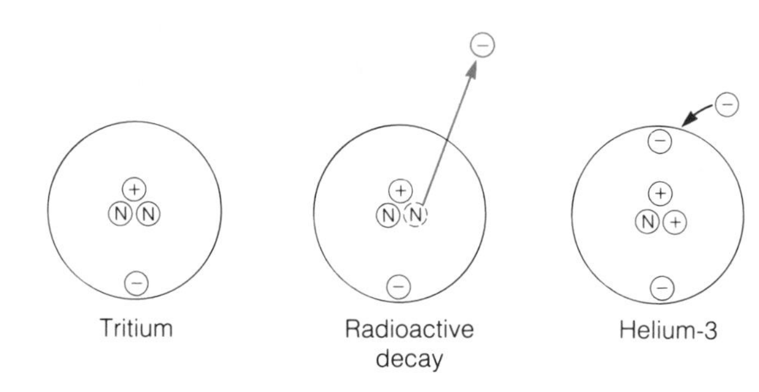

FIGURE 1-40
Decay of tritium. Tritium is an unstable isotope of hydrogen. During radioactive decay it emits an electron. The remaining atom contains two protons and one electron in the atomic nucleus. Capture of an electron from the surroundings reestablished balance between positive and negative charges, yielding an isotope of helium (helium-3).

ABLE 1-2
roperties of Four Radioactive Substances Frequently Used in Cell Research

Element	Symbol	Energy of β Particle Emitted	Half-Life	Use in Labeling
Tritium	^{3}H	0.018 MEV*	12.3 years	To Label Any Organic Molecule
Radioactive Carbon	^{14}C	0.180 MEV*	5570 years	To Label Any Organic Molecule
Radioactive Phosphorus	^{32}P	1.8 MEV*	14.3 days	To Label Nucleic Acids, Phospholipids, Phosphoproteins, ATP
Radioactive Sulfur	^{35}S	0.180 MEV*	87.2 days	To Label Proteins

Million Electron Volts

ally done with a scintillation counter. In scintillation ounting, radioactive cells, cell fractions, or cell extracts re mixed into a solution containing a molecule that vill scintillate, that is, emit light when it absorbs a β article. Each decay produces a pulse of light that can e counted. Furthermore, the intensity of the light ulse is proportional to the energy of the decay and can e used to identify the isotope that is decaying. This eature allows one to use two different isotopes simultaneously to label different substances (e.g., protein and DNA).

Alternatively, radioactivity in cells or macromolecules can be detected by a photographic film. Radioactivity causes exposure of the film in the same way that light does. This technique, called **autoradiography,** has the advantage of showing precisely where the radioactivity is located in the cell. The autoradiographic technique is described in Chapter 7.

ROBLEMS

1. What are the main structural parts of a bacterial cell like *E. coli,* and what is the function of each?
2. What important properties are shared by all cells on this planet?
3. What properties distinguish a prokaryotic cell from a eukaryotic cell?
4. Eukaryotic cells have more DNA than prokaryotic cells do. What is the significance of this difference?
5. The bacterium *E. coli* divides every 20 minutes in a rich nutrient medium and every 40 minutes in medium containing only glucose as a source of organic carbon. Why is cell proliferation more rapid in rich nutrient medium?
6. In what way are photosynthetic bacteria important to survival of animals, including humans?
7. What crucial roles do lysosomes play in the life of an amoeba? What role does the contractile vacuole play?
8. What are the main structural components of an animal cell, and what are their functions?

9. What additional structures does a plant cell have that an animal cell does not, and what are their functions?
10. Why do all cells require a continuous source of energy?
11. How many new cells does an adult human produce every 24 hours?
12. In what way do all heterotrophic cells depend on autotrophic cells for their existence?
13. How have cell fractionation and radioactive isotopes been crucial in the study of how cells work?
14. When *E. coli* are cultured in medium rich in glucose, they lack flagella. When grown in nutritionally poor media, they possess flagella. What does this indicate about the value of flagella?
15. The electron microscope reveals a very high concentration of ribosomes in the cells of some tissues. What does that suggest about the general function of those cells?

CHAPTER 2

The Molecules of Cells

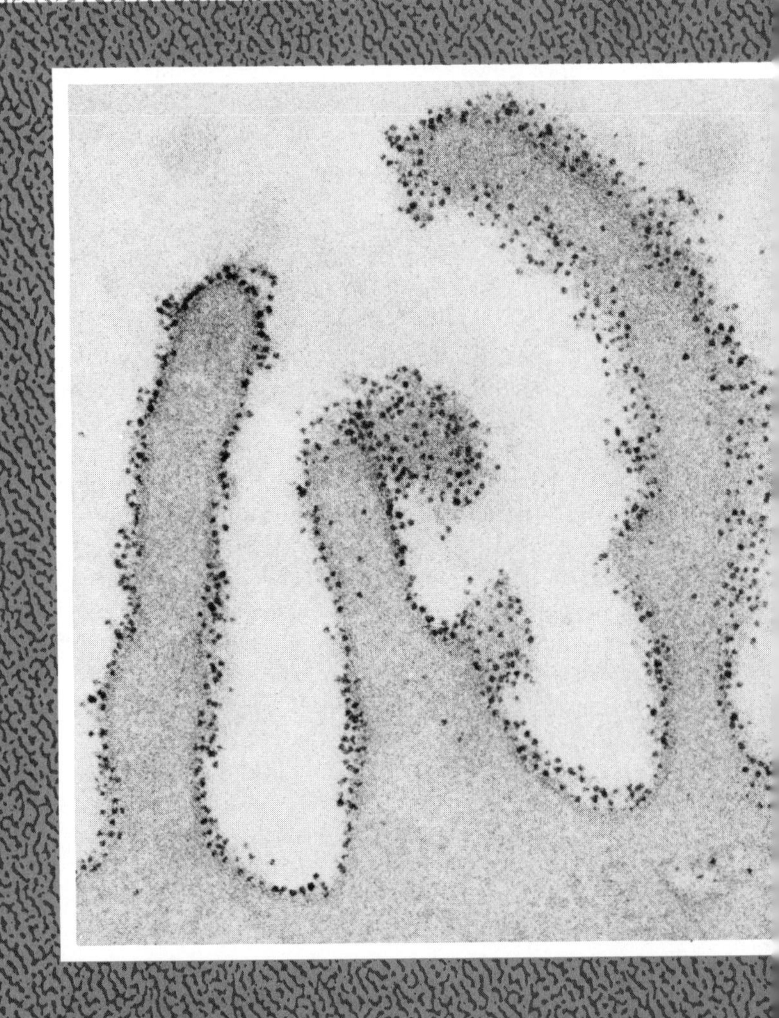

See Figure 2-35.

Many complex processes take place within a cell. They are based on a large number of different molecules and inorganic ions working in concert. For example, the degradation of sugars is performed by a group of enzymes and various other molecules. As a result of degradation, the sugar molecule is converted to carbon dioxide and water, and released energy is captured as chemical energy in a form that can be used to drive the various kinds of cell activities. The first part of the process of energy release from sugar is carried out by molecules present in the cytosol, and the second part is performed by molecules present in mitochondria. So too, the synthesis of proteins is carried out by the cooperative action of a different set of molecules organized into another structure, the ribosome.

Every cellular activity, including the formation of every structure in a cell, depends on the interaction of particular sets of molecules. In some cases, interacting molecules are dissolved in the cytosol, and cooperative action is achieved by frequent and extremely transient meetings of molecules. In other cases, particular molecules join together in relatively stable associations so that a structure is formed. For example, cell membranes are the result of the association of lipid molecules with each other and with certain kinds of proteins. Microtubules are formed by association of protein molecules called **tubulin.** A complete understanding of cell function and structure will ultimately require a detailed knowledge of the molecules that work cooperatively to carry out the many activities of the cell and to form structures.

This chapter provides a survey of the principal molecules and inorganic ions that make up cells. In subsequent chapters these various molecules are examined more closely in connection with particular functions and structures of the cell.

Even the smallest cells, the bacteria, are chemically very complex. *E. coli* contains 5000 or 6000 different kinds of organic molecules (nucleic acids, proteins, lipids, polysaccharides, coenzymes, sugars, amino acids, nucleotides, and others). In keeping with their greater functional and structural complexity, eukaryotic cells contain even more kinds of molecules, upwards of 10,000 in some kinds of cells. With so many different molecules, a complete explanation of cell function and structure at the chemical level would appear to be a hopelessly complex task. However, the chemical components of cells form a few large classes; hence the chemical analysis of cells is reduced to a manageable level. The three major kinds of components of the cell are water, inorganic ions, and organic molecules. Water is a single component, inorganic ions form a small group of components, and organic molecules account for most of the chemical complexity of cells. Organic molecules of cells fall, in turn, into a few well defined groups that allow order to be brought to their study.

WATER

Water is by far the most abundant substance in cells, accounting for 70 to 90 percent of cell weight. For example, frog eggs are about 70 percent water. However, the water content of a frog egg is less than in most cell types because so much of the weight of the egg consists of large quantities of stored proteins and lipids in the form of **yolk granules** or **platelets,** which are food reserves of the developing embryo. An amoeba or mammalian cell in culture is almost 90 percent water by weight.

Water is the liquid in which many of the ions and molecules of the cell are dissolved and in which the various cell structures and organelles are bathed. As the solvent in a cell, water interacts with the various inorganic and organic constituents. These interactions are possible because of the electrical polarity of water molecules. In a water molecule, two hydrogen atoms are covalently bonded to an oxygen atom (Figure 2-1(a) and (b)). The two hydrogen atoms form an angle of 104.5° with the oxygen atom. Each hydrogen atom has a slight positive charge on the side away from the oxygen atom, and the oxygen atom has a slight negative charge. Molecules, such as water, which have local positive and negative regions are called **polar molecules.**

In addition to its electrical interaction as a solvent with virtually all the other molecules of the cell, water also enters into many of the chemical reactions of the

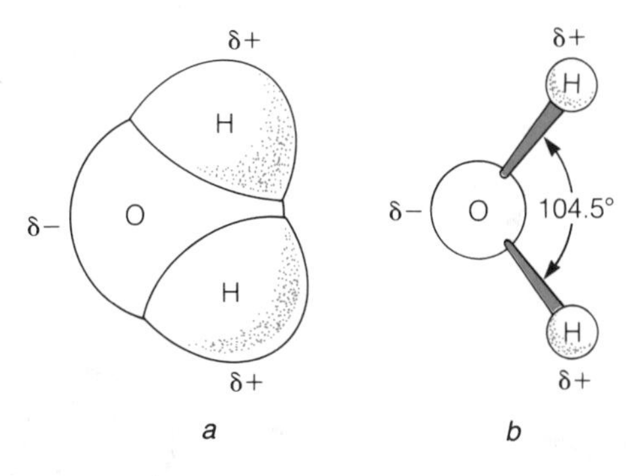

FIGURE 2-1
A water molecule consists of two hydrogen atoms and one oxygen atom. (a) A space-filling model of a water molecule, and (b) a ball and stick model showing the bond angle for the two hydrogen atoms. The symbols δ^- and δ^+ indicate slight negative and positive charges on the molecule.

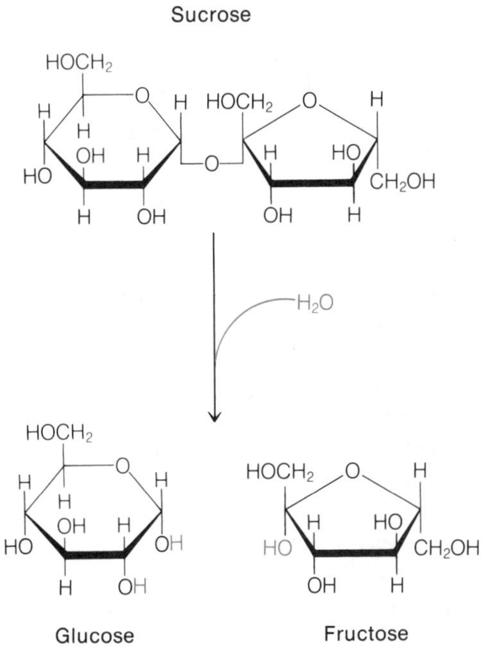

FIGURE 2-2
Sucrose is a disaccharide consisting of two 6-carbon sugar molecules, one glucose and one fructose molecule. Carbon atoms are present at all the angles in the sugar rings. Sucrose can be split into its component 6-carbon sugar molecules by hydrolytic cleavage, in which the atoms of one water molecule are divided between the glucose and fructose molecules.

cell. Organic molecules are commonly broken down by enzymatic addition of water. These reactions are known as **hydrolytic cleavages.** Figure 2-2 shows the hydrolytic cleavage of sucrose into fructose and glucose.

Many chemical reactions occur by the removal of water from organic molecules in a manner that allows two organic molecules to be joined (condensed) by a covalent bond to form a larger organic molecule. These are called **dehydration-condensation reactions.** For example, the formation of peptide bonds between amino acids in the synthesis of proteins is a dehydration-condensation reaction (Figure 2-3) that yields water as a by-product.

In some chemical reactions water is added to molecules to yield other molecules. For example, water can be added to carbon dioxide to make carbonic acid:

$$CO_2 + H_2O \rightleftharpoons H_2CO_3$$

The CO_2 produced in tissues of higher animals is released into the blood, where it is combined with water inside red blood cells. The reaction is reversed in the

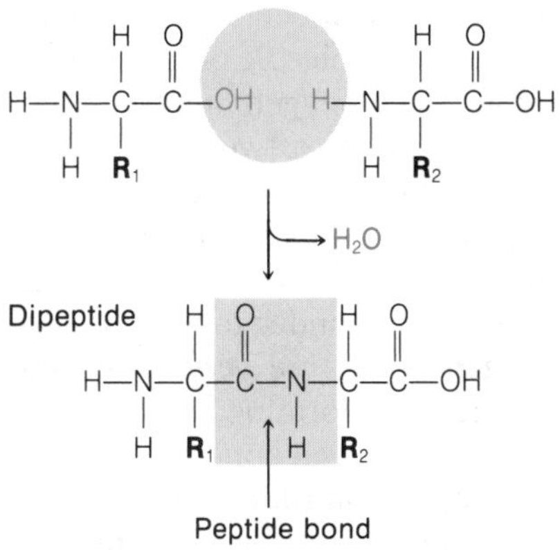

FIGURE 2-3
Formation of a peptide bond between two amino acids by elimination of water (shaded circle). The peptide bond is formed between the terminal carbon atom of one amino acid and the nitrogen atom of another amino acid. The atoms in the shaded square are called the peptide group.

$$H_2O \rightarrow \underbrace{2e^- + 2H^+}_{\text{Used for sugar synthesis}} + 1/2O_2 \leftarrow \text{Released into the atmosphere}$$

FIGURE 2-4
During photosynthesis water is split into electrons and hydrogen ions, which are used in photosynthetic synthesis of glucose, and oxygen, which is released to the atmosphere. A single water molecule yields one oxygen atom, which is half of an oxygen molecule (O_2).

lungs, where the CO_2 is released into the air that fills the lungs.

Another example of the involvement of water in cell activities is the splitting of water into electrons, H^+, and O_2 during photosynthesis (Figure 2-4). A final example is the production of water in the complete degradation of sugars (Figure 2-5).

Hydrogen Bonds

Because of their electrical polarity, water molecules are able to form weak bonds called **hydrogen bonds** with one another and with other polar molecules. A hydrogen bond can form between a slightly positively charged hydrogen atom belonging to one molecule and a slightly negatively charged atom, such as oxygen or nitrogen, in another molecule or in another part of the same molecule. Each oxygen atom in water can form two hydrogen bonds with two other water molecules (Figure 2-6). Each of the two hydrogen atoms of water can form a hydrogen bond with the oxygen atom in another molecule. Therefore, each water molecule tends to be hydrogen-bonded to four other water molecules. Hydrogen bonds are very weak and are constantly breaking and reforming. Despite their weak and transient nature, hydrogen bonds bring about some degree of structural organization among water molecules. Liquid water contains extensive clusters of molecules held together by hydrogen bonds.

$$C_6H_{12}O_6 + 6O_2 \rightarrow 6CO_2 + 6H_2O$$

FIGURE 2-5
Cells obtain energy by degrading sugars (using oxygen) to carbon dioxide and water.

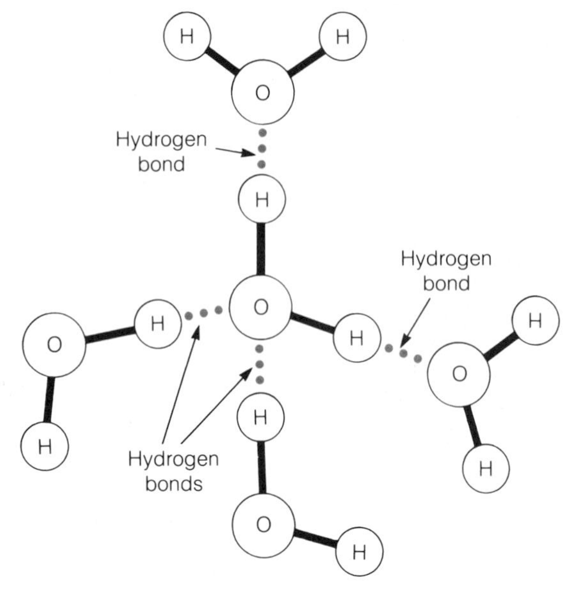

FIGURE 2-6
The weak electrical charges of water molecules allow the molecules to join transiently by hydrogen bonds. Hydrogen bonds are weak and constantly break and reform.

Hydrogen bonds occur not only among water molecules but also between water molecules and organic molecules. Water can form a hydrogen bond with an oxygen atom in an organic molecule (Figure 2-7(a)). Hydrogen bonds can also form between two organic molecules, for example, between an oxygen atom in one molecule and a hydrogen atom that is covalently joined to a nitrogen atom of another (Figure 2-7(b)). Similarly, hydrogen bonds are possible with a hydrogen atom between two oxygen atoms or between two nitrogen atoms (Figure 2-7(c) and (d)). A single hydrogen bond is too weak to provide stable bonding between two organic molecules, but two molecules may stick to each other if multiple hydrogen bonds are possible (Figure 2-7(d)).

Because of its polar nature, water is an excellent solvent for substances held together by electrostatic forces, that is, ionic substances such as inorganic salts. A crystal of sodium chloride is held together by the

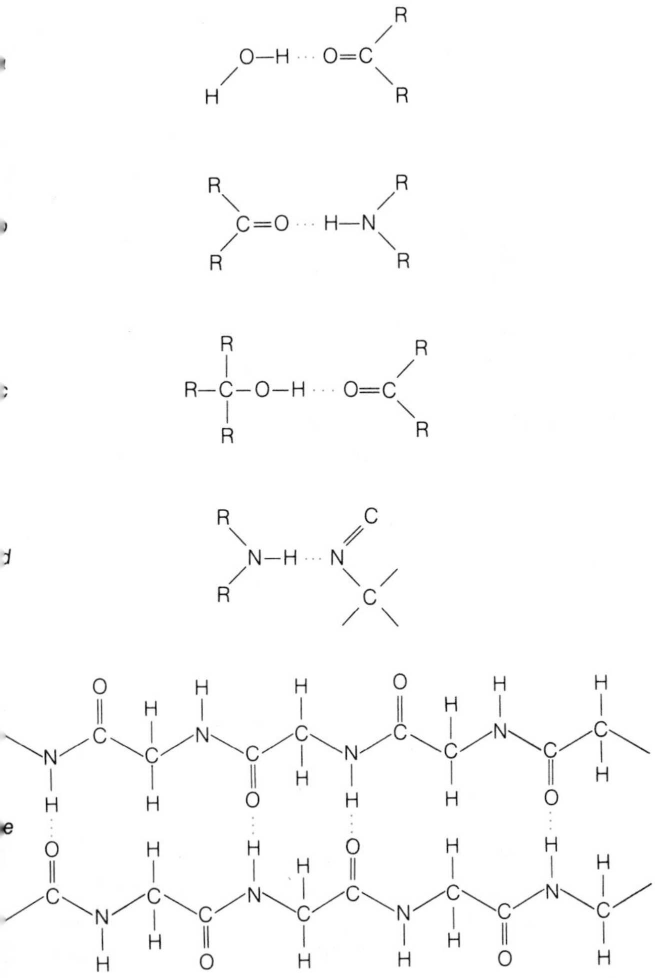

FIGURE 2-7
Hydrogen bonds between various kinds of molecules. (a) Between a water molecule and an oxygen atom in an organic molecule. (b) Between an oxygen and a nitrogen atom. (c) Between two oxygen atoms. (d) Between two nitrogen atoms. (e) Multiple hydrogen bonds between two large molecules.

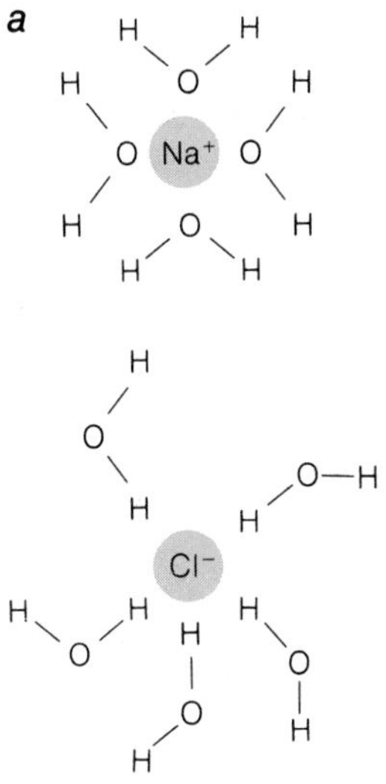

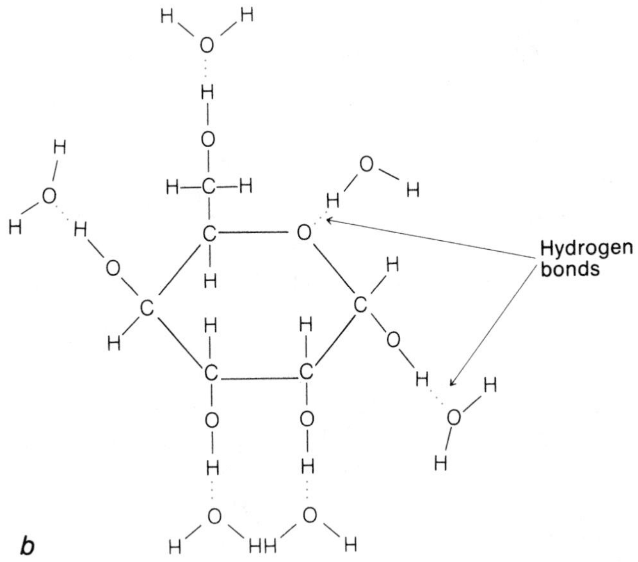

FIGURE 2-8
Interaction of water with ions and molecules. (a) The weak charges present on water molecules allow them to interact with negatively and positively charged ions. In this case NaCl is dissolved in water by formation of shells of water molecules around the Na^+ and Cl^- ions. (b) Glucose dissolves in water by formation of a cloud of water molecules held transiently by hydrogen bonds.

electrostatic attractions between positively charged sodium ions (Na^+) and negatively charged chloride ions (Cl^-). Sodium chloride readily dissolves in water because negatively charged oxygens of water molecules are electrostatically attracted to a Na^+ ion and positively charged hydrogens of water molecules are strongly attracted to a Cl^- ion. These electrostatic attractions result in dissociation of Na^+ from Cl^- with the formation of shells of water molecules around the two kinds of ions (Figure 2-8(a)).

Nonionic substances (substances that do not dissociate into ions in the presence of water) are also soluble in water if they are, like water, polar in nature, and therefore can form hydrogen bonds with water. Thus, sugars, ethyl alcohol, and a wide variety of other organic molecules readily dissolve in water (Figure 2-8(b)). Nonionic organic molecules that are essentially **nonpolar,** such as benzene and the fatty acid chains of lipid molecules, cannot interact with water molecules and therefore are insoluble in water.

Ionization of Water and pH

Water undergoes reversible ionization to yield H^+ and OH^- ions:

$$H_2O \rightleftharpoons H^+ + OH^-$$

In any one instant only a tiny fraction of the water molecules is in the ionized state (2 in 10^9 for pure water), and the concentrations of H^+ and OH^- ions are extremely low. Dissociation and reassociation are normally in equilibrium, which means that the rate of dissociation of H_2O into H^+ and OH^- ions is equal to the rate of reassociation of H^+ with OH^- ions to form H_2O. At equilibrium the concentration of H_2O molecules is very high, the concentrations of H^+ and OH^- ions are very low, and the concentrations of all three remain unchanged. Even in low concentrations, H^+ and OH^- ions have profound effects within the cell, affecting virtually all of the interactions among molecules that underlie cell activities. For example, the ability of enzymes to bind their substrates is highly dependent on the concentration of H^+ ions (Chapter 3).

The concentration of H^+ ions is often expressed in pH units. Because of the importance of H^+ ion concentration for the cell, the pH scale will be reviewed. The pH is defined as the $\log_{10}$ of the reciprocal of the molar concentration of H^+ ions. Molar concentration is indicated by brackets:

$$\text{pH} = \log_{10}(1/[H^+])$$

The concentration of H^+ ions in water at 25°C is 10^{-7} M. Therefore, at 25°C the pH of pure water is

$$\text{pH} = \log_{10}(1/10^{-7}) = 7.0$$

The pH of pure water (7.0) is defined as neutral pH because the concentrations of H^+ and OH^- ions are equal. However, the concentration of H^+ ions may be increased by addition of an acid such as HCl ($H^+ + Cl^-$) or decreased by the addition of a base such as NaOH ($Na^+ + OH^-$). As the H^+ ion concentration is increased by addition of HCl, the pH falls below 7.0. The addition of OH^- ions results in a momentary increase in the rate of association of OH^- with H^+ ions to form H_2O. Equilibrium is quickly reestablished, but at a decreased concentration of H^+ ions and therefore at a pH above 7.0. On the other hand, one mole of HCl mixed with water to make one liter of solution makes a one molar solution of HCl. Since HCl completely dissociates into H^+ and Cl^- ions, the result is a one molar solution of H^+ ions (1 M H^+). Therefore, the pH is

$$\text{pH} = \log_{10}(1/1) = 0$$

The pH of a 0.1 M solution of HCl is

$$\text{pH} = \log_{10}(1/0.1) = 1$$

The pH of a 0.01 M solution of HCl is

$$\text{pH} = \log_{10}(1/0.01) = 2$$

Weak Acids

Organic molecules that dissociate in water to yield H^+ ions are weak acids because, in contrast with many inorganic acids, they do not dissociate completely. Thus, acetic acid added to water partially dissociates, establishing an equilibrium between undissociated and dissociated molecules:

$$CH_3COOH \rightleftharpoons CH_3COO^- + H^+$$

The degree of dissociation is expressed as the **dissociation constant,** K' (the $'$ means that concentrations are given as molarities), which is equal to the product of the concentrations of the dissociated ions divided by the concentration of undissociated molecules. For acetic acid

$$K' = \frac{[CH_3COO^-][H^+]}{[CH_3COOH]}$$

Each organic acid has its own characteristic value r K', which can be determined by direct measurement f the amount of dissociated and undissociated compoents at equilibrium. For acetic acid

$$K' = \frac{[H^+][CH_3COO^-]}{[CH_3COOH]} = 1.74 \times 10^{-5}$$

Since the value of K' is known by measurement for large range of temperatures, the H^+ ion concentration pH) can be calculated for an acetic acid solution of any iven molarity and temperature. For a 1 M solution of cetic acid at 25°C

$$1.74 \times 10^{-5} = \frac{[H^+][CH_3COO^-]}{1 - [H^+]}$$

ince some acetic acid molecules will be dissociated, he concentration of undissociated molecules given in he denominator is not 1 M but 1 M minus the concenration of dissociated molecules; the concentration of lissociated molecules is equivalent to the concentration f either ion, $[CH_3COO^-]$ or $[H^+]$. In practice this corection of the denominator may be ignored because nly a minute fraction of the CH_3COOH molecules is in he dissociated state at equilibrium. (This is true for nost organic acids.) Hence, we may write

$$1.74 \times 10^{-5} = \frac{[H^+][CH_3COO^-]}{1\,M}$$

ince $[H^+] = [CH_3COO^-]$, then $[H^+] = \sqrt{1.74 \times 10^{-5}}$ $= 4.17 \times 10^{-3}$ M, and the pH of a 1 M solution of acetic cid at 25°C is

$$pH = \log_{10} \frac{1}{4.17 \times 10^{-3}} = 2.38$$

Thus, when the value of K' for an organic acid is known, the pH for a water solution of known molarity of the acid can be calculated.

Bases

Any substance that decreases the concentration of H^+ ions is called a **base.** Since pH is a measure of H^+ ion concentration, calculation of the increase in pH requires that the effect of addition of base on the H^+ ion concentration first be determined. This can be done because the product of the concentrations of H^+ and OH^- ions in any aqueous solution at any pH is a constant known as the **ion product** of water and represented by $\boldsymbol{K_w}$. At 25°C the value of K_w is 10^{-14}:

$$K_w = [H^+][OH^-] = 10^{-14}$$

For water at the neutral pH of 7.0,

$$K_w = [H^+][OH^-] = (10^{-7})(10^{-7}) = 10^{-14}$$

Since NaOH dissociates completely to Na^+ and OH^- ions in water, a 1 M solution of NaOH will be 1 M with respect to OH^- ions. $[H^+]$ can be calculated using the constant K_w. Thus,

$$[H^+] = \frac{K_w}{[OH^-]} = \frac{10^{-14}}{1} = 10^{-14}$$

and

$$pH = \log_{10} \frac{1}{10^{-14}} = 14.0$$

Similarly, the pH for a 0.1 M NaOH solution is calculated

$$[H^+] = \frac{K_w}{[OH^-]} = \frac{10^{-14}}{10^{-1}} = 10^{-13}$$

so

$$pH = \log_{10} \frac{1}{10^{-13}} = 13$$

Since the pH scale is logarithmic, a change of one pH unit, for example, from pH 7 to pH 6, represents a 10-fold increase in H^+ ion concentration (from 10^{-7} to 10^{-6} M) and a 10-fold decrease in OH^- ions (from 10^{-7} M to 10^{-8} M).

The pH of Biological Systems

The aqueous solution inside most eukaryotic cells is influenced by the many organic and inorganic ions that are present and is usually a few tenths of a pH unit

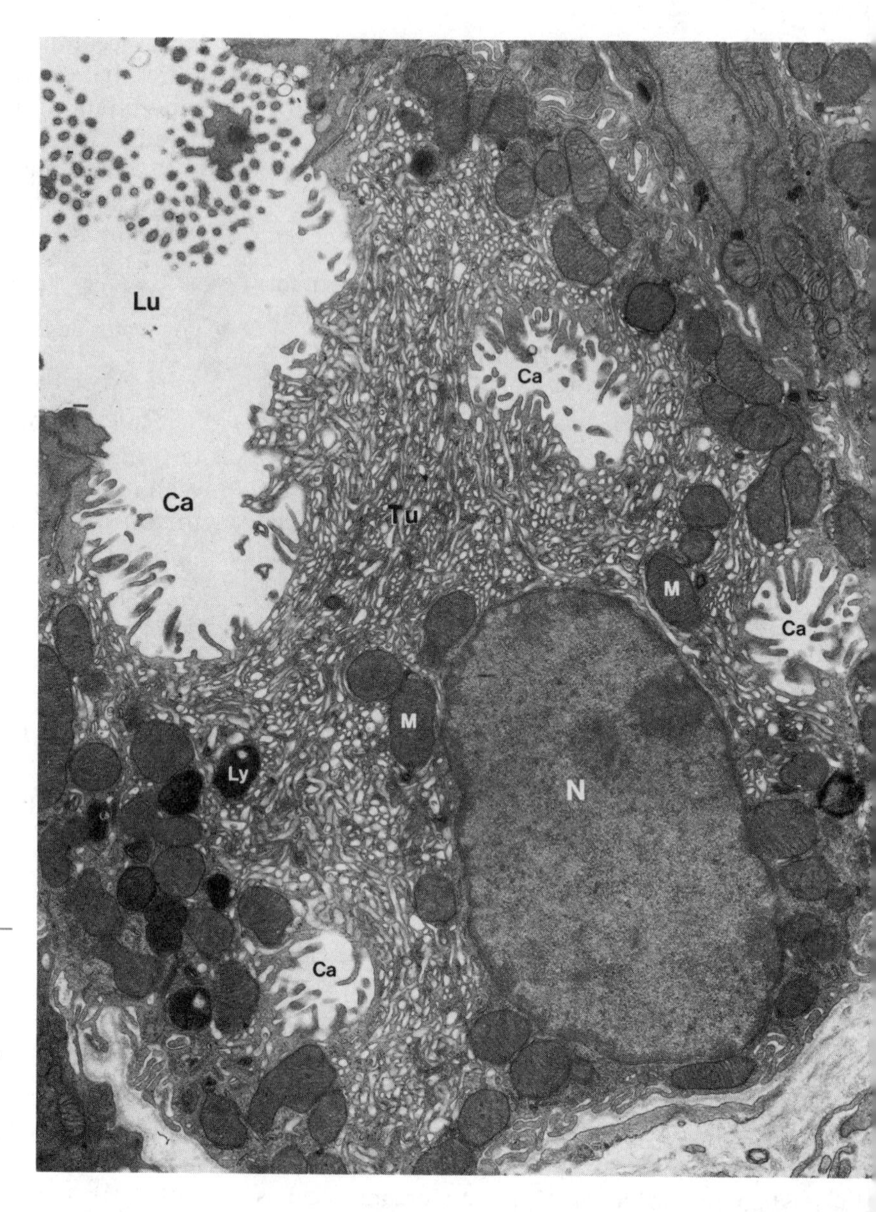

FIGURE 2-9
Electron micrograph of a parietal cell of the stomach. The cytoplasm contains an extensive system of membranous tubules (Tu), which are probably involved in secretion of acid. Acid may be secreted into canaliculi (canals) (Ca), which open to the cell surface and into the lumen (Lu) of the stomach. Ly = lysosome, M = mitochondria, N = nucleus. [Courtesy of Keith R. Porter.]

above pH 7.0. The pH inside organelles, such as a mitochondrion, a chloroplast, or a contractile vacuole, may be different from the pH in the cytosol. The formation of ATP in a mitochondrion involves the establishment of a pH difference on the two sides of the inner membrane of the mitochondrion (Chapter 4). The pH of blood plasma is slightly alkaline (7.4) and the pH of saliva is slightly acid (about 6.5). Certain kinds of cells can produce extremely acidic extracellular solutions by secretion of H^+ ions. **Parietal cells** in the epithelial lining of the stomach (Figure 2-9) secrete H^+ and Cl^- ions to form extracellular gastric juice in the stomach with a pH typically between 1.5 and 3.0. This extremely acidic solution in the stomach is important in ne digestion of food, destroying the cells in food and nitiating the breakdown of the three-dimensional ructure of most macromolecules, thereby making nem more susceptible to subsequent degradation by nzymes in the intestine. The acidity of the stomach lso serves as a barrier (although not always success-ully) to infection by the killing of bacteria, viruses, and ther parasitic organisms.

In summary, the large amount of water in cells is nportant in cell functions, because of its properties as solvent for inorganic and organic ions and molecules, ecause it enters into many chemical reactions, and ecause the H^+ and OH^- ions produced by dissociation f water have an enormous effect on the interactions mong molecules in the cell.

INORGANIC IONS

In most cells inorganic ions account for less than ne percent of cell weight. The cytosol component of ytoplasm contains a variety of inorganic ions that are equired in many cell functions and structures. The po-assium ion K^+ is a positively charged ion, or **cation**, resent in cells and is a required participant in many spects of cell metabolism. For example, a high intra-ellular concentration of K^+ ions is needed in protein ynthesis, and K^+ ions are required for maximal activity f some enzymes.

Na^+ ions are always present in some unicellular rganisms and in multicellular marine organisms. Iowever, depending on the culture conditions, Na^+ ons can be absent in many kinds of cells—for example, n some bacteria (e.g., *E. coli*), in some protozoa (e.g., *etrahymena*), and in plant cells in general. The plasma nembrane of many kinds of cells contains a molecular **ump** that transports K^+ ions from the extracellular nedium into the cell, at the same time extruding Na^+ ons from the cell (Chapter 5). The **Na^+-K^+ pump,** vhich requires Mg^{2+} ions for activity, is particularly vell developed in nerve and muscle cells. Gradients of K^+ and Na^+ ions across the plasma membrane are cre-ted by pumping the two ions in opposite directions. he gradients are the basis of an electrical potential across the plasma membrane that allows transmission of an impulse by either a nerve cell or muscle cell (Chapter 14).

The principal divalent cations (ions with two positive charges) of cells are Mg^{2+} and Ca^{2+}; these enter into almost every facet of cell structure and function. Among other things, the Ca^{2+} ion is important in the synthesis and breakdown of starch and for the contraction of muscle cells and motility of cells (Chapter 14). Ca^{2+} ions are necessary for the association of certain protein molecules with DNA, and in this regard play a major role in the folding and packing of chromosomes in eukaryotic cells. The stability of the ribosome, the organelle responsible for the synthesis of proteins, is dependent upon an appropriate concentration of Mg^{2+} ions in the cellular solution. A Mg^{2+} ion also forms a complex with a chlorophyll molecule and is essential for photosynthesis in all plant cells (Chapter 4). Various enzymes contain or require cations such as Ca^{2+}, Zn^{2+}, Mn^{2+}, Fe^{2+}, Co^{2+}, and Cu^{2+} in order to function. The phosphate ion (PO_4^{3-}), an **anion,** plays a central role in the energy metabolism of every cell, particularly in the synthesis of adenosine triphosphate (ATP), the molecule that drives most of the energy-requiring activities of a cell. Phosphate is also a major component of nucleic acid molecules and is an important covalent addition to some protein molecules.

The positive charges of all the monovalent and divalent cations in the cell are matched by an equivalent number of negative charges. These counter charges are contributed mainly by chloride ions (Cl^-), phosphate ions (PO_4^{3-}), bicarbonate ions (HCO_3^-), and negative charges of nucleic acids, proteins, and other organic molecules. A complete catalogue of the roles of inorganic ions in cells would not serve any purpose here; instead, it is more useful to discuss them as they enter into various aspects of cell function and structure described in subsequent chapters.

ORGANIC MOLECULES OF CELLS

The thousands of different intracellular organic molecules, which consist primarily of carbon, hydrogen, oxygen, and nitrogen, account for more than 90 percent of the weight of the cellular dry matter.

Much of the organic material in every cell consists of only three classes of very large molecules—proteins, nucleic acids, and carbohydrates. These are called **macromolecules** and are made up of a few kinds of small molecules known as **monomers.** Lipid molecules, another major constituent of the cell's organic material, are intermediate in size between macromolecules and small organic molecules. Most organic material not accounted for by the three types of macromolecules and the lipids consists primarily of (1) monomers that have not been assembled in macromolecules or lipids, (2) molecules that represent intermediate stages in the synthesis of monomers, (3) molecules that are intermediates in the digestion of sugar to carbon dioxide and water (Chapter 4), and (4) a relatively small number of molecules, such as certain vitamin derivatives, that serve mostly as helper molecules (coenzymes) for certain enzymes.

Molecular Weight of Organic Molecules

The weight of a molecule is the sum of the weights of all its atoms. The weights of atoms are given in **avograms.** One avogram is equal to 1.6×10^{-24} grams. The weights of organic molecules, particularly macromolecules, are useful to know, because they are a convenient measure of molecular size and complexity. Glucose ($C_6H_{12}O_6$) has a molecular weight of 180 (6 carbons = 72, 12 hydrogens = 12, 6 oxygens = 96) avograms. The molecular weights of proteins fall mostly in the range of ten thousand to several hundred thousand. DNA molecules are among the largest macromolecules. The DNA molecule in the chromosome of *E. coli* has a molecular weight (MW) of about 3.13×10^9, and the MW of individual DNA molecules in eukaryotic cells commonly exceeds 10^{10}.

There is no exact rule that defines when a molecule should be called a macromolecule, but generally, organic molecules with molecular weights greater than a few thousand are classified as macromolecules. Most lipid molecules have molecular weights of less than 1000 and hence do not fit into the somewhat arbitrar definition of a macromolecule.

Amino Acids and Proteins

Amino acids are the monomers from which pro teins are synthesized. It is possible to construct man kinds of amino acids, but only 20 different amino acic are used to make proteins. All biological amino acic have the basic skeleton

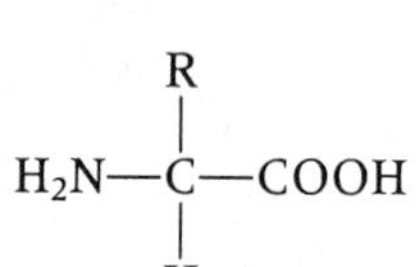

$$\begin{array}{c} R \\ | \\ H_2N-C-COOH \\ | \\ H \end{array}$$

The name *amino acid* derives from the presence of a amino group (NH_2) and a carboxyl group (COOH which can dissociate into COO^- and H^+ ions whe dissolved in water and hence is acidic because it adc H^+ ions to the solution. The 20 amino acids are define by the nature of the side group R. In glycine, the sim plest amino acid, R is a hydrogen atom:

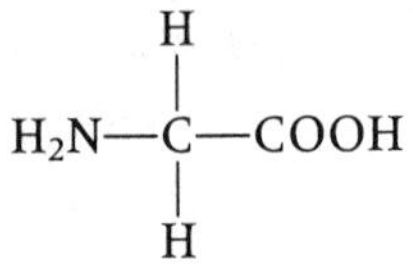

$$\begin{array}{c} H \\ | \\ H_2N-C-COOH \\ | \\ H \end{array}$$

In alanine, R is a CH_3 (**methyl**) group:

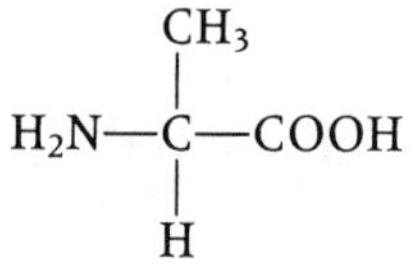

$$\begin{array}{c} CH_3 \\ | \\ H_2N-C-COOH \\ | \\ H \end{array}$$

The R groups (often called side chains or side group for 20 amino acids are shown in Figure 2-10. These sid

FIGURE 2-10
The 20 kinds of amino acids that make up proteins. Each amino acid is characterized by a different side group (R group).

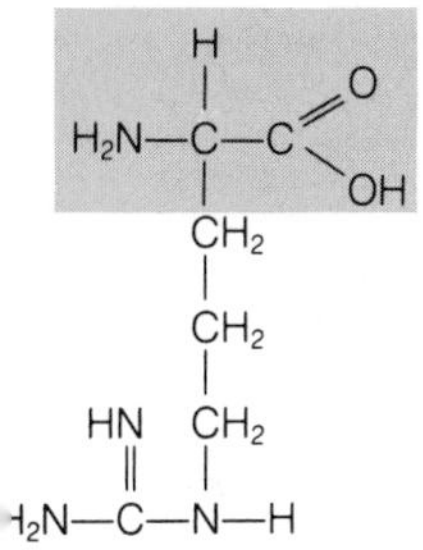

Arginine

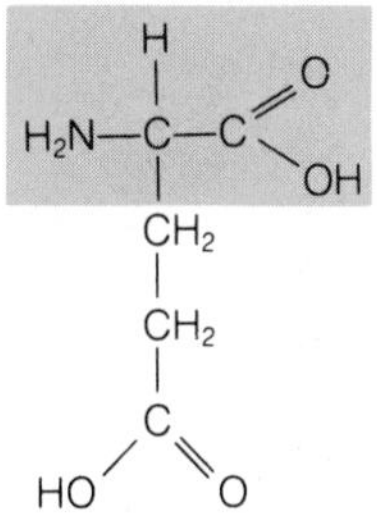

Glutamic acid

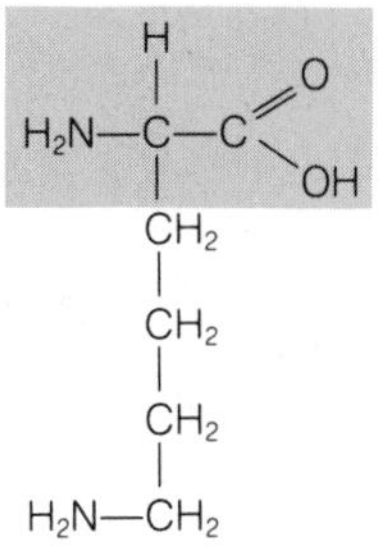

Lysine

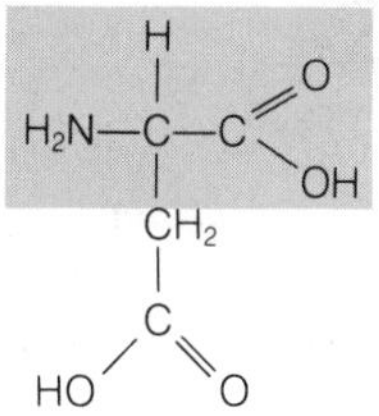

Aspartic acid

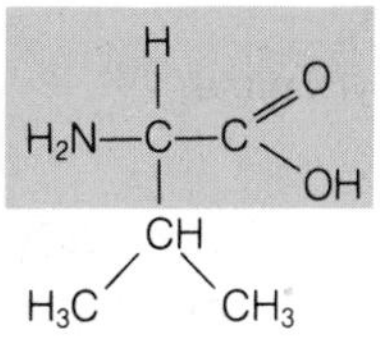

Valine

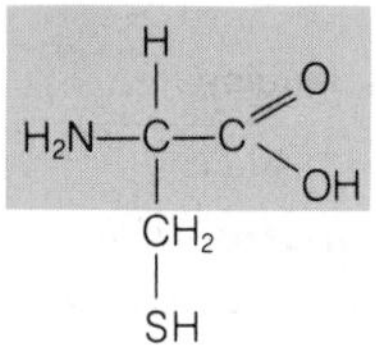

Cysteine

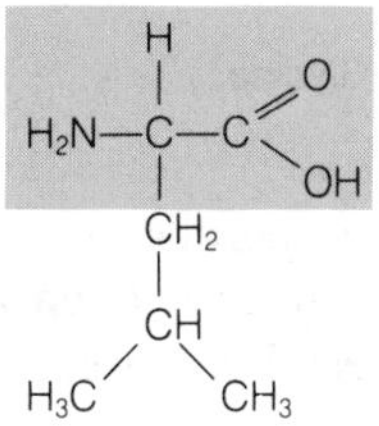

Leucine

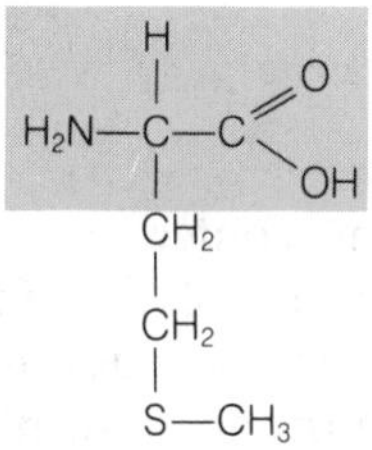

Methionine

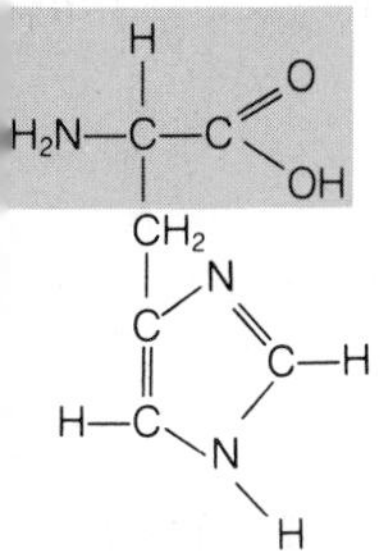

Histidine

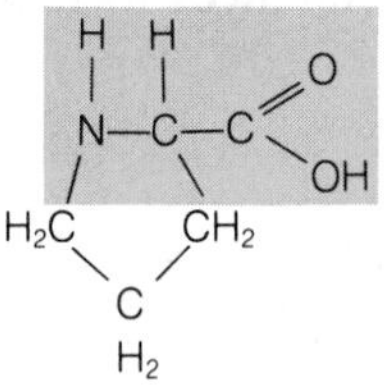

Proline

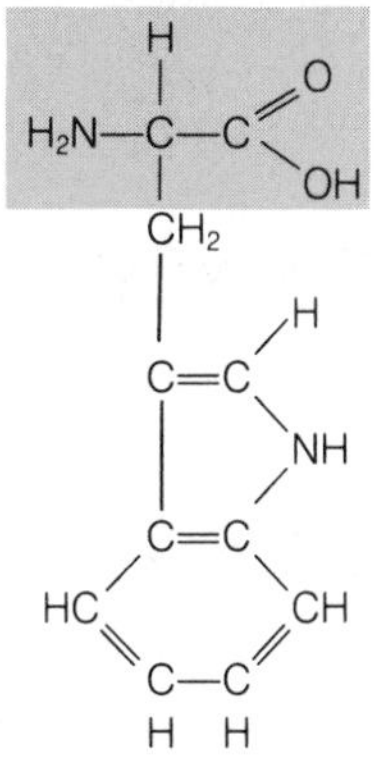

Tryptophan

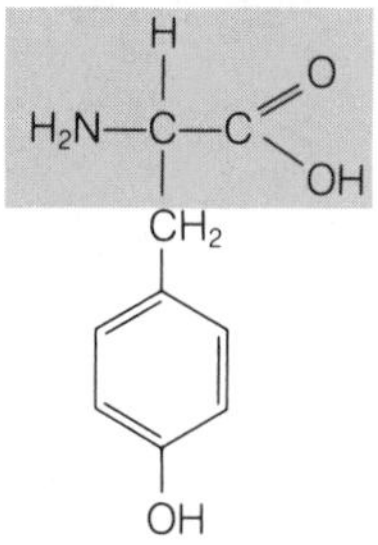

Tyrosine

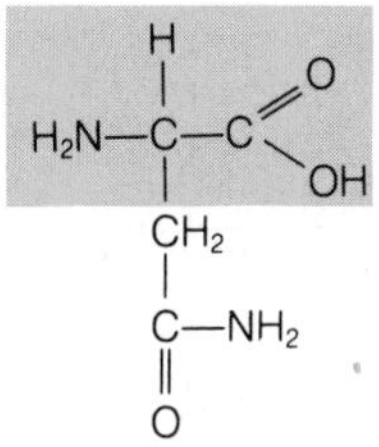

Asparagine

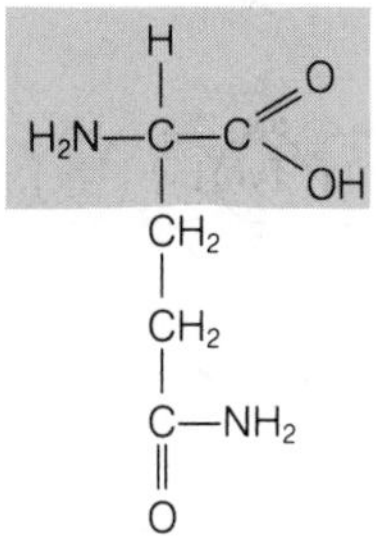

Glutamine

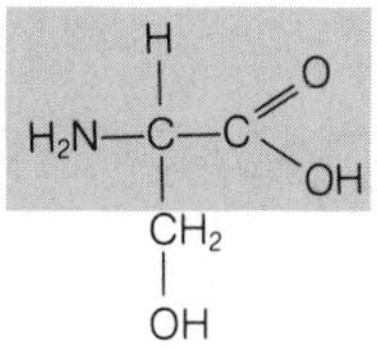

Serine

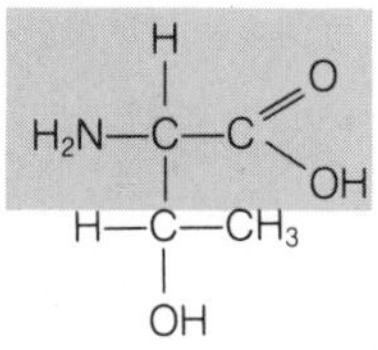

Threonine

→

FIGURE 2-10 (Cont.)

Alanine

Glycine

Isoleucine

Phenylalanine

chains confer specific chemical properties on each amino acid. Eight of the amino acids (**alanine, valine, leucine, isoleucine, proline, phenylalanine, tryptophan,** and **methionine**) have nonpolar R groups. Since these nonpolar R groups do not interact with water, these eight are classified as ***hydrophobic*** amino acids. Twelve of the amino acids have polar R groups. Since these groups interact with water, these 12 are called ***hydrophilic*** amino acids. Seven of these (**glycine, serine, threonine, cysteine, tyrosine, asparagine,** and **glutamine**) possess uncharged polar R groups. Five (**aspartic acid, glutamic acid, lysine, arginine,** and **histidine**) have charged polar R groups.

Amino acids are joined together in chains in various sequences to make macromolecules called **poly-**

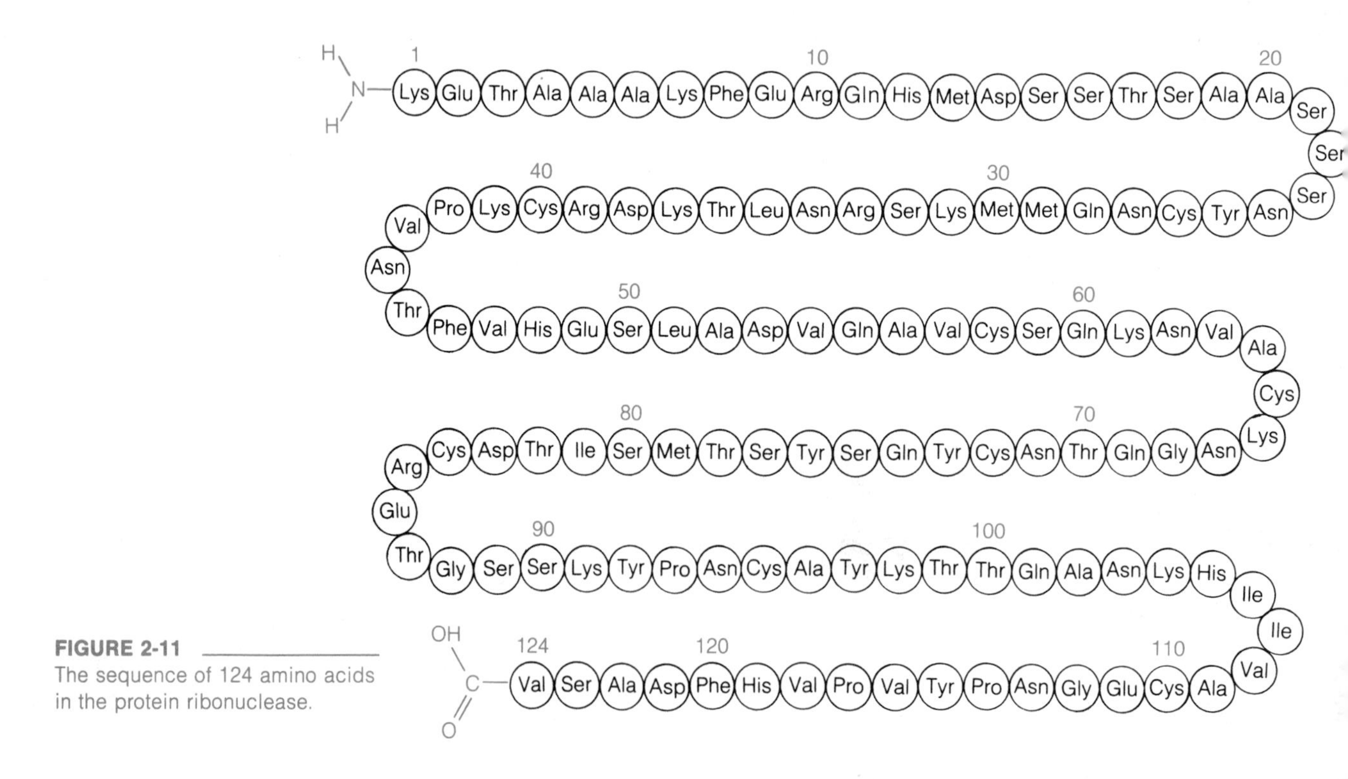

FIGURE 2-11
The sequence of 124 amino acids in the protein ribonuclease.

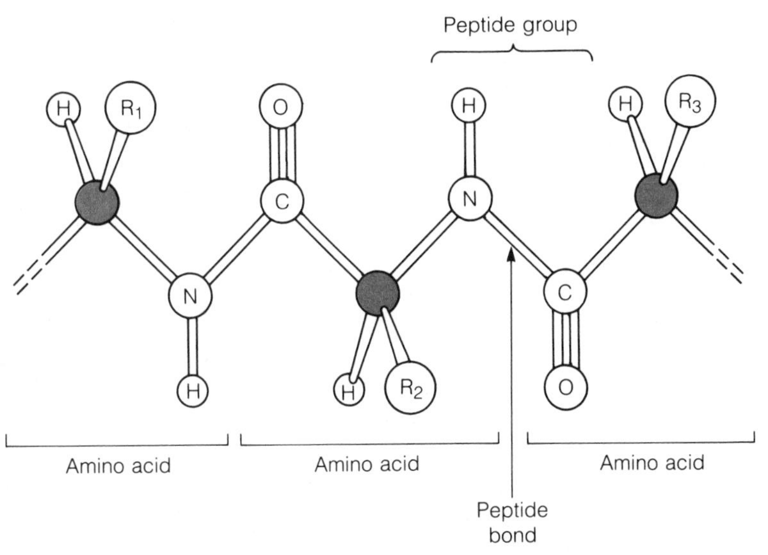

FIGURE 2-12
The arrangement of atoms in amino acids joined by peptide bonds.

peptides, which contain a few dozen to more than 1000 amino acids. Most polypeptides consist of a few hundred amino acids. Some **proteins** consist of a single folded polypeptide chain; an example is the enzyme ribonuclease (RNase) with a single polypeptide chain of 124 amino acids (Figure 2-11). Other proteins contain two or more chains, which may or may not be identical, held together by noncovalent bonds (see Figure 2-19).

In a polypeptide chain amino acids are joined by **peptide bonds.** A peptide bond is a covalent bond between the C of a carboxyl group of one amino acid and the N of the amino group of the next amino acid. To form such a C—N bond one molecule of water is removed (Figure 2-3). The four atoms involved in a peptide bond have bond angles that produce the structures shown in Figure 2-12.

The sequence of amino acids in the chain or chains of a protein is called the **primary structure** of the protein. Each chain has an amino group at one end (the **amino terminus**) and a carboxyl group at the other end (the **carboxyl terminus**) (Figure 2-11). Polypeptide chains are synthesized in the cell beginning with the amino acid that forms the amino terminus. Therefore, the amino terminus is designated as the start of the chain and the carboxyl terminus as the end. The amino acids are accordingly numbered 1, 2, 3, and so forth.

Rarely are polypeptide chains fully extended. They tend to collapse with one part of a chain sticking to another part of the same chain, particularly by the formation of hydrogen bonds between different parts of a chain. The C in a peptide bond has an O covalently attached to it, and the N in a peptide bond has an H covalently attached to it. The O has a slight negative charge, and the H has a slight positive charge, and therefore each can participate in the formation of a hydrogen bond. These hydrogen-bonding capabilities are important in determining the structure of a protein. A hydrogen bond may form between an O in one part of a peptide chain and an N in another part of the same chain. These hydrogen bond possibilities result in two forms of secondary structure in polypeptide chains, the **α-helix** and the **β-structure.**

The α-helix. A polypeptide chain may form a coil or helix in much the same way that a piece of thread can be wound into a coil. Coiling a polypeptide chain with 3.6 amino acids in a full turn of the helix permits the formation of intrachain hydrogen bonds between successive coils of the helix (Figure 2-13(a) and (b)). The

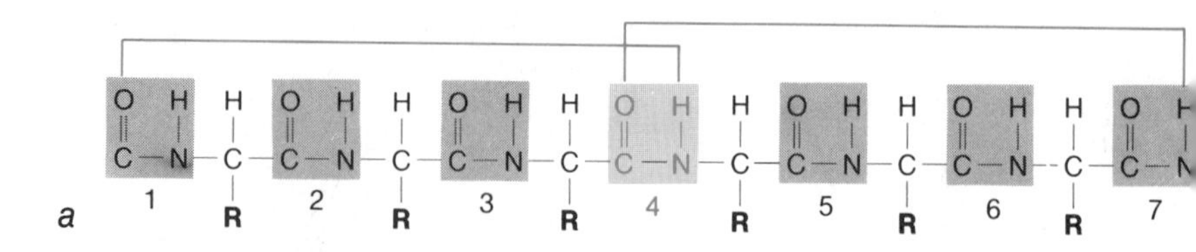

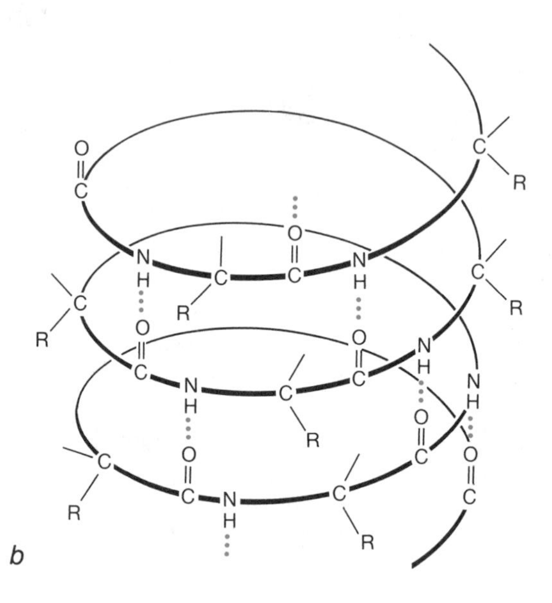

FIGURE 2-13
Hydrogen bonds in the α-helix. (a) Each peptide group forms a hydrogen bond with the fourth peptide group in each direction along the amino acid chain. (b) Coiling of an amino acid chain brings peptide groups into juxtapositions so that the hydrogen bonds shown in (a) can form. The multiple hydrogen bonds stabilize the helical configuration. Red dots indicate hydrogen bonds.

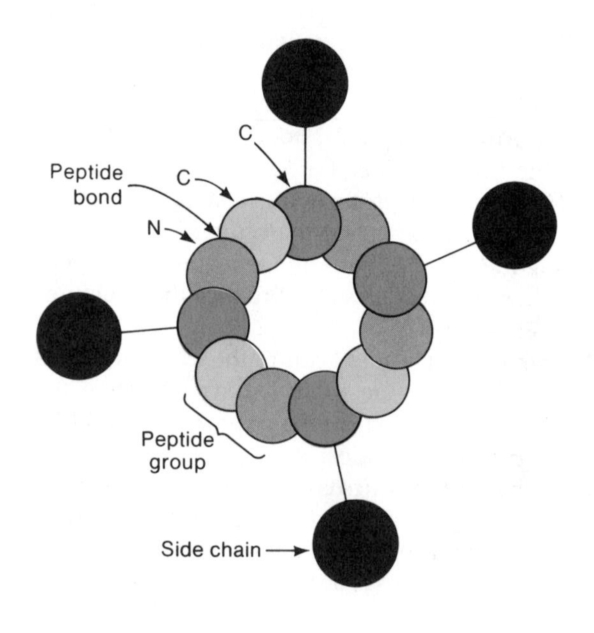

FIGURE 2-14
A schematic diagram of a cross section of one turn of an α-helix. Peptide bonds are between pink carbon atoms and gray nitrogen atoms. Side groups (represented by black circles) project outward from carbon atoms colored red. In a fully space-filling model the carbon and nitrogen atoms would fill most of the core of the helix.

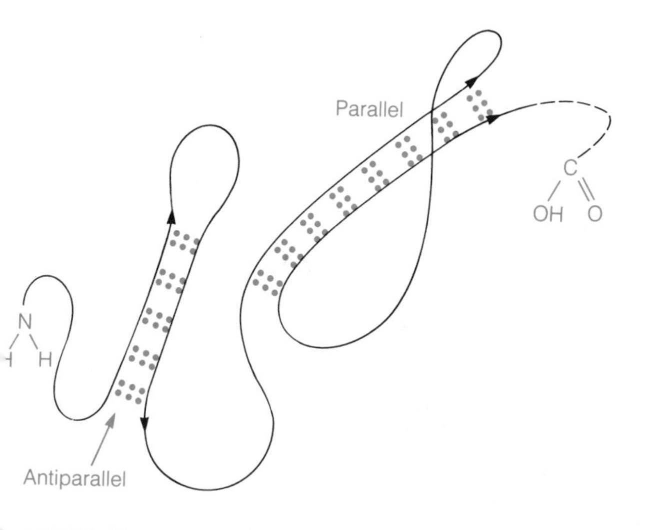

FIGURE 2-15
β-structures can be formed between stretches of a polypeptide chain that are antiparallel or parallel with respect to NH_2 to COOH polarity of the chain. Red dots indicate hydrogen bonds.

hydrogen bonds are more or less parallel to the axis of the helix and extend from the nitrogen of one peptide bond and the carbonyl oxygen (C=O) of the fourth amino acid down the chain.

FIGURE 2-16
β-structure formed between two stretches of a polypeptide chain in antiparallel orientation. Hydrogen bonds between the two sections of the polypeptide stabilize their association. Side groups (red circles) of the amino acids project below and above the plane of the β-structure.

In this form of folding of a chain, which is called an **α-helix,** every C and N atom in a peptide bond participates in a hydrogen bond. The R groups project outward from the helix (Figure 2-14). The size and charge (if any) of R groups strongly influence whether a stable α-helix will form. For example, the bulky R group of isoleucine (Figure 2-10) gets in the way of adjacent R groups so that the amino acids cannot occupy positions necessary for α-helical hydrogen bonding. The three-dimensional configuration conferred by such intrachain bonds is called the **secondary structure** of the polypeptide as distinguished from the **primary structure,** which is just the sequence of amino acids.

Portions of the polypeptide chains of many proteins are in the α-helical configuration, but usually this form of secondary structure extends for only 10 or 15 amino acids and is then interrupted by non-α-helical stretches. Keratin (the protein of hair and wool), major proteins in muscle cells, and hemoglobin (the protein of red blood cells) are examples of proteins that contain large amounts of α-helical structure in their polypeptides.

The β-structure. In contrast to the helical coil arrangement, a polypeptide chain may exist in a more stretched out form in which a polypeptide may fold back on itself (Figure 2-15) such that two parts of the chain lie side by side in a parallel or antiparallel arrangement. In this array the two portions can become hydrogen-bonded to one another, giving rise to a form of secondary structure called the **β-structure** (Figure 2-16). A portion of the polypeptide chain can form

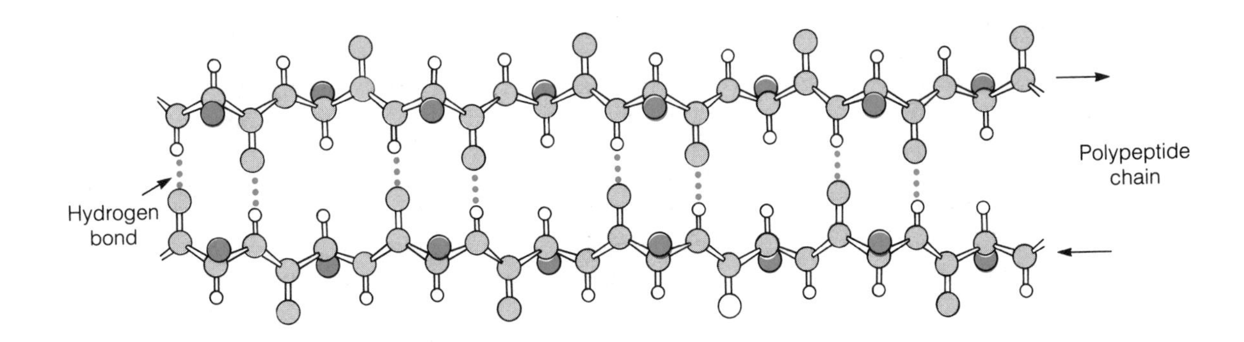

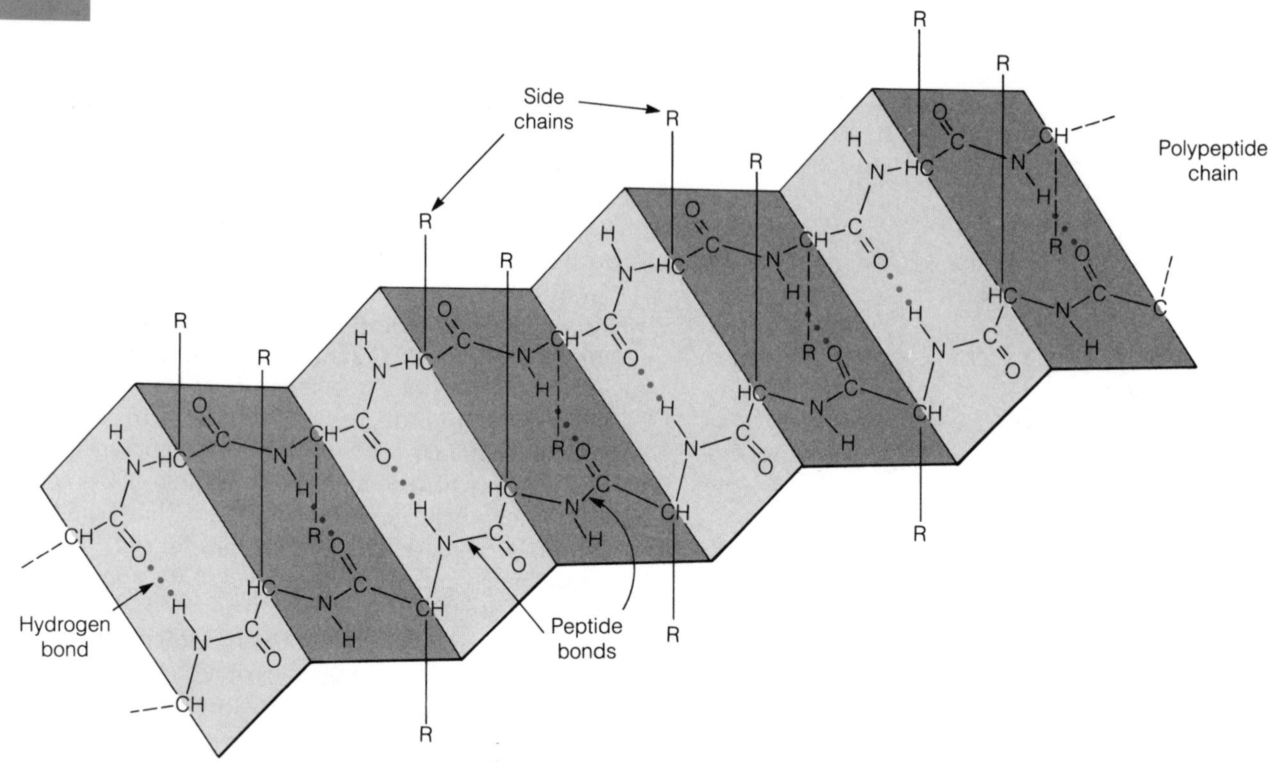

FIGURE 2-17
The plane formed by the association of stretches of a polypeptide chain in a β-structure is actually a pleated sheet because of the bond angles of the carbon atoms. In this structural representation the disposition of side groups above and below the pleated sheet is more easily perceived. Hydrogen bonds are shown in red.

hydrogen bonds with a portion of the chain lying to each side of it. This structural arrangement, with two or more portions of a chain binding to each other, is also called a **β-pleated sheet** because the bond angles in the parallel chains are in register so as to give the impression of a sheet with pleats (Figure 2-17). In the β-pleated sheet the R groups project above and below the plane of the sheet. Regions of β-structure are found in many proteins, but β-pleated sheets with three or more portions of a chain joined to each other are unusual. An example of a protein with extensive β-pleated sheets is the protein that makes up silk fibers. The sheen of silk is due to the reflective property of β-pleated sheets.

The β-structure and the α-helix constitute an intermediate degree of structural organization. A polypeptide with α-helical structures and β-structure undergoes further folding to produce another higher order of structural organization.

Tertiary structure of proteins. The primary structure of a protein is the sequence of amino acids in its polypeptide(s), and the secondary structure consists of regions of the polypeptide chain organized into α-helices and β-structures. A polypeptide chain containing such regions usually undergoes further folding. This folding is called **tertiary structure** (Figure 2-18(a), (b), and (c)). The R groups of amino acids, which project outward from α-helical regions and β-structures, play a major role in determining the tertiary structure of a protein. R groups influence folding through interactions with one another and with water. Hydrophobic R groups interact with one another because of their

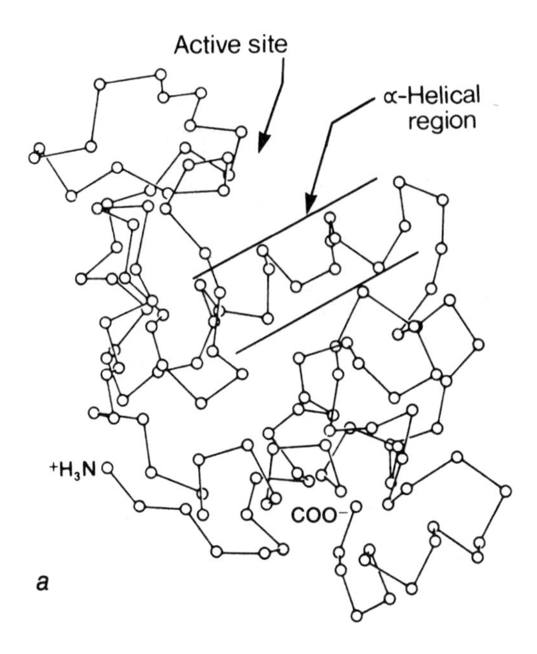

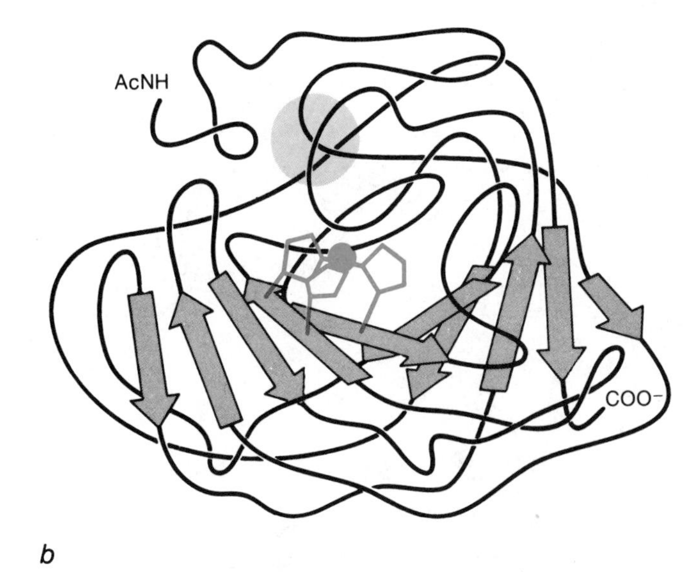

FIGURE 2-18
Three-dimensional structure of proteins. (a) Stick model of lysozyme (an enzyme that degrades certain polysaccharides). Each circle denotes an amino acid; side groups are not shown. The indicated region of α-helix forms one side of a cleft in which the active site of the enzyme is located. (b) A drawing of the tertiary structure of carbonic anhydrase. Regions of the amino acid chain that form β-structures are shown as arrows. A cluster of interactive hydrophobic amino acids occurs in the shaded red circle. Three histidines (red) and a zinc ion (red dot) form the active site. (c) Space-filling molecular model of lysozyme. The active site is in the cleft indicated by the arrow. (C atoms are black; H, white; N, gray; O, gray with slots.) [Courtesy of John A. Rupley.]

mutual noninteraction with water. Because of such **hydrophobic interactions** a polypeptide chain tends to fold such that the hydrophobic R groups are concentrated inside the folded structure out of contact with water. At the same time, chains tend to fold such that the hydrophilic R groups are at the surface of a protein, to maximize polar interactions with water. Thus, in the folding of a polypeptide chain, with its secondary structure, a balance is struck between the internalization of hydrophobic R groups and surface localization of hydrophilic R groups. The overall result is that a polypeptide with a particular sequence of amino acids will usually fold into a small number of configurations, or even a single, tertiary configuration.

Quaternary structure of proteins. Many proteins contain more than one polypeptide chain. The individual polypeptides fold to yield secondary and tertiary

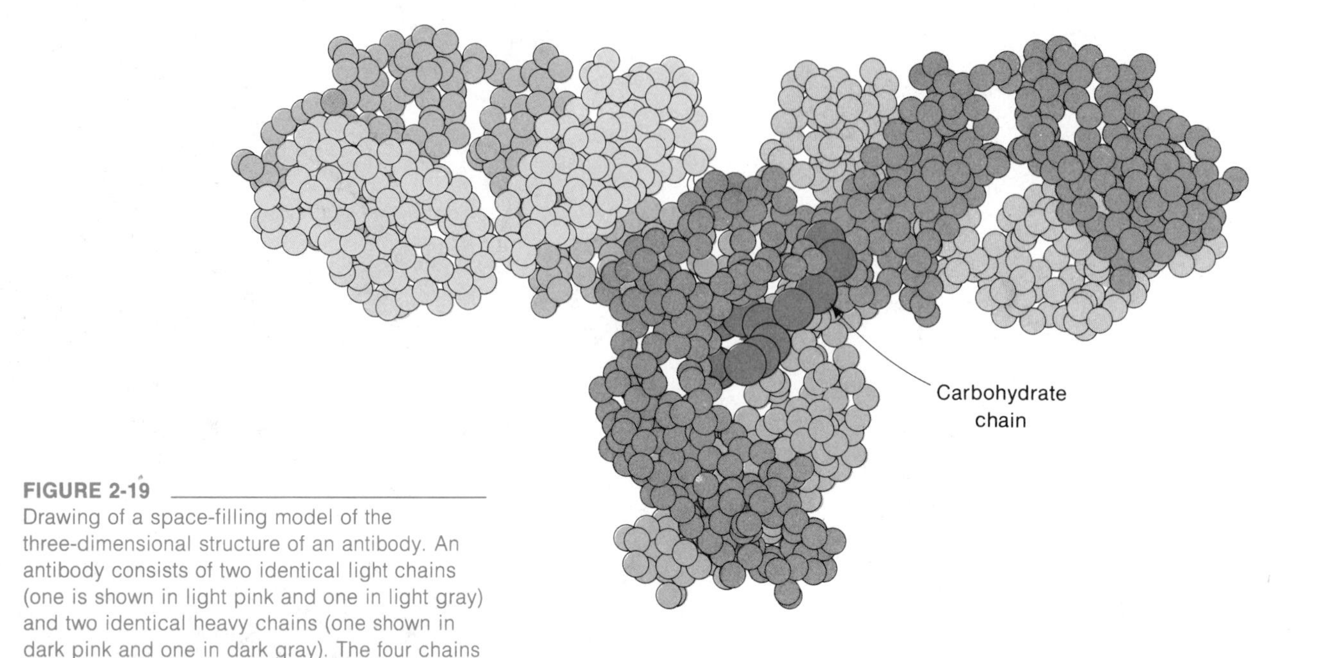

FIGURE 2-19
Drawing of a space-filling model of the three-dimensional structure of an antibody. An antibody consists of two identical light chains (one is shown in light pink and one in light gray) and two identical heavy chains (one shown in dark pink and one in dark gray). The four chains are held together in a quaternary structure by disulfide bridges. [After E. W. Silverton, M. A. Navia, and D. R. Davies. 1977. *Proc. Natl. Acad. Sci. USA*, 74: 5142.]

structure but also bind to one another in precise ways through hydrogen bonds, van der Waals forces (described later), ionic bonds, disulfide bridges (described later), and hydrophobic interactions. An *antibody* is a protein composed of four, folded polypeptide chains held together by several kinds of interactions and bonds, including disulfide bonds (Figure 2-19). The organization of two or more polypeptides into one unit is called the **quaternary structure** of a protein.

The orders of structure in a protein are summarized in Figure 2-20(a-e). The primary structure of a polypeptide determines the secondary, tertiary, and quaternary structure of a protein. Primary and secondary structure are fitted into higher level tertiary and quaternary structure, producing a three-dimensional arrangement of amino acids in space. The tertiary and quaternary structure give most proteins their functional properties.

Stabilization of tertiary structure by disulfide bonds. In many proteins tertiary structure is determined only by various kinds of weak interactions between amino acids, such as hydrogen bonds and hydrophobic interactions. The tertiary structure of some proteins, for example, ribonuclease (Figure 2-11), is further stabilized by covalent bonds between the **sulfhydryl groups** (SH groups) of the amino acid cysteine. The sulfur atom in the sulfhydryl group of cysteine is in a chemically reduced state, since it has a hydrogen atom bound to the sulfur in its side group (CH_2—SH). Two cysteines may be oxidized (removal of H atoms) and join covalently in a **disulfide bridge** (Figure 2-21).

The folding of some polypeptides brings cysteine residues into juxtaposition so that covalent disulfide bridges may form. These bridges further stabilize the tertiary folding of ribonuclease, in which the role of disulfide bridges in protein structure has been much studied. Ribonuclease contains eight cysteines in four disulfide bridges (Figure 2-22). The eight cysteines oc-

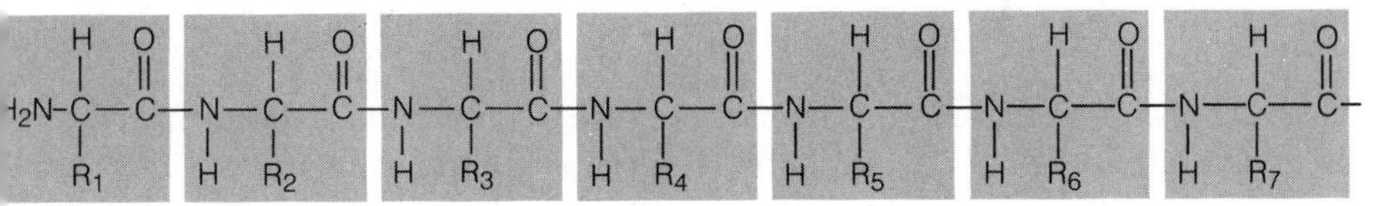

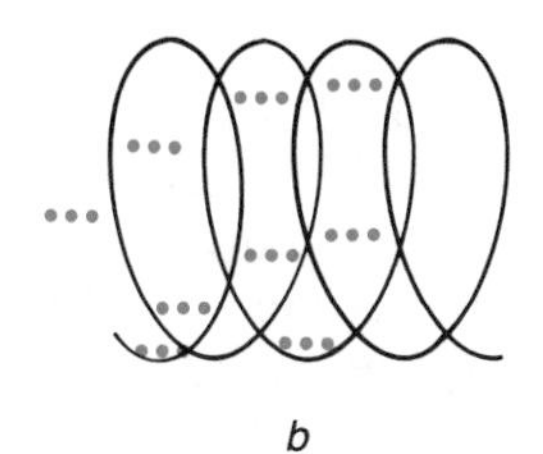

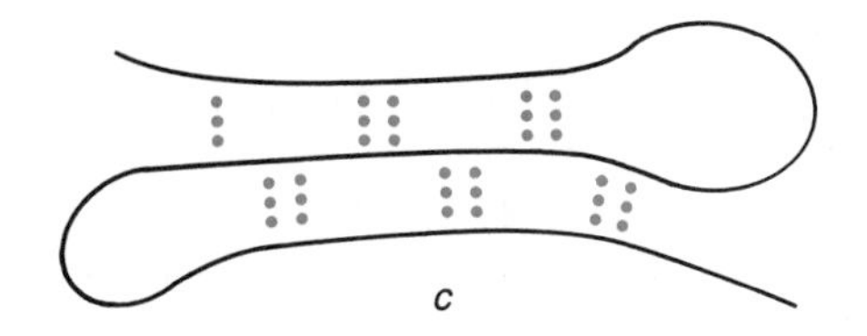

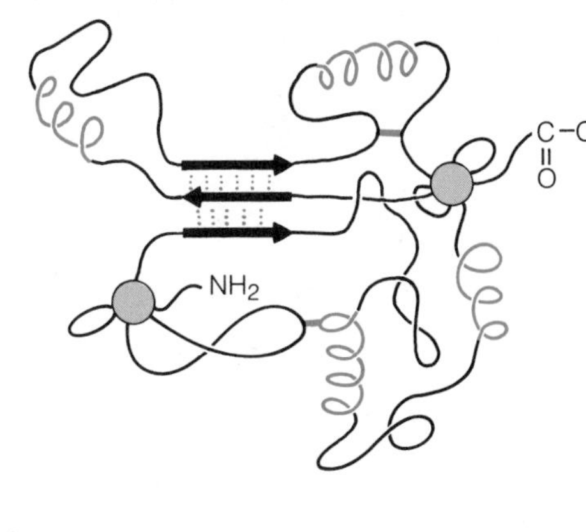

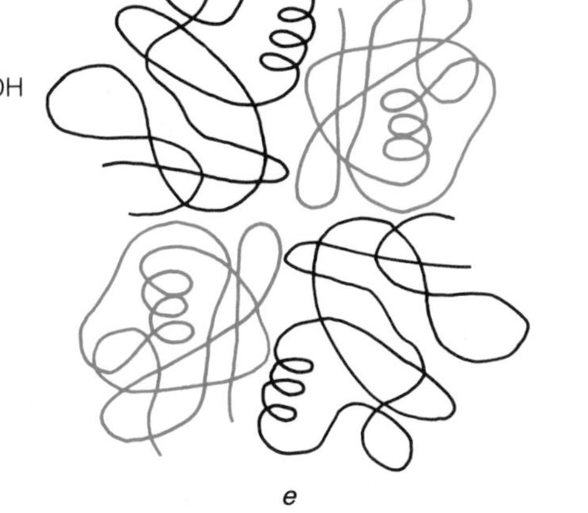

FIGURE 2-20
The structural orders of a protein. (a) Primary structure is the sequence of amino acids, joined by peptide bonds. The first seven amino acids (shaded boxes) of a polypeptide chain starting from the amino terminus are shown. (b) A region of a polypeptide in an α-helix. Hydrogen bonds join successive turns of the helix. (c) Antiparallel β-structure formed among three regions of a polypeptide chain. (d) A schematic diagram of a polypeptide folded into a specific form with various kinds of interactions among side groups of amino acids. Helical regions are in red. The heavy black arrows represent a region of β-structure. The shaded dots are clusters of hydrophobic side groups of amino acids, and the two heavy red bars represent disulfide bonds between side chains of cysteine molecules (see Figures 2-21 and 2-22). (e) Four polypeptide chains held together by weak bonds in a quaternary configuration in a protein. The two chains in red are identical to one another, and the two chains in black are identical to one another.

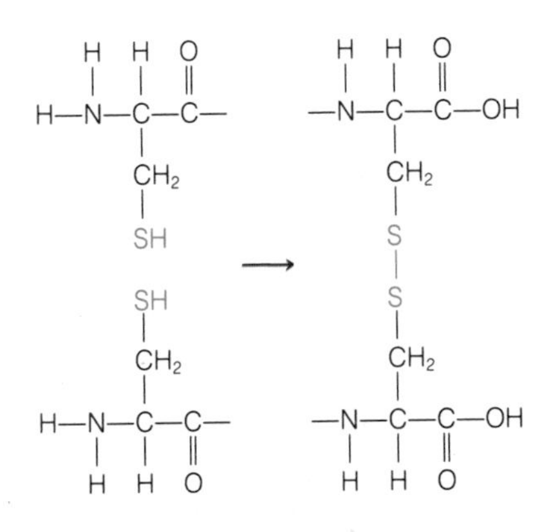

FIGURE 2-21
Two sulfur-containing amino acid cysteine molecules can form a disulfide bridge between different regions of a polypeptide chain, stabilizing the three dimensional form of a protein.

cupy positions 26, 40, 58, 65, 72, 84, 95, and 110 in the polypeptide chain. Folding of the chain into a functional protein brings each cysteine residue in juxtaposition with its correct partner (26–84, 40–95, 65–72, and 58–110) to form stabilizing disulfide bridges.

Denaturation of proteins. The tertiary and quaternary structure of proteins can easily be disrupted by conditions that overcome the weak forces on which polypeptide folding depends. Heat, high and low pH, detergents (which eliminate hydrophobic interactions), and agents that disrupt hydrogen bonds cause polypeptides to unfold. Such unfolding is called **denaturation;** it is invariably accompanied by loss of the normal biological function (e.g., enzyme activity) of the protein. Denatured proteins are usually not soluble in water, in part because denaturation exposes internal hydrophobic R groups. Hence, when proteins alread dissolved in water are heated, they may lose solubilit and precipitate. A familiar example is the denaturatio of proteins caused by heating milk; the denatured proteins precipitate and form a skin on the surface.

Functions of proteins. Almost all enzymes, whic catalyze the chemical reactions within a cell (Chapte 3), are proteins. Although the vast majority of enzyme are proteins, not all proteins are enzymes. Some proteins have no enzymatic activity and have other roles Collagen, a protein synthesized by the fibroblast cells o connective tissue and released into the extracellula space, has a structural function. After release from th cell, collagen precursor proteins assemble into fiber

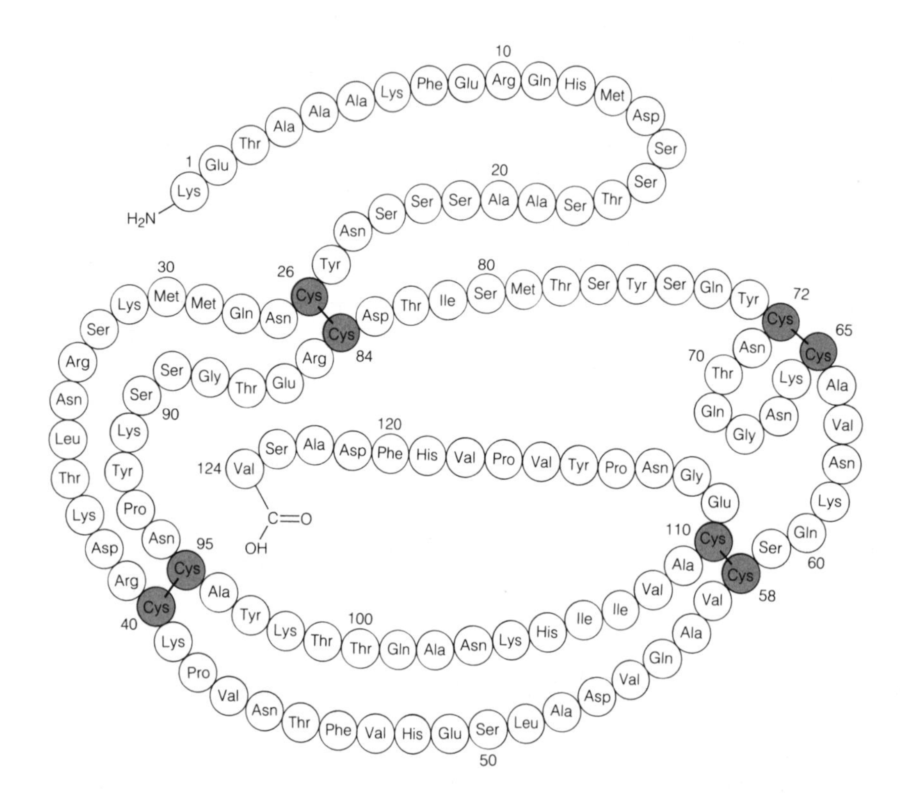

FIGURE 2-22
Schematic diagram of the protein ribonuclease, which contains 124 amino acids stabilized in a specific configuration b four disulfide bridges among eight cysteine molecules.

hat form tendons, ligaments, and the proteinaceous part of bone. Some nonenzymatic proteins, such as histones, have roles in the structure and function of chromosomes.

Even relatively simple cells, such as the bacteria, contain perhaps 2000 different proteins. The functions of less than half of these are known. Eukaryotic cells contain even greater numbers of different proteins, most of whose specific functions are not known. All of the proteins of a cell contribute to one or another aspect of cell function and structure. The solutions to many of the unsolved problems in cell biology will depend on identification of the functions, enzymatic or otherwise, of many of these proteins.

Proteins are distributed throughout the entire cell. Many proteins function in a soluble state in the cytosol and in the nuclear sap. Others are functional parts of structures and organelles like membranes, ribosomes, chromosomes, mitochondria, and so forth. Each structure or organelle in a cell contains a particular spectrum of proteins. Many proteins in eukaryotic cells occur in both the nucleus and cytoplasm, although some proteins are found only in either one or the other. In addition, many proteins move continually back and forth between the nucleus and the cytoplasm. The migration of at least some of these proteins constitutes a molecular communication between the nucleus and cytoplasm by which the two compartments of the cell influence the activities of each other and achieve coordination.

Turnover of proteins. The proteins within a cell are continually being synthesized and broken down to amino acids; that is, they undergo **turnover.** Turnover comes about in part because protein molecules are damaged by collision with other molecules. Damaged molecules have impaired functions. By means that we do not understand, damaged proteins inside the cell are recognized and selectively broken down to amino acids by degradative enzymes called **proteases.** The released amino acids are then reused in the continuous synthesis of new proteins. Most other molecules (RNA, lipids, carbohydrates, etc.) undergo similar turnover. DNA is an exception and does not undergo turnover. Instead, the cell has enzymes that recognize and repair sections of damaged DNA molecules, provided the damage is not too extensive.

In a mammalian cell the lifetimes of various kinds of protein molecules are typically several hours to a week or more. Because of this turnover, it is possible to label proteins with radioactivity by feeding to cells an amino acid containing radioactive carbon (^{14}C) or radioactive hydrogen (^{3}H or tritium); labeling occurs even in nongrowing cells. The ability to label proteins with radioactivity provides an important tool in the analysis of the mechanism of protein synthesis and the behavior of proteins within the cell.

Nucleotides and Nucleic Acids

Nucleic acids account for 0.5–1 percent of the dry weight of a cell. Cells contain two kinds of nucleic acid molecules, **ribonucleic acid (RNA)** and **deoxyribonucleic acid (DNA).** These two molecules differ in the chemical structure of their component nucleotides. The differences are slight, but important. RNA and DNA have quite different, although related, functions within the cell. A nucleic acid molecule is constructed of monomeric units called **nucleotides** joined together to form long chains. Nucleotides are composed of three

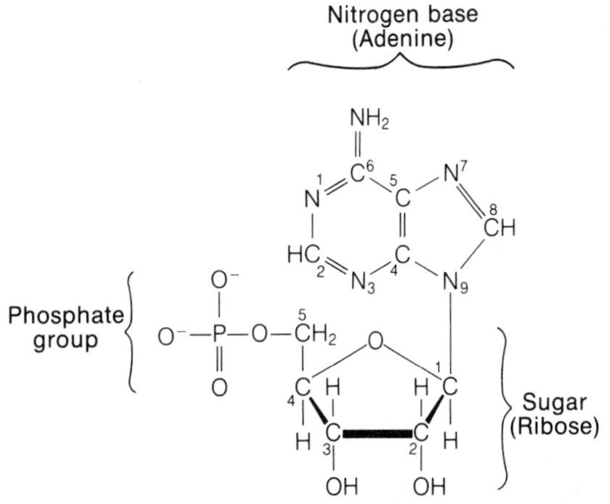

FIGURE 2-23
A nucleotide consists of a nitrogen base (in this case, adenine), a sugar molecule, and a phosphate group.

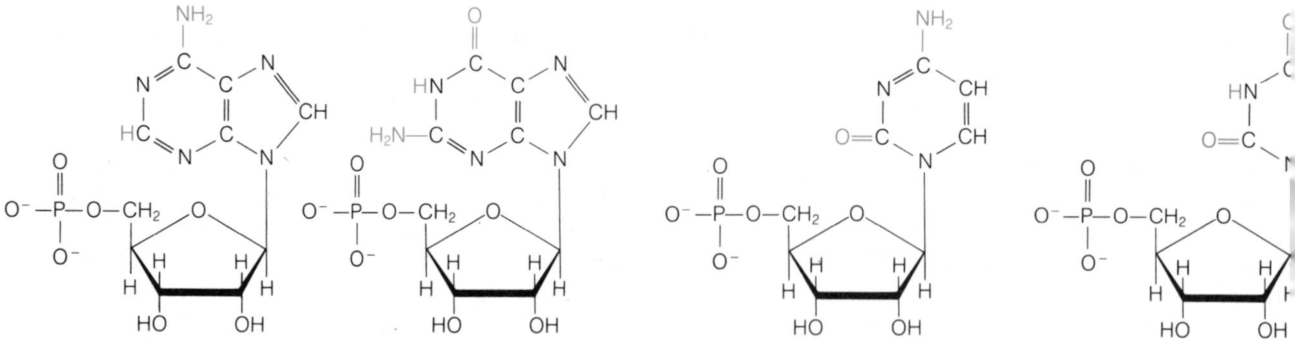

Adenosine monophosphate

Guanosine monophosphate

Cytidine monophosphate

Uridine monophosphate

FIGURE 2-24

The four different nucleotides that form RNA. The four nucleotides contain different nitrogen bases but are otherwise the same. The atoms colored red show the difference between the two purines (adenine and guanine) and between the two pyrimidines (cytosine and uracil).

parts: a nitrogen base, which may be a **purine** or a **pyrimidine,** a sugar molecule, and a phosphate group (PO_4) (Figure 2-23).

RNA is built from four kinds of nucleotides. Figure 2-24 shows the structure of the four nucleotides of RNA. They contain the four nitrogen bases, two purines (**adenine** and **guanine**), and two pyrimidines (**cytosine** and **uracil**). In each case, a ribose sugar is attached to the purine or pyrimidine (Figure 2-24), and a PO_4 group is in turn attached to the number 5 carbon atom of the ribose molecule. A purine or pyrimidine with just a ribose molecule attached is called a **nucleoside.** The four nucleosides of RNA are called **adenosine, guanosine, cytidine,** and **uridine.** Attachment of PO_4 to the sugar of a nucleoside makes a **nucleotide.** The four nucleotides of RNA are **adenylic acid, guanylic acid, cytidylic acid,** and **uridylic acid.** A nucleotide is sometimes also called a **nucleoside monophosphate,** that is, adenosine monophosphate (AMP), guanosine monophosphate (GMP), cytidine monophosphate (CMP), and uridine monophosphate (UMP).

In summary,

Purine or pyrimidine + ribose

Nucleoside + PO_4

Nucleotide (nucleoside monopho phate)

or specifically,

Adenine + ribose

Adenosine + PO_4

Adenylic acid (adenosine monophospha or AMP)

DNA is also built from four nucleotides (Figure 2-25), but is different from RNA in two important way (1) The pyrimidine uracil in RNA is represented in DN by the pyrimidine **thymine.** Uracil and thymine a very similar in structure; in thymine the pyrimidir ring has a methyl group (CH_3) attached to the fift atom in the ring (a carbon) and uracil has a hydroge atom instead (compare Figures 2-24 and 2-25). (2) R bose is present in RNA nucleotides and **deoxyribose** present in DNA nucleotides. Deoxyribose is a ribos molecule that has an H atom on the second carbo atom instead of an OH group (Figure 2-26).

The names of the components in the nucleotides RNA and DNA are summarized in Table 2-1. Attach ment of deoxyribose to the two purines (adenine an guanine) and to the two pyrimidines (cytosine and thy

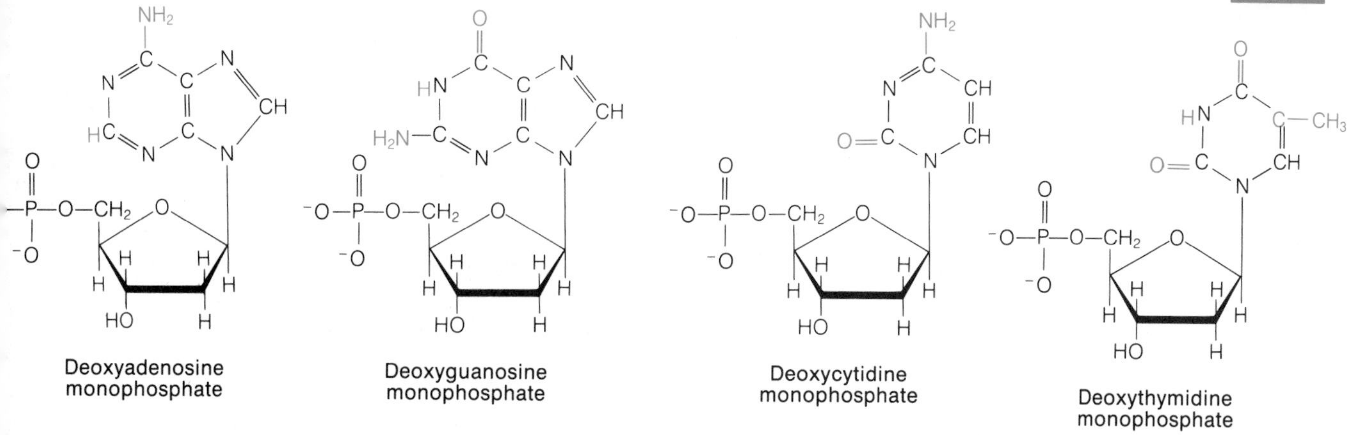

Deoxyadenosine monophosphate

Deoxyguanosine monophosphate

Deoxycytidine monophosphate

Deoxythymidine monophosphate

FIGURE 2-25
The four different nucleotides that form DNA. The four nucleotides contain different nitrogen bases but are otherwise the same. The atoms colored red show the difference between the two purines (adenine and guanine) and between the two pyrimidines (cytosine and thymine).

TABLE 2-1
Nucleic Acid Nomenclature

Base	Nucleoside[1]	Nucleotide[2]
	Purines (Pu)	
Adenine (A)	Adenosine (rA)	Adenylic acid, or adenosine monophosphate (AMP)
	Deoxyadenosine (dA)	Deoxyadenylic acid, or deoxyadenosine monophosphate (dAMP)
Guanine (G)	Guanosine (rG)	Guanylic acid, or guanosine monophosphate (GMP)
	Deoxyguanosine (dG)	Deoxyguanylic acid, or deoxyguanosine monophosphate (dGMP)
	Pyrimidines (Py)	
Cytosine (C)	Cytidine (rC)	Cytidylic acid, or cytidine monophosphate (CMP)
	Deoxycytidine (dC)	Deoxycytidylic acid, or deoxycytidine monophosphate (dCMP)
Thymine (T)	Thymidine[3] (dT)	Thymidylic acid, or thymidine monophosphate (TMP)
Uracil (U)	Uridine[4] (rU)	Uridylic acid, or uridine monophosphate (UMP)

[1] Note that the names of purine nucleosides end in -osine and the names of pyrimidine nucleosides end in -idine.
[2] Note that each nucleotide has two names for the same substance.
[3] Thymidine is the deoxy- form. The ribo- form, ribosylthymine, is not generally found in nucleic acids.
[4] Uridine is the ribo- form. Deoxyuridine is not commonly found, although deoxyuridylic acid is on the pathway for synthesis of thymidylic acid—i.e., deoxyuridylic acid is methylated to yield thymidylic acid.

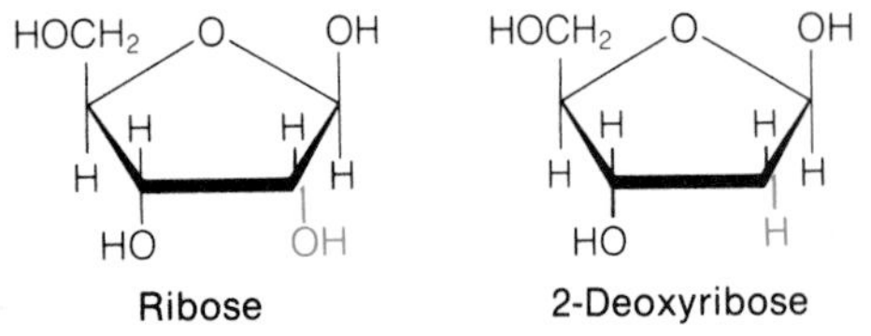

HOCH₂ O OH H H H H HO OH

Ribose

HOCH₂ O OH H H H H HO H

2-Deoxyribose

FIGURE 2-26

Ribose is the 5-carbon sugar in the nucleotides of RNA, and deoxyribose is the sugar in DNA. The difference between the two sugars is shown in red.

mine) yields the four nucleosides, but because the sugar is deoxyribose instead of ribose, the nucleosides are called **deoxynucleosides.** The four deoxynucleosides of DNA are **deoxyadenosine, deoxyguanosine, deoxycytidine,** and **thymidine.** Because thymidine occurs only in DNA, it always contains deoxyribose, and the deoxy- prefix is omitted.

A deoxynucleotide is a deoxynucleoside with a PO_4 group (deoxynucleotide = deoxynucleoside phosphate). The four deoxynucleotides or deoxynucleoside monophosphates are deoxyadenosine monophosphate (dAMP), deoxyguanosine monophosphate (dGMP), deoxycytidine monophosphate (dCMP), and thymidine monophosphate (TMP). The four deoxynucleotides also are called by the synonyms deoxyadenylic acid, deoxyguanylic acid, deoxycytidylic acid, and thymidylic acid.

In short, the four nucleotides of RNA consist of ribose attached to adenine, guanine, cytosine, and uracil and with phosphate (PO_4) attached to the ribose. Similarly, the four deoxynucleotides of DNA consist of deoxyribose attached to adenine, guanine, cytosine, and thymine with phosphate attached to deoxyribose.

In RNA and DNA the four nucleotide monomers are joined into long chains in which the phosphate of one nucleotide is linked to a carbon atom in the sugar of the next nucleotide. Thousands of nucleotides are joined to make one macromolecular chain of RNA or DNA. Details of how the joining occurs will be given in Chapter 7.

RNA occurs abundantly in the nucleus, where it is synthesized, and in the cytoplasm, where it plays a

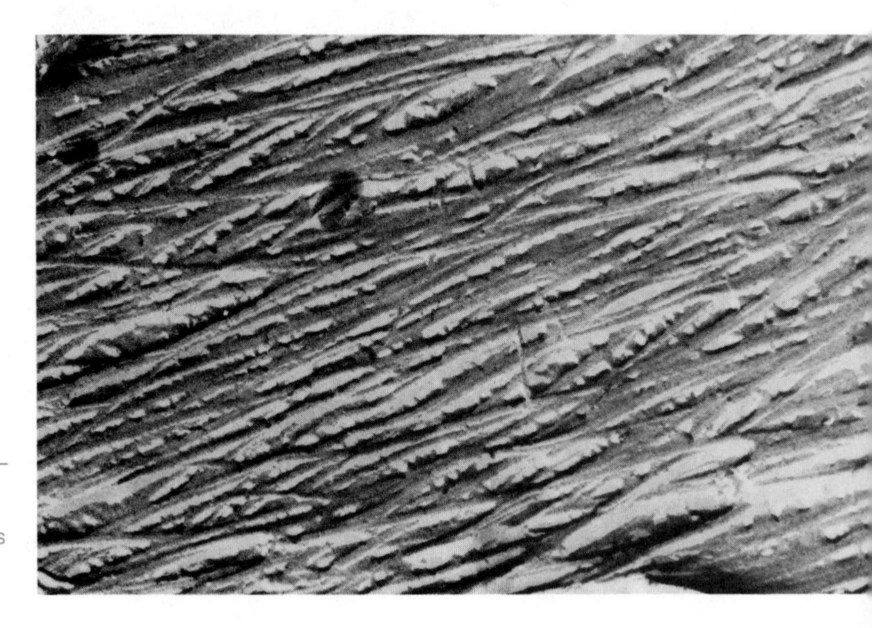

FIGURE 2-27

Electron micrograph of the cell wall of the alga *Closterium*. The cell was frozen and then split open by fracturing. The many fibers are made of cellulose. [Courtesy of Thomas H. Giddings, Jr.]

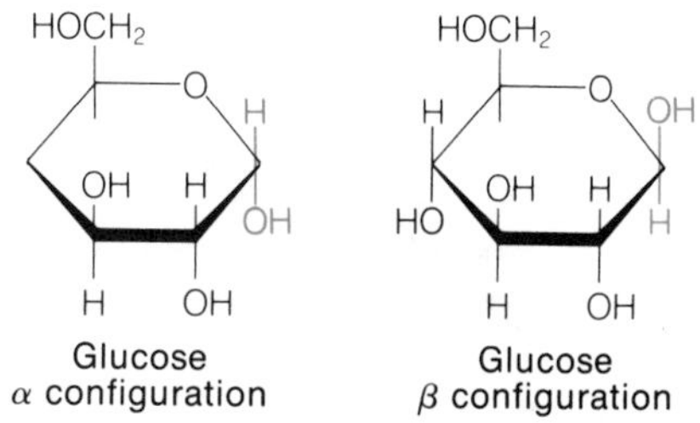

FIGURE 2-28

Two alternating forms of glucose defined by the positions of the H and OH groups (in red) attached to carbon number one.

central role in protein synthesis. DNA occurs in three locations in eukaryotic cells. Most of the DNA is contained in the chromosomes in the nucleus. Some DNA is also found in mitochondria and chloroplasts but it represents a minute fraction of the genes. Further details about the functions of the various kinds of RNA and DNA molecules in the nucleus and cytoplasm will be discussed in Chapters 7 through 10.

Sugars and Polysaccharides

By far the most important sugar in cells is **glucose.** Glucose occupies a central position in energy metabolism (Chapter 4) and is also the molecule from which other sugars, such as ribose and deoxyribose of nucleic acids, are derived.

Glucose molecules can be joined together in several different ways to yield long chains called **polysaccharides.** The three most common polysaccharides are **glycogen** (made and stored in animal cells), **starch** (made and stored in plant cells), and **cellulose.** Cellulose is the primary constituent of the rigid cell wall that surrounds most kinds of plant cells (Figure 2-27). Familiar forms of these cell walls are wood and cotton.

In starch, glycogen, and cellulose, the glucose molecules are joined through an oxygen atom between the number 1 carbon of one molecule and the number 4 carbon of another (see Figures 2-29, 2-30). Two types of 1–4 linkages are possible because glucose exists in two configurations, an α configuration and in a β configuration (Figure 2-28). The two configurations are spontaneously alternating forms of glucose in solution. In starch and glycogen, glucose molecules are joined with a glucose in the α configuration (Figure 2-29), forming an α 1–4 linkage. In cellulose, glucose molecules are joined with a glucose in the β configuration (Figure 2-30), forming a β 1–4 linkage.

In both starch and glycogen some of the glucose molecules form branches by an oxygen linkage between carbons number 1 and number 6 of two glucose

FIGURE 2-29

Glucose molecules in the α configuration are joined linearly between carbons number 1 and 4 (α 1-4 linkage, shown in red) in the formation of starch and glycogen.

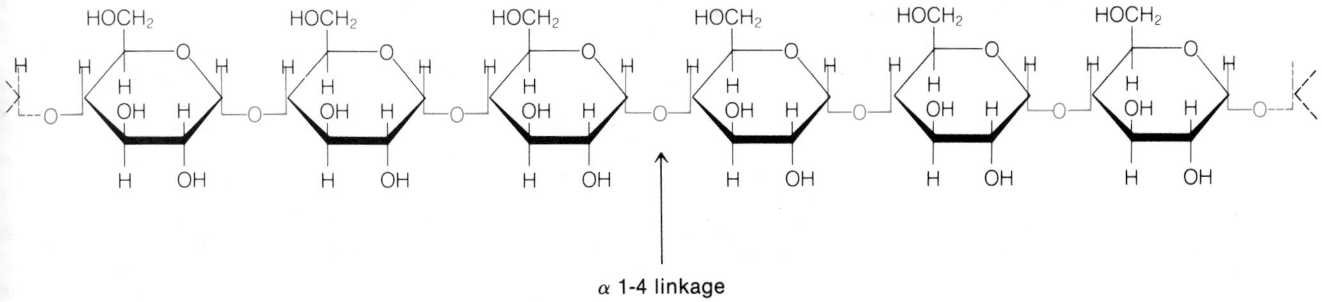

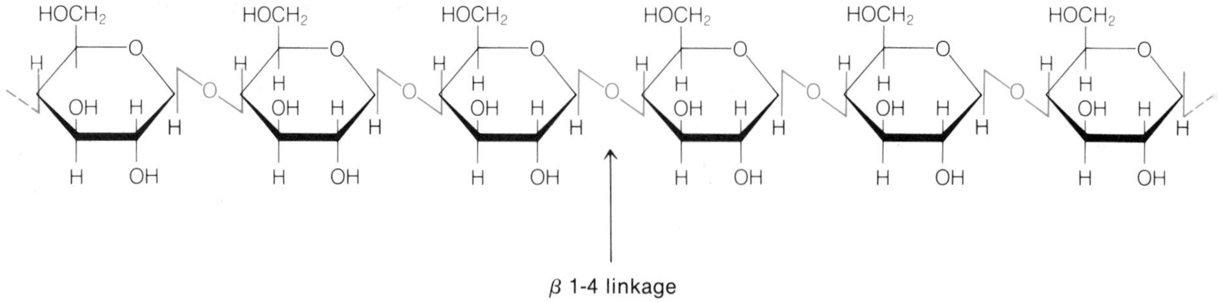

β 1-4 linkage

FIGURE 2-30
Glucose molecules in the β-configuration are joined linearly between carbons number 1 and number 4 (β 1-4 linkage, shown in red) in the synthesis of cellulose.

molecules (Figure 2-31). Glycogen has essentially the same glucose chain structure as starch except that branches occur much more frequently in glycogen. Both consist of hundreds of glucose molecules. A cellulose molecule contains thousands of glucose molecules but it is not branched.

In summary, polysaccharides serve for storage of glucose in macromolecular form (starch and glycogen), or in the construction of cell walls in plants (cellulose). The polysaccharides account for most of the organic material on this planet (e.g., wood is made of plant cell walls and contains a large amount of cellulose).

Fatty Acids and Lipids

A **fatty acid** is a long hydrocarbon chain with a carboxyl (COOH) group at one end (Figure 2-32). An even number of carbon atoms is present in naturally occurring chains because fatty acid chains are synthesized by the joining together of units containing two carbon atoms. Commonly, 16, 18, or 20 carbon atoms are present in fatty acid chains. The fatty acids are distinguished by the number of carbon atoms in their chains and by the number and location of double bonds between carbon atoms. For example, **stearic acid** has 18 carbon atoms. Each carbon is connected to the next

FIGURE 2-31
Two chains of glucose molecules (starch and glycogen) joined with each other through a branch linkage between carbons number 1 and number 6 (α 1-6 linkage, shown in red).

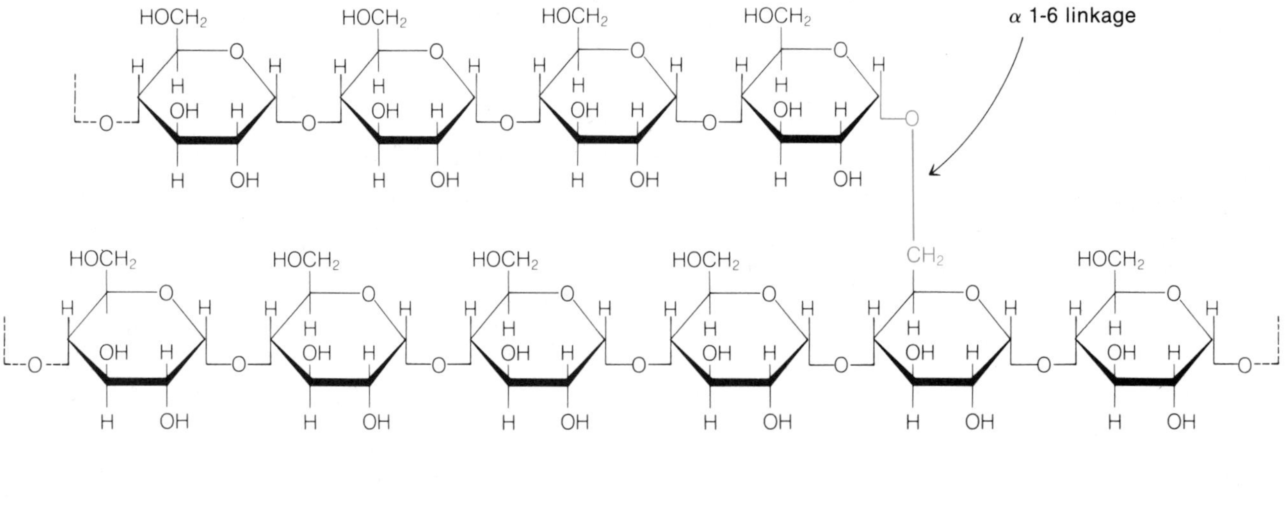

α 1-6 linkage

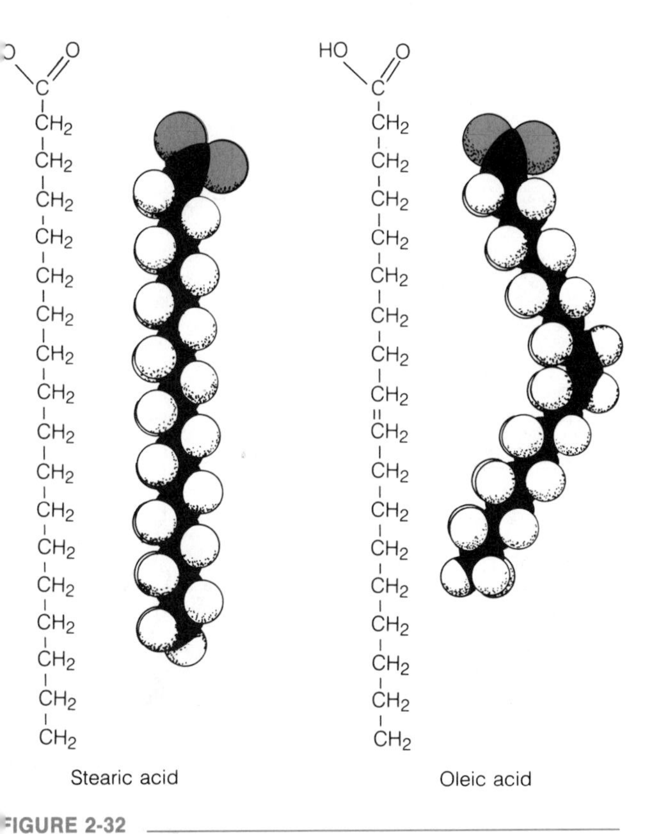

FIGURE 2-32
Diagrammatic and space-filling models of two fatty acids that occur in lipids. Each contains a backbone of 18 carbon atoms. Stearic acid is saturated; oleic acid is unsaturated, with a double bond between carbons number 9 and number 10. The double bond causes the carbon chain to be kinked.

by a single covalent bond (no double bonds), and the chain is straight (Figure 2-32). All but the first and last carbon atoms in stearic acid have two hydrogens; thus, it is saturated with hydrogen atoms and is a **saturated fatty acid.** Oleic acid also has 18 carbon atoms but contains a double bond between carbons number 9 and number 10 (Figure 2-32). Only one hydrogen is present at carbon atoms number 9 and number 10, so oleic acid is said to be an **unsaturated** fatty acid. A double bond puts a kink in the hydrocarbon chain and lowers the melting temperature. Stearic acid remains a solid until heated to 69°C, and oleic acid, with one double bond, melts at 13.4°C. An additional double bond in the molecule (linoleic acid) would lower the melting temperature to −5°C. Fatty acids are primary constituents of membranes, where they can be in either a solid or liquid state. The state has important consequences for membrane function and structure (see Chapter 5). Fatty acids are predominantly nonpolar and nonionic and hence are insoluble in water.

Fatty acid chains can be joined to a 3-carbon molecule, to form a lipid molecule. When the 3-carbon molecule is **glycerol phosphoric acid (phosphoglyceric acid),** the lipid formed is a **phosphoglyceride** (Figure 2-33). Most phosphoglycerides have an additional small molecule, usually **ethanolamine** or **choline,** attached to the phosphoric acid (Figure 2-34).

In summary, many lipids are built from four smaller molecules—two fatty acid chains, which form long hydrophobic tails, one molecule of glycerol phosphoric acid, and another molecule, which is usually choline or ethanolamine. The latter two components form the hydrophilic head. It is important to understand the structure of lipid molecules because their structure accounts for much of their behavior within the cell. The two fatty acid tails are insoluble in the aqueous solution (cytosol) of the cell, but the polar or ionic properties contributed by the two molecular components that make up the head of lipid molecules are soluble in water. This arrangement of insoluble and soluble parts is the key to the formation of the plasma membrane and the various internal membranes of the cell, as discussed in Chapter 5.

AGGREGATION OF MACROMOLECULES TO FORM CELL STRUCTURES

Some macromolecules occur individually in a soluble form in the cell, but the majority are joined together in various numbers and combinations by noncovalent bonds to form larger functional and structural units. Thus, some proteins contain only a single folded polypeptide chain, but many consist of two or more such chains or subunits. In some of these multisubunit mole-

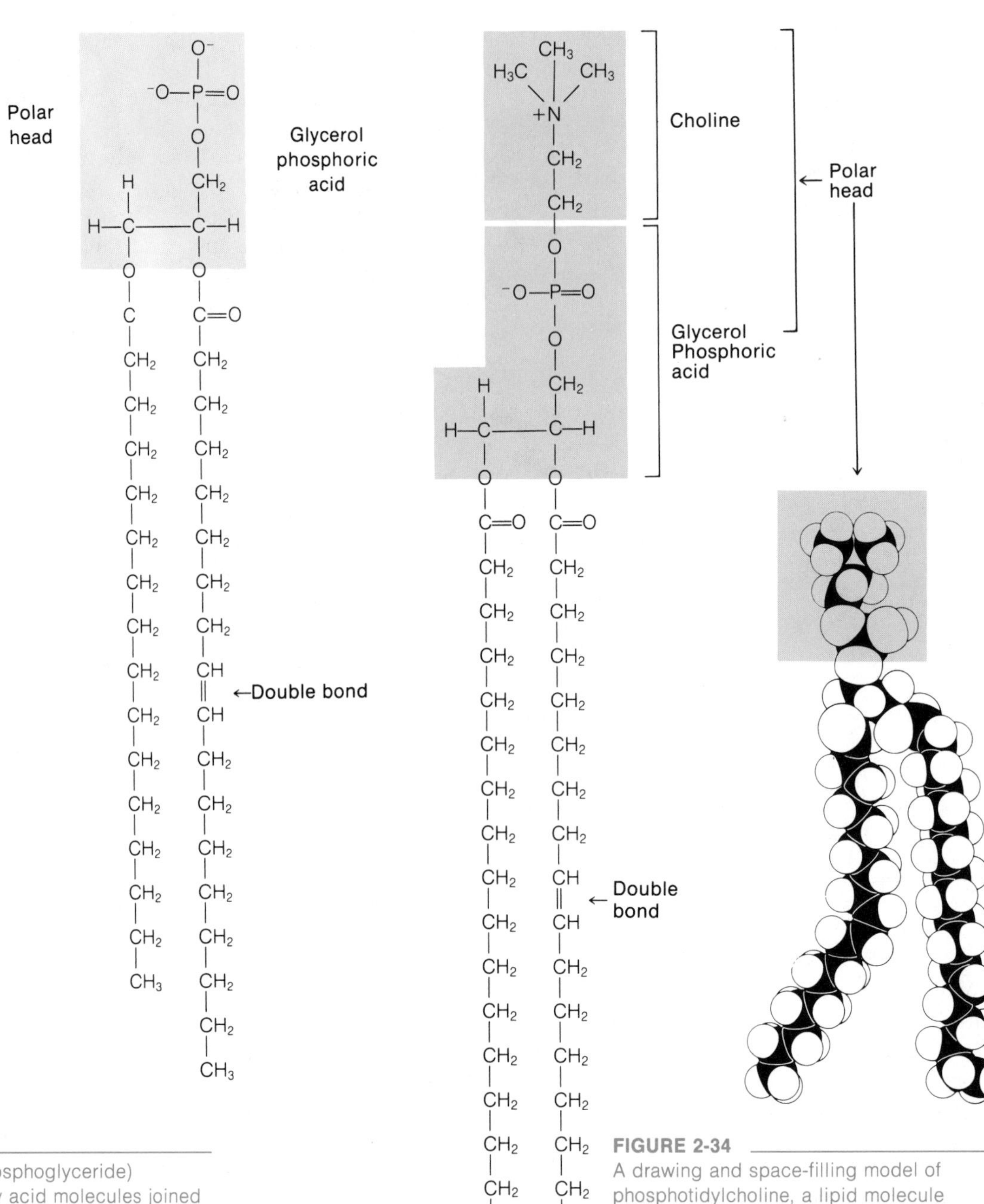

FIGURE 2-33

A lipid molecule (phosphoglyceride) consisting of two fatty acid molecules joined to glycerol phosphoric acid. The glycerol phosphoric acid forms a polar, or hydrophilic head, and the fatty acid chains form a hydrophobic tail. The chain to the right contains a double bond and would be kinked at that point.

FIGURE 2-34

A drawing and space-filling model of phosphotidylcholine, a lipid molecule composed of two fatty acid chains, glycerol phosphoric acid, and choline. The hydrophilic head is indicated by pink shading. Note the kink in one of the chains in the space-filling model; it corresponds to the double bond in the drawing to the left.

ules the polypeptide subunits are identical. **Ferritin,** n iron storage protein in liver cells, consists of 24 identical polypeptides, each with a molecular weight of approximately 18,500 and containing about 130 amino cids. The 24 subunits, each folded into a roughly pherical shape, aggregate in a specific pattern around a ore made up of about 2000 atoms of Fe^{3+}, making a article with a diameter of about 12 nm visible by electron microscopy (Figure 2-35). The Fe^{3+} stored in the ver can be transferred as ferritin through the blood to he bone marrow, where it is used in the synthesis of emoglobin in newly forming red blood cells.

Other polypeptides aggregate to form fibrous structures. Actin in mammals contains a chain of 374 amino cids that folds into a roughly spherical shape. Thousands of actin molecules join to form long microfilaments visible in cells by electron microscopy (Figure -11). Similarly, the flagellum of a bacterium is formed y aggregation of a single protein, **flagellin,** into 11 laments joined by lateral interactions to form a flagellum (Figure 1-3).

Aggregates are often mixtures of different molecules; for example, a eukaryotic ribosome consists of our different RNA molecules and about 80 different rotein molecules. All of these proteins are complexed with the RNA molecules or with each other in a specific attern to yield a structure (the ribosome) that functions in protein synthesis. Likewise, membranes are aggregates of lipid molecules and particular kinds of roteins (Chapter 5). Microtubules are specifically rranged aggregates of the protein tubulin (Chapter 6). n summary, all the fibers, tubules, particles, and membranes seen in a cell represent combinations of various inds of molecules, mostly macromolecules.

As a convenience, cell biologists classify as a ellular structure any macromolecular aggregate large nough to be seen clearly by electron microscopy. This efinition is arbitrary since there is really a continuum n size extending from aggregates of a few macromolecules, which may not be visible by electron microscopy, o those of many thousands of macromolecules, which may be visible even by light microscopy.

The sticking together of macromolecules to form tructures leads to several questions. What is the nature of the forces by which macromolecules bind to each other? What provides for the specificity that governs which macromolecules will bind to one another? And finally, what is achieved by aggregation of macromolecules into cellular structures? These questions are addressed in the next section.

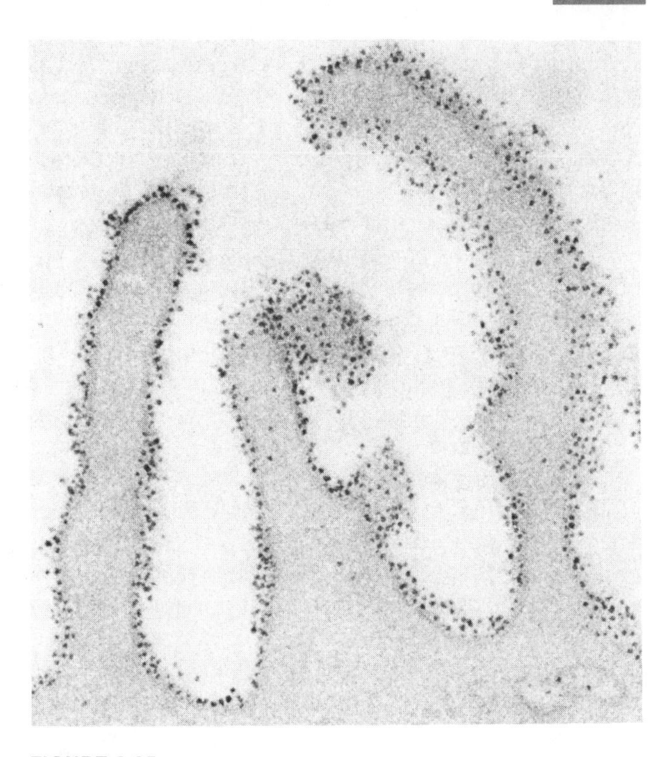

FIGURE 2-35
Electron micrograph of several hundred ferritin particles bound to the plasma membrane of a thyroid cell. [Courtesy of Volker Herzog.]

The Chemical Nature of the Interactions Between Macromolecules

Aggregates of macromolecules are generally not held together by covalent bonds. The forces by which assemblies are maintained are instead relatively weak. These forces are of four types: van der Waals forces, hydrogen bonds, ionic or salt linkages, and hydrophobic interactions.

Van der Waals forces exist between all molecules as a result of attractive forces between their atoms. The strength of the attractive force between two atoms is proportional to $1/r^6$, where r is the distance between the nuclei of the atoms. Thus, the attractive force becomes significant only when two atoms are very close to one another, that is, within 1 to 2 Å. When two atoms come so close that their outer shells of electrons overlap, they are powerfully repulsed from one another. The distance at which the attractive and repulsive forces equal one another for a particular atom defines the **van der Waals radius** of the atom. The attractive force is at a maximum when the repulsive force equals the attractive force, that is, when the atoms are at their closest.

The van der Waals radii differ from one kind of atom to another, and the shape of any molecule is defined by the van der Waals radii of its component atoms. Figure 2-36 shows the van der Waals shapes of two amino acids. Even when two atoms approach very closely such that the attractive force between them is near maximal, the force is still too weak to hold them together. However, if two molecules can approach each other such that *several* atoms in one molecule are close enough to several atoms in the other molecule to produce nearly maximal attractive force between them, then the two molecules will tend to stay together, as long as there are no other repulsive interactions. Thus, *if two molecules have complementary shapes, they can bind to each other by means of van der Waals forces.* The greater the number of atoms in the two molecules that can interact with near-maximal attractive force, the stronger the binding between the molecules.

A hydrogen bond consists of the attraction between a slightly positive hydrogen atom that is covalently attached to an oxygen or nitrogen atom and the slightly negative charge belonging to a nearby oxygen or nitrogen atom. To join two molecules together even weakly requires at least several hydrogen bonds. Earlier we saw an example of stable binding by means of multiple hydrogen bonds when two sections of a polypeptide chain form a β-structure (Figure 2-16).

Ionic bonds are formed by electrostatic attraction between oppositely charged atoms, for example the negatively charged phosphate groups of nucleic acids and the positively charged amino groups belonging to arginine and lysine residues in proteins. The histone proteins (found in eukaryotic chromosomes), which are rich in arginine and lysine, are complexed with DNA by just such phosphate-amino attraction (Figure 2-37).

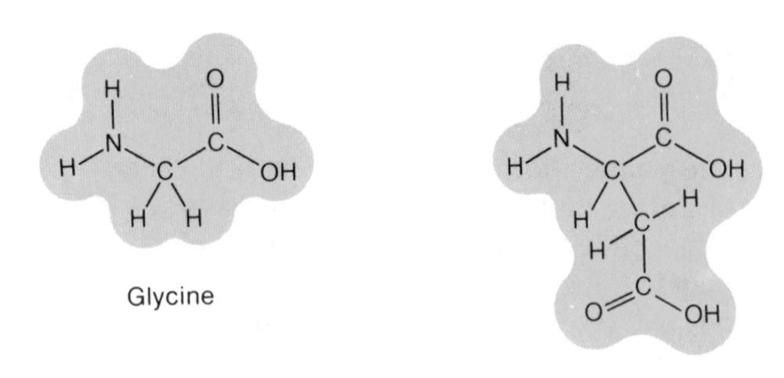

FIGURE 2-36
The shapes of two amino acids, glycine and aspartic acid, defined by the van der Waals radii of their atoms. For clarity the molecules are drawn in two dimensions but are actually globular 3-dimensional structures.

Hydrophobic interactions occur between nonpolar or hydrophobic portions of macromolecules. For thermodynamic reasons, molecules or parts of molecules that cannot interact with water (that is, they are hydrophobic) cannot stay in solution unless contact with water is minimized. Bringing such molecules or parts of molecules together reduces the number of surrounding water molecules per molecule of some substance, and this is the source of the hydrophobic interaction. An example is the aggregation of lipid molecules into structures called **micelles** in an aqueous environment (Figure 2-38(a) and (b)). In a micelle the hydrophobic tails of lipid molecules interact to form a hydrocarbon phase while the polar ends of the lipids remain in contact with water at the surface of the micelle. In addition to lipid–lipid interactions, some protein–protein and most protein–lipid interactions are hydrophobic. Formation of the hydrophobic interaction of lipids and proteins in the formation of membranes will be examined in detail in Chapter 5.

Specificity of Macromolecular Interactions

The rules that govern the specificity of macromolecular aggregation are all implied in the kinds of forces that hold macromolecules in stable associations. As noted already, to be effective a number of van der Waals bonds or hydrogen bonds must form in concert. Therefore, to bind through van der Waals forces or hydrogen bonds, macromolecules must have surface regions that are complementary in shape. The complementary shapes allow them to fit closely with one another so that multiple weak bonds can form and thereby create a sufficiently strong attraction to keep the two molecules together. Similarly, for associations to be stabilized by ionic bonds or hydrophobic interactions macromolecules must have shapes that allow them to fit together. Thus, the possibilities for establishment of stable aggregates of macromolecules are sharply limited, because it requires a high degree of complementarity of shape and chemical properties among the component macromolecules. In most cell structures specific associations are maintained by combinations of two or more of the four kinds of weak forces. For example, the structural arrangements of macromolecules in membranes probably involve all four types.

Knowledge of what holds macromolecules together in specific patterns at the same time provides a general understanding of how macromolecules are assembled into cellular structures. Once the individual macromolecules have been synthesized, specific interactions often guide their assembly in the correct pattern within a structure. This can take place in two ways: by integration of new macromolecules into existing structures or by self-assembly of the components to form

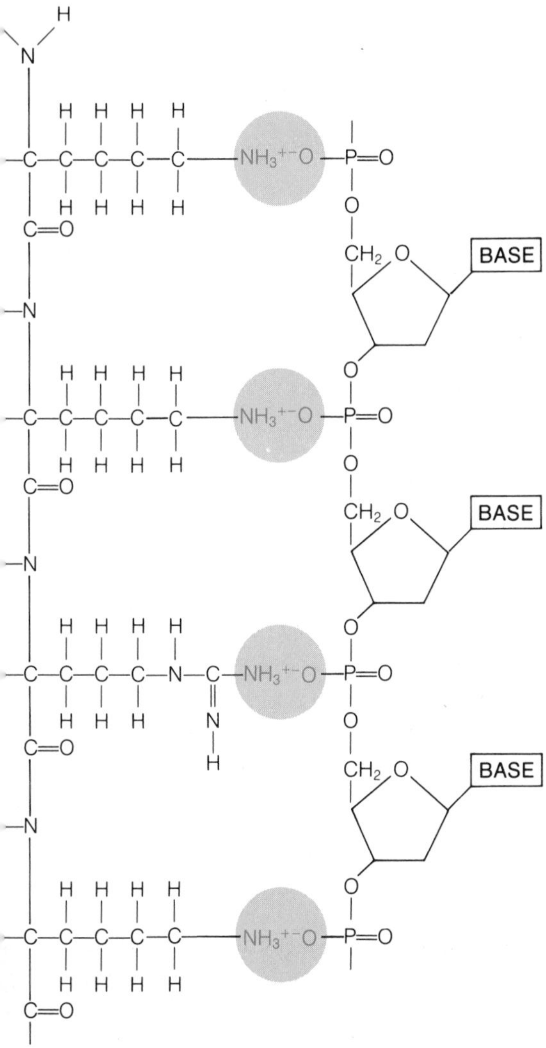

GURE 2-37
art of a polypeptide chain bound to part of a DNA chain by nic bonds between positively charged side groups of lysine d arginine and negatively charged phosphate groups of DNA.

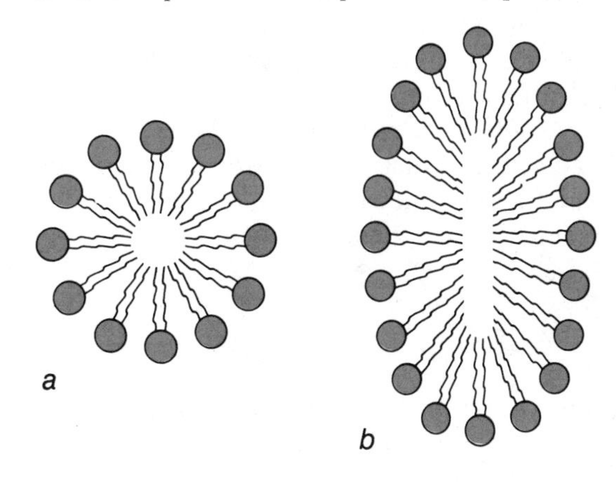

FIGURE 2-38
Diagrams of sections of (a) spherical and (b) ellipsoid micelles formed from phospholipid molecules.

independent new structures. For example, new molecular components are integrated into the plasma membrane and into the membranes of other structures in the course of growth of these cell parts. On the other hand each ribosome is produced individually by assembly *de novo* of the component macromolecules.

What is Achieved by the Assembly of Macromolecules into Aggregates?

Many functions of the cell can be achieved only through cell structures. These structures are all the outcome of the aggregation of macromolecules and smaller molecules in highly specific patterns. A muscle cell can contract because of functional cooperativity among millions of subunits, containing only a few kinds of proteins, into a contractile structure. The movement of chromosomes during mitosis requires microtubules, each of which is made up of many thousands of protein subunits; the unaggregated protein subunits are incapable of performing the functions accomplished by the aggregate. The regulation of the movement of materials in and out of a cell would be impossible without a functional cooperativity of protein and lipid molecules that is achieved by proper aggregation to make the plasma membrane. Further, the efficient operation of enzyme pathways often depends on the assembly of enzymes into patterns or sequences within a structure; examples are the enzymes that carry out respiration in the inner mitochondrial membrane and those that carry out photosynthesis in the internal membranes of the chloroplast (Chapter 4). One can view every cell structure as a group of macromolecules assembled into a specific pattern designed for the efficient achievement of a particular function. We may reasonably assert then that the structures of contemporary cells are the outcome of evolutionary selection for increased functional cooperation among macromolecules.

OTHER ORGANIC COMPOUNDS

Cells also contain some other kinds of organic molecules. Various cofactors and coenzymes are necessary for the function of many enzymes. An example of a coenzyme is **nicotinamide adenine dinucleotide (NAD^+).** How it enters into cellular activities is described in Chapter 4. The structure of NAD^+ is shown in Figure 2-39. The NAD^+ molecule is composed of two nucleotides, adenine nucleotide and nicotinamide nucleotide, joined through their phosphate groups to form a dinucleotide.

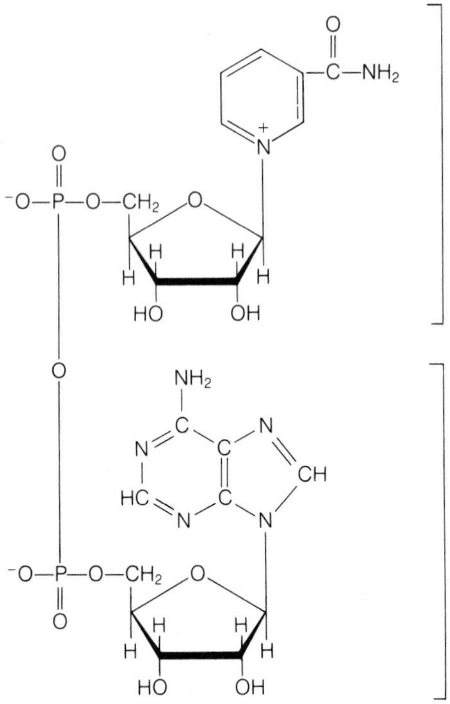

Nicotinamide monophosphate

Adenosine monophosphate

FIGURE 2-39
The structure of the coenzyme nicotinamide adenine dinucleotide (NAD^+).

The cells of humans and many other mammals cannot synthesize the nicotinamide part of NAD^+. Therefore, they must obtain it in the diet, where it is classified as a **vitamin.** Vitamins are organic compounds, needed in minute amounts for cell metabolism, that cannot be synthesized by the particular cell species and must be acquired from an outside source. Vitamins can be synthesized by cells of other species of organisms; for example, plants synthesize all the human vitamins. Fifteen different vitamins are required by humans. Additional examples of vitamins are **riboflavin** (vitamin B_2), which is required to make the coenzyme **flavine adenine dinucleotide,** and **pantothenic acid,** which is required to make **coenzyme A.** Both of these coenzymes are important in the transfer of energy from sugar to ATP (see Chapter 4).

The specific biochemical functions of some vitamins—for example, vitamins A, D, E, and K—are less well known, although their importance in cell functions is understood in a general way. Vitamin A is essential for development of many kinds of cells in mammals. A dietary deficiency of this vitamin leads to defects in bones, nervous tissue, skin, kidneys, gonads, and various glands. A lack of vitamin A also leads to night blindness, which is a loss of the ability of the retina's rod cells to absorb light (Chapter 15). The role of vitamin A in vision is the only one of several roles that is understood in molecular terms. Vitamin A is chemically modified to make **retinal;** retinal, in turn, is complexed with a particular protein to make **rhodopsin,** the light-absorbing pigment of rod cells in the retina and in some photosynthetic bacteria.

Vitamin D is believed to be converted in the kidneys into a hormone that is crucial for the uptake of Ca^{2+} ions from the intestine by intestinal epithelial cells and for deposition of Ca^{2+} ions in the formation of bones. Vitamin K is necessary for blood coagulation, but its precise molecular role is not well understood. A deficiency of vitamin E leads to sexual sterility and degeneration in many tissues in laboratory animals, but its molecular role(s) in cell metabolism remains obscure. Vitamin C (**ascorbic acid**) is essential for maintenance of connective tissue; a deficiency results in the disease called **scurvy.** Vitamin C is required by several mammals including humans and monkeys; cells of rats and most other mammals can synthesize ascorbic acid, and therefore it is not a vitamin in these species.

PROBLEMS

1. Describe the property of water that makes it a good solvent for ions and molecules in the cell.
2. Why are hydrolytic cleavage reactions and dehydration-condensation reactions so named?
3. Describe a hydrogen bond. If water molecules were incapable of hydrogen-bonding to one another, would water be liquid, gaseous, or solid at 25 °C?
4. What is meant by equilibrium in the dissociation and reassociation of water molecules?
5. Calculate the pH of a 0.0001 M solution of HCl and for a 0.0001 M solution of NaOH.
6. Some TV advertisers claim that aspirin causes gastric irritation because it is a weak acid. It has a value for K' that is less than that of acetic acid. Remembering that the fluid in the stomach has a pH between 1.5 and 3, comment on the validity of the claim.
7. What are the three main classes of organic molecules that account for most of the molecular complexity of cells?
8. Which amino acid is more soluble in water, tryptophan or lysine? Why?
9. Draw a peptide bond between these two amino acids.

$$\mathrm{H_2N{-}\overset{\displaystyle H}{\underset{\displaystyle R_1}{C}}{-}\overset{\displaystyle O}{\overset{\|}{C}}{-}OH} \qquad \mathrm{H_2N{-}\overset{\displaystyle H}{\underset{\displaystyle R_2}{C}}{-}\overset{\displaystyle O}{\overset{\|}{C}}{-}OH}$$

10. What are the two major forms of secondary structure in proteins? What kinds of bonds stabilize these structures?
11. How do the side groups of amino acids influence the tertiary structure of a protein?
12. The tertiary structure of many proteins dissolved in water is disrupted by heating above 80°C, but primary structure is unaffected. Why is this so?
13. What are the two primary chemical differences between RNA and DNA?
14. What are the three parts of a nucleotide or nucleoside monophosphate?
15. How are glucose molecules joined together to make starch, glycogen, and cellulose?
16. What is the difference between a saturated and an unsaturated fatty acid?
17. What kinds of bonds or forces are usually responsible for the aggregation of macromolecules into cellular structures?
18. What is achieved by aggregation of macromolecules into structures?

CHAPTER 3

Enzymes and Metabolism

See Figure 2-18c.

For the most part, the thousands of different small molecules and different macromolecules that compose a cell must be synthesized by the cell itself from simple precursor molecules. The sets of chemical reactions by which the syntheses of molecules are achieved is called **anabolism.** Cells also degrade molecules, a familiar example being the digestion of sugar to CO_2 and H_2O to generate energy in a usable form. In fact, a cell usually has the ability to degrade any organic molecule that it is capable of synthesizing; however, the processes of synthesis and degradation are usually not simply the reverse of one another, but instead occur by a different series of chemical steps. **Catabolism** is the totality of the chemical reactions by which cells degrade molecules. Catabolism consists of the degradation of carbohydrates, fatty acids, nucleic acids, and proteins into simpler molecules that then can be used either to provide energy or to serve as starting materials in the anabolic buildup of new molecules. For example, some cells degrade proteins acquired from the environment into free amino acids and then use them for synthesis of new and different protein molecules. Indeed, the primary source of amino acids for protein synthesis in humans is from the degradation of ingested proteins.

Together, synthesis (anabolism) and degradation (catabolism) of molecules constitute **metabolism.** Both processes usually occur in stepwise fashion in which molecules are progressively modified through a sequence of chemical reactions called a **metabolic pathway.** For example, glucose is degraded by a catabolic pathway composed of many chemical reactions that finally yield CO_2 and H_2O. Each reaction creates an intermediate molecule that is converted into the next intermediate molecule by another reaction, that is,

$$\text{Glucose} \rightarrow A \rightarrow B \rightarrow C \rightarrow D \rightarrow E \rightarrow . . . \rightarrow CO_2 + H_2O$$

Amino acids, nucleotides, and other molecules are synthesized by anabolic pathways starting from simpler organic molecules according to the general plan

$$X \rightarrow Y \rightarrow Z \rightarrow \text{product}$$

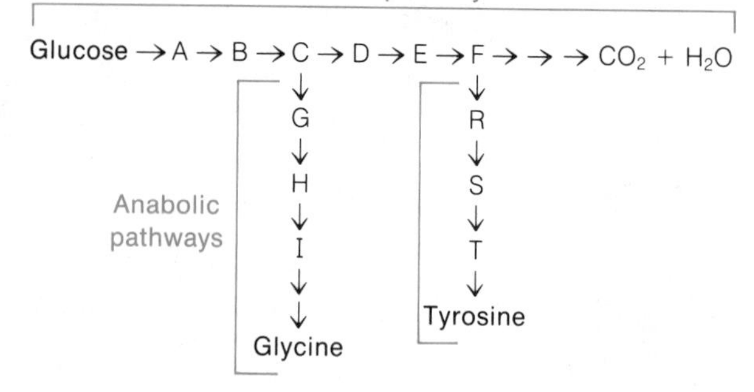

FIGURE 3-1
Several different intermediate molecules in the catabolic pathway of glucose degradation are starting points for anabolic pathways for synthesis of amino acids.

Catabolic and anabolic pathways frequently interconnect. For example, various intermediate molecules in the degradation of glucose are used as the start of anabolic pathways to produce amino acids (Figure 3-1).

Some chemical reactions within the cell are neither anabolic nor catabolic but make up such cell functions as uptake of nutrients from the environment (Chapter 5), cell movement (Chapter 14), conduction of impulses in nerve and muscle cells (Chapter 15), and export of specialized products (hormones, digestive enzymes, etc.) out of the cell (Chapter 5). A few chemical reactions will be described in this chapter to illustrate principles of cell metabolism. It is important to understand these illustrations in order that the principles can be firmly grasped, but it is not necessary to memorize the reactions.

CATALYSIS OF CHEMICAL REACTIONS BY ENZYMES

Many of the thousands of chemical reactions of cell metabolism can occur spontaneously; others must be driven by the input of energy. However, virtually all reactions that can proceed spontaneously do so at rates that are far too low to be useful to a cell. The catalytic action of a class of proteins called **enzymes,** increase the rates of the reactions, often by more than a millionfold. Almost all reactions in a cell are enzymatically accelerated. Some enzymes catalyze the breakdown of molecules, and others catalyze the synthesis of molecules from simpler starting materials. Like all true catalysts, an enzyme remains unchanged after catalyzing the conversion of one kind of molecule to another. A chemical reaction that would take hours or days without catalysis can occur in seconds when accelerated by a specific enzyme. For example, the enzyme **catalase** found in microbodies is one of the most potent catalysts known. It specifically accelerates the conversion of hydrogen peroxide (H_2O_2) to water and oxygen by a factor of 10^8. Acceleration of this reaction is important because H_2O_2, which is a by-product of other chemical reactions in the cell, can irreversibly damage many molecules; therefore, H_2O_2 must be quickly eliminated.

Enzymes do not change the equilibrium point of a reaction but only decrease the time it takes to reach equilibrium. For example, the breakdown of H_2O_2 to H_2O and O_2 is a reversible reaction in which two molecules of H_2O_2 can form two molecules of H_2O and one molecule of O_2, and vice versa:

$$2H_2O_2 \rightleftharpoons 2H_2O + O_2$$

In a concentrated solution of H_2O_2, the concentration of H_2O_2 falls very slowly as it breaks down into H_2O and O_2. As the concentration of H_2O_2 slowly decreases, the rate of conversion of H_2O_2 to H_2O and O_2 falls and eventually, when the concentration of H_2O_2 is extremely low, the rate becomes equal to the rate of H_2O_2 formation from H_2O and O_2, that is, equilibrium is reached. At equilibrium the concentrations of the reactants, H_2O_2, H_2O, and O_2, remain unchanged. The enzyme catalase accelerates the rate of reaction such that equilibrium is reached in seconds instead of months; however, at equilibrium the concentrations of the reactants are the same as they would be if equilibrium were reached without the enzyme. The main purpose of this chapter is to explain the common principles by which the thousands of different enzymes of a cell work to accelerate reactions.

SPECIFICITY OF ENZYMES

Enzymes are highly specific; for example, a particular enzyme catalyzes only a single reaction, and an enzyme usually recognizes only one specific molecule. Thus a cell contains thousands of different enzymes, and each catalyzes a particular reaction or group of reactions. The molecular basis of the specificity of an enzyme for a particular molecule depends on the structure of a region of the enzyme called the **active site,** the region at which catalysis occurs. The active site is always located either on the surface of the molecule or in a crevice on the surface. The molecule acted on catalytically by a particular enzyme is called the **substrate** of that enzyme. The specificity of an enzyme for its substrate resides in the ability of the active site to recognize and to bind its substrate reversibly. Some enzymes are extremely specific, acting on only a single kind of molecule. Other enzymes are less specific and catalyze changes in two or more different but structurally related molecules. For example, the enzyme hexokinase, which is present in the cells of almost all organisms, catalyzes the conversion of glucose to glucose 6-phosphate, a reaction in which a phosphate group is added to the number 6 carbon of glucose (Figure 3-2(a)). This reaction is one of the beginning steps in the catabolism of glucose (Chapter 4). The same hexokinase also catalyzes the addition of phosphate to other 6-carbon sugars (**hexoses**) such as **fructose** and **mannose** (Figure 3-2(b)).

Another enzyme, **glucokinase,** also catalyzes the phosphorylation of glucose. Glucokinase occurs primarily in liver cells and is important in lowering the level of glucose in the blood after a meal rich in sugar. The active site of glucokinase has a shape that makes the enzyme absolutely specific for glucose, and no other 6-carbon sugar is a substrate for glucokinase.

In summary, hexokinase and glucokinase illustrate three principles:

1. Some enzymes can catalyze a reaction with a class of compounds; for example, the addition of phosphate groups to several different sugars by hexokinase.

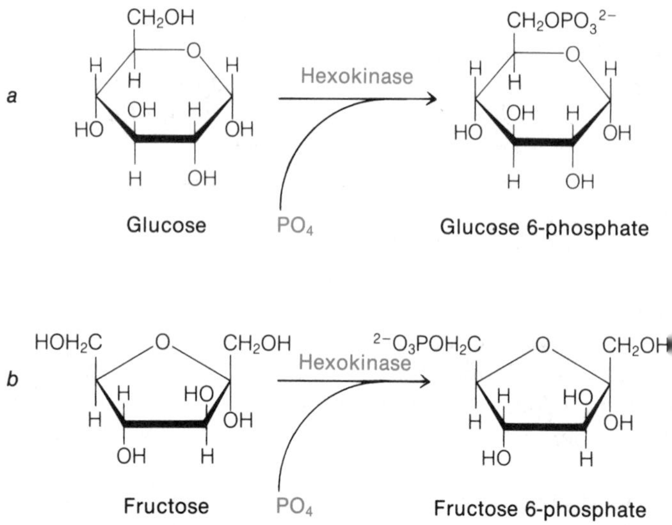

FIGURE 3-2 The enzyme hexokinase catalyzes phosphorylation of two structurally related, 6-carbon sugars. (a) Phosphorylation of glucose to glucose 6-phosphate. (b) Phosphorylation of fructose to fructose 6-phosphate.

2. Sometimes two different enzymes can catalyze the same reaction; for example, hexokinase and glucokinase both catalyze the addition of phosphate to glucose.
3. Two enzymes catalyzing the same reaction may have different ranges of substrates; for example, hexokinase works with glucose and several related sugars, but glucokinase works on glucose only.

Enzymes not only speed up enormously the rates of chemical reactions within a cell, but through their specificity they also provide an essential element of control of cell metabolism. Because many thousands of chemical reactions are theoretically possible within a cell, enzyme specificity provides the basis for selection of which reactions will, in effect, be allowed to occur at a significant rate, namely, those reactions useful to the cell. Occurrence of chemical reactions without such specific selection would create chaos and lead to destruction of a cell. Thus, the cell possesses the specific enzymes by which it selectively catalyzes those chemical reactions that are necessary or useful to carry out functions and to maintain cell structure.

In addition to the overall control of metabolism provided by enzyme specificity, cells have also evolved complex mechanisms by which the catalytic activity of particular enzymes can be increased or decreased. It is therefore possible for cells to adjust individually the rates at which many reactions are permitted to occur. Regulation of individual enzymes allows the cell to adjust various metabolic functions in response to changes in intracellular and extracellular conditions. Regulation of the *activities* of enzymes is presented in greater detail later in this chapter. A second way that cells can control rates of reactions is by adjusting individually the rates at which they *produce* the different kinds of enzymes. Regulation of enzyme production and of protein synthesis in general is discussed in Chapters 9 and 10.

MECHANISM OF ENZYME CATALYSIS

To appreciate the pivotal role that enzymes play in virtually all the reactions that make up cell metabolism, we need to know something about chemical reaction mechanisms. This is examined in this section.

Any chemical reaction in which a molecule S is converted into a product P occurs because during any moment some of the S molecules acquire enough energy from their surroundings to reach an activated condition called the **activated state (S^*)** (Figure 3-3). In is state the probability that a particular covalent bond ill be made or broken enabling S* to become P is eatly increased:

$$S \rightarrow S^* \rightarrow P$$

he rate at which S is converted to P is proportional to e fraction of S molecules that are in the activated ate. The more readily S molecules achieve the activated state, the more frequently S converts to P, that is, e more rapid the overall rate of the reaction. Viewed another way, the need to reach the activated state efore S can change to P represents an energy barrier at impedes the overall conversion of S to P.

There are two ways to accelerate a reaction. Energy ay be added by heating, thereby increasing the fraction of S molecules that reaches the activated state at ny instant of time. Raising the temperature by 10°C pproximately doubles the proportion of molecules hat are elevated to the activated state at any moment, oubling the rate of the reaction. Alternatively, the rate f reaction may be increased by adding a catalyst. A atalyst transiently combines with S molecules in a way o lower the energy needed for activation. Enzymes are catalysts and accelerate reactions by lowering the requirement to reach the activated state (Figure 3-3).

Neither way of increasing the rate of a reaction changes the final outcome. As S is converted to P, the concentration of S molecules decreases, and hence the rate of the conversion decreases. Likewise, as the concentration of P increases, the rate of the reverse reaction increases until finally the rate of the forward reaction (S → P) is equaled by the rate of the back reaction (P → S). At this point equilibrium has been reached, and the relative concentrations of S and P remain constant. Every chemical reaction is characterized by the relative concentrations of reactants (S) and products (P) present at equilibrium, and the concentrations of S and P in moles ([P] and [S]) are used to define the equilibrium constant for the particular reaction. A mole of a substance is equal to its molecular weight measured in grams:

$$K_{eq} = \frac{[P]}{[S]}$$

The ratio of [P] to [S] at equilibrium, that is, the value of K_{eq}, is the same whether or not the reaction is catalyzed by an enzyme. The only difference is that

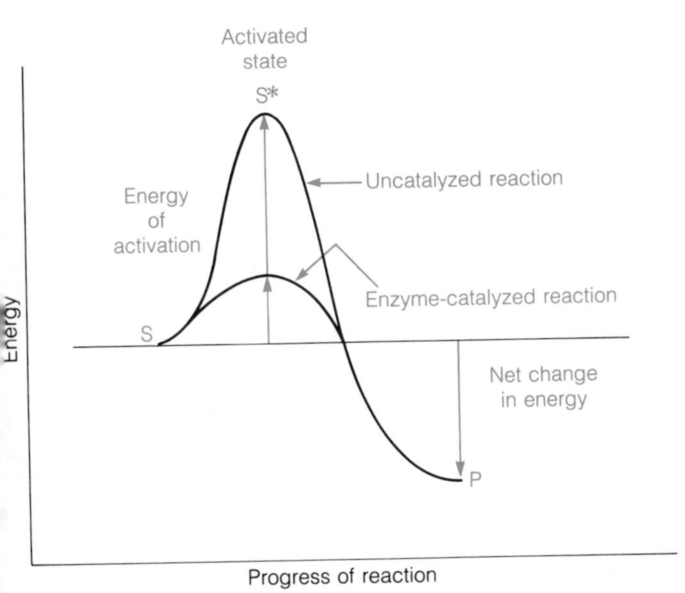

FIGURE 3-3
To undergo a chemical change a molecule S must acquire energy from its surroundings to reach an activated state S*. A molecule in the activated state may lose energy and revert to S or may undergo a reaction and convert to P, with an attendant loss of energy. An enzyme speeds up conversion of S to P by lowering the amount of energy needed to reach the activated state.

starting with all S molecules, the arrival at equilibrium might typically take days without catalysis but takes only a few minutes when catalyzed by an enzyme. The earlier example with H_2O_2 and catalase is a more extreme example of rapid catalysis.

All spontaneous reactions, such as the conversion of S to P in the discussion above, are accompanied by a release of energy. The released energy is referred to as **free energy** because it is available to do work. For example, the burning of wood is a chemical reaction in which carbohydrate molecules are converted to H_2O and CO_2. The heat given off (free energy) can be used to power a steam engine. In a cell the breakdown of carbohydrates releases free energy that may be used in the formation of new chemical bonds rather than released entirely as heat. The free energy released in a spontaneous reaction is indicated by the symbol ΔG. The release of energy in a reaction is indicated by a negative value for ΔG, that is, $-\Delta G$ or $\Delta G < 0$. Thus,

$$S \rightarrow P \quad \Delta G < 0$$

For some enzymes, lowering the activation energy is less important than simply providing binding sites that can bring together two molecules into a particular orientation that allows them to react with one another (Figure 3-4). The required orientation would be expected to occur for free molecules in solution only by chance, and hence rarely.

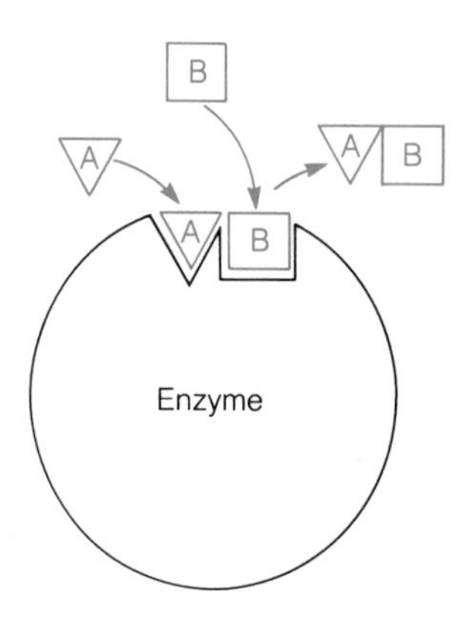

FIGURE 3-4
Some enzymes accelerate reactions by providing adjacent binding sites that orient the molecules in juxtapositions favoring a reaction between them.

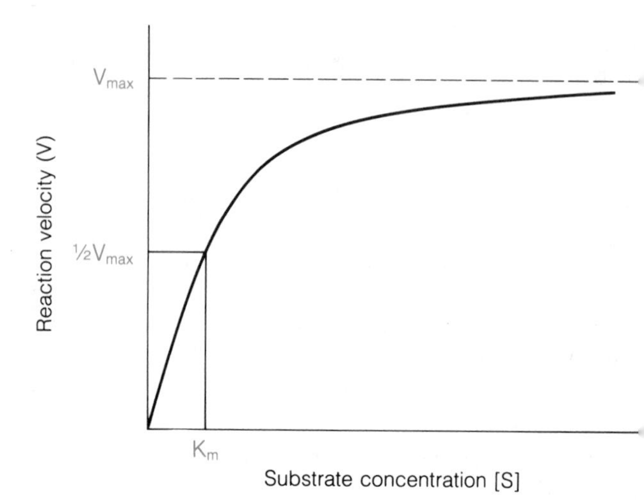

FIGURE 3-5
The curve shows the relation between the velocity of a reaction and the substrate concentration for an enzyme. As V_{max} is approached, increasing the concentration has little effect on the rate of enzyme catalysis. $1/2V_{max}$ is the substrate concentration (K_m) at which the reaction is half of its potential maximum (V_{max}).

ENZYME-CATALYZED REACTIONS

Different enzymes catalyze their respective reactions with different efficiencies. The efficiency of an enzyme depends on the way in which it interacts with its substrate.

The Enzyme-Substrate Complex

A most important insight about how an enzyme works was first obtained by observing how the velocity (V) of a simple reaction (e.g., $S \rightarrow P$) is related to the concentration of the substrate [S], when the amount of enzyme in the solution remains constant (Figure 3-5). *At low concentrations of the substrate* the velocity of the reaction $S \rightarrow P$ increases approximately linearly in proortion to the increase in the concentration of the sub- rate [S]. At higher and higher concentrations of sub- rate further increases in the substrate concentration ring about less and less of an increase of the velocity of ie reaction, until finally, a substrate concentration is ·ached at which the velocity of the reaction reached is arely affected. This velocity is denoted V_{max}. Further ıcreases in substrate concentrations are without signifi- ant effect on the velocity of the reaction because every nzyme molecule is working virtually at its maximum ite.

The existence of a maximum rate is critical in un- erstanding how an enzyme works. It led to the idea ıat in an enzymatically catalyzed reaction a substrate inds reversibly to an enzyme, forming a transient **en- yme-substrate complex:**

$$E + S \rightarrow ES$$

t V_{max} the substrate concentration is so high that vir- ıally all enzyme molecules are constantly in the ES orm.

While bound to an enzyme molecule, the substrate ıolecule may undergo a chemical reaction, either be- ause its requirement for activation energy has been owered by its binding to the enzyme or because it has een brought into proper juxtaposition with another ound substrate molecule. Thus, the substrate may be onverted to a product P, which is then released from he active site:

$$E + S \rightarrow ES \rightarrow E + P$$

Release of P from the enzyme regenerates free en- yme, which may repeat the cycle with another mole- ule. These events occur at rates that are difficult to omprehend. In most cases, an individual enzyme mol- cule forms ES complexes and releases product mole- ules a few hundred to many thousands of times per econd, depending on the particular enzyme.

The value of V_{max} and the amount of substrate ıeeded to saturate an enzyme differ from one kind of nzyme to another, depending on a number of factors. Among these factors are the strength of the binding etween enzyme and substrate and between enzyme and product, and the nature of the chemical reaction that is catalyzed.

The formation of an enzyme-substrate (ES) complex as an intermediate molecule in the enzymatic catalysis of any reaction was presented as a theory by Michaelis and Menten in 1913 to explain enzyme **saturation kinetics.** The existence of the ES complex has been validated in a variety of ways for many enzymes. For example, ES complexes have been directly detected by x-ray crystallographic study of the molecular structure of an enzyme saturated with its substrate. The existence of ES complexes has also been detected indirectly by observation of changes in physical properties of an enzyme. Such changes include solubility and stability of the enzyme's three-dimensional structure in the presence of the substrate of the enzyme.

The Michaelis-Menten Constant

Different enzymes work at different rates. Many accelerate a reaction 10^8- to 10^{10}-fold, and a few are even more powerful catalysts. A convenient way to characterize the catalytic power of any particular enzyme is the **Michaelis-Menten constant, K_m.** One can find formal derivation of an expression for K_m in almost any textbook on biochemistry. For our purposes it is sufficient to know that K_m is equal to the substrate concentration (moles/liter) at which the enzyme is working at half its maximal rate, that is, [S] at $\frac{1}{2} V_{max}$. This also means that when the substrate concentration is equal to K_m, the active sites on half of the enzyme molecules are occupied with substrate molecules, that is, half the enzyme molecules are in the complexed form ES. The lower the value of K_m (in moles/liter), the lower the concentration of S required to elicit one-half the maximum activity of the enzyme.

The values of K_m for different enzymes cover a broad range, usually falling between substrate concentrations of 10^{-1} and 10^{-6} M. In addition to giving the substrate concentration at which the velocity of the enzyme-catalyzed reaction equals $\frac{1}{2}V_{max}$, for many enzymes K_m is equal to the dissociation constant for the ES complex,

$$K_m = \frac{[E][S]}{[ES]}$$

and hence is a measure of the strength of the ES complex. A high K_m means weak binding between the enzyme and its substrate, and a low K_m means strong binding.

Table 3-1 lists K_m values for some enzymes. Hexokinase reaches $\frac{1}{2}V_{max}$ in the phosphorylation of glucose at 0.15 mM glucose, but achievement of $\frac{1}{2}V_{max}$ for phosphorylation of fructose by the same enzyme requires a 10-fold higher concentration of substrate (1.5 mM), that is, the active site of the enzyme has a higher affinity for glucose than for fructose. Both hexokinase and glucokinase catalyze the phosphorylation of glucose to glucose 6-phosphate. However, the value of K_m for hexokinase is 67 times smaller than it is for glucokinase, showing that glucose binds less readily to glucokinase than to hexokinase.

The V_{max} is important to know. It shows how well the enzyme catalyzes a reaction when the concentration of substrate is high enough so that all enzyme molecules exist as ES complexes, that is, when the enzyme molecules are saturated. When all the enzyme exists as ES, the rate of the reaction $ES \rightarrow E + P$ is at a maximum.

TABLE 3-1

K_m's for Some Enzymes

Enzyme	Substrate	K_m (mM
Catalase	H_2O_2	25.0
Hexokinase	Glucose	0.15
	Fructose	1.5
Glucokinase	Glucose	10.0
Carbonic anhydrase	HCO_3^-	9.0
Chymotrypsin	Glycyltyrosinamide	122.0
	N-benzoyltyrosinamide	2.5
Glutamate dehydrogenase	Glutamate	0.12
	α-ketoglutarate	2.0

Turnover Number or Molecular Activity of an Enzyme

At saturation, all of the enzyme molecules are in ES complexes; hence, the rate of the reaction (V_{max}) is determined by the rate of $ES \rightarrow E + P$ and not by the rate of $E + S \rightarrow ES$. A 10^{-6} M solution of carbonic anhydrase catalyzes the formation of 0.6M of H_2CO_3 per second when the enzyme is saturated with substrate:

$$CO_2 + H_2O \rightleftharpoons H_2CO_3$$

The rate constant for $ES \rightarrow E + P$ is therefore equal to

$$\frac{0.6M \text{ (product formed/sec)}}{10^{-6}M \text{ (enzyme concentration)}}$$

or 6×10^5 per second per enzyme molecule. In other words, a single enzyme molecule converts 600,000 molecules of CO_2 and H_2O into 600,000 H_2CO_3 molecules per second. This rate constant is also called the **turnover number** or the **molecular activity** of the enzyme. The turnover number for carbonic anhydras is one of the highest for any enzyme. Turnover num bers of most enzymes range between one and 10,00 per second. The enzyme DNA polymerase, which i animal cells has a turnover number of about 50, cata lyzes the replication of DNA molecules. That is, it add deoxynucleotides to a replicating DNA molecule at rate of 50 per second. To keep pace with all of the DN polymerase molecules, an animal cell must synthesiz deoxynucleotides needed to make DNA at a rate 400,000 per second.

Finally, in any unregulated metabolic pathway th overall rate of conversion of the starting molecule t the end product is determined by the enzyme in th pathway with the lowest turnover number.

COENZYMES

Many enzymes require **cofactors** or **coenzyme** in order to carry out their catalytic function. Coen zymes are organic molecules, smaller than proteins that in most cases serve as intermediate carriers c chemical groups or electrons. For example, the coen zyme **tetrahydrofolate** is a carrier of a methyl grou (CH_3) in many enzymatic reactions. A particularly im portant reaction that uses tetrahydrofolate occurs in th

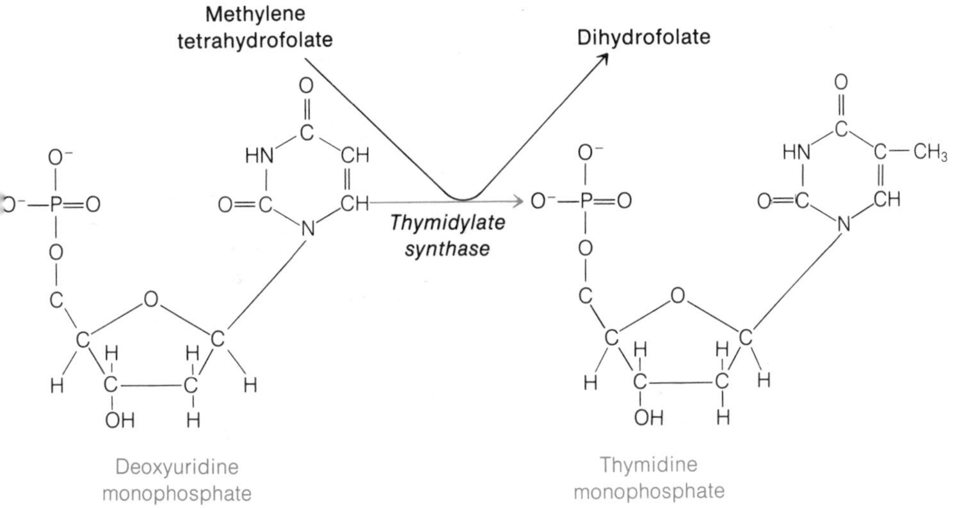

FIGURE 3-6
The coenzyme tetrahydrofolate carries a methyl group (as methylene), which it donates to deoxyuridine monophosphate (dUMP) under the catalytic action of thymidylate synthase. The products of this coupled reaction are thymidine monophosphate (TMP) and dihydrofolate.

enzyme pathway by which thymidine monophosphate (TMP) is synthesized from deoxyuridine monophosphate (dUMP) (Figure 3-6). In this reaction, which is catalyzed by the enzyme **thymidylate synthase,** a methyl group is carried to dUMP by methylene-tetrahydrofolate. TMP is one of the four nucleotides in DNA and is therefore essential for DNA synthesis.

Tetrahydrofolate is synthesized by mammalian cells from the vitamin **folic acid.** Mammals must obtain folic acid in the diet. When it is lacking, tetrahydrofolate cannot be synthesized, and reactions requiring this coenzyme, such as the formation of TMP, cannot take place. Thus, DNA synthesis is one of the cell activities affected by lack of folic acid.

The coenzyme nicotinamide adenine dinucleotide (NAD^+) functions as a carrier of electrons, working in concert with a number of enzymes that catalyze the transfer of electrons from one molecule to another. Electron transport is the central activity in the generation of biologically useful energy, for example, in the degradation of glucose to CO_2 and H_2O. Another coenzyme, flavin adenine dinucleotide (FAD), has the same function but works with a different group of enzymes. The functions of NAD^+, FAD, and several other coenzymes are discussed in more detail in connection with energy metabolism in the next chapter.

The major coenzymes, of which there are eight, are synthesized from vitamins. **Nicotinamide** (commonly known as **niacin,** a member of the group of B vitamins) is a building block for NAD^+. **Riboflavin (vitamin B_2)** forms a part of FAD. Other vitamins required for the synthesis of other coenzymes are **thiamine (B_1), pantothenic acid, pyridoxal phosphate (B_6), biotin, lipoic acid** (also called **thioctic acid),** and **vitamin B_{12}.** A deficiency of any one of these vitamins results in a lack of certain enzyme reactions. The lack of a vitamin-(coenzyme)-dependent reaction causes characteristic deficiencies in cell functions, which are in turn manifested as a vitamin-deficiency disease.

An enzyme, such as thymidylate synthase, that utilizes a coenzyme in a catalytic reaction possesses two binding sites—a specific site at which the coenzyme is bound and an active site for binding the substrate. The coenzyme-binding site must be in a particular location relative to the active site so that the bound coenzyme is properly positioned to accept or donate a chemical group or electrons under the catalytic influence of the active site (Figure 3-7).

THE NATURE OF THE ACTIVE SITE

Every enzyme has an **active site.** The active site has two functions: to bind the specific substrate molecule(s) of the enzyme and to catalyze a chemical

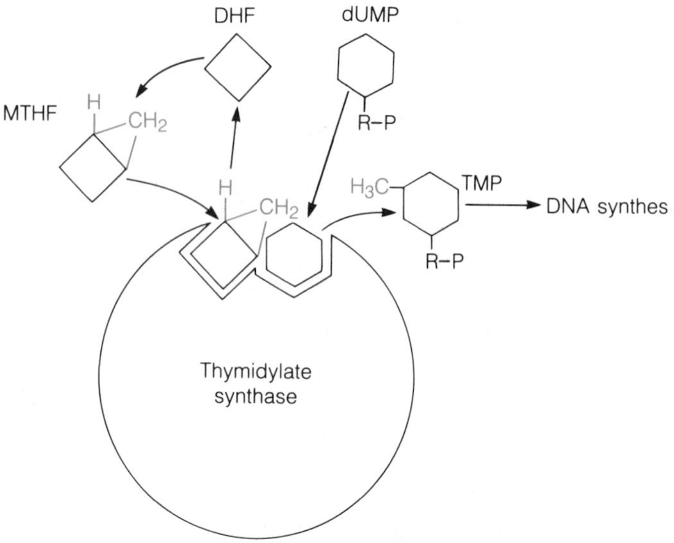

DHF dUMP

MTHF H CH_2

R–P

H CH_2 H_3C TMP → DNA synthes

R–P

Thymidylate synthase

FIGURE 3-7
The enzyme thymidylate synthase has two binding sites, one for the coenzyme methylene tetrahydrofolate (MTHF) and one for deoxyuridine monophosphate (dUMP). The enzyme catalyzes transfer of CH_3 from the coenzyme to dUMP to yield thymidine monophosphate.

MTHF = Methylene tetrahydrofolate
DHF = Dihydrofolate
dUMP = Deoxyuridine monophosphate
TMP = Thymidine monophosphate

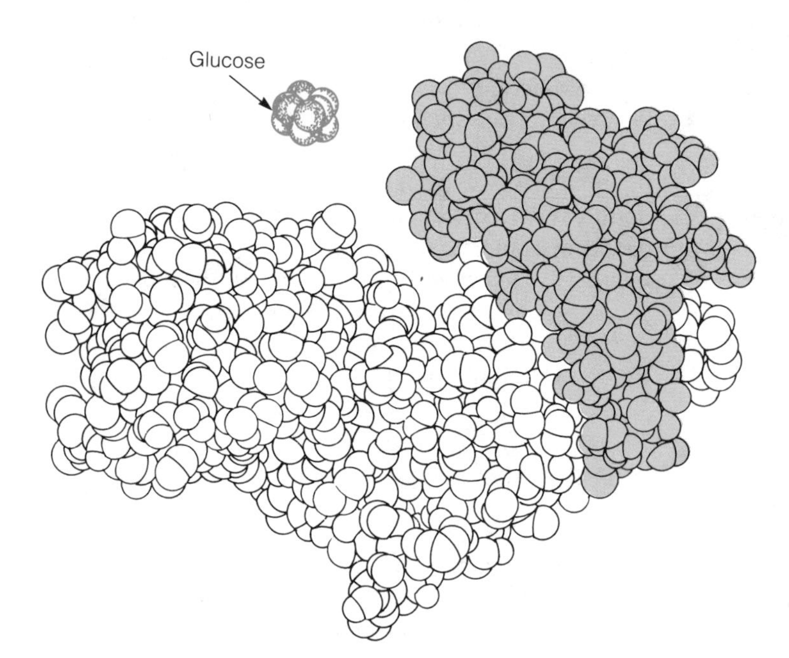

FIGURE 3-8
The active site of hexokinase has the form of a cleft. Glucose binds in the cleft and is phosphorylated to glucose 6-phosphate.

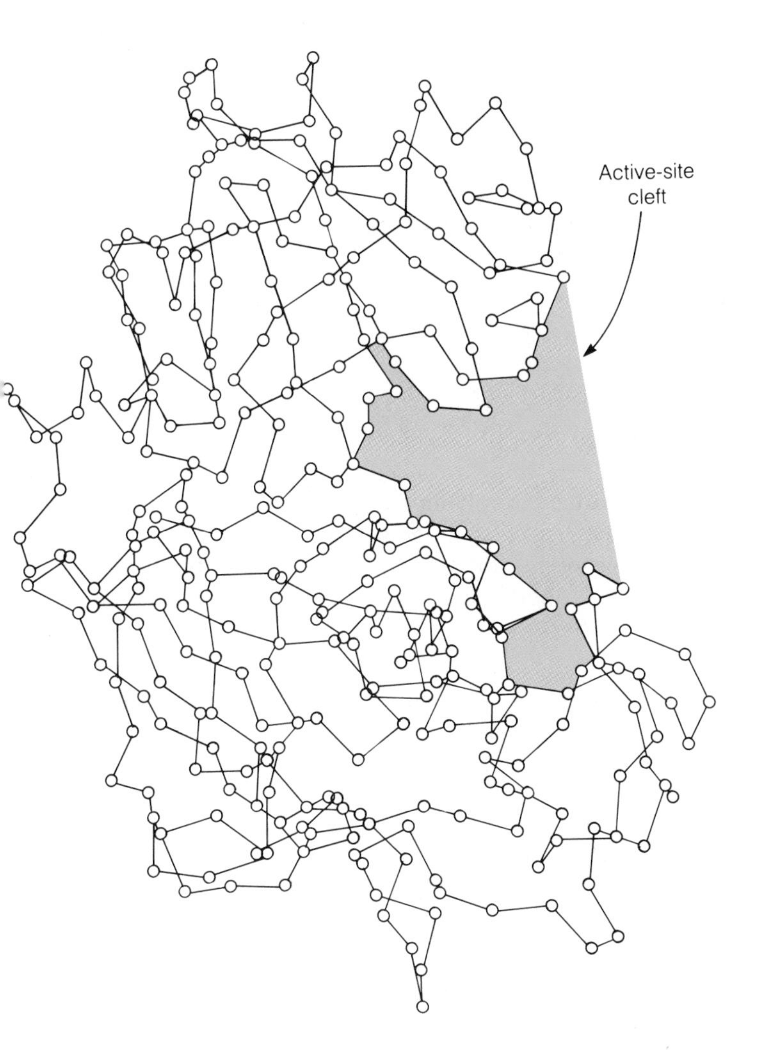

FIGURE 3-9 The active site of this protease is formed by side groups of several amino acids brought together by a complex pattern of folding of the polypeptide chain. [From E. Subramanian. 1978. *Trends in Biochem. Sci.* 3:2.]

change in the bound substrate molecule. To achieve these functions the active site must have a three-dimensional shape that is specific to the reaction and to the substrate.

Recall that almost all enzymes are proteins, and proteins are polypeptides containing 50 to 1000 amino acids. With few exceptions, polypeptides have no activity in a cell until they fold into particular three-dimensional shapes. The surface of the folded polypeptide has particular chemical properties that are determined by the overall pattern of amino acid side groups that are exposed at the surface. The polypeptide chain of an enzyme usually folds in such a way that the side groups of a small number of amino acids form a particular topography at one place on the enzyme surface, often a small cleft or groove of a highly specific shape (Figure 3-8). These clefts often are the active sites. Because of folding, the several amino acids that form the active site are not always adjacent to one another in the polypeptide chain but usually are brought together from widely different parts of the polypeptide chain (Figure 3-9).

The active site is formed by only a few amino acid

side chains that occupy just a few percent of the protein surface. Why then are enzymes so large? In part at least, the answer is that the amino acids of the active site are held in stable positions in space by the specific folding configuration of the rest of the polypeptide chain. This configuration is the result of many interactions among all of the amino acids in the protein.

Since the folding of a polypeptide chain into a specific three-dimensional configuration is the result of interactions between side chains of the amino acids (Chapter 2), substitution of one amino acid by another in the polypeptide chain can alter the folding pattern. This is indeed often the case. The magnitude of the alteration in folding pattern depends primarily on the difference in the chemical properties of the side chains of the original amino acid and the substituted amino acid. For example, the three-dimensional shape of a folded polypeptide depends to a great extent on the interactions among amino acids possessing hydrophobic side chains (valine, leucine, isoleucine, etc.). Because of their hydrophobic nature, these amino acids not only interact with one another but tend strongly to occupy positions inside the three-dimensional structure and away from water. The amino acids with hydrophilic side chains (glutamic acid, histidine, arginine, serine, etc.) strongly tend to occupy positions at the surface of the proteins. If a hydrophobic amino acid is replaced by another hydrophobic amino acid of roughly the same size (e.g., the substitution of leucine for valine), the effect of the replacement on folding of the polypeptide may be expected to be small. Similarly, if a hydrophilic amino acid is replaced by another hydrophilic amino acid with the same charge (e.g., arginine for lysine), the effect will again usually be small. However, if a hydrophilic amino acid is replaced by a hydrophobic amino acid (or vice versa), the resultant perturbation in folding is usually much larger. For example, the sixth amino acid in one of the two kinds of polypeptides in hemoglobin (β-globin) is normally occupied by the hydrophilic amino acid glutamic acid (Figure 3-10). In the genetically determined disease sickle cell anemia, this glutamic acid is replaced by the hydrophobic amino acid valine. The substitution, which is the result of a mutation in the gene for this particular polypeptide (Chapter 8), so alters the folding pattern of the polypeptide that it does not function properly in the transport of oxygen by hemoglobin in red blood cells.

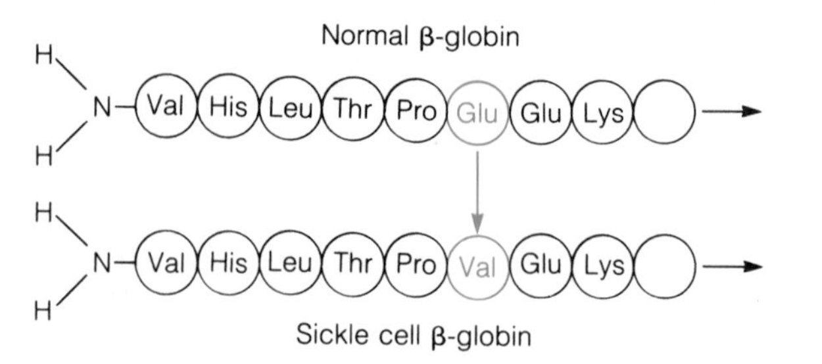

FIGURE 3-10
The sixth amino acid in the polypeptide chain of [Gb-globin is glutamic acid. In the genetically determined disease sickle cell anemia, the glutamic acid is replaced by valine.

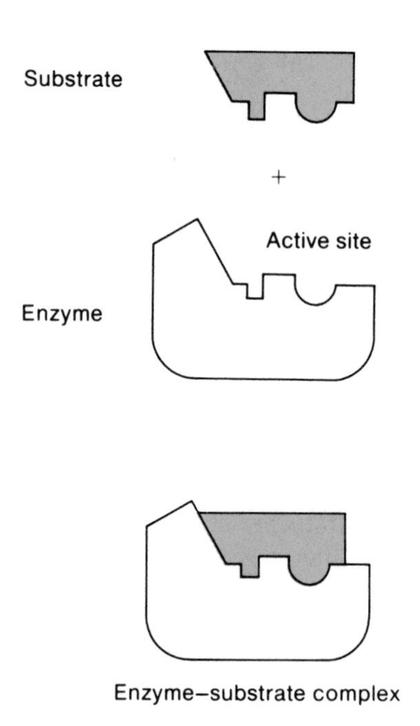

FIGURE 3-11
Specificity of binding of a substrate molecule to the active site of an enzyme is provided by structural complementarity between the molecule and the active site. This complementarity is likened to a lock and key.

Enzyme-Substrate Interaction

In order to fit into the active site of an enzyme, the ubstrate and the active site must have a matching, or omplementary, shape. This complementarity has been ikened to the matching of a key to a lock (Figure -11). This analogy applies particularly to those en-ymes that bring two substrate molecules into a relative rientation that allows them to react with each other. or other enzymes the interaction of an enzyme with its ubstrate is somewhat more complex, since in addition o binding, the activation energy is lowered. Efforts to xplain both the binding of substrate to an enzyme and atalysis of a chemical change in the substrate have led o the concept of **induced fit.**

In the induced-fit hypothesis both the substrate nd the active site are viewed as slightly flexible struc-ures. Although the substrate and the active site are complementary structures, the complementarity is not perfect. Binding of a substrate molecule to the active site is believed to induce a slight alteration in the shape of the active site toward closer complementarity with the substrate (Figure 3-12). The alteration in the shape of the active site distorts the tertiary structure of the protein. The tendency of a protein to return to its original configuration places a strain on a covalent bond in the substrate molecule bound in the active site. This strain is critical to the catalytic activity of the enzyme, for it causes a slight distortion in the substrate. This distortion lowers the activation energy of the substrate molecule by greatly increasing the probability that the bond will break. Breakage of the bond is followed by the release of the products and the return of the active site to its original configuration.

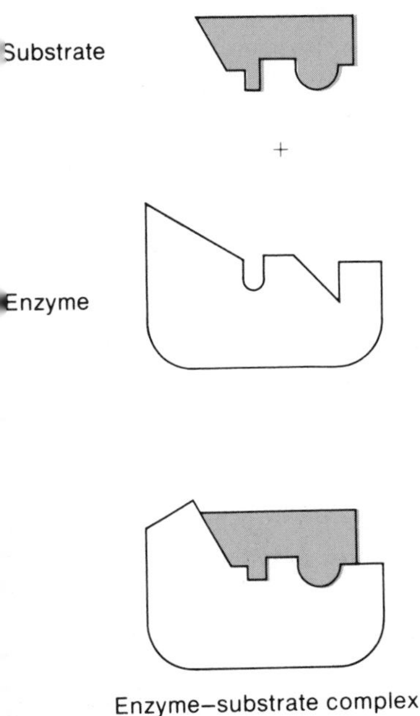

FIGURE 3-12
In the induced-fit concept of how an enzyme works, binding a substrate molecule to an enzyme induces a change in the shape of the active site so that the substrate molecule and active site become structurally complementary.

The induced-fit-strain model applies mostly to unimolecular reactions, like hydrolysis, in which a single substrate molecule is split in two with the addition of water (Chapter 2). The model is consistent with a large amount of experimental observations, although we have not yet obtained final proof of its validity.

Bonds Between Enzyme and Substrate

The binding of a substrate molecule to the active site occurs by weak bonds, which may be ionic bonds, hydrogen bonds, van der Waals bonds, or a combination of these. Thus, the bonds that hold a substrate molecule to its enzyme are weak compared with the covalent bonds that hold the atoms of a molecule together. When an enzymatic reaction utilizes a coenzyme, the surface of a protein outside but near the active site contains a specific site for binding of the coenzyme. Many proteins also contain specific sites for the binding of molecules that inhibit or enhance the enzymatic activity of the protein. These sites serve to regulate activity of the enzyme (discussed in the next section).

Many enzymes are composed of two or more polypeptide chains. In some cases the polypeptide chains in a single enzyme molecule are identical, and each possesses an active site. In other enzymes the polypeptides have different amino acid sequences, and the different polypeptides interact with each other by weak forces to form a single active site.

Finally, in some instances enzymes that work together to form a particular enzymatic pathway may bind to each other through specific physical interactions, presumably increasing the efficiency of the catalytic pathway in which they work. (An example is pyruvate dehydrogenase, discussed in Chapter 4.) These specific physical interactions come about by surface complementarity between two proteins at regions outside the area of the active site and may engage a considerable amount of the enzyme surface.

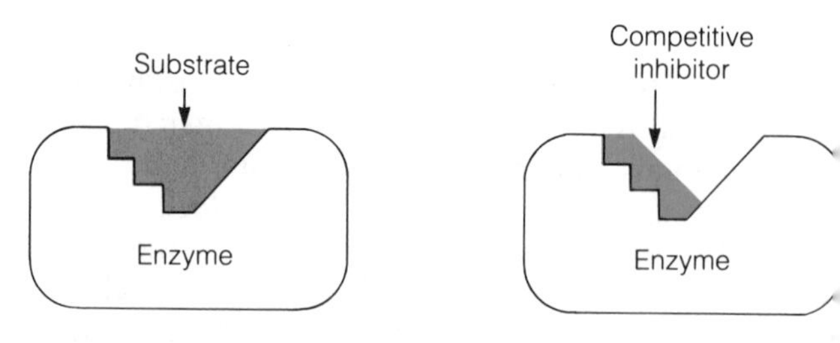

FIGURE 3-13
A competitive inhibitor may inhibit the catalytic action of an enzyme by binding to the active site and preventing binding of the normal substrate. The inhibitor is not altered by the enzyme.

INHIBITION OF ENZYMES

Inhibition of enzymes is the basis of a major mechanism by which metabolism is normally regulated in the cell. Inhibition occurs because the activity of an enzyme can be affected by the binding of substances other than the substrate. For example, an enzyme may be inhibited by an ion or molecule that binds to the protein outside the area of the active site and in so doing slightly shifts the tertiary or quaternary configuration of the enzyme. In turn, the shape of the active site changes. In other instances the inhibitor binds to the active site itself, thereby blocking access to the active site by substrate molecules. These inhibitions of an enzyme may be reversible or irreversible.

Reversible Inhibition

In reversible inhibition an inhibitor binds reversibly to the active site or to another location on the enzyme in a way that changes the shape of the active site. Reversible inhibition occurs normally in cells for many enzymes and is important in regulating the activity of enzymes (discussed later).

Reversible inhibition can be **competitive** or **noncompetitive.** In competitive inhibition a second substance with a structure similar to the normal substrate molecule may bind reversibly to the active site, forming an enzyme-inhibitor complex (EI), but the inhibitor is not acted upon by the enzyme (Figure 3-13):

$$E + I \rightleftharpoons EI$$

Since the inhibitor binds reversibly with the active site the substrate and the inhibitor compete for binding to the active site. The degree of inhibition depends on (1) the ratio of the binding strengths of the inhibitor and the substrate to the active site, and (2) the ratio of the concentration of inhibitor to substrate. The stronger the binding of the inhibitor relative to the binding of the substrate, the more enzyme molecules will at any moment be in a complex with the inhibitor, and hence the greater the inhibitory effect. Likewise, the higher the concentration of inhibitor relative to the concentration of the substrate, the greater the inhibition. This type of inhibition is called competitive because if the concentration of substrate is increased, the activity of the enzyme increases; the more substrate molecules present, the more likely substrate molecules will be to bind to free enzyme molecules. At a sufficiently high concentration of substrate the substrate monopolizes the enzyme, and competition by the inhibitor is virtually eliminated. Under these conditions the enzyme reaches its V_{max} despite the presence of some inhibitor molecules. Thus, the inhibitor does not change V_{max} but it raises the value of K_m.

One of the earliest and best studied cases of competitive inhibition is the inhibition of the enzyme succinate dehydrogenase (see Chapter 4 for the place of succinate dehydrogenase in cell metabolism). This enzyme catalyzes the removal of two hydrogen atoms from succinate to form fumarate. To carry out the catalysis the

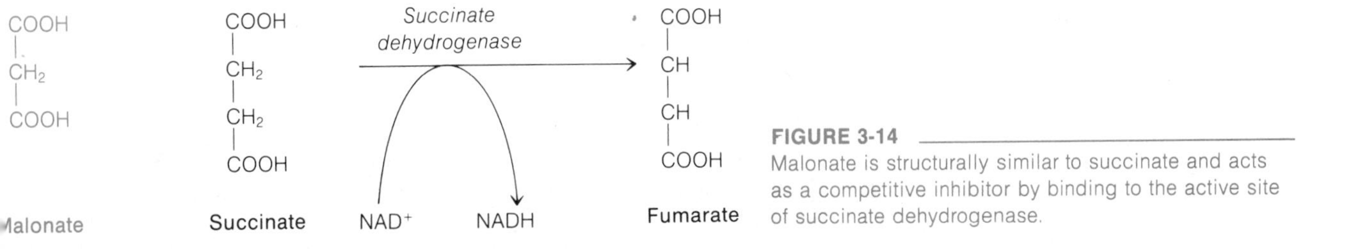

FIGURE 3-14
Malonate is structurally similar to succinate and acts as a competitive inhibitor by binding to the active site of succinate dehydrogenase.

enzyme requires NAD^+ as a coenzyme, coupling the removal of H_2 from succinate to the reduction of NAD^+ (Figure 3-14). Malonate has a structure similar to succinate; it has one less CH_2 group than succinate (Figure 3-14). The active site of succinate dehydrogenase reversibly binds malonate but does not catalyze a change in malonate. Therefore, in the presence of malonate the opportunity for succinate to bind to the enzyme is reduced.

In other cases of competitive inhibition, the inhibitor is the product of the reaction catalyzed by the enzyme. Enzymes usually catalyze relatively small chemical changes in their substrates. Hence, the chemical structure of the product of a reaction still bears a strong structural resemblance to the original substrate. In the case of succinate dehydrogenase the product, fumarate, is sufficiently similar to the structure of succinate to compete with succinate for the active site of the enzyme. Therefore, an increasing accumulation of fumarate in the cell (in mitochondria, where succinate dehydrogenase is located) results in a progressively increased inhibition of the enzyme and a decreased conversion of succinate to fumarate.

Many enzymes in the cell are subjected to competitive inhibition, and such inhibition plays an important role in the regulation of enzyme pathways (discussed below). Through regulation of enzyme pathways the concentrations of molecules within the cell can be controlled.

In noncompetitive inhibition the inhibitor does not compete with the substrate for binding at the active site of the enzyme. Instead noncompetitive inhibitors bind reversibly to the enzyme at a location *other than the active site*. Some noncompetitive inhibitors bind to the free enzyme to form an EI complex ($E + I \rightleftharpoons EI$), whereas others bind to the enzyme-substrate complex (ES):

$$E + S \rightleftharpoons ES + I \rightleftharpoons ESI$$

Binding of the inhibitor to the enzyme causes a slight shift in the three-dimensional structure of the enzyme. This distortion changes the shape of the active site and thereby reduces the binding affinity of the substrate for the active site or the catalytic activity of the active site. Similarly, in cases in which the inhibitor binds to ES, the dissociation of ES into E + P may be slowed, again presumably because of a slight change in the shape of the active site. In both cases, the turnover number of the enzyme is reduced in proportion to the strength of the reversible binding of the inhibitor to the enzyme and in proportion to the concentration of the inhibitor. Unlike competitive inhibition, however, noncompetitive inhibition results in a decrease in V_{max}. In this kind of inhibition, the K_m is unchanged. Properties of competitive and noncompetitive inhibition are summarized in Table 3-2.

Irreversible Inhibition

In irreversible inhibition binding of an inhibitor molecule or ion is so strong that dissociation is extremely slow, and virtually all of the enzyme molecules remain permanently inactivated. Irreversible inhibition of enzymes does not normally occur in cells but follows uptake of a poison from the environment. For example, **α-amanitin**, a toxic agent of many types of poisonous mushrooms (e.g., *Amanita*), inhibits the enzyme **RNA polymerase** nearly irreversibly. RNA polymerase cat-

TABLE 3-2

A Comparison of Competitive and Noncompetitive Inhibition of Enzymatic Activity

Competitive Inhibition	Noncompetitive Inhibition
1. Inhibitor molecules compete with substrate molecules for binding to the active site.	**1.** Inhibitor binds reversibly to a site other than the active site. Inhibitor may bind to free enzyme (E) or to an enzyme-substrate (ES) complex.
2. Binding of inhibitor does not necessarily shift the three-dimensional structure of the enzyme.	**2.** Binding of inhibitor changes the shape of the active site.
3. Turnover number of the enzyme is reduced in proportion to binding of the inhibitor to the enzyme (E).	**3.** Turnover number of the enzyme is reduced in proportion to binding of the inhibitor to the enzyme (E) or enzyme-substrate complex (ES).
4. Inhibitor raises the K_m for the enzyme.	**4.** The K_m for the enzyme is not changed.
5. Inhibitor does not change V_{max}.	**5.** Inhibitor decreases V_{max} of the enzyme.

alyzes the joining of nucleotides in the cell nucleus to form RNA molecules (see Chapter 8), and a cell cannot long survive without RNA synthesis.

REGULATION OF ENZYMATIC ACTIVITY BY FEEDBACK INHIBITION

The activity of almost every enzyme in a cell is regulated by feedback inhibition. In the simplest case the inhibition is competitive and occurs because a product P of a reaction may reversibly bind to the active site of an enzyme and act as a competitive inhibitor, $E + P \rightleftharpoons EP$. Thus, the product P regulates enzyme activity by acting as a feedback inhibitor.

Most enzymatic pathways are regulated by feedback inhibition of a different sort. In these cases the final product in an enzymatic pathway acts as a noncompetitive inhibitor that binds to the first enzyme (at a site outside the active site) in the pathway. For example, some species of cells can fulfill their requirement for **isoleucine** by synthesis from **threonine** by means of a pathway consisting of five enzymes (E_1, E_2, etc.) and four intermediate substrates (S_1, S_2, S_3, and S_4) (Figure 3-15). The first enzyme in the pathway is **threonine dehydratase**, which catalyzes the conversion of threonine to α-ketobutyrate. In addition to the active site, threonine dehydratase possesses a second

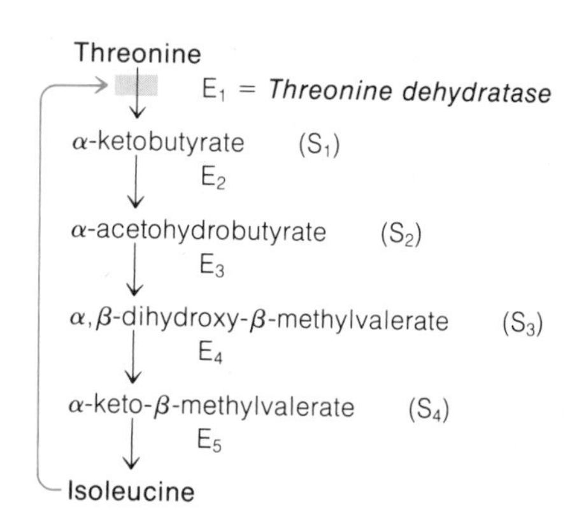

FIGURE 3-15

Isoleucine is synthesized from threonine in five steps, each catalyzed by a different enzyme. Isoleucine acts as a noncompetitive inhibitor of the first enzyme in the pathway, threonine dehydratase, by binding to a site other than the active site of the enzyme.

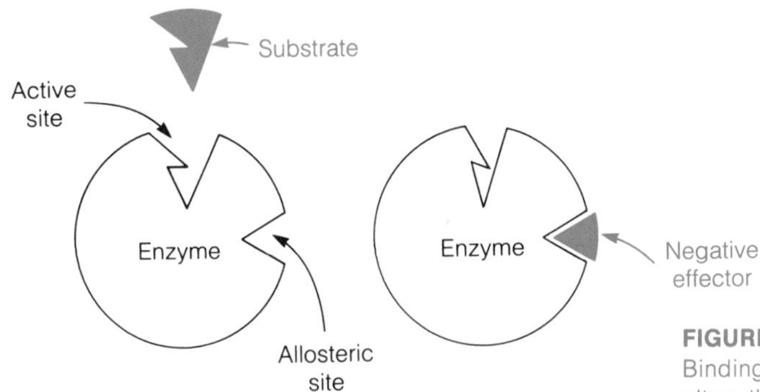

FIGURE 3-16
Binding of a negative effector to an allosteric site of an enzyme alters the shape of the active site, preventing binding of the substrate to the enzyme.

ite for reversible binding of isoleucine, the end product f the pathway. When isoleucine is bound, threonine ehydratase is inactivated. Such noncompetitive inhiition of an early enzyme (usually the first) in a pathway by the end product of the pathway is usually called **nd-product inhibition.**

Enzymes that are regulated by end-product inhibiion are called **allosteric** enzymes (from *allo* meaning ther and *steric* meaning space or structure). The site at which the end-product inhibitor binds (isoleucine in he example above) is called the **allosteric site.** A subtance that reversibly binds to the allosteric site is nown as an **effector** or **modulator** of the enzyme.

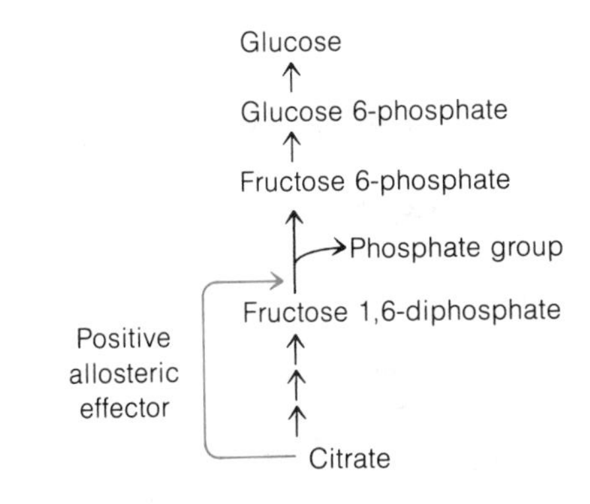

FIGURE 3-17
Citrate acts as a positive effector of the enzyme fructose 1,6-diphosphatase by binding to an allosteric site of the enzyme. Binding of the effector activates the enzyme by altering the shape of the active site.

In most known cases of allosteric control of enzymes, the effector is a **negative effector;** that is, binding of the effector inhibits the enzyme by slightly changing the three-dimensional structure of the protein, thereby altering the shape of the active site (Figure 3-16). In some cases, effectors enhance enzymatic activity and are called **positive effectors.**

REGULATION OF ENZYMATIC ACTIVITY BY POSITIVE EFFECTORS

In addition to degradation of glucose, cells are able to synthesize glucose, and do so under some circumstances. Citrate is a 6-carbon acid important in energy metabolism (see Chapter 4). Citrate can be converted to glucose through a series of enzymatically catalyzed reactions (Figure 3-17). The pathway becomes active when citrate accumulates in the cell. Citrate might accumulate for a number of reasons; the simplest would be the appearance of a large amount of citrate in the medium and its uptake by the cell. One way of regulating the pathway involves fructose diphosphatase, which splits fructose 1,6-diphosphate to fructose 6-phosphate and phosphate. Citrate is a positive effector of this enzyme. The accumulation of citrate leads to activation of fructose diphosphatase (the other enzymes in the pathway are normally present in active form whether or not citrate is present). This activation causes the pathway to function, which results in the excess citrate being converted to glucose. Glucose can be stored by polymerization into glycogen or starch (Chapters 2 and 4).

REGULATION OF ENZYME PATHWAYS

Allosteric interactions play a major role in the regulation of enzymatic pathways, and therefore of cell metabolism. The example given above of the five-step conversion of threonine to isoleucine can be expanded to give a beginning idea of the intricacy of the regulatory mechanisms in enzymatic pathways.

Threonine can itself be synthesized from another amino acid, aspartic acid, by a sequence of five enzymatically catalyzed steps. This conversion is allosterically controlled by end-product inhibition, in which threonine acts as a negative effector on the first enzyme in the pathway, aspartate kinase (Figure 3-18).

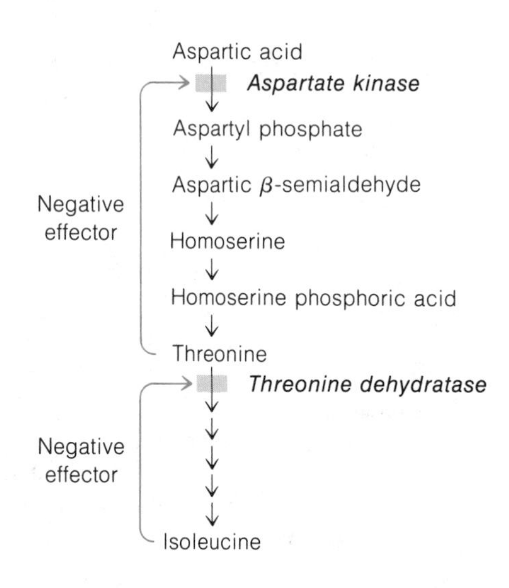

FIGURE 3-18
Aspartic acid can be converted to isoleucine through a pathway that is allosterically regulated at two points. The amino acid threonine, an intermediate in the pathway, acts as a negative effector of aspartate kinase, the first enzyme in the pathway. The end product of the pathway, isoleucine, is a negative effector of the enzyme (threonine dehydratase) that begins the conversion of threonine to isoleucine.

The consequences of the allosteric regulations i this system are the following: the buildup of the intra cellular pool of isoleucine results in feedback inhibitio of threonine dehydratase, enabling the pool c threonine in the cell to increase. The intracellular ir crease in threonine results in feedback inhibition of as partate kinase, leading, in turn, to the accumulation c aspartic acid. Aspartic acid is formed from glutamic aci and oxaloacetate, a reaction catalyzed by a trans aminase:

$$\begin{array}{c}\text{glutamic acid} + \text{oxaloacetate} \\ \downharpoonleft\upharpoonright \ (\text{transaminase}) \\ \alpha\text{-ketoglutarate} + \text{aspartic acid}\end{array}$$

If aspartic acid accumulates (because its conversion t threonine is shut off, and it is synthesized faster than is used for protein synthesis), then aspartic acid itse starts to work as a competitive inhibitor of the trans aminase. This leads to accumulation of glutamic acic the synthesis of which is regulated by another compl cated series of controls.

Interconnections of Enzyme Pathways

Almost every enzymatic pathway is directly inter connected with one or more other enzymatic path ways. The pathway that leads from aspartic acid t threonine to isoleucine connects with a pathway fo methionine synthesis. Methionine is synthesized fror homoserine, an intermediate in the pathway from as partate to threonine (Figure 3-19). Methionine is negative effector of the first enzyme of this branch, an the intracellular concentration of methionine regulate the amount of homoserine that enters the pathway t methionine. *In branch pathways the blockage is always a the first step in the branch,* a precise way to assure tha only the product present in excess is not made.

Various other intermediates in these pathways joi yet other pathways, creating an extraordinarily intri cate network of enzyme-catalyzed reactions. These in terconnections, along with multiple points of allosteri regulation, make up a finely tuned metabolic system i which each amino acid is finally produced in the righ amount to meet the needs of protein synthesis.

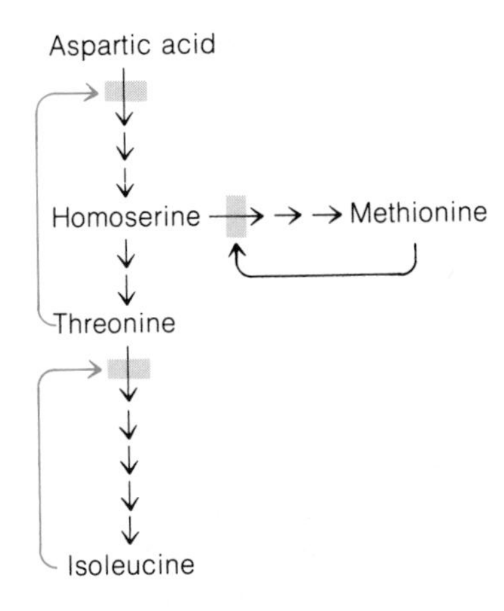

FIGURE 3-19
Another amino acid, methionine, can be synthesized from aspartic acid. The three amino acids threonine, isoleucine, and methionine are negative effectors of their own synthesis, inhibiting an earlier enzyme in their anabolic pathways from aspartic acid.

Aspartic acid is also the starting point for an allosterically regulated pathway for synthesis of uridine triphosphate (UTP) and cytidine triphosphate (CTP) (Figure 3-20), two of the four precursors needed for RNA synthesis. In the first step in this pathway aspartic acid is combined with carbamoyl phosphate to form *N*-carbamoylaspartic acid, a reaction catalyzed by an enzyme known as aspartate transcarbamoylase:

$$\text{carbamoyl phosphate} + \text{aspartic acid} \rightarrow N\text{-carbamoylaspartic acid} + \text{P}$$

N-carbamoylaspartic acid is converted through four additional steps to uridine monophosphate (UMP). Uridine monophosphate is phosphorylated to form uridine diphosphate (UDP) and then further phosphorylated to form UTP. UTP is used for RNA synthesis and in the formation of cytidine triphosphate (CTP). Cytidine triphosphate exerts end-product inhibition on the first enzyme in the pathway, aspartate transcarbamoylase, an allosteric enzyme, thereby controlling the rates of synthesis of both UTP and CTP. Aspartate transcarbamoylase was, in fact, the enzyme with which allosteric end-product inhibition was first discovered.

The aspartic acid–CTP pathway has yet other metabolic connections, one of which leads to DNA synthesis (Figure 3-21). These various pathways, starting with aspartic acid and leading to the synthesis of

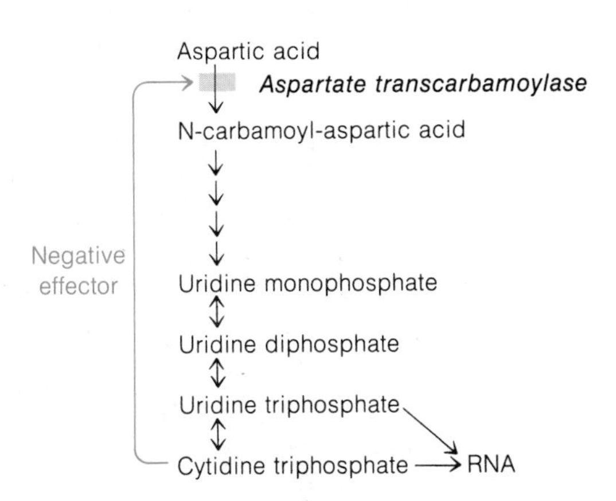

FIGURE 3-20
Aspartic acid is the starting point for synthesis of uridine triphosphate and cytidine triphosphate. Cytidine triphosphate is a negative effector of the first enzyme in this anabolic pathway.

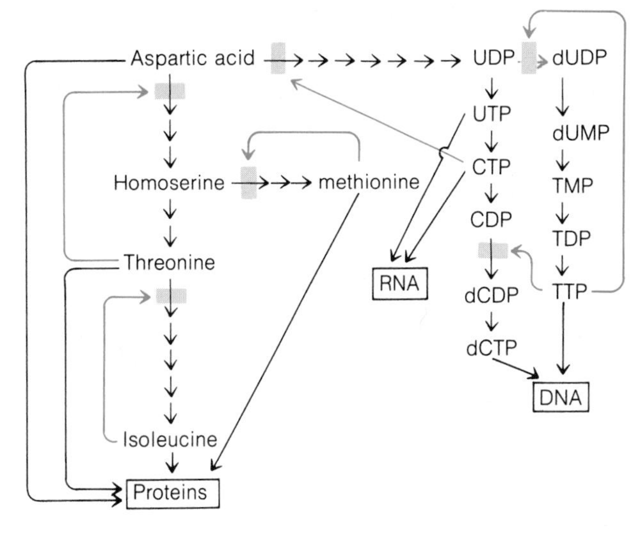

FIGURE 3-21
A summary of regulation of anabolic pathways starting with aspartic acid and leading to three amino acids, two nucleoside triphosphates (CTP and UTP) and two deoxynucleoside triphosphates (dCTP and TTP).

threonine, isoleucine, methionine, UTP, CTP, TTP, and dCTP may, on first encounter, seem terribly complex. However, these pathways and their regulatory mechanisms represent only a small segment of a far larger and more complex system of interconnecting pathways by which the syntheses of all the molecules in the cell are both regulated and integrated with one another. One must remember that the extraordinary complexity and regulatory efficiency of cell anabolism did not originate suddenly but were elaborated step-by-step over hundreds of millions of years of cellular evolution.

Because of its sparing effect on the anabolic demands within a cell, regulation by feedback inhibition is important for unicellular organisms. Uptake of a substance that may be available in the environment leads to closing down of the enzymatic pathway that synthesizes that substance. If isoleucine is present in the medium, then it is unnecessary to convert any threonine to isoleucine, and less aspartic acid need be converted to threonine.

In a fashion similar to the regulation of the syntheses shown in Figure 3-21, the synthesis of essentially all small molecules in cells is precisely regulated. A few other examples of allosteric regulation are discussed in connection with the metabolism of glucose (Chapter 4) and the production of deoxynucleotides for DNA synthesis (Chapter 12). Other mechanisms besides allostery regulate activities of certain enzymes. Regulation of many enzymes occurs by **covalent modification.** In covalent modification a particular chemical group is added to one or more amino acids, causing a shift in the tertiary structure of the protein and changing the properties of the active site. For example, phosphate groups are often covalently added to amino acid side groups. For some enzymes addition of phosphate completely inactivates the enzyme; for others addition of phosphate causes an inactive enzyme to gain catalytic activity. Covalent regulation of particular enzymes will be discussed in several subsequent chapters.

PROBLEMS

1. What are two ways to accelerate a chemical reaction?

2. Only those reactions that are useful to a cell are allowed to proceed at significant rates. How is this important selectivity achieved?

3. Some enzymes lower the activation energy of a reaction. How else do some enzymes catalyze a reaction, particularly between two molecules?

4. What was the first experimental observation to suggest that an enzyme and its substrate form a complex?

5. What does the value of the Michaelis-Menten constant, K_m, reveal about an enzyme?

6. In order for hexokinase to catalyze phosphorylation of fructose and glucose molecules at the same rate, what must the relative concentrations of the two substrates be?

7. What is the function of a coenzyme?

8. What is the role of some vitamins, including niacin and riboflavin, in cell metabolism?

9. How is an active site in an enzyme formed?

10. Why is a mutation that results in substitution of a hydrophilic amino acid by a hydrophobic one likely to have a greater effect on protein function than a mutation that substitutes a hydrophobic amino acid with another hydrophobic amino acid?

11. What physical property of an enzyme underlies its specificity for a given substrate?

12. What is the difference between the lock-and-key model and the induced-fit-strain model of enzyme catalysis?

13. The active site occupies only a small fraction of the surface of a protein enzyme. What are two other roles of the surface of a protein?

14. Explain how increasing the concentration of the normal substrate of an enzyme may decrease the effect of a competitive inhibitor.
15. How does binding of a competitive inhibitor to an enzyme differ from binding of a noncompetitive inhibitor?
16. How do reversible and irreversible inhibitors differ in the molecular mechanism of their actions?
17. What is the role of the allosteric site in regulating the activity of an enzyme?
18. Is an allosteric effector molecule that reduces enzyme activity a competitive or noncompetitive inhibitor? A reversible or irreversible inhibitor?
19. Why is it important to regulate enzyme pathways in a cell?
20. How else besides allostery is the activity of some enzymes regulated?

CHAPTER 4

Energy Flow and Metabolism

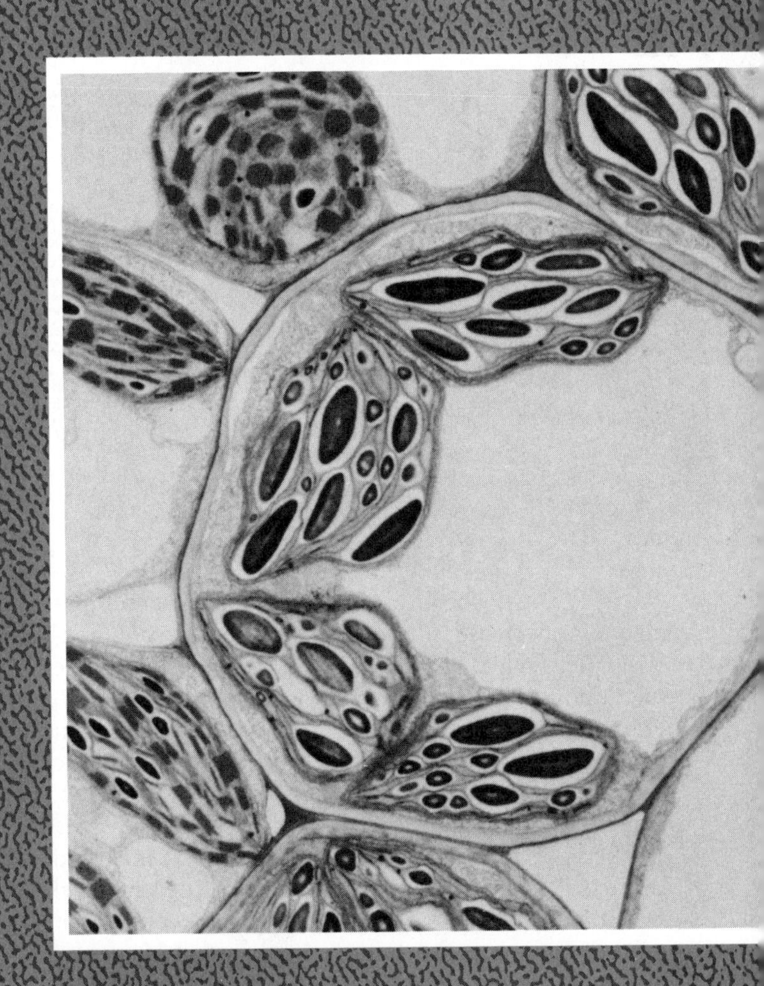

See Figure 4-14.

Practically all of the activities carried out by a cell demand energy. Therefore, a cell requires a constant source of energy from the outside. When the inflow of energy is cut off, internal reserves are depleted and activities necessarily slow down. One activity present in all cells is self-maintenance, in which many kinds of molecules must be continuously synthesized to replace those that have been damaged and degraded within the cell. Without this replacement capability, the integrity of structures and metabolic processes quickly deteriorate. Without an external source of energy most kinds of cells die within a few days. This chapter describes the means by which cells obtain energy from the environment.

CELL METABOLISM AND FREE ENERGY

Only those chemical reactions that are accompanied by a decrease in free energy can occur spontaneously. The released free energy is usually lost to the surroundings as heat. This loss is designated as $-\Delta G$. The most important chemical reactions in the cell that have a negative value of ΔG are those that make up the stepwise breakdown of sugars and fats to CO_2 and H_2O. On the other hand, a primary activity of cells is the synthesis of the thousands of molecules that make up cell machinery and structures. Most of the chemical reactions by which such synthesis is achieved cannot proceed spontaneously but require the input of energy; that is, these reactions have positive values of ΔG. A reaction with a $\Delta G > 0$ can proceed if it is **coupled** to another reaction in which ΔG is larger *and negative:*

$A \rightarrow B$ $\quad \Delta G_1 > 0$ (energetically unfavored)
$S \rightarrow P$ $\quad \Delta G_2 < 0$ (energetically favored)

A B / S P (crossed arrows) $\quad \Delta G_1 > 0 + \Delta G_2 < 0 = \Delta G < 0$ (energetically favored)

The crossed-arrow symbol means that the reactions are coupled. By coupling of the two reactions some of the energy normally lost as heat ($-\Delta G_2$) in $S \rightarrow P$ is transferred to the energy-requiring reaction $A \rightarrow B$ ($+\Delta G_1$). The coupled reaction is energetically favored because more energy is released by $S \rightarrow P$ than required for $A \rightarrow B$; that is, the coupled reaction still releases energy as heat ($\Delta G < 0$). Cells have evolved ways to couple reactions so that energy released by catabolic reactions can be used for energy-requiring anabolic reactions.

Photosynthetic organisms capture energy of sunlight for the synthesis of organic molecules from CO_2 and H_2O (Figure 4-1). In the process some of the energy is lost as heat. Almost all nonphotosynthetic organisms obtain energy by breaking down organic molecules, originally derived from photosynthetic organisms, to CO_2 and H_2O. The energy obtained is used to drive energy-requiring processes with loss of some of the energy as heat.

THE FLOW OF ENERGY IN CELLS

Ultimately, all cells obtain their energy from sunlight. Photosynthetic organisms (plants, algae, and photosynthetic bacteria) absorb energy in the form of light from the sun. This energy is channeled through coupled reactions into the synthesis of organic molecules, mostly carbohydrates, from CO_2 and H_2O. All other organisms obtain energy by breaking down (oxidizing) the organic molecules obtained through food chains that start with the photosynthetic organisms. The end products of the oxidation of organic molecules are CO_2 and H_2O, which are then available for recycling into photosynthesis. In this section we will examine the major pathways for energy flow.

THE ROLE OF ATP IN ENERGY FLOW

Energy derived from the breakdown of organic molecules such as carbohydrates is not used directly for the energy-requiring reactions in the cell but is first channeled into an energy-storing molecule, usually **adenosine triphosphate (ATP).**

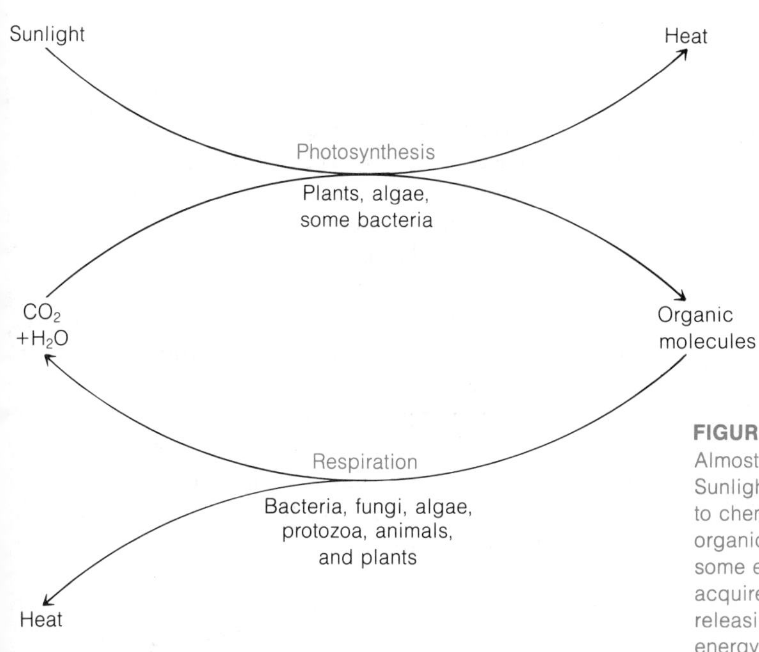

FIGURE 4-1
Almost all energy used in metabolism comes from sunlight. Sunlight is absorbed by photosynthetic organisms and converted to chemical energy, with the incorporation of carbon dioxide into organic molecules and release of oxygen. During the process some energy is lost as heat. Nonphotosynthetic organisms acquire chemical energy from photosynthetic organisms, releasing carbon dioxide and water. Once more, some of the energy is lost as heat.

The chemical structure of adenosine monophosphate was described in Chapter 2. In ATP three phosphate groups are attached in a row to the ribose portion of adenosine (Figure 4-2). A substantial amount of free energy is released when the bond between the second and third phosphate groups is hydrolyzed, yielding adenosine diphosphate (ADP) and inorganic phosphate (P_i):

$$\text{ATP} \rightarrow \text{ADP} + \text{P}_i \qquad \Delta G < 0$$

Similarly, hydrolysis of the bond between the first and second phosphate groups yields adenosine monophosphate (AMP), two phosphate groups, and a large amount of negative free energy:

$$\text{ATP} \rightarrow \text{AMP} + \text{P} \sim \text{P} \qquad \Delta G < 0$$

The bonds between the phosphate groups are called high-energy bonds, not because the bond energy is particularly high but because an unusually large amount of energy is *released* by hydrolysis. This is so because the four negatively charged oxygen atoms in ATP (Figure 4-2) are very close, and the repulsive force

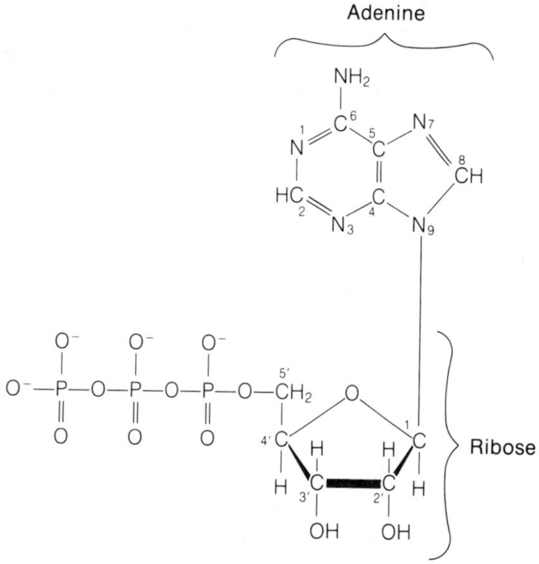

Adenine

Ribose

FIGURE 4-2
The structure of adenosine triphosphate (ATP). The molecule consists of adenine, ribose, and three phosphate groups.

between them is high. Hydrolysis of a bond reduces this repulsive force and releases free energy.

The presence of a bond with a large free energy of hydrolysis is indicated by the symbol ~. Thus, ATP may be written

$$AMP \sim P \sim P$$

and ADP as

$$AMP \sim P$$

The two phosphate groups released by hydrolysis of ATP to AMP are also joined by a high-energy bond and are written as P~P, which is known as **pyrophosphate.** Hydrolysis of pyrophosphate yields two individual phosphate groups, (orthophosphate), denoted P_i.

$$P \sim P \rightarrow 2P_i \qquad \Delta G < 0$$

Pyrophosphate is quickly hydrolyzed to two P_i in a cell through the catalytic action of the enzyme pyrophosphatase, with the release of free energy as heat.

ATP is synthesized from ADP by the addition of a phosphate group or from AMP by the stepwise addition of two phosphate groups. This requires, of course, the input of energy. The energy is obtained by coupling the energy-releasing breakdown of *fuel* molecules such as carbohydrates and fats to the energy-requiring synthesis of ATP in a coupled reaction,

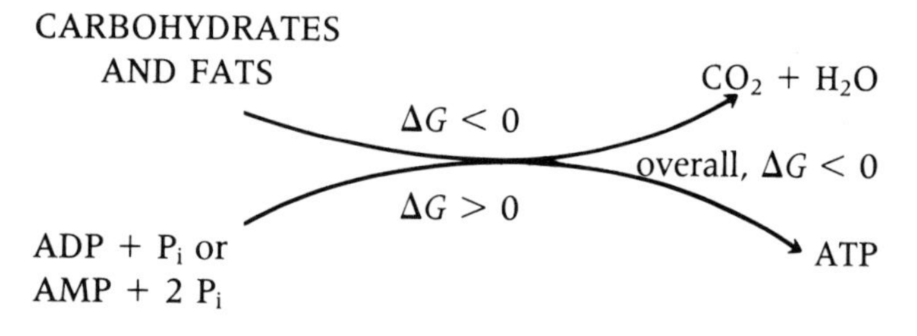

The overall coupled reaction still has $\Delta G < 0$, because more energy is released by breakdown of fuel molecules than used to synthesize ATP. The excess energy is lost in the cell as heat. The transfer of energy from fuel molecules to ATP is described in detail later in this chapter. The next section contains a brief overview of how ATP is used for cellular activities.

COUPLING OF ATP HYDROLYSIS TO ANABOLIC REACTIONS

When ATP is hydrolyzed to ADP or AMP, free energy is released primarily as heat. However, the energy of heat cannot be used efficiently to bring about chemical reactions that are energetically disfavored. In order for such energy to be used, hydrolysis of ATP must be coupled to such a reaction. Exactly what is meant by coupling? A simple example is the synthesis of uridine triphosphate (UTP) from uridine monophosphate (UMP) by the successive addition of two phosphate groups. Uridine triphosphate is one of the precursor molecules in the synthesis of RNA. These energy-requiring reactions are made possible by coupling the hydrolysis of two ATPs, with transfer of energy as well as the phosphate groups:

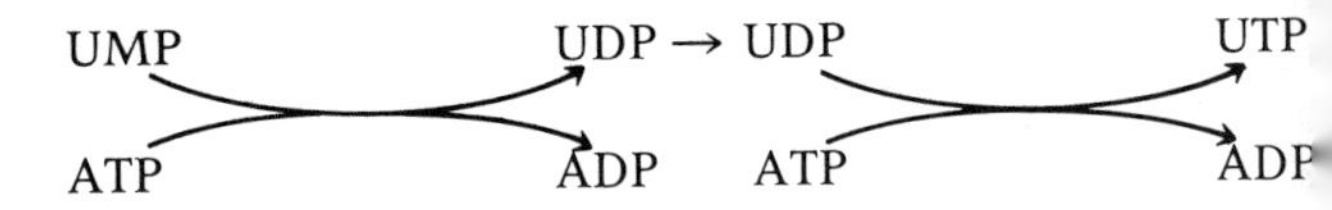

Many coupled reactions of this type occur in cells where not only energy but also a phosphate group is transferred from ATP to another molecule.

Another example of ATP utilization is the addition of a phosphate group (phosphorylation) to glucose, a reaction important in the breakdown of glucose:

Glucose → glucose–PO_4; ATP → ADP $\qquad \Delta G < 0$

In protein synthesis, amino acids are first *activated* in a coupled reaction with ATP in which the AMP part of ATP becomes covalently joined to the amino acid:

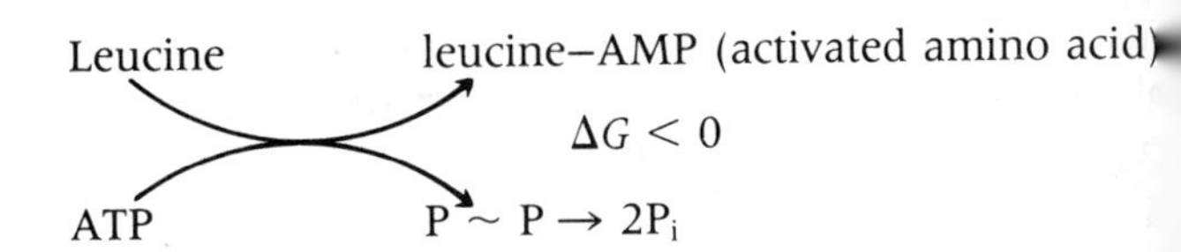

The activated amino acid (leucine = leu) can then

e added to a growing polypeptide chain with release f AMP:

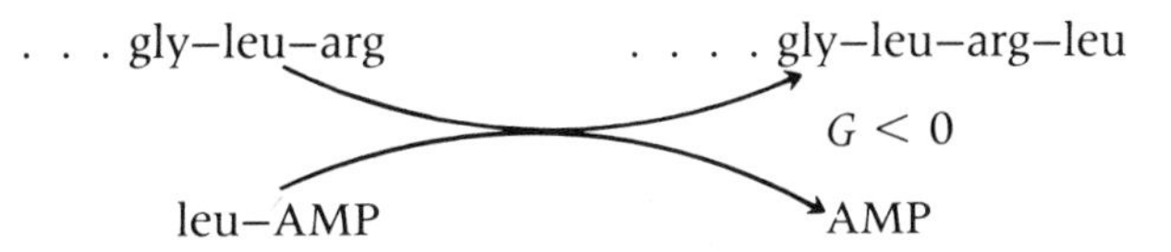

(The mechanism of amino acid addition to a poly- eptide chain is described in Chapter 8.)

The AMP produced in this reaction and the ADP roduced in the phosphorylation of glucose mentioned bove are available to be rephosphorylated to ATP in eactions that are coupled to the breakdown of carbo- ydrates and lipids.

The hydrolysis of ATP serves not only to make pos- ible hundreds of anabolic reactions in the cell, but also rovides the energy for almost all other kinds of work erformed by the cell. This work includes cell move- nent as in contraction of muscle cells, uptake of ions nd molecules from the environment, export of mate- ials from the cell, and generation of electrical energy.

Adenosine triphosphate occupies the pivotal posi- on in all energy-requiring processes that take place in ne cell. Thus, the means by which cells produce useful nergy in the form of ATP is a matter of major impor- ance. Much of this chapter deals with the various ways nat cells synthesize ATP.

PHOTOTROPHIC AND CHEMOTROPHIC CELLS

The synthesis of ATP requires the input of energy. ll energy in almost all cells ultimately originates from unlight through the process of photosynthesis in algae, lants, and photosynthetic bacteria. Photosynthetic or- anisms are consumed by nonphotosynthetic organ- sms, and chemical energy captured in organic mole- ules is passed to nonphotosynthetic organisms. Cells f algae, plants, and photosynthetic bacteria obtain en- rgy directly from light and are **phototrophs.** A pho- otroph synthesizes ATP using light energy captured in hotosynthesis. The ATP produced by photosynthesis elps to fuel the metabolism of phototrophic cells, including the synthesis of large amounts of sugars, particularly glucose. The glucose is in turn polymerized into the storage macromolecule starch, and into cellulose, which is a major constituent in the cell walls of plant cells and of many algae. The organic molecules produced (primarily glucose in the form of starch and cellulose) then serve as the source of biologically useful energy for the synthesis of ATP in nonphotosynthetic cells. These cells, (e.g., most bacterial species, protozoa, fungi, and animal cells) derive their energy from chemical digestion of organic molecules (primarily glucose) and are **chemotrophs.**

In summary, two principal ways by which cells obtain energy for the synthesis of ATP are (1) by capture of energy in sunlight (**photosynthesis**) and (2) by the breakdown of organic molecules (**catabolism**).

PHOTOSYNTHESIS

In photosynthesis energy in the form of light (photons) is captured and used to synthesize carbohydrates, usually glucose, from CO_2 and H_2O. ATP is also produced directly in photosynthesis, and this ATP provides energy for some of the steps leading to the production of glucose. The overall equation of photosynthesis is

$$6H_2O + 6CO_2 \xrightarrow{\text{light}} \rightarrow \rightarrow \rightarrow \underset{\text{glucose}}{C_6H_{12}O_6} + 6O_2$$

Six CO_2 and $6H_2O$ molecules are built into glucose ($C_6H_{12}O_6$). Of the 18 oxygen atoms in $6H_2O + 6CO_2$, 6 become incorporated into glucose and 12 are released as molecular oxygen ($6O_2$). Thus, photosynthesis yields two products essential for chemotrophs: glucose and oxygen.

In most photosynthetic cells the first step in photosynthesis is the absorption of energy, in the form of light, by the green pigment **chlorophyll** contained in membranes within chloroplasts. The chlorophyll molecule consists of a complicated ring structure containing a magnesium atom and a long hydrocarbon tail (Figure 4-3). Phototrophs contain accessory pigments, for example, **carotenoids** and **phycobilins,** whose func-

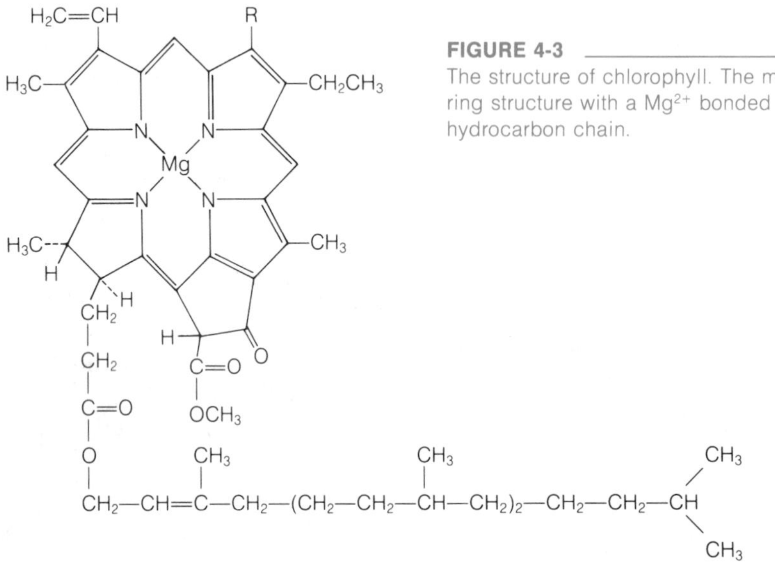

FIGURE 4-3

The structure of chlorophyll. The molecule consists of a complex ring structure with a Mg^{2+} bonded in the center and a long hydrocarbon chain.

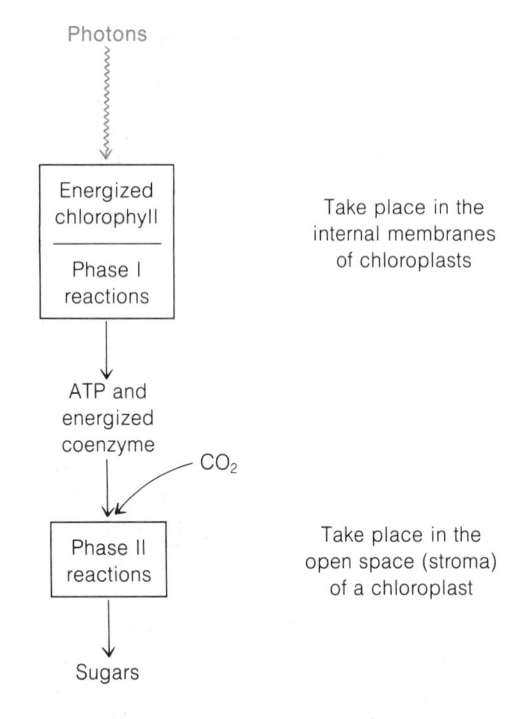

FIGURE 4-4

In reactions in phase I, energy captured as light by chloroplasts energizes an electron in chlorophyll. The energized electron is used to produce ATP and energized (reduced) coenzyme. These energy-rich products are used in phase II reactions to convert CO_2 and H_2O to sugars.

tion it is to expand the region of the light spectrur from which energy can be captured. These accessor pigments account, respectively, for the color of red an brown algae among eukaryotic phototrophs and for th bluish color of cyanobacteria. However, all of these or ganisms contain chlorophyll, and energy absorbed b accessory pigments is transferred to chlorophyll in earl stages of photosynthesis.

In halobacteria (*halo* = salt), light is not absorbe by chlorophyll but by **rhodopsin,** the light-absorbin purple pigment that is also present in the eyes of ani mals and that is the basis for sight. Rhodopsin is com posed of a protein and a smaller molecule made fror vitamin A. The discovery of rhodopsin-based synthesi of ATP in halobacteria raises some interesting ques tions. Did such a complicated molecule as rhodopsir specialized for absorption of light, arise twice in evolu tion, once in the evolution of the ATP synthetic mecha nism in halobacteria and again in the evolution of eye sight? Or, did the ability for rhodopsin synthesis i halobacteria find its way into the evolution of the light detecting mechanism that became the basis for the ani mal eye?

Photosynthesis in plants and algae is made up c two interconnected sets of reactions (Figure 4-4). Phas I reactions take place in the membranes inside a chloro plast. Phase II reactions take place in the open spac

stroma) inside a chloroplast. Phase I reactions form he first stage of photosynthesis. They begin with the apture of energy in the form of photons, which ener- izes an electron in chlorophyll. The subsequent reac- ions in phase I use the energized electron to make two roducts: ATP and energized coenzyme molecules. The eactions in phase I are often referred to as the "light eactions" because they are said to be dependent on ght. Only the capture of photons is a light-dependent eaction. The remaining phase I reactions proceed us- ng the light energy captured in chlorophyll but can ake place in the absence of light.

The reactions of phase II, which take place in the pen space or stroma inside the chloroplast, use the TP and energized coenzyme molecules produced by he reactions in phase I to convert CO_2 and H_2O into ugar molecules. The reactions of phase II are often alled the "dark reactions" because they are not di- ectly dependent on the presence of light and can pro- eed in light or darkness. The classification of the reac- ions in phase I as light reactions and in phase II as dark eactions is not strictly correct since only the first event n phase I is light-dependent. All other reactions are not hemselves dependent on light.

Phase I Reactions

Light is defined as the class of electromagnetic ra- iation whose wavelengths are visible to the human ye. The range of visible wavelengths extends from bout 400 nanometers (violet) to about 700 nanome- ers (red) (Figure 4-5). Above 700 nanometers is infra- ed radiation, and beyond the infrared are radiowaves. he ultraviolet range extends downward from 400 anometers.

Chlorophyll molecules absorb radiation in the lue-violet and yellow-red range of the visible part of the radiation spectrum, but do not absorb green light. Depending on the particular plant or alga, a number of accessory photosynthetic pigments (carotenoids, phy- cobilins, etc.) are present that absorb photons in still other parts of the visible spectrum. Light energy cap- tured by accessory light-gathering pigments is trans- ferred to chlorophyll for entry into photosynthesis.

The photon energy absorbed by a chlorophyll mol- ecule (or other photosynthetic pigment) is taken up by an electron in one of its atoms. Excited atoms are usu- ally unstable and extremely short-lived; the electron may give up the extra energy it has absorbed as electro- magnetic radiation (e.g., light, giving rise to fluores- cence) or, more often, as heat. The electron of chloro- phyll that is energized by the absorption of light has an entirely different fate. Instead of returning to its original state by giving off light or heat, the energized electron is transferred from chlorophyll through a chain of **elec- tron carriers.** Ultimately, the electrons are passed in pairs to the coenzyme nicotinamide adenine dinucleo- tide phosphate ($NADP^+$). $NADP^+$ differs from the coen- zyme NAD^+, described in Chapter 2, by the presence of an additional phosphate group. $NADP^+$ and NAD^+ each serve as a coenzyme for different sets of enzymes in different electron transfer pathways.

Phase I reactions are made up of two interacting photosystems, called **photosystem I** and **photosys- tem II,** and transfer of electrons through a series of carriers. The way in which the two photosystems work and how they are connected by electron carriers can be explained by following the fate of electrons energized in the two photosystems.

Photosystem I. Photosystem I (Figure 4-6) is made up of thousands of identical functional units in a single chloroplast. Each functional unit contains between 200

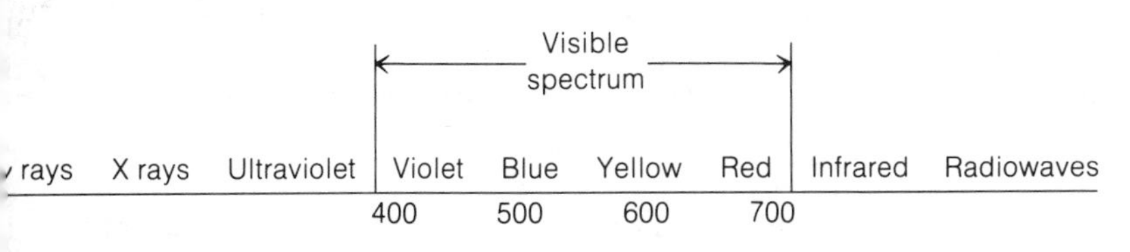

FIGURE 4-5
Wavelengths of electromagnetic radiations. The human eye perceives radiations with wavelengths of 400 to 700 nanometers.

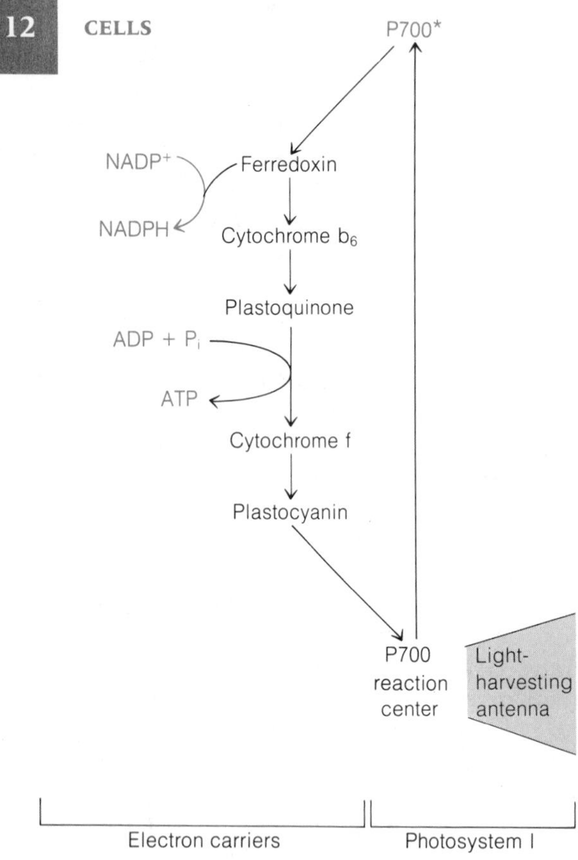

FIGURE 4-6
Photosystem I and electron carriers. The product is either NADPH or ATP.

and 400 chlorophyll molecules and accessory pigments. All but two of the chlorophyll molecules and all the accessory pigments in a single unit form a light-harvesting antenna that absorbs photons. From the antenna the absorbed energy is transferred to the **reaction center.** The reaction center consists of two chlorophyll molecules associated with a complex of proteins, which together are designated as **P700.** Transfer of energy energizes an electron in the reaction center (P700*). From the reaction center the energized electron can follow one of two routes: it can pass to a carrier and then to $NADP^+$ or through additional carriers back to the reaction center (Figure 4-6). In both routes the electron is passed to **ferredoxin,** an iron-containing protein that alternates between the ferric (Fe^{3+}) and the ferrous (Fe^{2+}) states as it shuttles elec trons from P700* either to coenzyme $NADP^+$ to yiel NADPH or to the carrier cytochrome b_6. Each NADP accepts two electrons from two ferredoxins. Two pro tons enter into the reduction of $NADP^+$ to NADPH balancing the negative charges of the electrons, bu they are energetically passive participants in the reac tions

$$NADP^+ \xrightarrow{2e^-} NADP^- \xrightarrow{2\,H^+} NADPH + H^+$$

The transfer of electrons from chlorophyll to $NADP^+$ i called **noncyclic electron flow.** The energy of th electrons carried by NADPH is used to bring about th reduction of CO_2 in the reactions of Phase II describe later.

Alternatively, an electron transferred from P700* t ferredoxin may be passed through the successive carri ers cytochrome b_6, cytochrome f, plastocyanin, an back to the reaction center (Figure 4-6). This is know as cyclic electron flow. During this flow, energy is give up by the electrons in a manner that brings about th synthesis of ATP from ADP and P_i. By taking the cycli route the transport of the energized electron is couple to the synthesis of ATP, and the electron is returned in deenergized condition to replace the electron originall given up by chlorophyll in the reaction center.

In summary, energized electrons from photosyster I can be transferred noncyclically to produce NADPH + H^+. Or electrons may travel cyclically, returning in deenergized state to chlorophyll in the reaction cente and driving the synthesis of ATP in the process. Th cyclic production of ATP is a form of photophosphory lation. Notice that photosystem I does not evolve O_2 and the energized electrons it generates are used t produce ATP only when they move cyclically. Whe electrons move noncyclically to $NADP^+$, they are with drawn from the system and must be replaced in th reaction center by a flow of electrons from photosyster II as described next.

Photosystem II. Like photosystem I, photosystem I in a single chloroplast is composed of thousands o

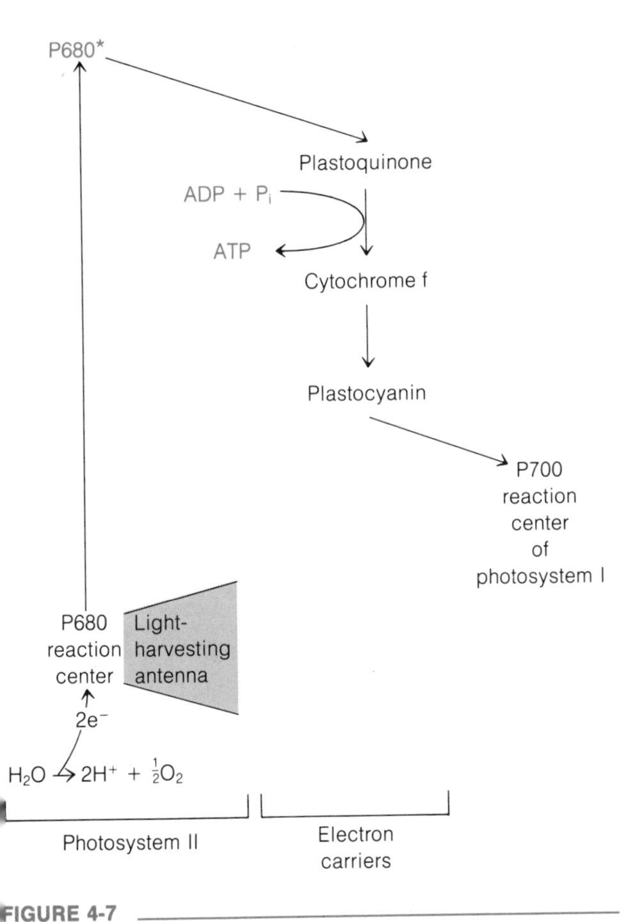

FIGURE 4-7
Photosystem II and electron carriers.

identical functional units. Each unit is made up of several hundred molecules of chlorophyll and accessory, light-capturing pigments that are part of a light harvesting antenna, plus a reaction center made up of two chlorophyll molecules bound to a complex of proteins. The reaction center of photosystem II is called **P680.** Each unit works as shown in Figure 4-7. An electron in one of the two chlorophyll molecules in the P680 reaction center is energized by the absorption of light by the antenna. The energized electron is transferred through a succession of carriers: plastoquinone, cytochrome f, plastocyanin, and finally to a chlorophyll molecule in a reaction center of photosystem I. These transfers are coupled to the phosphorylation of ADP to ATP.

Thus, electrons from photosystem II move by a noncyclic pathway in which energized electrons are passed from carrier to carrier and finally to chlorophyll in photosystem I, producing ATP molecules in the process. The flow of electrons results in replacement of electrons in chlorophyll of photosystem I, a fact that allows photosystem I to operate noncyclically to produce NADPH + H^+.

The transfer of electrons from photosystem II to photosystem I leaves chlorophyll in photosystem II in an electron-deficient state. However, the electron deficiencies in these chlorophyll molecules are filled by electrons obtained from splitting H_2O into $2H^+ + \frac{1}{2} O_2$:

$$H_2O \rightarrow 2H^+ + \frac{1}{2}O_2 + 2e^-$$

All the oxygen evolved in photosynthesis is released in photosystem II. The entire scheme of phase I reactions is summarized in Figure 4-8. The two photosystems and the electron carriers that make up the reactions of phase I generate both ATP and NADPH + H^+. These high-energy molecules are then used in the phase II reactions to reduce CO_2 to carbohydrates.

Phase II Reactions: Carbon Fixation

The ATP and NADPH + H^+ produced by the phase I reactions are used in the phase II reactions in the chloroplast as sources of energy and electrons to synthesize glucose from CO_2, a process called **carbon fixation.** Phase II reactions have been called the "dark reactions" because they are not directly dependent on light. However, most of the reactions of phase I are likewise not directly dependent on light and the designation of phase II reactions is a preferable name for the reactions of carbon fixation.

In the reactions of phase II, six CO_2 molecules are converted into a glucose molecule ($C_6H_{12}O_6$) using 6 ATP molecules and 12 pairs of electrons in 12 NADPH molecules (Figure 4-9). The key step in this process is the covalent joining of CO_2 to a 5-carbon molecule, **ribulose 1,5-bisphosphate,** to form a 6-carbon intermediate that is hydrolyzed into two 3-carbon mole-

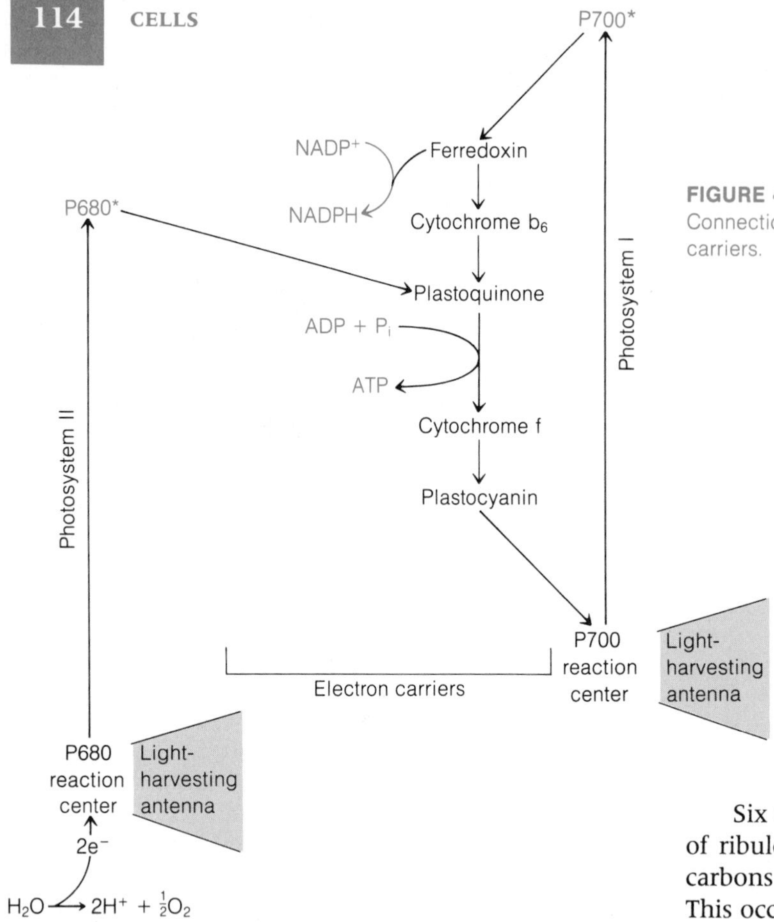

P700*
NADP+
Ferredoxin
NADPH
Cytochrome b_6
P680*
Plastoquinone
ADP + P_i
ATP
Cytochrome f
Plastocyanin
Photosystem I
Photosystem II
P700 reaction center
Light-harvesting antenna
Electron carriers
P680 reaction center
Light-harvesting antenna
$2e^-$
$H_2O \rightarrow 2H^+ + \frac{1}{2}O_2$

FIGURE 4-8
Connection of photosystem II with photosystem I by electron carriers.

cules (Figure 4-10). This reaction is catalyzed by an enzyme called **ribulose bisphosphate carboxylase.** This enzyme makes up about half of the total protein in chloroplasts and is probably the most abundant protein in the world.

Six CO_2 molecules must be joined to six molecules of ribulose 1,5-bisphosphate in order to gain the six carbons needed to make one 6-carbon sugar (glucose). This occurs in the cyclic process shown in Figure 4-11 which accomplishes two ends: first, a molecule of ribulose 1,5-bisphosphate that is used to fix a CO_2 molecule is regenerated so that the cycle can continue; and, second, for every six CO_2 molecules that enter the cycle two 3-carbon molecules that occur as intermediates in the cycle are drawn off to be subsequently combined to

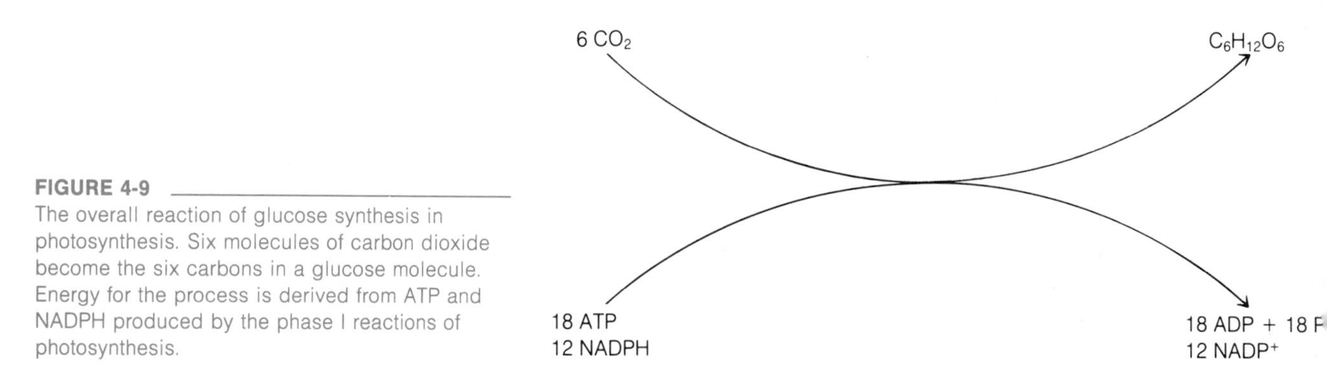

FIGURE 4-9
The overall reaction of glucose synthesis in photosynthesis. Six molecules of carbon dioxide become the six carbons in a glucose molecule. Energy for the process is derived from ATP and NADPH produced by the phase I reactions of photosynthesis.

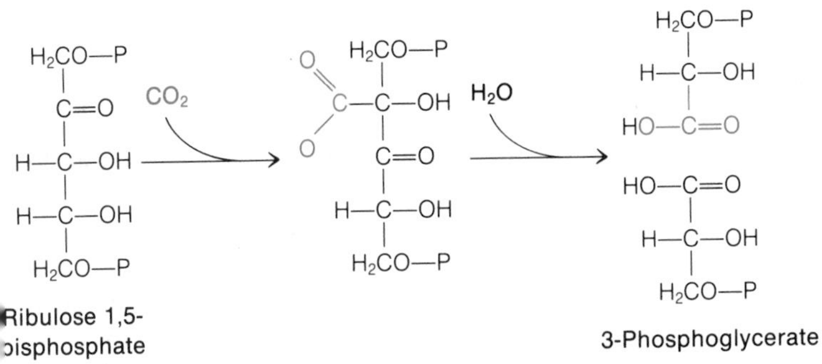

$H_2CO{-}P$ | $C{=}O$ | $H{-}C{-}OH$ | $H{-}C{-}OH$ | $H_2CO{-}P$ → (CO_2) → $O{=}C(O^-){-}C(OH)(H_2CO{-}P){-}C{=}O{-}H{-}C{-}OH{-}H_2CO{-}P$ → (H_2O) → two molecules of $H_2CO{-}P$ | $H{-}C{-}OH$ | $HO{-}C{=}O$

Ribulose 1,5-bisphosphate

3-Phosphoglycerate

FIGURE 4-10
Carbon dioxide is captured by addition to ribulose 1,5-bisphosphate in the first of the phase II reactions. The resulting 6-carbon molecule is immediately split into two 3-carbon molecules, with the addition of a molecule of water.

make glucose. The synthesis of glucose from the two 3-carbon molecules, glyceraldehyde 3-phosphate, occurs stepwise through several reactions. These same reactions occur, but in reverse direction, when glucose is broken down in any type of cell to obtain energy. The breakdown is described later in this chapter.

One CO_2 molecule is fixed with each turn of the CO_2 fixation cycle, and therefore the cycle must turn six times to make one glucose molecule. The overall outcome is that energy captured in the formation of 18 ATP and 12 NADPH molecules (from 12 $NADP^+$) in the phase I reactions is transferred into glucose in the phase II reactions. The transfer is not 100 percent efficient; although some of the energy goes into glucose, much of it is lost as heat.

CO_2 Availability and Photosynthesis

The atmosphere contains about 0.03 percent CO_2, and this is normally sufficient for photosynthesis; plants can continue photosynthesis as long as the atmosphere contains at least 0.005 percent CO_2. However, CO_2 enters a leaf primarily through pores in the epidermal layer of cells that make up the surface of the leaf (Figure 4-12). The pores, or **stomata**, can open and close. At high temperature and low humidity the pores remain almost closed in order to minimize water loss from the leaf. When the pores are almost closed, very little CO_2 can enter the leaf, and photosynthesis stops in many kinds of plants.

Ribulose bisphosphate carboxylase catalyzes the interaction of O_2 as well as CO_2 with ribulose 1,5-bisphosphate. The joining of O_2 to ribulose 1,5-bisphosphate, called **photorespiration,** is wasteful since it results in partial breakdown of the molecule with re-

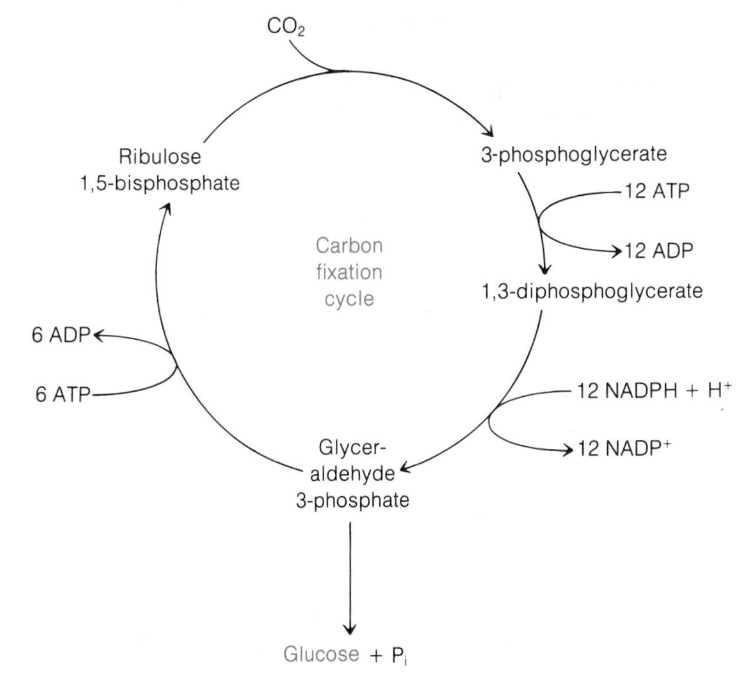

FIGURE 4-11
The carbon fixation cycle of the phase II reactions. One glucose molecule is produced for every six carbon dioxide molecules fixed by the cycle.

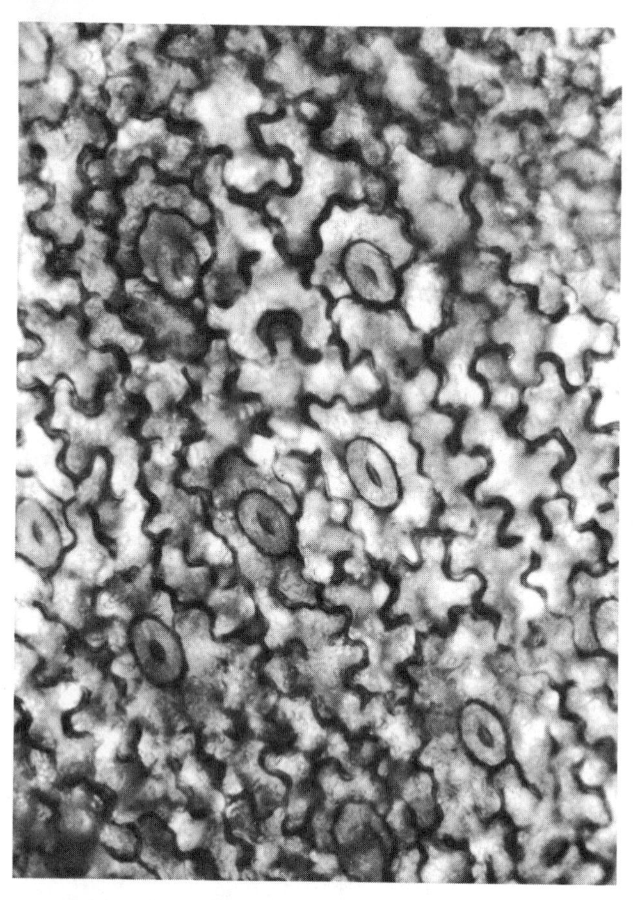

FIGURE 4-12
Light micrograph of the surface of a living duckweed leaf. The cells on the surface of the leaf (epidermis) have interlocking shapes that add strength to the layer. The oval structure with slit-shaped openings in the center are stomata formed by guard cells.

lease of one molecule of CO_2. In effect, photosynthesis runs backwards. However, the enzyme has a much stronger preference for CO_2, and photosynthesis continues as long as the CO_2 concentration remains above 0.005 percent even though the O_2 concentration may exceed 20 percent.

The problem of nearly closed stomata in a hot dry atmosphere is solved in certain kinds of plants, called **C_4 plants**, such as corn and sugarcane. In these plants CO_2 is concentrated in nonphotosynthetic cells, and the arrangement of cells in the leaf is different. In plants that require a cooler, more humid atmosphere, photosynthesis is carried out in **mesophyll cells** (Figure 4-13). In plants adapted to hot dry climates mesophyll cells, instead of carrying out photosynthesis, catalyze the joining of CO_2 to a 3-carbon molecule in a reaction that proceeds even when the CO_2 concentration drops below 0.005 percent. The resulting 4-carbon molecule is then shuttled into another kind of cell called the **bundle sheath cell** (Figure 4-14(a)), where it is broken down into the original 3-carbon molecule and CO_2, thereby raising the concentration of CO_2 in the bundle sheath cells (Figure 4-14(b)). In effect, mesophyll cells collect and pump CO_2 into bundle sheath cells, raising the CO_2 concentration and promoting photosynthesis and preventing the wasteful breakdown of ribulose 1,5-bisphosphate by joining to O_2 (**photorespiration**). Because fixed CO_2 is first detected in a 4-carbon molecule, these plants are called C_4 plants. Cells lacking this mechanism of CO_2 fixation are called C_3 plants, because fixed CO_2 is first detected in a 3-carbon molecule, 3-phosphoglycerate.

Photon Use in Photosynthesis

In photosystem I, two photons of light are needed to energize the two electrons required to reduce one $NADP^+$ to $NADPH + H^+$. Twelve NADPH are used to synthesize one glucose, and 24 photons must be absorbed for each glucose formed. However, the 24 electrons removed from chlorophyll in photosystem I must be replaced by 24 electrons from chlorophyll of photosystem II. The movement of the 24 electrons from photosystem II to photosystem I is accompanied by the generation of ATP, which is used to synthesize glucose.

Phototrophs Become Chemotrophs at Night

During daylight, photosynthetic organisms convert energy of photons into chemical energy in the form of ATP and NADPH. This energy can be used for all cell activities. During nighttime, phototrophic organisms

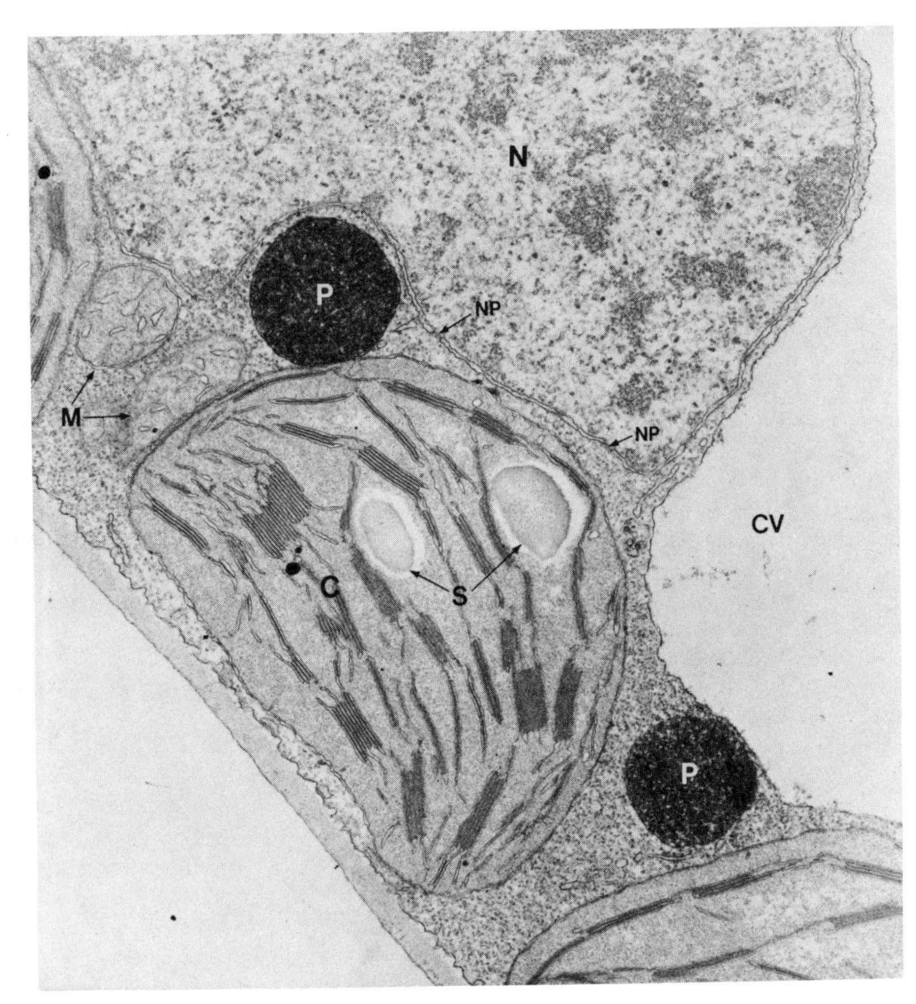

FIGURE 4-13
Electron micrograph of a mesophyll cell in a leaf of tobacco, a C_3 plant. The nucleus (N) is in the upper part of the picture. The two membranes of the nuclear envelope and nuclear pores (NP) are clearly visible. The chloroplast (C) contains grana and stromal thylakoids and starch granules (S). The dark round vesicles are peroxisomes (P) stained with a chemical that reveals the presence of catalase. (CV = central vacuole. M = mitochondria). [Courtesy of Eldon H. Newcomb, from S. E. Frederick, and E. H. Newcomb. 1969. *Journal of Cell Biology*, 43: 343.]

are deprived of a direct photosynthetic source of ATP and NADPH, and they must break down some of the glucose produced during the day to obtain energy; that is, they become chemotrophs. However, they capture more energy during daylight than they need for nighttime metabolism; some of the excess is used for cell growth and reproduction and some is stored as starch. Later sections in this chapter contain a description of how energy is obtained in the chemotrophic breakdown of organic molecules such as sugars.

The Structural Organization of Photosynthesis—The Chloroplast

In photosynthetic bacteria the chlorophyll and other light-absorbing pigments and the enzymes that carry out the phase I reactions are partially embedded in the plasma membrane and in infoldings of the plasma membrane called **lamellae** (Figure 1-7). These infoldings enormously increase the amount of membrane available for holding photosynthetic machinery

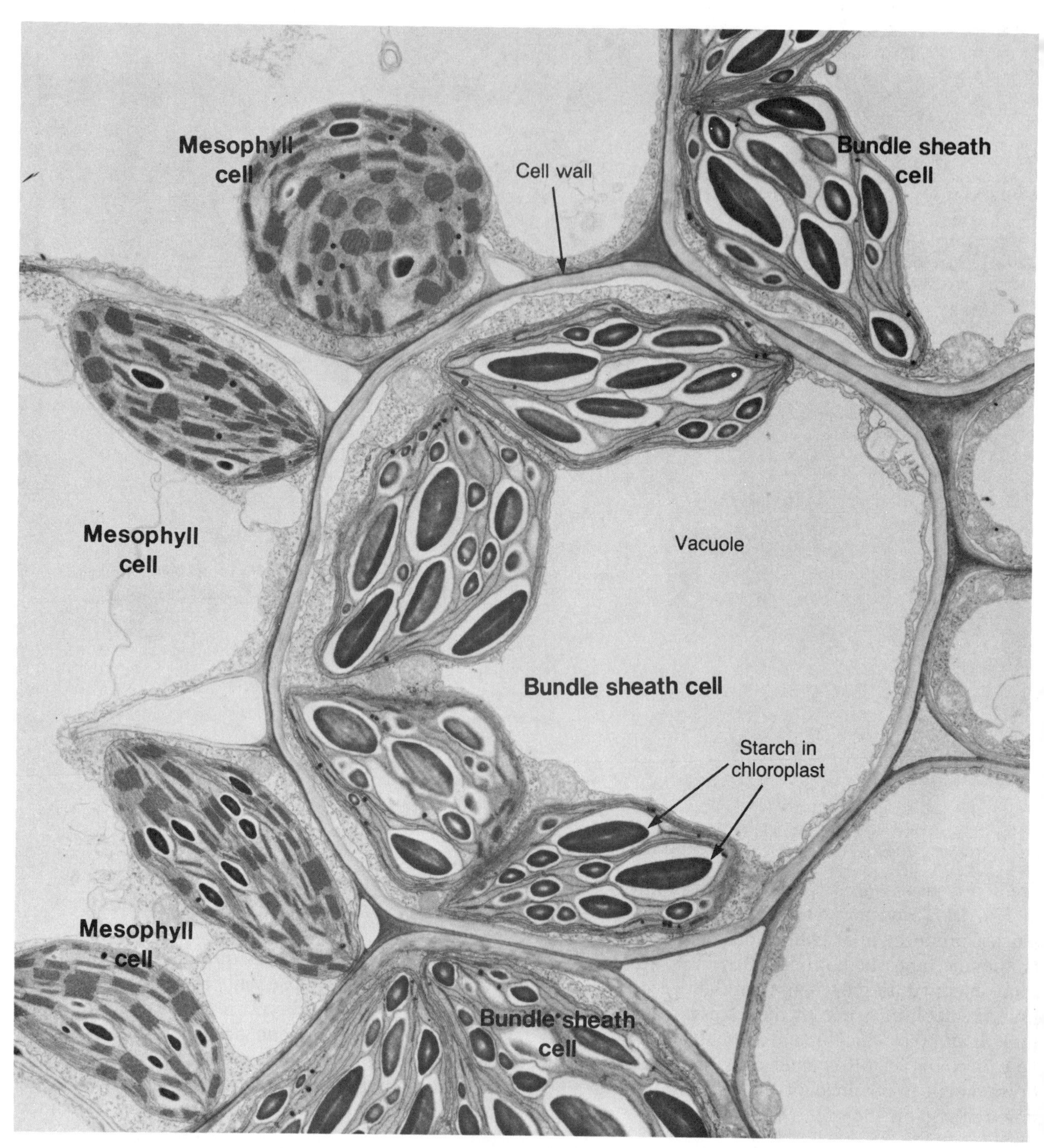
Mesophyll cell
Cell wall
Bundle sheath cell
Mesophyll cell
Vacuole
Bundle sheath cell
Starch in chloroplast
Mesophyll cell
Bundle sheath cell

a

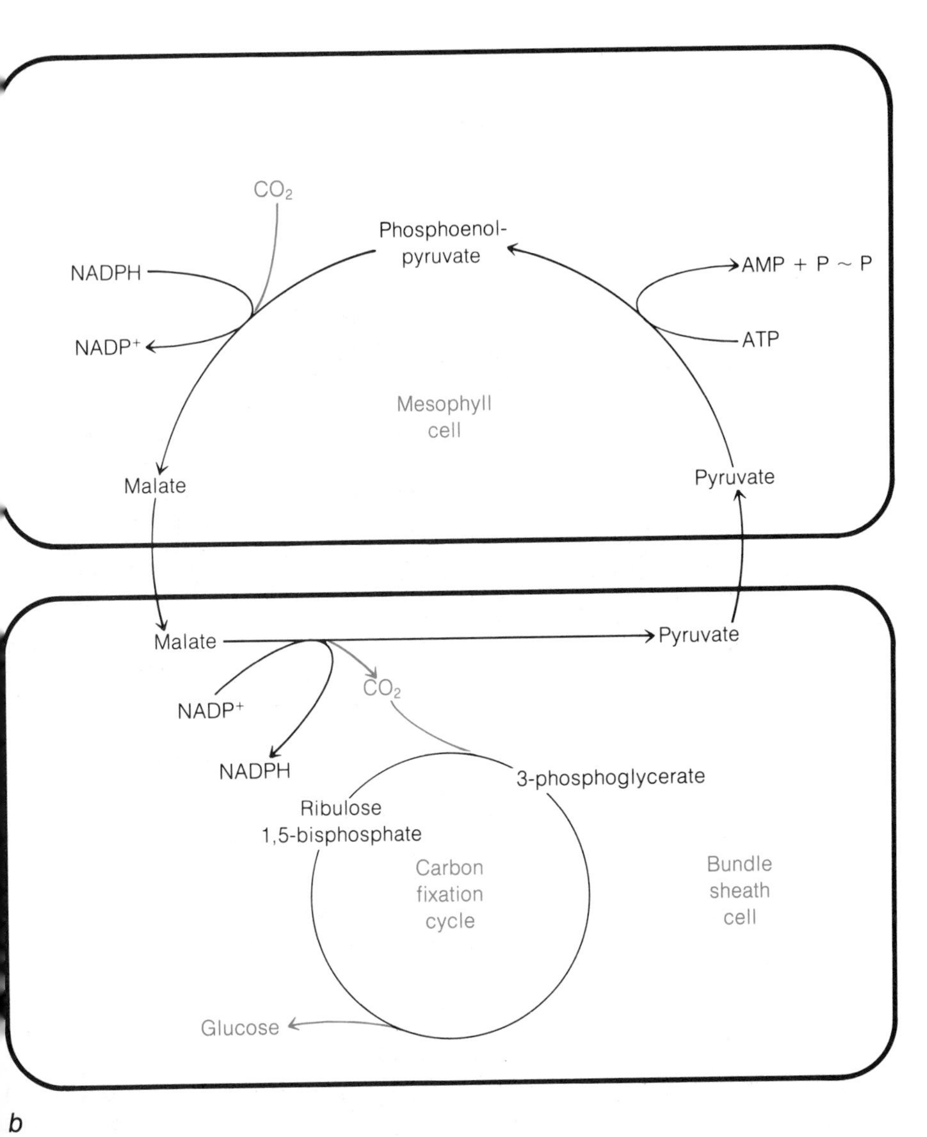

FIGURE 4-14
(a) Electron micrograph of mesophyll cells and bundle sheath cells in the leaf of corn *Zea mays*, a C_4 plant. Note the prominent starch granules (end products of photosynthesis) in the chloroplasts of bundle sheath cells. Nuclei are absent from this section. (b) In some plants mesophyll cells are specialized for capturing carbon dioxide in the synthesis of malate. Malate is transferred to bundle sheath cells, where it donates carbon dioxide to the carbon fixation cycle.
[(a) Courtesy of Eldon H. Newcomb.]

and hence are essential for high rates of photosynthesis. Phase II reactions take place in the cytosol. In plants and algae, photosynthesis is carried out in a cytoplasmic organelle specialized for the task, the **chloroplast.**

Chloroplasts in different species vary widely in the details of their internal structure, but they are all constructed on the same basic plan. The chloroplast is enclosed by a continuous envelope made of two membranes, an **outer membrane** and an **inner membrane** (Figure 4-15). These two membranes are separated by a narrow intermembrane space. The inner membrane encloses the stroma. The stroma contains a variety of dissolved molecules, including the enzymes that carry out the fixation of CO_2 into glucose (phase II reactions), as well as starch granules, ribosomes, and a

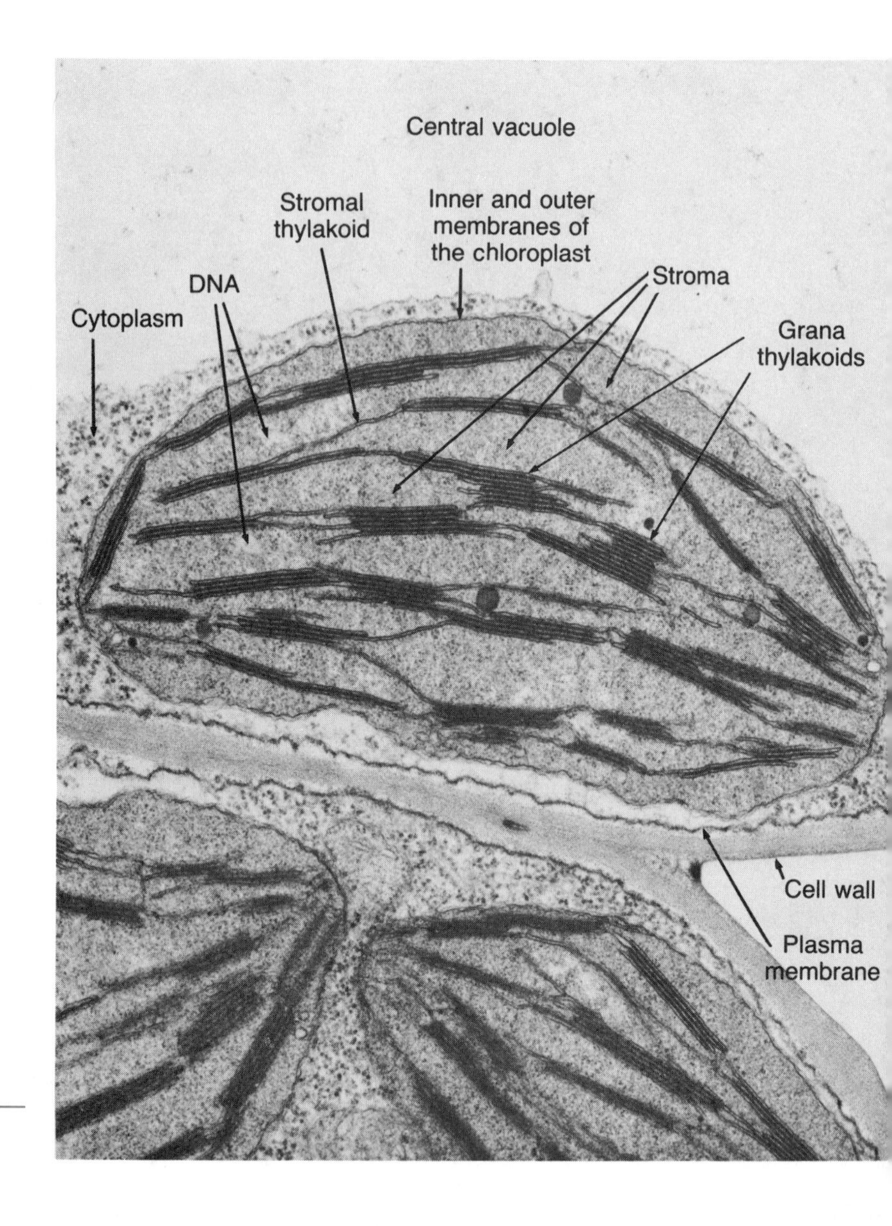

FIGURE 4-15
A chloroplast in a leaf cell of a tobacco plant. [Courtesy of L. Andrew Staehelin.]

small amount of DNA. Within the stromal compartment is a system of membranes (Figure 4-15) in which the chlorophyll molecules and accessory light-absorbing pigments are embedded. The stromal membranes also contain the enzymes for the phase I reactions of photosynthesis.

The numerous membranes in the stromal compartment are in the form of flattened sacs, called lamellae or **thylakoids,** arranged in parallel arrays (Figure 4-15). In many algae and in some kinds of plant cells the thylakoids are uniform in structure (Figure 4-16), but in other algae and plant cells the thylakoids occur

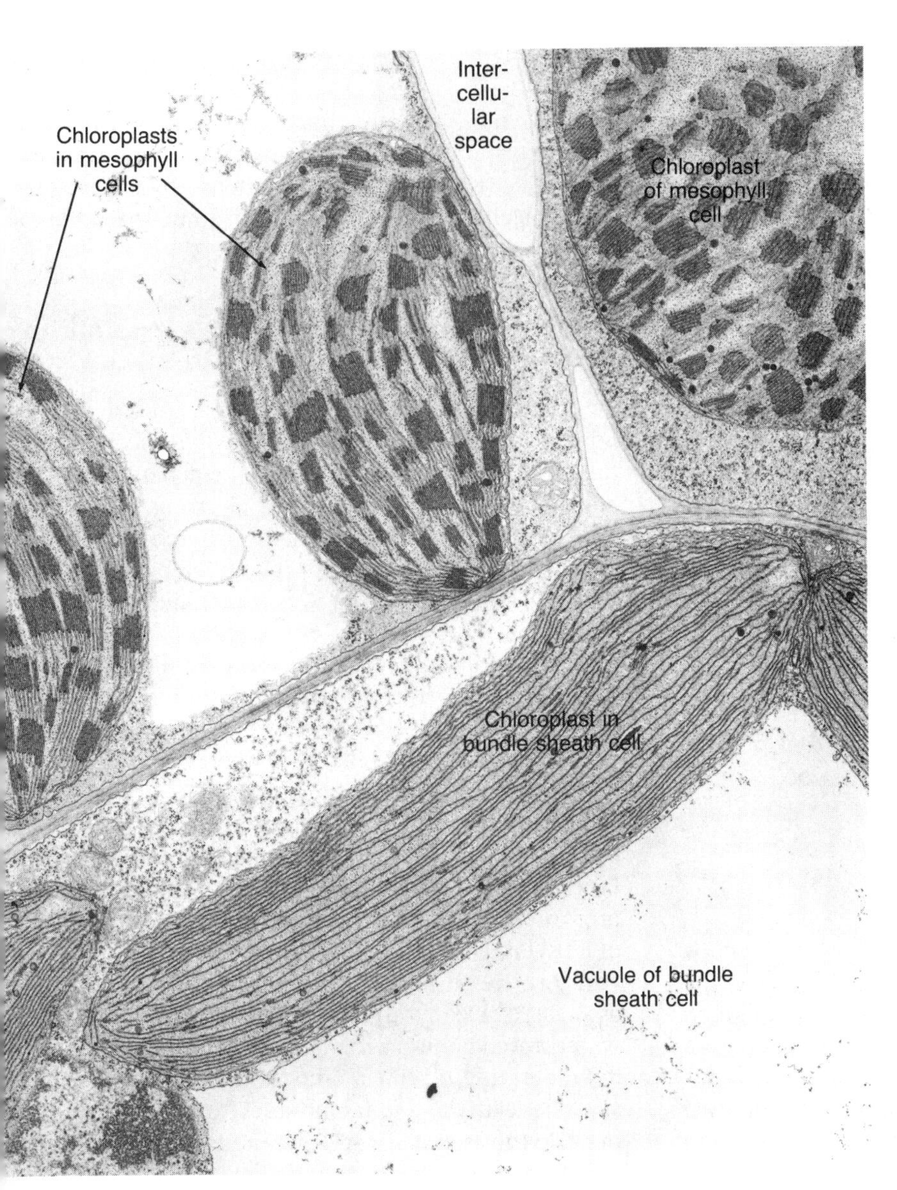

FIGURE 4-16 Electron micrograph of a section through a leaf of *Sorghum sudanense*, a C_4 plant. Note the presence of grana thylakoids and stromal thylakoids in mesophyll cells but only stromal thylakoids in the bundle sheath cell. [Electron micrograph by S. E. Frederick, courtesy of Eldon H. Newcomb.]

n two forms in a single chloroplast: (1) Some thylakoids are flattened sacs that have a disc or coin shape, and occur in uniform stacks called **grana.** These are called **grana thylakoids**. (2) Some thylakoids extend out of the grana and form bridges with thylakoids in other grana; these are called **stromal thylakoids.**

The Structural Basis of the Two Photosystems

It has been known for many years that the molecular machinery of the two photosystems is tightly bound to the thylakoid membranes. The development of the

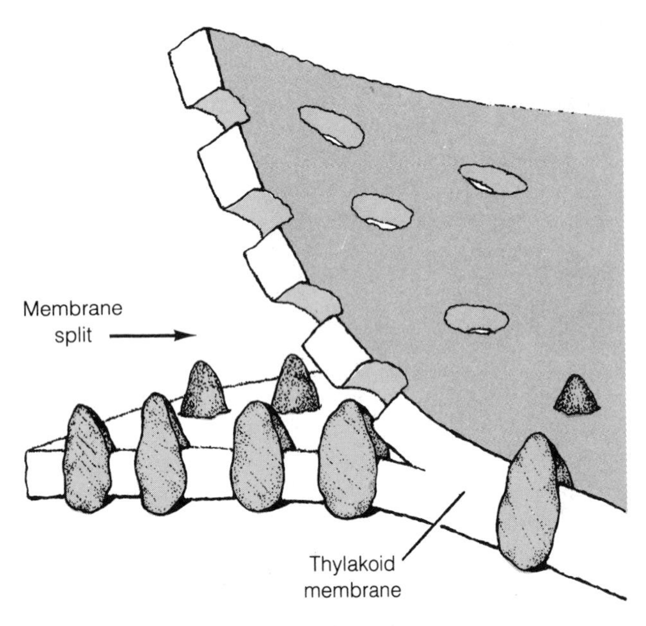

FIGURE 4-17
In the freeze-fracture technique cells are frozen and then fractured. Fracture planes frequently split membranes such as thylakoid membranes of chloroplasts, revealing particles embedded in the membranes.

freeze-fracture technique for electron microscopy has led to the identification of the structural basis of the two systems. In this technique cells are frozen and then mechanically fractured. The fracture plane through a cell often passes through the planes of the various membranes of the cell, splitting the membrane into two leaflets (Figure 4-17). This splitting occurs because in the frozen state the center of a membrane forms a plane of mechanical weakness.

By freeze-fracture it becomes possible to observe structural details inside the membrane. When a thylakoid membrane of a chloroplast is split into two leaflets, one can see a dense population of particles (Figure 4-18). Each particle is a functional unit of photosystem I, photosystem II, electron carriers, or an ATP synthesizing system. That is, each particle contains a complete set of the molecules that make up one or another com ponent of phase I reactions. Electrons are transferre from photosystem II to photosystem I by electron car rier molecules that move about within the thylakoi membrane.

The Mechanism of ATP Generation in the Phase I Reactions

The essential feature of the phase I reactions is th flow of energized electrons derived from chlorophy (and ultimately from water) and the generation of AT and NADPH. How this flow of electrons is translate into ATP synthesis is explained by the **chemiosmoti hypothesis.** The hypothesis is compelling for its inher ent logic and is supported by considerable experimenta evidence.

In the phase I reactions, energized electrons ar passed from carrier to carrier within the thylakoi membranes of the chloroplast, releasing energy durin these transfers. This release of energy is coupled to proton (H^+) pump in the thylakoid membranes tha transports protons from the stromal compartment t the small, narrow space enclosed by the membranes Proton pumping produces a thousand-fold difference i proton concentration (pH 5 vs. pH 8) on the two side of a thylakoid membrane. This **electrochemical pro ton gradient** can be used to drive a membrane associated ATP synthase to make ATP, much like th water stored behind a dam can be used to drive electric ity-generating turbines. Thus as the protons flow bac across the thylakoid membranes to the stroma, the provide the energy for ATP synthesis from ADP and P_i

ATPase is usually thought of as an enzyme tha hydrolyzes ATP:

$$\text{ATP} \rightarrow \text{ADP} + \text{P}_i \qquad \Delta G < 0$$

However, if the enzyme is coupled to a source of en ergy—in this case, a flow of proteins—the input o energy can result in a reversal of the reaction catalyze by ATPase, allowing the enzyme to act as ATP syn thase. The ΔG for the coupled reaction is, of course, stil

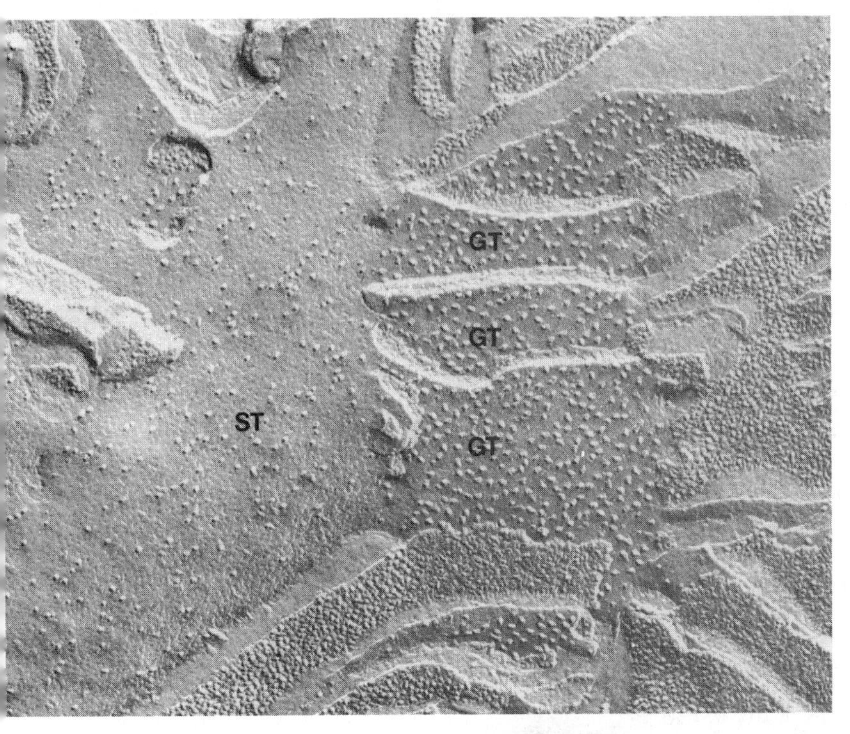

FIGURE 4-18

A portion of a chloroplast split by the freeze-fracture technique. Membranes of several grana thylakoids (GT) and a stromal thylakoid (ST) have been split revealing populations of particles. Each particle is an aggregate of proteins that carries out some part of photosynthesis, for example, photosystems I and II. [Courtesy of L. Andrew Staehelin.]

FIGURE 4-19

Light micrograph of living cells in the stem of a water fern. Each cell contains several hundred disc-shaped chloroplasts.

negative, that is, more energy is released by the proton flow than is needed to generate ATP.

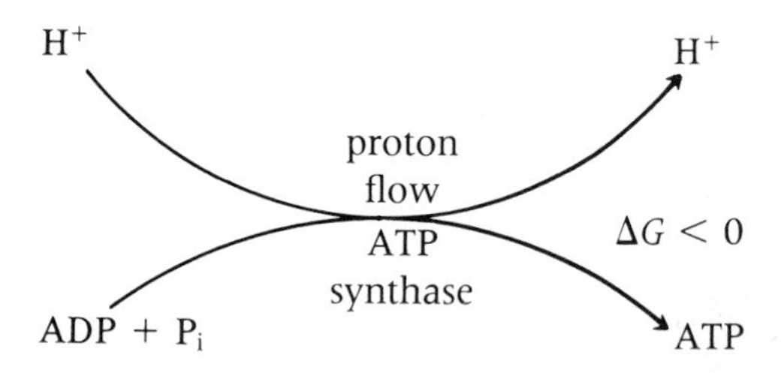

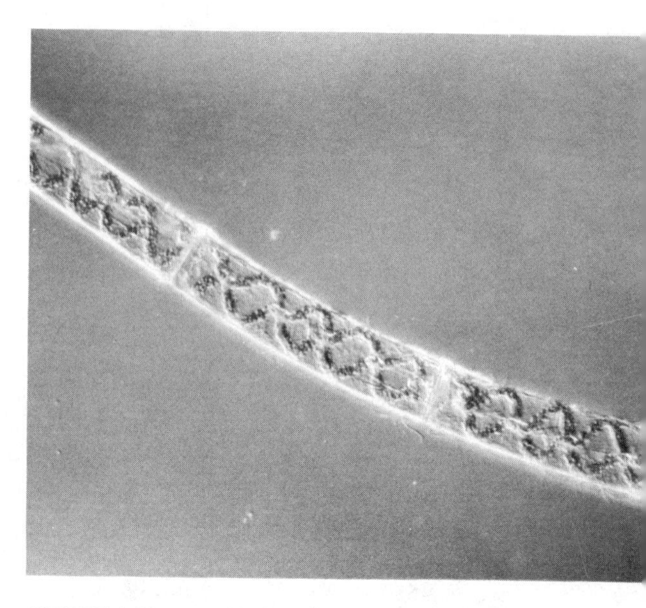

FIGURE 4-21
Light micrograph of three living *Spirogyra* algal cells. Each cell contains two long ribbon-shaped chloroplasts that spiral along the length of the cell.

Number, Size, and Form of Chloroplasts

Chloroplasts vary enormously in number, size, and form from one species of cell to another. Generally, in higher plants they have a discoid or ovoid form and range from 2 to 10 μm in their maximum dimension (Figure 4-19). Typically there are 20 to 40 chloroplasts in a plant cell. Some algae contain a single large chloroplast (Figure 4-20), while others have up to 30,000 globular-shaped chloroplasts. In some types of the filamentous algae, one or more long chloroplasts form a spiral just inside the cell surface. *Spirogyra,* shown in Figure 4-21, exemplifies this kind of alga.

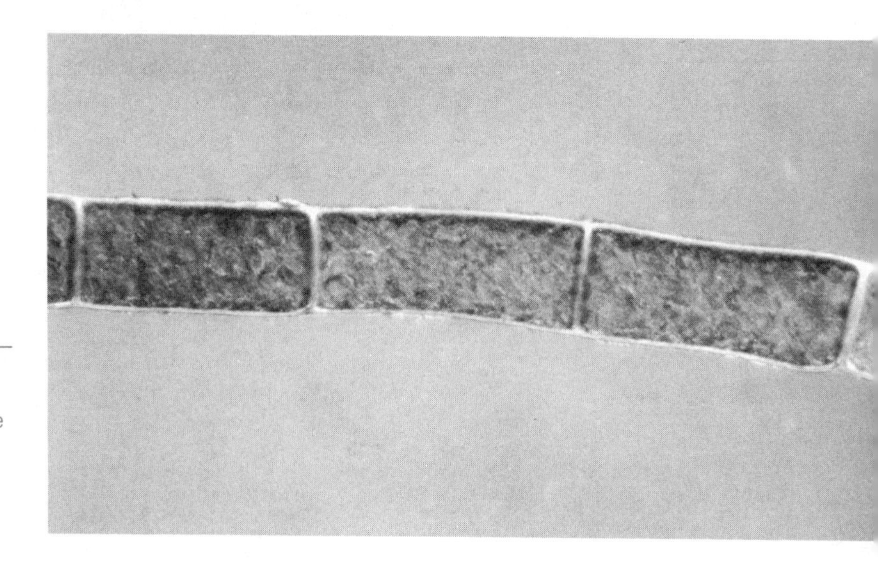

FIGURE 4-20
A light micrograph of living algal cells (*Mougeotia*). The cells are attached end-to-end, forming a long filament. A single large, flat chloroplast fills the center of the cylindrical cell. Other cell components (not visible) occupy space above and below the chloroplast.

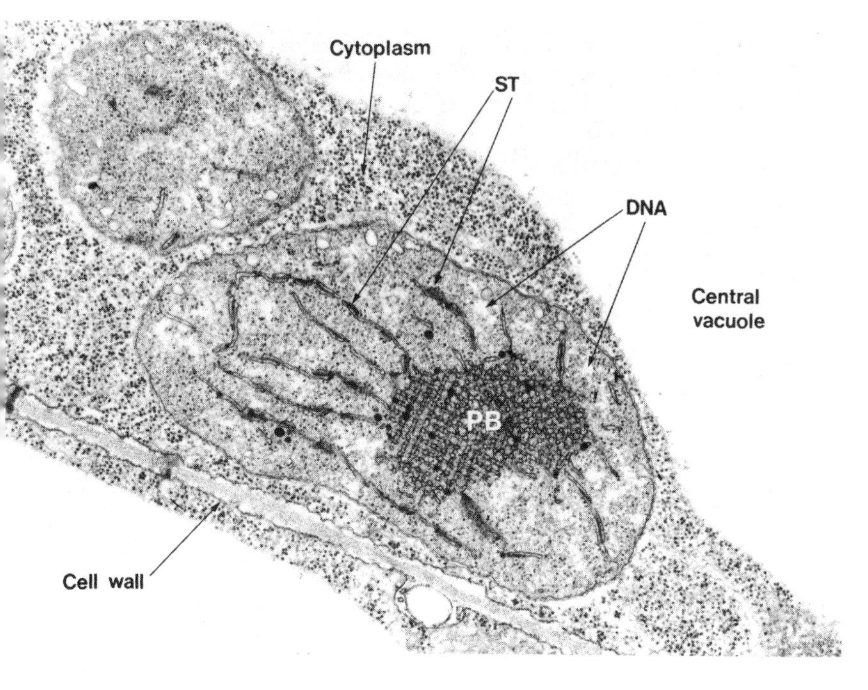

FIGURE 4-22

An electron micrograph of part of a leaf cell deprived of light for many days. Etioplasts have formed from chloroplasts. In the larger etioplast some stromal thylakoids (ST) are still present but grana thylakoids are absent. The prolamellar body (PB) is a storage depot of lipids. The light areas contain DNA. [Courtesy of L. Andrew Staehelin.]

Maintenance of Chloroplasts Requires Light

When plant cells are kept continuously in the dark for days, the thylakoid membranes gradually disappear, along with the ability of the chloroplast to carry out photosynthesis. This is readily observed as a disappearance of the green coloration and hence disappearance of the chlorophyll molecules. The loss of green coloration is called **etiolation.** Eventually the chloroplast recedes to an organelle called an **etioplast.** Etioplasts contain very few thylakoid membranes (Figure 4-22). Instead of thylakoid membranes, a lattice structure that appears to consist of tubules in regular array is present. This is usually called the **prolamellar body.** It is mostly a storage form of membrane lipids. Thus, an etioplast is delimited by an envelope consisting of an inner and outer membrane and has a stromal compartment that contains a prolamellar body, the chloroplast DNA, ribosomes, and the enzymes and other molecules required to carry out protein synthesis.

When a plant cell that has been etiolated by being kept in prolonged darkness is again exposed to light, thylakoid membranes begin to reappear within minutes. This is followed in a few hours by the synthesis of new chlorophyll and the other components of the photosynthetic machinery. Just how the thylakoid membranes re-form is not clear. Electron micrographs suggest that the material of the prolamellar body is reorganized to form the first thylakoid membranes; the membranes continue to increase long after the prolamellar body has disappeared, showing that thylakoid membranes, once formed, continue to grow.

These observations show that preventing chloroplast function by deprivation of light leads to the loss of those syntheses necessary to maintain the photosynthetic machinery and the structure of the thylakoid membrane. The molecular basis by which inhibition of photosynthesis by deprivation of light turns off synthesis (and hence maintenance) of the photosynthetic machinery is unknown. However, this interaction between function and maintenance of structure is not present in all species. Many kinds of algae maintain their chloroplasts in fully functional states without structural deterioration during prolonged darkness. In-

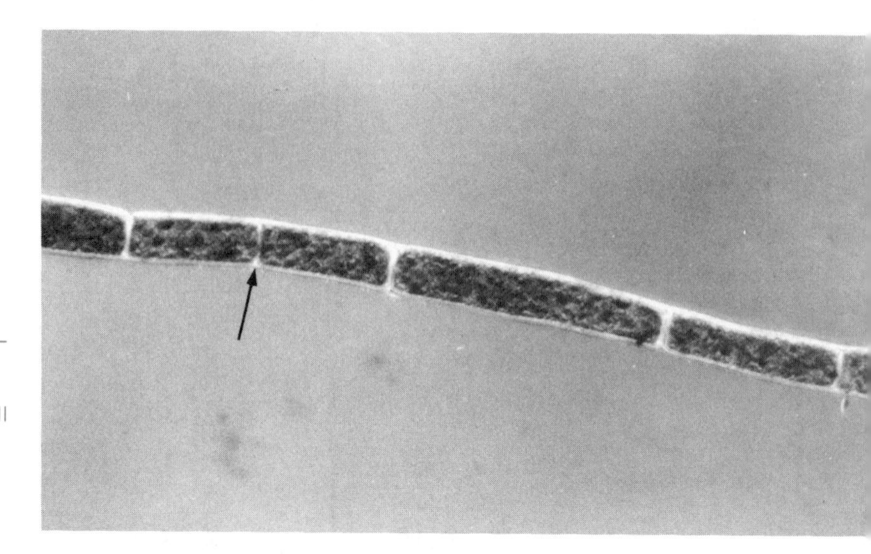

FIGURE 4-23
Light micrograph of several living algal cells (*Mougeotia*). One of the cells is just completing division and is forming a cell wall (arrow) between the two daughter cells. During division the large plate-like chloroplast has been cut in two by the formation of the cell wall.

deed, the chloroplasts grow and increase in number while kept in darkness. When algae are kept in darkness, they can no longer subsist phototrophically. They must assume a chemotrophic mode of existence, taking in organic molecules from their aqueous environment to obtain energy.

Formation of New Chloroplasts

Because chloroplasts are generally large organelles easily observed in living cells by light microscopy, it was discovered by microscopists in the nineteenth century that chloroplasts increase in number by fission. One can observe fission most easily in those algae containing a single, large chloroplast. *Mougeotia* is an example of such an alga. When cell division occurs, the chloroplast also divides into two in the same plane that divides the cell (Figure 4-23).

The regulation of the synthesis of macromolecular parts and the route of assembly of molecules into thylakoid membranes is important in new chloroplast formation. Specific interactions must exist among the molecular components of the thylakoid membranes that govern the relative proportions of the various molecules (e.g., chlorophyll and components of the electron transport chain assembled into the structure).

Chloroplasts, like mitochondria, are now recognized as *semiautonomous organelles.* Their DNA contains some of the genes necessary for chloroplast structure and function, although many of the genes that serve chloroplast structure and function are located in the cell nucleus. More will be said about chloroplast DNA in Chapter 11.

ATP PRODUCTION BY CHEMOTROPHY

Using the chemical energy acquired through photosynthesis, many phototrophic cells (cells of plants, algae, and photosynthetic bacteria) are able to synthesize all of the amino acids, nucleotides, sugars, vitamins, and other organic molecules needed for cell maintenance, growth, and reproduction. Almost all other forms of life are chemotrophic, acquiring chemical energy needed for anabolism and other cell activities by taking in organic molecules, especially sugars, and breaking them down into simpler molecules. The breakdown of sugars, principally glucose and fructose, occurs through a series of chemical steps in which a chain of intermediate molecules is created, ending with CO_2 and H_2O. At almost every step, part of the energy

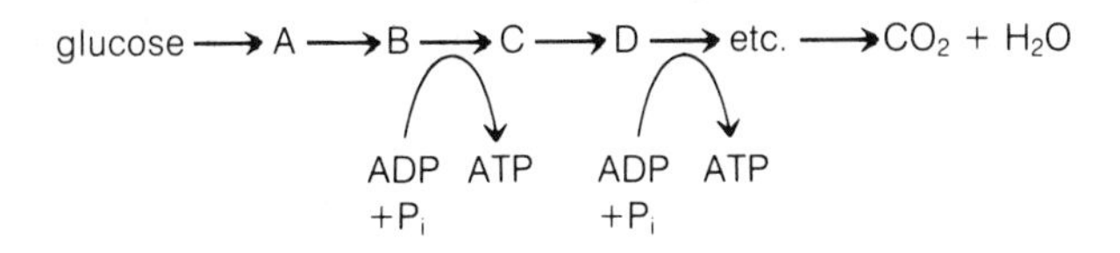

FIGURE 4-24
Glucose molecules are broken down step-by-step into carbon dioxide and water. At many of the individual steps some energy is released and coupled to the synthesis of ATP from ADP and P_i.

in the original sugar molecule is released. Some of this energy is captured in the cell by coupling particular breakdown steps to the synthesis of ATP from ADP and P_i (Figure 4-24).

Breakdown of other organic molecules to obtain energy, principally amino acids and fatty acids, joins the pathway at specific points that will be described later in the chapter.

Glucose is the principal sugar in the biological world. Breakdown of glucose occurs in two stages. The first stage, called **glycolysis,** consists of ten successive steps in which glucose, a 6-carbon molecule, is converted into two 3-carbon molecules (pyruvate), and two ATP molecules are synthesized from two ADP and two P_i (Figure 4-25). The second stage of glucose breakdown, called **cellular respiration,** converts the pyruvate molecules into CO_2 and H_2O through 15 steps, some of which are coupled to the synthesis of many molecules of ATP from ADP + P_i (Figure 4-26).

The glycolytic breakdown of glucose into two pyruvate molecules occurs in essentially all prokaryotes and eukaryotes. It does not require oxygen and is the means by which cells that live in the total absence of environmental O_2 are able to produce ATP. Cells that live without O_2 are known as **anaerobes.** There are two types: **obligate anaerobes,** primarily certain species of bacteria that live in the absence of O_2 and that are killed by O_2; and **facultative anaerobes,** which include many bacterial species, some fungi (e.g., yeast cells), some protozoa, and some invertebrate parasites. These various kinds of organisms can live in the absence *or* presence of O_2.

Facultative anaerobes use glycolysis to synthesize ATP when O_2 is not available and use both glycolysis and respiration when O_2 is available. When yeast cells are grown under anaerobic conditions, they live by glycolytic breakdown of glucose to pyruvate. However, these cells carry the process one step further; they convert the 3-carbon molecules to 2-carbon molecules, ethyl alcohol, a reaction that is the foundation of wine, beer, and liquor production. Some anaerobic bacteria convert pyruvate to another 3-carbon molecule, lactate. The formation of lactate is the principal event in the production of yogurt from milk by anaerobic bacteria. Muscle cells also produce lactate when the supply of O_2 is depleted during heavy use.

Finally, organisms known as **aerobes**, a group that includes many unicellular organisms and all animal cells, require O_2. Although they carry out glycolysis, aerobes are not able to subsist on glycolysis alone and must also carry out respiration, completing the diges-

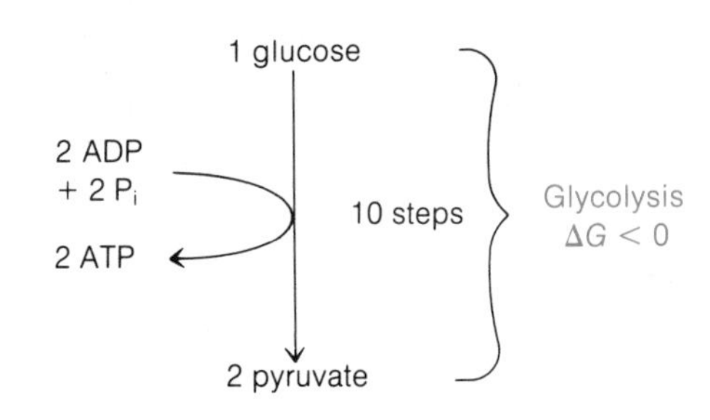

FIGURE 4-25
Glycolysis is the first phase of the breakdown of glucose, in which one glucose molecule yields two pyruvate molecules, and two ATP molecules are produced from ADP and P_i.

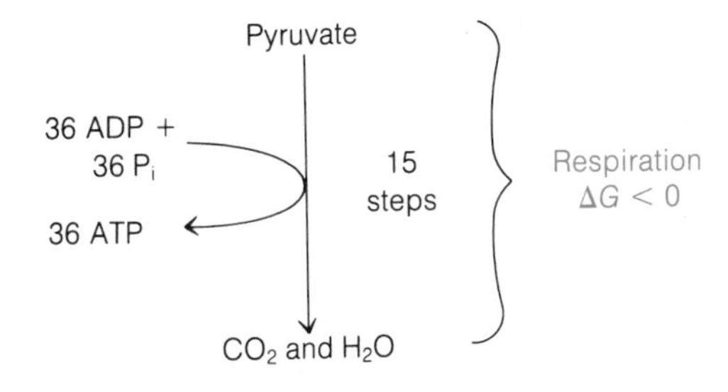

FIGURE 4-26
Respiration is the second phase of the breakdown of glucose, in which the pyruvate molecules produced in glycolysis are converted to carbon dioxide and water. Thirty-six ATP molecules are produced from 36 ADP and 36 P_i.

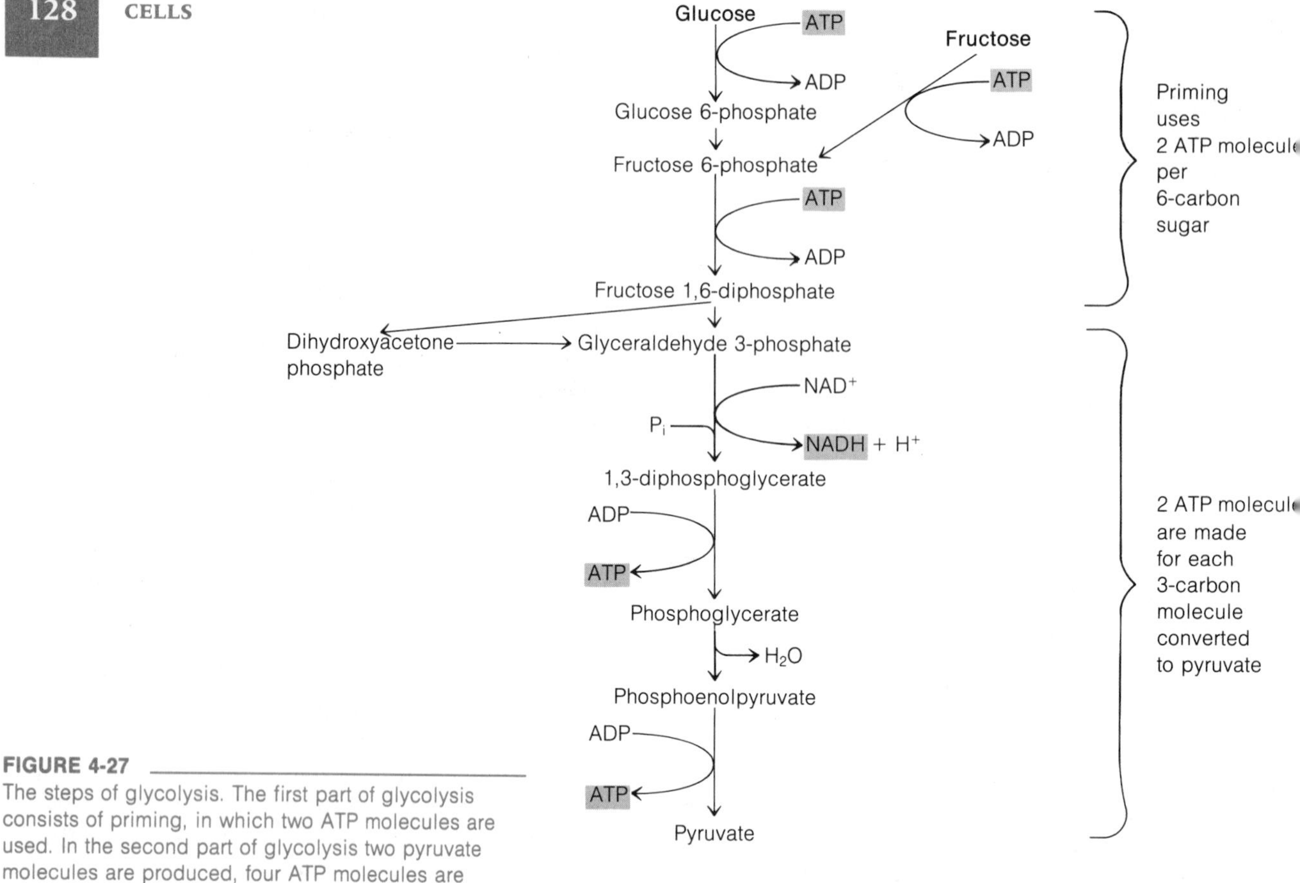

FIGURE 4-27
The steps of glycolysis. The first part of glycolysis consists of priming, in which two ATP molecules are used. In the second part of glycolysis two pyruvate molecules are produced, four ATP molecules are synthesized, and two NAD^+ are reduced to two NADH.

tion of glucose to CO_2 and H_2O. Phototrophs are also aerobes since, in addition to the ability for photosynthesis, these cells can digest sugar through glycolysis and respiration to produce ATP.

Glycolysis and cellular respiration are described successively in the next two sections. The objective in studying glycolysis and respiration is to understand the step-by-step disassembly of glucose into CO_2 and H_2O and the concomitant synthesis of ATP. Memorization is pointless if the logic of the overall process is not comprehended. In fact, there is little point in memorization of the individual steps.

Glycolysis

Figure 4-27 shows the reaction sequence by which glucose is converted to pyruvate with a net gain of two ATP molecules. The first half of glycolysis consists of priming reactions in which the sugar is phosphorylated twice, using two ATP molecules, and rearranged into the 6-carbon sugar fructose with two phosphate groups, fructose 1,6-diphosphate. Figure 4-27 shows also the entry of fructose into the glycolytic pathway. Fructose 1,6-diphosphate is split into two 3-carbon molecules, glyceraldehyde 3-phosphate (already encountered as an intermediate product in photosynthesis) and dihydroxyacetone phosphate. By a rearrangement of its atoms, dihydroxyacetone phosphate becomes glyceraldehyde 3-phosphate. Hence, glucose has now been converted into two molecules of glyceraldehyde 3-phosphate, and two molecules of ATP have been used up.

In the second half of glycolysis an inorganic phosphate group is added to each molecule of glyceral-

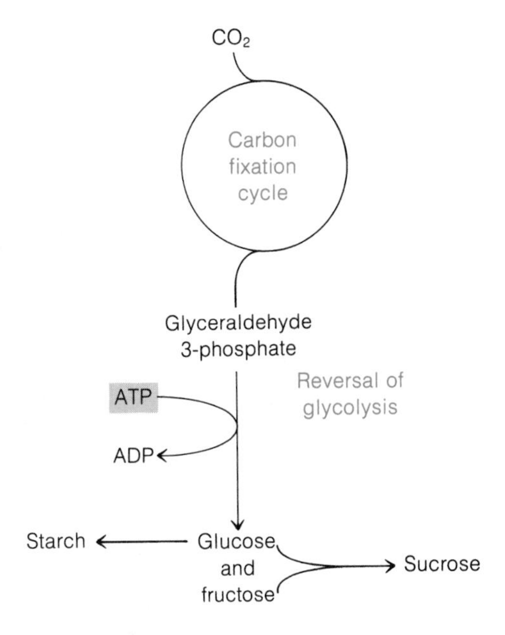

FIGURE 4-28
Carbon dioxide captured in the carbon fixation cycle of photosynthesis is converted into sugar molecules by a reversal of the first several steps of glycolysis. Sugars are converted into the disaccharide sucrose or polymerized into starch.

dehyde 3-phosphate to yield two molecules of 1,3-diphosphoglycerate. This reaction is coupled to the transfer of a pair of electrons from glyceraldehyde 3-phosphate to the coenzyme NAD^+, producing NADH + H^+. A phosphate group is next removed from each molecule of 1,3-diphosphoglycerate in reactions coupled to synthesis of two ATP molecules from ADP, leaving two 3-carbon molecules of pyruvate.

A total of four ATP molecules is produced from the two molecules of 1,3-diphosphoglycerate. However, two molecules of ATP were used in the first half of glycolysis in the phosphorylation of glucose. Therefore, glycolysis provides a net gain of two ATP molecules for each glucose converted to two pyruvates. Glycolysis also results in transfer of a pair of electrons to NAD^+. These electrons are important in the synthesis of ATP as described in the next section.

Recall also that glyceraldehyde 3-phosphate is produced in the phase II reactions (carbon fixation) of photosynthesis. The glyceraldehyde 3-phosphate is converted into glucose and fructose by a reversal of the glycolytic pathway. Since the reverse reactions are energetically disfavored, they must be coupled to the hydrolysis of some of the ATP produced in the phase I reactions of photosynthesis (Figure 4-28).

Glucose and fructose may be joined to form sucrose, a disaccharide produced by some photosynthetic plants, for example, sugarcane. Alternatively, glucose may be polymerized into starch, as in potatoes and maize.

The pyruvate molecules produced as end products of glycolysis may take one of three routes already mentioned earlier. Pyruvate can be converted to ethyl alcohol (by yeast growing anaerobically) or to lactate (by anaerobic bacteria), or it may serve as the starting molecule for the sequence of reactions that makes up cellular respiration (aerobic organisms), as described in the next section.

Cellular Respiration

When pure glucose is ignited and burned, it is completely converted to CO_2 and H_2O with the release of heat.

$$C_6H_{12}O_6 + 6O_2 \rightarrow 6CO_2 + 6H_2O \qquad \Delta G < 0$$

The breakdown of glucose in a cell leads to the same result. Six O_2 molecules are needed to catabolize each molecule of glucose, and six molecules of CO_2 and six of H_2O are the end products. For one mole of glucose (180 grams) 686 kilocalories are released. In a cell the breakdown occurs stepwise, with some energy released at each step. By coupling certain of the steps to synthesis of ATP from ADP, energy is captured. Glycolysis carries the breakdown to pyruvate, during which some energy is transferred into a net gain of two ATP molecules. Glycolysis, however, taps very little of the free energy in glucose; about 93 percent of this energy is still present in the two pyruvate molecules left after glycolysis. This energy is released in stepwise fashion by conversion of pyruvate to CO_2 and H_2O in **cellular respiration,** with the generation of 36 ATP molecules.

Most prokaryotes and essentially all eukaryotes possess the enzymatic pathways for cellular respiration; only obligate anaerobes lack them. Cellular respiration occurs in two stages. In the first stage, pyruvate is oxi-

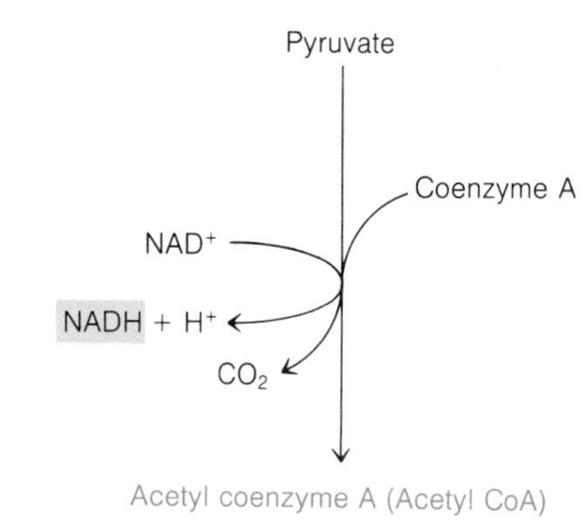

FIGURE 4-29
Pyruvate produced by glycolysis reacts with coenzyme A to yield acetyl coenzyme A (acetyl-CoA), carbon dioxide, and NADH + H^+.

dized in a sequence of reactions to CO_2 and H_2O, and electrons are transferred to coenzymes. In the second stage the passage of electrons along an electron transport chain is coupled to the synthesis of ATP from ADP and P_i in a manner similar to the transfer of electrons and generation of ATP that occurs in phase I reactions of photosynthesis.

The breakdown of pyruvate is shown in Figure 4-29. It begins with a complex coupled reaction in which two electrons of pyruvate are transferred to NAD^+, a molecule of CO_2 is released, and the remaining 2-carbon fragment, known as an acetyl group, is transferred to a coenzyme, **coenzyme A.** The acetyl-coenzyme-A complex is called **acetyl-CoA.** The 2-carbon acetyl group of acetyl-CoA is transferred to a 4-carbon molecule (oxaloacetate) to form a 6-carbon molecule (citric acid) and thereby initiates the **citric acid cycle** (also known as the Krebs cycle) (Figure 4-30). In the citric acid cycle citric acid is converted in a sequence of reactions back to the original 4-carbon molecule (oxaloacetate) by the splitting off of two CO_2 molecules and the transfer of four pairs of electrons to coenzymes. Three pairs are transferred to three NAD^+ molecules, and one pair is transferred to another coenzyme, flavine adenine dinucleotide (FAD). In addition, one of the reactions in the cycle is coupled to the phosphorylation of a molecule of guanosine diphosphate (GDP) to guanosine triphosphate (GTP). In turn, the conversion of GTP to GDP can be coupled to synthesis of ATP from ADP.

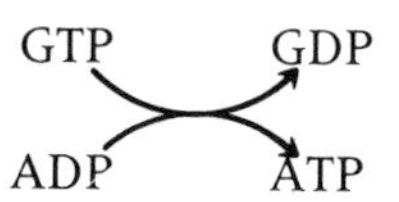

This is the only ATP molecule made directly from the citric acid cycle.

Overall, in the foregoing reactions (Figures 4-29, 4-30) all three carbons in pyruvate are released as CO_2, five pairs of electrons are transferred to coenzymes, and one ATP is generated. Remember that in glycolysis a single glucose molecule yields two pyruvate molecules and hence drives two turns of the citric acid cycle (Figure 4-31).

The stepwise breakdown of glucose to this point is summarized in Table 4-1. It yields six CO_2, four ATP, and results in 12 pairs of electrons transferred to NAD^+ and FAD to form NADH + H^+ and $FADH_2$.

In the second half of cellular respiration, called **oxidative phosphorylation,** each of the 12 pairs of electrons is transferred from the coenzymes along a se-

TABLE 4-1
Products from Various Stages of Glucose Catabolism

	Process	Coenzymes with Electron Pairs	ATP Molecules	CO_2
Produced from ONE GLUCOSE MOLECULE	glycolysis	2 NADH	2 ATP	
	pyruvate to acetyl-CoA	2 NADH		2 CO_2
	citric acid cycle	6 NADH 2 $FADH_2$	2 ATP	4 CO_2

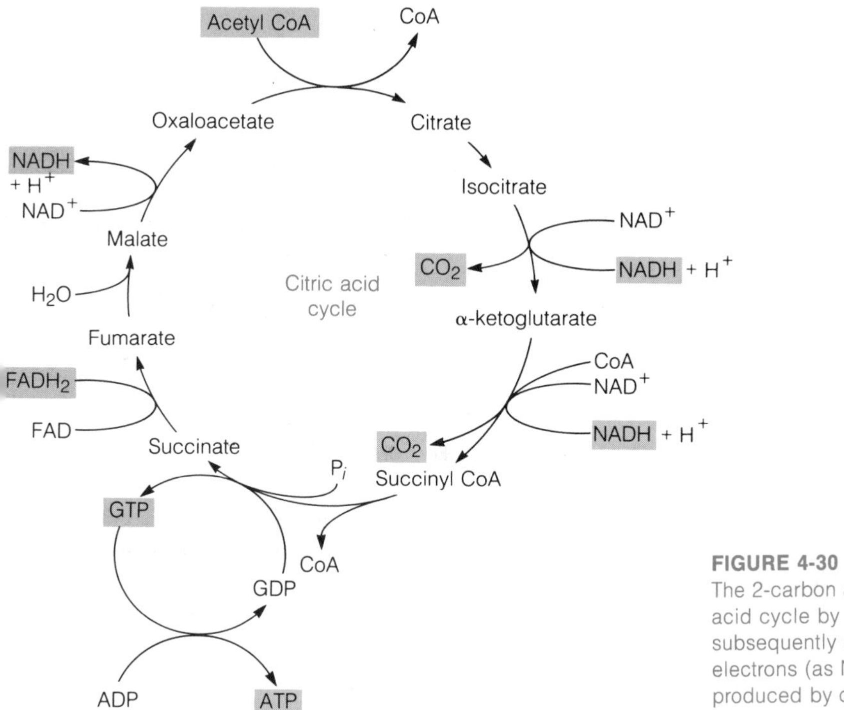

FIGURE 4-30 The 2-carbon acetyl group of acetyl-CoA enters the citric acid cycle by transfer to citric acid. The two carbons are subsequently released as carbon dioxide. Four pairs of electrons (as NADH + H^+ and $FADH_2$) and one ATP are produced by one turn of the citric acid cycle.

quence of carriers to oxygen (Figure 4-32). The ten pairs of electrons in 10NADH + $10H^+$ (Table 4-1) are first transferred to ten FAD to yield ten $FADH_2$ and ten NAD^+. For each pair of electrons transferred to FAD one ATP is produced from ADP and P_i. Therefore, a total of ten ATP molecules is formed at this point with the ten pairs of electrons derived from the breakdown of one glucose molecule. As a result of the electron transfers, the ten molecules of NADH are returned to NAD^+, which are then again available for receiving electrons as additional glucose is broken down. Since two molecules of $FADH_2$ were produced per glucose molecule in the citric acid cycle, 12 pairs of electrons are now present as $FADH_2$.

The pairs of electrons in $FADH_2$ are transferred to another carrier, ubiquinone. From ubiquinone the electrons are passed to the cytochrome b-c complex with the formation of 12 ATP from 12 ADP and 12 P_i, one ATP for each electron pair transferred. The electrons are next transferred to cytochrome c and then to cytochrome oxidase with the formation of 12 more ATP. From cytochrome oxidase, electrons are transferred to oxygen. One pair of electrons combines with one oxygen atom (½ O_2) and $2H^+$, yielding one molecule of water (H_2O). Only in the second half of cellular respiration (oxidative phosphorylation) is oxygen consumed, and it is to this purpose that oxygen taken in by an organism is directed.

In the transfer of electrons from glucose, first to coenzymes, then to electron carriers, and finally to water, one H^+ is present in some form for each electron transferred. In the case of NAD^+ two electrons and one H^+ are bound to give NADH; the second H^+ remains free in solution. In the case of FAD two electrons and two H^+ are bound, giving $FADH_2$. In subsequent electron transfers through the electron transport chain of

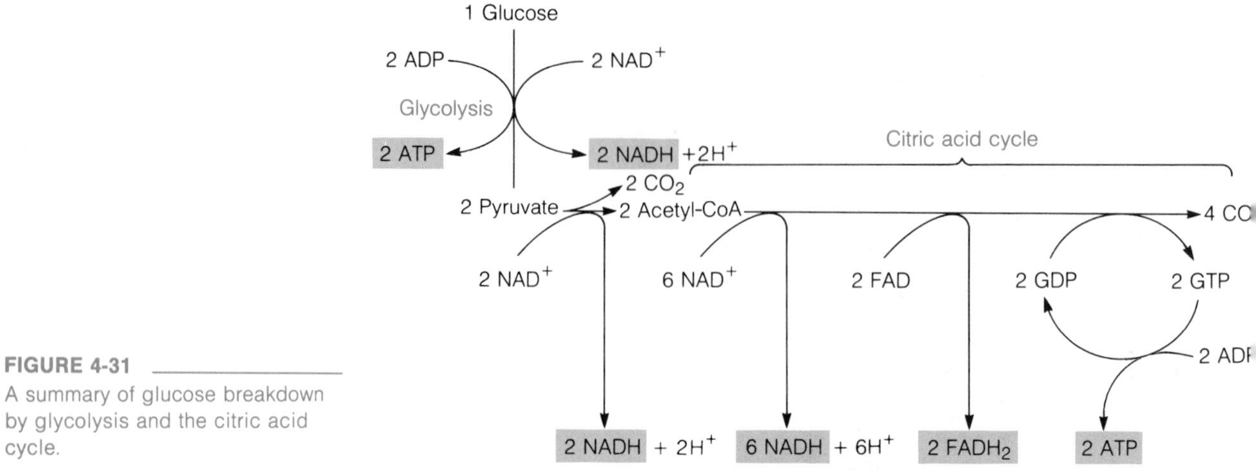

FIGURE 4-31
A summary of glucose breakdown by glycolysis and the citric acid cycle.

oxidative phosphorylation all the H^+ ions appear in solution. At the end of their journey the electrons derived from glucose are transferred to O_2 along with H^+ to make H_2O molecules.

Altogether, the transfer of 12 pairs of electrons, derived during the breakdown of one glucose, yield 10 + 12 + 12 = 34 ATP. Adding these 34 ATP to the two from glycolysis and the two from the citric acid cycle brings the total to 38 ATP for each glucose molecule broken down to CO_2 and H_2O.

An Overview of Glucose Breakdown

All the events in the breakdown of glucose to CO_2 and H_2O with the formation of ATP can now be summarized in one diagram with this sequence: glycolysis, the citric acid cycle, and oxidative phosphorylation (Figure 4-33). For each glucose molecule the products are two ATP and two pairs of electrons from glycolysis, two pairs of electrons from pyruvate breakdown, two ATP and eight pairs of electrons from the citric acid cycle, and 34 ATP from oxidative phosphorylation.

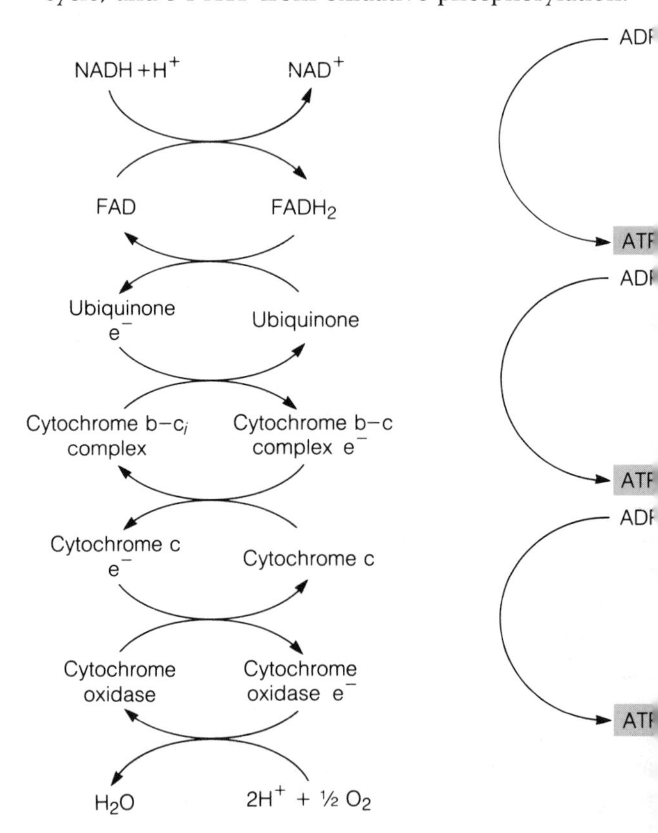

FIGURE 4-32
Electron pairs derived from glucose in glycolysis and the citric acid cycle are transferred to oxygen through a series of carriers. In transfer, energy is captured in the synthesis of ATP from ADP and P_i.

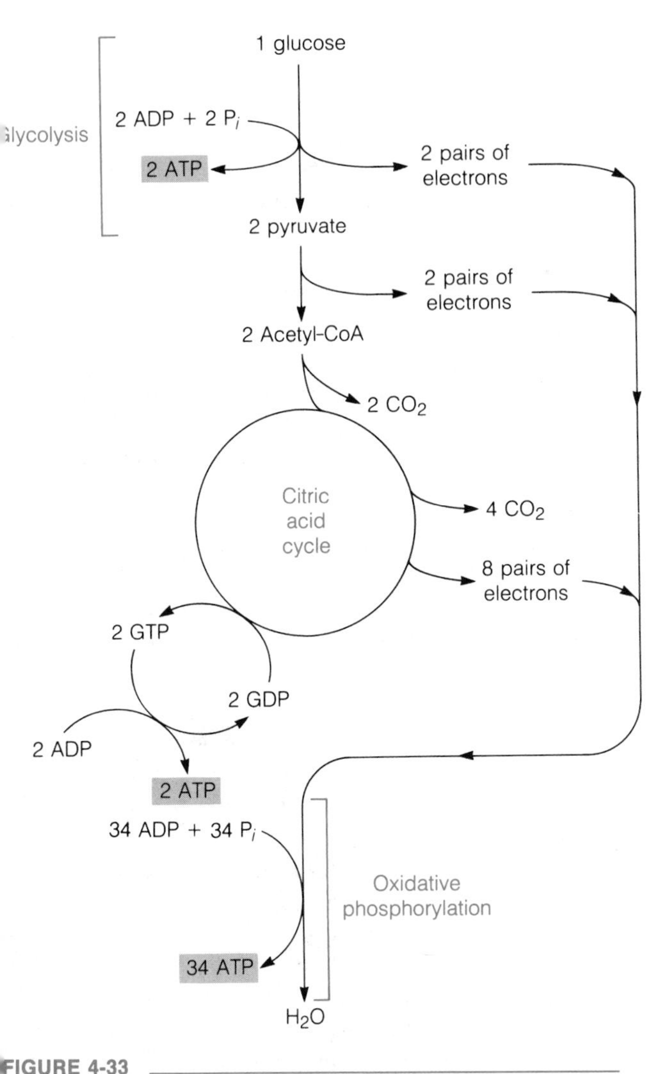

FIGURE 4-33
The step-by-step conversion of glucose to carbon dioxide and water by glycolysis, the citric acid cycle, and oxidative phosphorylation yields a net total of 38 ATP molecules.

The Efficiency of Energy Capture from Glucose Breakdown to CO_2 and H_2O

The breakdown of glucose to CO_2 and H_2O yields an amount of free energy (686 kilocalories) that could theoretically generate nearly 100 ATP molecules from ADP and P_i. In the many coupled reactions in glucose breakdown where energy is transferred, however, some energy is lost as heat, and only 38 ATP are actually formed. This represents an efficiency of energy capture of 38 percent. This may appear low, but it far exceeds the efficiency of any device humans have designed for energy utilization.

Energy Obtained from the Breakdown of Fats and Proteins

When carbohydrates are available, their breakdown is the principal source of energy. If the supply of carbohydrates is inadequate, fats may be broken down to yield energy. If both carbohydrates and fats are unavailable, even proteins can be utilized to generate ATP.

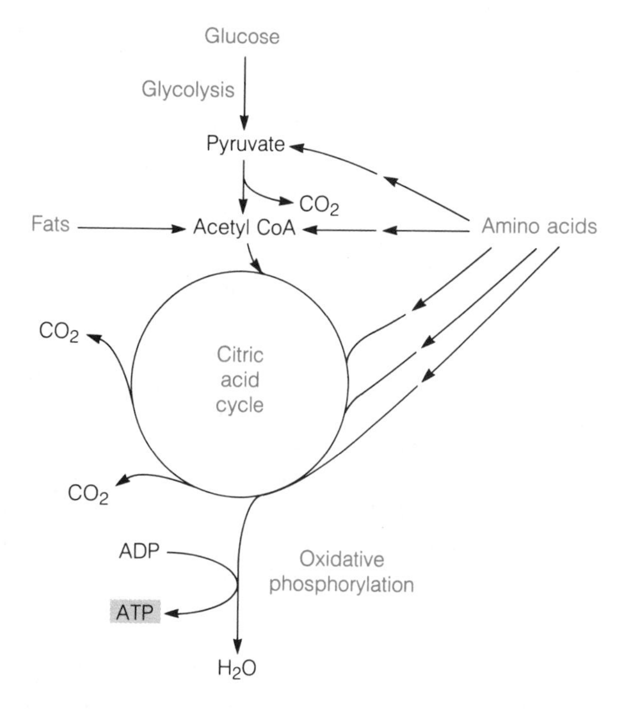

FIGURE 4-34
The carbon-containing parts of amino acids can enter into the pathway of glucose breakdown at various points and be converted to carbon dioxide, with the production of ATP.

Most of the energy in fats is contained in the long hydrocarbon chains of fatty acids. Through a sequence of reactions, 2-carbon fragments are sequentially removed from fatty acid chains to generate acetyl-CoA, which provides entry into the citric acid cycle and then oxidative phosphorylation (Figure 4-34).

Proteins enter energy metabolism by being broken down first into individual amino acids. Some kinds of amino acids are metabolized to pyruvate, some to acetyl-CoA, and still others to intermediates in the citric acid cycle (Figure 4-34). In each case molecules are produced that are intermediates in glucose metabolism, and energy is obtained by completing their breakdown to CO_2 and H_2O.

Control of ATP Synthesis and the Principle of Energy Charge

ATP is used as rapidly as it is synthesized to provide energy for cellular activities. To give an idea of the rate of hydrolysis and resynthesis of ATP, an adult human engaged in minimal physical activity must synthesize an amount of ATP equal to his own body weight every 24 hours. The total amount of ATP present in a 70 kg man is about 50 grams. Therefore, his entire body content of ATP must be converted to ADP and back to ATP about 1400 times every 24 hours.

To meet the needs of a cell the production of ATP is closely regulated at many points in glycolysis, the citric acid cycle, and oxidative phosphorylation. One example of regulation is in the flow of electrons in oxidative phosphorylation. Electrons can continue to flow if ADP is available to be converted to ATP. When ADP is phosphorylated to ATP faster than energy-requiring activities in the cell convert it back to ADP, the ADP will be in short supply, which results in a slowing of the flow of electrons. This, in turn, leads to increased loading of NAD^+ and FAD with electrons and a decreased availability of coenzymes for electron accepting, which slows both glycolysis and the citric acid cycle and results in a decreased rate of glucose breakdown. In addition, NAD^+, AMP, and ADP are positive allosteric effectors (Chapter 3) that increase the rate of catalysis by a number of enzymes in the glucose breakdown pathway. Conversely, NADH and ATP act as negative allosteric effectors, slowing down the rate of catalysis of the same enzymes.

The individual amounts of AMP, ADP, and ATP in a cell may fluctuate depending on how fast ATP is being generated and used up. However, the total amount of AMP + ADP + ATP remains practically constant. Each time a high-energy bond of ATP is used, an ADP or AMP is formed (ATP → ADP + P_i or ATP → AMP + 2 P_i). Or for each ATP synthesized, the amount of AMP or ADP decreases by one molecule. If all the molecules are present as ATP, then the energy system is full, and the energy charge is 1.0. If all the molecules are present as AMP, then the energy system is empty and the charge is 0. If half the molecules are AMP and half are ATP, or if all are ADP, the energy charge is 0.5. The equation for calculating energy charge is

$$\text{ENERGY CHARGE} = \frac{[\text{ATP}] + \frac{1}{2}[\text{ADP}]}{[\text{ATP}] + [\text{ADP}] + [\text{AMP}]}$$

Ordinarily, when ATP is being used at the same rate as it is being synthesized, the energy charge is close to 0.85. If the rate of ATP utilization increases, the energy charge decreases. A decreased energy charge, however, results in stimulation of the ATP-producing machinery through allosteric effects on key enzymes in energy metabolism. If the rate of ATP utilization decreases, then the energy charge may increase above 0.85, and the rate of ATP production is cut back. The system is controlled to maintain the energy charge within narrow limits, usually between 0.8 and 0.95. It is clear that the rate of ATP utilization regulates the rate of ATP production. In general, cells have the capacity to produce ATP at a considerably higher rate than usually demanded. If the demand for ATP exceeds the capacity to synthesize ATP, cells can respond by increasing their enzymatic machinery for glucose metabolism.

Compartmentalization of Energy Metabolism in the Cell

When the enzymes and coenzymes that carry out glycolysis are extracted from a cell, they are all found dissolved in the cytosol. In the breakdown of glucose each intermediate is apparently released from an enzyme as it is formed, and its conversion to the next intermediate depends on its chance collision with the active site of the appropriate enzyme until the final product, pyruvate, is formed. The concentration in the cytosol of enzymes, coenzymes, and intermediates is such that correct collisions occur frequently and glycolysis proceeds rapidly. Some recent experiments suggest that the enzymes of glycolysis may in fact be loosely bound together in functional groups. If such were the case, then transfer of intermediates from enzyme to enzyme might be more efficient than usually supposed. The breaking open of cells to examine the cytosol may disrupt the loose aggregates of glycolytic enzymes unless special methods are used to preserve them.

In aerobic prokaryotes, pyruvate is metabolized by enzymes of the citric acid cycle present in the cytosol. The electrons derived from the citric acid cycle then flow through the chain of electron carriers, which are present in the plasma membrane. In eukaryotes, pyruvate produced in glycolysis is taken into mitochondria, where breakdown to CO_2 and H_2O by the citric acid cycle occurs, followed by oxidative phosphorylation in the membranes of the mitochondria.

MITOCHONDRIA

Mitochondria were discovered by microscopists in the mid 1800s, but their role in energy metabolism was not discovered until the late 1940s. The first experiment consisted of breaking open rat liver cells and purifying cell parts by centrifugation (Chapter 1). Only that fraction containing mitochondria consumed O_2 and produced CO_2. Subsequent studies showed that 95 percent of the ATP of nonphotosynthetic eukaryotes is produced by the respiratory machinery of mitochondria, and these organelles are aptly called the *power plants* of the cell.

Mitochondria occur in various shapes and sizes even within the same cell; in some cells they are ellipsoidal and measure 1 μm × 2 μm. They also occur frequently as rods 1 or 2 μm in diameter and many micrometers in length (Figure 4-35). Time-lapse photography with a phase contrast microscope of chick cells in culture has shown that the rod-shaped mitochondria are in a continual state of motion, changing shape, forming branches, breaking up into smaller rods, and fusing again into larger structures. The significance of these activities is not understood, although we do know that mitochondria reproduce during cell growth by dividing.

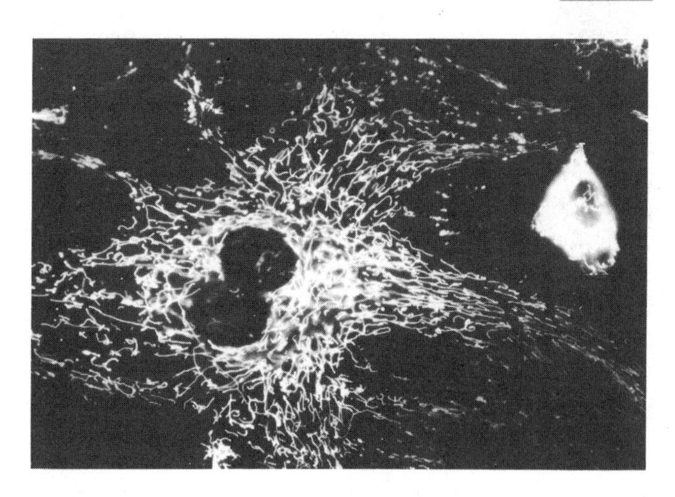

FIGURE 4-35
Light micrograph of a living tumor cell in culture taken from a gerbil. The living cell has been stained with a fluorescent dye rhodamine 123, which binds specifically to mitochondria. The mitochondria are threadlike and vary greatly in length. The nucleus is the central clear area. [From the *Proc. Natl. Acad. Sci.*, 77:990. Courtesy of Lincoln V. Johnson.]

The number of mitochondria varies greatly from one kind of cell to another. Some yeast cells contain one or a few highly branched mitochondria that extend through the cytoplasm (Figure 4-36). Liver cells (**hepatocytes**) contain about 1000 small ovoid mitochondria. These account for about 25 percent of the total volume of the hepatocyte. Large multinucleated amoebae contain about 500,000 small ovoid mitochondria. Red blood cells of vertebrates lose all their mitochondria during differentiation, providing a rare example of a vertebrate cell that must generate all of its ATP by glycolysis alone. However, the requirement for ATP in red blood cells is much reduced because of the total absence of anabolic processes.

Mitochondria are generally distributed throughout the cytoplasm but sometimes aggregate in specific locations. In ciliated protozoa such as *Tetrahymena* and *Paramecium*, the mitochondria are concentrated just under the cell surface, a favorable position for providing the large amount of ATP needed for the activity of the hundreds of cilia in these cells (Figure 4-37).

Structure of the Mitochondrion

The internal structure of mitochondria in hundreds of kinds of cells has been studied extensively by electron microscopy. Although the structural details may vary from one species to another, the basic organizational pattern is the same in all species.

The mitochondrion consists of an **outer membrane** and an **inner membrane** (Figure 4-38). The outer membrane completely encloses the organelle. The inner membrane is situated immediately inside the outer membrane, but it also extends by numerous folds called **cristae** into the internal region of the mitochondrion. The many cristae enormously increase the area of the inner membrane, which thereby provides space for a much larger number of the components responsible for oxidative phosphorylation. An extremely thin space (about 100 Å wide) separates the two membranes (Figure 4-38). One can enlarge this **intermembrane space** by placing isolated mitochondria in a sucrose solution. The sucrose penetrates the outer membrane but not the inner one. This creates an osmotic pressure (Chapter 5) difference between the intramembrane space and the internal compartment, which causes loss of water from the internal compart-

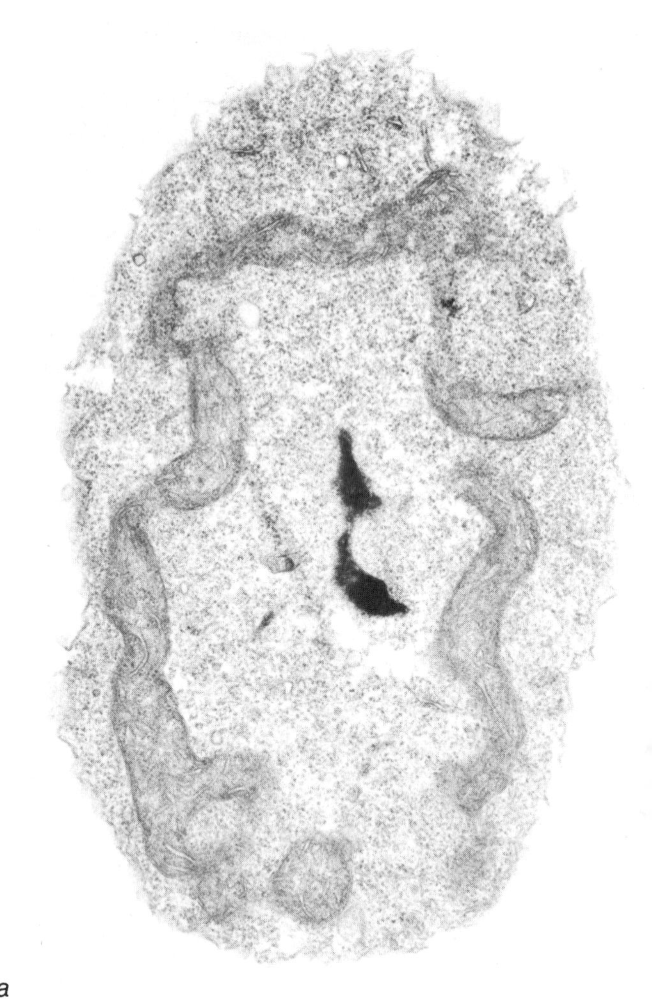

a

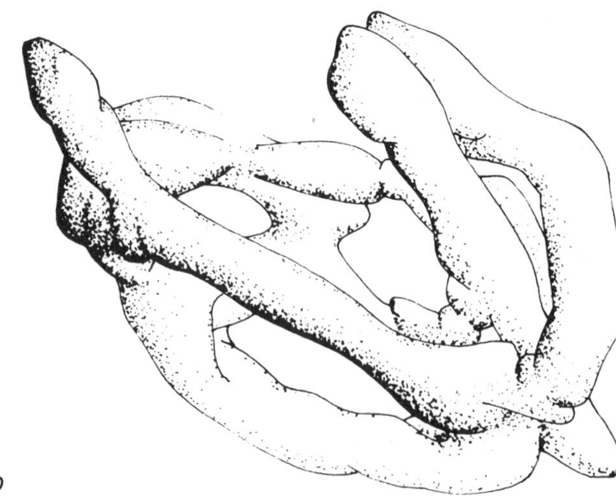

b

FIGURE 4-36

At some stages of their life cycle yeast cells contain a single, branched mitochondrion. (a) Electron micrograph of a yeast cell during germination from a spore. The mitochondrion leaves and reenters the plane of the section and its total continuity cannot be seen. (b) Drawing of the complete mitochondrion in a spore. [(a) Courtesy of Barbara Stevens.]

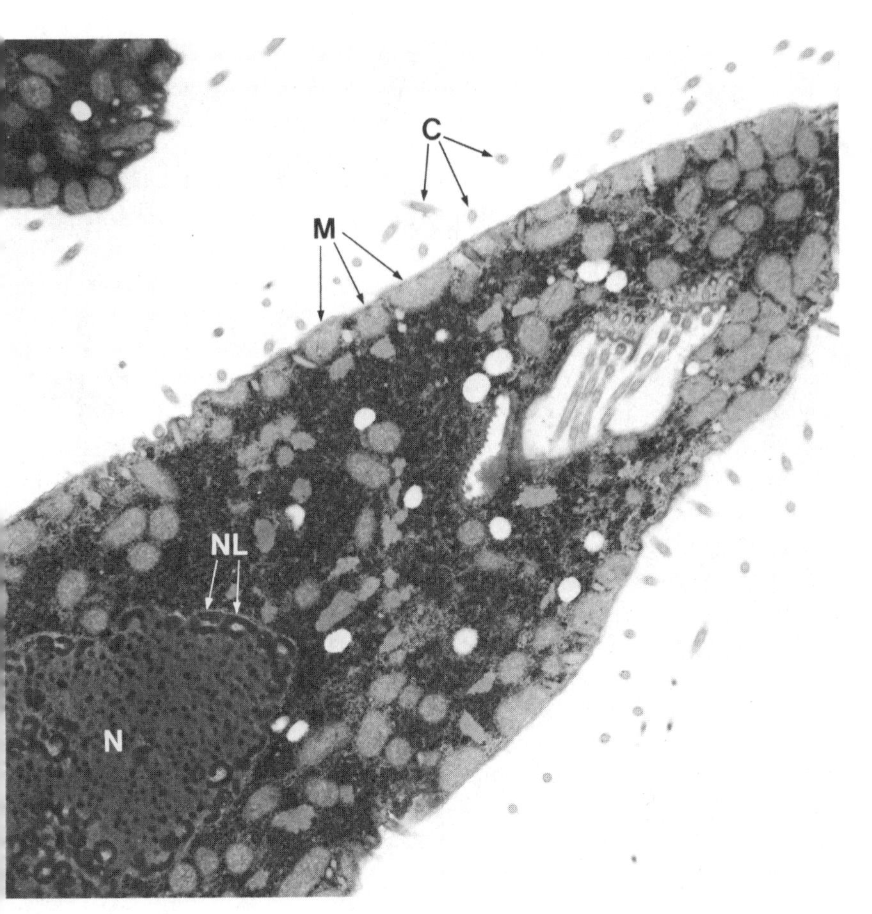

FIGURE 4-37
Electron micrograph of the ciliated protozoan *Tetrahymena* showing a high concentration of mitochondria (M) just below the cell surface. The nucleus (N), nucleoli (NL), and cilia (C) are also visible. [Courtesy of Ivan L. Cameron.]

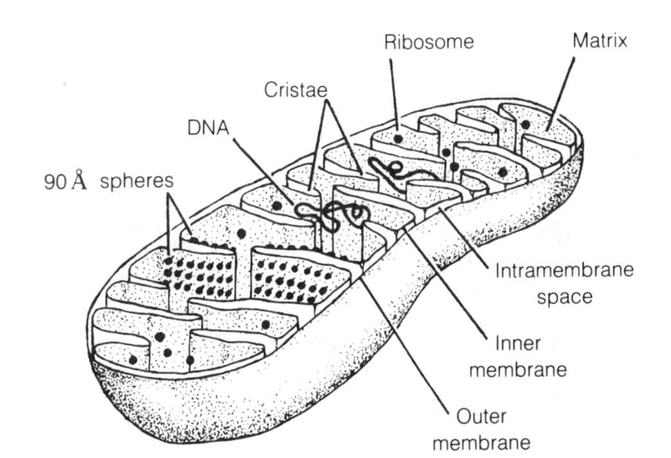

FIGURE 4-38
A three-dimensional schematic diagram of a mitochondrion.

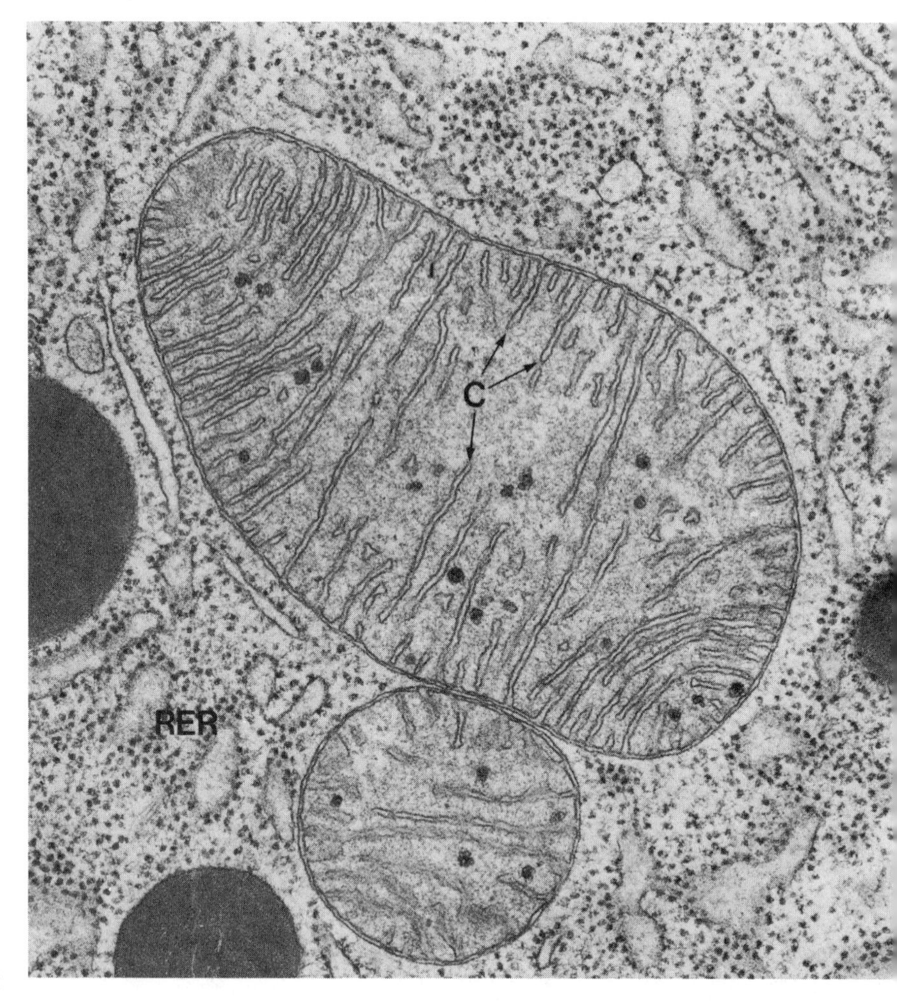

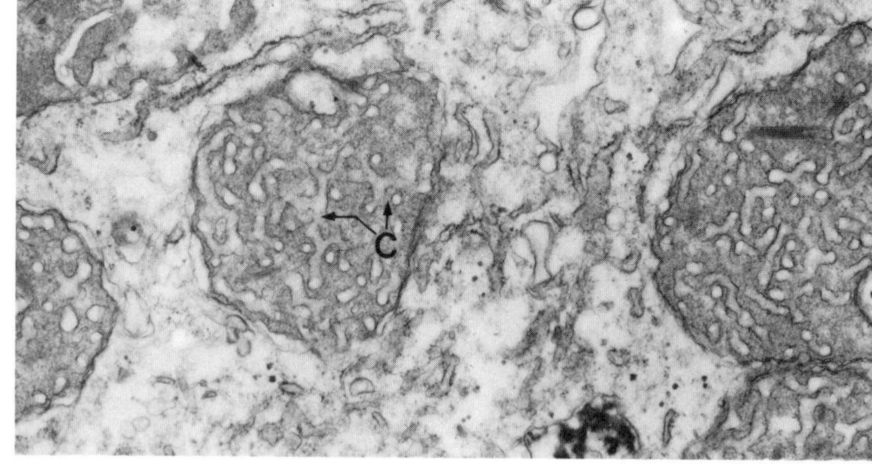

FIGURE 4-39

Electron micrographs of mitochondria illustrating two types of cristae. (a) Mitochondria with sheetlike cristae (C) in a pancreas cell of a bat. The mitochondria are surrounded by rough endoplasmic reticulum (RER). (b) Mitochondria in *Amoeba proteus* with tubular cristae (C). [(a) Courtesy of Keith R. Porter. (b) Courtesy of Charles J. Flickinger. From *Exptl. Cell Res.*, 60:225–236 (1970).]

nent enclosed by the inner membrane. The resulting hrinkage draws the inner membrane away from the uter membrane. This makes it easy to break the outer nembrane into pieces by gentle vibration with ultra-ound, without damaging the inner membrane. Differ-ntial centrifugation can separate the outer membrane ragments from the inner membrane component. The nalysis of the two membranes separated from each ther in bulk amounts has provided much of the infor-nation now available about the macromolecular com-osition and function of the outer and inner mem-ranes.

The outer and inner membranes create two com-artments in the mitochondrion. The outer compart-nent is the intermembrane space, which extends in-vard between the infolded membranes of the cristae. he compartment enclosed by the inner membrane is alled the **matrix.** It contains a variety of materials, ncluding enzymes of the citric acid cycle, ribosomes, nzymes and factors needed for protein synthesis, and a mall amount of DNA. The inner membrane is rather mpermeable, severely restricting the passage of mole-ules and ions between the matrix and the outer mito-hondrial compartment. This impermeability plays an mportant role in ATP production by the mitochon-rion.

Most of the differences in the internal structure mong mitochondria of different kinds of cells are in he form and number of the infoldings or cristae of the nner membrane. Often the cristae are sheetlike as in igure 4-39(a); in other cells the cristae are tubular nvaginations that appear circular in cross section (Fig-re 4-39(b)). The cristae in the mitochondria of proto-oa are usually tubular, although tubular cristae occur lso in some kinds of vertebrate cells, for example, cells f the ovary and adrenal gland. Mitochondria differ vith respect to the amount of cristae membranes. In eneral, mitochondria in cells with high rates of respi-ation (e.g., cardiac muscle cells and flight muscle cells f insects, shown in Figure 4-40) have a higher density f cristae than do mitochondria in cells with low rates f respiration (e.g., amoebae and liver cells).

The outer membrane is about 60 Å thick and is omposed of about 50 percent protein and 50 percent lipid. The arrangement of molecules in the outer membrane conforms to the molecular organization of membranes generally: there is a double or bilayer of lipid molecules with some protein molecules attached to the inner and outer surfaces of the membrane and other proteins embedded in the lipid bilayer (Chapter 5).

The inner membrane is also composed of lipid and protein but here the proportion of protein is much higher (80 percent). More than 60 different kinds of protein molecules are present in the inner membrane; most are part of the molecular machinery that carries out oxidative phosphorylation. Some of these proteins are organized into a complex that forms a 90 Å sphere on the inner surface of the inner membrane (Figure 4-41). The roles of the 90 Å spheres and of other proteins of the inner membrane in oxidative phosphorylation have been studied in the following way. Isolated mitochondria were disrupted, and the inner membranes were collected by differential centrifugation. The inner membranes were disrupted by exposure to ultrasound into membrane fragments (Figure 4-41). The fragments spontaneously formed sealed vesicles but in an inside-out fashion; the original outer surface of the inner membrane was now inside, and the original inner surface with its attached spherical particles was now outside. These vesicles retained the ability to carry out electron transport and phosphorylation of ADP + P_i to ATP.

The 90 Å spheres were dissolved away from the inner membrane vesicles to yield two fractions—the vesicles and a solution of proteins derived from the spheres. The vesicles were still able to carry out electron transport but could not synthesize ATP. The protein fraction derived from the spheres could not perform electron transport but could catalyze the hydrolysis of ATP to ADP and P_i, that is, this protein fraction contains an ATPase. Mixing of the proteins with the membranous vesicles resulted in reattachment of the proteins to the vesicles, concomitant re-formation of the spheres, and return of the ability to phosphorylate ADP to ATP. These and other experiments have shown that the inner membrane of the mitochondrion contains the electron transport chain, and that the spheres attached

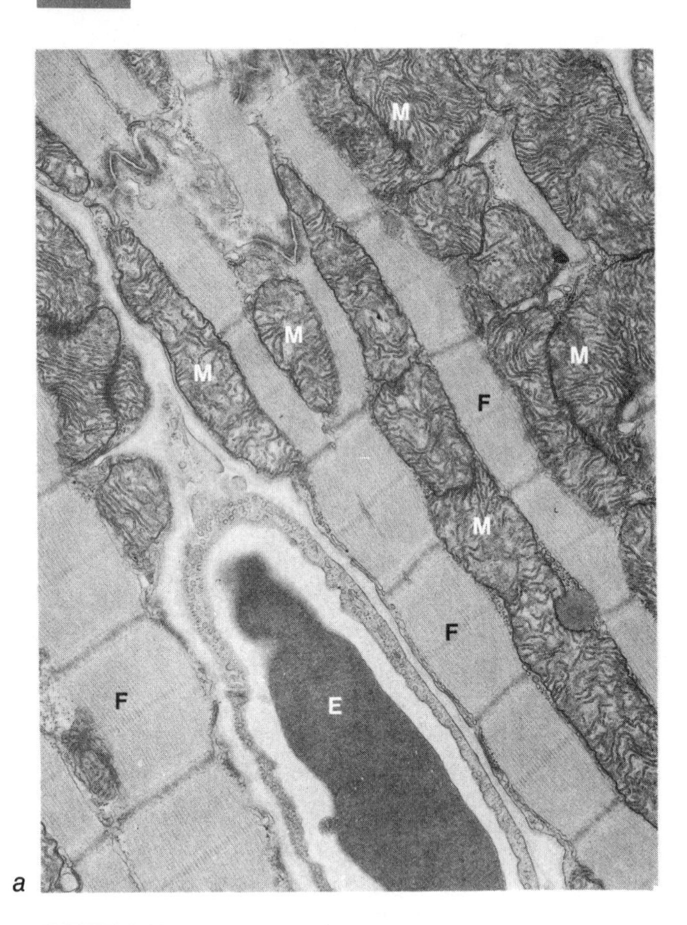

FIGURE 4-40

Electron micrographs of cells with large numbers of mitochondria. (a) Portions of two cardiac muscle cells of the bat. Many mitochondria (M) are present between the fibrous elements (F) that make up the contractile machinery of the muscle cell. A capillary containing an erythrocyte (E) is present between the two cells. (b) Electron micrograph of part of a muscle cell in a flight muscle of the moth *Manduca sexta*. The bundles of parallel filaments are the contractile machinery of the muscle cell (see Chapter 14). Mitochondria are packed in large numbers between the bundles of contractile filaments. [(a) Courtesy of Keith R. Porter. (b) Courtesy of Mary B. Rheuben.]

to the inner membrane contain the machinery for phosphorylation of ADP to ATP.

The ATPase activity that can be solubilized from the 90 Å spheres works in the reverse direction, that is, it catalyzes the synthesis of ATP from ADP and P_i when the enzyme is present in the sphere and coupled to the electron transport chain in the membrane. How this works is explained by the chemiosmotic coupling hypothesis already described briefly in connection with electron transport and ATP synthesis in photosynthesis.

arriers in the electron transport chain are oriented in ıe inner membrane of the mitochondrion such that ansfer of electrons from carrier to carrier causes umping of H^+ ions from the matrix into the inter- ıembrane space (Figure 4-42). The molecular mecha- ism of the pumping is still poorly understood. The esult of pumping is creation of a high concentration of $^+$ ions in the intermembrane space—creating a con- entration difference, which is used for driving the syn- ıesis of ATP. When the H^+ ions flow back across the ıner membrane through special channels, the ATPase ı the 90 Å spheres attached to the matrix side of the ıembrane catalyzes the synthesis of ATP from ADP nd P_i.

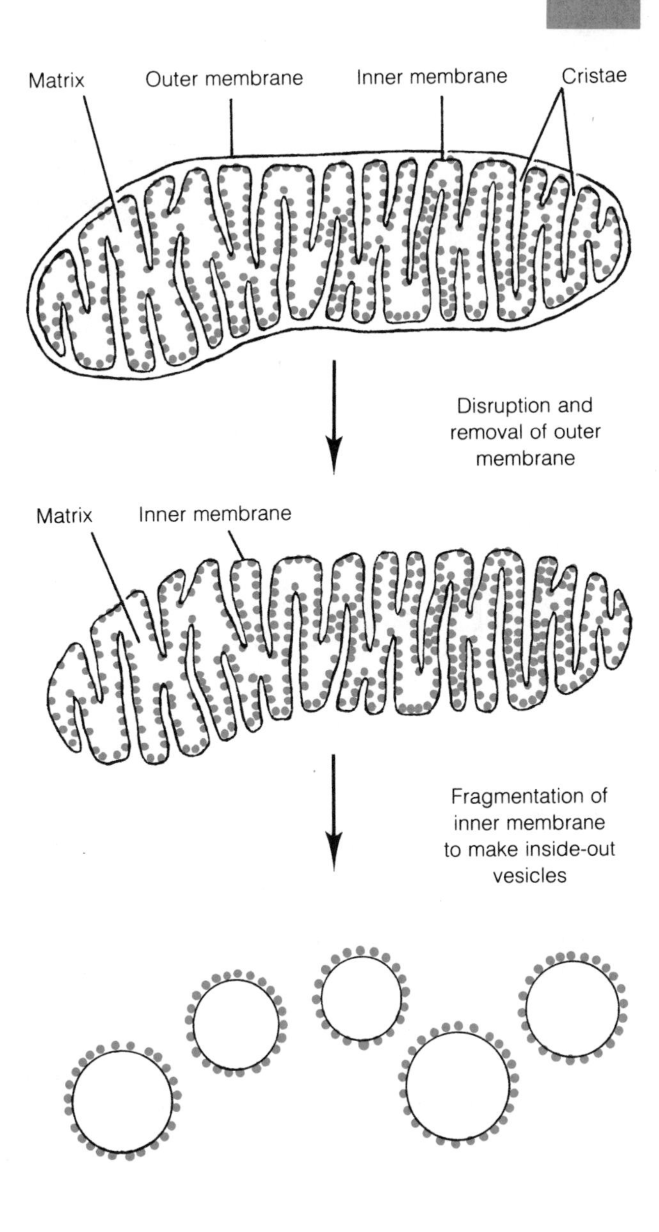

FIGURE 4-41
A drawing of a longitudinal section through a mitochondrion, removal of the outer mitochondrial membrane, and fragmentation of the inner mitochondrial membrane into inside-out spherical vesicles capable of oxidative phosphorylation.

Export of ATP from Mitochondria to the Cytosol

Although mitochondria require some ATP for their wn anabolism, almost all the ATP produced by respi- ation is exported for use in the cytosol and nucleus. he ATP produced by electron-flow phosphorylation is ccumulated in the matrix compartment of the mito- hondrion. The inner membrane is impermeable both ɔ ADP and ATP, and a special mechanism is present in ıe inner membrane to transport ADP into and ATP out f the mitochondrion. Both are moved by **facilitated liffusion** (Chapter 5). A carrier protein in the mem- rane binds ADP at the outside of the inner membrane nd carries it to the inside, releasing it in the matrix ompartment (Figure 4-42). The same carrier, in turn, arries ATP outward across the membrane. The trans- ort of ADP and ATP are interdependent: ATP is carried ut only if ADP is carried in. Thus, if ATP is not hydro- yzed to ADP in the cytosol and the nucleus, ATP can- ot be exported from the mitochondrion. Because facil- tated diffusion of ADP and ATP are linked, the process s called **facilitated exchange diffusion.**

Maintenance of Mitochondria

In general, the total mitochondrial mass in a partic- ılar cell appears to conform to the demands of cellular netabolism for energy. An extreme example of the cor- espondence in one direction is the erythrocyte, which

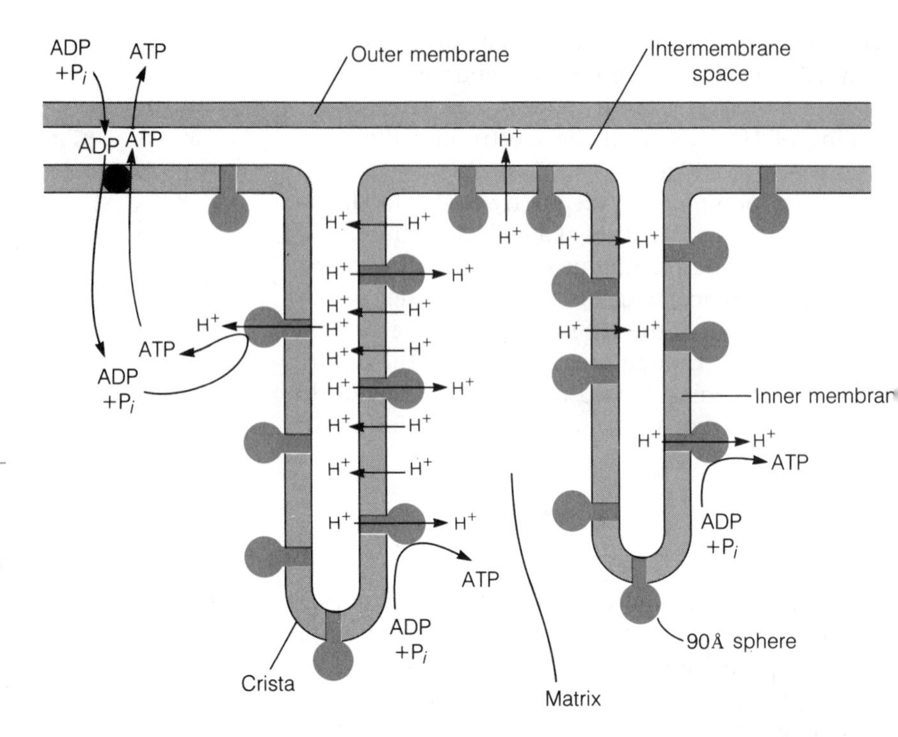

FIGURE 4-42
A schematic drawing of parts of the outer and inner membranes of a mitochondrion. H^+ ions are pumped into the intermembrane space. The flow of ^3H ions back across the inner membrane into the matrix is coupled to the synthesis of ATP in the 90 Å spheres. ADP is imported into and ATP is exported from the mitochondrion by a carrier.

carries out very little biosyntheses (e.g., it synthesizes no nucleic acids or proteins) and contains no mitochondria. An extreme example in the other direction is cardiac muscle cells, which contain a heavy concentration of mitochondria (Figure 4-40(a)). Flight muscle cells of insects, which are capable of sustained rapid contraction, also contain large numbers of mitochondria (Figure 4-40(b)). Moreover, these mitochondria are tightly packed with cristae, reflecting the presence of a large amount of electron transport machinery and an ability for a high rate of ATP production. Such observations indicate that the maintenance and probably the growth of mitochondria are tied to the cellular requirement for energy, although how the demand for ATP might turn on the synthesis of all the particular proteins and other components that constitute a mitochondrion is not known.

Just as maintenance of chloroplasts requires light, the maintenance of mitochondria also requires availability of oxygen, as illustrated in yeast cells. Yeast are facultative anaerobes. When yeast are grown in th absence of oxygen, the mitochondria decrease in siz and the cristae largely disappear. These changes ar accompanied by the disappearance of almost all cyto chrome molecules. The mitochondria, in fact, becom small, almost featureless vesicles incapable of oxidativ phosphorylation. When such yeast cells are provide with oxygen, the mitochondria develop cristae, and cy tochromes are synthesized. The mitochondria not onl regain a typical mitochondrial appearance, but also in crease in number.

Formation of New Mitochondria

Like chloroplasts, mitochondria contain DNA, al though less than do chloroplasts. The DNA of mamma lian mitochondria consists of 15,000 base pairs (Chap ter 7). The DNA codes for ribosomal rRNAs and tRNA that function in protein synthesis in the matrix com partment of the mitochondrion. The DNA also codes for a few proteins needed in oxidative phosphorylation. Most of the proteins of a mitochondrion are coded for by nuclear genes and are synthesized outside the mitochondrion on cytoplasmic ribosomes and then imported into the mitochondrion.

Like chloroplasts, mitochondria increase in number by binary fission. For this reason and because they contain DNA that codes for a limited number of components, both mitochondria and chloroplasts are called *semiautonomous organelles*.

In summary, photosynthetic cells obtain energy for synthesis of organic molecules from sunlight. In photosynthetic bacteria photosynthesis occurs in cytoplasmic membranes; in eukaryotes it takes place in chloroplasts. Nonphotosynthetic cells obtain energy by the step-by-step breakdown of organic molecules, usually glucose, coupling the released energy to synthesis of ATP. Breakdown of glucose occurs in two stages: glycolysis, which occurs in the cytosol and requires no O_2, and respiration, which takes place in mitochondria and is O_2-dependent.

PROBLEMS

1. What is achieved energetically by the coupling of reactions?
2. What is meant by the statement that all life on Earth is sustained by tiny electrical currents driven by the sun?
3. How is energy provided to drive an energy-requiring reaction like the formation of a peptide bond between two amino acids?
4. What is meant by the statement that ATP molecules are the currency of energy metabolism in a cell?
5. Photosynthesis yields *two* products that most chemotrophs need to survive. What are they?
6. What are the products of phase I reactions in photosynthesis? What is the fate of these products?
7. Photosystem I generates energized electrons that may be passed to $NADP^+$. How is the continuation of this function dependent on coupling of photosystem I to photosystem II?
8. What is the source of electrons and O_2 produced by the phase I reactions of photosynthesis?
9. Phase I reactions of photosynthesis yield energized electrons and ATP that are used in the phase II reactions. What product is yielded by phase II reactions?
10. What problem do plants face in hot, dry climates, and how do C_4 plants solve the problem?
11. Where do phase I and phase II reactions take place inside a chloroplast?
12. How do yeast cells deprived of O_2 obtain energy from glucose?
13. What are the products of glycolysis?
14. When the passage of electrons along the electron transport chain in respiration is blocked by a poison, the citric acid cycle cannot operate. Why?
15. Explain in words instead of an equation what is meant by the energy charge of a cell.
16. What are the two end products of cellular respiration?
17. Where do the reactions of the citric acid cycle and oxidative phosphorylation occur in a mitochondrion?
18. The same protein in a mitochondrion can work as an ATPase or as an ATP synthase (synthesizing ATP). What governs which activity will occur?
19. What is the function of the stomata in a plant leaf?

THE PLASMA MEMBRANE

Properties of the Plasma Membrane
The Lipid Nature of the Plasma Membrane

The fluid mosaic model of the plasma membrane—the lipid component
The fluid mosaic model—the proteins

DIFFUSION OF SUBSTANCES IN AND OUT OF CELLS

TRANSPORT ACROSS THE PLASMA MEMBRANE

Facilitated Diffusion or Passive Transport
Active Transport

Na^+-K^+ transport
Active transport of sugars and amino acids into animal cells
Active transport of amino acids and sugars into bacterial cells
Other systems of active transport

IMPORT OF MATERIALS BY ENDOCYTOSIS

Pinocytosis
Receptor-Mediated Endocytosis (Pinocytosis) of Specific Macromolecules
Phagocytosis

EXOCYTOSIS

Synthesis and Release of Digestive Enzymes by Pancreas Cells
Formation of the Surface of Intestinal Epithelium

CHAPTER 5

Cell and Environment

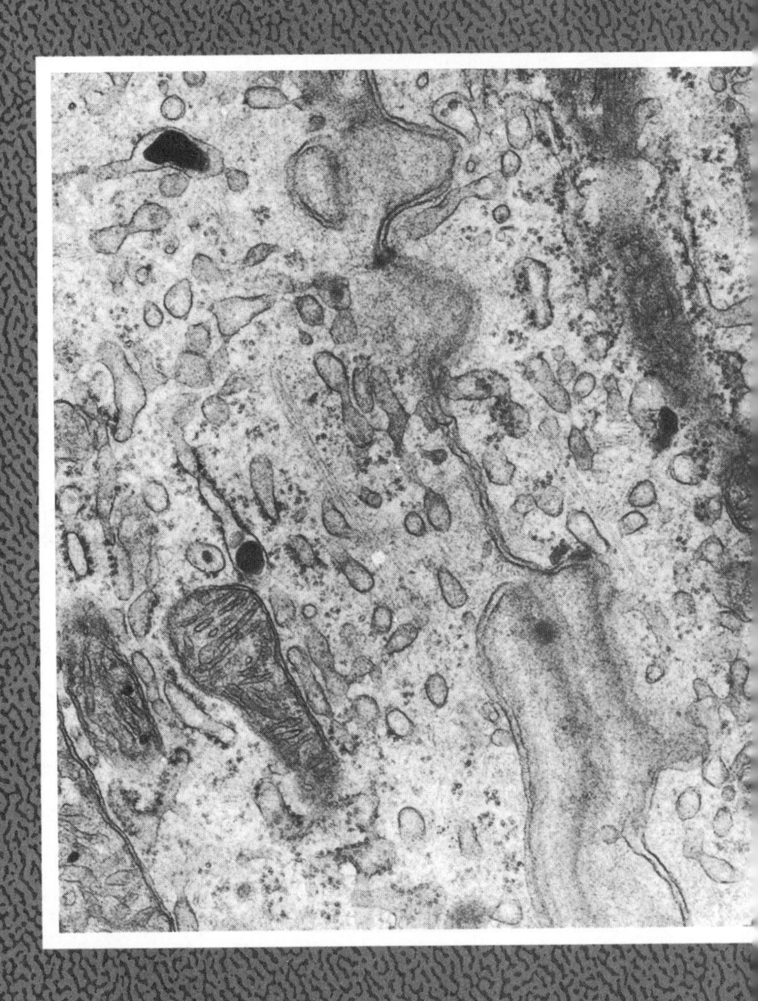

See Figure 5-36.

All cellular activities and functions depend on maintenance of the complex mixture of molecules and ions within a cell. Maintenance of cell composition depends not only on cellular metabolism, but on the uptake of molecules and ions from the surroundings and on the export of materials. Import into and export out of the cell are functions provided by the **plasma membrane,** a retentive barrier that encloses every cell (Figure 5-1).

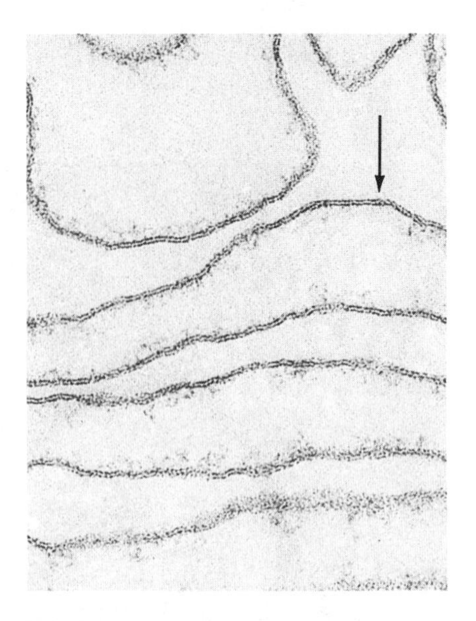

FIGURE 5-1
Electron micrograph of plasma membranes isolated from red blood cells. A single membrane is indicated by the arrow. [Courtesy of Vincent T. Marchesi.]

THE PLASMA MEMBRANE

A few substances, such as water and dissolved gases (CO_2, O_2), diffuse across the plasma membrane at high enough rates to meet the requirements of metabolism; however, most ions and molecules are transported in and out by molecular components present in the plasma membrane.

An understanding of how substances diffuse or are transported into or out of a cell requires a knowledge of the molecular structure of the plasma membrane. Therefore, the following section traces the development of our current understanding of the structure of the plasma membrane.

Properties of the Plasma Membrane

Observations of the nineteenth century established the presence of a barrier at the cell surface that allowed a relatively high rate of diffusion of water but more severely restricted the diffusion of all other materials, including organic molecules and inorganic ions. Because of the greater permeability to water and the much lower permeability to substances dissolved in the water (**solutes**), considerable water may be taken up or lost from a cell under certain conditions. The following experiment explains this net movement of water across the plasma membrane.

A container is separated into two compartments by a membrane that is freely permeable to water and solutes (Figure 5-2(a)). The two compartments are filled to the same level with a glucose solution. The dissolved glucose molecules and the water molecules are in constant motion, and both are able to move freely through the permeable membrane in both directions. No *net* movement of water or glucose takes place, and each compartment continues to hold the same amount of both substances. Consider the result of filling one compartment with the glucose solution and the other with pure water (Figure 5-2(b)). The water and glucose molecules can move freely across the permeable membrane. The glucose molecules distribute themselves uniformly throughout the water in both compartments. For this to happen, a net movement of glucose molecules must take place, finally resulting in the same concentration of glucose in the two compartments; that is, the glucose becomes uniformly distributed throughout the two compartments.

If the membrane separating the two compartments were freely permeable to water but impermeable to glucose, and both compartments were filled with a solution containing the same concentration of glucose (Figure 5-2(c)), no net movement of water from one compartment to the other would take place because the concentration of water is the same in both compartments. If one compartment were filled with glucose solution and the other with water, the concentration of water would be higher (since it is not diluted by the

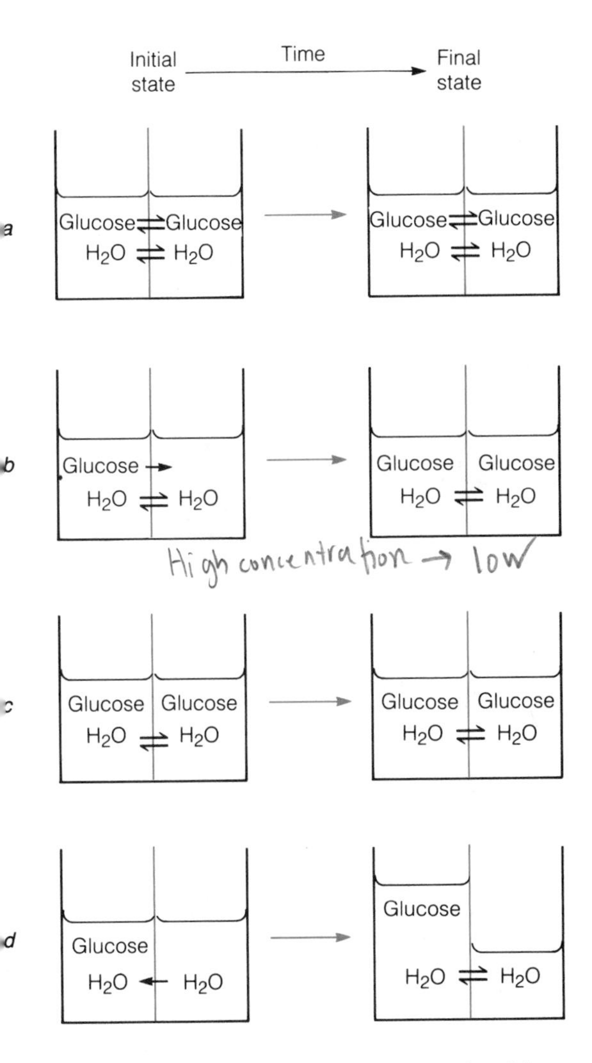

FIGURE 5-2
Diffusion of water and a solute through a membrane—generation of osmotic pressure.

(a) The two compartments, each containing a glucose solution, are separated by a membrane permeable to water and glucose molecules. Both move freely through the membrane but no *net* movement occurs.

(b) The two compartments, one containing a glucose solution and one pure water, are separated by a permeable membrane. Net diffusion of glucose occurs through the membrane until the concentration of glucose molecules is the same on each side of the membrane, that is, until equilibrium is reached.

(c) The two compartments, each containing the same concentration of glucose, are separated by a membrane that is permeable to water but impermeable to glucose. Water can move through the membrane but glucose cannot. No change takes place with time.

(d) The two compartments, one containing a glucose solution and one containing pure water, are separated by a membrane that is permeable to water but impermeable to glucose. Net diffusion of water occurs until the hydrostatic pressure equals the osmotic pressure. At equilibrium water molecules move through the membrane in both directions, but no further net movement of water takes place.

presence of glucose molecules) on one side of the membrane versus the other side (Figure 5-2(d)). Because any substance diffuses from a region of higher concentration to a region of lower concentration, a net movement of water into the glucose solution must occur. (Glucose would likewise move into the compartment containing pure water but is prevented from doing so by the membrane.) The net movement of water into the glucose-containing compartment is called **osmosis**, and it continues until the pressure created by the increase in height of the glucose-containing solution (hydrostatic pressure) is sufficient to counterbalance the **osmotic pressure** that drives water from a higher concentration to a lower one. If the hydrostatic pressure were decreased by removal of some of the glucose solution, osmotic pressure would again drive water through the membrane.

A solution with a higher water concentration is said to be **hypotonic** relative to a solution with a lower water concentration, which is **hypertonic.** In Figure 5-2(d) the pure water is hypotonic to the glucose solution, and conversely the glucose solution with its lower water concentration is hypertonic to pure water. If two solutions have the same water concentrations they are **isotonic.** Osmotic pressure is always present when nonisotonic solutions are separated by a membrane, as in Figure 5-2(d), that allows the passage of water but does not allow the passage of solute molecules.

A membrane that allows the passage of water but prevents diffusion of ions or molecules dissolved in the

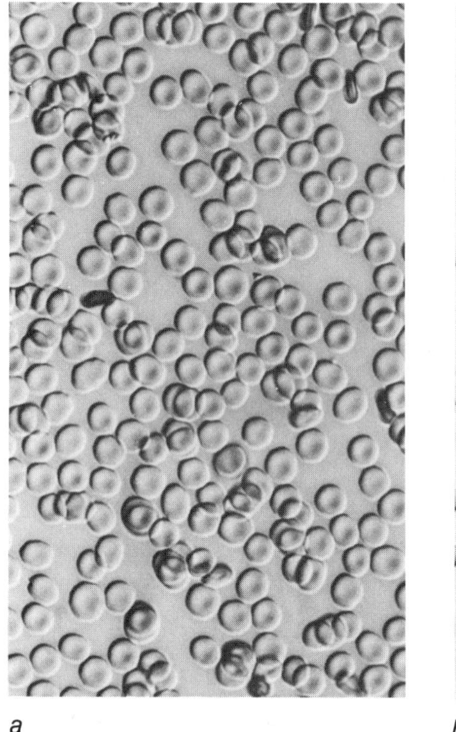

a

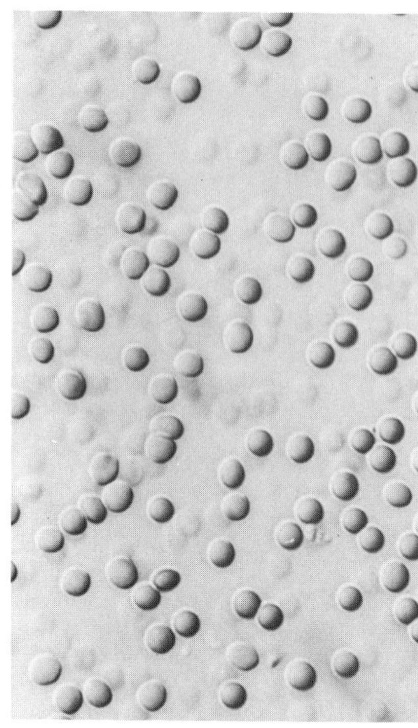

b

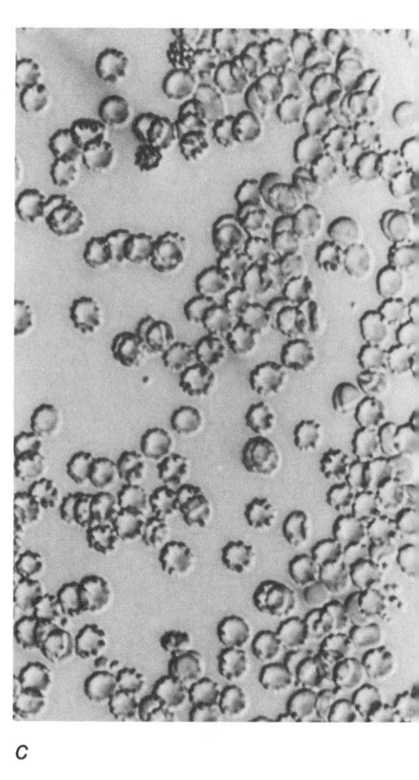

c

FIGURE 5-3
Swelling and shrinkage of red blood cells. (a) Light micrograph of red blood cells of normal shape in isotonic medium. The cells are biconcave discs. (b) Red blood cells swollen in hypotonic medium. The cells have changed from discs to spheres with the intake of water. Some cells have released their content of hemoglobin and appear as pale ghosts. (c) Red blood cells shrunken in hypertonic medium. The loss of water has caused the discs to become crenated.

water is said to be **semipermeable.** The plasma membrane approaches the properties of a semipermeable membrane, but is more appropriately called a **differentially permeable** membrane because of its relatively high permeability to water and low permeability to solutes such as glucose. In a manner similar to osmosis across a perfectly semipermeable membrane, osmotic movement of water can be induced across the differentially permeable plasma membrane by placing cells in hypotonic or hypertonic solutions. Placing a cell in a hypotonic solution creates a difference in osmotic pressure across the cell surface. This difference can be relieved only by equalizing the water concentrations on the two sides of the membrane by the osmotic uptake of water into the cell or by movement of solutes out of the cell into the solution. Since the plasma membrane has a higher permeability to water than to solutes, the result is an osmotic movement of water into the cell, with consequent cell swelling. For example, red blood cells are normally bathed in an isotonic solution, the fluid part of blood (Figure 5-3). When red blood cells are placed in a hypotonic solution, such as blood diluted with water, they swell and eventually burst. If red blood cells are placed in a hypertonic solution, for example, serum to which sucrose has been added, they shrink by the loss of water until the intracellular water concentration has decreased to the same level outside the cell.

Animal cells normally live under isotonic conditions because the medium surrounding the cells generally has the same total solute concentration as is

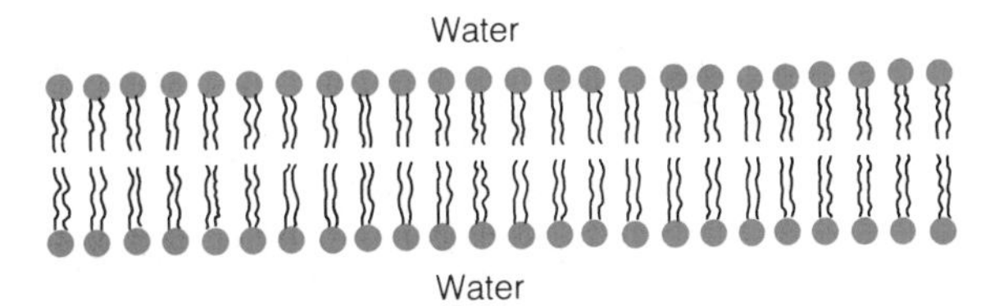

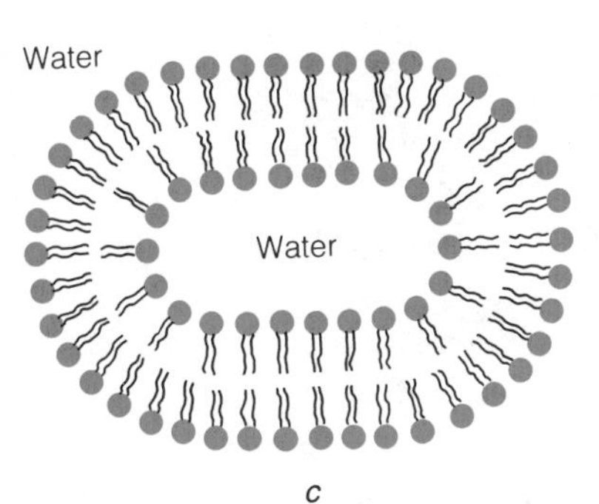

FIGURE 5-4
Aggregation of lipid molecules in water. (a) A spherical micelle. (b) A bilayer of lipid molecules with water on each side. (c) A vesicle formed from a bilayer of lipid molecules, with water on the inside and outside.

present in cells themselves. However, plants, algae, bacteria, and fungi often live in a hypotonic environment (freshwater ponds, streams, etc.). They do not swell because they possess strong cell walls. They are also protected against loss of solutes by a rather impermeable plasma membrane. For most invertebrates, algae, fungi, protozoa, and bacteria living in a marine environment, seawater is isotonic or nearly so. Protozoa living in fresh water are not protected from the hypotonic environment by a rigid cell wall, but avoid osmotic swelling by pumping excess water out as rapidly as it is taken in by means of a contractile vacuole (Figure 1-17).

The Lipid Nature of the Plasma Membrane

Studies of permeability to a variety of solutes led to the discovery in the 1890s that the greater the solubility of a substance in oil, the faster its penetration through the plasma membrane. This observation suggested that the diffusion barrier at the cell surface might consist of lipid molecules through which lipid-soluble substances might diffuse more rapidly than substances not soluble in lipid.

The lipid nature of the plasma membrane was demonstrated in 1925 by direct chemical analysis of plasma membranes purified from mammalian red blood cells. These cells are a particularly favorable source for preparation of pure plasma membranes because these cells contain no nucleus, no mitochondria, no endoplasmic reticulum, or no other membranous structures to complicate the analysis. To obtain plasma membranes, red blood cells were put in a hypotonic medium, which caused them to swell and rupture (Figure 5-3), releasing the internal contents of the cell, primarily hemoglobin. The membrane fragments could then be collected by centrifugation. In the original experiment the amount of lipid extracted from purified membranes was equal to an amount expected if the plasma membrane contained a **double layer** or **bilayer of lipid molecules** over the entire surface of the cell. This finding has been thoroughly validated in recent years by a variety of more precise analytical methods, and it is now generally accepted that the plasma membrane contains a bilayer of lipid molecules.

Two major kinds of lipid molecules make up plasma membranes in eukaryotes and prokaryotes: **phospholipids** and **glycolipids.** In addition, plasma membranes in eukaryotes (but usually not prokaryotes) also contain sterol molecules; in animal cells and in some other species the most common sterol is

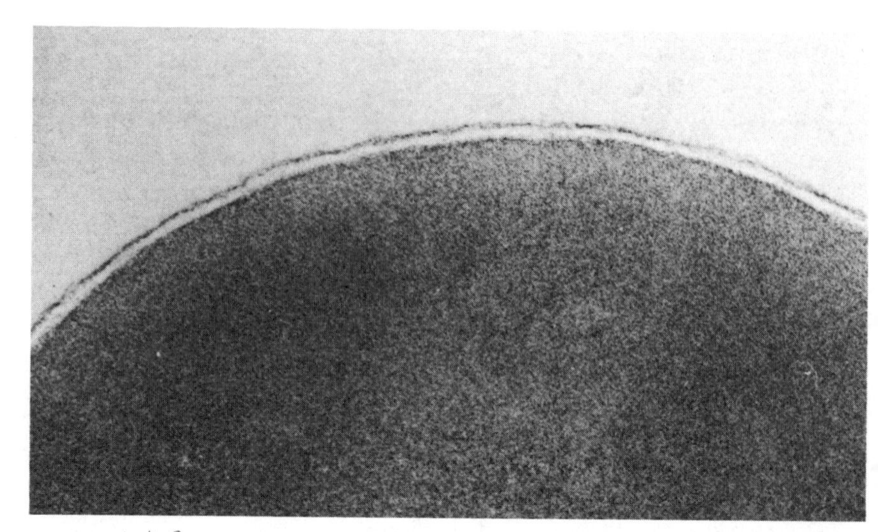

FIGURE 5-5
Electron micrograph of part of an erythrocyte showing the plasma membrane. [Courtesy of J. David Robertson, 1981].

cholesterol. The molecular structure of phospholipids was discussed in Chapter 2. Glycolipids have a similar structure, consisting of two fatty acids that form a hydrophobic tail, plus a hydrophilic head that lacks a phosphate group (present in phospholipids) but has one or more sugar groups. Such molecules, with one portion insoluble and one portion soluble in water, are said to be **amphipathic.**

Despite their hydrophilic heads, lipid molecules are insoluble in water because of their long hydrocarbon tails. When lipid molecules are mixed into water, they can form spherical or ellipsoidal aggregates or micelles (micelles were described in Chapter 1), with the hydrophobic tails buried inside the aggregate away from water and the hydrophilic heads at the surface, where they interact with water (Figure 5-4(a)). With the proper conditions the aggregates can take the form of a sheet made of two layers of lipid molecules (Figure 5-4(b)). The hydrocarbon tails of the two layers project inward toward each other, and the hydrophilic heads form the two surfaces of a bilayer exposed to water molecules. Such artificial lipid bilayers can spontaneously form vesicles, called **liposomes,** that enclose a portion of the aqueous medium (Figure 5-4(c)).

Thus, the internal portion of the bilayer is a hydrocarbon domain in which hydrophobic interactions sequester the hydrocarbon tails of the lipid molecules away from contact with water. The hydrophobic interactions and the interactions of the hydrophilic heads with each other and with water impart structural stability to the bilayer.

Analogous to the artificial vesicles just described, the lipids of the plasma membrane are arranged in a bilayer that completely encloses the cell. The evidence for this comes in part from a comparison of artificial lipid bilayers with the plasma membrane. The permeability characteristics of the two are strikingly similar. The plasma membrane (Figure 5-5), however, is not simply a pure lipid bilayer; it contains a variety of proteins, some of which function in the transport of substances across the lipid bilayer.

The fluid mosaic model of the plasma membrane—the lipid component. In 1971 the various facts that had been uncovered about the molecular structure of the plasma membrane made possible the assembly of the **fluid mosaic model.** The core of this model is a lipid bilayer with the lipid molecules in a fluid state.

The kinetic state of the lipid molecules in membrane bilayers has been studied by highly sophisticated, physical techniques. These analyses have shown that the hydrocarbon tails of the lipid molecules in a bilayer are ordered in the sense that they extend inward from

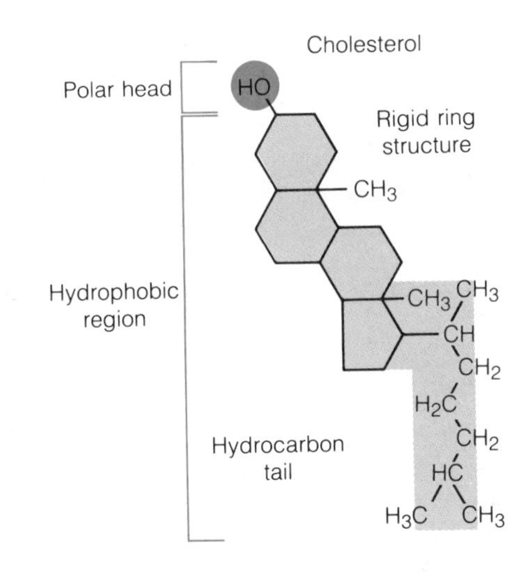

FIGURE 5-6
Structure of cholesterol, an amphipathic molecule.

the surface of the membrane, but they are not fixed in parallel arrays to form a crystalline-like structure. Rather, the hydrocarbon tails are in continuous motion, particularly rotating on their long axes, and form a liquid rather than a solid.

The state of the lipids in the plasma membrane is dependent on the temperature. In the temperature range in which a cell normally lives the lipid molecules are in the liquid state. If the environmental temperature of a cell is quickly lowered below the normal range, a point called the **transition temperature** is reached at which the hydrocarbon chains cease their movement and the lipids *freeze,* much as animal lipids such as melted butter solidify when the temperature is lowered. The transition temperature for a given kind of lipid molecule depends on a number of factors, including the number of unsaturated bonds in the two hydrocarbon chains. The greater the number of unsaturated bonds the lower the transition temperature.

Cholesterol or related sterol molecules are lipid components of the plasma membrane of eukaryotes. Cholesterol, like other lipids of the plasma membrane, is an *amphipathic* molecule: it consists of a group of ring structures and a hydrocarbon tail (Figure 5-6), both of which are hydrophobic. A hydroxyl group provides a slight water solubility to one end of the molecule. Because of their shape and minimal water solubility, cholesterol molecules cannot by themselves form a bilayer, but their amphipathic property causes them to orient in a bilayer of phospholipids and glycolipids (Figure 5-7). Cholesterol probably increases the mechanical strength of the membrane through its interactions with the hydrocarbon tails of the other lipids. These hydrophobic interactions reduce the fluidity of the lipid, raise the transition temperature, and generally lower membrane permeability.

Lowering the temperature of a cell below the transition point reduces the rate of diffusion of substances across the plasma membrane because lipid molecules in a fixed crystalline state form a less permeable structure. In a fixed state the lipids may also impair transport of substances such that the cell cannot survive. Thus, the transition temperature is a major factor in defining the lower limit of temperature at which a particular cell can

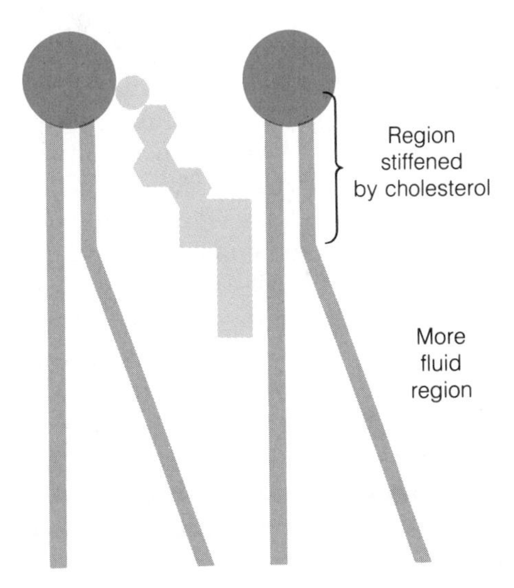

FIGURE 5-7
Drawing showing the arrangement of cholesterol between lipid molecules of a lipid bilayer. Cholesterol stiffens a lipid bilayer.

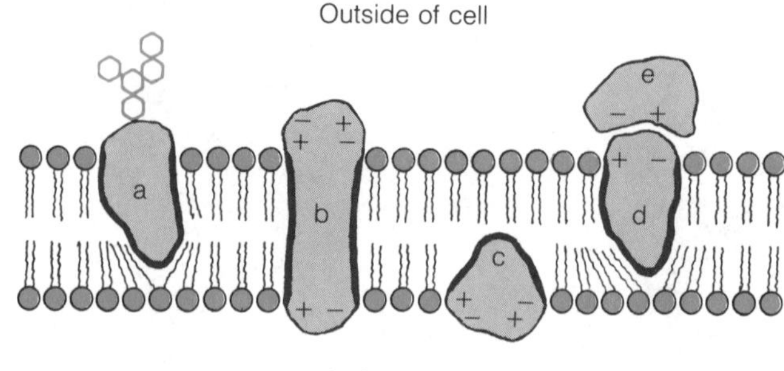

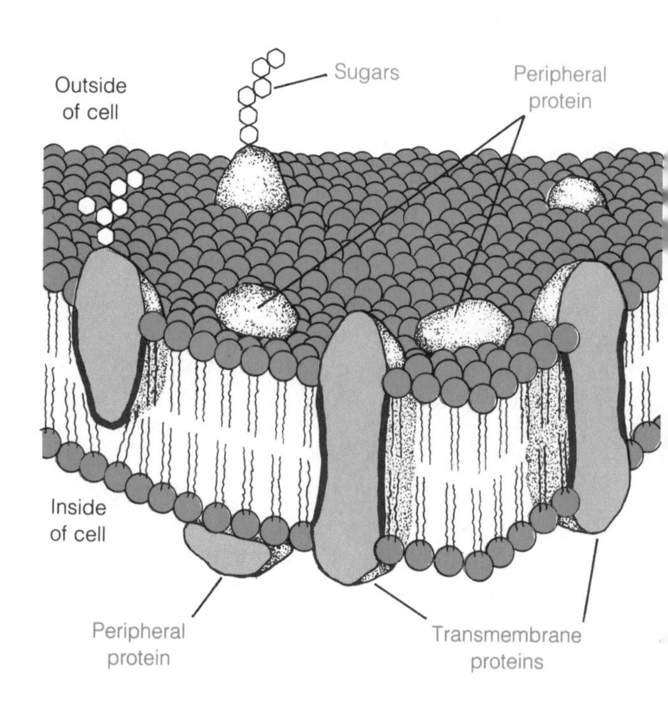

FIGURE 5-8 (a) Drawing of the fluid mosaic model of a membrane showing various dispositions of four integral proteins (a through d) and one peripheral protein (e). Heavy lines indicate hydrophobic surfaces of proteins embedded in the lipid bilayer. Protein a is a glycoprotein with sugar groups (red) projecting from the cell surface. Protein b is a transmembrane protein, with charged groups at both the inside and outside surfaces. Protein c projects from the inner surface of the membrane. Protein d has peripheral protein e bound to it by ionic bonds on the cell surface. (b) A schematic drawing in three dimensions of the fluid mosaic model of membrane structure. The peripheral proteins are shown attached to hydrophilic heads of lipid molecules. Peripheral proteins also attach to integral proteins at the membrane surfaces, as shown in (a).

survive. Some organisms are able to adapt to temperatures below the usual range. A major part of this adaptation is a lowering in the transition temperature of the plasma membrane. Immediately after the temperature is decreased the organisms shift synthesis of fatty acids to more unsaturated forms. These are incorporated into lipids that are added to the plasma membrane during growth. The continued incorporation of unsaturated fatty acids gradually lowers the transition temperature through replacement of existing lipids with lipids containing more unsaturated bonds in the hydrocarbon tails.

Because they are in a fluid state, the lipid molecules of the plasma membrane are more or less free to move laterally in the plane of the lipid bilayer. This means, of course, that the polar heads at the membrane surface are in constant motion. However, the flip-flop of individual lipid molecules from one layer to the other within the bilayer is severely restricted because of the difficulty of moving the hydrophilic head of a lipid molecule through the central hydrophobic region of the bilayer. Flip-flop occurs ten billion times more slowly than lateral diffusion of lipid molecules.

To summarize, the lipid bilayer of the plasma membrane is more like a fluid than a solid. Yet the bilayer is a relatively stable structure in an aqueous environment because of the exclusion of water from the central hydrocarbon region formed by the tails of the lipid molecules and because of the interactions of the hydrophilic heads with water at each surface. The fluid bilayer forms the permeability barrier at the cell surface. The proteins contained in the lipid bilayer are the key to how specific substances are transported across the barrier.

The fluid mosaic model—the proteins. Most membranes contain 50 to 70 percent protein and 30 to 50 percent lipid by weight. The proteins, which play major roles in membrane function and structure, are of two types: **peripheral** and **integral**. The peripheral proteins typically comprise about one-third of the total proteins of the plasma membrane. They are attached to the outside and inside surfaces of the plasma membrane by ionic bonds formed between negatively and positively charged groups of the proteins and charged groups on the surface of the bilayer (Figure 5-8). Therefore, peripheral proteins are easily dissociated from the membrane when washed with an aqueous salt solution. The ions of a salt solution (for example, Na^+ and Cl^- ions) competitively bind to charges on the proteins and the membrane surface and abolish ionic bonds between them.

Integral proteins, which make up the majority of membrane proteins, are within the lipid bilayer. These proteins are amphipathic, like the lipid molecules, possessing hydrophobic clusters of amino acids with hydrophobic R groups (leucine, valine, etc.) and water-soluble regions consisting of hydrophilic amino acids. The hydrophobic part of an integral protein usually takes the form of an α-helix in which the R groups are directed away from water and into the hydrophobic region of the lipid bilayer; the hydrophilic regions project from both surfaces of the bilayer where they interact with water (Figure 5-8).

Most integral proteins are **transmembrane proteins;** that is, they extend from one side of the membrane to the other with one hydrophilic region exposed at one surface, another hydrophilic region exposed at the other surface, and an intervening α-helical hydrophobic region within the lipid bilayer. Some integral proteins also have covalently attached fatty acids, which further stabilizes their arrangement in the membrane. Other proteins are anchored in the membrane with a hydrophobic region but possess only a single hydrophilic portion that projects from one or the other surface (Figure 5-8).

Because of their hydrophobic properties integral proteins are insoluble in salt solution and require a stronger treatment to extract them from membranes than is needed for peripheral proteins. The most common method employs detergents to solubilize integral proteins. Detergents such as sodium dodecyl sulfate (Figure 5-9(a)) are amphipathic molecules that aggregate with each other when mixed with water to form micelles. When plasma membranes are mixed with a detergent solution, the hydrophobic portion of the detergent molecules binds to the hydrophobic regions of integral proteins (Figure 5-9(b)). The hydrophilic heads of the detergent also interact with water, making the detergent-coated protein water-soluble. When the detergent is removed from the sample, the integral proteins aggregate and form a precipitate that is easily collected.

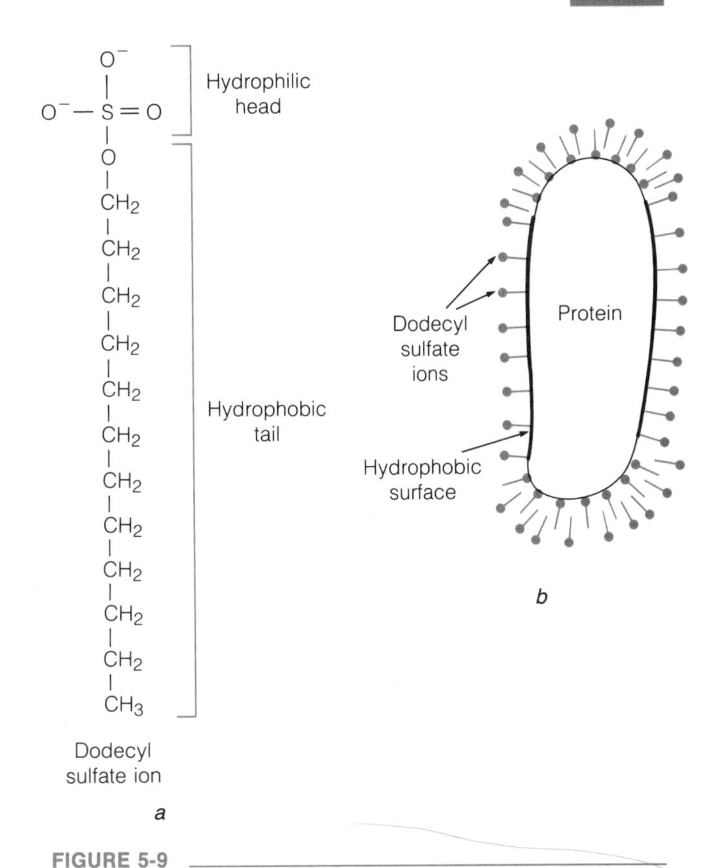

FIGURE 5-9
(a) Structure of the dodecyl sulfate ion, a detergent. Because of their amphipathic structure, dodecyl sulfate ions can solubilize in water protein molecules that have extensive hydrophobic surfaces. (b) The ions bind by their hydrophobic tails to the hydrophobic surface of protein molecules and bind by their hydrophilic heads to hydrophilic regions of protein. The effect is to surround the protein with a halo of dodecyl sulfate ions.

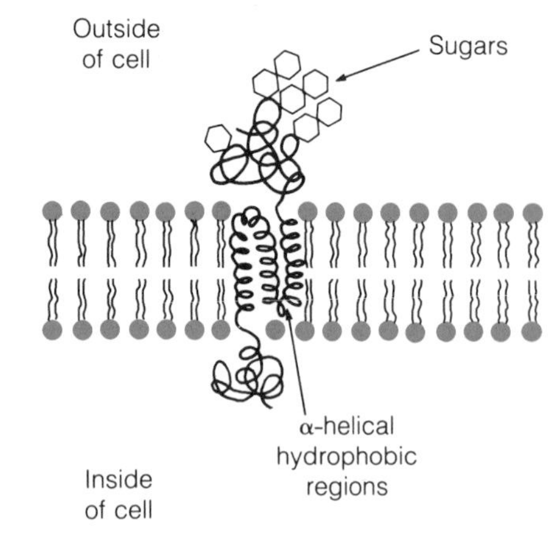

FIGURE 5-10
Schematic diagram of a transmembrane glycoprotein with three α-helical regions spanning the lipid bilayer of a membrane.

Many integral proteins have been purified from plasma membranes. Their amino acid sequences are precisely what one would predict for transmembrane proteins: two parts of the sequence contain a high proportion of hydrophilic amino acids separated by a third series of hydrophobic amino acids, usually 16 to 18, that forms an α-helix long enough to span a lipid bilayer. Some transmembrane proteins have two or more hydrophobic regions separating multiple hydrophilic regions. In such cases the polypeptide of a protein crosses the lipid bilayer two or more times and has two or more hydrophilic regions that project from the surface of the bilayer (Figure 5-10).

The extension of proteins through the lipid bilayer can be seen in freeze-fracture images of membranes. As described in Chapter 4, when a frozen cell is mechanically cracked, fracture planes frequently form between the two layers of lipid molecules of the bilayer (Figure 4-17). Particles are observed by electron microscopy to be scattered in intramembrane fracture planes (Figure 5-11), representing integral proteins anchored within the bilayer.

The hydrophilic portions of a transmembrane protein contain different amino acid sequences, reflecting

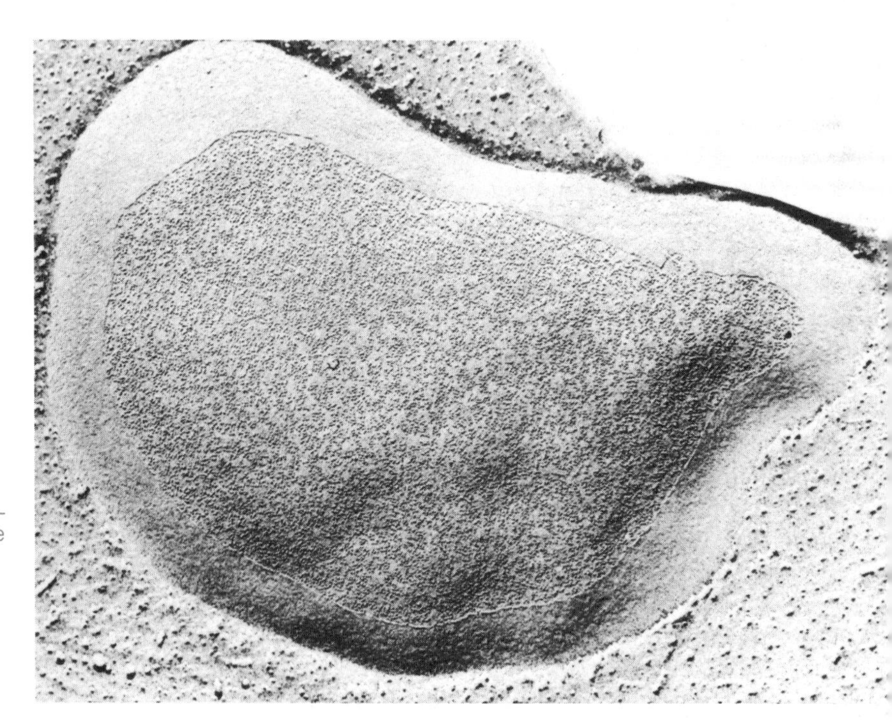

FIGURE 5-11
Electron micrograph of the plasma membrane of a red blood cell. By freezing and then fracturing the cell the plasma membrane has been split between the two lipid layers revealing a dense population of particles, which are integral membrane proteins. [Courtesy of Vincent T. Marchesi.]

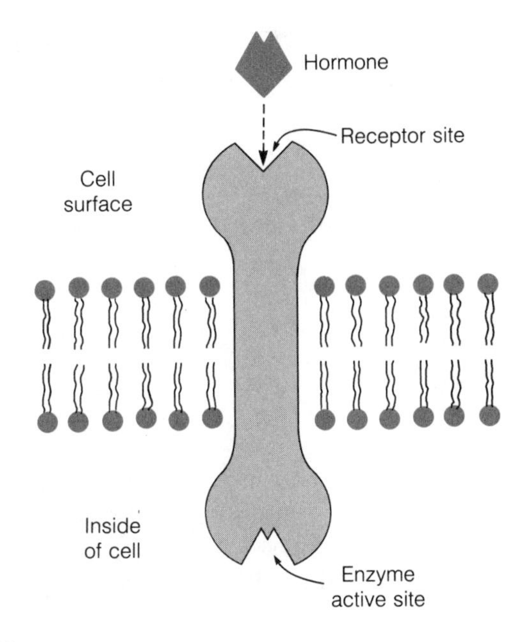

FIGURE 5-12
A transmembrane protein with a receptor site for a hormone molecule at the outer surface of the membrane and an enzymatic active site at the inner surface. Binding of the hormone shifts the structure of the protein so as to activate (or inactivate) the active site, a process called *signal transduction* across the membrane.

different functions for these domains of the proteins. The proteins are oriented in the membrane such that a particular domain always projects from the same surface. That is, one particular domain of a protein always faces the cytoplasmic side, and the other always occupies the outer surface of the cell. This has been shown by using a radioactive reagent that will bind to exposed parts of a protein but that cannot pass through the membrane. Binding to intact cells identifies the outside portion of a transmembrane protein. Furthermore, if cells are first broken open, the plasma membranes can be caused to reseal in an inside-out fashion. The portion of an integral protein that normally faces the cytoplasmic side of the membrane can then be identified with a radioactive agent that will bind to the protein.

The structural asymmetry of proteins with respect to the lipid bilayer reflects the asymmetry of membrane functions. For example, some transmembrane proteins are receptors for hormones or growth factors (see Chapter 12). They are usually glycoproteins with the sugar-containing region positioned at the outer surface of the plasma membrane. A particular configuration of the external domain provides for specific binding of a hormone or growth-factor molecule (Figure 5-12). In some cases binding of a hormone or growth factor activates enzymatic activity in the domain of the transmembrane protein located at the cytoplasmic surface of the membrane. The signal generated by binding of a hormone or growth factor molecule is transferred to the cytoplasmic domain of the protein, presumably through a slight shift in the tertiary configuration of the entire protein.

Another example of functional asymmetry of transmembrane proteins is the transport of various substances from the outside to the inside of the cell, but not the reverse. Other transmembrane proteins transport substances unidirectionally out of a cell.

Integral proteins are usually not held in fixed positions within the plane of the membrane, but, like the lipid molecules of the bilayer, are free to move laterally. This has been clearly shown by tracking the movement

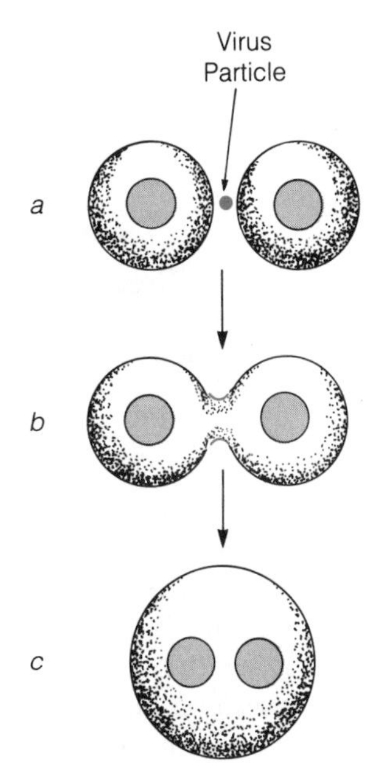

FIGURE 5-13
Drawing of the fusion of two animal cells induced by a Sendai virus. (a) Binding of a virus particle to two adjacent cells. (b) Fusion of the membrane of the virus with the membranes of both cells, creating a cytoplasmic bridge. (c) Binucleated cell resulting from fusion.

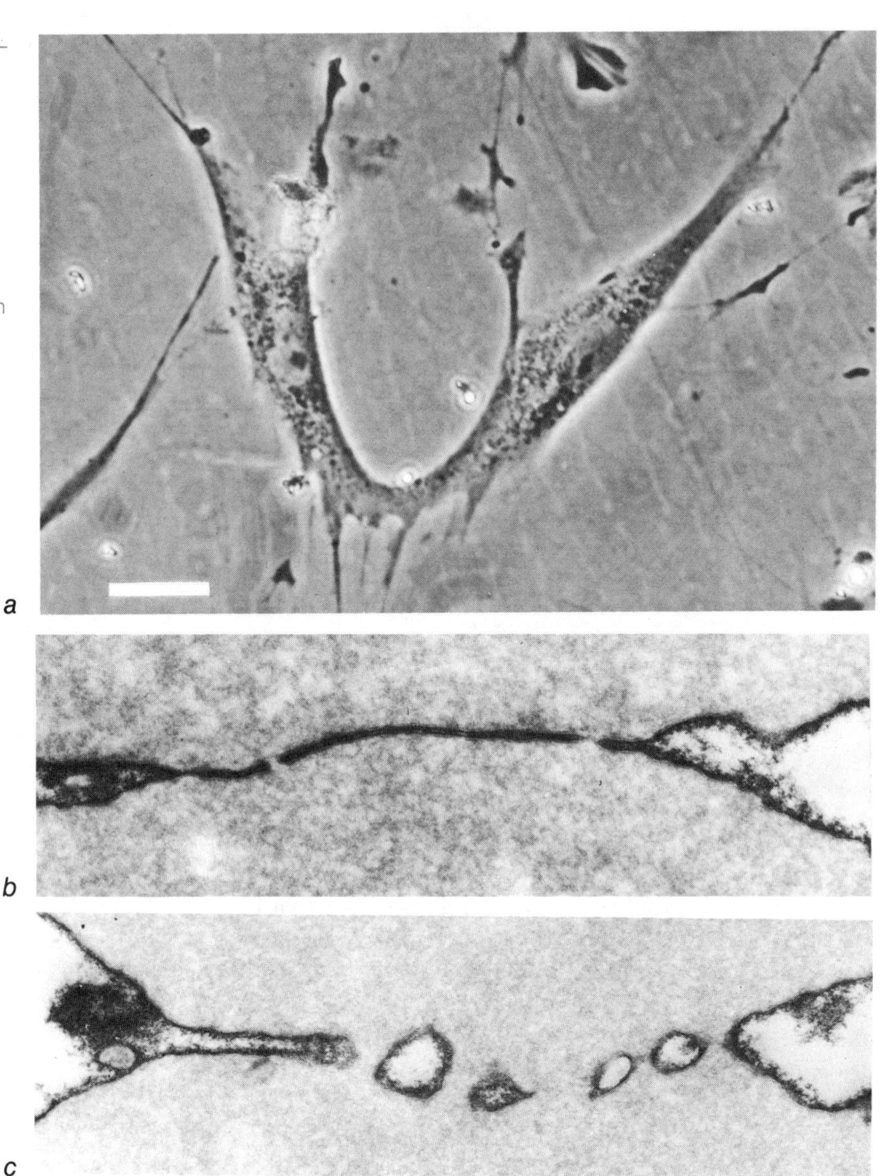

FIGURE 5-14
(a) Light micrograph of two human fibroblast skin cells in culture undergoing fusion as a result of treatment with polyethylene glycol. This micrograph was taken three hours after treatment. Eventually the two cells fuse into one cell body with two nuclei. The white bar is 30 μm long. (b) Electron micrograph of an early stage of fusion of two red blood cells mediated by Sendai virus. Two cytoplasmic bridges have formed. (c) Electron micrograph of a more advanced stage of fusion. [(a) Courtesy of G. Pontecorvo, Peter N. Riddle, Anne Hales and the Imperial Cancer Research Unit. (b and c) Courtesy of S. Knutton and The Company of Biologists Limited.]

of proteins and glycoproteins on the surface of the plasma membrane by immunological procedures. The experiment begins with the fusion of two cultured animal cells. Fusion can be accomplished with Sendai virus. This virus has a membranous envelope that fuses with the plasma membrane of a cell in order to gain entry into the cell. If the virus fuses with two closely adjacent cells simultaneously, the plasma membranes of the two cells fuse with each other (Figure 5-13(a-c)), making a small cytoplasmic bridge that widens to join the two cells into one binucleate cell. Polyethylene glycol is a more convenient fusing agent; it causes the lipid bilayers of two cells that are touching each other to fuse, creating a single cell from two cells (Figure 5-14).

In the original experiment, mouse cells were fused with human cells. Mouse and human cells have different glycoproteins in their plasma membranes. These glycoproteins served as marker antigens in an immunological test. The location of antigenic molecules could be determined by exposing the cells to antibody molecules directed against the antigens and labeled with a fluorescent dye. An antibody is a protein produced by lymphocytes that can bind to one specific antigen, for example, another protein (Chapter 15). Binding of the antibody to specific antigen molecules can be visualized as fluorescent spots by a fluorescence microscope. Antibodies to antigens on the mouse cell membrane were labeled with dye that produced fluorescence of one color, and antibodies to antigens in the human cell membrane were labeled with a dye that produced a different color of fluorescence (Figure 5-15). Treatment of the fused cells with both dyes simultaneously allowed the antigens to be located. Immediately after fusion, fused cells had a green fluorescent half and a red fluorescent half, indicating that the two groups of antigen molecules were separate (Figure 5-15). However, within 40 minutes both red and green fluorescence was seen all over the fused cell, indicating that the two kinds of antigens were completely intermixed. The most reasonable interpretation of this result is that the glycoproteins are free to move about within the plasma membrane.

The free lateral diffusion of membrane proteins has also been demonstrated in another way. The procedure is based on the fact that an antibody has two binding sites and hence can crosslink antigens (proteins in this case) into large aggregates (Figure 5-16(a) and (b)). The aggregation could be observed again by using a fluorescent antibody. Cells were exposed to the antibody. Initially, a weak fluorescence covered the cell because the antibody binds antigens all over the cell surface. However, within minutes the fluorescence became concentrated into intensely fluorescent patches, which indicated a clustering of the antigens. Such clustering can occur only if the membrane proteins are free to diffuse laterally in the plane of the membrane; in this way, they may encounter one another and become crosslinked into large aggregates by the antibodies.

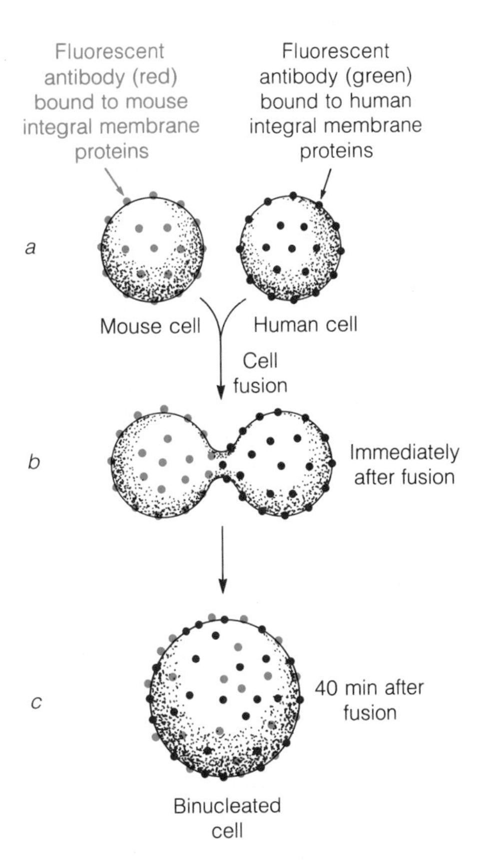

FIGURE 5-15
An experiment showing the mobility of integral membrane proteins. Fluorescently labeled antibody molecules were added to cells at various times in the experiment: (a) before cell fusion, (b) immediately after fusion, and (c) 40 minutes after fusion. The fluorescent antibodies show that the integral proteins mix rapidly.

The membranes of various intracellular structures contain many different integral proteins. In only a few cases do we know the functions of these proteins. For example, the proteins of electron-transport chains are located in the inner membranes of mitochondria and chloroplasts. Also, fatty acid chains synthesized in the cytosol are assembled into lipids by integral proteins on the cytosol side of the endoplasmic reticulum; the

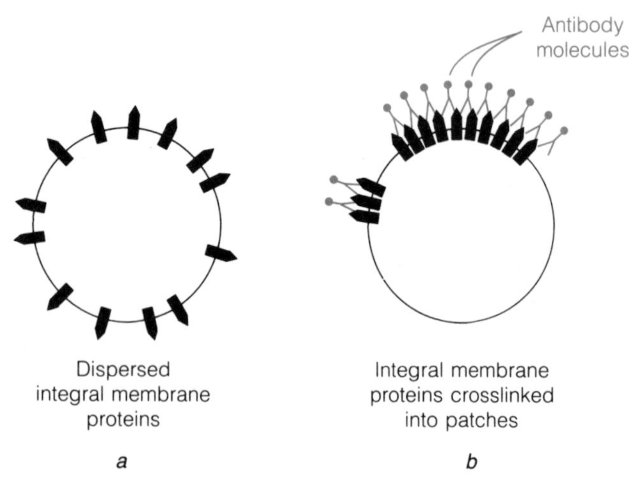

FIGURE 5-16
Diagram of integral proteins (a) dispersed throughout the plasma membrane, which (b) become crosslinked by antibody molecules, forming a protein patch.

newly produced lipids are inserted into the endoplasmic reticulum.

In summary, according to the fluid mosaic model, which has been extensively documented by experimental analysis, the plasma membrane consists of a lipid bilayer in which the lipid molecules are in a fluid state and free to diffuse laterally in the plane of the membrane. Integral proteins remain stably inserted in the bilayer through hydrophobic interactions with lipid molecules. In most cases, the integral proteins are transmembrane proteins that extend across the membrane, with hydrophilic domains projecting from both surfaces. Peripheral proteins are joined to the membrane surface by ionic bonding to heads of lipid molecules and to integral proteins. The peripheral proteins are also asymmetrically arranged. Some kinds are located on the outer surface of the plasma membrane, but most are attached to the inside surface of the membrane.

DIFFUSION OF SUBSTANCES IN AND OUT OF CELLS

Ions or molecules in any liquid or gas are in constant random motion. For example, water molecules are in constant motion; a molecule of water collides with other water molecules 10^{14} to 10^{15} times per second, a motion so rapid that it is difficult to imagine. When a lump of sugar is put into water, the sucrose molecules immediately start to dissolve. The dissolved sucrose molecules move randomly, colliding with water and other sucrose molecules. Eventually, such motion leads to an even distribution of all the sucrose molecules throughout the water. Thus, through diffusion *net* movement of a dissolved substance occurs from a region of high concentration to a region of low concentration. When the ions or molecules of the substance are evenly distributed, no further net movement occurs, although the random collisions and motion continue.

Some kinds of ions and molecules move back and forth across the plasma membrane by diffusion. If the concentration of a substance is higher on one side of the membrane than the other, net movement occurs until the concentration is equalized on the two sides of the membrane. Therefore, diffusion is the simplest means by which substances can move in or out of a cell. However, the plasma membrane is a permeability barrier that severely restricts diffusion, and not many substances important in cellular metabolism, other than water and the dissolved gases oxygen and carbon dioxide, leave and enter cells by diffusion at sufficiently high rates to meet the needs of the cell. Instead, transport mechanisms that ensure sufficient uptake or export of most substances are present in the plasma membrane. These transport systems are discussed in the subsequent section.

Passage of a substance into a cell by diffusion does not require expenditure of energy by the cell but does depend on a higher concentration of the substance in the surrounding fluid than in the cell. It also requires that the substance be able to pass through the plasma membrane. Such passage is easily measured by using molecules containing either radioactive hydrogen (tritium or ^3H) or radioactive carbon (^{14}C). For example, if cells are placed in a medium containing ^{14}C ethyl alcohol and at intervals cell samples are washed and assayed for radioactivity, the time course of alcohol uptake can be accurately determined.

Several properties of a substance affect its penetration by diffusion across the plasma membrane: (1) The rate of diffusion through the membrane depends on the molecular weight of a substance; the greater the molecular weight, the lower the rate of penetration. (2) Lipid solubility is important; the greater the solubility, the more easily the substance moves in or out of the cell. Ethanol and urea are not much different in size, but ethanol is 100 times more soluble in lipid than urea is and passes through the plasma membrane 100 times more rapidly. (3) Ionic charge is important; positively charged substances penetrate more easily than those with a negative charge. (4) Finally, molecular shape influences the rate of penetration; globular molecules diffuse across the membrane more rapidly than molecules with highly asymmetric shapes.

Even for substances that appear to enter the cell rapidly, the plasma membrane represents a formidable barrier. Water moves rapidly in and out of the cell, yet the free movement of water is reduced 100,000-fold by the plasma membrane. In general, two factors make diffusion inadequate as a means of moving substances other than water in and out of cells: (1) The difference in concentration of a substance inside and outside the cell is usually insufficient to drive the net movement of the substance across the membrane; (2) the plasma membrane severely retards diffusion of most ions and molecules, and other mechanisms are required to ensure an adequate rate of movement of substances both into and out of cells.

TRANSPORT ACROSS THE PLASMA MEMBRANE

The import and export of substances by cells is accomplished by transport systems within the plasma membrane that enormously increase the rate of movement of substances in a highly selective way. Each transport system is specific for a particular inorganic ion, organic molecule, or group of organic molecules important to cell metabolism. For example, some transport systems specifically act on a series of structurally similar sugars. Thus, the transport systems not only solve the problem of the permeability barrier at the cell surface, but they selectively transport substances useful in metabolism. This selectivity resides in the specific membrane components of each transport system. Transport is of two general types, passive and active, as discussed in the following two sections.

Facilitated Diffusion or Passive Transport

In **facilitated diffusion** the cell does not directly expend energy in transporting substances. This is also called **passive transport;** like diffusion, net movement of a substance by this mechanism occurs only when its concentration is higher on one side of the plasma membrane than the other.

The several characteristics that distinguish facilitated diffusion from simple diffusion are the following:

1. The rate of movement across the plasma membrane of substances at low concentrations by facilitated diffusion is much greater than it is in simple diffusion. This greater rate of movement is due to the presence in the plasma membrane of carrier systems that bind specific substances and accelerate their movement across the membrane. The carriers are integral proteins of the plasma membrane. We do not understand how these carriers function. A current hypothesis is that carrier proteins creating channels through the lipid bilayer facilitate diffusion. Figure 5-17 illustrates facilitated diffusion. The carrier system in facilitated diffusion is bidirectional; substrates are carried in either direction across the membrane. Net transportation occurs only when there is a concentration difference between the cell interior and the surrounding medium. In this sense facilitated diffusion is like simple diffusion. However, the carrier system facilitates transmembrane movement and greatly accelerates the rate at which the concentrations on the two sides of the membrane are equalized.
2. The second characteristic that distinguishes facilitated diffusion from simple diffusion is **saturation** of the facilitated diffusion mechanism. The rate of net movement or transport by simple diffusion is linearly proportional to the concentration difference. Thus, a plot (Figure 5-18) of rate of movement with increasing concentration difference pro-

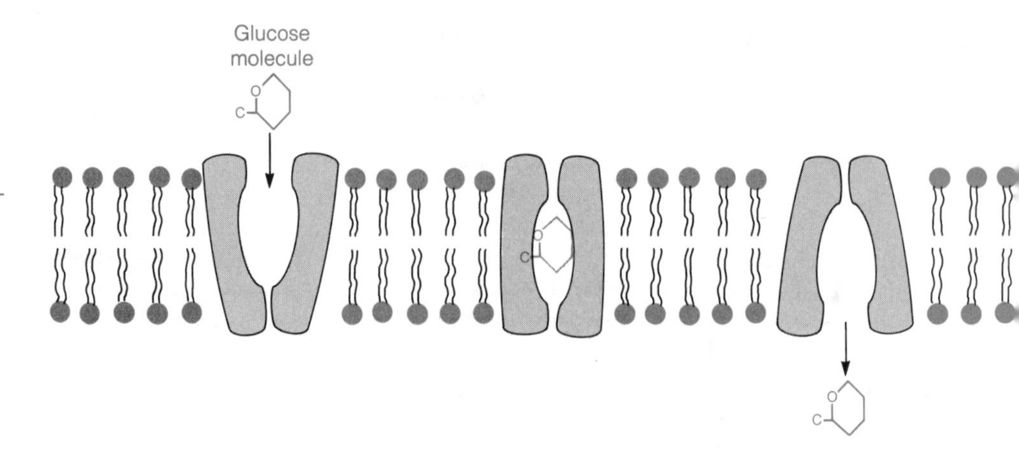

FIGURE 5-17
A general model of facilitated diffusion. A membrane protein specifically binds a substrate at one side of the plasma membrane and acts as a carrier. The bound substrate may then be released at the opposite side of the membrane.

duces a straight line. In contrast, for facilitated diffusion, plotting the rate against the concentration difference produces a hyperbolic curve (Figure 5-18) that approaches a maximum in the same way as an enzyme activity approaches a maximum in a Michaelis-Menten plot of enzyme activity against substrate concentration (Chapter 3). At low concentrations of a substance undergoing transport, most of the carrier molecules at any one instant are not complexed with the substrate. As the substrate concentration is increased, more and more carrier molecules form such a complex. At a sufficiently high concentration, all carrier molecules have a substrate molecule bound to them all of the time; that is, saturation of the carrier system has been reached, and a further increase in rate of transport is impossible, except by simple diffusion, which may be so slow as to be negligible.

3. The carrier in a particular facilitated diffusion system carries only a particular substance or several structurally related substances. The system that transports glucose does not transport an amino acid, but it may carry a number of sugars with a molecular structure similar to glucose, such as galactose and ribose. As would be predicted, chemically similar substances compete with one another for binding sites on the same carrier molecules, and each reduces the rate of transport of the other.

Facilitated diffusion systems can often be inhibited by chemical reagents. The latter bind tightly to certain groups on the side chains of amino acids, such as the sulfhydryl group of cysteine and the amino groups of arginine and lysine, providing evidence that the carriers are proteins. The inhibitors either reduce the affinity of

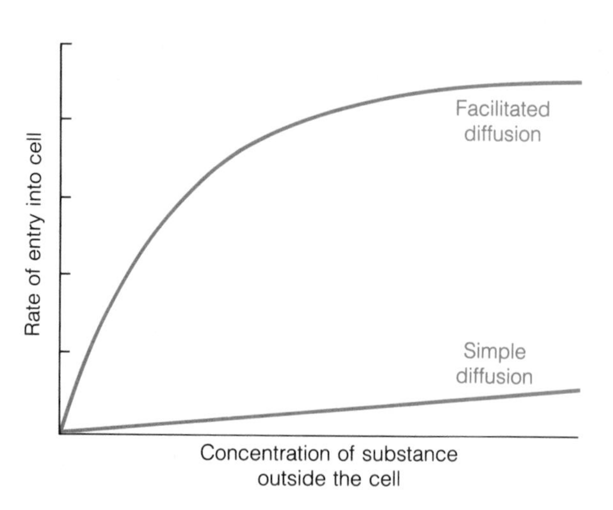

FIGURE 5-18
The rate of transport through the plasma membrane by facilitated diffusion is greater when the concentration difference on the two sides of the membrane is greater. However, unlike simple diffusion, facilitated diffusion has an upper limit that is reached when the carrier mechanism becomes saturated.

he carrier for its specific substrate or prevent the carrier from transporting bound substrate.

In summary, facilitated diffusion is driven by a concentration difference. Facilitated diffusion is accomplished by carrier proteins in the plasma membrane that greatly speed up movement across the relatively impermeable plasma membrane of substances useful in cell metabolism.

Active Transport

Import of a substance by facilitated diffusion is effective when the concentration of the particular substance is higher in the surroundings than in the cell. **Active transport** enables a cell to acquire specific molecules and ions even when they are present at very low external concentrations. For this reason active transport is a major means by which a cell maintains constancy in the concentrations of ions and nutrient molecules in its interior, even though the concentration of such substances in the surrounding medium may change over a wide range.

Active transport and facilitated diffusion are similar in most respects. For example, both processes are based on systems of carriers in the plasma membrane that have high specificity for their substrates. Also, the rate of active transport bears the same relation to concentration of substrate shown in Figure 5-18; that is, at higher concentrations the carrier molecules become saturated, and this limits the maximum rate of transport.

The essential difference between the two mechanisms is that the import (or export) of substances by active transport requires the cell's expenditure of energy to drive the process. The requirement of energy for active transport can be demonstrated by blocking the production of ATP with inhibitors such as fluoride, arsenate, or cyanide; addition of such substances eliminates transport of many materials. With input of energy, usually in the form of ATP, a carrier system can transport a substance that is present in the surrounding medium at a very low concentration into a cell that already contains a higher concentration, enabling much higher intracellular concentrations to be reached. For example, amino acids can still be actively accumulated by some cell types when the amino acid concentration is 300 times higher in the cell than in the surrounding fluid. In addition, unlike facilitated diffusion, active transport results in movement of molecules and ions unidirectionally across the plasma membrane. This is often called **pumping.** Although the direction of pumping is usually from outside into the cell, some pumps work in the opposite direction. The most important of these, as we shall see, is the Na^+ pump, which extrudes Na^+ ions from the cell.

Several different mechanisms of active transport are known, but none is completely understood. All systems appear to have three parts: binding of the substrate to be transported to a protein at the surface of the plasma membrane, transporting of the substrate across the plasma membrane, and an energy coupling by which metabolic energy is used to drive the transporting step unidirectionally against a concentration difference.

Na^+-K^+ transport. One of the most extensively studied transport systems is the **Na^+-K^+ pump** of animal cells, by which Na^+ ions are pumped out of the cell and K^+ ions are simultaneously pumped in, driven by the hydrolysis of ATP. The critical element in this system is a transmembrane protein that enzymatically hydrolyzes ATP (Figure 5-19). The active site of the enzyme is on the cytoplasmic side of the membrane and requires Na^+ ions for activity. The phosphate group split from ATP is not released but instead becomes covalently linked to the enzyme. These observations have been incorporated into the model of Na^+ and K^+ transmembrane transport shown in Figure 5-19. Ionic binding of three Na^+ ions to the protein activates the enzyme to split ATP and to gain a covalently attached phosphate group. The phosphate group alters the shape of the protein in a way that causes the three Na^+ ions to be released to the outside. In this configuration the protein can now bind two K^+ ions. The binding of two K^+ ions brings about dephosphorylation of the protein and restoration to its original form, with concomitant release of two K^+ ions inside the cell. All of these events occur extremely rapidly; a single ATPase protein can transport 300 Na^+ ions out and 200 K^+ ions into a cell

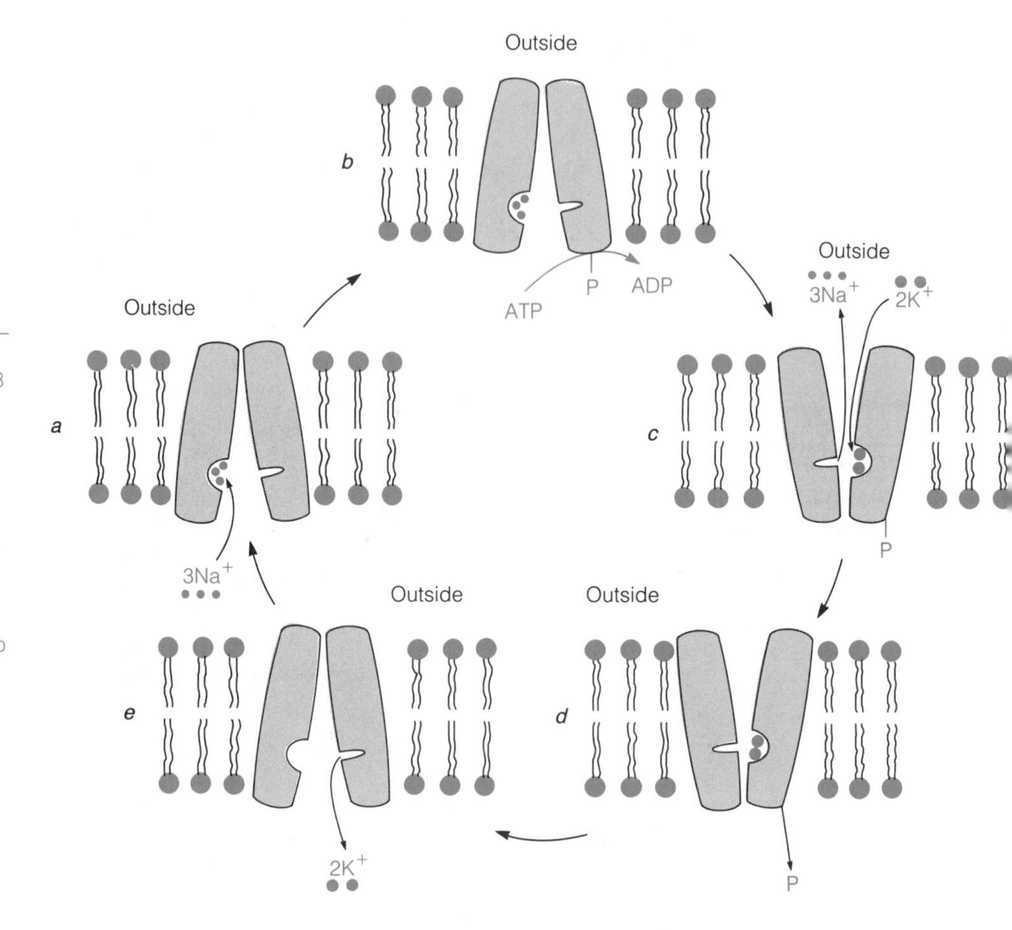

FIGURE 5-19 A schematic drawing of the Na^+-K^+ pump. (a) Binding of 3 Na^+ ions to the pump inside the cell is followed by (b) phosphorylation of one of the protein subunits that make up the pump. (c) Phosphorylation of the protein changes the conformation of the pump allowing the 3 Na^+ ions to be released outside, with concomitant binding of 2 K^+ to the pump outside. (d) Binding of 2 K^+ is followed by dephosphorylation of the pump protein, which (e) changes the conformation of the pump and allows the 2 K^+ ions to be released inside the cell. The pump is then ready to repeat the cycle.

per second. A single mammalian erythrocyte contains more than 100 of these ATPase molecules in its membrane, giving the cell a pumping capability of more than 30,000 Na^+ ions and 20,000 K^+ ions per second.

The Na^+-K^+ ATPase operates continuously, maintaining a lower concentration of Na^+ ions inside the cell than outside and a higher concentration of K^+ ions inside compared to the outside. Animal cell membranes also contain a protein that acts as a K^+ ion channel, allowing K^+ ions to leak out of the cell down the steep concentration difference created by the Na^+-K^+ ATPase. The inside of the cell therefore becomes electrically negative relative to the outside of the cell. As the electrical negativity increases with loss of more and more K^+ ions, it increasingly retards the loss of K^+ ions. Finally, an equilibrium is reached in which the rate of loss of K^+ ions through the K^+ ion-leak channels is equal to the rate of influx of K^+ ions maintained by the Na^+-K^+ ATPase. At equilibrium the combined concentration of Na^+ and K^+ ions is slightly higher outside the cell than inside. However, the concentration of anionic charges (-) is considerably greater inside the cell than outside, creating an electrical potential across the plasma membrane of 20 to 200 mV depending on the kind of cell. This membrane potential is the basis for conduction of impulses by nerve cells, as described in Chapter 15.

Active transport of sugars and amino acids into animal cells. The active transport of sugars and

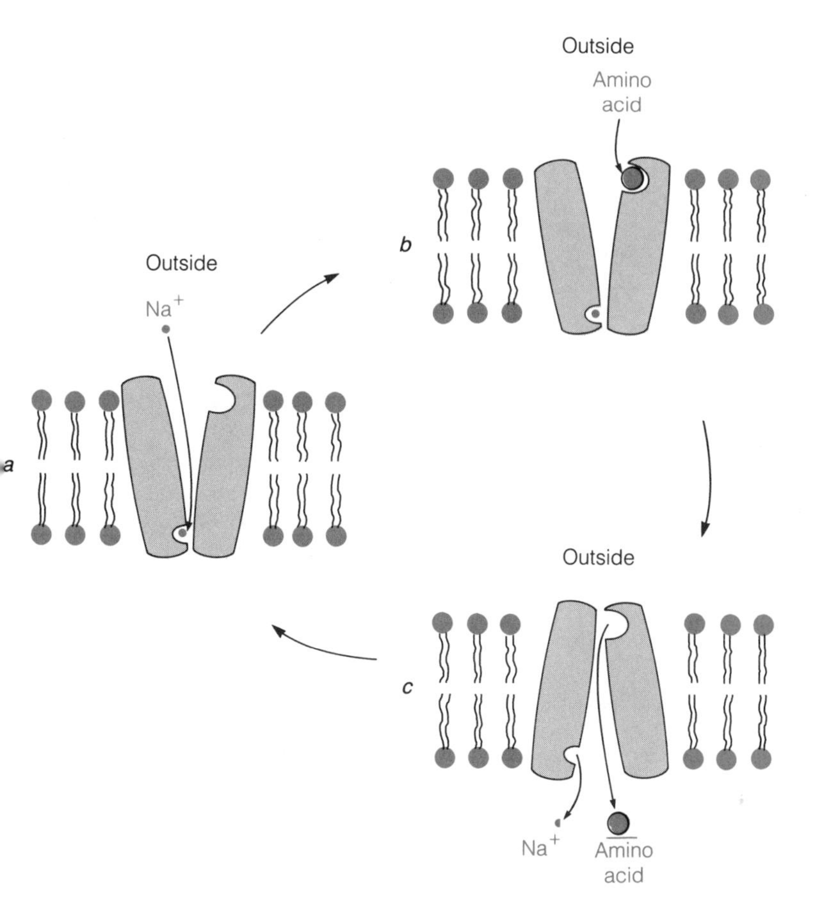

FIGURE 5-20
Active transport of an amino acid by coupling to the movement of Na^+ ions into the cell. Binding of both (a) a Na^+ ion and (b) an amino acid molecule at the cell surface is required to change the conformation of the carrier mechanism, which allows (c) the Na^+ ion and amino acid molecule to be released inside the cell.

amino acids into certain animal cells is driven by the hydrolysis of ATP but indirectly through the Na^+ ion concentration difference created by the Na^+-K^+ ATPase. The plasma membrane contains a carrier protein that binds both an Na^+ ion and an amino acid molecule at the outer cell surface (Figure 5-20(a–c)). Because the concentration of Na^+ ions is much higher outside the cell, the carrier is driven to transport Na^+ ions into the cell. The binding and transport of Na^+ ions can occur only if the amino acid is also bound and transported. Driven by the high concentration of Na^+ ions, an amino acid can be pumped inward even if the concentration of amino acid is higher inside the cell than outside.

Coupling the inward transport of Na^+ ions with sugars and amino acids is called **cotransport.** The plasma membrane of animal cells contains a family of proteins, each specific for a single substrate or a group of structurally similar substrates. The 20 different amino acids are transported by eight or nine different protein carriers that are specific for groups of amino acids.

Active transport of amino acids and sugars into bacterial cells. Many bacteria are able to synthesize all of the amino acids from glucose and inorganic salts. Nevertheless, these bacteria rapidly take in amino acids from the surrounding medium by active transport. In bacteria, H^+ ions are pumped out of the cell and a concentration difference is thus created. The influx of

H^+ ions, driven by the concentration difference, is coupled to the transport of amino acids in the same way that Na^+ ions are coupled to amino acid transport in animal cells; that is, H^+ ions and amino acids are cotransported.

Active transport of most sugars (e.g., lactose) in bacteria also occurs by cotransport with H^+ ions. The transport of other sugars is coupled to their phosphorylation. The best understood example of this mechanism, called **group transport,** is the transport of glucose (Figure 5-21). Four proteins are involved in the transfer of phosphate from phosphoenolpyruvate, an intermediate in glycolysis, to glucose as it crosses the plasma membrane. How the sugar is physically moved across the plasma membrane and how phosphorylation of the sugar contributes to its movement are not known. Once inside the cell the phosphorylated sugar cannot leak out because the strong negative charge of the phosphate group prevents its diffusion through the plasma membrane.

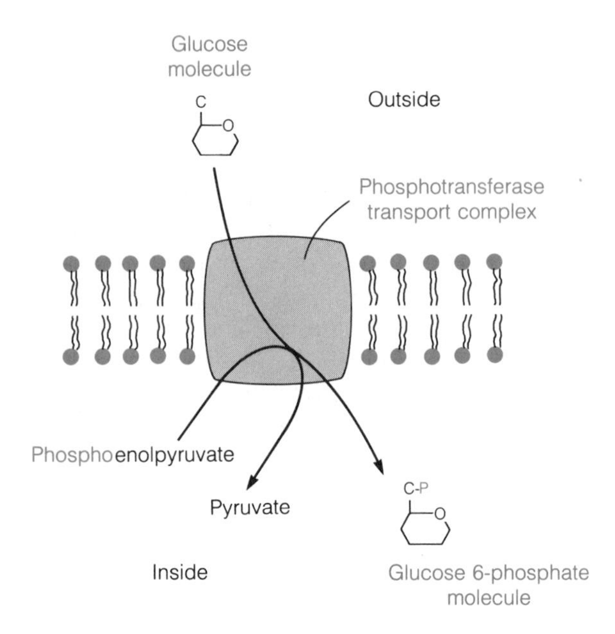

FIGURE 5-21
Group transport of glucose in bacteria. A glucose molecule is bound by a protein in the phosphotransferase complex in the plasma membrane. In a coupled reaction the glucose molecule is phosphorylated to glucose 6-phosphate and released inside the cell.

Other systems of active transport. A variety of other inorganic ions, purines, pyrimidines, nucleosides, vitamins, and other organic molecules are taken into cells by active transport. The plasma membranes of the cells of plant rootlets are particularly well equipped for extracting a wide range of inorganic ions that may occur in extremely low concentrations in the soil. Highly charged molecules, particularly negatively charged molecules, cannot penetrate the plasma membrane by diffusion and must be acquired by active intake. Pumps that ensure an adequate supply of PO_4^{3-} ions are present in all cell types.

Intake of Ca^{2+} ions by intestinal cells is mediated by a Ca^{2+}-transport protein whose presence is dependent on vitamin D. Animals with vitamin D deficiency are impaired in the active transport of Ca^{2+} ions by intestinal epithelium. This impairment leads to the disease ricketts, in which bone structure is weakened by the insufficiency of Ca^{2+} ions. The transport of Ca^{2+} ions across membranes, which is accomplished by a Ca^{2+}-dependent ATPase, is discussed in Chapter 14 in connection with muscle contraction. Cells of the kidney are specialized for transport of several substances, particularly Na^+ ions and water. This specialization enables the kidney to produce urine that is hypertonic relative to blood or tissue fluids. A final example is the active secretion of Cl^- ions by cells in the gills of marine fishes. The active export of Cl^- ions is followed by passive flow of Na^+ ions. By active transport of Cl^- ions out of cells, a marine fish is able to keep its internal NaCl concentration below that of seawater.

IMPORT OF MATERIALS BY ENDOCYTOSIS

Cells can take in materials from their environment by **endocytosis** (*endo* = within; *cytosis* = cellular), a mechanism entirely different from active transport by membrane pumps. Two general forms of endocytosis are pinocytosis and phagocytosis, both introduced in Chapter 1.

Pinocytosis

Pinocytosis occurs in eukaryotic cells that lack gid cell walls, and has been studied mostly in proto- a and animal cells. In the large *Amoeba proteus,* it can induced by a variety of dissolved substances, such as organic cations, amino acids, proteins, and viruses. ugars, polysaccharides, fatty acids, and nucleic acids e ineffective. Within two or three minutes after the ldition of an inducer to the medium, an amoeba rms many short **pseudopods.** Invagination of the ll surface occurs at the tip of the pseudopods (Figure -22). The invagination deepens to form a long tube- ke channel that penetrates deep into the cytoplasm. s a channel forms, it fills with medium that surrounds e cell. These channels, which form in a few seconds, e typically one to a few μm in diameter and 20 to 40 m long (Figure 5-22). When the channel is well rmed, pinocytic vesicles begin to pinch off rapidly one a time from the inner end. An individual channel mains for only 5 to 10 minutes and then completely breaks up into a string of pinocytic vesicles, which disperse into the cytoplasm. Hundreds of channels can form simultaneously, but all disappear within an hour after induction of pinocytosis. Usually no new pinocytic channels are formed for several hours even in the continuous presence of an inducer, possibly because some membrane component needed for the process has been exhausted from the surface by internalization with pinocytic vesicles. Pinocytosis is brought about in amoebae by binding of the inducer to the cell surface by means of positive charges on the inducer molecule and negative charges on the cell surface; the requirement for positive charges is the reason that sugars and anionic substances do not induce the process. We do not know how binding of inducer molecules leads to formation of pinocytic channels and vesicles.

Typically, an amoeba drinks an amount of fluid equal to one-third its own volume in one hour of pinocytosis. The excess water acquired by pinocytosis is excreted by the contractile vacuole. The amount of pinocytic water is in any case small compared to the amount

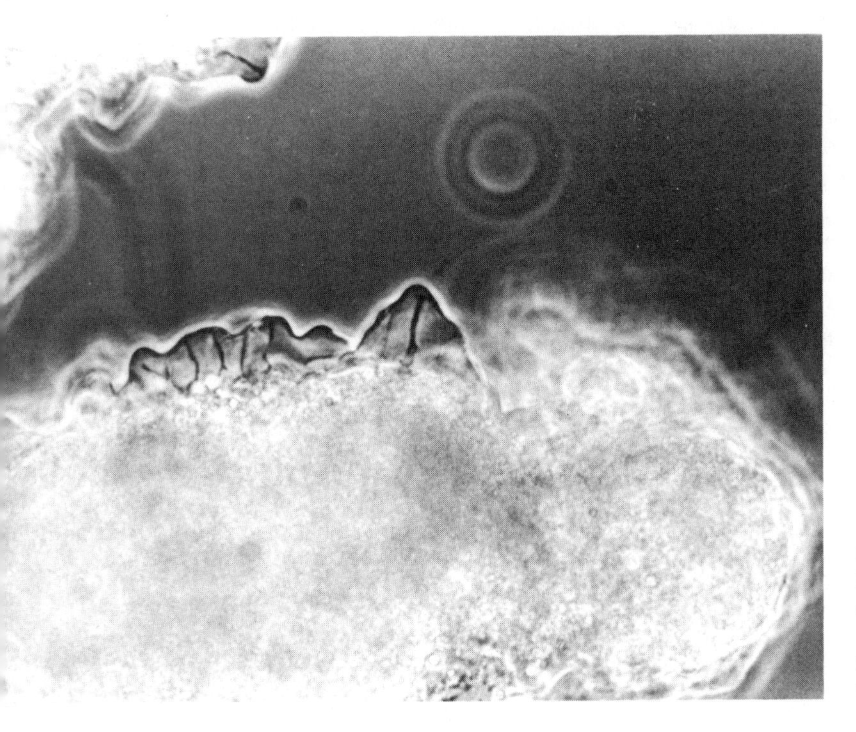

FIGURE 5-22
Light micrograph of pinocytosis in a living amoeba (*Amoeba proteus*). Several clear pseudopods extend from the body of the amoeba. Each of these contains a channel that extends deep into the cytoplasm.

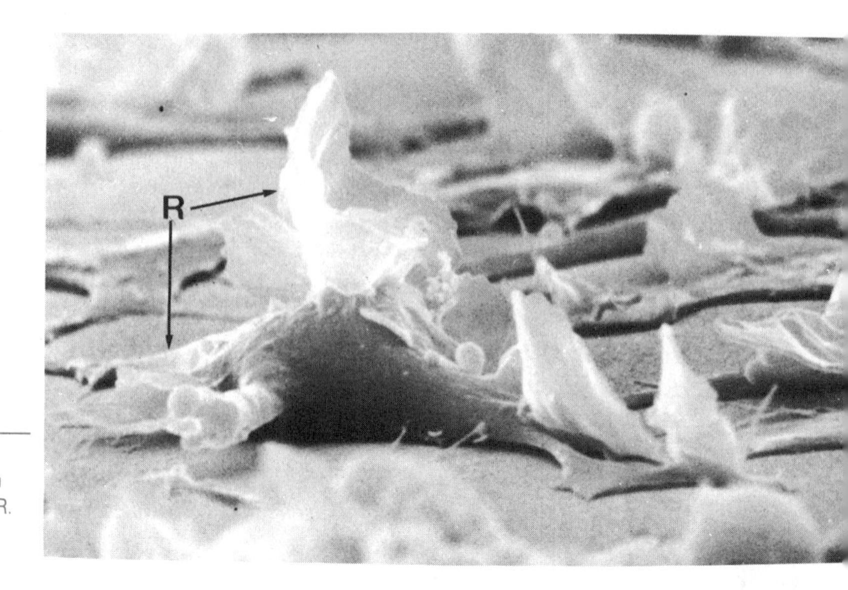

FIGURE 5-23
Scanning electron micrograph of a lymphoid leukemia cell showing extensive ruffles (R), in which pinocytosis occurs. [Courtesy of Keith R. Porter.]

of water continuously acquired by osmosis. The high rate of pinocytosis in amoebae also means that the plasma membrane and surface coat of the amoeba turn over rapidly.

Very shortly after formation, pinocytic vesicles fuse with lysosomal vesicles (Figures 1-16 and 5-25). The composite vesicle is called a **secondary lysosome** to distinguish it from the original or **primary lysosome.** The permeability of the secondary lysosome's membrane increases so that small molecules and ions quickly leak into the cytosol. Macromolecules taken in by pinocytosis are degraded into monomers within the vesicles by enzymes carried in the lysosomes.

In ciliated protozoa such as *Tetrahymena* and *Paramecium* pinocytic vesicles form directly at the cell surface without prior occurrence of channels, but only in a specialized region called the **oral apparatus.** Phagocytosis also takes place in the oral apparatus of ciliated protozoa.

Pinocytosis was first observed in animal cells in culture. It occurs at the leading edge of cells such as fibroblasts as they slowly move across the surface of the culture vessel. In animal cells motility and pinocytosis are too slow to be easily observed by simple microscopy but can be seen by **time-lapse photography.** With this technique, living cells are photographed at intervals, for example, every 15 seconds. When the se quence of photographs is observed at the normal pro jector speed of 16 frames per second, cellular motion appear accelerated, in this case by a factor of 240. With such an acceleration, motility and the formation ar movement of pinocytic vesicles can be detected. Unlik pinocytosis in large amoebae (Figure 5-22) pinocyt vesicles in small species of amoebae form by inwa folding of the plasma membrane (Figure 5-25), as i animal cells, and are so small that they are at the lim of visibility with a light microscope. However, they fu with one another quickly as they form. Before fusir with lysosomes the fused vesicles become readily vis ble just inside the cell surface. The vesicle formed b fusion of a pinocytic vesicle with a lysosome is called secondary lysosome.

In animal cells in culture, pinocytosis occurs at re gions of active motility on the cell surface called **ruffle** (Figures 5-23 and 5-24). Within the folds of these ru fles pinocytic vesicles are formed. Ruffles can form o any portion of the cell surface and often exist simulta neously at several places. In cultured animal cells or can determine the rate of drinking by counting th number of larger, fused vesicles as they form and b measuring their volumes by microscopy. In a few hou a cultured mammalian cell may take in an amount nutrient medium several times greater than its own volume. With the formation of each pinocytic vesicle, a cell also necessarily engulfs a portion of its own plasma membrane. New membrane must be added to the plasma membrane to replace internalized membrane. This replacement is believed to occur, in part, by eventual fusion of secondary lysosomes with the plasma membrane after digestion and release into the cytoplasm of the ingested substance (Figure 5-25).

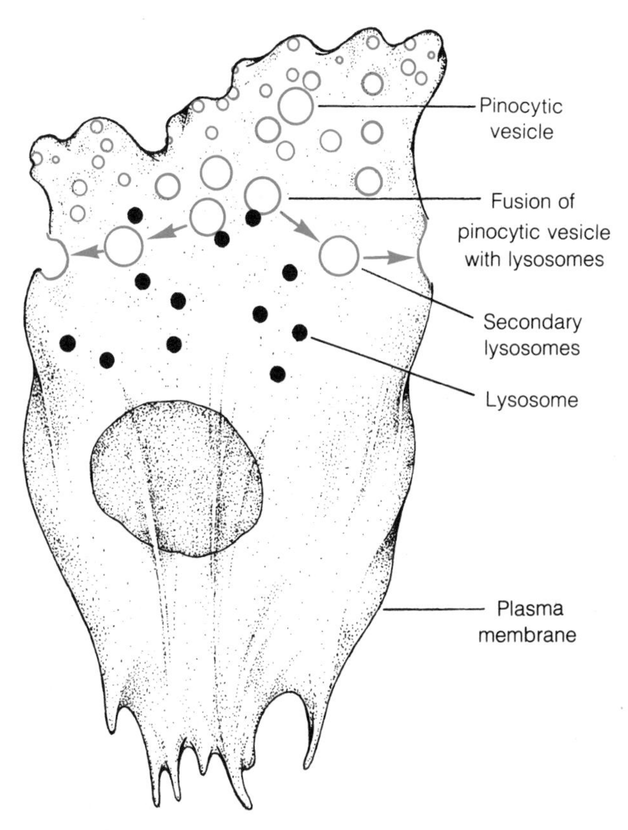

FIGURE 5-25
Membrane recycling in pinocytosis. Pinocytic vesicles fuse with each other to make larger vesicles that then fuse with lysosomes, producing secondary lysosomes. Secondary lysosomes eventually fuse with the cell surface, returning membrane components to the plasma membrane.

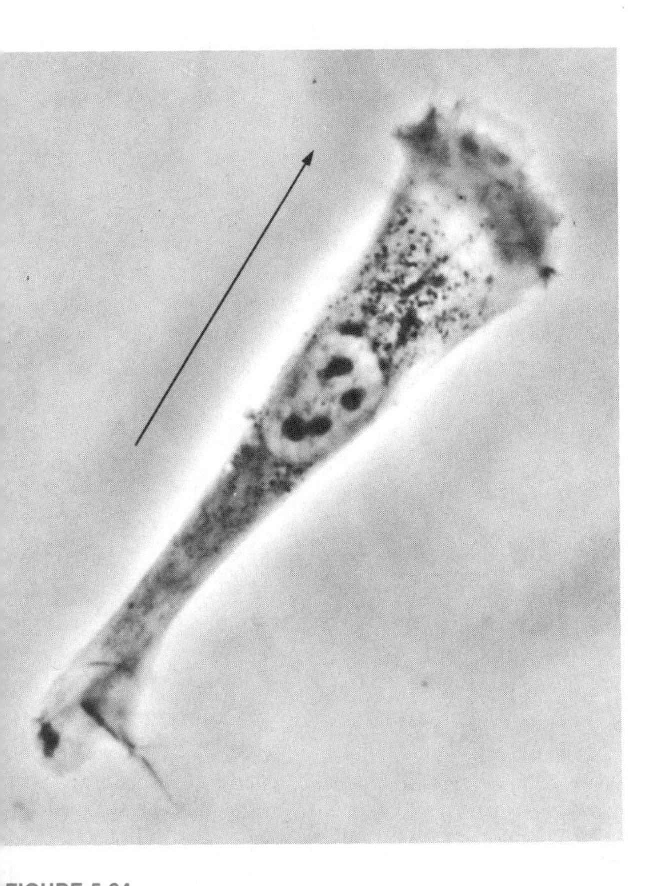

FIGURE 5-24
Light micrograph of a living animal cell in culture. The cell is moving in the direction indicated by the arrow. The leading edge consists of a series of ruffles, in which pinocytosis is occurring.

Receptor-Mediated Endocytosis (Pinocytosis) of Specific Macromolecules

Many of the proteins, especially glycoproteins, projecting from the surface of animal cells are specific cell **surface receptors** that bind particular molecules in the extracellular fluid. One of the best understood examples is the system for intake of cholesterol by animal cells (Figure 5-26). Most cholesterol is carried in the blood in a protein-cholesterol complex called **low-**

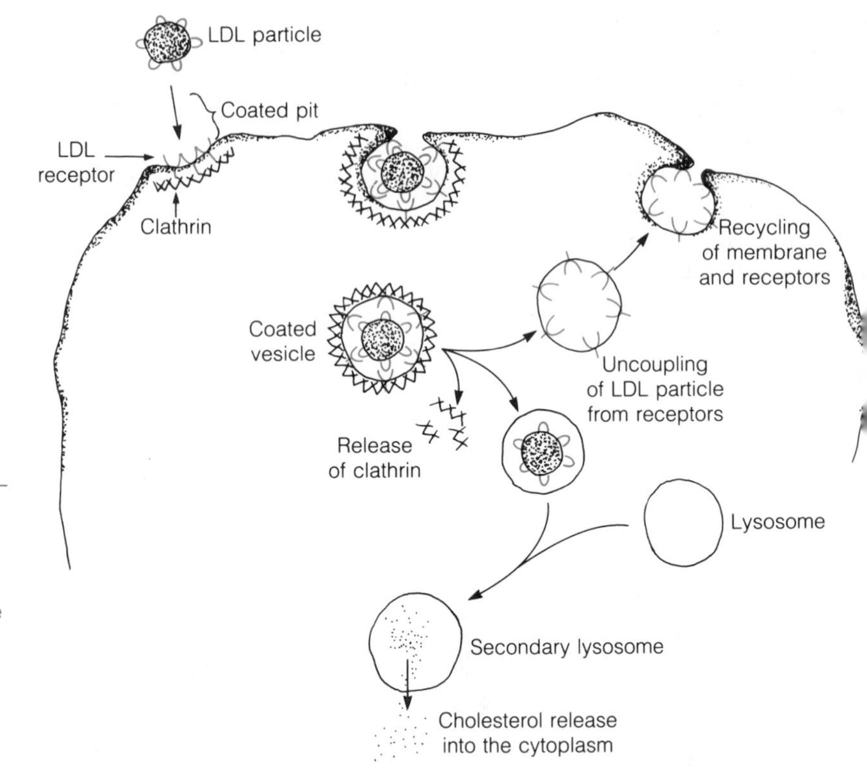

FIGURE 5-26
The uptake of cholesterol by receptor-mediated endocytosis in animal cells. Receptor proteins in coated pits bind LDL particles (containing cholesterol). Binding of particles induces pinocytosis. The receptors and plasma membrane are returned to the cell surface by vesicles. Vesicles containing LDL particles fuse with lysosomes, which brings about disassembly of the particles and release of cholesterol.

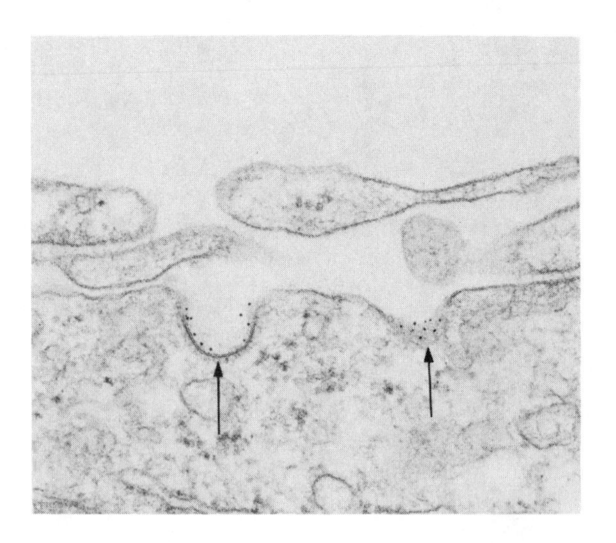

FIGURE 5-27
Electron micrograph of a portion of a liver cell (hepatocyte) with two coated pits in the plasma membrane (arrows). The black dots are gold particles that have been attached to a protein (asialoorosomucoid) in order to identify the location of the protein by electron microscopy. Receptors in the plasma membrane in the coated pits specifically bind the protein as the first step in receptor-mediated endocytosis. [Courtesy of Ann L. Hubbard.]

density lipoprotein. A single complex consists of a core of about 1500 cholesterol molecules encapsulated by a lipid bilayer containing several copies of a single large protein. Receptor proteins in the plasma membrane specifically bind the protein of the low-density lipoprotein complex, which is then brought into the cell by inward budding of the plasma membrane to form a small vesicle—in essence, a form of pinocytosis. In the cytoplasm, these vesicles fuse with lysosomes, which break up the low-density lipoprotein and release the cholesterol for use primarily in the formation of new plasma membrane. The receptor proteins are not destroyed in lysosomes but are quickly returned to the plasma membrane.

Receptor-mediated endocytosis occurs in specialized regions of the plasma membrane called **coated pits** (Figure 5-27), which make up a few percent of the cell surface in cultured animal cells. Coated pits appear in the electron microscope as small depressions with a coating of material lining the pit on the cytoplasmic face of the plasma membrane. The coating material is made up of proteins, one of which, called **clathrin,** forms a regular network. Receptor proteins become localized in the membrane of coated pits, hence the restriction of receptor-mediated endocytosis to coated pits. Coated pits presumably serve to concentrate receptor proteins in small regions of the plasma membrane, which may facilitate the endocytic process. The endocytic vesicles formed from coated pits are also initially coated and are called **coated vesicles** when internalized, but quickly lose their coats before fusing with lysosomes.

Phagocytosis

Phagocytosis, the engulfment of microscopically visible particles by cells, is the mechanism by which many protozoa obtain most of their nutrients (Figure 1-15). *Tetrahymena,* which is about 30,000 times larger than a bacterial cell, can obtain all of its nutrients by phagocytosis of bacteria and can ingest 20,000 bacteria per hour. A larger ciliate, *Euplotes,* can ingest about 20 *Tetrahymena* per hour as its sole source of food (Figure 5-28), although in nature it ingests a variety of bacteria, algae, fungi, and protozoa.

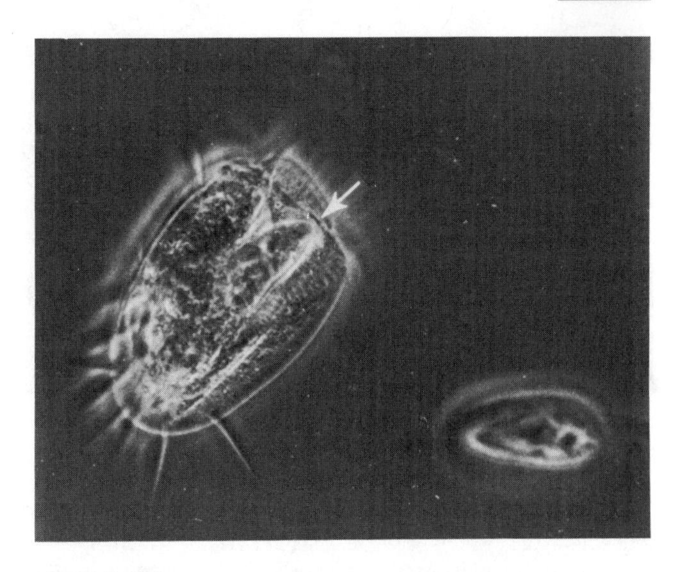

FIGURE 5-28
Light micrograph of a living ciliated protozoan *Euplotes* ingesting another ciliate (*Tetrahymena*) by phagocytosis. The *Tetrahymena* (arrow) has just become drawn into the oral apparatus of the *Euplotes* by action of cilia. In the oral apparatus the prey is enclosed in a vesicle and internalized, where it is killed by lysosomal enzymes.

Capture of other organisms by phagocytosis is a spectacular behavior of species of large amoebae (Figure 1-15). Phagocytosis of *Tetrahymena* by an amoeba provides an example of how the process works. To induce a phagocytic response the *Tetrahymena* must touch the surface of the amoeba. The touching may last less than a second. The amoeba responds with a forward flow of cytoplasm all around the area of contact (Figure 5-29). The effect is the forming of a cup that surrounds the *Tetrahymena.* The rim of the cup closes and the prey is trapped. A vesicle formed in this way usually continues to be connected to the outside through a narrow channel (Figure 1-15(c)). The vesicle next contracts around the food organism, squeezing out the medium taken in with the *Tetrahymena.* Finally, the connecting channel is closed off, and the food vacuole drifts into the interior. The entire process from ini-

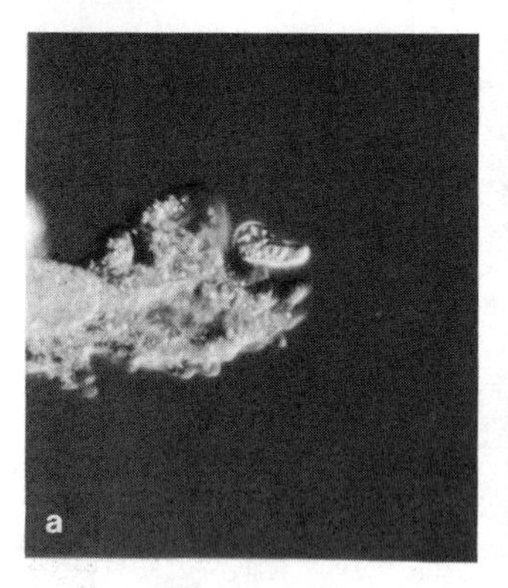

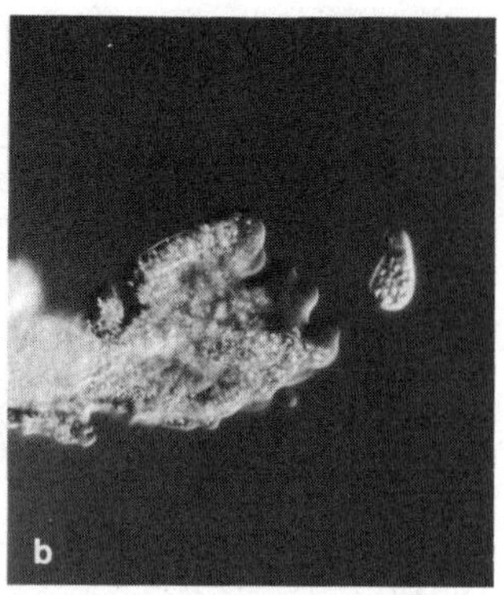

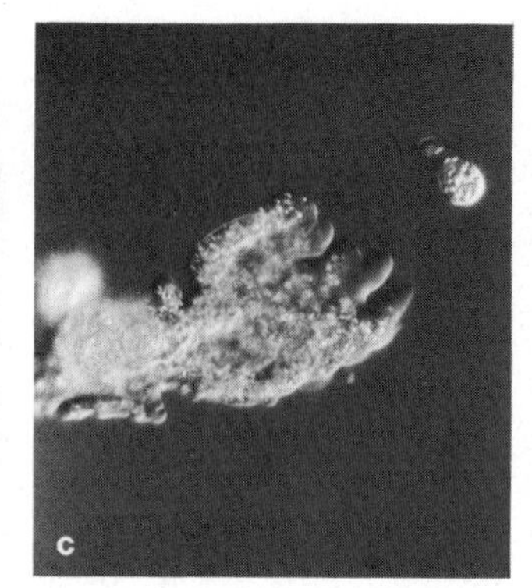

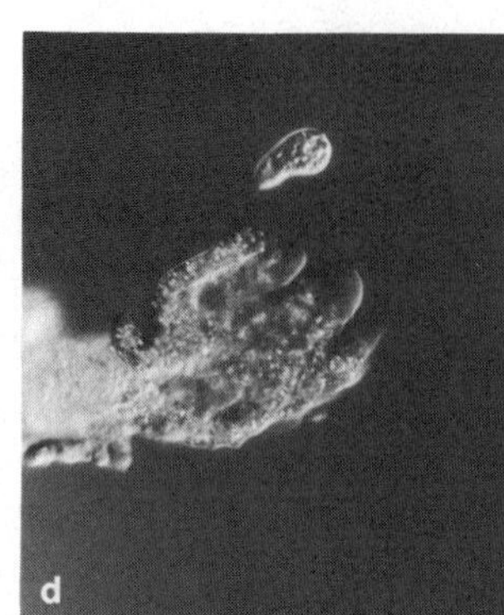

FIGURE 5-29
Response of an amoeba (*Amoeba proteus*) to touch by a ciliated protozoan (*Tetrahymena*). (a) A ciliate has touched the amoeba. (b) Two seconds later the ciliate has moved away but the amoeba responds at the point at which it was touched by extension of pseudopods. (c) Continuation of the response at four seconds. (d) At six seconds the ciliate has moved out of the path of the still extending pseudopods. By this type of response an amoeba is often able to make a renewed contact with a highly motile food organism and successfully ingests it by phagocytosis.

tial contact by the *Tetrahymena* to sealing of the food vacuole may occur in only 30 seconds. Within minutes after formation, the food vacuole fuses with one or more primary lysosomes, and the digestion process by lysosomal enzymes begins (Figure 1-16). Food organisms such as *Tetrahymena* are not totally digested, perhaps because of a lack of degradative enzymes for some molecules. The food vacuole ends its existence when the membrane of the vacuole fuses back into the plasma membrane, releasing undigested remnants to the outside. This export of residual material of food vacuoles is one form of exocytosis (see next section).

Scavenger cells that ingest foreign particles, including bacteria, occur commonly in all multicellular animals. Phagocytosis of bacteria by leukocytes is a part of the defense against bacterial infection among vertebrates (Figure 5-30).

Microscopic particles of asbestos, particles in cigarette smoke, particles in urban air, and so forth, are ingested by phagocytosis by several kinds of cells in the lungs. These indigestible materials may remain lodged in lung cells for months or years, often contributing to the development of lung diseases.

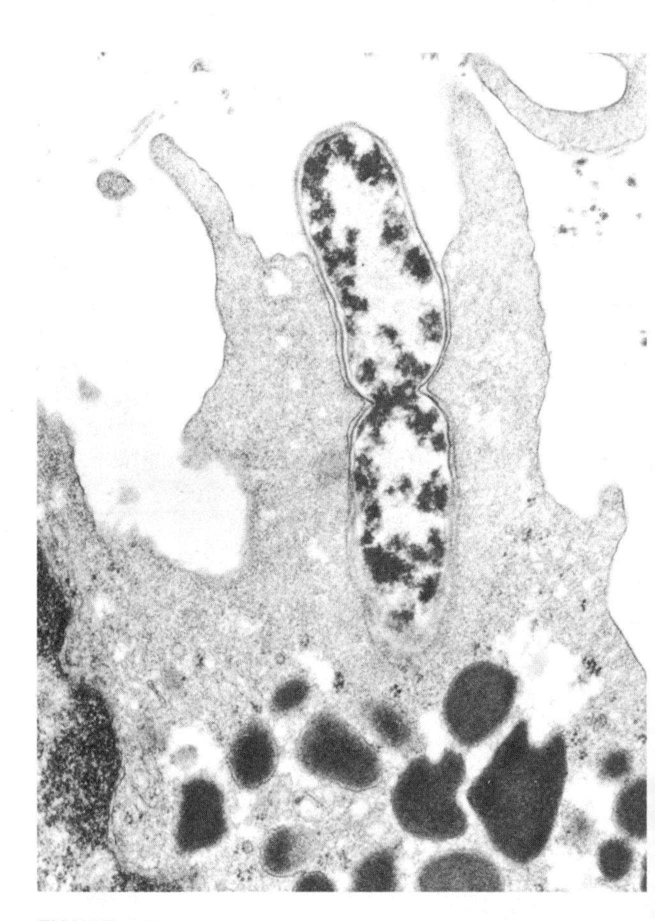

FIGURE 5-30
Electron micrograph of a dividing bacterium being engulfed by phagocytosis by a white blood cell called a neutrophil. [Courtesy of Dorothy Bainton.]

EXOCYTOSIS

Exocytosis is the export from cells of bulk nounts of materials. It includes such varied phenom- ıa as the bulk excretion of water by means of a con- actile vacuole, excretion of remnants from phagocytic acuoles, secretion of hormones by endocrine glands, lease of neurotransmitter molecules by nerve cells hapter 15), secretion of mucus into the gut by epithe- ıl cells of the intestine, secretion of saliva by salivary and cells, export of enzymes synthesized in pancreas lls for use in the intestine (see below), the secretion of aterials that form the extracellular coats of cells (in- uding the cell walls of plants, fungi, and algae), and e release of some kinds of viruses. Secretion of diges- ve enzymes is a particularly well understood example ıd is described in the next section.

Synthesis and Release of Digestive Enzymes by Pancreas Cells

The synthesis and secretion of digestive enzymes by ancreas cells are particularly amenable to study be- ıuse they occur at high rates. Pancreas cells synthesize ızymatically inactive precursors of several important igestive enzymes. The precursor enzymes chymotryp- nogen, trypsinogen, procarboxypeptidases A and B, nd proelastase are called **zymogens.** Zymogens are creted into small ducts that join to form the main **ancreatic duct**, which leads to the first portion of ıe small intestine, the **duodenum,** into whose lumen ıe pancreatic secretions are emptied.

The zymogens are activated in the duodenum by nzymatic removal of amino acids from the polypeptide hain. For example, trypsinogen is converted to the active proteolytic enzyme form, trypsin, by removal of six amino acids (a hexapeptide) by the enzyme entero- kinase, which is synthesized and secreted by epithelial cells of the intestinal lining. Chymotrypsinogen is con- verted to the active proteolytic form, chymotrypsin, by removal by trypsin of amino acids number 14, 15, 147, and 148 (Figure 5-31). This cuts the polypeptide chain into three smaller polypeptides. However, these remain joined to one another by disulfide bonds between cys- teine residues. (Monomers that are part of a macromol- ecule are called **residues.**) A crucial change is the re- moval of amino acids 14 and 15. This removal allows the polypeptide to refold in a way that creates an active site that now catalyzes the breakage of peptide bonds in proteins ingested in food.

The synthesis and secretion of zymogens by pan- creas cells has been followed by a combination of bio- chemical analysis and electron microscopic autoradi- ography. Before the experiment is explained, the tech- nique is described. Autoradiography is a technique for detection of radioactivity in cells or cell parts by photo- graphic film (Figure 5-32). Cells are given radioactive precursors that they may take up and use, usually for the synthesis of one or another macromolecule. For example, radioactive thymidine (thymidine containing ^3H or ^{14}C) is taken up by cells and incorporated into DNA. Similarly, the proteins of cells can be made radio- active when a radioactive amino acid is supplied. After the cells have been given sufficient time to incorporate the radioactive precursor into macromolecules, they are killed with a chemical that preserves their structure, for example, ethyl alcohol or formaldehyde. This pres- ervation of cell structure is called **fixation.** The fixed cells are washed to remove unincorporated radioactive precursor and are prepared for observation by micros-

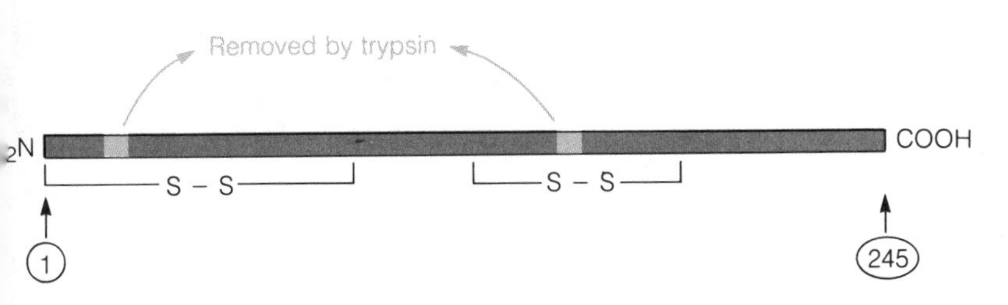

FIGURE 5-31
Chymotrypsinogen is converted to chymotrypsin by excision of amino acids 14 and 15 (left red block) and 147 and 148 (right red block). Excision is catalyzed by trypsin. The three polypeptide segments created by excision are held together by disulfide bonds.

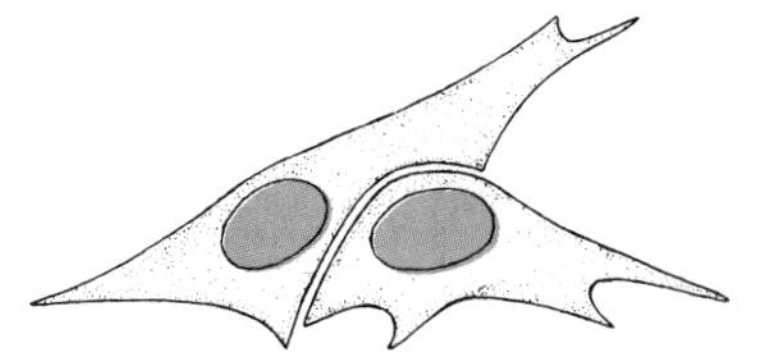

Living cells are incubated for 10 min. in medium containing ^3H thymidine.

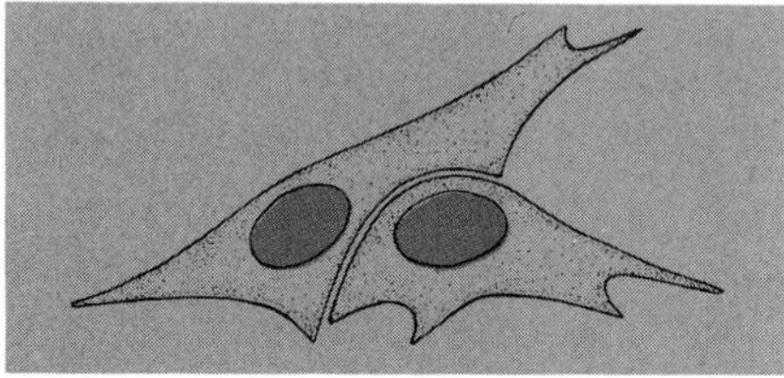

Cells are fixed and washed to remove ^3H thymidine not incorporated into DNA. The cells are then coated with a thin layer of photographic film and stored in the dark for several days to allow exposure of the film by radioactivity.

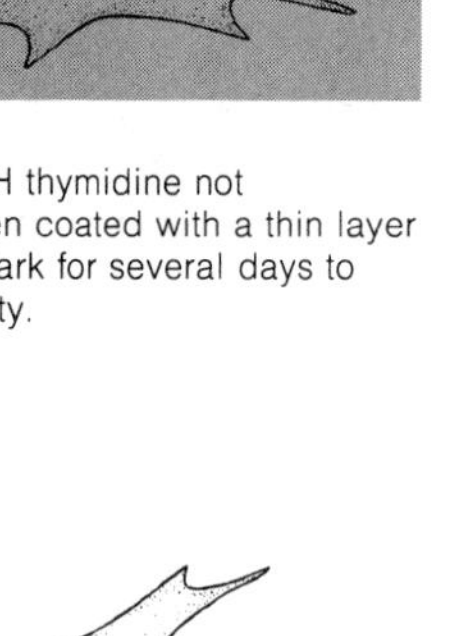

Photographic development produces silver grains in the film over the region of any cell that has incorporated radioactivity into macromolecules, in this case ^3H thymidine incorporated into DNA of one of the two cells.

FIGURE 5-32
Autoradiography of cells to show the location of radioactivity incorporated into macromolecules.

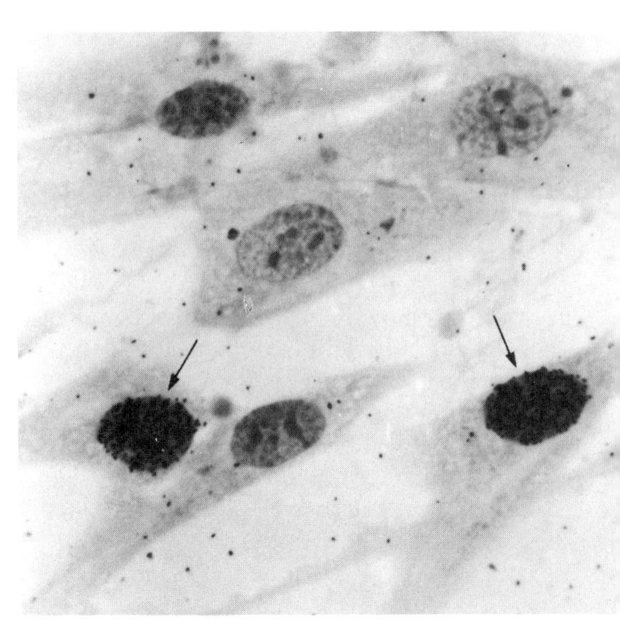

a

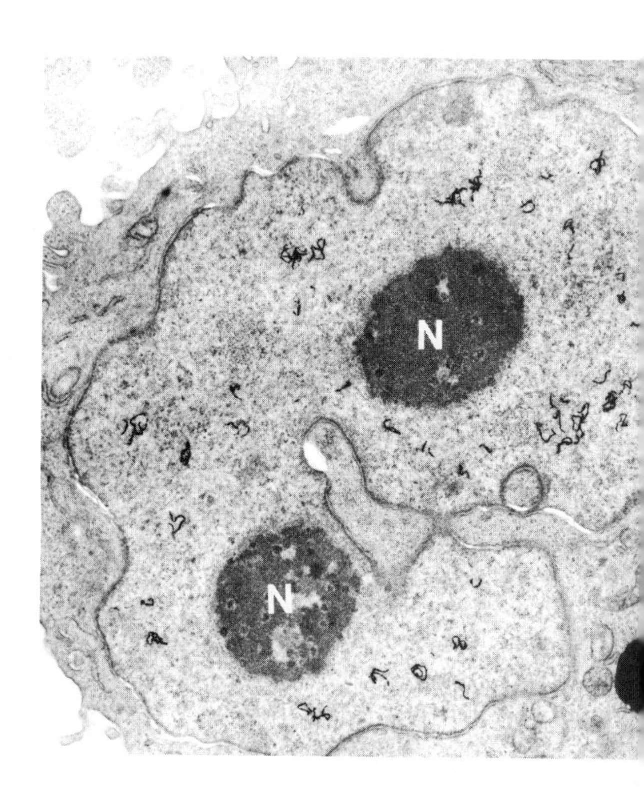

copy. Before examination by microscopy the cells are covered with a thin photographic film and stored in the dark for days or weeks. Radioactivity (usually β particles) present within the cell exposes the overlying photographic film. When the film is subsequently devel-

FIGURE 5-33

Autoradiographic images of cells whose DNA has been labeled with ^{3}H-thymidine. (a) Autoradiography with the light microscope. Of these six cultured cells, two (arrows) have incorporated ^{3}H-thymidine into their DNA as indicated by the silver grains in the overlying photographic film. (b) Autoradiography with the electron microscope. The DNA of the irregularly shaped nucleus [with two nucleoli (N)] has been labeled with ^{3}H-thymidine. The silver grains appear as short, black curled threads in the photographic film overlying the thin section of the cell. There are fewer silver grains than in part (a). Note that in both (a) and (b) the silver grains appear only over nuclei. [(b) Courtesy of Gary E. Wise. From *Exptl. Cell Res.*, 81:63–72 (1973).]

oped, it contains an image showing where radioactivity is present in the underlying cell (Figure 5-33(a)). The image consists of a pattern of silver grains exposed by the radioactive decay. Autoradiography can also be done on thin sections of cells observed by electron microscopy (Figure 5-33(b)). The location of radioactivity in a cell can be more closely defined by electron microscopy autoradiography because of the higher resolving power of electron microscopy.

In an experiment designed to follow synthesis and secretion of zymogen, a radioactive amino acid was injected into the pancreas of a guinea pig to label proteins undergoing synthesis. Most of the proteins synthesized in pancreas cells are zymogens, so the radioactive amino acid labels primarily zymogen proteins. Three minutes after injection of the amino acid almost all of the incorporated radioactivity was found by electron microscope autoradiography to be present in the rough ER (Figure 5-34), which indicates that zymogen proteins are synthesized on the rough ER.

The three-minute labeling period was followed by injection of a large amount of nonradioactive amino acid, effectively diluting out radioactive amino acid that had not yet been used for protein synthesis. The injection of the nonradioactive amino acid is called a **chase,** and its purpose is to reduce to a negligible level further incorporation of radioactivity into proteins undergoing synthesis. By using a chase it was possible to follow the movement of the radioactive proteins made in the three-minute period. Seventeen minutes after the chase was started, most of the radioactive protein was found in rough ER near the Golgi complex. By 117 minutes the radioactive protein was localized in zymogen storage vesicles called **zymogen granules** (Figure 5-34). The identity of zymogen granules was established by biochemical analysis of the contents of granules isolated in bulk from pancreas cells. The zymogen-labeled granules gradually migrated to the surface of the cell that faces a duct (*apex* of the cell). The contents of zymogen granules are then released from the cell into the duct by exocytosis. Exocytosis occurs by fusion of the membrane surrounding the zymogen granule with the plasma membrane, thus releasing the granule to the lumen of the duct.

These experiments led to the scheme shown in Figure 5-35. Zymogens are synthesized in the rough ER, and from there the finished proteins move to the Golgi complex, where they are concentrated and packaged in zymogen granules. The zymogen granules then migrate to the cell surface and release their contents into a collecting duct. Release of zymogen granules at the cell surface adds substantial membrane to the plasma membrane. Addition of membrane is believed to be balanced by withdrawal of membrane from the apical surface of the cell by formation of small vesicles, perhaps at coated pits. These vesicles, in turn, may fuse with membranes of the Golgi complex to complete the recycling of membrane material.

Formation of the Surface of Intestinal Epithelium

Epithelial cells that line the intestinal tract are specialized for absorption of food molecules from the intestinal lumen. Part of the specialization consists of a large number of tiny cytoplasmic extensions called **microvilli** that project into the lumen of the intestine (Figure 5-36). The microvilli are so numerous and so regular in their arrangement that they give the epithelial surface a distinctive appearance that has given rise to the descriptive term **brush border.** Because of the microvilli these epithelial cells have an enormous surface area available for absorption of materials from the intestinal lumen. In addition, the lumenal surface of

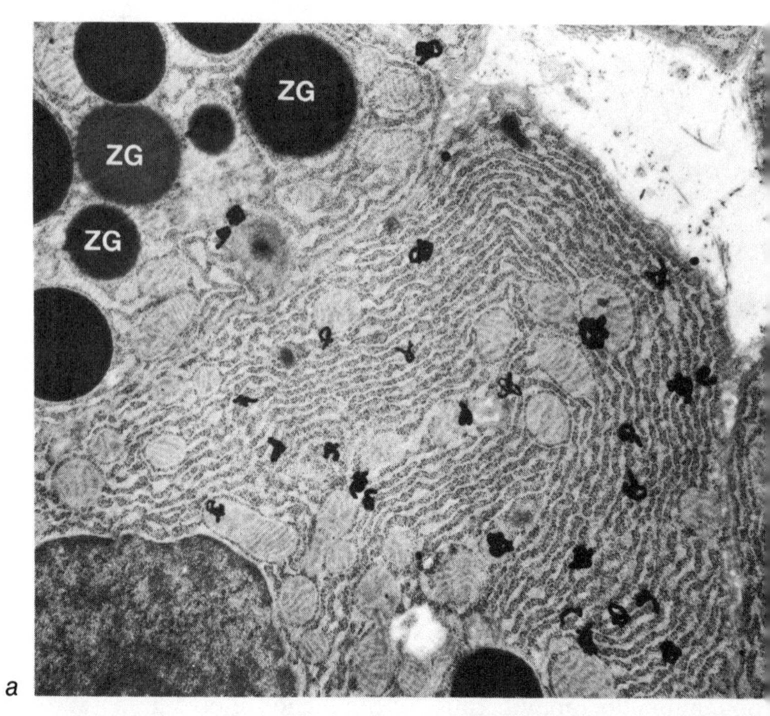

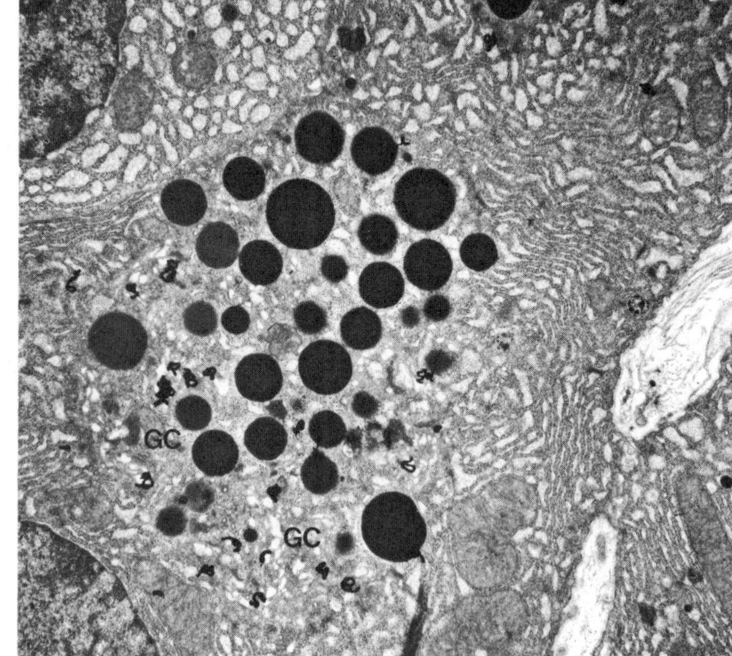

FIGURE 5-34

An experiment using electron microscope autoradiography to follow synthesis and secretion of zymogen by pancreas cells. (a) An autoradiograph of a pancreas cell three minutes after the start of labeling of proteins with ^{3}H-leucine. The silver grains are located almost exclusively over the rough ER, and none are over zymogen granules (ZG). (b) Seven minutes after the end of labeling of proteins with ^{3}H-leucine most of the radioactive proteins are associated with the Golgi

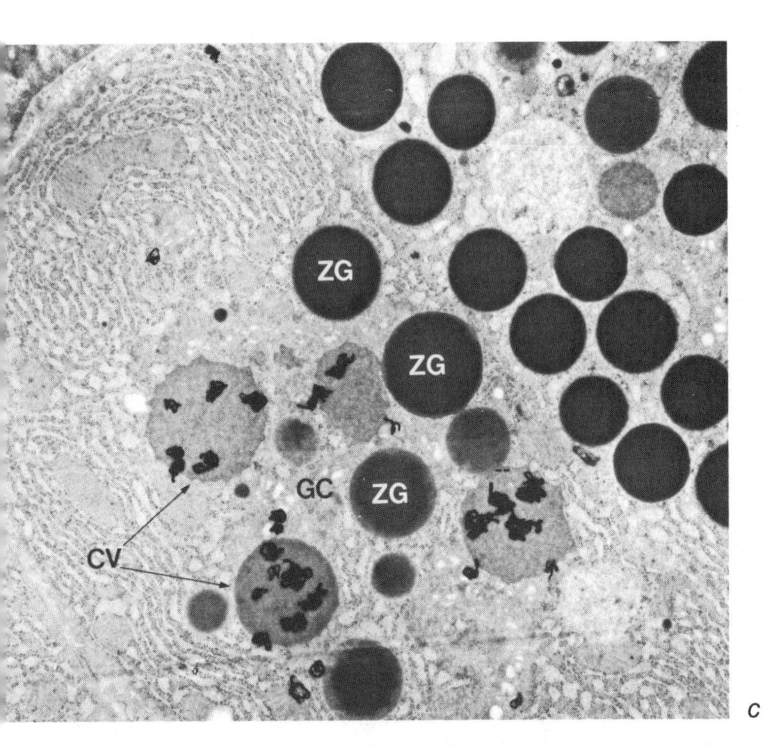

c

d

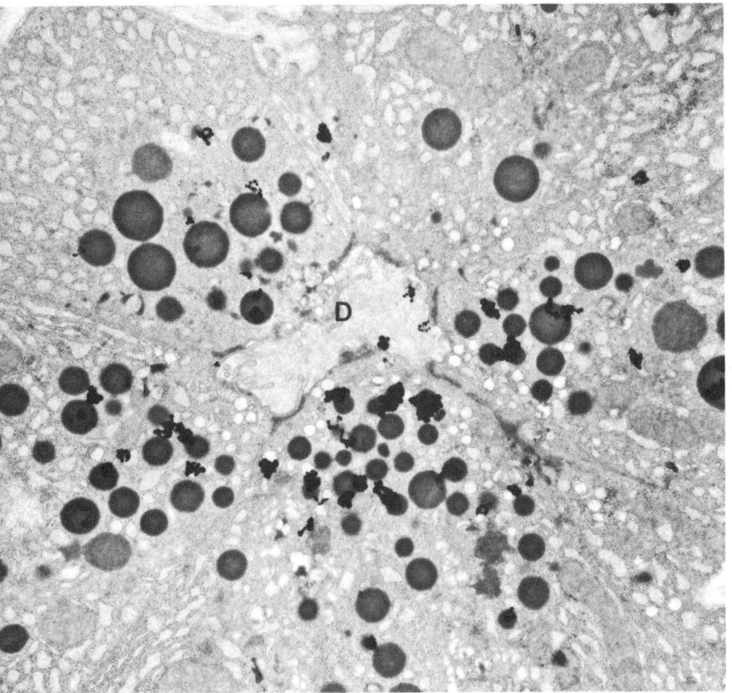

complex (GC), with little remaining in the rough ER. (c) Portion of a pancreas cell 37 minutes after the end of labeling of proteins. The radioactive proteins are concentrated in condensing vacuoles (CV) of the Golgi complex. The zymogen granules are unlabeled. (d) Several pancreas cells nearly two hours after the labeling of proteins with ^{3}H-leucine. The radioactive proteins are now located primarily in zymogen granules adjacent to the lumen of a duct (D). [Courtesy of James D. Jamieson. Reproduced from *The Journal of Cell Biology*, 34:597–615 (1967) by copyright permission of the Rockefeller University Press.]

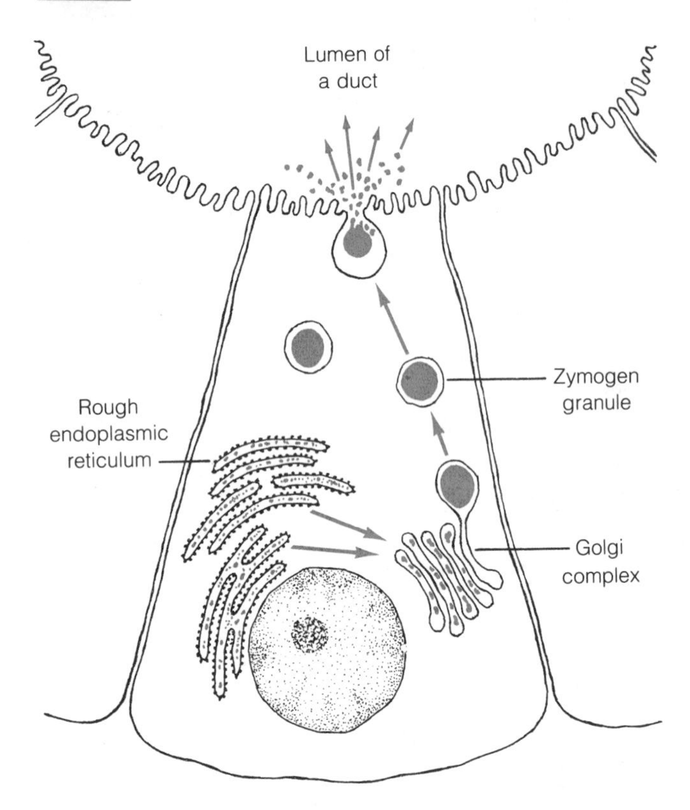

FIGURE 5-35
Schematic diagram of exocytosis of zymogens by pancreas cells. Zymogens are synthesized on ribosomes of rough endoplasmic reticulum (ER) and transported to the Golgi complex. In the Golgi complex, zymogens are concentrated and packaged into zymogen granules. The membrane surrounding zymogen granules fuses with the plasma membrane, releasing the zymogens into a pancreatic duct leading to the small intestine.

these cells possesses a surface coat that is sufficiently prominent to be seen easily by electron microscopy (Figure 5-37). This surface coat, which appears as a network of fine fibrils, presumably has a role in absorption of materials, but its precise function has not yet been identified.

The prominent surface coat of intestinal epithelium provides a good opportunity to study the intracellular events leading to formation of surface coats in general.

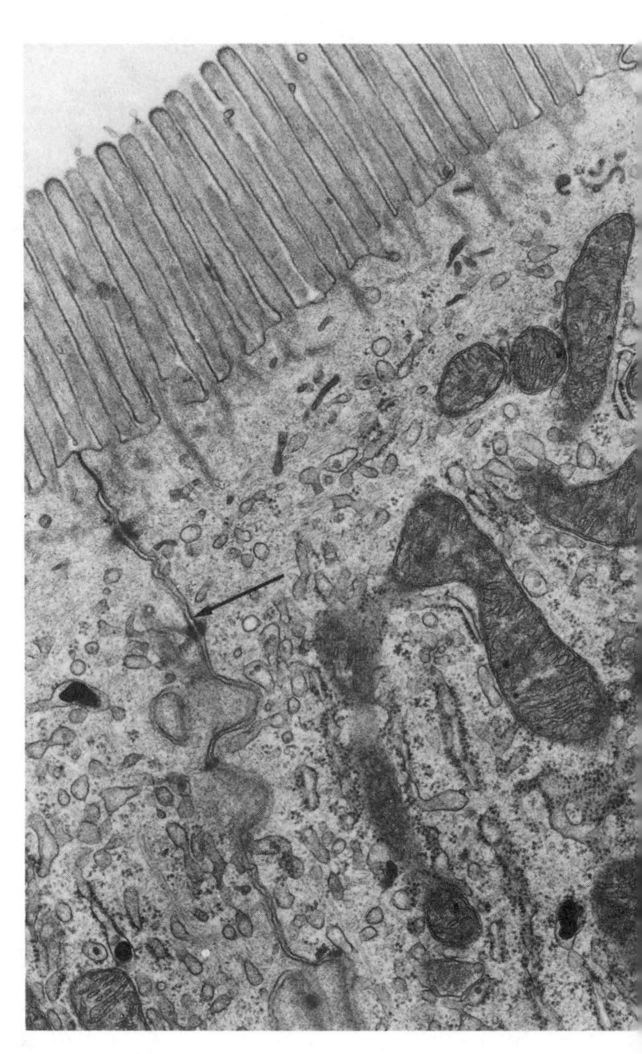

FIGURE 5-36
Electron micrograph of a portion of two epithelial cells in the rat intestine showing some of the many microvilli at the lumenal surface of the cell. Each microvillus contains a bundle of microfilaments. The arrow indicates the border between the two cells. [Courtesy of Keith R. Porter.]

The main molecular components of the surface coat are glycoproteins. The protein portions of the glycoproteins are synthesized on the ribosomes of the rough ER, passed into the cisternae of the ER, and collected in the Golgi complex (Figure 5-38) in a manner similar to synthesis and collection of zymogens in pancreas cells.

One of the sugars of the surface glycoproteins is **L-fucose,** a 5-carbon sugar. Fucose is used by the cell almost exclusively in the synthesis of glycoproteins of the surface coat and perhaps to some extent in the synthesis of glycoproteins of mitochondrial membranes. Therefore, radioactive fucose and electron microscope autoradiography have been used to follow the production of glycoproteins and their transport to the cell surface. Ten minutes after injection of ^{3}H-fucose into a mouse most of the silver grains produced by ^{3}H-fucose in autoradiographs of intestinal epithelial cells are associated with Golgi complexes (Table 5-1). Since in the preparation of the tissue any ^{3}H-fucose that has not become part of glycoproteins has been removed, the radioactivity detected in the Golgi complexes indicates the presence of newly formed glycoprotein in that organelle. Since most of the ^{3}H-fucose appears first in the Golgi complexes, it is reasonable to conclude that the Golgi complex is the site of addition of this sugar to proteins to form glycoproteins.

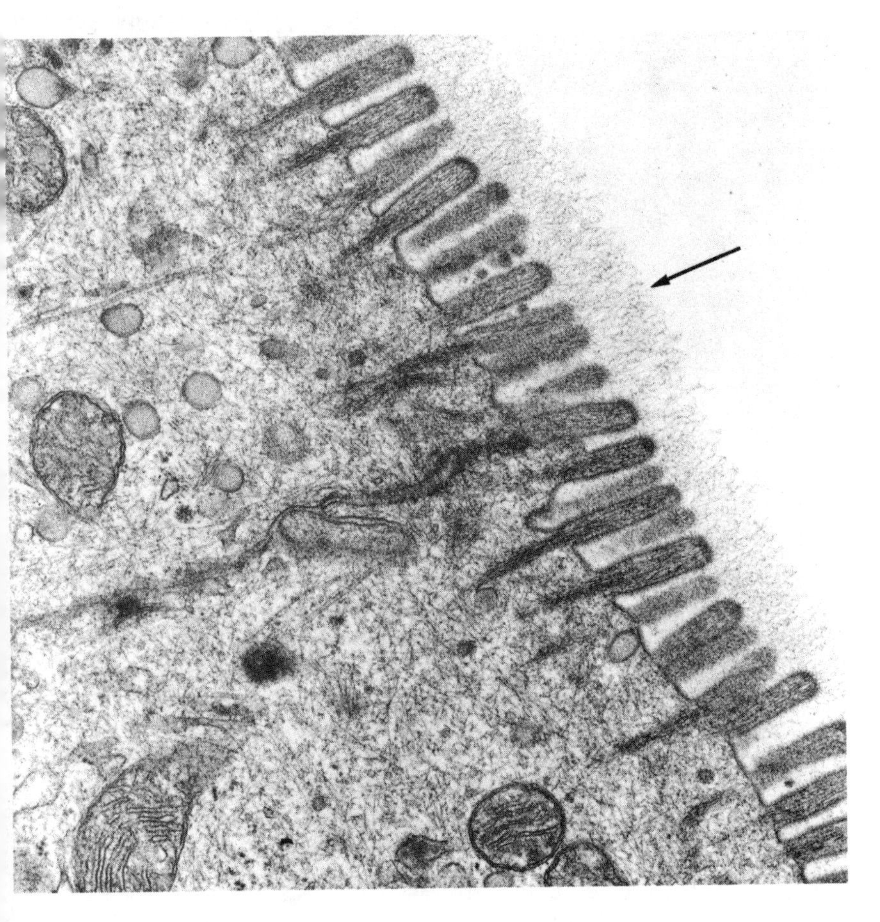

FIGURE 5-37
Electron micrograph of microvilli on an intestinal epithelial cell. In this cell the extracellular coat of fine filaments (arrow) has been retained. [Courtesy of Charles P. Leblond.]

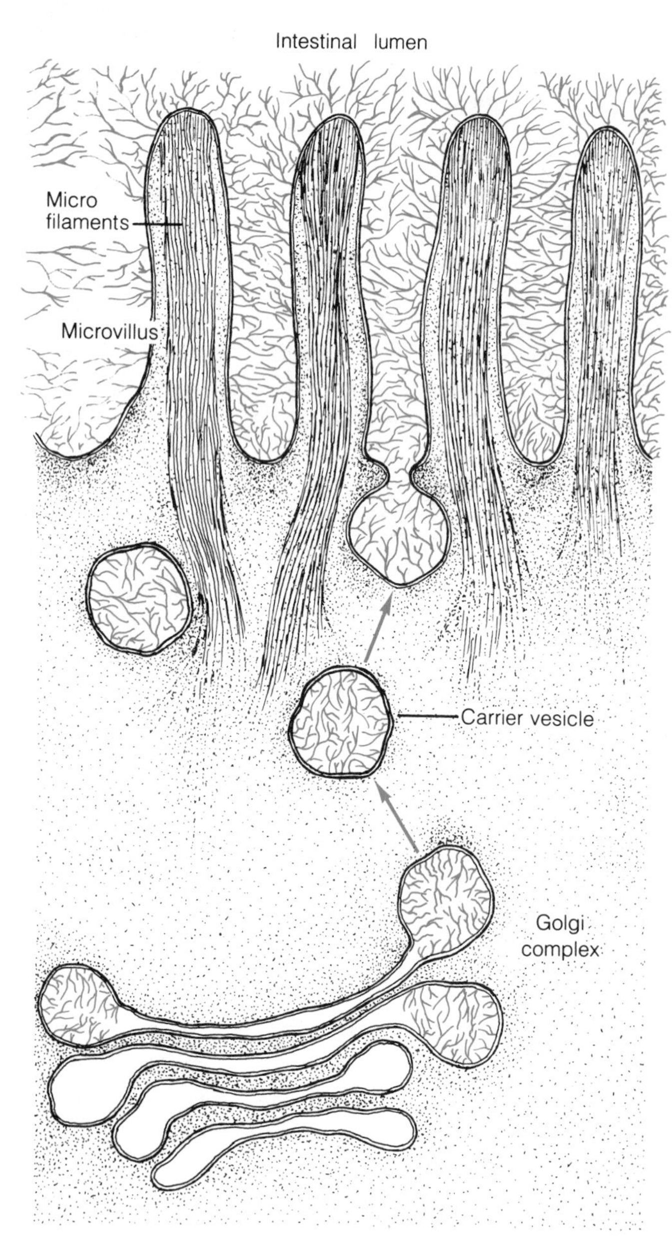

FIGURE 5-38
Glycoproteins destined to become integral proteins of the plasma membrane are packaged into vesicles in the Golgi complex. The vesicles migrate to the surface of the cell, in this case the brush border of an intestinal epithelial cell, and fuse with the plasma membrane.

Ten minutes after injection of ^{3}H-fucose little radioactivity is found associated with the surface coat. Twenty minutes after injection, by which time ^{3}H-fucose has largely disappeared from the blood, the percent of total radioactivity has declined in Golgi complexes and increased in small vesicles that bud from the Golgi complex (Table 5-1).

With longer times after injection of ^{3}H-fucose, radioactivity declines to a low amount in the Golgi complexes and small vesicles, and most of the radioactivity is in the surface coat. The early appearance of ^{3}H-fucose in mitochondria (Table 5-1) and its subsequent decline may reflect rapid turnover of glycoproteins in mitochondrial membranes.

The changing distribution with time of incorporated ^{3}H-fucose indicates that sugars are added to proteins that have been collected in Golgi complexes from the rough ER. The newly formed glycoproteins are then packaged in small vesicles that migrate to the lumenal surface of the epithelial cell and release their contents into the surface by fusion of the membrane around the vesicle with the plasma membrane. This route of exocytosis conforms with electron microscopic observations of the vesicles. They often appear near the cell surface and contain a fuzzy, fibrous material that closely resembles the material of the surface coat.

The Golgi complex of membranes is important not only in the collection, packaging, and export of protein but also for distribution of proteins within the cell. The functions of the Golgi complex are discussed in more detail in Chapter 8.

TABLE 5-1

Grain Counts Made in Autoradiographs of Intestinal Epithelial Cells at Five Different Times After Injection of ^{3}H-fucose in Mice [From J. E. Michaels, and C. P. Leblond. 1976. *J. de Microscopie et de Biologie Cellulaire*, 25:243.]

		10 min	20 min	30 min	1 hr	4 hr
Average Grain Counts in Autoradiographs	Golgi complexes	65.4	22.2	16.6	8.3	9.5
	Rough endoplasmic reticulum	9.3	4.7	8.1	12.0	3.3
	Mitochondria	13.1	34.5	21.0	13.6	12.7
	Vesicles	8.0	29.9	31.7	18.6	6.4
	Plasma membrane	4.3	8.7	22.6	47.7	68.1

PROBLEMS

1. How do protozoa living in fresh water cope with the hypotonic condition of their environment?
2. Describe the main features of the fluid mosaic model of plasma membrane structure.
3. Why does an animal cell swell when placed in a hypotonic medium?
4. Why are detergents needed to dissolve integral membrane proteins in water?
5. What observation by electron microscopy confirms the existence of transmembrane integral proteins?
6. What is one function of transmembrane proteins?
7. How was it demonstrated experimentally that integral membrane protein molecules diffuse laterally in the plasma membrane?
8. What properties of a molecule govern how rapidly it passes through the plasma membrane?
9. What is a major way in which the mechanisms of facilitated diffusion and active transport resemble one another? What are two ways in which they differ?
10. Na^+-K^+ ATPase transports Na^+ ions out of many kinds of eukaryotic cells. How does this activity contribute to active transport of sugars and amino acids into the cell?
11. How does pinocytosis differ from phagocytosis?
12. How was a radioactive amino acid used to show the pathway of secretion of zymogens by pancreas cells?
13. What is the fate of LDL receptors in the intake of cholesterol by receptor-mediated endocytosis?
14. What purpose is served by photographic film in the autoradiographic technique?

MICROTUBULES

Molecular Structure of Microtubules
Organization of Microtubules
Molecular Biology of Microtubule Assembly and Disassembly
Interaction of Microtubules with Other Organelles

MICROFILAMENTS

Assembly and Disassembly of Microfilaments
Actin-Binding Proteins

Proteins that bind to G-Actin
Proteins that cap the ends of microfilaments
Proteins that cap and sever microfilaments
Proteins that crosslink microfilaments

Interaction of Microfilaments with Other Cellular Components
Association of Myosin with Microfilaments
Interaction of Microfilaments with the Plasma Membrane

INTERMEDIATE FILAMENTS

Types of Intermediate Filaments
Construction of Intermediate Filaments
Function of Intermediate Filaments

C H A P T E R 6

Cellular Organization: The Cytoskeleton

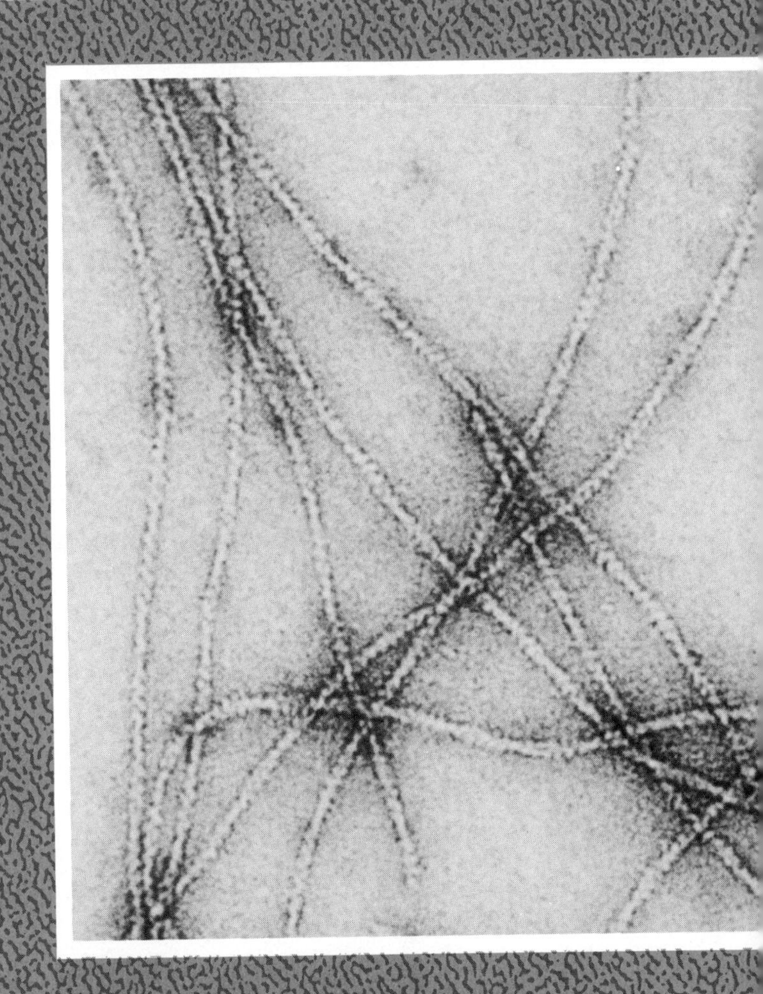

See Figure 6-11.

Cells contain a variety of structures, each of which is the basis of a particular function or group of functions. For example, in eukaryotic cells mitochondria generate ATP by oxidative phosphorylation, chloroplasts carry out photosynthesis, and ribosomes carry out protein synthesis. Also, lysosomes serve as storage vesicles for degradative enzymes, the plasma membrane controls the flow of materials in and out of the cell and mediates interaction of a cell with its environment, and the nucleus contains the genetic material and provides RNA for protein synthesis in the cytoplasm. All cellular structures represent the organization of different sets of molecules into highly ordered, functional aggregates. *Structures simply reflect the evolution of efficient ways for molecules to interact to achieve a particular cellular activity.*

The functions of all the molecules in the cell, whether they are organized into structures or not, are ultimately integrated with each other through molecular interactions, the bare outlines of which are only just beginning to be understood. For example, the rate of ATP production by mitochondria is, on the one hand, tied to production of pyruvate by the glycolytic enzymes in the cytosol. On the other hand, the production rate is governed by the demand for ATP by other parts of the cell—ATP for synthesis of proteins by ribosomes, as a precursor for synthesis of RNA by the nucleus, for beating of cilia and flagella, for amoeboid movement, for synthesis of lipids by the endoplasmic reticulum, and for many other energy-requiring activities. Similarly, the synthesis of proteins destined for secretion from a cell requires a cooperative action involving nuclear genes, cytoplasmic ribosomes, the endoplasmic reticulum, the Golgi membranes, and the plasma membrane. The production of cytoplasmic ribosomes requires cooperative function of many nuclear genes and transport of maturing ribosomes from the nucleus to the cytoplasm. The *rate* of production of ribosomes is determined minute-by-minute by the total demand for production of proteins. In general, increases in individual parts of the cell are regulated by functional demands on the individual parts. Overall, the function of each cell component is tied by molecular signals to the functions of all the other parts. In the end, it is such total functional integration of cellular components that accounts for global activities of cells such as growth and reproduction, differentiation movement, secretion, conduction of nerve impulses adaptation to environmental changes, and so forth.

Organelles and other structures in a cell often seem to lack particular positional relationships. However, closer study reveals some rules about the structural organization of a cell. The plasma membrane is always at the periphery of the cell, although it is a highly flexible structure that accommodates changes in cell shape in cells not constrained by a rigid cell wall. The nucleus almost always occupies a particular position in a given kind of cell, often somewhere near the center in animal cells, but often displaced to the periphery by the central vacuole in plant cells. Mitochondria are generally scattered throughout the cytoplasm, but in some cells they conform to specific patterns. In ciliated protozoa, mitochondria form an irregular layer just below the cell surface, an arrangement that may be important in supplying cilia with ATP. The contractile vacuole in protozoa is surrounded by a layer of mitochondria. In animal cells the Golgi complex is usually positioned close to the nucleus in an exact orientation (Chapter 8). In sperm cells the mitochondria, nucleus, and flagellum are arranged to form a cell streamlined for swimming. Nerve cells have cytoplasmic extensions that serve for the conduction of nerve impulses. The contractile machinery of muscle cells is precisely arranged to achieve maximum, efficient contraction.

Many examples of structural organization of cells and of the functional significance of such organization are cited in other chapters. This chapter is concerned with a structure that has a major role in cell organization, namely the **cytoskeleton.** Since the last century microscopists had debated the presence of a system of microfibers in the cytoplasm, that is, a cytoskeleton. Only in the last decade was its existence finally proven, via newer methods of light and electron microscopy combined with molecular analyses. Both the nucleus and cytoplasm have skeletal frameworks. In the cytoplasm it is called the cytoskeleton; the nuclear counterpart is known as the **nuclear matrix.** They appear to lack direct connections with each other, being sepa-ated by the nuclear envelope. The nuclear matrix is discussed later in connection with chromosome structure in Chapter 11. The present chapter deals with architectural organization of the cytoplasm by the cytoskeleton and the interaction of the cytoskeleton with other cell structures.

The cytoskeleton consists of three main fibrous elements—**microtubules, microfilaments** and **intermediate filaments**—and various proteins associated with them. These fibrous elements connect with other cell structures; for example, microfilaments connect to integral proteins of the plasma membrane. In addition, in the electron microscope the three main fibrous elements of the cytoskeleton are seen to interconnect with each other by means of fine protein strands, which have not yet been chemically characterized. The cytoskeleton, and its extensive matrix of interconnecting strands, creates a three-dimensional architectural framework throughout the cytoplasm.

The term cytoskeleton is perhaps misleading since skeleton carries the connotation of a permanent, rigid structure. However, the cytoskeleton often undergoes rapid reshaping, and parts of it are in a constant state of assembly and disassembly. The cytoskeleton gives a cell its three-dimensional form, and the cytoskeleton changes as cell shape changes. For example, during cell division it is completely restructured to serve in the mitotic distribution of chromosomes. The cytoskeleton provides a framework for transport of vesicles and particles within the cytoplasm, for cell movement (Chapter 14), and for general organization of organelles within the cytoplasm.

MICROTUBULES

Microtubules are hollow rods of various lengths, sometimes many μm long, with a diameter of 30 nm. They are important in determining overall cell form. They occur singly or in loose bundles throughout the cytoplasm of virtually all eukaryotic cells and are present in the nuclei of a few kinds of cells. They are also the main structural component of cilia and flagella and participate in the generation of force of ciliary and flagellar beating (Chapter 14). The distribution of microtubules in the cell can be seen in the light microscope (Figure 6-1) when the fluorescent antibody technique is used. A simple form of this technique is as follows: **Tubulin,** the protein monomer from which microtubules are constructed, is injected into an animal such as a rabbit or goat (Figure 6-2). Some lymphocytes in the animal respond by synthesizing and releasing into the blood a large amount of antibody molecules that can bind specifically to the injected tubulin. The antibody to tubulin binds to tubulin but to no other protein. Antibodies to tubulin are separated from the blood and covalently joined to a dye that fluoresces when excited by light. The dye-antibody complexes (**fluorescent antibodies**) are allowed to bind to tubulin in microtubules in fixed cells or are injected into living cells. One such fluorescent dye is **fluoroscein,** which is excited by blue light and emits green light. Illumination of cells stained with fluoroscein-labeled antibodies with blue light reveals the pattern of microtubules (Figure 6-1), which emit green light. The fluorescent antibody also binds to tubulin molecules not assembled into microtubules (**free tubulin**). Free tu-

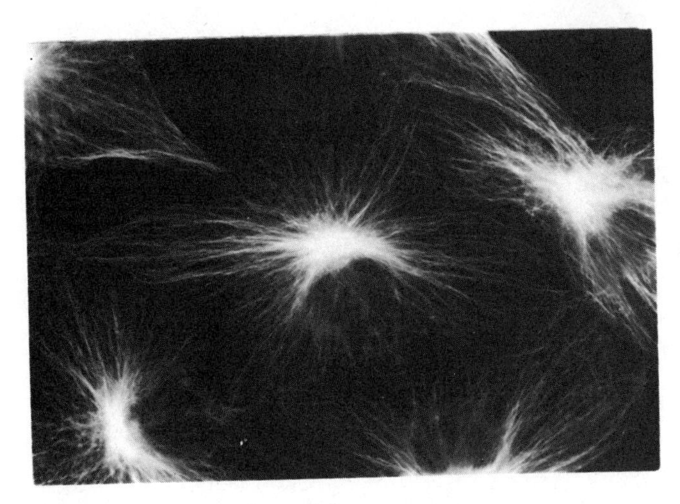

FIGURE 6-1
Light micrograph of cultured animal cells stained by the fluorescent antibody technique to show microtubules. Microtubules radiate out from the centrosome next to the nucleus (not visible) in the center of the cell. [Courtesy of J. Richard McIntosh.]

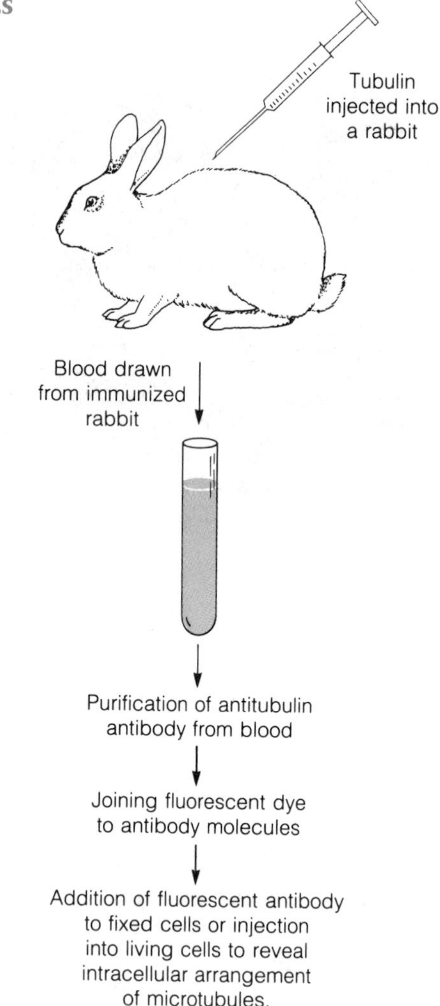
Tubulin
injected into
a rabbit
Blood drawn
from immunized
rabbit
Purification of antitubulin
antibody from blood
Joining fluorescent dye
to antibody molecules
Addition of fluorescent antibody
to fixed cells or injection
into living cells to reveal
intracellular arrangement
of microtubules.

FIGURE 6-2

The fluorescent antibody technique for visualizing the intracellular location of specific macromolecules, in this case tubulin.

bulin, which accounts for about 50 percent of the total tubulin in some cells, is evenly dispersed in the cytosol and gives rise to a faint background of fluorescent antibody staining that does not interfere with observation of the intensely fluorescent microtubules.

Changes in cell shape are accompanied by remodeling of the cytoskeleton; the microtubules throughout the cytoplasm are in a constant process of disassembly and reassembly. This dynamic nature of microtubules is exploited during remodeling of cell shape. A dramatic example is the essentially total disassembly of the mi crotubule network of an interphase cell when the ce enters mitosis and reassembly of the tubulin to form th microtubules of the mitotic spindle (Chapter 12). Disas sembly of microtubules is accompanied by disassembl of microfilaments and the disruption of the organiza tion of intermediate filaments, hence, by a global reor ganization of the cytoskeleton. As a result, the cell lose its interphase shape and becomes spherical (Figure 6 3). In animal cells, microfilaments are assembled into contractile ring just inside the plasma membrane; con traction of the ring divides the cell into two. With th completion of cell division the microtubules of th spindle are disassembled, and the microtubules an other cytoskeletal elements are reassembled into thei interphase forms.

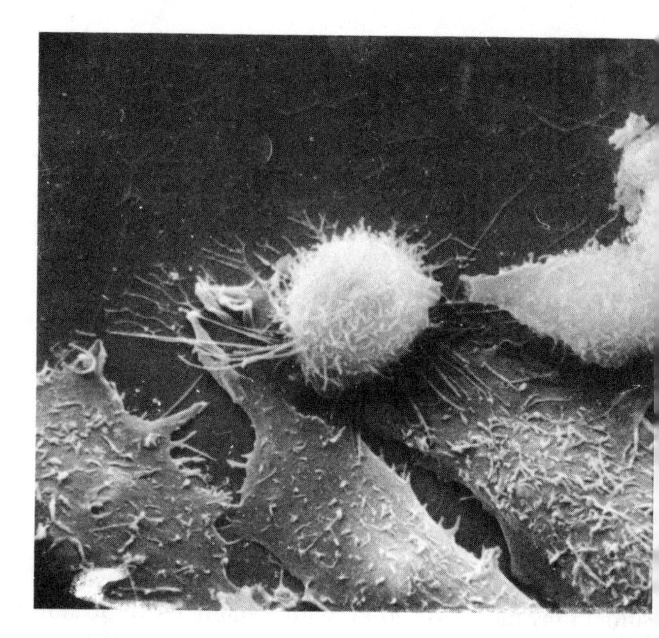

FIGURE 6-3

Scanning electron micrograph of hamster cells in culture. The three flat cells are in interphase. The cell in the center has entered mitosis (late prophase or metaphase) and become spherical. Many filamentous cytoplasmic projections cover its surface. Some of these projections are long and attach to the surface of the culture vessel and adjacent interphase cells. [Courtesy of Keith R. Porter.]

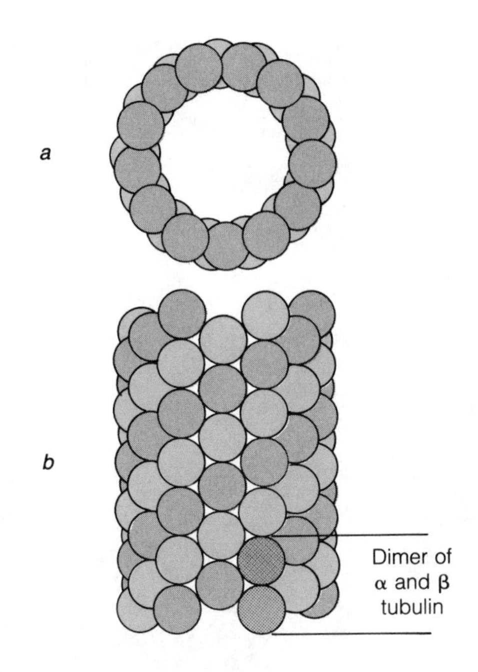

FIGURE 6-4
Drawing of a microtubule. (a) Cross-sectional view of the 13 protofilaments. (b) Longitudinal view.

Molecular Structure of Microtubules

Microtubules are built by assembly of equal amounts of two kinds of tubulin molecules, called **α-tubulin** and **β-tubulin,** each with a molecular weight of 50,000. One α-tubulin molecule binds noncovalently with one β-tubulin molecule to form a dimer of 100,000. Dimers bind to one another noncovalently to form a long, straight protofilament (Figure 6-4) about 8 nm in diameter. Thirteen protofilaments are arranged side-by-side in a circular pattern within a microtubule. The 13 protofilaments form the microtubular wall, about 8 nm thick, and enclose a lumen about 14 nm in diameter. Thus, in the formation of a microtubule each tubulin molecule participates in three levels of binding: α-tubulin binds to β-tubulin to form a dimer; dimers bind to each other to form protofilaments; and the dimers in one protofilament bind to dimers in adjacent protofilaments to form a tubule. A fourth set of interactions occurs in which tubulin molecules assembled into microtubules bind certain other proteins, called **microtubule-associated proteins** or **MAPS** (see next section).

Organization of Microtubules

In animal cells in interphase the microtubules extend outward in a starlike pattern from a body adjacent to the nucleus called the **microtubule center** or **centrosome.** This can be seen clearly if the microtubules are first caused to disassemble into free tubulin within a cell and then allowed to reassemble. Microtubules can be disassembled by exposing cells briefly to cold or by treatment with certain drugs. Cold prevents the tubulin–tubulin interactions that are the basis of microtubule structure. The poisonous drug **colchicine** binds tightly to tubulin and prevents its assembly into microtubules. Since microtubules of the cytoskeleton are normally in a dynamic state, with individual tubulin molecules constantly entering and leaving microtubules (discussed later in the chapter), colchicine treatment quickly leads to disassembly. In recovering from cold or drug treatment, tubulin reassembles into microtubules starting at the centrosome (Figure 6-5). The microtubules grow outward from the centrosome and eventually extend through the entire cytoplasm.

The centrosome is an ill-defined body composed of material that appears amorphous in the electron microscope (Figure 6-6). In some species the centrosome contains granules of unknown composition and significance. In most kinds of animal cells a centrosome contains two **centrioles** arranged at a right angle to each other. As described in Chapter 1, each centriole is made up of triplets of microtubules (Figure 1-27) surrounded by the amorphous material of the centrosome. During mitosis the centrosome separates into two, and the two daughter centrosomes reorganize the microtubules into the bipolar mitotic spindle, which plays the central role in the mitotic separation of daughter chromosomes (Chapter 12). Long before the centrosome splits into two at the start of mitosis, the centrioles in a pair separate slightly. A new centriole then forms at a right angle to each old centriole to give rise to two pairs. Later,

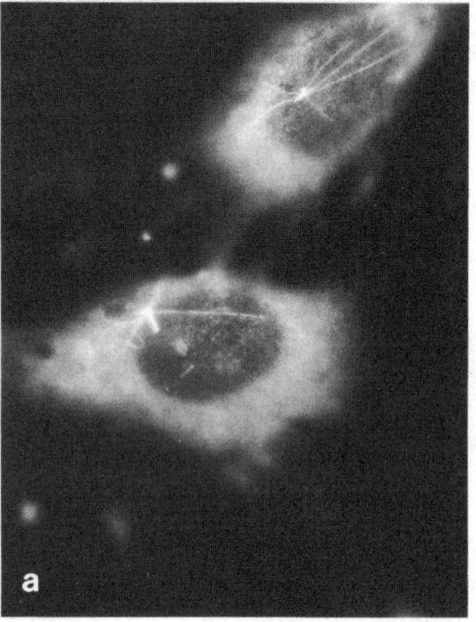

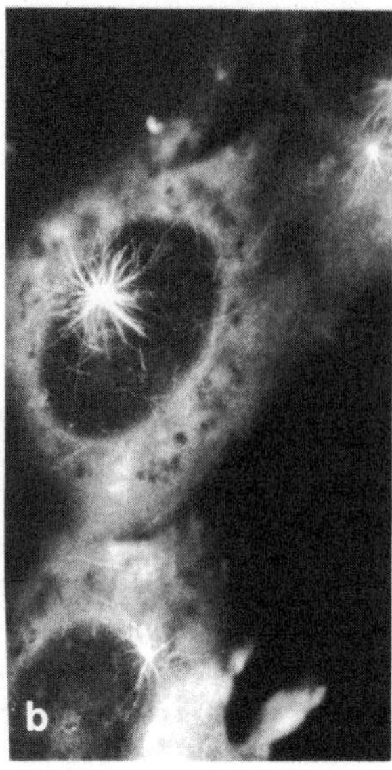

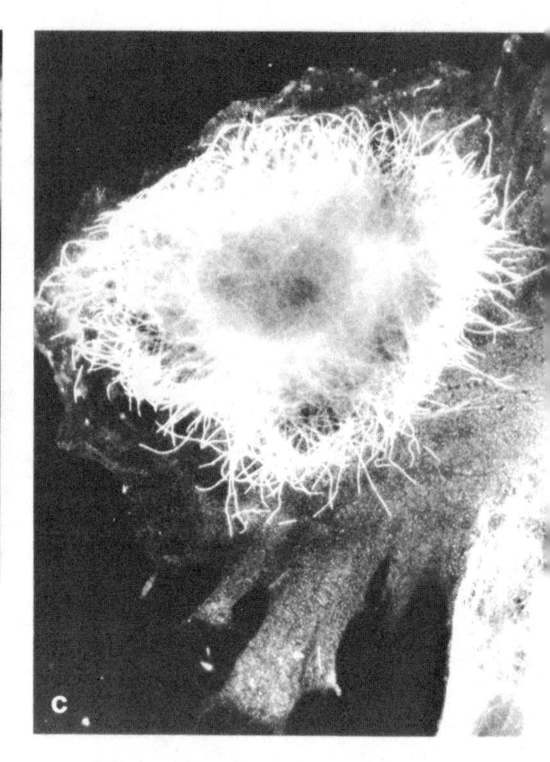

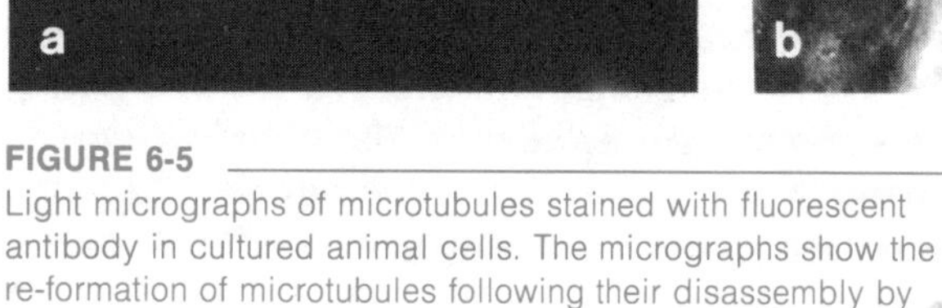

FIGURE 6-5
Light micrographs of microtubules stained with fluorescent antibody in cultured animal cells. The micrographs show the re-formation of microtubules following their disassembly by treatment of living cells with colcemid for one hour. (a) Two cells stained to show microtubules immediately after colcemid treatment. Only a few microtubules remain, radiating from the centrosome. (b) Thirty minutes after removal of colcemid many microtubules have formed and radiate from the centrosome. (c) Fifty minutes after treatment recovery is complete, and a dense population of microtubules extends throughout the cell from the centrosome. [Courtesy of Mary Osborn. From M. Osborn, and K. Weber. 1976. *Proc. Natl. Acad. Sci. USA*, 73: 867.]

when the centrosome splits, each daughter centrosome receives a pair of centrioles.

A centriole is structurally similar, if not identical, to the **basal body** of a cilium or a flagellum (Figure 6-7). Basal bodies serve as nucleating structures in the formation and organization of microtubules into cilia and flagella (Chapter 14). Therefore, they may be considered as microtubule-organizing centers for the formation of cilia and flagella. Thus, the basal body serves in the skeletal organization of a highly localized protrusion of the cytoplasm, namely, the cilium or flagellum (these structures are covered by the plasma membrane), while the centrosome functions in global organization of the microtubules of the cytoskeleton.

Strong evidence for the functional relatedness o centrioles and basal bodies is the fact that the basa body of the flagellum in the sperm cell of some specie becomes a centriole, taking up a position in the centro some in an egg cell after fertilization. Furthermore, ba sal bodies isolated from the flagellated alga *Chlamydo monas* or from the ciliated protozoan *Tetrahymena* an injected into frog eggs cause the organization of star shaped arrays of microtubules resembling those orga nized by centrosomes during reassembly of microtu bules after cold or colchicine treatment.

Centrioles are absent from the centrosomes of few species of animals, from the centrosomes of man unicellular eukaryotes, and from plant cells in general Indeed, although mouse cells usually contain centri oles, they are absent from the centrosomes during th first several divisions of a fertilized egg. At that poin they presumably form *de novo* and become associate with the centrosome. Such observations indicate tha centrioles are not necessary for organization of the mi totic spindle by the centrosome.

Molecular Biology of Microtubule Assembly and Disassembly

When cells are mechanically disrupted for the purpose of isolating cell parts, most organelles and structures remain intact. However, the cytoskeleton is lost, largely because microtubules and microfilaments disaggregate into their protein monomers and the network of intermediate filaments collapses. A major advance in the study of the cytoskeleton came in the 1970s, when conditions were worked out by which microtubules could be induced to form from tubulin purified from homogenized brain tissue. Nerve cells contain unusually large numbers of microtubules, so brain tissue is a rich source of tubulin.

Tubulin isolated from warm-blooded animals spontaneously assembles into microtubules at 35°C in the presence of GTP, Mg^{2+} ions, and a Ca^{2+} ion chelator called **EGTA** ([ethylene-bis(oxyethylenenitriolo)]tetracetic acid). EGTA tightly binds Ca^{2+} ions in the cell homogenate and thereby prevents the interaction of Ca^{2+} ions with tubulin. Interaction of Ca^{2+} ions with tubulin prevents formation of microtubules, perhaps by interfering with the role of Mg^{2+} ions in the sticking of one dimer to another. Binding of GTP to tubulin probably alters the tertiary shape of dimers in a way that allows one dimer to bind to the next. As a free dimer binds to a dimer at the end of a microtubule, the GTP is hydrolyzed to GDP + P_i by a GTPase activity of the tubulin dimer itself. The GDP remains bound to assembled dimers. Although GTP is essential for dimer addition to microtubules, the hydrolysis of bound GTP to GDP is not essential; evidence for this is that analogues of GTP that bind to tubulin but that tubulin cannot hydrolyze to the corresponding GDP analogs still promote assembly of tubulin into microtubules. However, both the binding and hydrolysis of GTP are

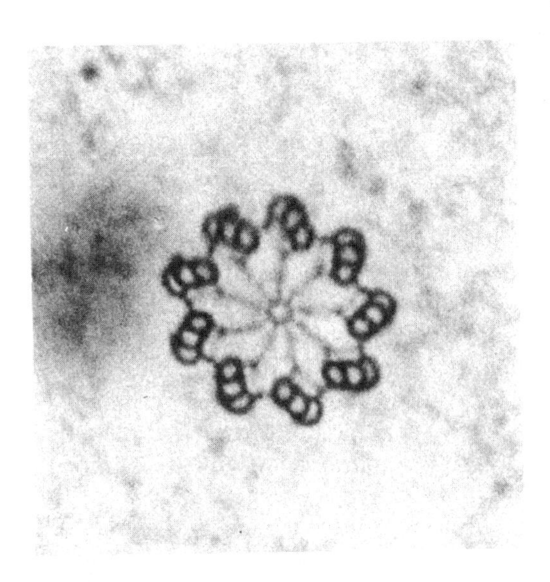

FIGURE 6-7
Electron micrograph of a cross section of the basal body of the flagellum in the flagellated protozoan *Trichonympha*. Nine microtubule triplets are connected to a central tubular structure by nine radial spokes. [Courtesy of Ian R. Gibbons. Reproduced from the *Journal of Biophysical and Biochemical Cytology*, 7:697–716 (1960) by copyright permission of the Rockefeller University Press.]

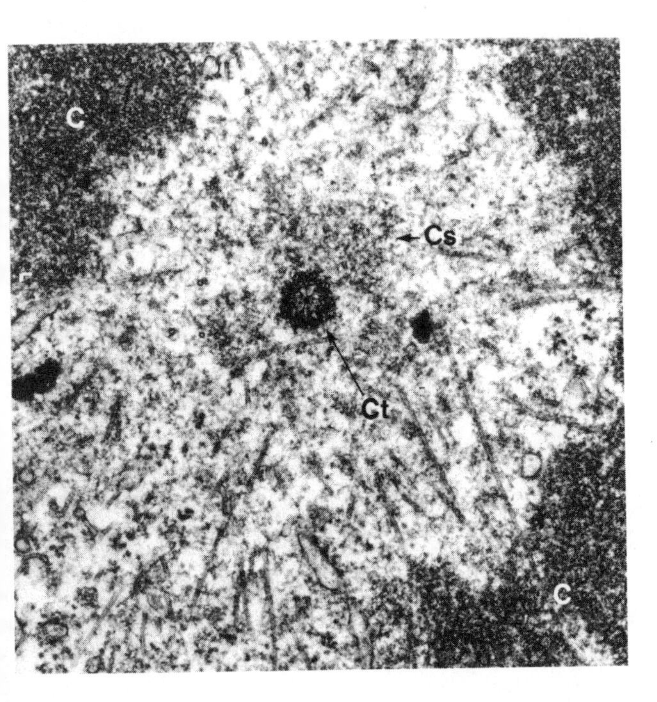

FIGURE 6-6
Electron micrograph of a cross section of a centriole (Ct) in a dividing animal cell. The centriole is surrounded by amorphous material that constitutes the centrosome (Cs). Microtubules radiate from the amorphous material. C = chromosomes. [Courtesy of J. Richard McIntosh.]

probably important in the regulation of microtubule assembly and disassembly (described later in this chapter). Once formed in a cell-free homogenate, microtubules can be separated from other cell components by centrifugation. The microtubules can then be disassembled into tubulin dimers, for example, by lowering the temperature from 35°C to 4°C. Reassembly can be induced again by raising the temperature to 35°C.

With several cycles consisting of assembly, washing the microtubules free of contaminants, and then disassembly, a tubulin preparation is obtained that contains α- and β-tubulin and several other proteins that bind to microtubules. Two of these microtubule-associated proteins (MAPs) are present in a constant proportion to tubulin. They have high molecular weights and are large enough to be observed in the electron microscope bound to the microtubule surface (Figure 6-8). Purified tubulin, free of MAPs, can aggregate into microtubules under conditions that do not occur in the cell (very high concentration of free tubulin or high Mg^{2+} ion concentration). However, the presence of MAPs promotes microtubule formation under conditions that approach those of the cell. Tubulin from cold-blooded animals, for example, from Arctic fish with body temperatures as low as −3°C, forms microtubules that are stabilized at low temperatures. Stabilization is achieved by microtubule-associated proteins that are slightly different from those found in organisms with higher body temperatures.

When the temperature of a solution of tubulin prepared from a warm-blooded animal is raised from 4°C to 35°C, formation of microtubules begins after a lag period. The lag period is the time required for **nucleation** events to occur. Nucleation consists of assembly of some of the tubulin dimers into small rings. Experiments suggest that the rings are short, coiled-up protofilaments, but we are not certain of their structure. The current hypothesis is that the rings uncoil, and the short linear protofilaments join in a side-by-side interaction to create a sheet that then folds into a short microtubule with 13 protofilaments (Figure 6-9). The short microtubules then serve as nucleation structures for addition of free dimers onto the ends. Separate experiments have shown that microtubules elongate by addition of dimers to ends and not by insertion of dimers into the structure. However, end-to-end joining of microtubules occurs, which rapidly creates long microtubules. Formation of nucleation structures proceeds slowly but once nucleation structures have formed, tubulin dimers are added rapidly—about 100 dimers per second—which gives a growth rate for a microtubule of

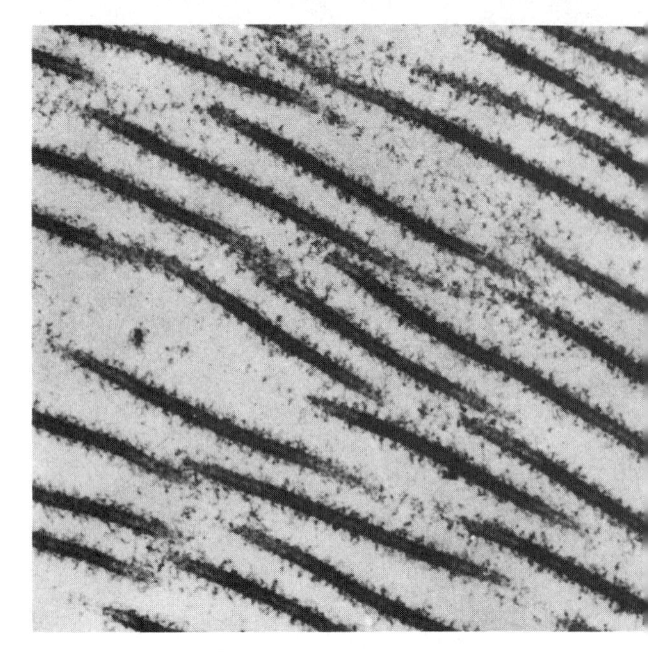

FIGURE 6-8
Electron micrograph of microtubules cut in a longitudinal section. Microtubule-associated protein-2 (MAP-2) molecules are attached to the microtubules, giving them a fuzzy appearance. [Reprinted from the *Annuals* of The New York Academy of Sciences, Volume 466:288. Courtesy of Helen Kim, Lester I. Binder, and Joel L. Rosenbaum.]

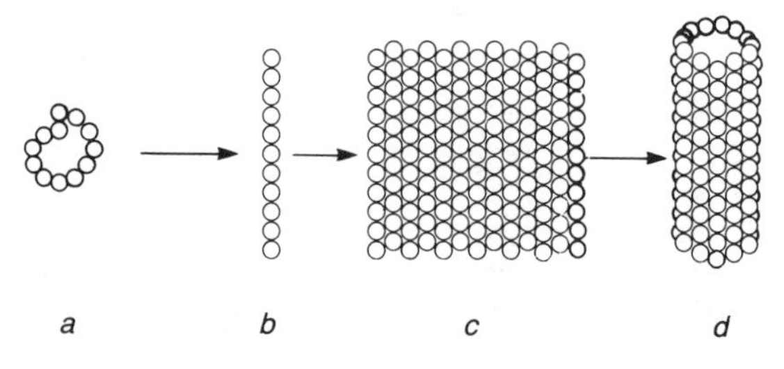

FIGURE 6-9
Nucleation of a microtubule assembly. (a) Formation of a small ring of tubulin molecules. (b) Rings unwind into straight protofilaments. (c) Thirteen protofilaments form a sheet. (d) The protofilament sheet rolls up into a microtubule that elongates by addition of tubulin molecules.

3.6 μm per minute. The lag period preceding microtubule assembly can be eliminated by adding to a tubulin solution short fragments of microtubules, which serve as nucleation structures. MAPs also shorten the lag period by accelerating the formation of rings. This may be a role for MAPs in the cell. Also, MAPs tend to stabilize microtubules against dissociation into free tubulin dimers.

The addition of tubulin dimers to the ends of microtubules is a reversible event. In a cell-free system, as long as the concentration of free tubulin dimers is high, net addition takes place at both ends, and the microtubules grow. At progressively lower concentrations of free tubulin dimers, the rate of addition of dimers decreases until it equals the rate of subtraction of dimers, that is, the system is in equilibrium. However, a closer examination of a cell-free system shows that events are more complicated than suggested by measuring the overall rate of tubulin assembly into microtubules. Addition of dimers occurs more readily at one end of a microtubule, called the (+) end, than the other, the (−) end (Figure 6-10). As a result, as the concentration of free tubulin dimers is decreased, a point is reached where the rate of addition of dimers to the (−) end

FIGURE 6-10
Both addition and subtraction of tubulin dimers to and from microtubules occur continuously in a cell-free system.

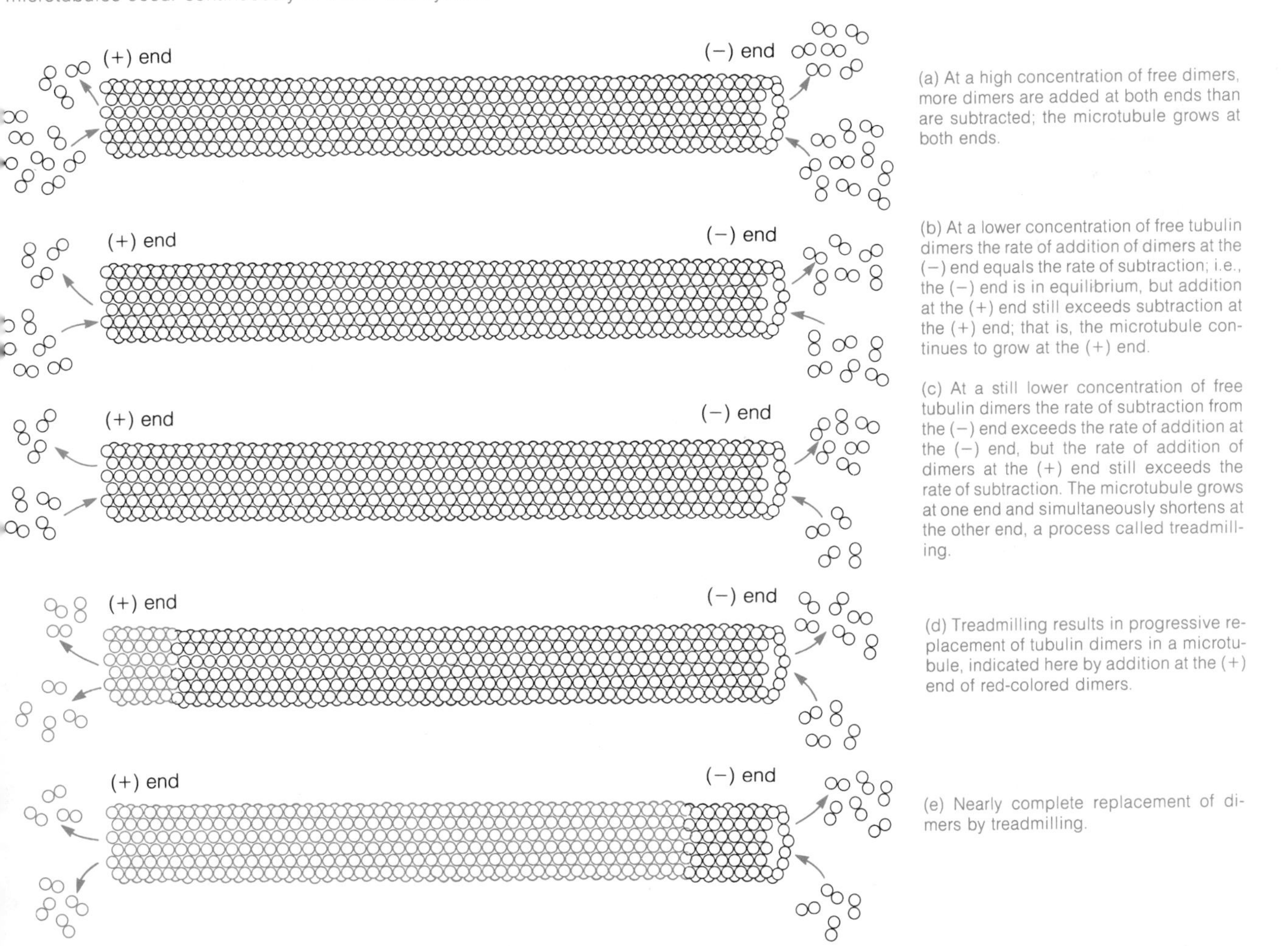

(a) At a high concentration of free dimers, more dimers are added at both ends than are subtracted; the microtubule grows at both ends.

(b) At a lower concentration of free tubulin dimers the rate of addition of dimers at the (−) end equals the rate of subtraction; i.e., the (−) end is in equilibrium, but addition at the (+) end still exceeds subtraction at the (+) end; that is, the microtubule continues to grow at the (+) end.

(c) At a still lower concentration of free tubulin dimers the rate of subtraction from the (−) end exceeds the rate of addition at the (−) end, but the rate of addition of dimers at the (+) end still exceeds the rate of subtraction. The microtubule grows at one end and simultaneously shortens at the other end, a process called treadmilling.

(d) Treadmilling results in progressive replacement of tubulin dimers in a microtubule, indicated here by addition at the (+) end of red-colored dimers.

(e) Nearly complete replacement of dimers by treadmilling.

equals the rate of subtraction at that end (equilibrium is reached at the (−) end). Therefore, growth of the microtubule at the (−) end stops (Figure 6-10(b)). When equilibrium is reached at the (−) end, tubulin dimers are still undergoing net addition to the (+) end, and the microtubule continues to grow at that end. With a further decrease in tubulin concentration, a net loss of tubulin starts to occur at the (−) end while net addition is still occurring at some lower rate at the (+) end (Figure 6-10(c, d, e)). At such a tubulin concentration the microtubule grows at one end and shortens at the other end—a condition known as **treadmilling.** With a further decrease in the concentration of tubulin dimers, growth at the (+) end stops, and the microtubule ultimately disappears because of loss of dimers from the (−) end. At even lower concentrations of tubulin, dissolution occurs at both ends.

These experiments make three points:

1. A microtubule is a dynamic structure, undergoing constant addition and subtraction of tubulin dimers at both ends.
2. Microtubules are different at both ends, i.e., they possess polarity. **Polarity** means that the tubulin dimers present different configurations at the two ends to account for different abilities to bind additional dimers.
3. Over a certain range of tubulin concentration in a cell-free system a microtubule can continue to grow at one end while diminishing in length at the other—that is, treadmilling can occur.

A reasonable expectation is that all the microtubules in a cell-free preparation, which are exposed to the same concentration of free tubulin dimers, should behave in the same way—lengthening, shortening, or retaining a constant length, depending on the concentration of free tubulin. However, direct observations of individual microtubules suggest that some grow while others shorten in the same preparation. The following hypothesis, which has an important bearing on microtubule behavior in the intact cell, has been proposed to explain this seemingly enigmatic behavior. Microtubules ending with dimers to which GTP is bound (GTP-tubulin dimers) may be preferentially stabilized because GTP-dimers bind more strongly than GDP-tubulin dimers. If GTP-tubulin dimers are added to the end of a microtubule faster than the GTP is hydrolyzed to GDP, some GTP-tubulin dimers will accumulate at the end, which may stabilize the end against loss of dimers. As a result, growth of the microtubule is favored. However, if the number of GTP-tubulin dimers at the end should decrease below some critical value because of a random, momentary increase in hydrolysis of GTP or decrease in GTP-tubulin dimer addition, or both, the microtubule will now have GDP-tubulin dimers at its end. The lower stability of binding of GDP-tubulin dimers to each other will result in more rapid dissociation of dimers, and the net loss of dimers will be favored. The loss will continue in a progressive manner unless, by random chance, events are reversed and the microtubule again acquires sufficient GTP-tubulin dimers at its end to be stabilized against loss of dimers.

Simultaneous growth and shortening of microtubules occur in an individual living cell. In a fibroblast stained with fluorescent antibody to tubulin, the total array of microtubules seen at any instant gives the impression of a stable population. In fact, injected tubulin to which a fluorescent dye has been covalently linked shows that the average half-life of microtubules is less than 10 minutes during interphase in a cultured animal cell and less than 30 seconds for the microtubules of the mitotic spindle. *Individual microtubules are constantly being formed and entirely disassembled.* Growth of new microtubules starts at the centrosome, to which they remain anchored by their (−) ends. Possibly, some microtubules may also initiate free in the cytosol; however, such microtubules will disappear more quickly than microtubules anchored in the centrosome because, at the concentration of free tubulin in the cell, net loss probably will always occur at free (−) ends that are not anchored in the centrosome. Anchorage in the centrosome prevents loss of dimers from (−) ends. Thus, the treadmilling seen in a cell-free system may not occur for the bulk of microtubules in the cell, namely, those anchored in the centrosome, but may favor dissolution of unanchored microtubules compared to anchored ones.

Regulation of microtubule assembly and disassembly in an intact cell is still poorly understood. The centrosome serves to nucleate microtubule assembly, but probably does not influence the exchange of dimers at the distant (+) ends. Other factors may modulate microtubule length and stability. For example, a small percentage of microtubules does not undergo rapid turnover. These persist for hours, compared to a half-life of less than 10 minutes for the bulk of microtubules. These few microtubules are possibly stabilized by specific capping proteins that bind to (+) ends and at least transiently prevent disassembly. Ultimately, even these more long-lived microtubules disassemble.

Microtubules in cilia and flagella have their (−) ends anchored in basal bodies. Growth occurs by addition of tubulin dimers to the (+) end, which extends the cilium or flagellum. The cilia or flagella in any particular cell species are always the same length. We do not know how length of microtubules is so precisely regulated in these structures. In addition, the microtubules in cilia and flagella, in contrast to those in the main body of the cytoplasm, do not undergo turnover or disassembly. Rather, they are highly stable, perhaps because they are capped in some way at their (+) ends. Remarkably, when cilia or flagella are mechanically torn from a cell, they are in most cases quickly regenerated from the pool of soluble tubulin present in the cell.

Interaction of Microtubules with Other Organelles

Microtubules interact with a number of other cell components; for example many microtubules terminate with their (+) ends very near the plasma membrane, and microtubules appear to be attached along their lengths to mitochondria. The significance of these interactions is unclear. Small cytoplasmic vesicles are attached to and transported along microtubules, a process particularly evident in axons of nerve cells. Also, during mitosis some microtubules are attached at their (+) ends to the chromosomes. These chromosomal attachments are essential for mitotic movement of chromosomes. Microtubules are important in maintaining the organization of the Golgi complex. Treatment of interphase cells with drugs that cause disassembly of microtubules results in dispersion of the vesicular sacs that make up the Golgi complex, which disrupts exocytosis of proteins. Finally, microtubules are joined to microfilaments and intermediate filaments by a network of proteinaceous strands called the **microtrabecular lattice** or **matrix.**

MICROFILAMENTS

Microfilaments are present in all eukaryotic cells. They are an important constituent of the cytoskeleton and form part of the contractile machinery both in muscle and nonmuscle cells (Chapter 14). Microfilaments are long, rodlike structures about 9.5 nm in diameter formed by joining together molecules of **actin** (Figure 6-11), a protein with a molecular weight of 42,000. It is one of the most conserved proteins in evolution in terms of amino acid sequence, tertiary structure, and function; thus, the actin of amoeba and yeast closely resembles the actin of mammals.

Microfilaments, or **actin filaments**, as they are also called, are often arranged in parallel in bundles, in which other kinds of protein molecules form cross-links among the microfilaments. Microfilaments also occur in meshworks, in which they appear in random orientations with respect to one another. Skeletal and cardiac muscle cells have extraordinarily well developed cytoskeletons, which are responsible for contraction. In such cells actin filaments are arranged in precise patterns. Actin filaments combine with fibers composed of the protein **myosin,** another major part of the muscle cell's contractile machinery (Chapter 14). In nonmuscle cells, bundles and meshworks of microfilaments are also associated with the protein myosin and participate in contractile processes that generate amoeboid movement and cytoplasmic streaming (Chapter 14). In some kinds of animal cells in culture, bundles of microfilaments are especially prominent and have been given the name **stress fibers** (Figure 6-12). These connect to the inner side of the plasma membrane in areas where the cell is in contact with the

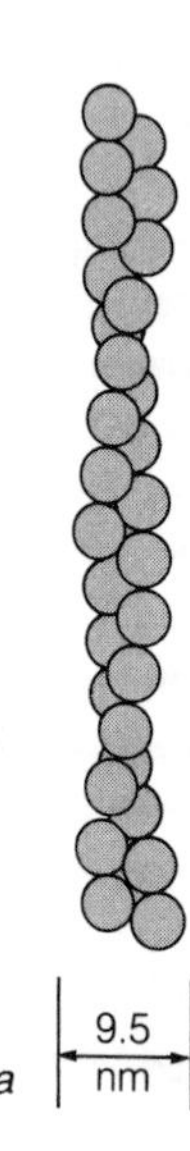

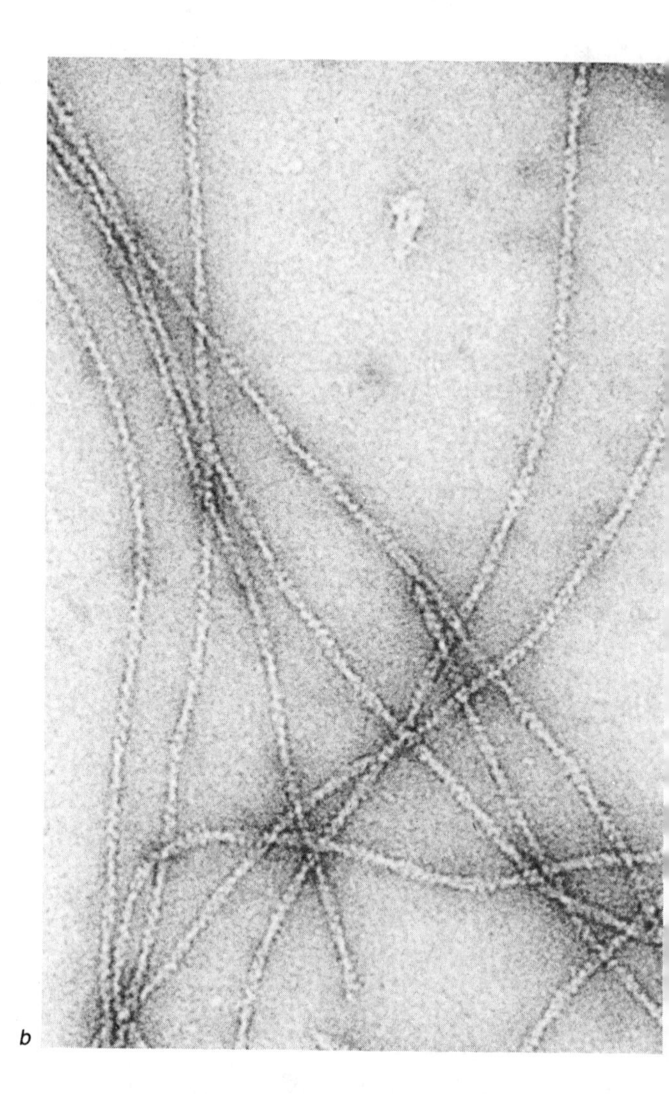

FIGURE 6-11
Actin microfilament. (a) Model consisting of actin molecules arranged as two intertwined helices. (b) Electron micrograph of actin microfilaments isolated from blood platelets. [(b) Courtesy of Roger Craig.]

surface of the culture vessel. The muscle cell proteins myosin and **α-actinin** are found at regular intervals along stress fibers. Stress fibers isolated from cells contract in the presence of ATP, which is consistent with a role in cell movement. The contractile role of actin filaments in cytokinesis is discussed in Chapter 12. Bundles of actin filaments also occur in epithelial cells, where they are associated with specialized surface structures called **belt desmosomes,** by which cells are joined to one another (Chapter 15). Actin filaments are particularly abundant in the microvilli that make up the brush border on the lumenal surface of epithelial cells of the intestine (Figure 6-13; see also Chapter 15).

Assembly and Disassembly of Microfilaments

Actin is a slightly elongated, globular molecule with several sites for binding to other actin molecules and several sites for binding to a variety of kinds of **actin-associated proteins.** In its unbound form it is called **G-actin** (globular actin); when assembled into microfilaments, it is called **F-actin** (fibrous actin). As

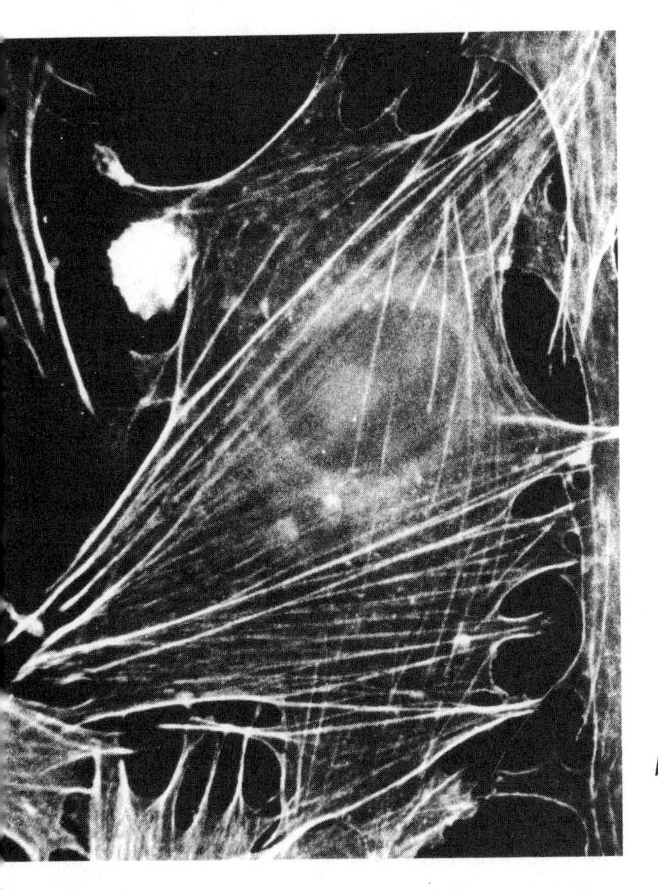

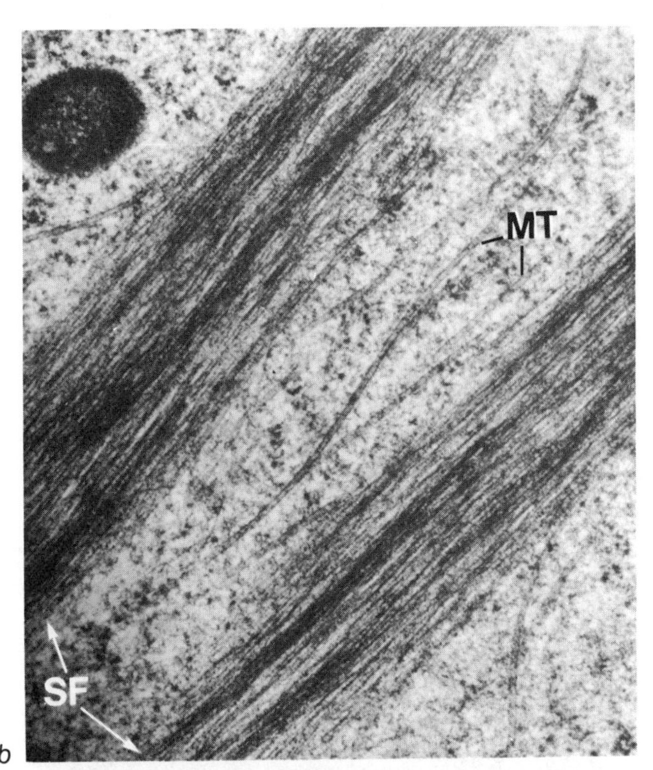

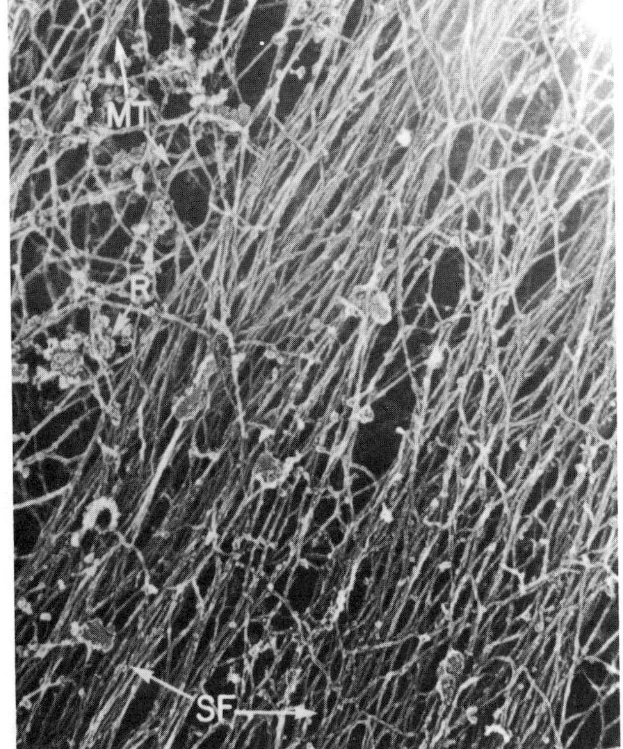

FIGURE 6-12
Microfilament stress fibers (SF) seen by three kinds of microscopy. (a) Light micrograph of stress fibers in a cultured animal cell stained with fluorescent dye. (b) Electron micrograph of two stress fibers in a fibroblast. A few microtubules (MT) are present between the two stress fibers. (c) Electron micrograph of a stress fiber in a fibroblast that was rapidly frozen and then freeze-dried to reveal microfilament stress fibers and microtubules (MT). R = ribosome clusters. [(a) Courtesy of K. G. Murti. (b and c) Courtesy of Marc W. Kirschner. Reproduced from *The Journal of Cell Biology*, 86:212–234 (1980) by copyright permission of the Rockefeller University Press.]

in the assembly of tubulin into microtubules, the assembly of G-actin into microfilaments is reversible. Hydrophobic interactions and ionic bonds are responsible for the noncovalent bonding of one actin molecule to another. The precise arrangement of actin molecules in microfilaments is not yet known. The model in Figure

6-11, suggested by several kinds of observations, shows the actin molecules arranged as a helix.

Actin was first purified from muscle cells, where it is particularly plentiful, but it is abundant in most kinds of nonmuscle cells; it has been isolated from a wide range of eukaryotes from yeast to mammals. Assembly and disassembly of microfilaments is readily achieved in a cell-free system, and one can easily monitor assembly and disassembly by measuring the viscosity of the solution. Formation of microfilaments from G-actin is accompanied by an increase in viscosity, and disassembly is accompanied by a decrease. Formation of microfilaments in a cell-free system is preceded by a lag period, in which small units form and act as nucleation structures. The small units consist of a few G-actin monomers, are unstable, and generally dissociate before forming a larger aggregate. When an aggregate containing a critical number of actin molecules is assembled, binding becomes more stable, continued addition of G-actin monomers becomes more probable than dissociation, and a microfilament is rapidly formed (Figure 6-14).

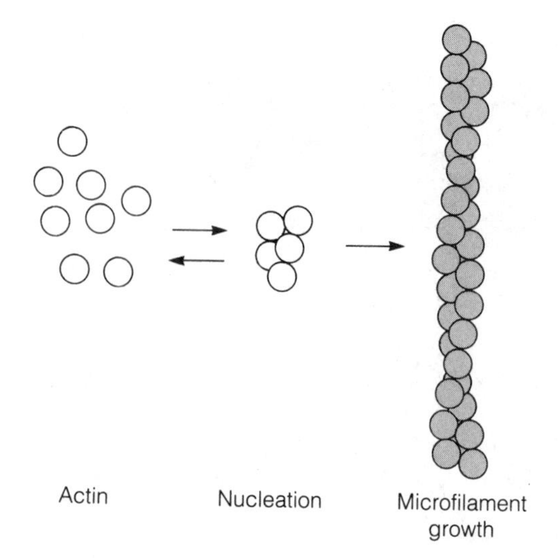

FIGURE 6-14
Formation of microfilaments from purified G-actin. Small, unstable aggregates that readily dissociate into G-actin act as nucleation structures. Once a nucleation structure exceeds a critical size, G-actin-binding becomes more stable, and the microfilament grows.

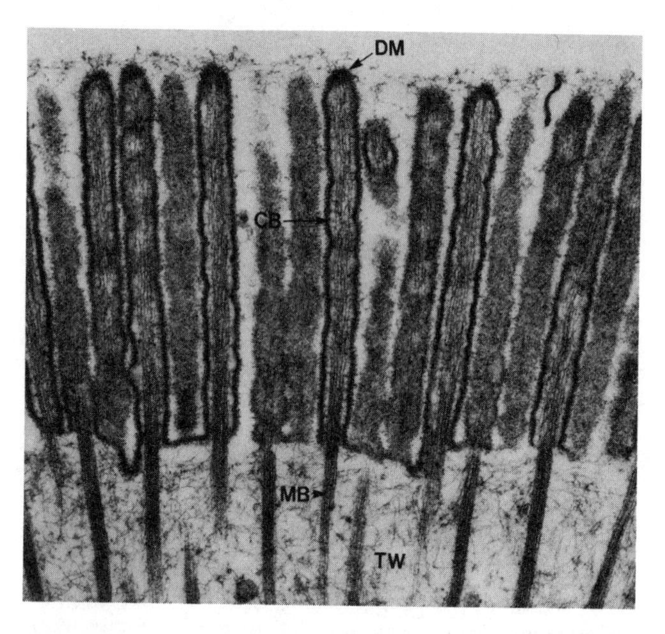

FIGURE 6-13
Electron micrograph of a portion of the brush border of an intestinal epithelial cell. Each microvillus contains a bundle of actin microfilaments (MB) that extends into the body of the cell into a terminal web (TW) of filaments. The actin bundles are attached by fine cross bridges (CB) to the plasma membrane along the length of a microvillus and a dense matrix (DM) of material at the mircrovillar tip. [Courtesy of Mark S. Mooseker. Reproduced from *The Journal of Cell Biology*, 67:725–743 (1975) by copyright permission of the Rockefeller University Press.]

Formation of microfilaments resembles production of microtubules in several respects. Assembly of G-actin in a cell-free system requires the presence of ATP and inorganic ions such as Ca^{2+} or Mg^{2+}. ATP binds to G-actin and is hydrolyzed to ADP + P_i in association with assembly into F-actin. Actin is not an ATPase, but it does stimulate an ATPase in the cytosol. ATP-actin binds more stably to the end of a microfilament than does ADP-actin. Therefore, if the hydrolysis of ATP to ADP is delayed, assembly of further ATP-actin is favored, and the microfilament continues to grow. However, when ADP-actin is present at the end, dissociation is favored and the microfilament shrinks progressively. ADP-actin released from microfilaments is regenerated to ATP-actin by exchange of ADP with ATP. Therefore, cycles of assembly and disassembly are an energy-consuming process.

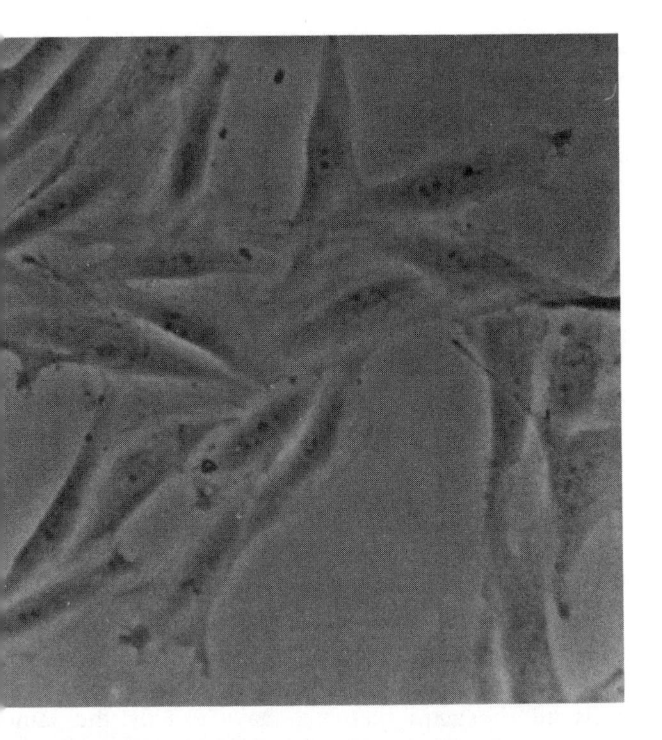

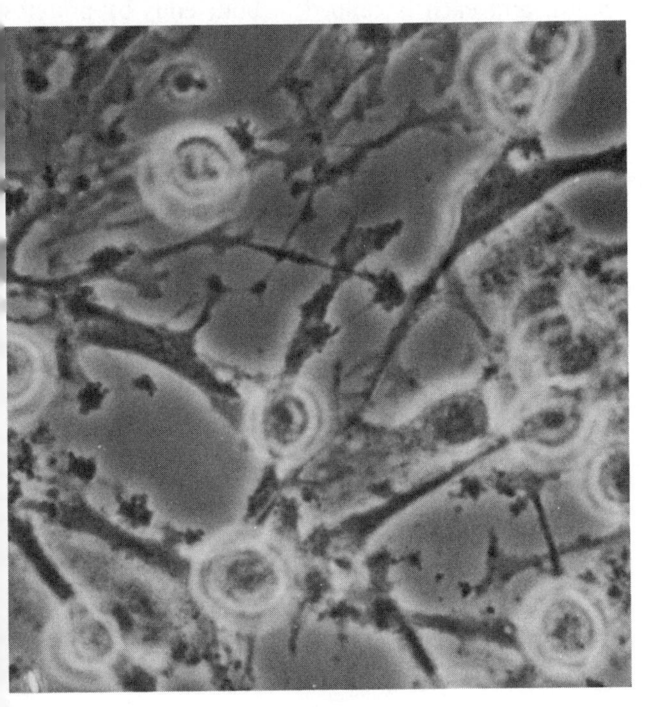

FIGURE 6-15
Effect of the mold toxin cytochalasin on animal cells.
(a) Photomicrograph of untreated cells in culture.
(b) Photomicrograph of cells incubated with cytochalasin for several minutes. The cells have a shrivelled appearance and their nuclei protrude upward from the main body of the cells and are out of the focus of the photograph. [Courtesy of Jerry W. Shay.]

Like microtubules, microfilaments are polar. Actin can be added 5- to 10-fold more rapidly to one end, designated the (+) end, than to the other, the (−) end. The rates of addition decline as the concentration of free ATP-actin decreases, and, at a critical concentration, ATP-actin may continue to be added to the (+) end while ADP-actin begins to dissociate from the (−) end; that is, treadmilling may occur in a cell-free system just as it does for microtubules. What role, if any, treadmilling has in an intact cell is uncertain, but as in the case of microtubules, the treadmilling phenomenon may reflect the ability of the cell to disassemble rapidly spurious microfilaments that are not anchored or capped in some way at their (−) ends.

Experiments show that several substances interfere with assembly and disassembly of microfilaments, affecting cell form and behavior. **Phalloidin,** a poison produced by the mushroom *Amanita phalloides,* binds strongly to microfilaments and inhibits their disassembly. It blocks the movement of amoebae and animal cells in culture. This means that disassembly of microfilaments is a necessary part of the motility mechanism in these cells. **Cytochalasins,** a family of chemically related substances produced by several species of molds, bind to the (+) ends of microfilaments, preventing their elongation. This leads to a drastic change in cell shape (Figure 6-15), reflecting partial collapse of the cytoskeleton. Cytochalasins inhibit movement of cultured animal cells and cytokinesis, but do not block mitosis. Cells that reach cell division can complete mitosis in a normal way, but because cytokinesis is prevented, a binucleated cell is produced. From these and other experiments we have learned that assembly of microfilaments is necessary for migratory movement of cells and for the operation of the contractile ring that pinches an animal cell into two at division.

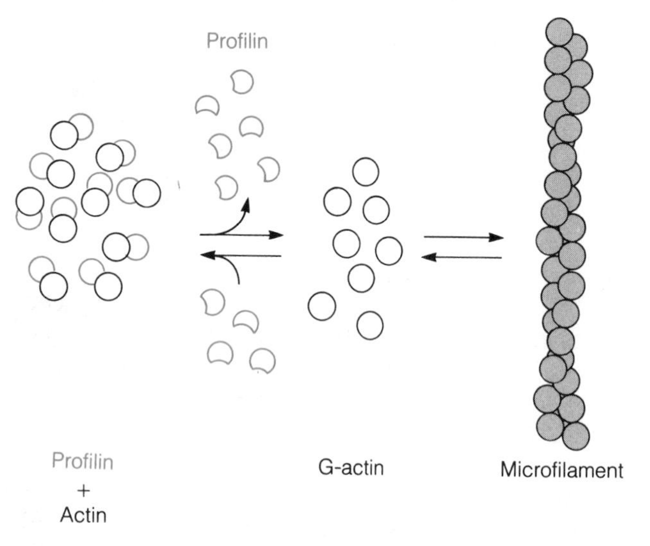

FIGURE 6-16
Profilin binds to G-actin and blocks addition of G-actin to microfilaments. Profilin, in effect, reduces the number of G-actin molecules available for assembly into microfilaments, and therefore its action contributes to disassembly of microfilaments.

Actin-Binding Proteins

More than 25 different proteins have been identified that bind reversibly to actin and affect the assembly, disassembly, form, and functions of microfilaments in different ways. Most of these actin-binding proteins are believed to have roles in regulating the microfilament component of the cytoskeleton. Others cross-link microfilaments to microtubules or other structures. Most actin-binding proteins fall into one of several classes, described below.

Proteins that bind to G-actin. Several kinds of proteins bind to G-actin and inhibit its assembly in microfilaments (Figure 6-16). An intensively studied example is a low molecular weight (16,000) protein called **profilin.** Profilin has been detected in many eukaryotes and has been isolated from various mammalian tissue, the slime mold *Physarum,* and the small soil amoeba *Acanthamoeba.* One profilin molecule binds reversibly to one molecule of G-actin. The addition of profilin to purified actin greatly retards both formation of nucleation structures and elongation of microfilaments from nucleation structures. In the cell, profilin almost certainly has an important role in controlling assembly of microfilaments by regulating the availability of G-actin. Since the binding of profilin to G-actin is reversible, the amount of G-actin available for assembly in the cell is presumably governed by a mechanism regulating binding of profilin to G-actin, but how binding is regulated is not known.

Proteins that cap the ends of microfilaments. Several kinds of proteins bind to the actin molecules at the ends of microfilaments and form caps that greatly retard addition and loss of actin molecules (Figure 6-17). Different capping proteins bind specifically either to the (+) end or (−) end. Most of those identified so far in various eukaryotic cells cap the (+) end and resemble in this respect cytochalasins, which bind only to the (+) ends. It is very likely that capping proteins play a role in controlling the elongation and disassembly of microfilaments. For example, in skeletal muscle cells all the actin filaments have exactly the same length, and each is capped at both ends by proteins specific for the (+) and (−) ends. Little is known about

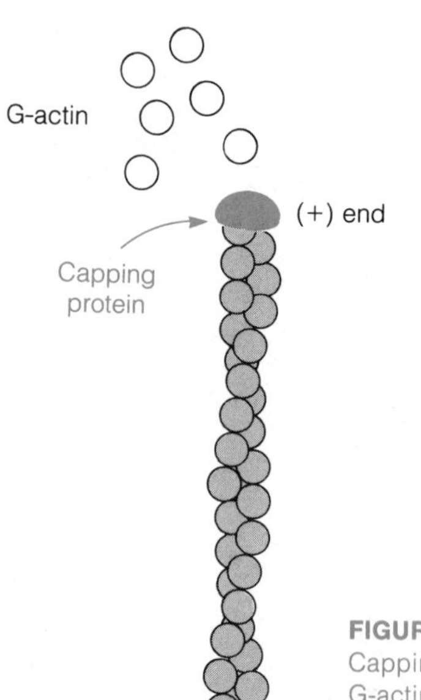

FIGURE 6-17
Capping protein blocks addition of G-actin to a microfilament, in this case at the (+) end.

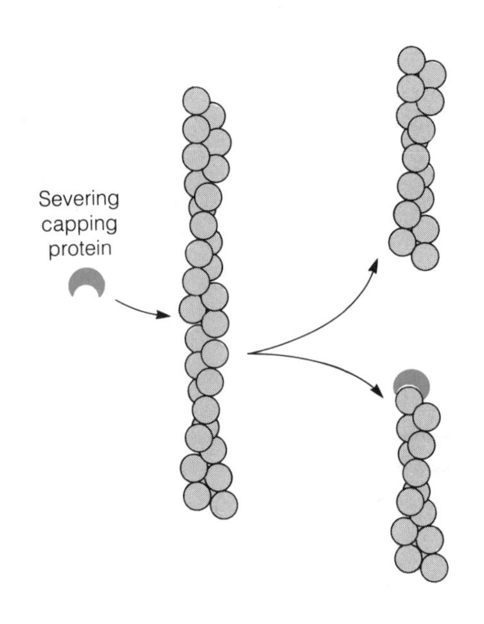

FIGURE 6-18
Severing-capping proteins break microfilaments and block growth at the new (+) end created by severing.

regulation of binding of capping proteins to microfilaments. Capping proteins may also serve to anchor microfilaments to other cell structures, for example, the plasma membrane (discussed later in this chapter).

Proteins that cap and sever microfilaments. Some capping proteins also sever microfilaments, breaking them into shorter structures. **Gelsolin** is a protein that can cap (+) ends of microfilaments, but it also binds to the sides of microfilaments and disrupts the noncovalent bonds between two successive actin molecules in the chain, thereby breaking the microfilament at that point (Figure 6-18). In the process of severing a microfilament, the gelsolin molecule binds to the new (+) end created by the breakage and inhibits assembly or disassembly of that microfilament fragment; that is, it acts as a cap.

The fragmentation of microfilaments allows for very rapid shortening of microfilaments and could have a major role in restructuring microfilaments during cell movement. The capping and severing activities of gelsolin are affected differentially by the concentration of Ca^{2+} ions. At a low concentration of Ca^{2+} ions, capping occurs but severing does not. At a high concentration of Ca^{2+} ions, capping action is increased and severing is induced. Other proteins with capping and severing activity have been isolated from a variety of eukaryotes, including slime molds, sea urchins, chickens, and mammals.

Proteins that cross-link microfilaments. A number of different proteins, isolated from various eukaryotic species, can cross-link microfilaments. To cross-link microfilaments, such a protein must have at least two binding sites for F-actin. Two general kinds of three-dimensional structures are produced by cross-linking of microfilaments: microfilaments aligned in parallel to form **bundles** and microfilaments arranged in various orientations to form **meshworks.**

By fluorescent antibody staining and by electron microscopy one can easily observe bundles of microfilaments, such as stress fibers, in intact cells (Figure 6-12). Bundles can be formed in cell-free systems and identified by various optical methods and electron microscopy. In general all the microfilaments in a bundle have their (+) ends pointing in the same direction.

Cross-linking proteins isolated from sea urchin eggs, pig brain, and starfish sperm are monomers with two binding sites for F-actin. These proteins bind at intervals of 13 nm along adjacent microfilaments (at every 4 to 5 F-actin molecules) to produce tightly packed bundles in which the microfilaments are about 9 nm apart (Figure 6-19(a)). An example is the protein **fimbrin,** which tightly crosslinks microfilaments in the bundles present in microvilli (Figure 6-13). Still other cross-linking proteins from other organisms produce bundles with different spacings between microfilaments.

Microfilaments are cross-linked into networks by other kinds of proteins. One of these, called **actin-binding protein,** is a dimer composed of two large subunits. The dimer is a rod 160 nm long and about 4 nm in diameter. An F-actin binding site is present near each end. It cross-links microfilaments, holding them in *perpendicular* orientations with respect to one another (Figure 6-19(b)). Such orthogonal arrays of mi-

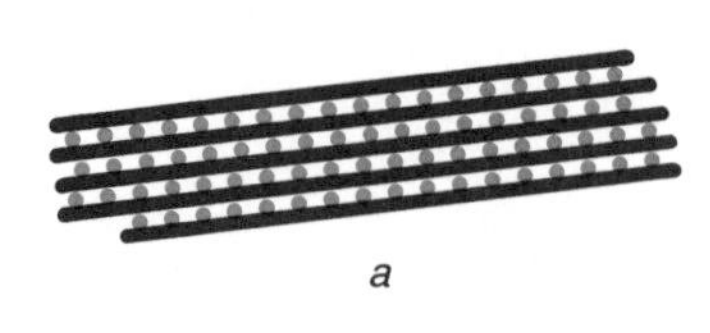

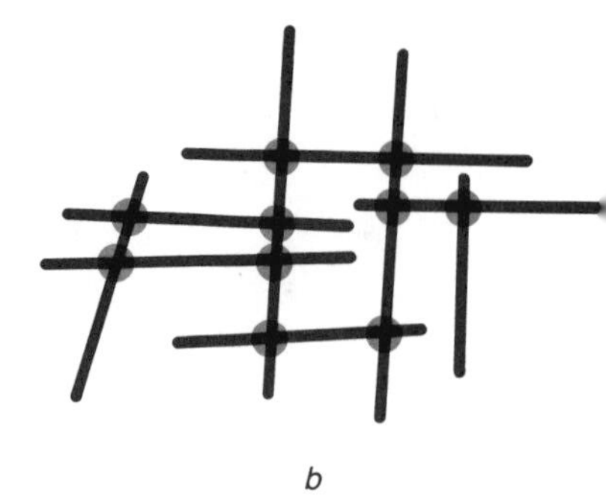

FIGURE 6-19
Cross-linking of actin microfilaments. (a) Microfilaments cross-linked into a bundle by the protein fimbrin (red). (b) Cross-linking of microfilaments into an orthogonal array by actin-binding protein (red).

crofilaments, cross-linked by actin-binding protein, are present in the cortical layer of cytoplasm just inside the plasma membrane of macrophages. The microfilament network is responsible for the gelled state of the cortex. Myosin is present in microfilament networks. Interaction of myosin and microfilaments is believed to cause contraction of the cortical gel, which may be the force-generating process in amoeboid movement (Chapter 14).

Interaction of Microfilaments with Other Cellular Components

Microfilaments are a major part of the cytoskeleton in eukaryotic cells. Much evidence also suggests that microfilaments are contractile elements and function like cellular muscles as well as cytoskeleton in nonmuscle cells. Two interactions with other cellular components are important for the contractile function of microfilaments: interaction with myosin filaments and interaction with the plasma membrane, as discussed next. Strands of material, probably proteins, appear in the electron microscope to connect microfilaments with microtubules, and possibly with intermediate filaments and various other organelles. These connections have not yet been well defined and are not understood, except as a general mechanism of structural integration of the cytoskeleton.

Association of Myosin with Microfilaments

In skeletal muscle cells microfilaments are arranged in orderly bundles (Chapter 14). The microfilaments interact with filaments made of myosin, a major protein of the contractile machinery of muscle cells. The myosin filaments, called **thick filaments** in muscle cells, are arranged parallel to the actin filaments (Figure 6-20). Contraction of a muscle cell is achieved by transient binding of microfilaments to myosin thick

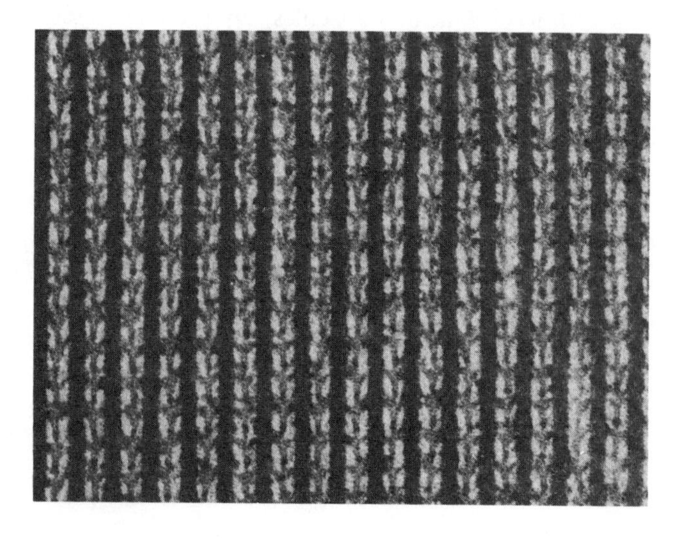

FIGURE 6-20
Electron micrograph of a small area of a skeletal muscle cell showing a highly ordered arrangement of myosin thick filaments and actin thin filaments. Crossbridges between the thin and thick filaments give the thin filaments the appearance of rows of chevrons. [Courtesy of Roger Craig.]

filaments and sliding of the two kinds of filaments along each other; during sliding they pull along each other, producing contraction. The process is described in detail in Chapter 14.

Muscle cells are specialized for contraction and are packed with microfilaments and myosin filaments, which are organized into a highly ordered, cytoskeletal pattern. Other types of cells are not specialized for contraction but do require contractile machinery for changing cell shape and migrating by the amoeboid-like motion commonly observed in macrophages, fibroblasts, and slime molds. The contractions that underlie these movements in nonmuscle cells are, as in muscle cells, almost certainly produced by interaction of microfilaments with myosin. Myosin molecules are associated both with bundles of microfilaments and microfilaments organized in networks. However, myosin filaments in nonmuscle cells are difficult to visualize, probably because they are short and not nearly as plentiful as in muscle. Fluorescent-antibody labeling has shown that myosin molecules are present at regular intervals along microfilaments in stress fibers. However, the details of the arrangement between microfilaments and myosin filaments and of how they interact to produce contraction are still being worked out.

Interaction of Microfilaments with the Plasma Membrane

Bundles of microfilaments such as stress fibers are usually seen terminating at the plasma membrane, which implies some relationship of microfilament ends to the plasma membrane. The molecular nature of this interaction is not yet clear. Microfilaments do not bind directly to integral proteins at the cytoplasmic surface of the plasma membrane. Instead several kinds of proteins together form an intermediate linkage (Figure 6-21). The microfilament-associated protein, α-actinin (present also in muscle cells), and proteins called **vinculin** and **talin** are found where microfilament bundles terminate at the plasma membrane. It is believed that these proteins form a connection between the microfilaments and one or more kinds of transmembrane glycoproteins of the plasma membrane.

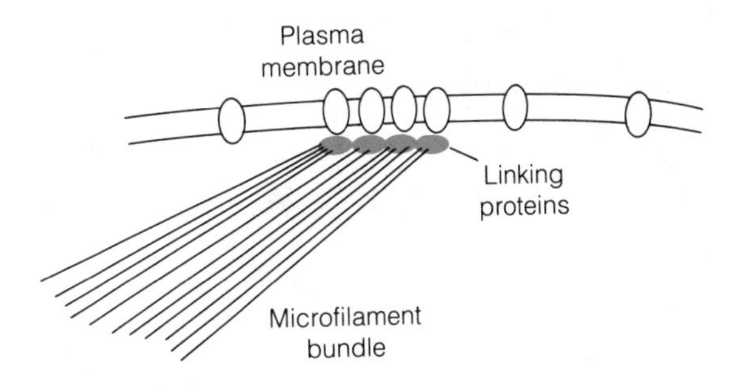

FIGURE 6-21
Schematic diagram of linking proteins (e.g., vinculin and talin) connecting ends of microfilaments to integral proteins in the plasma membrane.

In cultured cells, some of the transmembrane glycoproteins projecting from the outer surface of the plasma membrane attach the cell to the surface of the culture vessel. The glycoproteins do not stick directly to the surface. Instead they stick to proteins secreted by the cells into the medium. These proteins attach to the surface of the culture vessel, forming an **extracellular matrix** (Figure 6-22). Two major proteins of the extracellular matrix are **fibronectin** and **laminin.** Transmembrane glycoproteins near the end of the microfilament bundle then stick to the extracellular matrix. The points of the cell's attachment to the substratum are called **adhesion plaques,** and these plaques are numerous. Fibronectin and laminin are also principal components of the extracellular matrix in tissues and have major roles in anchoring cells and providing a molecular surface for migrations of cells, particularly during development.

A clear example of the relationship between a microfilament bundle and an adhesion plaque is seen in long, thin cytoplasmic extensions from cultured cells called **microspikes** (Figure 6-23). A microspike terminates within an adhesion plaque that attaches it to the surface of the culture vessel. A microspike contains a bundle of microfilaments, which provides structural stability. The microfilament bundle terminates at the tip of the microspike, where the adhesion plaque is located.

The bundle of microfilaments in a microvillus of an intestinal epithelial cell terminates at the plasma mem-

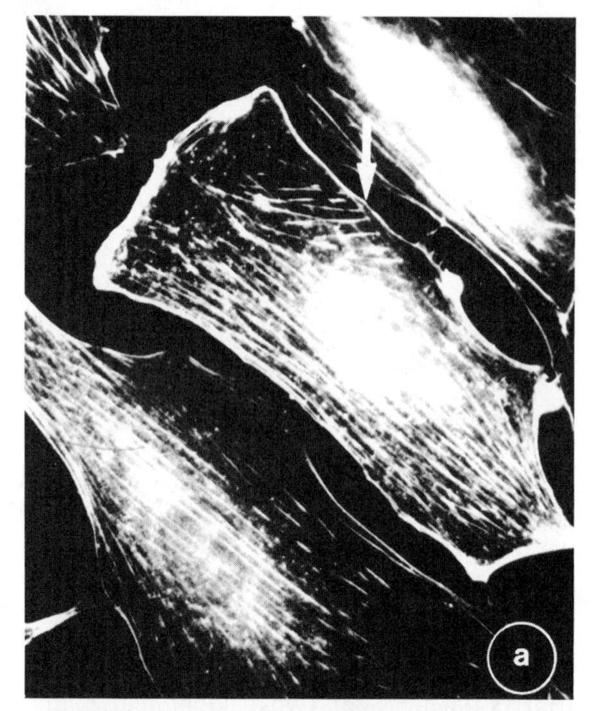

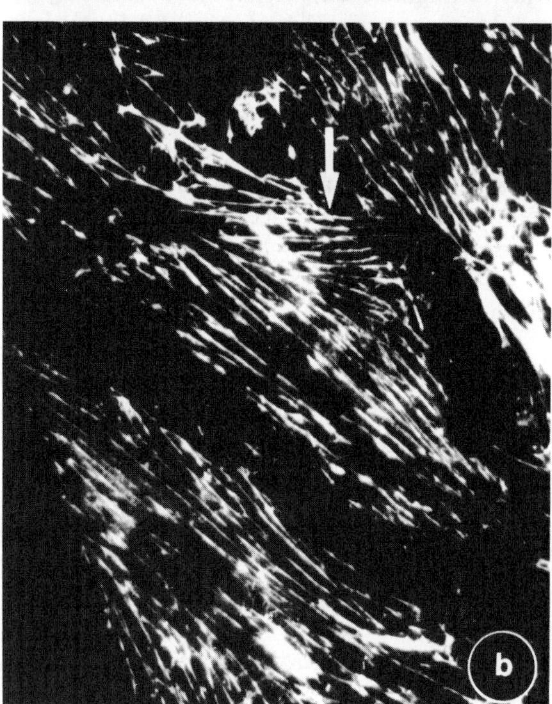

FIGURE 6-22

Light micrograph of fibronectin and actin stained by the fluorescent antibody technique. (a) Bundles of fibronectin fibers outside three rat fibroblasts in culture are attached to the surface of the culture vessel. (b) Microfilament bundles inside the three cells. The orientation of microfilament bundles inside the cells corresponds to the orientation of fibronectin bundles in the extracellular matrix. In some areas (arrows) the fibronectin bundles extend beyond the edge of the cell. [Courtesy of Richard O. Hynes; R. O. Hynes, and A. T. Destree 1978. *Cell,* 15:875.]

brane at the tip of the microvillus (Figure 6-13). Here the microfilaments are attached to the plasma membrane presumably through other proteins. The tip of a microvillus projects into the intestinal lumen, and no adhesion plaque is present. The other end of the microfilament bundle terminates in the cytoplasm in a network of microfilaments called the **terminal web,** below the base of the microvillus.

FIGURE 6-23

Scanning electron micrograph of a neuroblastoma cell (nerve cell tumor) with many microspikes. The long microspikes at the base of the cell anchor it to the surface of the culture vessel. [Courtesy of Keith R. Porter.]

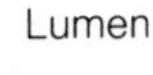

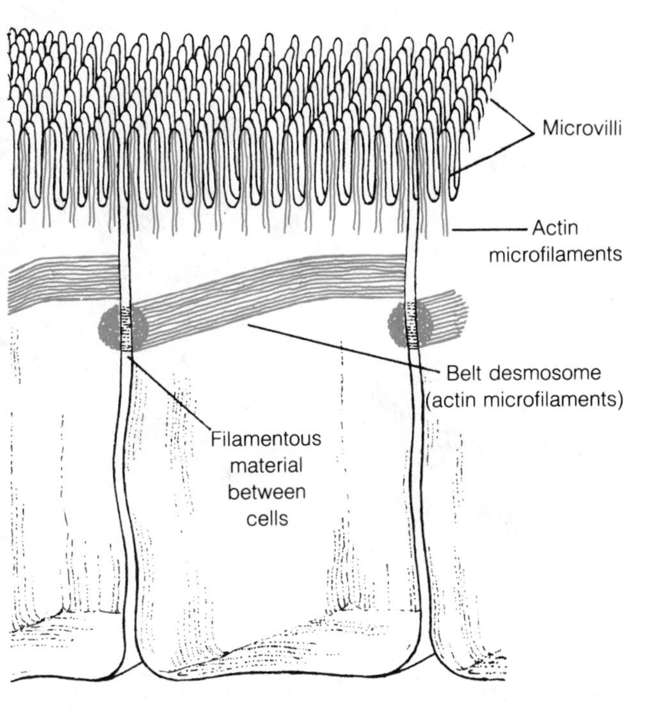

FIGURE 6-24
Schematic drawing of intestinal epithelial cells. Belt desmosomes (red bands) consist of bundles of actin microfilaments that form a continuous belt around the inner surface of the cell. Adjacent cells are connected by filamentous material in the region of belt desmosomes. Actin microfilaments fill the microvilli and extend down into the cytoplasm. Nuclei and other organelles have been omitted from the drawing.

Microfilaments that terminate at the plasma membrane are all arranged with (+) ends at the plasma membrane and (−) ends extending into the cytoplasm. The (−) ends are capped with proteins and are possibly joined to other structures.

In skeletal muscle cells the (+) ends of microfilaments are anchored. This anchorage is essential in producing contraction when microfilaments interact with myosin (Chapter 14). Since (+) ends of microfilaments in nonmuscle cells are anchored to the plasma membrane, they are in a position to exert an inward pulling force on the plasma membrane as they interact with short myosin filaments. Just how this might contribute to cellular movement is not clear. Since stress fibers are in effect anchored at least at one end by adhesion plaques, these fibers could exert a pulling force on a cell toward the adhesion plaque. The sides of microfilaments can also attach at multiple sites along their length to the plasma membrane. This can be seen in microvilli (Figure 6-13); thin strands of material, made of protein, project from the surface of microfilaments and appear to attach to the plasma membrane.

The bundles of microfilaments that form the contractile ring of cytokinesis must be joined along their length to the plasma membrane, since contraction of the ring progressively pulls the plasma membrane into the cleavage furrow as the cell is pinched into two. Also, epithelial cells, which form sheets of cells that cover the surfaces of tissues—for example, the epithelial cells that line the intestines—adhere to each other around their entire circumference by junctions called belt desmosomes (Figure 6-24). Just inside the plasma membrane where desmosomes occur is a substantial band of microfilaments; these form a continuous ring that appears to be connected, probably through other proteins, to the plasma membrane. The function of the microfilament ring presumably is to add reinforcement to the cell–cell binding (see Chapter 15).

INTERMEDIATE FILAMENTS

The third kind of fibers of the cytoskeleton is **intermediate filaments.** They are present in most eukaryotic cells, although they are far more prominent in animal cells than in other cell types and have been studied most intensively in animal cells (Figure 6-25). Intermediate filaments are long, straight or slightly bent rods with diameters between 7 and 12 nm.

Intermediate filaments are protein fibers occurring in bundles or networks that could provide considerable physical strength to the cytoskeleton. They are far more stable than microtubules and microfilaments. They are not disrupted by treatments that extract almost all other proteins from a cell, including tubulin and actin, for example, by treatment with a concentrated salt solu-

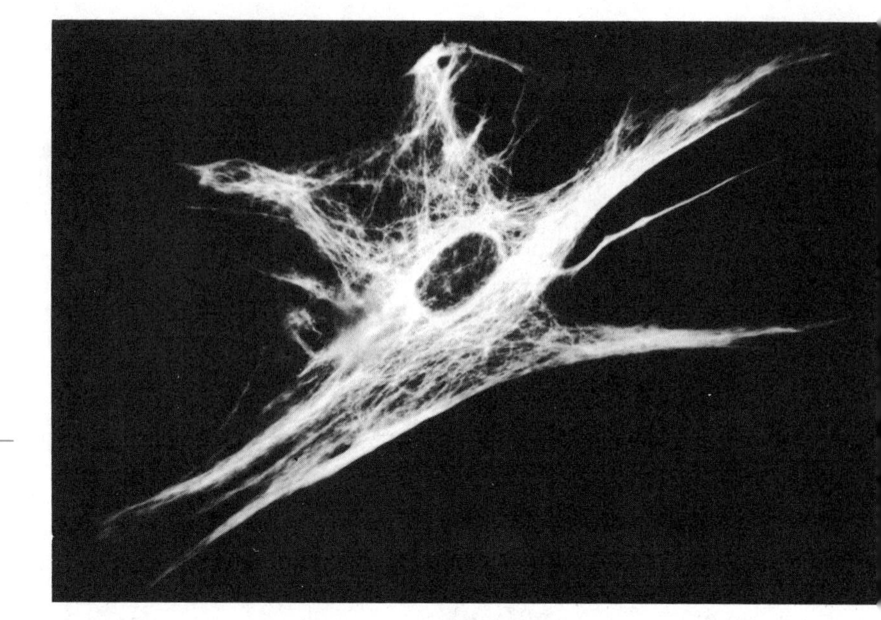

FIGURE 6-25
Light micrograph of a cell stained by the fluorescent antibody technique showing the distribution of intermediate filaments throughout the cytoplasm. Intermediate filaments are not present in the nucleus. [Courtesy of Michael W. Klymkowsky.]

tion or with detergent. Intermediate filaments are particularly prominent in cells that normally are subjected to mechanical stress, such as muscle cells. Nerve axons, which are long and thin, contain bundles of intermediate filaments, which give the axon great resistance to mechanical breakage. In a sheet of epithelial cells, each cell, in addition to being joined to its neighbors by belt desmosomes, is connected by **spot desmosomes.** These are specialized areas at the cell surface at which cells in a tissue can strongly attach to one other (Figure 6-26; also Chapter 15). The many spot desmosomes on the surface of an individual epithelial cell are connected to one another inside the cell by a network of intermediate filaments. Thus, the network of intermediate filaments in one cell connects through spot desmosomes with the networks in its immediate neighbors. In effect this interconnecting produces a continuous network throughout the epithelial sheet that gives the sheet mechanical strength.

An extreme example of mechanical strength conferred by intermediate filaments occurs in epithelial cells that form the epidermis of the skin in mammals (Figure 6-27). Only the cells in the basal layer of the epidermis reproduce. When a cell in the basal layer

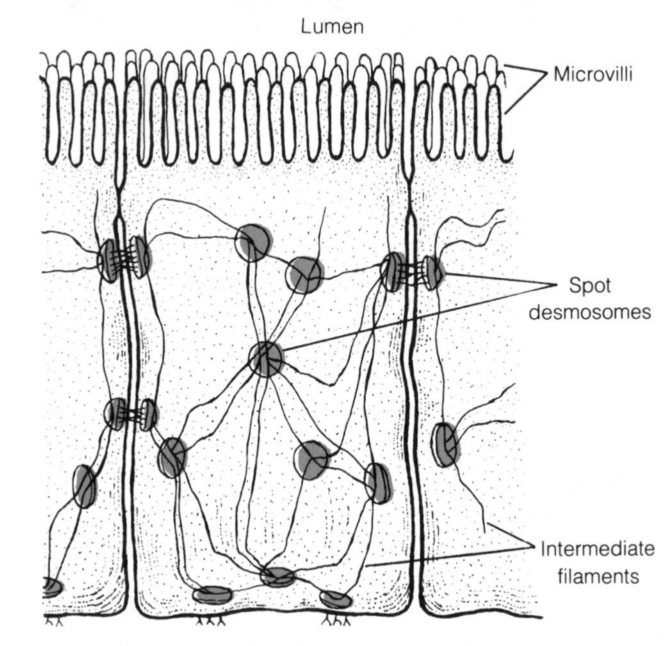

FIGURE 6-26
Epithelial cells of the small intestine. Spot desmosomes (red) are connected inside the cell by a system of intermediate filaments. Adjacent cells are connected by filamentous material at spot desmosomes.

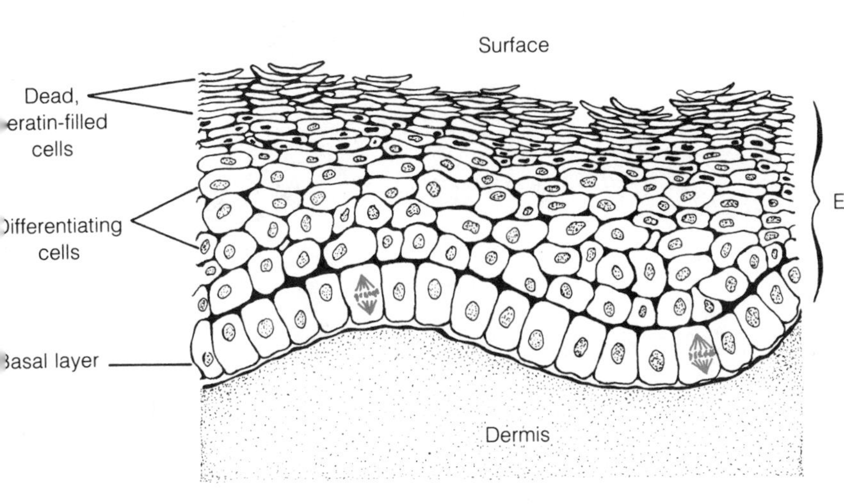

FIGURE 6-27 Drawing of animal epidermis. Cells in the basal layer reproduce and provide cells to the layer of differentiating cells. The differentiating cells become filled with keratin and form the outer layer of tough, flattened cells at the surface that slough off.

divides, one of the daughters moves into the layer just above and begins to differentiate and flatten. Differentiation is primarily the production of intermediate filaments made of the protein keratin. As the differentiating cells are pushed toward the surface of the skin by cell reproduction in the basal layer, they become more and more flat and packed with dense bundles of intermediate filaments. In the process they lose their nuclei and cytoplasmic organelles; for these cells an end point of differentiation is death. Many spot desmosomes bind the keratin-filled cells together into a multilayered structure. This structural arrangement gives skin great resistance to physical stress and chemicals, protects cells of the basal layer and below against ultraviolet radiation, and prevents loss of water from the body. At the surface of the epidermis, dead cells, which are little more than extremely flattened discs of keratin, are continuously sloughed from the surface.

Types of Intermediate Filaments

Five major classes of intermediate filaments can be distinguished on the basis of their distributions in different cell types and on the basis of the proteins of which they are made.

1. **Keratin filaments** are made of proteins called keratins or cytokeratins. Keratin filaments are present in epithelial cells or cells of epithelial origin.
2. **Desmin filaments,** composed of a protein called desmin, are found primarily in smooth, skeletal, and cardiac muscle cells.
3. **Vimentin filaments,** made of the protein vimentin, occur in many kinds of differentiating animal cells and animal cells grown in culture.
4. **Glial filaments,** consisting of the protein called **glial fibrillary acidic protein,** occur only in glial cells of the central nervous system. There are several kinds of glial cells, each performing a supportive function for nerve cells. Most of the volume of the brain is occupied by glial cells.
5. **Neurofilaments** are detected only in nerve cells.

In most cells only one type of intermediate filament is present, but some cells have more than one type. For example, both desmin and vimentin are present in muscle cells. On the other hand, there are also a few cell types (e.g., some nerve cells) that appear not to contain any type of intermediate filament protein.

Construction of Intermediate Filaments

About 30 different kinds of cytokeratins have been identified in keratin filaments. Nineteen are known for

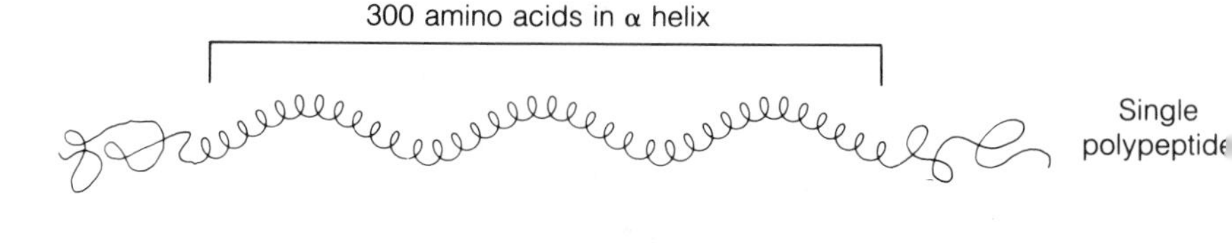

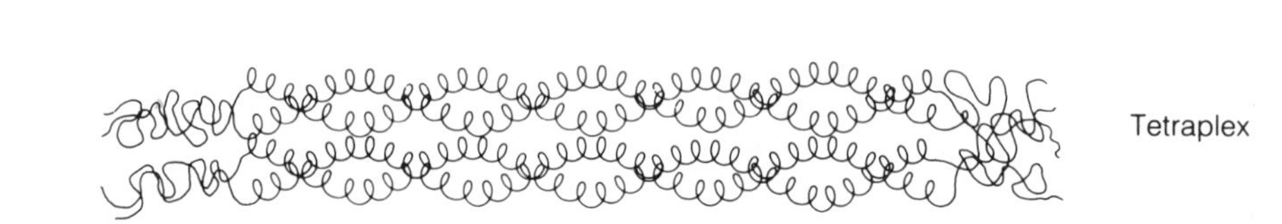

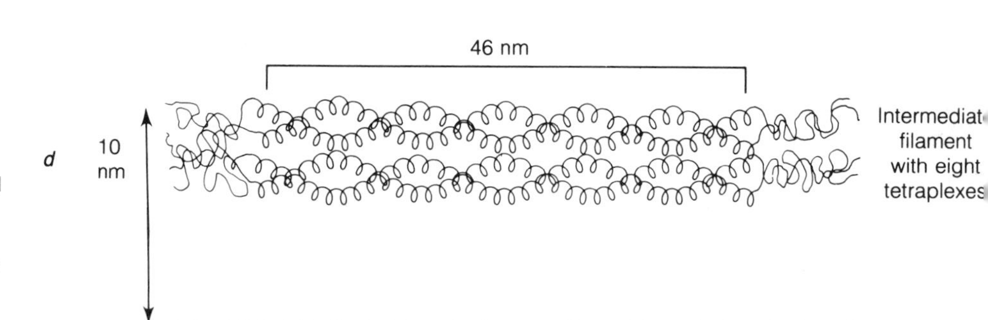

FIGURE 6-28
Structure of an intermediate filament. (a) Polypeptides of an intermediate filament contain a central region of α-helix with 310 amino acids. (b) The α-helical regions of polypeptides coil around one another to make a duplex coiled coil. (c) Two coiled coils associated to form a tetraplex. (d) Eight tetraplexes combine to form a 46 nm section of an intermediate filament.

various types of human cells, with several different kinds occurring in one cell type. Eight kinds of cytokeratins are present in human hair. For vimentin, desmin, and glial fibrillary acidic proteins only a single type of protein has been identified. Three different proteins are joined in a complex to form a neurofilament. Regions of vimentin and desmin molecules have similar amino acid sequences, suggesting that they represent evolutionary variants of a single, ancient protein. Vimentin and desmin are similarly related to proteins called **lamins** that make up a major part of the nuclear matrix (Chapter 11).

The molecular weight of proteins of intermediate filaments ranges from 40,000 (one of the cytokeratins) to about 135,000 (one of the neurofilament proteins). Despite the differences in sizes of their component proteins, all intermediate filaments have the same rod-shaped morphology, although they differ somewhat in diameter.

The protein molecules of intermediate filaments are all built on a common plan. In each case the central section of about 300 amino acids forms an α-helix (Figure 6-28). The two end domains, which are folded in various ways, differ in size and account for the large range of molecular weights of the different proteins. In formation of a filament the central α-helical regions of two proteins coil around each other, and two such coils pair to form a **tetraplex.** Eight tetraplexes align side-by-side to form the basic rod-shaped unit of the filament (Figure 6-28). The proteins within one unit join end-to-end with proteins in successive units to produce a filament. The disposition and function of the end domains of the individual proteins in formation of a filament are not known. However, the end domains are thought to give the different intermediate filaments different properties. A protein called **fillagrin** cross-links cytokeratin-type intermediate filaments into bundles in skin cells.

Unlike microtubules and microfilaments, intermediate filaments are not in a dynamic state. There is no cytoplasmic pool of unassembled proteins for intermediate filaments, and disassembly and reassembly of protein subunits does not occur.

Function of Intermediate Filaments

The function of intermediate filaments is poorly understood. As the most insoluble part of the cytoskeleton, intermediate filaments appear to provide structural stability for the cell, but this has been difficult to verify experimentally. In epidermis the structural role of the dense bundles of keratin filaments seems clear, but the function of intermediate filaments in other kinds of cells is uncertain. Injection of antibodies to proteins of intermediate filaments causes severe disruption or collapse of the network of intermediate filaments, but this disruption has no effect on the organization of microtubules or microfilaments and does not affect cell shape, cell motility, or cell growth and division. Disruption of microtubules with colchicine results in collapse of the network of intermediate filaments, which implies that organization of intermediate filaments is dependent on microtubules and may be linked to them. These observations suggest that intermediate filaments may provide the cell with mechanical strength reinforcement, while cell shape and movement are determined by microtubules and microfilaments.

PROBLEMS

1. What are the three fibrous elements of the cytoskeleton?
2. How is the distribution of microtubules seen in a whole cell using a light microscope?
3. What is the role of the centrosome in formation of the cytoskeleton?
4. Describe the step-by-step self-assembly of a microtubule from tubulin molecules.
5. What is meant by treadmilling of microtubules?
6. How is GTP thought to influence the growth of microtubules?
7. What two observations with poisons suggest that assembly and disassembly of microfilaments are important in amoeboid movement?
8. What are two functions of fibronectin and laminin?
9. What is thought to be the function of intermediate filaments?
10. Compare the assembly and disassembly of microtubules and microfilaments with the assembly and disassembly of intermediate filaments.

THE GENETIC MATERIAL

The Transformation Experiment
The T2 Bacteriophage Experiment
Metabolic Stability of DNA

THE STRUCTURE OF DNA

REPLICATION OF THE DNA DOUBLE HELIX

Proof that DNA Replicates Semiconservatively
The Mechanism of DNA Replication

CHROMOSOMES AND CELL DIVISION

The Cell Cycle
Mitosis
The Karyotype of a Cell
Tracing the Replication and Mitotic Distribution of DNA
Production of Germ Cells: Meiosis
Fertilization

MENDELIAN PRINCIPLES OF INHERITANCE

Rule of Segregation
Rule of Independent Assortment

GENE LINKAGE, CROSSING OVER, AND CHROMOSOME MAPPING

Gene Linkage
The Cytology of Chromosome Pairing and Crossing Over
Mapping of Genes along Chromosomes
Determining the Locations of Genes in Human Chromosomes
Sexual Reproduction in Unicellular Eukaryotes

Yeast
Protozoa

CYTOPLASMIC INHERITANCE

Inheritance Through Chloroplasts
Mitochondrial Inheritance
Cytoplasmic Inheritance in *Paramecium*

CHAPTER 7

Cell Heredity

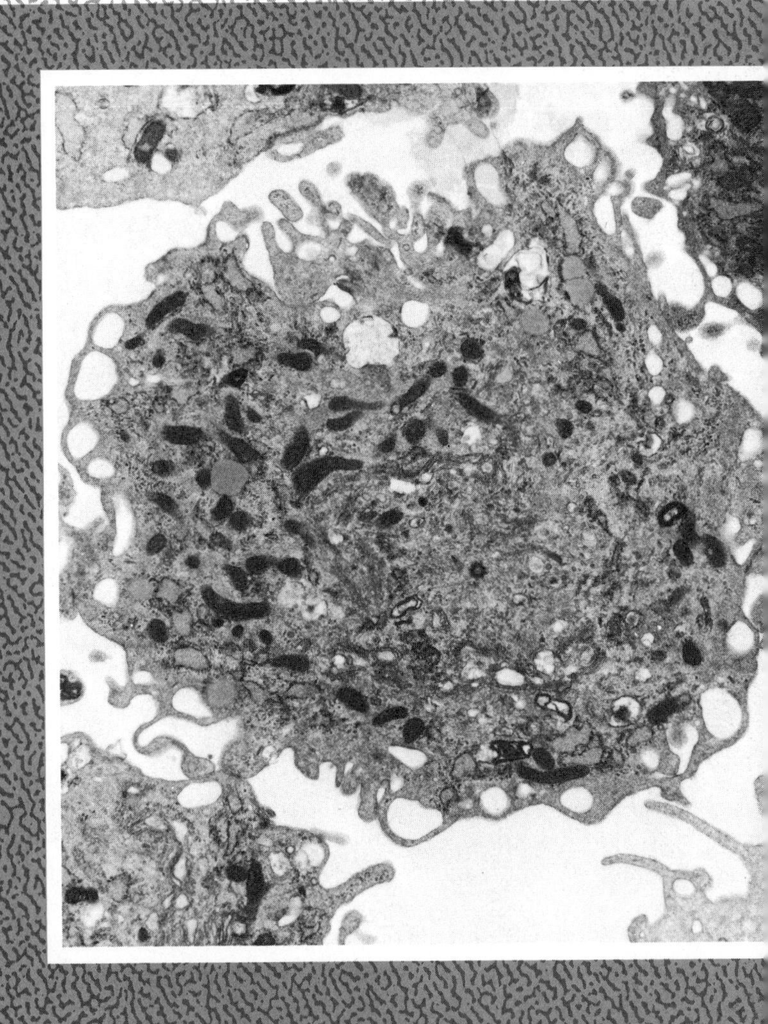

See Figure 7-40.

THE GENETIC MATERIAL

The discovery in the mid-nineteenth century that all cells come from preexisting cells through the process of cell division introduced in a general way the concept of inheritance from parental cell to daughter cell. Realization that this inheritance is based on genes carried by chromosomes came many years later, at the beginning of the twentieth century.

Chromosomes are complex structures consisting of DNA and many kinds of proteins. For a long time it was supposed that genes were made of chromosomal proteins, and DNA was hypothesized to be a structural polymer that helped to hold the chromosome together. A primary source of DNA for chemical study was the calf thymus gland, and DNA was for a while called thymus nucleic acid; this is the origin of the name of the base, thymine, which is present in DNA but not in RNA. RNA was easily obtained in large quantities from yeast cells and hence was for a while called yeast nucleic acid.

The Transformation Experiment

Proof that DNA is the genetic material in cells came from several experiments performed in 1944 and 1952. The first of these experiments was based on a 1928 study of inheritance of the ability of the bacterium *Pneumococcus* to cause pneumonia in mice. Some strains of *Pneumococcus* are highly virulent and some are nonvirulent. A virulent bacterium has a polysaccharide coat that gives bacterial colonies growing on nutrient agar a smooth surface (Figure 7-1). A nonvirulent bacterium lacks the polysaccharide coat, and colonies grow with a rough surface. The presence of the polysaccharide coat is a genetic trait inherited from bacterium to bacterium through cell division. When injected into a mouse, the virulent strain is protected by its polysaccharide coat from attack by white blood cells. Therefore, the bacteria multiply, cause pneumonia, and eventually kill the mouse (Figure 7-2). Nonvirulent bacteria injected into a mouse are destroyed by lymphocytes, and the animal is protected from disease. In a critical series of experiments mice were injected with

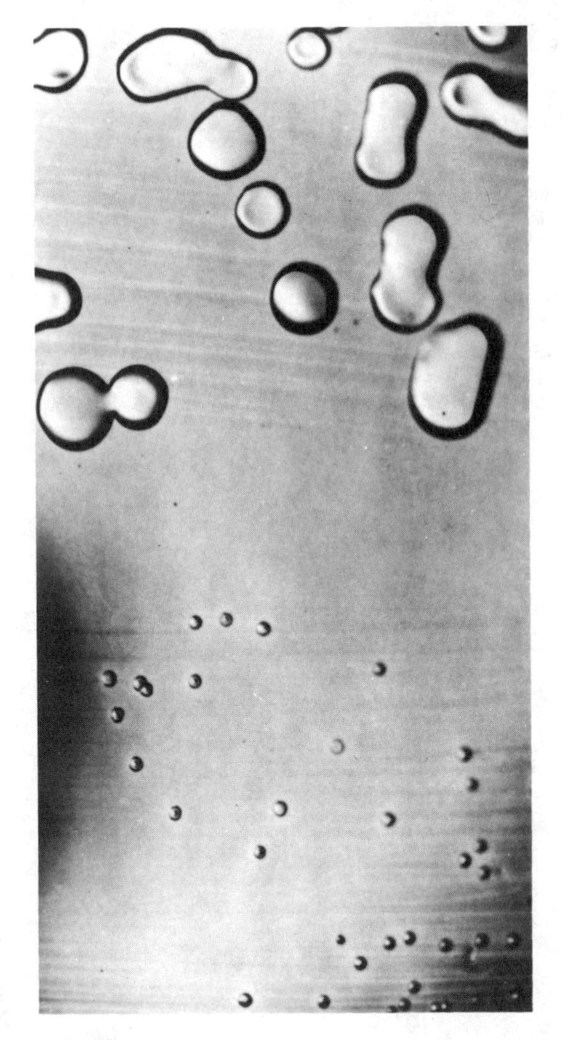

FIGURE 7-1
Transformation of avirulent R pneumococci (small colonies, below) to virulent S pneumococci (large colonies, above) by DNA from heat-killed virulent pneumococci. [Courtesy of Maclyn McCarty. Reproduced from the *Journal of Experimental Medicine*, 79:137–158 (1944) by copyright permission of the Rockefeller University Press.]

virulent *Pneumococcus* that had been killed by heating to 60°C. Since the bacteria could not proliferate, no disease resulted. In another trial heat-killed, virulent bacteria *plus* nonvirulent, live bacteria were injected together into mice. Unexpectedly, the animals died and virulent bacteria (with polysaccharide coats) were isolated from the dead animals. It appeared that in some way the live nonvirulent bacteria had acquired from the dead virulent bacteria the genetically prescribed

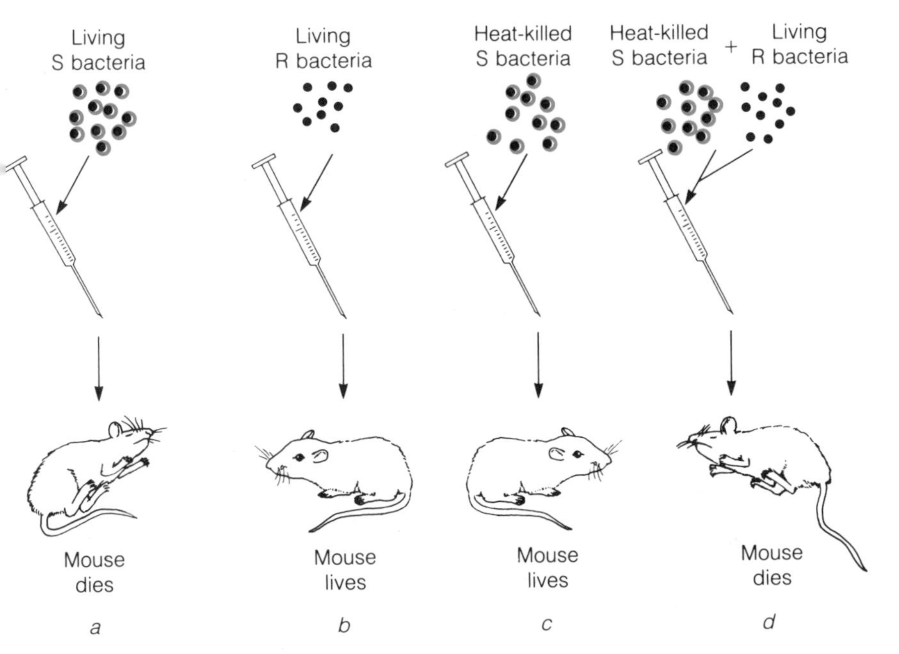

FIGURE 7-2
The pneumococcus transformation experiment. (a) Mouse injected with living S (virulent) bacteria dies of pneumonia. (b) Mouse injected with living R (avirulent) bacteria remains healthy. (c) Mouse injected with heat-killed S (virulent) bacteria remains healthy. (d) Mouse injected with both heat-killed S (virulent) bacteria and living R (avirulent) bacteria dies of pneumonia.

ability to synthesize a polysaccharide coat. In 1944 such transformation of an inherited (genetic) trait was achieved by addition of an *extract* of virulent bacteria to a culture of living nonvirulent bacteria (Figure 7-3). Hence, it was clear that some chemical material, called a **transforming principle,** was present in the bacterial extract and was responsible for conferring a new genetic trait on nonvirulent bacteria.

The chemical nature of the transforming principle was identified when it was partially purified from an extract of virulent bacteria. Treatment of these extracts with enzymes that attack either protein, RNA, or DNA showed that only extracts in which DNA was retained possessed transforming principle. DNA was then isolated from virulent bacteria, added to nonvirulent bacteria, and about one out of 1000 of these bacteria was transformed by this DNA to the virulent type. The change was permanent, since virulence was transmitted to all progeny bacteria during subsequent cell divisions.

These experiments did not convince all biologists that DNA was the genetic material; some continued to adhere to the hypothesis that genes were made of proteins. Chemical analyses of DNA had suggested that the polymer was made up of a regular repeating arrangement of four nucleotides, one each of adenine, thymine, guanine, and cytosine—the tetranucleotide hypothesis. Such a regular chemical structure could not

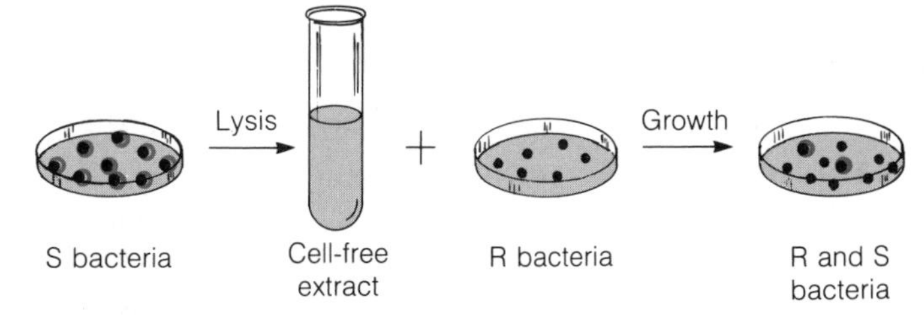

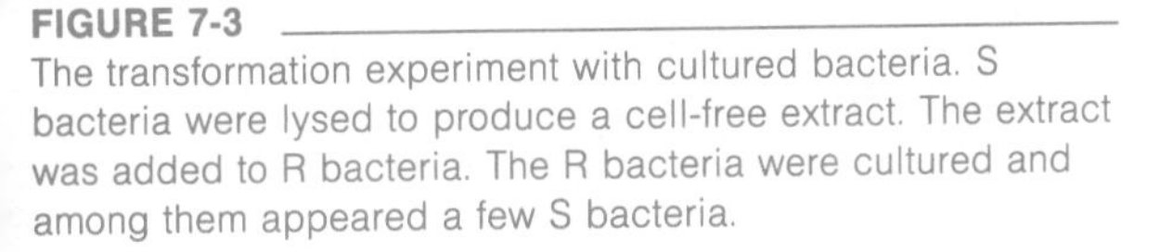

FIGURE 7-3
The transformation experiment with cultured bacteria. S bacteria were lysed to produce a cell-free extract. The extract was added to R bacteria. The R bacteria were cultured and among them appeared a few S bacteria.

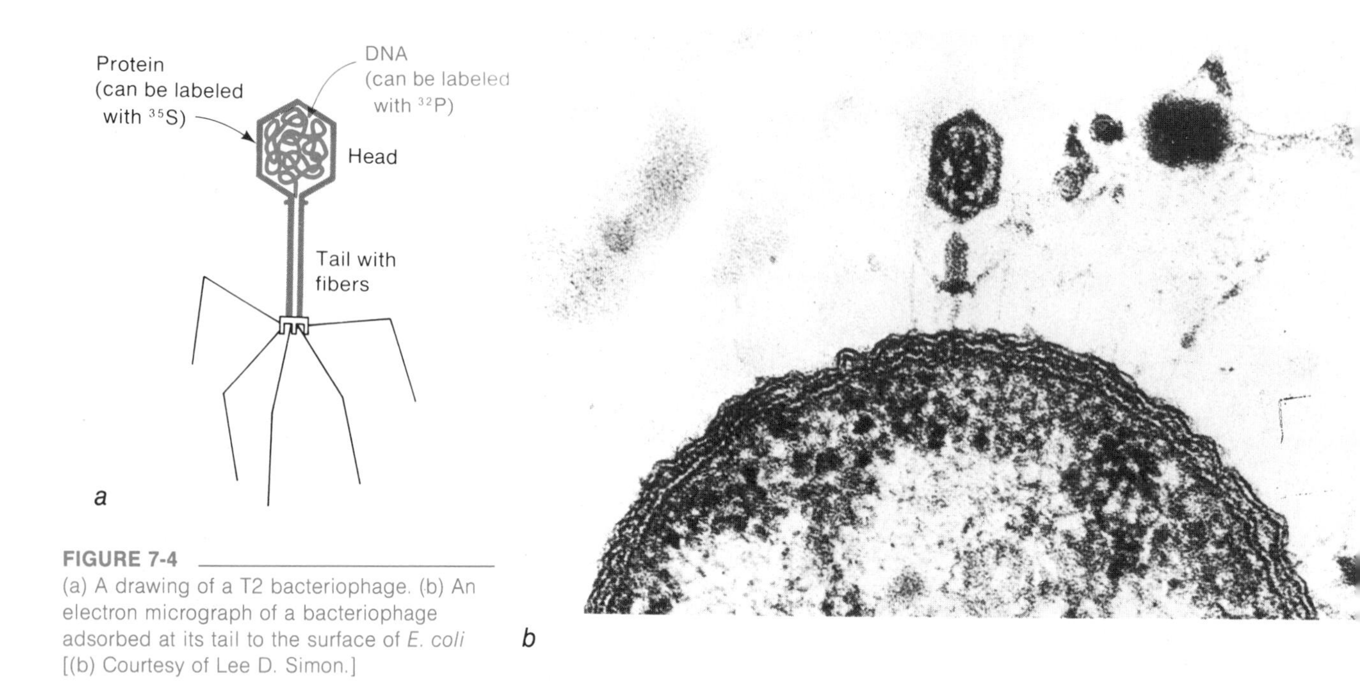

FIGURE 7-4
(a) A drawing of a T2 bacteriophage. (b) An electron micrograph of a bacteriophage adsorbed at its tail to the surface of *E. coli* [(b) Courtesy of Lee D. Simon.]

possibly contain the information needed to make many different genes. Improved methods for determining base composition and examination of many organisms showed that the four nucleotides are rarely found in a 1:1:1:1 ratio. These data were inconsistent with the hypothesis that DNA is a repeating tetranucleotide and indicated that the sequences of the four nucleotides could be enormously variable, providing a complexity of chemical structure that could meet the requirements for the genetic material.

The concept of DNA as the genetic material received confirmation from a 1952 experiment with a bacterial virus.

The T2 Bacteriophage Experiment

An experiment in 1952 with the bacterial virus **T2 bacteriophage** showed that genes must be made of DNA. This virus consists of proteins and DNA. Proteins of the virus were labeled with radioactive sulfur (^{35}S), which is incorporated into the two sulfur-containing amino acids, methionine and cysteine, by growing virus-infected *E. coli* in the presence of $^{35}SO_4$. In a separate experiment viral DNA was labeled with radioactive phosphorus (^{32}P) by growing the virus-infected bacteria in medium containing $^{32}PO_4$. When bacteriophages (or **phages**) containing ^{35}S were used to infect *E. coli*, most of the ^{35}S did not enter the cells, showing that little viral protein had entered the cells. By contrast, when ^{32}P-labeled viruses were used, most of the ^{32}P entered the cells. This indicated that only phage DNA was necessary to produce an infection in which new virus particles with a full complement of viral proteins were produced. Hence, the DNA had to be the genetic material.

Details of the experiment are as follows: a T2 phage has a long tail by which it attaches to the surface of a bacterial cell (Figure 7-4). Attached phages can be detached from the bacterial surface by violent agitation of the medium in an ordinary kitchen blender. Hence this experiment is sometimes called the "blender experiment." In the experiment, phages labeled in their proteins with ^{35}S were allowed to attach to bacteria. The bacteria, with attached phages, were collected by centrifugation to separate them from unattached phages. The bacteria were resuspended in fresh medium and agitated vigorously in a blender to shear the phages from the bacterial surface. The bacteria were again collected by centrifugation, which left the detached phages in suspension. Most of the ^{35}S was found with the suspended phages. When the entire procedure was repeated with phages labeled in their DNA with ^{32}P, little of the ^{32}P appeared in the final suspension and most was at the bottom of the centrifuge tube. When the bacteria at the tube bottom were resuspended and incubated in nutrient medium, they produced progeny phages, showing that they had been successfully infected. Thus, the ability of a bacterium to produce new phages was associated predominantly with the transfer of DNA into the bacterium rather than proteins.

Metabolic Stability of DNA

When DNA is labeled with ^{32}P during synthesis, it does not, at a later time, lose any radioactivity through DNA breakdown. By contrast, radioactivity incorporated in cellular RNA or proteins gradually disappears, because these macromolecules undergo constant breakdown and resynthesis. Such a high degree of metabolic stability of DNA is a predicted property of the genetic material, since breakdown would result in permanent loss of genetic information and eventual death of the cell. The absence of detectable metabolic turnover of DNA was another feature that supported the hypothesis that DNA is the genetic material.

THE STRUCTURE OF DNA

DNA consists of two very long polymeric chains that twist around each other to form a double helix. Each chain is made up of four monomeric units. Two are the pyrimidine deoxynucleotides, thymidine monophosphate (TMP) and deoxycytidine monophosphate (dCMP). The other two are the purine deoxynucleotides, deoxyadenosine monophosphate (dAMP) and deoxyguanosine monophosphate (dGMP) (Figure 2-25). Each deoxynucleotide consists of a nitrogen base (thymine, cytosine, adenine, or guanine) linked to a deoxyribose (a sugar) that is linked to a phosphate group, as described in Chapter 2.

Deoxynucleotides of the four types are joined together to form a long chain. The linkage between successive nucleotides is called a **phosphodiester bond,** in which the phosphate group of one deoxynucleotide is linked to the deoxyribose part of the next deoxynucleotide (Figure 7-5). This is called a 3′,5′-phosphodiester bond because the phosphate group on the 3′ carbon of deoxyribose becomes linked to the 5′ carbon of the next nucleotide by covalent bonds called an ester linkage. In this way thousands or even millions of the four deoxynucleotides are linked together to form an enormously long polymeric chain.

The two polymeric chains in DNA interact to form a stable double helix in which a thymine base in one chain is always paired to an adenine base in the other chain, and a guanine in one chain is always paired to a

FIGURE 7-5
Deoxynucleotides in DNA are joined in long chains by 3,5-phosphodiester linkages.

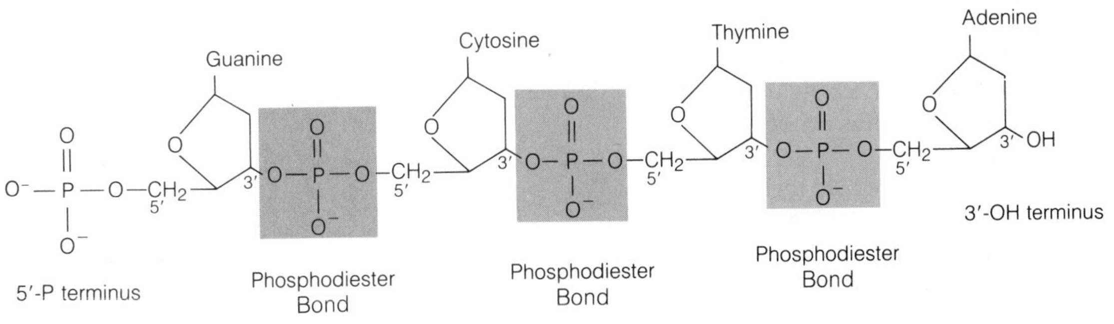

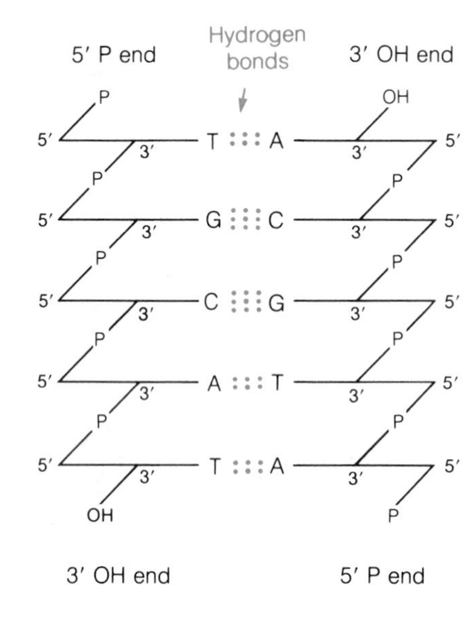

FIGURE 7-6
Two deoxynucleotide chains form a double helix in which thymine (T) in one chain is always matched with adenine (A) in the other chain, and guanine (G) is always matched with cytosine (C).

cytosine (Figure 7-6). Thus, for all cells the number of thymines is always equal to the number of adenines, and the number of guanines is equal to the number of cytosines, that is,

$$[A] = [T]$$

$$[G] = [C]$$

in which [ ] denotes mole fraction.

However, the number of AT pairs in a double helix of DNA can and usually does differ from the number of GC pairs. The ciliated protozoan *Tetrahymena* has one of the most AT-rich DNAs known, with about 80 percent AT base pairs and 20 percent GC base pairs. Human DNA contains about 60 percent AT and 40 percent GC base pairs, and *E. coli* DNA contains about 48 percent AT and 52 percent GC pairs.

The chains of the double helix are held together by two hydrogen bonds between each adenine and thymine and three hydrogen bonds between each guanine and cytosine (Figure 7-7). On the average, the DNA molecule in a human chromosome consists of about 1.3×10^8 base pairs. Since 60 percent are AT pairs and 40 percent are GC pairs, the two chains are joined by about 3.1×10^8 hydrogen bonds.

The ring portions of the purine and pyrimidine bases are flat structures whose planes are at right angles to the long axis in the double helix (Figure 7-8). The successive bases within each chain are stacked upon one another, with only a 12° rotation between them. Thus a double helix contains two complementary stacks of bases that wind around each other. The two kinds of weak forces, the hydrogen bonds between successive bases in opposite chains and hydrophobic interactions between bases within each chain, stabilize the double-helical structure.

A DNA chain has polarity because a phosphodiester linkage is between the 3′ carbon of one sugar and

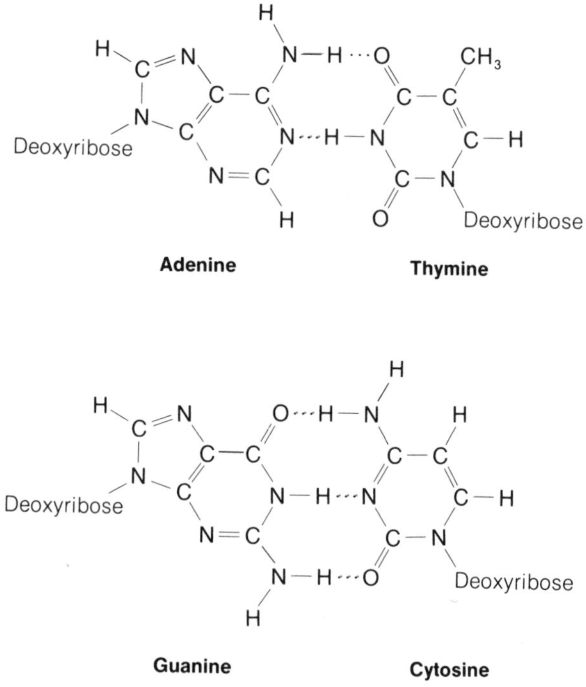

Deoxyribose — Adenine ··· Thymine — Deoxyribose

Adenine **Thymine**

Deoxyribose — Guanine ··· Cytosine — Deoxyribose

Guanine **Cytosine**

FIGURE 7-7
In a DNA double helix adenine binds by two hydrogen bonds to thymine, and guanine binds by three hydrogen bonds to cytosine.

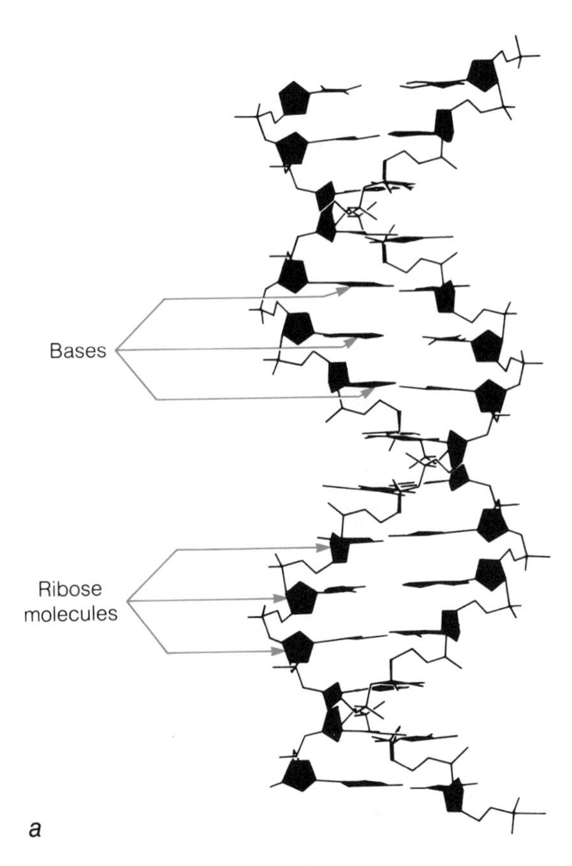

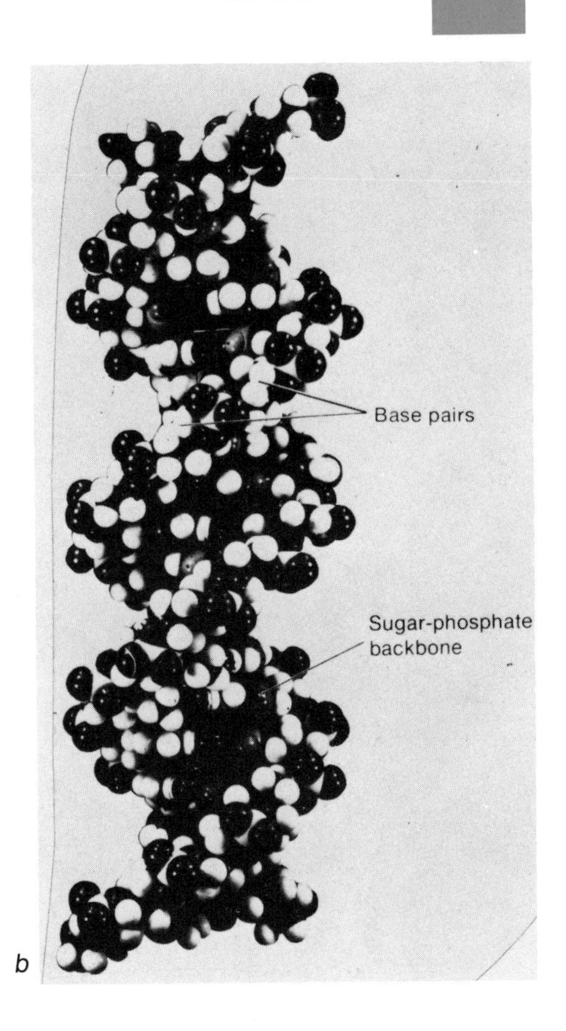

FIGURE 7-8
Structure of the DNA double helix. (a) A skeletal model. The bases, whose planes are at nearly right angles to the long axis of the double helix, are seen on edge between the chains. (b) A space filling model of a DNA double helix. The sugar-phosphate backbones of the two chains wind around the outside of the molecule. The bases are largely but not completely buried within.

the 5′ carbon of the next sugar. Thus, the ends of the chain are different: a deoxyribose 3′-OH group occurs at one end, and a 5′ carbon bearing a PO_4 group is at the other end (Figure 7-6). In a double-stranded structure (a duplex) the two chains have opposite polarities: at each end of a duplex one chain bears a 3′-OH and the other has a 5′-PO_4.

REPLICATION OF THE DNA DOUBLE HELIX

The discovery of precise base-pairing between the two chains of the double helix immediately suggests the mechanism by which DNA could replicate to produce two identical daughter double helices from one parental double helix.

Because adenine (A) is always paired with thymine (T) and guanine (G) is always paired with cytosine (C), the sequences of the two chains of the double helix are complementary to each other. For example, if a section of one chain is made up of nucleotides with the sequence of bases 5′-ATGGCGTTGA-3′, then the corre-

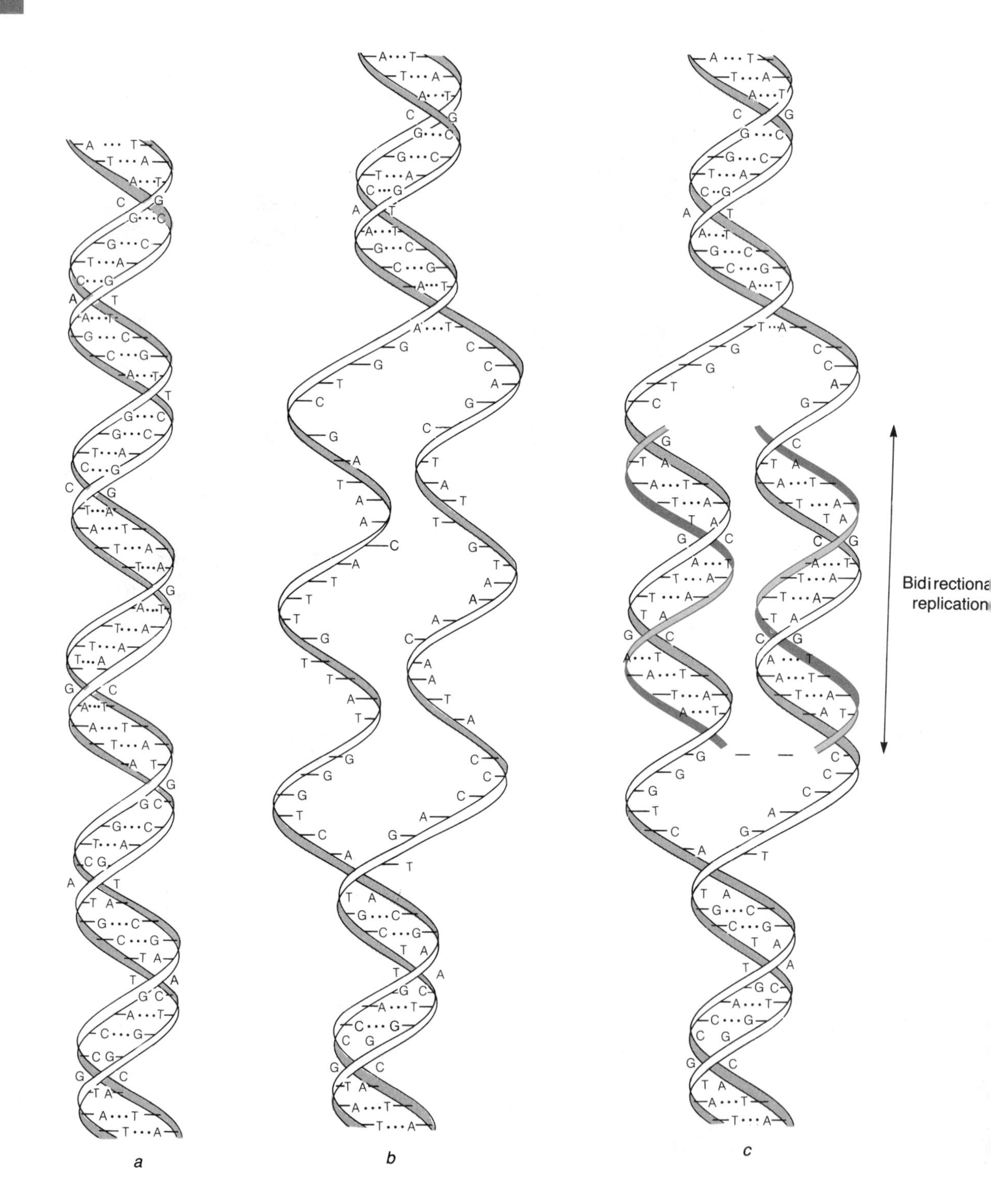
Bidirectiona
replication
a
b
c

FIGURE 7-9
Replication of DNA. (a) DNA double helix before replication. (b) Opening of the double helix at the start of replication. (c) Progression of replication bidirectionally along the parental double helix.

ponding section of the complementary chain will read 3′-TACCGCAACT-5′, forming a section of double helix or duplex with the composition,

5′-ATGGCGTTGA-3′
3′-TACCGCAACT-5′

This complementarity of the two chains suggested how the molecule might guide its own replication by a mechanism known as **semiconservative replication** of DNA. The first step in semiconservative replication is the separation of the two complementary chains from each other at the point in the DNA duplex where replication is to begin. The two chains are held together by hydrogen bonds, and these weak bonds are relatively easy to break. Cells contain a protein that specifically attaches to a DNA duplex in such a way that the two complementary chains become separated from each other over a short region (Figure 7-9(a)).

When localized disruption of the duplex has taken place, each of the two chains becomes free to serve as a **template** for the polymerization of a new complementary chain (Figure 7-9(b)). The enzyme DNA polymerase selects nucleotides that can pair with the bases in the DNA and joins successive deoxynucleotides together with phosphodiester bonds. Thus, wherever an A, T, G, or C occurs in the existing (parental) chain, a complementary, base-pairing deoxynucleotide will be added to the new chain (Figure 7-9(c)). The growth of new chains by copying the base sequence of the parental chains continues by extending the unzippering of the duplex and adding complementary deoxynucleotides to the two new chains, one growing on each parental chain.

The overall outcome of the replication process is that two daughter DNA duplexes are produced that are each precisely identical to the parental duplex in the sequence of base pairs. Since successive sequences of deoxynucleotides of various lengths along the DNA duplex constitute genes, the replication of the DNA is equivalent to making two identical sets of genes from a single parental set. Although the original duplex is not conserved, the two original chains are conserved, each chain becoming a part of one of the two daughter duplexes; hence, the original duplex is in a sense halved or semiconserved in the replication process.

Proof that DNA Replicates Semiconservatively

Although the logic of semiconservative replication was extremely appealing, experiments were needed to prove that DNA does, in fact, reproduce in this way. Direct proof of semiconservative replication in bacteria was obtained in 1958 by use of a technique for measuring the density of DNA. *E. coli* cells were grown in medium containing the nonradioactive, heavy isotope of nitrogen, ^{15}N, instead of the usual lighter isotope ^{14}N. Growth continued for many generations and, for all practical purposes, all of the DNA was labeled with ^{15}N.

DNA containing ^{15}N is more dense than ^{14}N-containing DNA, and the increased density can be measured by centrifuging it in a density gradient made with CsCl. Because the Cs^+ ion is much denser than water, it sediments toward the bottom of a tube during centrifugation. However, diffusion prevents complete sedimentation and a gradient of Cs^+ ions forms, with the greatest concentration of ions and therefore the greatest density of solution at the bottom of the tube and the least concentration of ions and least density of solution at the top. When DNA is present in the solution, it moves during centrifugation to that part of the density gradient at which its density equals that in the gradient (Figure 7-10(a)). Thus, DNA containing ^{14}N would separate from ^{15}N-containing DNA in such a gradient because the two DNAs have different densities by about 0.015 g per cm^3.

When *E. coli* DNA containing only ^{15}N atoms is centrifuged in a CsCl solution, all of the DNA forms a single band at the position of the gradient with a density matching its own density (Figure 7-10(a)). If bacterial cells that have been grown in ^{15}N-containing me-

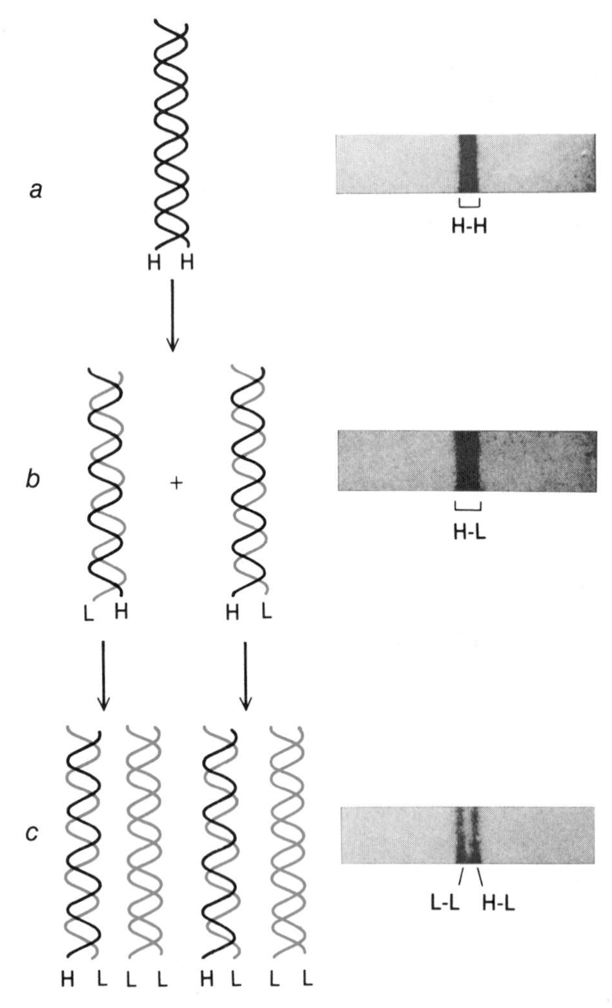

FIGURE 7-10
Experiment showing semiconservative replication of DNA. (a) DNA duplexes from cells grown in ^{15}N-containing medium. Both chains contain ^{15}N and are heavy (H-H). In a CsCl gradient (right) the DNA equilibrates at a position corresponding to the density of H-H DNA. (b) After one replication in ^{14}N-containing medium both DNA duplexes contain one H (^{15}N) and one light or L (^{14}N) chain (H-L). The DNA is less dense than in (a) and equilibrates in a CsCl gradient at a correspondingly lower density (shifted to the left). (c) After two successive replications in ^{14}N-containing medium two duplexes contain H and L chains and two duplexes contain only L chains. The L-L duplexes equilibrate in a CsCl gradient at a position of less density than H-L duplexes.

dium are transferred to ^{14}N-containing medium and allowed to grow in that medium, the density of the total DNA decreases as new DNA is synthesized. With semiconservative replication this decrease follows a specific, predictable pattern. During the first replication after the switch from ^{15}N to ^{14}N medium, two new DNA duplexes form from each old one. Each of the new duplexes have one old DNA chain containing ^{15}N and a new chain containing only ^{14}N. Therefore, each daughter duplex is a ^{15}N/^{14}N hybrid and has a density that is the average of the densities of pure ^{15}N DNA and pure ^{14}N DNA. Such intermediate density for DNA after one replication is precisely what is observed (Figure 7-10(b)). As predicted, subsequent replication of the hybrid DNA duplexes in the bacteria yields daughter duplexes, half of which are hybrid and half of which have a density reflecting the presence of ^{14}N in both chains (Figure 7-10(c)). After the third replication one fourth of the DNA in the cell population is still hybrid and three-fourths contain only ^{14}N. As cell proliferation continues, the fraction of DNA that is hybrid gradually declines to an undetectable level.

This density experiment, first done with bacteria, has been repeated with both plant and animal cells with the same results. All data indicate that semiconservative DNA replication is a universal phenomenon.

The Mechanism of DNA Replication

Polymerization of new nucleotide chains along the template (parental) chains in semiconservative replication of DNA is catalyzed by an enzyme called **DNA polymerase.** The monomeric units in this reaction are the deoxynucleoside *tri*phosphates dTTP, dCTP, dATP and dGTP, which are made by successive addition of two phosphate groups (donated by ATP) to the monophosphates, for example,

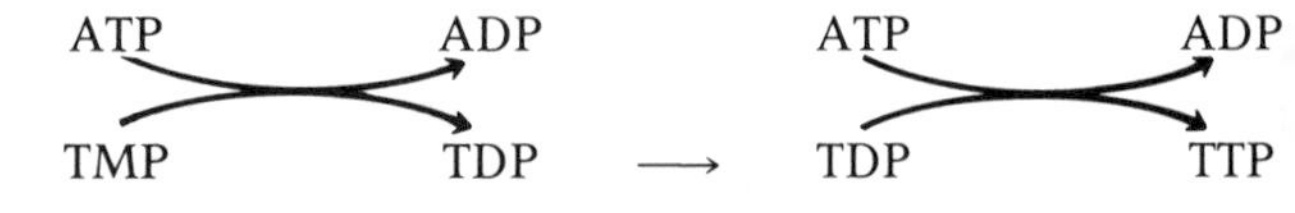

In building a deoxynucleotide chain, DNA polymerase catalyzes the joining of the deoxyribose-linked

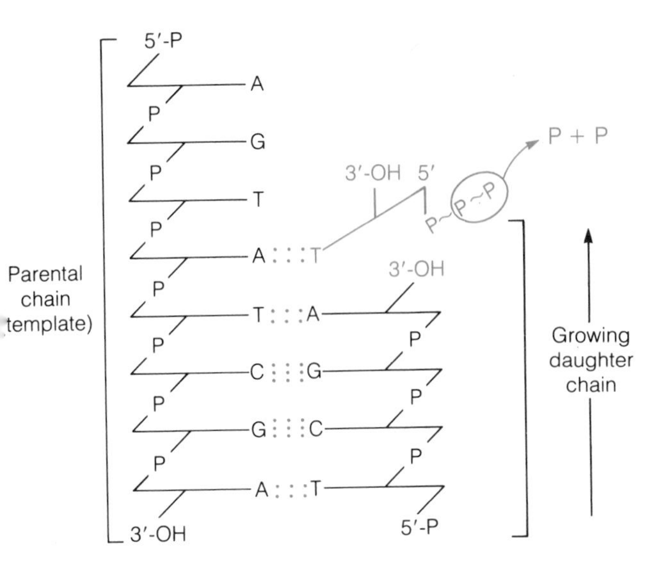

FIGURE 7-11
Deoxynucleoside molecules are added to the 3′-OH end of a growing DNA daughter chain. P ~ P is cleaved from a nucleoside triphosphate as it is joined to the chain.

phosphate group of a deoxynucleoside triphosphate to the OH group attached to the 3′ carbon of the deoxyribose of the terminal nucleotide in the growing chain (Figure 7-11). In the reaction the two terminal phosphate groups are removed from the nucleotide as a pyrophosphate (PP_i) group. The 3′-5′ linkage is made by reaction with an OH group attached to the 3′ carbon. The PP_i released in the reaction is rapidly split into two phosphate groups by the enzyme pyrophosphatase. With the elimination of PP_i, DNA polymerase cannot catalyze the back reaction, that is, cleavage of deoxynucleotides from the end of the chain with reformation of deoxynucleoside triphosphates, because one of the necessary substrates (PP_i) is missing.

DNA polymerase can catalyze only the unidirectional extension of a nucleotide chain, that is, it catalyzes additions of deoxynucleotides to the 3′-OH end of a chain but not to the 5′-PO_4 end. No enzyme exists that can elongate a polynucleotide chain by adding nucleotides to the 5′ end.

DNA polymerase can catalyze additions of nucleotides to the 3′-OH end of an existing chain, but cannot initiate a new chain from individual deoxynucleotides. Thus, initiation requires an existing polynucleotide chain (a **primer**) in order to add deoxynucleotides to a chain. In the laboratory the primer can be a short segment of either DNA or RNA, but in cells the primer is a short RNA chain. The primer is synthesized in the following way: first, a localized separation of the two chains of the double helix occurs within a DNA molecule, not at the end of the molecule (Figures 7-9, 7-12(a,b)). DNA molecules in bacteria, many viruses, in chloroplasts, plasmids, and in mitochondria of most cell species are circular and have no ends anyway. In eukaryotic cell nuclei the DNA molecules are extremely long linear molecules, and replication starts within these molecules. Opening the double helix creates short single-stranded regions that can serve as templates for replication (Figure 7-12(b)). An enzyme called **primase** catalyzes the formation of a short RNA chain from nucleoside triphosphates. Unlike DNA polymerase, primase can initiate a polynucleotide chain. The sequence of the primer so formed is dictated by the DNA template (Figure 7-12(c)) following the base-pairing rules, with the substitution of U for T in RNA. In other words A, T, C, and G in DNA prescribe U, A, G, and C, respectively in the RNA. The RNA primer remains base-paired to the DNA template; once the primer is made, a DNA polymerase can take over from the primase and extend the chain by addition of deoxynucleotides (Figure 7-12(d)).

An RNA primer is formed on each DNA parental chain in the opened region of the double helix, and new complementary daughter chains are synthesized

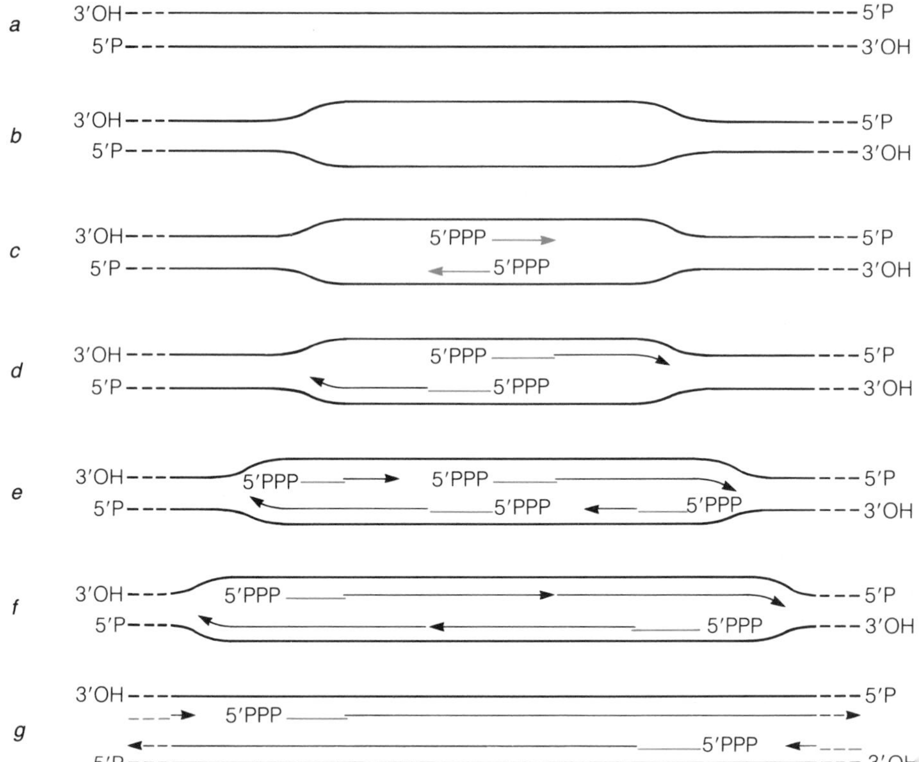

FIGURE 7-12 Replication of DNA. (a) Parental double helix. (b) The double helix opens, creating single-stranded parental chains that serve as templates. (c) Short RNA primers (red) are synthesized on the DNA templates by RNA primase. (d) RNA primers are extended as DNA daughter chains by DNA polymerase. (e) Extension of daughter chains continues and more parental DNA is progressively opened up. New RNA primers are formed and then extended as DNA chains to backfill along unreplicated parts of parental chains. (f) Original RNA primers are removed as backfilling chains are extended. (g) The 3′-OH end of backfilling daughter chains are joined to the 5′-P end of the next section of daughter chain, completing replication of this portion of the DNA duplex.

on each parental chain simultaneously. Because the two parental chains have opposite polarities, the RNA primers formed on the parental chains have opposite polarities. Since DNA polymerase catalyzes additions of deoxynucleotides only at 3′-OH ends, the two daughter chains grow in opposite directions (Figure 7-12(d)).

As the daughter strands are extended from their 3′-OH ends, the double helix progressively unzippers in both directions ahead of the growing ends of the new complementary daughter chains. The point of helix unwinding and growth of a 3′-OH end of a chain is called a **replication fork.** The progression of each replication fork leaves behind it a new double helix formed from one of the parental chains, while the other parental chain remains unreplicated because deoxynucleotides cannot be added to the 5′-P end of the RNA primer (Figure 7-12(d)). Replication of the other parental chain is achieved in the following way: when a portion of single-stranded parental DNA has accumulated behind a replication fork, a new RNA primer is formed on the parental chain, which allows a DNA polymerase molecule to extend the chain in the backward direction from the replication fork (Figure 7-12(e)). The process is repeated as the replication fork progresses along the DNA and leaves behind a region of single-stranded parental DNA. The sections of new DNA chains made in the backward direction are a few hundred to a thousand nucleotides long and are called **Okazaki fragments** after the biochemist who discovered them.

As an Okazaki piece grows in the backward direction, it meets the 5′ end of the RNA primer of the previous Okazaki fragment. When this encounter occurs, an exonuclease activity of the DNA polymerase successively removes RNA nucleotides from the RNA primer, simultaneously adding deoxynucleotides to the backward-growing DNA chain. Ultimately the entire RNA primer is removed, leaving the 3′-OH end of one Okazaki fragment abutting the 5′-P end of a deoxynucleotide in the previous Okazaki fragment (Figure 7-12(f)). The two ends are then joined in the usual phosphodiester linkage by an enzyme called **DNA ligase,** which can join a 3′-OH of a chain to a 5′ monophosphate of a nucleotide chain (Figure 7-12(g)). In effect, repeated formation of Okazaki fragments, removal of RNA primers, and ligation of the DNA sections achieve formation of a new, continuous daughter chain on one parental chain while the other daughter chain forms continuously on the corresponding section of parental chain immediately opposite. Since initiation of the discontinuous strand (the one formed from Okazaki fragments) cannot begin until the continuously formed strand has opened the helix, synthesis of the discontinuous strand lags behind that of the continuous strand. Thus, the terms **leading strand** and **lagging strand** are commonly used for the continuous and discontinuous strand, respectively.

One problem in DNA replication has not been considered. Separation of chains at the replication fork requires unwinding of the double helix. The double helix contains one complete turn per ten base pairs. Therefore, for each ten base pairs unwound and separated at the replication fork the double helix ahead of the fork must undergo one rotation in the opposite direction. In *E. coli* a replication fork moves through 60,000 base pairs (6000 turns of the helix) per minute, requiring the double helix ahead of the fork to rotate at a rate of 6000 turns per minute. In eukaryotes the rate is about ten times less. Rotation of the long regions of double helix ahead of the replication fork is impossible because the DNA is totally constrained by the many protein molecules attached to it (Chapters 9 through 11). Without rotation, torsion in the double helix ahead of the replication fork would prevent unwinding and prevent DNA replication. One way the problem is solved is by an enzyme known as **topoisomerase,** which creates transient breaks in one chain (transient single-stranded nicks) a short distance ahead of the replication fork. Each break allows the covalent bond in the other chain directly opposite to the nick to act as a swivel, which in turn permits the double helix immediately ahead of the replication fork to rotate freely (Figure 7-13). Single-stranded nicks are rapidly created and repaired by topoisomerase ahead of the progressing replication fork, preventing the occurrence of torsion. In replicating circular DNA, the product is a pair of molecules linked like the links in a chain and is called a **catenane.** Topoisomerase is also responsible for separating the circles.

CHROMOSOMES AND CELL DIVISION

Chromosomes consist of DNA and many kinds of proteins. The DNA carries the genetic information. The proteins have a variety of functions in packaging of the very long DNA molecules into compact form (Chapter 11) and in the genetic functions of the DNA (Chapters 9 and 10). For the present purpose of understanding the general principle of cell heredity it is necessary to know only that DNA is the genetic material, that it is contained in chromosomes, that it replicates semiconservatively to produce two identical daughter double helices from one parental helix, and finally that the daughter double helices are distributed to the two daughter cells at cell division.

The Cell Cycle

DNA replication and distribution of the replicated DNA at cell division are fundamental events in the reproduction of cells (described in detail in Chapter 12). Here we consider them in the context of cell heredity in the cell cycle diagrammed in Figure 7-14. Prior to each cell division the DNA double helix in each chromosome replicates. The daughter helices form the two daughter chromosomes, which are distributed to daughter cells at cell division, providing each daughter cell with a full set of chromosomes and therefore a full complement of genetic material. The process is repeated cyclically with an alternation of DNA replication and distribution to daughter cells through successive cell reproductions.

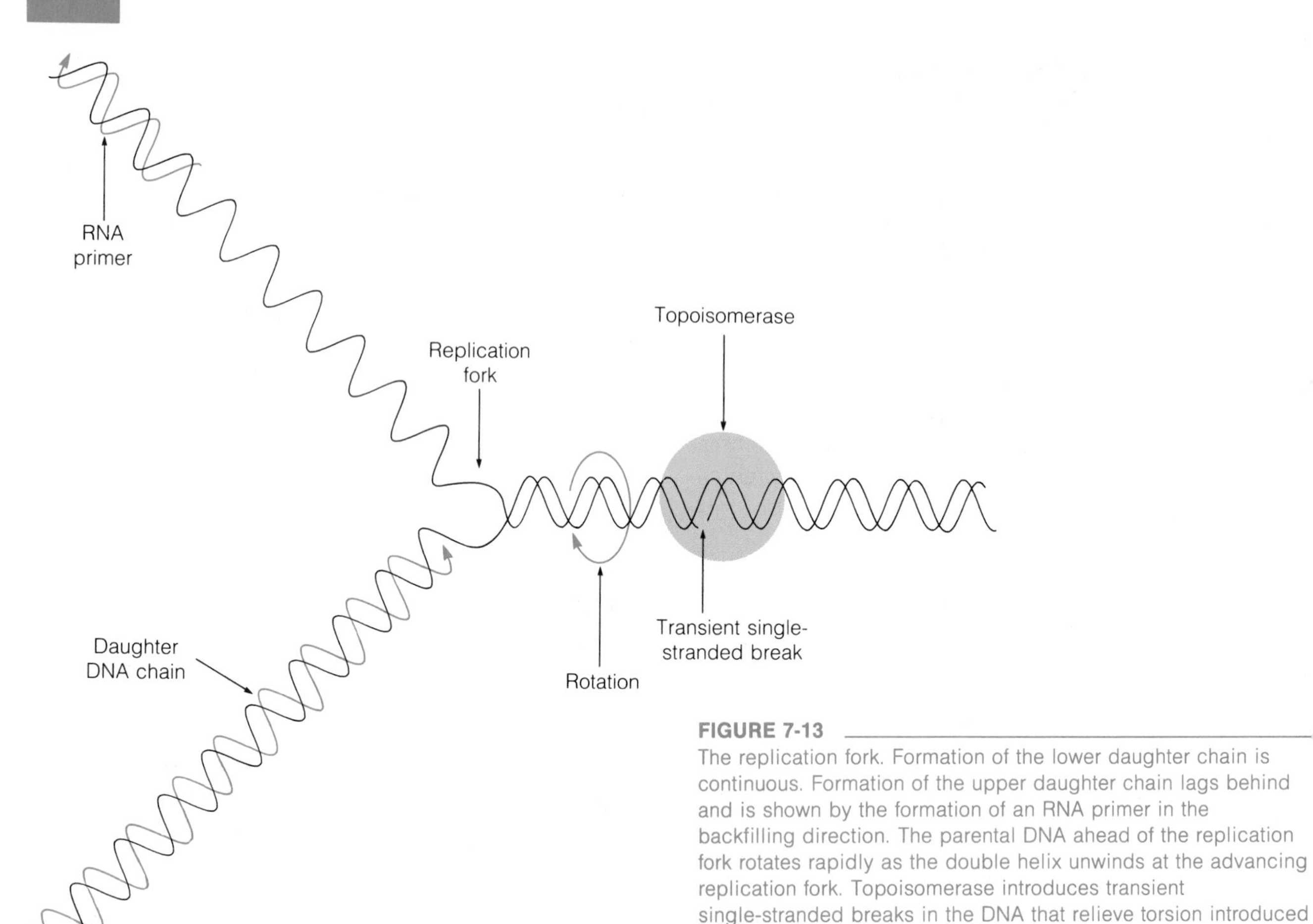

FIGURE 7-13
The replication fork. Formation of the lower daughter chain is continuous. Formation of the upper daughter chain lags behind and is shown by the formation of an RNA primer in the backfilling direction. The parental DNA ahead of the replication fork rotates rapidly as the double helix unwinds at the advancing replication fork. Topoisomerase introduces transient single-stranded breaks in the DNA that relieve torsion introduced by rotation of the parental double helix.

Mitosis

Mitosis is the process in eukaryotic cells by which chromosomes are distributed at cell division (Figure 7-15). It consists of condensation of the duplicated chromosomes during **prophase,** movement of the chromosomes to the central region of the cell at **metaphase,** separation of the duplicated parts of each chro-

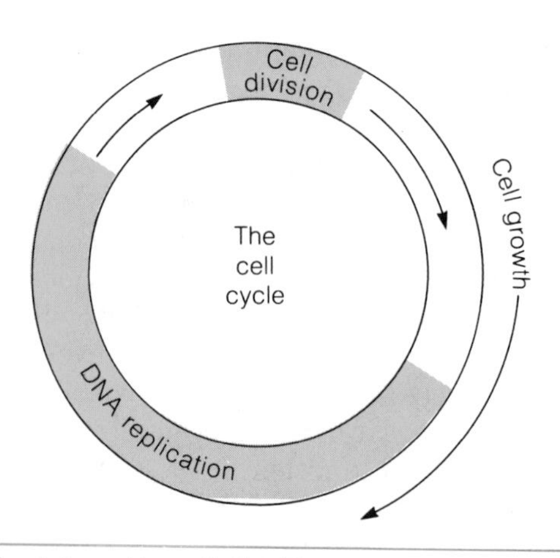

FIGURE 7-14
The cell reproductive cycle consists of cell growth and DNA replication followed by cell division.

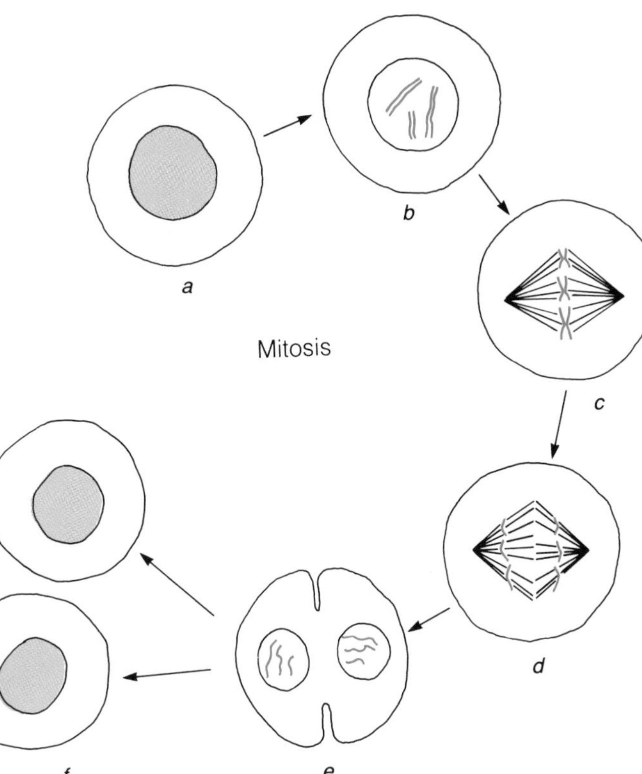

FIGURE 7-15 Events of mitosis in an animal cell. (a) A cell in interphase. The chromosomes are decondensed and not individually discernible. (b) Prophase. The chromosomes are condensing. (c) Metaphase. The fully condensed, doubled chromosomes are aligned in the center of the cell on the mitotic spindle (composed chiefly of microtubules), and the nuclear envelope has disappeared. (d) Anaphase. The doubled chromosomes have each split and daughter chromosomes are moving to opposite sides of the cell. (e) Telophase. The chromosomes are undergoing decondensation and nuclear envelopes have reformed around groups of daughter chromosomes. Cytokinesis (division of the cytoplasm) has begun. (f) Two daughter cells in interphase.

mosome to separate sides of the cell in **anaphase,** and formation of daughter nuclei by decondensation of the chromosomes and formation of a nuclear envelope in **telophase.** At the end of telophase the cell undergoes **cytokinesis,** forming two daughter cells.

The number of chromosomes is constant in the cells of any given species, and therefore the number of different DNA double helices is constant. However, chromosomes are exceedingly long threads in the cell nucleus that are difficult to detect during most of the cell cycle. After the DNA has replicated and just before a eukaryotic cell divides, each chromosome shortens enormously by complex coiling and folding and becomes many times thicker. Through this process of condensation each chromosome becomes visible as a distinct, elongated structure (Figure 7-16). Each condensed chromosome is seen to be a double structure, because DNA replication has occurred prior to condensation.

The stage of chromosome condensation, called prophase, is the first of four periods that make up mitosis (Figure 7-15(b)). In mammalian cells prophase condensation typically takes 20 to 30 minutes. In other species, for example, in amphibians and in many kinds of plants, prophase condensation may take one to several hours. In the latter part of prophase, when chromosome condensation has reached an advanced state, the nuclear envelope and nucleoli in most kinds of cells disappear.

With the completion of prophase condensation, the cell enters the second stage of mitosis, metaphase. In metaphase the chromosomes become arranged in a plane usually in the middle of the cell, which later becomes the plane through which the cell is cleaved into two (Figure 7-15(c)). At metaphase the duplicated nature of each chromosome is clear. The two duplicated parts, called **chromatids,** have separated from each other except at a single point of connection at the

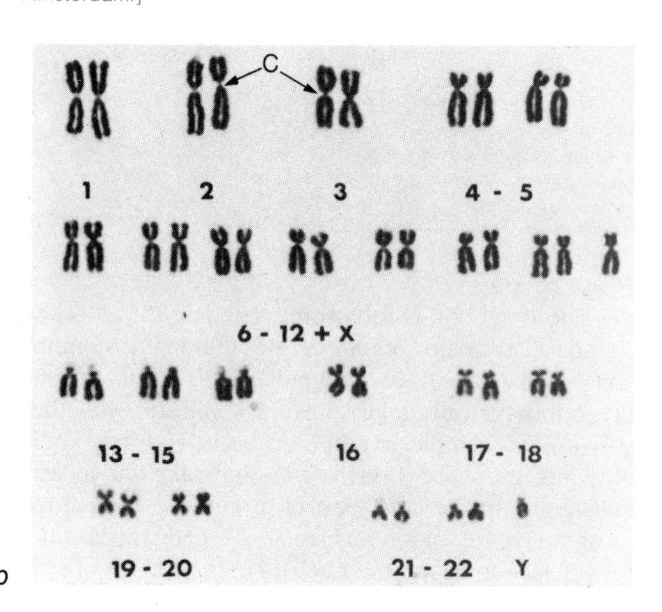

FIGURE 7-16
Light micrograph of human male metaphase chromosomes. (a) Chromosomes as they appear when spread from a dividing cell. (b) Chromosomes arranged in pairs. Some chromosomes cannot be distinguished from one another without special staining techniques. For example, the chromosomes in pairs 4 and 5 cannot be distinguished from each other and those in pairs 6 through 12 plus the X chromosome are all almost identical in size and position of the centromere (C). (c) Human male metaphase chromosomes treated and stained to reveal a pattern called G banding. Each chromosome can be identified by its unique banding pattern (see Chapter 11 for more about banding). [Courtesy of A. T. Sumner. From C. J. Bostock and A. T. Sumner. *The Eukaryotic Chromosome* (1978) North Holland, Amsterdam.]

centromere (Figure 7-16). The end of metaphase and the beginning of the third stage of mitosis, anaphase, is marked by the separation of centromeres of the two chromatids. From the point of separation onward the two chromatids are called **daughter chromosomes.** In anaphase the two daughter chromosomes move away from each other, each going to opposite sides of the cell (Figure 7-15(d)). This happens more or less simultaneously for all of the chromosomes, so anaphase consists of the synchronous movement in opposite directions of two groups of daughter chromosomes.

The orderly arranging of chromosomes in metaphase and the subsequent orderly distribution of daughter chromosomes in anaphase are achieved by a special apparatus called the **mitotic spindle,** which is composed of microtubules and other elements. How the spindle achieves the precise movements of chromosomes in mitosis is not completely understood. Models of how the mitotic spindle works are described in Chapter 12.

The fourth and final stage of mitosis, telophase, begins as the two clusters of daughter chromosomes finish moving away from each other (Figure 7-15(e)). At this point the chromosomes begin to decondense, returning to the form of highly dispersed, fine threads (Figure 7-15(f)). As decondensation proceeds, a nuclear envelope appears around each group, forming two daughter nuclei at opposite sides of the cell. Different species of cells follow different time schedules for mitosis, but typically the entire process is completed in 30 to 60 minutes.

About this time, division of the cytoplasm, cytokinesis, begins by a splitting of the cell through the region between the two groups of daughter chromosomes in the plane formerly occupied by the metaphase chromosomes. In animal cells, cell splitting is accomplished by a deepening, circular furrow that cleaves the cell in two (Figure 7-15(e)). The contraction-like process underlying the formation of the cleavage furrow involves actin, one of the proteins that achieves contraction of muscle cells. In cytokinesis, actin filaments and myosin are present in a contractile ring in the cytoplasmic region of the furrow. In plant cells, the cell becomes split by rapid construction of a membrane and wall in the plane between the two groups of daughter chromosomes. The Golgi complex plays a major role in assembling materials from which the membrane is built (Chapter 12).

The behavior of chromosomes in mitosis makes complete sense because we know that chromosomes are composed of hereditary units known as **genes.** In the duplication and mitotic distribution of chromosomes we see how a parent cell endows its two daughters with a full set of hereditary characteristics.

The Karyotype of a Cell

The different chromosomes in a cell can be identified by their lengths and by the locations of their centromeres. The centromeres are always located at the same position in a given chromosome; in different chromosomes this location may be anywhere from either end to the middle. Four characteristics define the **karyotype** of the cell: the number of chromosomes, their individual sizes, the location of their centromeres as seen during metaphase of mitosis, and the banding pattern of individual chromosomes. Banding patterns in chromosomes produced by various staining procedures reveal underlying differences in the structure of chromosomes (Chapter 11) and allow otherwise similarly sized chromosomes to be clearly distinguished from one another. In Figure 7-16 is a set of metaphase chromosomes from a human male cell prepared by breaking open the cell and spreading them on a glass slide. In almost all normal human cells the genes are carried in 23 different chromosomes. There are two copies of each chromosome, so the total number is 46. The two chromosomes in a pair are genetically homologous to one another and are called **homologues.** In a human female cell there are 22 pairs of chromosomes, called **autosomes,** which are by convention numbered 1 through 22 on the basis of decreasing size. The 23rd pair of chromosomes is the **sex chromosomes** called the **X chromosomes.** The sex chromosomes carry genes important in determining sex as well as many other genes unrelated to sex. Thus, human female cells contain two complete sets of genes. A cell with two of each chromosome is said to be **diploid.**

Human male cells are also diploid, containing the same 22 pairs of autosomes and a pair of sex chromosomes. The sex chromosomes form the 23rd pair, but they do not match; one is an X chromosome, as in a female, and the other is a much smaller one called the **Y chromosome.** There are many fewer genes in the Y chromosomes and those that are present are different from those carried by the X. Some genes of a Y chromosome are, like some genes of the X, important in determining sex characteristics.

An important exception to diploidy in human cells occurs in cancer cells. Most kinds of cancer cells have more than 46 chromosomes, some of which cannot be matched into pairs. Such cells are said to be **aneuploid.** Aneuploidy is connected in some way with the change in behavior of a normal cell when it is converted into a cancer cell (Chapter 13), but the connection is poorly understood.

Another important exception to diploidy is sperm and egg cells. These contain only one of each chromosome and are said to be **haploid.** The haploid chromosome number in human egg and sperm cells is 23.

The cells of animals and many plants and unicellular eukaryotes are diploid, with the exceptions noted above. Some kinds of plants are made of haploid cells. Some unicellular organisms, such as budding yeast, are commonly haploid. The significance of a haploid chromosome number versus a diploid number is discussed later in this chapter.

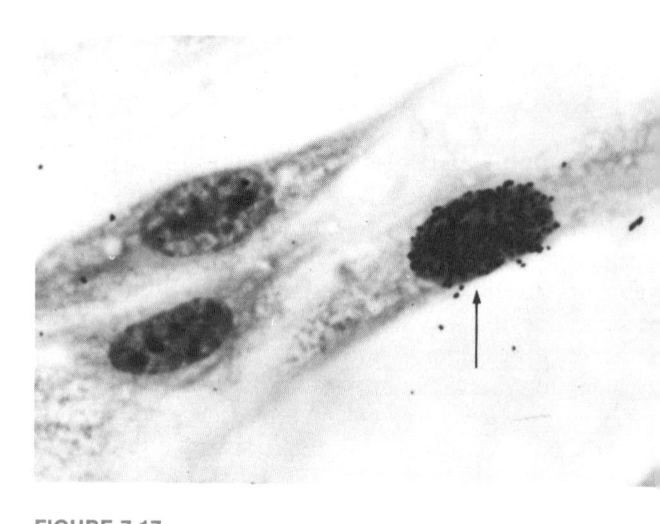

FIGURE 7-17

A light microscope autoradiograph of a cell (arrow) that has incorporated ^{3}H-thymidine into its DNA. The other two cells did not make DNA during the experiment and therefore their DNA did not become labeled with ^{3}H-thymidine.

Tracing the Replication and Mitotic Distribution of DNA

A variety of experiments (Chapter 11) has shown that a chromosome contains a single DNA molecule. The length of the DNA molecule varies from one chromosome to another but is constant for any given chromosome. Among the 23 different chromosomes in a human cell, the lengths of the DNA molecules range from about 0.5 cm in the smallest chromosomes to 7 cm in the longest.

The semiconservative replication of the DNA molecule in a chromosome and the subsequent distribution of the two daughter double helices into two daughter chromosomes at mitosis can be visualized by way of a radioactive tracer and the technique of autoradiography. In this experiment cells are given thymidine in which one of the hydrogen atoms has been replaced by the radioactive isotope of hydrogen—tritium (^{3}H). Most species of cells take up thymidine; once inside, the thymidine is converted to thymidine triphosphate (dTTP), and the nucleotide is then incorporated into new deoxynucleotide chains as they are synthesized. With the completion of DNA replication the original double helix in a chromosome is now present as two identical double helices, each composed of one of the original deoxynucleotide chains and a new chain labeled with radioactivity.

The presence of radioactivity (^{3}H-thymidine) in the new DNA duplexes can be detected by autoradiography. Figure 7-17 shows an autoradiograph of a cultured animal cell in interphase that has incorporated ^{3}H-thymidine into DNA. A large number of silver grains is present in the film directly over the cell nucleus, indicating the presence of radioactivity in the newly synthesized DNA.

When these chromosomes condense into the metaphase form, each is observed to consist of two chromatids still held together at their centromeres, as in Figure 7-16. Since the cell was incubated in ^{3}H-thymidine during the immediately preceding period of DNA replication, the condensed metaphase chromosomes contain radioactivity. Semiconservative replication of the DNA double helix in the presence of ^{3}H-thymidine should produce two helices, which condense to form the two chromatids at metaphase. Each chromatid should contain a radioactive chain in its DNA duplex, and autoradiography confirms this (Figure 7-18(a)). The number of silver grains over the two chromatids is about the same; this means that each contains about the same amount of radioactivity since the number of silver grains formed is proportional to the amount of radioactivity present.

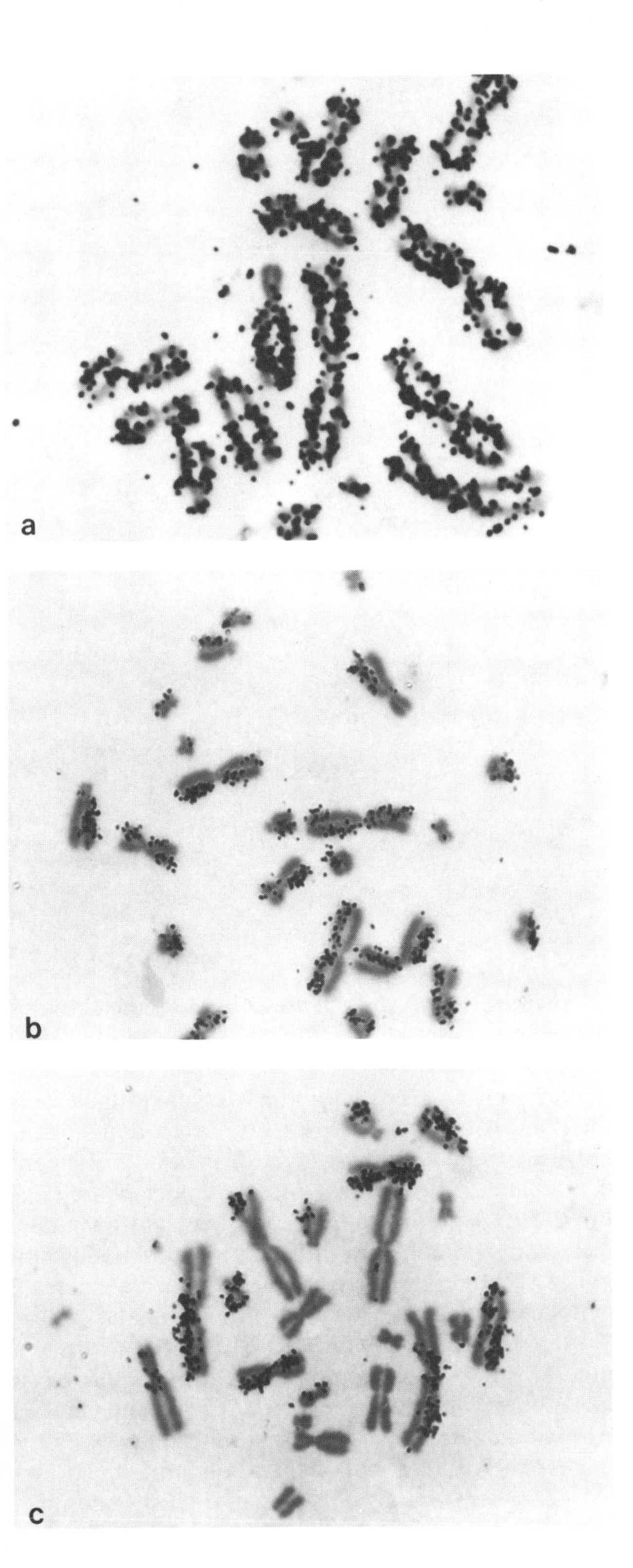

FIGURE 7-18
Autoradiography showing the distribution of radioactivity in chromosomes at metaphase after labeling DNA with ^{3}H-thymidine. (a) At the first metaphase after labeling of DNA both chromatids of each chromosome are labeled. (b) At the second metaphase after labeling one chromatid of each chromosome is labeled and one is unlabeled. (c) At the third metaphase about half the chromosomes have one labeled and one unlabeled chromatid. The remaining chromosomes are unlabeled.

If, after incorporation of ^{3}H-thymidine into DNA, a cell is allowed to divide, then each DNA double helix in the two daughter cells should contain one radioactively labeled chain and one unlabeled chain. When such cells replicate their DNA in preparation for the next mitosis, but in the absence of ^{3}H-thymidine (cells make their own nonradioactive thymidine), then, by the rules of semiconservative replication, each chain of a double helix will appear in one of the two daughter double helices. Condensation of the duplicated chromosomes at the ensuing metaphase (second metaphase after labeling of DNA) should yield a chromosome in which one chromatid is radioactive and one is not, which is precisely what is found (Figure 7-18(b)). Assuming that for the total complement of chromosomes the distribution of the radioactive chromatids to the two daughter cells during anaphase is random, then half of the chromosomes in each daughter cell should contain a double helix with one radioactive chain. When these chromosomes replicate and reappear in condensed form at the next mitosis (third mitosis after labeling of DNA), half of them are found to contain one radioactive chromatid, as expected, and the other half are totally nonradioactive (Figure 7-18(c)).

Production of Germ Cells: Meiosis

The distribution of DNA in chromosomes by mitosis explains how hereditary features governed by genes are passed from parent cell to daughter cells at cell division. Inheritance of genetically determined features from a parental organism to its offspring in sexual reproduction among plants and animals is also based on chromosome distribution at cell division, but involves

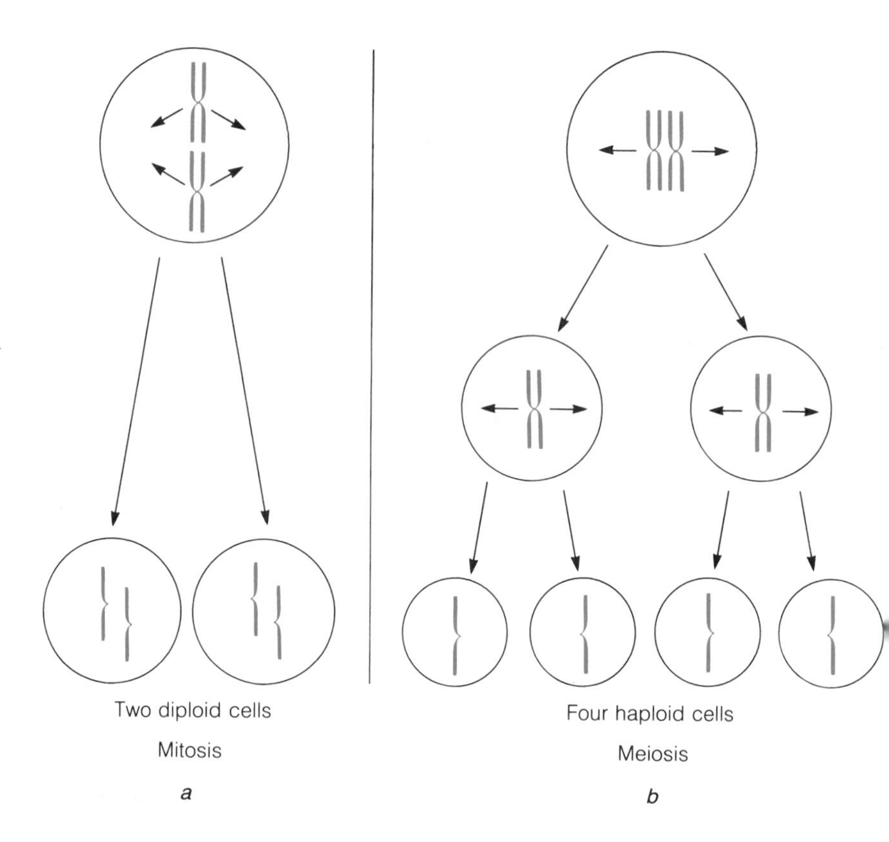

FIGURE 7-19
Mitosis and meiosis. (a) In mitosis all of the chromosomes line up individually at metaphase. In anaphase, the chromatids in each chromosome separate into daughter chromosomes, and cell division produces two diploid cells. (b) In the first meiotic division each pair of chromosomes lines up side by side instead of independently as in mitosis. In anaphase chromosomes of a pair move to opposite sides of the cell, and cell division produces two haploid cells (although each chromosome is still composed of two chromatids). In the second meiotic division the chromatids in each chromosome separate and move to opposite sides of the cell, yielding four haploid granddaughter cells.

some additional steps because two parents participate in making a single cell, the fertilized egg.

The first step in sexual reproduction is the formation of germ cells—egg and sperm cells. Egg and sperm cells are called **gametes** to distinguish them from cells that form the body of a multicellular organism, the **somatic cells**. Gametes are haploid cells that are formed from diploid cells by the process of meiosis in the gonads of plants and animals. When a diploid cell undergoes mitosis, the two homologous chromosomes of each pair behave independently, that is, each splits into two daughter chromosomes as metaphase proceeds into anaphase. As a result, each daughter cell receives a copy of the two homologues of every chromosome, and therefore each daughter cell is diploid like the parent cell. **Meiosis** is an alternate form of chromosome distribution consisting of two consecutive cell divisions. In the first of these divisions the chromosomes of the diploid cell condense as usual, but they align differently at metaphase than in mitosis. The two homologous chromosomes in each pair (in male cells the X and Y chromosomes are not strictly homologous but they form a pair), instead of behaving independently, line up next to one another at metaphase (Figure 7-19). When metaphase proceeds into anaphase, one entire homologue with its two chromatids still joined at the centromere is carried to one daughter cell and the other homologue to the other daughter (Figure 7-19). In short, the key difference between meiosis and mitosis is that in meiosis homologues are paired at metaphase and are distributed to the two daughter cells without having undergone splitting at their centromeres. The result of this first meiotic division is that each daughter cell has one copy of each chromosome, ıat is, it has a haploid number of chromosomes, but ach of these chromosomes is still in the duplicated tate (Figure 7-19).

The first meiotic division of the nucleus is followed y cytokinesis, and the two daughter cells almost imıediately begin the second meiotic division, usually efore the chromosomes have even undergone full deondensation from the first division. No DNA repliation occurs between the first and second meiotic livision (each chromosome still consists of two chroıatids). At metaphase of the second division the hapoid set of chromosomes in each daughter become aranged in the center of the cell in the same way they do or mitosis. The two chromatids in each chromosome hen separate by splitting at the centromere and move o opposite sides of the cell (Figure 7-19). Cytokinesis ollows, producing by the two meiotic divisions of one riginal diploid cell four daughter cells, each with a ıaploid set of unduplicated chromosomes. The four ıaploid cells produced in the male gonad (testis) then lifferentiate into four male gametes (sperm). In the emale gonad (ovary) in animals three of the four hapoid cells produced degenerate, and the fourth matures nto an egg cell (ovum).

Fertilization

Fertilization consists of the fusion of a male gamete with a female gamete. Within the fusion product, called a **zygote,** the two haploid nuclei fuse, forming a single nucleus with a diploid set of chromosomes. The DNA in these chromosomes then replicates, and the fertilized ovum enters the first mitotic cell division as the start of the development of a new multicellular individual. Thus, *meiosis produces haploid cells from diploid cells in the gonads, and fertilization reestablishes a diploid number of chromosomes by fusion of two haploid nuclei,* one from each parent, in the zygote.

Knowledge of the diploid state of somatic cells, the production of haploid cells by meiosis, and the re-formation by fertilization of a diploid somatic cell to start development make the principles of inheritance from parental organisms to offspring readily understandable.

MENDELIAN PRINCIPLES OF INHERITANCE

Humans had practiced genetics in the breeding of domesticated plants and animals for several thousand years without knowing anything about mitosis, meiosis, fertilization, or genes. In 1865 the Austrian botanist Gregor Johann Mendel proposed a set of principles from his breeding experiments with peas. These principles explain heredity among plants and animals. The significance of Mendel's results were not appreciated at the time. In 1900 plant breeders performed experiments similar to Mendel's and confirmed Mendel's principles about inheritance, marking the formal beginning of the science of genetics. The principles of inheritance described below were soon interpreted in terms of the behavior of chromosomes during mitosis and meiosis.

Rule of Segregation

Mendel studied seven inherited characteristics of pea plants: flower color, seed shape, seed color, pod shape, pod color, location of flowers on the stem, and length of the stem. He first determined that each of his plants bred "true;" that is, when a plant with a particular characteristic—for example, red flowers—was crossed with a plant of the same strain, all of the offspring had the same trait, that is, red flowers. Without such true-breeding plants his experiments would have failed, or at least would not have yielded results that could have been interpreted at that time. He made crosses between different true-breeding strains and observed the pattern of inheritance of the particular traits. When pollen (equivalent to sperm cells) from plants with red flowers was used to fertilize the egg cells of plants breeding true for white flowers, all of the resulting seeds grew into plants with red flowers (Figure 7-20). When plants breeding true for wrinkled seeds were crossed with plants that produced smooth seeds, all of the offspring plants had only smooth seeds. In experiments of this kind the parental plants are called the **parental** or **P generation.** The offspring from breeding two parental plants are called the $\mathbf{F_1}$ for **first**

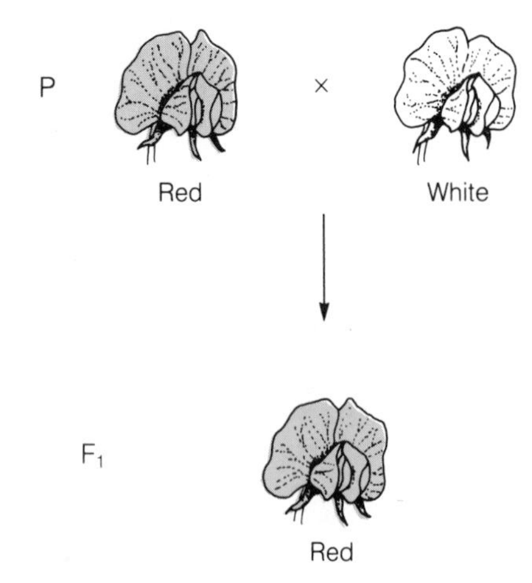

FIGURE 7-20
A cross between plants that breed true for red flowers and plants that breed true for white flowers produces only red-flowering plants in the F_1 generation.

filial generation. In these experiments Mendel demonstrated that one form of a trait dominated over the other form in the F_1. For example, red flower color is **dominant** over white, smoothness of seeds is dominant over wrinkledness, yellow seed color is dominant over green, and so on. Alternatively, one form of a trait is **recessive** to another; for example, white flower color is recessive to red flower color, wrinkled is recessive to smooth, and so forth. In short, the form of a particular trait could be dominant or recessive with respect to another form of that trait. In a cross between a true-breeding parent carrying the dominant form of a trait and a second true-breeding parent with the recessive trait, all progeny have the dominant trait.

Mendel next crossed members of the F_1 generation with one another to produce the next generation, called the $\mathbf{F_2}$ generation. For each of the seven traits that he examined he found that 25 percent of the F_2 plants displayed the recessive form of the trait and 75 percent showed the dominant form (Figure 7-21). For example, in the case where one of the original parents had red flowers and one had white, 25 percent of the F_2 plants had white flowers and 75 percent had red flowers. When F_2 plants with a recessive trait (white flowers) were bred with one another, all of the next generation, the F_3 plants, bred true for the recessive trait (white flowers). The F_2 plants with the dominant trait were bred with one another and discovered to be of two types. One-third of the red-flowered F_2 plants bred true and produced only progeny with the domi-

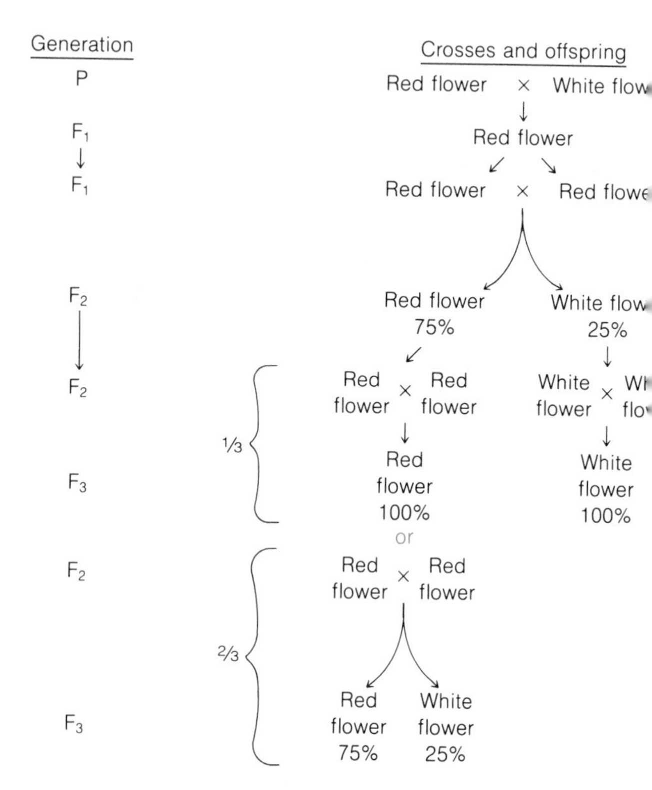

FIGURE 7-21
A cross between a true-breeding plant with red flowers and a true-breeding plant with white flowers (P generation) produces only red-flowering plants in the F_1 generation. A cross between two such F_1 plants produces an F_2 generation of 75 percent red-flowering plants and 25 percent white-flowering plants. All of the white-flowering plants in the F_2 generation breed true when self-pollinated, i.e., yield only white-flowering plants in the F_3 generation. One-third of the red-flowering plants in the F_2 generation breed true when self-pollinated, i.e., produce only red-flowering plants in the F_3 generation. When self-pollinated, two-thirds of the red-flowering plants in the F_2 generation yield 75 percent red-flowering plants and 25 percent white-flowering plants in the F_3 generation.

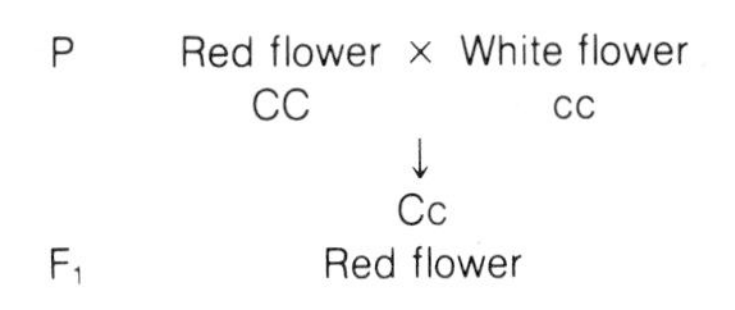

FIGURE 7-22 True breeding red flowers carry two copies of a dominant factor C, and white flowers carry two copies of a recessive factor c. All F_1 offspring from a cross between them have one C and one c factor and are red-flowering.

nant trait (red flowers). The other two-thirds gave a mixture in the F_3 generation, 75 percent with red flowers and 25 percent with the white flowers.

Mendel concluded from these results that flower color and the other six traits were each determined by pairs of factors, one from the male parent and one from the female. Each purebred parent contains two identical factors for each trait. For example, a true-breeding plant with red flowers contains two factors (*CC*) that determine red coloring (a capital indicates dominance of a factor), and a purebred plant for white flowers contains the pair *cc* for white coloring (a lowercase letter indicates recessiveness of a factor). In crosses between them every F_1 plant receives one factor from each parent, one dominant and one recessive, and is *Cc* (Figure 7-22). Since *C* is dominant over *c*, every F_1 plant has red flowers. The form of the trait displayed by an organism is called its **phenotype;** for example, the red-flower phenotype in pea plants. The pair of factors controlling phenotype are genes, one carried by one chromosome of a pair of homologues and the other carried by the other homologue in a diploid cell. The pair of genes specifying a phenotypic trait together constitute the **genotype** for that trait. The genotype for the red flowers in pea plants could be either *CC* or *Cc*, but the genotype for white flowers can only be *cc*. When the two genes for a trait in a cell or organism are identical, as in *CC* or *cc*, the cell or organism is said to be **homozygous** for that trait. When the two genes are different, as in *Cc*, the cell or organism is said to be **heterozygous** for that genetic trait.

In a cross between a *CC* plant (homozygous for red flowers) and a *cc* plant (homozygous for white flowers) one plant produces haploid gametes containing *C* and the other produces gametes containing *c*. All of the zygotes produced (F_1 generation) must have the *Cc* genotype (heterozygous) and develop into red-flowering plants. Half the gametes produced by an F_1 plant with a *Cc* genotype will contain *C* and half will contain *c*. When F_1 plants were crossed with one another (to produce the F_2 generation), Mendel found a ratio of three red-flowering plants to one white-flowering plant. To Mendel this meant that two factors, one from each parent, govern the appearance of a given trait. We now know that a single copy of a factor, or a gene, occurs in gametic cells. Mendel recognized that four combinations of factors were possible when two of the above F_1 plants were crossed. Explained in modern terminology a *c*-containing gamete may combine with either another *c*-containing gamete or a *C*-containing gamete, and a *C*-containing gamete may likewise combine with either type.

The two forms of a particular gene, in this case *C* and *c*, are called **alleles** of a single gene. A gene may exist in more than two allelic forms, but a given diploid cell and usually any diploid organism carries two alleles of a gene. The two alleles in a cell may be the same, for example, *CC*, or they may be different, *Cc*. To restate the definition, an allele is one of two or more alternative forms of a gene that occupy corresponding sites on homologous chromosomes.

The simplest way to determine the genotypes of the offspring of a cross is to use a **Punnett square.** In Figure 7-23 the genotypes of the two parents in a cross heterozygous for a gene are designated *Cc* and *Cc*. Each parent can produce gametes in equal numbers with the *C* or the *c* allele. These possibilities are aligned on the Punnett square for each parent. The genotypes resulting from a mating are indicated in the subdivisions of the square. Among the offspring the genotypes *cc*, *Cc*, and *CC* will occur in a ratio of 1:2:1. Since *Cc* and *CC* give the same phenotype, for example, red flowers in peas, the red:white ratio (phenotypic ratio) among the offspring will be 3 : 1.

Experiments of this type, performed before the chromosomal basis of heredity was known, show that the two copies (alleles) of a gene governing a pheno-

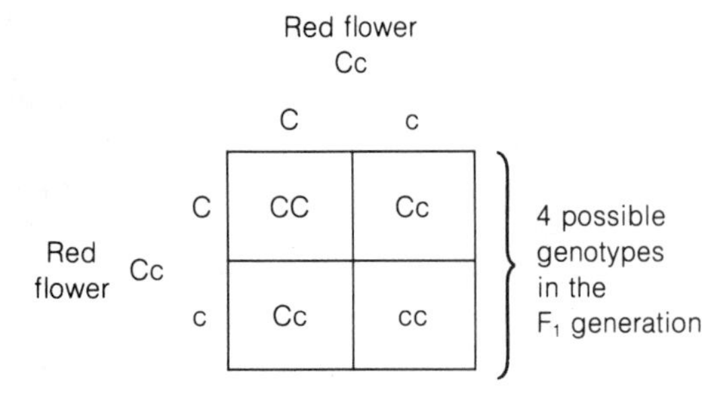

FIGURE 7-23
The Punnett square. Red-flowering plants that are heterozygous for the gene that determines flower color have the genotype Cc. Such a plant produces an equal number of C-containing and c-containing gametes. A cross between two such plants yields four possible combinations in F_1 plants. F_1 plants with one or two Cs are red, and plants with only c are white.

typic trait are transmitted through the gametic cells separately and independently of each other. Each is carried on a separate chromosome, that is, on the two homologues. This rule of **segregation** for the two versions of a gene (in this case *C* and *c*) is known as **Mendel's first law.**

Rule of Independent Assortment

Mendel also determined the pattern of inheritance when two different traits were involved simultaneously. He crossed a purebred plant with yellow, smooth seeds (*YYSS*) with a purebred plant with green, wrinkled seeds (*yyss*) (Figure 7-24(a)). Yellow and smooth are dominant over green and wrinkled, and hence, all of the F_1 had yellow, smooth seeds (*YySs*). Mendel then crossed F_1 plants with one another, and found in the F_2 generation the original parental phenotypes—yellow, smooth and green, wrinkled—plus two new phenotypes—yellow, wrinkled and green, smooth. He interpreted the presence of the new combinations (yellow, wrinkled and green, smooth) to mean that the factor determining one trait (seed color) had been inherited through the gametes independently of the factor determining the other trait (seed texture). The entire experiment is explained as follows: the genotype of the yellow, smooth parent was *YYSS*, and the genotype for the green, wrinkled parent was *yyss*; also, only one kind of gamete is possible for each parent, namely, *YS* and *ys* (Figure 7-24(a)). The F_1 genotype *YySs*, however, can give rise to four kinds of gametes—*YS*, *Ys*, *yS*, and *ys* (Figure 7-24(b)). When two of these gametes unite at fertilization to yield the F_2 generation, 16 combinations are possible, as shown in Figure 7-24. Each square contains an F_2 genotype formed by

a

Yellow-smooth YYSS (gamete YS) × Green-rough yyss (gamete ys)

	YS
ys	YySs

F_1 phenotype yellow-smooth

b

Yellow-smooth YySs × Yellow-smooth YySs

	YS	Ys	yS	ys
YS	YYSS	YYSs	YySS	YySs
Ys	YYSs	YYss	YySs	Yyss
yS	YySS	YySs	yySS	yySs
ys	YySs	Yyss	yySs	yyss

Phenotype	Ratio
Yellow-smooth	9
Yellow-rough	3
Green-smooth	3
Green-rough	1

FIGURE 7-24
(a) A cross between a true breeding plant producing yellow-smooth seeds (YYSS) and a true breeding plant producing green-rough seeds (yyss) yields F_1 plants that produce only yellow-smooth seeds (YySs). (b) A cross between the F_1 heterozygous plants (YySs) produces F_2 plants with nine different genotypes (among the 16 possible combinations) that generate four different phenotypes in a ratio of 9:3:3:1.

fusion of two F_1 gametes. All those squares that contain at least one *Y* are plants that will have the yellow-seed phenotype, and all those containing at least one *S* are plants that will have the smooth-seed phenotype. Squares with *yy* are plants with green seeds, and squares with *ss* are plants with wrinkled seeds. Thus, among the 16 genotypes there are four different phenotypes, and these occur in a ratio of 9:3:3:1—9 yellow, smooth, 3 yellow, wrinkled, 3 green, smooth and 1 green, wrinkled.

These results show that the two different kinds of traits, seed color and seed texture, are assorted in gametes and inherited independently of each other; that is, the genes for the two traits are carried by different chromosomes. This **independent assortment** of traits is called **Mendel's second law.**

When Mendel's laws were rediscovered, they were quickly correlated with the behavior of chromosomes in meiosis, which was unknown in Mendel's day. Shortly after the affirmation of Mendel's work, it was observed that each cell in an organism contains two copies of each chromosome (diploidy) except for the germ cells, which contain only one of each kind of chromosome (haploidy). That observation led to the proposal that the pairs of factors that determine the traits studied by Mendel are carried on pairs of chromosomes. Thus, the pair of factors (alleles) determining seed color are carried by one pair of chromosomes, the pair of factors determining seed texture by another pair of chromosomes, and so on.

One additional example of the correlation between the phenotype of an organism and the behavior of chromosomes during meiosis was uncovered in 1905, two years after the discovery of diploidy and haploidy. In females, all chromosomes occur in homologous pairs because, in addition to the autosomes, they have two X chromosomes forming an XX pair. Each haploid egg cell formed by meiosis must contain an X chromosome. In males of most species the autosomes form homologous pairs, but the X and Y, which are different in form, also make a pair during meiosis. Cytologists confirmed that two kinds of sperm cells must be formed, one carrying a Y chromosome and one carrying an X chromosome. If the X-carrying sperm fertilizes an egg, then the *genotype* of the offspring will be XX, and the offspring will be a female. If a Y sperm fertilizes an egg, the *genotype* will be XY and the offspring will be a male.

By this discovery the inheritance of sexual phenotype became clearly traceable to the X and Y chromosomes, providing the first of many subsequent proofs of the hypothesis that chromosomes carry the heredity factors or genes. Chromosomes carry genes not only in meiosis and in inheritance from an organism to its offspring, but in inheritance from a parent cell to its daughter cells through the process of mitosis.

GENE LINKAGE, CROSSING OVER, AND CHROMOSOME MAPPING

Extension of Mendel's experiments to other phenotypic traits in a variety of organisms showed that not all traits assort independently, but instead some are inherited in groups. In this section we will examine this phenomenon.

Gene Linkage

Traits that are inherited together are said to be **linked,** and they form a **linkage** group. A linkage group is defined by genetic analysis as traits that assort together, and its physical basis is that the genes in the group are contained in a single chromosome. For example, fruit flies of the genus *Drosophila* normally breed true for red eyes (female genotype *RR*) and for wing length (female genotype *NN*). These genes can occur in mutated form. Female flies homozygous for a mutation in the *R* gene have the genotype *rr* and have vermilion (a shade of red) eye color. Female flies homozygous for a mutation in the *N* gene have the genotype *nn* and have miniature wings. Both mutations are recessive, so that female flies heterozygous for either trait (*Rr* or *Nn*) have normal phenotypes. Usually, a recessive allele of a gene is a mutated version of the dominant allele, but there are important exceptions. Sometimes mutated alleles are dominant, and in some cases traits produced by the normal gene and its mutated form may both appear in the same cell (this phenomenon will be discussed later in this chapter).

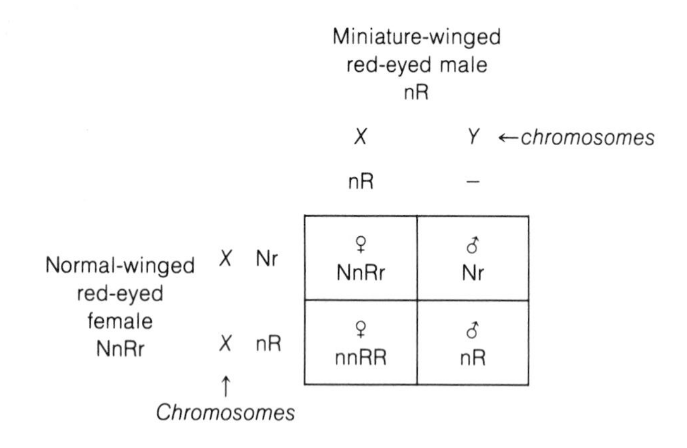

FIGURE 7-25
A cross between fruit flies to observe the inheritance of two genes carried in the X chromosome.

By breeding experiments, the gene for eye color (*R* or *r*) and the gene for wing size (*N* or *n*) were shown to be carried by the same chromosome; that is, they are linked. The eye-color gene and the gene for wing size in *Drosophila* are carried on the X chromosome. Consider a normal-winged, red-eyed female heterozygous for both traits that has the combination *Nr* on one X chromosome and the combination *nR* on the other X chromosome. This female is mated with a miniature-winged, red-eyed male whose single X chromosome carries the genes (*nR*). The F_1 females receive an X chromosome from each parent, but F_1 males receive their X chromosome from the female parent only (Figure 7-25). Half the F_1 females have the genotype *NnRr* (normal-winged, red-eyed) and half are *nnRR* (miniature-winged, red-eyed). Half of the F_1 males have the genotype *Nr* (normal-winged, vermilion-eyed) and half are *nR* (miniature-winged, red-eyed). Occasionally among the males two kinds of exceptions occur. These exceptions are normal-winged, red-eyed (*NR*) and miniature-winged, vermilion-eyed (*nr*) individuals. The original X chromosomes in the female parent were *Nr* and *nR*. Therefore, the X chromosome in the exceptional F_1 males represent new combinations, namely *NR* and *nr*.

The new combinations represent breakdown in old linkages between the alleles of two genes and the formation of new combinations during meiosis. In meiosis homologous chromosomes are already paired with one another by the time chromosome condensation is first detectable in the prophase of the first meiotic division. Meiotic pairing is called **synapsis.** A synapsed pair of homologous chromosomes is called a **bivalent** (Figure 7-26(a)). Cytological observations show that symmetrical exchanges occasionally occur between a chromatid belonging to one chromosome and a chromatid of its synapsed homologue. This phenomenon of exchanges between chromatids of homologous chromosomes is called **crossing over.** The key event in crossing over is the breakage of the DNA duplexes at the same points in the two chromatids undergoing crossing over and a rejoining of the DNA duplex in one chro-

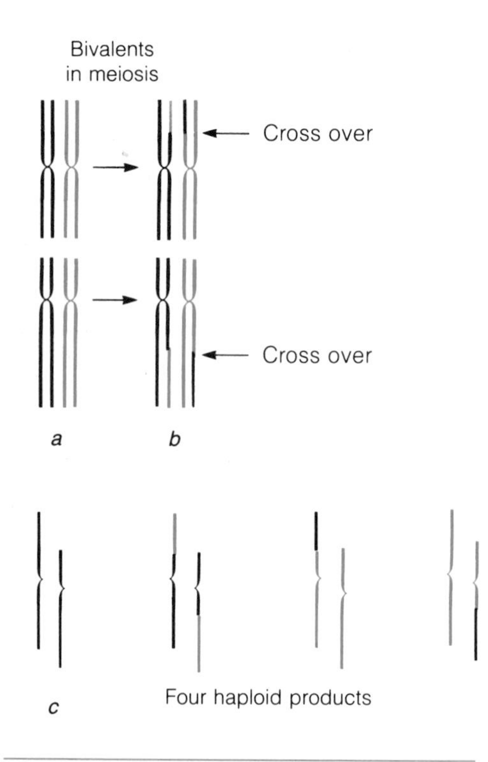

FIGURE 7-26
Crossing over illustrated in two different chromosomes distinguishable by the location of their centromeres.
(a) Homologous chromosomes align with each other into pairs to form bivalents during the first nuclear division in meiosis.
(b) Crossing over has occurred between the two bivalents.
(c) Four haploid gametes formed at the end of meiosis.

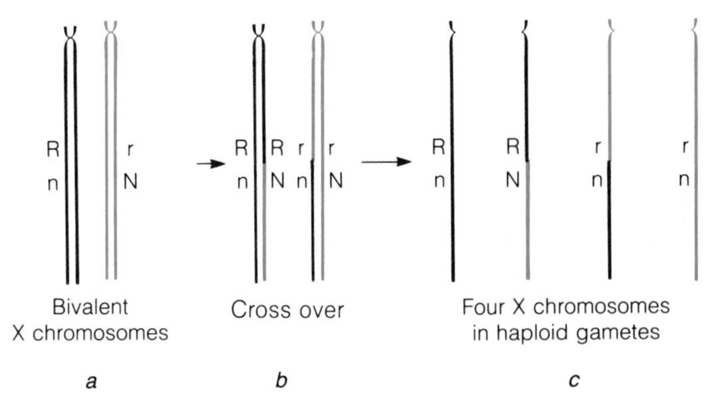

FIGURE 7-27
(a) A bivalent formed by two X chromosomes in a female fruit fly. (b) Crossing over between gene for eye color (R or r) and for wing size (N or n). (c) As a result of crossing over four kinds of X chromosomes appear in the four haploid nuclei at the end of meiosis.

matid with the DNA in the other (Figure 7-26(b)). The physical event of crossing over between the two X chromosomes in *Drosophila* accounts for the appearance of new linkage combinations for genes governing wing length and eye color (Figure 7-27).

In sexual reproduction F_1 offspring are created that contain new mixtures of the alleles of genes by virtue of receiving half of their chromosomes from one parent and half from the other (chromosomal mixing). When germ cells are subsequently formed in the mature F_1 organism, an additional mixing of genes takes place through exchanging of parts by crossing over of the two original parental chromosomes (subchromosomal mixing). Gene mixing creates new combinations of alleles of genes, which in turn results in offspring with new combinations of phenotypic traits.

The Cytology of Chromosome Pairing and Crossing Over

Crossing over occurs during prophase (prophase I) of the first of the two nuclear divisions of meiosis. Five stages can be recognized in prophase I; these are defined by the state of chromosome condensation, homologue pairing, and crossing over.

The beginning of chromosome condensation is called the **leptotene** stage (Figure 7-28). The chromosomes in leptotene are so extended that individual chromosomes are still difficult to discern. As condensation continues and the chromosomes become more distinct the nucleus enters the **zygotene** stage. In this stage the joining or synapsis of homologous chromosomes into pairs is apparent. The pairing between homologous chromosomes into bivalents is reflected by the formation of a structural arrangement called the **synaptonemal complex,** seen by electron microscopy (Figure 7-29). The synaptonemal complex is a tripartite structure consisting of a central region flanked by two lateral elements. The central region has a central protein axis and a regular array of protein threads that extend out to the lateral elements. The central region varies from a few hundred to 1200 Å in diameter, depending on the species. The two lateral elements are a few hundred Å in diameter, and each represents one of the two homologous chromosomes. In the lateral element the two chromatids in a homologue are tightly attached to each other and cannot be discerned individually. However, the DNA protein fibers that make up the two chromatids loop out repeatedly all along the outer edge of the lateral element.

The two homologous chromosomes are held in close alignment with each other by the synaptonemal complex, but remain separated by the width of the central region. Crossing over between a chromatid of one homologue and a chromatid in the other must occur across the central region. This crossing over is probably mediated by a particle called a **recombination nodule** composed of proteins that catalyze the breakage and rejoining of DNA between two chromatids. Recombination nodules occur randomly at widely spaced, irregular intervals along the synaptonemal complex. What governs their positioning or precisely how they

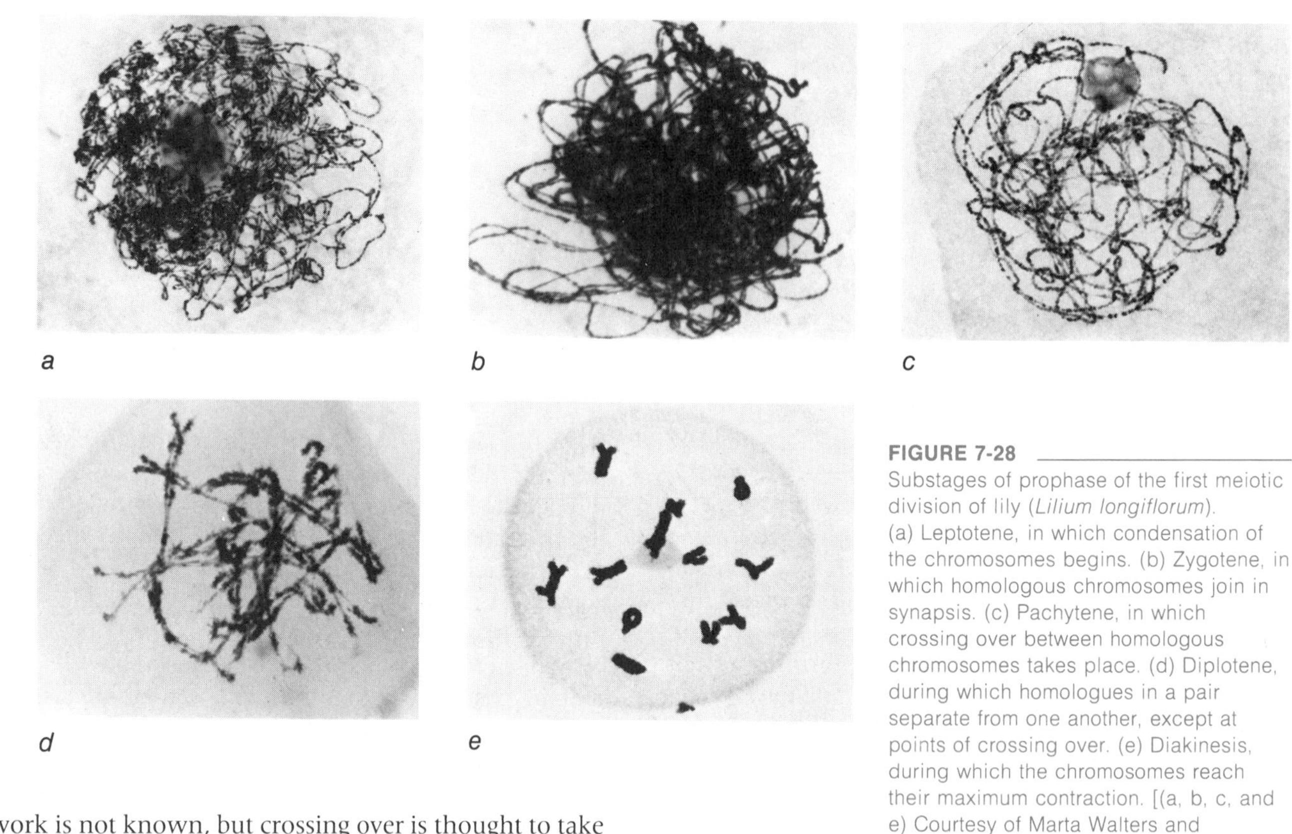

FIGURE 7-28
Substages of prophase of the first meiotic division of lily (*Lilium longiflorum*). (a) Leptotene, in which condensation of the chromosomes begins. (b) Zygotene, in which homologous chromosomes join in synapsis. (c) Pachytene, in which crossing over between homologous chromosomes takes place. (d) Diplotene, during which homologues in a pair separate from one another, except at points of crossing over. (e) Diakinesis, during which the chromosomes reach their maximum contraction. [(a, b, c, and e) Courtesy of Marta Walters and (d) courtesy of Herbert Stern.]

work is not known, but crossing over is thought to take place at the location of each nodule.

Condensation of the zygotene chromosome pairs continues, leading to the **pachytene** stage (Figure 7-28). It is probably in this stage that recombination nodules bring about crossing over between the homologues of a pair. Next, the paired chromosome partners begin to separate, and the **diplotene** stage begins. In addition, the two chromatids in each chromosome, which up to now have been tightly joined, begin to separate. Each pair of homologues is now seen to consist of four threads, which are the four chromatids; this structure is called a **tetrad.** The crossovers between chromatids that occurred earlier are now apparent as points at which the pairs of homologues remain joined (Figure 7-30). Points are called **chiasmata** (sing.: **chiasma**).

During the diplotene stage the still incompletely condensed chromosomes consist of darker staining beads or blocks, called **chromomeres,** connected by thinner **interchromomeric regions** (Figure 7-28). A given chromosome always shows the same beaded pattern, and therefore homologous chromosomes match each other bead for bead. During the diplotene stage in some species, **lateral loops** form that project out from the chromomeres of the two chromatids in each homologue. These lateral loops are particularly large in some animals with large chromosomes, such as amphibians (Figure 7-31). The fuzzy appearance produced by loops led early microscopists to liken them to a brush that was used for cleaning the glass chimneys of oil lamps, and hence such diplotene chromosomes came to be called **lampbrush chromosomes.** In some species, chromosomes may remain arrested in the diplotene stage for many months (e.g., amphibians) or years

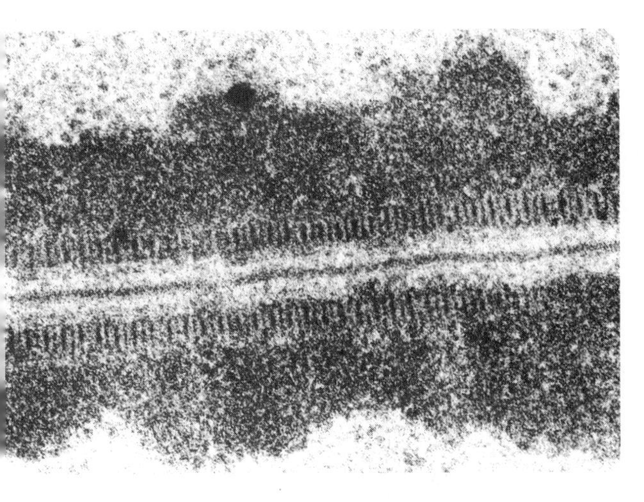

FIGURE 7-29
Electron micrograph of a short segment of a synaptonemal complex formed by synapsed homologous chromosomes in the first meiotic prophase in the fungus *Neotiella rutilans*. [Courtesy of D. von Wettstein.]

(e.g., humans). The lampbrush chromosomes of amphibians have been especially useful to the analysis of chromosome structure and organization (Chapter 11).

The final stage of prophase of the first meiotic division is **diakinesis.** By this stage the chromosomes are highly condensed (Figure 7-28). The end of diakinesis and beginning of metaphase is marked by the breakdown of the nucleoli and nuclear envelope. The first and second meiotic divisions are then completed, as described earlier.

Mapping of Genes along Chromosomes

Crossing over provides a way to determine the relative positions of genes along a chromosome. Crossing over usually occurs randomly along the length of a chromosome. Therefore, the farther apart two genes are positioned along a chromosome, the more likely a crossover event will occur between them at meiosis. Thus, the frequency of crossing over between two genes is usually a direct measure of the relative distance between those genes. However, certain regions of chromosomes sometimes undergo higher or lower rates of crossing over for reasons that are not yet understood. The frequency of crossing over between two genes is determined by mating experiments of the type described above for the eye-color and wing-size genes in *Drosophila* and counting the proportion of F_1 organisms that have a phenotype reflecting a crossover.

The method for determining the relative positions of genes is simple in principle. Starting with three genes *a*, *b*, and *c* the distances between *a* and *b*, between *b* and *c*, and between *a* and *c* can each be measured by the frequency of crossing over between any pair. Suppose that the crossover frequency between *a* and *c*, is one and one-half times the frequency between *a* and *b*

FIGURE 7-30
A pair of homologous chromosomes in the diplotene stage in a male germ cell of the salamander *Oedipina poelzi*. Two chiasmata are present where chromatids of the homologous chromosomes have undergone crossing over. (a) Light micrograph. (b) Interpretive drawing. [From F. W. Stahl. 1964. *The Mechanics of Inheritance*. Prentice-Hall, Inc.; courtesy of James Kezer.]

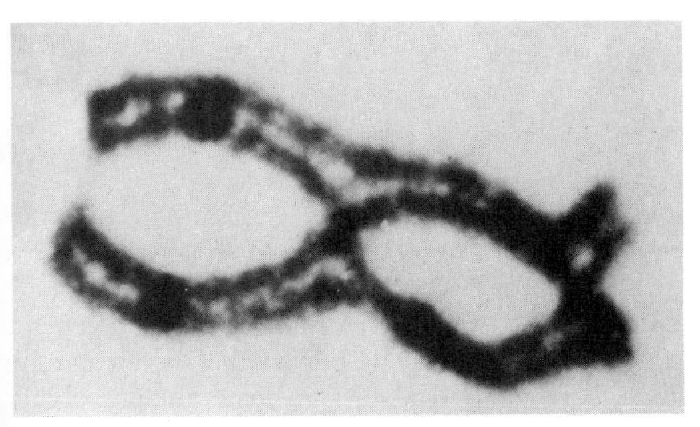

a

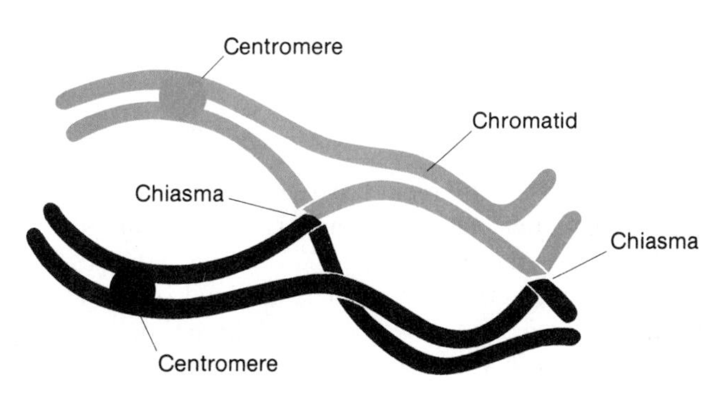

b

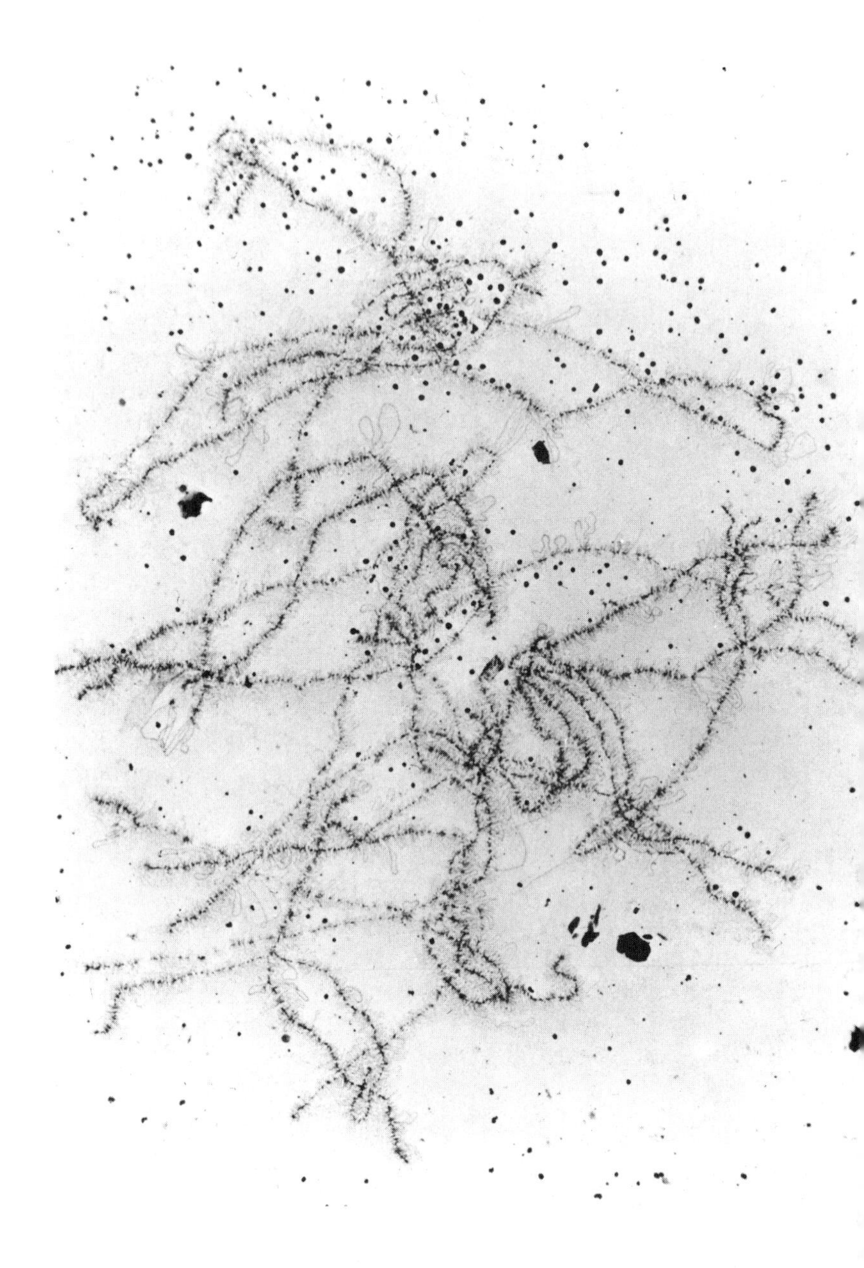

FIGURE 7-31
Light micrograph of a set of lampbrush chromosomes isolated from an oocyte of the newt *Notophthalmus viridescens*. The fuzzy appearance is due to hundreds of pairs of lateral loops; some of the loops are large and easily seen. [Courtesy of James Kezer.]

and three times the frequency between *b* and *c*. The crossover frequency between *a* and *b* is therefore about twice that between *b* and *c* (Figure 7-32). This indicates that the distance between *a* and *c* is one and one-half times the distance between *a* and *b*, and three times the distance between *b* and *c*. These distance relationships can be satisfied only if the three genes are in the order *a*—*b*—*c* in the chromosome with the relative distances shown in Figure 7-32.

In assessing the relative distances between genes a correction must be made for the fact that the greater the distance between two genes, the greater the probability

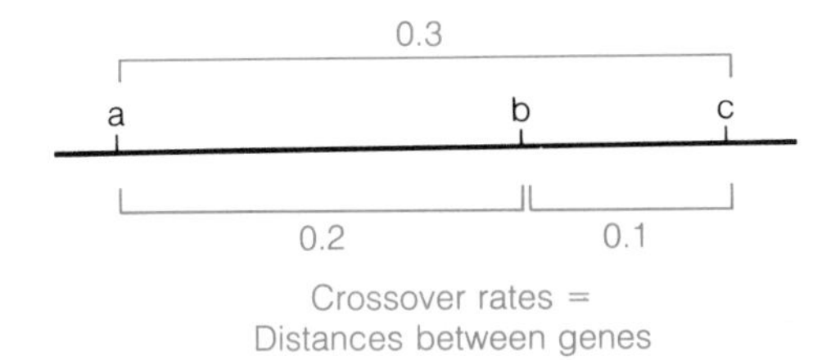

FIGURE 7-32
Crossover rates between pairs of genes *ac*, *ab*, and *bc* are measures of the physical distance between the genes in a chromosome.

hat *two* crossovers will occur during the same meiosis between these genes. Two crossovers will reestablish the original linkage, and the experimentally determined crossover frequency between the two genes will be less than the true frequency. Hence, the two genes will be a little farther apart than the genetic analysis indicates. A correction can be introduced by calculating the probability with which two crossovers will occur. By measuring crossover distances the locations of hundreds of genes have been mapped for the chromosomes of *Drosophila* and many other organisms (Figure 7-33).

Determining the Locations of Genes in Human Chromosomes

For humans mapping of genes in chromosomes by the technique just described above is essentially impossible because of the long generation time and because it is not feasible to do controlled matings between individuals of particular genotypes. Considerable progress has been made both in assignment of genes to particular chromosomes and determining locations along chromosomes using several other techniques, one of which is described here.

Cultured cells of two species can be fused together, as described earlier in Chapter 5, to produce a **hybrid cell.** Initially, the hybrid cell is binucleate, but at the first mitosis after fusion a single mitotic apparatus is formed that distributes the chromosomes of the two nuclei to the daughter cells. Each daughter cell then forms a single nucleus containing the chromosomes of both species. Thus, a mouse and a human cell can be fused to produce a hybrid cell whose nucleus contains both mouse and human chromosomes. During subsequent proliferation of mouse-human hybrid cells the human chromosomes (not those of the mouse) are gradually lost, so the hybrid cell progeny contain a full set of mouse chromosomes but only one or a few human chromosomes. Eventually, all the human chromosomes may disappear. The reason for the preferential loss of chromosomes of one species is not known.

The first step for assignment of genes to chromosomes is to make hybrids. We can do this using a mutant mouse cell, for example, one that has lost the gene for the enzyme **thymidine kinase** (TK). Without thymidine kinase the cell is unable to convert thymidine to TMP, and this can be demonstrated by the failure of such a cell to incorporate radioactive thymidine (^{3}H-thymidine) into DNA. Instead, the cell derives all of its TMP by synthesis from dUMP. When a mouse cell lacking thymidine kinase (TK$^-$) is fused to a human cell containing the gene for thymidine kinase (TK$^+$), the resulting hybrid cell can incorporate ^{3}H-thymidine into DNA because of the presence of the human gene for TK.

As the hybrid cell grows and divides, it loses human chromosomes. If it should lose the human chromosome containing the gene for TK, the hybrid cell will lose the ability to use thymidine for DNA synthesis. With the use of the inhibitor **methotrexate** it is possible to favor the proliferation of hybrid cells that continue to contain the human gene for TK. Methotrexate inhibits the synthesis of TMP from dUMP, which forces a cell to obtain all its TMP by phosphorylation of thymidine. Since TK catalyzes this reaction, only cells that contain TK can proliferate. Hence, any mouse-human hybrid that loses the human chromosome with the TK gene stops reproducing. Culturing hybrid cells in the presence of methotrexate therefore selectively allows proliferation of hybrid cells that retain the human TK gene. Under such conditions hybrid cells are eventually obtained that contain a full complement of mouse

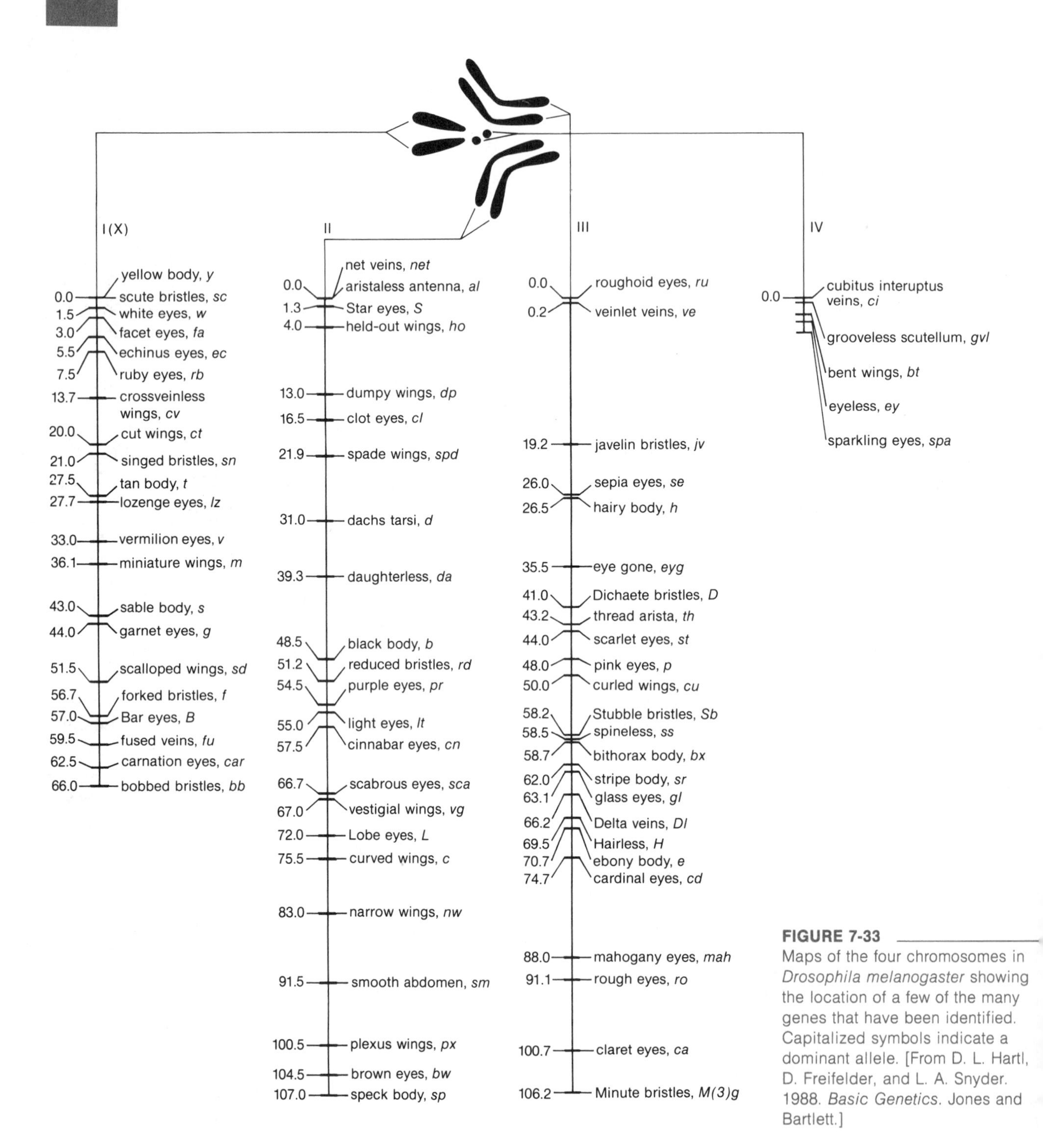

FIGURE 7-33 Maps of the four chromosomes in *Drosophila melanogaster* showing the location of a few of the many genes that have been identified. Capitalized symbols indicate a dominant allele. [From D. L. Hartl, D. Freifelder, and L. A. Snyder. 1988. *Basic Genetics*. Jones and Bartlett.]

chromosomes, but only human chromosome number 17. Since such hybrids must also contain human TK to grow, it follows that the human gene for TK must be located in human chromosome 17.

Using this and related procedures several hundred human genes have been assigned to the 22 autosomes and X and Y chromosomes. Other techniques have allowed the approximate location of many of these genes within their chromosomes. The total number of different genes carried by all the human chromosomes is unknown; most estimates place the number between 50,000 and 100,000.

Sexual Reproduction in Unicellular Eukaryotes

Sexual reproduction is not limited to multicellular plants and animals but occurs commonly among unicellular eukaryotes (**protists**). The details of sexual reproduction among many protists—such as certain species of yeast, alga, and protozoa—are well understood. Sexual reproduction among some protists, such as the many species of amoebae, has never been detected in spite of considerable study. It can be argued effectively that new combinations of gene alleles, and hence new combinations of phenotypic traits, generated by sexual reproduction are essential for any protist to compete successfully for resources in the microenvironment. Sexual reproduction of an organism like an amoeba could easily escape detection under laboratory conditions, as it did for many years in the much studied slime mold *Physarum*.

Yeast. The budding yeast cell has been used extensively to study many problems in cell biology, molecular biology, and genetics, and its life cycle is particularly well understood. The life cycle has a haploid phase and a diploid phase connected by sexual events. In the haploid phase (haploid chromosome number = 17) the budding yeast reproduces with a typical cell cycle in which DNA replication and cell growth alternate with cell division. In almost all protists cell growth is apparent as enlargement of a cell that is then cleaved into two at cell division. In budding yeast, growth is apparent as a bud that forms at the surface of the cell and gradually increases in size (Figure 1-10). The mother cell and its bud remain connected by a cytoplasmic bridge, and anabolic activities in both mother cell and bud contribute to bud growth. The DNA in the 17 chromosomes in the nucleus of the mother cell replicates in preparation for mitosis. When the bud has nearly reached the size of the mother cell, mitosis takes place at the bridge between the two. Mother cell and bud each receive a haploid set of chromosomes through the mitotic division, and cytokinesis consists of a closing off of the bridge between the two cells, followed by physical separation. Both mother and daughter enter the next cell cycle, each forming a bud soon after cytokinesis. Haploid yeast cells may proliferate in this manner indefinitely.

Budding yeasts contain two genes, a and α, that determine the mating phenotype. Only one or the other gene can be expressed in a given yeast cell. If the a gene is expressed, the cell acquires an a mating phenotype. If the α gene is expressed, the cell has an α mating phenotype. The a and α cells can mate with each other; an a cell cannot mate with another a cell, and an α cell cannot mate with another α. Mating consists of fusion of an a cell with an α cell, followed by fusion of the two haploid nuclei. The result is a single diploid cell. The new diploid cell can then reproduce by budding, with a cell cycle identical to that of the haploid (Figure 7-34).

When the growth and division of diploid cells are prevented, as happens when the cells are deprived of an essential nutrient, the cells undergo meiosis. As usual, the first meiotic division reduces the chromosome number from diploid to haploid, and the second meiotic division divides the duplicated chromosomes into unduplicated ones. Each of the four resulting haploid cells becomes a **spore,** which is a nongrowing form of the cell having a nearly zero rate of metabolism. Spores have tough walls and are extremely resistant to drying, heating, freezing, and other environmental stresses. They may remain as dormant spores for years. When a spore is placed in conditions favorable for growth, it reactivates (**germinates**) and resumes proliferation as a haploid cell.

Protozoa. Ciliated protozoa are found in large numbers in all bodies of fresh and salt water through-

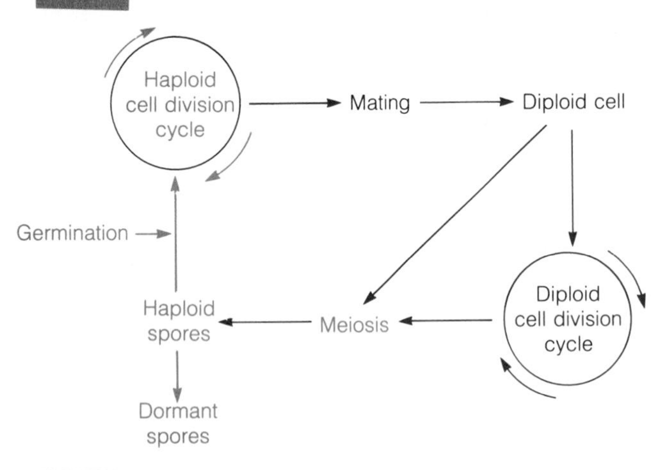

FIGURE 7-34
The life cycle of budding yeast. Haploid cells can mate to produce diploid cells. Diploid cells can undergo meiosis or reproduce (diploid cell division cycle) and subsequently undergo meiosis. Meiosis produces four haploid spores, which can remain dormant for a long time, or if nutrient conditions are favorable, germinate into haploid cells that reproduce (haploid cell division cycle).

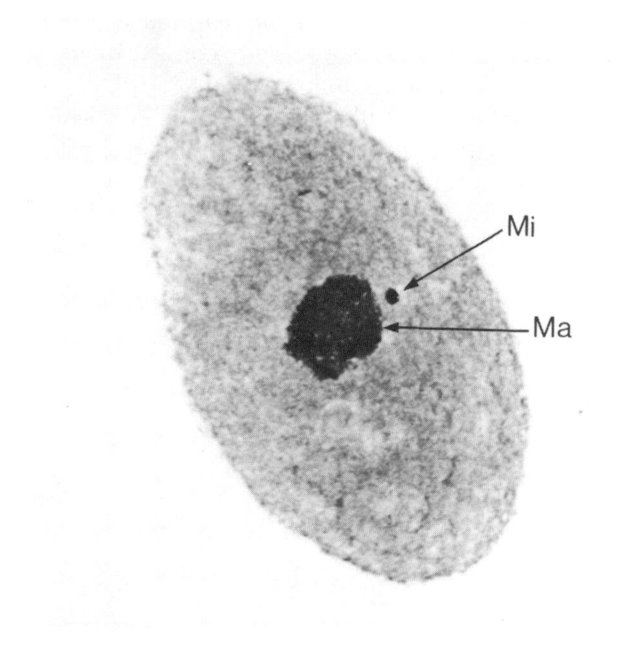

FIGURE 7-35
Light micrograph of the ciliate *Tetrahymena*. Mi = micronucleus. Ma = macronucleus. [Courtesy of Joseph G. Gall.]

out the world. Many thousands of species have been described and named, but many more remain unknown. Over 50 new species are discovered each year by protozoologists.

Ciliates are generally very large cells with two nuclei of different sizes (Figure 7-35). The smaller nucleus, the **micronucleus,** is usually diploid. The larger nucleus, the **macronucleus,** contains a much larger amount of DNA that is not organized into usual chromosomes. Only the genes in the macronucleus govern the phenotype of the cell; the micronuclear genes remain silent.

Ciliates such as *Paramecium* and *Tetrahymena* mate by the sticking together of two cells and the formation of a cytoplasmic bridge between them (Figure 7-36). Variations in the behavior of micronuclei in mating occur in different species, but the principles are always the same. Mating in *Tetrahymena* is described here.

The joining of two cells in mating is accompanied by meiosis of the micronucleus to produce four haploid micronuclei in each cell of a mating pair (Figure 7-37). (Meiosis in this case is not accompanied by cell division.) Three of the four haploid micronuclei are destroyed. The remaining haploid micronucleus divides by mitosis so that a cell in a mating pair of cells has two genetically identical haploid micronuclei. The two mating cells each exchange single haploid micronuclei through the cytoplasmic bridge. The migratory haploid micronucleus fuses with the haploid micronuclei in the cell into which it has migrated. This formation of a new diploid micronucleus constitutes fertilization in each cell. The new diploid micronucleus divides mitotically twice without cell division to yield four diploid micronuclei. The mating cells now separate and are called **exconjugants.** The four diploid micronuclei in each exconjugant have the following fates: two of them develop into two new macronuclei, and the old macronucleus disintegrates. One of the new diploid micronuclei disintegrates. The fourth becomes the new diploid micronucleus. Development of the two new macronuclei occurs by many replications of the DNA and by some reorganization in DNA as described in Chapter 11. After all of these nuclear events an exconjugant cell now has two new macronuclei and one diploid micronu-

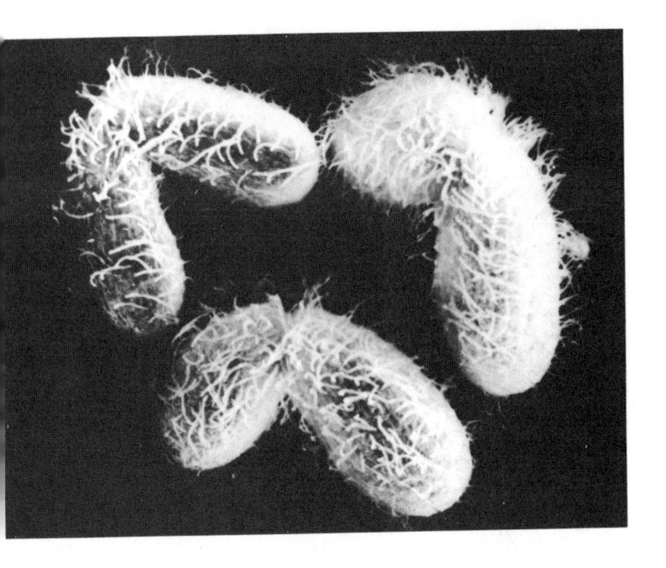

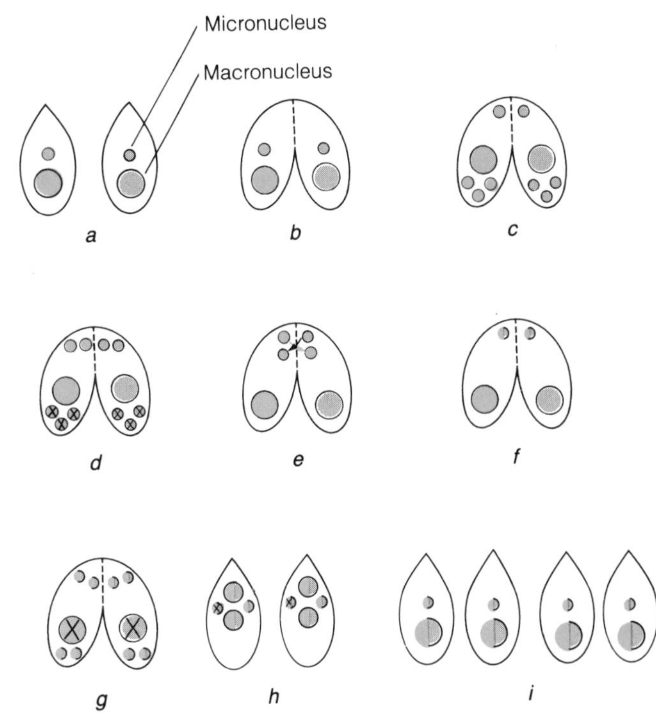

FIGURE 7-37
Mating in *Tetrahymena*. (a) Two cells of opposite mating type. (b) Joining of cells by a cytoplasmic bridge. (c) Meiosis in each cell yielding four haploid products. (d) Three haploid micronuclei disintegrate and one divides by mitosis. (e) Cells exchange haploid micronuclei. (f) Fusion of exchanged haploid micronucleus with resident haploid micronucleus in each cell to produce new diploid micronuclei (red-gray). (g) Two divisions of new diploid micronuclei yielding four diploid micronuclei in each cell. The macronuclei degenerate. (h) Mating cells separate into exconjugants. Two diploid micronuclei develop into two new macronuclei (red-gray). The third diploid micronucleus degenerates; the fourth diploid micronucleus is retained. (i) The two exconjugant cells divide producing a total of four daughter cells that resume proliferation.

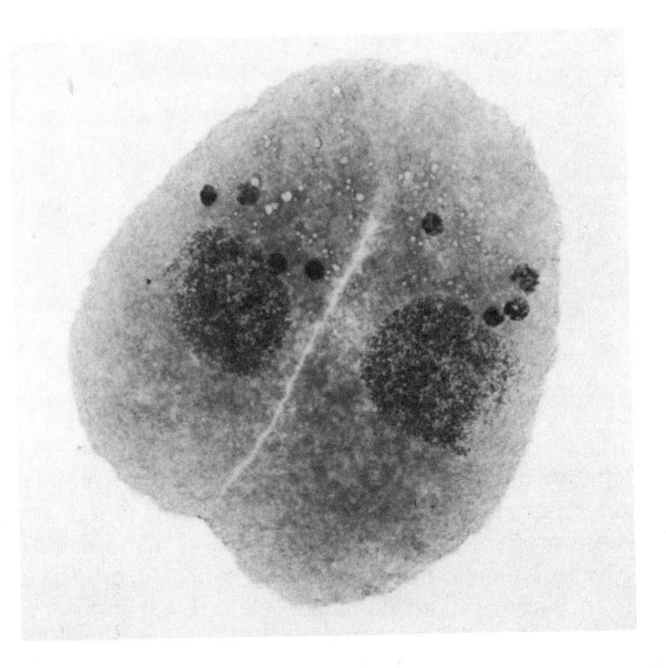

FIGURE 7-36
(a) Scanning electron micrograph of three mating pairs of *Tetrahymena*. The cells have joined at their anterior ends and formed a cytoplasmic bridge, through which haploid micronuclei will be exchanged. (b) Light micrograph of a pair of mating *Tetrahymena*. A cytoplasmic bridge and four haploid micronuclei just produced by meiosis in each cell are visible. [(a) From G. Shih and R. Kessel. 1982. *Living Images*. Jones and Bartlett. (b) Courtesy of Joseph G. Gall.]

cleus. The exconjugant divides, with mitosis of the micronucleus but without division of the macronuclei. The daughter cells, each possessing a single diploid micronucleus and a single macronucleus, now resume proliferation. The phenotype of an exconjugant cell is determined by the genes that it has received from its two parents and that are present in its macronucleus.

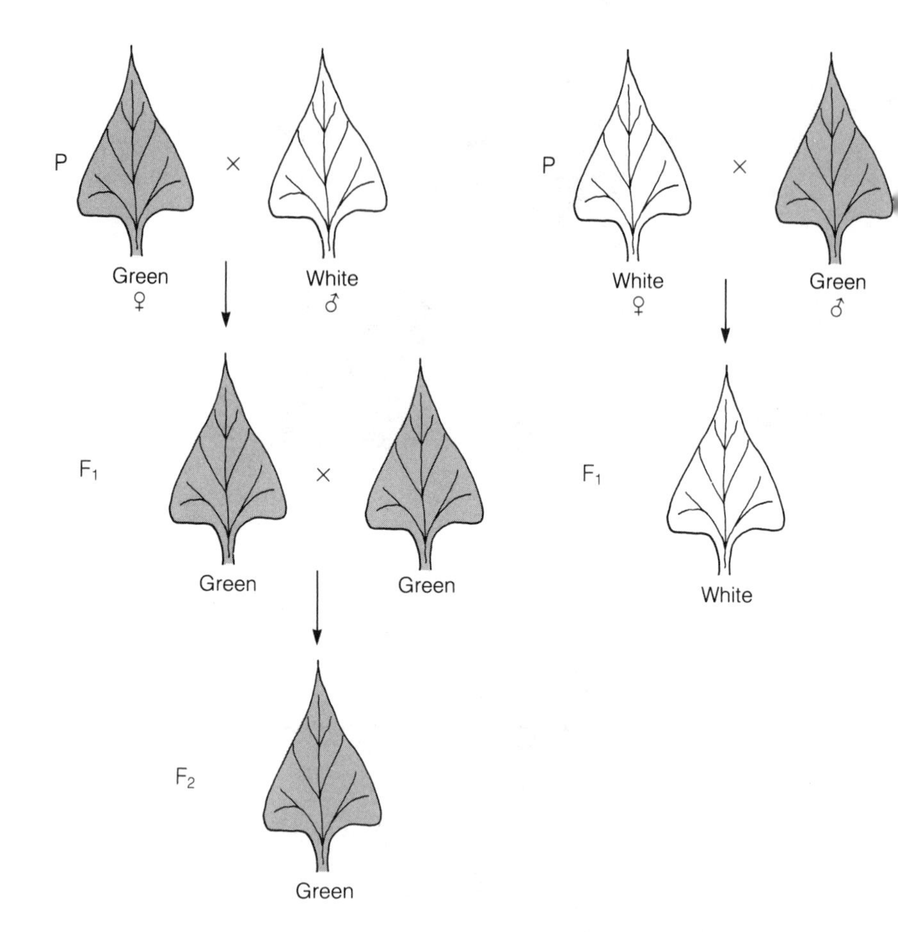

FIGURE 7-38
Leaf color in four o'clocks is inherited through chloroplasts. Only female gametes carry chloroplasts; therefore the phenotype of all of the offspring of a cross is the same as the phenotype of the maternal parent.

Other groups of eukaryotes show other variations in arrangements and timing of meiosis, mating, fertilization, and the like, but the overall principles are always the same and accomplish the same ends: formation of new combinations of alleles in chromosomes by crossing over during meiosis, and formation of new combinations of chromosomes by the fusion of haploid nuclei.

CYTOPLASMIC INHERITANCE

Almost all (99 percent) of the proteins and hence most of the functional and structural properties of a cell are prescribed by genes in the cell nucleus. A few proteins are determined by genes carried in the DNA molecules present in mitochondria of all eukaryotes and more proteins by the DNA molecules in the chloroplasts of plants and algae. Inheritance of properties from parent cell to daughter cell through mitosis or meiosis of the nucleus in unicellular and multicellular eukaryotes is said to be Mendelian inheritance, because it conforms to the laws first defined by Mendel's breeding experiments with plants and later extended (linkage and crossing over) by other researchers. Inheritance of a few cell properties takes place in a non-Mendelian fashion by factors carried in the cytoplasm, for example, mitochondrial and chloroplast DNA molecules.

Inheritance Through Chloroplasts

Early in the century, shortly after the first rules of Mendelian inheritance had been established, examples of non-Mendelian inheritance in plants were discovered. In *Mirabilis,* the "four-o'clock," the leaves in a single plant can be green or white, or a mixture of green and white. Flowers on a green-leaved branch (female) pollinated by flowers (male) of a white-leaved branch produce only F_1 offspring plants that are green-leaved (Figure 7-38). Matings (crosses) between F_1 offspring produce only green-leaved plants in the F_2 generation. In the reciprocal type of cross, in which flowers on a white-leaved branch (female) are fertilized with pollen (male) from a green-leaved branch, seeds are produced that germinate into F_1-generation plants that are all white. In short, in any genetic cross the offspring inherit the leaf color of the **maternal** parent. Such maternal inheritance is due to genes carried in chloroplasts. Pollen cells in *Mirabilis* do not possess chloroplasts; all chloroplasts are inherited through the egg cytoplasm. Flowers on white-leaved branches carry defective chloroplasts that do not produce chlorophyll and can produce only eggs with colorless chloroplasts; flowers on green-leaved branches contain green chloroplasts and produce eggs with green chloroplasts. The totally white plants produced in such an experiment do not survive long after seed germination because without chlorophyll they are incapable of photosynthesis.

Other cases of cytoplasmic inheritance involving chloroplasts have been found. The discovery of DNA in chloroplasts in the early 1960s provided an explanation of chloroplast inheritance. Some of the properties of chloroplasts are governed by genes carried in chloroplast DNA. Other properties of chloroplasts are inherited in Mendelian fashion because they are governed by genes carried in the cell nucleus. Chloroplast function and structure are prescribed cooperatively by genes in the organelle itself and by genes in the cell nucleus.

Mitochondrial Inheritance

Mitochondria contain a small amount of DNA, which codes for a few of the proteins found in mitochondria. Genes in this DNA are the basis for cytoplasmic inheritance of a limited number of traits.

Inheritance of resistance to the drug **chloramphenicol** in mouse cells is an example of mitochondrial inheritance. Chloramphenicol (also called chloromycetin) is an antibiotic that inhibits protein synthesis in bacteria, which is the basis of its action in treating bacterial infections in humans and other mammals. It also inhibits protein synthesis in mitochondria, but does not inhibit protein synthesis in the rest of the cytoplasm. Inhibition of mitochondrial protein synthesis causes side effects of the antibiotic in humans, and it must be used cautiously. Overall, bacteria are more sensitive to chloramphenicol than are eukaryotic cells, and it can be used effectively. The difference in sensitivity between mitochondrial protein synthesis and protein synthesis in the rest of the cytoplasm is understood in considerable detail but need not be discussed for present purposes. Mitochondrial protein synthesis resembles bacterial protein synthesis in important respects (Chapter 8) that account for their similar sensitivities to chloramphenicol.

Mouse cells grown in the presence of chloramphenicol stop growing and eventually die because of inhibition of mitochondrial protein synthesis necessary to maintain mitochondrial function and structure. On rare occasions a cell becomes insensitive to chloramphenicol and grows normally in the presence of the drug. Such resistance is thereafter inherited through cell division, giving rise to a permanently resistant strain.

Cytoplasmic inheritance of chloramphenicol resistance was shown in the following way: the nuclei of chloramphenicol-resistant cells were removed by treating cells with the drug **cytochalasin B** and centrifuging them. Cytochalasin B binds to G-actin, which leads to disassembly of microfilaments. This weakens the cytoskeleton such that the nucleus, which is denser than the cytoplasm, can be pulled out of the cell by centrifugal force (Figure 7-39). This leaves an enucleated cell, which is called a **cytoplast** (Figure 7-40). Cytoplasts derived from chloramphenicol-resistant mouse cells were fused with normal-type (called **wild-type**) nucleated (chloramphenicol-sensitive) cells (Figure 7-41),

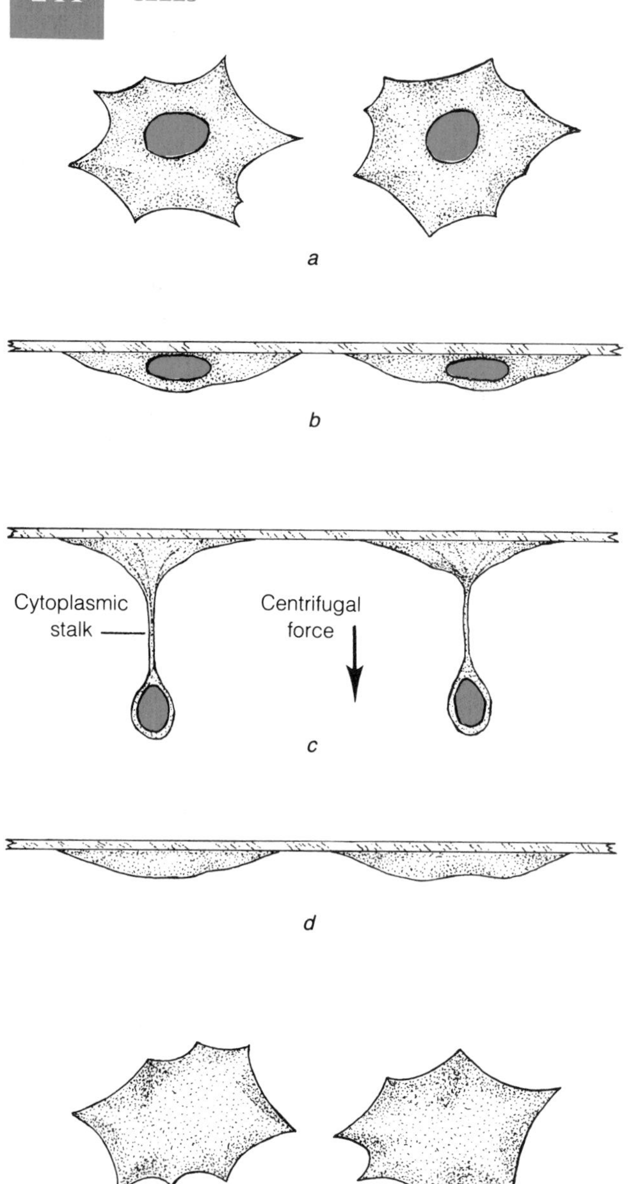

FIGURE 7-39

Enucleation of cultured animal cells. (a) Intact cells attached to a coverglass viewed from above. (b) Same cells in a side view. (c) Centrifugation in the presence of cytochalasin. Nuclei have been drawn away from the cell and are attached by a cytoplasmic stalk. (d) Stalks have been severed, releasing nuclei from the cells, and retracted into the body of the cell. (e) Enucleated cells viewed from above.

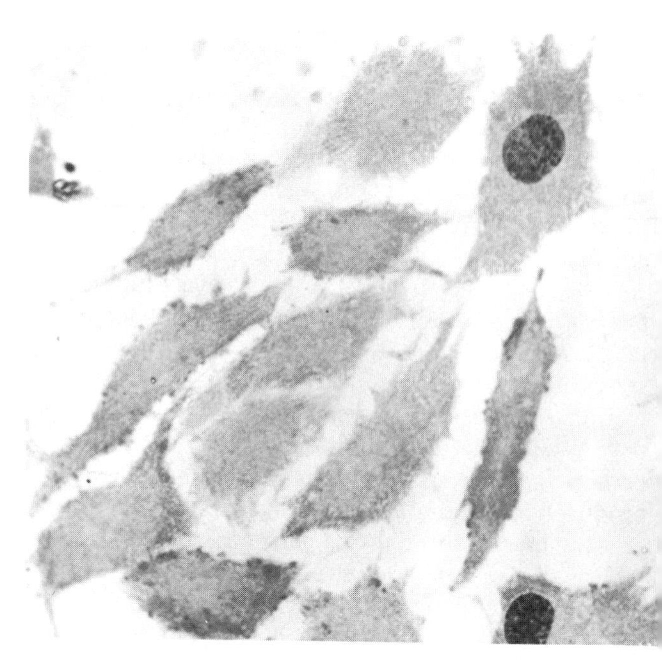

FIGURE 7-40

(a) Light microscope photograph of a group of animal cells in culture enucleated by the cytochalasin-centrifugation technique. Three cells have retained their nuclei. (b) Electron micrograph of an enucleated animal cell. [(b) Courtesy of Jerry W. Shay.]

creating a cytoplasmic hybrid. Such cytoplasmic hybrids are resistant to chloramphenicol and transmit resistance to progeny through cell division. Thus, resistance has been transferred to a sensitive cell by adding to it the cytoplasm of a resistant cell. This cytoplasmically inherited property is due to acquisition of chloramphenicol resistance possessed by mitochondria of the resistant cell type.

Mitochondrial inheritance was first discovered in yeast cells. Mutant yeast cells were found that were unable to carry out aerobic respiration. Such cells grow slowly because of the limited supply of ATP, derived from glycolysis. Because they make small colonies on agar plates, such cells are called **petites.** The normal or wild-type, are called **grandes.** In sexual crosses, the petite characteristic is inherited in non-Mendelian fashion. When a wild-type yeast cell is crossed with some kinds of petite mutants, all the F_1, F_2, and so on, are

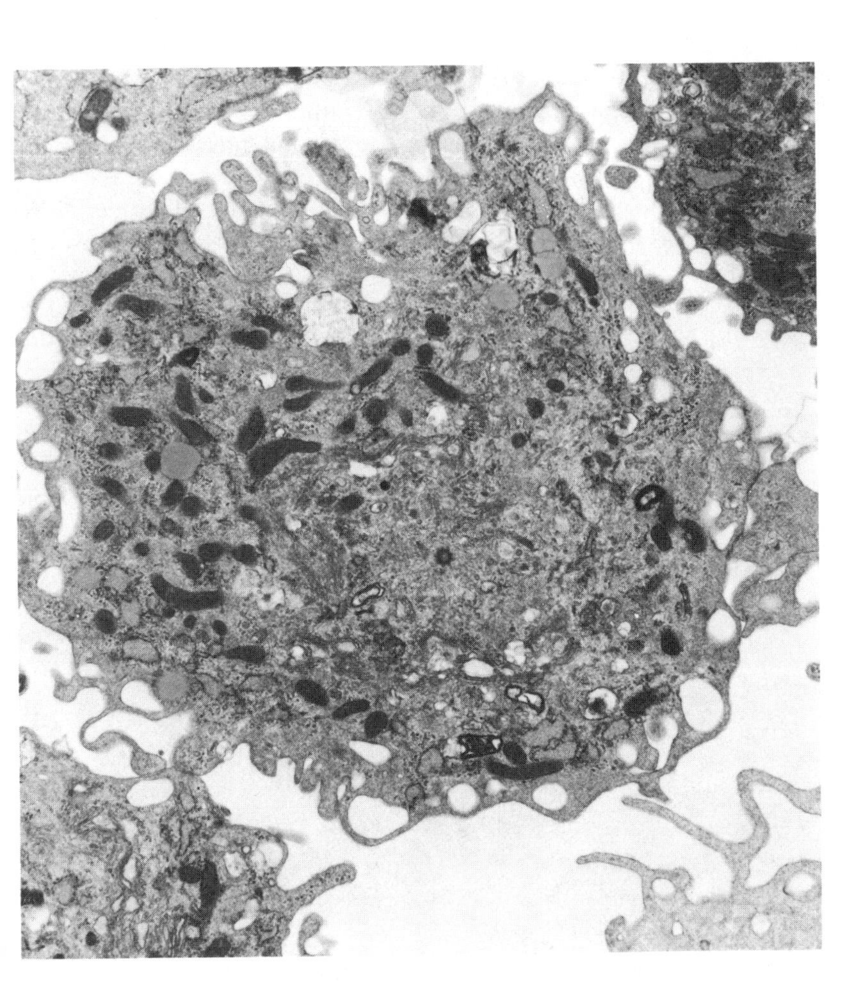

wild-type (Figure 7-42) demonstrating that the characteristic is based on cytoplasmically inherited factors.

Petites are defective in aerobic respiration, and it was reasonable to suppose that the mitochondria were the cytoplasmic factors responsible for inheritance of the petite phenotype. When mitochondrial DNA was discovered in the early 1960s, it was not long before it was shown that the petite phenotype is due to a deletion in the mitochondrial DNA. In some petite strains the amount of DNA deleted is small, and in others the entire chromosome is absent in all mitochondria. In mating between some kinds of petite mutants and a wild-type, the normal mitochondria from the wild-type yeast grow and reproduce more rapidly than the defective ones, and eventually come to be the only mitochondrial type in such cells.

Not all petite colonies of yeast are due to loss of mitochondrial DNA. Mutations in certain nuclear genes (detected by their Mendelian-type inheritance) cause defects in aerobic respiration in mitochondria, resulting

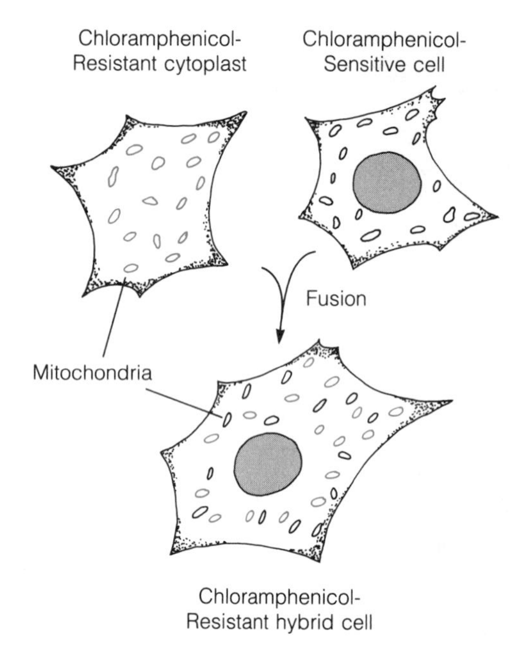

FIGURE 7-41
Fusion of a cytoplast derived from a chloramphenicol-resistant cell with a chloramphenicol-sensitive cell yields a hybrid that remains permanently chloramphenicol-resistant during subsequent proliferation.

in slow cell growth. The function and structure of mitochondria are supported cooperatively by mitochondrial and nuclear genes.

Cytoplasmic Inheritance in *Paramecium*

A completely different form of cytoplasmic inheritance occurs in the ciliated protozoan *Paramecium*. This organism swims by means of cilia that are arranged in a precise pattern at the cell surface. The pattern can be altered by microsurgery, reversing the orientation of some of the cilia so that they then beat in the opposite direction to normal cilia. This abnormal ciliary pattern is inherited by daughter cells from the parent cell through cell division. The experiment demonstrates the cytoplasmic inheritance of a particular structural arrangement, namely, the organizational pattern of cilia. The pattern of cilia is in fact governed by structures in the cortex of the cell; the cortex is the cytoplasmic region that is in a gel state rather than a liquid and found immediately inside the plasma membrane of many kinds of cells.

Whether cytoplasmic inheritance of structural organization is peculiar to ciliated protozoa or involves other structures in other kinds of cells is not known. Nevertheless it shows that a structural property *can* be inherited independently of the genes in the nucleus, mitochondria, or chloroplasts.

The inheritance of ciliary pattern differs fundamentally from other forms of cytoplasmic inheritance because it is apparently not mediated by DNA. Searches for DNA that might be responsible for this cytoplasmic inheritance have centered on the basal bodies of cilia. Basal bodies are centrioles, and although several claims of identification of DNA in basal bodies and centrioles have been made, the evidence at present is predominantly against it.

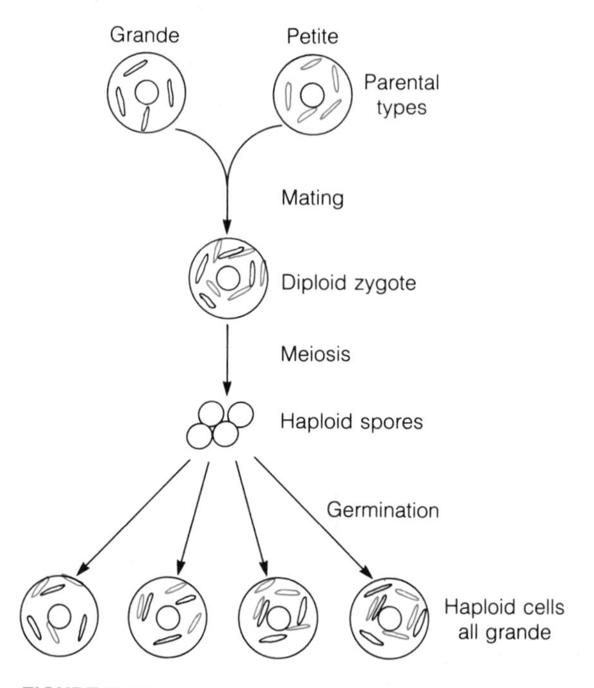

FIGURE 7-42
Mitochondrial inheritance in yeast. All the haploid progeny of a mating between grande (normal) and petite cells are grande.

PROBLEMS

1. In the DNA of a ciliated protozoan named *Oxytricha,* 22 percent of the bases are thymines. What percent are guanines?
2. Of the three studies (the pneumococcus experiment, the phage-blender experiment, and the stability of DNA), which seems to provide the most convincing evidence that DNA is the genetic material? Which seems to provide the least convincing evidence? Explain.
3. How many hydrogen bonds hold the two deoxynucleotide chains together in a DNA molecule that is 10^6 base pairs long and in which 20 percent of the bases are adenine?
4. What is meant by the statement that the deoxynucleotide chains in DNA have opposite polarities?
5. In the experiment with *E. coli* that demonstrates semiconservative replication of DNA what would the DNA density pattern (Figure 7-10) be like if the cells were switched back to ^{15}N medium for one generation after having grown for two generations on ^{14}N medium?
6. To which end of a nucleotide chain does DNA polymerase catalyze the addition of deoxynucleotides?
7. What function do RNA primers serve in DNA replication?
8. What two functions do topoisomerases perform in DNA replication?
9. Describe what happens to the chromosomes in each of the four stages of mitosis.
10. What characteristics are used to define the karyotype of a species?
11. In a multicellular organism what kinds of normal cells are haploid?
12. What is the difference in the way chromosomes are arranged in metaphase in mitosis and in the first meiotic division?
13. What purpose is served by meiosis?
14. What phenotypes would occur in the offspring of a mating of a pea plant with yellow-smooth seeds with a heterozygous genotype for the two traits and a pea plant with green-wrinkled seeds? What would be the ratios of the phenotypes?
15. What phenotypes would occur in the female offspring of a mating between a vermilion-eyed, miniature-winged female fruit fly and a vermilion-eyed, normal-winged male? Remember, these traits are carried in the X chromosome. Would crossing over affect the outcome in the F_1 generation?
16. Suppose the crossover frequency between linked genes d and e is X, between linked genes e and f is 4X, and between genes f and d is 3X. What is the order and relative spacing of the genes in a chromosome?
17. Which two organelles are responsible for cytoplasmic inheritance among eukaryotes?

TRANSFER OF GENETIC INFORMATION FROM DNA TO PROTEINS

Protein Synthesis and DNA
The Role of RNA in Protein Synthesis
Transcription of DNA Nucleotide Sequences into RNA Nucleotide Sequences
The Chemical Structure of RNA
The Concept of Messenger RNA (mRNA)
RNA Polymerase and the Promoter in Prokaryotes
Termination of Transcription in Prokaryotes
Transcription from One Strand of DNA
Translation of RNA Nucleotide Sequences into Proteins

THE GENETIC CODE

Cell-Free Synthesis of Polypeptides
Degeneracy of the Genetic Code
The Near Universality of the Genetic Code
Mutations and the Triplet Code
Reverse Mutations and Suppressor Mutations
Intragenic suppression
Intergenic suppression

TRANSLATION

The Initiation Phase
The Elongation Phase
Termination of Translation
Polysomes
Coupled Transcription and Translation
Agents that Block Translation

ADDITIONAL FEATURES OF PRODUCTION OF mRNA IN EUKARYOTES

The Three RNA Polymerases in Eukaryotes
The 5′ Cap of Eukaryotic mRNA
The 3′-Poly(A) Tail of Eukaryotic mRNA Molecules
Cutting and Splicing of RNA in Eukaryotes
Excision of Introns
Self-Splicing of Primary Transcripts
The Function of Introns
The Primary Transcripts in the Nucleus

FORMATION OF RIBOSOMES

The Nucleolus
Multiplicity of rRNA Genes
Synthesis of rRNA Precursors in Eukaryotes
Self-Assembly of Ribosomes from Component Macromolecules

GENE EXPRESSION AND CELL PROPERTIES

TARGETING OF PROTEINS TO COMPARTMENTS AND STRUCTURES IN THE CELL

Proteins that Remain in the Cytosol
Targeting of Proteins to the Cytoskeleton
Targeting of Proteins to Mitochondria
Targeting of Proteins to Chloroplasts
Targeting of Proteins to the Nucleus
Targeting of Proteins to the Endoplasmic Reticulum, the Golgi Complex, and Lysosomes

CHAPTER 8

Function of Genes

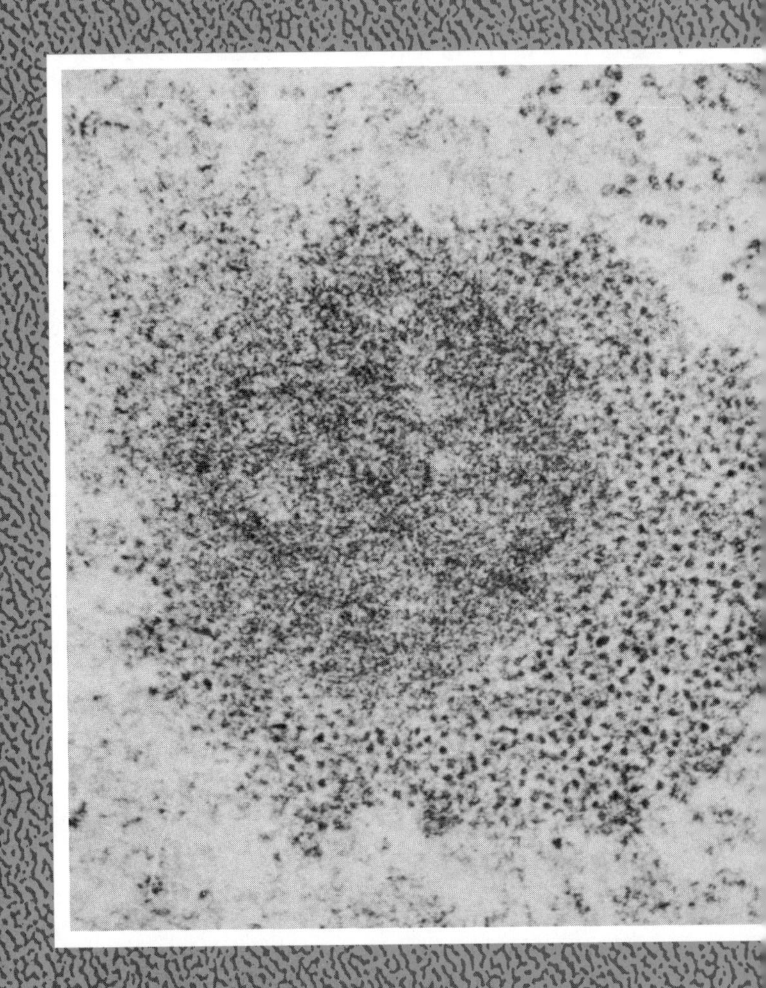

See Figure 8-36.

Two major discoveries in genetics and cell biology were introduced in Chapter 7. First, inheritance of traits from parental cell to daughter cells and therefore from parental organism to offspring is based on genes carried in chromosomes. Second, genes are contained in very long DNA molecules that are distributed to daughter cells by mitosis and meiosis. The establishment of these facts brought into focus new, pressing questions about cell function and structure to which cell biologists turned their attention. A primary enigma was how DNA contained in chromosomes in the nucleus exerts its genetic influence on cell metabolism. Proteins were recognized as the determinants of cell metabolism through their enzymatic properties, and hence as the determinants of the phenotypic traits of the cell. Does a direct relation exist between DNA and proteins, and if so, what is the molecular nature of the relation? Answers to this and related questions came from a series of major experimental discoveries in the late 1950s and the 1960s, which included explanation of the genetic code of DNA, the role of RNA in the transfer of genetic information from DNA to proteins, and the general mechanism of protein synthesis. These are the topics of this chapter.

TRANSFER OF GENETIC INFORMATION FROM DNA TO PROTEINS

Genetic information contained in DNA is transferred first to certain RNA molecules. The information in these molecules is then converted to amino acid sequences. The various transfer steps will be described in this section.

Protein Synthesis and DNA

For a few years after DNA had been proven to be the genetic material, the possibility was considered that DNA had an immediate role in protein synthesis. This idea implied that proteins are synthesized in the cell nucleus because the DNA is contained in chromosomes and is restricted to the nucleus. The demonstration of the presence of substantial amounts of proteins in the nucleus was compatible with the idea. One could test the hypothesis directly by determining whether proteins are synthesized in the nucleus or cytoplasm.

Radioactive amino acids added to the growth medium of cells are taken into cells within seconds, and they label proteins undergoing synthesis. Cells examined a few minutes later by autoradiography show that radioactive proteins are present both in the nucleus and in the cytoplasm. The experiment does not tell the location of protein synthesis because some proteins move rapidly back and forth between the nucleus and cytoplasm. However, if enucleated cells are created by removal of the nucleus, radioactive amino acids are incorporated in cytoplasmic proteins, showing that proteins can be synthesized in the absence of nuclear DNA (Figure 8-1). Therefore, it is clear that protein synthesis does not require direct participation of DNA. Indeed, subsequent kinetic experiments have shown that no protein synthesis occurs in the nucleus. *All nuclear proteins are synthesized in the cytoplasm and subsequently migrate into the nucleus.*

The Role of RNA in Protein Synthesis

The idea that RNA might participate in protein synthesis came from the discovery in the late 1930s that cells particularly active in protein synthesis are rich in RNA. For example, cells of the liver produce large amounts of serum proteins, and cells of the pancreas synthesize large quantities of digestive enzymes for use in the intestines. Cells of both organs stain intensely when treated with dyes that bind to RNA and, by chemical extraction, yield large amounts of RNA. However, direct proof of RNA involvement in protein synthesis did not come for many more years.

Experiments done in the late 1950s proved several important points about RNA.

1. Incorporation of radioactive precursors such as ^{3}H-cytidine and ^{3}H-uridine into RNA of individual cells can be readily detected by autoradiography (Figure 8-2). However, if the nucleus is first removed from the cell, no detectable incorporation of

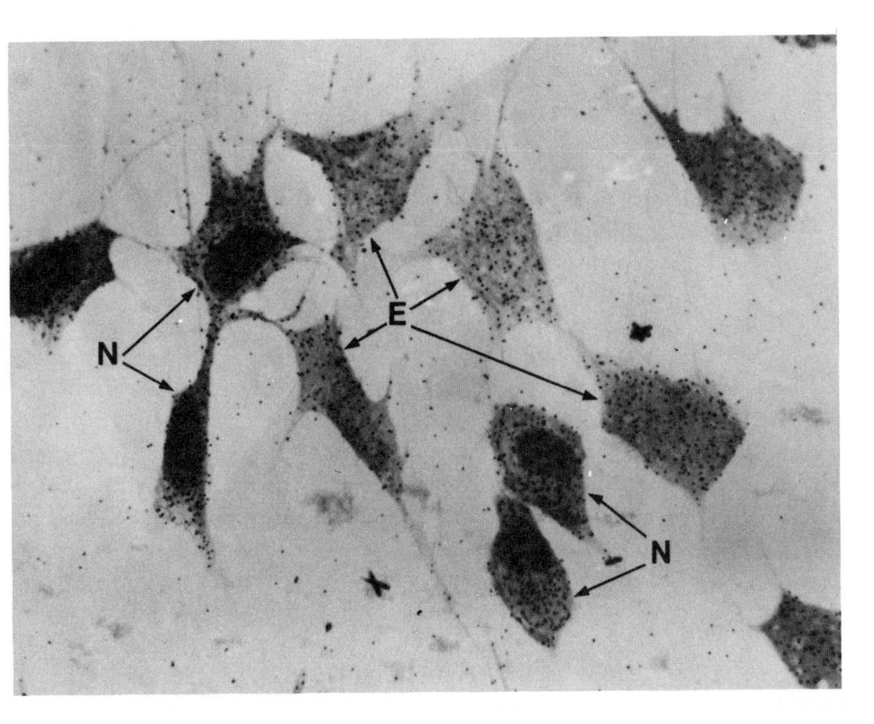

FIGURE 8-1 Autoradiograph of a group of enucleated (E) and nucleated (N) animal cells in culture that have been incubated with tritium-containing amino acids. The incorporation of radioactivity into proteins by enucleated cells is as intense as it is by nucleated cells.

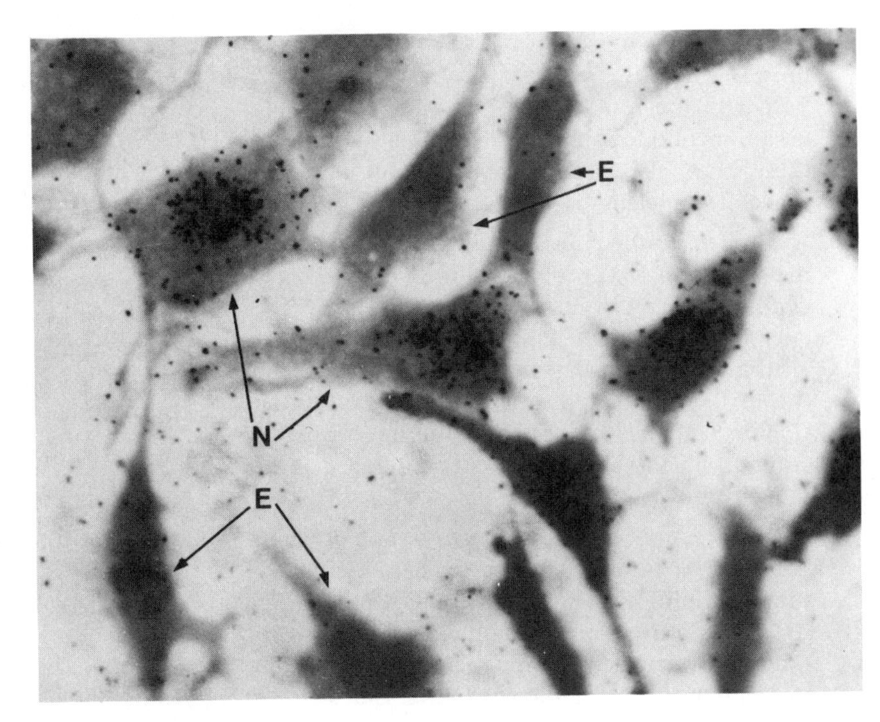

FIGURE 8-2 Autoradiograph of several enucleated (E) and nucleated (N) cells incubated for ten minutes with ^{3}H-uridine to label RNA. Only nucleated cells show silver grains, indicating that RNA synthesis has occurred and the silver grains are located predominantly over nuclei.

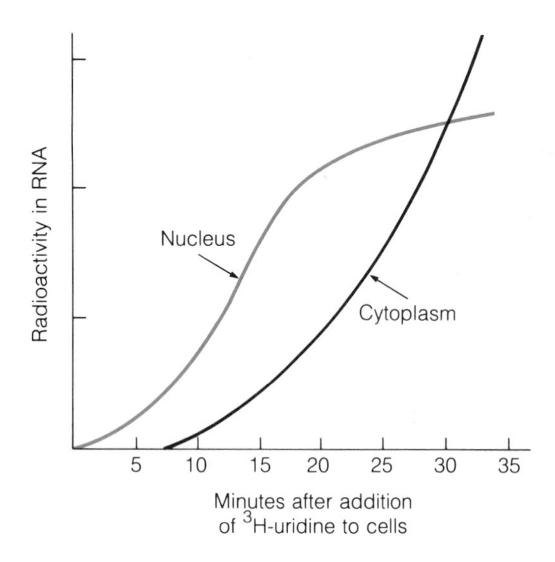

FIGURE 8-3
Time course of ^{3}H-uridine into nuclear and cytoplasmic RNAs in mammalian cells.

precursors into RNA takes place. A reasonable interpretation of the result is that RNA synthesis does not take place in the cytoplasm and, therefore, it must occur in the nucleus. Contrast this observation with the result of the experiment of similar design mentioned above showing that protein synthesis does take place in the cytoplasm in the absence of a nucleus.

2. A corollary to the enucleation experiment was the measurement of incorporation of ^{3}H-uridine or ^{3}H-cytidine in RNA of intact cells. When the ^{3}H precursor is added to the culture medium, essentially all the radioactivity incorporated into the cells in the first several minutes is found in the nucleus (Figure 8-3). The results of the experiment are consistent with the idea that RNA synthesized in the nucleus moves into the cytoplasm.

3. When a cell that has been cultured for several minutes with ^{3}H-uridine is transferred to medium lacking ^{3}H-uridine, labeling of nuclear RNA stops. If the cell is grown for 30 minutes or longer in the nonradioactive medium and then examined by autoradiography, little radioactive RNA is detected in the nucleus, but it is now found in the cytoplasm (Figure 8-4). This confirms that newly synthesized RNA migrates from the nucleus to the cytoplasm.

In short, the enucleation experiment and the time course study of the location of radioactive RNA lead to the conclusion that RNA is synthesized in the nucleus but not in the cytoplasm and that cytoplasmic RNA comes from the nucleus.

Subsequent experiments have shown that a very small amount of RNA is synthesized in mitochondria and chloroplasts. This RNA represents less than one percent of total cellular RNA and would not have been detected in the autoradiographic experiments just described. Considerable RNA synthesis occurs in the cytoplasm of cells infected with some kinds of RNA viruses, such as polio virus, and after infection with certain DNA viruses, such as smallpox virus. These are exceptional situations. In normal, virus-free cells more than 99.0 percent of RNA synthesis takes place in the nucleus. Discovery that RNA is made in the nucleus and migrates to the cytoplasm supported the idea that RNA might be the intermediate between DNA in the nucleus and protein synthesis in the cytoplasm.

Transcription of DNA Nucleotide Sequences into RNA Nucleotide Sequences

For RNA to serve as the intermediary between the DNA in the nucleus and the synthesis of proteins in the cytoplasm, genetic information in the form of sequences of deoxynucleotides in DNA must be transferred into ribonucleotide sequences in RNA and then into amino acid sequences in polypeptides. The first step, synthesis of RNA directed by DNA, is called **transcription.** The second step, namely the synthesis of polypeptide chains under the guidance of RNA is called **translation.**

The Chemical Structure of RNA

Like DNA, RNA is a long polymer consisting of hundreds of nucleotides of four kinds joined to each

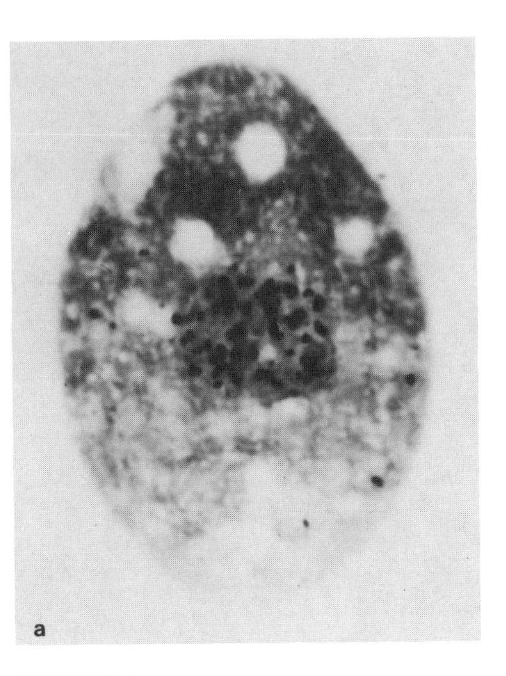

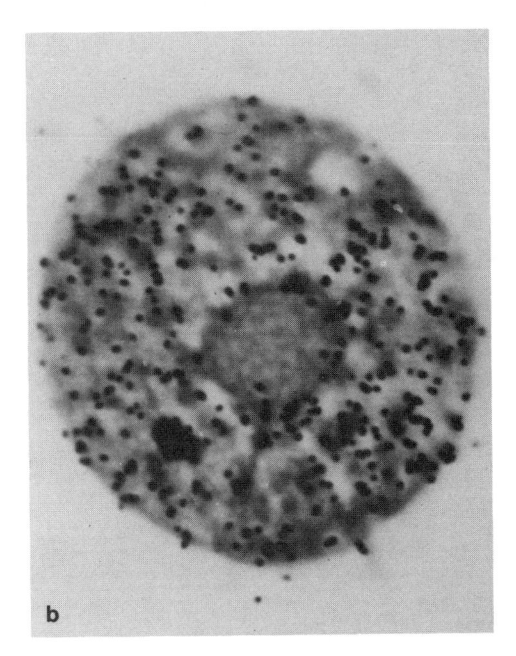

FIGURE 8-4
Experiment with the ciliate *Tetrahymena* showing that RNA is synthesized in the nucleus and moves to the cytoplasm. (a) Autoradiograph of a cell incubated with ^{3}H-cytidine for 15 minutes. Almost all the silver grains are localized over the nucleus. (b) Autoradiograph of a cell incubated with ^{3}H-cytidine for 12 minutes and then incubated with nonradioactive cytidine for 88 minutes. Almost all the radioactive RNA has moved from the nucleus to the cytoplasm.

other by phosphodiester linkages (Figure 8-5). RNA is chemically different from DNA in three ways:

1. In RNA the pyrimidine uracil is present instead of thymine. Thus, the four bases in RNA are uracil, cytosine, adenine, and guanine. Uracil is structurally similar to thymine. In thymine a methyl group (CH_3) is attached to the 5′ position in the pyrimidine ring. In uracil a hydrogen atom is present instead of a methyl group (Figures 2-24 and 2-25).
2. In RNA the sugar part is ribose instead of deoxyribose. In ribose an OH group is present on the 2′ carbon. In deoxyribose a hydrogen atom is present instead of an OH group (Figure 2-26).

FIGURE 8-5
Structure of RNA.

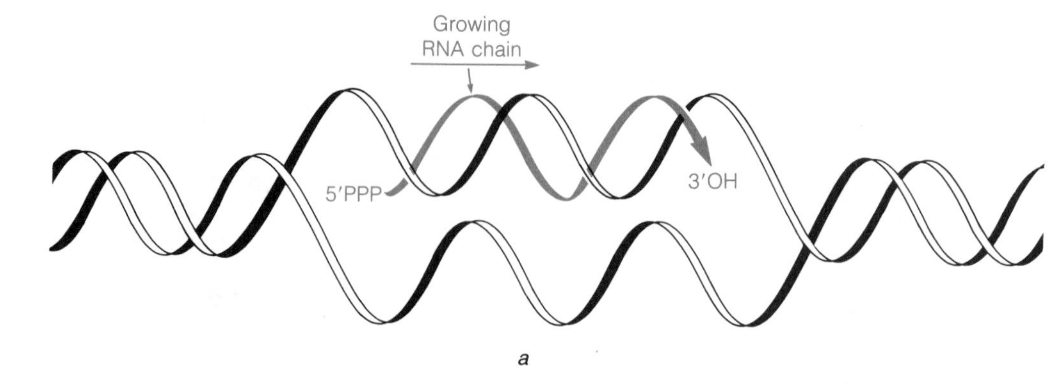

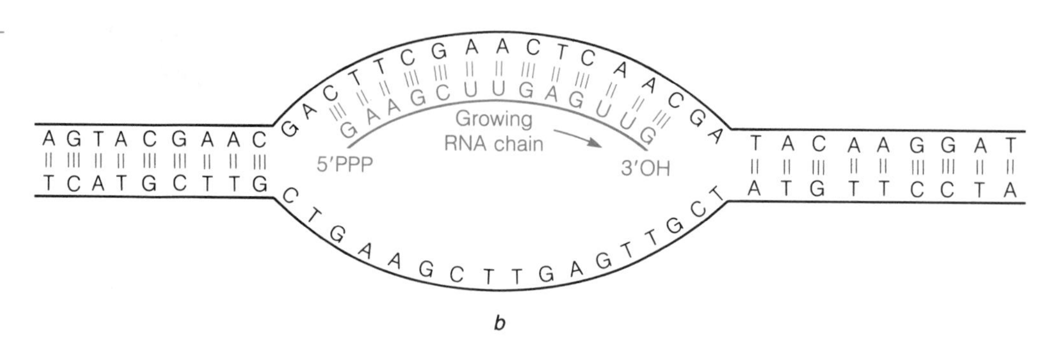

FIGURE 8-6 RNA chain forming along one strand of an opened DNA duplex. (a) In a helical configuration. (b) A more diagrammatic representation showing base pairing between the growing RNA chain and the DNA template strand. No RNA is made along the other DNA strand in a given gene.

3. RNA is usually a single-stranded polymer in contrast to the double-strandedness of the DNA double helix. Since RNA is not double-stranded, the base ratios of uracil to adenine and cytosine to guanine need not be and usually are not equal to one, as they are in the DNA double helix.

The Concept of Messenger RNA (mRNA)

The structure of RNA suggests in part how RNA might serve as a messenger between DNA and proteins. The genetic information in DNA is encoded in its sequence of deoxynucleotides. RNA has a similar potential for containing genetic information in its nucleotide sequences; therefore, it seemed possible that the transfer of information could occur by the synthesis of RNA on a DNA template by the base-pairing rules that operate in DNA replication. This has proven to be exactly what happens. Thymine, adenine, cytosine, and guanine in a DNA chain prescribe adenine, uracil, guanine, and cytosine, respectively, in RNA (Figure 8-6). To serve as a template for RNA synthesis the DNA duplex must open, as it does in DNA replication, so that free ribonucleotides may align by base pairing with the DNA template. By about 1960 the idea of RNA acting as an intermediate between DNA and synthesis of proteins had been formalized in the **messenger RNA (mRNA)** hypothesis.

The mRNA hypothesis had to accommodate a further fact. Although many kinds of proteins are synthesized continuously by the cell, the synthesis of some may be turned on and off under certain circumstances. Many examples of this are known and some are cited in Chapters 9 and 10. One example is **β-galactosidase** in *E. coli,* an enzyme that catalyzes the splitting of the disaccharide lactose into the simple sugars glucose and galactose. When *E. coli* are grown in glucose medium, they synthesize almost no β-galactosidase. When grown in medium where lactose is the only sugar, *E. coli* synthesizes large amounts of the enzyme. Thus, the synthesis of this protein can be switched on and off by changing the nutrient medium. It was proposed that mRNA molecules serve for the synthesis of a limited number of copies of a polypeptide chain, after which they are degraded. In this way the synthesis of a protein could be regulated by controlling the synthesis of mRNA from the corresponding gene. If synthesis of the mRNA for a particular gene were turned off, the synthesis of the corresponding protein would be turned off as rapidly as the existing mRNA molecules were degraded. To maintain continuous synthesis of a protein, such as the enzymes of the glycolytic pathway, would require continuous transcription of the appropriate genes into mRNAs. This also fits with the observation that RNA undergoes turnover, that is, it is continuously synthesized and degraded.

Since their introduction around 1960, the concepts of mRNA and control of protein synthesis through control of mRNA synthesis have been validated by a wealth of experimental evidence, although it is now known that protein synthesis can be controlled in other ways (Chapters 9 and 10).

RNA Polymerase and the Promoter in Prokaryotes

A crucial step in the proof that RNA synthesis takes place on DNA templates was the discovery of **RNA polymerase,** an enzyme that catalyzes the polymerization of ribonucleotides to form RNA. The principles of transcription by which DNA is copied into RNA are the same for both prokaryotes and eukaryotes, but important differences are present, as discussed in detail later in the chapter.

For transcription to occur, RNA polymerase must first bind to DNA. This occurs at a specific sequence of deoxynucleotides in DNA, called a **promoter,** located a short distance in front of the DNA sequence to be transcribed (Figure 8-7). The base pair at which transcription actually starts is designated as number 1. By convention the DNA is usually portrayed in an orientation such that transcription then proceeds rightward in a *downstream* direction along the DNA sequence that makes up the gene. The base pairs in this DNA are numbered successively 1, 2, 3, 4, 5, and so on, to the final base pair, at which transcription is terminated. The base pairs in the leftward direction (*upstream*) from the base pair at which transcription starts are numbered −1, −2, −3, −4, −5, and so forth. The promoter in bacteria consists of two sections of base pairs in this leftward, or upstream, region. The first section, called the **Pribnow box,** consists of six base pairs of DNA extending upstream from about base pair −8 (Figure 8-7). The term **box** stems from the practice of drawing a box around a short sequence in order to make it easy to distinguish. The second part of the promoter is located further upstream, extending to about base pair −42.

An RNA polymerase molecule binds to these two sites in DNA and brings about a local unwinding of the

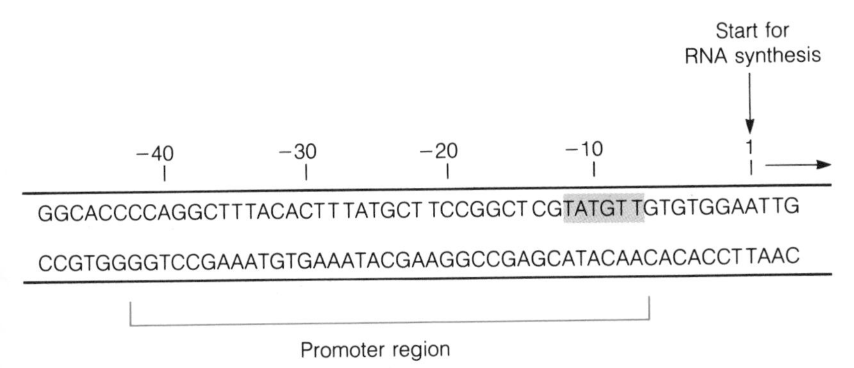

FIGURE 8-7
The promoter region for binding of RNA polymerase upstream from a gene in a bacterial chromosome. The Pribnow box of the promoter is shaded pink.

DNA double helix, creating stretches of single-stranded DNA. Transcription then begins on one of these templates at deoxynucleotide number 1, eight bases downstream from the Pribnow box, following the rules of base complementarity, in which the bases A, T, G, and C in DNA specify U, A, C, and G, respectively, in RNA. The first base in DNA to be transcribed is almost always a pyrimidine and usually a T. Therefore, the first base in the RNA transcript is usually A. The RNA polymerase then moves along the DNA, opening further regions of double helix and catalyzing the successive addition of ribonucleotides to the growing RNA chain under the guidance of the DNA template.

Behind RNA polymerase, the growing, single-stranded RNA chain peels off from its DNA template, and the DNA template rejoins its complementary chain to reform the double helix (Figure 8-8). The unwound region of the DNA duplex is called the **transcription bubble** and is 17 base pairs long in *E. coli.* Successive RNA polymerase molecules may attach one after the other at the promoter and move down the DNA, each catalyzing synthesis of an RNA transcript in a separate transcription bubble.

Like DNA polymerase, RNA polymerase catalyzes the growth of a nucleotide chain only in the $5' \rightarrow 3'$ direction, adding nucleotides to the 3′OH attached to the ribose of the last nucleotide on the chain. The immediate precursors in RNA synthesis are the ribonucleoside triphosphates UTP, CTP, GTP, and ATP. As a triphosphate precursor is added to the chain by the joining of the 3′OH of the growing chain to the 5′P of the next nucleotide, pyrophosphate (P~P) is released. Unlike DNA polymerase, RNA polymerase does not require a primer polynucleotide in order to make 3′-5′ phosphodiester linkages. Thus, RNA polymerase starts the formation of RNA transcripts with a single ribonucleoside triphosphate.

Termination of Transcription in Prokaryotes

A specific sequence at the end of the gene signals RNA polymerase to stop transcribing. Termination is different in prokaryotes and eukaryotes. In prokaryotes

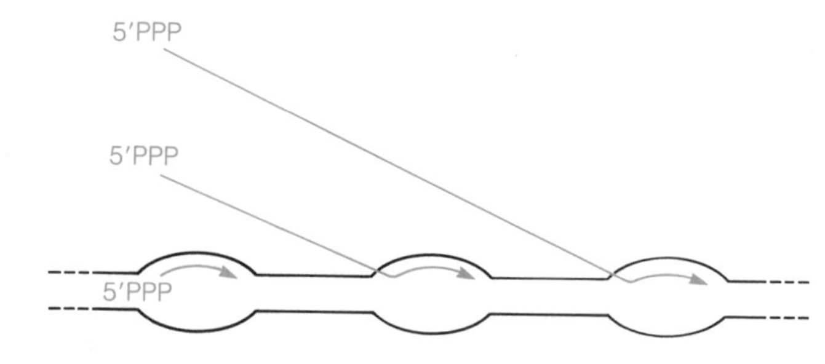

FIGURE 8-8
An RNA molecule peels away from its DNA template strand, and the DNA duplex reforms behind the region of RNA polymerization. Here successive RNA chains are being formed from the DNA of a single gene.

the termination signal consists of a sequence of eight or nine base pairs in DNA that is repeated, with several base pairs interposed between the two repeats (Figure 8-9(a)). The two repeats are in inverse order. Immediately after these repeats are four or more AT base pairs that are transcribed as a sequence of U residues in the RNA transcript. The inverted repeats, after transcription into RNA, make it possible for the RNA to fold back on itself, forming a short hairpin region of RNA double helix intramolecularly as shown in Figure 8-9(b). The hairpin in RNA is believed to signal RNA polymerase to stop transcription within a few bases after the hairpin. The U residues on the end of the RNA transcript may possibly participate in termination.

The termination of transcription is accompanied by release of the newly formed RNA transcript and the RNA polymerase from the DNA template (Figure 8-9(c)). The termination of transcription of some genes in bacteria is helped by a special terminator protein called **Rho,** but the molecular mechanism of Rho action is poorly understood.

Transcription from One Strand of DNA

A gene contains information in the form of a sequence of bases in the DNA double helix. Ordinarily, only one of the two strands of the DNA is transcribed, yielding an mRNA transcript with a specific nucleotide sequence that subsequently directs the assembly of a specific sequence of amino acids to make a polypeptide.

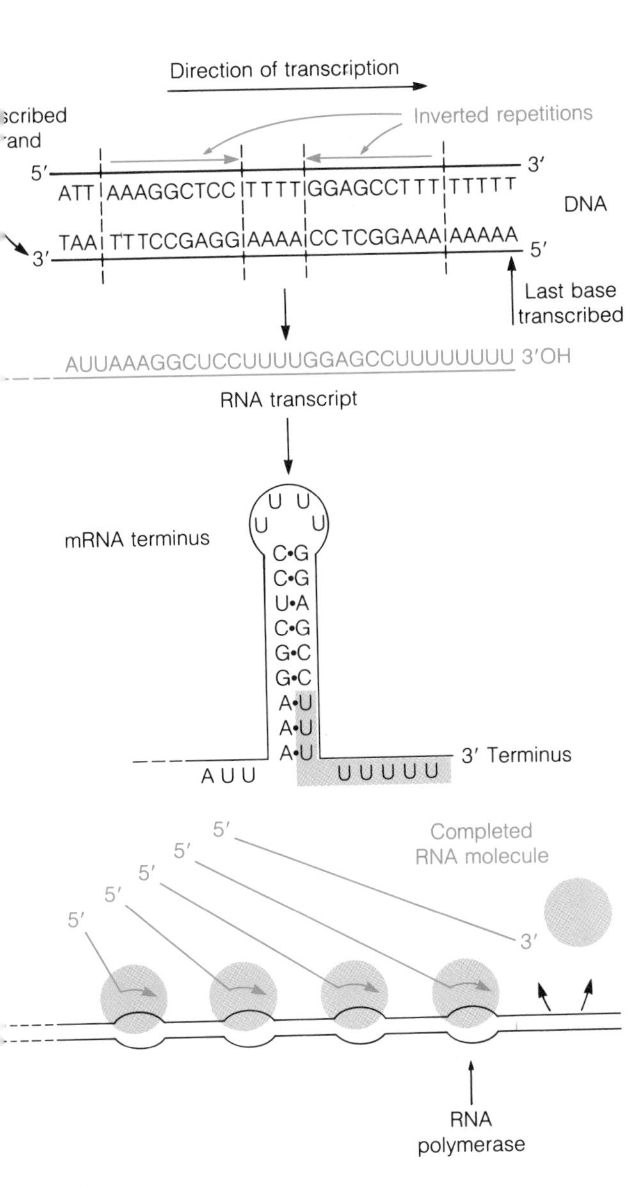

FIGURE 8-9
Termination of mRNA synthesis. (a) A termination signal in the DNA terminates synthesis of an mRNA molecule. (b) An mRNA molecule forms a hairpin by intramolecular base pairing. The hairpin may trigger release of RNA polymerase. (c) At termination RNA polymerase and the completed mRNA are released from the DNA template.

If the opposite DNA strand were used as a template, a transcript with an entirely different nucleotide sequence would be made, which would direct the synthesis of an entirely different sequence of amino acids in a polypeptide. The function of a polypeptide depends on a precise sequence of amino acids. It is unlikely that both strands of DNA could direct, through mRNA transcripts, two functional polypeptide chains. The improbability of this will become clearer with later discussion of polypeptide synthesis. The promoter determines the orientation of binding of RNA polymerase to the DNA double helix, and this orientation, in turn, determines which DNA chain will be transcribed. The transcribed DNA chain is sometimes referred to as the **sense strand,** and the complementary, untranscribed strand is sometimes called the **nonsense strand.**

Sequences of base pairs define successive genes along a DNA molecule. Each gene has a promoter that determines the direction of transcription and defines the transcribed strand and a terminator that defines its end. However, successive genes are not all arranged such that the transcribed strand of one gene is the continuation of the transcribed strand of an adjacent gene. The transcribed strand may switch from one strand of the DNA double helix to the other in successive genes. When switching occurs the direction of transcription must be reversed as diagrammed in Figure 8-10 because RNA polymerase catalyzes RNA synthesis only in the 3′OH → 5′P direction along a DNA template that has the opposite polarity.

Translation of RNA Nucleotide Sequences into Proteins

Experiments done in the 1950s had shown that RNA is synthesized in the nucleus and then migrates to the cytoplasm. During this same period studies on protein synthesis were beginning to show how RNA was involved in the formation of proteins. The primary experiment was to inject radioactive amino acids into rats and, a short time later, to remove the liver, homogenize the cells, separate the homogenate by differential centrifugation into fractions containing different cell parts (Chapter 1), and finally to determine the fraction in which the radioactive amino acids had been incorporated into proteins. The liver was chosen because it is a large organ and has a high rate of protein synthesis.

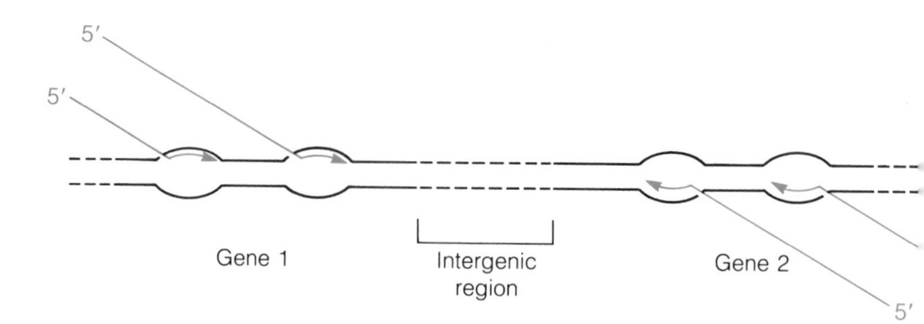

FIGURE 8-10
The template DNA strand copied into RNA in two successive genes can be the same DNA strand or can be different, as shown here. When switching of this type occurs, the RNA molecules are synthesized in opposite directions because the two DNA strands have opposite chemical polarities. The region between the genes is larger than shown here.

Liver from a rat injected with radioactive amino acids was fractionated. Most of the radioactivity was found in the microsomes (endoplasmic reticulum fragments), provided that only a few minutes had elapsed between injection with the radioactive amino acids and fractionation of the liver. If killing of the animal was delayed for several hours, every cell fraction contained radioactivity. This observation indicates that protein synthesis in the liver cells occurs in the endoplasmic reticulum and that newly synthesized proteins subsequently migrate to other organelles and compartments in the cell.

The microsomal fraction of liver cells contains most of the ribosomes; this is so because most ribosomes are attached to the endoplasmic reticulum in a liver cell. It was soon discovered that the fraction of ribosome-containing microsomes isolated from liver cells could itself incorporate amino acids into proteins if some of the soluble fraction (cytosol) and ATP were added to the microsomal fraction. This was the first demonstration of protein synthesis in a cell extract.

Ribosomes were shown to be the organelles that carry out protein synthesis by detaching them from microsomal membranes of liver cells with a detergent, purifying them by centrifugation, and measuring their ability to promote protein synthesis, using radioactive amino acids to follow the process. Messenger RNA is also needed to direct the joining together of amino acids in specific sequences. In early experiments with purified ribosomes, the mRNA requirement was met by mRNA molecules that remained bound to ribosomes during their purification. In addition to ribosomes and mRNA, two components from the cytosol are necessary: amino acid activating enzymes called **aminoacyl-tRNA synthetases** and a class of small RNA molecules known as **transfer RNAs** or **tRNAs**. The tRNA molecules vary in length between 76 and 83 nucleotides. Finally, energy is required to drive the entire process of protein synthesis.

Thus, the list of components needed to synthesize polypeptide chains is as follows:

mRNA
ribosomes
amino acids
aminoacyl-tRNA synthetases
tRNA molecules
energy (ATP)

The task now is to see how each of these components is used to make a polypeptide.

Particularly important in protein synthesis is the mechanism by which mRNA guides the sequence of amino acids during polypeptide synthesis. Attempts were made to explain how mRNA could serve as a template to direct assembly of amino acid chains by binding directly to amino acids, in an analogous fashion to the directing of mRNA synthesis by a DNA template. These attempted explanations proved futile. The solution was found in the tRNA molecules. These molecules, along with aminoacyl-tRNA synthetase, serve as interpretors of the mRNA.

The first step in protein synthesis is the energetic activation of amino acids with ATP. Activation of an amino acid is achieved by being joined covalently to a tRNA molecule, with the hydrolysis of an ATP to AMP

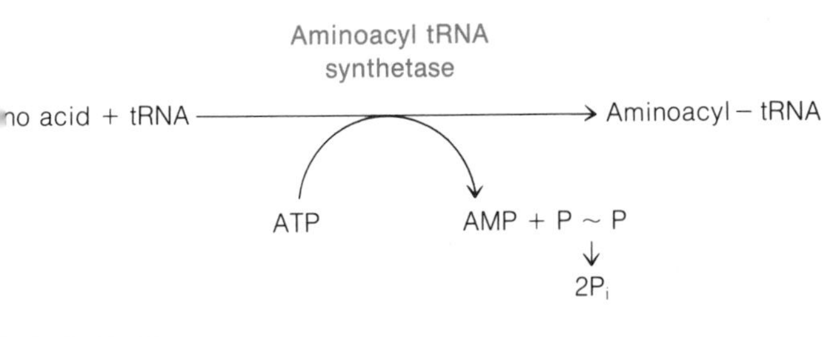

FIGURE 8-11
Activation of an amino acid by covalent joining of the amino acid to a tRNA molecule in a reaction coupled to splitting of ATP.

and PP_i. The reaction is catalyzed by one of the aminoacyl-tRNA synthetases, and the aminoacyl-tRNA product is called **aminoacyl-tRNA** (Figure 8-11).

The amino acid in an aminoacyl-tRNA is in an activated state and can be transferred from the tRNA to a growing polypeptide chain with the formation of a peptide bond (Figure 8-12). The transfer of the amino acid from aminoacyl-tRNA to the growing end of the polypeptide chain releases the tRNA, which can accept another amino acid in a repeat of the activation process.

The mRNA determines the order in which amino acids are transferred from tRNA molecules to a polypeptide chain. The assembly of amino acids into protein under the guidance of mRNA is called mRNA translation. The key to the translation of mRNA is in the structure and function of tRNA and the aminoacyl-tRNA synthetases. First, each of the 20 different amino acids has its own particular tRNAs to which it is joined

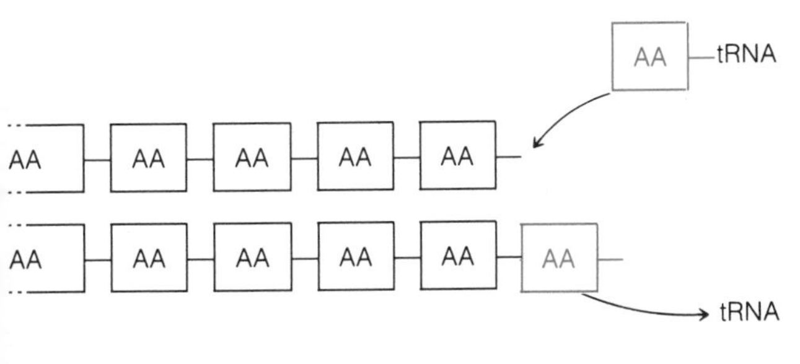

FIGURE 8-12
The overall reaction in which an activated amino acid joined to a tRNA molecule is added to the carboxyl end of a growing polypeptide chain.

in the formation of aminoacyl-tRNA. The specificity between amino acids and tRNA molecules is complete: glycine can be joined only to a glycine-accepting tRNA, alanine only to an alanine-accepting tRNA, and so forth. All of the tRNA molecules are close to the same size and have nucleotide sequences in common, but important differences in sequences identify one tRNA as glycine tRNA, another as alanine tRNA, and so on, for all 20 amino acids.

Second, a specific aminoacyl-tRNA synthetase serves to catalyze the joining of a particular amino acid to its tRNA. Thus glycyl-tRNA synthetase joins glycine to its particular tRNA, alanyl-tRNA synthetase joins alanine to its tRNA, and so forth for all 20 amino acids.

The high degree of specificity between a tRNA and its corresponding synthetase resides in a recognition between a site on the synthetase protein and specific structural features of its tRNA partner. All tRNA molecules fold up into roughly the same three-dimensional structure through formation of intramolecular double helices by pairing of bases in one part of the molecule with bases in another part (Figure 8-13). In tRNA, U residues hydrogen bond with A residues and C with G. Four short regions of double helix are formed by four sets of complementary sequences. The four regions of double helix give a tRNA molecule the overall three-dimensional configuration (tertiary structure) shown in Figure 8-14. As diagrammed in Figure 8-13, but difficult to see in the three-dimensional model (Figure 8-14), formation of the four double helices gives rise to three loops and a **terminal stem** containing the 5′P and 3′OH ends of the tRNA. Each loop and the terminal stem have specific roles in tRNA function. In the terminal stem the 3′OH end accepts the amino acid and therefore is also called the **acceptor stem.**

Small differences in sequence among the different tRNAs give rise to small differences in tertiary structure that ensure that a given tRNA species interacts only with its own synthetase. The interaction is such that the active site of the synthetase is precisely positioned to add an amino acid to the acceptor stem. A feature of the active site is that it recognizes only one of the 20 different amino acids. In short, the synthetase has two specific interactions. It binds specifically to a particular

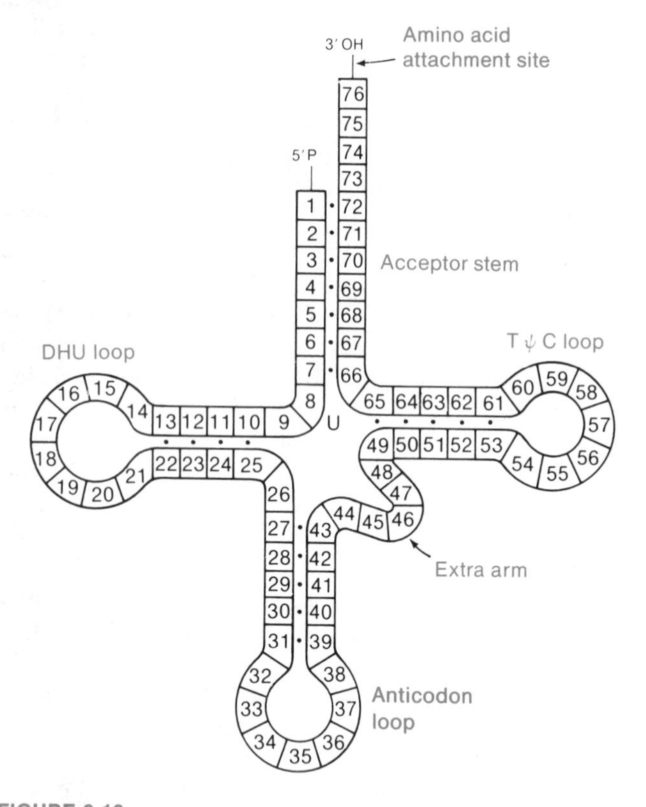

FIGURE 8-13
The folded structure of a tRNA molecule containing 76 nucleotides. Folding gives rise to three loops, an extra arm, and an acceptor stem.

tRNA molecule, and it binds a particular amino acid at its active site. These two interactions ensure that a given tRNA becomes charged only with its specific amino acid.

The different tRNA molecules, each charged with its own particular amino acid, transfer their amino acids to a growing polypeptide chain in an order dictated by mRNA in the following way. The anticodon loop in tRNA molecules consists of seven unpaired nucleotides (Figure 8-15). The three central nucleotides (bases) form a recognition site between the tRNA, charged with its amino acid on the acceptor stem, and an mRNA molecule. The recognition is in the form of complementary base pairing of the three bases in the anticodon loop of tRNA and three bases in mRNA. Each successive set of three bases in mRNA interacts with a tRNA with a complementary set of three bases in its anticodon loop. In this way each successive set of three bases in mRNA selects, through tRNA molecules, the order in which amino acids are added to a polypeptide chain. After the tRNA has bound to mRNA, its amino acid is transferred to the end of a polypeptide chain. The three bases in mRNA that in this way specify an amino acid are called a **codon triplet** or simply a **codon.** The three complementary bases in tRNA are called the **anticodon.** The final component in protein synthesis, the ribosome, serves two major roles. (1) It provides binding sites for mRNA and tRNA molecules that hold them in the stable orientations necessary for correct initiation and progression of polypeptide synthesis. (2) It contributes actively to the formation of peptide bonds in a growing polypeptide. These functions are discussed in a later section, which will present the integration of the steps in polypeptide synthesis, beginning with formation of the **initiation complex.** First it is necessary to discuss the genetic code.

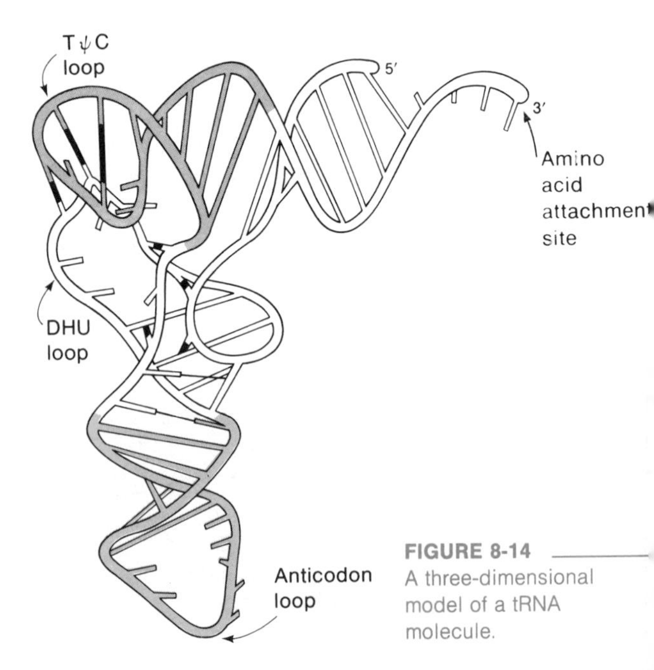

FIGURE 8-14
A three-dimensional model of a tRNA molecule.

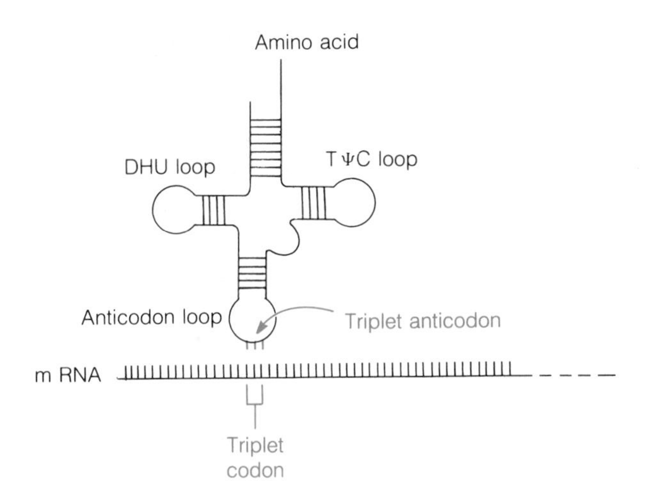

FIGURE 8-15
tRNA molecules translate the nucleotide sequence of an mRNA molecule by base pairing of three bases (an anticodon triplet) in the anticodon loop with three bases (a triplet codon) in mRNA.

THE GENETIC CODE

From the time that DNA had been shown to be the genetic material, it was assumed that a sequence of nucleotides in the DNA gene specified the sequence of amino acids in the gene product, a protein. A wealth of indirect observations supported this idea, and it was virtually established as fact with the proof of mRNA in the early 1960s as the intermediate carrier of genetic information between DNA and the assembly of proteins.

The essential question of how the four-letter alphabet of DNA (A, T, G, and C) corresponded to the 20-letter alphabet of amino acids remained to be solved. It was assumed that more than two nucleotides had to be used to specify a single amino acid because only 16 (4^2) different combinations of two nucleotides are possible, but 20 amino acids must be coded for. The number of possible different combinations containing three nucleotides is 64 (4^3), which is more than needed to code for 20 amino acids. The number of combinations of four nucleotides is 256 (4^4), which is far more than the number needed. Based on these numbers it seemed reasonable to assume that the codon for each amino acid consists of three nucleotides arranged in a specific order, and indeed certain genetic experiments gave strong evidence for a triplet code. Identification of the codons began in 1961 via cell-free systems of polypeptide synthesis. Some of these experiments also provided definitive proof of the triplet nature of the code.

Cell-Free Synthesis of Polypeptides

Studies in 1955 had shown that ribosomes purified from liver could carry out incorporation of amino acids into polypeptides for a short while when the ribosomes were mixed with ATP, tRNA, and aminoacyl-tRNA synthetases. In about 1961 it was discovered that the limited incorporation of amino acids was dependent on the small amount of mRNA that had remained bound to the ribosomes during their isolation. When purified mRNA was added to the protein-synthesizing system, the incorporation of amino acids was greatly enhanced.

One of the first demonstrations of the faithfulness of translation was done with an mRNA from the **bacterial virus f2** in a protein-synthesizing mixture prepared with ribosomes from *E. coli.* The viral mRNA, which codes for the coat protein of the virus, was accurately translated into complete coat protein molecules with the *in vitro* system. Similar success was also achieved with other kinds of mRNA molecules, including the synthesis of hemoglobin with mRNA and ribosomes obtained from immature red blood cells or **reticulocytes.**

The method of cell-free protein synthesis was exploited to decipher the genetic code by using synthetic polynucleotides of known base composition as mRNAs and then examining the polypeptides that were synthesized. Initially, the synthetic mRNAs were prepared with **polynucleotide phosphorylase,** an enzyme that catalyzes the formation of polynucleotide chains from nucleoside diphosphates in a purified cell-free system without DNA templates. A polynucleotide made from uridine diphosphate, and therefore containing only Us in a long series, was used as an mRNA for *in vitro* protein synthesis. The only amino acid incorporated was phenylalanine, and a polypeptide of pure phenylalanine was produced (Figure 8-16(a)). This result suggested that UUU was the **triplet** that codes

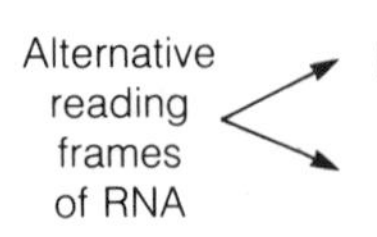

FIGURE 8-16
Synthesis of polypeptides from synthetic RNA molecules. (a) Poly (U) codes for a polypeptide containing phenylalanine. (b) Poly (U) to which a G has been added codes for polyphenylalanine terminated by leucine. (c) Poly (U) with many G monomers added codes for a string of phenylalanines, followed by a single leucine *or* tryptophan (depending on the reading frame of the RNA), followed by a string of glycines.

for phenylalanine. Similar experiments indicated that the triplet CCC codes for proline and AAA codes for lysine.

This basic experiment was extended in several ways. For example, if a single guanine was added to the end of a polyuridylic acid chain (poly(U)), polyphenylalanine terminated by a leucine was sometimes formed in an *in vitro* protein-synthesizing system (Figure 8-16(b)). Therefore, UUG must be a leucine codon. If more guanines were added to poly(U), glycine was sometimes found at the end of polyphenylalanine indicating that GGG codes for glycine. However, occasionally tryptophan was found, which showed that UGG is a codon for tryptophan (Figure 8-16(c)). Other mixed polymers were synthesized, for example, composed of two kinds of triplets ...ACA CAC ACA CAC... The resulting polypeptide consists of alternating threonine and histidine residues. Other experiments had proven that the codon for threonine contained two As and one C, although the order of nucleotides was not known. Therefore, ACA must be a threonine codon and CAC must be a histidine codon.

Finally, various synthetic trinucleotides of known sequence were bound to ribosomes in place of mRNA. Binding of a particular trinucleotide to ribosomes brought about binding of specific aminoacyl-tRNAs. The binding of UUU was accompanied by binding of phenylalanyl-tRNA to ribosomes; no other aminoacyl-tRNA would bind. The trinucleotide GUU caused binding of valyl-tRNA, GUC caused binding of alanyl-tRNA, and so on. This technique also demonstrated that the polarity of triplets was crucial to their translation. Thus, the triplet GUU with a $5' \rightarrow 3'$ polarity of the phosphodiester linkages bound valyl-tRNA, and 5′UUG3′ (the same triplet with the opposite polarity) bound leucyl-tRNA.

With this work the assignment of all 64 possible codons was completed (Table 8-1). Sixty-one codons were assigned to amino acids; three codons (UAA, UAG, and UGA) do not have amino acid assignments but instead signal the termination of translation.

Degeneracy of the Genetic Code

Among other things, the deciphering of the genetic code showed that the code is **redundant** or **degenerate**; that is, most of the 20 amino acids are prescribed by more than one codon. The degeneracy is, moreover, uneven. For example, there are six codon triplets for leucine, four for proline, three for isoleucine, and two for tyrosine. Only two amino acids, methionine and tryptophan, have single codons. The words degenerate or redundant do not imply that the code is inaccurate, only that more than one codon can specify a particular amino acid.

As seen in Table 8-1, redundancy of the code usually involves the third base in codons. For example, alanine is specified by GCU, GCC, and GCA. All three codons are recognized by the same alanyl-tRNA. The first base in the anticodon in alanyl-tRNA pairs with the third base (the variable base) in the codons for alanine because codon and anticodon have opposite

TABLE 8-1
The "Universal" Genetic Code

First Position (5′ End)	Second Position				Third Position (3′ End)
	U	C	A	G	
U	Phe	Ser	Tyr	Cys	U
	Phe	Ser	Tyr	Cys	C
	Leu	Ser	Stop	Stop	A
	Leu	Ser	Stop	Trp	G
C	Leu	Pro	His	Arg	U
	Leu	Pro	His	Arg	C
	Leu	Pro	Gln	Arg	A
	Leu	Pro	Gln	Arg	G
A	Ile	Thr	Asn	Ser	U
	Ile	Thr	Asn	Ser	C
	Ile	Thr	Lys	Arg	A
	Met	Thr	Lys	Arg	G
G	Val	Ala	Asp	Gly	U
	Val	Ala	Asp	Gly	C
	Val	Ala	Glu	Gly	A
	Val	Ala	Glu	Gly	G

Note: The boxed codons are used for initiation. GUG is very rare.

polarities. The first base in the anticodon is modified in such a way that it can hydrogen-bond with U, C, or A. The modified base is inosine (I) and is formed by removal of an amino group from adenosine. The anticodon in this alanyl-tRNA is therefore 3′CGI5′, and it can pair with any of the particular three codons for alanine wherever they occur in mRNA,

CODON	5′GCU3′	5′GCC3′	5′GCA3′
ANTICODON	3′CGI5′	3′CGI5′	3′CGI5′

By convention the bases in codons and anticodons are read in the 5′ → 3′ direction and accordingly numbered 1, 2, and 3 with that directionality. Alanine is also coded by a fourth codon, GCG. A second kind of alanyl-tRNA, one containing the anticodon 3′CGU5′, which translates GCG to alanine during protein synthesis, is present in the cell.

Wobble refers to a flexibility in the orientation of the first base in the anticodon (in this case inosine) such that it can hydrogen-bond with more than one kind of base in the third position of the complementary codon. The significance of wobble is that a single tRNA, if it contains inosine in the first position of its anticodon, can recognize three different codons. Hence, one tRNA can work with three codons in mRNA. Wobble does not decrease the accuracy of the code, because the three codons recognized by an inosine-containing tRNA are not recognized by any other tRNA.

Wobble is not restricted to tRNAs containing inosine. A tRNA with G in the first-position of the anticodon can pair with two codons, one with U and one with C in the third position of the codon. Likewise, a first-position U in an anticodon can pair with a third-position A or G. The possibilities for base pairing between the third-position bases in codons and first bases in anticodons are summarized in Table 8-2. Adenine never appears in the first position of anticodons because it is always deaminated to form inosine (I).

In some cases codons differing only in the third-position code for different amino acids. For example, UUU and UUC code for phenylalanine, and UUA and UUG code for valine. In such cases the third-position is occupied by a pyrimidine for one amino acid (U or C for phenylalanine) and by a purine for the other amino acid (A or G for valine). The anticodon in phenylalanyl-tRNA is 5′GAA3′; the first-position G can pair with either C or U in the third position of the two codons. The anticodon in the valyl-tRNA is 5′UAA3′; the first-position U can pair with the third-position G or A.

TABLE 8-2
The Wobble Pairings

Third-Position Codon Base	First-Position Anticodon Base
A	U, I
G	C, U
U	G, I
C	G, I

TABLE 8-3
Exceptions to Universality of the Genetic Code

Organism	Codon Difference
Mycoplasma capricolum	UGA = tryptophan instead of stop
Mammalian mitochondrial DNA	UGA = tryptophan instead of stop AGA and AGG = stop rather than arginine AUA and AUU are initiation codons, as well as AUG AUA codes for methionine instead of isoleucine
Ciliated protozoa	UAA and UAG = glutamine instead of stop

The Near Universality of the Genetic Code

Assignment of the various triplet codons to amino acids has been checked in many widely divergent organisms, including viruses, bacteria, protozoa, algae, plants, insects, amphibians, birds, and mammals. With three exceptions the codon assignments to amino acids have proven to be the same; that is, the genetic code is nearly universal among all viruses, prokaryotes, and eukaryotes. This near-universality is evidence that all viruses, prokaryotes, and eukaryotes had a single common evolutionary origin.

The three exceptions known to date occur in mRNA transcribed from the DNA in the bacterium *Mycoplasma capricolum,* in mitochondria, and in mRNA transcribed from nuclear DNA in ciliated protozoa (Table 8-3). In *Mycoplasma capricolum* UGA is a codon for tryptophan instead of acting as a stop codon. In the DNA of mitochondria in human cells UGA codes for tryptophan instead of acting as a stop codon, AGA and AGG are stop codons instead of arginine codons, AUA codes for methionine instead of isoleucine. Further differences occur in mitochondrial DNA of other organisms. In the DNA of ciliated protozoa, UAA and UAG code for glutamine instead of acting as stop codons. In ciliates the only stop codon is UGA.

Mutations and the Triplet Code

Once the genetic code had been deciphered, it became possible to explain mutations in very specific terms. Changing one or more nucleotides in a codon so that it now stands for a different amino acid is called a **missense mutation.** For example, more than 100 mutations in human hemoglobin have been found that consist of a single substitution in the amino acid sequence in one of the two polypeptide chains of hemoglobin. The single amino acid change impairs the function of the hemoglobin molecule. These mutations are caused by a single base substitution (mutation) in the codon for the changed amino acid. The disease known as sickle-cell anemia in humans is due to a change in the sixth amino acid in the β polypeptide chain of hemoglobin from glutamic acid to valine. This is accounted for by changing the middle base in a codon for glutamic acid (GAA or GAG) from A to U, making a codon (GUA or GUG) that translates into valine.

A base change that alters a codon from specifying an amino acid to specifying termination of translation is called a **nonsense mutation.** For example, changing a lysine codon AAA or AAG to UAA or UAG by substituting U for A in the first position creates a termination signal at a position formerly designating lysine. Nonsense mutations therefore cause the translation of the particular mRNA to be prematurely terminated, forming an incomplete polypeptide chain. The length of the incomplete chain depends on where in the gene the mutation has occurred. Such truncated polypeptides usually have no function.

The deletion or insertion, usually of a single base pair, in a gene constitutes a **frameshift mutation.** The addition or loss of a base in a codon necessarily shifts the reading frame of all of the triplet codons coming after the change. An example of part of an mRNA sequence and the sequence of amino acids into which it translates is shown in Figure 8-17(a). If the first C (in the histidine codon CAU) were deleted by removing a

a	met	val	his	lys	tyr	arg	ser
	AUG	GUU	CAU	AAA	UAU	CGU	AGU
			↓ C missing				
b	met	val	ile	asn	ile	val	val
	AUG	GUU	AUA	AAU	AUC	GUA	GU-

FIGURE 8-17
Frameshift mutation as a result of deletion of one base pair in DNA. (a) Normal mRNA. (b) Deletion of a CG base pair in DNA results in an mRNA lacking a C, resulting in a change in the downstream reading frame.

GC base pair in the gene, the message would be modified and translated differently (Figure 8-17(b)). As a result of the deletion of this GC base pair, the reading frame will be shifted such that all amino acids from the third one onward are changed, and an almost entirely different polypeptide will be coded for by the mutated gene. Moreover, the polypeptide is likely to be truncated by the occurrence of a stop codon in the new reading frame.

Reverse Mutations and Suppressor Mutations

A mutation can be reversed by a second mutation. Such a change is called a **reversion** or a **back mutation,** while the initial mutation is a **forward mutation.** Back mutations are rare compared with forward mutations. Any one of many nucleotide changes in a gene can constitute a forward mutation, but only certain nucleotide changes can restore the activity of the mutant protein. Especially rare is a back mutation that restores the original nucleotide sequence. More commonly, the effect of a mutation is reversed by a second mutation occurring elsewhere in the same gene (intragenic suppression) or even in a different gene altogether (intergenic suppression). These two modes of reversion are described in the following subsections.

Intragenic suppression. A frameshift caused by deletion of a base pair can sometimes be corrected by addition of an extra base pair at some other point in the gene (the insertion may be on either side of the deletion). The shorter the distance along the gene between the deletion and the insertion, the fewer amino acid changes will occur before the correct reading frame is reestablished. The fewer the amino acid changes, the greater is the probability that the polypeptide produced will still show some of its normal biological activity when folded into its tertiary configuration.

In a reciprocal manner, a frameshift mutation resulting from insertion of an extra base pair can be reversed by a subsequent deletion of a base pair somewhere else in the gene (Figure 8-18). Such double mutation within a single gene that restores at least partial gene function is one type of **intragenic suppression.**

Intragenic suppression also can occur for missense mutations. A single amino acid change resulting from a missense mutation can inactivate the protein in which it occurs by interfering with folding of the polypeptide chain into the tertiary configuration necessary for it to function. A second missense mutation at another site in the same gene can, in certain circumstances, cause amino acid change that allows the polypeptide chain to fold into a tertiary shape that approaches normal. In other words the second mutation may partially cancel the effect of the first and restore some degree of function to the protein. An example of this kind of intragenic suppression occurs in the *trpA* gene of *E. coli.* This gene codes for the A polypeptide, one of the two different polypeptides that combine to form the enzyme **tryptophan synthetase,** which catalyzes the final

a	Normal	met	val	his	lys	tyr	arg	ser	—
		AUG	GUU	CAU	AAA	UAU	CGU	AGU	---
				↓ C missing					
b	Mutant	met	val	ile	asn	ile	val	val	—
		AUG	GUU	AUA	AAU	AUC	GUA	GU-	---
					↓ A added				
c	Revertant	met	val	ile	lys	tyr	arg	ser	—
		AUG	GUU	AUA	AAA	UAU	CGU	AGU	---

FIGURE 8-18
An example of intragenic suppression. (a) Normal mRNA. (b) mRNA copied from the same gene after a frameshift mutation causing C to be missing from the third codon. (c) Restoration of the original reading frame by addition of a TA base pair to the gene. The polypeptide in the revertant has a single amino acid difference (ile instead of his), which may restore partial or full function to the protein.

step in tryptophan synthesis. The A polypeptide in *E. coli* consists of 267 amino acids. A mutation that changes amino acid number 210 from glycine to glutamic acid inactivates the protein by changing the folding of the polypeptide chain so that the active site cannot be formed. A second missense mutation in the same gene that changes amino acid number 174 from tyrosine to cysteine partially restores enzyme activity. Presumably, the side groups of amino acids 210 and 174 are adjacent to each other in the properly folded polypeptide chain. Glycine (210) has a small, uncharged side chain; a change to glutamic acid, with a bulky side chain, could readily disrupt folding. The presence of the bulky charged group of glutamic acid can be partially compensated for by substitution of tyrosine, with a bulky side group, to cysteine, which has a much smaller and weakly charged side group. Otherwise stated, if the amino acid in position 210 has a bulky side group, the amino acid in position 174 must have a small side group, and vice versa.

Intergenic suppression. In **intergenic suppression** of a mutation the effect of the mutation is partially or completely reversed by a mutation in a different gene. Genes that reverse a mutational effect are called **suppressor genes.** One of the ways that intergenic suppression can work is by altering the reading of a mutated codon in another gene. A single suppressor gene can correct misreading in a number of different mutated genes.

The special case of suppression of nonsense mutations gives some insight into how a suppressor gene works. Mutations can occur in genes that code for tRNA molecules. A mutation can change the anticodon triplet of tRNA such that the tRNA now recognizes a termination codon (UGA, UAA, or UAG) instead of its normal codon. When this happens, an amino acid is added to the polypeptide chain where termination would normally take place (Figure 8-19). Hence, if a nonsense mutation occurred in a gene causing the premature termination of a growing polypeptide, the mutational effect will be reversed by the mutated tRNA (called a **suppressor tRNA**) that can insert an amino acid at the position of the nonsense mutation.

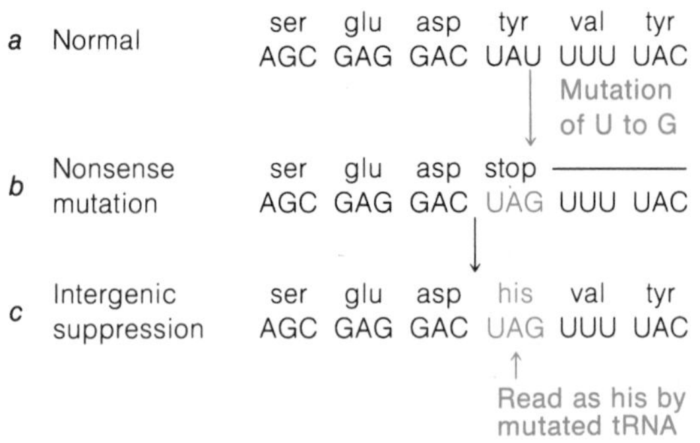

FIGURE 8-19
An example of intergenic suppression. (a) Normal mRNA. (b) mRNA after a nonsense mutation. (c) Intergenic suppression as a result of mutation in a $tRNA^{his}$ gene that changes the anticodon to pair with UAG.

Two other consequences of a suppressor mutation in tRNA would be predicted. First, all polypeptide chains normally terminated by that termination codon now recognized by the mutated tRNA would continue to grow beyond the normal termination point. This does not happen often enough to kill the cell because normal termination is frequently signaled by two different termination codons in succession in an mRNA molecule. If one fails, termination is ensured by the presence of the second. However, improper termination occurs often enough that such cells usually grow more slowly.

Second, a mutation of a tRNA to a nonsense suppressor would leave the cell without a means of inserting the amino acid served by that tRNA into all polypeptide chains. This, in fact, does not happen because the cell usually possesses at least two genes that code for the particular tRNA; mutation in one to a nonsense suppressor leaves the other to function normally in polypeptide synthesis.

TRANSLATION

Translation can conveniently be considered in three phases:

1. An initiation phase, in which mRNA and the first tRNA charged with its amino acid bind to a ribosome;

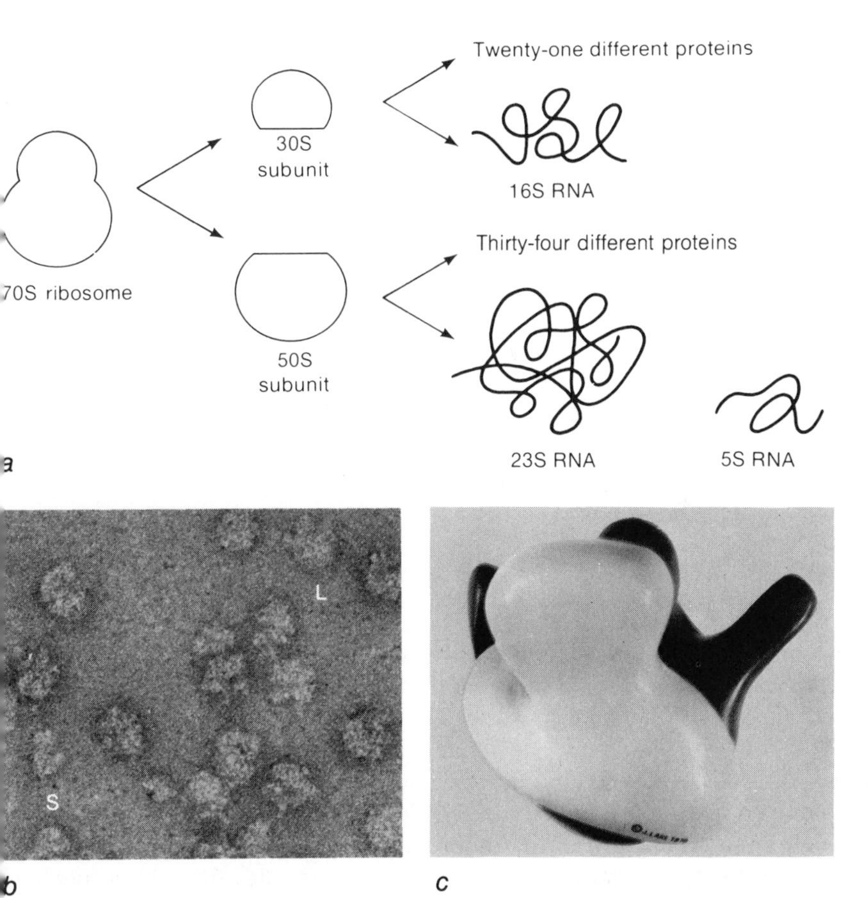

FIGURE 8-20
Structure of a 70S ribosome of a bacterial cell. (a) rRNA and protein components of the ribosome. (b) Electron micrograph of ribosomes, including a large (L) and small (S) subunit. (c) A model of a ribosome. The 30S subunit is white and the 50S subunit is red. [(a) Courtesy of David Freifelder. (b and c) By permission of James Lake.]

2. An elongation phase, in which the ribosome moves along the mRNA three nucleotides at a time as each triplet codon is translated into an amino acid that is added to the growing polypeptide chain; and
3. A termination phase, in which the ribosome comes to a stop signal in the mRNA and the completed polypeptide chain, the mRNA, and the ribosome all dissociate from one another.

The Initiation Phase

The formation of peptide bonds takes place on ribosomes; for this to happen the mRNA and the first aminoacyl-tRNA must bind to a ribosome. The formation of this 3-component complex (mRNA, aminoacyl-tRNA, and ribosome) occurs in a very specific way. Each ribosome consists of two particles or subunits of different sizes that in bacteria are called the **30S subunit** and the **50S subunit** (Figure 8-20), names derived from their rates of sedimentation in a centrifugal field (S = sedimentation constant). The 30S particle consists of a ribosomal RNA (rRNA) molecule of about 1500 nucleotides (16S rRNA), which is folded and complexed with 21 different protein molecules. The 50S particle consists of two RNA molecules, 5S rRNA (120 nucleotides) and 23S rRNA (about 3000 nucleotides), which are folded and complexed with 34 different proteins. The two particles together form a **70S ribosome.** Eukaryotic ribosomes are larger, having larger rRNA molecules and more proteins: one 40S and one 60S particle join to make an **80S ribosome.** The

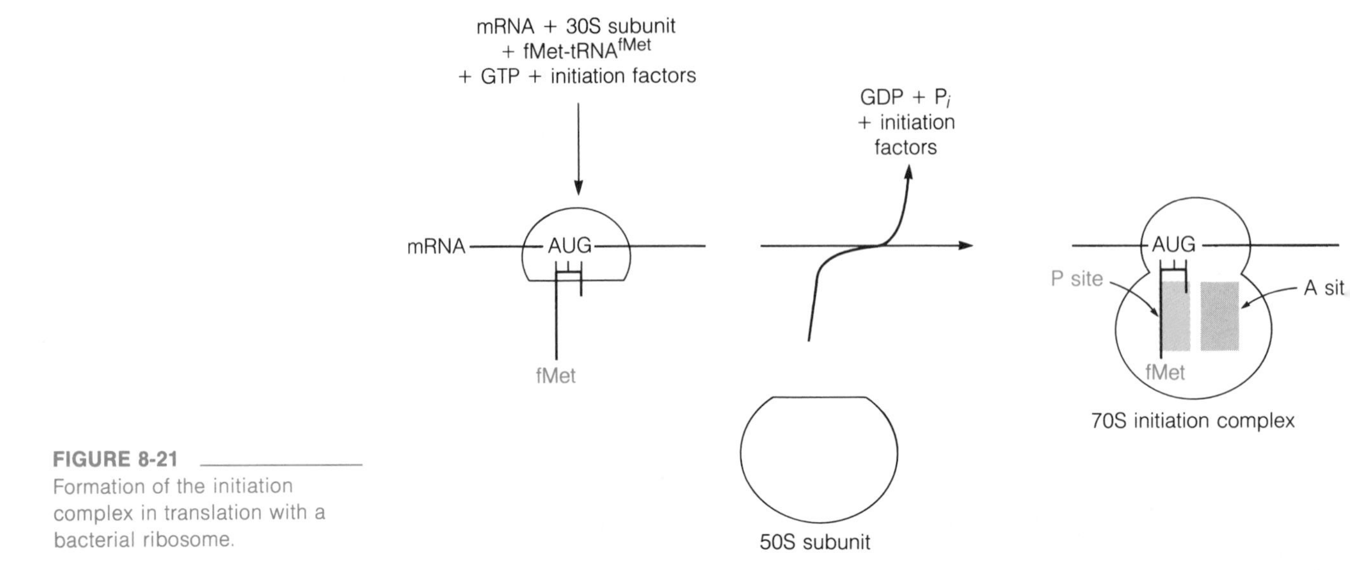

FIGURE 8-21
Formation of the initiation complex in translation with a bacterial ribosome.

RNA and protein components of the ribosome are all arranged in a specific way to form binding sites for mRNA and tRNA molecules and to catalyze several steps in polypeptide synthesis.

The first step in initiation of translation is the binding of a special **initiator tRNA** molecule charged with a modified form of methionine, formylmethionine in prokaryotes, and methionine in eukaryotes to a site on the small ribosomal subunit called the P site (Figure 8-21). The initiator tRNA has the anticodon that recognizes the codon AUG in mRNA. Initiation is catalyzed by several proteins called **initiation factors** present in the cytosol. One of these, initiation factor 2 (IF-2) catalyzes the loading of the initiator tRNA molecule onto the small ribosomal subunit. After binding of the initiator tRNA molecule, the small ribosomal subunit binds to an mRNA molecule near its 5′ end so that the first AUG codon (the **start codon**) in the mRNA pairs with the anticodon in the initiator tRNA (Figure 8-21). The AUG codon at which translation starts is usually 20 to 30 nucleotides inward from the 5′ end of the mRNA. In prokaryotes these nucleotides form an untranslated **leader** sequence that is essential in the initial binding of mRNA to the small ribosomal subunit. Depending on the particular mRNA molecule, the leader contains a stretch of five to ten nucleotides (the **Shine-Dalgarno sequence**) that base-pairs to a complementary sequence in the rRNA molecule of the small subunit. This pairing facilitates the correct aligning of the AUG start codon with the anticodon of the bound initiator tRNA. In eukaryotes the leader sequences of mRNA molecules do not contain such a complementary sequence. Instead, the 5′ ends of a eukaryotic mRNA molecule bind to the small subunit and then moves along the subunit until the first AUG codon arrives at the P site (occupied by an initiator methionyl-tRNA) and remains there until translation begins. In the final step of initiation in both prokaryotes and eukaryotes, the large ribosomal subunit joins the small subunit, and the initiation complex is complete. The complex consists of an intact ribosome (70S in prokaryotes, 80S in eukaryotes), mRNA, and the initiating tRNA in the P site.

The Elongation Phase

In addition to the P site occupied by the initiator tRNA in the initiation complex, the ribosome contains a second binding site for a charged tRNA called the **A site.** The particular tRNA that binds to the A site is determined by the codon adjacent (on the 3′ side) to he initiator AUG codon in the mRNA. For example, if he next codon were UUA (that is, the sequence is 'AUGUUA), the A site would be occupied by phenylaanyl-tRNA. Once the A site is occupied, the two amino cids on the two tRNA molecules are joined y a peptide bond. The bond between the formylnethionine, or methionine, and the initiator tRNA is roken, but that linking the second amino acid and its RNA in the A site remains (Figure 8-22). The reaction s catalyzed by an enzyme of the large subunit called **eptidyl transferase.**

Loss of the amino acid from the initiator tRNA eaves that tRNA uncharged, and consequently it dissoiates from the P site and becomes available for re-charging in the cytosol. The two amino acids, linked by a peptide bond (a **dipeptide**), remain attached to the tRNA still in the A site. This tRNA, with the two amino acids still attached to it, then moves into the unoccupied P site, and the ribosome simultaneously moves three nucleotides along the mRNA to align the next (the third) codon to the now unoccupied A site (this movement is called **translocation**). The tRNA now in the P site has the growing polypeptide chain (consisting at this point of two amino acids) attached to it (Figure 8-22), and it is called the **peptidyl-tRNA.**

The vacant A site then becomes occupied by another tRNA charged with its amino acid; again the particular tRNA and therefore the next amino acid is determined by the codon then aligned with the A site. The dipeptide attached to the tRNA molecule in the P site is then linked to amino acid number 3 by a peptide bond. This removes the dipeptide from the tRNA in the P site, and the uncharged tRNA molecule in the P site dissociates from the ribosome. The tRNA with the tripeptide is

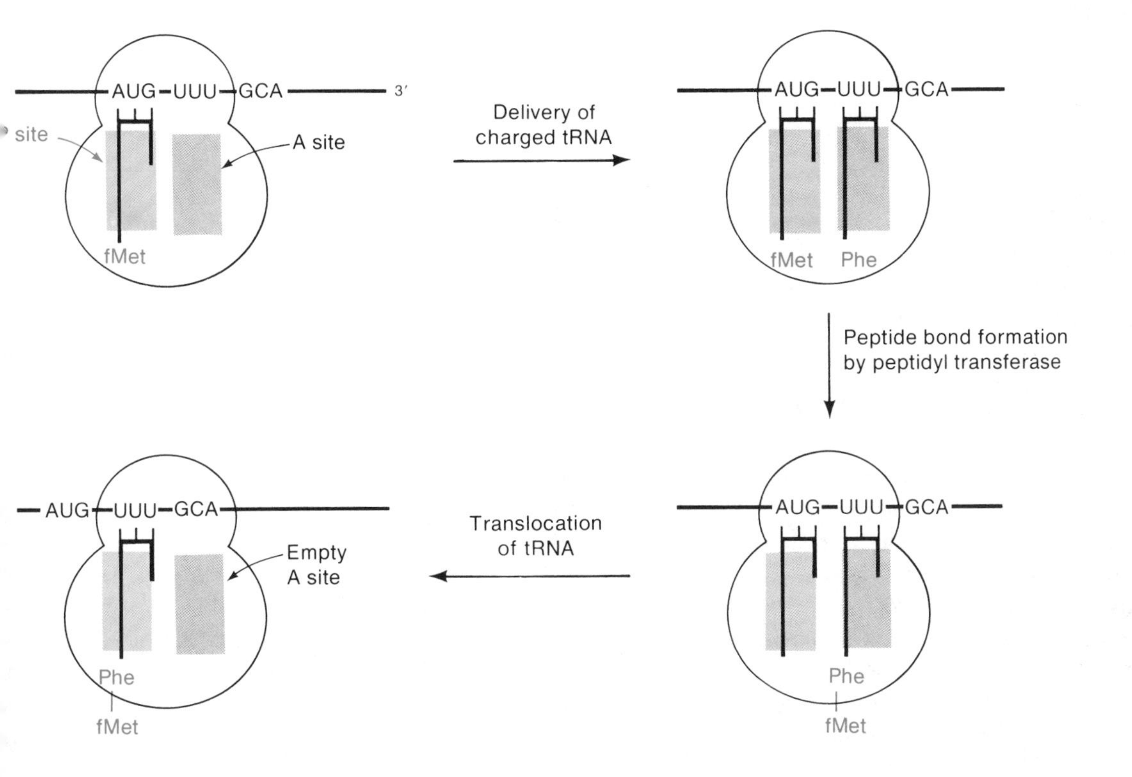

IGURE 8-22
longation phase of polypeptide synthesis. (a) Initiation complex. b) Binding of a charged tRNA molecule to the A site. c) Formation of a peptide bond. (d) Translocation of tRNA from he A site to the P site.

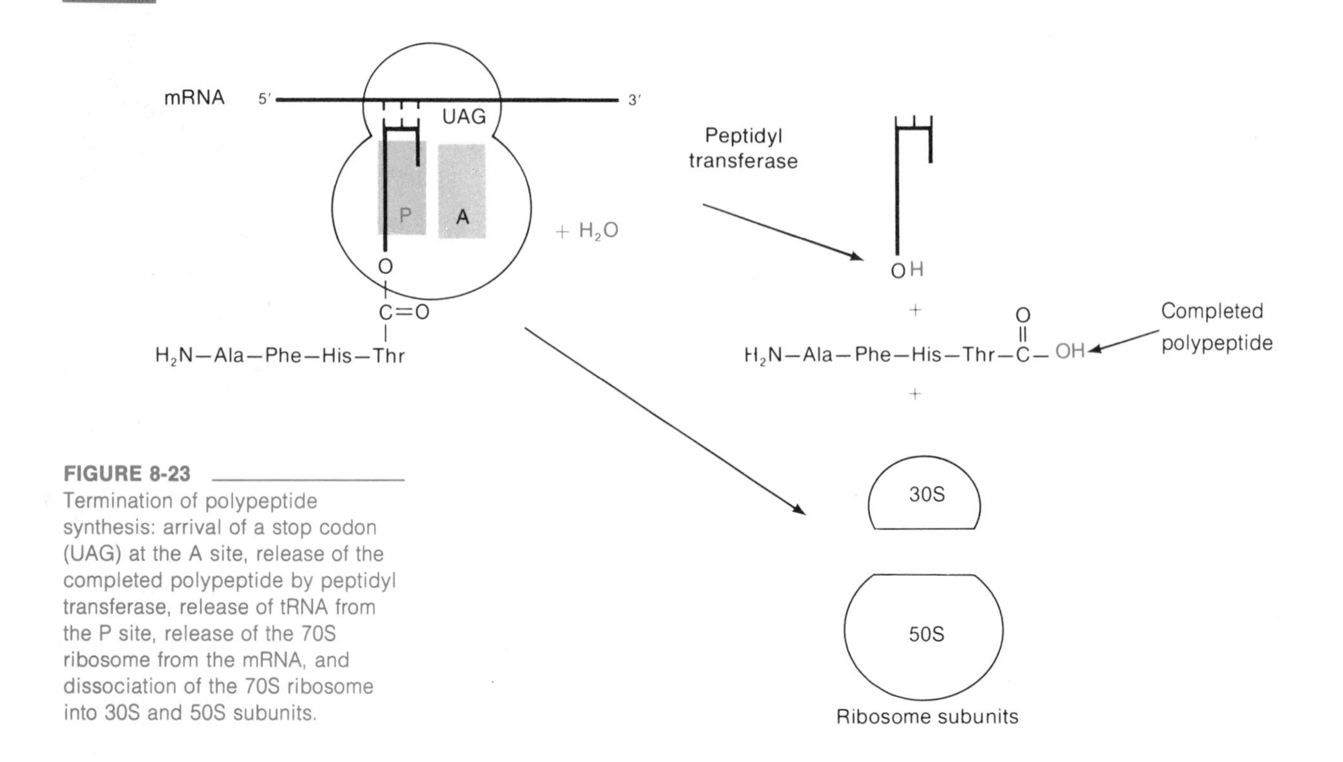

FIGURE 8-23
Termination of polypeptide synthesis: arrival of a stop codon (UAG) at the A site, release of the completed polypeptide by peptidyl transferase, release of tRNA from the P site, release of the 70S ribosome from the mRNA, and dissociation of the 70S ribosome into 30S and 50S subunits.

shifted to the P site, the mRNA moves three nucleotides to bring the next codon to the A site, and a new charged tRNA enters the A site. In short, each time a triplet codon of mRNA moves into position at the A site, a charged tRNA molecule enters the A site, and its amino acid is linked to the growing polypeptide chain. The tRNA in the A site moves to the P site, leaving a free A site for the next tRNA molecule. The cycle is repeated until all of the codons have been translated into amino acids of the polypeptide chain.

The complexity of the elongation process may leave the impression that translation is slow. However, the various steps in synthesis of a polypeptide chain are catalyzed by several proteins called **elongation factors.** In bacteria, amino acids are added to growing polypeptide chains at a rate of about 20 per second. Synthesis of an average-sized polypeptide (about 300 amino acids) takes only about 15 seconds. Althoug protein synthesis in eukaryotes follows the same princi ples as in prokaryotes, some of the steps in eukaryote are more complex, and translation is somewhat slowe in eukaryotes, primarily because formation of the initi ation complex takes longer.

Termination of Translation

The key event in termination of the synthesis of polypeptide chain is the arrival of a **termination co don** in mRNA at the A site (Figure 8-23). Any one o three different codons—UAA, UAG, UGA—signals ter mination, because no tRNA molecules possess antico dons for them. Thus, when one of these codons ir mRNA arrives at the A site, no tRNA is available tc occupy the A site, and elongation of the polypeptid

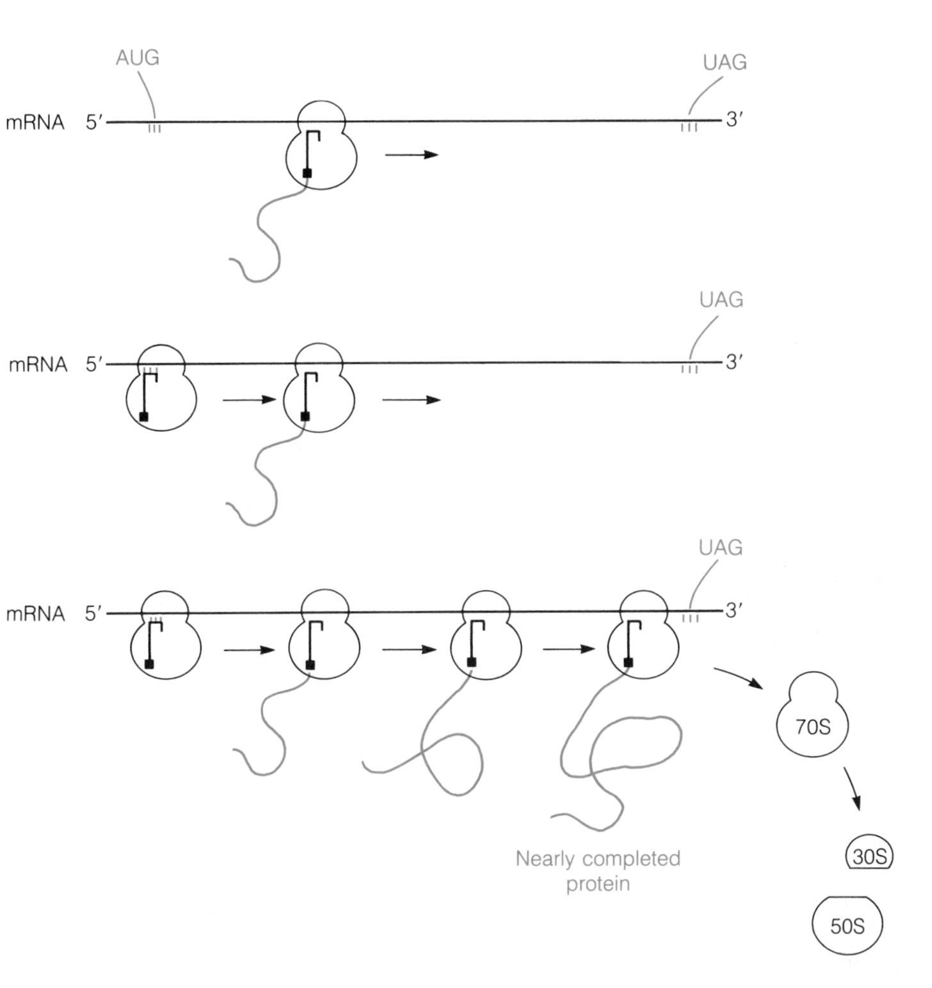

FIGURE 8-24
Successive translations of an mRNA molecule, forming a polysome.

stops. Release of the polypeptide from the tRNA molecule in the P site is accomplished by proteins called **release factors**. Following release of the completed polypeptide the ribosome dissociates into its two subunits, which are then available to start a new round of polypeptide synthesis.

Polysomes

After about 25 amino acids have been joined together in a growing polypeptide chain, the mRNA molecule has moved along the ribosome so that its end and the AUG initiation codon have become free of the ribosome (Figure 8-24). The 5′ end of the mRNA molecule can then form an initiation complex with a second ribosome, and synthesis of another polypeptide chain begins. The process is repeated until the mRNA is attached to ribosomes along its entire length, with one ribosome present for each successive group of about 80 nucleotides. This translation complex of multiple ribosomes attached to one mRNA is called a **polysome** (Figure 8-25).

In eukaryotes mRNA molecules typically have translation regions about 900 nucleotides long (300 triplets coding for 300 amino acids). A polysome formed with such an mRNA molecule contains about 11 ribosomes. In polysomes multiple copies of the same kind of polypeptide chain are produced by simultaneous translation of a single mRNA molecule, which greatly enhances the rate of protein synthesis.

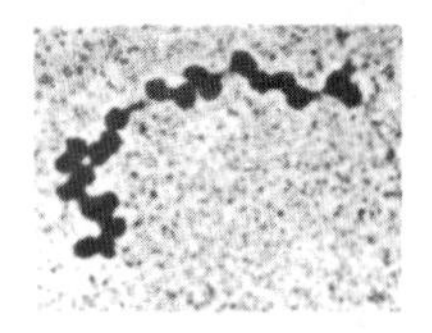

FIGURE 8-25
Electron micrograph of a polysome isolated from *E. coli*. [Courtesy of Barbara A. Hamkalo.]

Coupled Transcription and Translation

In prokaryotes there is no nuclear envelope separating the DNA from ribosomes in the cytoplasm. As a result mRNA molecules can begin to be translated before their own synthesis is complete. A DNA template is transcribed in the $3' \rightarrow 5'$ direction producing an RNA transcript with the opposite polarity, $5' \rightarrow 3'$. Therefore, mRNA molecules peeling away from DNA have 5′ ends. Translation of an mRNA molecule begins near its 5′ end. Ribosomes may attach to the 5′ end of an mRNA molecule and begin translation while the downstream part of the mRNA is still being transcribed from its DNA template. Figure 8-26 is an electron micrograph of such coupled transcription-translation made up of DNA-RNA-protein complexes isolated from *E. coli*.

Coupled transcription-translation cannot occur in eukaryotes because transcription of DNA occurs in the

FIGURE 8-26
Coupled transcription-translation. (a) Transcription of a section of DNA of *E. coli* with concurrent translation of the partially completed mRNA molecules. The string of dark particles are ribosomes adhering to the mRNA. (b) A drawing of structures in part (a). The mRNA is red and coated with ribosomes. The red dots on the DNA are RNA polymerase. The dashed arrows show the distances that each RNA polymerase has traveled from the transcription initiation site. Arrows 1, 2, and 3 have the same length as mRNA molecules 1, 2, 3. mRNA 4 is shorter than arrow 4, probably because its 5′ end (free end) has been partially digested by RNase. No arrows are included for some of the lower mRNA molecules in (a). [(a) From O. L. Miller, Barbara A. Hamkalo, and C. A. Thomas, *Science*, 169, 392. (b) Courtesy of David Freifelder.]

a

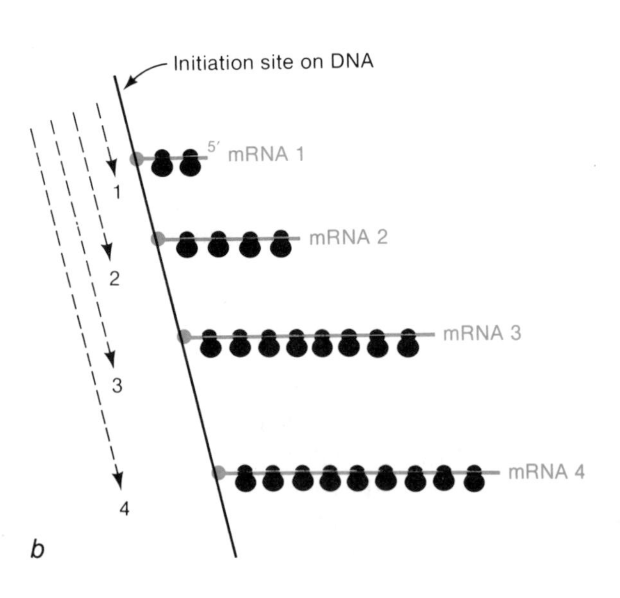

nucleus and is physically separated by the nuclear envelope from translation of mRNA on ribosomes in the cytoplasm. In eukaryotes completed mRNA molecules are packaged in the nucleus and then transported to the cytoplasm through nuclear pores for subsequent translation. However, coupled transcription-translation possibly occurs in mitochondria and chloroplasts of eukaryotic cells. The DNA and ribosomes within these organelles are not physically separated from one another.

Agents that Block Translation

A variety of agents, including many antibiotics, prevents growth of bacteria because they block one or another step in protein synthesis. They are useful in treating bacterial infections because they are nontoxic to eukaryotic cells either because they cannot penetrate the plasma membrane of eukaryotic cells or because they do not bind to eukaryotic ribosomes. **Streptomycin** and **neomycin** bind to one of the proteins in the 30S bacterial ribosomal subunit and block the binding of the initiator tRNA to the P site. **Chloromycetin** (chloramphenicol) inhibits the peptidyl transferase of bacterial ribosomes. **Tetracycline** inhibits binding of charged tRNA molecules to the 30S subunit in bacteria. **Erythromycin** binds to free 50S subunits and prevents attachment to 30S subunits to form 70S particles. Some of these antibiotics do affect ribosomes in mitochondria (e.g., tetracycline), causing toxic side effects but do not affect the ribosomes in the cytosol. Their toxic action is much greater against bacteria than against mitochondria; otherwise they could not be used in medical treatment.

Some bacteria synthesize toxins that account for their disease-producing effects. The bacterium *Corynebacterium diphtheriae*, which causes diphtheria, produces an enzyme that blocks the elongation phase of translation in eukaryotes. The inhibition of protein synthesis is the main cause of disease symptoms. **Cycloheximide** is a drug that inhibits peptidyl transferase in eukaryotes. It is used in cancer patients to kill cancer cells (chemotherapy), but is limited in its usefulness because it also affects protein synthesis in normal cells as well as cancer cells.

ADDITIONAL FEATURES OF PRODUCTION OF mRNA IN EUKARYOTES

The basic mechanism of transcription in eukaryotes is much the same as in prokaryotes. However, several features of genes and RNA transcripts are different in eukaryotes. These features, some of which are important in regulation of gene expression, are the following:

1. All eukaryotes possess three kinds of RNA polymerase. The three polymerases transcribe different classes of genes.
2. The 5′ ends of mRNA molecules are modified in the nucleus after transcription by addition of a **cap.**
3. The 3′ ends of most kinds of mRNA molecules are modified in the nucleus by addition of up to 200 adenylic acid monomers, forming **polyadenylated** or **poly(A) tails.**
4. The RNA molecule that is directly transcribed from DNA, the **primary transcript,** is longer—usually much longer—than the final mRNA that is translated in the cytoplasm. Generally, transcription proceeds well beyond the end of the gene and part of the 3′ end is removed prior to addition of a poly(A) tail. In addition, after capping at the 5′ end and polyadenylation at the 3′ end, regions of the RNA transcript, called **intervening sequences** or **introns,** are cut out of the RNA transcript and destroyed within the nucleus. The preserved fragments are spliced together to make the final mRNA molecule with a cap and a poly(A) tail.
5. Some primary transcripts contain the sequences of several genes and are processed into individual mRNA molecules, each encoding a single polypeptide. Many bacterial mRNA molecules represent several genes, but these mRNA molecules are not split into separate molecules for each gene sequence. Instead, the gene sequences present in a

single mRNA molecule are translated individually into different polypeptides.

6. Transcription and translation in eukaryotes are not coupled, as they can be in prokaryotes. The nuclear envelope separates sites of transcription in the nucleus from sites of translation on cytoplasmic ribosomes.
7. Messenger RNA molecules are packaged into particles in the nucleus by complexing with specific packaging proteins and then move into the cytoplasm, where they are at least partially unpackaged prior to translation.
8. Many kinds of mRNA molecules are long-lived in the cytoplasm, lasting for days, weeks, and occasionally months. The lifetime of mRNA molecules in prokaryotes is measured in minutes or hours.

Most of these differences are now described in more detail.

The Three RNA Polymerases in Eukaryotes

The three enzymes are designated RNA polymerase I, II, and III. Each catalyzes transcription of different types of genes. **Polymerase I** transcribes only those genes encoding rRNA. It is largely localized in the nucleolus, where genes that encode rRNA are located. **Polymerase II** transcribes genes that encode mRNA molecules and is present throughout the nucleus. **Polymerase III** is also present throughout the nucleus. It transcribes genes encoding tRNA, one of the rRNA molecules (5S rRNA) and a group of small RNA molecules called **snRNA molecules** (sn = small nuclear), discussed later. All three polymerases are large proteins with molecular weights of about 500,000. They are made up of two large and several small subunit polypeptides. Presumably one or more subunits in each kind of polymerase is responsible for recognition of the different classes of genes that the polymerase transcribes. All three polymerase types catalyze the same reaction as catalyzed by the single RNA polymerase in prokaryotic cells, the formation of 3′-5′-phosphodiester bonds in RNA. Eukaryotic polymerases catalyze addition of 50 to 100 nucleotides per second to growing RNA chains. The three enzymes can be separated from one another in cell extracts by chemical fractionation procedures.

The significance of the presence of three different kinds of RNA polymerase, each specific for transcription of a different set of genes, is not well understood. The three polymerases may reflect three different mechanisms for regulating the expression of three major groups of genes. Some experimental evidence suggests that regulation of the genes transcribed by polymerase III (genes encoding tRNAs, 5S rRNA, and snRNAs) may be quite different from regulation of genes transcribed into mRNA and rRNA.

The 5′ Cap of Eukaryotic mRNA

Capping of mRNA occurs in the nucleus and is the first step in the processing of the primary RNA transcript. A cap is added to the 5′ end of an RNA molecule even before transcription from the gene is completed (Figure 8-27(a)). The cap consists of a terminal guanylic acid in which the N in position 7 of the purine ring is methylated (addition of a CH_3 group) (Figure 8-27(b)). Three phosphate groups join the 5′ carbon of the ribose part of 7-methylguanylic acid to the 5′ carbon in the sugar of the next nucleotide, which may be adenylic acid or guanylic acid. The second nucleotide is joined in a usual 3′-5′-phosphodiester linkage to the third nucleotide, which may be either adenylic or guanylic acid. In animal cells the second nucleotide of the cap is modified by addition of a methyl group to the ribose. In other eukaryotes the third nucleotide may also be methylated.

The transcript is said to be capped because it has no free 5′ end (Figure 8-27). The cap is thought to protect the 5′ end of mRNA from degradation by ribonuclease. Experimentally produced mRNA molecules lacking a cap are quickly degraded after injection into a cell, with little or no translation having taken place from them. In the cytoplasm one or more **cap-binding proteins** attach to the cap and enhance formation of an initiation complex. The details of how cap-binding proteins func-

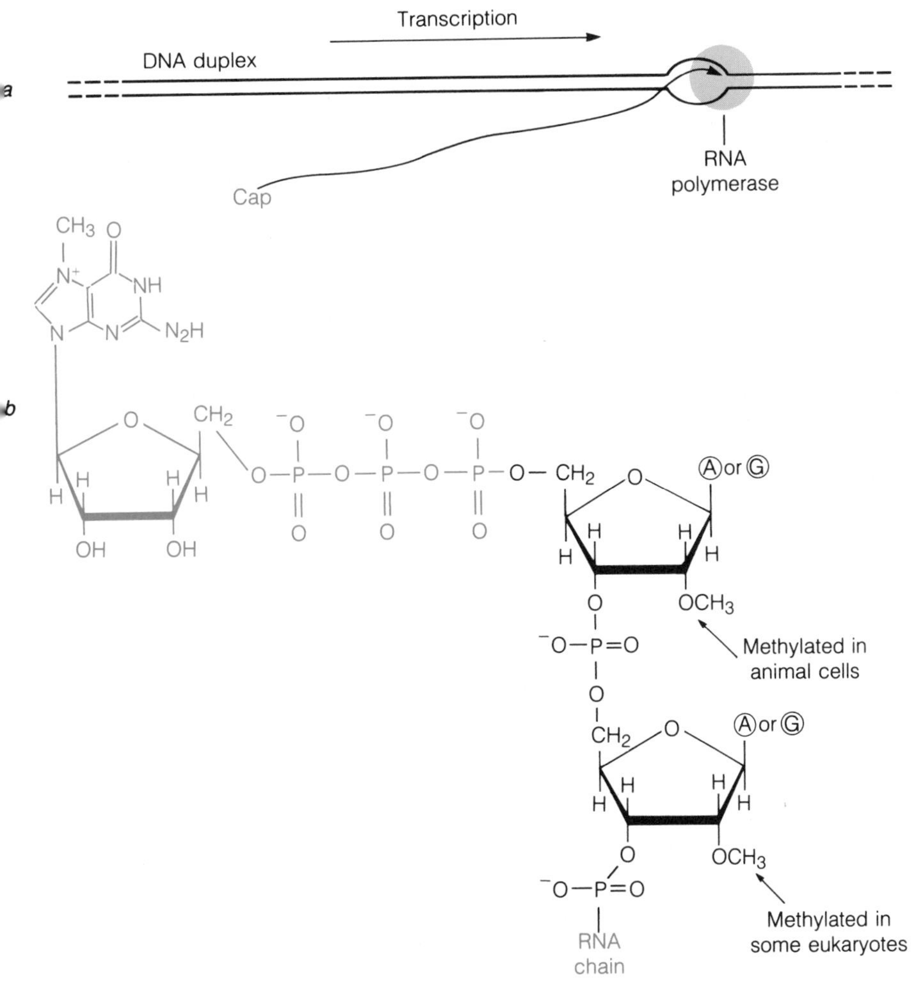

FIGURE 8-27 Cap at the 5′ end of mRNA in eukaryotes. (a) A cap is added to the 5′ end of mRNA before its synthesis is completed. (b) The three deoxynucleotides that constitute a cap.

tion is not known. Caps might have a role in regulating translation, but there is as yet not much firm evidence to support the idea.

The 3′-Poly(A) Tail of Eukaryotic mRNA Molecules

After transcription is complete, a string of 20 to 200 adenylic acid monomers is added to the 3′ end of most eukaryotic mRNA molecules in the nucleus by the enzyme **poly(A) polymerase** (Figure 8-28). Monomers are frequently called residues, when they are part of a macromolecule. Poly(A) tails are not usually added to the end of the primary transcript. Instead, transcription proceeds beyond the eventual site of poly(A) addition, and a section of the 3′ end of the transcript is removed in the nucleus before poly(A) addition. Addition of poly(A) is signaled by the sequence AAUAAA found 10 to 25 bases upstream from the poly(A) addition site. What governs the length of the poly(A) tail is not clear. It is likely that the poly(A) tails are in fact rather heterogeneous in length. At least one kind of mRNA molecule lacks poly(A) tails, namely, those encoding the **histone proteins** (Chapter 11). The function of

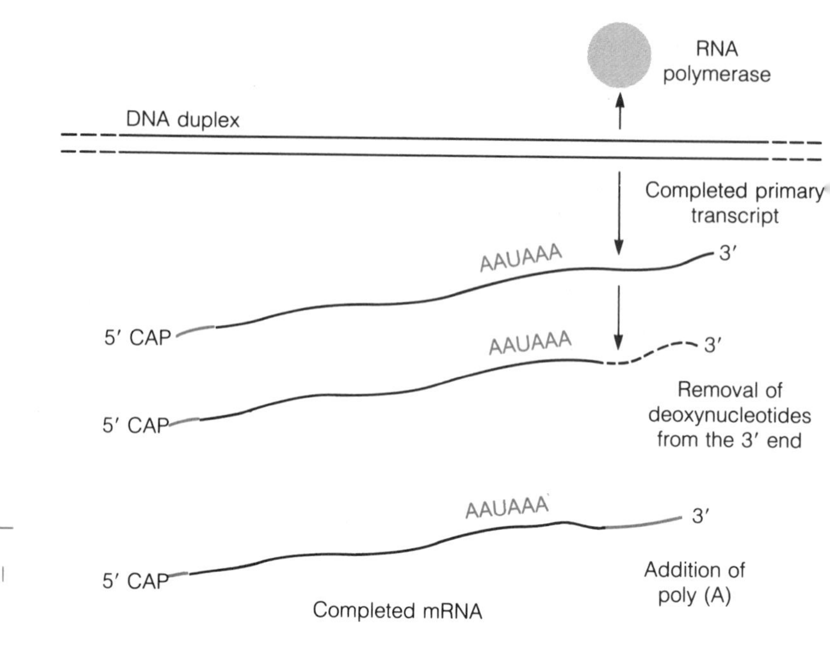

FIGURE 8-28 Removal of deoxynucleotides from the 3′ end of a primary transcript and addition of the 3′ poly(A) tail by poly(A) polymerase.

the poly(A) tail on an mRNA molecule is not well understood. It appears to increase the stability of the mRNA. The tail gradually shortens during successive translations of the mRNA in the cytoplasm. When the poly(A) has been totally removed, the mRNA molecule is degraded. In this sense poly(A) tails may contribute to regulation of gene expression by helping to govern the life span of mRNA molecules. The few types of mRNA molecules that lack poly(A) tails are transported to the cytoplasm and translated normally, so poly(A) tails cannot be essential for mRNA function.

Cutting and Splicing of RNA in Eukaryotes

In prokaryotes the DNA sequence that codes for a polypeptide chain is continuous and uninterrupted. For example, DNA polymerase in *E. coli* is an unusually large protein with a molecular weight of 109,000 and consisting of a polypeptide chain of 975 amino acids. The polypeptide chain is encoded by an uninterrupted region of 2925 base pairs of DNA in the chromosome. In eukaryotes, the region of DNA that codes for a polypeptide chain is, for most genes, interrupted within the gene by one or more noncoding DNA sequences called intervening sequences or introns. Examples are the genes that code for β-globin in mammals, one of the two polypeptide chains of hemoglobin. The coding sequence of this gene in most organisms is interrupted by two introns (Figure 8-29). In the gene encoding human β-globin these are 130 and 850 bp long. The two introns divide the coding sequence into three regions.

The gene for β-globin in rabbits, mice, and humans contains a large and a small intron. The introns are slightly different in size from species to species but are located in the same relative position within the gene or its primary transcript. The coding regions, **exons,** and introns are all transcribed as one continuous primary transcript (Figure 8-30). After addition of a cap and poly(A) tail, the transcript is cut at the junctions between the exons and the introns. The two intron segments are destroyed, and the three coding regions are spliced together in the nucleus to yield an mRNA molecule with a cap and a poly(A) tail. After transfer to the cytoplasm, the mRNA is translated into a polypeptide of 146 amino acids.

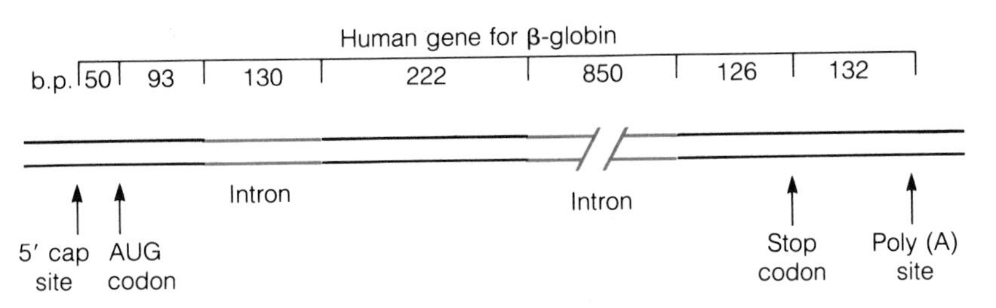

FIGURE 8-29
Diagram of the human gene encoding β-globin. The primary transcript begins at the 5′ cap site and extends beyond the poly(A) addition site. The number of base pairs in various segments of the gene are given. A 50-base-pair leader extends between the cap site from the start of translation at the AUG site. Two introns interrupt the protein coding region. One hundred thirty-two base pairs form a trailer between the stop codon for translation and the signal (AAUAAA) for poly(A) addition.

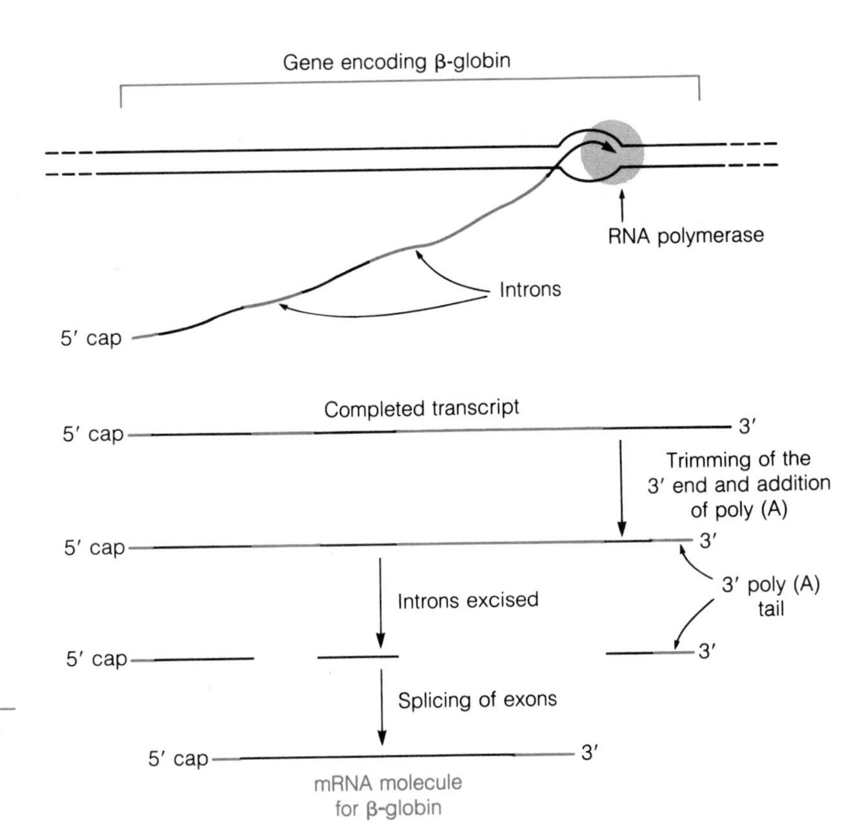

FIGURE 8-30
The gene encoding human β-globin contains two introns, which are removed from the primary transcript. The exons are spliced to yield a completed mRNA molecule.

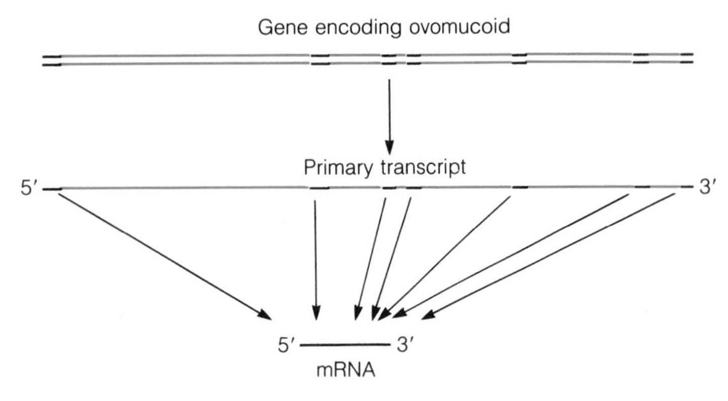

FIGURE 8-31
The ovomucoid gene in the chicken contains six introns (red) and seven exons (black). The completed mRNA is only about one-seventh as long as the gene.

The gene in chickens encoding the protein ovomucoid is made up of six introns and seven exons. Ovomucoid is a major protein of egg white and is synthesized in large amounts by cells of the oviduct. The organization of the gene is shown in Figure 8-31. The six introns account for 85 to 90 percent of the primary RNA transcript from the ovomucoid gene. These divide the coding region, which is about 750 base pairs long. Subsequent cutting and splicing of the primary transcript of the gene yields an mRNA molecule that is about one-seventh as long as the primary transcript.

The $\alpha 2$-collagen gene in chicken cells is an extreme example of division of a gene by introns. Collagens are proteins synthesized and secreted by fibroblasts and are the main proteins of connective tissue. Tendons, for example, are composed almost entirely of collagen. There are 52 introns in the $\alpha 2$-collagen gene of the chicken, and these range in size from 100 to 3000 base pairs (Figure 8-32). They separate the coding region into 53 exons. The primary transcript is 40 kilobases long and is cut and spliced in the nucleus to form an mRNA molecule less than one-tenth as long.

All eukaryotes have many copies of the genes for rRNA. In the frog *Xenopus* some of these contain introns, and others do not. In one strain of the ciliate *Tetrahymena* an intron is present in all of its rRNA genes, but in a closely related strain no introns are present in any of the rRNA genes. The rat has two genes for insulin; one has two introns, and the second gene has only one. Some tRNA genes in yeast contain introns, but in other organisms introns are absent from these tRNA genes. Most protein-coding genes in animals contain introns but at least a few genes do not.

Excision of Introns

The mechanism of cutting and splicing of transcripts to remove introns and generate mRNA molecules is partly understood. Cutting and splicing must be precise to the level of single nucleotides. If it were not, then in translation of the spliced exons the triplet reading frame in the mRNA would be distorted in the same way that it is distorted by a frameshift mutation. In most cases a nearly identical sequence two to six bases long occurs at each end of an intron (Figure 8-33). Introns generally begin with GU and end with AG (the

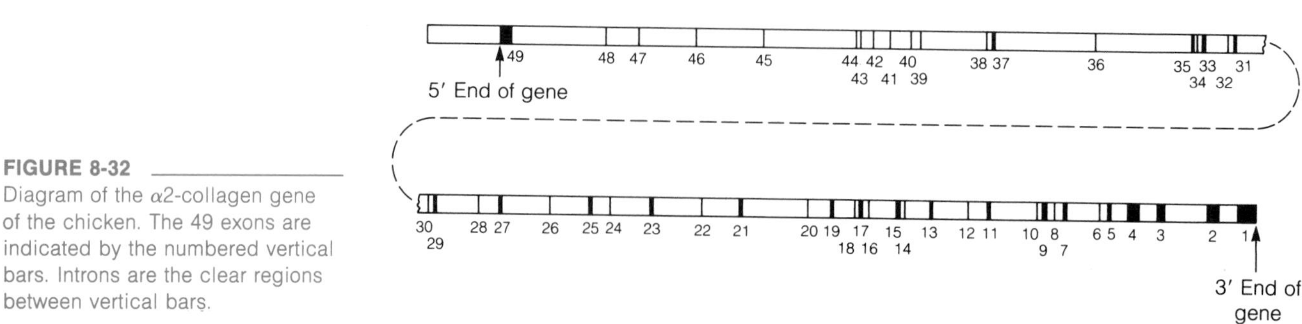

FIGURE 8-32
Diagram of the $\alpha 2$-collagen gene of the chicken. The 49 exons are indicated by the numbered vertical bars. Introns are the clear regions between vertical bars.

Exon Intron Exon

AGGUAAGU PyNCAGGN

FIGURE 8-33
Common features of the base sequence at exon-intron junctions. Py = pyrimidine; N = any nucleotide. Introns generally begin with GU and end with AG.

start and end of an intron are defined by the direction of transcription of the gene).

These sequence regularities have a crucial role in excision of introns, and alteration of one of these sequences by mutation can disrupt normal splicing. For example, normal blood clotting requires a series of different proteins synthesized by liver cells and released into the blood by exocytosis. One of these proteins, called **factor IX,** is defective in one form of the genetic disease known as **hemophilia B.** The defect in factor IX can be any one of several types. One of these is the result of a mutation changing a single nucleotide at the splice junction site for removal of one of the introns from the mRNA for factor IX. The mutation has changed the obligatory GU at the start of the intron to UU (CA to AA in the transcribed strand of the DNA). The change disrupts splicing and leads to synthesis of an abnormal factor IX, and hemophilia.

Removal of an intron occurs in a fixed sequence of steps (Figure 8-34). First, a cut is made at the start of the intron. This creates a 3′OH end on the exon and a 5′P terminus on the intron. The 5′P terminus becomes covalently joined to the OH on the 2′ carbon (forming a 2′-5′-phosphodiester bond) of a nucleotide in a region 18 to 37 nucleotides upstream from the other end (the right end) of the intron, creating a lariat configuration. Next a cut is made at the right terminus of the intron, releasing the lariat, and the two free ends of the transcript are spliced together. The free lariats are eventually degraded to single nucleotides. The significance of this order of steps is not known; presumably the enzymes (at least four) involved in cutting and splicing have properties that require that they work in a fixed order.

Another element involved in excision of introns is a small, nuclear RNA molecule, snRNA, mentioned earlier in the chapter. The nuclei of eukaryotes contain a small family of different kinds of snRNA molecules. Like rRNA and tRNA they are not translated. The snRNA molecules combine with particular proteins to form snRNA ribonucleoprotein particles in the nucleus. An snRNA molecule known as U1 (165 nucleotides) contains sequences that are complementary to the sequences at intron boundaries. It has been demonstrated experimentally that ribonucleoprotein particles containing U1 RNA participate in the cutting and splicing

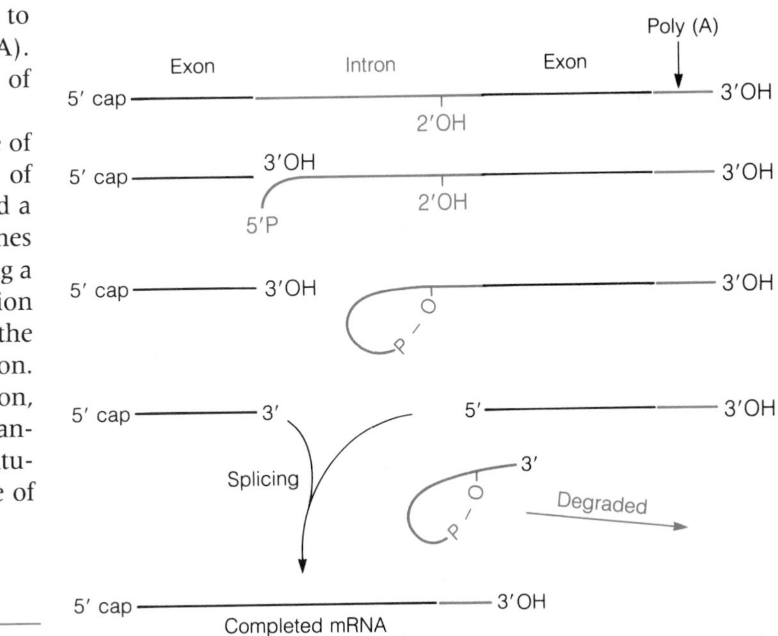

FIGURE 8-34
Removal of an intron sequence from a pre-mRNA molecule.

of primary transcripts. Base-pairing between U1 RNA and boundary sequences of an intron may put the intron into a configuration that allows enzymes to carry out proper excision.

Self-Splicing of Primary Transcripts

Removal of introns from primary transcripts is usually mediated by enzymes as just described. An alternate cutting-splicing mechanism has been found for rRNA primary transcripts in the ciliate *Tetrahymena* and in the slime mold *Physarum,* and for primary transcripts of both rRNA and mRNA in the mitochondria of yeast. These transcripts contain short complementary sequences that allow intramolecular base-pairing, which folds the molecule into a specific configuration. In this configuration the RNA transcript carries out its own cutting and splicing correctly without catalysis by protein enzymes. The RNA is *autocatalytic,* catalyzing its own processing, and is called a **ribozyme.**

The Function of Introns

What purpose introns serve is still poorly understood. They occur in many but not all eukaryotic genes. As noted earlier, introns may occur in some copies of rRNA genes and be absent in other copies in the same cell. Introns are generally absent in genes of prokaryotes, although they have been found in two tRNA genes among bacteria known as *Archaebacteria* and in a protein-coding gene in T4 bacteriophage. Introns vary greatly in size, most falling in the range of 40 to 5000 bp. By DNA cloning techniques a gene can be isolated from a cell, and the size of the intron can be varied experimentally. As long as the short repeat sequences at the borders between intron and exons are unchanged, the intron, altered in length, is correctly excised from RNA transcripts when the gene is reintroduced into a cell. As might have been predicted from this experiment, introns show much more divergence in sequence than do coding sequences, without affecting expression of the gene. Although the sequence and size of an intron can vary, the location of introns within a gene remains fixed. For example, the two introns present in the β-globin genes in human, rabbit, and mouse cells differ somewhat in size and sequence, but they interrupt the coding sequences (exons) at precisely the same nucleotide positions in the genes of all three species.

Some believe that the intron serves to carry out **exon shuffling.** Proteins often contain structually different domains. These include domains of α-helices and β-structure, domains of hydrophobic amino acids, and domains that are responsible for binding various small molecules that modulate the activity of the enzyme, for example, allosteric effectors (Chapter 3). An exon might represent a sequence of nucleotides coding for a particular domain so that the exons present in a gene could represent the kinds and order of domains in the protein. The exons are thought to have been brought together during evolution by a process of exon shuffling, in which new combinations and arrangements of domains form the basis for creation of new genes encoding proteins with new structural and functional properties. The introns in this hypothesis represent *neutral* noncoding, joining regions by which exons might be shuffled. Shuffling of exons could provide a means for rapid evolution of new genes.

Exon shuffling does indeed occur in lymphocytes (see later in this chapter), in which exons coding for discrete domains of antibody proteins are joined in many combinations to produce an enormous diversity of antibodies. However, for most proteins, like β-globin, there is no clearly apparent correspondence between exons and major, functional domains in protein molecules. Also, coding regions for readily definable structural and functional domains in proteins are often not separated by introns in the genes, although it might be that the introns were lost subsequent to the joining of the domains. Overall, the hypothesis that introns separate functional domains, although still under consideration, does not appear to fit with many intron-exon patterns.

A clear relationship between exons and introns is observable in genes coding for collagen. The collagen molecule contains a regularly repeating pattern in its amino acid sequence consisting of gly-x-y-gly-x-y-gly-

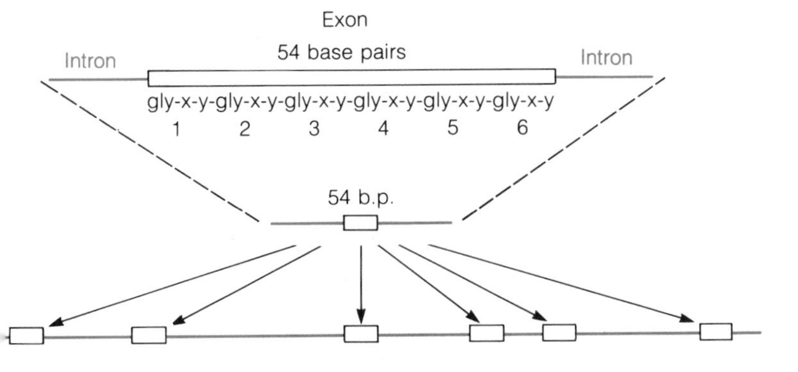

FIGURE 8-35 Diagram of how the collagen gene might have evolved from an original exon of 54 base pairs encoding six repeats of the amino acid sequence gly-x-y. Repeated replication of the exon and joining of units together with inclusion of variable amounts of intron sequence would produce the pattern of exons of 54 base pairs joined by introns of variable length.

x-y, and so forth, which is important in the assembly of collagen molecules into fibers. Every third amino acid is glycine, and x and y can be any of a variety of amino acids. The fundamental repeat, gly-x-y is encoded by 9 base pairs. Some of the 50 exons in the collagen gene consist of 54 bp, encoding six repeats of gly-x-y. Other exons range in size from 45 to 108 bp by multiples of 9 bp (encoding 5 to 12 repeats of gly-x-y). This suggests that the collagen gene arose by replication and stringing together copies of an original unit of 54 bp encoding 6 repeats of gly-x-y (Figure 8-35). The stringing together occurred by means of adjacent, noncoding sequences that formed introns as the gene lengthened in evolution. Exons shorter or longer than 54 bp presumably arose by subtracting or adding units of 9 bp coding for gly-x-y. A similar regularity of pattern of exons and introns is present in the gene encoding the protein α-fetoprotein. However, proteins in general do not have regular repeating patterns of amino acids, and there is no regular repeating pattern of exons and introns in their genes.

It must also be considered that introns are remnants of some past phenomenon in the evolution of genes and have no contemporary function. In this sense they may represent *junk* DNA (Chapter 11). The idea of no function is supported by an experiment with the actin gene of yeast. Yeast cells contain only a single actin gene, which has one intron. By genetic engineering the intron was removed from the gene and the intronless gene was used to replace the single, resident actin gene in haploid yeast cells. The amount of actin in yeast is important for viability; too much or too little of this protein inhibits cell growth. Cells containing the intronless gene appeared normal and grew at the normal rate. Therefore, the intron in the actin gene appears to have no function, although it is still possible that the intron has a nonessential function that could not be detected by measuring growth rate.

The Primary Transcripts in the Nucleus

Because of their size heterogeneity and nuclear location, newly synthesized, or primary, transcripts have long been called **heterogeneous nuclear RNA** or **HnRNA.** HnRNA molecules are, on average, several thousand bases longer than mRNA molecules in the cytoplasm. These observations, first made in the 1960s, are now explained by the occurrence of introns and by the removal of nucleotides from the 3′ end of primary transcripts. Excision of introns from HnRNA molecules and 3′ trimming drastically reduce their average size to that of mRNA molecules. In addition, the processing of HnRNA (capping, polyadenylation, and excision of introns) explains the earlier observation of a delay of about 30 minutes (in animal cells) between synthesis of a primary transcript and its appearance as an mRNA molecule in the cytoplasm. As it separates from its DNA template, the completed portion of a primary transcript becomes complexed with protein molecules, forming

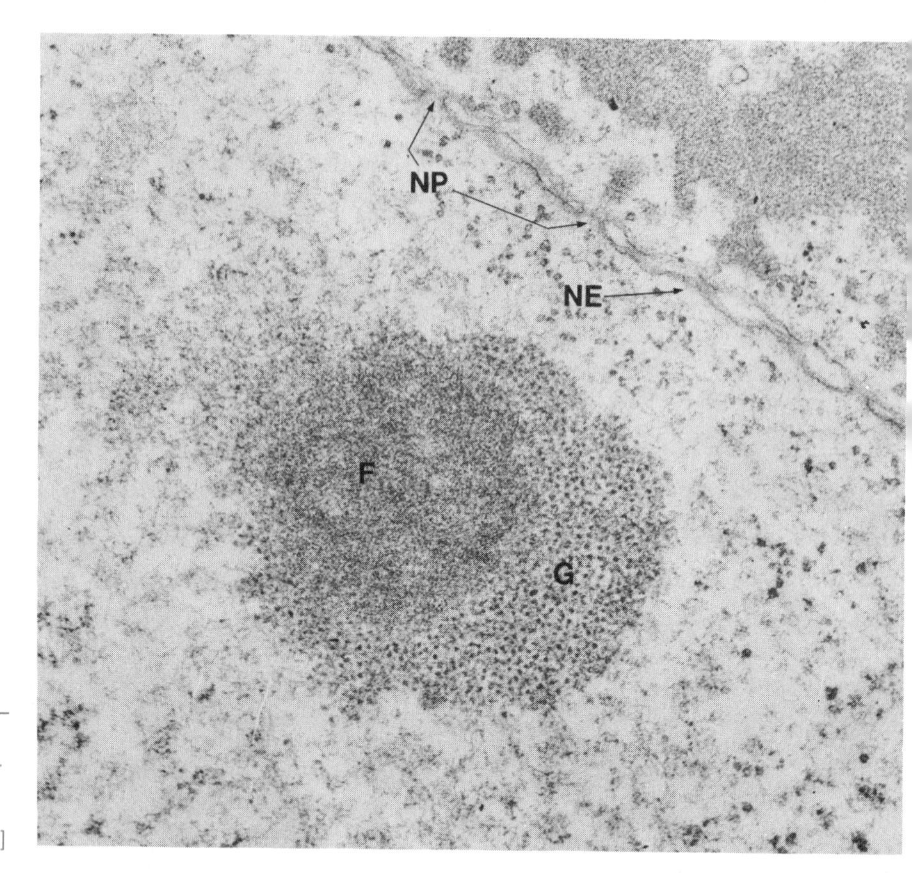

FIGURE 8-36
Electron micrograph of a nucleolus in an oocyte of the green frog (*Rana clamitans*). Granular (G) and fibrous (F) regions are visible. NE = nuclear envelope. NP = nuclear pore. [Courtesy of O. L. Miller, Jr.]

ribonucleoprotein (RNP) particles. At least six different kinds of proteins are involved. These proteins aggregate with each other as well as with an HnRNA molecule in a highly specific configuration, much like the formation of a ribosomal subunit from ribosomal proteins and rRNA. Processing of HnRNA then occurs within RNP particles. The resulting mRNA molecule is also complexed with protein molecules and moves into the cytoplasm as an RNP particle. In subsequent translation the mRNA becomes dissociated from the *carrier* proteins, which probably return to the nucleus and participate in a new cycle of RNA packaging and transport to the cytoplasm.

FORMATION OF RIBOSOMES

Ribosomes are cytoplasmic organelles whose exclusive function is the synthesis of proteins. Ribosomes are rather stable structures but nevertheless do undergo turnover. In nonreproducing cells active in protein synthesis, such as liver hepatocytes and pancreas cells, new ribosomes are continuously produced to replace old ones that presumably have suffered damage to one or more of their macromolecular constituents. In cells that are reproducing, ribosome production is one of the main events that make up cell growth. The formation of a ribosome from its individual components is one of the better understood examples of self-assembly of a cell structure.

The Nucleolus

In eukaryotic cells the large and small subunits of ribosomes are assembled within a discrete organelle, the nucleolus, within the nucleus. Diploid cells sometimes have two nucleoli, although these often fuse to form one. Such fusion probably has no functional significance but simply occurs if the nucleoli are closely adjacent to one another. The nuclei of large amoeba and ciliated protozoa contain tens or even hundreds of nucleoli. Nucleoli are never found in prokaryotes, where ribosome assembly takes place in the cytosol.

Nucleoli form at specific regions, called **nucleolar organizers,** on one or more chromosomes. The nucleolar organizer is a set of DNA sequences that code for the synthesis of the two larger rRNA molecules (28S and 18S) and one of the two smaller rRNA molecules (5.8S) that occur in ribosomes. The other small rRNA molecule (5S) is encoded by genes in other parts of the chromosomes. Nucleoli usually have two regions readily discernible in the electron microscope, a fibrous region and a granular region (Figure 8-36). The fibrous region contains the nucleolar organizer DNA of the chromosome, rRNA newly transcribed from the DNA, and ribosomal subunits in various stages of assembly. Ribosomal proteins are synthesized in the cytoplasm and imported into the nucleolus for assembly into ribosomal subunits. The granular region consists of nearly completed ribosomal subunits, less distinct in shape and slightly smaller than the mature ribosomal subunits found in the cytoplasm. The ribosomal subunits in the granular region migrate to the nuclear envelope, pass through the pores in the envelope, and enter the cytoplasm. When the nascent ribosomal subunits reach the cytoplasm, they undergo a poorly understood maturation step and become available for protein synthesis.

Multiplicity of rRNA Genes

The genes for rRNA synthesis are one of several that have been found to be present in multiple copies. In bacteria there are usually five to ten copies (seven in *E. coli*) of each gene for the 23S and 16S molecules. These genes occur in sets containing one gene for 23S rRNA and one for 16S rRNA. The two rRNA genes are separated by a gene that encodes a tRNA molecule. A single, large transcript is produced from each gene set and is then cleaved to yield one 23S rRNA, one 16S rRNA, and one tRNA molecule. The primary transcript is larger than the sum of the three rRNAs, and a substantial part is removed and degraded during processing. The purpose (if there is one) of this seemingly wasteful arrangement is not known.

In eukaryotes the multiplicity of genes coding for rRNA is much greater. Human cells have around 200 rRNA genes per haploid set of chromosomes. *Drosophila* has about 160 copies, and *Xenopus* has about 600 rRNA gene copies. The genes for 28S, 18S, and 5.8S rRNA occur in sets arranged in tandem in the DNA of the nucleolar organizer. In some mammalian species the several hundred copies of rRNA genes are all contained in a single organizer. The multiple genes for 5S rRNA are not located in the nucleolar organizer but occur in groups at other sites on the chromosomes.

The multiplicity of genes for rRNA allows the production of ribosomes to keep pace with cell growth during cell reproduction (Chapter 12). In nongrowing cells, even those actively engaged in protein synthesis, probably only a fraction of the rRNA genes is utilized.

In human cells the 200 sets of rRNA genes are distributed among five chromosomes (numbers 13, 14, 15, 21, and 22). In each case the 40 or so sets of genes for rRNA are located at one end of the chromosome. In the interphase nucleus all five regions containing rRNA genes come together and form a single nucleolus. When a cell enters mitotic prophase, the nuclear envelope disassembles into small fragments of membrane, the nucleolus disperses, synthesis of rRNA stops altogether, and the genes for rRNA in the chromosomes separate and become condensed into the five respective chromosomes. At the end of mitosis, these events are reversed. The chromosomes decondense, the nuclear envelope reforms, rRNA synthesis resumes, and nucleoli begin to form around the ends of the five chromosomes. Within a short while the five ends come together toward the center of the nucleus and the several nucleoli fuse into one large nucleolus.

Synthesis of rRNA Precursors in Eukaryotes

As in bacteria, 28S rRNA and 18S rRNA in eukaryotes are transcribed from **rDNA genes** as a single primary transcript. In mammalian cells the precursor is a very large molecule, about 13,000 nucleotides long. Within this molecule are the nucleotide sequences for one each of 28S, 18S, and 5.8S rRNA molecules ar-

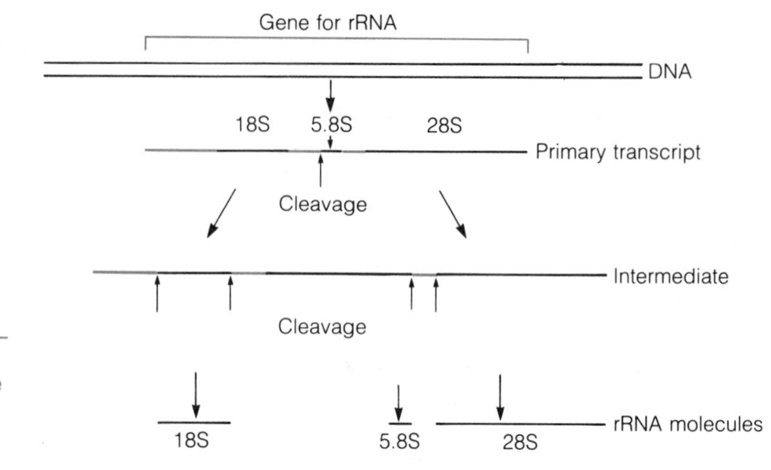

FIGURE 8-37
Processing the primary transcript encoded by a mammalian rRNA gene. Processing consists of cleavage of the molecule, release of the three rRNA molecules, and degradation of the three segments colored red.

ranged as shown in Figure 8-37. These sequences are each about 5100, 2100, and 150 nucleotides long, accounting for about 7350 of the 13,000 nucleotides in the precursor. In the nucleolus the precursor is enzymatically cleaved into the three rRNA products, with breakdown of about 5650 nucleotides of precursor sequence into individual nucleotides. Part of the sequences destroyed during posttranscriptional processing form spacers between the three rRNA sequences (and between corresponding regions of DNA in the chromosome) and are called **transcribed spacers.** A portion of the precursor located at the 5′ end of the molecule is also degraded. The significance of these expendable sequences is poorly understood. As discussed earlier, mRNA and tRNA sequences are also synthesized as larger molecules that are subsequently reduced in size by processing. The pattern of rRNA synthesis is the same in other eukaryotes, although the amount of expendable sequence in the precursor varies.

The length of double-helix DNA transcribed into an rRNA precursor is called a **ribosomal DNA (rDNA) transcription unit.** As noted above, the sets of rRNA genes, or rDNA transcription units, occur in multiple copies. These are arranged in tandem along the DNA molecule of the chromosome with each unit separated from the next by a DNA spacer that is not transcribed

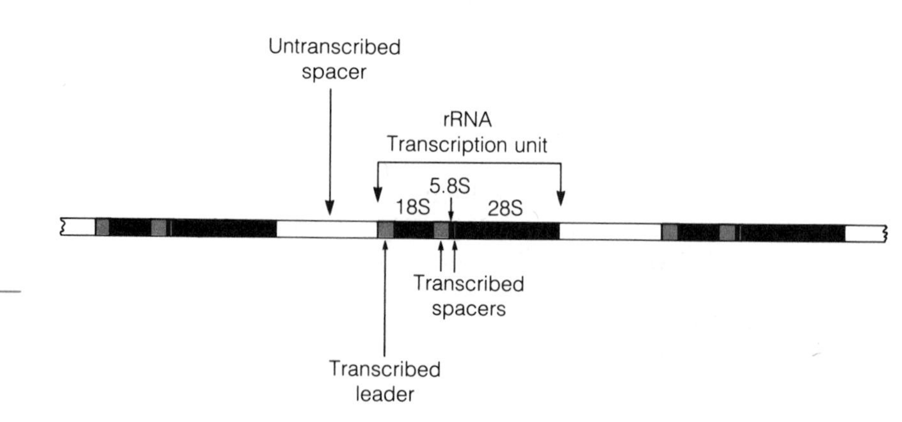

FIGURE 8-38
Transcription units for rRNA arranged in tandem in a chromosome of a frog. Transcription units are separated by untranscribed spacers of unknown function.

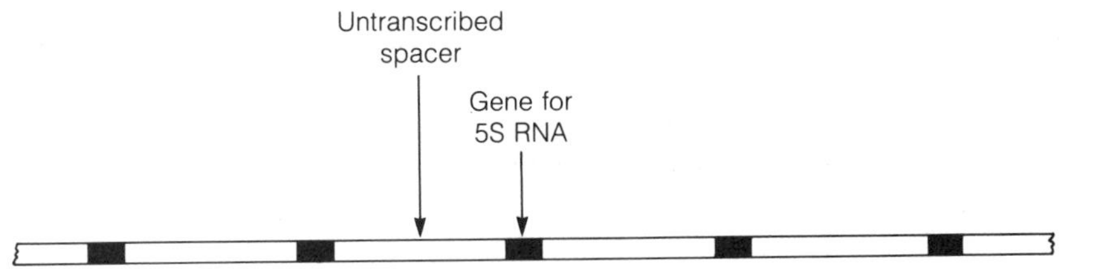

FIGURE 8-39
Tandem arrangement of genes for 5S rRNA in a frog. The genes are separated by untranscribed spacers that are several times larger than the genes.

(Figure 8-38). **Untranscribed spacer** DNA occurs between genes in most eukaryotes (Chapter 11).

The fourth kind of rRNA molecule, the 5S RNA (120 nucleotides long) found in the large ribosomal subunit, is coded for by thousands of copies of a gene located elsewhere on one or more chromosomes. In human cells the 24,000 genes for 5S RNA are all grouped in a tandem array near one end of chromosome number 1. The individual transcription units are separated by long untranscribed spacers (Figure 8-39).

Self-Assembly of Ribosomes from Component Macromolecules

The assembly of rRNA molecules and proteins into 30S and 50S subunits of bacterial ribosomes has been achieved in cell-free preparations starting with purified rRNA and protein components. Assembly of the 30S subunit begins with the noncovalent binding of several ribosomal proteins to the 16S rRNA molecule. Other proteins then bind to the initial complex in an orderly sequence until all 21 proteins have assembled in a specific pattern and the subunit is complete. Omission of a single protein from a reassembly mixture interferes with proper assembly and, for most proteins, results in a functionally defective subunit. Similarly, 5S rRNA and 23S rRNA become complexed in an orderly sequence with the 34 different proteins of the 50S subunit. The reconstituted subunits are the same in their physical properties as the subunits obtained from cells, and they are active in cell-free protein synthesis. These experiments provide a particularly clear example of the self-assembly of a cell structure or organelle requiring no guiding mechanism beyond that contained in the shape and properties of the component macromolecules themselves.

GENE EXPRESSION AND CELL PROPERTIES

Genes are by themselves like inert tapes of information, which do not directly participate in cell metabolism. An earlier part of this chapter dealt with the means by which the sequence of deoxynucleotides in a gene prescribes the construction of a product that is the active representative of the gene in cell metabolism. For almost all genes, this means the synthesis of mRNA molecules whose role is to be translated into the final gene products, polypeptide chains. A small proportion of the genes in a cell encode rRNA, tRNA, snRNA, and a few other RNA molecules, which are not translated but directly carry out various functions. All of these gene products—polypeptides, rRNA, tRNA and snRNA—are directly responsible for the structure and activity of a cell.

Some protein and RNA molecules aggregate and form structures with specific functions, for example, ribosomes. Other proteins become part of the plasma membrane, where they function in active transport. Still others become part of one or another organelle, contributing to organellar function and structure. Many proteins, such as those that make up the glycolytic pathway, do not form stable structures but remain in a soluble state in the cytosol. The role of many proteins is to catalyze synthesis of additional kinds of molecules used for other functional and structural parts of a cell. Overall, the properties of a cell are a direct reflection both of the amounts and of the kinds of gene products that the cell contains.

Cells of one species differ from those of another because many of their genes are different, specifying different proteins. In general, the longer ago in evolution that two species diverged from each other, that is, the greater the evolutionary distance between them, the fewer genes, and therefore the fewer proteins, they possess in common. Almost none of the proteins of *E. coli* are the same as those found in an amoeba, and almost none of those in an amoeba are the same proteins in a human cell. However, most of the genes in a chimpanzee are very similar to or identical to genes in a human.

Gene differences among species fall into two broad categories. First, some gene differences reflect distinct differences in cell functions. For example, human cells possess genes encoding hemoglobin, but cells of amoebae do not. Yeast cells have genes that encode proteins needed to make spores, but human cells and some amoebae do not make spores and lack the genes to do so. These are the gene differences that clearly make one species different from another. Second, genes may differ even though they encode essentially identical functions. For example, all prokaryotic and eukaryotic cell species have the enzymes for glycolysis. However, during evolution amino acid changes have occurred in these proteins that are compatible with maintenance of the original enzyme activities and allosteric properties of these proteins. Amino acid changes are permissible if they do not distort the active site or sites of a protein. This means, in effect, maintenance of at least some part of the same tertiary configuration of the protein, which can be done with different primary sequences of amino acids. Thus, the glucose kinases in *E. coli* and a human cell are almost totally unrelated in amino acid sequence, but both catalyze addition of phosphate to glucose because they have similar tertiary forms and active sites. The greater the evolutionary distance, the more likely such gene differences will be present, as for example between *E. coli* cells and human cells. The more closely related two species are in evolution, the smaller the differences in genes encoding a common function. The genes encoding glycolytic enzymes in humans and other mammals are almost identical in nucleotide sequence, but differ considerably from those in yeast cells or amoebae. On the other hand, the amino acid sequences of some other proteins have changed little in evolution. Genes coding for the large polypeptide subunits of RNA polymerase are remarkably alike for the enzyme in *E. coli,* yeast, and human cells. This likeness suggests that the function of RNA polymerase is stringently dependent on one particular sequence of amino acids and cannot be achieved with any other set. Similarly, genes encoding tubulin proteins that make up microtubules are similar in ciliated protozoa, algae, yeast, fruit flies, chickens, and humans.

Still, it is the unshared genes that give rise to the most distinct differences among species. *E. coli* does not have a gene for tubulin, but human cells and amoeba do. *E. coli* has genes coding for enzymes that catalyze synthesis of its cell wall; an amoeba or human cell lack those genes. Plant cells have genes coding for enzymes for photosynthesis; nonphotosynthetic cells lack those genes.

TARGETING OF PROTEINS TO COMPARTMENTS AND STRUCTURES IN THE CELL

The last step in gene expression, the one by which a gene finally exerts its influence on a cell, is the integration of the gene product into the structure and metabolic activities of the cell. For rRNA molecules and ribosomal proteins this integration means assembly into a ribosome and participation of ribosomes in protein synthesis. For tRNA molecules it means transport from the nucleus to the cytosol and participation in translation. Each protein molecule is delivered from the polysome on which it is synthesized to some part of the cell where it carries out its function. Some principles that direct protein traffic in the cell have been worked out. Most protein molecules contain information in their structure that functions to guide them to appropriate locations in the cell.

Proteins that Remain in the Cytosol

Many kinds of protein molecules are released from polysomes into the cytosol. Most of these become distributed to cell organelles and structures. Some remain in the cytosol—for example, enzymes of the glycolytic pathway and enzymes engaged in synthesis of amino acids, purine, pyrimidines, and amino acyl-tRNA synthetases. Even these proteins contain information in their tertiary and quaternary structure that promotes loose or transitory associations into multimolecular complexes. For example, enzymes that catalyze synthesis of pyrimidines can be isolated from the cytosol as a functional aggregate. Enzymes of the glycolytic pathway probably also form loose aggregates that readily fall apart when a cell is lysed. Some evidence also suggests that glycolytic enzymes may be loosely bound, perhaps as aggregates, to the cytoskeleton.

Targeting of Proteins to the Cytoskeleton

The α- and β-tubulins are examples of proteins whose surface configuration guides their assembly into structures (Chapter 6). Microtubule-associated proteins (MAPs) bind to microtubules in patterns dictated by specific surface properties of the microtubules and the MAPs. Similarly, actin molecules bind to one another to form microfilaments, which, in turn, bind actin-associated proteins. These are two-way processes in which the protein molecules continuously exchange between the cytosol and microtubules and microfilaments as the cytoskeleton is continuously reshaped in many kinds of cells. Proteins of intermediate filaments are also thought to undergo self-assembly guided by the surface properties of the protein molecules. In contrast to that of microtubules and microfilaments, assembly of the protein molecules into intermediate filaments yields permanent or static structures.

The assembly of cytoskeletal elements illustrates one principle that regulates traffic of some proteins in the cell, namely, that the tertiary folding of a protein produces specific surface characteristics that target the proteins to particular destinations. Different mechanisms regulate the traffic patterns of other proteins.

Targeting of Proteins to Mitochondria

Mitochondria contain several hundred different proteins. Almost all of these are encoded by genes in the nucleus, synthesized on polysomes in the cytosol, and imported into mitochondria. Some, such as enzymes of the tricarboxylic acid cycle, RNA and DNA polymerases, and proteins of mitochondrial ribosomes, are delivered to the matrix compartment of mitochondria. Others are sequestered in the space between the inner and outer membranes, or inserted as integral proteins in the outer membrane. Still others, such as, those that carry out oxidative phosphorylation, are assembled into functional aggregates on and in the inner membrane.

The deoxynucleotide sequences of more than 20 nuclear genes that encode mitochondrial proteins have been determined. The amino acid sequences predicted by the gene sequences are, in most cases, longer than the actual amino acid sequences of the proteins as they are found in the mitochondria. The extra length is accounted for by 20 to 30 amino acids on the amino-terminal end of the protein. These amino acids constitute a **leader sequence** that directs the protein to the mitochondrion and targets the protein for importation into the correct mitochondrial compartment (Figure 8-40). After arrival of a protein at its correct location, the

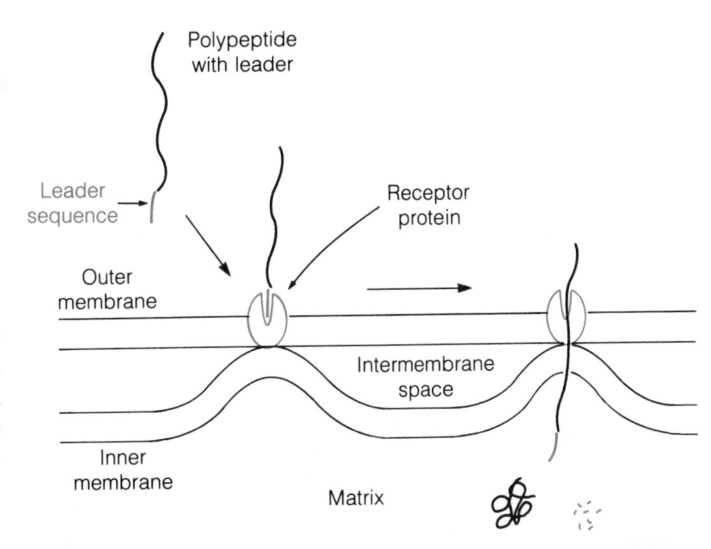

FIGURE 8-40
Import of a polypeptide across the inner and outer mitochondrial membranes into the matrix compartments is mediated by a receptor protein, possibly at specialized sites where the inner and outer membranes adhere to one another. Once inside the matrix compartment the leader is cleaved into amino acids, and the polypeptide folds into its functional tertiary conformation.

leader is clipped off. Some evidence for this comes from an experiment in which a leader was added to a protein not normally imported into mitochondria: addition of a leader caused the nonmitochondrial protein to enter the mitochondrion.

The leader of a mitochondrial protein is believed to recognize a receptor protein on the surface of the mitochondrion. Some experiments suggest that different kinds of receptors may be present, each specific for a different leader or group of similar leaders.

The leader on proteins destined for the matrix compartment, once attached to a receptor, directs transport of the protein across the two mitochondrial membranes into the matrix. The details of how this happens are not known, although an electrochemical proton gradient across the inner membrane is required. Transport is believed to occur at points in the mitochondria where the inner and outer membranes are closely joined to one another (Figure 8-40). In the matrix the leader is cleaved from the newly imported protein molecule by a protease, generating the mature functional form of the protein. Leaders on other proteins apparently direct them to their destinations in the inner mitochondrial membrane or intermembrane space and then are cleaved from their proteins.

A few mitochondrial proteins, including particularly those of the outer mitochondrial membrane, do not have cleavable leaders, but are targeted to their destinations by amino acid sequences that are retained in the amino-terminal end or in other parts of the protein.

Targeting of Proteins to Chloroplasts

The majority of the hundreds of proteins present in chloroplasts is encoded by nuclear genes and transported into chloroplasts after completion of synthesis in the cytosol. The principles of targeting and import are similar to those by which mitochondria acquire proteins. Chloroplast proteins are also synthesized in a precursor form with extra amino acids on the amino-terminal end. For example, the small subunit of the CO_2-fixing enzyme, ribulose-l,5-bisphosphate carboxylase (Chapter 4) is synthesized on cytosolic polysomes as a precursor with 57 extra amino acids on the amino-terminal end of the polypeptide. These amino acids are cleaved from the molecule when it enters the chloroplast, and the molecule becomes a functional enzyme. The segment of the polypeptide containing the extra amino acids, which is comparable to the leader of mitochondrial precursor proteins, is called the **transit peptide.** Protein molecules from which the transit peptide has been deleted do not enter chloroplasts but remain in the cytosol. Addition of a transit peptide to a protein that normally remains in the cytosol causes it to enter a chloroplast.

Like mitochondria, chloroplasts appear to have receptor proteins to which precursor proteins bind and which then facilitate import of the precursor. Isolated, purified chloroplasts efficiently bind the precursor protein of ribulose-1,5-bisphosphate carboxylase; pretreatment of chloroplasts with a protease abolishes their ability to bind the enzyme, which is evidence that receptors are proteins.

Targeting of Proteins to the Nucleus

The nucleus contains hundreds of different proteins, all of which are synthesized in the cytoplasm. Some of these constantly shuttle back and forth between the cytoplasm and nucleus, probably by way of the pores in the nuclear envelope. Some shuttling proteins bind to mRNA molecules in the nucleus to form mRNA-protein particles, which move to the cytoplasm through nuclear pores. These proteins return to the nucleus after dissociation of mRNA-protein particles in the cytoplasm. Newly synthesized ribosomal proteins migrate from the cytoplasm to the nucleolus, where they are assembled with rRNA molecules to form the large and small ribosomal subunits. These shuttle only once, that is, they migrate to the cytoplasm and remain there. Other shuttling proteins may be involved in regulation of gene expression. For example, in cells that are targets of steroid hormones, certain cytoplasmic proteins are steroid receptors that are thought to migrate into the nucleus when complexed with steroids and activate transcription of particular genes. The shuttling of some proteins between nucleus and cytoplasm may have no significance but may simply represent diffusion between the two compartments. For example, G-actin apparently diffuses back and forth freely, simply because it is a soluble protein small enough to pass readily through pores in the nuclear envelope (Chapter 11). The specific function of most of the shuttling proteins is not known.

Many other proteins, for example, RNA and DNA polymerases, remain in the nucleus during interphase, are released into the cytoplasm during mitosis, and then quickly return to the daughter nuclei when mitosis is completed. Still other proteins, primarily histones, enter the nucleus as soon as they are synthesized, and remain permanently associated with DNA throughout the life of the cell.

The mechanisms that are responsible for the extensive protein traffic between the cytoplasm and the nucleus and for retention of certain proteins in the nucleolus are partially understood. Some smaller proteins may be able to diffuse through nuclear pores and then be held in the nucleus by binding to nuclear chromatin or to the nucleolus. Other proteins contain signal sequences in their structure that cause them to be specifically imported into the nucleus. Clear evidence for this comes from the study of a virus called SV40 (Chapter 13). A virus reproduces by entering a cell and using the metabolic machinery of the cell to synthesize viral molecules, which are then assembled into new virus particles within the cell and then released, often by lysis of the cell. The SV40 virus contains a gene that encodes a protein known as the **large-T antigen.** This viral protein is synthesized in the cytoplasm of a cell infected with SV40 virus. The protein then migrates into the nucleus, where it accumulates. Mutation studies have identified a sequence of amino acids in large-T antigen that acts as a nuclear localization signal. The sequence consists of eight amino acids beginning at amino acid 123 from the amino-terminal end of the protein. Mutations that change an amino acid in the signal sequence result in failure of the large-T antigen to accumulate in the nucleus. Joining of the 8-amino acid signal sequence to a cytoplasmic protein that normally does not enter the nucleus causes that protein to accumulate in the nucleus.

Evidence for a signal sequence that targets a ribosomal protein to the nucleus has been obtained in yeast cells. Joining of the 21 amino acids on the amino-terminal end of the yeast ribosomal protein L3 to a β-galactosidase molecule from *E. coli,* causes the β-galactosidase molecule to become localized in the yeast nucleus. Several percent of the nonhistone proteins in animal cell nuclei consists of **nucleoplasmin,** a protein of unknown function. When nucleoplasmin is injected into the cytoplasm, it quickly accumulates in the nucleus. If part of the carboxyl terminus is first removed with trypsin, nucleoplasmin fails to enter the nucleus.

Signal sequences that target proteins to the nucleus, unlike cleavable signal sequences for proteins targeted to mitochondria, chloroplasts, and the endoplasmic reticulum (see below), are permanent parts of the proteins. Permanent signal sequences would cause nuclear proteins released to the cytoplasm during mitosis to reaccumulate in the reforming nuclei at the end of mitosis, that is, a signal sequence may be used repeatedly to direct a protein into the nucleus.

Targeting of Proteins to the Endoplasmic Reticulum, the Golgi Complex, and Lysosomes

Proteins destined for exocytosis, for insertion into the plasma membrane, for peroxisomes, or for lysosomes are synthesized by ribosomes attached to the rough endoplasmic reticulum (ER). The rough ER transfers these proteins to the Golgi complex, where they are sorted and sent to their destinations, as considered briefly in Chapter 5. The synthesis of proteins was discussed earlier in this chapter. Targeting of proteins via the rough ER and Golgi complex can now be discussed in more detail.

The synthesis of proteins targeted to the rough ER begins on ribosomes in the cytosol not attached to membranes of the rough ER (Figure 8-41). These proteins contain a **signal sequence** of amino acids that is recognized by a **signal recognition particle (SRP).** An SRP consists of six polypeptides and a small RNA molecule. The signal sequence is usually 16 to 26 amino acids long and is located at the amino terminus

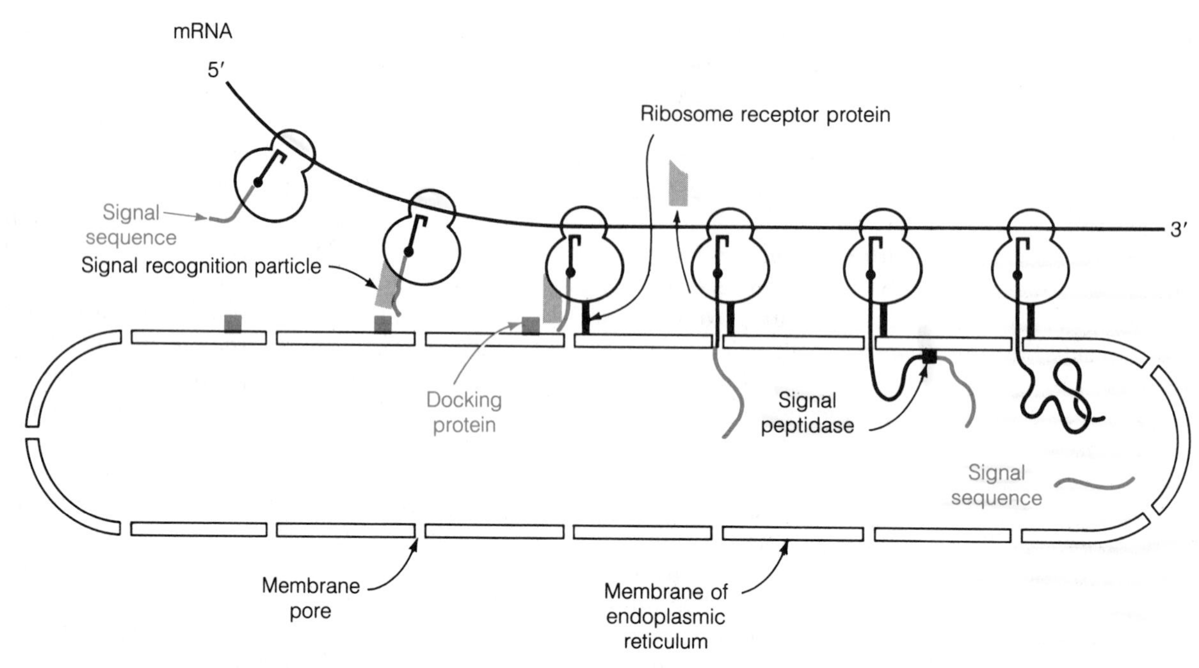

FIGURE 8-41
Translation-translocation coupling of proteins destined for the endoplasmic reticulum. Pores are thought to be present in the membrane of the endoplasmic reticulum that allow polypeptides to pass through the membrane as they are synthesized. When protein synthesis is completed, the protein passes into the lumen of the endoplasmic reticulum. The signal sequence is removed by a signal peptidase.

of some proteins and in an internal region of others. As soon as the signal sequence has been synthesized, before completion of the entire polypeptide, it is recognized by an SRP, which binds to the ribosome with its partially completed polypeptide. Binding of an SRP particle stops further translation. In the case of polypeptides with a signal sequence at the amino terminus, only about 80 amino acids, including the signal sequence, are translated from the mRNA molecule when synthesis stops. When translation is blocked, the release into the cytosol of proteins destined for exocytosis, for inclusion in lysosomes, or for insertion into the plasma membrane is prevented.

An SRP that has bound to a ribosome is recognized by an **SRP receptor,** also called a **docking protein,** which is an integral protein of the rough ER membrane. Binding of the SRP-ribosome complex to the SRP receptor is followed by binding of the ribosome to the membrane by means of a ribosome receptor protein of the ER. Signal sequences contain a region of nonpolar amino acids, which facilitates insertion of the signal sequence into the membrane. Binding of the ribosome and insertion of the signal sequence are accompanied by release of the SRP. Release of the SRP allows translation to resume, and as the polypeptide elongates, it is translocated across the membrane. This is known as **translation-translocation coupling.** As additional ribosomes successively engage the end of the same mRNA molecule, they become bound to the membrane of the rough ER. With completion of translation each ribosome dissociates into its two subunits and is released from the membrane.

The details of how polypeptides are translocated across the membrane are not known. In general, in the case of secreted polypeptides, such as antibodies (Chapter 15) and digestive enzymes (Chapter 5), the signal sequence is cleaved from the polypeptide by an enzyme called the **signal peptidase** and is released into the lumen of the rough ER. Proteins destined for

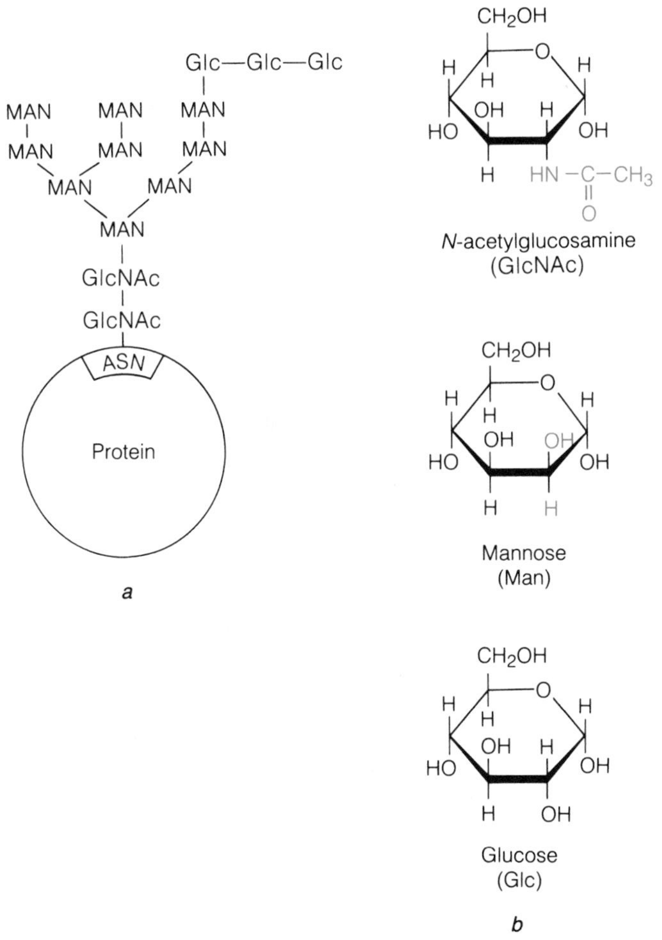

Glc—Glc—Glc

MAN MAN MAN

MAN MAN MAN

MAN MAN

MAN

GlcNAc

GlcNAc

ASN

Protein

a

CH_2OH

N-acetylglucosamine (GlcNAc)

Mannose (Man)

Glucose (Glc)

b

FIGURE 8-42

Glycosylation of proteins in the lumen of the endoplasmic reticulum. (a) Addition of two molecules of *N*-acetylglucosamine (GlcNAc), nine molecules of mannose (Man), and three molecules of glucose (Glc) to the side group of asparagine (ASN). (b) Structure of GlcNAc, Man, and Glc. GlcNAc and Man differ from glucose by the atoms shown in red.

lysosomes are probably treated in a similar fashion. In the case of polypeptides with internal signal sequences, no cleavage occurs and the polypeptide apparently remains attached to the membrane.

The newly completed polypeptides in the lumen of the ER and attached to the ER membrane are glycosylated by enzymes called **glycosyltransferases,** which add fourteen 6-carbon sugar molecules to the side chain of a single asparagine residue (Figure 8-42). Two of the sugars are *N*-acetylglucosamines (GlcNAc), nine are mannose molecules, and three are glucose molecules. The three glucose molecules and one mannose molecule are removed from every glycoprotein while they are still in the ER.

The newly synthesized glycoproteins are carried to the Golgi complex by vesicles that bud from the ER and fuse with the flattened sac on one side of the Golgi complex, which is defined by this activity as the *cis* face (Figure 8-43). The *cis* face is usually oriented toward the nucleus.

The Golgi complex contains at least three kinds of sacs defined by different functions. Within the cisternae of the one or two sacs on the *cis* face (***cis* sacs**), phosphate groups are added to mannose molecules of those glycoproteins that are destined for lysosomes. These proteins are not modified further. Phosphorylation serves as a specific signal to direct proteins to lysosomes. Sacs in the middle part of the Golgi stack are called ***medial* sacs.** In the cisternae of the *medial* sacs, enzymes called **mannosidases** remove more mannose molecules; transferase enzymes then add GlcNAc molecules to glycoproteins destined for insertion in the plasma membrane or for secretion (exocytosis). In the one or two sacs on the opposite side of the Golgi complex—the ***trans* sacs**—galactose molecules are added. It is in the *trans* sacs that sialic acid molecules (sialic acid is a modified 6-carbon sugar) are probably added also. Glycoproteins are apparently passed from *cis* to *medial* to *trans* sacs by vesicles that bud from one sac and fuse to the next.

Within the Golgi sacs proteins are sorted according to destination by poorly understood mechanisms. Sorting is first detected by budding of vesicles carrying different populations of glycoproteins from the *trans* face of the Golgi complex. One kind of vesicle fuses with lysosomes, delivering about 50 different degradative enzymes (hydrolases) to lysosomes. Another type of vesicle fuses with the plasma membrane, delivering glycoproteins still inserted in membrane (those with internal signal sequences). A third type of vesicle carries glycoproteins to secretory storage granules, which subsequently fuse with the plasma membrane to release their contents outside the cell.

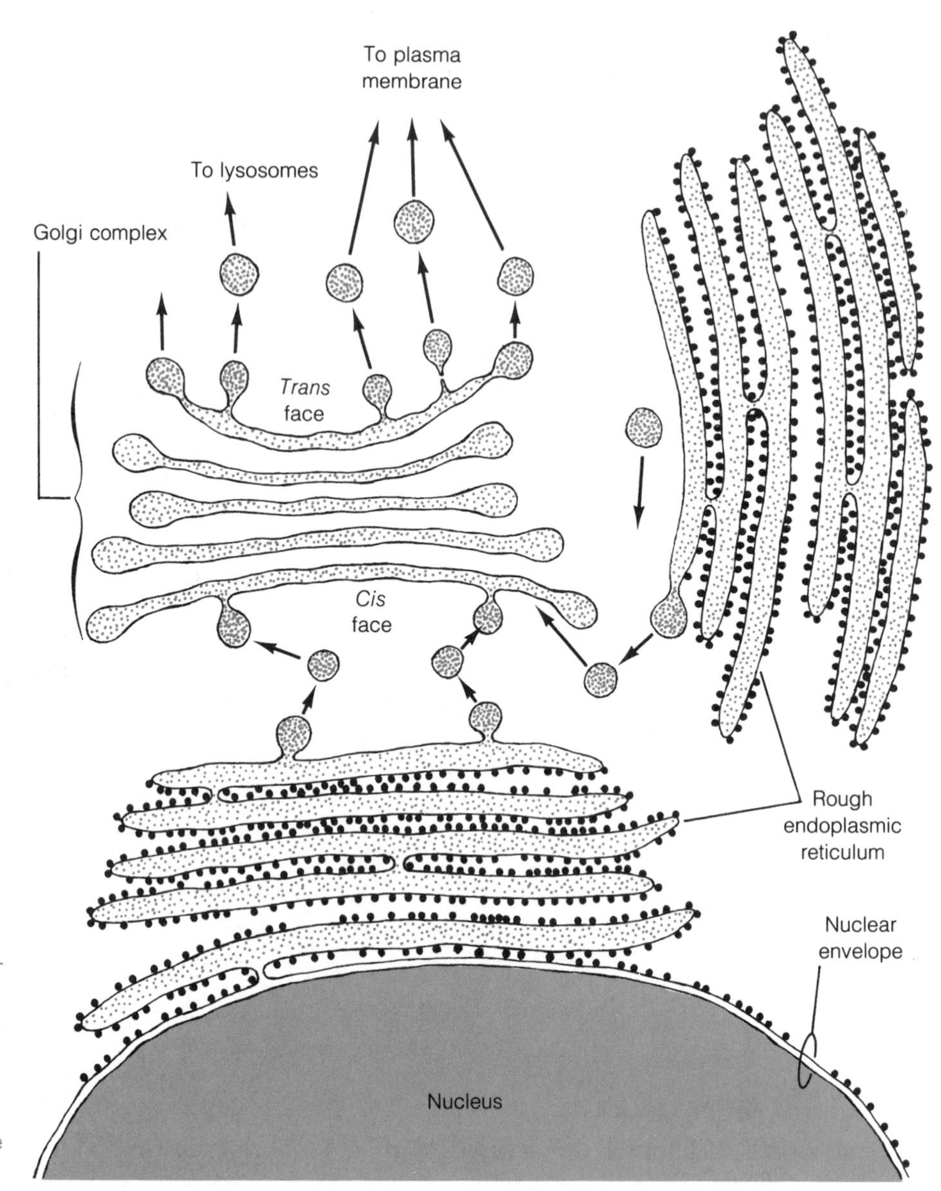

FIGURE 8-43
Glycoprotein-containing vesicles from the rough endoplasmic reticulum fuse with a membrane sac at the *cis* face of the Golgi complex. The glycoproteins are modified in the Golgi complex. Vesicles budded from the *trans* face carry proteins to the plasma membrane and lysosomes.

PROBLEMS

1. What experiment first showed that protein synthesis does not require the direct participation of DNA?
2. How was it proven that RNA is synthesized in the nucleus and moves into the cytoplasm?
3. What is the function of the promoter of a gene in prokaryotes?
4. Two adjacent genes may be encoded by opposite strands of the DNA duplex. When these genes are transcribed, why do the RNA polymerase molecules travel along the DNA in opposite directions?
5. Which two kinds of macromolecules are needed to decode the nucleotide sequence of an mRNA molecule into the amino acid sequence of a polypeptide?
6. How does the ribosome contribute to formation of a polypeptide?
7. Describe the experiment that gave amino acid assignments to 61 of the 64 possible nucleotide triplets in RNA.
8. What is the significance of the wobble phenomenon?
9. Define missense mutation, nonsense mutation, and frameshift mutation.
10. The effect of a mutation can be suppressed by a second mutation. What are the two main types of suppression and how do they work?
11. What are the functions of the A and P sites on a ribosome?
12. How do mRNA molecules in eukaryotes differ from those in prokaryotes?
13. What are two major differences between most primary transcripts and mRNA molecules in a eukaryote?
14. Why is it important that excision of introns be precise to the level of a single nucleotide?
15. Where in a eukaryotic cell do rRNAs and ribosomal proteins assemble into ribosomal subunits?
16. Most mitochondrial proteins are synthesized outside the mitochondrion. How are they selectively guided to their proper mitochondrial destination?
17. Describe how a newly synthesized polypeptide destined for the plasma membrane is guided along its route. How is it modified along the way?

CHAPTER 9

Regulation of Gene Expression in Prokaryotes

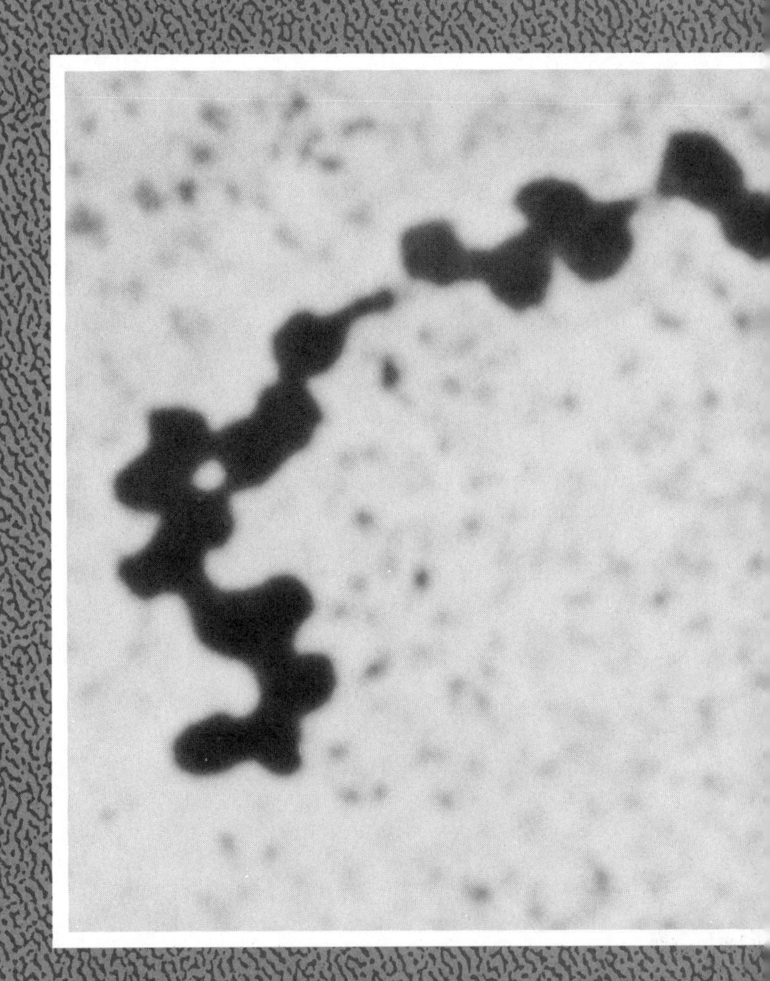

See Figure 8-25.

Most genes are expressed through the protein products that they encode. A few are expressed as rRNA, tRNA, and several kinds of small RNA molecules. Each of the thousands of different gene products is needed in a particular amount in accordance with the structural and functional properties of the cell. Hence, different gene products are synthesized in a cell in different amounts, sometimes vastly different. For example, enzymes that catalyze the steps in glycolysis and in respiration are generally abundant, while other enzymes are present in lesser amounts. Different enzymes also vary greatly in catalytic power (turnover number); weaker enzymes sometimes must be present in larger amounts than stronger ones so that they do not create unwanted barriers in metabolic pathways. However, barriers are important in controlling metabolic pathways, and these are achieved by two general mechanisms: (1) by regulation of the *activity* of a particular enzyme in a pathway, as discussed in Chapter 3, and (2) by regulation of the *absolute amount* of a particular enzyme that is synthesized, which is a main subject of this chapter. Cells are efficient: they produce an amount of an individual gene product, such as an enzyme, sufficient to meet functional needs and normally do not waste resources by overproduction.

The need for a particular gene product may change with alterations in the environment of a cell. *E. coli* growing in medium containing tryptophan possesses extremely low amounts of the enzymes needed to catalyze the synthesis of tryptophan. Synthesis of such gene products would be wasteful when tryptophan is readily available in the medium. When the tryptophan molecules in the medium are used up, *E. coli* immediately synthesizes the enzymes needed to make tryptophan. Thus, the levels of these gene products within a cell vary according to the need to make tryptophan. In prokaryotic and eukaryotic cells the need for various products often changes dramatically. Cells accommodate such changes by regulating the expression of the relevant genes. A growing cell produces new ribosomes at a much higher rate than a nongrowing cell, which needs only to replace ribosomes that become defective. All eukaryotic cells subjected to an abnormally high temperature for a short time respond by curtailing drastically synthesis of most proteins and by turning up sharply the synthesis of a few kinds of proteins, called **heat shock** or **stress proteins.** Yeast cells and some bacterial species respond to starvation by forming spores, a process that requires greatly increased expression of at least 80 genes. Thus, by regulating gene expression cells can (1) produce a balance of gene products to meet structural and functional needs, and (2) change the rate of synthesis of particular gene products to bring about changes in structure and function.

The expressions of most, if not all, genes in prokaryotes and eukaryotes are regulated. The major manifestations of regulation and the molecular mechanisms by which regulation is achieved are different enough in prokaryotes and eukaryotes to require separate discussion. Regulation of gene expression in prokaryotes, which is understood well, is discussed in this chapter. Regulation of eukaryotic genes, currently an extremely active area of research in cell biology, is covered in the next chapter.

MECHANISMS OF REGULATION

Regulation of the expression of a gene is achieved in prokaryotes in two principal ways: (1) by regulating of the number of mRNA molecules transcribed per unit time from a gene; and (2) controlling translation of polypeptides from mRNA molecules. An important part of translational control is that of regulating how many times a given mRNA molecule is translated. Most mRNA molecules in prokaryotes have life spans of only a few minutes and hence are translated a limited number of times before destruction. Without continued synthesis of its mRNA, the synthesis of a particular protein is quickly cut off.

In a sense, regulation of gene expression can also be achieved by regulating the activity of the final gene product, for example, covalent modification of protein products or feedback effects in enzyme pathways, as discussed in Chapter 3. Regulation of gene expression as discussed in this and the next chapter is defined as regulation of the abundance of gene products and does not include regulation of the activity of products, such as enzymes.

Regulation of gene expression in bacteria is achieved primarily through regulation of the rate of initiation of transcription. The rate of initiation is regulated in two senses, extrinsic and intrinsic. **Intrinsic regulation** refers to different rates of transcription of one gene compared with another as a result of inherent properties of the genes themselves. In intrinsic regulation, transcription is governed primarily by how strongly RNA polymerase binds to the promoter of a gene, which, in turn, governs how often transcription of the gene occurs. The strength of binding of RNA polymerase varies according to the deoxynucleotide sequence of the promoter region. Substitution of a single base pair in the promoter region can markedly increase or decrease polymerase binding. Genes with **weak promoters** are normally transcribed at intrinsically low rates. These are usually genes whose products are needed in the cell in low amounts, and a weak promoter therefore provides economy in expression of that gene. Genes with **strong promoters** have higher intrinsic rates of transcription. Intrinsic regulation is the result of long-term evolution in the strength of various promoters. It has no role in changing the rates of synthesis of particular proteins in response to short term changes in cellular metabolism.

In **extrinsic regulation** the rate at which a particular gene is transcribed may be turned up or down, even to the point that transcription may be almost totally stopped, by factors that are outside, that is extrinsic, to the gene itself. Turning transcription of individual genes up or down is commonly imposed on genes with strong promoters, although such active regulation may also come into play in turning down transcription of some genes that normally have weak promoters.

Expression of many genes is turned up or down as the needs of the cell change, for example, as a result of changes in the environment. The many genes needed to code for the enzymes that make up the pathways for synthesis of the 20 amino acids are present in the chromosome of *E. coli.* When one or another amino acid is available in the medium, enzymes that catalyze synthesis of the amino acid are not needed, and the expression of the genes for the enzymes is markedly turned down. Likewise, the use of different sugars, as sources of energy, requires different enzymes to channel them into the glycolytic pathway. The rates of synthesis of these enzymes may be regulated according to availability of the different sugars that the enzymes act on. A well understood example is the enzyme β-galactosidase in *E. coli.* β-galactosidase splits the disaccharide sugar lactose into two monosaccharides, galactose and glucose. When *E. coli* is grown in a medium containing glycerol as a carbon and energy source, the cells do not synthesize β-galactosidase. If the cells are switched to a medium containing lactose as the sole carbon and energy source, within one to two minutes they begin to synthesize the lactose-cleaving enzyme β-galactosidase. If switched back to lactose-free glycerol medium, the synthesis of β-galactosidase quickly ceases, and gradually the concentration of β-galactosidase per cell declines. The early work on the gene for β-galactosidase, which has subsequently been expanded and elaborated with many other genes, established a conceptual framework for understanding active control of gene expression in prokaryotes, as described in the next section.

Rates of transcription of at least a few genes appear not to be actively regulated. Examples are the genes that code for the enzymes of the glycolytic pathway in *E. coli.* These genes are transcribed and translated into the various enzymes (each at its own characteristic rate) even when glucose or some other sugar is not available to fuel glycolysis. *E. coli* will grow using energy sources such as acetate or succinate, which are fed directly into the citric acid cycle (Chapter 4). Even though this eliminates glycolysis, the glycolytic enzymes continue to be synthesized at rates determined only by promoter strength, no matter how long the cells are cultured in a medium containing a carbon source that cannot enter the glycolytic pathway. This is in contrast with some eukaryotes, for example, yeast cells, in which the concentration of glycolytic enzymes declines markedly when no appropriate sugars are present, implying regulation of expression of these genes in yeast.

THE VALUE OF GENE REGULATION

Transcription and translation consume 50 percent of the ATP produced by *E. coli*. Therefore, regulation of these processes is a significant means for a cell to maximize the efficiency of utilization of resources. For example, a cell whose gene for β-galactosidase is regulated would presumably have a competitive edge over a cell whose gene for β-galactosidase was unregulated.

The advantage provided by regulation of gene expression can be tested directly. When tryptophan is present in the medium, transcription of the five genes required for tryptophan synthesis is turned off in *E. coli*. The nucleotides, amino acids, and ATP that would be used to transcribe and translate the tryptophan genes are therefore available for other activities in the cell. The importance of this saving is demonstrated by inoculating medium containing tryptophan with two strains of *E. coli* differing in only one respect. One strain has the ability to regulate expression of the tryptophan genes; in the other, a mutation has rendered the cell incapable of turning off expression of the genes so that it continues to produce the enzymes for tryptophan biosynthesis even though tryptophan is present in the medium. When a culture is started with equal numbers of cells of each type, a few days later most of the cells in the culture will be the regulated type; they have outgrown the unregulated type. Clearly, the saving of resources by repression of the tryptophan genes allows the regulated strain to proliferate faster. Consider how this could happen. Suppose a cell saves one percent of its resources by repressing genes and as a result can reproduce one percent faster than a mutant in which expression of the same genes is unregulated. When the mutant cell has gone through 200 generations (less than three days in a rich medium), the normal cell will have gone through 202 generations. The two additional generations provide an additional quadrupling in cell number, and the normal cell type will outnumber the mutant by 4 : 1. In less than nine days (600 generations) the ratio will be 64 : 1. Thus we see that a mutant cell is at a competitive disadvantage and will eventually disappear in a normal, competitive situation.

THE OPERON CONCEPT

The regulation of almost all gene expression in prokaryotes occurs at the level of gene transcription. For many genes **repressor molecules** play the key role. Most repressors are members of a particular class of protein molecules that are capable of binding to genes (DNA) in a way that prevents transcription of the gene. A given repressor usually binds strongly to a particular site adjacent to its target gene. Specificity of binding is based on structural complementarity between a protein molecule and a particular sequence of deoxynucleotides. This high degree of specificity of binding between a gene and its repressor is the principal means (but not the only one) by which the regulation of one gene is achieved independently of another. Genes whose rates of transcription are governed by repressors are divided into two classes, **inducible** and **repressible,** depending on whether they are involved in catabolic pathways (inducible) or anabolic pathways (repressible). This distinction is discussed in detail later in the chapter. Rates of transcription of at least a few other genes are **autoregulated.** That is, the product of the gene may itself act as a repressor and therefore become the key element in regulating the rate of transcription or translation of that gene. An example of autoregulation is the transcription of genes encoding rRNA, as discussed later in the chapter.

A repressor blocks transcription of its target gene by binding to a specific short sequence of nucleotides, called the **operator,** positioned for many genes between the promoter and the coding region of the gene (Figure 9-1). Binding of a repressor molecule to the operator sequence prevents RNA polymerase from binding to the promoter and proceeding to the coding part of the gene, and thereby prevents transcription. Repressor proteins are encoded by genes that are separate from the genes being regulated (Figure 9-1). Genes coding for repressors are **regulatory genes.** Such genes are continuously transcribed and translated and are said to be **constitutive.** Only a few copies of repressor proteins are needed per cell since their function is to block transcription by binding to one or a few sites on the prokaryotic chromosomes. Because repressor

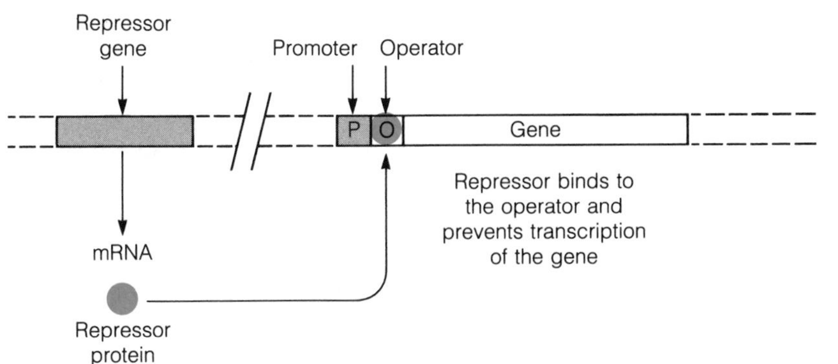

FIGURE 9-1
The components of a bacterial operon. The repressor gene need not be located near the operon that it helps regulate.

molecules are constitutively synthesized and always present in a bacterium, a mechanism is needed to modulate the interaction of repressor molecules with the operators of their target genes. Otherwise, the expression of the target genes would remain permanently inhibited by the repressor molecules. Such modulation is achieved by small molecules called **inducers** and **corepressors** that affect the binding of repressor molecules to their operator targets, as described in the next section.

The elements that have just been described, namely a gene with its promoter and operator sequences, together constitute a unit of genetic operation or an **operon.** Regulation of transcription of the gene or genes in an operon is now considered in detail, using examples that introduce different features of operon control.

Regulation of Catabolic and Anabolic Pathways

The regulation of catabolic pathways (e.g., catabolism of sugars leading to glycolysis) is different from regulation of anabolic pathways (e.g., synthesis of amino acids) in one crucial respect: the enzymes required to catabolize some substances are generally produced only when the particular substance is *available* to the cell. The enzymes of anabolic pathways are produced only when the substance synthesized by the pathway is *unavailable* to the cell. This difference makes for efficient use of resources.

The *lac* Operon in *E. coli*

It is economical for a cell to be able to restrict synthesis of catabolic enzymes to times when the substrates to be degraded are available. The mechanisms by which this is achieved are illustrated by the ***lac* operon.**

The *lac* operon contains three genes (Figure 9-2) encoding enzymes needed to use lactose, which must be split into glucose and galactose before it can fuel

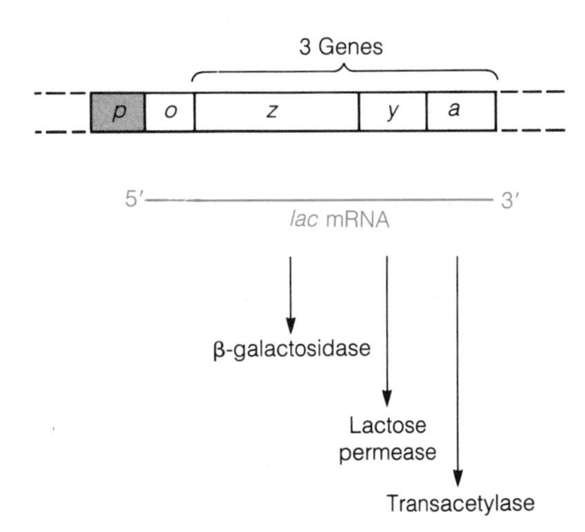

FIGURE 9-2
The three genes in the *lac* operon are transcribed into a single mRNA molecule that is, in turn, translated into three proteins.

FIGURE 9-3
Hydrolytic cleavage of lactose to galactose and glucose by β-galactosidase.

glycolysis (Figure 9-3). Lactose is found in milk and nowhere else. The genes of the *lac* operon were the first to be studied intensively with respect to regulation of gene expression. One of the three genes, designated the *z* gene, codes for β-galactosidase mentioned above. β-galactosidase splits lactose into glucose and galactose. The second gene, the *y* gene, encodes a protein in the plasma membrane that functions specifically in the transport of lactose into the cell. Without this protein, called **lactose permease,** lactose cannot enter the cell rapidly enough to be useful. A third gene, the *a* gene, is controlled coordinately with the *z* and *y* genes, but its role in lactose metabolism is a bit complicated and need not concern us here. The three genes *z*, *y*, and *a*, are adjacent to one another in the chromosome of *E. coli* and are under the control of a single promoter and a single operator (Figure 9-2). RNA polymerase binds to the promoter and transcribes a single, continuous mRNA molecule for all three genes. Because the mRNA molecule contains the transcripts of more than one gene, it is called a **polycistronic (polygenic) mRNA.** A **cistron** corresponds to a segment of DNA coding for one polypeptide chain plus the start and stop signals required for proper translation. A **monocistronic**

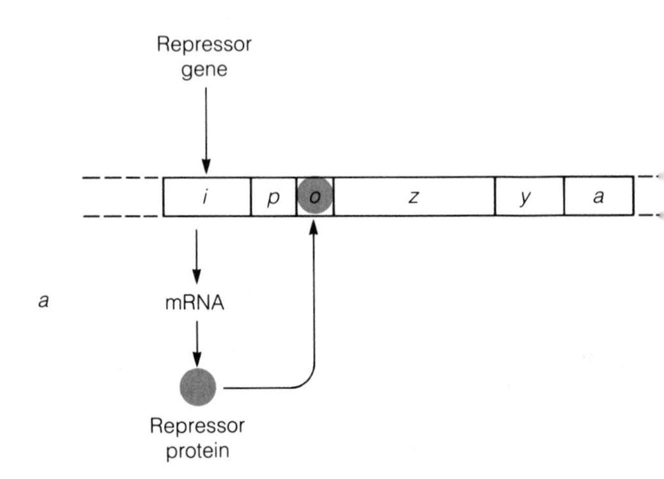

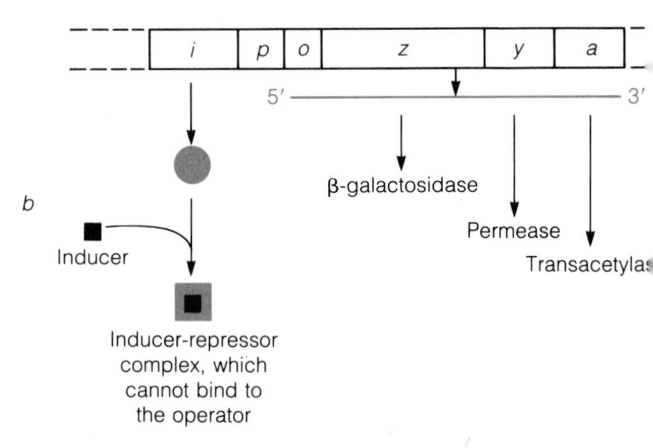

FIGURE 9-4
The *lac* operon. (a) The repressor protein encoded by the repressor gene binds to the operator and prevents transcription when lactose is not present. (b) Lactose (allo-lactose, see text) acts as an inducer by binding to repressor protein molecules. With the inducer bound to it, repressor protein cannot bind to the operator, which allows transcription to occur.

mRNA contains a sequence encoding only one polypeptide.

Another important gene in the *lac* operon is the *i* gene, (*i* for inhibitor), which is located some distance ahead of the promoter for the *z, y,* and *a* genes. This gene encodes the *lac* repressor. When *E. coli* is grown in the absence of lactose, this repressor binds to the operator and prevents transcription of the *z, y,* and *a* genes (Figure 9-4(a)). The *i* gene is constantly transcribed but at a very low rate. About 10 to 20 *lac* repressor molecules are always present in the cell. Binding of the repressor protein to the operator is reversible, although in the absence of lactose the *lac* operator has a repressor bound to it most of the time. Occasionally, the operator may be unoccupied by a repressor molecule long enough (a fraction of a second) for RNA polymerase to pass through and initiate transcription of an mRNA molecule encoding all three genes. Thus, a few molecules of β-galactosidase and lactose permease are present even in a cell not growing in the presence of lactose. These few molecules are insufficient to metabolize lactose fast enough to meet cell needs, but turn out to be important for reasons discussed shortly.

When *E. coli* growing in the absence of lactose, for example, with glycerol as an energy and carbon source, are transferred into medium containing only lactose, repressor molecules are prevented from binding to the operator of the *lac* operon. Transcription and translation occur within one to two minutes, providing the cell with vastly increased amounts of permease to transport lactose into the cell and β-galactosidase to split the lactose that has been taken in. Prevention of binding of the repressor to the operator is brought about by lactose as follows (Figure 9-4(b)): the *lac* repressor possesses a site that is specific for the binding of **allo-lactose,** an alternate form of lactose in which glucose and galactose are linked in a slightly different way. Allo-lactose is produced from lactose by the catalytic action of β-galactosidase. Thus, the active site of β-galactosidase has two effects on lactose: interconversion of lactose and allo-lactose (lactose ⇌ allo-lactose) and the splitting of lactose to glucose and galactose (Figure 9-5). The binding of allo-lactose to the repressor prevents the repressor from binding to the operator.

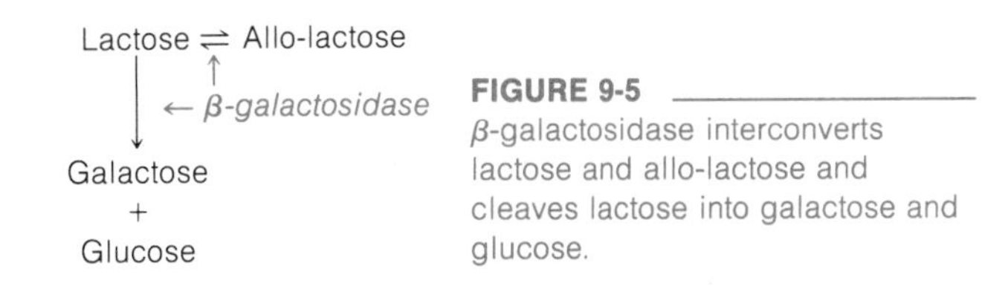

FIGURE 9-5
β-galactosidase interconverts lactose and allo-lactose and cleaves lactose into galactose and glucose.

Hence, the immediate inducer of the *lac* operon is allo-lactose.

Here the few molecules of lactose permease and β-galactosidase already present in a cell at the time the cells are switched from lactose-free to lactose-containing medium are important. The few permease molecules are sufficient to transport a small amount of lactose into the cell. The small amount of β-galactosidase present converts some of the lactose to allo-lactose. The allo-lactose rapidly binds to the 10 to 20 *lac* repressor molecules, including the repressor molecule that at the moment is bound to the operator of the *lac* operon. The repressor molecule is an allosteric protein whose three-dimensional shape is slightly altered by the binding of allo-lactose. This shift in configuration changes the shape of the site by which the repressor binds to the operator. As a result, the repressor molecule already bound to the operator loses its strong affinity for the operator and dissociates from it. Removal of the repressor then allows the three genes of the operon to be transcribed, which leads to synthesis of lactose permease and β-galactosidase and catabolism of lactose. All this happens very quickly; within the first one to two minutes after exposure to lactose, thousands of β-galactosidase molecules are synthesized in each cell. As long as the repressor molecules remain complexed with allo-lactose molecules, the genes of the *lac* operon continue to be transcribed and translated. If, for whatever reason, the lactose concentration in the cell falls, allo-lactose, which is converted to lactose in accordance with the chemical equilibrium between the two molecules, also falls. The decrease in allo-lactose concentration results in free repressor molecules that then bind to the operator and turn off the *lac* operon.

Because it induces the synthesis of permease and β-galactosidase, allo-lactose is called an **inducer,** and

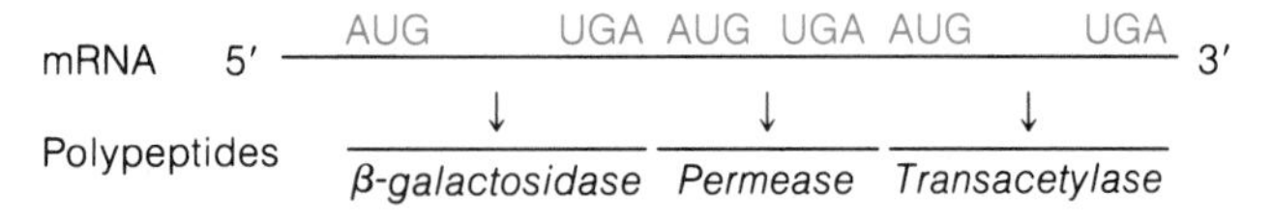

FIGURE 9-6
The mRNA encoded by the *lac* operon contains start (AUG) and stop (UGA) codons that signal the translation of three separate polypeptides.

the enzymes are said to be **inducible enzymes.** The transcription of a variety of genes encoding catabolic enzymes is controlled in prokaryotic cells by inducer-repressor mechanisms very similar to that described for the *lac* operon.

Translational Regulation of Expression of the *lac* Operon

It makes sense that the genes coding for enzymes needed for lactose metabolism should be coordinately controlled at the transcriptional level. An important refinement of this coordination is imposed at the translational level. The three gene transcripts contained in single polycistronic mRNA molecules transcribed from the *lac* operon are each translated individually into the three protein products (Figure 9-2). In each case translation begins with an AUG codon and is terminated by a stop codon (Figure 9-6). For reasons not well understood the probability of initiating translation decreases with each successive initiation codon such that the segment encoding β-galactosidase (*z* gene) is translated more often than that for lactose permease (*y* gene), which is translated more often than the transcript for galactoside transacetylase (*a* gene). In addition, the polycistronic mRNA begins to break down after a few minutes through digestion by a ribonuclease. Digestion begins preferentially at the 3′OH end of the mRNA, that is, the end of the third gene (*a* gene), allowing translation of the upstream segments to continue a bit longer, producing more products of the *a* and *y* genes. As a result of these events, the ratios of the three enzymes, β-galactosidase, permease, and transacetylase, in the fully induced cell are 10 : 5 : 2. These ratios are believed to reflect the level of need of the three enzyme activities in a cell growing with lactose as its energy and carbon source.

CONVERSION OF AN INDUCIBLE GENE TO A CONSTITUTIVE GENE BY MUTATION

In the absence of inducer, inducible genes remain in the repressed state (Figure 9-1). The repressed state requires proper repressor proteins and a normal operator. Mutations can occur in a repressor gene such that the protein loses affinity for the operator, whether or not the repressor has an inducer bound to it (Figure 9-7). Likewise, a mutation in the operator sequence may decrease the affinity of the repressor for the operator. As a result of either kind of mutation, the genes controlled by the repressor are converted from inducible genes to constitutive ones; that is, in the absence of a proper binding of repressor to the operator the genes remain in a state of permanent, continuous transcription.

Inducibility of genes such as those in the *lac* operon increases the efficiency of operation of the cell. Although inducible genes require that a cell carry *extra* genes (coding for repressor molecules), the cost of replicating, transcribing, and translating such genes is far outweighed by the conservation in resources that is achieved by turning down the expression of other genes when they are not needed.

CYCLIC AMP AND CATABOLITE REPRESSION OF THE *lac* OPERON

Glucose is a more efficient source of energy than is lactose. The use of lactose requires that the cell make several additional enzymes to cleave lactose to galactose and glucose and to metabolize galactose, which are unnecessary expenses if glucose is directly available for glycolysis. Indeed, when glucose and lactose are both present, glucose is used in preference to lactose in the

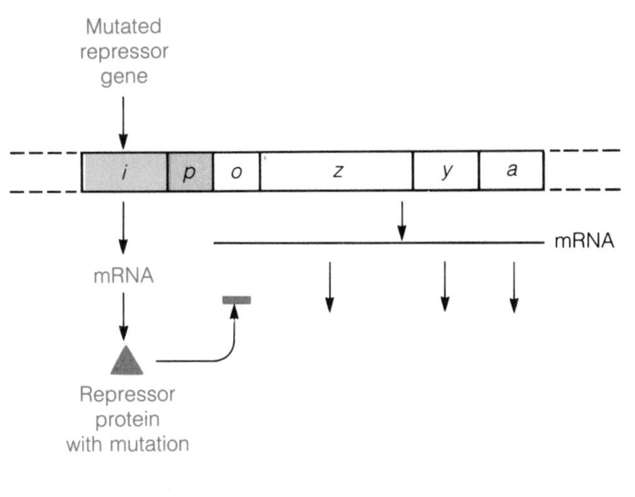

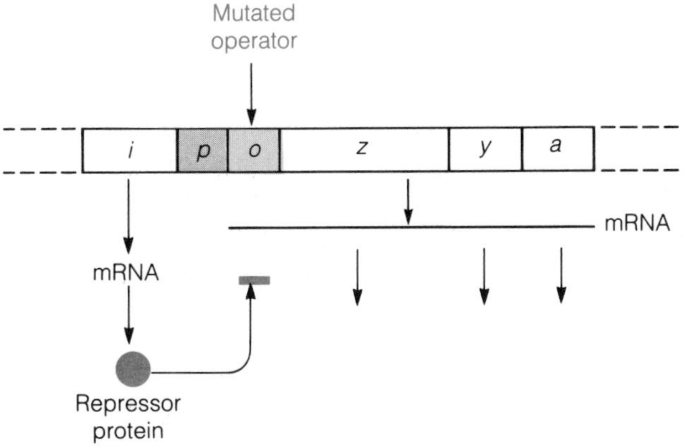

FIGURE 9-7
Mutational conversion of the *lac* operon from an inducible to a constitutive state. (a) A mutation in the repressor gene may result in inability of the repressor protein to bind to the operator. (b) A mutation in the operator may reduce the binding of the repressor protein by the operator.

following way. Transcription of the genes in the *lac* operon is not only modulated by a repressor protein that binds to the operator, but is also controlled by a second protein called **cyclic AMP receptor protein (CRP)**, which binds in the promoter region of the *lac* operon. In contrast to the repressor protein, which binds to the operator and inhibits transcription, CRP greatly stimulates transcription of the operon by binding to its promoter (Figure 9-8) and by doing so, enhancing the binding of RNA polymerase. Just as the binding of the repressor to the operator is governed by lactose, the binding of CRP in the promoter region is governed by glucose. When glucose is plentiful, CRP does not bind and expression of the *lac* operon remains low *even if lactose is present* (Figure 9-9(a)). When glucose is absent in the cell, CRP binds in the promoter region and greatly enhances transcription, but only if lactose is also present to prevent the repressor from binding to the operator (Figure 9-9(b)). CRP cannot enhance transcription if repressor is still blocking transcription at the operator (Figure 9-9(c)).

Glucose does not directly modulate binding of CRP in the promoter region of the *lac* operon but instead works through **cyclic AMP (cAMP).** A cAMP molecule is an AMP in which the phosphate group is covalently attached to both the 5′ and 3′ carbons of the ribose instead of just the 5′ carbon, creating a ring (Figure 9-10); it is formed from ATP by the catalytic action of the enzyme **adenylate cyclase.** The cyclizing between the 3′ and 5′ carbons through a phosphate allows cAMP to bind to particular proteins where AMP

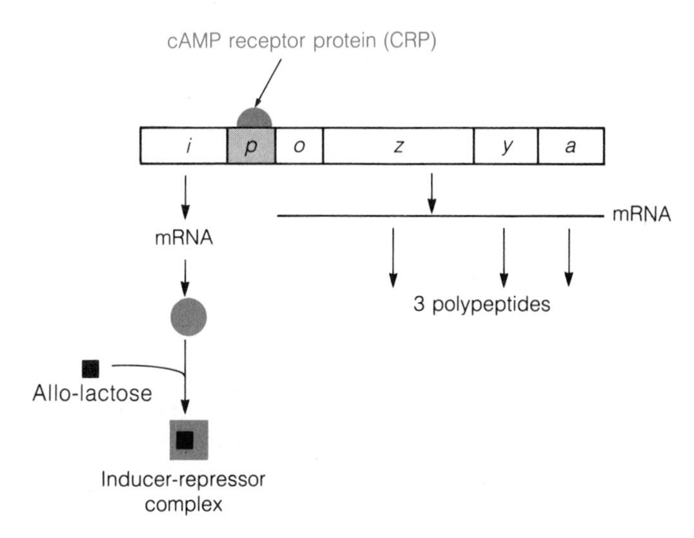

FIGURE 9-8
Binding of cyclic AMP receptor protein (CRP) to the promoter of the *lac* operon enhances the binding of RNA polymerase to the promoter, which results in expression of the three genes.

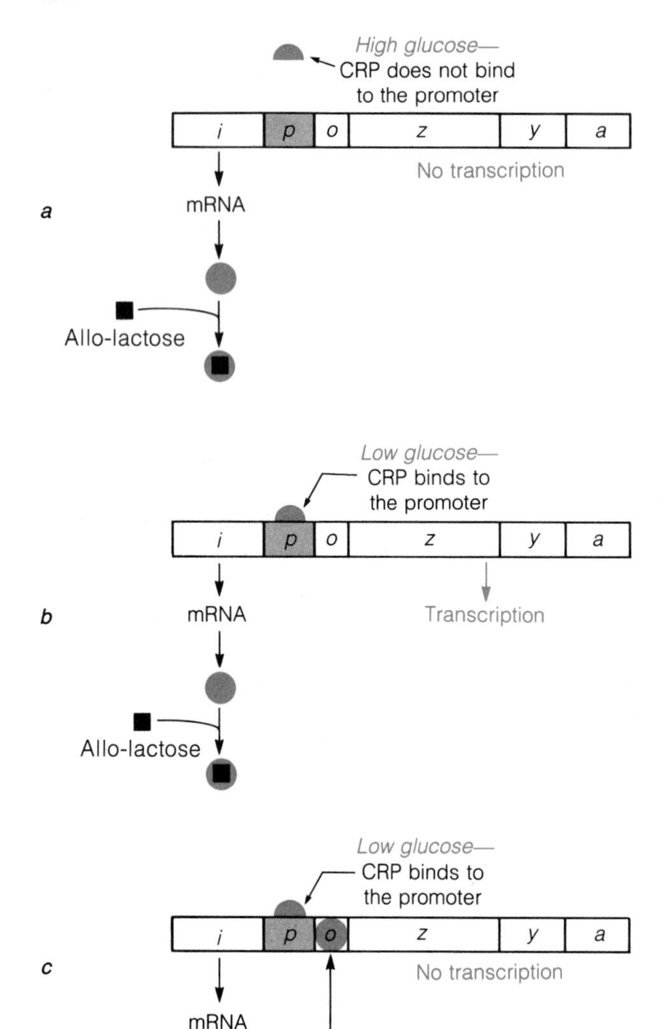

FIGURE 9-9
Effect of glucose concentration on the binding of CRP to the promoter. (a) At a high concentration of glucose CRP does not bind to the promoter, and transcription of the *lac* operon does not occur. (b) At a low concentration of glucose CRP binds to the promoter, and if lactose is present, transcription occurs. (c) At a low concentration of glucose CRP binds to the promoter, but in the absence of lactose, transcription is blocked by binding of the repressor protein to the operator.

cannot bind, for example, to CRP. In order for CRP to bind to the promoter of the *lac* operon it must have cAMP bound to it (Figure 9-11). When glucose is plentiful in the cell, the concentration of cAMP remains low, and CRP does not bind to the promoter. The mechanism by which glucose affects the concentration of cAMP is not fully understood, but when glucose is low, adenylate cyclase is activated to produce more cAMP. Then cAMP binds to CRP, and the cAMP-CRP complex binds to the promoter of the *lac* operon, stimulating binding of RNA polymerase. This results in more transcription, provided lactose is present to prevent binding of the repressor to the operator. Cyclic AMP is involved in modulating other operons that are responsible for carbohydrate metabolism in bacteria. In eukaryotes cAMP also plays an important role in regulating transcription of certain genes.

Many inducible genes that code for enzymes in degradative pathways are controlled in bacteria by mechanisms that are illustrated in principle by the above discussion of the *lac* operon. Expression of still other genes, particularly genes coding for enzymes involved in anabolic pathways, is regulated by a different mechanism, as will now be discussed.

REGULATION OF GENES INVOLVED IN ANABOLIC PATHWAYS

In prokaryotes the expression of many genes coding for anabolic enzymes is turned up and down ac-

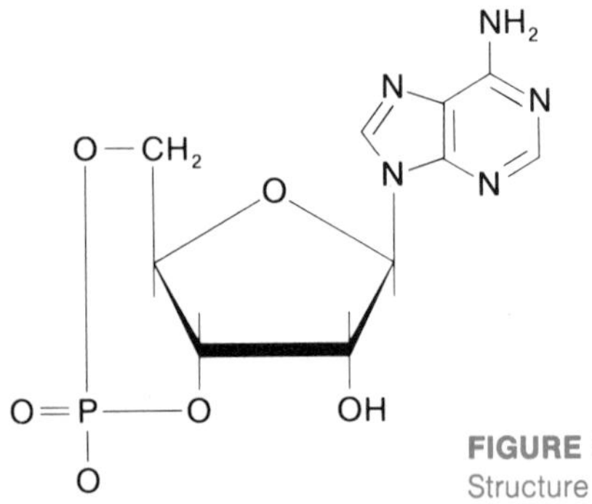

FIGURE 9-10
Structure of cyclic AMP.

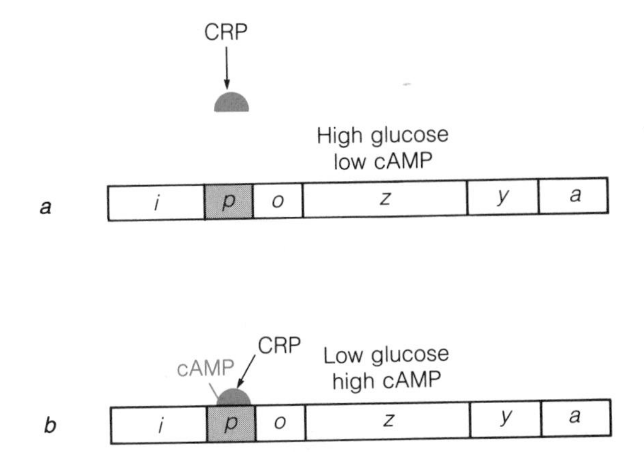

FIGURE 9-11
(a) At a high glucose concentration the intracellular concentration of cAMP is low. cAMP is not bound to CRP, and CRP, in turn, cannot bind to the *lac* promoter. (b) At a low glucose concentration the concentration of cAMP in the cell is increased. cAMP becomes bound to CRP, which enables CRP to bind to the *lac* promoter.

cording to need. This can be seen by comparing the enzyme compositions of cells grown in different media. *E. coli* can be grown on a simple minimal medium containing a carbon and energy source such as glucose, NH_4^+ as a source of reduced nitrogen, and several inorganic ions. In this medium the bacterium contains the enzyme pathways to synthesize the 20 amino acids for protein synthesis, the five nucleotides for nucleic acid synthesis, and all the other organic molecules of the cell. However, if a useful molecule, such as an amino acid, becomes available in the medium, the enzyme pathway that catalyzes its synthesis in the cell becomes superfluous. For example, if the amino acid histidine is added to the minimal medium, the nine enzymes that make up the pathway for the synthesis of histidine, starting from glucose and NH_4^+, soon disappear from the cell because histidine inhibits transcription of the nine genes coding for these enzymes. Genes encoding enzymes involved in the synthesis of other amino acids are similarly regulated in relation to availability of the particular amino acid. The principle of such regulation will now be described using genes for tryptophan synthesis as the example.

Arrangement of Sequences in the Tryptophan Operon

The synthesis of tryptophan occurs in several steps, each catalyzed by a different enzyme. The five proteins are encoded by five genes arranged in an operon (Figure 9-12). These genes, called *trpE, trpD, trpC, trpB,* and *trpA* are transcribed into a single mRNA molecule. The five polypeptides are then translated individually from the mRNA molecule. The first gene, *trpE,* is preceded by an operator sequence and a promoter sequence. In addition, a region called the **leader-attenuator region** is present between *trpE* and the operator-promoter region (Figure 9-12). The repressor gene for the operon *trpR* is located in a distant region of the chromosome, unlike the repressor gene for the *lac* operon, which is just upstream from the *lac* promoter. A gene encoding a repressor protein need not be closely linked to its target operon, since the repressor protein is a diffusible product.

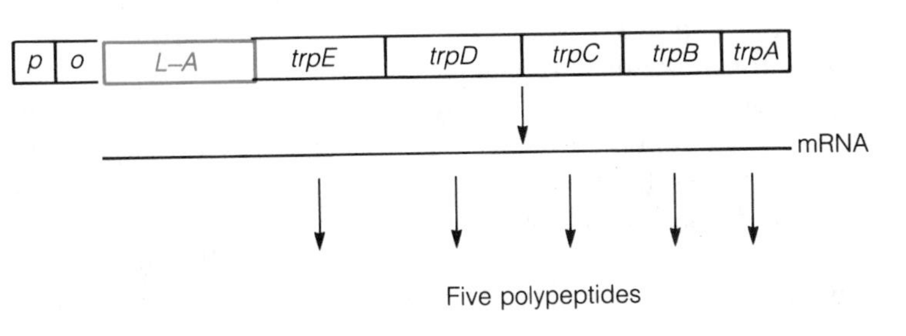

FIGURE 9-12
The *trp* operon in *E. coli.* A single mRNA is transcribed from the five genes and translated into five separate polypeptides. A leader-attenuator segment (L-A) is present between the operator and the first gene.

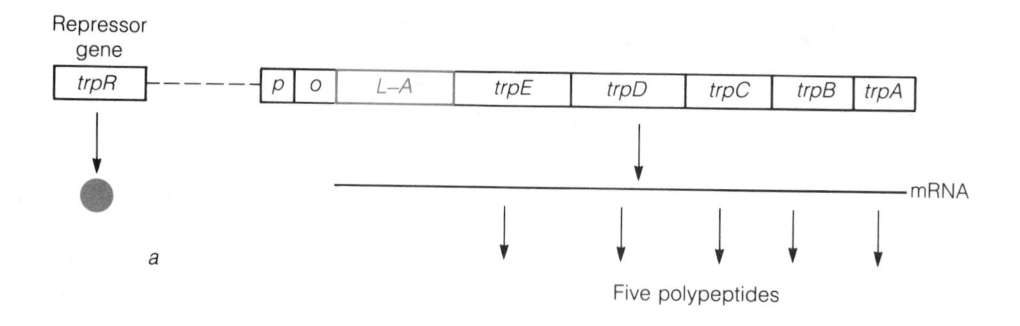

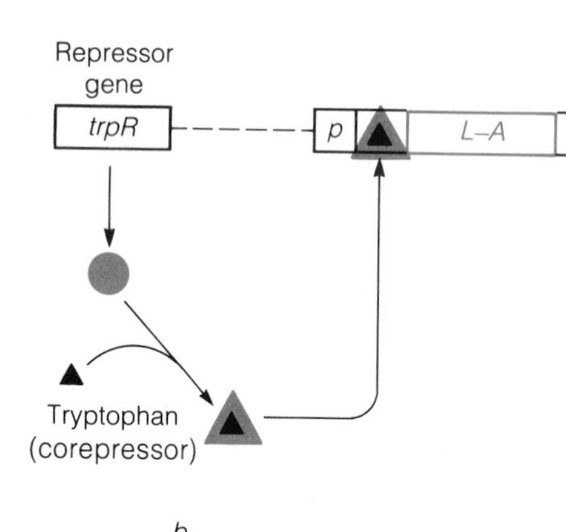

FIGURE 9-13
Regulation of the *trp* operon by repressor protein. (a) The repressor gene (*trp*R) encodes a repressor protein, which, in the absence of tryptophan, cannot bind to the operator; the operon is transcribed. (b) When tryptophan is present, it binds to the repressor. The tryptophan-repressor complex binds to the operator and prevents transcription.

Repressor Control of the *trp* Operon

The repressor gene, *trpR,* codes for a protein that has a binding site for the operator of the *trp* operon. The repressor also has a binding site for tryptophan and binds to the operator only if tryptophan is present (Figure 9-13). Tryptophan binds to the repressor, which enables the repressor to bind to the operator. Therefore, if tryptophan is available for uptake from the medium, it binds to the repressor, which then binds to the operator and prevents transcription of the five genes required for tryptophan synthesis. If tryptophan is absent in the medium, the repressor cannot form a complex with tryptophan and remains dissociated from the operator. This allows transcription of the operon and production of the five proteins that catalyze the steps of tryptophan synthesis.

The *trp* repressor and operator form a mechanism that regulates the synthesis of tryptophan up and down in a graded fashion according to how much tryptophan is present in the medium. The regulatory mechanism just described for anabolic pathways differs in an important respect from the mechanism in an inducible operon for a catabolic pathway, such as the *lac* operon. In the *lac* operon an inducer (allo-lactose) *prevents* binding of the repressor to the operator. In the *trp* operon tryptophan *enhances* the binding of repressor to the operator. For that reason tryptophan is called a **corepressor.**

In addition to the variable switch using a repressor, corepressor, and operator, the *trp* operon is also controlled by a second mechanism called **attenuation,** described as follows.

Attenuation of Transcription

As noted above and illustrated in Figure 9-12, a leader-attenuator region is interposed between the operator-promoter and the first gene of the *trp* operon. The attenuator-leader forms the basis of a regulatory mechanism that is sensitive to the prevailing concentration of tryptophan. Transcription of the *trp* operon begins one base pair down from the right end of the operator (Figure 9-13(a)) and proceeds rightward through the 162 base pairs that make up the leader-attenuator region before entering the *trpE* gene. Within the first part of the transcript, representing the leftward por-

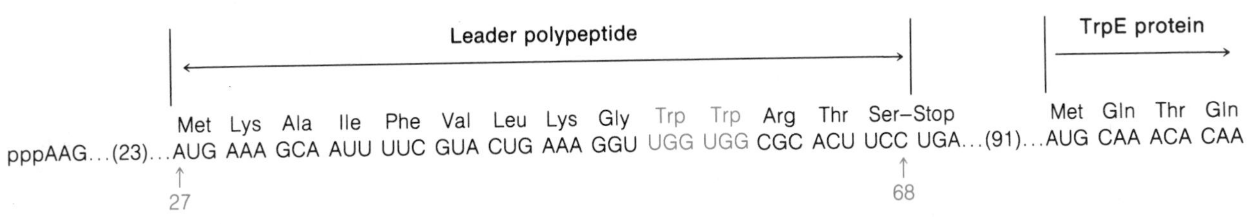

FIGURE 9-14 The leader-attenuator region of the transcript of the *trp* operon. This part of the transcript is 162 bases long. The 14 amino acids of the leader peptide are encoded by bases 27 through 68. Ninety-one bases connect the UGA stop codon from the translation start codon (AUG) of the first gene (*trpE*.)

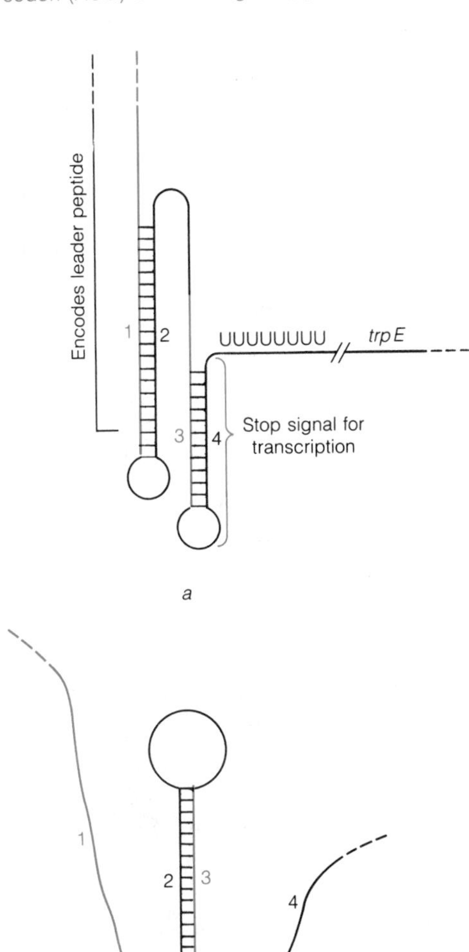

tion of the leader-attenuator region of the operon, is an AUG translation start codon. Some base pairs later (but before the *trpE* gene) this codon is followed by a translation terminator, UGA (Figure 9-14). Translation of this region forms a polypeptide of 14 amino acids, called the **leader polypeptide.** This polypeptide contains two adjacent tryptophan molecules, which are thought to have special significance.

The leader-attenuator region has four segments, denoted 1, 2, 3, and 4, that can base-pair with one another (intramolecular base-pairing) in two kinds of configurations in the single-stranded RNA transcript (Figure 9-15(a)). One configuration is segment 1 base-paired with segment 2 and segment 3 with segment 4; another is segment 2 base-paired with segment 3. Base-pairing of segment 3 with segment 4 forms a stem-and-loop that is immediately followed by seven Us in the mRNA transcript. Thus, base-pairing of segments 3 and 4 constitutes a termination signal for transcription, as described in Chapter 8. Therefore, if segments 3 and 4 base-pair with each other, transcription terminates at this point, and the five genes in the operon are not transcribed. Base-pairing of segment 2 with segment 3 in the RNA transcript does not form a termination signal *and* prevents formation of a termination signal by preventing base-pairing between segments 3 and 4 (Figure 9-15(b)). In that case transcription continues through the leader and the five genes of the operon.

FIGURE 9-15 Transcript of the leader-attenuator segment of the *trp* operon in *E. coli.* (a) Segments 1 and 2 can base-pair with each other, and segments 3 and 4 can base-pair. Base-pairing of segments 3 and 4 forms a termination signal for transcription, preventing transcription of the five genes in the operon. (b) Base-pairing between segments 2 and 3 prevents formation of a termination signal between 3 and 4, and transcription can continue through the operon.

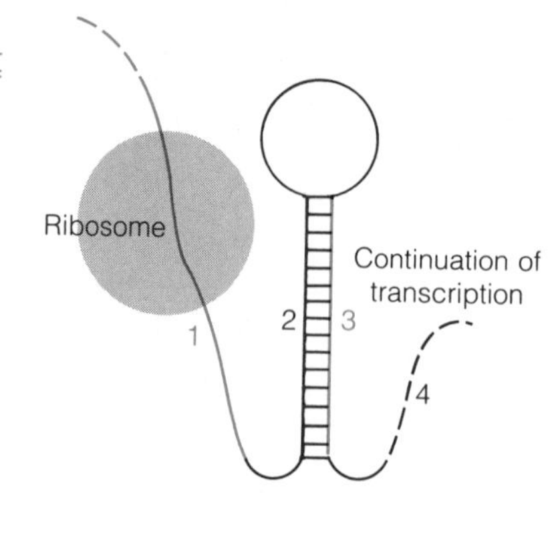

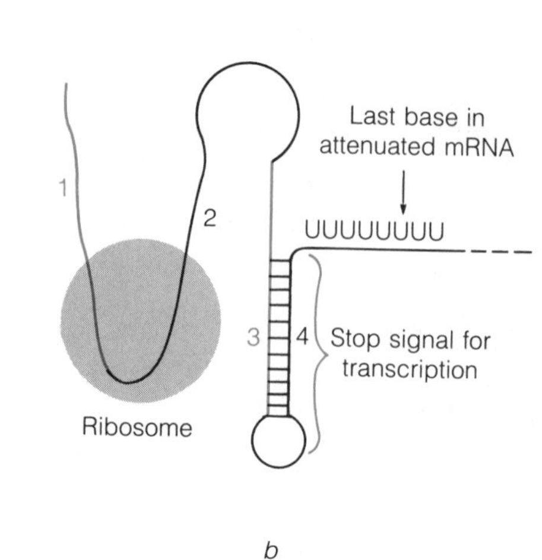

FIGURE 9-16
Diagram of attenuation in the *trp* operon of *E. coli*. (a) When the concentration of tryptophan is low, the ribosome stalls in segment 1, preventing base-pairing between segments 1 and 2 and allowing base-pairing between segments 2 and 3. No stop signal for transcription is formed by pairing of 3 with 4, and transcription continues through the five genes of the operon. (b) When the concentration of tryptophan is high, the ribosome does not stall in segment 1, but continues into segment 2, preventing base-pairing between segments 2 and 3. Segments 3 and 4 base-pair, forming a stop signal for transcription.

Tryptophan regulates whether termination occurs in the leader-attenuator sequence. In prokaryotes transcription and translation are coupled; translation of mRNA begins shortly after its 5′ end dissociates from the DNA template—well before synthesis of the transcript has been completed. Events are believed to proceed as depicted in Figure 9-16. A ribosome attaches to the 5′ end of the transcript and translation of the leader polypeptide is initiated and proceeds into segment 1 of the mRNA. If the level of tryptophan in the cell is low, the ribosome stalls transiently in segment 1 at the two tryptophan codons because tRNA charged with tryptophan ($tRNA^{trp}$) is in short supply. In holding segment 1 the ribosome prevents base-pairing of segment 1 with segment 2 of the transcript. Segment 2 is therefore free to base-pair with segment 3, which in turn prevents base-pairing of segment 3 with segment 4. Failure of the latter segments to base-pair prevents termination in the leader-attenuator. When the potential transcription-termination signal, consisting of base-pairing between segments 3 and 4 is not formed, the RNA polymerase continues transcription into the coding region of the genes, producing the polycistronic mRNA for the five genes. Thus, a low tryptophan concentration favors synthesis of complete mRNA, with subsequent translation to yield the five enzymes of the biosynthetic pathway for tryptophan.

When the concentration of tryptophan in the cell is high, a ribosome translates the leader region without stalling in segment 1 ($tRNA^{trp}$ is plentiful) and reaches segment 2 without delay (Figure 9-16(b)). This removes the possibility of base-pairing between segment 2 and 3, and leaves segment 3 free to base-pair with segment 4, forming a transcription-termination signal, which causes premature termination and failure to make a transcript of the five genes. The portion of the leader-attenuator region in which a termination signal can form is called the **attenuator** since it terminates transcription prematurely, that is, it attenuates transcription. Thus, the frequency of premature termination depends on how much tryptophan is available in the cell. The leader-attenuator region is the basis of a sensitive mechanism for regulating transcription of the *trp* operon and maintaining a precise level of tryptophan in the cell whenever the amount of tryptophan available in the medium is too low to meet requirements for general protein synthesis.

The transcription of many genes encoding enzymes that function in amino acid synthesis is regulated by attenuators, presumably by the mechanism just described for the *trp* operon. The histidine operon in the bacterium *Salmonella typhimurium* encodes nine enzymes for the biosynthetic pathway for histidine. This operon is not regulated by a repressor protein at all but is regulated entirely through an attenuator mechanism. The operons encoding enzymes for synthesis of threonine, leucine, isoleucine, valine, and phenylalanine in various bacteria are all regulated similarly to the *trp* operon.

REGULATION OF COMPLEX CELL ACTIVITIES

The chromosome of *E. coli* contains about 1000 operons. Principles for individual regulation of these operons are well understood. By mechanisms not so well understood the regulation of individual operons is in turn integrated with regulation of other functionally related operons. The regulation of such groups of operons are in turn regulated by unknown mechanisms within a single cell to achieve a global regulation of complex cell activities. For example, regulation of energy metabolism is integrated with anabolic pathways, and these are in turn coordinated with the biosynthesis of structures such as membranes, ribosomes, cell wall, and flagella. The ultimate expression of integration is the coordinated increase in all cell elements during growth and reproduction of a cell. In addition, the entire complex of events that make up cell growth and reproduction responds in a coordinated way to changes in growth conditions. In nutrient medium with an inefficiently used carbon source, such as proline, *E. coli* doubles in size and divides about once every 20 hours at 37 °C. When cells growing in proline-containing medium are switched to a rich nutrient medium containing glucose, amino acids, purines, pyrimidines, vitamins, and fatty acids, the cells respond rapidly with coordinated rates of production of all cellular constituents increased about 60-fold. In rich medium the cell then doubles in size and divides about every 20 minutes.

The mechanisms by which all phases of cell metabolism are integrated, and how total cell metabolism is globally regulated are poorly understood. However, an example of recent progress in this direction is the regulation of the biosynthesis of a main prokaryotic cell organelle, the ribosome.

REGULATION OF RIBOSOME PRODUCTION

Prokaryotic ribosomes are assembled from three different rRNA molecules and 55 different proteins (Chapter 8). Genes for rRNA molecules are encoded in seven different operons of the chromosome in *E. coli.* Each operon contains a coding region for each of the rRNA molecules (16S, 23S, and 5S) transcribed as a unit and processed into separate molecules. The genes for the 55 proteins are organized into 12 operons located at many places throughout the chromosome. Each ribosome possesses one copy of each of the RNA molecules and one copy of 53 of the 55 proteins. Four copies of the other two proteins are present per ribosome. Expression of all of the rRNA genes and genes for ribosomal proteins (**r-proteins**) is regulated such that all the gene products are synthesized in balanced amounts. Expression is further regulated in relation to nutrient conditions. In rich nutrient medium 15 to 20 ribosomes are completed per second (a cell generation time of 20 minutes with 20,000 per cell at the start of the cell cycle). Under less favorable growth conditions ribosome production and cell growth are both proportionally lower.

Ribosomal RNA synthesis in bacteria is controlled at the transcriptional level, but the mechanism is somewhat complicated (Figure 9-17). Experimental introduction into a cell of extra genes coding for rRNA leads to a transient two-fold increase in 30S and 50S subunits; however, the number of ribosomes engaged in translation remains the same. Shortly afterwards such overproduction of free ribosomal subunits represses synthesis of both rRNA and tRNA molecules. If genes for rRNA are introduced that are defective such that their rRNA transcripts cannot be assembled into ribosomal subunits, repression of transcription of rRNA and tRNA molecules from the normal genes in the chromosome does not occur. Thus, ribosomal subunits, not rRNA molecules, regulate (through a route not yet discovered) expression of genes for rRNA and tRNA. Such regulation by the structure in which rRNA is a major component is a form of autoregulation.

Production of ribosomal proteins is also autoregu-

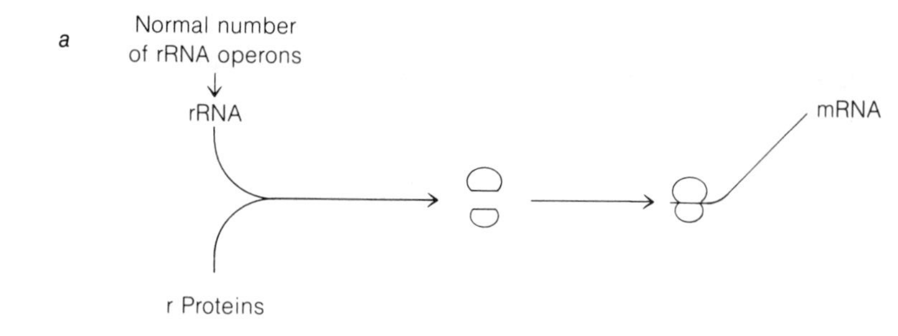

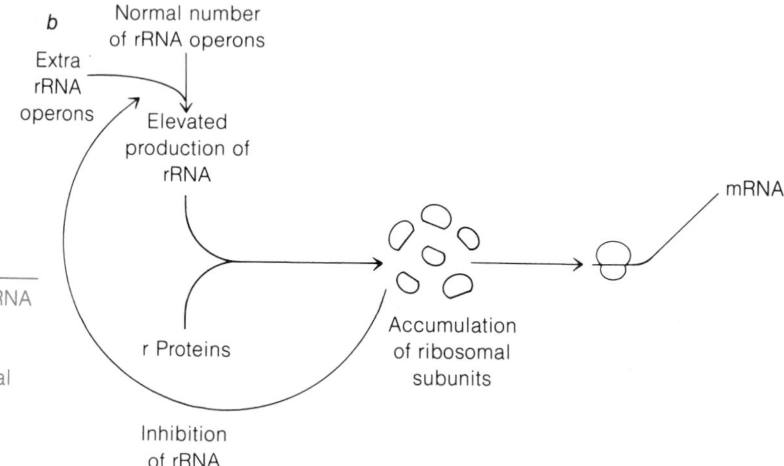

FIGURE 9-17
Diagram of an experiment showing autoregulation of rRNA synthesis. (a) rRNA and r-proteins are assembled into ribosomal subunits, which are used in translation of mRNA. (b) Addition of extra rRNA genes to the bacterial cell results in elevated production of rRNA and accumulation of ribosomal subunits. By an unknown mechanism the accumulation of ribosomal subunits reduces the synthesis of rRNA.

lated. Introduction into a cell of extra genes encoding certain r-proteins, such as L4 of the small ribosomal subunit, brings about overproduction of this protein. In turn, overproduction of L4 causes repression of synthesis of several other r-proteins encoded in the same polycistronic mRNA that contains the L4 gene. Presumably, expression of the normally present L4 gene will also be repressed, but the experimental design precluded detection of such an effect.

Unlike the regulation of genes for rRNA just described and for other prokaryotic genes described earlier, regulation of expression of r-protein genes occurs both at the transcriptional and translational levels (Figure 9-18). Introduction of extra r-protein genes is followed by a proportional increase in mRNA synthesis for the particular r-proteins. However, production of the r-proteins themselves does not increase, showing that the increase in the number of mRNA molecules is compensated for by less translation of the individual mRNA molecules. Initiation of translation per mRNA

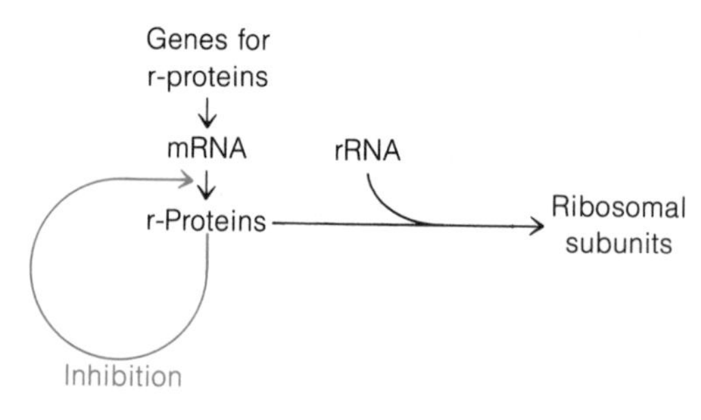

FIGURE 9-18
The rate of synthesis of r-proteins is at least in part autoregulated by r-proteins themselves. The r-proteins not assembled into ribosomal subunits inhibit translation of mRNA for r-proteins.

molecule is less frequent, and regulation of translation apparently occurs by a feedback mechanism in which r-proteins not assembled into ribosomes act as specific inhibitors of the translation of mRNA molecules for r-proteins. In this example, autoregulation of gene expression is achieved at the translational level.

The rate of production of r-proteins is proportional to the rate of rRNA production under a variety of growth conditions. When the growth rate increases, for example as a result of transfer of the cells to a richer nutrient medium, the syntheses of rRNA and r-proteins increase in the same proportion. This coordination strongly implies a regulatory interconnection for the two types of ribosomal components, but the molecular nature of such an interconnection is not known. Finally, operons containing genes for r-proteins also contain genes other than those encoding r-proteins—for example, genes for some of the subunits of RNA polymerase and for various enzymes that function in protein synthesis—which indicates that the rate of ribosome production is integrated with the rate of production of enzymes involved in gene expression.

PROBLEMS

1. Why is it advantageous for cells to be able to regulate individually the synthesis of different kinds of proteins?
2. What are the two principal mechanisms by which prokaryotes regulate expression of a gene?
3. Why is it important that the binding of a repressor molecule to an operator be reversible?
4. Describe what happens to the *lac* operon when cells growing in glycerol-containing medium are switched to medium containing only lactose.
5. Why is it important that *E. coli* growing in medium without lactose continues to produce a few β-galactosidase molecules?
6. Explain the mechanism by which the *lac* operon is kept inactive in *E. coli* growing in medium containing both lactose and glucose.
7. What is the principal difference in the mechanism of regulation between genes encoding catabolic enzymes and genes encoding anabolic enzymes?
8. How does the concentration of tryptophan in a bacterial cell regulate whether termination of transcription occurs in the leader-attenuator sequence of the *trp* operon?
9. Synthesis of rRNA is autoregulated. What is the regulating element?
10. Describe the experiment showing that autoregulation of r-proteins can occur at the translational level.

CHAPTER 10

Regulation of Gene Expression in Eukaryotes

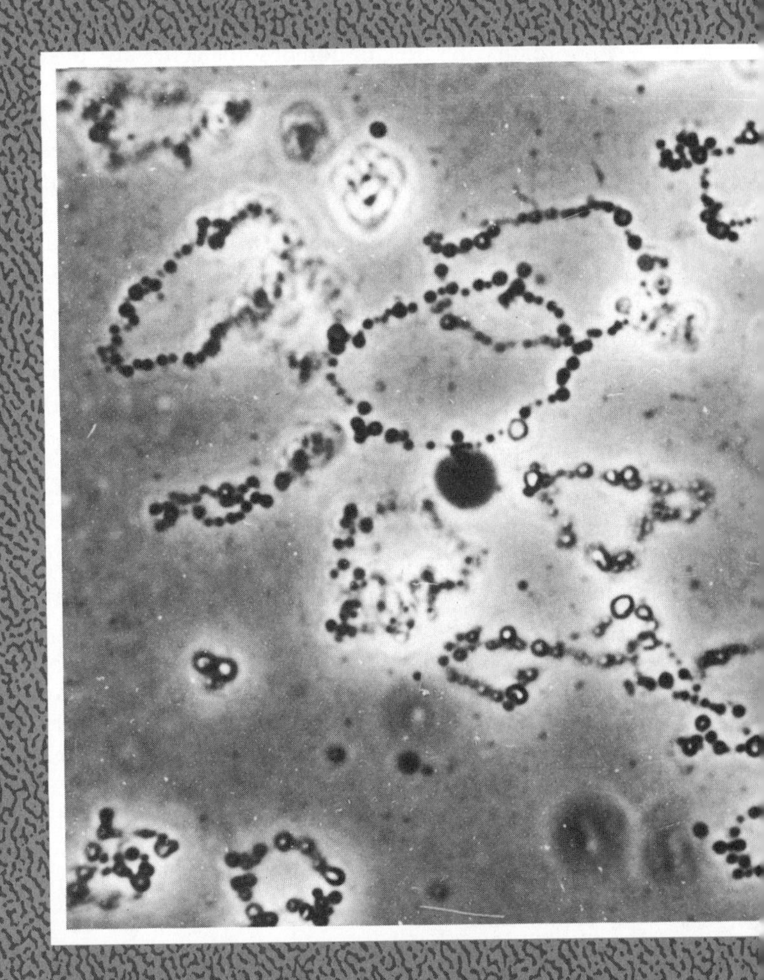

See Figure 10-29.

Gene expression in eukaryotes is regulated primarily by regulation of transcription, as it is in prokaryotes. Regulation of some genes also occurs at the translational level in both groups of organisms. In eukaryotes gene expression is also sometimes regulated by a third mechanism: alternative modes of processing of primary RNA transcripts within the nucleus. The molecular mechanisms for regulation of transcription are for the most part different in the two groups of organisms. It is not clear why they have evolved away from one another in this regard. Eukaryotes possess more genes than do prokaryotes and have more complicated life-styles and life cycles, which might necessitate different molecular mechanisms for regulation of transcription.

The regulatory systems of prokaryotes are heavily geared toward maximum efficiency in utilization of resources to achieve maximum rates of cell reproduction. This is also true to a great extent for unicellular eukaryotes like fungi, algae, and protozoa, although these kinds of cells are considerably more complex in their functions and structures than prokaryotes are. In contrast, the metabolism of cells of multicellular organisms is generally not directed toward maximum rates of cell multiplication. Rather, the evolution of multicellular organisms from unicellular organisms depends on the evolution of mechanisms that efficiently restrict and control cell multiplication. *In the cells of multicellular organisms regulation of gene expression is geared to bring about structural and functional specializations of cells to carry out different tasks in the various tissues in the organism.* The clearest examples are seen in the development of a multicellular organism from a single undifferentiated cell, the fertilized ovum, into a complex adult with 200 or more structurally and functionally different cell types. We now know that the majority of the 200+ different cell types in an animal contains the same set of nucleotide sequences in DNA and hence the same store of genetic information. The wide differences in structure and function among cells, for example between a muscle cell and a liver cell, arise through extensive differences in which genes are expressed in these two cell types. In muscle cells genes encoding proteins involved in contraction are heavily expressed, and genes encoding the blood protein albumin and some liver cell enzymes are not expressed. In liver cells there is little expression of genes encoding proteins for muscle contraction, and genes encoding blood albumin and various liver-specific enzymes are heavily expressed. In short, cell differentiation is generally achieved through regulation of gene expression.

Gene regulation is under intensive study in a variety of unicellular and multicellular eukaryotes. In no case has a regulatory mechanism been completely elucidated at the molecular level, but the outlines and many details are understood. One thing is clear: regulation is achieved in several ways.

TRANSCRIPTION IN EUKARYOTES

Transcription in eukaryotes is more complex than in prokaryotes. The essential features are described in this section.

Promoter Sequences

Most promoters are present upstream from the start point of transcription of eukaryotic genes. Although the eukaryotic promoters do essentially what promoters of prokaryotic genes do, that is, initiate transcription of the adjacent gene, the nature of the promoters and the mechanism of their workings are quite different.

Promoters of genes that encode proteins (genes transcribed by RNA polymerase II) contain two or more sequences that are recognized by specific proteins called **transcription factors.** Binding of the transcription factors permits RNA polymerase II and several other proteins to form an initiation complex on the DNA and to begin transcription from a specific start site. RNA polymerase II itself does not recognize any of the promoter sequences, and it cannot initiate transcription at the normal start site without guidance provided by binding of transcription factors to promoter sequences. As far as we know, each gene or set of related genes uses a unique transcription factor. The number of transcription factors must be far less than the number of genes.

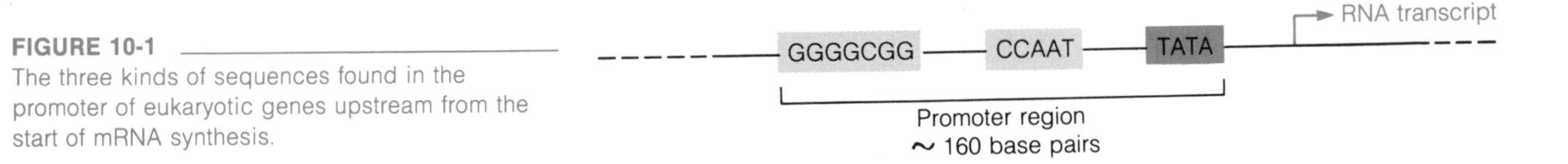

FIGURE 10-1 The three kinds of sequences found in the promoter of eukaryotic genes upstream from the start of mRNA synthesis.

Three different kinds of sequences have been identified in eukaryotic promoters. All occur in a specific order, usually within the 160-bp region immediately upstream from the nucleotide at which transcription starts (Figure 10-1).

1. **The TATA Box.** The sequence 5′TATAA3′, called the TATA box (pronounced *tah-tah)* is found 25 to 30 bp upstream from the start site of transcription of most but not all genes. The TATA box does not provide differential activation of one gene versus another. Rather, the TATA box binds a specific transcription factor (a protein) required for initiation of transcription at the correct nucleotide. Most mutations in the TATA box do not reduce transcription from the adjacent gene, but cause initiation to occur randomly in the 5′ region of the gene instead of at the normal start site. Thus, the TATA box with its bound transcription factor directs RNA polymerase to the initiation site.
2. **The CAAT Box.** The CAAT box (pronounced *cat*) consists of the sequence 5′CCAAT3′ and is usually located about 50 bp upstream from the TATA box (Figure 10-1). Many genes have a CAAT box in their promoters, but some do not. Binding of a specific transcription factor (a protein) to the CAAT box fulfills another general requirement for transcription (in addition to the TATA box with its transcription factor), but how it promotes transcription is not known.
3. **The GC Box.** The GC box most frequently consists of the sequence 5′GGGGCGG3′ but slight variations are present in the promoters of some genes. The GC box is located upstream from the CAAT box, usually between 80 and 160 bp from the start of transcription. The GC box may be present in a single copy or in several copies. Although they occur in the promoters of many genes and have been found particularly in the promoters of genes of viruses that infect animal cells, GC boxes appear not to occur as widely as TATA and CAAT boxes. The GC box binds yet another transcription factor (again a protein) that has been partially purified and given the designation SP-1. Binding of SP-1 to the GC box (or boxes) promotes transcription, but the molecular mechanism is unknown.

Enhancer Sequences

In addition to the three kinds of sequences that occur in what is now generally accepted as the promoter, other sequences called **enhancers** greatly affect rates of transcription (Figure 10-2). Enhancers are usually between 50 and 200 bp long and differ from the promoter in several respects:

1. An enhancer is equally effective in either orientation of its sequence relative to the coding sequence it affects. In other words, experimental removal of the enhancer and reinsertion in the same position but in the opposite orientation does not alter its enhancing activity for the adjacent coding sequence.

FIGURE 10-2 Enhancers occur upstream of the promoter or downstream from the 3′ end of the transcribed segment of a gene.

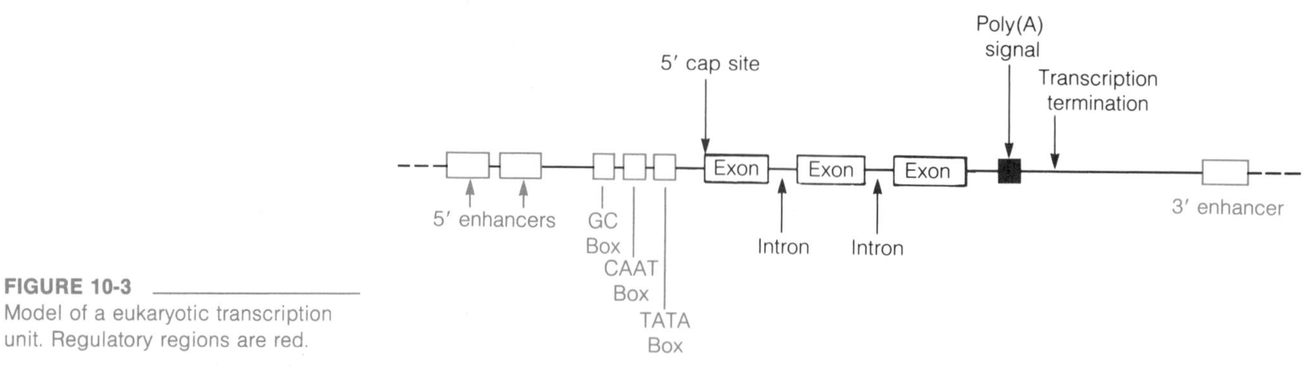

FIGURE 10-3
Model of a eukaryotic transcription unit. Regulatory regions are red.

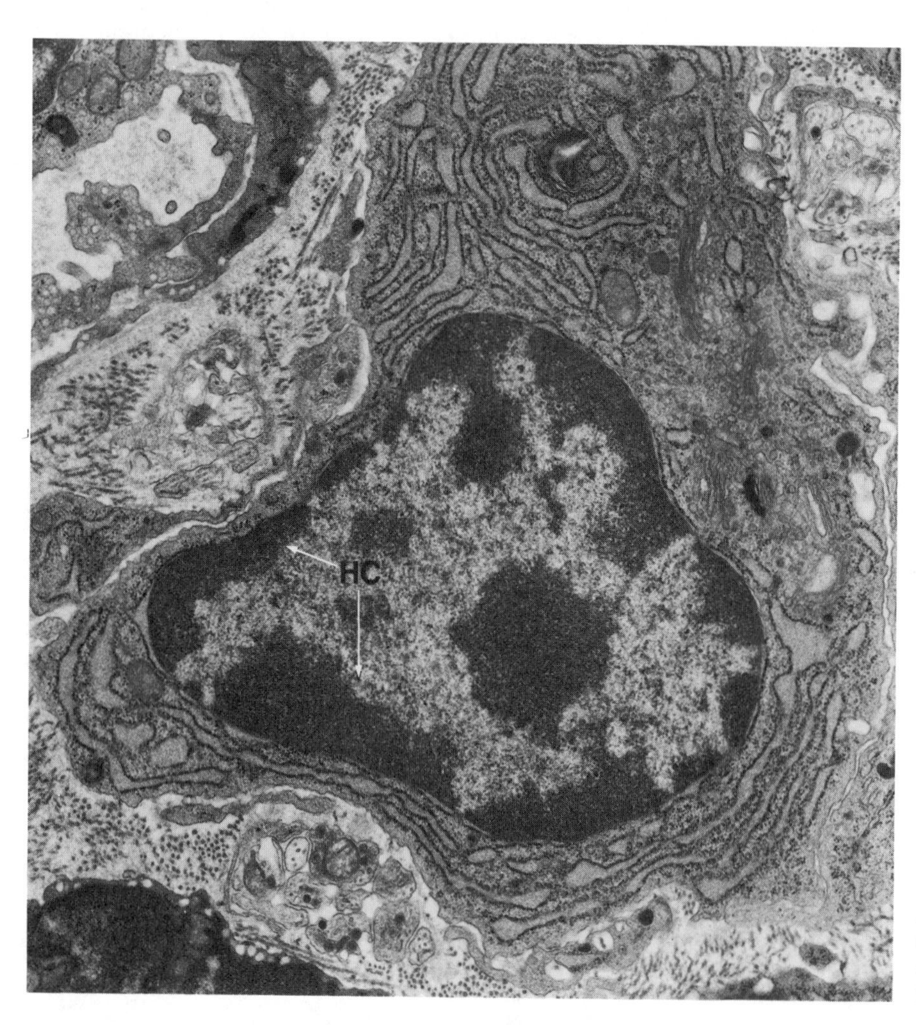

FIGURE 10-4
Electron micrograph of a plasma cell (antibody-producing cell) in rat intestine. A large amount of the DNA is present as heterochromatin (HC), much of which is closely adhered to the nuclear envelope. [Courtesy of Keith R. Porter.]

2. Not all genes have enhancer sequences, but when one is present, it may occupy a position up to 1000 bp upstream from the promoter or downstream from the 3′ end of the gene. This means that enhancers work over long distances to increase the frequency of initiation of transcription.
3. Two or more enhancer sequences may be present in tandem.

Enhancers were first discovered near coding sequences in viral chromosomes and have subsequently been found associated with many cellular genes, for example, genes coding for immunoglobins (antibody polypeptides). An enhancer increases the frequency of initiation of transcription of its associated gene up to several hundred-fold, but how it functions is not known. One possibility is that it allows chromatin structure to be loosened (see next section), which, in turn, may increase access of transcription factors to the promoter. This idea fits with the observation that an enhancer works in both orientations.

A Generalized Transcription Unit in Eukaryotes

The sequences present in a eukaryotic transcription unit are summarized in Figure 10-3. The TATA box, CAAT box, or GC box is absent for some genes. Enhancer elements may not be associated with all genes. For some genes more than one enhancer sequence may be present.

CHROMATIN AND THE REGULATION OF TRANSCRIPTION

The DNA packed into highly compacted or condensed chromatin, called **heterochromatin,** remains transcriptionally inactive. More loosely packed DNA, called **euchromatin,** is transcriptionally active. The packing of DNA into heterochromatin and euchromatin is achieved by proteins that bind to DNA. The blocks of heterochromatin seen by light microscopy, or better by electron microscopy (Figure 10-4), are large enough to contain amounts of DNA corresponding to hundreds or even thousands of genes. The loosening of the packing in heterochromatin to form euchromatin does not occur readily, and the presence of heterochromatin apparently reflects a mechanism for maintaining large segments of chromosomes in an inactive state. In certain kinds of insects whole haploid sets of chromosomes are heterochromatic and are genetically inactive. In the male mealy bug, but not in the female, the entire set of chromosomes inherited from the male parent is heterochromatic and inactive in transcription. The chromosome set received from the maternal parent is euchromatic and provides genetic information in somatic cells. In females both sets of chromosomes are euchromatic.

En bloc inactivation of genes occurs in female mammals by heterochromatic packing of one of the two X chromosomes early in embryogenesis. An example of the consequence of this is observable in the coat color patches of **calico** cats. The X chromosome of cats carries a gene that affects coat color. If a female that is homozygous (*BB*) for the X-linked gene for black coat color is crossed with a male whose X chromosome carries a gene for yellow coat color (*b*), all the female offspring (*Bb*) have random patches of black and yellow fur; all the males are black (*B*). Females are calico because early in development one X chromosome is inactivated at random in each cell of the female offspring; in some cells the X chromosome derived from the paternal parent is inactivated and in other cells the maternally derived X chromosome becomes inactivated. When a given X chromosome is inactivated in a cell, all the descendants of that cell have the same inactive X chromosome. Therefore, female skin cells with the paternal X chromosome inactivated form black patches, and cells with the maternal X chromosome inactivated form yellow patches. If both chromosomes were active in a heterozygous female (*Bb*), its skin would be uniformly black.

Similarly, in humans the gene for glucose 6-phosphate dehydrogenase is located in the X chromosome. A mutation that changes the protein slightly without altering its enzyme activity occurs in some humans. Consequently, if a female offspring receives from her mother an X chromosome that carries a slightly altered gene for glucose 6-phosphate dehydrogenase and from her father an X chromosome with a normal gene, she will possess two populations of cells, one with an altered enzyme (the paternal X is inactive) and the other with a normal enzyme (the maternal X is inactive).

The inactivated X chromosome in a female cell of a human or other mammal appears in an interphase cell as a distinct body of heterochromatin called the **Barr body** (Figure 10-5). Because male cells carry a single X chromosome, which is not inactivated, they lack Barr bodies. In mice, humans, and most other mammals either the paternal X or the maternal X may become inactivated; the process appears to be random. In marsupials, such as the kangaroo, it is nonrandom; the X chromosome received from the paternal parent is the one that is always inactivated. How the inactivation mechanism specifically recognizes the paternal X chromosome in marsupials (or the entire paternal set of chromosomes in mealy bugs) is not known.

Thus, although male cells contain only one X chromosome and female cells contain two, male and female cells are equivalent with respect to X chromosome activity since only one is fully active genetically. The inactivation of the second X chromosome is a form of dosage compensation—the dose of genes in the second X chromosome is compensated for by inactivation; thus, male and female cells are equal with regard to the number of active X chromosomes. Presumably, two active X chromosomes in females would be deleterious, although, in contrast, two active homologues for all the autosomal chromosomes is in fact necessary for normal development. An exception to the rule of only one X chromosome being active occurs in the cells of very early female embryos. Both X chromosomes are active, and such activity is essential for normal female development. Later, one of the X chromosomes is inactivated in every cell of the embryo.

Some parts of autosomal chromosomes remain permanently and apparently irreversibly condensed as heterochromatin. Such chromatin is called **constitutive heterochromatin.** An example is the permanently condensed heterochromatin near the centromere in chromosomes of many kinds of eukaryotes (Chapter 11). Other heterochromatic regions may convert back to the euchromatic state under certain normal conditions and these are referred to as **facultative heterochromatin.** An example is the switch in very early development in female mammals, in which one of the X chromosomes becomes heterochromatic (the Barr body). Around the time of meiosis in oocytes, the heterochromatic X chromosome returns to the euchromatic state.

Gene inactivation also occurs coordinately with chromatin condensation in the differentiation of two kinds of cells in animals. In most mammals (camels are an exception) the final step in differentiation of the erythrocyte is ejection of the nucleus to create an anucleated cell, but in other vertebrates the nucleus is retained. In the red blood cells of birds and amphibians, for example, the chromatin becomes tightly condensed and all transcription stops. The cells are, in effect, functionally anucleated. Similarly, in the differentiation of sperm cells the chromatin condenses and all transcription stops. The condensed sperm chromatin then reverts to euchromatin after fertilization.

Although euchromatin is less condensed than heterochromatin, it may have to be loosened further for transcription to occur. This idea is supported by many observations that show that the DNA of genes engaged in transcription is much more susceptible to degradation by DNase (i.e., is **hypersensitive**) than is DNA of inactive genes in isolated nuclei or isolated chromatin. This greater accessibility of active genes to DNase suggests that loosening of the packing in euchromatin is necessary for transcription to occur. DNA segments upstream from the transcription initiation point of an active gene are particularly sensitive to DNase. In some cases at least, these correspond to sequences in the promoter region. Furthermore, this apparent loosening of chromatin appears to be necessary but not sufficient for transcription. For example, genes that are transcribed only during a limited part of interphase of the cell cycle remain hypersensitive to DNase even in parts of the cycle when they are not actively transcribed.

These observations lead to a two-step hypothesis of gene activation in which the first step consists of loosening of chromatin upstream from the coding region of

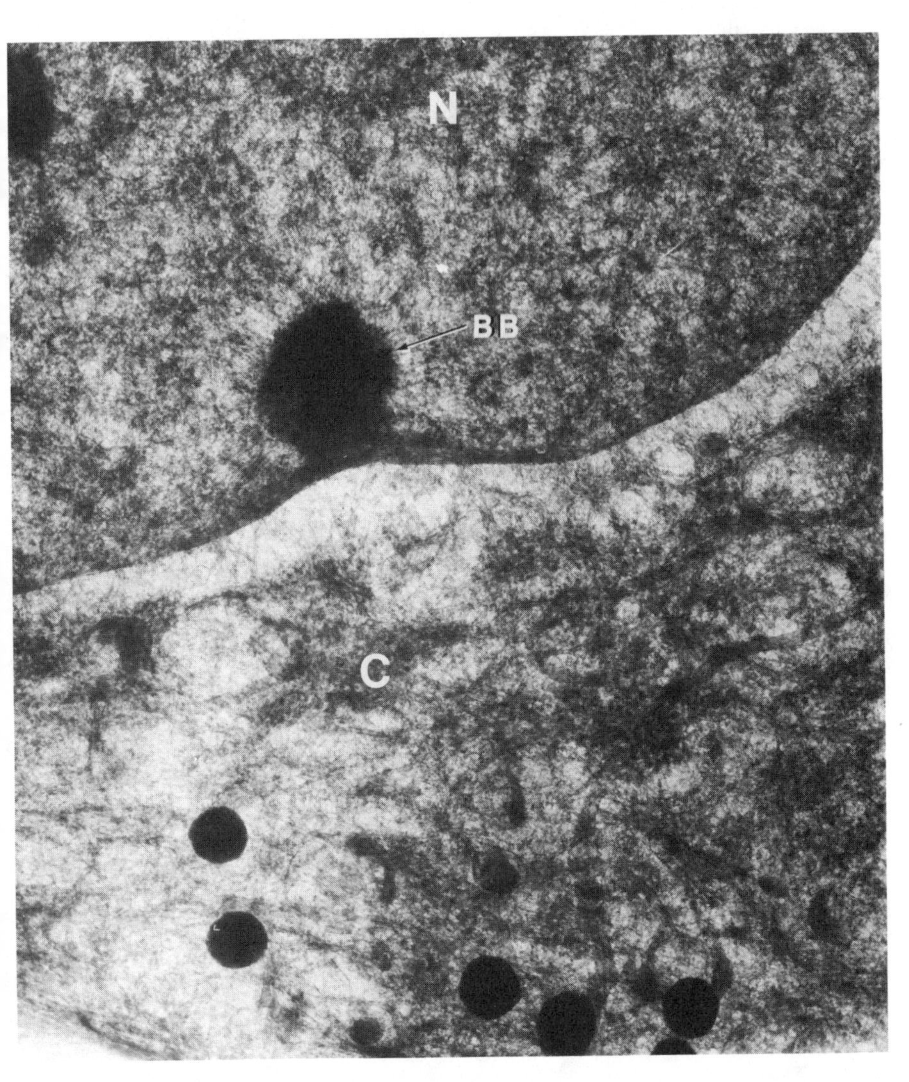

FIGURE 10-5
Electron micrograph of a Barr body (BB) in a human fibroblast. This is part of a whole cell viewed with the high voltage electron microscrope (Boulder, Colorado). N = nucleus, C = cytoplasm. [Courtesy of Stanley M. Gartler and Karen A. Dyer.]

a gene. This occurs in a gene-specific manner, but almost nothing is known about the signals that recognize and lead to loosening of the chromatin of the gene chosen for activation. The second step is also not well understood but apparently consists of the interaction of protein transcription factors with the promoter, which then brings about initiation of transcription. This general model is consistent with many experimental observations but is at present no more than a guiding framework; detailed molecular evidence is still lacking on several crucial points, such as on the postulated signals that loosen chromatin.

The following sections discuss a variety of experimental systems for studying regulation of gene expression. Many of the observations support the general model just outlined, and others are at least consistent with it. Considering a variety of systems shows the broad scope of regulation of gene expression in eukaryotes and gives a better idea of what is and is not known about mechanisms of regulation.

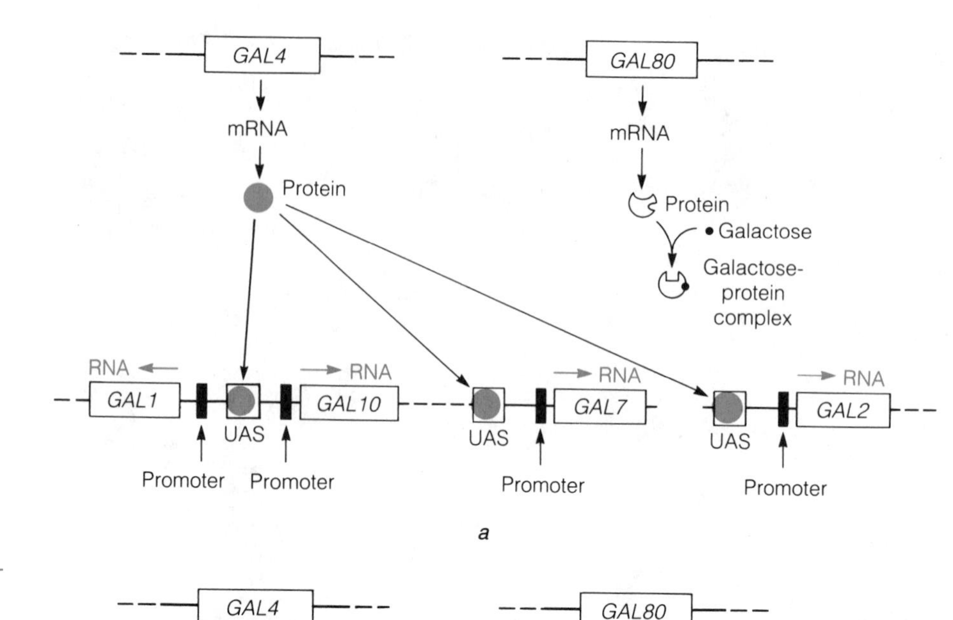

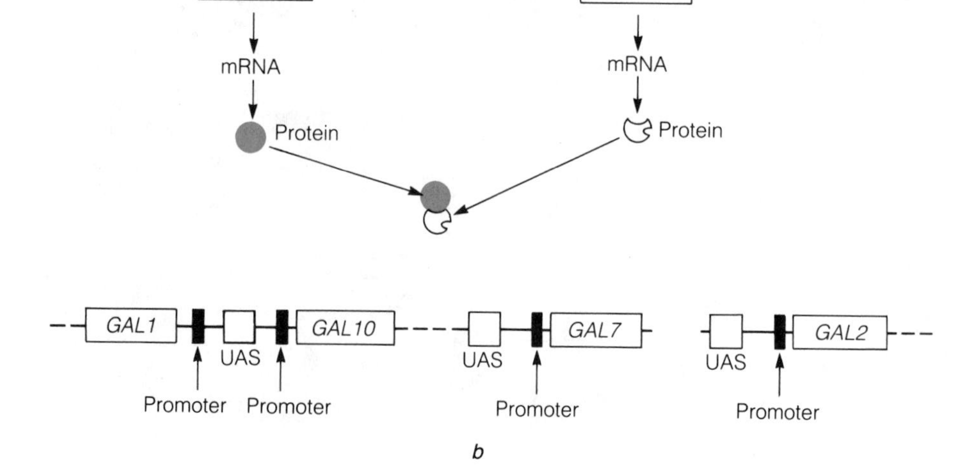

FIGURE 10-6
Scheme for regulation of genes needed for metabolism of galactose in yeast. (a) The *GAL4* gene encodes a protein that activates transcription of four genes (*GAL1, 2, 7, and 10*) by binding to upstream activation sites (UASs). *GAL1 and 10* have a single, common UAS that is positioned between them; these two genes are transcribed in opposite directions. Galactose binds to the protein encoded by *GAL80* and prevents this protein from interacting with the protein encoded by *GAL4*. (b) In the absence of galactose the *GAL80* protein interacts with the *GAL4* protein and prevents it from activating transcription of the four genes *GAL1*, *2*, *7*, and *10*.

REGULATION OF GENE EXPRESSION IN YEAST CELLS

Regulation of gene expression in unicellular eukaryotes such as yeast makes possible adaptation to the environment, as it does in prokaryotes, although the mechanisms are different from those of the prokaryotes. Regulation provides for maximum efficiency in utilization of resources by closing down or activating anabolic and catabolic pathways in relation to the supply of various nutrients, particularly carbon sources and amino acids. Many cases of regulation of gene expression in yeast have been studied. Three examples are described here: galactose utilization, synthesis of amino acids, and metabolism of ethyl alcohol.

Genes for Galactose Catabolism in Yeast Cells

In order for yeast cells to use galactose for glycolysis, four genes must be expressed, *GAL1, 2, 7,* and *GAL10*. The protein product of *GAL2* functions in the transport of galactose into the cell. *GAL1, 7,* and *GAL10* are required to convert galactose to glucose-1-P, which then enters the glycolytic pathway. Galactose in the medium induces a 5000-fold increase in the transcription of these genes in cells that were previously growing with ethanol as a source of energy. Also, glucose represses induction of the *GAL* genes. Therefore, in medium containing both galactose and glucose the four genes remain repressed. Two regulatory genes, *GAL4* and *GAL80,* control the four genes for galactose metabolism. Experimental observations support the following model of how galactose, glucose, and the products of the six genes (*GAL1, 2, 4, 7, 10,* and *GAL80*) interact to regulate metabolism of galactose (Figure 10-6).

The protein encoded by *GAL4* is necessary for expression of *GAL1, 2, 7,* and *GAL10,* four genes needed for galactose metabolism. The *GAL4* gene is expressed constitutively, and its encoded protein activates transcription of the four regulated genes by binding to a region about 250 bp upstream from the initiation site for transcription of each gene. The genes *GAL1, 7,* and *GAL10* are adjacent on a single chromosome and although regulated coordinately, they are transcribed separately. The other *GAL* genes are scattered at other sites on various chromosomes. The upstream region of *GAL1, 2, 7,* and *GAL10* to which the *GAL4* protein binds consists of a 17-bp sequence repeated four times in tandem. Such a sequence in yeast, located far upstream from the initiation site for transcription, is called an **upstream activation site,** or **UAS.** The UAS is present in addition to promoter sequences already discussed (e.g., TATA).

GAL80 is a regulatory gene that is also expressed constitutively. In the absence of galactose the protein encoded by *GAL80* blocks the binding of the *GAL4* protein to the UASs of the four genes, and the synthesis of the enzymes for galactose metabolism remains at an uninduced level. If galactose is present, it binds to the *GAL80* protein, which in turn prevents *GAL80* protein from blocking the action of *GAL4* protein. As a result, transcription of the genes for the galactose-metabolizing enzymes is induced. Glucose represses induction of transcription by interacting directly or indirectly with *GAL4* protein to prevent its binding to the UASs.

The galactose system in yeast, which also involves genes other than the six mentioned, is still incompletely understood. However, it illustrates the following several points about regulation of transcription:

1. Two kinds of regulatory genes are present, one encoding a protein that stimulates transcription (positive regulation) and the other encoding proteins that repress transcription (negative regulation).
2. The 5′ upstream regions of genes contain sequences that are recognized by protein transcription factors that act as positive regulators.
3. The action of the activator protein may be blocked by binding of a repressor protein to it.
4. Transcriptional events of several genes can be coordinately regulated because they possess the same recognition site for a regulatory protein. This is probably a general mechanism for coordinate control of functionally related genes in eukaryotes.
5. Regulation can be modulated by catabolite repression, e.g., it can be repressed by glucose.

Genes for Amino Acid Synthesis in Yeast

Expression of genes for synthesis of at least some amino acids in yeast is regulated separately, amino acid by amino acid. This process is called **specific control.** It provides for separate turning on of genes needed to synthesize a specific amino acid when that amino acid is not available in the environment. The molecular mechanism for such individual control is not completely understood. The 50 or so genes required for synthesis of at least the majority of amino acids are also regulated as a group, a mechanism known as **general control.** In general control, the lack of a single kind of amino acid triggers the activation of all 50 genes and establishment of many anabolic pathways for amino acids. A key gene in general control is *GCN4* (*g*eneral *c*ontrol of *n*itrogen metabolism). It regulates expression of other genes by the mechanism shown in Figure 10-7. *GCN4* is expressed constitutively; transcripts from the gene remain at a constant level whether or not yeast cells are starved for one or more essential amino acids. However, translation of the mRNA for *GCN4* is regulated. If cells are starved for any single amino acid, translation of mRNA for *GCN4* increases 100-fold. The protein encoded by *GCN4* is a positive transcriptional factor that turns on the transcription of many genes, probably all 50 genes involved in synthesis of amino acids, and not just genes for the amino acid that is lacking in the medium. Cells with a mutation in the *GCN4* gene fail to synthesize the enzymes for amino acid anabolic pathways. The protein encoded by *GCN4* binds to the sequence 5′TGACTC3′ found upstream of the TATA box for 15 genes examined so far that code for enzymes for amino acid synthesis.

The mechanism of general control therefore involves both translational control (i.e., of the *GCN4* gene) and transcriptional control. How translational control is achieved is not known, but hints come from the structure of the *GCN4* mRNA molecule. This mRNA molecule has an unusually long leader consisting of 577 nucleotides that are not translated into the final protein product. However, the leader contains four translational start codons (AUG), each of which is followed a few codons later by a translational stop codon. Thus, if translation begins at any one of the four start codons, it is soon aborted. The unsolved mystery is how amino acid starvation directs initiation of translation away from these four start codons to a fifth and *real* AUG that results in synthesis of the *GCN4* protein.

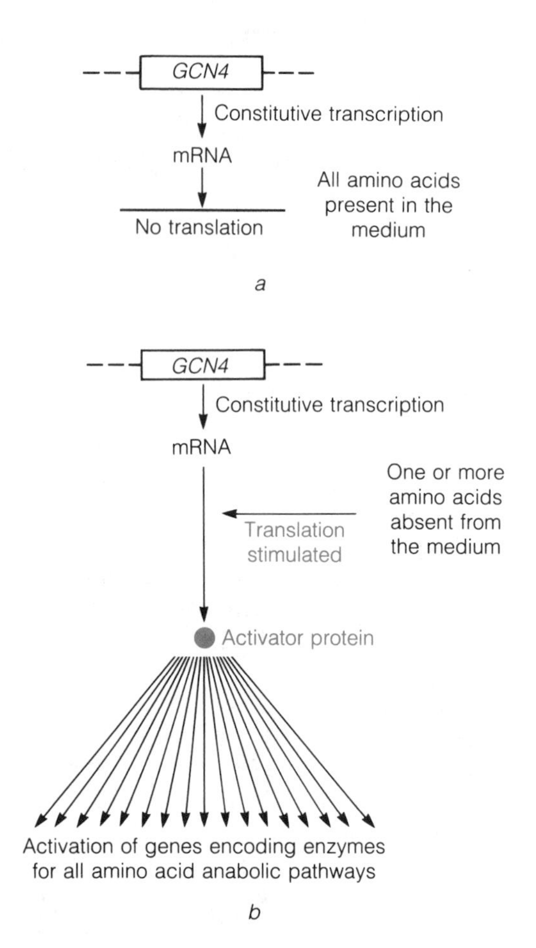

FIGURE 10-7
General regulation of genes encoding enzymes for amino acid synthesis in yeast. (a) Gene *GCN4* is transcribed constitutively. When all amino acids are available in the medium, translation of the mRNA is inhibited. (b) In the absence of a single amino acid (or more than one), translation of the mRNA of *GCN4* is stimulated, resulting in synthesis of 50 or so enzymes that make up the anabolic pathways for all amino acids.

Genes for Ethanol Metabolism in Yeast

Yeast nuclei possess two genes that encode the enzyme **alcohol dehydrogenase (ADH).** ADH1 is encoded by the gene *ADC1* and is involved with **production** of alcohol during fermentation (Chapter 4). *ADC1* is transcribed constitutively, so ADH1 is always present in the cell. ADH2 is encoded by gene *ADR2* and is involved in **utilization** of alcohol as a source of energy. Transcription of the *ADR2* gene is repressed in the presence of glucose and derepressed in the presence of ethanol, if glucose is not present. When yeast cells are switched from glucose to ethanol, the level of ADH2 increases 100-fold. Thus, gene *ADC1* is unregulated (constitutively expressed) and gene *ADR2* is regulated.

The two genes, as they exist in chromatin, have been compared with respect to their sensitivities to digestion by DNase. Under conditions of repression, gene *ADR2* (repressible gene) is much less sensitive to DNase than *ADC1* (constitutively expressed gene). Under conditions of derepression the two genes are equally sensitive to DNase. In addition, hypersensitive sites appear to be present upstream from the point of transcription initiation in both genes under conditions of derepression. The experiment with ADH in yeast fits with the hypothesis that a change in chromatin configuration is of general importance in regulation of genes in unicellular eukaryotes.

REGULATION OF GENE EXPRESSION DURING THE CELL CYCLE

In cells that are proliferating, progress through the cell cycle requires different amounts of certain gene products at particular times. For example when DNA replication starts, the transcription of genes encoding the enzymes dihydrofolate reductase and thymidine kinase increase sharply, as do those genes encoding histone proteins.

The principal preparation for cell division is the replication of DNA during interphase of the cell. Replication of DNA requires production of several enzymes that catalyze the synthesis of the precursors for DNA synthesis—the four deoxynucleoside triphosphates. One of these enzymes is **dihydrofolate reductase,** which catalyzes the synthesis of methylene tetrahydrofolate from folate, a reaction mentioned in Chapter 3 (Figure 3-6). Methylene tetrahydrofolate is the coenzyme for **thymidine monophosphate synthase,** the enzyme that catalyzes synthesis of **thymidine monophosphate (TMP).** In the synthesis of TMP the coenzyme donates a methyl group. In order for TMP synthesis to continue, new coenzyme molecules with methyl groups must be continuously synthesized. Therefore, dihydrofolate reductase is essential for the continuous synthesis of TMP, and TMP is required for continuation of DNA replication. Without dihydrofolate reductase, DNA replication stops.

In cells that are not replicating DNA, the level of dihydrofolate reductase is very low or undetectable. When DNA replication begins, the transcription of the gene encoding the enzyme increases many-fold, and subsequently falls when DNA replication is over. The gene encoding the enzyme must be activated again at the next entry of the two daughter cells into DNA replication. Regulation of expression of the gene is ultimately connected to initiation of DNA replication, but nothing is known about the signals that turn on transcription of the gene coordinately with DNA replication.

In a similar fashion transcription of genes encoding the five histone proteins bound to DNA in chromatin (Chapter 11) are turned up sharply at the start of DNA replication (Chapter 12). Histone proteins are then synthesized and become complexed with the steadily increasing amount of DNA in the nucleus.

Genes encoding for histones are present in multiple copies, presumably because the demand for rapid histone production during DNA replication cannot be met by a single gene for each of the five histones. Histone genes represent one of the very few kinds of protein-coding genes that are present in multiple identical copies. Some other protein-coding genes, such as actin genes, are also multiple, but the genes in such a family represent slightly different versions of the gene. Most vertebrate species have about 40 sets of histone genes per genome with one copy of each of the five histone genes per set. In sea urchins there are about 1000 sets of the five histone genes. The genes within a set are separated from one another by untranscribed DNA spacers (Figure 10-8). Sets of histone genes are repeated in tandem, with untranscribed DNA spacers between sets. Each gene in a set is transcribed separately. Transcription of the different genes in the various sets is regulated coordinately.

The signal that activates transcription of histone genes has not been identified, but a change in chromatin configuration is involved. In proliferating animal cells a region upstream from the start of the gene for H4 histone has been found to be sensitive to DNase even

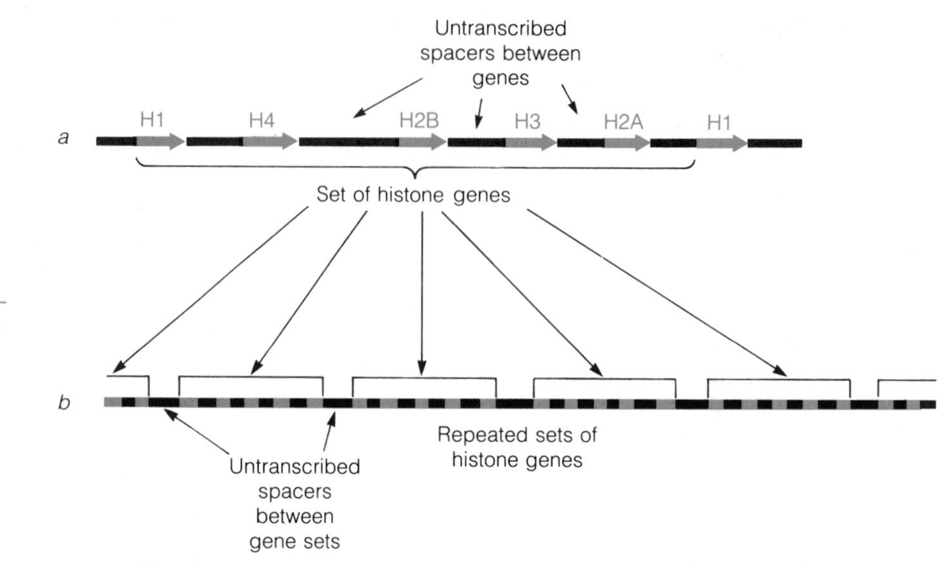

FIGURE 10-8
(a) Set of five histone genes in the sea urchin. The genes (red) are separated by untranscribed spacers (black) and all are transcribed in the same direction in this case. In other organisms genes in a single set are transcribed in opposite directions. (b) Tandem sets of histone genes are separated by untranscribed spacers.

before DNA replication starts, suggesting that the first step in gene activation, namely, chromatin loosening, has already taken place. When DNA replication starts, the size of the region sensitive to DNase increases, implying expansion of the region of chromatin loosening. In contrast, the gene encoding β-globin of hemoglobin, which has no role in the cell cycle and which is not expressed in these cycling cells, remains insensitive to DNase throughout the cycle.

Transcription of histone genes must be turned up every time DNA replication starts, which is about every 20 hours in the experiment just cited. Transcription of histone genes is turned down when DNA replication is not occurring. The structure of the chromatin is altered cyclically in rhythm with DNA replication; however, as long as cells continue to reproduce, chromatin in which histone genes reside apparently does not condense to the extent found for genes permanently in an inactive state.

The regulation of expression of histone genes occurs not only at the transcriptional level, as just described, but also at a posttranscriptional level. If DNA synthesis is blocked experimentally with a drug, histone genes continue to be transcribed, but all histone mRNA molecules are rapidly destroyed, and histone protein synthesis stops. If the block to DNA synthesis is removed, the histone mRNA molecules are again stabilized, and histone synthesis resumes. The expression of the histone genes is, therefore, also regulated through differential stabilization of the mRNA, a kind of translational control. The molecular mechanism by which destruction of histone mRNA is regulated is unknown. Experiments in which the H4 histone gene has been isolated and experimentally modified have shown that the regulation of the turnover of histone mRNA is dependent on intactness of at least some part of the second half of the gene (toward the 3′ end) but does not require the presence of any of the 5′ half of the mRNA molecule. This finding may be related to other observations suggesting that the poly(A) tail at the 3′ end of mRNA molecules in eukaryotes has a role in mRNA stability.

AUTOREGULATION OF GENE EXPRESSION

The expression of some genes in eukaryotes is regulated by the proteins they encode. The phenomenon, called autoregulation, has been studied in more depth in prokaryotes (e.g., autoregulation of rRNA and r-proteins discussed in Chapter 9). However, there are a few known cases of autoregulation in eukaryotes.

The protein tubulin, from which microtubules are built and which all eukaryotic cells possess, is synthesized under autoregulatory control. The first experimental evidence came from treatment of cells with drugs (e.g., colchicine or vinblastine) that prevent assembly of tubulin into microtubules. This inhibition results in a raising of the level of free tubulin in the cytoplasm, which is quickly followed by rapid destruction of mRNA molecules coding for tubulin. In a subsequent experiment, purified tubulin was injected with a micropipette into individual (cultured) mammalian cells. This increase in intracellular tubulin by injection rapidly suppressed synthesis of tubulin by the cell. This observation showed that the synthesis of tubulin is regulated by unassembled tubulin acting as a repressor. The autoregulation is possibly achieved in part at a post-transcriptional level, since transcription of the tubulin gene appears unaffected in cells treated by an agent blocking assembly of tubulin into microtubules.

REGULATION OF HEAT SHOCK GENES

All cells have a particular upper limit of temperature at which they can survive. For *E. coli* and most mammalian cells this temperature is about 42°C. For some kinds of yeast cells and for some species of *Tetrahymena* the upper temperature limit is 32°C, and for chick cells it is 43°C. All types of cells can survive a short treatment at a temperature above their lethal limit. When subjected to such a heat shock, all kinds of cells, from bacterial to mammalian, immediately respond by synthesizing large amounts of several proteins, called **heat-shock proteins (hsps).** *E. coli* synthesizes 17 hsps, mammalian cells about ten, *Drosophila* eight, *Xenopus* two, and so on. In many kinds of cells, the response includes virtually complete closing down of synthesis of all other proteins. When cells are returned to a sublethal temperature, the synthesis of heat-shock proteins rapidly decreases and synthesis of the normal spectrum of proteins resumes.

Some of the hsps are among the most highly conserved in evolution. In one hsp, with a molecular weight of 70,000 and hence called hsp70, 50 percent of the amino acids in the polypeptide chain are the same in *E. coli* and *Drosophila*. This remarkable sequence homology for a protein in a prokaryote and a eukaryote implies that heat-shock proteins participate in fundamental cell processes. However, almost nothing is known about them. Synthesis of hsps is also induced by other kinds of stress, such as exposure to ethanol, ultraviolet light, viral infection, and certain toxic chemicals. Hence, a better name for these proteins is **stress proteins,** which they are sometimes called.

Expression of hsps is a remarkable example of transcriptional regulation. Studies in *Drosophila* in which segments of DNA have been deleted show that regulation is mediated by a 16-bp sequence upstream from the coding region of the heat-shock gene. Activation of heat-shock genes is accompanied by binding of a protein transcription factor within the region between 40 and 99 bp upstream from the TATA box; presumably the 16-bp sequence identified by deletion is the specific binding site in this region. Experimental joining of the 5′ flanking region of the heat-shock gene from *Drosophila* to a gene not normally activated by temperature converts it to an hsp capable of responding to heat shock in a mammalian cell. Thus, the molecular signal(s) functioning in the heat-shock response in a mammalian cell can recognize the regulatory site from a *Drosophila* gene. Finally, examination of several different heat-shock genes in the same species has revealed the same or a similar sequence (called a **consensus sequence**) in the 5′ flanking regions in each case. Such a consensus sequence is the means by which the multiple separate genes are coordinately turned on by stress.

REGULATION OF GENE EXPRESSION BY CYCLIC AMP (cAMP)

Cyclic AMP has an important role in increasing transcription of certain operons in bacteria. High levels of cAMP stimulate binding of RNA polymerase to the promoters of these operons. Transcription of other bacterial genes is inhibited by cAMP (although no example was given in Chapter 9). The transcription of many genes in eukaryotes (e.g., genes encoding tyrosine aminotransferase, phosphoenol pyruvate carboxykinase, and the hormone prolactin) is increased when the concentration of intracellular cAMP rises. The transcription of other genes (e.g., those encoding aldolase and discoidin I) is decreased by high levels of cAMP.

Cyclic AMP is believed to modulate transcription in eukaryotes through cAMP-dependent protein kinases. Protein kinases are enzymes that catalyze addition of phosphate groups to particular threonine, serine, and tyrosine residues in certain proteins. The phosphate groups are derived from ATP. Phosphorylation of proteins usually profoundly changes the activity of a protein. For example, phosphorylation inactivates some enzymes and activates others. Furthermore, some protein kinases are activated by cAMP (cAMP-dependent kinases), whereas others are unaffected. An increase in cAMP in mammalian cells activates two kinds of cAMP-dependent kinases, which then phosphorylate a specific set of proteins. According to one hypothesis, some of these are regulatory proteins that increase transcription of certain genes and decrease transcription of other genes. Phosphorylation of regulatory proteins may increase or decrease their actions depending on the gene involved. The effect of increased cAMP, through phosphorylation of proteins by cAMP-dependent protein kinases, is reversed when the level of cAMP falls. This occurs by dephosphorylation of proteins by the action of the enzyme phosphatase.

REGULATION OF GENE EXPRESSION IN PLANT CELLS BY LIGHT

Many kinds of plant cells lose their chlorophyll and become white (etiolated) when kept in complete darkness for many days. Etiolation involves loss of enzymes that catalyze the steps in chlorophyll synthesis as well as many other enzymes in chloroplasts. Exposure of etiolated plants to light is followed within a few hours

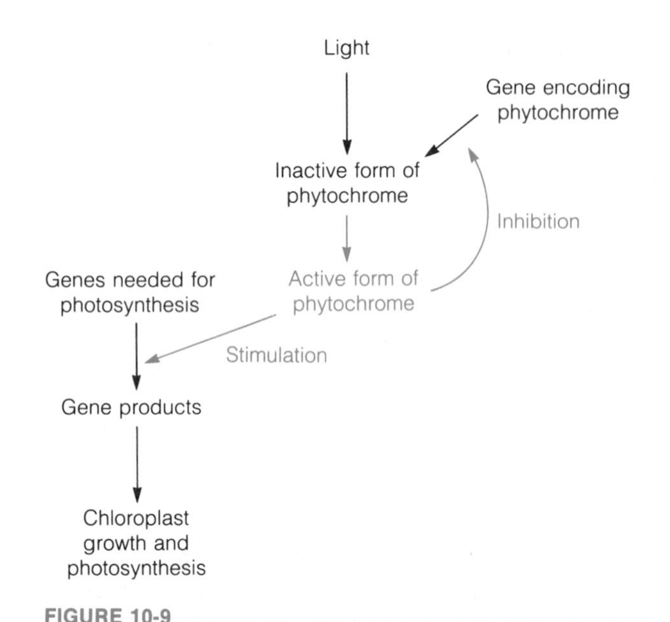

FIGURE 10-9
Light converts phytochrome into an activated form. Activated phytochrome stimulates expression of genes whose products are required for chloroplast growth photosynthesis. Activated phytochrome also autoregulates its own production by inhibiting expression of the phytochrome gene.

by appearance of chlorophyll molecules, synthesis of more than 60 enzymes involved in photosynthesis, synthesis of rRNA in chloroplasts, and overall growth of chloroplasts. This remarkable induction of synthesis of gene products by light has been studied in many kinds of plants, and although differences occur, an overall generalized pattern of regulation has emerged (Figure 10-9).

Syntheses of the polypeptides in the small and large subunits of ribulose bisphosphate carboxylase, a key enzyme in the conversion of CO_2 into carbohydrate (Chapter 4) have been investigated in detail. The small subunit protein of ribulose bisphosphate carboxylase is encoded by a nuclear gene and is synthesized in the cytoplasm. From the cytoplasm the protein is transported into chloroplasts. Regulation occurs at the transcriptional level and is mediated by another molecule called **phytochrome.** This is a protein molecule with a light-absorbing pigment molecule covalently bound to it. In the absence of light, phytochrome exists in an inactive form (Figure 10-9). Exposure to light converts it to an active form. The activated form is essential for induction of transcription of the gene encoding the small subunit protein of ribulose bisphosphate carboxylase and of other genes.

The large subunit of ribulose bisphosphate carboxylase is encoded by a gene in the chloroplast DNA molecule and is synthesized inside the chloroplast. Induction of transcription in the chloroplast by light also involves phytochrome although other factors participate, and the effect of phytochrome may be indirect. In any case the outcome is that genes encoding the two subunits of an enzyme required for photosynthetic assimilation of CO_2 are not transcribed in the absence of light and are turned on when light is present.

Phytochrome also mediates negative regulation of at least two genes. One of these genes encodes the phytochrome protein itself. In plants kept in the dark the phytochrome gene is transcribed, and phytochrome molecules are synthesized. On exposure to light, phytochrome molecules are activated and act to repress the transcription of the phytochrome gene, a clear example of autoregulation of a gene. Presumably, as the cell grows, the concentration of phytochrome will fall, which could lead to decreased transcription of genes essential for photosynthesis. However, a fall in phytochrome concentration would also result in derepression of the gene encoding phytochrome, establishing a more or less steady state of phytochrome concentration.

Light also regulates gene expression at the translational level in the simple multicellular alga *Volvox*. The reproductive life cycle of *Volvox* can be controlled by continuous alternation of 32 hours of light and 16 hours of dark. Cells in the dark period contain essentially the same population of mRNA molecules as cells in the light period. Yet the shift from dark to light results in a dramatic increase in total translation and differentially enhanced synthesis of several polypeptides. The increases occur even if photosynthesis is specificially blocked with drugs, and hence are not due to increased availability of ATP generated by photosynthesis. The molecular mechanism by which light exercises this translational regulation of gene expression is unknown.

REGULATION OF GENE EXPRESSION IN CELL DIFFERENTIATION

The development of multicellular organisms begins with a fertilized ovum. Following a short period of rapid cell multiplication, groups of cells form various tissues specialized to carry out particular functions. In an adult animal there are about 200 readily distinguishable cell types. The genome of each cell remains essentially constant, and cells differ from one another because they express different genes. A first indication of constancy of the genome was provided by measurements of the DNA content of individual cells. Most cells of various tissues contain a diploid amount of DNA. A few percent of cells are tetraploid, and germ cells that have completed meiosis are haploid. Because almost all cells contain a diploid amount of DNA, there can be no gross changes in DNA in association with differentiation. More compelling proof that the genome remains constant during differentiation has been shown by two kinds of experiments, one with plant cells and one with transplantation of nuclei into frog eggs.

Totipotency of Plant Cells

The cells of an adult plant can be dissociated, and cells cultured on nutrient medium. Individual cells multiply and form **clones,** groups of cells originating from a single cell. Ultimately, growth of clones forms a mass called a **callus** (Figure 10-10) from which a complete, mature plant may form. This experiment, which has been carried out with a single cell from a carrot, from the stem of a tobacco plant, and single cells from several other plants, proves that individual cells in differentiated tissues contain all the genes needed to specify the different kinds of cells in an adult plant. Differentiation must be the result of regulation of gene expression rather than some permanent modification of the genome. Theoretically, it might be possible to do

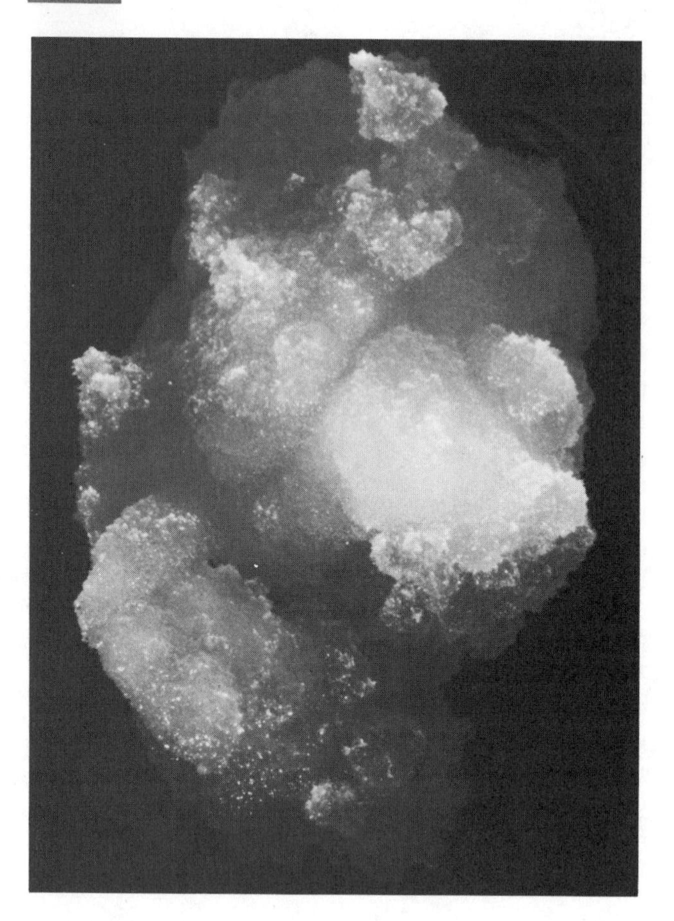

FIGURE 10-10
Callus of soybean. The callus started as a single cultured cell and has grown into a mass of many thousands of cells. [Courtesy of Kathleen J. Danna.]

the same kind of experiment with animal cells, producing a complete adult animal from a single cell. However, it has not yet been achieved, even for simple animals.

Totipotency Demonstrated by Nuclear Transplantation

The first successful transplantation of a nucleus from one cell to another was performed 50 years ago with the large, free-living amoeba, *Amoeba proteus* (Chapter 1). Transplanting a nucleus into a cell as large as a frog egg is easy using a micromanipulator to control a micropipette. Such a procedure has been used to show that the nucleus from a differentiated cell has a full set of genes. Transplantation of a nucleus from an intestinal epithelial cell from a tadpole into an egg whose own nucleus has been destroyed has resulted in development of a tadpole and then an adult frog (Figure 10-11).

Nuclei taken from an adult frog instead of a tadpole can also support development, for example, transplantation of an erythrocyte nucleus into an egg. This represents an extreme case because an erythrocyte is a highly differentiated cell that has ceased to reproduce and, moreover, all transcription in its nucleus has been shut down. After transplantation to an enucleated egg,

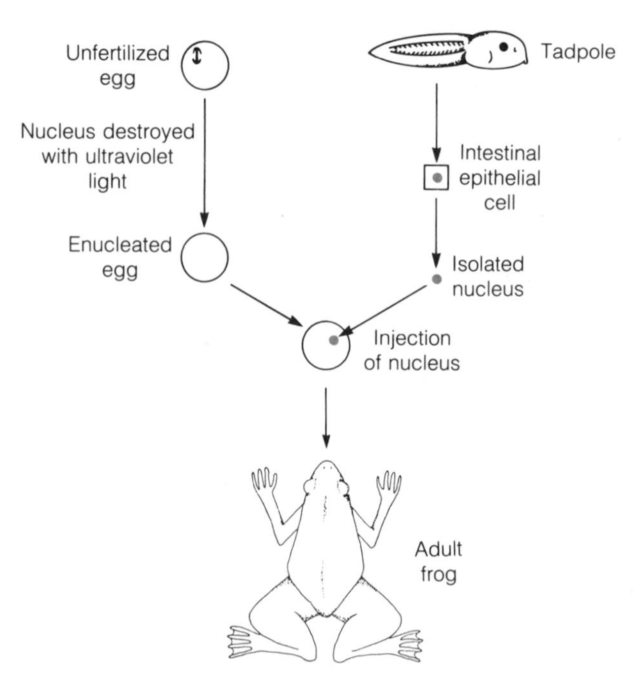

FIGURE 10-11
Nuclear transplantation to show totipotency of a nucleus from a differentiated cell. A nucleus from an intestinal epithelial cell of a tadpole injected into an enucleated egg can, in some cases, develop into an adult frog. The egg is enucleated by destroying its nucleus (engaged in meiosis) with ultraviolet light.

transcription and DNA replication are reactivated in the erythrocyte nucleus. Development advances to the tadpole stage, but tadpoles raised using the nucleus from an adult frog (as opposed to a nucleus from another tadpole) do not metamorphose into frogs, for a reason that is not understood. Nevertheless, in both cases it is clear that the nucleus from a differentiated cell contains all the genes needed to produce all of the other kinds of differentiated cells in the animal. It may be argued that cells from an adult may lack one or more genes necessary for conversion of a tadpole to a frog.

Reactivation of an erythrocyte nucleus by transplantation into an egg is instructive in another context. An old argument in developmental biology is whether or not cell differentiation is reversible. Normally in an animal, completely differentiated cells such as muscle cells, erythrocytes, nerve cells, and others, do not dedifferentiate into unspecialized cells, and completely differentiated normal cells do not dedifferentiate in culture. It has been argued from such observations that dedifferentiation cannot occur because the transcription pattern of genes has been irreversibly fixed. Erythrocyte transplantation shows that this is not so. Given an appropriate cytoplasmic environment the differentiated state of the nucleus can be reversed.

Regulation of Translation of mRNA in an Egg Cell

During the formation of an egg cell, transcription and translation take place at substantial rates. Later, in a mature egg cell, the rates of mRNA production and protein synthesis are much diminished. Immediately upon fertilization transcription and translation again increase. Protein synthesis is typically stimulated more than 20-fold as a result of two effects. First, the efficiency of the translation machinery increases several fold as measured by more rapid transit of ribosomes along mRNA molecules. Second, mRNA molecules synthesized in the egg nucleus before fertilization and stored in the cell are mobilized for translation. Indeed, these mRNA molecules are sufficient for early development. The egg can proceed to a point at which several thousand cells are present and the beginnings of cell differentiation are recognizable in the presence of the drug actinomycin D, which inhibits transcription. Thus, mRNA molecules stored in the egg, called **masked mRNA,** consist of transcripts of those genes whose products are needed for early development. In the eggs of the sea urchin, an organism much used to study development, the stored mRNA molecules correspond to six percent of the sequences in DNA. These mRNA molecules therefore represent many thousands of genes.

Thus, transcription of genes important in early development occurs in an egg cell during its formation. The primary transcripts are processed (capping, polyadenylation, and intron excision) and complexed with proteins, which presumably prevent translation. How the inhibition of translation is reversed at fertilization is unknown. However, the phenomenon of masked mRNA molecules represents a prime example of the regulation of gene expression at the level of translation.

CELL DIFFERENTIATION

Experiments described earlier showed that the nuclei of adult, differentiated cells of plants and animals contain a full set of the genes essential for development and differentiation. This means that differentiation is not based on selective elimination of some genes and selective retention of others. In general, differentiation is achieved by selective turning on and turning off of the expression of particular genes, although the origin and nature of the signals that do this are still poorly understood. Several mechanisms that could contribute to regulation of gene expression in differentiation have already been discussed. These are:

1. Regulation of transcription
2. Regulation of RNA processing in the nucleus
3. Regulation of translation of mRNA

Regulation of the mRNA life span and the phenomenon of masked mRNA stored in egg cells are examples of regulation of translation. One of the clearest cases of regulation of mRNA life span in the cytoplasm is mRNA for histone proteins during the cell cycle.

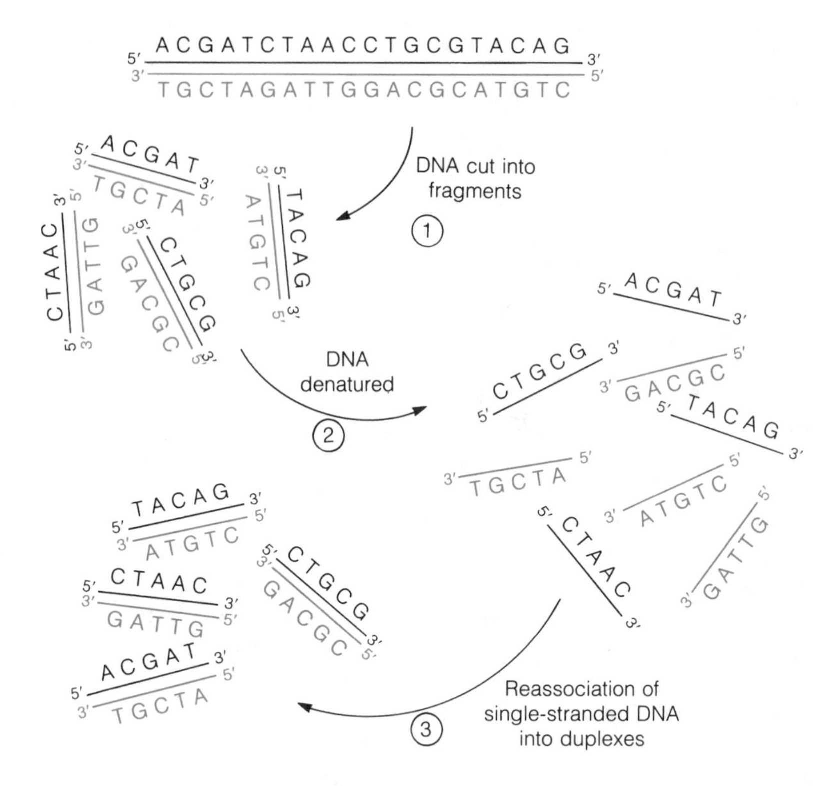

FIGURE 10-12 Denaturation of duplex DNA and reassociation of single strands into duplexes. (1) Duplex DNA cut into short fragments. (2) Duplex fragments denatured into single strands by heat. (3) Reassociation of single strands into duplexes after cooling. In an actual experiment the duplex DNA is cut into fragments of several hundred base pairs.

When DNA synthesis is completed, histone synthesis is no longer needed and existing mRNA molecules for histones are rapidly destroyed. The degree to which regulation of mRNA life span contributes to cell differentiation is unclear, as is also the case for the role of RNA processing in differentiation. Regulation of gene expression in cell differentiation is usually exercised at the level of transcription.

Differences in mRNA Among Different Kinds of Cells in an Organism

If differentiation of cells is the result of transcription of different sets of genes in different kinds of cells, then there should be differences in the mRNA composition of these cells. This has, indeed, been demonstrated by use of nucleic acid hybridization techniques to compare populations of mRNA molecules among various adult cell types.

Nucleic acid hybridization is done as follows: When DNA in solution is heated, the hydrogen bonds between the two strands are broken and the strands separate, yielding single-stranded DNA (Figure 10-12). This is called **denaturation** or melting of double helix DNA. If the solution of single-stranded DNA is cooled, complementary chains (strands) eventually collide and reassociate with one another by means of base-pairing.

Every primary transcript in a cell is a sequence that is a complement of the DNA chain from which it was originally transcribed. Therefore, if purified DNA is denatured and allowed to react with RNA purified from the same kind of cell, the RNA molecules can form hybrid double helices with the DNA (Figure 10-13). Two competing reactions actually take place because single-stranded DNA can reassociate either with its complementary chain of DNA or with a complementary RNA chain. By using a large excess of RNA and selecting a particular temperature, formation of RNA/

DNA hybrids is strongly favored. Primary transcripts (nuclear RNA) hybridize to more of the DNA than does mRNA (cytoplasmic RNA) because (1) primary transcripts contain intron sequences, some of which are very long, that are spliced out in the formation of mRNA molecules, and (2) many primary transcripts are totally destroyed in the nucleus and do not yield mRNA molecules for transport to the cytoplasm. Some of the duplex DNA is not transcribed and cannot form DNA/RNA hybrids.

All cell types in an organism are expected to have many of the same mRNA molecules, namely, those encoding enzymes involved in energy metabolism, proteins of the cytoskeleton, proteins of ribosomes, and other proteins required for general maintenance of cell function and structure. Beyond these there are differences. For example, about 4.5 percent of total DNA forms hybrids with the nuclear RNA from mouse liver cells. This means that in liver cells only about 4.5 percent of the DNA is transcribed. Similarly, the RNA mol-

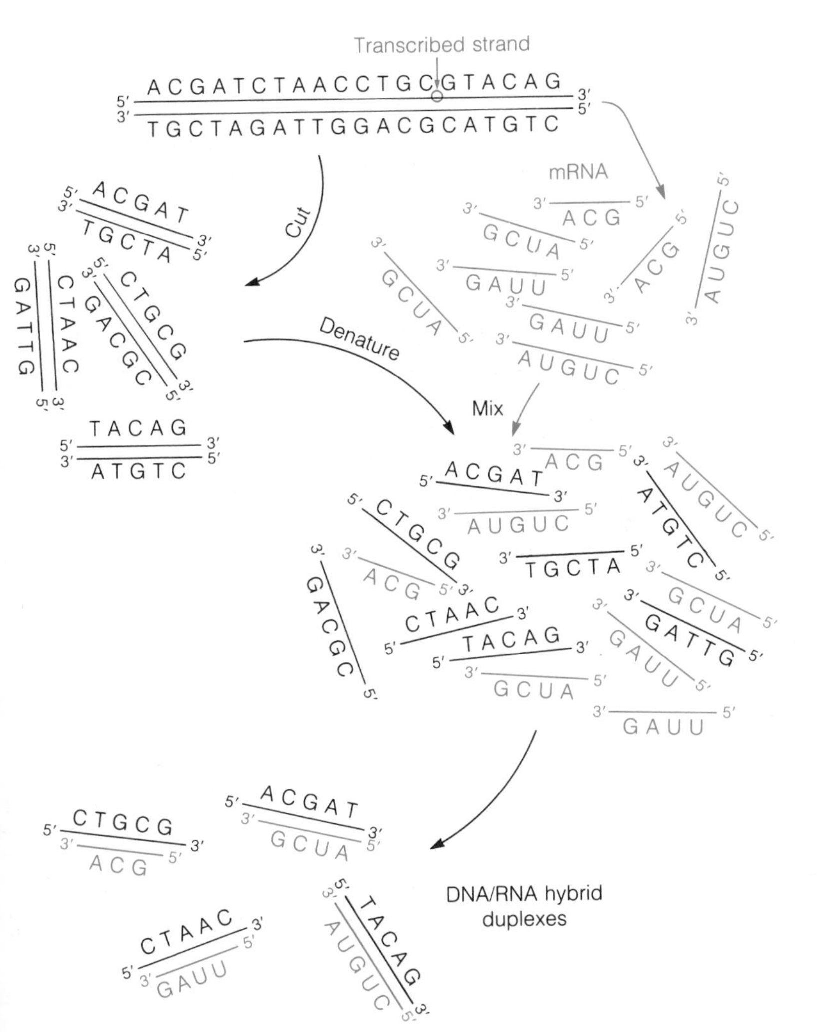

FIGURE 10-13

Formation of DNA/RNA hybrid duplexes. Duplex DNA is cut into fragments, denatured, and mixed with mRNA. mRNA molecules base-pair with complementary DNA molecules and form DNA/RNA hybrid duplexes. Many of the single-stranded DNA and RNA molecules in the mixture cannot form duplexes because of the absence of a complementary single-stranded molecule.

ecules that are present in nuclei of kidney cells represent transcription of about 4.0 percent of the total DNA. If the RNA molecules in kidney cells are the same as those in liver, that is, transcribed from the same group of genes, then a mixture of liver and kidney RNAs should hybridize with 4.5 percent of the DNA. If the RNA molecules in kidney cells are completely different from those in liver (that is, transcribed from a completely different set of genes), then a mixture of liver and kidney RNAs will hybridize with $(4.0 + 4.5) = 8.5$ percent of the DNA. The actual result is intermediate: a mixture of RNA molecules from the two kinds of cells hybridizes with 7.5 percent of the DNA, which means that some of the RNA molecules in the two cell types are the same but most are different. Similarly, liver cells contain kinds of RNAs not found in muscle or brain cells, muscle cells contain RNAs that are undetectable in liver and brain cells, and other kinds of RNA molecules are found only in brain cells.

A fundamental question is how these differences in RNA populations arise. One of the most thoroughly studied examples is the activation of expression of hemoglobin genes in development and during differentiation of erythrocytes.

Activation of Hemoglobin Genes in Development

Hemoglobin consists of four polypeptide subunits, two α-globin subunits and two β-globin subunits. The α- and β-globins in mammals are different in amino acid sequences but have sufficient similarities to suggest that they may have evolved from a common ancestral type. This idea is strongly supported by the studies of α- and β-globin in many vertebrates, which indicate that the α- and β-globin genes began to diverge from one another about 500 million years ago. Divergence led to the two distinctly different kinds of subunits that, when combined as a tetramer (containing four subunits) of two α- and two β-globins, form an extraordinary efficient carrier of O_2.

In mammals there are several different forms of α- and β-subunits differing by one to several amino acids and reflecting further gene evolution. The main forms in the α family are α_2 and α_1, and the main forms in the β family are ε, γ_G, γ_A, δ, and β. The different forms are expressed at different stages of human development (Figure 10-14). Within three weeks after fertilization the first red blood cells form, and the heart begins to pump blood. The hemoglobin in the first blood cells and continuing up to about 8 weeks is made up of ζ_1 and ζ_2 chains (α family) and two ε chains (β family). At eight weeks transcription of these genes is turned off and formerly silent genes for α_1- and α_2-globins (α family) and γ_G- and γ_A-globins (β family) are transcribed in newly differentiating blood cells. At about 30 weeks after fertilization the β-globin gene begins to be transcribed. By the time of birth, transcription of γ_G and γ_A genes has been almost completely shut off, and most hemoglobin molecules produced from that point on

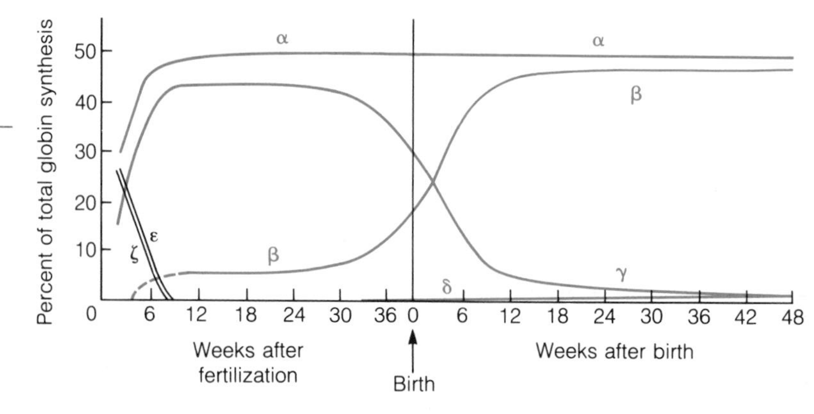

FIGURE 10-14
Hemoglobin switching in development. Hemoglobin synthesis begins at about three weeks after fertilization. Synthesis of ζ- and ε-globin ceases by eight weeks. α-globin increases to 50 percent at about 20 weeks and remains at 50 percent through life. γ-globin is dominant during fetal development but is replaced by β-globin. A small amount of δ-globin is synthesized beginning around birth. [From S. Karlsson, and A. W. Nienhuis. 1985. *Annual Review of Biochemistry*, 54: 1075.]

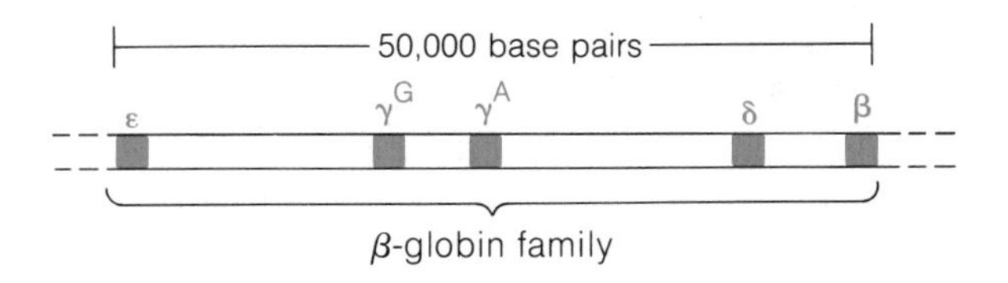

FIGURE 10-15
The five members of the β-globin family occur within a region of 50,000 base pairs in human chromosome number 11. Most of the 50,000 base pairs form spacers of unknown function between genes.

consist of two α and two β chains with no further switching during the life of the individual. However, around the time of birth transcription of the δ-globin gene (β family) begins, so that about two percent of the total hemoglobin in an adult is made up of two α and two δ chains.

The genes of the α family (α_1, and α_2) are located as a cluster in chromosome 16, and the five genes of the β family (ε, γ_G, γ_A, δ, and β) form a cluster in chromosome 11 of humans. The arrangement of the β-globin family of genes in DNA is shown in Figure 10-15. The physical order is the same as the order in which transcription is activated in development except that transcription of the δ-globin gene begins after transcription of the β-globin gene. We do not know whether this arrangement is a coincidence or reflects something about the transcription switching mechanism during development. The segment of DNA containing the five β-globin family genes is 50,000 bp in length. Only a small fraction codes for transcripts, and most of the DNA forms *spacers* between transcription units (Chapter 11).

The developmental succession of expression of hemoglobin genes is sometimes modified in the group of genetic diseases called **thalassemia**. In β-thalassemia, insufficient β-globin is produced because of mutations, such as a mutation in the TATA box of the β-globin gene, or deletions in the promoter. The disease varies in severity depending on the degree of impairment of synthesis of β-globin. In some severe cases a deletion removes the entire β-globin gene and extends upstream into the gene for δ-globin. The disease occurs only if β-globin genes are affected in both chromosome homologues. One normal copy of the gene is sufficient for full hemoglobin production. Persons with defects in both chromosomes suffer from anemia, and depending on the severity of the defects, require lifelong blood transfusions.

In some cases of thalassemia, transcription of γ-globin genes, which are normally switched off at birth, continues into adulthood. This continuation of transcription significantly, but not completely, ameliorates the anemia due to lack of β-globin. It is possible that the deletion in such cases has removed sequences in the region between the δ- and β-globin genes that are required for switching off transcription of the γ-globin gene.

There are two major questions about the regulation of expression of the β-globin gene family in development. Why are these genes expressed only in differentiating red blood cells (erythroid cells), and how is switching from one member of the gene family to another signaled in development? When introduced into the mouse genome, an extra adult mouse β-globin gene, including 1300 bp of the 5′ flanking region, is transcribed only in erythroid cells and undergoes proper switching during development. This means that whatever sequence is needed to evoke expression of the β-globin gene in erythroid cells and whatever sequence is recognized by the signal that brings about switching during development is located somewhere in the stretch of DNA that includes the coding region of the adult β-globin gene and the 1300 bp upstream from the coding region. Nothing is known about the nature of the switching signal. When erythroid cells synthesizing fetal hemoglobin in a sheep fetus were transplanted into an adult animal, they continued to transcribe the fetal globin genes for many days before finally switching to transcription of adult β-globin genes. This delay suggests that switching is not due to some extracellular signal in an adult animal, but rather indicates that erythroid cells follow an intracellularly programmed course of developmental switching.

Like eukaryotic genes in general, the β-globin gene has a TATA box and a CAAT box at the usual locations upstream from the start of transcription. Presumably these come into play only after the chromatin has been loosened by factors that regulate transcription. In adult erythroid cells but not in other cell types the region 5′ to the transcribed region of the β-globin gene is hypersensitive to DNase. In human fetal erythroid cells the 5′ regions of the γ_G-, γ_A-, δ-, and β-globin genes are all hypersensitive to DNase although the γ_A- and γ_G-globin genes are by far predominantly expressed. Thus, some additional regulatory element beyond the chromatin loosening reflected by DNase hypersensitivity must be necessary to turn up expression of the δ- and β-globin genes toward the end of fetal life. In adult erythroid cells δ- and β-globin genes remain hypersensitive to DNase. However, the γ_A- and γ_G-globin genes lose their hypersensitivity, which suggests that the permanent switching off of these genes is accompanied by the tighter packing of DNA into chromatin in their 5′ regions.

Developmental Switching of Other Genes

In addition to the expression of globin genes in erythroid cells hundreds of other genes encoding differentiation-specific functions must be expressed in the approximately 200 different cell types in an animal. One set of genes studied in detail is that encoding actin. Like the family of β-globin genes, actin genes are present as a family. The different members of the family differ slightly in the amino acid sequences they encode. The different forms of actin are transcribed in a temporally and spatially regulated manner during development. Expression of the six actin genes of *Drosophila* has been well characterized. Messenger RNA molecules for the six actins show differences in accumulation during development. The six genes can be grouped into three sets of two each, according to the temporal patterns of appearance of their products during development. Two of the actin genes encode cytoplasmic actins that are synthesized predominantly during early development and are probably components of the cytoskeleton. These are the only actins found in permanent cell cultures of *Drosophila,* cultures in which the cells actively proliferate and show no overt differentiation. Transcripts of the second set of two actin genes become abundant during later embryogenesis and larval growth, when larval musculature is formed. The actins encoded by these genes become part of the contractile machinery in larval muscle (Chapter 14). The last two actin genes are expressed later in development, after metamorphosis of the larva into a fly, and specify actins in muscles of the adult.

Similarly, the chicken contains at least three actin genes that undergo developmental switching. The α-actin gene encodes muscle actin and the β- and γ-genes encode actins found in all nonmuscle cells. In undifferentiated cells (**myoblasts**) that give rise to muscle, most of the actin mRNA molecules code for β- and γ-actin and a few percent code for α-actin. As differentiation proceeds, the concentration of mRNA for α-actin increases several hundred-fold, and mRNA molecules for β- and γ-actins gradually decline to undetectability.

In developmental switching of the α-amylase gene in the mouse, the underlying mechanism seems to be different. The mouse genome contains a single gene encoding the enzyme α-amylase, which is synthesized in the parotid gland (a salivary gland) and secreted into the oral cavity, where it catalyzes the breakdown of starch and glycogen to glucose. Therefore α-amylase is an important digestive enzyme. Expression of the gene for α-amylase is controlled by two promoters, a weak promoter located close to the coding region of the gene and a strong promoter farther upstream. Shortly after birth, transcription is initiated by the weak promoter. As development of the mouse proceeds, more and more cells switch to use of the strong promoter, and by adulthood only the strong promoter is active. A consequence of unknown significance is that mRNA molecules in the adult have a longer leader sequence, because the strong promoter specifies an initiation site that is farther upstream.

A similar example occurs for the gene encoding alcohol dehydrogenase in *Drosophila.* This enzyme is encoded by a single gene with two promoters. In the larval stage a promoter just upstream from the coding region of the gene is used. In the adult fly, switching to a promoter farther upstream occurs, and results in a longer transcript being produced. We do not know the significance of developmental switching of promoters

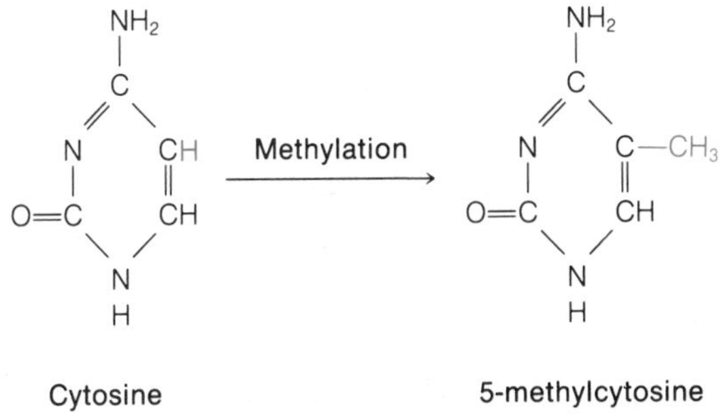

FIGURE 10-16

Some cytosines in the DNA of vertebrates and higher plants are methylated to form 5-methylcytosine.

for the α-amylase gene in mouse and the alcohol dehydrogenase gene in *Drosophila.* (More discussion about alternative transcripts from single genes is presented in a later section.) Using promoters of different strengths would be one way to provide two rates of transcription. In addition, a change in length of the leader sequence could conceivably affect efficiency of translation.

GENE EXPRESSION AND METHYLATION OF DNA

In the DNA of vertebrates and higher plants some of the cytosine bases have a methyl group attached to carbon 5 (Figure 10-16). The fraction of cytosine residues with methyl groups ranges from 0.7 percent to 8.0 percent in different species of animals and can also vary from tissue to tissue in a single species. In some higher plants 50 percent of the cytosine residues are methylated. Methylation of DNA takes place on cytosine residues already in DNA, not by incorporation of 5-methylcytosine (C^{me}) during DNA replication. Methyl groups are added to cytosine by the catalytic action of **methyltransferase.** More than 90 percent of the cytosine residues that are methylated are followed by a guanine in the same chain (5′-C^{me}G-3′), and the cytosine in the 5′-CG-3′ of the opposite strand is also methylated giving the form

$$\begin{matrix} 5'C^{me}G\ \ 3' \\ 3'G\ \ C^{me}5' \end{matrix}$$

When the DNA replicates, the new complementary chains are initially unmethylated. The presence of C^{me}G in a parental chain is recognized by methyltransferase, which methylates the cytosine in the complementary chain. In this way the pattern of methylcytosines in the DNA is maintained through cell reproduction.

Methylation of cytosine residues in the DNA 5′ to the coding region of some genes may have a role in regulating expression of that gene. Evidence has come from the following kind of experiment. Animal cells in culture can take up DNA from the medium under appropriate conditions. Some of the DNA may become stably integrated into chromosomes through breakage and rejoining of chromosomal DNA. Other DNA fragments may be taken into the nucleus in a metastable state, presumably not integrated into chromosomal DNA. The experimental procedure of DNA transfer is known as **DNA transformation.** Both stably and metastably incorporated DNA can express genetic functions, provided that genes have remained intact during DNA transformation. Mammalian cells transformed with methylated γ-globin genes and simultaneously with unmethylated β-globin genes express the β-globin genes but not the γ-globin genes, suggesting that methylation of the γ-globin genes suppresses their expression. When an unmethylated γ-globin gene is introduced, it is transcribed. In fact, when most of the coding region is methylated, but a region extending from 760 bp upstream to 100 bp downstream of the 5′ end of the coding region of the gene remains unmethylated, the gene is still transcribed. This result fits with the hypothesis that methylation of certain cytosines in the vicinity of the 5′ end of the gene are crucial for preventing transcription.

Another example is transcription of genes for **vitellogenin,** a protein precursor that is posttranslationally split into two proteins, phosvitin and lipovitellin. Both proteins are present in large amounts in the yolk of avian and amphibian eggs. Transcription of the vitellogenin genes in chickens is activated in liver cells by the presence of the hormone estrogen in the blood (discussed in the next section). Activation of transcription is associated specifically with loss of methylation of a cytosine residue about 600 bp upstream from the transcription initiation site. Other sites in the 5′ flanking region remain methylated and do not affect transcription.

In differentiated cells of mammals about 70 percent of the total CG sequences in total DNA are methylated. However, only 20 to 30 percent of CG sequences are methylated in that part of chromatin engaged in transcription, consistent with the idea that demethylation may be necessary for transcription. Also, when **5-azacytosine** is incorporated into DNA, activation of genes sometimes follows. 5-azacytosine is an analog of cytosine that contains a nitrogen atom in position 5 of the pyrimidine ring instead of a carbon atom (Figure 10-17). The nitrogen cannot be methylated, and incorporation of 5-azacytosine in place of cytosine during DNA replication results in reduced methylation of the DNA and activation of some formerly silent genes. In one of the first reports of the gene-activating ability of 5-azacytosine, mouse fibroblasts were converted to muscle cells as a result of the incorporation of 5-azacytosine into DNA. This conversion is likely to require a complex change in the pattern of gene transcription.

5-methylcytosine 5-azacytosine

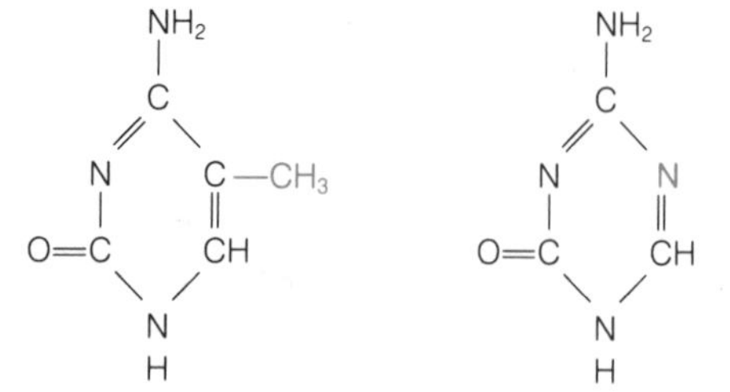

FIGURE 10-17
5-methylcytosine has a methyl group attached to the number 5 carbon. 5-azacytosine has a nitrogen atom at position 5 in the ring and cannot be methylated.

The DNA in an inactive X chromosome is in some way structurally modified, as shown by the following experiment. The enzyme hypoxanthine phosphoribosyl transferase (HPRT) is encoded by a gene in the X chromosome in humans. Fibroblasts taken from a female with a mutation in the gene in one of the X chromosomes (heterozygous for the *HPRT* gene) were grown in culture as clones. In some clones HPRT was present because the normal gene was carried by the euchromatic X chromosome and the mutated gene was carried by the inactive X chromosome. Other clones lacked HPRT because the X chromosome carrying the normal gene was inactive. DNA purified from each of the two kinds of clones was used to transform mouse cells permanently lacking the enzyme. Only DNA from human cells with the normal *HPRT* gene in the active chromosome resulted in appearance of the enzyme in the mouse cell. DNA purified from human cells with the normal *HPRT* gene in an inactive X chromosome failed to stimulate mouse cells to make the enzyme. This means that the DNA sequence of the *HPRT* gene in the inactive X chromosome must have been different (modified) compared with the same DNA sequence in the active X chromosome.

Whatever the modification of DNA that occurs in transcriptional inactivation of an X chromosome, it is reversed during formation of oocytes. A current hypothesis is that X-chromosome inactivation (and heterochromatin formation) involve methylation of cytosine residues, a potentially reversible modification. The idea fits with the inactivating effect of methylation discussed earlier. Supporting evidence comes from an experiment with cells lacking the enzyme HPRT but carrying the gene in an inactive X chromosome. When such cells incorporate 5-azacytosine into their DNA, the enzyme begins to be produced in a few percent of the cells. The experiment is interpreted to mean that reduced methylation of DNA resulting from incorporation of 5-azacytosine allows the formerly inactive gene for *HPRT* to be expressed. Another possible mechanism is that reduced methylation activates a gene in another chromosome that is possibly a protein that regulates expression of the *HPRT* gene.

Constitutive heterochromatin contains a higher percentage of methylated cytosines in DNA than does euchromatin. However, using antibodies that bind specifically to DNA containing 5-methylcytosine, no difference in 5-methylcytosine content in active and inactive X chromosomes can be detected. One possibility is that gene inactivation in the X chromosome is achieved by an increase in methylation too small to be detected by antibody binding.

Some observations contradict the hypothesis that methylation is important in regulation of transcription. Some inactive genes are unmethylated. At least one kind of gene, that coding for rRNA in the frog *Xenopus*, is actively transcribed when injected into the nucleus of an oocyte of *Xenopus*. Even though this gene is methylated at all 19 CG sites in the promoter region, transcription takes place.

In short, it seems likely that methylation of DNA contributes to the regulatory mechanism for some genes in higher eukaryotes, but certainly not for all genes. Finally, activation and inactivation of transcription by demethylation and methylation cannot be a regulatory mechanism that applies generally to eukaryotes. Invertebrates have very low levels of methylation in their DNAs, and the DNA of *Drosophila* and some unicellular eukaryotes totally lacks 5-methylcytosine.

REGULATION OF GENE EXPRESSION BY HORMONES

Within multicellular organisms cells in one specific tissue may modify cellular activities in another specific tissue by means of a chemical messenger called a **hormone.** Some hormones, like insulin and glucagon, modulate enzyme activities in target cells. Insulin is a small protein synthesized by islet cells (β cells) of the pancreas and released into the blood. It becomes bound to specific receptor proteins in the plasma membranes of cells in several target tissues, particularly liver cells (hepatocytes), muscle cells, and fat cells (**adipocytes**). The binding of insulin to receptor proteins stimulates synthesis of glycogen in muscle cells and hepatocytes, accelerates glycolysis in liver cells, and promotes transport of glucose into muscle cells and adipocytes.

The ingestion of carbohydrates by an animal increases the sugar level in the blood. It also stimulates the β cells of the pancreas to synthesize and release insulin, which, by the various actions on target tissues just cited, brings about the lowering of the sugar level in the blood. By these various actions it helps maintain a steady level of glucose in the blood. Glucagon is a small polypeptide hormone secreted by α cells of the pancreas in response to a low blood-sugar level. Glucagon's target is the hepatocyte, in which it inhibits glycogen synthesis and stimulates breakdown of glycogen to glucose. The hepatocytes release the glucose, raising the sugar level in the blood. In addition to such regulation there is a neurologically induced, emergency response in which the hormones **norepinephrine** and **epinephrine** (adrenalin) secreted by the adrenal gland and sympathetic nerve endings stimulate the breakdown of glycogen to glucose in muscle cells and hepatocytes, stimulate the secretion of glucagon, inhibit the secretion of insulin, and promote the release of stored fat molecules (triacylglycerides) from adipocytes into the blood. These changes form the physiological basis of the adrenalin-stimulated flight/fight reaction of the body whereby energy reserves are mobilized to meet a sharp demand for energy by body tissues. There are many other hormones that, in various ways, coordinate and integrate the activities of cells in different tissues. Some of these are discussed in other chapters.

For purposes of this chapter the most important hormones are the steroids (e.g., estrogens, progesterone, androgens, and glucocorticoids), which are synthesized from cholesterol. Rather than regulating enzyme activity, these hormones modulate transcription of particular genes in target cells. Steroids do not bind to cell-surface receptors, but because they are hydrophobic molecules, readily pass through the lipid bilayer of the plasma membrane. Target cells contain receptor proteins in the cytoplasm and/or the nucleus, and steroid molecules bind to these to form hormone-receptor complexes. Binding of a hormone molecule presumably changes the configuration of the receptor, allowing it to bind to chromatin containing specific target genes. Binding of the hormone-receptor complex activates transcription of the target genes, but not much is known about the molecular details.

In mammals the estrogen **β-estradiol** is synthesized by cells of the ovary, and its principal targets are the mammary gland and the uterus. In the uterus activation of transcription of particular genes leads, among other things, to proliferation of uterine epithelial cells.

In birds estrogen activates transcription of the gene encoding **ovalbumin** in cells of the oviduct. Ovalbumin is the principal protein of egg white, and its production is regulated by estrogen as the egg yolk passes through the oviduct and receives its envelope of ovalbumin. In both birds and amphibians estrogen also activates transcription in the liver of genes encoding the protein **vitellogenin.** Vitellogenin is released into the blood by hepatocytes and is then taken up by pinocytosis by growing oocytes in the ovaries. If the estrogen concentration in the blood falls, transcription of the gene for vitellogenin is quickly turned off.

In birds the synthesis of ovalbumin in the oviduct and vitellogenin in the liver is coordinated by a variety of effects of estrogen secreted by the ovary. Estrogen also induces synthesis of its own receptor protein, but whether this is a case of transcriptional or post-transcriptional control is not known. In *Xenopus*, estrogen causes stabilization of mRNA for vitellogenin. In the presence of estrogen the mRNA has a half-life of three weeks; in its absence the half-life falls to 16 hours.

In plants, activities of different cell populations are regulated by a variety of hormones. One class of plant hormones is the **auxins.** The principal auxin is indole-3-acetic acid, which is synthesized from tryptophan. This hormone regulates cell reproduction and cell elongation in certain tissues. At least some of its regulatory effects are at the transcriptional level, activating several genes in target tissues. Target tissues contain receptor proteins to which auxin binds. Presumably, it is the auxin-receptor complex that mediates gene activation.

It is clear that hormones have major roles in regulating transcription of genes. One of the problems in unraveling this hormone action is to identify the molecular pathways by which the hormone turns on or increases transcription of specific genes. The simplest hypothesis is that the hormone-receptor complex recognizes transcription regulation sites of specific genes, or perhaps specific proteins bound to these sites. Another mechanism we know very little about is that which turns on the synthesis of receptor proteins in particular cells, thereby making them specific targets of particular hormones.

CYTOLOGICAL OBSERVATION OF GENE ACTIVATION IN POLYTENE CHROMOSOMES

Ordinarily, it is impossible to see discrete chromosomes in the cell nucleus with either the light microscope or the electron microscope. Usually, the only time in the life of the cell when chromosomes are visible as discrete structures is during meiosis or mitosis. During mitosis the extremely long thread that is the chromosome (Chapters 7 and 11) coils up to form a compact structure that is easily visible by light or electron microscopy. In chromosomes that are condensed into the mitotic form all gene transcription (RNA synthesis) ceases (Chapter 12). When mitotic chromosomes decondense after mitosis is over, transcription resumes. In any case, the study of mitotic chromosomes has taught us almost nothing about gene activity and its regulation.

There is an exception to the invisibility of the interphase (nonmitotic) chromosomes. In certain cells of the larvae of many flies and midges (the *Diptera*) and some other insects, the interphase chromosomes duplicate and reduplicate many times without an accompanying nuclear division (Figure 10-18(a)). The multiple copies of a single chromosome, numbering up to several thousand identical threads, are aligned in parallel, forming a structure that is easily seen in the light microscope (Figure 10-18). These are called **polytene chromosomes** (*poly* = many; *tene* = thread). Polytene chromosomes are genetically active, interphase chromosomes. They also occur in certain plant cells and in some ciliated protozoans but have been most extensively studied in larvae of *Drosophila* and of the midge, *Chironomus.* Polytene chromosomes do not enter mitosis, and therefore, cells that contain them are incapable of reproduction.

A striking feature of polytene chromosomes is their distinct banding pattern. A given chromosome, for example, chromosome 2 in *Drosophila,* always shows the same banding pattern. Genetic studies have shown that a particular gene is always found in the same band. Other studies also suggest that most bands contain only a single type of gene, but at least a few bands have more than one gene. Any given gene is, of course, present

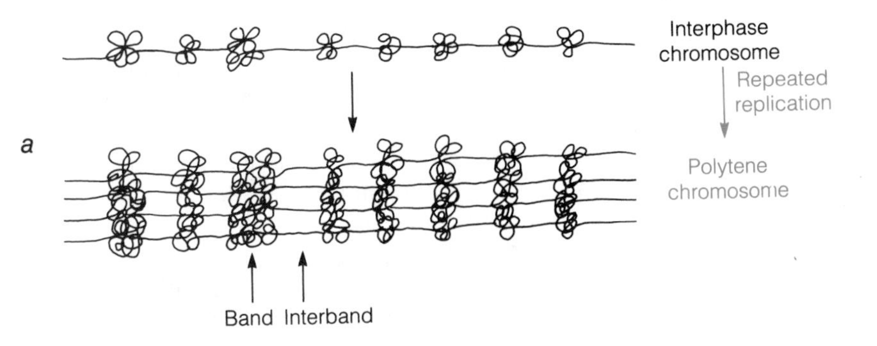

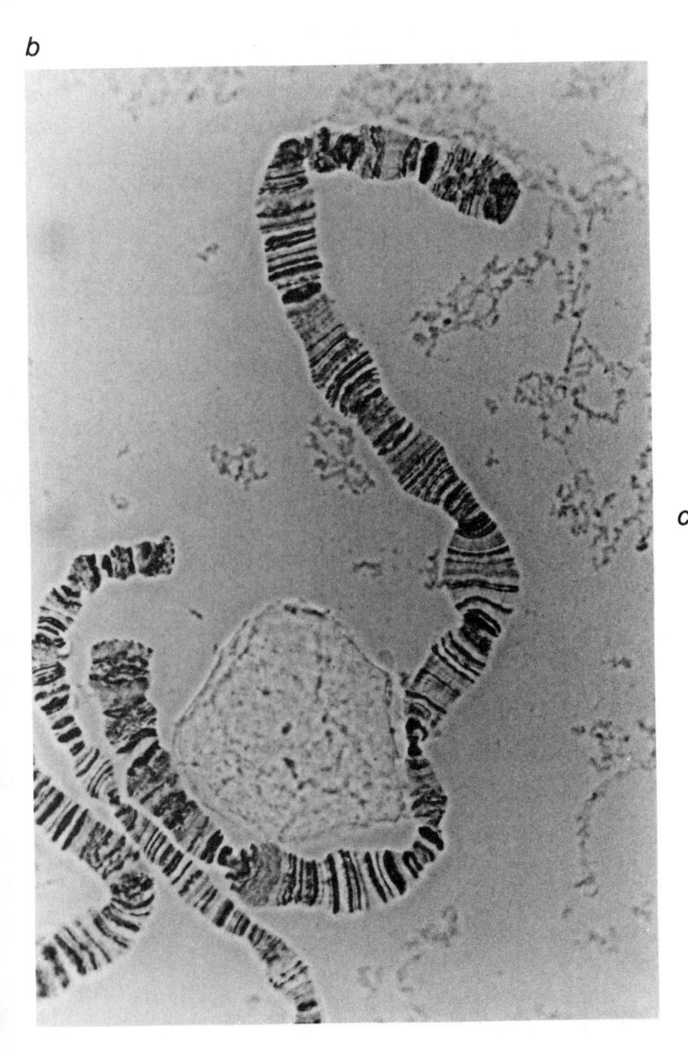

FIGURE 10-18

(a) Polytene chromosomes form by repeated replication of the DNA duplex in an interphase chromosome. In this simplified diagram only four copies of the DNA-histone thread are shown in the polytene chromosome; over one thousand such threads are present in the chromosome in (b). (b) Light micrograph of one polytene chromosome and portions of two others isolated from the salivary gland of a fruit fly larva (*Drosophila silvarentis*). (c) Polytene chromosomes from the larval salivary gland of *Drosophila*. The chromosomes are attached to one another at heterochromatin-containing regions around centromeres. The left (L) and right (R) arms of chromosomes 2 and 3 are indicated. The centromere of the X chromosome is very near one end and, therefore, only one arm is present. Chromosome 4 is very short, with the centromere at one end. [(b) Courtesy of Jong Sik Yoon. (c) By permission of George Lefevre.]

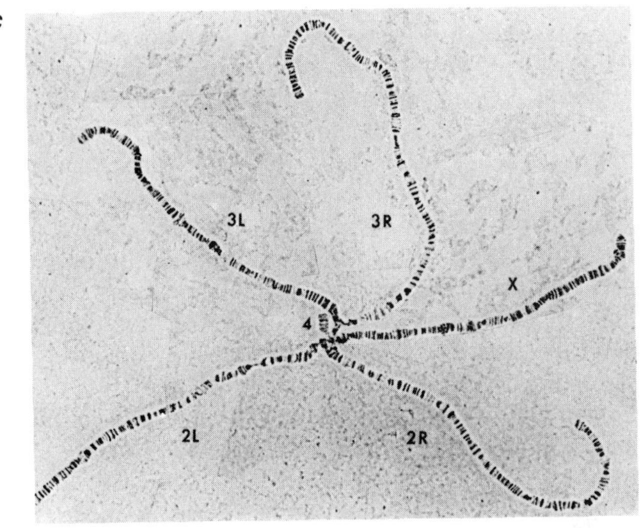

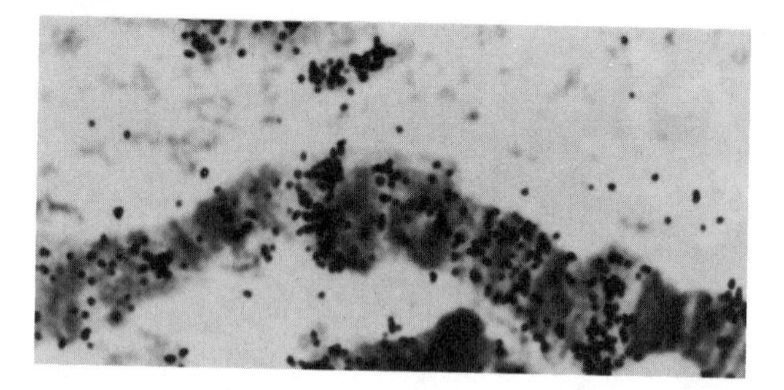

FIGURE 10-19
Autoradiograph by light microscopy of a portion of a polytene chromosome *Drosophila* incubated with ^{3}H-uridine. Tritiated uridine has been incorporated into RNA in many bands of the chromosome. [Courtesy of Jose J. Bonner.]

once in each of the multiple threads that make up the polytene chromosome. *Drosophila* has four different chromosomes and these contain altogether about 5000 bands. (It is important to understand that the bands seen in polytene chromosomes *do not* correspond to the far fewer bands detected in mitotic chromosomes by special staining techniques. The bands of polytene and mitotic chromosomes are discussed further in Chapter 11.)

RNA synthesis occurs in many of the bands along the polytene chromosome, reflecting the transcription of genes. This can be demonstrated by autoradiographic detection of the incorporation of radioactive precursors (for example, ^{3}H-uridine) (Figure 10-19). Genes in certain bands become especially active in RNA synthesis, apparently in response to demand for a large amount of specific gene products. When this occurs, the tightly packed threads in a band unravel and bulge out, forming a pufflike structure (Figure 10-20(a)). Puffing of a band is accompanied by development of hypersensitivity of the DNA to degradation by DNase. Thus, the unraveling that occurs in puffing is reminiscent of the loosening of chromatin underlying the appearance of hypersensitivity to DNase that accompanies activation of genes discussed earlier. Again, ^{3}H-uridine incorporation and autoradiography can be used to show that puffed bands are sites of particularly intense RNA synthesis (Figure 10-20(b)). Particularly large puffs in the midge *Chironomus* are called **Balbiani rings** after the cytologist who first described them.

Different tissues in the larva of *Drosophila,* such as salivary gland, Malpighian tubules (kidney), and intestine, show different puffing patterns. These patterns change as the larva develops toward the next phase of development, the pupal stage. Certain puffs are seen in all tissues at all stages of larval development. These presumably represent expression of genes coding for functions that are needed continuously in all tissues. Some puffs occur in all tissues but only at a certain stage of development, for example, when the larva molts. Other puffs occur only in certain tissues. In *Drosophila* over 100 bands undergo puff formation and re-

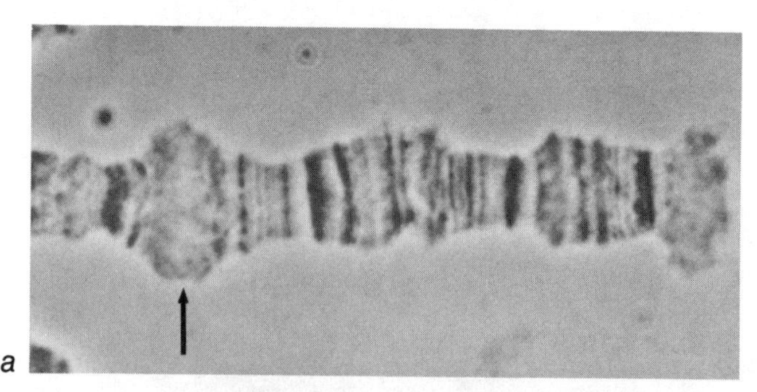

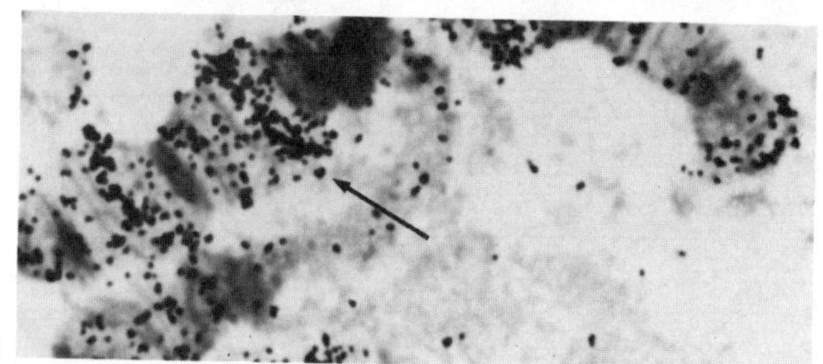

FIGURE 10-20
(a) Light micrograph of a polytene chromosome (end of the long arm of the third chromosome) isolated from a *Drosophila* salivary gland cell. The chromosome contains a prominent puff (arrow). (b) Autoradiograph by light microscopy of part of a polytene chromosome of *Drosophila* incubated with ^{3}H-uridine. Tritiated uridine incorporation into RNA is particularly intense in a puff (arrow). [(a) Courtesy of Joseph G. Gall. (b) Courtesy of Jose J. Bonner.]

gression in a well defined temporal sequence during larval development.

Activation of heat-shock genes, discussed earlier in the chapter, is also easily observed in polytene chromosomes. In fact, heat-shock genes were originally discovered by the formation of a specific set of puffs in response to heat shock applied to larvae of *Drosophila*.

The puffing of some bands is under hormonal control. Molting of the larva is normally induced by the insect steroid hormone, **ecdysone,** which is synthesized in a gland at particular times in development and released into the **hemolymph** (insect equivalent of blood). Ecdysone, like steroid hormones in vertebrates, binds to a receptor protein in target cells, and the ecdysone-receptor complex binds to specific sites on the polytene chromosomes, inducing transcription at those sites. What triggers synthesis of ecdysone, and hence molting, is not well understood, but it is certain that molting requires highly specific changes in gene expression. Cell biologists have studied in detail the cause-and-effect relation between ecdysone and activation of genes by injecting ecdysone into larvae to induce molting and then watching for changes in the puffing pattern in the polytene chromosomes. Ecdysone induces puffing in three chromosome bands within minutes of its injection in larvae. As expected, the formation of the three puffs reflects intense RNA synthesis, as shown by incorporation of ^{3}H-uridine into RNA in the puffs detected by autoradiography.

Five other bands undergo puffing four to six hours after injection of ecdysone. These delayed puffs are not directly induced by ecdysone but come about through a train of events triggered in the cell by ecdysone. This is shown in experiments with the drug cycloheximide, a specific inhibitor of the peptidyl transferase of eukaryotic ribosomes, which blocks translation of mRNA into polypeptides but does not directly affect RNA synthesis. When ecdysone and cycloheximide are simultaneously injected into a larva, the first three puffs appear on schedule. This means that the induction of the first three puffs by ecdysone does not require synthesis of new proteins. However, injection of cycloheximide prevents formation of the five other puffs that would normally appear four to six hours later. This means that protein synthesis is necessary for induction of RNA synthesis in the late puffs. A plausible but unproven explanation is that one or more of the first three puffs codes for a protein that is instrumental in the induction of the later puffs (Figure 10-21). That is, when synthesis of the protein encoded by a gene in one of the early puffs is blocked by cycloheximide, the later puffs cannot be induced.

Cytoplasmic Modulation of Nuclear Gene Expression

At least some of the factors that regulate transcription in eukaryotes are proteins. These may not be confined to the nucleus but move back and forth between nucleus and cytoplasm (e.g., steroid hormone receptors). Experiments with cell fusion demonstrate directly the presence in the cytoplasm of factors that profoundly affect transcription in the nucleus.

Fusion of a chicken erythrocyte with a mammalian cell growing in culture creates a **heterokaryon** (*hetero* = other or different; *karyon* = nucleus), a cell with two nuclei. Originally only the nucleus of the mammalian cell is active in RNA transcription because the chicken erythrocyte nucleus contains only heterochromatin, and hence all of its transcription has been shut down. Shortly after cell fusion, the erythrocyte nucleus begins to enlarge, much of the heterochromatin is converted to euchromatin, and RNA synthesis resumes (Figure 10-22). Much the same happens when a frog erythrocyte nucleus is injected into a frog oocyte, an experiment described earlier. Cytoplasmic factors contributed by the cytoplasm of the mammalian cells are thought to be responsible for reactivation of the inert nucleus of the erythrocyte. To eliminate any doubt about the cytoplasmic origin of the factors, an erythrocyte was fused with a mammalian cell from which the nucleus had previously been removed (a cytoplast) by the cytochalasin-centrifugation technique (Chapter 8); the erythrocyte nucleus was also activated in this case. Nucleoli, which were initially absent from the erythrocyte nucleus, were formed, and many newly synthesized gene products encoded by the chicken genome appeared in

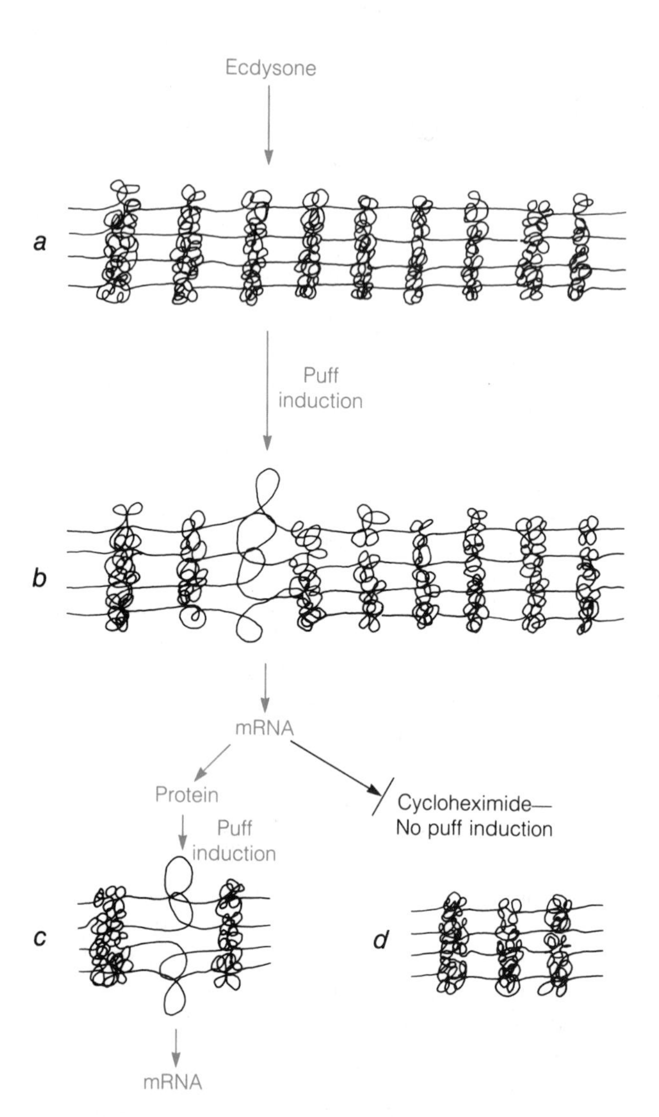

FIGURE 10-21
Schematic diagram of puff induction by ecdysone in a polytene chromosome of *Drosophila*. (a) Polytene chromosome before puff formation. (b) Induction of puffing of a band by ecdysone. mRNA transcription occurs in the puff and the mRNA is translated into a protein product. (c) Induction of a second puff through the action of the protein encoded by the DNA in the first puff. (d) Block of induction of the second puff by an inhibitor (cycloheximide) of protein synthesis.

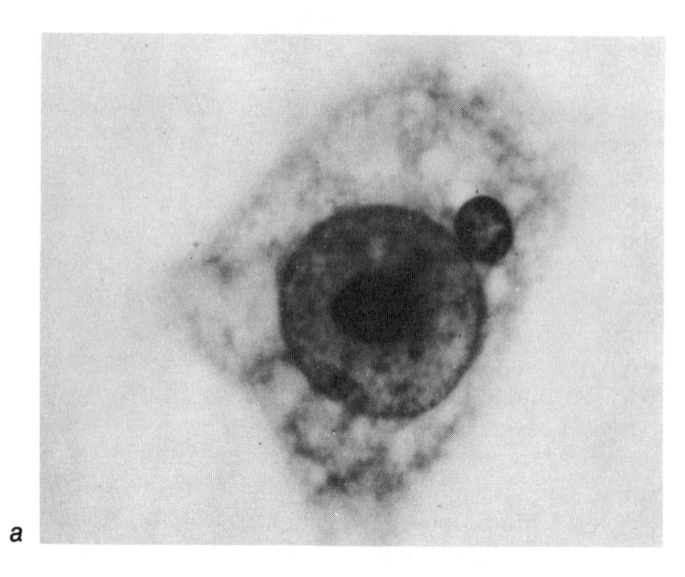

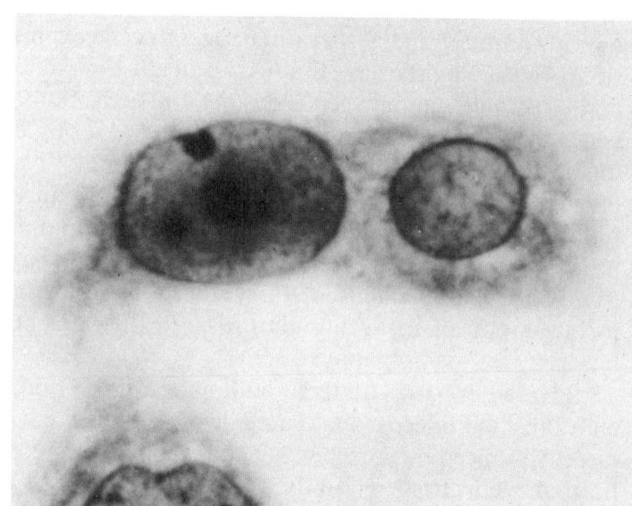

FIGURE 10-22
Reactivation of an erythrocyte nucleus. (a) Light micrograph of a binucleated cell produced by fusing a chicken erythrocyte with a HeLa cell. One day after fusion the erythrocyte nucleus is still small. (b) A day later the erythrocyte nucleus (right) has increased in volume many fold and has resumed RNA synthesis. [Courtesy of Henry Harris.]

the cytoplasm. These products included chicken-specific α- and β-globin, proteins of the plasma membrane, and enzymes. Since the mammalian cell lacked a nucleus, the signals for the synthesis must have come from the cytoplasm.

In a similar experiment a normal human diploid cell from an amnion (the sac surrounding a fetus) was fused with a muscle cell of a mouse. In the resulting heterokaryon, transcription of genes encoding muscle-specific proteins occurred within 24 hours in the human nucleus. These muscle-specific genes, not normally expressed in an amnion cell, were activated by a factor(s) present in the cytoplasm of the mouse muscle cell.

For unknown reasons, fusion of a chicken erythrocyte with an intact mammalian cell or cytoplast does not yield a cell that will proliferate. However, fusion of two animal cells, both of which are independently capable of proliferation, can produce progeny capable of proliferation. The two parental cells can be of the same type, (**homotypic fusion**) or of different types (**heterotypic fusion**). The product of a heterotypic fusion is a **cell hybrid.** Initially, fusion creates a binucleated cell. When the binucleated cell divides, the nuclear envelopes break down and the chromosomes in the two nuclei condense coordinately and become arranged in a single mitotic spindle. The chromosomes are then distributed into two telophase groups, each of which forms a nucleus that contains a set of chromosomes from each original nucleus. If two diploid cells are fused, then each daughter nucleus formed in the first division after fusion is tetraploid. Hybrid cells can be made by fusing two cells in culture from the same species, for example a fibroblast with a skin cell or liver cell. Cell fusions of this type give some insight about regulation of gene expression. When parental cells expressing one or more different genes are fused, in the resulting hybrid cell expression of the particular gene or genes may be extinguished. One possible explanation is that the parental cell not originally expressing the gene contains a factor that turns off expression of that gene in the other parental genome.

A potentially more interesting result is the turning on of new genes in hybrids, for example, in fusion of a human lymphocyte with a liver cell from a mouse. The mouse liver cell produces the protein albumin, which is a liver-specific function. Hybrid cells produce not only mouse albumin but also human albumin. (The two albumins differ sufficiently to be distinguishable when separated by gel electrophoresis.) This means that the gene for albumin, which human lymphocytes do not express, was turned on by a factor present in the mouse liver cell. Since the nuclei in a binucleated cell are separated by cytoplasm, the lymphocyte nucleus must have either acquired the factor from cytoplasm contributed by the liver cell, or at least the transfer was mediated by the cytoplasm.

In a similar experiment, fusion of a human amnion cell with a mouse liver cell led to activation of human genes encoding four serum proteins synthesized by liver cells but not by amnion cells (albumin, transferrin, ceruloplasmin, and α-1 antitrypsin). It will be enormously important for understanding gene regulation to identify the factors responsible for these activations of specific genes.

ALTERNATIVE TRANSCRIPTS FROM SINGLE GENES

In the earlier section on **Developmental Switching of Other Genes** two examples of synthesis of alternative transcripts from a single gene were described for the α-amylase gene in the mouse salivary gland and the gene for alcohol dehydrogenase in *Drosophila.* The significance of alternative transcripts is not well understood, but presumably the use of alternative promoters, producing transcripts with different length leader sequences, is important in regulating gene expression, at the translational level.

Alternative transcripts are also sometimes produced from a single gene by use of different termination sites for transcription at the 3′ end of the gene. An example of alternative termination occurs for the gene that encodes **vimentin,** the protein from which intermediate filaments of the cytoskeleton are formed in fibroblasts and some other types of animal cells. The vimentin gene in the chicken has polyadenylation sites

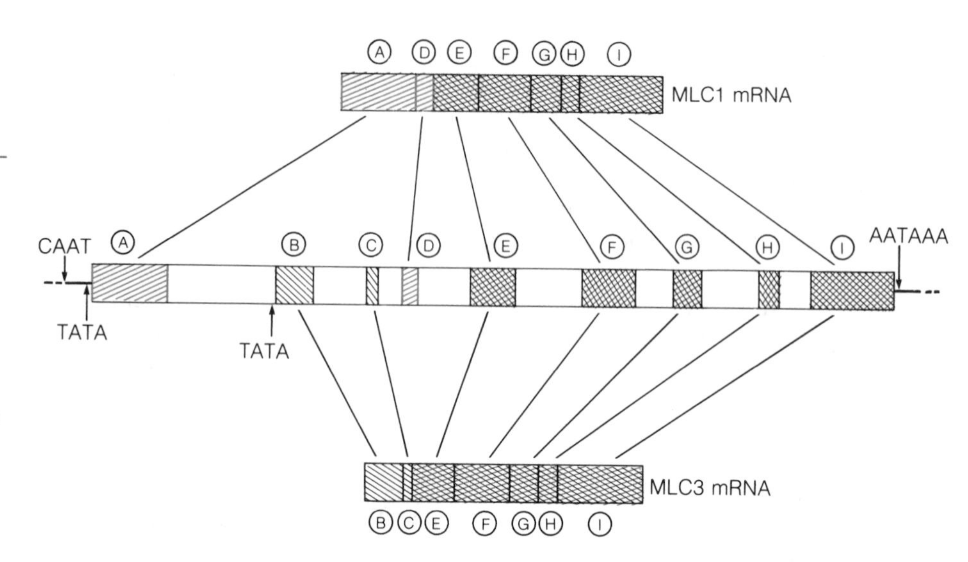

FIGURE 10-23 The gene for myosin light chains in the mouse. A single gene encodes two myosin proteins, MLC1 and MLC2. The transcript (not shown) for MLC1 is copied from the full length of the gene and contains seven exons that are spliced into an mRNA molecule. The transcript (not shown) for MLC2 is shorter, starting at the TATA box at the end of the first intron and contains seven introns that are spliced into an mRNA molecule. Circled letters indicate the nine total exons for both mRNA molecules.

at 249 and 532 bp downstream from the stop codon for translation. Two types of mRNA molecules are produced from primary transcripts, one almost 300 bases longer at the 3′ end than the other. Both mRNA types contain the same protein-coding sequence and yield the same protein. Both forms of mRNA molecules are found in fibroblasts, cells of the eye lens, and several other types, but erythroid cells contain predominantly the shorter form of mRNA. This suggests that the synthesis of two forms of vimentin mRNA is somehow related to cell differentiation, but the significance of this alternate mode of transcription is not known.

The protein myosin (Chapter 14) is a major component of the contractile apparatus in muscle cells. It consists of two identical, long polypeptides (heavy chains) and four shorter polypeptides (light chains). In the mouse genome two of the light chains, myosin light chains 1 and 3 (MLC1 and MLC3), are encoded by a single gene from which two different transcripts are produced (Figure 10-23). The primary transcript for MLC1 is initiated at the 5′ end of the gene, where TATA and CAAT boxes are present 28 bp and 70 bp upstream from the site of initiation of transcription (the cap site). Seven exons (A, D, E, F, G, H, and I in Figure 10-23) are spliced together (removal of six introns) to generate the mRNA molecule for MLC1. The first exon (A) contains the leader sequence of the mRNA and the first part of the coding region for the polypeptide. The sixth exon (H) contains the sequence encoding the end of the polypeptide and the first part of the trailer sequence of the mRNA. The seventh exon (I) completes the trailer; the sequence AATAAA at the end of the trailer signals the addition of a poly(A) tail.

The primary transcript for MLC3 is initiated almost 10,000 bp downstream from the initiation site for the primary transcript for MLC1. Therefore, the initiation site for MLC3 is within what constitutes the first intron in the MLC1 transcript (Figure 10-23). The initiation site for MLC3 is preceded by a TATA box 34 bp upstream from the cap site, but no CAAT box is present. The sequence that constitutes the first intron for MLC1 contains the first and second exons (B and C) of MLC3. The second exon (D) of MLC1 is part of the first intron of the MLC3 transcript. Beginning with exon E both transcripts are identical in exon and intron composition to the ends of the transcripts. Thus, two mRNA molecules are produced from two different length transcripts of the same gene. Each mRNA specifies a different amino acid sequence for the first part of the two polypeptides (MLC1 and MLC3). Most of the amino acid sequence (specified by exons E, F, G, H, and I) is identical in the two proteins. MLC1 accumulates earlier in fetal development than MLC3. How developmental timing, differential transcription, and differential splicing are achieved is not known.

ALTERNATIVE PROCESSING OF PRIMARY TRANSCRIPTS

In addition to synthesis of alternative transcripts, a single transcript can be processed in different ways, producing distinct mRNA molecules that code for similar but different proteins. Alternative processing of transcripts allows two or more proteins to be encoded by a single gene. In all cases studied so far the proteins share extensive regions of amino acid sequence but differ in certain regions of their polypeptide chains.

As an example, the human genome contains a single gene encoding three different forms of the protein fibronectin. Fibronectins are large glycoproteins that have major roles in cell-to-cell adhesion in tissues, cell migration, and maintenance of cell shape. The three fibronectins are translated from three types of mRNA molecules produced by alternative splicing of a single primary transcript (Figure 10-24). A single intron is spliced out by use of the same splicing site at the 5′ end of the intron, but three different 3′ splice sites are used. This causes removal of successively larger segments of the primary transcript, producing three different mRNA molecules. These mRNA molecules have the potential to code for three versions of fibronectin that differ in length because of omission of sections of amino acid sequences in two of them. All other regions of amino acid sequence are the same in the three proteins.

The significance of the alternative splicing of fibronectin transcripts is probably related to multiple domains. Fibronectin contains a number of domains (regions of amino acid sequence) that contribute the

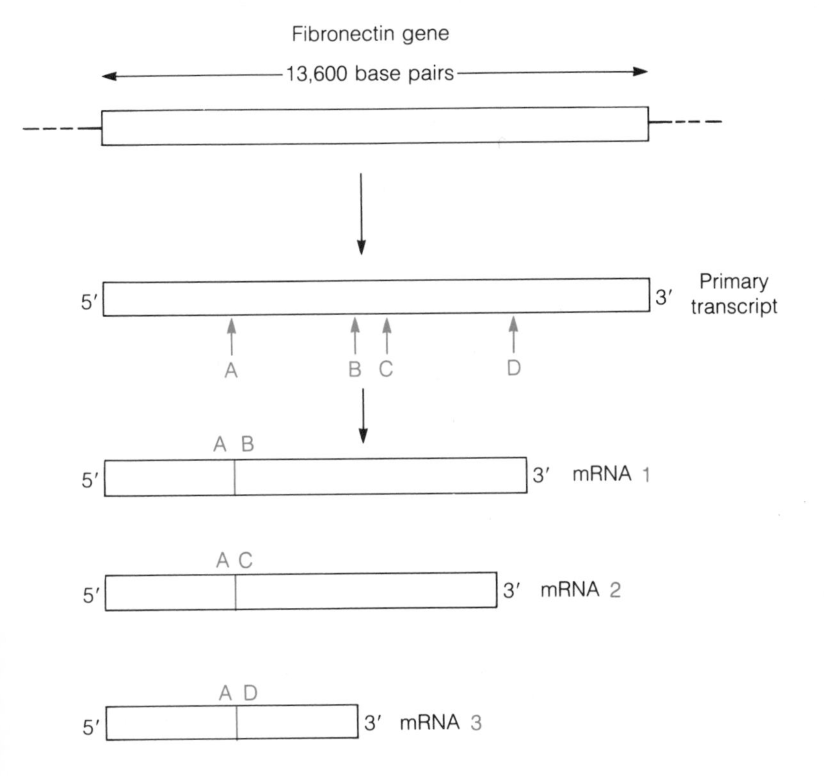

FIGURE 10-24
Probable structure of the fibronectin gene in the rat. Splicing of the primary transcript between A and B, C, or D creates three different mRNA molecules encoding three fibronectin proteins.

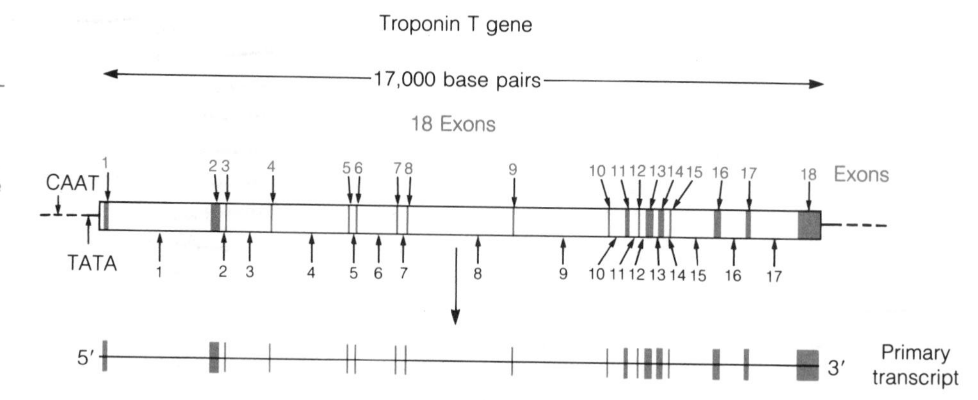

FIGURE 10-25
Organization of 18 exons and 17 introns in the gene encoding the muscle protein troponin T in the rat. Exons account for six percent of the total sequence and introns for 94 percent. Splicing of different combinations of exons yields at least ten different mRNA molecules encoding 10 different proteins. [Redrawn from R. E. Breitbart et al. 1985. *Cell*, 41: 67.]

various properties of the molecule. Such properties include binding of fibronectin to itself, binding of **heparin** (a polysaccharide in blood), binding to **fibrin** (a protein important in blood clotting), and others. The omission of a portion of amino acid sequence in two of the three versions of fibronectin would presumably alter one or more of these binding sites and alter functional properties of the protein.

The list of genes that codes for transcripts that are subject to alternative processing is currently growing rapidly. Several proteins that make up the contractile apparatus of muscle cells are recently discovered examples. An extreme case of alternative processing occurs for the transcript of the single gene that encodes **troponin T,** a muscle protein. The troponin T gene in the rat spans a region of 17,000 bp and is made up of 18 exons and 17 introns (Figure 10-25). The 18 exons account for only six percent of the sequence in the gene. Exon 1 consists exclusively of an untranslated sequence upstream from the protein coding region and therefore represents a spliced leader. Exon 2 includes part of the leader and extends into the protein coding region. Exons 3 through 17 represent coding regions and exon 18 includes the end of the translated region and extends through the 3′ trailer. At least ten versions of troponin T (**isoforms**) are coded by the single gene by alternative RNA splicing. The various isoforms occur in a developmentally regulated and tissue-specific pattern. All isoforms of troponin T contain the same sequence of amino acids from residue 44 through 228 (a constant domain) by splicing of exons 10 through 15. An interaction of this region of the protein with two other muscle proteins is important in regulating muscle contraction (Chapter 14). The amino acid sequence from residues 229 through 242 is variable because it is encoded by either exon 16 or 17, depending on which one is retained in RNA processing. The region from amino acid 11 through 36 is even more variable in sequence and length because it is encoded by various combinations of exons 4 through 8.

Presumably, different isoforms of troponin T are synthesized in different groups of muscle cells and are necessary for achieving subtle differences in the contractile apparatus in different muscle tissues. Why alternative splicing of a transcript from a single gene has evolved instead of a set of different genes regulated in a tissue specific pattern is not understood. As in other cases of alternative splicing, little is known about regulation of splicing of the primary transcript for troponin T.

STRUCTURE AND TRANSCRIPTION OF IMMUNOGLOBULIN GENES

Immunoglobulins (antibodies) are the major proteins of the immune system. They bind to foreign molecules (antigens) and, by forming an antibody-antigen complex, cause them to be inactivated by one of

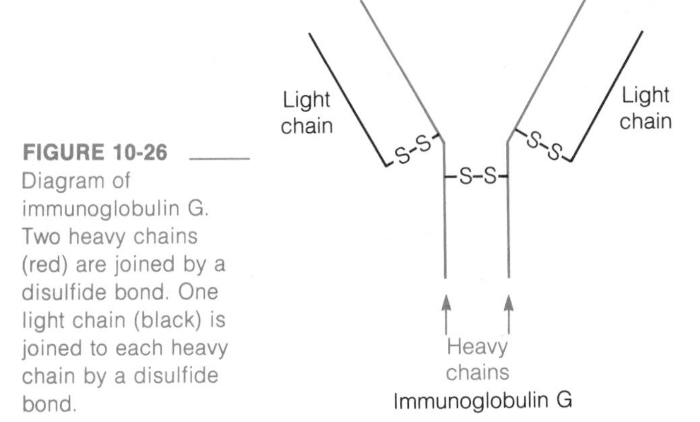

FIGURE 10-26 Diagram of immunoglobulin G. Two heavy chains (red) are joined by a disulfide bond. One light chain (black) is joined to each heavy chain by a disulfide bond.

several mechanisms. As is discussed in Chapter 15, there are five classes of immunoglobulins (Ig), among which **immunoglobulin G (IgG)** is the best understood.

IgG contains two identical long polypeptide chains (**heavy** or **H chains**) and two identical short chains (**light** or **L chains**). The short chains are held to the long chains and the long chains are held to each other by disulfide bonds formed between the SH group of cysteine residues (Figure 10-26). An L and an H chain together form an antigen binding site, so a single IgG molecule possesses two identical antigen binding sites. There are many thousands of types of IgG molecules, each specific for a different antigen. This diversity of IgG molecules resides in both the L and H chains. Each L and H chain is composed of a **variable region** (V) and a **constant region** (C).

The C regions of all L chains (C_L) are identical in amino acid sequence in all types of IgG molecules. Likewise, the C regions of all H chains (C_H) are identical, but different from C_L regions. The V regions of light chains (V_L) differ in amino acid sequence from one IgG type to another, as do the V regions of heavy chains (V_H).

A mammal can produce more than 10^7 different antibody molecules capable of recognizing and binding to more than 10^7 different antigen molecules. Therefore, the genome is able to encode more than 10^7 different combinations of L and H chains. The genome of a species does not contain a separate gene for each L and each H chain, but rather the genes are assembled from banks of sequences in different combinations to yield the diversity of IgG genes.

A single lymphocyte is programmed to produce only a single kind of antibody molecule. The process of assembly of a gene to encode that specific antibody is fairly well understood. Within the IgG class of immunoglobulins there are two kinds of L chains, called **kappa** and **gamma** chains. The mechanisms for generating the two kinds of L chains are very similar (Figure 10-27). Three *DNA segments* are brought together in the assembly of a gene for an L chain: a V segment, a J segment, and a C segment. There are about 300 different V segments in the mammalian genome arranged tandemly in a single chromosome. Some distance from the cluster of V segments is a cluster of four different J segments, followed a short distance along the DNA molecule by a C segment. Each V segment possesses a leader region with sequences that signal initiation of transcription. By comparing the arrangement of V, J, and C segments in embryonic cells (before immunoglobulin genes are assembled from V, J, and C segments) with the arrangement of V, J, and C segments in a series of lymphocyte clones, each clone producing one kind of immunoglobulin, the following picture of gene assembly has emerged. Any one of the 300 different V segments may be joined to any one of the four different J segments by removal of the intervening DNA. For example, in Figure 10-27, V_{200} is joined to J_3 by removal of the intervening segment of DNA (containing segments V_{201} through V_{300}, the VJ spacer, and segments J_1 and J_2), creating a transcribable unit or gene. Transcription is initiated at signals in the leader region of the V_{200} segment, proceeds through the J_3 and J_4 segments, through the JC spacer, and finally through the C segment to produce a primary transcript (Figure 10-27). Processing of the primary transcript removes all the RNA sequence corresponding to the region between the end of J_3 and the C segment (removal of VJ spacer, J_4, and JC spacer) to yield an mRNA molecule consisting of $V_{201}J_3C$. The VJ segment codes for the V region and the C segment for the C region of the immunoglobulin.

The 300 V segments and four J segments in the genome can be joined in 1200 different combinations.

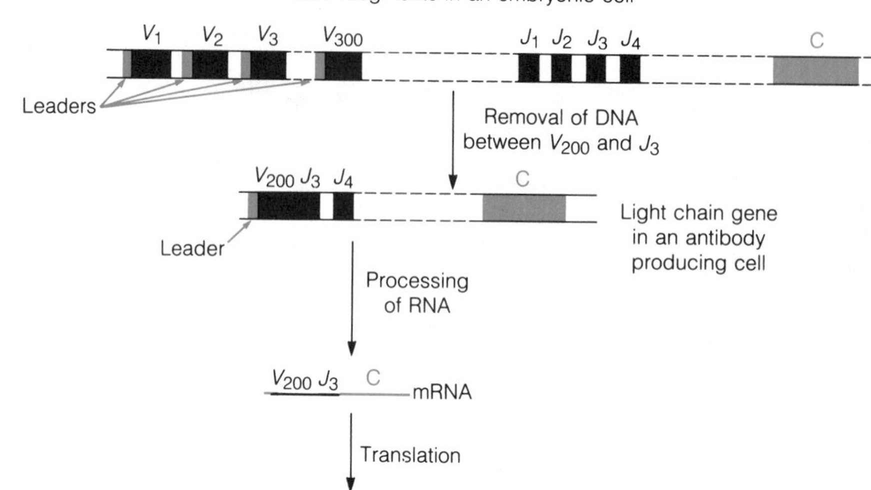

FIGURE 10-27 Formation of a gene encoding an L chain. A V segment is joined to a J segment by removal of the intervening DNA. In this case DNA encoding V_{201} through V_{300} and J_1 and J_2 are removed, then V_{200} is spliced to J_3. Transcript of the spliced gene begins in the leader of the V_{200} segment and proceeds to the end of the C segment. Sequences from the end of the J_3 segment to the start of the C segment are spliced out of the transcript. [Courtesy of David Freifelder.]

In addition, splicing of the V and J segments in DNA occurs within a triplet codon at any site in that codon (Figure 10-28), creating one of four possible codons. Because the genetic code is degenerate, on the average in VJ junctions only 2.5 different amino acids are specified. This amino acid variability increases the number of possible variable regions to 3000 (2.5 × 1200).

Genes for H chains are assembled in a similar way, with the exception that a fourth short segment of DNA, a **D** segment, contributes to formation of the V region. More than 5000 different VDJ combinations are possible for H chains. Since an IgG molecule contains two identical H chains and two identical L chains, about $3000 \times 5000 = 1.5 \times 10^7$ different IgG molecules are possible. In addition the DNA encoding the V regions of H and L chains contains hypervariable segments, in which mutations are frequent, increasing further the number of possible L and H chains. The number of possible IgG molecules is more than enough to account for the diversity of antibodies in an animal species.

GENE AMPLIFICATION

In general the large amounts of a particular gene product required in the life of a cell can be produced by transcription and translation of the two copies of a gene in a diploid cell. This is because expression of a gene is expanded at two points—transcription, which yields multiple coding sequences (mRNA molecules) from a single gene, and translation, which yields multiple polypeptides from a single mRNA (Figure 11-30). However, single copies of certain genes are inadequate to meet the need for the products of those genes.

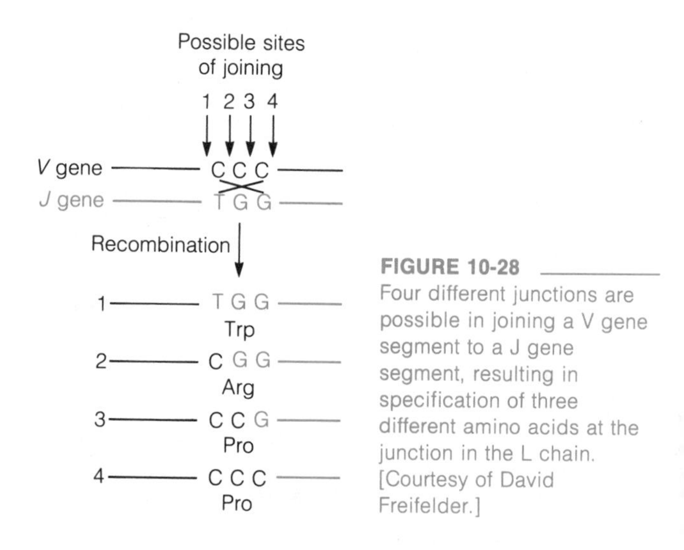

FIGURE 10-28 Four different junctions are possible in joining a V gene segment to a J gene segment, resulting in specification of three different amino acids at the junction in the L chain. [Courtesy of David Freifelder.]

Multiple Copies of Genes for rRNA

Several hundred copies of the genes encoding rRNA molecules are present in each genome. The second step in amplification of gene expression, namely translation, does not occur for rRNA genes, which is presumably compensated for by the presence of multiple gene copies. Maintenance of a large ribosome population, particularly in growing cells, requires a high rate of production of synthesis of rRNA that can be met only with multiple gene copies.

Three rRNA molecules, 28S, 18S, and 5.8S, are transcribed from a transcription unit into a single transcript (Chapter 8). These transcription units are arranged in large tandem arrays. The transcription units are separated by untranscribed spacer regions of DNA, usually hundreds of bp long. The nucleolus, in which rRNA transcripts are processed and ribosomal subunits are assembled, is organized around the section of a chromosome containing the multiple copies of the genes encoding rRNA. Genes for 5S rRNA range in copy number from a few hundred (500 in *Drosophila*) to many thousands (about 24,000 in *Xenopus*). They too are arranged in tandem with untranscribed spacers between successive genes. Genes for 5S rRNA are usually located in other regions of the genome away from nu-

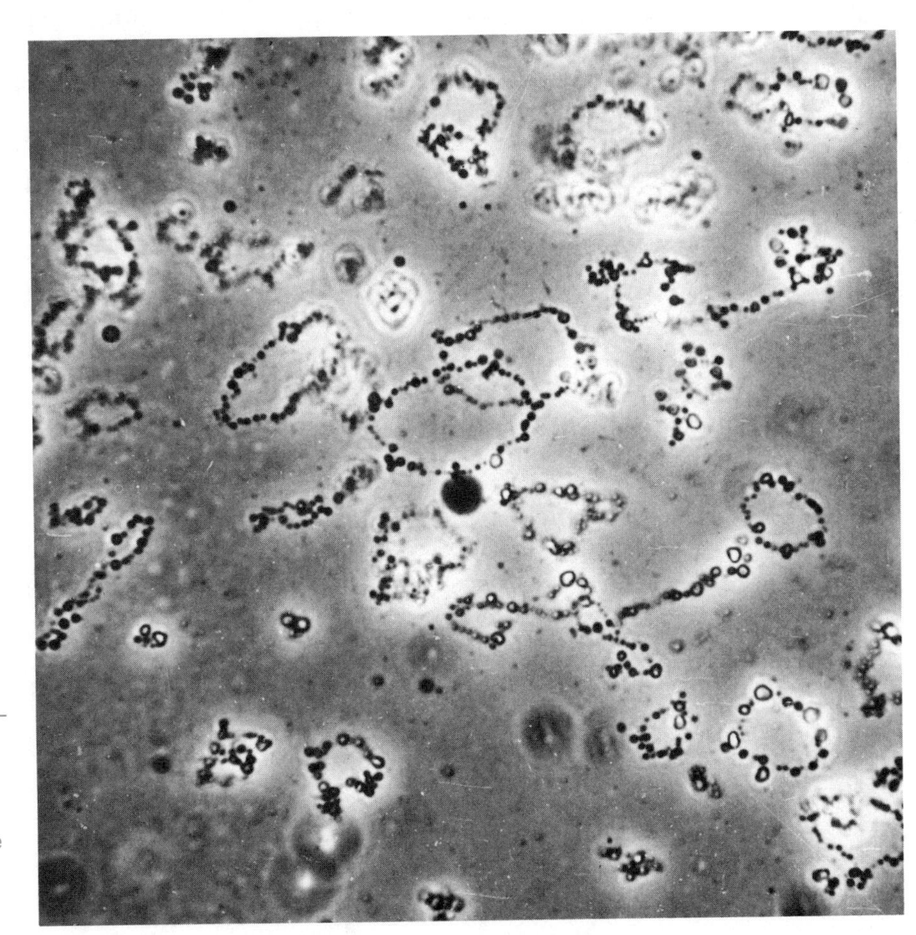

FIGURE 10-29
Light micrograph of rings released from nucleoli in a salamander oocyte (*Triturus pyrrhogaster*). Each ring contains a DNA molecule with multiple genes encoding rRNA molecules. The beaded appearance is due to proteins and perhaps rRNA precursors still attached to the DNA. [Courtesy of O. L. Miller, Jr.]

cleoli, often on different chromosomes than 28S, 18S, and 5.8S genes.

Amplification of Genes for rRNA

In some species of animals, particularly those that produce large oocytes, genes for rRNA undergo transient amplification, which in a broad sense represents a type of regulation of gene expression. By increasing the copy number of rRNA genes, the synthesis of rRNA can be sharply increased to meet the demand for a vast quantity of ribosomes in a large growing cell like an oocyte. Amplification is particularly prominent in amphibian oocytes. The genome of *Xenopus* contains about 500 copies of the transcription unit for rRNA molecules. This increases 4000-fold to about 2×10^6 copies in the egg. These gene copies support the production of 10^{12} ribosomes during oocyte growth.

The extra copies of genes for rRNA are present in tandem arrays in circular, extrachromosomal molecules of DNA (Figure 10-29). Nucleoli form around these DNA molecules so that several hundred nucleoli are present. After fertilization the task of producing ribosomes is again taken over entirely by the chromosomal copies of the genes.

Amplification of the Gene for Chorion Protein in *Drosophila*

The chorion is a tough, thick envelope that surrounds the mature oocytes of some species, providing mechanical protection. The chorion around *Drosophila* eggs consists of many kinds of proteins arranged in complex layers and synthesized over a period of five hours by follicle cells surrounding the oocyte in the ovary. At least three different genes encoding three proteins of the chorion undergo ten-fold amplification in the ovary before the onset of synthesis of chorion proteins. The amplification of chorion genes is the only example discovered so far for increasing transcriptional potential of protein-coding genes to meet a sudden demand for large quantities of specific proteins.

PROBLEMS

1. One kind of transcription factor binds to the TATA box of eukaryotic genes. How is this important for transcription?
2. What three kinds of short promoter sequences are often found in the 5′ upstream region of eukaryotic genes?
3. How do enhancer sequences differ from promoter sequences?
4. Draw a generalized eukaryotic transcription unit with its various regulatory sequences.
5. Why does the calico coat pattern occur only in female cats?
6. What is the experiment that shows that loosening of chromatin is involved in activation of transcription of a gene?
7. How does the protein encoded by *GAL80* block expression of genes necessary for galactose metabolism?
8. Give an example of a kind of gene whose expression is regulated in relation to the cell cycle.
9. Describe an experiment showing that production of tubulin is autoregulated.
10. What might be the value of regulation of gene expression by light in plant cells?
11. Describe two experiments showing that differentiated cells in multicellular plants and animals contain a full set of genes.
12. Describe an experimental observation showing that differentiation of cells in a multicellular organism is achieved by regulation of transcription.
13. Give an example of regulation of gene expression by a hormone.
14. How is it known that puffing of a band in a polytene chromosome represents activation of gene expression?
15. What is the probable significance of alternative splicing of primary transcripts of the fibronectin gene?
16. Describe the general principle that underlies the synthesis of 10^7 different antibodies from a few hundred gene segments.

PROKARYOTIC CHROMOSOMES

Chromosome Size
Circularity of Prokaryotic Chromosomes
Structural Organization of Bacterial Chromosomes
Replication of Bacterial Chromosomes

- Initiation of DNA replication
- Bidirectional replication
- Termination of replication

EUKARYOTIC CHROMOSOMES

Chromosome Numbers in Eukaryotes
Amount of DNA in Eukaryotes
The Size of DNA Molecules in Eukaryotes

- Length of DNA molecules in *Drosophila*
- Length of DNA molecules in mammalian cells
- Length of DNA molecules in salamanders

Organization of Deoxynucleotide Sequences in Eukaryotic DNA

- Repetitive DNA sequences
- Telomeric DNA sequences
- Autonomously replicating sequences
- Centromeric sequences
- Pseudogenes
- Junk DNA

The Essential Sequences of a Chromosome
Total Complexity of DNA Sequences in Genomes
The C-Value Paradox
Estimates of Gene Numbers, the S-Value Paradox, and Spacer DNA

- Gene number and spacers in *Drosophila*
- Gene number and spacers in mammals
- Gene number and spacers in ciliated protozoa

Gene Families
Packing of DNA in Chromosomes

- Nucleosomes and chromatin
- Organization of chromatin
- The nuclear envelope
- Structure of condensed chromosomes

Banding in Mitotic Chromosomes
Replication of Chromosomes

- Replication units
- Bidirectional replication
- Termination between replication units
- Size and number of replication units
- Rate of replication

DNA MOLECULES IN MITOCHONDRIA

DNA MOLECULES IN CHLOROPLASTS

CHAPTER 11

Nuclear Structure and Function: Chromosomes

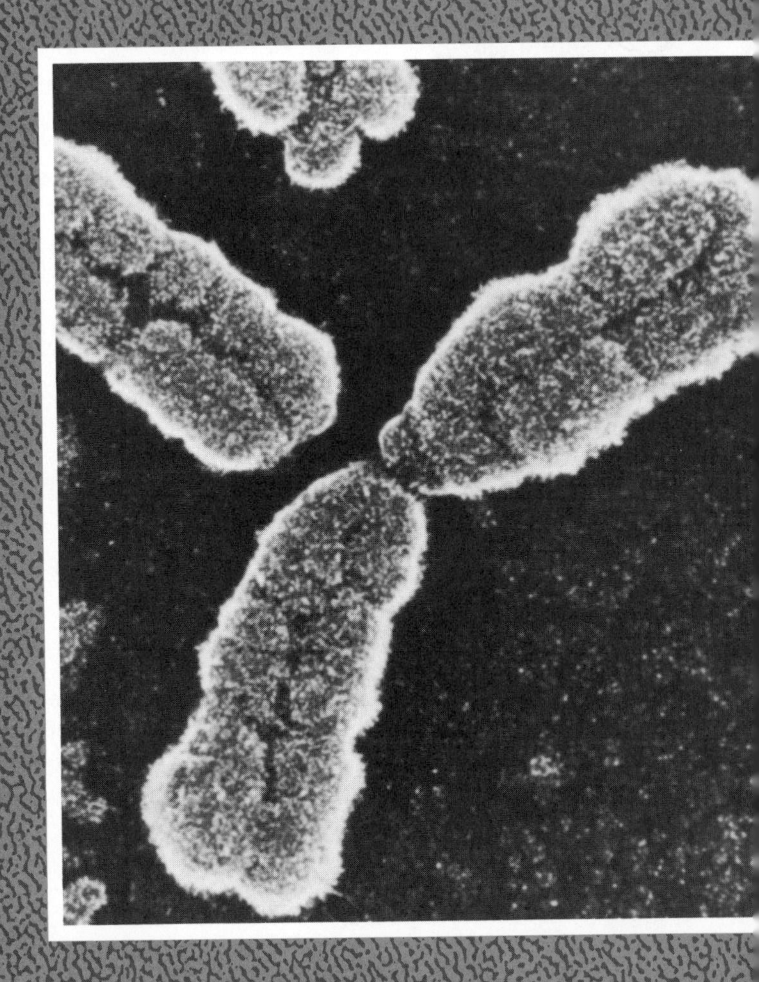

See Figure 11-44.

Chromosomes as the physical basis for the genetic properties of cells and for inheritance from parent cells to daughters have been discussed in a variety of contexts in earlier chapters. This chapter deals with the molecular structure of chromosomes and with the replication of chromosomes in preparation for cell division.

The primary property of chromosomes, namely the carrying of genes in DNA molecules, is the same in the chromosomes of prokaryotes and eukaryotes. However, prokaryotic and eukaryotic chromosomes differ in both structure and mode of replication in several fundamental ways, and it is convenient to discuss the two types of chromosomes separately. In spite of the differences, the principles of structure and replication already uncovered for prokaryotic chromosomes serve as a basis from which to analyze the much larger and structurally complex chromosomes of eukaryotes.

PROKARYOTIC CHROMOSOMES

A bacterium contains only one chromosome, which encodes all of the genes needed for the life of the cell. Some bacteria contain a very small molecule of DNA called a **plasmid,** which encodes only a few genes. Some plasmids are valuable to the cell because they confer antibiotic resistance; however, in general, plasmids are dispensable.

Chromosome Size

Several methods have been developed to measure either the length or molecular weight of the DNA in an intact prokaryotic chromosome. Each of these methods requires chemical extraction of unbroken DNA from cells. Extracting unbroken DNA is difficult because thin molecules easily break by mechanical scission or shear and DNases are present that degrade DNA. The DNA double helix is a stiff molecule that is easily broken by shearing forces created during stirring, pouring, or pipetting of a DNA solution. The longer the DNA molecule, the greater the probability of shearing. With simple precautions molecules with molecular weights as large as 10^8 (150,000 bp) can be isolated without shearing. However, enzymatic breakage can be inflicted by DNases that are activated when a cell is lysed. Because of their small size the DNA molecules of plasmids are the easiest to isolate (Figure 11-1).

Early estimates of the size of bacterial DNA were low because of shearing and enzyme degradation during preparation. These problems have been solved, and we now know that the sizes of DNA molecules of chromosomes in various bacterial species cover a 15-fold range from 750,000 bp to a little over 10,000,000 bp. Table 11-1 gives the sizes of DNA molecules, measured by various methods, for some bacterial species, a bacteriophage, and a plasmid. Why the DNA molecules of some species of bacteria are so much larger than others is not fully understood. Some bacteria have more genes, giving that cell greater functional versatility. However, it is unlikely that differences in gene number account for more than a part of the differences in DNA amounts.

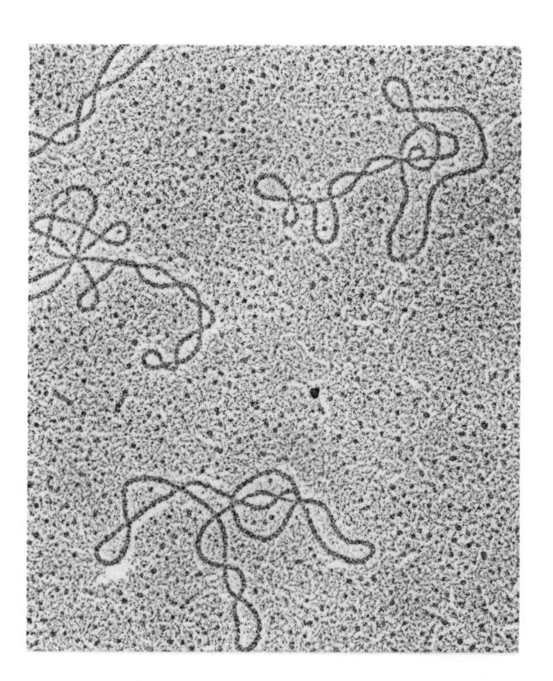

FIGURE 11-1
Circular DNA molecules of a bacterial plasmid. The circles are twisted because the DNA has coiled upon itself. [Courtesy of David Freifelder.]

TABLE 11-1
Sizes of DNA Molecules in Some Bacteria

	Molecular Weight	Number of Base Pairs
Plasmid (in *E. coli*)	5.0×10^6	7.50×10^3
Bacteriophage T4	110.0×10^6	0.165×10^6
Mycoplasma pneumoniae	0.50×10^9	0.75×10^6
Acholeplasma laidlawaii	1.0×10^9	1.50×10^6
Haemophilus influenzae	1.01×10^9	1.52×10^6
Neisseriaceae gonorrhoea	1.28×10^9	1.92×10^6
Staphylococcus aureus	1.43×10^9	2.15×10^6
Streptococcus faecalis	1.47×10^9	2.21×10^6
Bacillus cereus	2.60×10^9	3.90×10^6
E. coli	3.13×10^9	4.70×10^6
Salmonella pullorum	2.83×10^9	4.25×10^6
Serratia marcesans	5.56×10^9	8.34×10^6
Pseudomonas aeruginosa	6.96×10^9	10.44×10^6

One of the first accurate estimates of the molecular weight of DNA in the chromosome in *E. coli* was made by autoradiography. Cells were grown for two generations with ^{3}H-thymidine to label chromosomes throughout their lengths; then the DNA was extracted very carefully and spread out on a surface for autoradiography. From measurements of the length of autoradiographic images (Figure 11-2), the DNA was estimated to be at least 1100 μm long, which corresponds to 3,300,000 bp or 3300 kilobase pairs (kbp). For interconversion of units of size for DNA it is only necessary to remember the 1-2-3 rule:

1 μm of DNA duplex = **2** $\times 10^6$ molecular weight
= **3** kbp

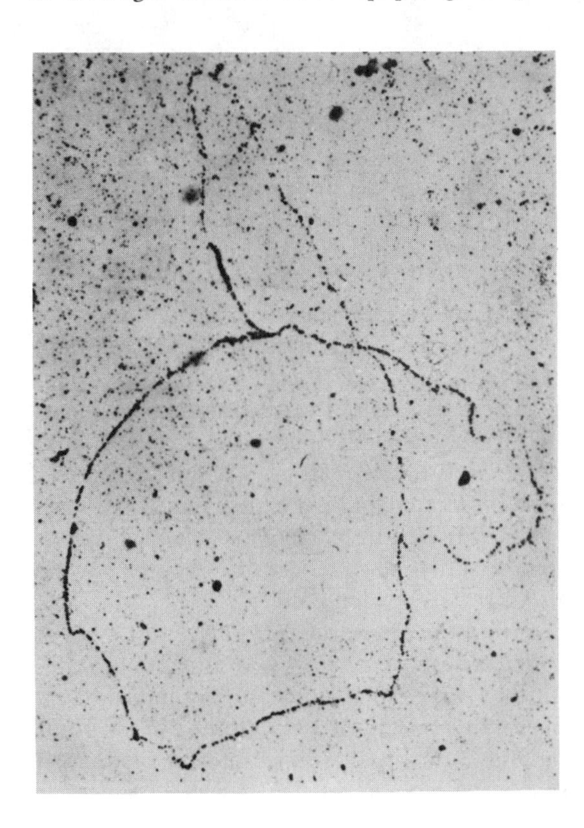

FIGURE 11-2
Autoradiograph of an intact replicating chromosome of *E. coli* that was labeled by incorporation of ^{3}H-thymidine. The circular DNA molecule was undergoing replication at the time of isolation and therefore has a partially doubled structure. [From John Cairns. 1963. *Cold Spring Harbor Symp. Quant. Biol.*, 28: 44.]

Three kbp have an absolute mass of 3.1×10^{-6} picograms. One picogram (pg) equals 10^{-12} g. More recent measurements by several methods have shown that the chromosome of *E. coli* is about 4700 kbp.

The sizes of smaller molecules, such as those of bacteriophages and plasmids, can often be measured directly by electron microscopy. In general, however, the large molecules of bacterial chromosomes cannot be measured by electron microscopy because the field of view in an electron microscope is small compared with the length of the DNA. An exception is the DNA molecule in the wall-less bacteria, the *Mycoplasma*. These are the smallest kinds of bacteria, some measuring only 0.3 μm in diameter. The intact DNA molecule extracted from *Mycoplasma hominis* measured by electron microscopy is about 250 μm long (Figure 11-3), which corresponds to 750 kbp. The mycoplasma contain the smallest cell chromosomes known. Assuming as a rough estimate 1000 bp per gene (coding se-

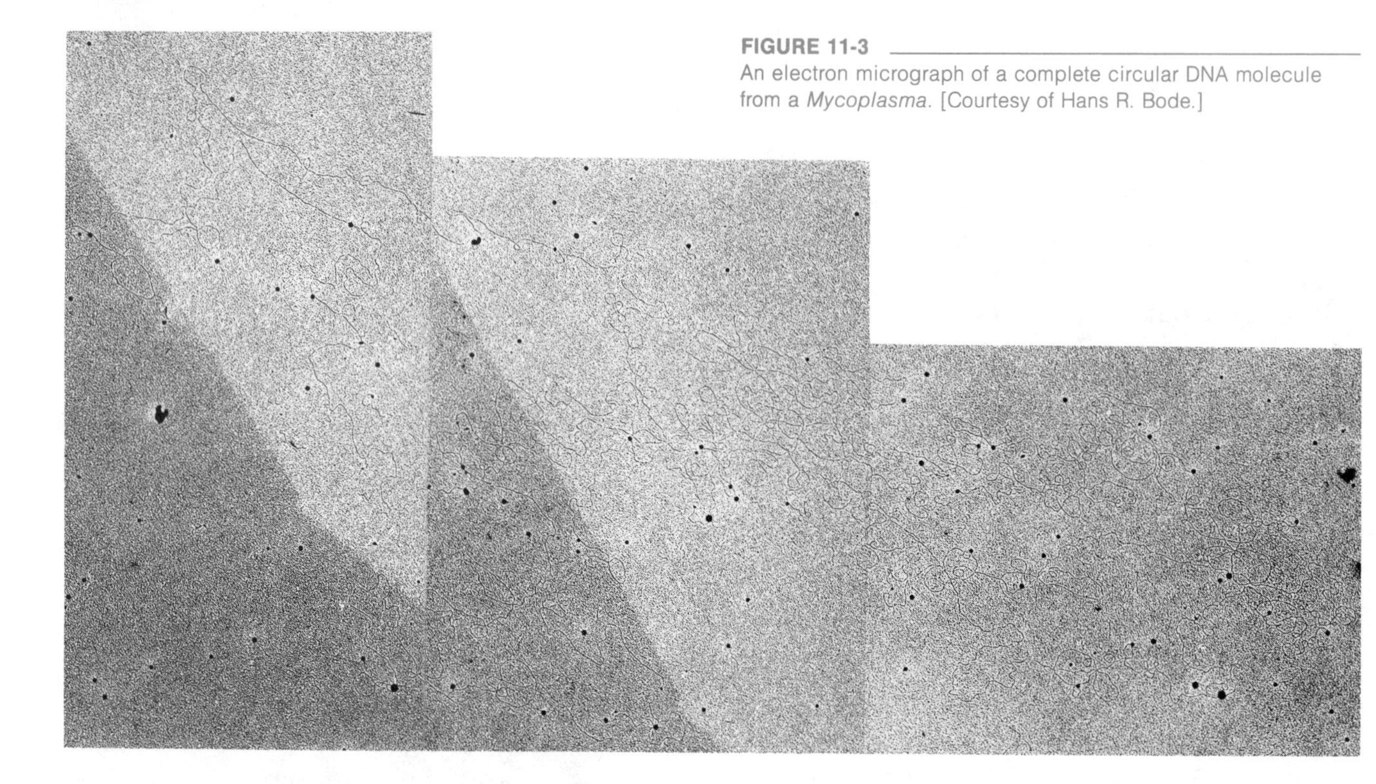

FIGURE 11-3
An electron micrograph of a complete circular DNA molecule from a *Mycoplasma*. [Courtesy of Hans R. Bode.]

quence plus regulatory sequences) and assuming that all of the DNA is used for genes, the mycoplasma have an upper limit of 750 genes. About 350 kinds of protein molecules have been identified in mycoplasma. Some proteins may have escaped detection, so the number of protein-encoding genes may be somewhat higher than 350. In addition, there are genes encoding rRNA and tRNA molecules. Therefore, the true number of genes in mycoplasma is somewhere between about 400 and 750. Because of its relatively small size the DNA molecule in mycoplasma might be the first cellular chromosome for which the entire nucleotide sequence will be determined. From the sequence virtually every gene and its product will be identifiable.

The chromosome in *E. coli* is over five times larger than the one in mycoplasma (4700 vs 750 kbp) and could encode 4700 genes. Genetic studies indicate that the number may be 1500 to 2000. Compare this number to the 50,000 to 100,000 different genes estimated to be present in a human cell.

Circularity of Prokaryotic Chromosomes

The DNA molecule of all bacterial chromosomes and plasmids is a closed circle. Circularity was first suggested by genetic maps constructed for the chromosome of *E. coli*. One could explain the positional relationship among various genes in the chromosome by assuming that the chromosome was continuous, that is, was a circle. But there were other explanations for the positional relationships among genes. Proof of circularity came from physical studies of bacterial chromosomes. The first physical evidence of circularity was obtained from the autoradiographic study of *E. coli* chromosomes, shown earlier in Figure 11-2. This particular chromosome is in an intermediate stage of replication as reflected by the doubling of part of the circle. DNA molecules isolated from a variety of plasmids and prokaryotic viruses have been shown by electron microscopy to be closed circles (Figure 11-1). The chro-

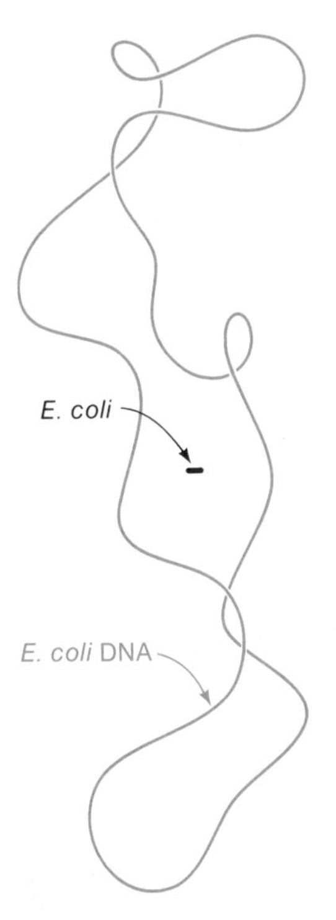

FIGURE 11-4
Schematic drawing showing the relative sizes of an *E. coli* cell and its DNA molecule, drawn to the same scale except for the width of the DNA molecule, which is enlarged in width approximately 10^6 times. [Courtesy of David Freifelder.]

mosome of *M. hominis* (750 kbp) is the largest chromosome shown to be circular by electron microscopy (Figure 11-3). The significance of circularity of chromosomes is not known, but it is important in chromosome replication (discussed later).

Structural Organization of Bacterial Chromosomes

Bacterial cells measure no more than a few micrometers in their maximum dimension but contain a DNA molecule hundreds of micrometers in length. *E. coli* is a cylinder about one μm in diameter and two μm long and contains a circular DNA molecule about 1560 μm long (Figure 11-4). The DNA molecule fits into the bacterial cell by being compactly folded in a way that does not interfere with transcription or replication. The folding is obvious in an electron micrograph of a bacterium. The DNA, complexed with protein molecules, forms a dense fibrous mass (Figure 11-5), the nucleoid. Removal of the tough cell wall by digestion with the enzyme lysozyme allows the cell to be gently lysed. The nucleoid can then be separated from the cytoplasm and studied by electron microscopy (Figure 11-6).

The DNA is held in the compacted state in the nucleoid by several kinds of proteins. Some of these proteins aggregate to form particles that in turn bind to DNA by means of ionic bonds between positively charged side groups of amino acids, particularly the amino group ($—NH_3^+$) of arginine, and the negatively charged phosphate groups of the DNA. The DNA coils around successive protein aggregates, producing a beaded string in which the extended DNA molecule is drawn into a shortened configuration. The beaded string is itself drawn into a more compacted state, probably by complexing with other proteins, to form the nucleoid.

FIGURE 11-5
A dividing *E. coli* cell. The daughter nucleoids in each cell occupy the lightly stained areas. [Courtesy of Nanne Nanninga. From *Molecular Cytology of Escherichia coli* (1985) N. Nanninga, ed. Academic Press, London.]

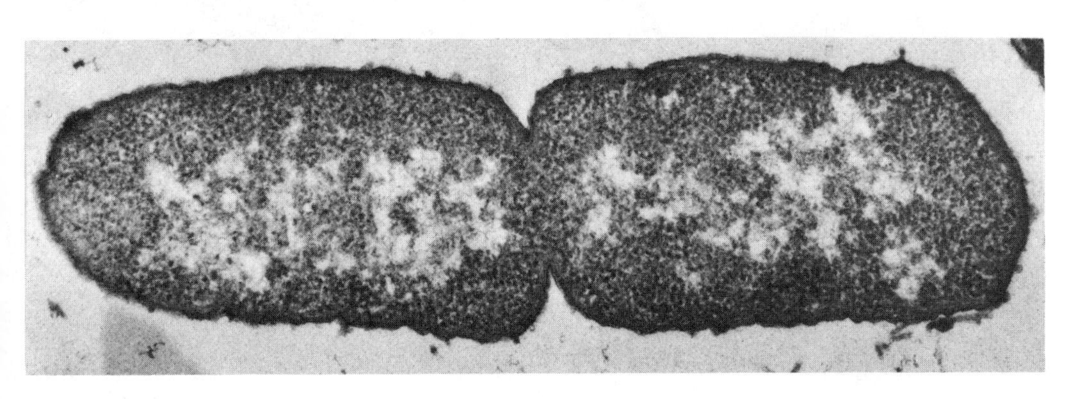

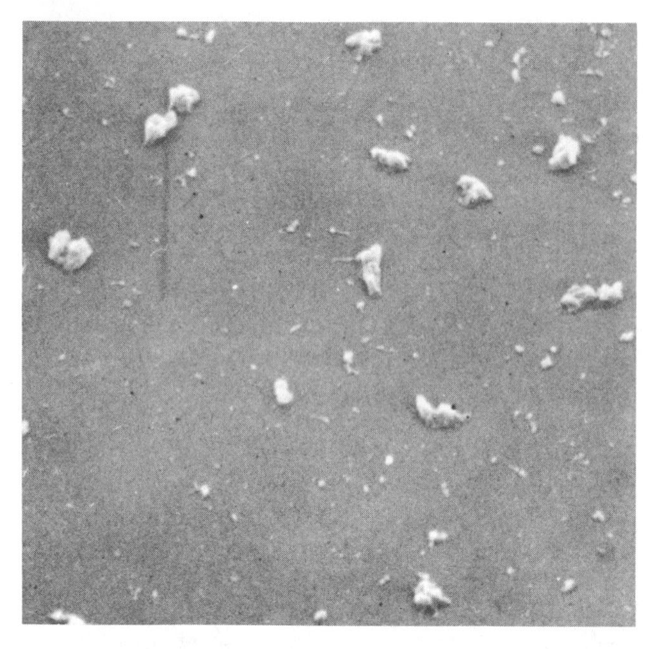

FIGURE 11-6
Scanning electron micrograph of individual nucleoids isolated from *E. coli.* [Courtesy of David E. Pettijohn. From *Molecular Cytology of Escherichia coli* (1985) N. Nanninga, ed. Academic Press, London.]

The beaded string, which in its fully extended state would be several hundred micrometers in length, is organized into loops called **domains,** with all the loops anchored at their two ends to one central core. This organization becomes clear when certain of the proteins, including those that aggregate to form the beads, are removed from an isolated nucleoid. Loops of DNA, each representing a domain containing many genes, extend out from the central protein core of the nucleoid (Figure 11-7). Because the chromosome is a continuous, circular DNA molecule, one loop must be joined to the next within the central core. This scheme of the organization of the bacterial chromosome in a nucleoid is shown in the drawing in Figure 11-8.

The organization of DNA molecules into loops or domains is thought to be important in transcription. The DNA double helix within a loop is **underwound.** To visualize what this looks like an analogy with a piece of string is useful (Figure 11-9). Some kinds of string consist of two threads wound around each other to form a double helix although the demonstration will work whatever the number of threads (Figure 11-9(a)). When the two ends of the string are held and then twisted in a direction intended to unwind the two threads, the string coils upon itself, that is, becomes supercoiled (Figure 11-9(b)). If one end of the string is released, the string is free to rotate on its own axis and the supercoiling is released. Supercoiling induced by underwinding a double helix is called **negative supercoiling.** Supercoiling can also be produced by overwinding a double helix; such supercoiling is said to be positive. Negative supercoiling is shown for the cir-

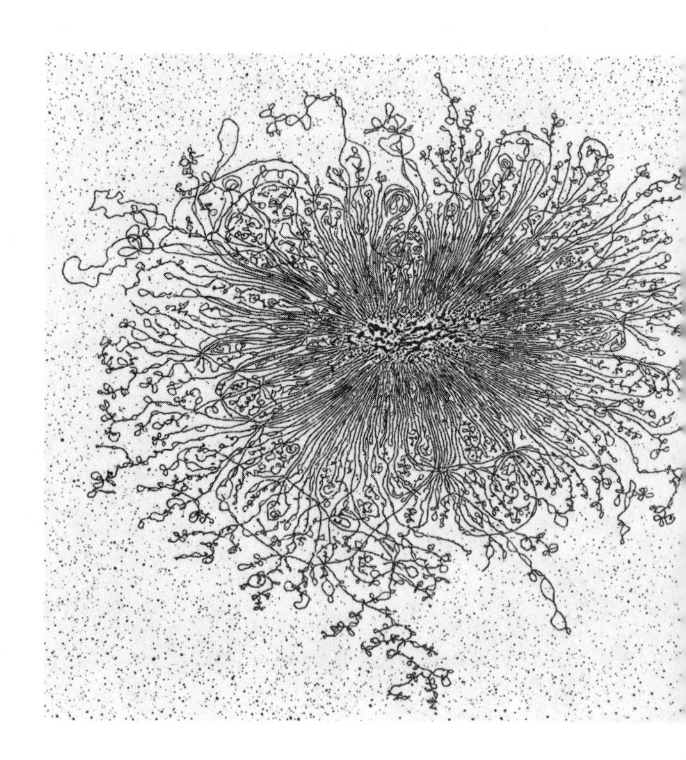

FIGURE 11-7
Electron micrograph of an *E. coli* chromosome partially released from a cell and spread over a wide area. Many loops extend out from the central region. [Bluegenes #1. All rights reserved by Designergenes Posters Ltd., P.O. Box 100, Del Mar, CA 92014-0100, from which posters, postcards, and shirts are available.]

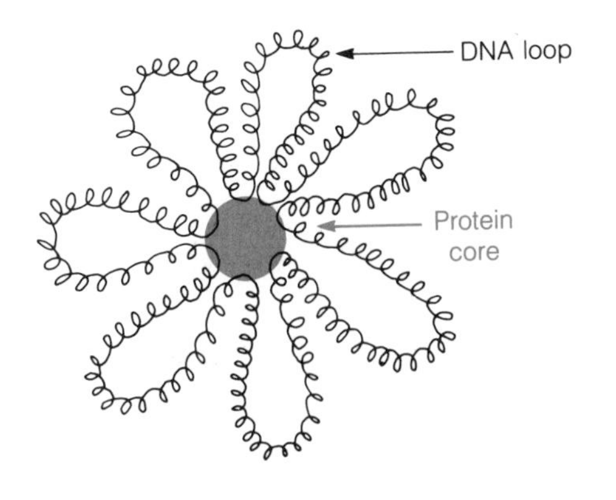

FIGURE 11-8
Schematic diagram of the organization of a bacterial nucleoid. A continuous circular molecule of DNA is organized into loops or domains that are anchored in a central core of protein. The DNA in each loop is coiled. Only seven of the approximately 50 loops are shown.

cular duplex molecule of a bacterial virus in Figure 11-10. Because the molecule has no free ends the supercoiling is stable. If a break is introduced into the phosphodiester backbone of the duplex, the bond opposite the break (in the other chain) will act as a swivel and allow the supercoiling to be released.

The portion of the bacterial DNA double helix in a loop is underwound and is therefore negatively supercoiled. The DNA is prevented from rotating and releasing the supercoiling because the two ends of a loop are immobilized by attachment to proteins in the central core of the nucleoid. Twisting of the DNA double helix to produce negative supercoiling is catalyzed by an enzyme called **DNA gyrase.** The proteins that bind to DNA to form the beaded structure of the chromosome then stabilize the supercoiled configuration of individual loops.

Negative supercoiling tends to destabilize the base-pair binding between the two chains in the DNA double helix, allowing the chains to separate in localized regions. This separation is believed to facilitate transcription by making the coding strand of the DNA double helix more accessible to RNA polymerase.

Replication of Bacterial Chromosomes

Replication of the chromosome is a central event in the reproduction of a cell. During cell division the two identical DNA molecules produced by replication are distributed to the two new daughter cells. After cell division the DNA molecule in each daughter cell begins to replicate in preparation for the next cell division. What triggers DNA replication is not known but it is coordinated with the growth of the cell (Chapter 12).

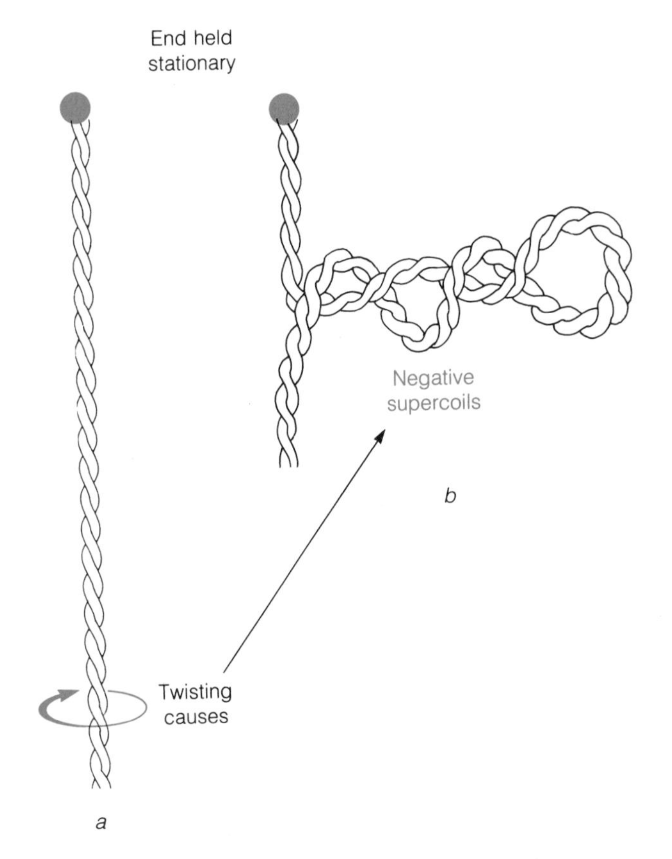

FIGURE 11-9
Supercoiling in a piece of string. (a) A piece of string immobilized at the top end and twisted on its axis at the bottom end in a direction opposite to the coiling of the threads in the string. (b) Twisting causes the string to writhe upon itself, forming negative supercoils. In the same way, twisting a DNA duplex in a direction opposite to the coil of the duplex causes superhelical twisting of the duplex.

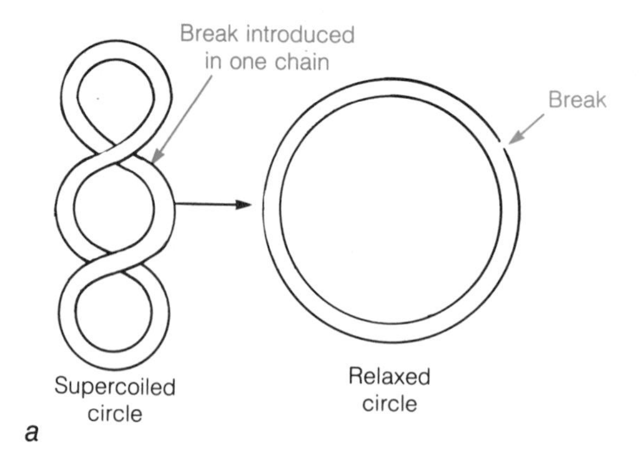

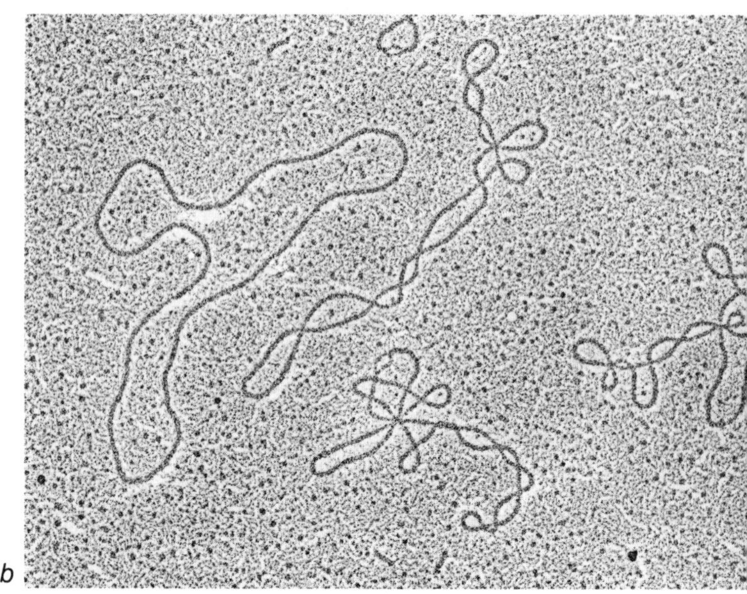

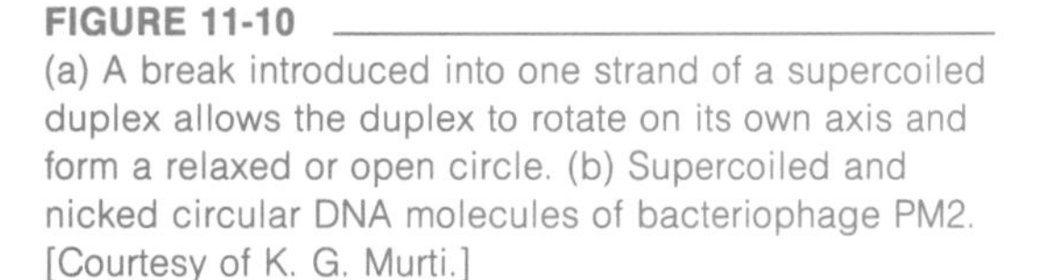

FIGURE 11-10
(a) A break introduced into one strand of a supercoiled duplex allows the duplex to rotate on its own axis and form a relaxed or open circle. (b) Supercoiled and nicked circular DNA molecules of bacteriophage PM2. [Courtesy of K. G. Murti.]

Initiation of DNA replication. Initiation of replication in *E. coli* begins at a special region of the DNA molecule called the **origin** or ***ori.*** The origin has been defined by mutational analysis as a 245-bp stretch of the chromosome. That is, mutations in which individual pairs are changed within the 245-bp region decrease the efficiency of initiation of replication. Base changes in adjacent regions are without effect on initiation. *Ori* contains several short, repeated sequences that are probably recognized by proteins required for initiating DNA replication. Three of these proteins are encoded by genes designated *DnaA, DnaB,* and *DnaC.* Their roles in initiation are only partially understood. The first step in initiation is binding of the protein encoded by *DnaA* to the origin, and hence the DnaA protein has been named the **initiator protein.** What controls the timing of binding (and therefore the timing of initiation) or how it causes initiation are not known. The function of the protein encoded by the gene *DnaC* is also essential, but precisely what it does is less clear.

The gene *DnaB* encodes a protein that is required to promote unwinding of the double helix (a **helicase**). Unwinding gives a special RNA polymerase called **primase** access to the two strands as templates for replication. The RNA primase synthesizes a short RNA primer, as described in Chapter 7 and as shown in Figure 11-11. DNA polymerase then takes over from RNA primase, continuing to extend the new chain with deoxynucleotides.

The unwinding of the double helix promoted by the DnaB helicase causes rotation of the double helix. This rotation is accomplished without inducing supercoiling by the breaking of one chain (nicking), which allows the bond opposite the nick in the other chain to act as a swivel that releases the torsion created by unwinding. Swiveling is done by an enzyme called **topoisomerase**, which continuously introduces and reseals nicks ahead of replicating templates to permit progressive unwinding as replication proceeds (Figure 11-11). Another protein binds to single-stranded DNA and is called **single-stranded binding protein** or **SSB.** This protein holds the two strands of DNA in the single-stranded state ahead of DNA polymerase as it moves along the single-stranded template. Once replication has been initiated with an RNA primer, DNA polymerase works continuously on one template strand but works discontinuously on the opposite strand by the backfilling mechanism described in Chapter 7.

The first region of the chromosome to replicate is the origin (Figure 11-11), yielding origins in two partial daughter chromosomes. Whatever condition in the cell caused the parental origin to be activated, it has dissipated, since the two new origins are not used to initiate new rounds of replication until the cell has grown, divided, and again met conditions needed for initiation. A molecular definition of the requirements for initiation of DNA replication and the molecular relation of initiation to cell growth remains one of the most important unsolved problems in cell biology.

Bidirectional replication. Figure 11-11 shows the synthesis of an initiation RNA primer on each strand of the newly opened DNA double helix in the origin giving rise to bidirectional replication. By the rules of polarity of the chains in a double helix (Chapter 7), two RNA primers are synthesized in a 5′ → 3′ direction, one on each parental template, with nucleotides being added at the end with the free 3′OH group. DNA polymerase replaces primase and extends the RNA chain with deoxynucleotides in the same direction as the primase (Figure 11-11). The continuous extension of RNA initiation primers and then DNA chains leaves a single-stranded parental chain immediately opposite. These are converted to double-helix DNA by additional RNA primers made as the parental double helix unwinds (Chapter 7 and Figure 11-11). These RNA primers are extended as DNA chains, the RNA primers are removed by exonuclease, and the individual segments of DNA are joined by DNA ligase to complete the double helix. Because filling in replication with Okazaki fragments necessarily comes after the synthesis of the continuous chain, called the **leading strand,** the backfilled strand is called the **lagging strand.**

The overall process as shown in Figures 11-11 and 7-12 leads to bidirectional replication of the chromosomal DNA, with two replication forks moving in opposite directions from the origin. Autoradiography provided direct visualization of bidirectional replication in the following experiment. An auxotrophic mutant of *E. coli* that cannot synthesize arginine or thymine was

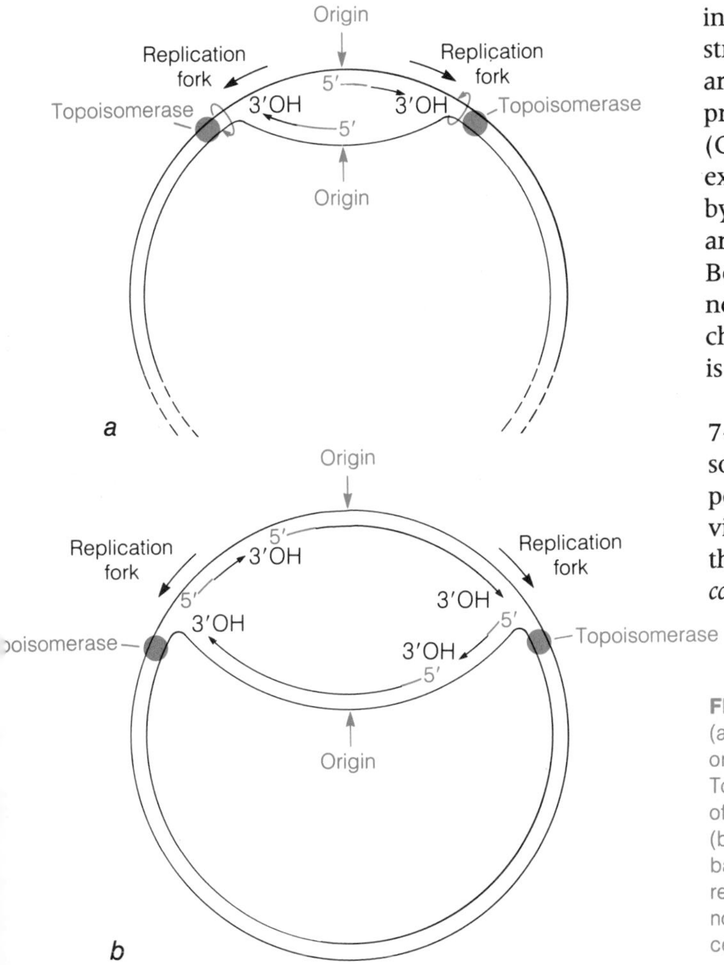

FIGURE 11-11
(a) Replication begins with unwinding of the chromosome at the origin and synthesis of primer RNA on each parental strand. Topoisomerase continuously relieves torsion in the duplex ahead of each replication fork as parental DNA continues to unwind. (b) Synthesis of additional RNA primers occurs in the 5′ to 3′ backfilling along each parental strand. The very beginning of replication forms origins in the daughter chromosomes that are not activated until reinitiation conditions are established in the cell.

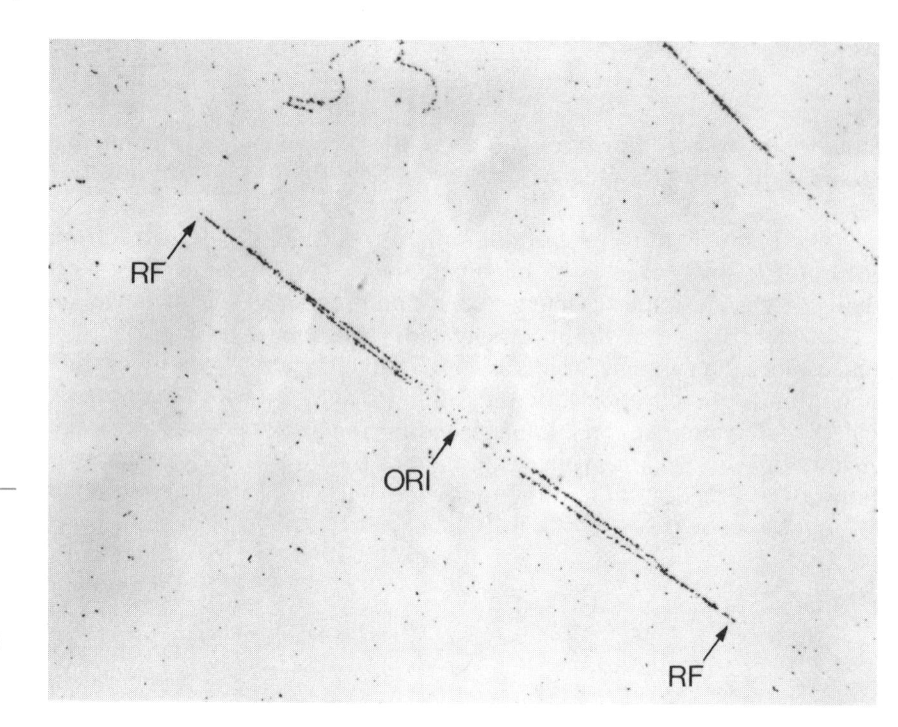

FIGURE 11-12
Autoradiograph showing bidirectional replication of the *E. coli* chromosome. The origin of replication (ORI) is at the center of the autoradiographic pattern and the replication forks (RF) at the two extremities. Unreplicated parental DNA extends outward from the two replication forks but is unlabeled and therefore is not detected.

grown in the presence of arginine and thymine and was then transferred to nutrient medium lacking arginine and incubated for several hours. Without arginine there is no net synthesis of proteins. In the absence of protein synthesis DNA replication cannot be initiated, and all the bacteria become arrested in the cell cycle at a point shortly before initiation of replication. Addition of arginine to the medium allows protein synthesis to resume, following which all the bacteria begin to replicate their chromosomes at nearly the same time, that is, the millions of cells in the culture have been synchronized with respect to initiation of DNA replication.

At the moment of addition of arginine to the arginine-starved culture ^{3}H-thymine of low specific activity was also added to the culture. (Low specific activity means that a small fraction of the thymine molecules possess a tritium atom.) When the cells begin replication, the incorporation of the low-specific-activity ^{3}H-thymine results in a low level of labeling of the newly synthesized DNA chains, and a weak autoradiographic pattern will be produced by the DNA. A minute or two after replication had begun, ^{3}H-thymi*dine* of high specific activity (roughly one ^{3}H per thymidine) was added to the culture. Two things are important about the ^{3}H-thymidine. First, thymidine is a deoxynucleoside and is incorporated into DNA preferentially to thymine, the free pyrimidine base. Second, the addition of ^{3}H-thymidine at high specific activity results in an immediate shift to incorporation of an intense level of radioactivity at replication forks and a resulting intense autoradiograph from the DNA.

After a minute or two of incorporation of ^{3}H-thymidine, the bacteria were cooled to stop further DNA replication, treated with lysozyme to digest the cell wall, and a drop of the culture was lysed with a drop of ionic detergent on a microscope slide. The detergent stripped the proteins from DNA. The drop of culture-detergent mixture was spread, which caused DNA to be extended out onto the microscope slide. Autoradiography then revealed the pattern of radioactive regions in DNA molecules, as shown in Figure 11-12. Each molecule produced a central light autoradiograph for the two partial daughter segments (^{3}H-thymine incorporation). The light pattern represented the first DNA to replicate after initiation. The intense autoradiographic patterns represented subsequent synthesis and showed that replication had proceeded bidirectionally from the origin (the origin is at the center of the light autoradiograph). The two partial daughter molecules were joined at the replication forks at the outer extremities of the autoradiographic patterns. The unreplicated, parental DNA ahead of the replication fork was not radioactive and hence not seen in the autoradiograph.

Autoradiography shows the overall pattern of replication but does not have the resolving power to reveal the finer molecular events such as backfilling with Okazaki fragments, which occurs within the first micrometer behind the replication fork.

During replication, proteins that bind to DNA to produce the folding pattern of the chromosome, as well as proteins involved in transcription (RNA polymerase, repressor molecules, etc.) are probably transiently displaced from the DNA as the replication forks move through it. New proteins that make up the replication enzymatic machinery as well as single-stranded binding protein (SSB) bind to the DNA in the immediate region of the replication fork and migrate with it. The SSB binds only to single-stranded DNA and stabilizes the lagging parental strand prior to its replication. Behind the fork the newly replicated segments of daughter double helices become folded and packed by binding with packing proteins and may resume transcription with binding of RNA polymerase and other transcription factors.

All of these events—release of packing proteins, displacement of transcription factors, opening the double helix, and replication—occur extremely rapidly. At its maximum rate a replication fork moves through 980 bp of DNA per second.

Termination of replication. The two replication forks, traveling in opposite directions, eventually meet (in 40 minutes at the maximum rate of DNA replication) on the opposite side of the circular chromosome. Figure 11-13 shows an autoradiograph of two replication forks approaching each other less than a minute before termination. Termination apparently does not occur at a specific deoxynucleotide sequence. Instead, in the termination region the rate of fork travel is enormously slowed (by an unknown means), and forks al-

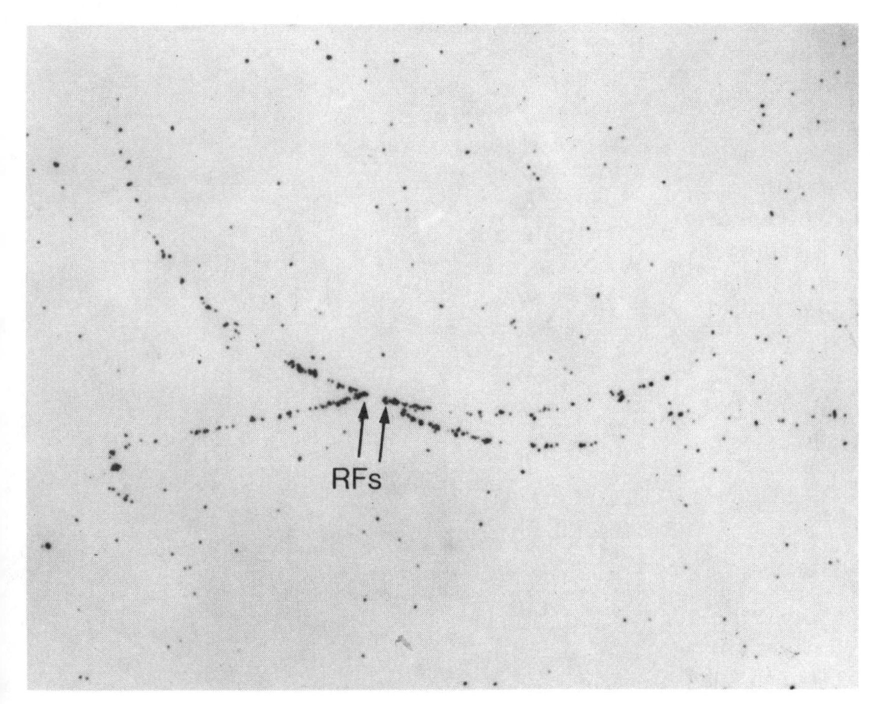

FIGURE 11-13

Autoradiograph of termination of replication of an *E. coli* chromosome. The two replication forks (RFs) have nearly met. A short segment of unreplicated, parental DNA (unlabeled, not visible) separates them. Portions of the two daughter chromosomes extend back from the replication forks.

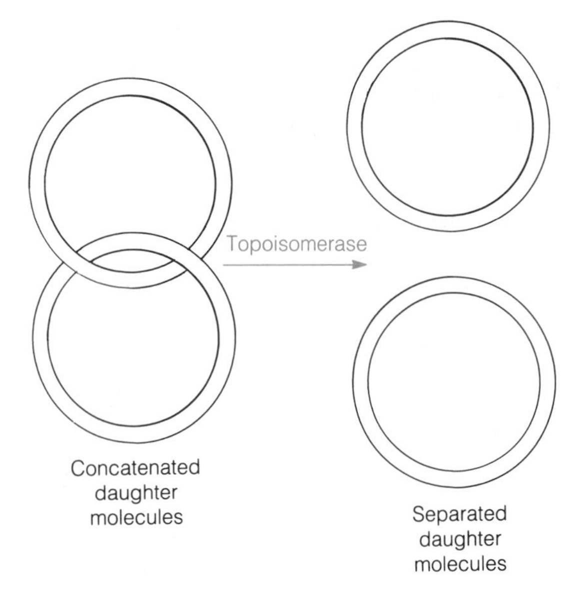

FIGURE 11-14
At the end of replication of a circular duplex the two daughter duplexes are concatenated. The two molecules can be separated by topoisomerse, which can break and reseal DNA chains.

ways meet in roughly the same region of the chromosome. The two replication forks abolish each other on meeting. At this point the two circular daughter molecules are complete, but are still interlooped with each other in concatenated fashion. The final step in replication is separation of the loops. Separation is catalyzed by a topoisomerase, which creates a transient double-stranded break in one of the daughter double helices, allowing the other to slip through (Figure 11-14), and then reseals the double helix.

The connection between the termination of DNA replication and subsequent cell division is discussed in Chapter 12.

EUKARYOTIC CHROMOSOMES

The structure and replication of eukaryotic genomes are more complicated than in prokaryotes. Although there are important similarities between the two groups of organisms, there are also major differences. Three are given here and others listed later.

1. In all prokaryotic organisms the entire necessary set of genes (genome) is carried in one chromosome. In contrast, almost all eukaryotic species distribute their genome over a number of chromosomes, ranging up to hundreds in some species.
2. The aggregate molecular weight of the DNA in a haploid set of chromosomes in a eukaryotic cell is a few times to tens of thousands of times greater than the molecular weight of the DNA in the largest

TABLE 11-2
Haploid Chromosome Numbers and Total DNA Content in Some Eukaryotes

Organism	Chromosome Number	Total Haploid DNA Content in pg (C-value)
Ant (*Myrmecia pilosula*)	1	—
Haplopapus	2	—
Tetrahymena	5	0.23
Budding Yeast	17	0.016
Nematode (roundworm) *Caenorhabditis elegans*	6	0.09
Chinese Hamster	11	3.3
Golden Hamster	22	3.7
Dog	39	2.8
Human	23	3.1
Salamander (*Amphiuma means*)	12	100.0
Lungfish (*Protopterus*)	17	50.0
Lily (*Lilium*)	12	50.0
Turtle (*Chelonia*)	28	2.7
Boa Constrictor (*Constrictor*)	18	1.8
Frog (*Rana pipiens*)	13	7.0
Chicken (*Gallus domestica*)	39	1.2
Alga (*Euglena gracilis*)	45	3.0
Fruit Fly (*Drosophila melanogaster*)	4	0.17

prokaryotic chromosome known. The greater amount of DNA underlies the greater genetic complexity of eukaryotes, although much of the DNA in most eukaryotes has no known function.

3. The chromosome in a prokaryotic cell is always a covalently closed, circular DNA molecule. Each nuclear chromosome in a eukaryotic cell contains a single, linear molecule.

Chromosome Numbers in Eukaryotes

The smallest haploid chromosomal complement in a eukaryote occurs in a species of ant—the number is one. The plant *Haplopapus* has two chromosomes, and all of the thousands of other eukaryotes examined so far possess more than two chromosomes. Table 11-2 lists the haploid chromosome numbers for a variety of eukaryotes, and Figure 11-15 shows mitotic chromosomes of several organisms. The chromosomal numbers show no regular pattern within groups of plants or animals. Different species of mammals have numbers that range from four in the Indian deer to 23 in the human to 39 in the dog. Mammalian species with fewer chromosomes have larger chromosomes. Some protozoa have as few as five chromosomes (e.g., *Tetrahymena*); others have hundreds (*Amoeba proteus*). Large differences occur among plants, algae, amphibians, fishes, reptiles, and so forth, although in most species the chromosome number is less than 100. The significance of such divergent chromosome numbers, which occur even among closely related species, is not known.

a

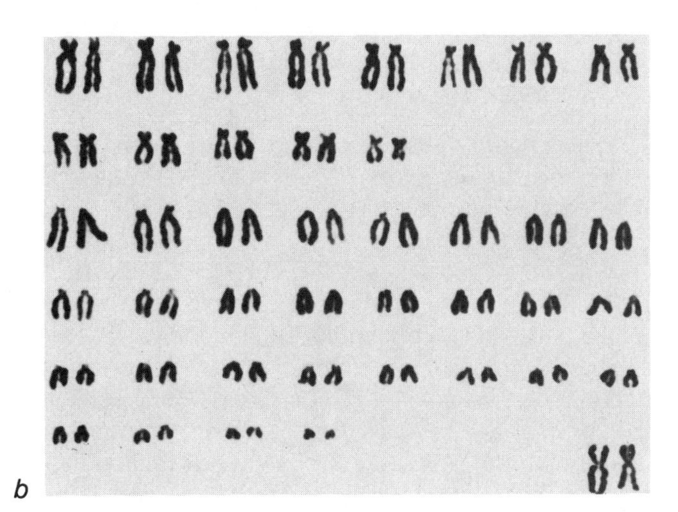

b

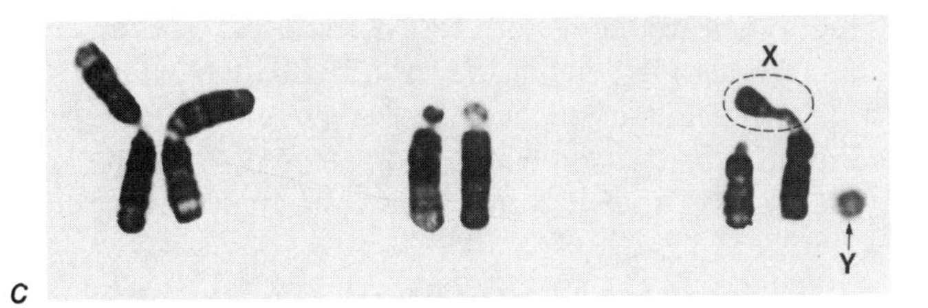

c

FIGURE 11-15
Karyotypes of three mammals. (a) Chromosomes of a female Dorcas gazelle (*Gazella dorcas*); the diploid chromosome number is 30. The pair of X chromosomes is in the lower right corner. (b) Chromosomes of a female black rhinoceros (*Diceros bicornis*); the diploid number is 84, one of the highest numbers among mammals. The pair of X chromosomes is in the lower right corner. (c) Chromosomes of a male Indian muntjak deer (*Muntiacus muntjac*). The diploid chromosome number in males is seven and in females it is six, the lowest numbers among mammals. The X chromosome is normally attached to the third autosome, an unusual feature among mammals. The Y chromosome is very small. Although differing in chromosome number, all three of these mammals have about the same total amount of DNA in their genomes. Hence, the chromosomes in the rhinoceros are small, those in the gazelle are intermediate, and those in the muntjac are large. [Courtesy of Kurt Benirschke.]

Amount of DNA in Eukaryotes

The total amount of DNA in a haploid set of chromosomes in a species is called the genomic value or C-value of the species. The C-values for eukaryotes exceed the amount of DNA in the largest prokaryotic chromosomes by a few times (yeast) to several million times (some salamanders). This greater amount of DNA underlies the greater genetic complexity of eukaryotic species. C-values differ greatly among eukaryotes (Table 11-2). For example, the C-value for the human is 3.1 pg, for the lungfish 50.0 pg., and for *Amphiuma means* 100 pg. Even *Amoeba proteus* has many times more DNA than a human. It might be assumed that more DNA in the genome means a greater number of different genes. However, it can hardly be supposed that salamanders have more kinds of genes than humans. The resolution of this apparent contradiction between DNA amount and expected gene number, called the **C-value paradox,** is discussed later.

The Size of DNA Molecules in Eukaryotes

The amount of DNA per chromosome differs over a wide range in eukaryotes. The budding yeast has the smallest chromosomes discovered so far. The C-value of 1.5×10^7 base pairs of DNA is distributed among 17 chromosomes in a haploid set. The sizes of the DNA molecules in these chromosomes have been determined partly by electrophoretic migration in a gel. The smallest chromosome contains a DNA molecule of 260,000 base pairs (87 μm), or about 1/16 as long as the chromosome in *E. coli.* The others extend upwards in size with the largest one well over one million base pairs (over 333 μm in length).

There is no conceptual difficulty in accepting the length measurements for yeast chromosomes; they are all considerably smaller than the DNA molecule in *E. coli.* Other eukaryotes possess much more DNA per chromosome, in some the equivalent of DNA molecules several meters long. Such long molecules seem improbable, and for years it was debated whether a eukaryotic chromosome contains a single, long DNA molecule or multiple short ones.

TABLE 11-3

Average Lengths of DNA Molecules in the Chromosomes of Some Eukaryotes

Organism	Average Length of DNA Molecules
Yeast (*Saccharomyces cerevisiae*)	0.3 mm
Unicellular Alga (*Chlamydomonas*)	1.33 mm
Moss (*Sphagnum*)	2.33 mm
Fruit fly (*Drosophila*)	14.0 mm
Chicken (*Gallus*)	9.8 mm
Pike (*Esox*)	3.1 cm
Human	4.0 cm
Shark (*Scyllium*)	4.7 cm
Amoeba (*Amoeba proteus*)	5.3 cm
Mouse (*Mus*)	5.3 cm
Cat (*Felis*)	6.2 cm
Pine tree (*Pinus*)	9.7 cm
Frog (*Rana*)	18.3 cm
Lungfish (*Protopterus*)	98.3 cm
Onion (*Allium*)	1.1 meters
Salamander (*Amphiuma*)	2.7 meters
Lily (*Trillium*)	4.3 meters

The most direct way to determine whether the DNA in a chromosome is present in one or many molecules is to purify the DNA and measure its size directly by some physical technique. This has been done for the relatively short molecules in yeast, but is virtually impossible for much longer molecules because of the ease with which they are broken, especially during purification. Indirect methods for determining molecular length had to be devised as described below.

The lengths of the DNA molecules in chromosomes of various species cover a wide range. Table 11-3 gives the average length of DNA molecules for some eukaryotic species. Why organisms package their genes in such vastly different sizes of DNA molecules is not known.

Length of DNA molecules in *Drosophila*. One method for measuring DNA length exploits an elastic property of DNA. Molecules of DNA free in solution coil randomly, much as a long piece of string coils when immersed in water. When a DNA molecule is pulled into a straight configuration and then released, it recoils back into a random configuration. The time taken for a straight molecule to relax is proportional to its length. Measuring the relaxation time provides an indirect but accurate measure of molecular length. A great advantage of the method is that the DNA does not need to be purified to be measured, which eliminates manipulations of DNA that inevitably break it. Cells are lysed with an ionic detergent, which strips all the proteins from DNA. The elasticity of DNA is measured directly in the detergent solution. When applied to cultured cells of *Drosophila*, the largest DNA molecule gives a molecular weight of 41×10^9, which corresponds to a length of 20.5 mm. The total amount of DNA in the largest chromosome was previously known from optical measurements to be 41×10^9, and the elasticity measurement proves that all this DNA is contained in a single molecule.

Length of DNA molecules in mammalian cells. Estimates of molecular length can be obtained for mammalian chromosomes with the autoradiographic technique previously described for visualizing the DNA of *E. coli*. Cultured human cells are first grown for at least one cell cycle with ^{3}H-thymidine, which completely labels the DNA molecules throughout their lengths. The cells are lysed in a drop of dilute ionic detergent on a microscope slide. Then the drop is gently spread over the slide, which draws the DNA into a long, straight form. Subsequent autoradiography shows the presence of tracks of silver grains several centimeters long, reflecting the presence of DNA molecules of such length. A human diploid cell contains 6.2 pg of DNA and 46 chromosomes, which means on average an equivalent of 4 cm of DNA per chromosome. The autoradiographic visualization shows that all the DNA in a chromosome is contained in a single molecule.

Length of DNA molecules in salamanders. The genomes and chromosomes of salamanders are among the largest known. Only the African lungfish, and a few species of lilies have comparable amounts of DNA per genome and per chromosome. A haploid set of 12 chromosomes in the salamander *Amphiuma means* (misnamed the Congo eel) contains 100 pg of DNA (Table 11-2). Therefore, on the average each chromosome contains the equivalent of 2.7 meters of DNA. According to experimental evidence, some of which is cited below, all 2.7 meters of DNA are present in a single, extremely long molecule.

Most studies of DNA in salamanders have been done on chromosomes in meiosis. In salamanders the formation of oocytes lasts many months. During this time the chromosomes are arrested in the prophase stage of the first meiotic division. Because the chromosomes are partially condensed, they are easily seen by light microscopy (Figure 11-16). A single lampbrush chromosome actually contains two homologous chromosomes that previously were tightly joined in meiotic pairing and subsequently have largely separated from each other. The chromosomes are now in the diplotene stage of the first meiotic prophase described in Chapter 7. The two homologues have separated except at several joined points called chiasmata (*sing.* = chiasma). Each chiasma is the cytological manifestation of a crossover event that took place earlier in prophase. Each homologue underwent duplication (DNA replication) in the premeiotic interphase and consists of two chromatids. Crossing over has occurred between a chromatid in one homologue and a chromatid in the other, which holds the chromosomes together at chiasmata.

The fuzzy appearance of the two homologues in a lampbrush chromosome is the result of threads that loop out from its two chromatids. The loops occur in pairs—wherever a loop occurs in one chromatid, it is

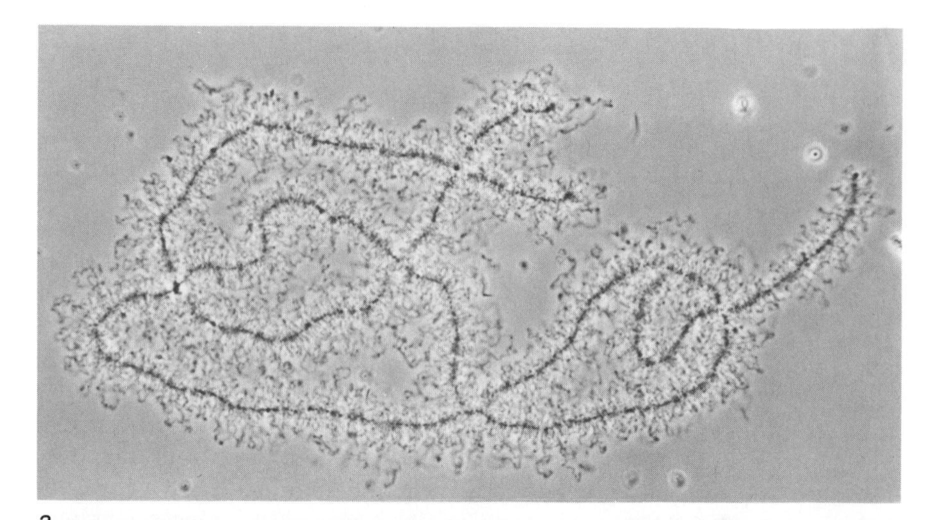

a

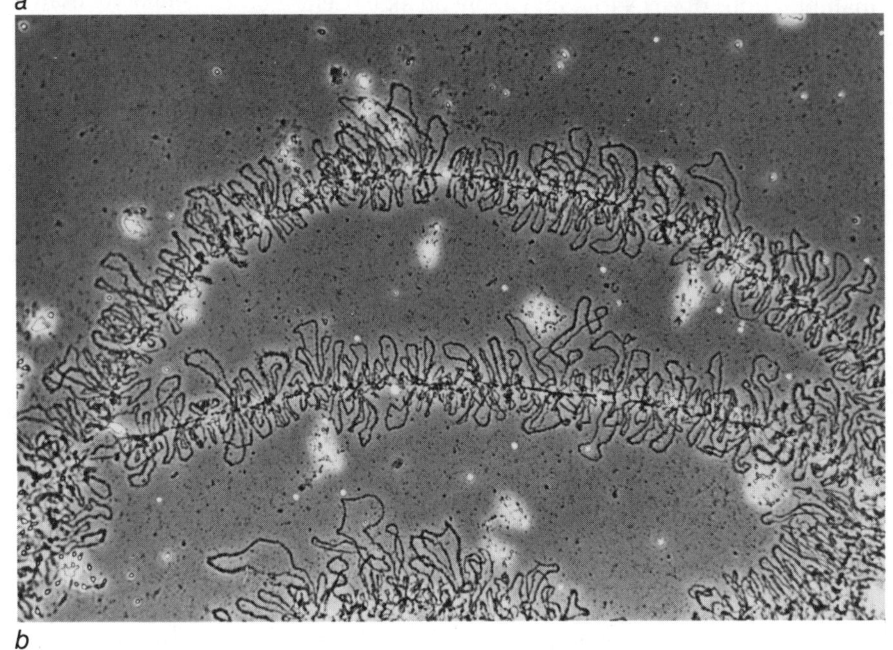

b

FIGURE 11-16 (a) Light micrograph of lampbrush chromosomes isolated from an oocyte of the newt *Notophthalmus viridescens*. Two homologues are present and still joined by several chiasmata. (b) Portions of two lampbrush chromosomes that have been flattened on a surface by centrifugation. The loops now all lie in the same plane and are more clearly visible than in the chromosomes in (a). [Courtesy of Joseph G. Gall.]

matched by a loop of the same size in the other chromatid (Figure 11-17). The loops are made up of DNA with attached ribonucleoproteins and are sites of transcription. The axis of the chromosome consists of a string of fine chromatin granules called **chromomeres.** One or more pairs of loops extend from each chromomere. In this arrangement each chromatid consists of a string of chromomeres with attached loops. Since the two chromatids are tightly apposed to each other, the chromosome axis is probably formed by fused pairs of chromomeres. The molecular structure of chromomeres and loops was deciphered by a combination of experiments. First, removal of proteins and RNA from lampbrush chromosomes with enzymes does not cause breakage of the chromosome. Treatment with DNase causes fragmentation. This suggests that the linear continuity of the chromosome is not dependent on protein or RNA but does depend on continuous intact DNA. When protein and RNA are stripped from loops, only a DNA thread is left. Measurements of the diameter of the thread seen in loops by electron microscopy show that the thread is a single DNA double helix. A comparison of the rates of breakage with DNase of loops versus the main axis of the chromosome shows that each loop contains one DNA duplex whereas the main axis must contain two DNA double helices, one for each chromatid.

These and other observations showed that each chromatid contains a single DNA double helix that extends through the entire length of the chromatid (Figure 11-17). Coiling and folding of the DNA gives rise to the successive chromomeres. Successive chromomeres are connected by segments of the DNA molecule that loop out from one chromomere and continue into the next. In short, one DNA molecule traverses the entire length of a chromatid. Proteins are responsible for folding the DNA in a specific pattern of successive chromomeres and loops.

The partial meiotic condensation of lampbrush chromosomes makes visible a structural organization along the chromosome that may be present in interphase chromosomes of eukaryotes in general, although

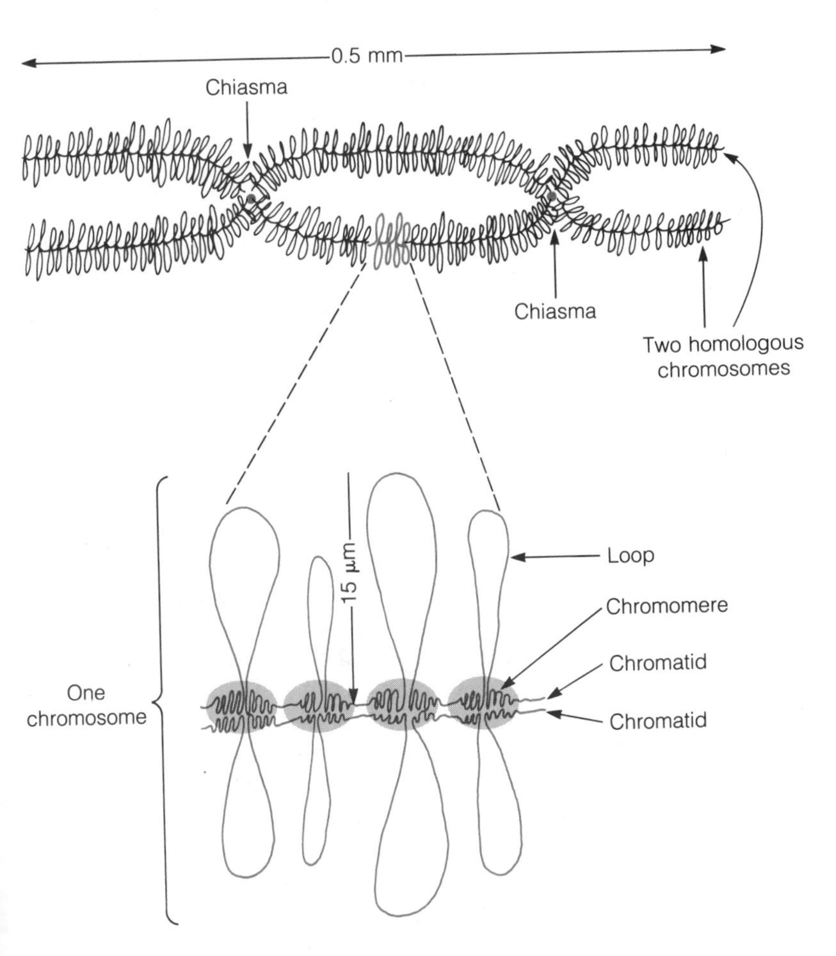

FIGURE 11-17 Structure of a lampbrush chromosome. (a) A lampbrush chromosome is, in fact, two homologous chromosomes joined by chiasmata, in this case two chiasmata. The two homologues have identical sets of hundreds of loops. (b) An enlarged portion of one homologue showing that it consists of two chromatids. Each chromatid consists of a continuous DNA-protein fiber that is periodically folded to form chromomeres. The DNA-protein fiber extends out from each chromomere to form a loop. Any given loop is the same size in both chromatids.

probably in a looser form. This presumed structural organization is not discernible because the chromosomes are more extended and intertwined with one another within the nucleus. Direct support for this view is provided by polytene chromosomes, described in Chapter 10. Polytene chromosomes are true interphase chromosomes. Their linear organization is visible because many identical copies of the DNA double helix, with attached proteins, are held together in exact parallel register, producing a clearly discernible chromosome. Each band of a polytene chromosome may correspond to a chromomeric granule and its loop.

Organization of Deoxynucleotide Sequences in Eukaryotic DNA

The DNA molecule of a prokaryotic chromosome is largely occupied by a succession of genes. A gene is defined here as a coding sequence and sequences upstream and downstream that are concerned with controlling the expression of the coding sequence. No genes have yet been found in some segments of the chromosome in *E. coli,* but most of the DNA consists of identifiable genes.

Eukaryotic DNA molecules are usually larger than prokaryotic molecules and possess a number of features not observed in prokaryotes. Earlier in the chapter three distinguishing features were mentioned to introduce eukaryotic chromosomes. Additional characteristics are the following:

1. The DNA of most eukaryotic genomes contains nongene sequences that are repeated many times (**repetitive DNA**).
2. A few kinds of eukaryotic genes are present in multiple copies.
3. Eukaryotic DNA contains imperfect, apparently nonfunctional copies of some genes called **pseudogenes.**
4. The linear DNA molecules in eukaryotic chromosomes terminate in special sequences called **telomeres,** except for mitochondrial and chloroplast chromosomes, which are usually circular and therefore have no ends.
5. Eukaryotic chromosomes possess **centromeres,** structures by which chromosomes are distributed at mitosis and meiosis. The centromere has its basis in a specific sequence in the DNA molecule.
6. Prokaryotic DNA molecules contain a single origin of replication. The linear molecules in eukaryotes have many origins dispersed along their lengths.
7. Prokaryotic chromosomes are complexed with DNA-binding proteins that package the DNA molecule in successive units. Eukaryotic DNA is bound to a special class of proteins, the histones, forming repeating units called **nucleosomes.**
8. Most genes in eukaryotes are interrupted at a number of sites by introns (Chapter 8). With a few exceptions introns are absent from prokaryotic genes.
9. Much of the DNA in most eukaryotic genomes forms long spacers between successive genes or groups of genes in a molecule. Spacers between genes in prokaryotes are very short.

Repetitive DNA sequences. All eukaryotes contain repetitive sequences. A small part of the repetitive sequences represents a few kinds of genes that are present in multiple copies, but most of the repetitive sequences have no known function, genetic or otherwise. The absolute amount of repetitive sequences tends to be greater in organisms with larger genomes. In yeast cells only a few percent of the total DNA consists of repetitive sequences. Typically in mammals 30 percent to 40 percent of the DNA is made up of repetitive sequences belonging to several families. The remaining 60 percent to 70 percent is made up of sequences that occur only once per haploid genome. These are called **unique sequences** and contain most of the genes as well as a large amount of DNA of unknown function.

Within a family of repetitive sequences all the sequences are identical or nearly identical with one another and differ from sequences in any other family. Families are classified according to the number of copies of the repeat. Families containing two to 100 copies are **slightly repetitive.** Families with 100 to several

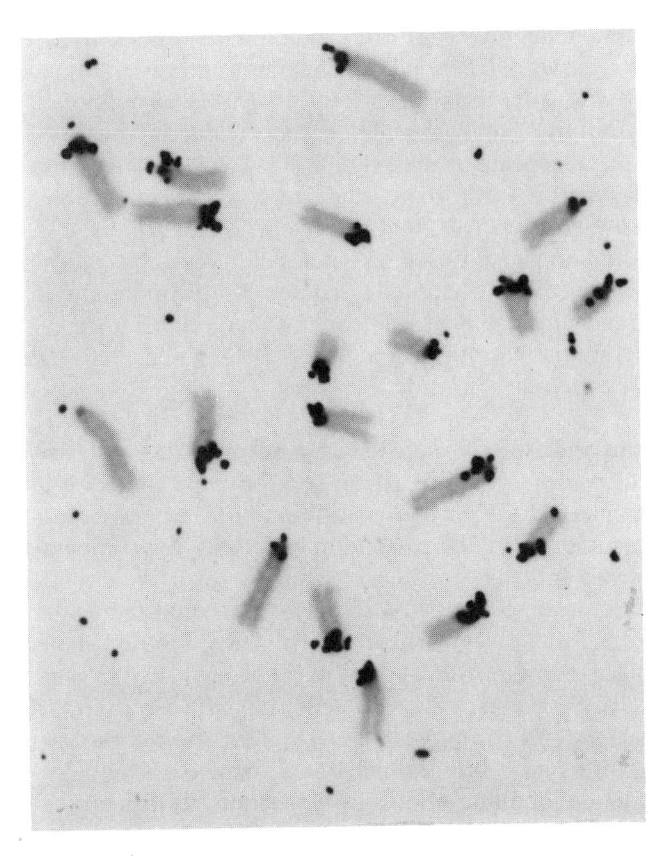

FIGURE 11-18
Autoradiograph of chromosomes from a culture of mouse cells after hybridization of tritium-labeled RNA prepared by transcribing purified mouse satellite DNA. The RNA has bound to its complementary sequences in the DNA in the chromosomes, showing that mouse satellite DNA is localized in the heterochromatin at the centromere. [Courtesy of Mary Lou Pardue.]

thousand copies are **middle repetitive.** Families with several thousand to several million repetitive sequences are called **highly repetitive.** Some of the families of slightly or middle repetitive DNA are, in fact, multiple copies of certain genes. For example, genes encoding actin are repeated several times in most eukaryotes and therefore make a family of slightly repetitive sequences. Genes encoding histone proteins (discussed later) are usually present in a few hundred copies and represent middle repetitive DNA. However, the functions, if any, of most families of middle repetitive sequences are unknown.

An example of a highly repetitive sequence occurs in the centromeric region of mouse chromosomes (Figure 11-18). This sequence is about 300 base pairs long, is repeated about 10^6 times per genome, and accounts for 10 percent of the DNA in the mouse genome. It occurs in blocks of thousands of copies in tandem arrays. Its location in the centromeric region of chromosomes implies a centromere-related function, but its function is not known. Blocks of highly repetitive sequences are packaged in a tightly condensed heterochromatin, which is in accord with the observed absence of transcription (Chapter 10). Blocks of highly repetitive sequences often differ in overall base composition from the rest of the DNA in the genome. If the repetitive DNA has a higher-than-average content of AT base pairs compared with the rest of the DNA, it has a lower-than-average density in a CsCl solution. GC-rich repetitive DNA has a higher average density. When total mouse DNA is cut into segments of a few hundred base pairs and centrifuged to equilibrium in a solution of CsCl (Chapter 7), the highly repetitive DNA forms a small band adjacent to the main band of DNA because it has a higher average AT content and therefore a lower density than the main DNA (Figure 11-19). This, in fact, was how highly repetitive DNA was first discovered, and it is sometimes referred to as **satellite DNA** because it forms a satellite peak in a density profile.

The individual copies of sequences in families of all three classes of repetitive sequences (slightly, middle, and highly repetitive) can occur in blocks as just described for the family of highly repetitive sequences in the mouse genome, or the individual copies can be dispersed apparently randomly among genes throughout the genome. Middle repetitive DNA is also usually distributed in the genome in between genes. An example of a highly repetitive sequence that is dispersed is the **Alu family** in the human genome. The repetitive sequence in the Alu family is about 300 base pairs long, more than 900,000 copies are present, and they account for five percent of human DNA. One or more copies have been found not far from all genes isolated. At least some Alu sequences are transcribed, forming

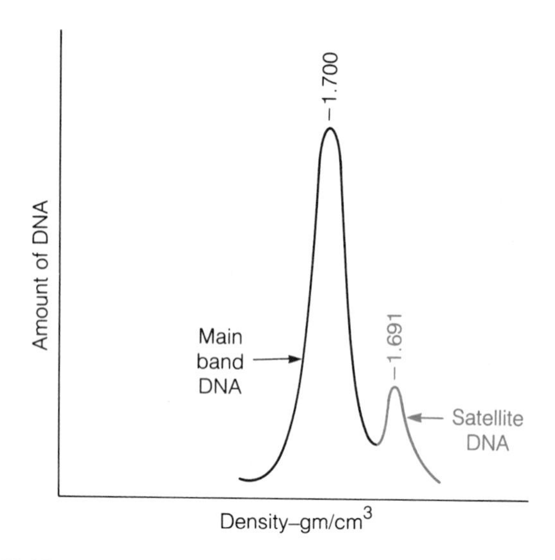

FIGURE 11-19
Profile of mouse DNA centrifuged to equilibrium in a CsCl gradient. The DNA separates by density into two components, main-band DNA at a density of 1.700 gm/cm³ and a band of satellite DNA at a density of 1.691 gm/cm³.

part of the primary transcript of some genes. The Alu segments of primary transcripts are removed and destroyed in processing of the transcript in the formation of an mRNA molecule. Like all repetitive sequences the function of the Alu sequences or their transcripts is not known.

Closely related species sometimes have greatly differing amounts of repetitive DNA. For example, the genome of *Drosophila melanogaster* contains about 15 percent of highly repetitive sequences, *D. virilis* has 40 percent, and *D. cyrtoloma* has 60 percent. In one species of the kangaroo rat 60 percent of the DNA is repetitive; in a closely related species less than ten percent is repetitive, and the total amount of DNA in the genome is correspondingly less.

Telomeric DNA sequences. The linear DNA molecules of eukaryotes have special sequences at their ends called **telomeric sequences.** The sequence of telomeric DNA has been determined in a number of lower eukaryotes. Although there are differences, all telomeric sequences conform to a common pattern. In *Tetrahymena* and *Paramecium*, telomeres consist of up to 70 repeats of the sequence 5′CCCCAA3′. In a different group of ciliates (the hypotrichs) telomeres consist of several repeats of the sequence 5′CCCCAAAA3′. In a flagellated protozoan the repeated sequence in telomeres is 5′CCCTAA3′, in yeast it is more variable, consisting of 5′CCACA3′ and slight variations of the sequence. The telomeric sequences are important in conferring stability on linear DNA molecules, since experimentally introduced linear DNA molecules that lack them are degraded.

Autonomously replicating sequences. An **autonomously replicating sequence** or **ARS** is so named because it confers autonomous replicative capacity on a circular plasmid in yeast cells. It is comparable in function to the *ori* sequence in a prokaryotic chromosome. The ARS in the yeast plasmid can be deleted, and a segment of foreign DNA inserted in its place to test for ARS activity of the foreign DNA in yeast cells. Segments with ARS activity have been found in this way in chromosomal DNA of yeast and other organisms including human DNA, and these may be sequences defining origins of replication in their normal locations.

The ARSs in yeast chromosomes differ somewhat from one another in sequence, but all those isolated so far have an 11-base-pair, AT-rich sequence with at least 200 base pairs flanking each side, which are necessary for normal initiation of replication. An ARS presumably is recognized by a specific protein that triggers the initiation of DNA replication, but such a protein has not yet been identified in eukaryotes.

Centromeric sequences. The proper distribution of chromosomes in eukaryotes during mitosis and meiosis is achieved by attachment of chromosomes to microtubules in the spindle apparatus in the dividing cell (Chapter 12). The attachment in most species occurs at a region in the chromosome called the centromere (Figure 11-20). The centromere is also the point at which the two chromatids remain attached to each other in metaphase. Separation of the two chromatids at the centromere marks the transition from metaphase to anaphase; from that instant on, the chromatids are called **daughter** chromosomes. The attachment of microtubules occurs at a part of the centromere called the **kinetochore,** which in the electron microscope looks like a plate on each chromatid (Figure 11-21).

The centromere is defined by a sequence of deoxynucleotides. Special proteins probably bind to the centromeric sequence and give the centromere its functional properties and form the kinetochore. Centromeric sequences from several chromosomes (**CEN sequences**) have been studied closely in yeast. They are quite similar but not identical in sequence, consisting of 80 to 90 base pairs of 93 to 94 percent AT base pairs next to a highly conserved 11-base-pair sequence. A CEN sequence can be removed from a chromosome and reinserted in the opposite orientation, and it works just as efficiently. CEN sequences, with slightly different composition, can be interchanged among yeast chromosomes without any loss of function.

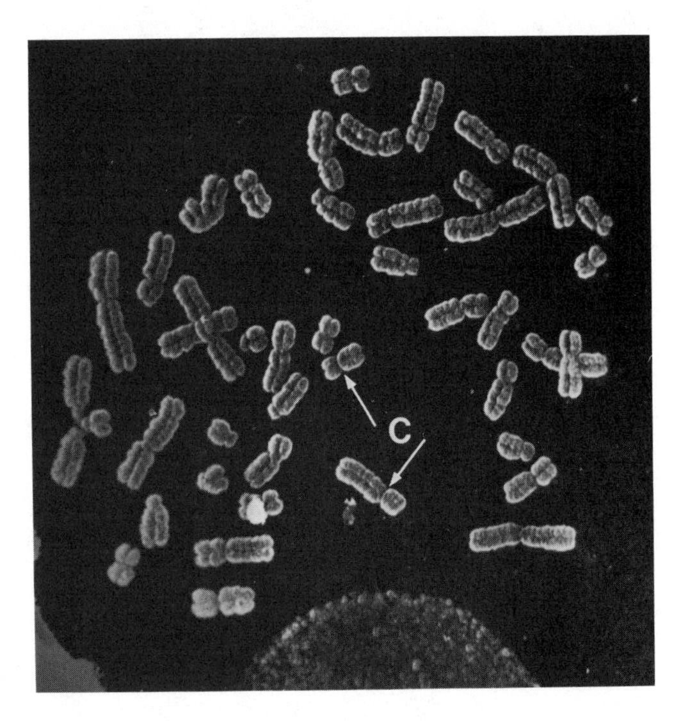

FIGURE 11-20
Scanning electron micrograph of a set of human chromosomes isolated from a cell in metaphase. The two chromatids in each chromosome are held together at their centromere (C). [Courtesy of Christine J. Harrison and The Company of Biologists Limited.]

Pseudogenes. Pseudogenes are nonfunctional, imperfect copies of normal genes. They occur commonly in higher eukaryotes such as mammals, but only a few have been found in lower eukaryotes such as yeast and *Drosophila.* In mammals pseudogenes account for as much as one percent of the total DNA. Formation of a pseudogene is a rare event in evolution, but once formed a pseudogene remains even though it is nonfunctional. For example, pseudogene $\psi\eta$ is a defective gene for β-globin of hemoglobin in mammals. A detailed evolutionary history of pseudogene $\psi\eta$ has been constructed by the study of the globin gene family in various animals, and it is clear that pseudogene $\psi\eta$ arose more than 140 million years ago, long before the evolution of modern mammals.

Pseudogenes arise by two mechanisms. Some, such as $\psi\eta$, originated by chance duplication of a normal functional gene. Duplication of a gene can occur by unequal crossing over during meiosis such that one chromatid receives both copies of a gene and the other chromatid receives no copy (Figure 11-22). Gene duplication can also occur through a mistake in DNA replication. A segment of DNA is replicated twice, and the extra copy becomes integrated into a chromosome by a poorly understood mechanism of breakage and rejoining. The result is the presence of two functional copies of a gene. Duplication of a gene may be followed sometime later in evolution by mutation of one of the copies into a nonfunctional form, creating a pseudogene. Defects observed in pseudogenes include abnormal initiation and termination codons, missense, nonsense, and frameshift mutations, altered promoters, and abnormal splicing in transcripts.

In the second mechanism pseudogenes are formed from mRNA molecules. Certain RNA viruses, called **retroviruses,** carry a gene encoding an enzyme that can synthesize double-stranded DNA using a single-stranded RNA chain as a template instead of DNA (Figure 11-23; Chapter 13). The enzyme is called **reverse**

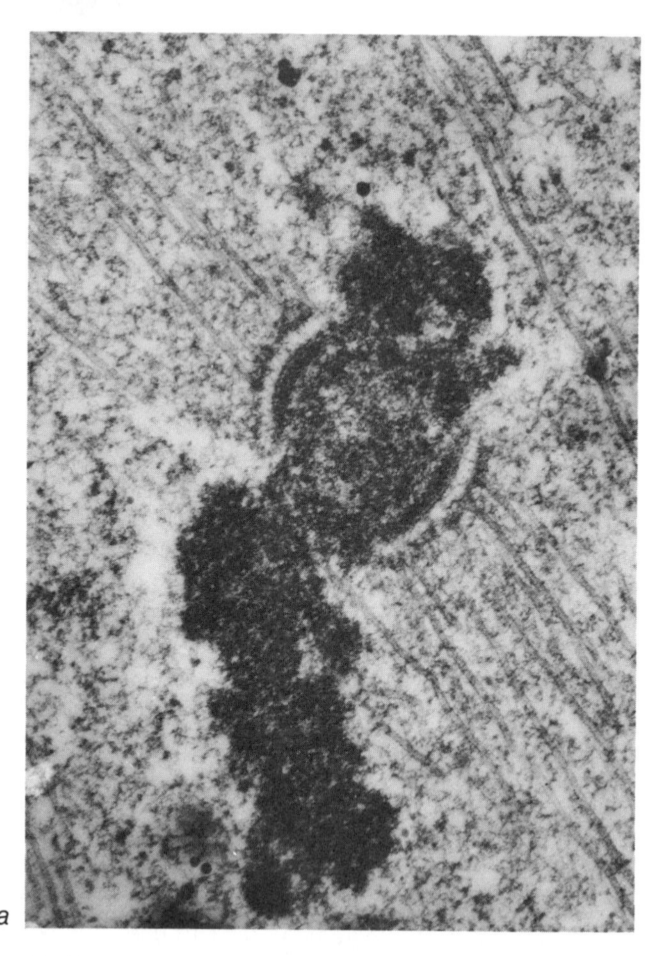

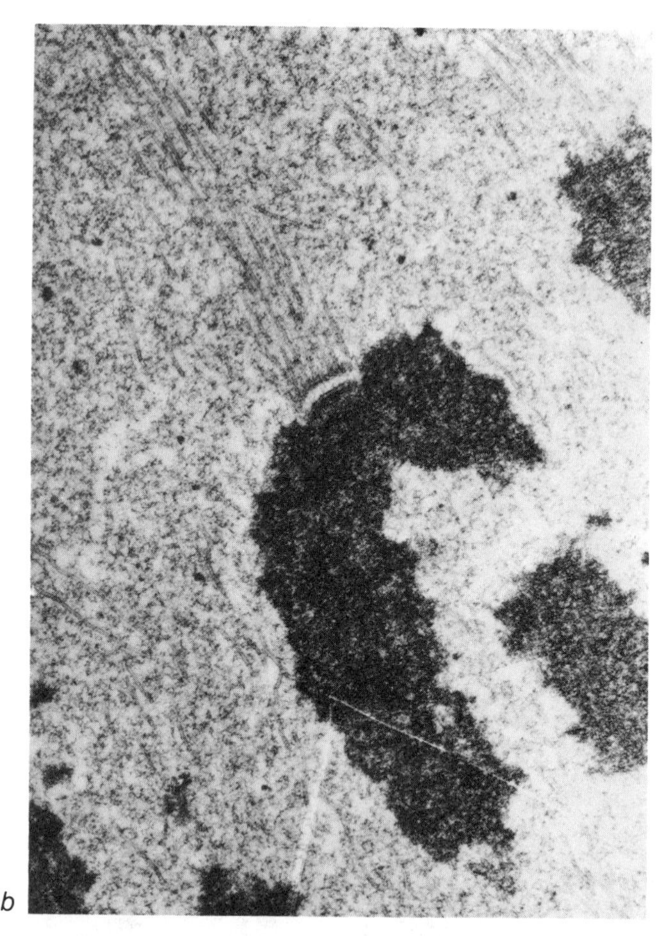

FIGURE 11-21
Electron micrographs of kinetochores. (a) Metaphase chromosome with a platelike kinetochore on each chromatid. Microtubules from opposite poles of the cell are attached to the kinetochores. Most of the chromosome is out of the plane of the section. (b) Kinetochore on a daughter chromosome in anaphase. The chromosome is moving in the direction of the microtubules. [Courtesy of Jeremy Pickett-Heaps.]

transcriptase. The resulting segment of DNA double helix, called **cDNA**, may become integrated into chromosomal DNA by breakage and resealing. The integrated cDNA, being a copy of mRNA, lacks a promoter and other sequences necessary for transcription and hence cannot be transcribed. Pseudogenes formed from mRNA lack the introns present in most normal genes, and are therefore easily distinguished from their normal, intron-containing counterparts.

Sequence analysis suggests that some families of dispersed repetitive DNA sequences are, in fact, pseudogenes of tRNA molecules, snRNA molecules, or other small non-mRNA molecules that are plentiful in the cell and therefore more likely to have been used by chance to make a cDNA. Thousands of nonfunctional, pseudogene copies of snRNA genes are found in mammals. The human genome contains about 20 copies of the gene encoding actin; many of these are nonfunctional pseudogenes.

Junk DNA. Pseudogenes, other dispersed repetitive sequences not clearly identified as pseudogenes, and perhaps other nonrepetitive sequences with no known function may all be defective relics of normal genes. For that reason they are sometimes called **junk DNA.** If such DNA sequences contribute nothing to the functioning of a cell, that is, are genetically neutral, it is not clear why they are not in the course of time lost through chance deletion. The cost of carrying junk DNA, for example, in the expense to the cell of its replication, may be so trivial that organisms containing it do not incur a survival disadvantage.

The Essential Sequences of a Chromosome

Many kinds of sequences have just been listed; some are essential for chromosome structure and function, and some are not. The minimal sequences in a chromosome are genes, telomeres, ARSs, and a centromere (Figure 11-24). So far, research has not identified functions for most repetitive sequences (not including normal gene families) and pseudogenes, and they are presumably nonessential.

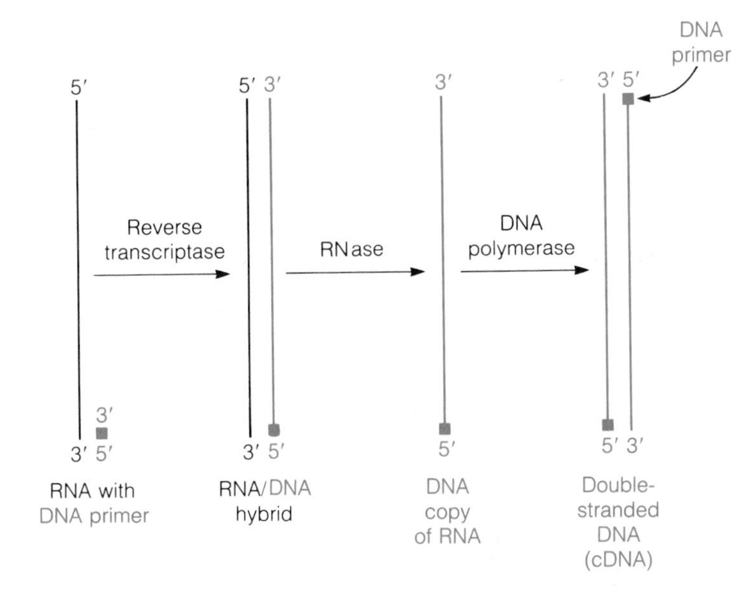

FIGURE 11-23
Synthesis of a double-stranded DNA molecule starting with a single-stranded RNA template. Reverse transcriptase can use single-stranded RNA as a template, but like DNA polymerase it requires a DNA or RNA primer. In this case a short fragment of DNA is the primer. The RNA/DNA hybrid synthesized by reverse transcriptase is reduced to a single-stranded DNA molecule by RNase. DNA polymerase converts the single-stranded DNA into double-stranded DNA (cDNA), which requires another primer.

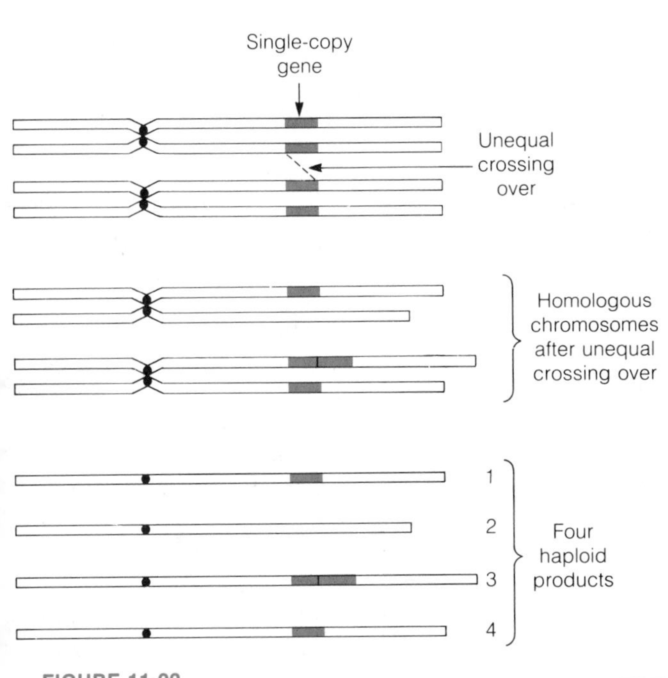

FIGURE 11-22
Duplication of a gene in a chromosome by unequal crossing over. An organism receiving the third haploid product at fertilization would contain a duplicate copy of the gene in one of its chromosomes.

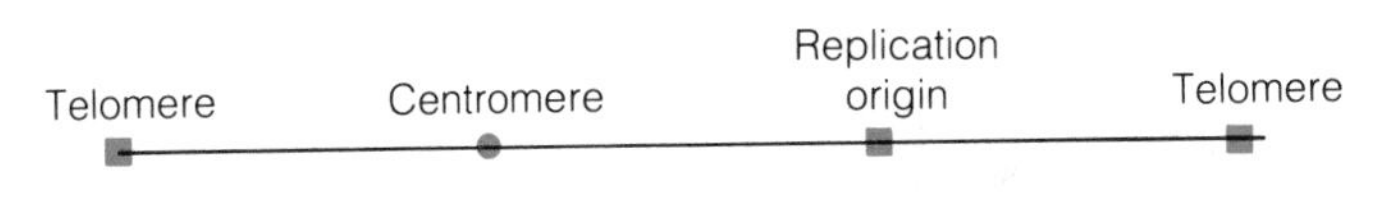

FIGURE 11-24
Essential sequences in a eukaryotic chromosome are those that are present in the two telomeres, the centromere, and an origin of replication. Multiple origins of replication are normally present, as well as genes and spacers between genes that give the chromosome a length required for reliable segregation at mitosis.

The sequences required for chromosome function have been defined experimentally in yeast cells in the following way: a linear piece of DNA containing an ARS and a few genes was combined with a CEN sequence, and telomeres were joined to the ends. The molecule, containing about 10^4 bp, was introduced into yeast cells. Although the molecule persisted through several cell divisions, it was lost because it was not properly distributed to daughter cells by mitosis. Apparently, the CEN sequence did not function adequately in such a short DNA molecule. A major difference between this artificially constructed chromosome and a normal yeast chromosome is length. The smallest yeast chromosome contains 2.6×10^5 bp, compared to the 10^4 bp in the artificial chromosome. The artificial chromosome was therefore enlarged to 5×10^4 bp using DNA from a bacteriophage. Bacteriophage DNA is functionless in yeast and was used to act as a filler. Increasing size in this way markedly improved the mitotic stability of the artificial chromosome and even allowed it to behave normally in meiosis, but it was still not as stable as the larger normal chromosomes in yeast. A normal yeast chromosome is accidentally lost with a frequency of once per 10^4 to 10^5 cell divisions. Smaller chromosomes are lost more frequently than larger ones. These experiments imply that a DNA molecule, in addition to possessing ARS, telomeres, genes, and a CEN sequence must have some minimum size in order to be distributed properly in mitosis and meiosis. The apparent size requirement is met in yeast almost exclusively by genes, with little or no contribution from repetitive sequences, pseudogenes, or junk DNA.

Total Complexity of DNA Sequences in Genomes

The total complexity of sequences, as opposed to total DNA, is the sum of all the different sequences in a genome. In prokaryotes total sequence complexity is essentially equivalent to the DNA content of the genome since almost all the DNA consists of single-copy, or unique sequences. In eukaryotes, total sequence complexity is less than the total DNA content because of the presence of repetitive DNA. Repetitive DNA accounts for a significant part of the total DNA but because it is composed of the same sequence repeated over and over, it contributes disproportionally less than unique-sequence DNA to total sequence complexity. For example, about 25 percent of the DNA in the mouse genome is repetitive and 75 percent is unique. Of the 25 percent repetitive DNA, 10 percent is a highly repetitive sequence of 300 bp (the satellite DNA), and 15 percent is middle repetitive DNA. The total DNA content in the mouse genome is about 3×10^9 bp. The unique-sequence DNA (75 percent of total or 2.25×10^9 bp) accounts for virtually 100 percent of the sequence complexity. The highly repetitive sequence accounts for 10 percent of the total DNA but only 0.000013 percent of the total complexity (300 bp/2.25×10^9 bp). The several families of middle repetitive DNA account altogether for less than one percent of the total sequence complexity in the mouse genome.

The sequence complexity of DNA is determined by shearing it into fragments and dissociating the two chains of the double helix by denaturation; then, the rate at which they reassociate is measured as a function of DNA concentration expressed as nucleotides per liter. The greater the complexity of DNA, the more slowly it reassociates. Consider, for example, two DNA molecules each consisting of 1000 bp. Suppose one of the molecules consists entirely of unique sequences and the other consists of a 100-base-pair sequence repeated identically ten times. Suppose the same number of molecules of 1000 bp is dissolved in a salt solution of the same volume, cut in segments about 100 bp long, denatured, and then allowed to renature. The rate of reassociation is governed by how often complementary chains encounter each other in solution. The concentration of chains of particular sequences is ten times higher for the repetitive DNA than for the unique sequence DNA; therefore the repetitive DNA will reassociate much faster than the unique sequence DNA. Likewise, when the total DNA of a genome is cut into lengths of a few hundred bp and dissociated, each kind of sequence reassociates at a rate that is proportional to its concentration; highly repetitive sequences reassociate more rapidly, followed by reassociation of middle

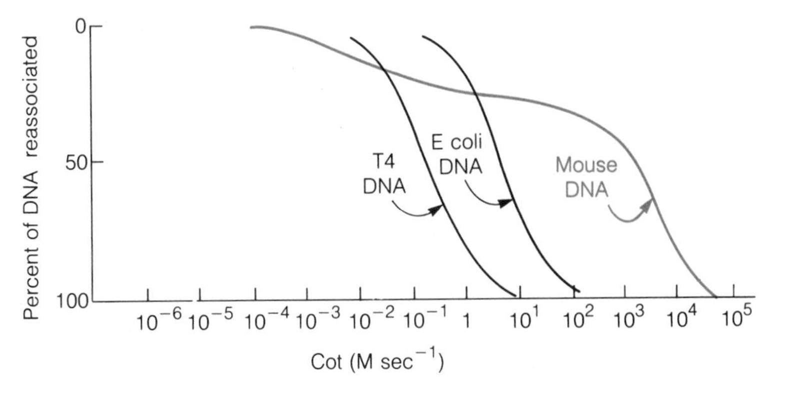

FIGURE 11-25
Reassociation (Cot) curves for bacteriophage T4 DNA, *E. coli* DNA, and mouse DNA.

repetitive sequences, and then by reassociation of the unique sequences.

Figure 11-25 shows reassociation curves for DNAs of T4 bacteriophage, *E. coli,* and the mouse. Since T4 and *E. coli* DNA do not contain repetitive sequences, they reassociate as uniform curves reflecting rates expected for genomes of their respective sizes. The genome of *E. coli* DNA is about 25 times larger than DNA of T4. If both DNAs are prepared at the same concentration (in nucleotides per liter), the DNA of *E. coli* reassociates more slowly than T4 DNA (because the number of DNA molecules is less). Mouse DNA shows a more complicated curve, the first part of which reflects reassociation of highly repetitive sequences (satellite DNA), followed by several families of intermediate repetitive DNA and finally unique sequence DNA.

A useful way to record reassociation events is to plot the initial molar concentration of nucleotides (C_o) in the DNA times the time (t) in seconds that reassociation has been occurring against UV absorption; a decline in UV absorption occurs as DNA reassociates. Such a plot of initial $C_o \times t$ (Cot) against UV absorption is shown in Figure 11-25. The UV absorption of totally dissociated DNA at $t = 0$ is defined as 100 percent. The value of Cot when half the DNA has reassociated, $Cot_{1/2}$, is used as a measure of the sequence complexity of the DNA in nucleotides. Notice in Figure 11-25 that the $Cot_{1/2}$ for *E. coli* DNA is about 25 times greater than for T4 DNA, and the $Cot_{1/2}$ for the **unique fraction** of mouse DNA is about 600 times greater than for *E. coli* DNA.

The measure of sequence complexity of genomes, or **S-value** (S for sequence), is in some ways a more useful value than is total DNA since total DNA can differ by wide margins from species to species because of large differences in the amounts of repetitive sequences. The DNA content of mammals covers a two-fold range, but all have about the same complexity, which is about 600 times the complexity of *E. coli* DNA. This does not mean that mammals possess 600 times more genes than *E. coli* because most of the DNA sequences in mammals do not encode genes.

The C-Value Paradox

The C-value paradox mentioned earlier occurred because of wide differences in DNA content among species. Comparing S-values among organisms instead of comparing C-values might be expected to show a closer correspondence. After all, S-values are measures of the amount of unique sequence complexity, and they avoid the wide differences in total DNA that stem from differences in the content of repetitive DNA. The correspondence is, indeed, closer when S-values are used. All mammals have close to the same structural and functional complexity, and although C-values vary over a two-fold range, they all have very close to the same S-value. However, among eukaryotes a lack of correspondence between S-values is still often observed. For example, salamanders have several times more unique-sequence DNA than do mammals. The solution to this S-value paradox is that a large amount of unique-sequence DNA does not encode genes, as described in the next section.

Estimates of Gene Numbers, the S-Value Paradox, and Spacer DNA

The precise number of genes is known for a few of the simpler viruses, but it is not known for any prokaryotic or eukaryotic cell. Assuming 1000 bp for one gene, mycoplasma contains a maximum of 750 genes. The chromosome of *E. coli* contains 4.7×10^6 bp, enough DNA for 4700 genes, although the number is certainly much less than that. Eukaryotes contain more genes than prokaryotes, but at present only rough estimates are possible.

Gene number and spacers in *Drosophila*. *Drosophila melanogaster* contains 1.65×10^8 bp of DNA, of which 85 percent or 1.40×10^8 bp is unique sequence DNA. Assuming a somewhat larger average size for the coding regions of genes in *Drosophila*, such as 1200 bp, 1.40×10^8 bp could contain 117,000 genes. Estimates based on other grounds are considerably less. Genetic evidence suggests that one band of a polytene chromosome corresponds to one gene, although some bands may contain more than one gene. There are a little over 5000 bands total in the four polytene chromosomes, so estimates of gene number are 5000 to 7000. At 1200 bp per gene there is a 17- to 23-fold excess of unique sequence DNA.

A clue about the disposition of the excess unique-sequence DNA is apparent from the study of the polytene chromosomes in *Drosophila*. On the average, bands in a polytene chromosome contain segments of DNA consisting of about 25,000 bp of unique sequence. Using the estimate of 1200 bp for a gene, with one gene per band, about five percent of the DNA in a band is used to encode a gene. Some part of the remaining 95 percent (23,800 bp) can be accounted for by introns and pseudogenes. However, the function of the bulk of the unique-sequence DNA cannot be accounted for, and at present it must simply be regarded as spacer DNA between successive genes.

Gene number and spacers in mammals. In mammals the amount of unique sequence is sufficient to encode over 1.5×10^6 genes. A number of genetic considerations suggest a number 15 to 30 times less, or between 50,000 and 100,000 genes. Most eukaryotic genes contain introns, sometimes many of them, and some are very large (Chapter 8). Even including introns as parts of genes does not resolve the S-value paradox. The solution is, again, probably the presence of large spacers of unique sequences between successive genes. For example, a 50,000-bp segment of DNA in humans contains the five members in the family of the β-globin genes (Chapter 8) and the $\psi\eta$ pseudogene for β-globin. Each gene consists of about 2/3 intron sequences and 1/3 coding sequences. However, the six genes, including their introns and known regulatory sequences upstream and downstream from the genes, account for only about 20 percent of the 50,000-bp segment (Figure 10-15). The rest of the DNA, most of which consists of unique sequences, contains no identifiable genes and appears to exist as spacers between members of the globin gene family.

Gene number and spacers in ciliated protozoa. In ciliated protozoa it is clear that most of the unique-sequence DNA in the eukaryotic chromosomes exists as spacers between genes and has no known function. Ciliates were introduced briefly in Chapter 7. A ciliate contains a small, diploid micronucleus and a large, DNA-rich macronucleus (Figure 11-26). The micronucleus does not produce RNA, and it functions during cell mating as described in Chapter 7. The macronucleus makes all the RNA needed for cell maintenance and reproduction.

The major activity of ciliates is growth and division but occasionally two cells mate (conjugate). During conjugation the micronucleus undergoes meiosis, and the cells exchange haploid micronuclei and then separate, as described in Chapter 7. Immediately after separation the macronucleus is destroyed, and a new macronucleus is formed from a micronucleus. Formation of a new macronucleus is a complex process that need not be described here. In one group of ciliates known as hypotrichs, the first stage in macronuclear development is the formation of polytene chromosomes by multiple rounds of DNA replication (Figure 11-27). The process takes two days. As soon as they have formed, the polytene chromosomes are cut into short segments, each segment corresponding to a band of a polytene chromosome (Figure 11-28). A band is believed to represent a single gene, and the cutting up of the polytene chromosomes is, therefore, a gene-by-gene process. Cutting produces segments of DNA that contain on the average over 40,000 bp. These are rapidly reduced to an average of 2200 bp by cleavage of most of each DNA segment to nucleotides. Each of the short molecules that is left represents a single gene. Short telomeric sequences are then added to the ends of each gene. Finally, all the genes replicate many times to produce a

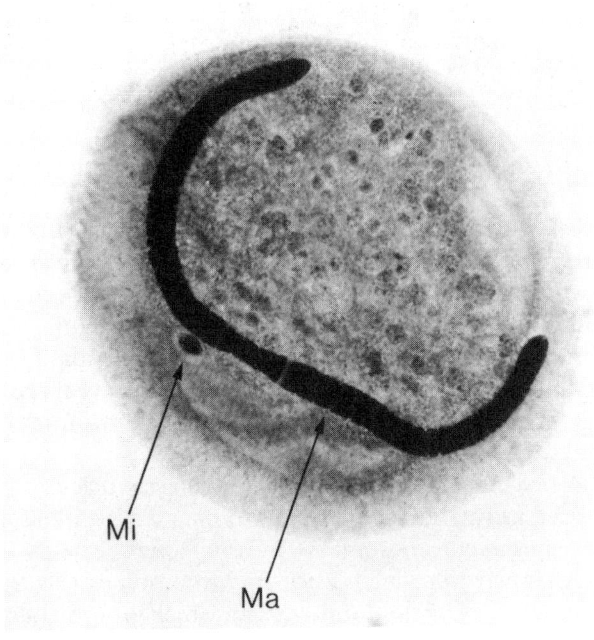

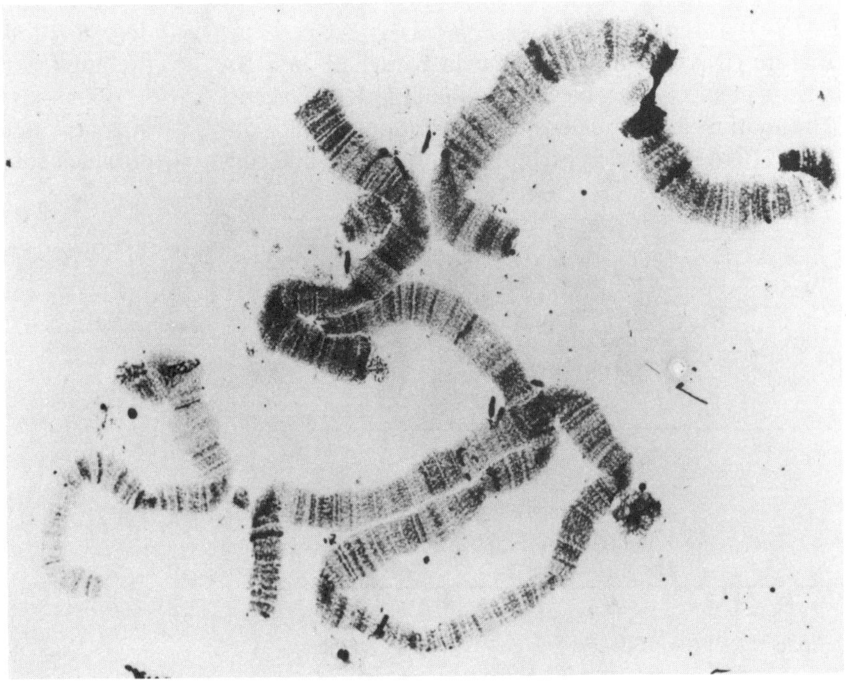

FIGURE 11-26
Light micrograph of the ciliated protozoan *Euplotes eurystomus*. The macronucleus (Ma) is C-shaped and darkly stained. The micronucleus (Mi) is in metaphase, and the cell is about to divide.

FIGURE 11-27
Electron micrograph of polytene chromosomes isolated from a developing macronucleus of the ciliate *Oxytricha nova*. [Courtesy of K. G. Murti.]

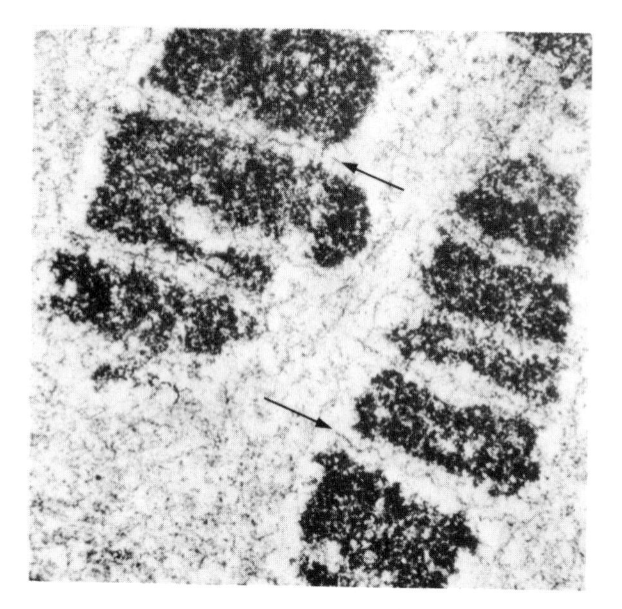

FIGURE 11-28
Electron micrograph of a section through polytene chromosomes in the developing macronucleus of a ciliate. Septa (arrows) have formed in all interbands, transecting the chromosomes band-by-band. [Courtesy of K. G. Murti.]

mature DNA-rich macronucleus in which all the DNA occurs in short molecules, each encoding a single gene. The mature macronucleus contains about 20,000 different DNA molecules, each present in about 1000 copies. These genes, representing only five percent of the sequence complexity in the chromosomes of the original micronucleus, are transcribed to provide all the mRNA, tRNA, rRNA, and other RNAs needed by the cell.

In summary, the development of the macronucleus consists of excision of all the genes from the chromosomes and destruction of most of the sequences in the chromosomes. Why polytene chromosomes are formed as an intermediate stage and many of the molecular details of how genes are excised and supplied with telomeres are still not known. However, the unusual manipulations of the genome by these ciliates reveal how gene and nongene sequences are organized in chromosomes.

The gene-sized molecules in the macronucleus can be used as hybridization probes to map the position of the genes in chromosomal DNA using long segments of DNA cloned from micronuclear chromosomes (Figure 11-29). Genes are found singly or in small groups along the DNA molecule separated by long spacers of nongene DNA. It is the spacer DNA, accounting for 95 percent of the total deoxynucleotide sequences, that is degraded following the gene-by-gene segmentation of the polytene chromosomes.

This extreme elimination of all nonessential sequences to form the functional nucleus presumably provides some advantage to these organisms. Ciliates

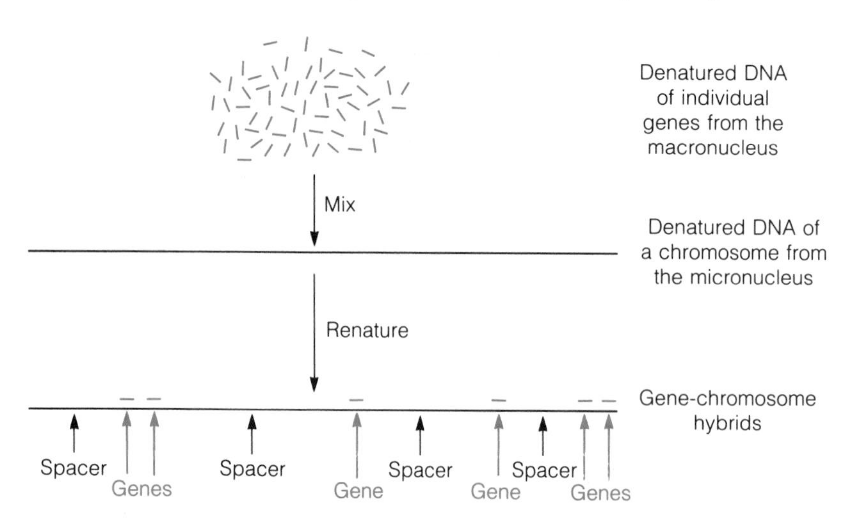

FIGURE 11-29
Organization of genes in a chromosome. Small DNA molecules containing single genes from the macronucleus are denatured and mixed with denatured DNA of chromosomes from the micronucleus. After renaturation, gene-sized molecules are found hybridized to chromosomal DNA in a highly dispersed pattern, with large spacers between genes or small groups of genes.

may require the many copies of each gene in the macronucleus to support their very large cell size. By elimination of all nonessential sequences, a high copy number for each gene can be achieved with a minimum amount of DNA and a minimum expansion of nuclear size.

The several observations on the DNA of *Drosophila*, mammals, and ciliates are consistent in suggesting a general model for eukaryotes, in which most of the DNA sequences form spacers of unknown significance between successive genes. Large differences in DNA sequence complexities (S-values) among eukaryotes, such as the high values in salamanders and lilies, are probably due to large differences in the size of spacers between genes.

Gene Families

Most genes in both prokaryotes and eukaryotes are present in single copies per genome. Although a few kinds of genes exist in multiple copies, in general single copies of genes are sufficient, because expression of genes that encode protein products is amplified at two points. First, many mRNA copies are transcribed from a single gene, and second, each of the many mRNA copies is translated into many copies of the protein (Figure 11-30). For most genes these amplifications at the transcriptional and translational levels are sufficient to meet the needs of a cell. Even during differentiation of cells, when a large amount of a gene product may be required, single copies of genes are sufficient. For example, an enormous number of hemoglobin molecules is produced in the formation of a red blood cell. Single copies of genes encoding α- and β-globin are sufficient.

For genes whose final products are RNA molecules, like genes encoding tRNAs and rRNAs, expression is by transcription alone. For these genes, as well as for a few kinds of genes that encode proteins, the rate of production of product cannot be achieved with a single copy of the gene per genome because of limits on rates of transcription (10 to 50 mRNA transcripts per gene per minute and 3 to 20 polypeptides per mRNA molecule per minute).

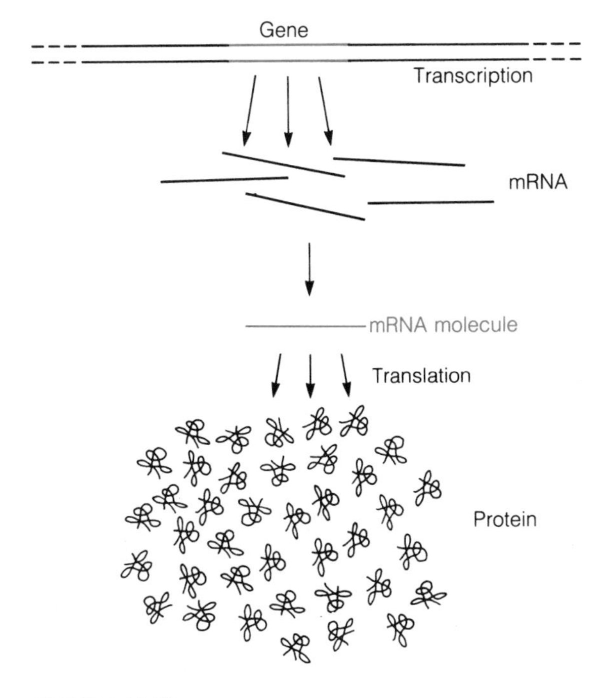

FIGURE 11-30
Expression of protein-encoding genes is amplified by multiple transcriptions, with accumulation of mRNA molecules, and by multiple translations of each mRNA molecule, with accumulation of protein molecules.

Almost all eukaryotes have multiple copies of the sequences encoding the four rRNAs. Sequences for three of the rRNAs (28S, 18S, and 5.8S in mammals) are grouped in a single transcription unit of DNA and are transcribed as a single precursor rRNA that is post-transcriptionally processed into three final products as described in Chapter 8. Transcription units for the three rRNAs are repeated in tandem in the chromosome in many copies (Chapter 10). In yeast cells the transcription unit is repeated 100 to 120 times in the arrangement shown in Figure 11-31 (see also Figure 10-8 for histone genes in the sea urchin). Successive transcription units are separated by about 2000 bp. Within the 2000 bp is a gene of about 120 bp that codes for the fourth rRNA, 5S rRNA. The rest of the 2000 bp is spacer. In other eukaryotes the 5S genes occur elsewhere in the genome as separate tandem arrays. In

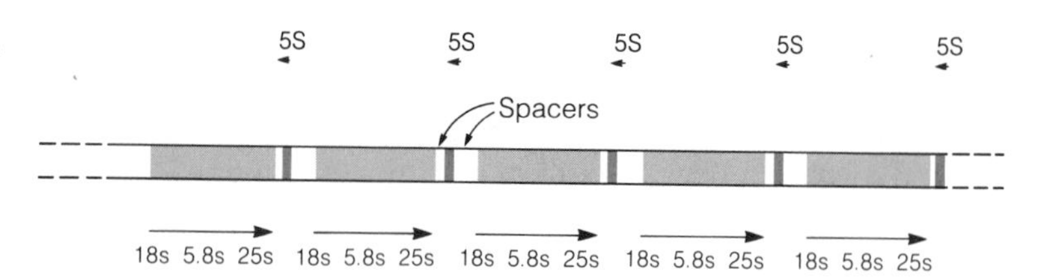

FIGURE 11-31 Tandem arrangement of genes encoding 5S, 18S, 5.8S, and 25S rRNA in yeast. The genes encoding 5S rRNA are transcribed separately and in the opposite direction from the genes encoding 18S, 5.8S, and 25S rRNA sequences. Arrows indicate the direction of transcription.

these eukaryotes the spacers between the transcription units for the three rRNAs contain no coding function, and no part of them is transcribed.

The size of spacers between rRNA transcription units ranges from about 2000 bp in some protists to 50,000 bp in some higher eukaryotes. There are 450 rRNA transcription units per haploid genome in *Xenopus,* arranged in tandem. Each is 7600 bp long and separated from adjacent units by untranscribed spacers of several thousands of base pairs (Figure 8-38). The transcription of rRNA genes and their separation by spacers can be observed directly by electron microscopy in gently lysed and dispersed nuclei (Figure 11-32).

The hundreds of genes for rRNA form the **nucleolar organizer,** the core of the nucleolus. In the nucleolus rRNA transcripts are processed into three ribosomal RNAs and combined with the fourth rRNA (5S) and with ribosomal proteins, which are synthesized in the cytoplasm and imported into the nucleus, to form the large and small ribosomal subunits. These are then exported to the cytoplasm. Thus, the nucleolus is the structural manifestation of rRNA transcription, rRNA processing, and assembly of ribosomal subunits.

Genes encoding 5S rRNA are arranged in multiple tandem clusters, usually in several different chromosomes and do not form part of the nucleolus. In *Xenopus* there are 24,000 copies of 5S genes that are separated by untranscribed spacers (Figure 8-39). The genes are 120 bp long and are separated by spacers of 600 bp. Genes encoding tRNA molecules are typically repeated about 200 times and are separated by spacers that are, on the average, 10 times longer than a tRNA gene itself.

In growing oocytes of amphibians and some insects even the many copies of genes encoding 18S, 28S, and 5.8S rRNAs are insufficient to meet the need for ribosome production. In these cells the genes for these rRNAs are extensively amplified further (Chapter 10).

Histones are proteins needed in the packing of long DNA molecules in the nucleus (see next section). There are five types of histones and 20 to several hundred copies of each gene are present. In most eukaryotes histone genes are arranged in groups, with one copy of each of the five genes per group (Figure 10-8). In sea urchin cells there are about 1000 such groups repeated in tandem. The five genes within a group are transcribed individually and separated by spacers. The groups in turn are separated from one another by spacers.

The spacers that occur between copies of repeated genes are usually similar or identical in sequence for a particular gene in a particular species. Thus, all the spacers between 5S rRNA genes are the same in *Xenopus laevis.* In the closely related species, *X. mulleri,* the 5S genes are identical to those in *X. laevis* but the spacers are three times longer and have a different sequence from the spacers in *X. laevis.* In general, spacers change in length and sequence far more rapidly during evolution than the genes they separate. Whatever the function of spacers, that function is compatible with large changes in spacer composition.

The significance of spacers is unknown. However, it is presumed that spacers between repeated genes represent the same phenomenon and have the same significance as spacers between single copy genes.

Packing of DNA in Chromosomes

The length of the DNA molecules in eukaryotic chromosomes exceeds by many thousands the diameter of the nuclei in which they are contained. Extreme examples are the DNA molecules of some salamanders. Twelve DNA molecules, each several meters long, are

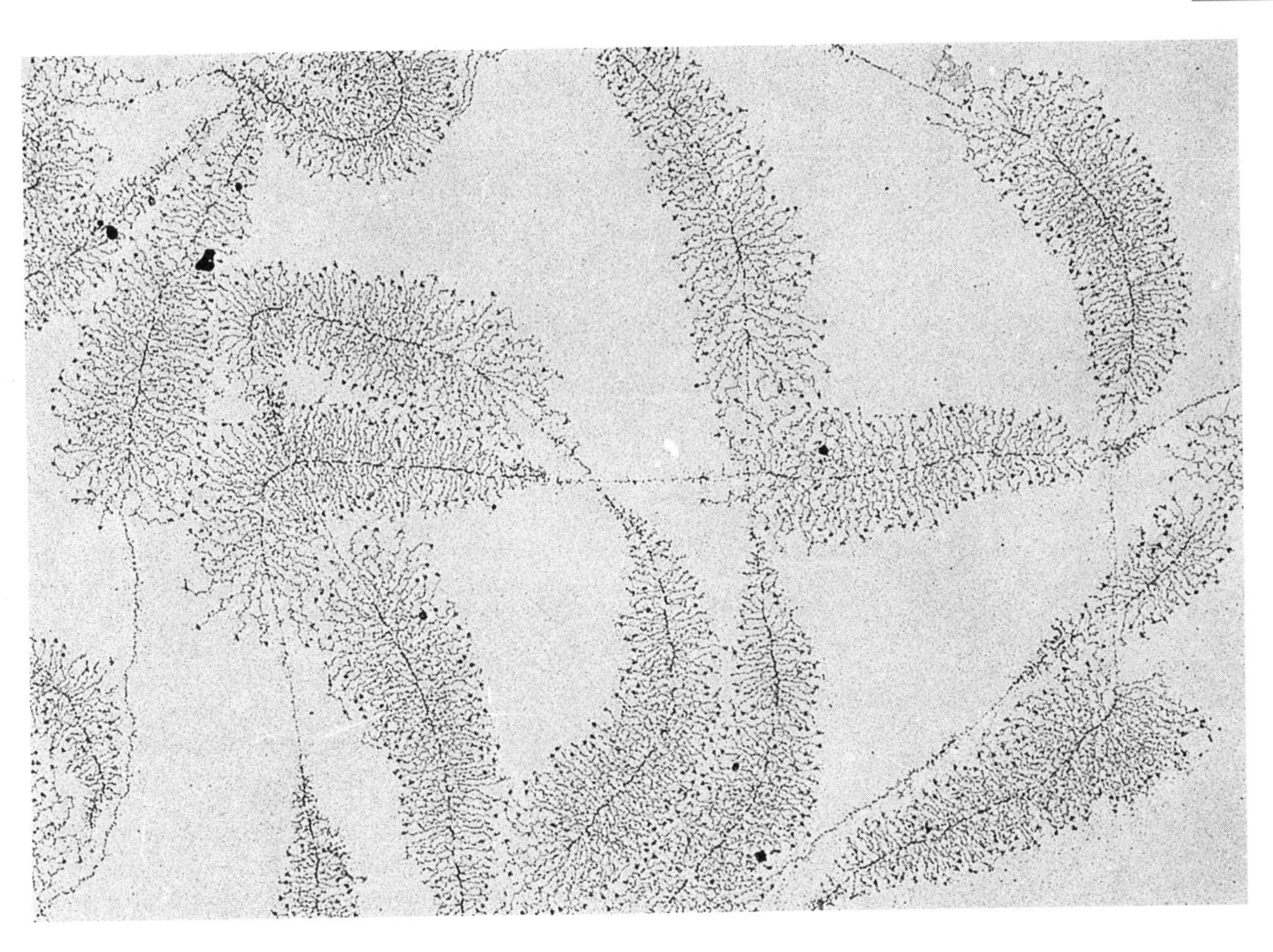

FIGURE 11-32
Electron micrograph of a portion of the DNA of the salamander *Triturus viridescens* containing tandemly repeated genes for ribosomal RNA in the process of being transcribed. The thin strands forming the featherlike arrays are rRNA molecules in various stages of completion, as indicated by the gradient of lengths. The axis of the feather is the DNA transcription unit for rRNA. Regions between transcription units (feathers) are untranscribed spacers. [Courtesy of O. L. Miller, Jr.]

contained in a spherical nucleus less than 20 μm in diameter in *Amphiuma*. In humans a total of 200 cm of DNA in the 46 chromosomes of a diploid cell are contained in a nucleus with a diameter of 6 or 7 μm. By coiling and packing, the extremely long DNA molecules can fit into the small nuclear volume. Certain nuclear proteins, described below and in Chapter 10, bring about the coiling and packing process.

Nucleosomes and chromatin. The DNA of eukaryotes is complexed with proteins that, in effect, package the DNA in stretches and hold it in specific configurations within the chromatin. The arrangement of proteins and DNA in chromatin is partially understood. Also, it is clear that the particular configuration of DNA in chromatin is very important in the regulation of transcription (Chapter 10). In general DNA packaged tightly with protein in chromatin is not transcribed; loosening of the chromatin is necessary, but

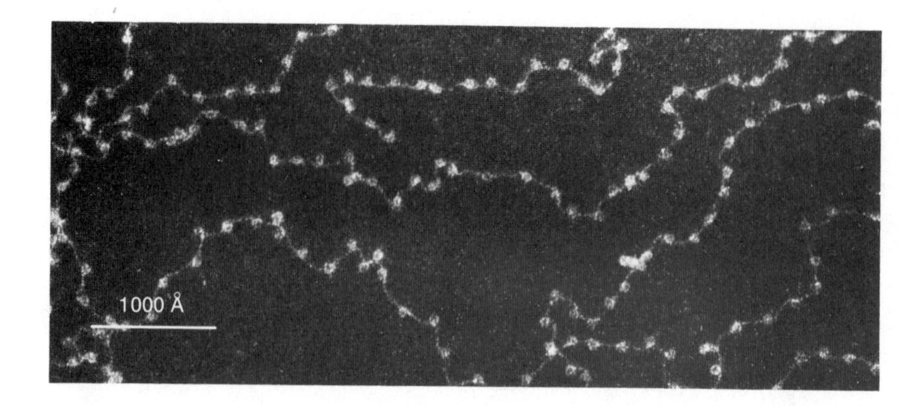

FIGURE 11-33 Electron micrograph of beadlike particles in chromatin. [Courtesy of Ada L. Olins and Donald E. Olins.]

alone is insufficient to bring about transcription. For transcription to occur other proteins must interact with the DNA sequence upstream from the coding region of a protein-encoding gene.

The packing of DNA in chromatin occurs in several successive orders, each of which accomplishes a further compaction of the DNA. The primary level of packing has been thoroughly elucidated by biochemical and physical analyses and electron microscope observations on chromatin isolated from interphase nuclei. It has been known for a long time that DNA is complexed with a family of five histone proteins. Histones contain large numbers of the basic amino acids lysine and arginine, with R groups that terminate in amino groups ($—NH_2$). In the pH range in the cell the amino groups acquire a proton (H^+) and are therefore charged positively ($—NH_3^+$). The large number of positive charges of arginine and lysine residues in histones form ionic bonds with the negative charges of the phosphates in the DNA strands. In the last decade the molecular details of how histones are complexed with DNA and form the first level of chromatin organization have been worked out.

Brief treatment with deoxyribonuclease (DNase) breaks chromatin into small particles, each composed of histones complexed with a fragment of DNA. After the histones are removed, the DNA consists of fragments of about 200 bp and multiples of 200 bp (i.e., 400, 600, etc.). If the enzymatic digestion is longer, all of the DNA is in units of 200 bp, and the multiple-size fragments are not found. These and other experiments show that the long, continuous DNA molecule in a chromosome is complexed with histones in 200-bp stretches. Each stretch is complexed with histones so as to form a particle. In corollary observations by electron microscopy, chromatin that has loosened and spread out is found to consist of long strings of particles (Figure 11-33). Brief treatment with DNase breaks the string into units containing one to several particles (Figure 11-34), and longer digestion reduces the string entirely to individual particles. Each particle can be dissociated into a 200-bp fragment of DNA and a particle consisting of eight histone molecules.

The particles of chromatin, or **nucleosomes,** are joined in long strings because the DNA molecule is continuous from one nucleosome to the next. The region of DNA linking nucleosomes is more accessible to DNase than DNA within nucleosomes. Hence, with properly gauged digestion chromatin is converted into single, physically separate nucleosomes.

Each of these nucleosomes is made up of a 200-bp fragment of DNA and nine histone molecules—one molecule of histone H1, and two molecules each of the other four histones H2A, H2B, H3, and H4. If the nucleosome preparation is digested further with DNase, the DNA fragment is gradually reduced in size until it reaches 146 bp and histone H1 is released. Further action of DNase is blocked because of a tight association between the remaining eight histone molecules and DNA. The structure that remains is called the **core par-**

ticle. The eight histone molecules (two each of H2A, H2B, H3, and H4) in a core particle are joined with each other in a specific pattern that forms a disc (Figure 11-35(a)). The DNA molecule makes almost two turns around the disc and then continues to the next core particle (Figure 11-35(b)). The DNA extending from one core particle to the next is called **linker DNA.** While the amount of DNA wrapped around the histone aggregate is constant (146 bp), the amount of DNA in the linker is variable from one kind of cell to another or from species to species. On the average, linkers are a little over 50-bp long, which, added to the DNA around the core particle (146 bp), accounts for the 200-bp fragments observed after brief DNase digestion of chromatin.

An H1 histone molecule is attached to each stretch of linker DNA (Figure 11-35); hence it is released when the linker DNA is digested with DNase. H1 histone plays a role in holding adjacent nucleosomes in tight association and in folding or twisting of the string of nucleosomes into the next order of DNA packing.

The packing of DNA into nucleosomes condenses the DNA about seven-fold, forming on the average for a human DNA molecule (four cm long) a thread of 600,000 nucleosomes that is about ten nm (1 nm = 10^{-9} meter) in diameter and 6000 μm long. Relative to the diameter of the nucleus (6 to 7 μm), the string of nucleosomes is still enormously long.

In the next order of DNA packing, the ten-nm nucleosome thread is folded into a fiber about 30 nm in

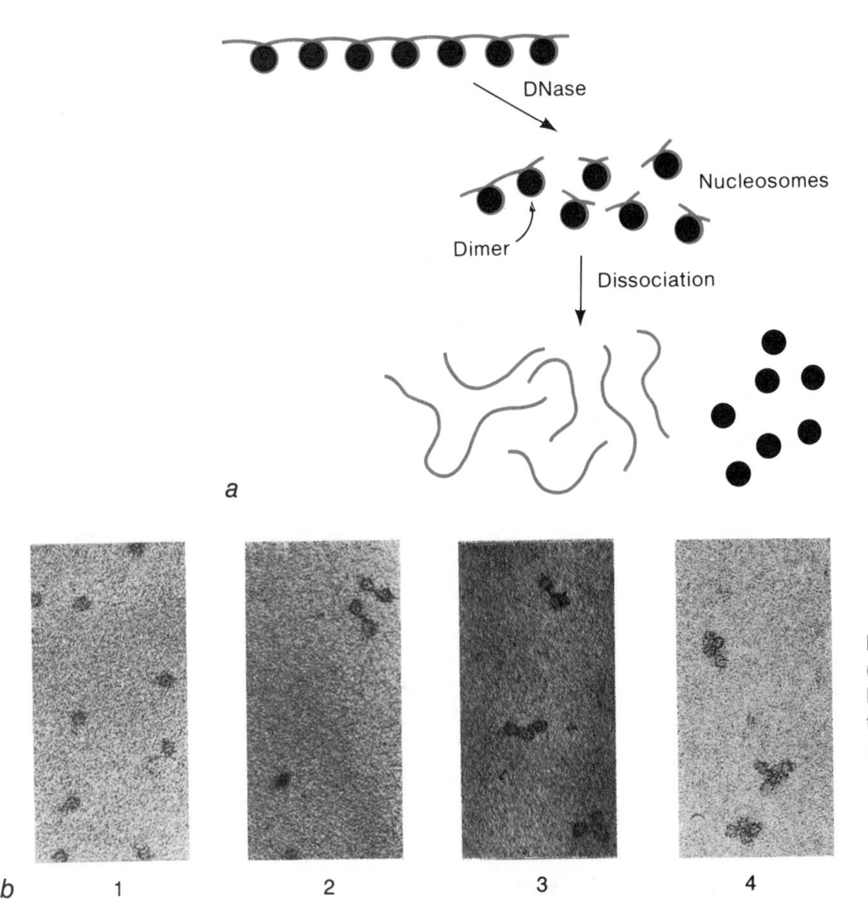

FIGURE 11-34
(a) Strings of nucleosomes digested with DNase form single nucleosomes, dimers, trimers, etc. (b) Electron micrographs of incompletely digested strings of nucleosomes, showing (1) monomers, (2) dimers, (3) trimers and (4) tetramers. [From J. T. Finch, M. Noll, and R. D. Kornberg. 1975. *Proc. Natl. Acad. Sci.*, 72:3321.]

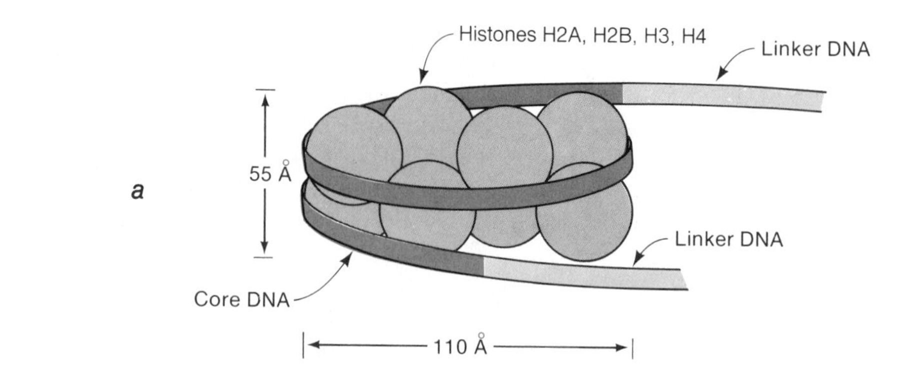

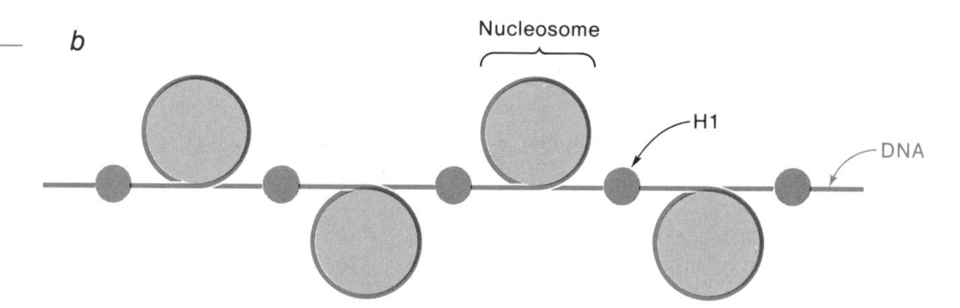

FIGURE 11-35
(a) Diagram of a nucleosome core particle. The DNA molecule makes 1 3/4 turns (consisting of 146 base pairs) around a histone octomer (two molecules each of histones H2A, H2B, H3, and H4). The DNA continues through linker DNA into adjacent nucleosomes. (b) A string of nucleosomes, with histone H1 bound to the linker DNA between individual nucleosomes.

diameter. According to the most widely accepted model the nucleosome thread is arranged as a hollow helix or solenoid, generating a thicker structure (Figure 11-36). The solenoid is believed to account for the thread of 30 nm in diameter seen by electron microscopy of chromosomes (Figure 11-37). The thicker filament formed in the solenoidal model is about seven times shorter than the extended nucleosome filament or a little over 800 μm for an average human chromosome. Observations by electron microscopy suggest that at the next higher level of packaging the solenoidal filaments are arranged as loops, containing 5 to 10 $\times$ 10^4 bp of DNA in an animal cell, extending out from a central axis composed of protein (Figure 11-38) and resembling the arrangement seen in lampbrush chromosomes (Figures 11-16 and 11-17).

The three orders of packing of DNA (nucleosomes, solenoids, and central axis with loops) accomplished by histones and other proteins is thought to produce the chromatin observed in the interphase nucleus. This chromatin contains the DNA that is transcribed and forms euchromatin. A portion of the chromatin undergoes additional packing to yield the more tightly condensed form, heterochromatin. The molecular arrangements in this mode of packing are not known. The significance of euchromatin vs heterochromatin is not

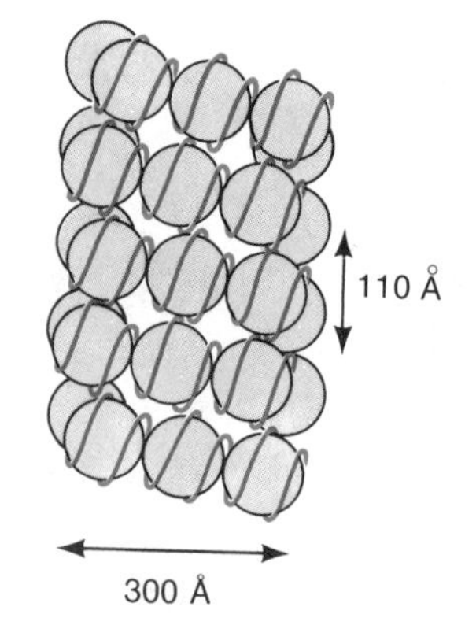

FIGURE 11-36
A solenoidal model of the packing of nucleosomes, yielding a fiber about 300 Å (30 nm) in diameter. The DNA double helix (red) is wound around each nucleosome. [After J. T. Finch, and A. Klug. 1976. *Proc. Natl. Acad. Sci.* 73: 1900.]

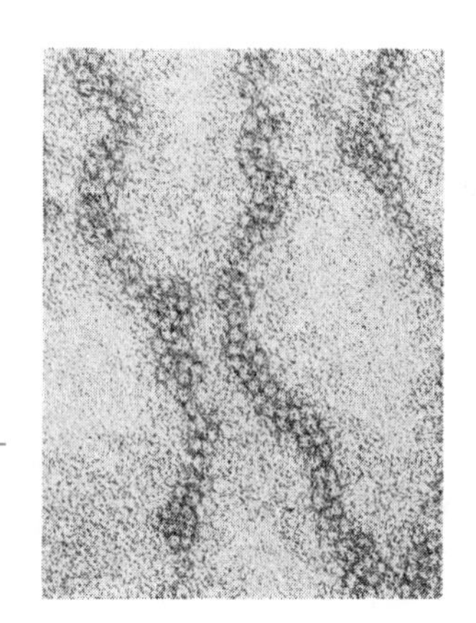

FIGURE 11-37 Electron micrograph of the 30 nm thread in mouse chromosomes. [Courtesy of Barbara A. Hamkalo.]

completely understood, but DNA present in heterochromatin is generally not transcribed (Chapter 10).

Organization of chromatin. The contents of the nucleus appear to form a gel with no movement of the euchromatin, heterochromatin, or nucleolus. There is, for example, no Brownian movement (vibratory movements caused by molecular collisions) of intranuclear structures in living cells viewed by light microscopy. By comparison, the cytoplasm is largely a solution in which organelles like mitochondria and lysosomes visible in the light microscope are in constant Brownian movement. The immobilized state of interphase chromosomes also contrasts sharply with the constant Brownian movement activity and shuffling movements of chromosomes in mitosis.

Those parts of interphase chromosomes that are condensed into heterochromatin are often closely associated with the inside of the nuclear envelope (Figure 11-39). A striking example is the inactivated X chromosome in a female mammalian cell (Chapter 10). The entire chromosome is packed as a discrete mass of heterochromatin, the Barr body, that is also positioned at the nuclear envelope. The inside of the nuclear envelope is lined with a thin layer of protein, the **nuclear lamina** (Figure 11-40). The lamina may serve to hold

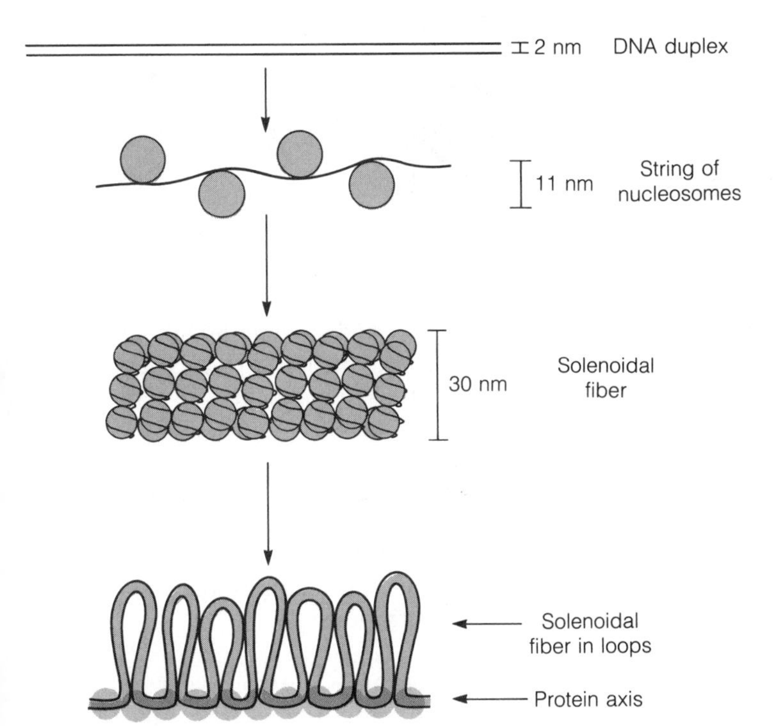

FIGURE 11-38 Model showing three orders of packing of DNA in chromatin: (1) The DNA duplex is wound around particles of histone to form a string of nucleosomes 11 nm in diameter. (2) Nucleosomes are packed into a solenoidal fiber 30 nm in diameter. (3) A solenoidal fiber is arranged in loops, with loops anchored into an axis composed of proteins.

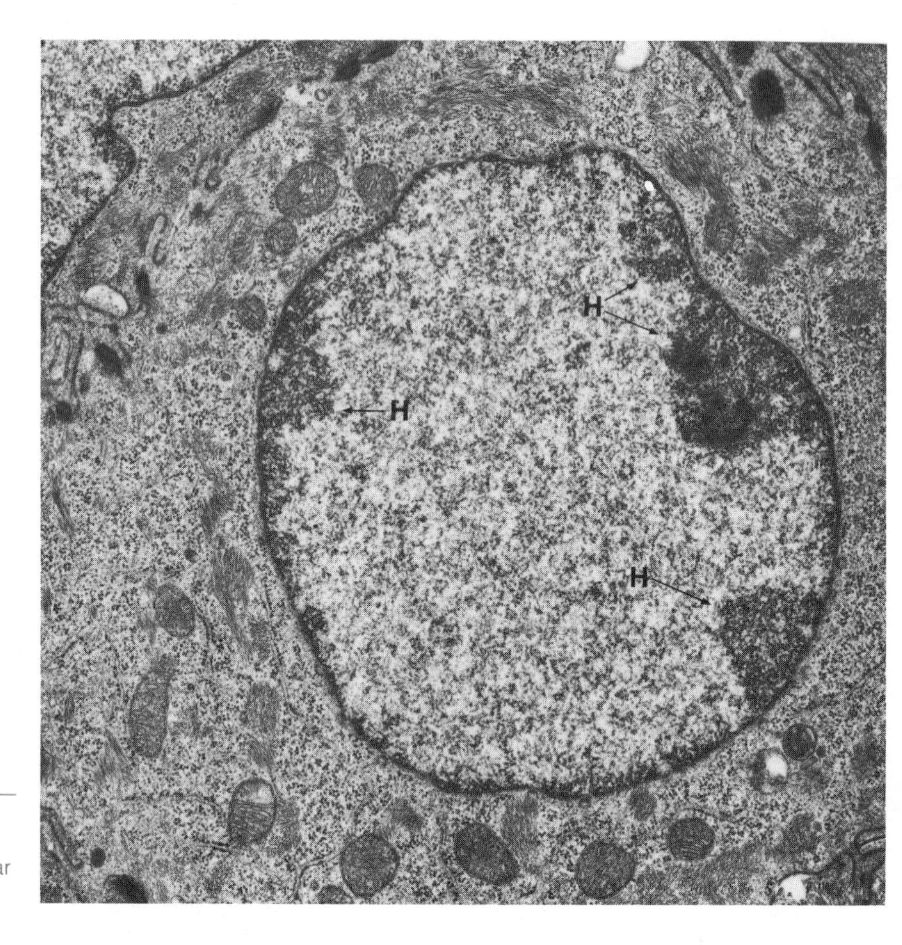

FIGURE 11-39
Electron micrograph of a taste bud cell of a mouse. Heterochromatic parts of chromosomes (H) are attached to the nuclear envelope. [Courtesy of John C. Kinnamon.]

both euchromatin and heterochromatin in a fixed orientation.

The global structural organization of the interphase nucleus is partially understood. A variety of observations suggest that the 30-nm fiber formed by packing of nucleosomes into solenoids (Figure 11-38) is organized into a series of loops much like the loops seen in lampbrush chromosomes (Figures 11-16 and 11-17). Although less clearly discernible, such a loop arrangement appears to be present in polytene chromosomes observed by electron microscopy. Unlike lampbrush chromosomes, in which the loops and central core might reflect early meiotic condensation, polytene chromosomes are interphase chromosomes. The many copies of the 30-nm fiber present in a polytene chromosome obscure observation of the individual fibers, but the overall appearance is compatible with the model in Figure 11-17. Loops, which contain 50,000 to 100,000 base pairs of DNA, are anchored to a central axis of protein. This protein axis is attached at points to the nuclear lamina. Together the protein axis and lamina constitute a nuclear matrix that remains after removal (with solutions of detergent and high salt) of the nuclear envelope and many intranuclear proteins, including histones. Each loop is thought to correspond to a functional domain of the chromosome that perhaps contains one to several genes.

The nuclear envelope. The nuclear envelope encloses the interphase chromosomes within the nucleus, thereby physically separating the transcription of genes and the processing of RNA from protein synthesis and other functions of the cell (Figure 11-41). The envelope consists of a double membrane with an interval spacing of 20 to 40 nm between them, the **perinuclear space.** The outer membrane is continuous at irregular and infrequent intervals with the endoplasmic reticulum (ER) and is often considered to be part of the ER (Figure 11-42). Indeed, the cytoplasmic face of the outer membrane is often studded with ribosomes and resembles the rough ER. The perinuclear space is therefore continuous with the lumen of the ER.

Attached to the nuclear face of the inner membrane is the nuclear lamina mentioned earlier (Figure 11-40). The lamina in mammalian cells is 100 to 300 nm thick and consists of a complex of three proteins, designated **lamins A, B,** and **C.** Lamins A and C appear to be the products of a single gene and are related in amino acid sequence to the proteins of intermediate filament families, suggesting a relationship of the lamina with the cytoskeleton. Lamin B has an affinity for the inner nuclear membrane and may be responsible for the tight binding of the lamina to the membrane. Lamins A and C have a binding affinity for chromatin

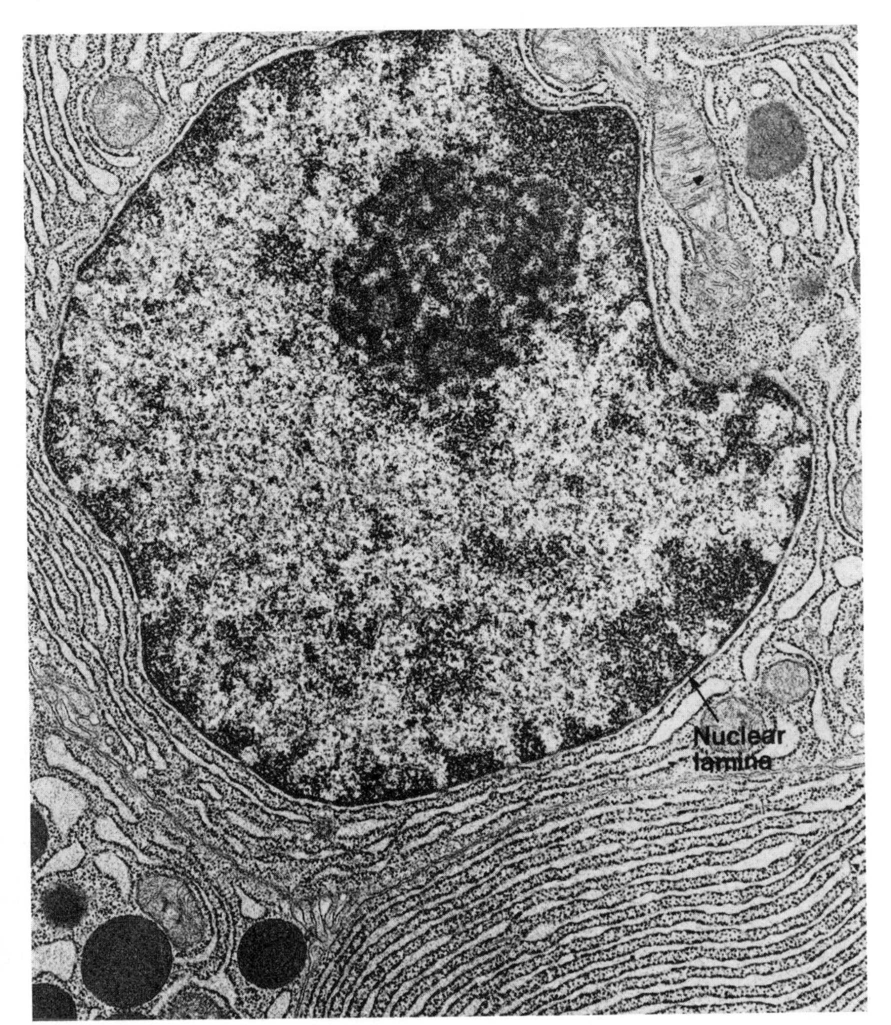

FIGURE 11-40
Electron micrograph of the nucleus in a pancreas cell. The outer membrane of the nuclear envelope (studded with ribosomes) and the perinuclear space (clear) are easily identified. The nuclear lamina (arrow) is the dark line along the inner border of the inner membrane of the nuclear envelope. [Courtesy of Susumu Ito.]

and therefore probably mediate the adhesion of chromatin to the nuclear envelope.

The two membranes of the nuclear envelope are continuous with one another as they form the walls of the many nuclear pores that are present in the nuclear envelope (Figure 11-42). The inner and outer openings of a pore and the interior of the pore contain eight granules, arranged radially and surrounding a central granule that is sometimes present in the opening (Figure 11-43). The granules, together with the pore, are called the **nuclear pore complex.** The entire complex is about 100 nm in diameter. The particles of the pore complex together with some ill-defined fibrous material appear to occlude most of the pore opening. Nuclear pores are nevertheless believed to be channels by which materials move between nucleus and cytoplasm. When molecules with molecular weights of less than 50,000 are injected into the cytoplasm, they quickly equilibrate between the nucleus and cytoplasm. Larger molecules enter the nucleus slowly. From such experiments it has been estimated that the effective diameter of the opening of the pore is about nine nm.

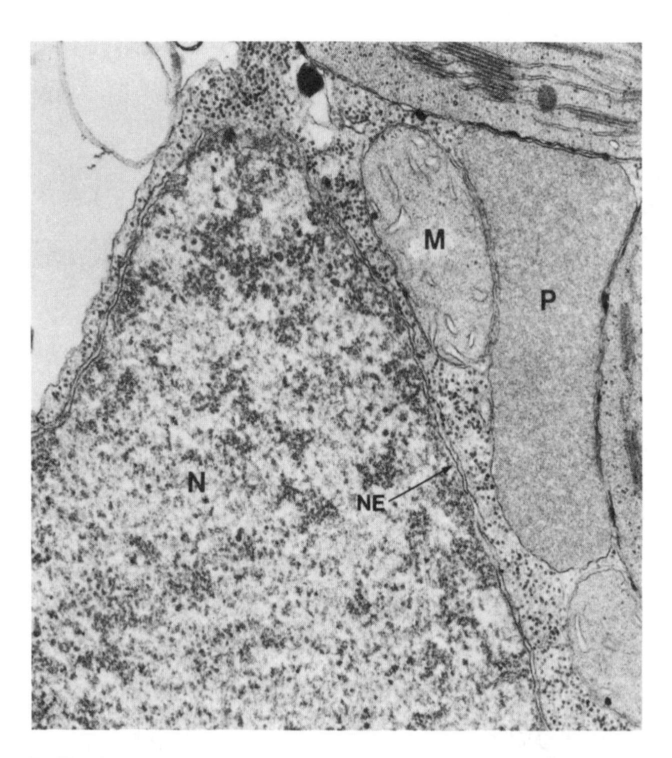

FIGURE 11-41
Electron micrograph of a portion of a tobacco leaf cell illustrating the two membranes of the nuclear envelope (NE) and the narrow perinuclear space between the two membranes. N = nucleus, M = mitochondrion, and P = peroxisome. [Electron micrograph by S. E. Frederick, courtesy of Eldon H. Newcomb.]

Molecular traffic through nuclear pores is intense. For example, during DNA replication hundreds of thousands of histone molecules synthesized in the cytoplasm enter the nucleus. Ribosome production in the nucleolus requires the continuous importation of great numbers of the approximately 80 different kinds of ribosomal proteins. Experiments (Chapter 8) suggest that histones and other proteins may be actively transported through the pores. In addition, large numbers of mRNA-protein complexes and new ribosomal subunits flow out of the nucleus every minute. These substances are too large to be accommodated by the nine-nm opening of a pore, and some means of increasing the effective opening to facilitate their passage must occur. The particle sometimes observed in the center of a nuclear pore may represent material caught in transit between nucleus and cytoplasm.

During mitosis in most kinds of cells the nuclear envelope breaks into small fragments that disperse in the cytoplasm. These come together again to reform nuclear envelopes around daughter nuclei in telophase. The specific reaggregation of these fragments shows that the membranes of the nuclear envelope must have properties that distinguish them from membranes of the endoplasmic reticulum. The re-formation of the nuclear envelope is closely coupled with the re-formation of the nuclear lamina on the nuclear surface of the inner membrane.

Structure of condensed chromosomes. As a cell enters prophase of mitosis or meiosis the interphase chromatin becomes tightly condensed by further packing, and individual chromosomes become recognizable. An early stage of such packing is represented by the lampbrush chromosome (Figure 11-17), in which the central axis has probably shortened and thickened compared to the interphase state. The central axis is, in

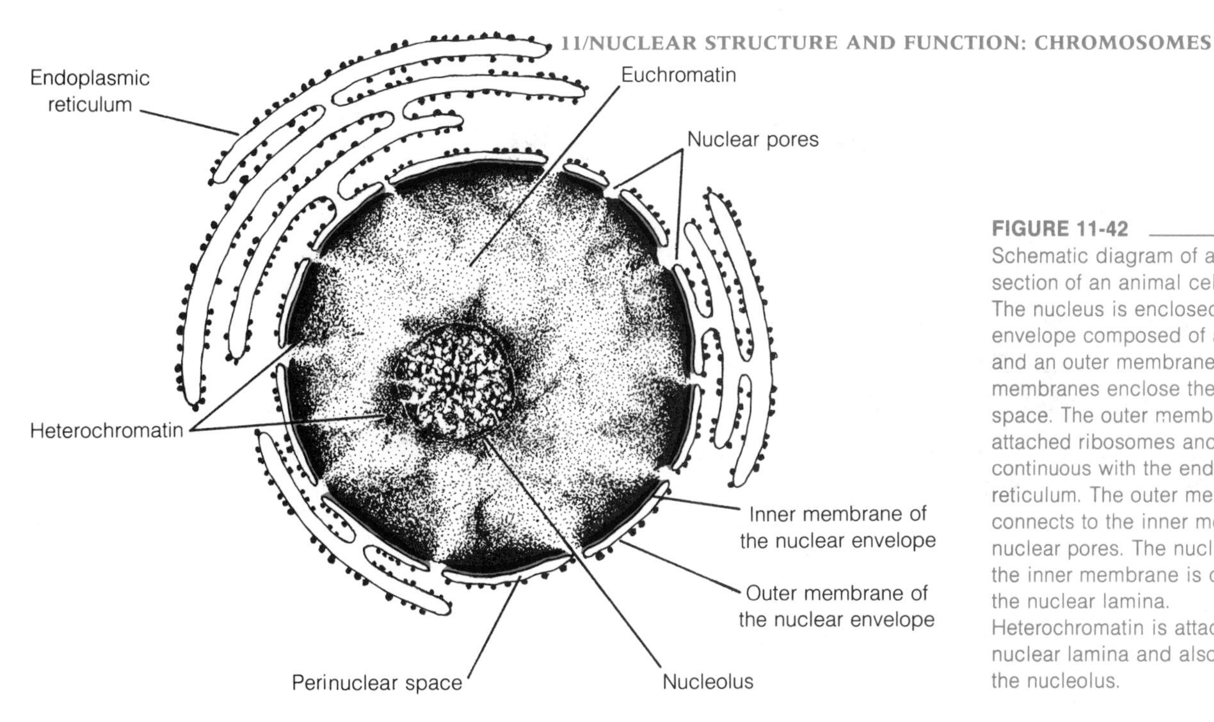

FIGURE 11-42 Schematic diagram of a cross section of an animal cell nucleus. The nucleus is enclosed by an envelope composed of an inner and an outer membrane. The two membranes enclose the perinuclear space. The outer membrane has attached ribosomes and is continuous with the endoplasmic reticulum. The outer membrane connects to the inner membrane at nuclear pores. The nuclear face of the inner membrane is coated with the nuclear lamina. Heterochromatin is attached to the nuclear lamina and also surrounds the nucleolus.

FIGURE 11-43 The nuclear pore complex. (a) A schematic diagram of cross section of a nuclear pore complex showing radially arranged granules at the edge of the pore and a central granule, which is not always present. (b) A surface view of a nuclear pore complex showing the eight radially arranged granules and the central granule. (c) Electron micrograph of nuclear pores in the nuclear envelope of a cultured mouse cell observed by freeze-fracture. The fracture plane is partly through the outer membrane (OM) and partly through the inner membrane (IM). The resolution is insufficient to see radial and central granules distinctly. [(c) Courtesy of L. Andrew Staehelin.]

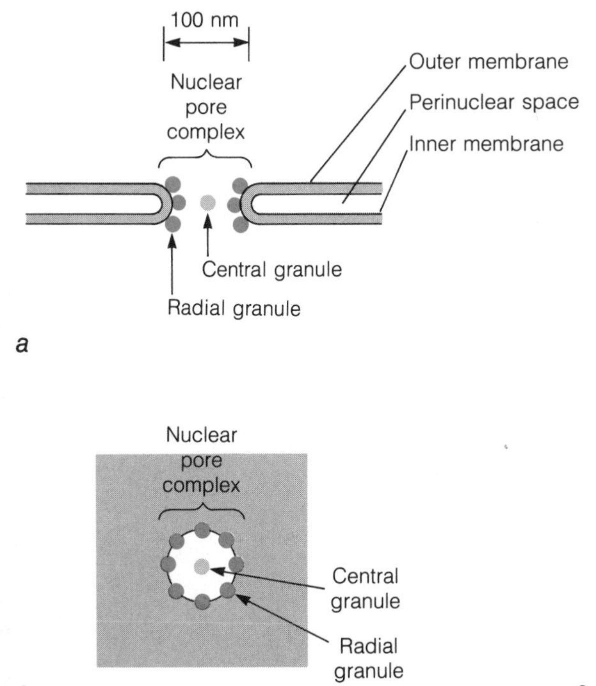

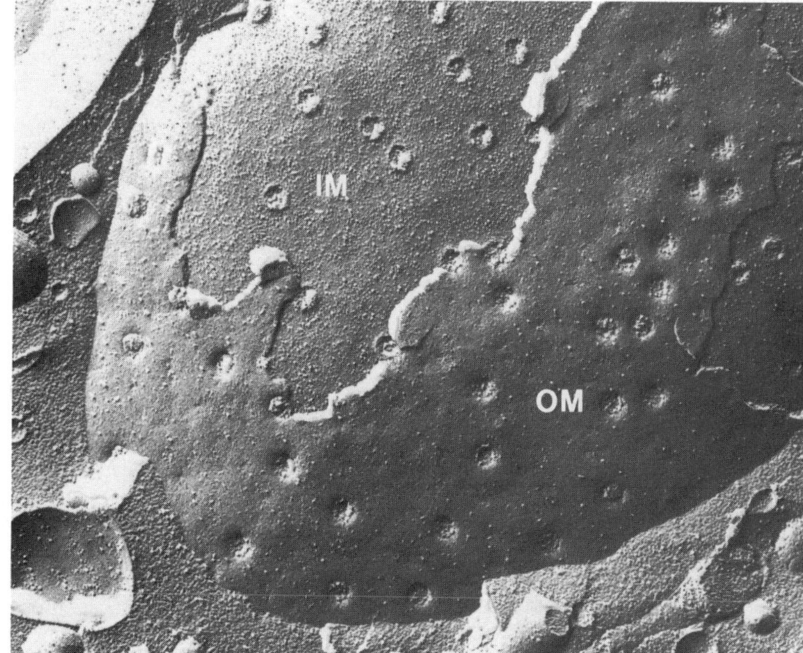

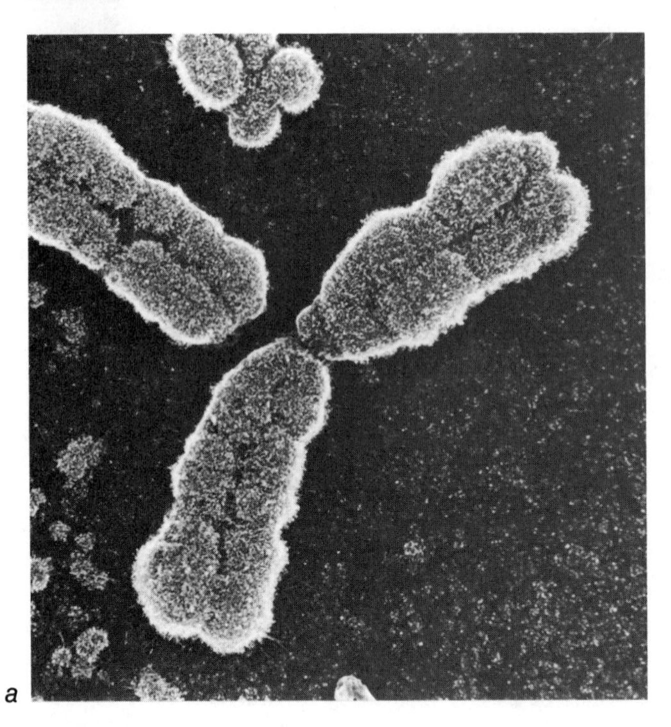

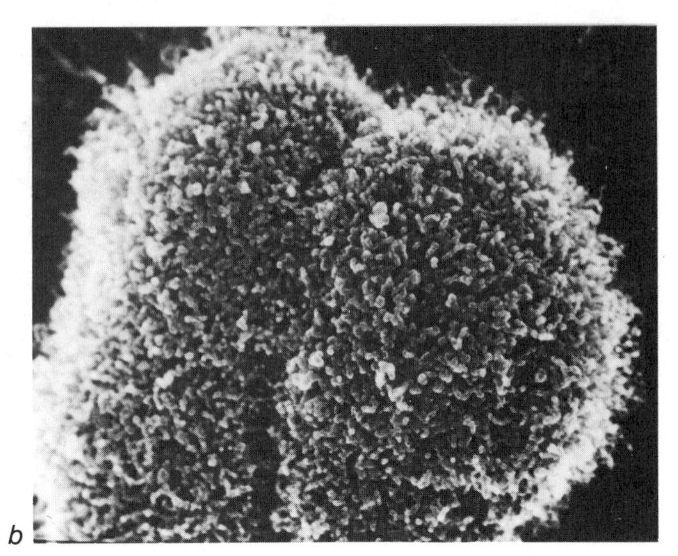

FIGURE 11-44
(a) Scanning electron micrograph of a fully condensed metaphase chromosome. (b) A portion of a chromosome at a higher magnification, showing individual DNA-protein fiber loops. [Courtesy of R. T. Johnson, A. M. Mullinger and The Company of Biologists Limited.]

turn, folded or coiled to produce the fully condensed chromatids of a chromosome (Figure 11-44). The condensation of chromosomes is achieved by proteins. The synthesis of a principal protein involved in condensation begins just around the start of prophase (Chapter 12), but details about the events of condensation are still lacking.

Removal of histones and DNA from metaphase chromosomes leaves a residual structure that is still recognizable as a chromosome. The proteins that are left are believed to form the scaffold onto which the DNA-histone components are organized and packed into the condensed mitotic or meiotic form. The chromosome scaffold and the nuclear matrix of the interphase nucleus may well be forms of the same structure, one designed to produce the compact mitotic chromosomes, the other responsible for the organization of the transcriptionally active interphase chromosome.

Banding in Mitotic Chromosomes

Mitotic chromosomes are not uniform in structure throughout their lengths; a considerable amount of substructure is revealed by a variety of staining techniques. The substructure takes the form of several kinds of transverse bands. Harsh treatments that extract DNA differentially from regions of a chromosome leave certain regions intact. In subsequent staining, these regions, known as C-bands, show up more intensely (Figure 11-45). C-bands in general correspond to DNA that is packed as heterochromatin in the interphase nucleus. In metaphase chromosomes, this DNA appears to be more tightly packed than DNA that occurs as euchromatin during interphase. C-banding occurs predominantly in the centromeric region of mammalian chromosomes (Figure 11-45), where heterochromatin containing highly repetitive DNA is known to be located.

Three other staining techniques produce R-, Q-, and G-bands, (Figure 11-46). The different patterns of C-, R-, Q-, and G-banding are constant in a species and allow each chromosome to be identified unequivocally. Figure 11-47 shows the banding map for the chromosomes of the human genome. Banding in mitotic chromosomes is an entirely different phenomenon from the banding seen in polytene chromosomes. The latter are interphase chromosomes and their banding should not be confused with the banding in mitotic chromosomes. Banding patterns in mitotic chromosomes are useful because they allow segments of chromosomes that have been exchanged between chromosomes (**translocations**) to be precisely identified. Translocations can be induced by radiation and sometimes they occur without known cause. They occur frequently in cancer cells, and often they have important consequences for the cell.

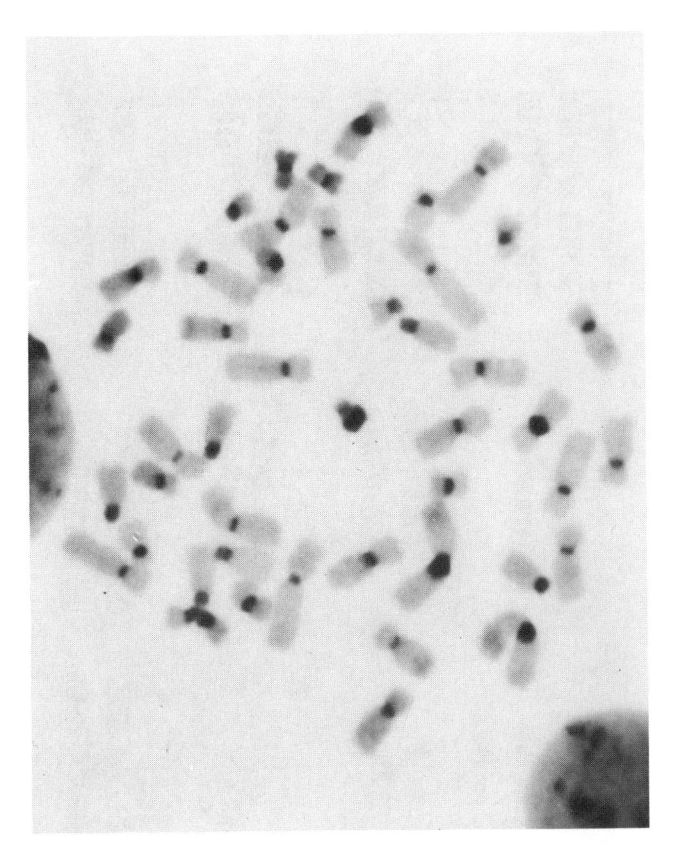

FIGURE 11-45
Light micrograph of human metaphase chromosomes treated and stained to show C bands, which are located in the centromeric region. [Courtesy of A. T. Sumner. From C. J. Bostock and A. T. Sumner. *The Eukaryotic Chromosome* (1978) North Holland, Amsterdam.]

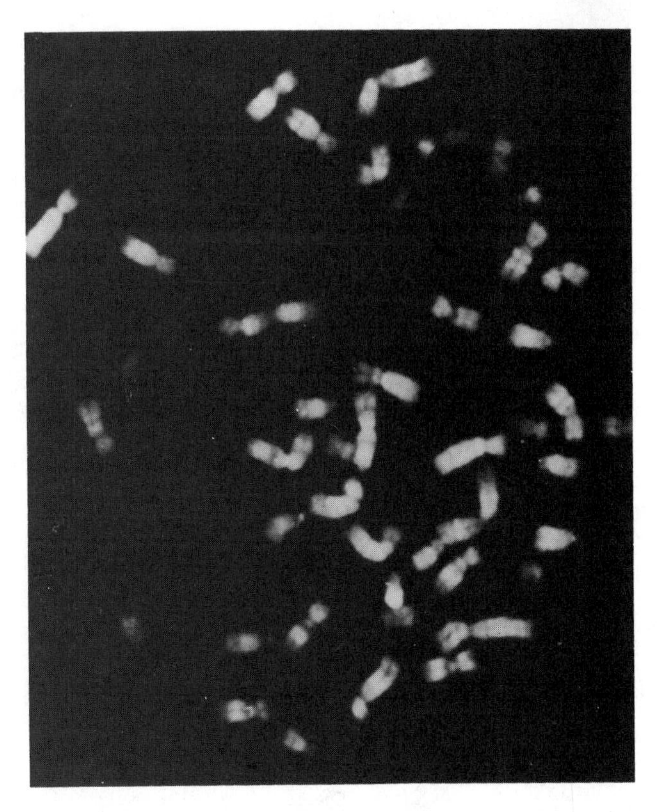

FIGURE 11-46
Light micrograph of human metaphase chromosomes treated and stained to show Q bands. [Courtesy of A. T. Sumner. From C. J. Bostock and A. T. Sumner. *The Eukaryotic Chromosome* (1978) North Holland, Amsterdam.]

Replication of Chromosomes

In most eukaryotes the DNA molecules of chromosomes are enormously long, raising the question of how replication is organized. Prokaryotic chromosomes, which are shorter and circular, replicate beginning at a single origin; that is, they are single replication units. In contrast, the linear DNA molecule of a eukaryotic chromosome contains many points of initiation of replication; that is, it contains many replication units.

Replication units. The multiple replication units of eukaryotic chromosomes are readily seen by autoradiography or by electron microscopy. The autoradio-

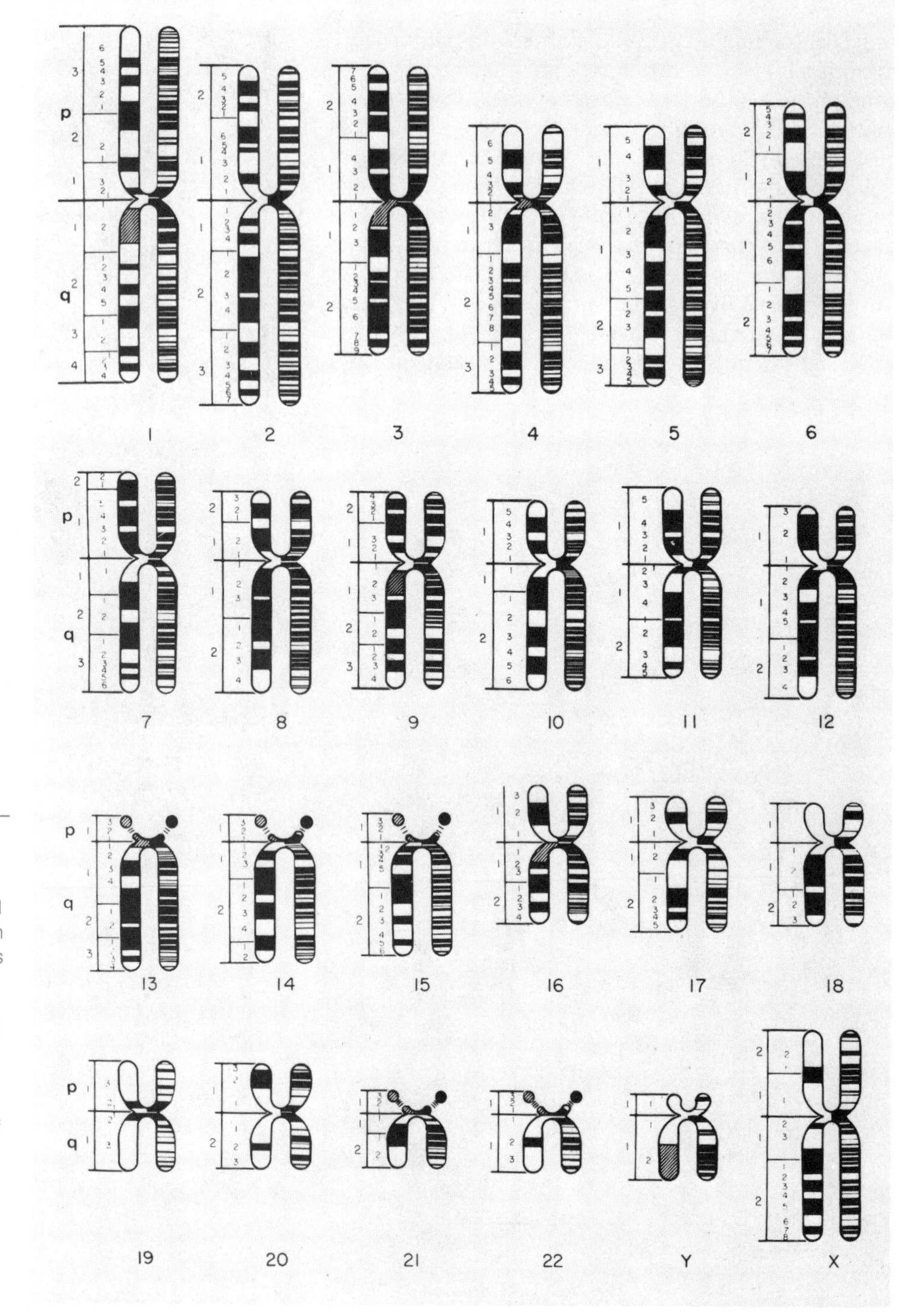

FIGURE 11-47
A drawing of the 22 human autosomes and the X and Y chromosomes showing the bands that have been revealed by several kinds of staining. The chromatid on the left in each chromosome shows the bands seen in metaphase chromosomes. The more refined banding pattern in each chromatid on the right was determined using prophase chromosomes but is displayed here on a metaphase chromosome. The greater length of incompletely condensed prophase chromosomes allows a finer analysis of banding structure. [Courtesy of Jorge J. Yunis. Reprinted from *Science*, 191:1268–1270. 1976. Copyright 1976 by the AAAS.]

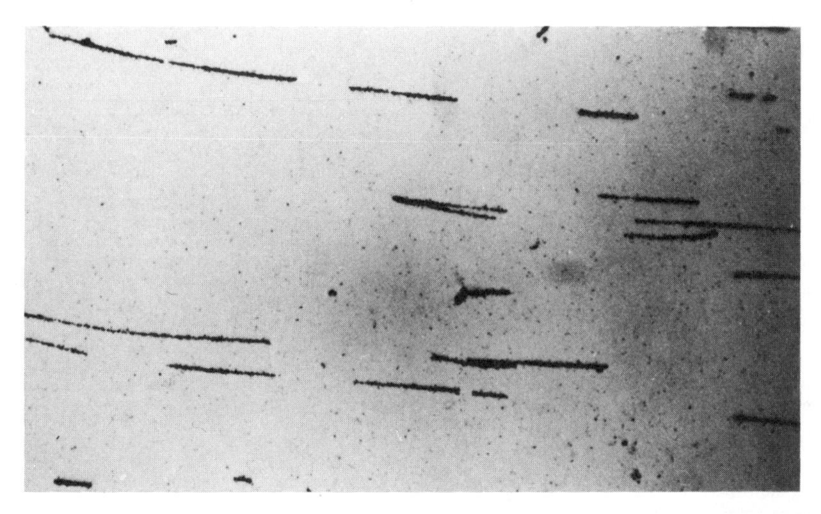

FIGURE 11-48 Autoradiograph of DNA isolated from a cultured hamster cell labeled with ^{3}H-thymidine for 30 minutes. The interrupted patterns of autoradiographic silver grains show that DNA replication was occurring at multiple places along a DNA molecule. [Courtesy of Joel A. Huberman. From *J. Mol. Biol.*, 75:5–12 (1973).]

graphic experiment is simple. Cultured cells are incubated with ^{3}H-thymidine for a few minutes during DNA replication. Cells are lysed with a drop of a solution of an ionic detergent on a microscope slide. The detergent lyses the nuclei and removes essentially all proteins from the DNA. The drop of lysed material is spread evenly over the entire microscope slide. This draws the DNA molecules out in long, linear stretches. Subsequent autoradiography shows that DNA replication was occurring at many places along individual molecules during the short incubation with ^{3}H-thymidine (Figure 11-48). Each stretch of radioactivity represents a replication unit. Replication can also be seen directly by electron microscopic observation of isolated DNA molecules (Figure 11-49). Each segment of the molecule that is doubled represents a separate replication unit.

In prokaryotic chromosomes origins of replication are defined by specific deoxynucleotide sequences. The strongest evidence that origins are similarly defined in eukaryotes comes from the study of yeast chromosomes. These contain ARSs discussed earlier in this chapter that confer on yeast plasmids the ability to replicate. ARSs that function as apparent specific origins of replication in yeast plasmids have been identified in the DNA molecules from a variety of eukaryotes. Conclusive evidence that these sequences function as origins

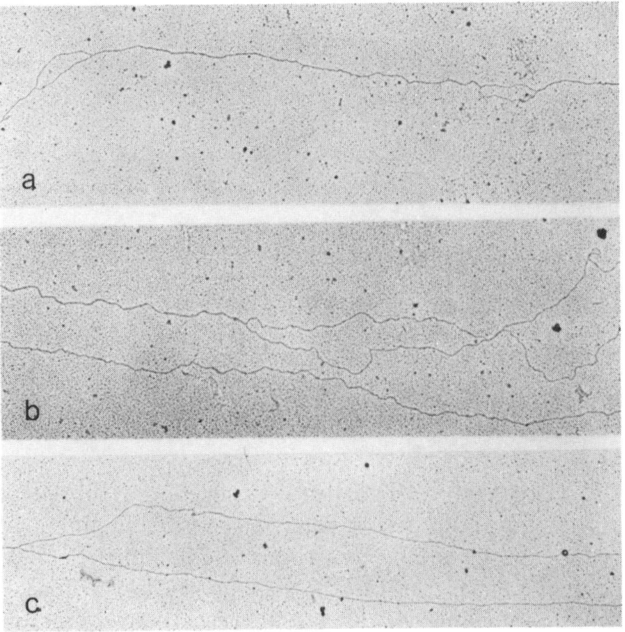

FIGURE 11-49 Electron micrographs of DNA molecules isolated from cultured animal cells in the S period. (a) Portion of a molecule with two regions partially replicated. (b) Portions of two molecules, one replicating and the other not replicating. The replicating molecule has two large regions that replicated but which are separated from each other by a short segment of unreplicated, parental DNA. (c) Portion of a molecule that has almost completely replicated. A short segment of unreplicated, parental DNA is present at the far left. [Courtesy of K. G. Murti.]

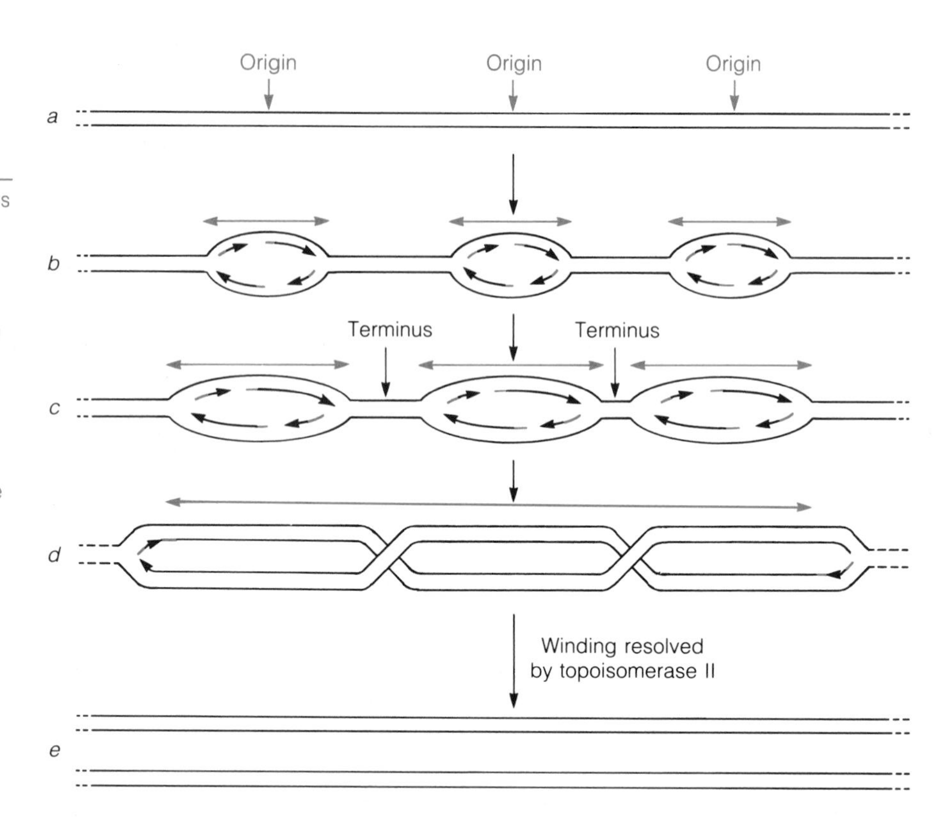

FIGURE 11-50 Replication of successive replication units in a eukaryotic chromosome. (a) An unreplicated DNA duplex with three origins of replication. (b) Some minutes after initiation of replication at the three origins. Replication forks are progressing bidirectionally from each origin.
(c) Replication forks from adjacent replication units are nearing termini between units. (d) Replication forks have met between adjacent replication units and disappeared. Replication forks at the right and left ends are approaching adjacent replication forks that are not shown. At termini the completed portions of daughter duplexes are wound around each other in each replication unit.
(e) The completed segments of daughter molecules are unwound from each other by introduction of transient double-stranded breaks by topoisomerase, allowing one duplex to separate from the other.

of replication in the chromosomes of these eukaryotes is still lacking.

Bidirectional replication. Replication is initiated at the single point within each replication unit and proceeds bidirectionally from that point much as in prokaryotes (Figure 11-50). Such points are labeled as origins in Figure 11-50, but whether they represent specific sequences has not been unequivocally proven. Bidirectional replication was discovered by the following autoradiographic experiment: fluorodeoxyuridine, which inhibits DNA replication (Chapter 12), was used to block entry into the S period. A culture of cells was thus synchronized. Cells in the G2, M, and G1 periods of the cell cycle are unaffected by the drug, proceed to the G1/S transition point, and stop. Cells already in S are killed by fluorodeoxyuridine. After enough time was allowed for cells to be synchronized at the G1/S border, the inhibitor was washed away and fresh nutrient medium containing ^{3}H-thymidine was added. As a result, replication was initiated, with incorporation of ^{3}H-thymidine. After a few minutes of replication, nonradioactive thymidine was added to the medium. The thymidine was taken up by cells and phosphorylated to TTP, which became part of the pool of TTP used for DNA replication. Over several minutes nonradioactive TTP replaced ^{3}H-TTP in the pool of TTP in the cell. As a consequence the amount of radioactivity incorporated into DNA progressively diminished. Some minutes after addition of nonradioactive thymidine, cells were lysed; autoradiography of DNA molecules showed the expected tandem arrays of replication units. However, each autoradiographic image was darkest in its central region and faded to imperceptibility at its two extremities (Figure 11-51). The fading images reflect the travel of replication forks during the decline of ^{3}H-thymidine

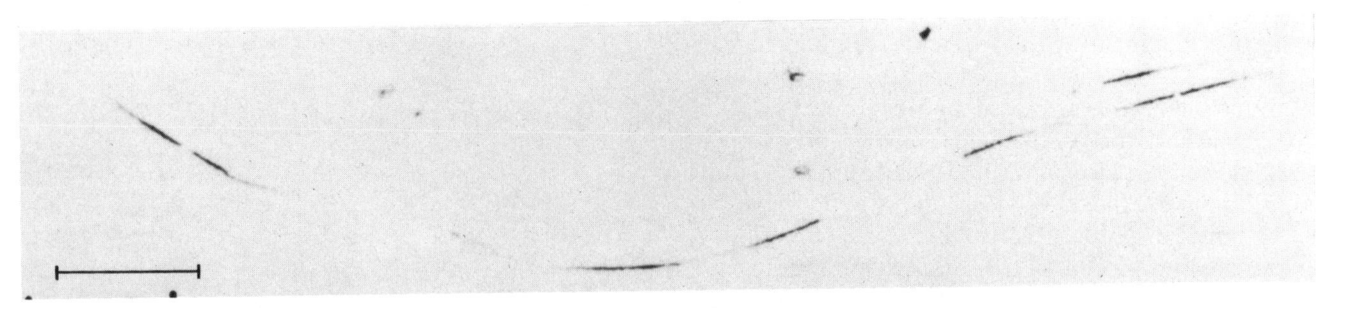

FIGURE 11-51 Autoradiographic patterns produced by a DNA molecule isolated from cultured hamster cells labeled with ^{3}H-thymidine for 30 minutes followed by incubation in medium containing nonradioactive thymidine for 45 minutes. The gradual fading from an intense autoradiographic pattern to a progressively lighter one shows the direction of travel of a replication fork. In several of the labeled segments a heavily labeled segment fades in both directions, showing bidirectional replication. The bar indicates 50 μm. [Courtesy of Joel A. Huberman. From *J. Mol. Biol.*, 75:5–12 (1973).]

triphosphate in the cell and show that replication has proceeded bidirectionally from a single point (origin) in each replication unit.

Termination between replication units. Each replication fork eventually meets a replication fork traveling toward it in the next adjacent replication unit (Figure 11-50). Meeting of forks completes replication of that particular segment of the chromosome, and the replication machinery dissociates from the DNA. The terminus between adjacent replication units is defined by the meeting place of replication forks rather than by a specific sequence in the DNA.

The meeting of forks and completion of replication between replication units leaves the two daughter molecules linked by one interloop between them for each replication unit (Figure 11-50). These interloops would be difficult to eliminate by untwisting of the two duplexes at the end of replication. An average of more than 1300 interloops are generated in a human chromosome. Loops are removed by the action of topoisomerase, which breaks the chains in the daughter double helices, allows the daughters to separate and then reseals the breaks. Mutant yeast cells lacking the proper topoisomerase cannot remove interloopings between the two daughter molecules. As a result, the two chromatids in a chromosome cannot separate from each other at the ensuing mitosis.

Size and number of replication units. An estimate of the size of replication units is obtained by measuring the distance between the centers of successive origins either in autoradiographic or electron microscope displays of DNA. Replication units in mammalian cells range from 15 to 100 μm (45,000 to 300,000 bp) with an average of 30 μm (90,000 bp). The four cm of DNA in an average mammalian chromosome therefore contain over 1300 replication units. A haploid set of mammalian chromosomes contains about 33,000 replication units.

Rate of replication. The rate of travel of replication forks has been estimated by several methods, among them autoradiography. The simplest procedure is to synchronize cells by blocking them at the G1/S border as previously described, then releasing the block and simultaneously adding ^{3}H-thymidine. DNA isolated at intervals thereafter produces progressively longer autoradiographic images, which shows that in mammalian cells replication forks travel about one μm per minute (3000 bases per minute). Since the average size of a replication unit is 30 μm and each unit has two forks, the time to replicate an average-sized unit is 15 minutes.

DNA MOLECULES IN MITOCHONDRIA

The mitochondria of all eukaryotes contain DNA molecules (mtDNA). These encode a few of the macromolecules that make up mitochondria, although all but a few of the 100 or more different kinds of polypeptides in mitochondria are encoded by nuclear genes. The polypeptide products of these nuclear genes are synthesized in the cytoplasm and subsequently imported into mitochondria (Chapter 8).

Among protozoa, yeasts, and other unicellular eukaryotes mitochondrial DNA molecules range in size from 15,000 to 75,000 bp. In a few species the molecules are linear, but in most they are circular. There are multiple identical copies of mtDNA molecules in each mitochondrion, and all molecules in all the mitochondria have the same coding functions.

Among unicellular eukaryotes mtDNA of the budding yeast has been studied intensively. It is a circular molecule containing 75,000 bp, and there are ten to 50 DNA molecules altogether in all the mitochondria of a single cell. These mtDNA molecules represent 5 to 25 percent of the total cellular DNA but only 0.5 percent of the total sequence complexity of the total DNA. These are unusually high percentages, and are due to the unusually large size of yeast mtDNA (75,000 bp) and the small size of the yeast nuclear genome (1.5×10^7 bp). In all other eukaryotes studied so far, mtDNA represents a far smaller percentage. In mammalian cells mitochondrial DNA makes up less than one percent of the total cellular DNA and about 0.0005 percent of sequence complexity of the total DNA in the cell.

Molecules of mtDNA in invertebrate and vertebrate animals are closed circles as in yeast but are much smaller, ranging in size from 13,000 to 18,000 bp. In mammalian mitochondria, mtDNA molecules contain about 16,000 bp (Figure 11-52). The entire sequence of 16,569 bp in human DNA has been determined. There are five to ten molecules per mitochondrion and 100 to several thousand mitochondria per cell, depending on the size and type of cell. Fibroblasts have about 100, liver cells about 1000, and heart muscle cells many thousands.

In plants, mtDNA molecules are much larger and more variable in size than those in animals. Even within a single organism the lengths of mtDNA molecules are variable, and some molecules are circular and some are linear. These variations are apparently not the result of DNA breakage during preparation and are not understood. Much less is known about the coding functions of mtDNA in plants compared with yeast and animal mtDNA (see below), but evidence so far suggests that plant mtDNA contains about the same number of different kinds of genes as other eukaryotes.

Mitochondrial DNA molecules are not complexed with nuclear-type histones to form nucleosomes. Histone-like molecules are present instead, and the mtDNA-protein complexes form nucleoid bodies in the matrix compartment. Mitochondria reproduce by fission, and the multiple mtDNA molecules, which are

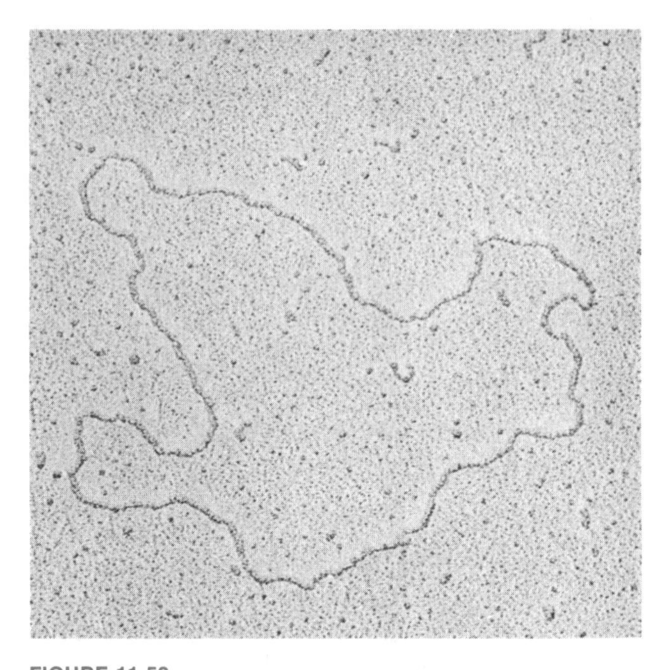

FIGURE 11-52
Electron micrograph of a circular DNA molecule isolated from a human mitochondrion. [Courtesy of David A. Clayton.]

FIGURE 11-53
Electron micrograph of a circular DNA molecule isolated from a chloroplast of Romaine lettuce. The several small circles are DNA molecules from a bacteriophage, included as size standards. [Courtesy of Richard Kolodner and Krishna K. Tewari.]

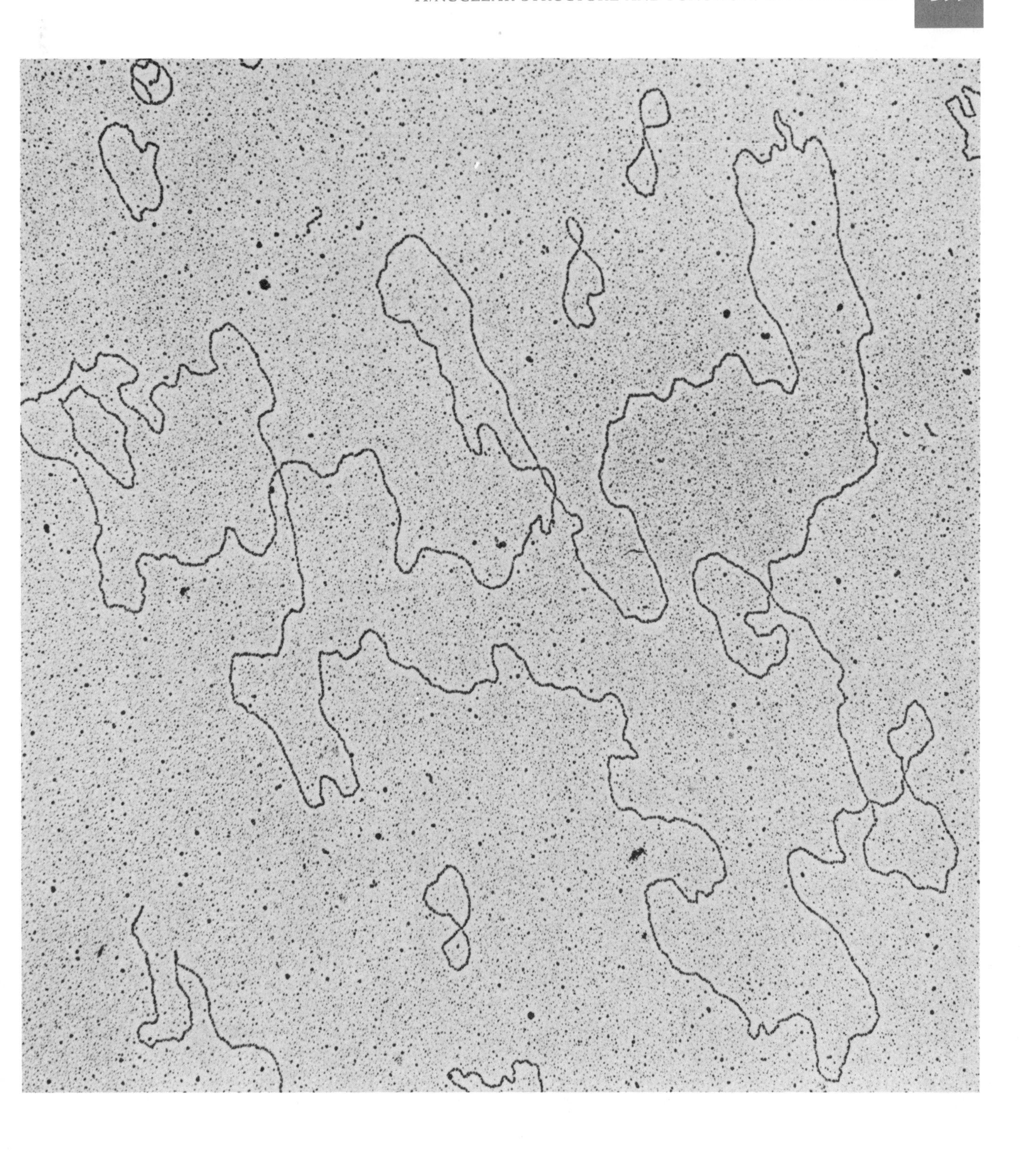

located in the matrix of mitochondria, are distributed randomly between daughter mitochondria. In some kinds of cells, for example, chick cells in culture, and probably in cells in general, mitochondria fuse and separate often, so that the population of mtDNA molecules are dynamically shared among the mitochondria within a single cell.

A yeast mtDNA molecule is five times longer than a mammalian mtDNA molecule but both encode almost an identical set of genes. Most of yeast mtDNA has no known coding or other function. MtDNA molecules in both organisms encode three of the subunit polypeptides of cytochrome *c* and seven or eight other polypeptides. Both encode the two major rRNA molecules for large and small subunits of ribosomes. These rRNAs are smaller than those encoded by nuclear genes, and mitochondrial ribosomes are smaller than cytoplasmic ribosomes. Mammalian mtDNA encodes 22 kinds of tRNA molecules and yeast encodes about 25.

Thus, mtDNA encodes some of the molecules needed to make up the translation machinery in the mitochondrial matrix. However, most elements, like most ribosomal protein and aminoacyl-tRNA synthetases, are encoded by nuclear genes, synthesized in the cytoplasm, and imported into the mitochondrion. The mitochondrial translation machinery is limited in use to the several kinds of mRNA molecules transcribed from mtDNA. No mRNA molecules are imported from the cytoplasm. Like bacterial mRNA molecules, mitochondrial mRNAs are not capped at their 5′ ends. They have short, 3′-poly(A) tails. Transcription of human mtDNA is unusual. One long transcript is produced from *each* strand of the DNA, in contrast to transcription from only *one* strand in nuclear DNA. The two transcripts are full-length copies of the two DNA strands. Both transcripts are processed, each yielding some of the rRNA, tRNA, and mRNA products. The coding segments for some genes are in one strand of the DNA and coding segments for others are in the opposite DNA strand.

Remarkably, the genetic code of mtDNA varies slightly from the nearly universal code present in the nucleus of eukaryotes, the nucleoid of prokaryotes, and viruses. In fungi, which includes yeast, and in animal cells, the codon UGA functions as a codon for tryptophan rather than as a stop codon for translation. In the mtDNA of some organisms AUA codes for methionine rather than for isoleucine. AGA and AGG code for stop rather than arginine.

During the cell cycle the total number of mtDNA molecules doubles, keeping pace with a doubling in mitochondrial number. Replication is catalyzed by a mitochondrial-specific polymerase called DNA polymerase γ, which is encoded by a nuclear gene. The mtDNA molecules replicate asynchronously with each other. Some molecules may replicate twice during a given cell cycle and others not at all. The time of replication is spread over the entire interphase and bears no relationship to the nuclear S period.

At least some segments of mtDNA are represented by nearly identical sequences at scattered locations in the nuclear genome. For example, the 16S rRNA sequence of human mtDNA is also present in nuclear DNA. The significance of mitochondrial sequences in the nucleus is not known.

DNA MOLECULES IN CHLOROPLASTS

Chloroplast DNA molecules (chlDNA) range in size from 33,000 bp to more than 2×10^6 bp (Figure 11-53). In higher plants chlDNA molecules consist of about 1.5×10^5 bp. The number of molecules per chloroplast ranges from 20 to more than 100, depending on the species, and these are distributed in multiple nucleoid-like bodies in the stromal compartment of the chloroplast (Figure 4-15).

More than 80 polypeptides are encoded by chlDNA, as well as rRNAs for large and small ribosomal subunits and tRNAs. Many other chloroplast proteins are encoded by nuclear genes, are synthesized in the cytoplasm, and are imported into the chloroplast. For example, the small subunit polypeptide of the enzyme ribulose-l,5-bisphosphate carboxylase is encoded by a nuclear gene; the large subunit polypeptide is encoded by chlDNA. This enzyme occupies a pivotal position in the conversion of CO_2 to carbohydrate during photosynthesis. Messenger RNA molecules transcribed from chlDNA are translated by the chloroplast translation machinery in the stromal compartment of chloroplasts.

Like mtDNA molecules chlDNA molecules double in number during the cell cycle and are distributed randomly between the daughters during chloroplast fission. Histones are not associated with chlDNA.

PROBLEMS

1. If the DNA molecule in a bacterium has a molecular weight of 4.0×10^9, how many base pairs does it contain?
2. What is the difference between negative and positive supercoiling of DNA?
3. Why does DNA replication begin with the synthesis of RNA primers?
4. DNA polymerase can add nucleotides only to the 3′OH end of a nucleotide or deoxynucleotide chain. How does this complicate DNA replication and how is the complication overcome?
5. How is it known that replication of the chromosome in *E. coli* is bidirectional from the origin?
6. If the bidirectional replication of the chromosome in *E. coli* takes 40 minutes, how many deoxynucleotides are added to a growing chain per second?
7. What are eight differences between the genomes of prokaryotes and eukaryotes?
8. What is satellite DNA?
9. What are the minimum DNA sequence components in a eukaryotic chromosome?
10. How is it known that some pseudogenes arise through reverse transcription of mRNA followed by insertion of the DNA into a chromosome?
11. How is the total complexity of deoxynucleotide sequences in the genome of a species measured?
12. What is the likely explanation of the S-value paradox?
13. What has been learned about the arrangement of genes in chromosomes from the study of ciliates of the hypotrich type?
14. How do histones contribute to the packing of DNA?
15. How was the location of satellite DNA in mouse chromosomes determined?
16. In general among eukaryotes how much of the total cellular DNA is present in the mitochondria?

REPRODUCTION OF PROKARYOTIC CELLS
REPRODUCTION OF EUKARYOTIC CELLS

DNA Replication in the Cell Cycle
Flow Microfluorimetry (FMF)
Lengths of the G1, S, G2, and M Periods
Progression through the Cell Cycle

The G1 period
The initiation of DNA replication
Progression through the S period
The G2 period

The Events of Mitosis
Cytokinesis

Cytokinesis in animal cells
Cytokinesis in plant cells
Distribution of cytoplasmic components

SYNCHRONIZATION OF THE CELL CYCLE

Synchronization of Prokaryotic Cells
Synchronous Eukaryotic Cells

Natural synchrony
Chemically induced synchrony
Mechanical selection of synchronous cells

Decay of Synchrony in Cell Populations

CELL-CYCLE GENES
REGULATION OF REPRODUCTION OF CULTURED CELLS

Specific Regulation of the Cell Cycle in Yeast
Proliferation of Vertebrate Cells in Culture

Density-dependent inhibition of cell reproduction
Anchorage dependence of normal cells
Serum growth factors
Finite life span of normal cells in culture
G0 state

REGULATION OF CELL REPRODUCTION IN ORGANISMS

Stimulation of Cell Reproduction by Hormones and Growth Factors
Regulation of Cell Reproduction by Inhibitors

An inhibitor of epithelial cells in culture
Autoregulation of liver cell reproduction

CHAPTER 12

Cell Reproduction

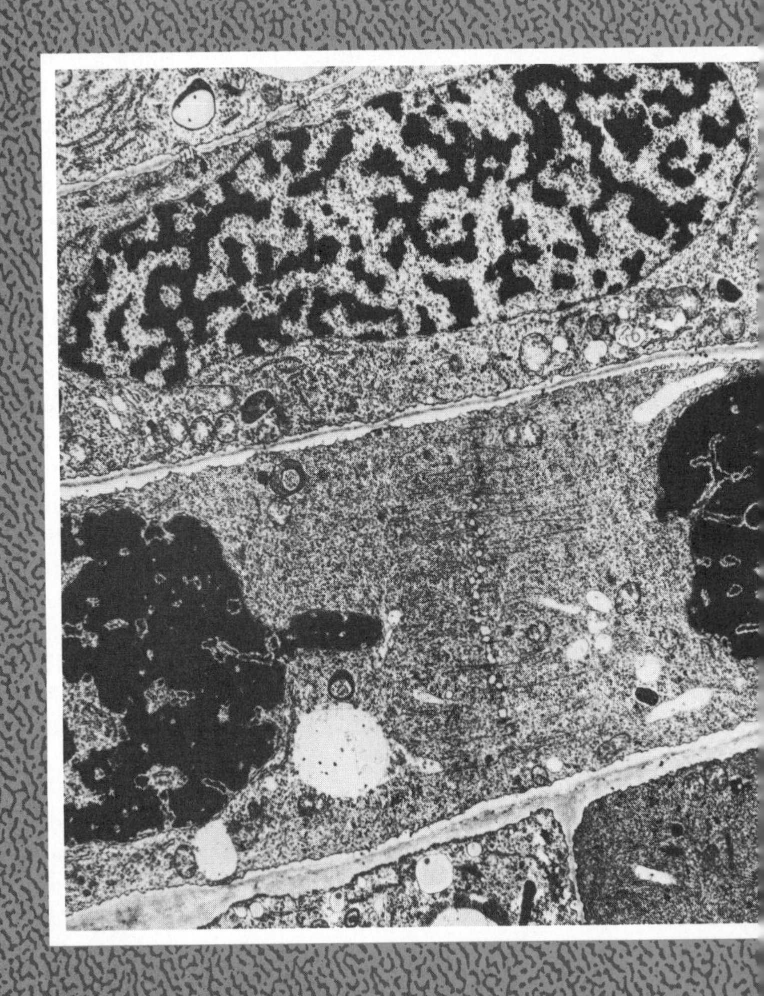

See Figure 12-17.

The ability to grow and reproduce is a fundamental attribute of cells—one on which the survival of every species of unicellular and multicellular organism is absolutely dependent. The study of cell reproduction is important in many areas of cell biology, including the development and functioning of every organism. A human, for example, begins existence as a single cell, the fertilized egg, and grows by cell reproduction to more than one hundred trillion cells (10^{14}) in adulthood. To maintain the functions of various tissues, about one trillion (10^{12}) cell divisions occur in an adult human every 24 hours, or 10^7 cell divisions per second. The precursor cells in the bone marrow that differentiate into red blood cells produce 2.5×10^6 new cells per second. Blood contains fewer white blood cells, but their life spans are measured in days instead of months, and they must be replaced by a similarly high rate of cell division. Epithelial cells that line the intestinal tract and cells in the generative layer of the skin also reproduce at high rates to replace dead cells. In some other tissues, such as those of the liver, kidneys, lungs, and bones, the rate of cell reproduction is much lower. Two major kinds of cells—skeletal muscle cells and nerve cells—do not reproduce in adults and cannot be replaced.

Reproduction of a cell is a momentous accomplishment. It starts with growth, which is the coordinated increase in the amounts of the thousands of different kinds of molecules that make up all of the parts and functions of a cell, and culminates with partitioning of cytoplasm and chromosomes into two daughter cells by cell division. Despite its complexity, cell reproduction is generally a mistake-free process, having been highly polished by several billion (10^9) years of evolution.

An essential prerequisite for cell reproduction is chromosome replication. Chromosome replication must be coordinated with cell growth, and both must be coordinated with cell division. How this is achieved is one of the major areas of study in cell biology.

Cell reproduction is different enough in prokaryotes and eukaryotic cells to warrant separate discussions. The following section summarizes main features of the reproductive process in prokaryotes.

REPRODUCTION OF PROKARYOTIC CELLS

A bacterial cell reproduces by increasing its size, simultaneously replicating its chromosome, and then dividing into two cells, each of which receives one of the daughter chromosomes. Growth from division to division consists of a doubling in the amounts of all components—ribosomes, plasma membrane, cell wall, enzymes, water, and other molecules. A new daughter

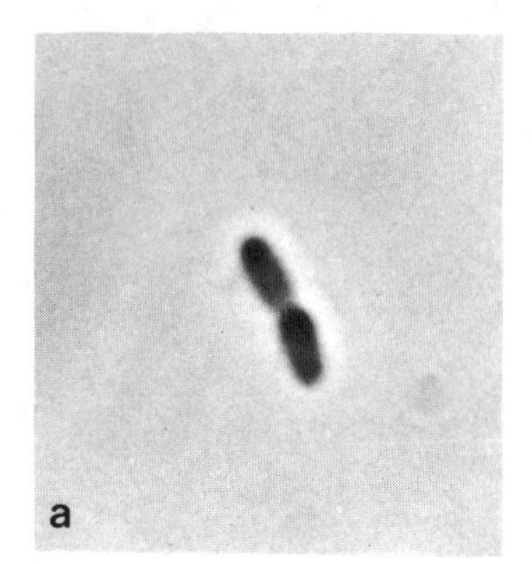

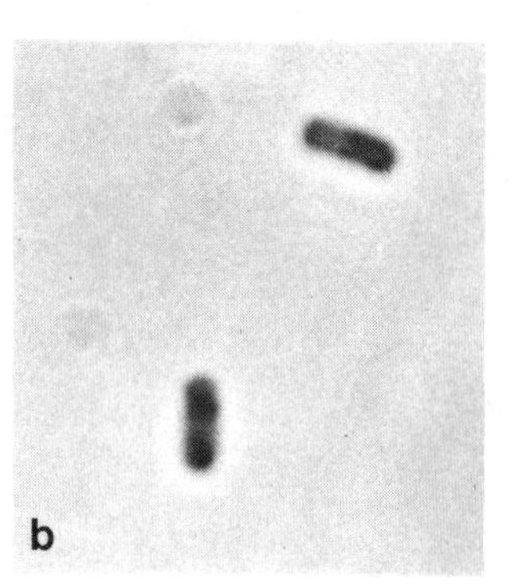

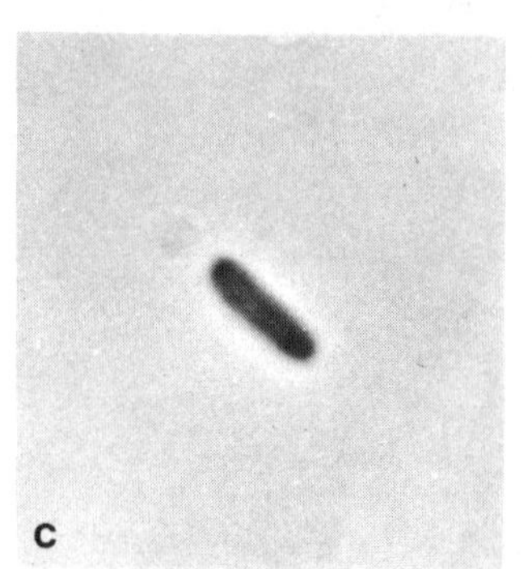

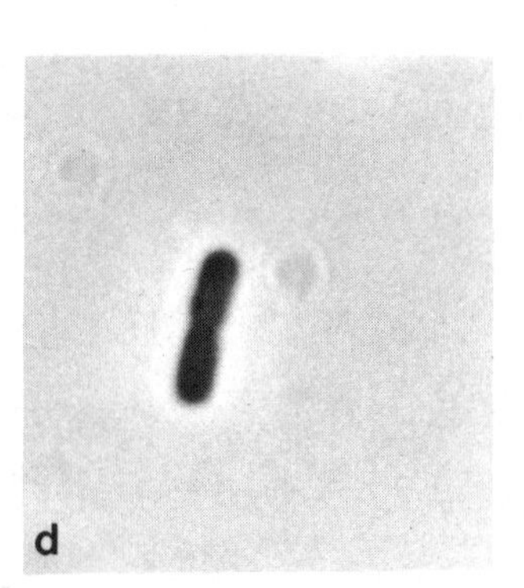

FIGURE 12-1
Light micrographs showing growth of *E. coli*. (a) A cell in division. (b) Eight minutes later the two daughter cells from the division in (a) have moved apart. (c) One of the daughter cells 24 minutes after division. (d) Division of the cell in (c) 42 minutes after the previous division.

cell of *E. coli,* for example, begins its independent existence immediately after cell division; it is then a cylinder one μm in diameter and two μm in length. The diameter remains constant, and the cell doubles its size by growing to four μm in length (Figure 12-1). During this growth the number of ribosomes increases from 20,000 to 40,000 with a concomitant doubling in the cell's capacity for protein synthesis. In a rapidly growing bacterial cell a major portion of the cell's ATP is used for production of ribosomes.

Cell growth, chromosome replication, and cell division are coordinated as the cell moves through its cycle, but the molecular basis for the coordination is still poorly understood. Measurements of cell size suggest that initiation of DNA replication is triggered when a cell reaches a threshold size, which is well below that reached just prior to division. The DNA replicates as the cell continues to grow. Division occurs at a fixed time interval after completion of DNA replication, by which time the cell has also completed its growth to a doubled size.

A connection between cell growth and initiation of DNA replication has been demonstrated in a variety of experiments. If protein synthesis is blocked with an inhibitory drug like puromycin or by excluding from the nutrient medium an amino acid that the cell cannot synthesize, initiation of DNA replication is prevented. Blocking growth in other ways, for example, by depriving a cell of sources of nitrogen, phosphate ions, or energy (glucose and organic molecules), produces the same result. Thus, if ATP production in *E. coli* falls because of a lack of energy-yielding carbon molecules in the medium, ribosome production stops, cell growth ceases, and DNA replication is not initiated.

In prokaryotes, cell division consists of a separation of the two daughter chromosomes into two distinct nucleoids, then the migration of the nucleoids toward the two ends of the cell, followed by an inward growth of the plasma membrane and cell wall to form a septum across the cell between the nucleoids (Figure 12-2). Subsequently, a split forms in the plane of the septum wall, and the two daughter cells separate.

The rate of growth of the individual cell, and therefore the rate of cell reproduction, depends on nutrient

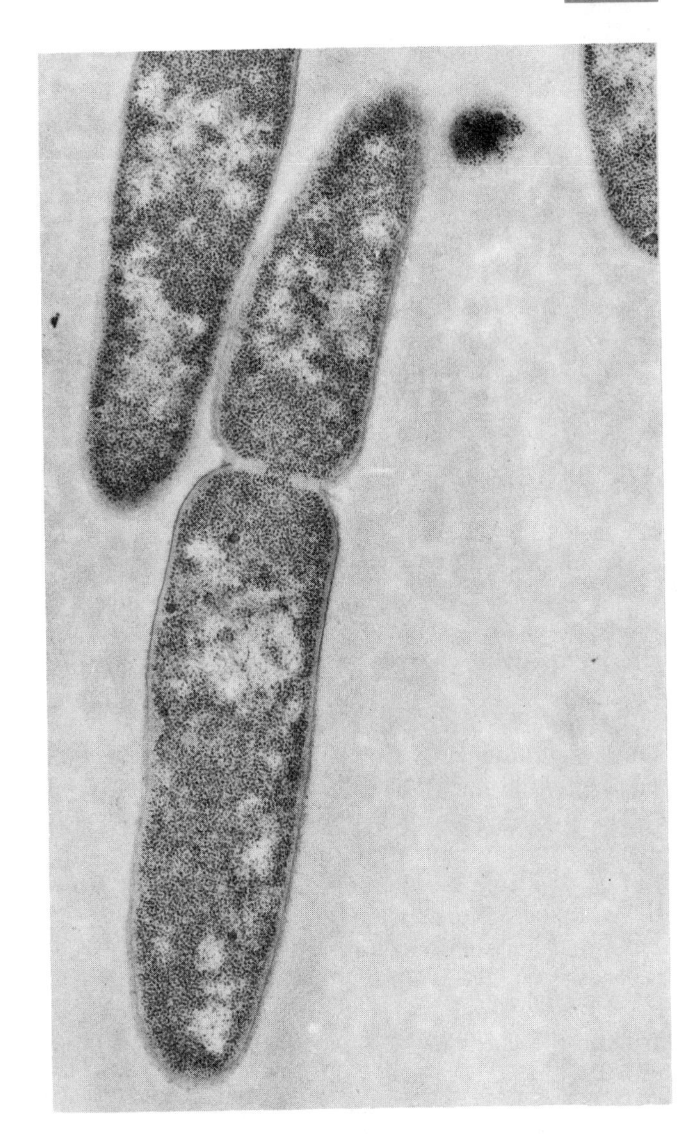

FIGURE 12-2
Electron micrograph of cell division in *E. coli.* The section includes only part of one of the two incipient daughter cells. Division is nearly complete. An opening, filled with ribosomes, still remains in the gap in the septum. [Courtesy of J. A. Hobot.]

conditions and temperature. *E. coli* can grow in a medium consisting of a simple organic molecule such as glucose as a source of energy and carbon and various inorganic ions (Chapter 1). In nutrient medium with glucose as the carbon and energy source, a cell doubles in size and divides at 37°C every 40 minutes on the average. In medium rich in amino acids, purines, pyrimidines, and other organic molecules cells grow faster, and the division time (the generation time) is reduced to 20 minutes. This is a remarkably high rate of cell proliferation. A single bacterium reproducing with a generation time of 20 minutes produces a population of 2^{24} or over ten million cells in 8 hours. As nutrients in a culture are used up, individual cell growth slows and then stops. Cell reproduction ceases, the number of cells remains stationary for a time, and eventually cells begin to die if not transferred into fresh medium.

It is often convenient to record the proliferative behavior of cells in culture with a growth curve, as shown in Figure 12-3. Growth of a culture is initiated by inoculating cells from a dense culture into fresh medium (time 0). Proliferation begins after a lag of some minutes, during which cells adjust to the fresh medium. Following the **lag phase** the cells multiply exponentially. For this reason the log of cell number plotted against time gives a straight line. The generation time can be read from the slope of the line as the time taken for the population to double. This period of rapid growth, called the **log phase** or **exponential phase,** is followed by decelerating growth and then a **stationary phase,** in which the cell number does not change. Finally, the culture enters the **decline phase** as the cells begin to die.

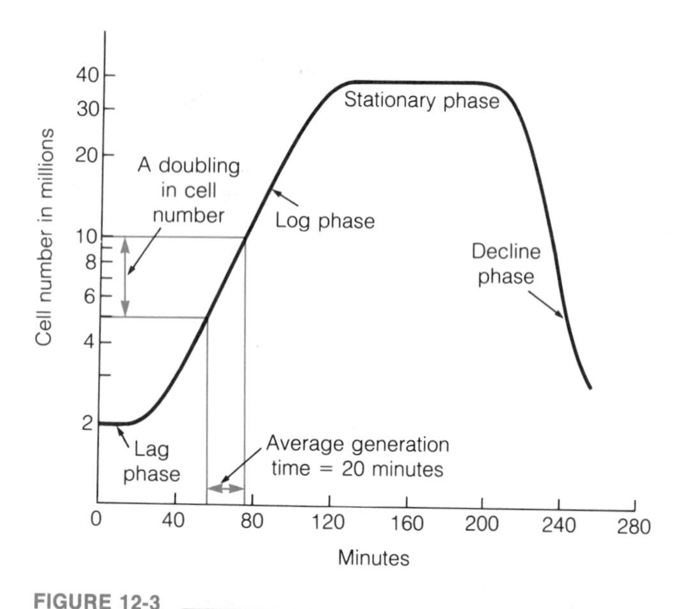

FIGURE 12-3
Growth curve for a bacterial cell population in rich nutrient medium. The logarithm of cell number is plotted against time in minutes. The average generation time can be read from the slope of the curve during log phase, i.e., a doubling in cell number occurs in 20 minutes, which is the average generation time.

REPRODUCTION OF EUKARYOTIC CELLS

The simplest description of cell reproduction is a cycle of growth and division. The cell grows during the period between cell divisions, the interphase, and then divides; the daughter cells repeat the cycle, and the number of cells increases exponentially. This description of the cell cycle is expanded by marking out the period of DNA replication within interphase. Because the timing of DNA replication in the cell cycle is crucial to the understanding of cell reproduction and its regulation, methods for determining timing are described next.

DNA Replication in the Cell Cycle

The most sensitive method for determining when DNA replication takes place is by autoradiographic detection of incorporation of ^{3}H-thymidine. Since thymidine is used exclusively for DNA synthesis, only those cells engaged in replication incorporate ^{3}H-thymidine and become radioactive. At any instant the cells in a growing population will be in all different parts of the cell cycle, that is, the cell cycles will be asynchronous. Most cells will be in one or another part of interphase; others will be in various stages of division. When ^{3}H-thymidine is given to the cells for a few minutes and the cells are then processed for autoradiography, those that are replicating their DNA (in the S period) will produce an autoradiographic image (Figure 12-4). This autoradiograph shows that (1) cells in division do not incorporate ^{3}H-thymidine and hence cells in division are not replicating DNA and (2) only some of the interphase cells are replicating DNA, and therefore DNA replication, or the S period, occupies only part of interphase. Where in interphase replication occurs is shown by the following experiment.

An asynchronously reproducing culture of animal cells is given ^{3}H-thymidine for a few minutes to label those cells that are in the S period at the time. Instead of processing the cells for autoradiography as in Figure 12-4, the cells are given fresh medium containing nonradioactive thymidine and no ^{3}H-thymidine so that no further labeling of DNA takes place. The cells are allowed to continue cycling, but from now on any cell that enters the S period will not become labeled. Further, cells that incorporated ^{3}H-thymidine during the short exposure and then leave the S period, having completed replication, will still be labeled in their DNA.

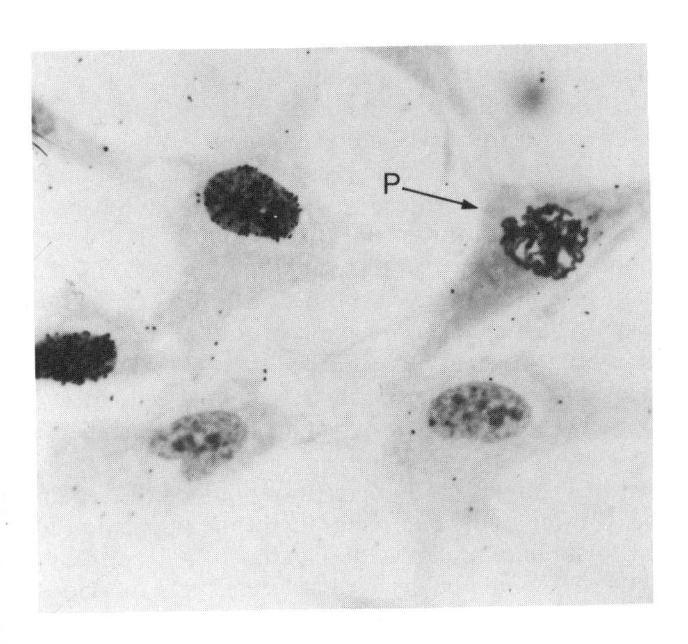

FIGURE 12-4
Autoradiograph of cells incubated with ^{3}H-thymidine for ten minutes. Two interphase cells are labeled and two are not. One cell (P) is in prophase of mitosis and is unlabeled.

One hour after removal of ^{3}H-thymidine from the medium a sample of the cells is examined by autoradiography. Particular attention is given to cells in mitosis, and none of these are found to be radioactive—only interphase cells show label. This means that one hour is not long enough for any of the cells that were in the S period at the time of ^{3}H-thymidine labeling to reach mitosis, that is, DNA replication must end at least one hour before mitosis. A sample of cells examined two hours after removal of ^{3}H-thymidine from the medium gives the same result, which means that DNA replication must end at least two hours before mitosis. In a sample of cells examined at three hours after removal of ^{3}H-thymidine from the medium, 50 percent of the cells in mitosis are found by autoradiography to be radioactive. These cells must have been in the last minutes of the S period when the ^{3}H-thymidine was available. In the subsequent three hours they finished DNA replication and progressed into mitosis. Half of the mitotic cells in the three-hour sample are not radioactive and must have already finished replication when the ^{3}H-thymidine was available. This means that cells progress from the end of the S period to mitosis at slightly variable rates: some take a little longer than three hours, some take a little less time. The average is calculated as the time required for the fraction of mitotic cells that is labeled to reach 50 percent, in this case, three hours.

In a sample of cells taken at four hours after the short exposure to ^{3}H-thymidine almost 100 percent of cells in mitosis are labeled, which means that almost all cells were replicating their DNA four hours before mitosis. Plotting the percentage of labeled mitotic cells against time after exposure to ^{3}H-thymidine on a graph shows the relationship of DNA replication to mitosis (Figure 12-5). In samples of cells taken at longer times after exposure to ^{3}H-thymidine all the mitotic cells are labeled, showing that DNA replication takes place continuously over many hours of interphase. Eventually, in cells taken at 11 hours after ^{3}H-thymidine exposure, some of the mitotic cells are not labeled. These are cells that had not yet entered DNA replication when ^{3}H-

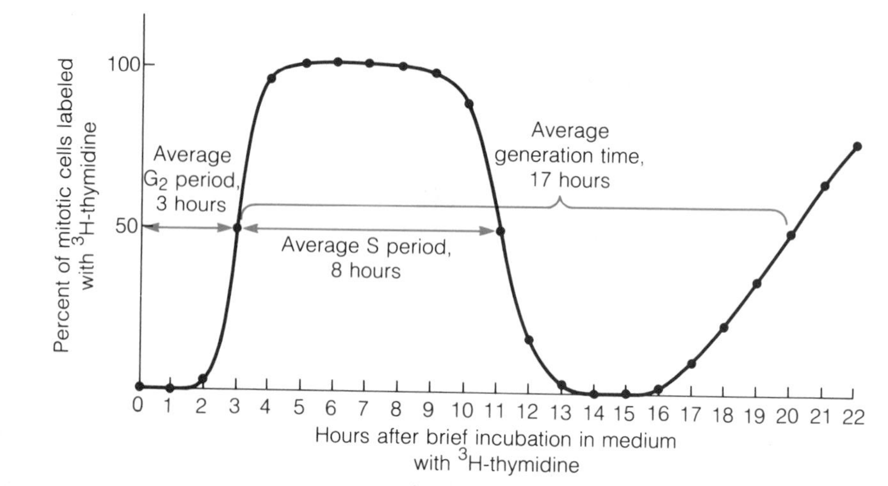

FIGURE 12-5
The labeled mitosis method for determining the length of the G2 period, the S period, and the generation time.

thymidine was available, that is, cells in the prereplicative part of interphase. The point on the curve in Figure 12-5 at which the percentage of mitotic cells that is labeled falls to 50 percent marks the average time preceding mitosis when DNA replication started, or at 11 hours in Figure 12-5. In cell samples taken at longer times none of the mitotic cells are labeled because none had reached DNA replication at the time ^{3}H-thymidine was available. Eventually, in cell samples examined at longer and longer times, labeled mitotic cells again appear. These are cells that were in DNA replication at the time of availability to ^{3}H-thymidine and have had time to reach mitosis, divide, transit the entire next interphase, and reach the next mitosis.

The information in the graph in Figure 12-5 can be used to construct a diagram showing the arrangement of events in the cell cycle (Figure 12-6). Replication of DNA occurs over an eight-hour interval, the S period, that ends about three hours before mitosis. The three-hour interval between the end of the S period and the start of mitosis is called the **G2 period** (G for *g*ap of time). The interval between mitosis and the next initiation of DNA replication is called the **G1 period.** Cell division takes about one hour in animal cells (determined by microscopic observation) and is designated as M (*m*itosis) or D (*d*ivision).

In Figure 12-5, 50 percent of mitotic cells are labeled three hours after exposure of the population to ^{3}H-thymidine. If allowed to continue, the labeled mitotic cells would be detected as labeled cells in mitosis at the end of the next cell cycle. They are responsible for

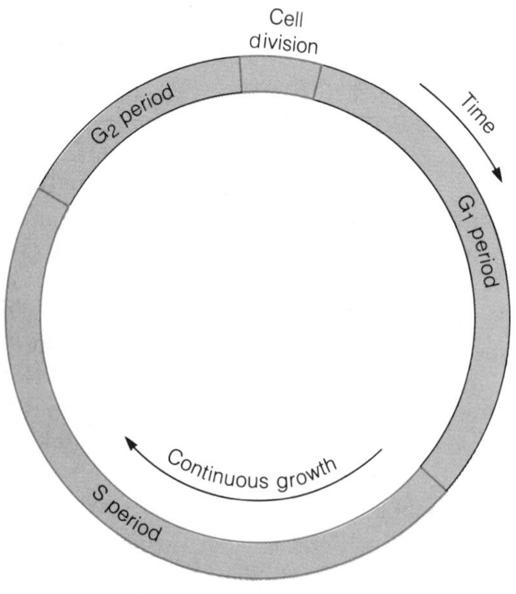

Cell division
Time
G_1 period
S period
G_2 period
Continuous growth

FIGURE 12-6
The cell cycle and its four major sections, the G1, S, G2 and division periods.

the second rise in labeled mitotic cells in Figure 12-5, reaching 50 percent labeling at 20 hours. The cells reached the first mitosis on average three hours after labeling and reached the second mitosis on average at 20 hours after labeling (second rise of the curve in Figure 12-5). Therefore, the time from one mitosis to the next, or the average generation time, is $20 - 3 = 17$ hours.

The data in Figure 12-5 reveal the generation time (GT), the length of the G2 period, and the length of the S period. The time taken for mitosis (one hour) is known from microscopic study. The length of G1 is then calculated: $G1 = GT - G2 - S - M$, in this case five hours, as shown in Figure 12-6.

This analysis shows that the cell cycle consists of four successive periods that are defined by the occurrence of DNA replication and mitosis. The analysis of these four periods and their interrelations forms the basis of much current research on how cell reproduction is achieved and how it is regulated.

Flow Microfluorimetry (FMF)

Another way to analyze the cell cycle is to measure the amount of DNA in individual cells in an asynchronously growing population. In a completely asynchronous population, at any instant cells are distributed randomly in all parts of the cell cycle. Some cells in the population have not started replication and contain the 2*C* (diploid) value of DNA (Chapter 11). The *C* value is the quantity of DNA in a cell with a haploid set of unreplicated chromosomes. Therefore, 2*C* is the amount of DNA in a diploid cell before DNA replication starts, that is, a diploid cell in G1. Some cells in the growing population contain a 4*C* amount of DNA, which is the amount expected in a cell in the G2 period or in mitosis before completion of cell division. The rest of the cells have DNA contents between the 2*C* and 4*C* amounts; these are in the process of replicating DNA.

The quantity of DNA in a large number of individual cells can be quickly determined by staining the cells with a dye that binds specifically to DNA. The amount of dye bound in a cell is then proportional to the amount of DNA in the cell. The kinds of dyes used for this technique are fluorescent: they emit light of one color when excited by light of another color. The intensity of emitted fluorescent light is proportional to the amount of dye that is present. Therefore, when a cell whose DNA has been stained with a fluorescent dye is exposed to a light beam of the right color, it fluoresces with a light intensity that is proportional to its DNA content.

In the technique called **flow microfluorimetry,** cells are fixed with alcohol or another fixative and then stained with a fluorescent dye. The cells are suspended in a salt solution and forced to pass rapidly in single file through a fine capillary tube positioned in the path of a laser beam. As each cell passes through the capillary, the fluorescent dye bound to its DNA is excited to emit light in all directions. The intensity of emitted light is measured by a photoelectric cell positioned out of the path of the laser beam. Cells can be forced through the capillary at a rate of several thousand per second. The light emitted by each one is measured separately and recorded electronically by the flow microfluorimeter (FMF).

Another component of the flow microfluorimeter compiles the individual cell measurements in a display called a histogram, in which the number of cells with a given amount of DNA is plotted against the DNA amount. The histogram in Figure 12-7 shows the DNA contents of tens of thousands of cells for an asynchronously growing population. A peak in cell number is present at the 2*C* amount of DNA, representing cells in G1. Another peak, at the 4*C* amount of DNA, represents cells in G2 and mitosis. Cells with intermediate amounts of DNA are S-period cells.

If the cells in a growing population are distributed randomly throughout the cell cycle, which is the case for a population that is growing completely asynchronously, then the percentage of cells in any given segment of the cell cycle is proportional to the duration of that segment. Thus, the percentage of cells in G1 is proportional to the length of the G1 period, the percentage of cells in the S period is proportional to the length of the S period, etc. In FMF analysis the percentages of cells in the G1, S, and G2 + M periods can be calculated by resolving the histogram into the three

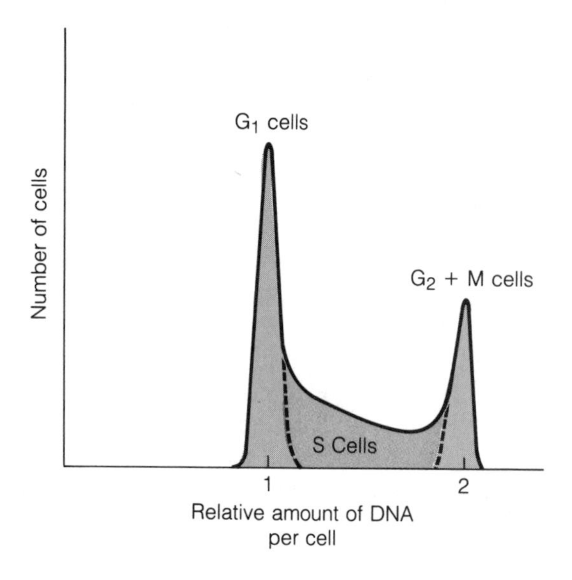

FIGURE 12-7
The DNA content per cell for a population of cells determined by flow microfluorimetry. The dashed lines represent resolution of the cell population into three components, G1 cells, S period cells, and cells in G2 and M.

corresponding component curves (Figure 12-7). The areas under the three component curves give the relative percentages of G1, S, and G2 + M cells. However, because there are always about twice as many cells entering G1 as there are cells leaving G2 in a logarithmically expanding population, with intermediate numbers of cells between the two extremes, a correction factor is applied in order to obtain accurate measures of the sizes of the G1, S, and G2 + M populations.

When the percentages of G1, S, and G2 + M cells have been determined, the relative durations of G1, S, and G2 + M are also known. If the average generation time for cells in the population is also known, for example, from a population growth curve, then the absolute length of each period can be determined. Since M usually lasts one hour, the length of G2 is simple to calculate. For example, if 40 percent of the cells are in G1, 40 percent in S, and 10 percent in G2 + M, and the generation time is 20 hours, then, as an approximation, G1 = 8 hours, S = 8 hours, G2 = 3 hours, and M = 1 hour. To make these estimates a bit more accurate, corrections can be introduced to account for the distribution of ages of cells in the cycle, as mentioned above.

Analysis of the cell cycle by FMF is not quite as accurate as analysis by ^{3}H-thymidine and autoradiography because incorporation of small amounts of ^{3}H-thymidine can be detected more readily than small increases in DNA content. However, the FMF method has the advantages of speed and convenience in analysis of cultured cells. For analysis of cell tissues in a multicellular organism the autoradiographic method ordinarily has to be used since the FMF requires that cells be suspended in a solution.

Lengths of the G1, S, G2, and M Periods

For mammalian cells, both in culture and in an organism, the lengths of the S, G2, and M periods tend to be constant in all cells. The S period lasts about 8 hours, G2 about 3 hours, and M about one hour. The length of G1 is far more variable from cell to cell. For cultured mammalian cells it usually ranges from 3 to 12 hours, depending on the species and on the type of cell. Within an organism, the length of G1 differs much more from one cell type to another. In adult tissues G1 is usually much longer than in cultured cells. For example, in the epithelial cells that line the digestive tract, which are among the most rapidly reproducing cells in the body, the length of G1 ranges from a few hours in the small intestine to more than 100 hours in the esophagus. The most rapidly reproducing cells in animals generally are those of early cleavage-stage embryos; in these the G1 period is totally absent, and such cells proceed from mitosis directly into the next S period.

In short, the differences in the length of G1 give rise to the main differences in cell cycle times and therefore determine differences in rates of cell reproduction. More about the length of G1 is included later in this chapter in connection with regulation of cell reproduction in normal cells and in Chapter 13 in connection with cancer cells.

The cell cycles for cells in other animals and in plants are similar to those in mammals. In some rapidly reproducing, unicellular organisms the generation times are much shorter (e.g., 1.5 hours in yeast), but the arrangement of G1, S, G2, and M is much the same. In others, such as *Amoeba proteus* and *Physarum*, the cycles are different. The G1 period is totally absent and G2 is the longest part of the cycle.

Progression through the Cell Cycle

The cell progresses through a sequence of stages to accomplish its reproduction. This section briefly describes the progression.

The Gl period. The completion of cell division marks the end of one cell cycle and the start of the next for the two daughter cells. With the few exceptions already noted above, the cycle begins with entry into the G1 period. During G1 a cell grows continuously, as it does during S and G2. It is logical to suppose that during G1 the cell is preparing in some way for entry into DNA replication, but no such preparations have yet been specifically identified. Enzymes important in DNA replication, such as DNA polymerase and those that catalyze synthesis of deoxynucleoside triphosphates, are not accumulated during G1; instead the numbers of these molecules crucial for DNA replication begin to increase around the start of the S period. The G1 period has been intensively studied, but aside from growth almost nothing is known about the molecular events in G1 that contribute to the progression of a cell toward the S period.

The G1 period has a particular importance in that the rate of cell reproduction is controlled by regulation of transit through G1. Rapidly reproducing cells proceed through G1 quickly. With slower rates of reproduction, cells remain in G1 for longer times. Cells that temporarily or permanently stop proliferating, as happens in many tissues of multicellular organisms, remain in G1. Regulation is believed to be achieved by blocking a particular molecular event required to allow progression toward DNA replication. Hence it is important to determine how a cell progresses through the G1 period. More will be said later about arrest of cells in the G1 period and regulation of cell reproduction.

The initiation of DNA replication. The G1 period ends and the S period starts when the first replication units are triggered into replication. This is an important transition point in the cell cycle. Delaying or preventing a cell from reaching this transition is the means by which cell reproduction is regulated in multicellular organisms and in many unicellular eukaryotes. Entry of a cell into the S period commits the cell to another cell division.

How DNA replication is triggered is not known. Synthesis of enzymes that catalyze synthesis of deoxynucleoside triphosphates (dNTPs) and synthesis of DNA polymerase increase sharply at the start of the S period. However, neither the synthesis of these enzymes nor production of dNTPs initiates replication. Instead these are simply coordinated with initiation by unknown molecular mechanisms.

Initiation of replication is in some way linked to cell growth. For example, if growth is stopped by depriving the cell of an essential nutrient or by inhibiting protein synthesis with a drug, DNA replication is not initiated. Further, a variety of observations suggest that initiation is triggered when cells reach a critical size. On the other hand, the apparent association between cell size and the initiation of DNA replication can be easily disrupted, that is, small cells can be made to initiate replication.

The start of the S period is marked by the activation of transcription of a number of genes. Some of these encode enzymes concerned with DNA replication, and their activation leads to sharp increases in synthesis of enzymes, as just noted above. For example, the transcription of at least four genes encoding enzymes in the synthetic pathway for thymidine triphosphate (TTP) is activated (Figure 12-8). Also, transcription of genes encoding histones is turned on at or slightly before the start of S (Chapter 10). Identification of the signal that coordinates the activation of the various genes with initiation of replication of DNA is an important problem.

Progression through the S period. Each DNA molecule (chromosome) in the eukaryotic nucleus contains many replicating units. A human diploid cell con-

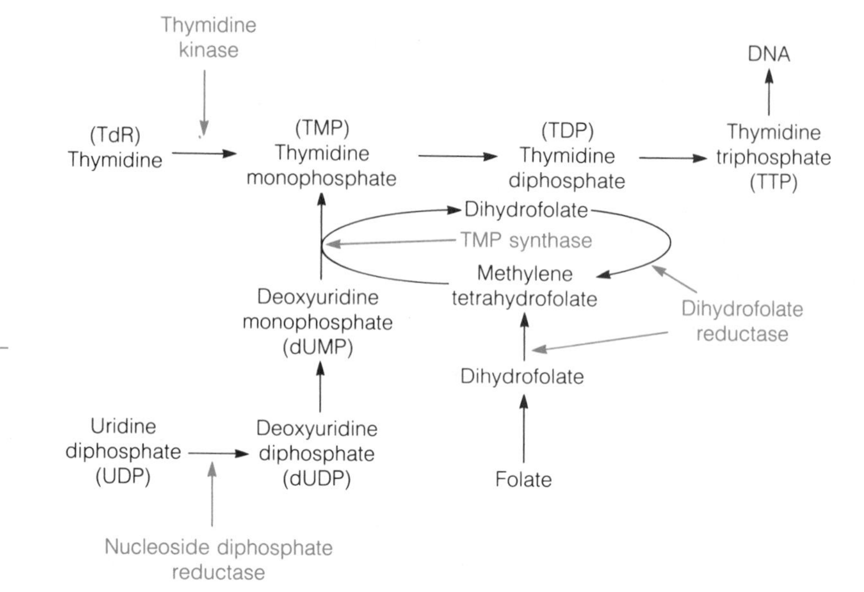

FIGURE 12-8
Enzyme pathways involved in the synthesis of thymidine triphosphate. Four enzymes (red) are turned on in the cell cycle at the transition from the G1 period to the S period and are turned off when DNA replication has been completed.

tains about 70,000 replicating units, or an average of about 1300 in the DNA molecule of each of the 46 chromosomes. The replicating units are not all initiated simultaneously at the start of the S period, but rather initiations are staggered throughout the S period. In a mammalian cell, replication units are on the average 30 μm long and each replicates in about 15 minutes. DNA replication in the nucleus is continuous over the eight-hour S period, and therefore replication units must be initiated almost continuously over that interval.

The firing of replication units follows, at least to some degree, a regular sequential order. Replicating units that initiate at the start of one S period also initiate at the start of the subsequent S period. Replicating units that initiate late in one S period also initiate late in the next S period. Order is also reflected by the replication of DNA in euchromatin in the first part of the S period and replication of DNA in heterochromatin in later S. This fits with two other experimental findings: (l) highly repetitious sequences, which are present in heterochromatin, generally replicate late in the S period, and (2) DNA that is actively transcribed is replicated in the first part of S, and DNA that is transcriptionally inactive (perhaps in heterochromatin) is replicated late in S. Thus, so-called housekeeping genes, which encode proteins that are synthesized continuously in all cells, appear to replicate early. The timing of replication of genes that encode proteins that give a cell differentiated properties varies according to cell type. Thus, in cells that are differentiating into erythrocytes and still reproducing, genes encoding α- and β-globins are replicated early. However the same genes are replicated late in cells in which transcription of the genes encoding α- and β-globins is turned off.

If protein synthesis is blocked in a cell in the S period, DNA replication quickly stops. This has been interpreted to mean either that histone synthesis is required for DNA replication to continue or that specific initiator proteins must be synthesized to trigger successive initiations in replication units. Beyond this, little is known about the control of initiation of the 70,000 replication units in a mammalian cell. The mechanism that controls the succession of initiations of replication units also provides that *every* replication unit is replicated and that normally no unit replicates more than once in a single S period.

The G2 period. The G2 period links the end of DNA replication with the initiation of mitosis, but not much is known about the molecular events that occur in this period. Inhibition of RNA or protein synthesis blocks transit through G2, so synthesis of new RNA and protein is necessary. Several proteins have been found to appear in G2 and then to disappear as soon as mitosis is over. One of the proteins synthesized only in the G2 period is known as **maturation promoting factor (MPF).** It brings about condensation of interphase chromosomes into the mitotic form. The existence of MPF was first demonstrated by a cell fusion experiment. When a mammalian cell in mitosis is fused with an interphase cell, the nuclear envelope of the interphase nucleus disappears and the chromosomes are caused to condense into the mitotic form, a phenomenon called **premature chromosome condensation** (Figure 12-9). If the interphase cell used in this fusion experiment happens to be in the G1 period, then the prematurely condensed chromosomes appear containing a single chromatid each. Fusion of a mitotic cell with a cell in the S period produces chromosomes that have a fragmented or ragged appearance, consistent with what might be expected for chromosomes with DNA replication taking place at many points along their lengths. Fusion with a G2 cell produces condensed chromosomes containing two chromatids each, as expected.

Maturation promoting factor was later identified in mature oocytes of *Xenopus,* cells that remain arrested in metaphase of meiosis under the influence of MPF until fertilization occurs. Injection of MPF into cells of *Xeno-*

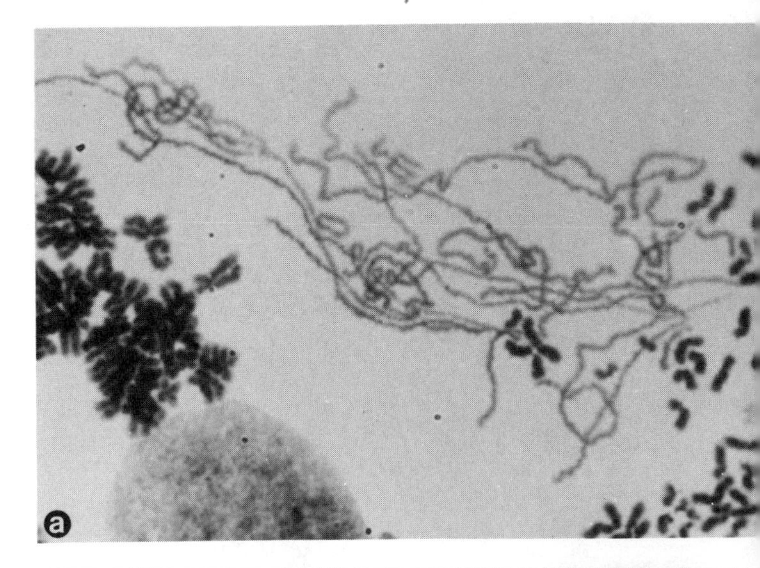

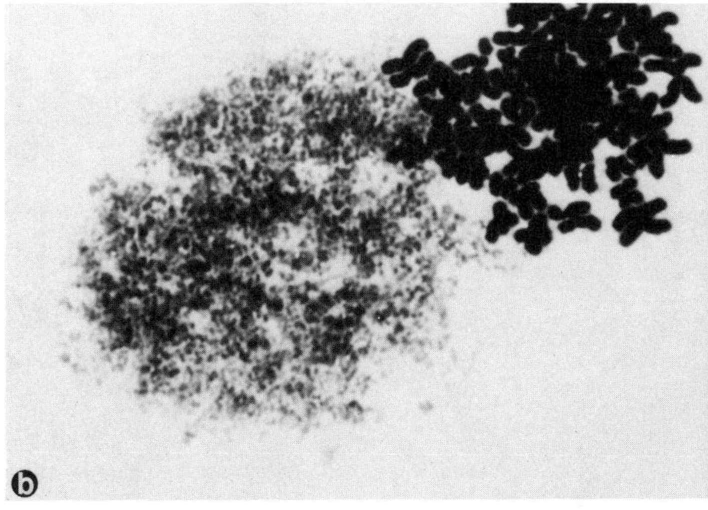

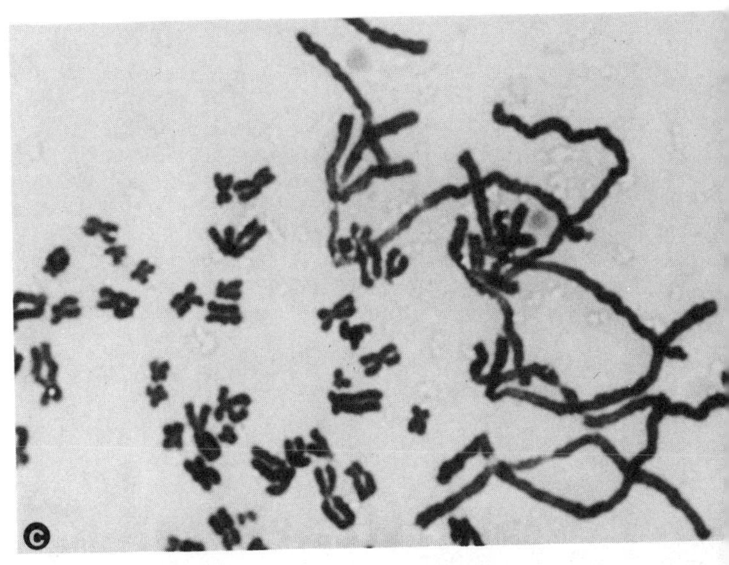

FIGURE 12-9
Premature chromosome condensation induced by fusing a cultured cell in mitosis with an interphase cell. (a) Fusion of a mitotic cell with a cell in the G1 period. The mitotic chromosomes (each with two chromatids) are short and darkly stained. The prematurely condensed chromosomes are not fully condensed. They contain only one chromatid each. (b) Fusion of a mitotic cell with a cell in the S period. The condensed chromosomes in the S period cell are fragmented and indistinct. (c) Fusion of a mitotic cell with a cell in the G2 period. The chromosomes in the G2 period are not fully condensed but two chromatids in each chromosome are visible. [Courtesy of Potu N. Rao.]

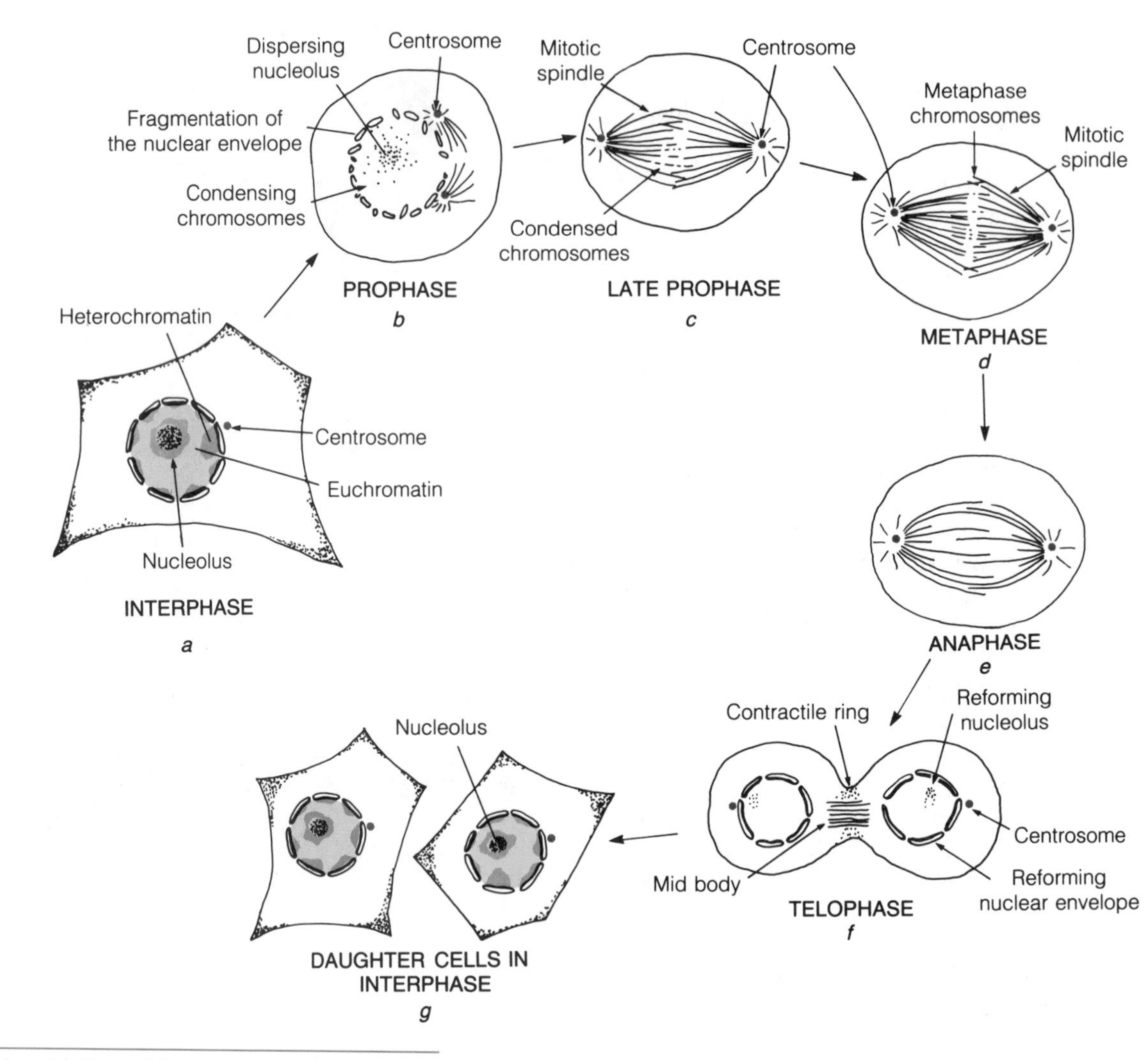

FIGURE 12-10
Events of cell division. (a) The cell in interphase before cell division. (b) Prophase. The chromosomes are partially condensed, the nucleolus is beginning to disperse, the nuclear envelope is fragmenting, and the mitotic spindle has begun to form between the separating halves of the centrosome. (c) Late prophase. The chromosomes are highly condensed and the mitotic spindle has formed. (d) Metaphase. The condensed chromosomes have aligned in the center of the cell. (e) Anaphase. The chromatids have separated into daughter chromosomes and are moving toward the centrosomes. (f) Telophase and cytokinesis. The chromosomes are decondensing, and the nucleolus and nuclear envelope are re-forming. Microtubular remnants form the mid-body. The contractile ring of actin filaments draws the cleavage furrow inward. (g) Daughter cells in interphase.

pus embryos that had been experimentally arrested at the end of the S period causes dissolution of the nuclear envelope and rapid mitotic condensation of the interphase chromosomes. MPF has also been identified in yeast cells in mitosis, in mammalian oocytes in meiosis, and in mammalian somatic cells. For example, injection of a semipurified MPF from mitotic mammalian cells into immature oocytes of *Xenopus* causes breakdown of the nuclear envelope and premature chromosome condensation.

Isolated interphase nuclei incubated in an MPF-containing extract from *Xenopus* oocytes undergo breakdown of the nuclear envelope and chromosome condensation. This experimental system makes feasible a molecular analysis of MPF action and the mechanism controlling chromosome condensation. For example, within 15 minutes after addition of MPF, two of the major proteins (lamins) of the nuclear lamina (lamins A and C) become hyperphosphorylated. This is followed by breakup of the nuclear lamina, presumably caused by hyperphosphorylation of lamins, and dissolution of the nuclear envelope 30 minutes later.

In the continued presence of MPF, cells remain in metaphase. An MPF-inactivating agent has been identified in oocytes of *Xenopus*. This agent appears in metaphase and inactivates MPF, which releases the cell from metaphase and allows it to complete mitosis (or meiosis) and enter interphase once again.

Blockage of cells in the G2 period occasionally occurs at a low frequency in somatic cells of plants and animals. For example, up to a few percent of the cells in the skin of mammals are arrested in the G2 period. When the skin is wounded, these cells are immediately released from the G2 block and enter mitosis. There are too few cells blocked in G2 to argue that they represent an important, rapid initial component in wound healing, and their significance remains unknown. They may occur because of a temporary failure of MPF to appear and trigger entry into mitosis.

The Events of Mitosis

The events of mitosis are summarized in Figure 12-10. Mitosis begins with prophase. This period of chromosome condensation is induced by MPF. It is the longest part of mitosis, lasting about 30 minutes in mammalian cells. Late in prophase the nuclear lamina and nuclear envelope disaggregate, and many nuclear components mix into the cytoplasm. With the breakdown of the lamina and nuclear envelope most of the nonhistone proteins of the nucleus and incompletely processed RNA transcripts are released to the cytoplasm. During prophase, synthesis of mRNA and rRNA slows and finally stops (synthesis of tRNA continues at a reduced rate) (Figure 12-11). In most kinds of cells the cessation of rRNA synthesis is accompanied by the complete dissolution of nucleoli. RNA synthesis stops (Figure 12-12) presumably because the tight condensation of chromatin excludes access of transcription factors and RNA polymerase to the DNA. The situation is analogous to the effect resulting from packing of heterochromatin in interphase nuclei; transcription does not occur in heterochromatin.

In late prophase, about the time of breakdown of the nuclear envelope and dispersion of the lamina, the rate of protein synthesis in the cytoplasm begins to decrease. It falls to 30 percent of its G2 value by the time

Changes at late prophase

Condensation of chromatin into mitotic chromosomes
Dissolution of the nuclear envelope
Dissolution of the nuclear lamina
Cessation of rRNA and mRNA synthesis
Dispersion of most nuclear proteins into the cytoplasm
Restructuring of the cytoskeleton and
Formation of the mitotic spindle
Decrease in protein synthesis

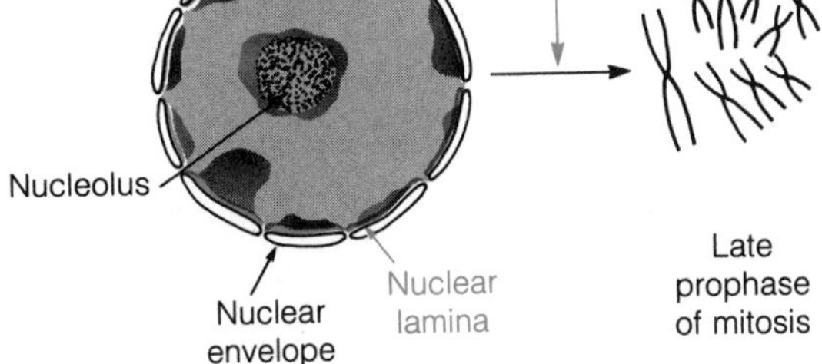

FIGURE 12-11
Transit of a cell from interphase to mitosis is accompanied by dramatic changes in cell function and structure. All the changes are reversed at reentry into interphase after mitosis is over.

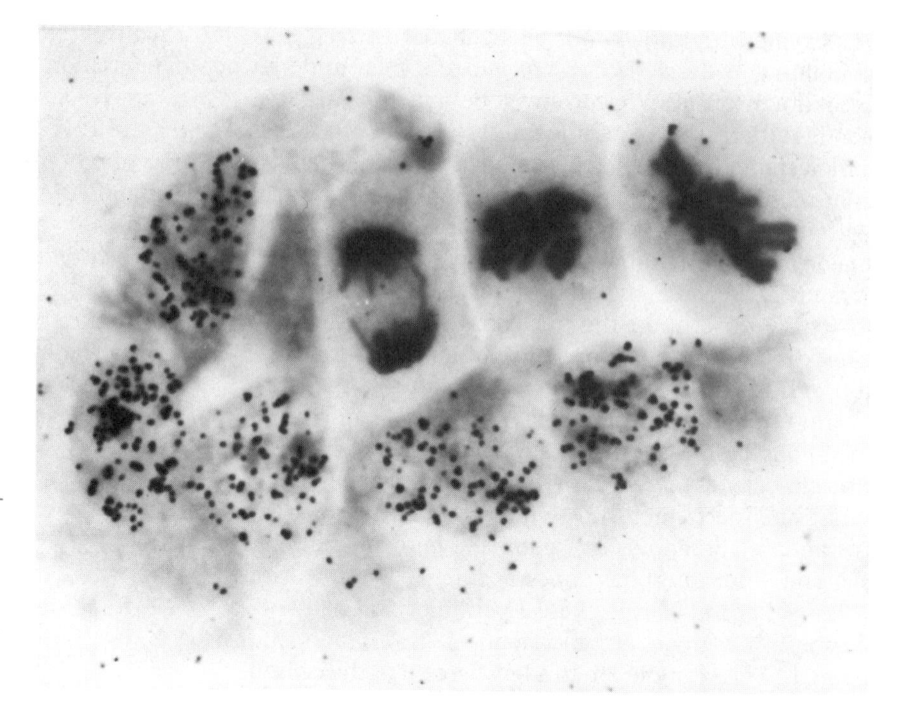

FIGURE 12-12 Autoradiograph of cultured animal cells incubated with ^{3}H-uridine for ten minutes. The five interphase cells have incorporated ^{3}H-uridine into RNA; no radioactive RNA has yet moved into the cytoplasm. The three cells in mitosis have not incorporated any ^{3}H-uridine into RNA.

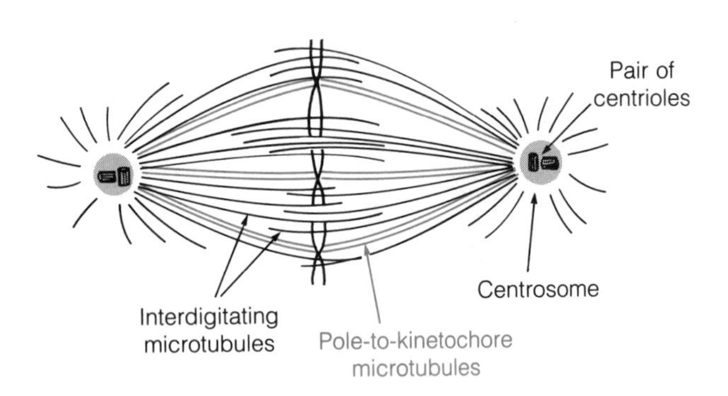

FIGURE 12-13 Schematic drawing of the mitotic spindle in an animal cell. Microtubules extend from the centrosomes toward the centrally aligned chromosomes. Two populations of microtubules are present. Centrosomal-kinetochore microtubules (red) and microtubules from opposite centrosomes that interdigitate in the central part of the spindle. A pair of centrioles is present in each centrosome.

metaphase occurs. The decrease in protein synthesis is not due to a decrease in the amount of mRNA, through normal turnover, since the life spans of almost all kinds of mRNA molecules in eukaryotes are much longer than the duration of mitosis. Some experiments suggest that protein synthesis slows because the ribosomes become incompetent to carry out protein synthesis. This may be a result of proteins that become bound to ribosomes, perhaps nuclear proteins released when the nuclear envelope breaks down.

A transition period between prophase and metaphase, called **prometaphase,** is characterized by a shuffling motion of the condensed chromosomes in which they are brought into alignment in a plane in the center of the cell, **the metaphase plate.** How the central alignment is accomplished is not understood, but these events require the **mitotic spindle.**

The mitotic spindle forms in late prophase and prometaphase; it consists of microtubules and other components (Figures 12-10, 12-13). The microtubules are assembled from tubulin molecules released by dissassembly of the cytoskeleton of the interphase cell. New tubulin synthesis occurs continuously in the cell but is not required for mitosis. A concomitant of this disassembly in most kinds of animal cells and in some unicellular eukaryotes like amoebae growing on a surface is a drastic change from a flat irregular shape to a spherical form (Figure 12-14). In animal tissues, shape changes are less pronounced because of physical constraints provided by surrounding cells. The cell wall of plant cells and fungi prevents changes in cell shape during mitosis.

The microtubules of the spindle originate from opposite poles of the cell and interdigitate in the central region of the cell (Figure 12-13). An amorphous material occupies some of the space between microtubules. The two poles from which the microtubules emanate are called the **centrosomes.** The molecular nature of centrosomes has not been determined. Pure centrosomes have not yet been isolated in large enough quantities to do biochemical analyses.

The center of a centrosome in animal cells is occupied by a pair of centrioles (Figure 12-13). Plant cells and many unicellular eukaryotes, *Amoeba proteus* for example, lack centrioles in their centrosomes. Even in animal cells the centrioles appear not to contribute to the function of the centrosome. Although cells of *Drosophila* normally have centrioles in their centrosomes, a culture of *Drosophila* cells has been obtained in which centrioles are absent and mitosis proceeds in the usual way. The presence of centrioles in centrosomes may simply represent a role of centrosomes in distributing centrioles equivalently to daughter cells. Centrioles serve as nucleating structures for assembly of microtubules in the formation of cilia and flagella (Chapter 14).

Centrosomes act as nucleating centers for assembly of tubulin into microtubules in the formation of the

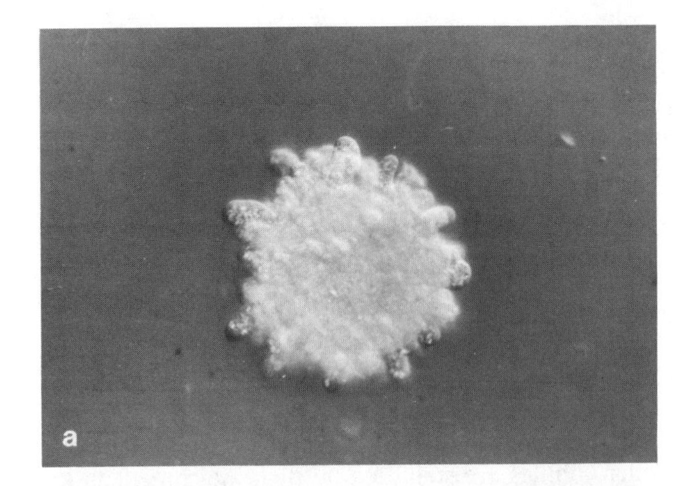

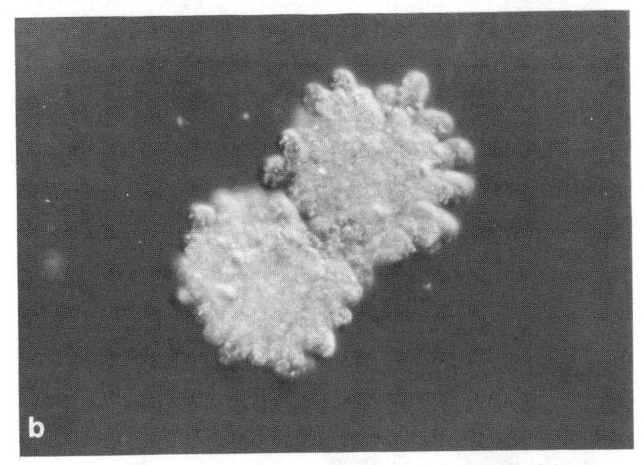

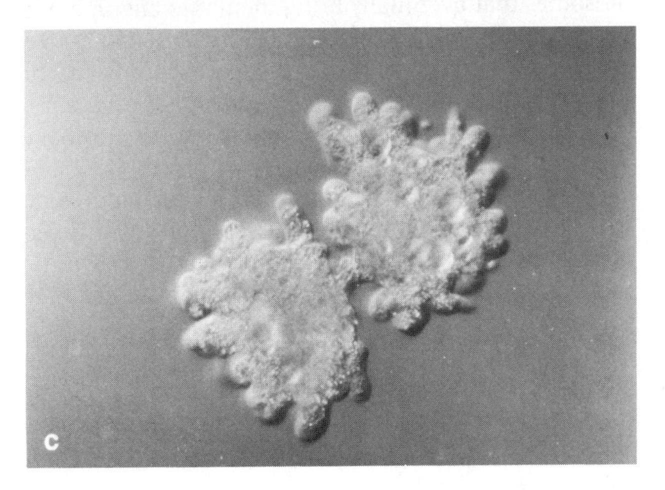

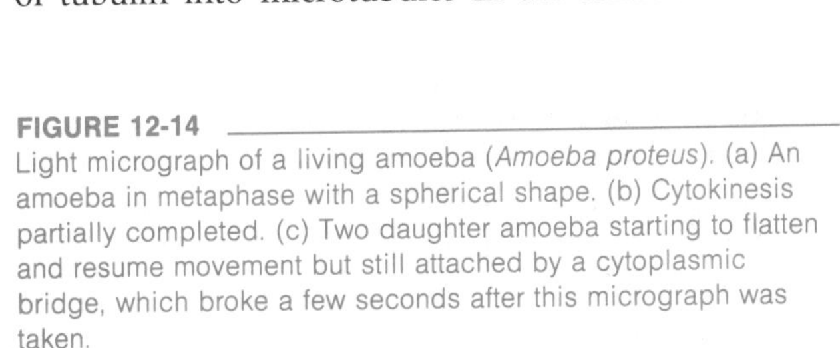

FIGURE 12-14
Light micrograph of a living amoeba (*Amoeba proteus*). (a) An amoeba in metaphase with a spherical shape. (b) Cytokinesis partially completed. (c) Two daughter amoeba starting to flatten and resume movement but still attached by a cytoplasmic bridge, which broke a few seconds after this micrograph was taken.

mitotic spindle. Tubulin molecules are continuously added to the free (distal) ends of the microtubules, which grow toward the opposite centrosome. As the growing microtubules invade the central region of the cell in late prophase, some of them terminate by attachment to the chromosomes at their kinetochores. Therefore, two types of microtubules are identifiable in the mitotic spindle (Figure 12-13): (1) interdigitating microtubules that grow out from a centrosome and continue through the central region of the cell, and (2) centrosome-to-kinetochore microtubules that connect kinetochores to centrosomes. Interdigitating microtubules form, disassemble, and reform continuously during mitosis (Chapter 6). Centrosome-to-kinetochore microtubules are stabilized by their attachment to kinetochores; attachment appears to act as a cap for microtubules (Chapter 6).

Both chromatids in a chromosome have a kinetochore. The two kinetochores in a pair of chromatids face outward on opposite sides of the chromosome (Figure 12-15). The two chromatids remain attached to each other at the region on their inner sides from kinetochores. This attachment region plus the kinetochores together comprise the centromere.

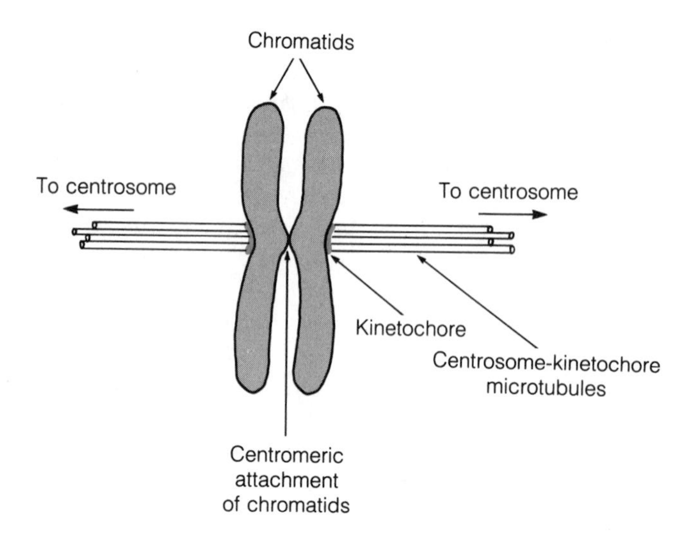

FIGURE 12-15
The plate-like kinetochore on a chromatid is composed of proteins and serves as an attachment site for microtutubles from a centrosome. Chromatids remain attached to each other at the centromere; detachment marks the beginning of anaphase.

Microtubules emanating from one centrosome attach to the kinetochore of one chromatid, and microtubules from the other centrosome attach to the kinetochore in the other chromatid. These attachments are essential for the back-and-forth motions of the chromosomes that eventually bring them into alignment in the metaphase plate. The completion of alignment defines arrival at metaphase. At this time, because of the microtubule attachments, one chromatid of each chromosome faces one centrosome and the other chromatid faces the opposite centrosome.

In somatic cells the chromosomes remain poised in mitotic metaphase for only a few minutes. However, during *meiosis* in many species of animals, chromosomes may remain in metaphase for months and even years. Mitotic metaphase ends and anaphase begins abruptly with the separation of the two chromatids in all the chromosomes simultaneously. The nature of the connection between chromatids, how the connection is broken, and the signal that brings it about are not known. Anaphase is defined as the stage in which the two groups of daughter chromosomes move to opposite poles; it takes only a few minutes (Figures 12-10, 12-16). Like the shuffling movement of condensed chromosomes during prometaphase, the movement of chromatids in anaphase is dependent on the centrosome-to-kinetochore microtubules. The movement of chromatids, which have now become daughter chromosomes, has been intensively studied for decades; nevertheless, the molecular mechanism is still unknown. A protein, named *kinesin,* that appears to act as a *motor* to propel cytoplasmic particles along microtubules in nerve axons (Chapter 14) has been isolated from the mitotic spindle of animal cells. Its association with microtubules in the spindle suggests the intriguing possibility that kinesin may serve to connect the kinetochore to microtubules and may somehow provide the motor that propels chromosomes on microtubules in the spindle. An explanation of how daughter chromosomes move must also account for the fact that the centrosome-to-kinetochore microtubules shorten in correspondence with poleward movement of the daughter chromosomes.

When the two groups of daughter chromosomes reach the two poles of the cell, telophase begins. In telophase the chromosomes decondense, and a nuclear envelope and nuclear lamina re-form around them. The envelope and lamina are reassembled from the components released by the earlier breakdown of the envelope and lamina in prophase. Dephosphorylation of lamins accompanies their assembly into the lamina. Simultaneously, the several other changes that occurred in prophase are reversed. The nucleolus re-forms around the nucleolar organizers of chromosomes, RNA synthesis resumes, and proteins released from the nucleus into the cytoplasm at the time of breakdown of the nuclear envelope in prophase quickly reaccumulate in the re-forming daughter nuclei. The mitotic spindle is disassembled, and the free tubulin molecules are reassembled into the cytoskeleton of the interphase cell. With completion of these telophase events, the nuclei enter the G1 period of the next cell cycle.

Cytokinesis

Cytokinesis is the division of the body of the cell to form two daughter cells. Cytokinesis together with mitosis comprises cell division. The term mitosis is sometimes used incorrectly as a synonym for cell division; mitosis refers only to nuclear division.

Cytokinesis is coordinated in time and physical orientation with mitosis. It begins in late anaphase or early telophase, and division of the cell occurs in a plane at a right angle to the long axis of the spindle, midway between the separated groups of daughter chromosomes. In cells that lack rigid walls (animal

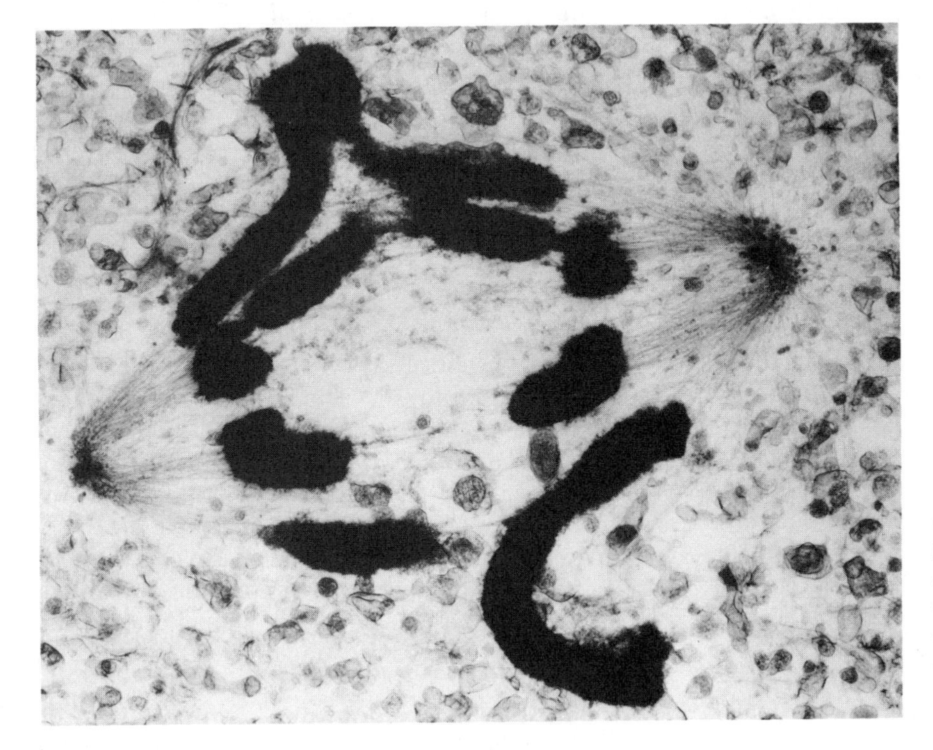

FIGURE 12-16
Electron micrograph of a section through an animal cell in anaphase. Microtubules extending from centrosomes to chromosomes are prominent. Centrioles are present in the centrosomes but are not clearly discernible. [Courtesy of J. Richard McIntosh.]

cells, protozoa, some algae) cytokinesis differs from the comparable process in cells with rigid walls (plant cells, some algae, and most fungi). The two mechanisms of cytokinesis are described in the following two sections, typified by animal cells and plant cells.

Cytokinesis in animal cells. Cytokinesis in animal cells is a contractile process. The contractile mechanism is contained in a **contractile ring** located just inside the plasma membrane (Figure 12-10). The ring forms around the circumference of the cell in the region where the edge of the metaphase plate used to be. The ring consists predominantly of a bundle of microfilaments assembled from actin. Also present in the contractile ring is myosin, one of the principal proteins that make up the contractile machinery in muscle cells (Chapter 14). An interaction between actin and myosin, accompanied by hydrolysis of ATP to ADP and P_i (see Chapter 14 for details) generates a contractile force that draws the contractile ring inward, forming a furrow in the cell surface and pinching the cell in two. Part of the evidence that myosin is involved is the inhibition of cytokinesis (but not of mitosis) when antibody molecules that specifically bind to myosin are injected into single animal cells. To accomplish cytokinesis the contractile ring must be attached to the inside of the plasma membrane. As the ring contracts, it retains the same thickness, which means that actin and myosin are released from the ring as it draws inward. The contractile ring reduces the connection between the two daughter cells to a narrow bridge that contains a compressed bundle of interdigitating microtubules of the mitotic spindle. This compact bundle, called the **mid-body,** persists as the last remnant of the spindle (Figure 12-10) and then disappears as the two cells finally break apart from each other. In animal cells the main part of cytokinesis occupies only a few minutes; the narrow bridge with its mid-body may persist for an hour or more in some kinds of cells. In some protozoa, such as the small amoeba *Acanthamoeba,* the contractile ring contracts extremely rapidly, and all of cytokinesis takes place in 40 to 50 seconds.

Even before cytokinesis is completed in animal cells, interphase cytoskeletons are assembling in the incipient daughter cells. As a consequence, the daughter cells flatten (if growing on a surface) and resume the amoeboid motion that stopped when the parental cell became spherical as it entered mitosis.

Cytokinesis in plant cells. Plant cells do not divide by means of a contractile ring. Instead they assemble plasma membranes and cell walls between the two daughter cells in the plane of the metaphase plate. The events of this assembly process have been defined in detail by observations in the light and electron microscopes.

When the daughter chromosomes have left the metaphase plate as a result of their anaphase movements, membranous vesicles derived from the Golgi complex accumulate in the plane where the metaphase plate used to be (Figure 12-17). The vesicles are drawn into the plane from both sides of the spindle, apparently by the action of microtubules. The small vesicles fuse to form a large, flat vesicle that grows perpendicular to the spindle until it reaches the plasma membrane. The accumulated vesicular material is called the **phragmoplast** (*phragma* = enclosure, *plast* = forming). By their fusion the vesicles of the phragmoplast form a double plasma membrane. The two plasma membranes fuse peripherally with the main plasma membrane of the cell, thereby forming two separate daughter cells. Thereafter, a double cell wall is formed between the two plasma membranes by the daughter cells.

Although mitosis and cytokinesis usually occur together, this is not always the case. For example, in *Drosophila* eggs early development consists of many successive and synchronous mitoses in the fertilized egg without accompanying cytokinesis. This creates a single cytoplasmic mass with several thousand nuclei. At a precise point in development the nuclei migrate to positions close to the surface, and plasma membranes form to create simultaneously several thousand uninucleated cells. Occasionally even in animal cells cytokinesis may fail to occur, creating binucleated cells.

Distribution of cytoplasmic components. In most instances cytokinesis divides a cell into roughly

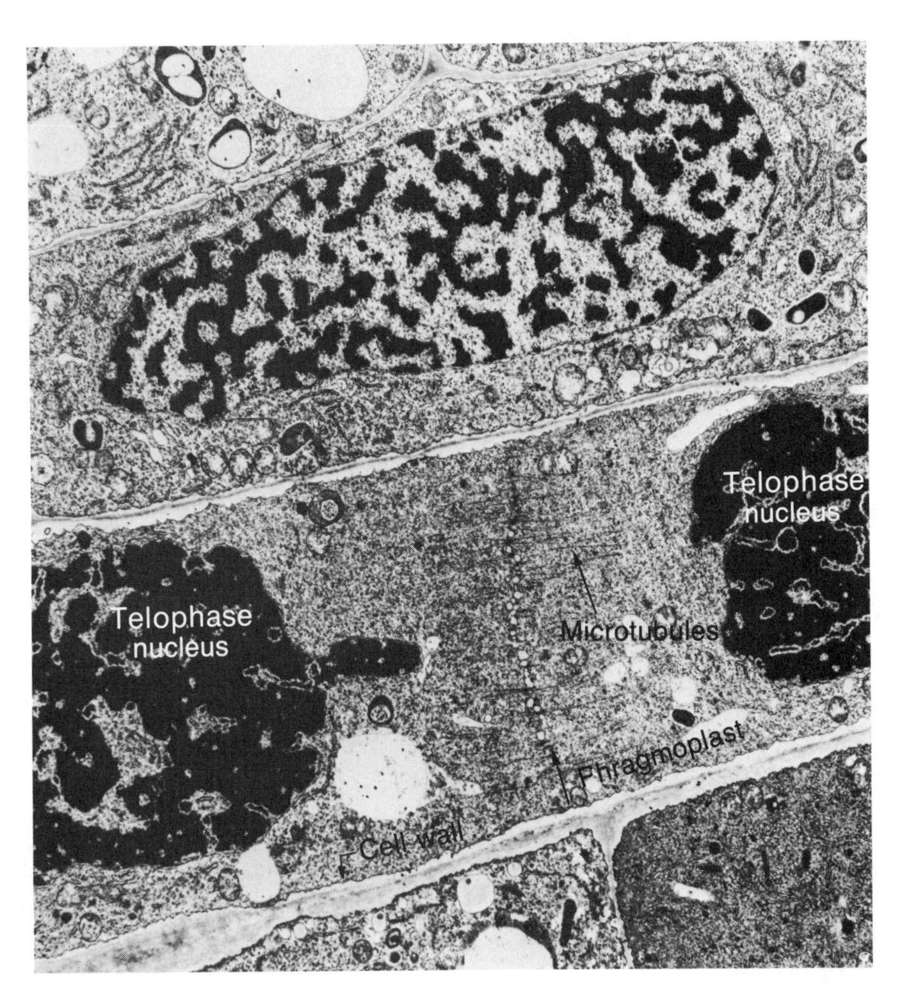

FIGURE 12-17 Electron micrograph of wheat plant cells. The lower cell is finishing mitosis; the two nuclei are in telophase. A phragmoplast has formed between the telophase nuclei and marks where a cell wall will form. Microtubules left from the mitotic spindle are visible. [Courtesy of Jeremy Pickett-Heaps.]

equal-sized daughters. Exceptions occur in early cleavage stages of embryos of many species, where cytokinesis produces large and small daughter cells in precisely defined patterns. Even in the usual case of equal division, cytokinesis often does not divide a cell into two exactly equal parts; one cell receives slightly more of the cytoplasm than the other. This is of no major consequence since most cytoplasmic organelles like mitochondria and ribosomes are present in many copies and each cell always receives some of these. Failure of a daughter cell to receive at least one mitochondrion (and at least one chloroplast for plant and algal cells) would likely be lethal since these DNA-containing organelles form by fission of preexisting organelles and cannot form *de novo*. A cell lacking a Golgi complex or a contractile vacuole can form new ones *de novo*.

SYNCHRONIZATION OF THE CELL CYCLE

The study of cell reproduction is based predominantly on measurements of the interrelated molecular events that occur during the cell cycle. Measurements on individual cells are sometimes possible, but usually they are difficult or impossible. Large populations of cells in which all the cells traverse the cycle in step with one another, that is, in synchrony, must be used for most kinds of molecular studies.

In some systems synchrony of the cell cycle in a population occurs naturally, a notable example being the cleavage of egg cells fertilized at the same time. Alternatively, a number of methods have been developed for obtaining synchronous masses of cells from populations in which the cells are originally completely out of synchrony with one another.

Synchronization of Prokaryotic Cells

The simplest way to synchronize the cycles of bacterial cells like *E. coli* in a culture is to allow a population to enter the stationary phase of culture growth. As a result of depletion of one or another nutrient in the medium, cell growth is blocked, and all the cells arrest in the cell cycle between cell division and initiation of DNA replication. When the cells are removed from the exhausted medium, for example by sedimentation by centrifugation, and introduced into fresh medium, they all initiate DNA replication over an interval of many minutes and subsequently divide somewhat synchronously. The cells may retain a trace of synchrony for one to two cycles, but quickly become asynchronous because of variations in generation times of the individual cells.

The only method for obtaining highly synchronous populations of *E. coli* is to collect proliferating cells on a membrane filter. Nutrient medium is passed through the filter slowly and continuously so that the medium emerges from the side of the filter on which the bacteria have been collected. Initially, loosely attached cells are washed off, but subsequently only new daughter cells are released, and these can be collected as a synchronous population (Figure 12-18). Using a series of filters, up to 5×10^8 synchronous cells per minute can be collected repeatedly. The increase in cell number for bacteria synchronized in this way is shown in Figure 12-19. By the first division after collection of synchronous cells, the degree of synchrony has seriously deteriorated, as indicated by the spread in the doubling of the cell population over a 15-minute interval.

Synchronous populations of cells make possible measurements of the transcription of particular genes, synthesis of particular proteins, or other activities in relation to progression of the cell cycle. Synchronous cells were used in the experiment described in Chapter 11 for determining that DNA replication is initiated bidirectionally in the chromosome of *E. coli.*

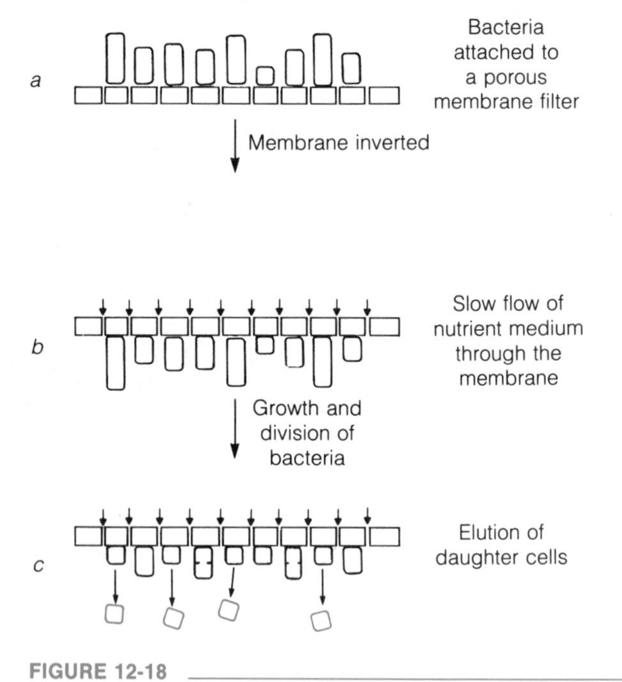

FIGURE 12-18
Synchronous bacteria obtained by the membrane filter method. (a) Bacteria are centrifuged onto a porous membrane filter, to which they remain attached. (b) Nutrient medium is allowed to flow slowly through the filter. (c) The cells grow and divide, and daughter cells are released into the medium and collected, for example, by centrifugation of the medium.

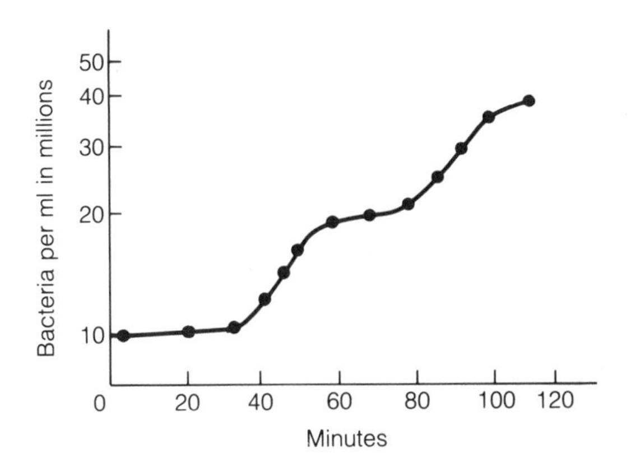

FIGURE 12-19

Cell increase in synchronous bacteria obtained by the membrane filter method shown in Figure 12-18. The original high degree of synchrony at 0 time has deteriorated considerably by the first cell doubling and further decreased by the second cell doubling.

Synchronous Eukaryotic Cells

Many strategies have been developed for obtaining synchrony of the cell cycle in eukaryotes. These methods fall into three categories: natural synchrony, chemically induced synchronization, and mechanical selection to obtain synchronous cells.

Natural synchrony. Fertilization of an egg cell is followed in many species by extremely precise synchrony of division of the cells of the early embryo (**blastomeres**). In amphibians, synchrony lasts for more than ten cell cycles. These cycles are rapid and lack G1 and G2 periods. They become asynchronous suddenly at a particular cell division, for example, division 12 in *Xenopus* embryos, at which time the embryo contains 2^{12} (4096 cells). In the thirteenth cycle both G1 and G2 periods appear, and asynchrony is already apparent at the thirteenth division. Within a few more cycles the cells become completely asynchronous.

The divisions in embryos follow an exact time schedule, so groups of eggs fertilized at the same time undergo subsequent cell divisions with each other (Figure 12-20). Fertilization of eggs therefore provides large numbers of cells reproducing with synchronous cycles. An example of the usefulness of such material is the study of the role of MPF in controlling initiation of mitotic condensation of chromosomes discussed earlier in the chapter.

Natural synchrony also occurs in the plasmodial slime molds. A **plasmodium** is a single large compartment of cytoplasm containing multiple nuclei. The slime mold *Physarum* grows as a flat cytoplasmic mass

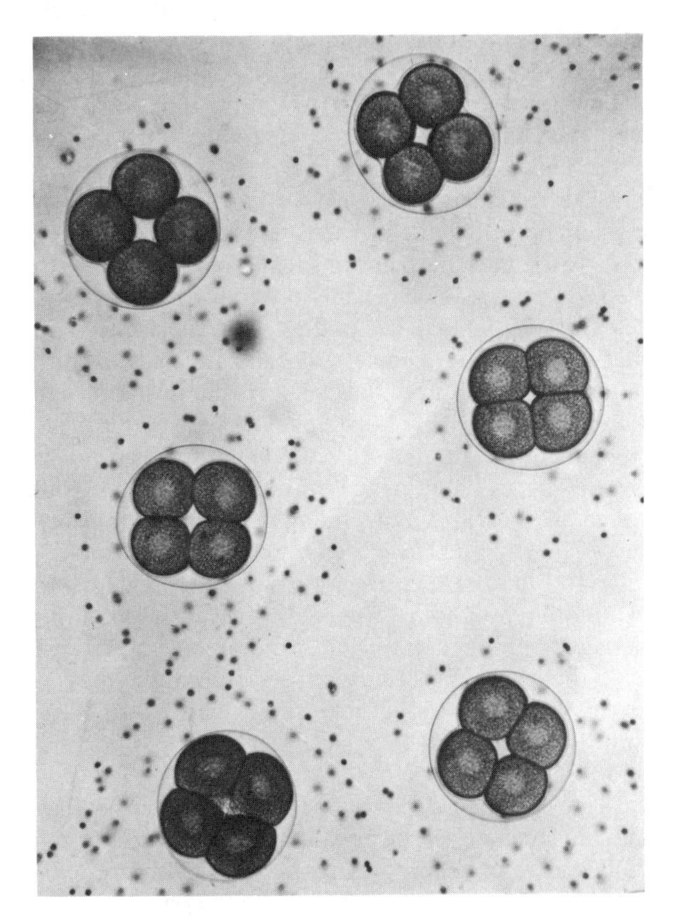

FIGURE 12-20

Light micrograph of dividing eggs of the sanddollar (*Dendraster excentricus*). Each embryo is in the four cell stage. Nuclei are visible as light round areas in the center of each cell. The embryos are enclosed in a tough protective membrane (the chorion) and are surrounded by a jelly coat. [Courtesy of Victor D. Vacquier.]

that can be many centimeters in diameter and contain tens of thousands of nuclei. As the plasmodium grows (without cytokinesis) all the nuclei proceed through G1, S, G2, and M in close to perfect synchrony. *Physarum* has been used extensively to study the nuclear events of the cell cycle.

Another type of natural synchrony occurs in organisms in which cell division is controlled by a biological timing mechanism known as a **circadian clock** (*circa* = about; *diem* = a day). In these organisms cell division in a population of cells (or population of unicellular microorganisms) occurs within an interval of a few hours at a particular time, often during the middle of the night.

The circadian clock, whose biochemical mechanism is completely unknown, is set to *local time* through a photoreceptor that transmits temporal information to the clock by sensing when it is day and when it is night. The clock is then coupled to the cell division process so that cells that have completed G2 and are ready to enter mitosis must wait until the clock reaches its permissive *temporal window* before dividing.

Synchronization of cell division by circadian clocks is just one example of the widespread occurrence of circadian (daily) rhythms among eukaryotic organisms. Circadian rhythms have been observed in both unicellular and multicellular organisms in such functions as cell motility, phototaxis, photosynthesis, and bioluminescence (production of light by an organism), as well as in the levels of specific enzymes, substrates, and cofactors and in metabolic processes such as DNA replication, RNA and protein synthesis, fatty acid synthesis, and respiration.

The circadian clock serves to couple the physiology and biochemistry of the organism to fluctuations in the daily environmental cycle, particularly light and temperature. The adaptive value of the biological oscillation is clear in the case of rhythms such as photosynthesis and bioluminescence. In many organisms there is a daily rhythm in the synthesis of components of the photosynthetic machinery that results in a rhythm in photosynthetic capacity with its maximum in the middle of the day. On the other hand rhythms in bioluminescence show a maximum in the middle of the night as a result, at least in part, of a circadian rhythm in the synthesis of luciferase, the key enzyme in the light-producing chemical reaction in the cell. The adaptive significance of other rhythms such as cell division are not as obvious, although some ecologists have suggested that different species could occupy different *temporal niches* by having their cell cycles phased differentially throughout the day.

Chemically induced synchrony. Blocking cells at a single point in the cell cycle with an inhibitor or by deprivation of an essential nutrient induces synchrony by chemical means. An example was introduced earlier in the blocking of initiation of DNA replication with fluorodeoxyuridine (FUdR), an analog of thymidine. Fluorodeoxyuridine is converted by the cell to FUdR monophosphate (FUdRP) by thymidine kinase, just as thymidine is converted to thymidine monophosphate (TMP) by the same enzyme. FUdRP is a powerful inhibitor of the enzyme TMP synthase. The pathway for TMP synthesis is shown in detail in Figure 12-8 because it is important for understanding the FUdR effect on the cell cycle of animal cells as well as the effects of two other inhibitors. The action of these inhibitors has additional significance because two of them are used as chemotherapeutic agents for combatting cancer by preventing DNA replication in cancer cells.

As shown in Figure 12-8, TMP is phosphorylated to thymidine diphosphate (TDP) and then to thymidine triphosphate (TTP) in two successive steps. Thymidine triphosphate is required for replication of DNA. A block in its synthesis prevents cells from entering DNA replication and stops DNA replication of cells already in the S period when the block is imposed.

Thymidine monophosphate can be synthesized from two precursors, thymidine and deoxyuridine monophosphate (dUMP). Neither prokaryotes nor eukaryotes can synthesize thymidine from simpler precursors. Thymidine is only produced by the breakdown of TMP, for example, when DNA is degraded. For this reason thymidine kinase is called a "salvage" enzyme because it allows thymidine produced by degradation processes to be salvaged.

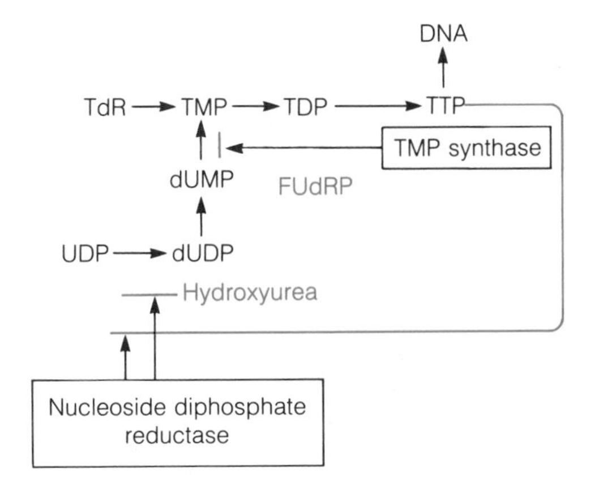

FIGURE 12-21
Three ways to inhibit DNA synthesis by inhibiting the TTP biosynthetic pathway. Such inhibition of DNA synthesis is used to synchronize animal cells in culture.

Ordinarily, thymidine is not available to cells in sufficient quantity to support DNA replication. Cells depend instead on synthesis of TMP from dUMP. All prokaryotes and eukaryotes produce dUMP by dephosphorylation of dUDP (Figure 12-8); dUDP is synthesized from uridine diphosphate (UDP), a process catalyzed by nucleoside diphosphate reductase in which the ribose of UDP is converted to deoxyribose. In turn, UDP can be synthesized by most kinds of cells from simpler precursors.

There are several steps at which synthesis of TMP can be blocked by inhibitors, thereby preventing DNA replication and stopping cell reproduction. Inhibition at two of these steps is useful for synchronizing cells. Inhibition by FUdRP has already been mentioned. It binds tightly to the active site of TMP synthase and prevents the enzyme from catalyzing the addition of a methyl group to dUMP to form TMP (Figure 12-21). As a result, cells reaching the end of the G1 period are prevented from entering the S period. Cells already in the S period when FUdRP appears are stopped in S and are killed. We do not understand why cells die when blocked part way through the S period. Inhibition of TMP synthesis by FUdRP is easily overcome by adding thymidine to the medium. Thymidine is converted to TTP and all the cells blocked at the end of the G1 period by FUdRP initiate DNA replication synchronously.

The second useful inhibitor is hydroxyurea (HU) (Figure 12-21). It inhibits the action of nucleoside diphosphate reductase, the enzyme that catalyzes both the synthesis of dUDP from UDP and synthesis of deoxycytidine diphosphate (dCDP) from cytidine diphosphate (CDP). Without dUDP and dCDP a cell cannot make TTP and deoxycytidine triphosphate (dCTP), and DNA cannot replicate. The result is accumulation of cells at the G1/S border. Washing of cells in fresh medium removes HU from the cells, relieves the inhibition, and allows cells to enter DNA replication in synchrony.

Thymidine can also be used to synchronize animal cells in culture in the following way (Figure 12-22). Several enzymes involved in production of deoxynucleoside triphosphates (dNTPs) are regulated such that levels of dNTPs adequate to support DNA replication without overproduction are maintained. Nucleoside diphosphate reductase is allosterically inhibited (Chapter 3) by TTP (Figure 12-21). When the level of TTP in the cell rises, it inhibits the reductase and prevents the synthesis of both dCTP and TTP. Thymidine added to the medium of a culture of animal cells is readily taken in and phosphorylated to TTP in an essentially unregulated manner. If a high concentration of thymidine is added to medium (e.g., 2 mM), then the resulting high concentration of TTP in the cell inhibits nucleoside diphosphate reductase. This stops synthesis of TTP, which is of no consequence since the cell is already flooded with TTP. However, synthesis of dCTP is also inhibited, and without it DNA cannot replicate. The inhibition of the enzyme is not total, and cells continue to produce a trickle of dCTP. As a result cells reaching the end of G1 enter the S period but replicate DNA extremely slowly and accumulate in the very first part of the S period (Figure 12-22). Cells already in S when a high concen-

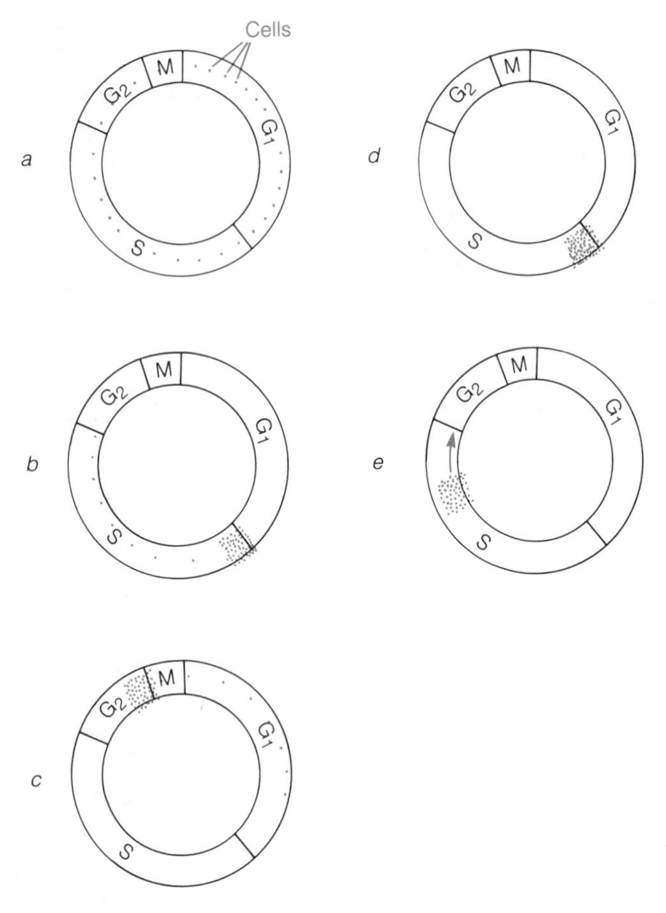

FIGURE 12-22
The double-thymidine-block method for synchronizing animal cells in culture. (a) Cells distributed throughout the cell cycle. (b) Treatment with 2 mM thymidine for 60 percent of a generation time accumulates all cells in S, with most in early S. (c) Release of the thymidine block allows all cells to progress into G2, M, and G1. (d) A second block with 2 mM thymidine accumulates all cells in early S. (e) Release of the second thymidine block allows the cells to proceed synchronously through the cycle.

tration of thymidine is added to the medium continue progress through DNA replication, but at a very slow rate. Cells in G2, M, or G1 are not affected in their rate of progress until they reach S.

If the cells are exposed to a high concentration of thymidine for an interval equal to the sum of the lengths of G2, M, and G1, then all cells in these periods reach the S period and become tightly bunched in early S. Cells already in DNA replication remain scattered throughout the S period and progress very slowly. When the medium with the high thymidine concentration is replaced with normal medium, the intracellular concentration of TTP falls rapidly, relieving the inhibition of synthesis of dCTP. Incubation of the cells in the fresh medium for a time equal to the length of the S period allows all the cells (theoretically 100 percent of the cells are in S) to complete replication. The cells originally scattered at various positions in S will now occupy positions in various parts of G2, D, and G1. The large fraction of cells originally arrested in early S will reach the G2 period. Thus, all the cells in the culture are in G2, D, or G1. When a high concentration of thymidine is again added to the culture, the entire population of cells will accumulate in early S.

This foregoing technique, called the **double-thymidine-block method** of synchronization, makes it possible to impose an experimentally useful degree of cell-cycle synchrony via a normal metabolic regulatory mechanism instead of a toxic drug. The double-thymidine-block method avoids some of the uncertainty that attends the use of drugs. The latter may have undetected effects perturbing the very processes that the researcher is attempting to measure in the cell cycle.

Mechanical selection of synchronous cells. Synchronous animal cells can also be obtained by mechanical separation of cells according to their positions in the cell cycle. One such method is based on selection of animal cells in mitosis. Animal cells generally attach to the surfaces of culture vessels and form monolayers (Figure 12-23). In attaching they flatten and therefore have low profiles. During mitosis the cells disassemble microtubules (composed of tubulin) of the cytoskeleton and assemble mitotic spindles. The cells lose their flat shape, become spherical, and remain tethered to the surface only by thin strands of cytoplasm (Figure 12-24). With a spherical shape mitotic animal cells protrude up into the medium farther than interphase cells do and, in contrast to the flat interphase cells, are only loosely attached. Shaking the culture vessel causes the

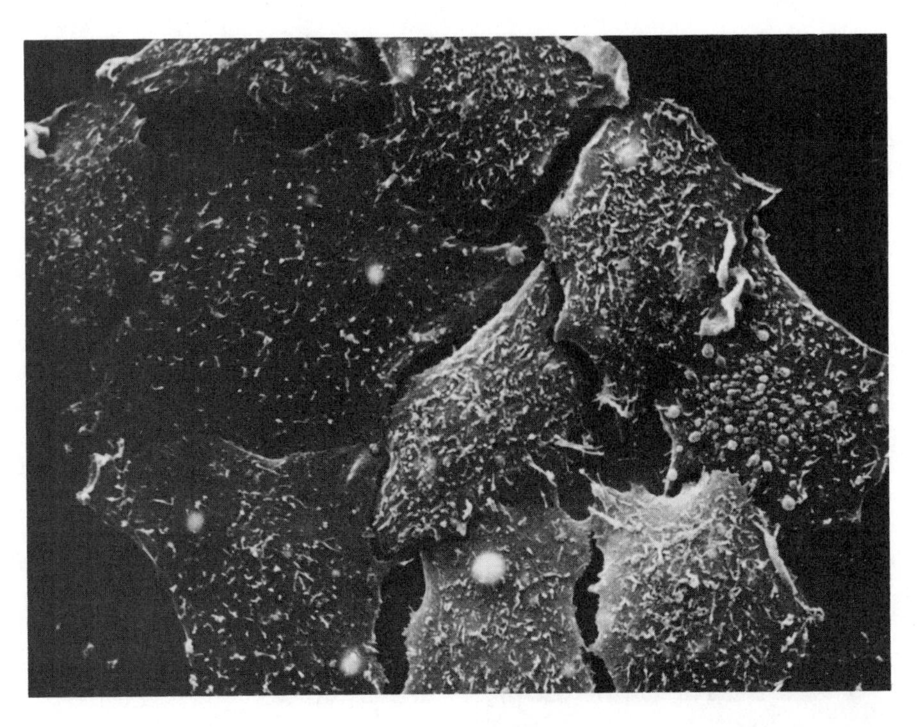

FIGURE 12-23
Scanning electron micrograph of part of a monolayer of cultured animal cells in interphase. The cells are flattened and firmly attached to the surface of the culture vessel.

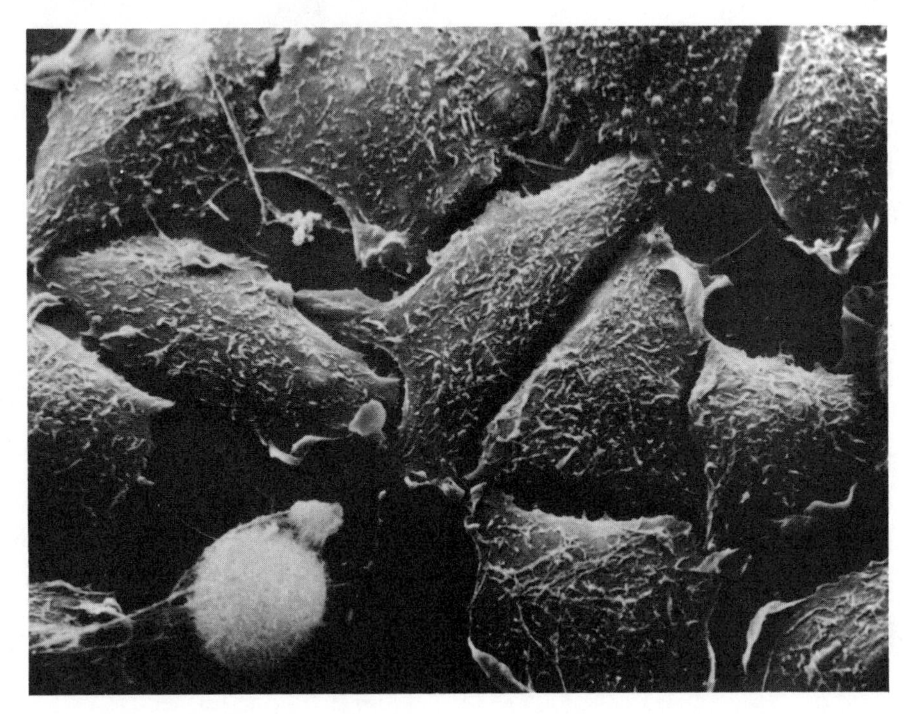

FIGURE 12-24
Scanning electron micrograph of cultured animal cells. The spherical cell in the lower left is in mitosis. The flattened cells are in interphase.

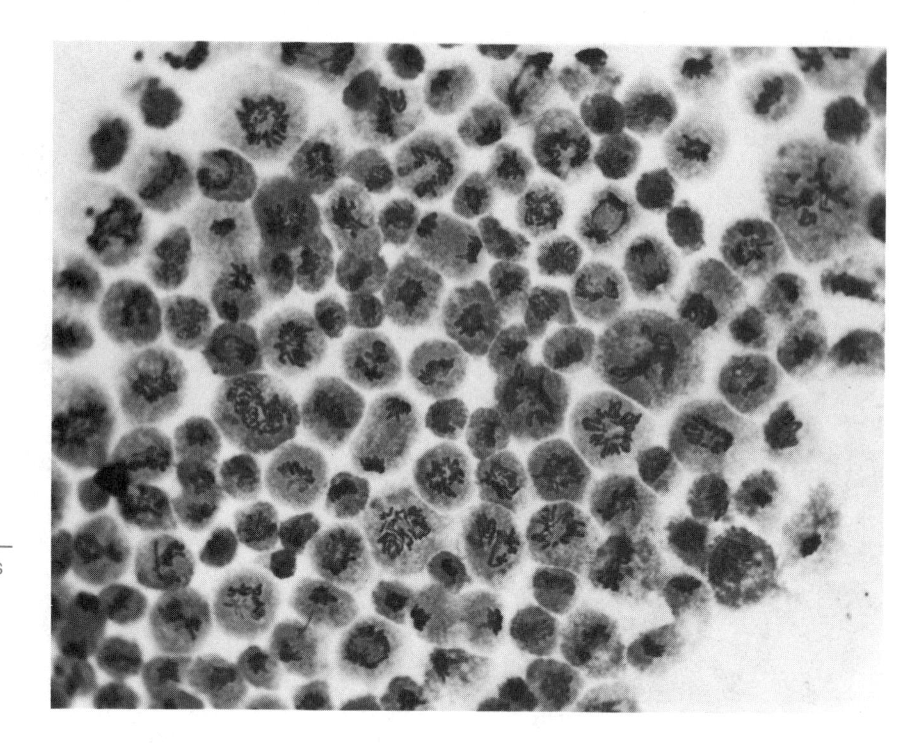

FIGURE 12-25
Light microscope photograph of mitotic cells collected by shaking a culture vessel. The flow of medium dislodges mitotic cells selectively from the surface of the culture vessel. They can then be concentrated by centrifugation of the medium.

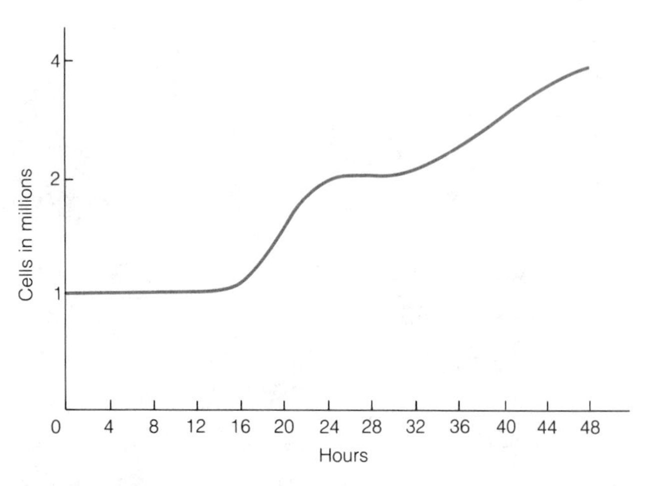

FIGURE 12-26
Increase in cell number over two cell cycles in a population of synchronous mammalian cells obtained by mitotic selection. The experiment starts with one million new daughter cells, which proceed through interphase and divide into two million cells over an 8-hour interval. Subsequently four million cells are formed over a 16-hour interval with only a trace of synchrony remaining.

medium to flow across the monolayer of cells, and the mitotic cells are preferentially dislodged. The medium is removed from the culture, and the mitotic cells are collected by sedimentation in a centrifuge and cultured. By this procedure a highly synchronous population of cells, initially all in prometaphase, metaphase, anaphase, and early telophase (Figure 12-25), is obtained that enters G1 synchronously and can be used for studies of the cell cycle. The original culture that yielded the mitotic cells can be incubated with fresh medium and shaken again, as more cells reach mitosis, to provide multiple synchronous populations. In addition, the double-thymidine-block method can be used for initial synchronization, and the mitotic selection method applied when the synchronized cells arrive at mitosis, a procedure that greatly increases the yield of mitotic cells.

Decay of Synchrony in Cell Populations

The best synchronization methods depend on creating synchronous populations of cells at the G1/S border or at mitosis. The usefulness of these methods is limited by the rapid disappearance of synchrony in the population in subsequent cell cycles. Figure 12-26 shows the increase in cell number in a culture initiated with mammalian cells selected at mitosis. The cells begin to traverse the cycle in almost perfect synchrony. When they arrive at the next division, they are not so tightly in step with one another, doubling in number over an interval of several hours. The next doubling in cell number occurs with only a trace of the original synchrony, that is, the synchrony has decayed during traverse of the cycle. The third doubling in cell number occurs with no synchrony evident: increase in cell number is uniform with time and the curve is perfectly straight. Essentially the same result is obtained when cells are synchronized at the G1/S border by blocking DNA replication.

The decay in synchrony is essentially the same in every type of prokaryotic and eukaryotic cell. It occurs because of variation in the rate with which individual cells traverse the cell cycle. A distribution of generation times is shown for several hundred individual *Tetrahymena* in Figure 12-27. All of these cells descended from a single starting cell, and hence they are genetically identical. The cells were grown in the same nutrient medium. Thus, the variation in generation times does not stem either from genetic difference or differences in nutrient availability or in other culture conditions. Generation times ranged from 84 to 149 minutes with an average of 111 minutes. When the cells with the shortest and longest generation times were used to start two new clones, the distribution of generation times and the average generation time in the descendants of these two cells were the same as in the original clone in Figure 12-27. Hence, shorter or longer generation times are not inherited properties, but instead represent random fluctuations in the behavior of individual cells from one cell cycle to another.

Random variation in generation times is due almost entirely to variation in the length of the G1 period. The combined lengths of the S, G2, and D periods is constant under a given set of culture conditions—in mammalian cells typically about 12 hours. The variation in the length of G1 is easily measured by incorporation of ^{3}H-thymidine in a mammalian cell population synchronized by mitotic selection. The cells complete division, traverse G1, and begin to incorporate ^{3}H-thymidine when they reach the S period. Entry into the S period can be monitored by sampling the population at regular intervals and determining by autoradiography the percentage of cells that is labeled (**labeling index**). Figure 12-28 shows the course of entry of cells into the S period. The first cells reached S between one and two hours after mitosis. The labeling index rose as more and more cells completed G1 and reached nearly 100 percent at about 10 hours, by which time the first cells that reached S were now leaving S and entering the G2 period. Thus, the length of G1 varied from one to almost 10 hours in this experiment, with an average of about 6.5 hours. This variation in G1 length almost completely accounts for the variation in generation times among individual cells and the rapid decline in synchrony in an initially synchronous population of cells.

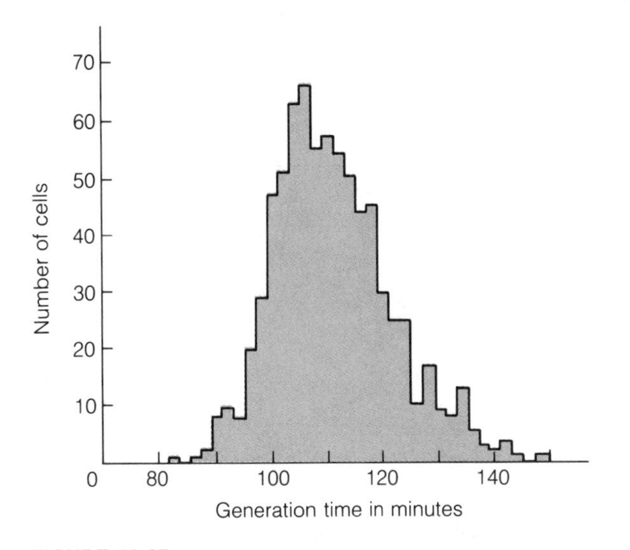

FIGURE 12-27
Distribution of generation times for *Tetrahymena*.

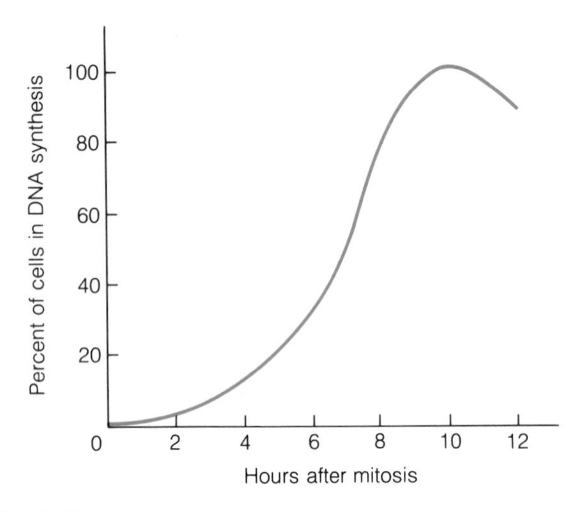

FIGURE 12-28
Entry into DNA synthesis of cells in a synchronous population obtained by mitotic selection. The curve shows the variability of the G1 period of different cells.

Variation in the length of G1 periods for individual cells occurs in a population in which the *average* G1 length and therefore the *average generation time* are constant. However, different kinds of cells in a multicellular organism usually have distinct and different average generation times. Such differences in average generation time are due to differences in the *average* length of the G1 periods in the different populations of cells. These phenomena, variations in individual G1 periods in a homogeneous population of cells (Figure 12-28) and differences in average G1 periods among distinct cell populations, should be carefully distinguished. Differences in average G1 periods represent regulation of reproductive rates of cell populations, as discussed below.

CELL-CYCLE GENES

Progress through the cycle is marked by changing activities in the cell, primarily centered around DNA replication and cell division. These activities require expression of a particular set of genes, called **cell-cycle genes.** For example, the initiation of DNA replication is accompanied by the turning on of genes encoding histones and enzymes for synthesis of deoxynucleoside triphosphates. The initiation of chromosome condensation at the start of mitosis is dependent on expression of the gene encoding MPF. Other gene products are required for progress through other parts of the cycle, but little is known about these genes or the action of their products. One of the major unsolved questions is how DNA replication is initiated. Is it triggered by expression of a particular gene and, if so, what regulates expression of the gene?

More is known about cell-cycle genes in the budding yeast than in any other cell type (Figure 1-10 and Chapter 7). About 50 different genes have been identified by **conditional mutations.** A conditional mutation is one that affects the activity of the gene product under one culture condition but not another. Almost all conditional mutations discovered so far are temperature-sensitive, or **Ts mutations.** In most of these the mutation is expressed and results in a mutant phenotype at a temperature high in the normal temperature range for the cell (Figure 12-29). Such Ts mutations are said to be heat-sensitive. A few Ts mutations are expressed as a mutant phenotype at a lower temperature and are normal at a higher temperature; these are cold-sensitive mutations. In most cases, Ts mutations are the result of an amino acid substitution that changes the stability of the three-dimensional configuration of the gene's protein product, making it nonfunctional above a certain temperature (in the case of a heat-sensitive mutation).

A heat-sensitive mutation in a cell-cycle gene causes the cell to arrest at a point in the cycle at which the gene product is essential for progression. At a lower temperature the gene product is active, so the cell cycle can proceed normally. The 50 cell-cycle genes identified in yeast by Ts mutations are crucial for progress at many points in the cycle. Some cell-cycle genes are required for initiation of DNA replication, some for progress through the S period, others function at various steps in mitosis, and still others are essential for cytokinesis. The analysis of the function of cell-cycle genes

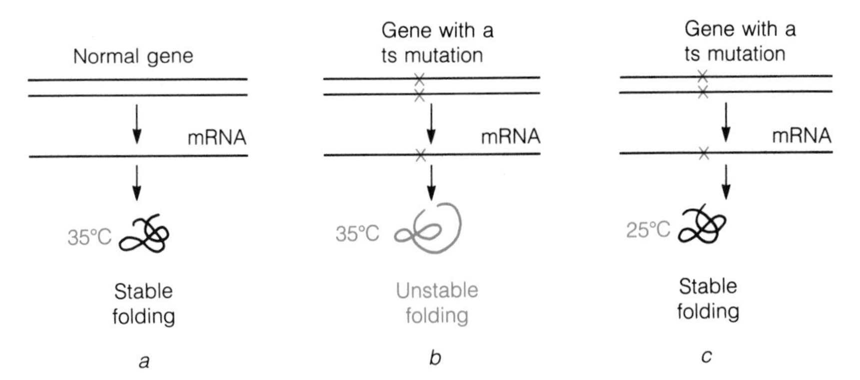

FIGURE 12-29
Effect of a Ts mutation. (a) Normal version of a gene and its stably folded polypeptide product in a cell at 35°C. (b) A mutated version of the gene. The polypeptide product does not fold stably at 35°C. (c) The same mutated gene as in (b). At the lower temperature of 25°C the polypeptide folds stably.

will eventually help elucidate how the cell is carried forward through its cycle.

Important examples of cell cycle genes identified by Ts mutations in yeast are six involved in control of entry into DNA replication. When any one of these genes suffers a Ts mutation, the cell cultured at 37°C is blocked at a point in the cycle located shortly before the start of the S period (Figure 12-30). The same cell cultured at 23°C is not blocked. The point in the cycle identified by these six mutated genes is called *Start* in yeast cells. When normal yeast cells at 37°C (or mutant yeast cells at 23°C) are starved for an essential nutrient, for example, phosphate, they arrest at Start. Similarly, prior to mating, yeast cells are arrested at Start by mating hormone (described a little later in this chapter). The products encoded by the six Start genes identified by Ts mutation probably form a molecular switch that controls entry into the S period. One of these genes, called CDC-28 (CDC = cell division cycle), encodes a protein kinase, suggesting that operation of the switch involves phosphorylation of proteins.

REGULATION OF REPRODUCTION OF CULTURED CELLS

All normal prokaryotic and eukaryotic cells regulate their own reproduction in relation to nutrient conditions. When nutrients are insufficient to allow growth, the cells become blocked at a particular point in the cycle, usually in the G1 period. The cycle in a few organisms, such as *Amoeba proteus* and *Physarum*, lacks a G1 period, and these cells arrest in G2. Cells arrested in G1 (or G2) may survive for weeks, months, or even years in that state without loss of viability. Cells experi-

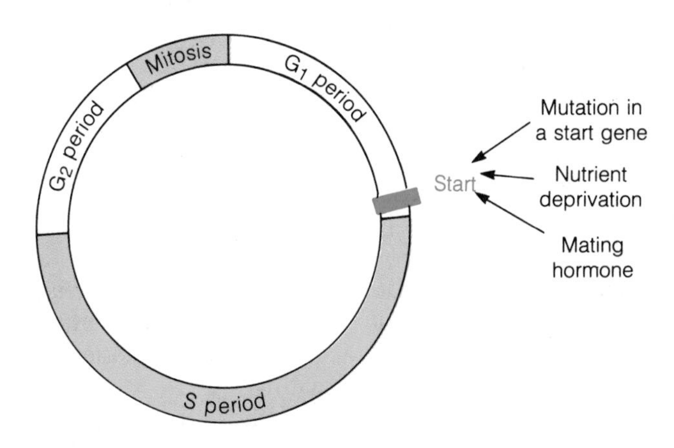

FIGURE 12-30
The Start switch in the cell cycle of the budding yeast is affected by mutation in a Start gene, nutrient deprivation, and mating hormone.

mentally arrested in S with drugs quickly lose viability, usually within hours. Apparently, the normal regulatory mechanism ensures that the cell is arrested in a particular metabolic state in which it may remain without jeopardy, but what defines this state is unknown. Nor is it known why cells die when arrested in S for more than a short time.

In addition to nonspecific regulation by nutrient availability, the cell cycle may also be regulated by specific signals. Specific regulation occurs in some unicellular organisms, and an example in yeast is discussed in the next section. Furthermore, evolution of specific regulation of cell reproduction was essential in the evolution of multicellular plants and animals, since unrestrained cell reproduction would be disastrous for a multicellular organism. What has been learned about regulation of cells in tissues during development of multicellular organisms and in adults is discussed in much of the remainder of this chapter.

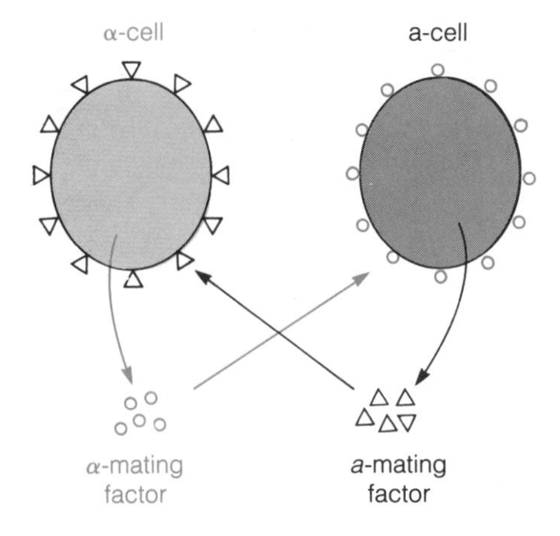

FIGURE 12-31
Mating in budding type yeast. An α-mating type cell secretes α-mating factor, which binds to receptors on the membrane of an a-type cell. Reciprocally, an a-type cell secretes a-mating factor, which binds to receptors on the membrane of a α-type cell. Binding of mating factors arrests the cell in G1.

Specific Regulation of the Cell Cycle in Yeast

Yeast, like most types of unicellular eukaryotes, can mate. Mating occurs between opposite mating types in a species. The mating type of a cell is determined by expression of a mating type gene. In the budding yeast cells are either mating type *a* or *α*.

Yeast cells proliferate as haploid cells (Figure 7-34). In mating, an *a* mating type cell fuses with an *α* type, producing a diploid cell. The diploid cell can multiply if nutrient conditions are adequate or, if starved, undergoes meiosis and forms four haploid spores. The spores are metabolically inert, protected by a tough wall, and highly resistant to heat, toxic chemicals, and drying. Spores remain viable for years. Under favorable nutrient conditions they germinate and again proliferate as haploid cells until the next mating.

Mating of yeast cells occurs only between cells that are in the G1 period. The two mating types cause each other to arrest in G1 by secreting a small polypeptide, a mating hormone, that arrests cells of the opposite mating type at a specific point in G1 called Start. A 12-amino-acid polypeptide, called ***α*-factor,** is secreted by *α*-type cells. This polypeptide binds to specific receptors on the surface of a-type cells and causes them to arrest at Start in G1 (Figure 12-31). The *α*-factor has no effect on *α*-type cells. Cells of *a* mating type secrete a polypeptide of ten amino acids—***a* factor**—that causes *α*-type cells to arrest at Start, but which has no effect on a-type cells. Thus, the two mating types signal each other to arrest at Start, following which they mate by fusing to make a diploid cell. The fused cells cease to produce mating hormones and also lose sensitivity to them and are released from Start. The diploid cell proliferates or undergoes meiosis depending on nutrient conditions.

The mating phenomenon in yeast is a simple model of how one cell may specifically affect the cycle of another cell, arresting it at a specific point in G1. The Start point in yeast is positioned 10 to 20 minutes before the initiation of DNA replication and is an on/off switch that controls entry into the S period. The Start switch can be operated by a number of signals. Presence of the a-type mating hormone or deficiency in any one of a variety of nutrients in the environment puts the switch in the *off* position. As long as nutritional conditions are adequate for cell growth and mating hormone is not present, the switch can be turned to *on*, and the cell completes its cycle.

Proliferation of Vertebrate Cells in Culture

Many kinds of normal cells from various tissues of animals proliferate in culture when supplied with amino acids, carbohydrates, vitamins, inorganic ions, and serum. Serum is blood from which all the cells and clotting factors have been removed. Serum provides several hormones and growth factors (discussed later in this chapter). Although the study of cultured cells has uncovered important facts about regulation of cell reproduction, the environment provided to cells in culture vessels is different from what is provided in intact tissues in ways that dramatically affect regulation of the cell cycle. In tissues in the body, cell reproduction is precisely regulated by mechanisms that are still poorly understood. The study of regulation in intact animals is difficult because there are so many influences on cell reproduction that complicate design and interpretation of experiments. These variables can be eliminated or controlled with cultured cells, but in cell cultures only remnants of normal control mechanisms continue to be present, and cells proliferate with much less restraint than in tissues. Development of culture methods that approximate conditions experienced by cells in intact tissues is a slow and difficult task.

Density-dependent inhibition of cell reproduction. When normal cells are placed in a culture, they attach to and flatten onto the surface inside the culture vessel. As the attached cells multiply, they form colonies that expand and fuse with one another until the entire surface of the culture vessel becomes covered with a complete, single layer of cells (a **confluent monolayer**) (Figure 12-23). When the monolayer is complete, cell reproduction stops, even though the nutrient medium is still capable of supporting more cell proliferation. The arrest of cell reproduction in the completed monolayer is called **contact inhibition of growth** or **density-dependent inhibition of growth** because the cells appear to exert an inhibition on one another when sufficiently crowded in a monolayer. The inhibited cells stop in the G1 period, where they may remain for many days without adverse effect. When the cells are removed from the culture and seeded into new culture vessels, they resume progress through the cell cycle. Density-dependent inhibition observed in cultured cells is generally assumed to reflect one mechanism by which cell reproduction is regulated in intact tissues. How the inhibition in culture works is not known, but the following four observations provide a better picture of the phenomenon:

1. Before the monolayer becomes confluent, addition of ^{3}H-thymidine to a culture shows which cells are still reproducing in cell colonies. Figure 12-32 is an autoradiograph of cells incubated with ^{3}H-thymidine for 24 hours. All the cells at the periphery of the colony have replicated DNA during the 24-hour interval, and many have divided. None of the cells in the interior of the monolayer have replicated DNA; they are, in fact, arrested in the G1 period. Clearly, crowding of cells in the interior of the colony somehow causes an inhibition that is not present at the periphery of the colony.
2. When fresh serum is added to a complete monolayer of cells, a large fraction of the cells is released from G1 arrest and divides once before again becoming arrested in G1 in an even more crowded state. This release suggests that the cells have exhausted essential growth factors provided by the serum. However, when the medium is withdrawn from a culture with a complete cell monolayer in which all cells are arrested in G1 and that medium is seeded with a low density of cells, the new cells proliferate to a complete monolayer. This proliferation shows that arrest in the first culture could not have been due to depletion of growth factors in the medium.
3. When a monolayer is *wounded* by drawing a glass needle across it and scraping cells from the surface, the cells immediately adjacent to the free surface are released from G1, move into the unoccupied

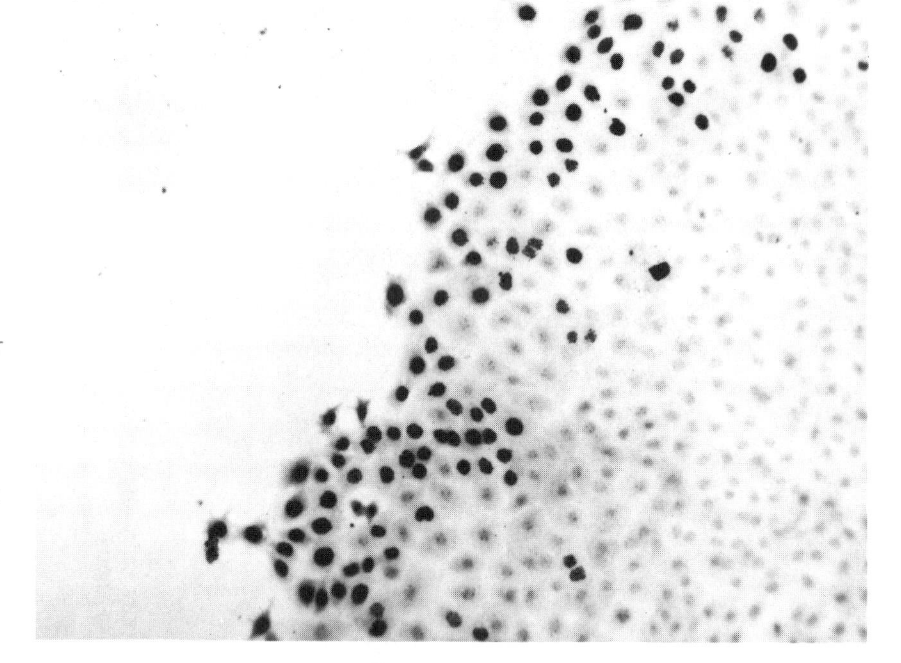

FIGURE 12-32
Autoradiograph of a colony of animal cells in culture illustrating density-dependent inhibition of cell reproduction. The cells were incubated with ^3H-thymidine for 24 hours. Almost all the cells at the edge of the colony have incorporated ^3H-thymidine, showing that they are still traversing the cell cycle and dividing. Almost all the cells in the inner part of the colony failed to incorporate ^3H-thymidine because they are arrested in G1 by density-dependent inhibition.

area, and proliferate until the wound has been repaired with a monolayer of new cells. This suggests but does not prove that relieving cell-cell contacts relieves the G1 inhibition.

4. Plasma membranes purified from cells in a G1-arrested monolayer inhibit cell reproduction when added to a culture with cells still proliferating. Plasma membranes from proliferating cells do not have this inhibitory effect, which supports the hypothesis that the inhibition of one cell by another is mediated by inhibitors appearing in the plasma membrane as cells become crowded.

Anchorage dependence of normal cells. Normal cells in culture survive and proliferate only when attached to the glass or plastic surface of a culture vessel. Normal cells secrete a large glycoprotein called fibronectin that sticks to surfaces of both the cells and the culture vessel and promotes attachment of the cells to the culture vessel. Fibronectin is believed to play a role in cell-cell adhesion in tissues and in cell migration occurring during development. Normal cells prevented from settling and attaching to the surface of the culture vessel when the medium is gently stirred die within a day or two without dividing. The requirement for attachment is called **anchorage dependence.**

Serum growth factors. Serum is required for the proliferation of normal cells in culture. Typically, serum makes up 10 to 15 percent of the nutrient medium; the other 85 to 90 percent is a solution of nutrients. If reproducing cells are transferred to medium containing less serum, for example 5 percent, they stop proliferating by arresting in the G1 period.

Serum contains growth factors that are necessary for cell reproduction. These growth factors are proteins. Two principal growth factors that when acting together fulfill the growth factor requirements of many normal cells in culture are **epidermal growth factor (EGF)** and **platelet-derived growth factor (PDGF).** Platelets are formed in bone marrow and are released into the blood. They are packets of cytoplasm that pinch off from giant cells in the bone marrow called **megakaryocytes.** Platelets carry factors essential for clotting of blood. When a blood vessel is damaged, platelets lyse and release clotting factors, preventing bleeding from the wound. Platelet-derived growth factor is also released from the platelets and is probably important in stimulating cell reproduction needed for wound healing.

Insulin in serum also acts as a growth factor for some kinds of normal cells. A variety of other growth factors are known that stimulate reproduction of specific cell types. For example, **erythropoietin** is a protein growth factor synthesized in the kidney. It stimulates reproduction of cells in the bone marrow that differentiate into red blood cells. **Interleukins** are growth factors that stimulate reproduction of certain kinds of white blood cells.

The effect of growth factors is mediated by means of receptors in the plasma membrane that specifically bind particular growth factors. For example, cultured normal cells that require EGF and PDGF have EGF and PDGF receptors. A receptor is a transmembrane glycoprotein that binds a signaling molecule like a growth factor or a hormone molecule to a binding site at the outer surface of the plasma membrane. In the case of EGF receptor, binding of EGF at the cell surface alters the receptor protein and activates an enzymatic site in the part of the protein at the inner surface of the plasma membrane (the cytoplasmic side) (Figure 12-33). The enzyme portion of the receptor is a protein kinase that phosphorylates particular proteins. The specific target proteins in the cytoplasm have not yet been identified, and the connection between the activated EGF receptor (receptor with EGF bound to it) and the regulation of cell reproduction is not understood in molecular terms.

Finite life span of normal cells in culture. Cells taken from normal tissue and put in culture have a limited life span. Typically, they are capable of 50 to 100 doublings (50 to 100 cell cycles) over a period of several months, and then die. The phenomenon is called **cell senescence.** Cells taken from a child will undergo more doublings than cells taken from an old adult; this might seem to support the idea that part of aging of a human may be due to loss of reproductive potential of cells. However, even the cells of an individual very old at the time of death still have the capacity to divide many times in culture.

Several theories have been proposed to explain why normal cells in culture are capable of only a limited number of reproductions. Most researchers believe that normal cells follow an intrinsic or intracellularly determined program (as opposed to extrinsic effects of the environment) that governs the number of reproductions of a cell. Such a program might consist either of the turning on or turning off of the transcription of certain genes, resulting in the closing down of reproduction. Whatever the mechanism, it works by arresting the cell permanently and irreversibly in the G1 period.

In the cultivation of cells the increase in cell number is logarithmic, and cells soon accumulate beyond the need for experiments and beyond what can practically be maintained. For example, during the first ten population doublings of cells newly explanted from a tissue, the cell number increases by 2^{10} (1000-fold). A convenient tactic is to store excess cells by freezing them in liquid nitrogen (−196°C). They can be thawed and placed in culture whenever cells are needed, thus the more arduous procedure of explanting new cells

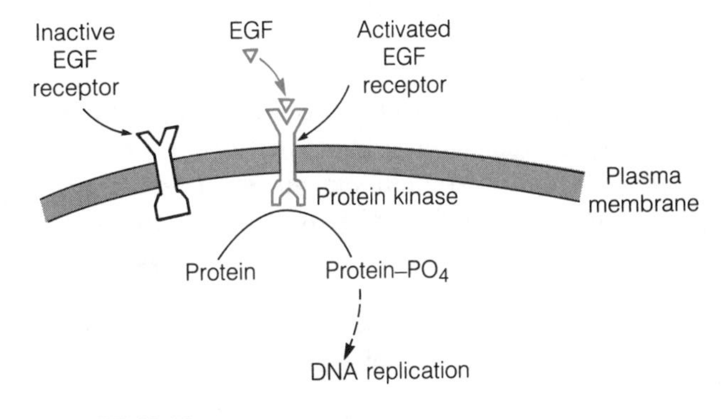

FIGURE 12-33
The binding of epidermal growth factor (EGF) to the receptor at the outer surface of the membrane activates the protein kinase site on the cytoplasmic domain of the receptor.

from tissues can be avoided. Storage in the frozen state does not decrease or increase the reproductive potential of cells; those frozen after ten population doublings and stored for a year or more undergo the same total number of population doublings, when thawed, as cells never frozen.

G0 state. Cells arrested in G1 for whatever reason, for example nutrient deprivation, lack of growth factors, or density-dependent inhibition, undergo metabolic changes described as entry into quiescence or a **G zero** or **G0 state.** In G0, rates of RNA and protein synthesis are decreased.

One consequence of entry into G0 is a delay of some hours in the return of a cell to its reproductive cycle when the condition causing G1 arrest is reversed. Typically, 12 or more hours are required for cells to begin DNA replication after they are released from G0. Some of this time may correspond to part of the G1 period, but most of it is taken up by reversal of metabolic changes that occurred during transition into quiescence. For example, cells that cycle with an average G1 period of four hours require 12 or more hours to reach the S period after release from quiescence.

REGULATION OF CELL REPRODUCTION IN ORGANISMS

Observation of the behavior of cells in culture provides some clues about regulation of reproduction of normal cells, but regulation is more complex in the tissues of intact organisms. First, regulation is different from one cell type to another. Reproductive rates range from very high in some tissues to zero in others. Second, reproductive rates for many cell types are flexible, for example, transiently increasing in response to injury to a tissue.

In general, three broad classes of cells are definable in animals by their rates of reproduction:

1. Some cell types reproduce continuously at high rates. Primary examples are cells in bone marrow and the lymphatic system that give rise to red and white blood cells, epithelial cells that line the alimentary canal, cells in the basal layer of the epidermis of skin, cells in hair follicles, male germ cells, and the epithelium of the uterus during the reproductive years in women.
2. Cells of some tissues reproduce at very low rates, for example, cells of liver, bone and other connective tissues, kidney, smooth muscle, pancreas, ovary, lung, and adrenal glands. Rates for some of these increase enormously in response to injury of a tissue, such as liver and bone, or in response to hormonal stimulation, such as smooth muscle of the uterus.
3. A few kinds of cells permanently lose the capacity for reproduction when they become differentiated. Examples are nerve cells, skeletal muscle cells, cardiac muscle cells, cells of the outer layers of epidermis, and mature red blood cells (which lack nuclei).

Differential rates of cell reproduction are achieved by arrest of cells in G1 for different lengths of time. Cells arrested for more than a few hours probably enter a G0 state comparable to that observed in cultured cells. Most bone marrow cells are only briefly arrested in G1. Liver cells in a rat divide on average less than once a year, and are almost constantly in G0. In the G0 state they perform the many differentiated functions of liver cells, such as breakdown of toxic substances like alcohol and other drugs that may be in the blood, production of bile, synthesis of certain blood proteins, and so forth. Nerve and skeletal muscle cells remain permanently in G0.

Stimulation of Cell Reproduction by Hormones and Growth Factors

Differential rates of cell reproduction are achieved by a variety of mechanisms. One of the better understood mechanisms is stimulation of cell reproduction by hormones and growth factors. For example, as described earlier, release of PDGF by platelets in the vicinity of a wound stimulates reproduction of cells as a part of wound repair. The rate of reproduction of bone marrow cells that differentiate into erythrocytes is stimulated by the hormone erythropoietin, which is synthesized by kidney cells at a rate determined by the O_2 level of the blood. When the O_2 level falls, the production of erythropoietin increases in the kidney, which causes an increased rate of erythrocyte production in the bone marrow. The increase in erythrocytes raises the capacity of blood to carry O_2. This regulatory mechanism not only maintains a rate of production of erythrocytes adequate to replace erythrocytes that wear out and die; the mechanism also steps up production of erythrocytes in response to loss of blood through bleeding and in response to the lower amount of O_2 in the atmosphere at a high altitude.

Reproduction of epithelial cells in the uterus is at least partially regulated by estrogen. Reproduction of male germ cells is at least partially regulated by the male hormone testosterone.

Regulation of Cell Reproduction by Inhibitors

In addition to stimulation of reproduction of specific target cells by growth factors and hormones, cell reproduction is also controlled by inhibitors.

An inhibitor of epithelial cells in culture. One identified inhibitor is a small protein molecule produced by epithelial cells cultured from monkey kidney. The inhibitor, known as **TGF-β,** is effective at an extremely low concentration (ng per ml of medium) and blocks the reproduction of epithelial cells that produce it, but does not block fibroblasts. It is presumed that the inhibitor is part of the mechanism by which reproduction of cells is controlled in the intact animal.

The blocking of reproduction of cells by an inhibitor they themselves synthesize is a form of **autoregulation.** Autoregulation of reproduction of liver cells is one of the better understood examples of this mechanism.

Autoregulation of liver cell reproduction. Practically all the cells in an adult mammalian liver are in the G0 state. The cell death rate in liver is very low and the rate of reproduction of liver cells is also correspondingly low. When part of the liver is removed by surgery, the remaining liver cells are triggered out of G0, reproduce, and replace the cells removed by surgery. Up to 75 percent of the liver can be removed from a rat, and the remaining 25 percent will reproduce to regenerate a full-sized liver in just a few days.

PROBLEMS

1. *E. coli* cells in rich medium divide every 20 minutes. Starting with one cell at 9 AM on Monday and assuming a sufficiently large culture vessel with an inexhaustible supply of nutrients, how many cells will be present at 9 AM on Tuesday?
2. Describe the experiment that shows that DNA replication does not occur during mitosis.
3. A culture of animal cells in which the cell cycles were asynchronous was incubated with ^{3}H-thymidine for ten minutes. Autoradiography showed that 50 percent of the cells were labeled. If the cell cycle time (generation time) was 16 hours, how long was the S period?
4. What is the particular importance of the G1 period of the cell cycle?
5. Maturation promoting factor induces mitotic condensation of chromosomes. How was it discovered?
6. What effect does mitosis have on transcription?

7. What is the major molecular component of the mitotic spindle?
8. What is the function of the kinetochore?
9. Compare cytokinesis in plant cells with cytokinesis in animal cells.
10. What is the role of the Golgi complex in plant cell cytokinesis?
11. How is a high concentration of thymidine used to synchronize the cell cycle in a population of cells?
12. Explain how Ts mutations have been used to identify cell-cycle genes in yeast cells.
13. How do opposite mating type cells of yeast prepare each other for mating?
14. Describe four growth-related characteristics of normal animal cells in culture.
15. What role do some transmembrane proteins provide in regulation of reproduction of cultured animal cells?

CHAPTER 13

The Cell Biology of Cancer

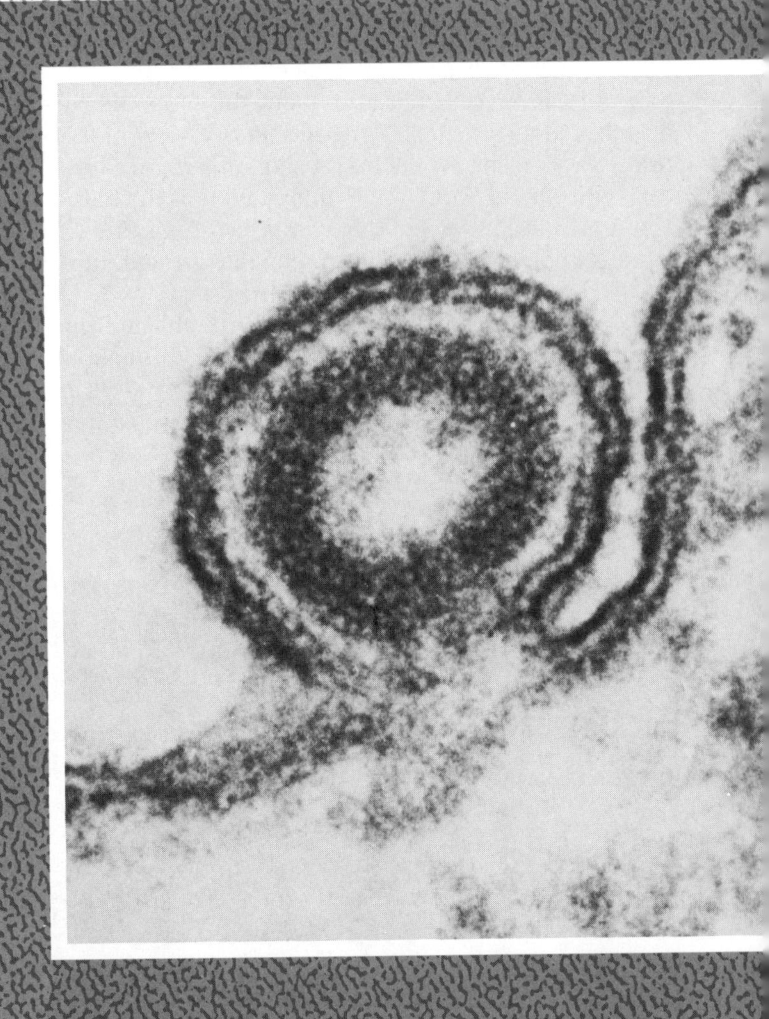

See Figure 13-14(a).

The study of cancer cells has been a major part of cell biology for the past 40 years. From this study much has been learned not only about the properties of cancer cells but also about the properties of normal cells. This chapter presents an overview of this research, focusing on the origin and properties of cancer cells. In addition some information is included about practical aspects of cancer as a disease, including what is known about its causes, how it is treated, how it kills, and how it may be prevented.

Cancer is an ancient disease. It has afflicted our ancestors throughout human history and throughout human evolution. Egyptian medical tracts 3500 years old describe diseases today recognizable as cancer. More direct evidence of cancer's antiquity is seen in distinctly scarred mummies and skeletons of ancient cancer victims (Figure 13-1).

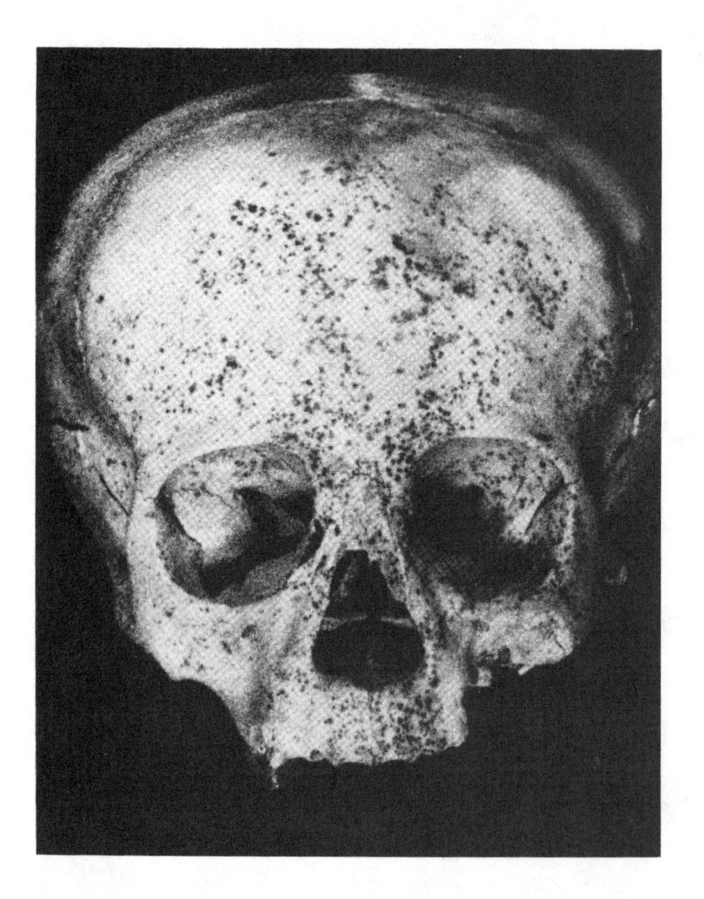

Cancer has always been a much feared disease but, until recently, was seldom encountered. In the last century or so, the situation has changed dramatically (Figure 13-2). Cancer now ranks second only to heart disease as the major cause of death in the United States and Europe and is similarly devastating in the rest of the world. During 1987, in the United States alone more than 965,000 people developed cancer and about 483,000 died of the disease. At least one out of every four Americans alive today will develop at least one cancer; nearly 25 percent of all Americans will be killed by cancer. Reasons for the substantial rise in the incidence of cancer are now understood. First, many years of exposure to cancer-causing agents (**carcinogens**) are usually required to produce cancer. The average life span of Americans and Europeans has increased dramatically during the last 100 years, and therefore many more people now live long enough to develop cancer. Half of the people who die of cancer are over the age of 65. Second, modern life-styles and the contemporary environment result in longer exposures to more kinds of carcinogens and in greater amounts. Carcinogens, how they convert normal cells to cancer cells, and how cancer cells differ from normal cells are discussed in this chapter.

CANCER CELLS AND DISEASE

Cancer, like all organic diseases, results from the misfunctioning of one of the many cells that make up the body. Cancers generally begin when a single normal cell is converted into a cancer cell, a conversion that is now partially understood. Each descendant of the single cancer cell is also a cancer cell and all, in turn, produce more cancer cells. Whatever the nature of the conversion, it is transmitted from one cell generation to the next at each cell division.

FIGURE 13-1
Skull of a pre-Columbian Inca Indian afflicted with cancer many centuries ago. The small dark spots are scars left by melanoma, which begins in a single site and spreads through the body. [From O. Orteaga, and G. T. Pack. 1966. *Cancer*, 19: 607. Print courtesy of Michael B. Shimkin.]

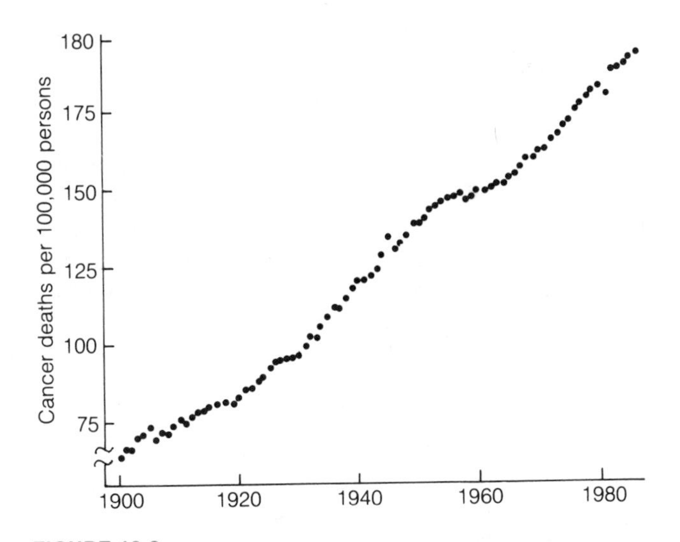

FIGURE 13-2
Total deaths from cancer per 100,000 persons in the United States from 1900 to 1986.

The misfunctions characteristic of cancer cells are fundamentally different from the misfunctions that occur in other diseases. In almost all other diseases, the misfunction stems from the injury or death of normal cells. Bacteria that cause tuberculosis, for example, destroy lung cells, and bacteria that cause diphtheria produce a poison that kills cells of many tissues by blocking protein synthesis. Similarly, polio viruses destroy nerve cells, and hepatitis viruses destroy liver cells. Cancer cells, by contrast, are not injured, nor are they dying. They are, by most criteria, remarkably healthy. Cancer cells have three important properties that underlie the nature of the diseases they cause:

1. *Regulation of reproduction is defective in cancer cells.* In an adult organism the reproduction of normal cells is tightly geared to replace cells that are lost. An insidious property of cancer cells is that they reproduce incessantly rather than responding to signals that turn off the reproduction of their normal counterparts.
2. *Cancer cells in an organism are intrinsically immortal.* Many kinds of normal cells in an intact animal have limited life spans (lymphocytes, intestinal epithelial cells, and epidermal cells of skin, are clear examples). Normal cells with limited life spans give rise to cancer cells with unlimited life spans. Cancer cells do sometimes die, for example because of lack of nutrients in cancerous masses inadequately provided with blood vessels. (As a cancer increases in size, blood vessels grow into it, but the process is sometimes inefficient so that cells deeper in the cancer die).
3. *Differentiation of cancer cells is defective.* Many kinds of normal cells in an organism undergo turnover. Cells die and are replaced by cell division. In some tissues this occurs by reproduction of undifferentiated stem cells. For example, the cells in the outer layer of epidermis die and are shed from the surface of skin (Figure 6-27). New cells are continuously supplied by reproduction of cells in a layer of undifferentiated cells located at the inner base of the epidermis. This is called the **basal layer** or **germinal cells.** Undifferentiated, reproducing cells that supply new cells to a tissue are called **stem cells** (Chapter 15).

On the average when a stem cell in the basal layer of epidermis divides, one of the daughter cells is forced out of the basal layer. Continuous reproduction of the stem cells causes cells in the overlying region to be gradually driven toward the surface of the skin. When a cell leaves the basal layer it becomes permanently blocked in G1 and starts to differentiate. Differentiation is predominantly the synthesis of the protein keratin, which is organized into intermediate filaments. Cells packed with keratin filaments give the outer layer of skin its tough, protective character. The keratin-packed cells die and eventually are sloughed from the surface of the skin. The entire course of events from exit of a cell from the basal layer, through differentiation, death, and sloughing from the surface occurs in a matter of days.

Cancer arises in skin when a stem cell in the basal layer undergoes a particular genetic change, in most cases caused by ultraviolet light of sunlight. Cells that leave the basal layer continue to reproduce instead of becoming permanently arrested in G1. They may undergo some degree of differentiation but become blocked in an immature state.

Undifferentiated stem cells are clearly evident in some other tissues, such as intestinal epithelium and bone marrow. In some other renewing tissues stem cells are not readily identifiable. For example, renewal of hepatocytes in liver appears to occur by division of fully differentiated hepatocytes instead of undifferentiated stem cells. However, in liver cancer the cancer cells may resemble hepatocytes, but they are incompletely differentiated and do not function normally.

A cancer begins with a single cell that has become defective in the three characteristics just described as a result of one or another change in its genetic makeup. The genetic changes that produce cancer cells are only partially understood; they appear to consist of mutations or changes in expression of genes involved in the control of the reproduction, differentiation, and life span of a cell.

A typical human cancer cell growing in culture divides about once every 24 hours. At that rate, a single cancer cell would generate over 10^9 descendants in 30 days, a mass that would weigh less than one gram. In another two weeks, the population would increase to nearly 10^{13} cells and weigh several kilograms. Fortunately, cancer cells rarely reproduce nearly so rapidly in a body. Normally, a cancer requires months or even years to give rise to a cancer mass that produces disease symptoms. Cancer represents an accumulation of cells, but cancer cells do not usually reproduce at particularly high rates. Some types of normal cells, for example blood-forming cells of the bone marrow and epithelial cells that line the intestines, reproduce much more rapidly than almost all kinds of cancer cells. However, because of their immortality and failure to stop reproducing, cancer cells accumulate. Accumulation of these functionally defective cells results in disease.

Spread of Cancer Cells

The immortality of cancer cells, added to their incessant reproduction, results in inexorable accumulation of incompletely differentiated, functionally useless cells that kill the organism by interfering with the functioning of normal cells. The malignancy is enhanced by invasion of the cancer cells into adjacent normal tissue and by the spread of cancer cells to distant sites, where they establish new centers of cancer growth. Spreading occurs by detachment of cancer cells from the primary growth and travel in the blood and the lymphatic system (Chapter 15), a process called **metastasis.** In laboratory mice, breast cancers that had grown to about 3 grams were found to release several million cancer cells a day into the blood. However, few of these cells survived to establish secondary cancer growths, or **metastases,** elsewhere in the animal. Even when particularly malignant cancer cells were injected into the blood of mice, only one in 1000 of those was still alive two weeks later. The immune system probably destroys many wandering cancer cells, but other, unknown protective mechanisms may also be at work. Metastases are ultimately established, often in liver, bone, lungs, brain, and lymph nodes. Metastases greatly complicate the problem of treatment and are often responsible for death.

Immediate Causes of Death from Cancer

How cancer cells kill an individual cannot always be precisely determined, but the general mechanisms are known. About 25 percent of all cancer deaths are the result of interference with the function of an essential organ by the growing mass of cancer cells; commonly the function of lungs, liver, brain, or kidneys is impaired. Ten percent of cancer deaths result from starvation. About seven percent of deaths are the result of hemorrhage, a frequent cause of death in leukemia. Almost 50 percent of all cancer deaths come from infection. Bacteria, molds, and other infectious organisms that are ordinarily destroyed by the immune system of a healthy individual can form fatal infections in cancer victims. The lowered resistance to infection results partly from the generally weak condition of an individual with advanced cancer but is usually more the result of an impaired immune system. By mechanisms that are poorly understood, many cancers severely suppress the immune system.

Treatment of Cancer

Without treatment, virtually all cancers are sooner or later fatal. The objective in the treatment of cancer is elimination of all cancer cells from an individual. Elimination of every cancer cell is important because cancer can regrow from a single surviving cancer cell. The three traditional methods of treatment are to remove cancer cells by surgery, to kill cancer cells with radiation, or to kill or prevent proliferation of cancer cells with chemotherapeutic drugs or hormones. These treatments are often used in combination. Surgical removal of cancerous tissue is frequently followed by radiation therapy or chemotherapy with the objective of killing any cancer cells not eliminated by the surgery.

All current strategies for treating cancer are limited by the fact that, by the time the disease is discovered clinically, most cancers already consist of billions of cells and usually have already metastasized to other sites. To remove, inhibit, or destroy every last cell of a cancer is difficult and usually impossible with presently available methods. Development of new ways to treat cancer is centered on manipulation of components of the immune system to kill cancer cells that may be disseminated in various parts of the body. The immunological approach has not yet yielded a primary therapy. A potentially powerful form of immunotherapy is to use monoclonal antibodies (Chapter 15) to deliver cytotoxic agents specifically to cancer cells. This immunological approach is based on the putative presence of antigens on the surfaces of cancer cells that are not present on normal cells. For example, in a pioneering test, 104 patients with liver cancer were injected with antibodies tagged with radioactive iodine (^{131}I). The antibodies were specific for an antigen on the cancer cell in the liver. Fifty of the 104 patients showed decreases in tumor size, and seven appeared to be cured. Liver cancer is ordinarily fatal. In general, the difficulty with the antibody approach has been in finding antigens that are present on cancer cells but not on any type of normal cell. Without such specificity antibodies

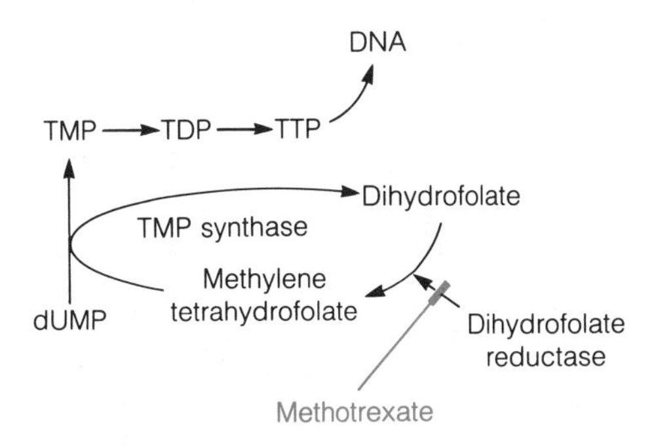

FIGURE 13-3
Methotrexate is a cancer chemotherapeutic drug that blocks DNA replication by inhibiting the enzyme dihydrofolate reductase.

will also bind to one or another kind of normal cell, producing unacceptable side effects of treatment.

About 35 drugs are now in use to kill cancer cells. Among those frequently used are methotrexate, fluorouracil, and hydroxyurea. Fluorouracil is converted to fluorodeoxyuridine (FUdR) in the cell. These three drugs block DNA replication as shown in Figures 12-21 and 13-3. Cells in the S phase are killed by such drugs. Other drugs inhibit other cellular reactions or damage cells in other ways, particularly cells that are proliferating. Unfortunately, none of the drugs is specific for cancer cells and therefore kill normal cells, particularly proliferating cells in the bone marrow that differentiate into blood cells and stem cells in the intestinal epithelium. Nevertheless, chemotherapy is useful in treatment of a limited number of kinds of cancers. One of the most effective uses of chemotherapy consists of treatment of acute childhood leukemia with methotrexate and x-irradiation. Well over half of treated children with this disease are now cured.

Unfortunately, all chemotherapeutic agents are toxic and many are mutagenic. Occasionally cancer patients are killed by chemotherapy. Generally people treated for cancer with chemotherapeutic drugs have an elevated risk of ultimately developing a second cancer induced by the mutagenic action of the drugs used to kill the first cancer.

The overall effectiveness of current methods of treatment can be judged from the mortality figures.

Twenty-three percent of all deaths in the United States in 1987 were due to cancer. Currently, nearly 75 percent of those who develop cancer ultimately die from the disease and the overall cure rate has not changed in the last 30 years. The improvements in cure rate for acute childhood leukemia and a few other cancers do not change the overall cure rate because these cancers account for less than one percent of all cancers. A major improvement in treatment is reflected in a prolonged survival time (but not cure) for many cancer patients.

ORIGIN OF CANCER

The conversion of a normal cell to a cancerous one occurs in at least two steps, called **initiation** and **promotion.** Initiation consists of mutation in or rearrangement of one, or possibly more than one, gene belonging to a class of genes known as **protooncogenes,** which are discussed in a later section. Promotion is less well understood but consists of a second change, not necessarily mutational, that converts an initiated cell into a cancer cell. This two-step process is in part responsible for the long delay, usually many years, between the start of exposure to carcinogens and development of a cancer. A third, even less well defined process called **progression** may occur in which cancer cells undergo further changes that increase their invasiveness and rates of reproduction and metastasis.

Some cancer-causing agents, X rays for example, can act both as initiators and promoters and are called **complete carcinogens.** Other agents, such as saccharin, appear to work only as promoters.

Causes of Cancer Cell Formation

The causes of cancer cell formation fall into three classes: radiation, chemicals, and viruses. In humans, chemicals are responsible for most cancer. Only several percent of all cancers are caused by radiation and viruses.

Radiation. It has been known for 80 years that radiation can cause cancer, but only gradually have we learned to appreciate how sensitive cells are to complete transformation by radiation. The increasing use of radiation for medical and industrial purposes has been paralleled by increasing rates of cancer in certain professions. Radiologists, for example, used to suffer a seven- to ten-fold higher rate of cancer than physicians in general because of the chronic exposure to low levels of X rays. Protective measures have since led to prevention of these excess cancers in radiologists. Uranium miners have a higher rate of lung cancer than other kinds of miners not in contact with radioactivity. The survivors of the atomic bombs dropped on Hiroshima and Nagasaki in World War II have higher rates of cancer. It has been estimated that as many as 39,000 people in Europe, outside the Soviet Union, may eventually die of cancer induced by radioactive fallout from the failure of the nuclear reactor at Chernobyl, USSR.

There are many examples of cancers induced by X rays and radioactivity. Tissues particularly susceptible to these radiations are breast, bone marrow, and thyroid gland, although radiation can cause cancer in virtually any tissue. X rays and radioactivity cause cell transformation by mutation or chromosomal rearrangements of genes involved in regulation of cell proliferation and differentiation.

Ultraviolet light is mutagenic, which probably accounts for a major part of its action in causing skin cancer (Figure 13-4). However, overexposure to ultraviolet light also causes suppression of the immune system, and the immune system has a major role in preventing the development of cancer. For example, the success rate in transplantation of cancer cells from one animal to another in the laboratory is greatly increased by suppressing the immune system in recipient animals by irradiating them with ultraviolet light before transplantation.

Altogether in the United States radiations from various sources, but mostly ultraviolet light, probably account for about three percent of all cancer deaths.

Chemicals. Several methods are used in trying to identify chemical carcinogens. First, observation of human populations sometimes reveals carcinogens. Such

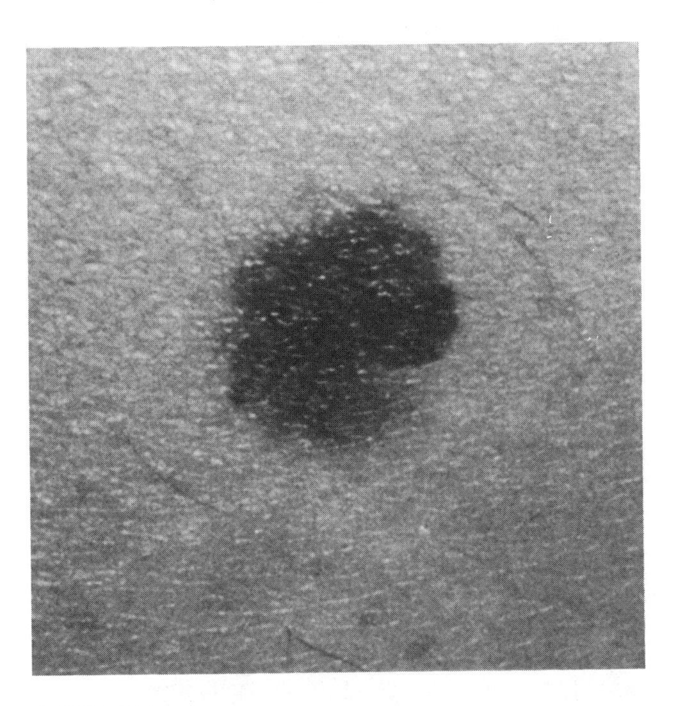

FIGURE 13-4
A human melanoma measuring about two cm in diameter. By the time a melanoma has reached this size, it often has metastasized. [Courtesy of William A. Robinson.]

was the case in the identification of tobacco smoke as the cause of most lung cancer and benzene as a cause of some cases of leukemia in adults. Second, a chemical suspected of being a carcinogen is administered to animals, and the subsequent cancer incidence is compared with the incidence in animals not exposed to the test chemical. Such tests take years and are expensive. Third, since cancer is the result of a genetic change in a cell and since most carcinogens have proven to be mutagens, chemicals are tested for their mutagenic properties. Any chemical that proves to be a mutagen becomes a suspected carcinogen.

Various methods for testing whether a substance is mutagenic have been developed. The oldest and most widely used, called the Ames test after its originator, is based on mutation in the bacterium *Salmonella typhimurium*. Strains of *S. typhimurium* have been developed that carry mutations that make the cell incapable of synthesizing histidine. Such histidine-minus mutants proliferate in medium containing histidine but not in medium that lacks histidine. Histidine-minus mutants may regain the ability to synthesize histidine as a result of a second mutation, called a back mutation. Bacteria that sustain such back mutations can proliferate in medium lacking histidine. A substance being tested for mutagenicity is added to a culture of histidine-minus bacteria in a medium lacking histidine. The number of bacteria that regain the ability to proliferate in the histidine-minus medium gives a measure of the mutagenic potency of the test substance. An example of the outcome of such a test is shown in Figure 13-5. The Ames test is extraordinarily sensitive, takes only a few days, and is inexpensive. Any substance that is mutagenic in the Ames test becomes a suspected carcinogen; however, other evidence, for example animal testing, is required to prove carcinogenicity. Short-term tests like the Ames test provide a quick, inexpensive means of identifying chemicals that should be subject to the far more expensive and laborious tests for carcinogenicity with animals.

Most human cancer is caused by chemicals. These chemical carcinogens occur primarily in air, water, and food. The major source of airborne carcinogens is tobacco smoke, which accounts for about one-third of all cancer deaths in the United States. Tobacco smoke contains over 6000 different kinds of organic molecules, some of which are listed in Table 13-1. About 100 of these are known mutagens and about 30 of these have been tested and shown to cause cancer in animals. Tobacco smoke also contains minute amounts of radioactive heavy metals absorbed from the soil by tobacco plants. These emit α radiation and include, particularly thorium (^{228}Th), radium (^{226}Ra), and lead (^{210}Pb), which are thought to contribute to the carcinogenicity of tobacco. Carcinogens in tobacco smoke are not only absorbed by cells of the respiratory tract, causing oral, throat, and lung cancer, but also enter the bloodstream and are distributed throughout the body. Cigarette smokers also suffer from higher risks of esophageal, bladder, kidney, and pancreatic cancers. Pipe and cigar smokers have the same elevated rate of oral cancer as cigarette smokers and have an elevated risk of lung

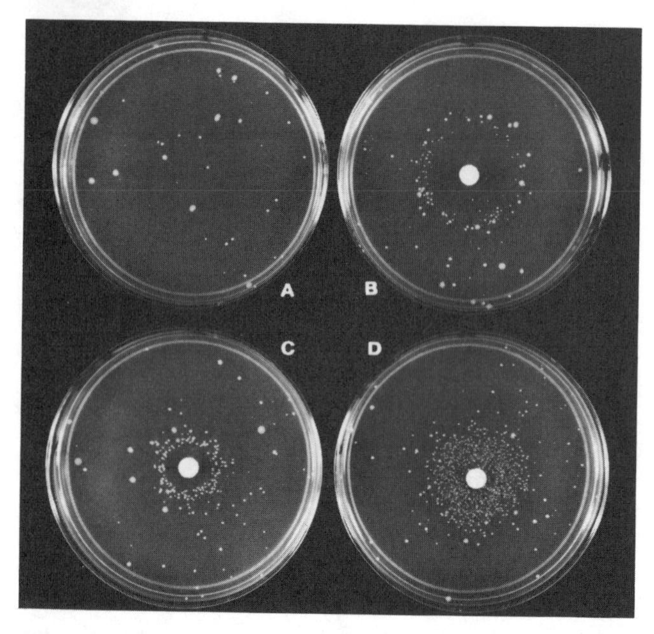

FIGURE 13-5

An example of the Ames test for detecting mutagens. Each petri plate contains a layer containing about 10^9 bacteria (*Salmonella*) on the surface of nutrient agar lacking histidine. Substances to be tested for mutagenic action were applied to a disc of filter paper in the center of the plates. (A) Control plate with no test substance. Each white spot is a clone of cells that grew from a single bacterium that had undergone a back mutation that restored the ability to synthesize histidine. The number of colonies is a measure of the background mutation rate. (B) One microgram of the Japanese food additive furylfuramide was added to the filter paper. Diffusion of the substance outward from the disc resulted in a ring of colonies representing bacteria mutated by the food additive. (C) One microgram of the mold carcinogen aflatoxin was added to the filter paper disc. (D) Ten micrograms of 2-aminofluorene were added to the filter paper disc. Plates C and D also contained a rat liver microsomal activation system. Microsomes contain enzymes that convert premutagens (and precarcinogens) to mutagens (and carcinogens). Furylfuramide (plate B) does not need activation. [Courtesy of Bruce N. Ames and Joyce McCann.]

TABLE 13-1

Some of the 6000 Chemicals in Cigarette Smoke. Asterisks (*) Mark Proven Carcinogens

Acetaldehyde	Fluorene
Acetone	Formaldehyde
Acetylene	Hexane
Acrolein	Hydrazine
Aminostilbene*	Indeno[1,2,3-*cd*]pyrene*
Ammonia	Indole
Arsenic*	Isoprene
Benz[*a*]anthracene*	Methane
Benz[*a*]pyrene*	Methanol
Benzene*	Methylcarbazole
Benzo[*b*]fluoranthene*	5-Methylchrysene*
Benzo[*c*]phenanthrene*	Methylfluoranthene*
Benzo[*j*]fluoranthene	Methylindole
Cadmium*	β-Naphthylamine*
Carbazole	Nickel compounds*
Carbon dioxide	Nicotine
Carbon monoxide	Nitric oxide
Chrysene*	Nitrobenzene
Cresols	Nitroethane
Crotonaldehyde	Nitromethane
Cyanide	*N*-Nitrosodimethylamine*
DDT	*N*-Nitrosomethylethyl-amine*
Dibenz[*a,c*]anthracene*	*N*-Nitrosodiethylamine*
Dibenzo[*a,e*]fluoranthene*	Nitrosonornicotine*
Dibenz[*a,b*]acridine*	*N*-Nitrosonanabasine*
Dibenz[*a,j*]acridine*	*N*-Nitrosopiperdine*
Dibenzo[*c,g*]carbazone*	*N*-Nitrosopyrrolidine*
N-Dibutylnitrosamine*	Phenol
Dichlorostilbene	Polonium-210*
2,3-Dimethylchrysene*	Propene
Dimethylphenol	Pyridine
Ethane	Sulfur dioxide
Ethanol	Toluene
Ethylphenol	Vinyl acetate
Fluoranthene	

cancer, but not as elevated as in cigarette smokers. Users of smokeless tobacco like chewing tobacco or snuff have an elevated risk of oral cancer. Altogether, tobacco currently causes about 160,000 cancer deaths per year in the United States. Lung cancer is the most common cancer caused by tobacco smoke. Treatments are not effective. About ten percent of those diagnosed with lung cancer survive five years. The discontinuance of use of tobacco has been clearly shown to be followed by a decrease in occurrence in lung cancer (Figure

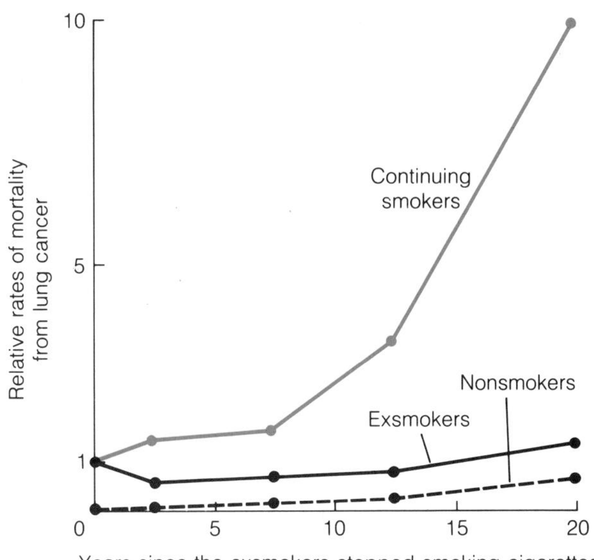

FIGURE 13-6
Relative rates of lung cancer mortality among nonsmokers, exsmokers, and continuing smokers of cigarettes for a 20-year period.

13-6). Industry and motor vehicles are minor sources of airborne carcinogens compared to tobacco smoke.

Essentially all municipal water supplies in the United States are contaminated with minute amounts of mutagens and proven carcinogens. A study of cancer mortality in the 473 largest cities in the United States suggests that as much as one percent of human cancer deaths (about 5000 out of a total of 483,000 cancer deaths in 1987) derive from this source. Regular intake of alcoholic beverages carries a higher risk of esophageal and perhaps bladder cancer. Even moderate use of alcohol increases the risk of breast cancer. Tea, including herbal teas, and coffee contain mutagenic substances, but whether they cause cancer in humans is still an unsettled question.

According to various estimates, between 30 and 50 percent of cancers in the United States are caused by agents in food. Some foods contain mutagens and carcinogens naturally. Examples are bracken fern (several chemicals), mushrooms (hydrazines), black pepper (piperine), alfalfa sprouts (canavanine), and cottonseed oil (gossypol). Some foods are contaminated with carcinogens known as mycotoxins produced by mold that can grow on the food during harvesting and storage. A serious example is the mycotoxin called **aflatoxin** produced by the mold *Aspergillus flavus,* which commonly grows on raw peanuts, corn, grains, and nuts. Aflatoxin is soluble in lipid and is readily absorbed in the lipid component of these foods. In laboratory animals a diet containing just several parts per million of aflatoxin results in an elevated rate of cancer. Aflatoxins are at least partially responsible for the high rate of liver cancer in Africa.

Other carcinogens are added to foods during processing, especially as a result of pickling, smoking, and fermenting. Chemical reactions that yield mutagens occur in many foods, particularly meats, during cooking at temperatures of 150°C and above. The most important source of carcinogens in food is probably fat. Animals fed diets rich in fat have higher rates of cancer than animals fed diets low in fat (Figure 13-7). In human populations throughout the world mortality from cancers of the colon, breast, prostate, and pancreas are correlated with the amount of fat in the diet (Figure

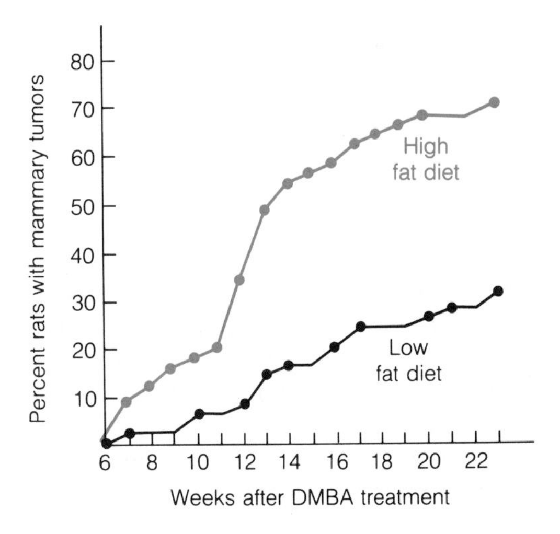

FIGURE 13-7
Rats given the carcinogen dimethylbenzanthracene (DMBA) develop mammary tumors more often when given a high-fat diet than when given a low-fat diet.

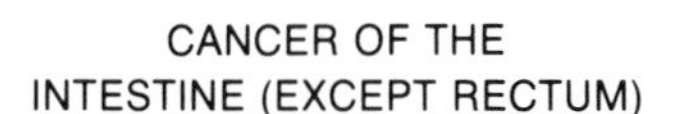

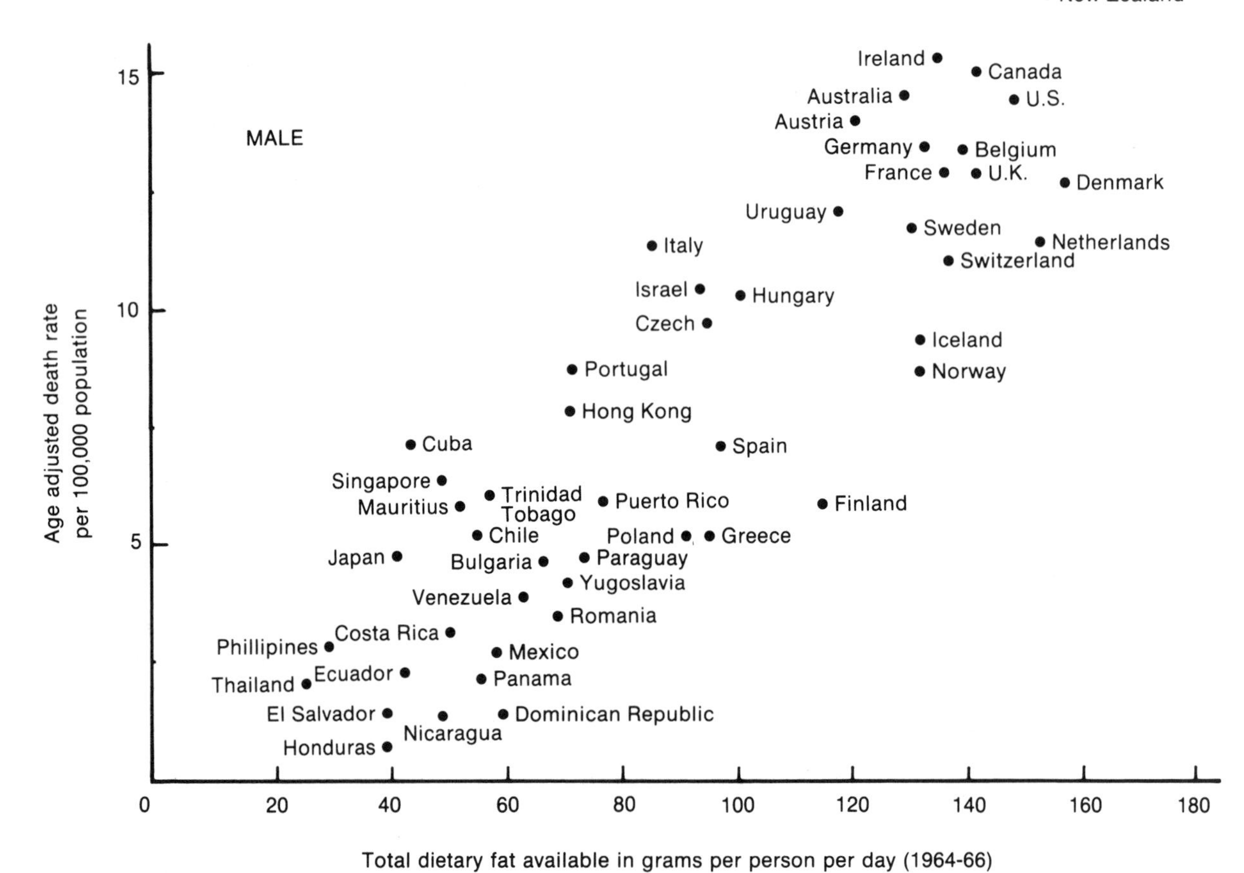

FIGURE 13-8
An epidemiological study in humans showing a worldwide correlation of daily fat consumption with intestinal cancer.

13-8). Several theories have been proposed to explain how fat might contribute to cancer. For example, the oxidation of fat molecules in a cell results in formation of substances called **free radicals.** Free radicals possess an unpaired electron, which makes them extremely reactive and mutagenic. Free radicals have extremely short lives, usually measured in fractions of a millisecond.

Cancer incidence in laboratory animals is related to total caloric intake. For example, animals placed on a restricted diet but with no change in the quality of the food develop fewer cancers, commonly 50 percent fewer.

Several occupations entail exposure to particular carcinogens. Workers engaged in the manufacture of some kinds of dyes have high rates of bladder cancer. Wood workers and leather workers have higher rates of nasal cancer as a result of breathing wood and leather dust. Asbestos workers have a higher risk of lung cancer.

Oncogenic viruses. **Oncogenic viruses** (*onco* = tumor; *genic* = forming) have been shown to cause at least a few kinds of human cancers: adult T-cell leukemia, some liver cancers, and Burkitt's lymphoma. Viruses have been implicated but not proven as causative agents in cancers of the uterine cervix, the nasopharynx, and a few other tissues. However, overall, viruses appear to account for a very small fraction of human cancers. Hundreds of different oncogenic viruses have been identified in animals, primarily in birds and mammals. Usually a virus that causes cancer in one species of animal is harmless in another, although there are exceptions. For example, Rous sarcoma virus can cause cancer in at least several kinds of birds and in mice.

An enormous amount of research has been and is being done on how oncogenic viruses cause cancer, using particularly the transformation of normal cells in culture to cancer cells by oncogenic viruses. This research has led to major new insights not only about transformation by viruses, but also by chemicals and radiation. Only the barest outlines of this work are included here and make up the latter part of this chapter.

REPRODUCTIVE PROPERTIES OF TRANSFORMED ANIMAL CELLS IN CULTURE

Certain properties of normal animal cells in culture (Chapter 12) are different from those in transformed cells. These differences are summarized in Table 13-2. The conversion of a normal cell to a cancer cell can be achieved in culture with chemical carcinogens, radiation, or oncogenic viruses in a process called **transformation.** The six properties listed in Table 13-2 are discussed in the following sections.

TABLE 13-2
Comparison of Growth Characteristics of Normal and Transformed Cells in Culture

Property	Normal Cells	Transformed Cells
Density-dependent inhibition	Present	Absent
Anchorage dependence	Present	Absent
Requirement for serum growth factors	High	Low or absent
Life span	Finite	Infinite
G1 arrest and entry into G0	Present	Absent
Karyotype	Diploid	Aneuploid

Absence of Density-Dependent Inhibition of Reproduction

Normal cells in culture grow as a monolayer, and only cells at the periphery of the monolayer reproduce. This can be shown by adding ^{3}H-thymidine to the culture medium and detecting DNA replication by autoradiography (Figure 12-32). Cells at the periphery of a colony become labeled; cells in the center of the colony are arrested in G1 by density-dependent inhibition. The behavior of transformed cells is very different. Feeding of ^{3}H-thymidine for 24 hours results in labeling of all cells throughout a colony (Figure 13-9(a)). In contrast to normal cells, transformed cells in the central part of a colony are not subject to density-dependent inhibition of the cell cycle. When a transformed cell arises in a monolayer of normal cells in a culture, it continues to proliferate, and progeny pile up on one another into irregular layers (Figure 13-9(b)).

Absence of Anchorage Dependence

Normal cells must attach to a surface in order to proliferate. Indeed, when kept in suspension in nutrient medium, they die within a few days. Transformed cells can usually be adapted to proliferate in suspension culture, that is, they are **anchorage independent.** This property is commonly induced so that transformed cells can be identified by the following method. Cells are seeded in nutrient medium that has been made into a loose gel by the addition of agar. The cells remain stationary in the agar gel and cannot attach to a surface. Normal cells fail to proliferate, and transformed cells form clones visible without a microscope, which can be removed and cultured by standard methods.

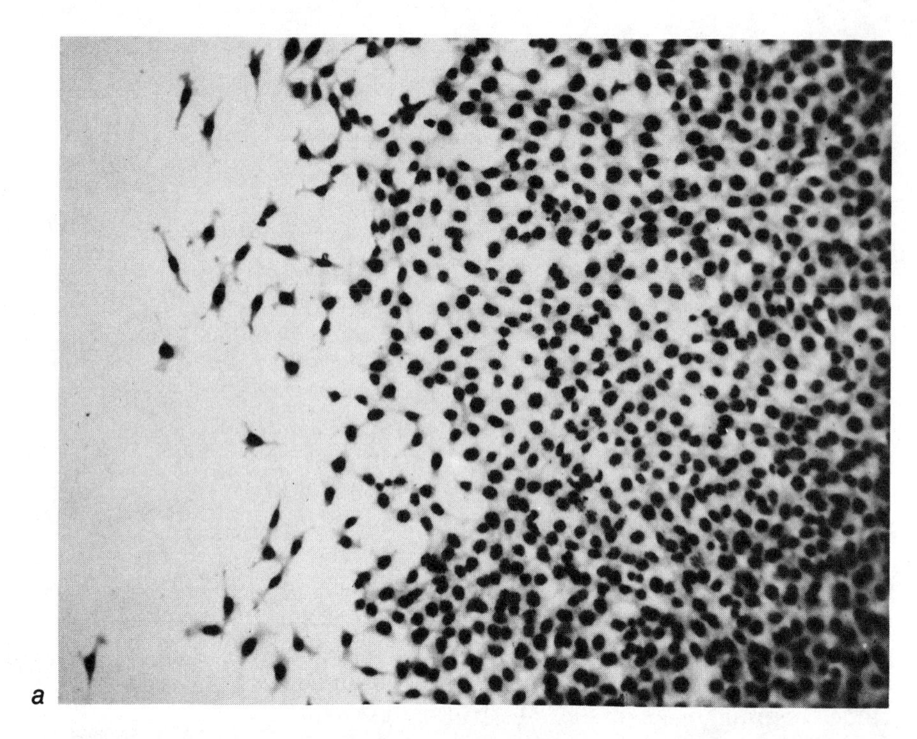

a

b

FIGURE 13-9
(a) A colony of transformed animal cells in culture incubated with ^{3}H-thymidine for 24 hours. All the cells have incorporated radioactivity into their DNA, showing that all cells, even those well inside the monolayer, are reproducing. This lack of density-dependent inhibition is in contrast to the behavior of normal cells shown in Figure 12-32. (b) Light micrograph of a monolayer of normal animal cells in which a transformed cell has arisen. The normal cells remain arrested in G0, but the transformed cell has proliferated, giving rise to an irregular mass of piled-up cells. [(b) Courtesy of Robert E. Scott.]

Lack of Requirement for Serum Growth Factors

Normal cells require serum growth factors, which are provided by nutrient media containing 10 to 15 percent serum. When the serum component is reduced to one percent, normal cells become arrested in the G1 period, but many transformed cells continue to proliferate. Transformed cells are independent of growth factors in the medium because they synthesize their own growth factors or undergo other changes that obviate the need for growth factors.

Immortality of Transformed Cells

Normal mammalian cells have finite life spans in culture; typically they divide between 50 and 100 times and die. In culture, transformed cells proliferate indefinitely, like their cancer counterparts in an organism. Some have divided thousands of times with no reduction in vitality or reproductive rate.

Absence of the G0 State

Most kinds of transformed cells are unable to enter the G0 state. When subjected to a treatment that arrests normal cells in G1 and causes entry into G0, for example deprivation of an essential nutrient, transformed cells cease proliferation and soon die.

Aneuploidy

Normal cells in culture have diploid sets of chromosomes, that is, they are **euploid.** Transformed cells have an abnormal number of chromosomes, that is, they are **aneuploid.** Typically, the chromosome number in an aneuploid cell is greater than the diploid number. For example, the chromosome number in the human carcinoma cell known as HeLa typically ranges from 55 to over 100. The HeLa cell was derived from a uterine (cervical) cancer in the early 1950s, and cultures of this cell have been used in research since that time in hundreds of laboratories throughout the world. It is named for *He*nrietta *La*cks, who died from the cancer.

Cancer cells in an animal are also usually aneuploid. In addition, in aneuploid cells extensive exchanges of segments between chromosomes occur so that normal chromosomes as observed in a diploid cell are not identifiable. Such exchanges are called **translocations.** The first step in development of aneuploidy is duplication of chromosomes in a diploid cell without cell division. The result is a cell with four copies of each chromosome, or a **tetraploid** set of chromosomes.

TABLE 13-3
Classes of Oncogenic Viruses

Viruses with Double-Stranded DNA	Viruses with Single-Stranded RNA (Retroviruses)
Papovaviruses (polyoma, SV40, BK, JC, and HD)	B-type viruses (several types)
Adenoviruses (many kinds)	C-type viruses (many types)
Herpesviruses (many kinds)	Human T-cell lymphotrophic viruses (many types)
Cytomegaloviruses (many kinds)	
Hepatitis B virus	

Within the tetraploid cell some chromosomes are lost and others undergo translocations over a number of cell cycles. The chromosomal makeup of an aneuploid cell is often unstable; cells with different numbers and forms of chromosomes are continuously produced.

Cells in culture sometimes undergo only partial transformation. An example is the 3T3 cell, which was derived from a mouse embryo. These cells exhibit density-dependent inhibition, require serum growth factors, can enter the G0 state, and are anchorage-dependent, all characteristics of normal cells. However, they have the transformed characteristics of immortality and aneuploidy. Generally, for a cultured cell to become cancerous, that is, to grow as a cancer when implanted into an animal, it must be fully transformed, that is, it must possess all six of the characteristics listed in Table 13-2.

ONCOGENIC VIRUSES, ONCOGENES, AND PROTOONCOGENES

Cancer-causing viruses are probably responsible for no more than a few percent of human cancers. However, the study of how such viruses work has provided major understanding of genetic changes that occur in the induction of cancer by chemicals and radiation. Therefore, oncogenic viruses are discussed here in some detail.

Viruses are inert particles that become active only after they have entered cells. They parasitize the metabolic machinery of cells, using ATP produced in mitochondria, using the cell's ribosomes and enzymes for protein synthesis, and so forth. Their activities are directed toward their own reproduction. Thousands of virus particles can be produced in one cell originally infected with only one or a few virus particles. Depending on the type of virus, they are released from the host cell by lysis of the cell or individually through the plasma membrane.

The minimum components of a virus are a genome of nucleic acid and a surrounding coat made of many copies of one or a few kinds of protein molecules. Many viruses are more complex, containing an additional outer membrane derived from the plasma membrane of a cell during exit from the cell. The genome of some of the viruses that invade eukaryotes is composed of double-stranded DNA; others have chromosomes composed of single-stranded DNA, or single- or double-stranded RNA. The chromosomes of oncogenic viruses are either double-stranded DNA or single-stranded RNA.

Oncogenic viruses fall into several classes. Those with chromosomes of double-stranded DNA are the papovaviruses, herpesviruses, adenoviruses, cytomegaloviruses, and hepatitis B virus. Those with single-stranded RNA chromosomes are called **retroviruses.**

Oncogenic Viruses with Double-Stranded DNA

These viruses belong to the five groups listed in Table 13-3. Most of the viruses in these groups cause cancer in animals and several are known or suspected to cause cancer in humans.

Papovaviruses. Two well studied oncogenic viruses of the papova group are the **polyoma** virus (*poly* = many; *oma* = tumors) and the simian virus 40 (**SV40**). The polyoma virus was so named because it can cause more than ten different kinds of cancer when injected into newborn mice. It causes neither cancer nor any other disease in adult mice. Viral transformation of normal cells in culture into cancer cells was discovered with polyoma virus. There is no evidence that it causes disease or cancer in humans.

The SV40 virus was discovered in 1960 during the production of Salk vaccine against poliomyelitis virus. Polio vaccines were prepared from poliovirus grown in cultured cells derived from monkey kidneys. Monkey cells were used because of the possibility that human cells might contain undetected, harmful human viruses that would become included in the polio vaccine. In 1960, SV40 virus was isolated from monkey cells and detected in polio vaccine produced with these cells; by this time several million people had been injected with polio vaccine containing SV40 virus. By 1962 it was clear that SV40 virus is oncogenic in rodents, causing tumors with a high efficiency when injected into baby hamsters. Fortunately, SV40 virus is not known to have caused tumors or disease in humans, and the inadvertent injection of humans with SV40 virus has had no detectable consequences. Under carefully arranged conditions, SV40 can transform human cells growing in culture.

Several other papovaviruses, known as BK, JC, and HD, have been isolated from a few human cancers, but it is not known whether they have a causative role in those cancers. Injection of JC virus causes brain cancer in monkeys. The presence of a virus in human cancer cells does not prove that the virus has a role in causation of the cancer. Viruses are sporadically observed in normal cells, producing no discernible pathological effect. To prove that a virus causes cancer requires that injection of the virus into animals consistently induces cancer.

Adenoviruses. **Adenoviruses** were discovered in 1953, and in 1962 a type of adenovirus obtained from the throat of humans was found to induce cancer when injected into baby hamsters. Adenovirus infections are very common in humans, and more than a dozen of the many varieties of human adenoviruses have been found to be oncogenic in mice, rats, and hamsters. The human adenoviruses are not known to be oncogenic in humans, but do cause transient respiratory and gastrointestinal diseases. They provide another illustration that a virus may have different effects in different species. Adenoviruses capable of causing cancer in rodents have also been obtained from cows, monkeys, and birds. More than 50 different adenoviruses, many oncogenic in animals, have been discovered.

Herpesviruses. The **herpesviruses** form a large, ubiquitous group of DNA viruses, members of which occur commonly in frogs, chickens, turkeys, guinea pigs, rabbits, monkeys, humans, and other species. In frogs, kidney cancer caused by a herpesvirus is common. *Herpesvirus saimiri,* recovered from the squirrel monkey *Saimiri sciureus,* does not cause cancer or other ill effects in the squirrel monkey but induces a highly malignant **lymphoma** (cancer of lymph tissues) when injected into marmosets. In guinea pigs a herpesvirus causes at least one kind of leukemia. Herpesvirus in chickens causes a highly contagious form of **leukemia,** called **Marek's disease,** which was common enough to be a serious economic problem. This leukemia is now prevented with an antiviral vaccine.

Herpesviruses are strongly implicated in several types of malignancies in humans. A kind of lymphoma (**Burkitt's lymphoma**) is common among children in certain parts of Africa. This lymphoma is almost certainly caused by a herpesvirus called the **Epstein-Barr virus (EB virus),** named after its discoverers. It is able to transform cultured human cells, adding to the evidence that it is the causative agent in Burkitt's lymphoma. The EB virus also causes **infectious mononucleosis,** a transient disease characterized by overproduction of lymphocytes. Why the virus can take two courses in humans is partially understood. Children in Africa who are also infected with the malarial parasite (a protozoan) have weakened immune systems. Immunological defense against EB virus is therefore probably impaired, although why this impairment leads to virus-induced lymphoma is not known. A normally functioning immune system apparently limits the virus to a transient lymphoma-like disease, infectious mononucleosis.

Herpes simplex virus II (also called **Herpes genitalis**) has been implicated as a causative agent in cervical cancer of the human uterus, but the evidence is not conclusive. Infection with herpes simplex II is now the most rapidly increasing venereal disease throughout the world.

Cytomegaloviruses. Many adult humans carry **cytomegaloviruses** with no apparent ill effect, that is, it remains in a latent state. It may be transmitted from one adult to another through saliva or sexual intercourse. The virus is transmitted from an infected mother to a fetus, for example, during passage of the fetus through the birth canal and can cause severe abnormalities in postnatal development. It sometimes causes abortion. Some observations have implicated cytomegalovirus as a causative agent in human prostatic and bladder cancers.

Hepatitis B virus. This virus causes a severe form of hepatitis. In some individuals the hepatitis is followed by liver cancer (hepatocarcinoma). In Africa, hepatitis B virus, in conjunction with aflatoxin consumed as a contaminant of nuts and grains, is responsible for the very high rate of liver cancer in that part of the world.

Oncogenic Viruses with Single-Stranded RNA—Retroviruses

These RNA viruses, known as **retroviruses** for a reason discussed later, are of three types, B-type, C-type, and human T-cell lymphotropic viruses.

B-Type viruses. These are also called milk viruses because they have been found in the milk of some mammals (mouse, rat, monkey, and human). The **B-type virus** was one of the first viruses proven to cause cancer in animals. It causes breast cancer in mice, but evidence that the human B-type milk virus causes human cancer is weak.

C-Type viruses. There are many **C-type viruses,** infecting a wide variety of animals. They cause leuke-

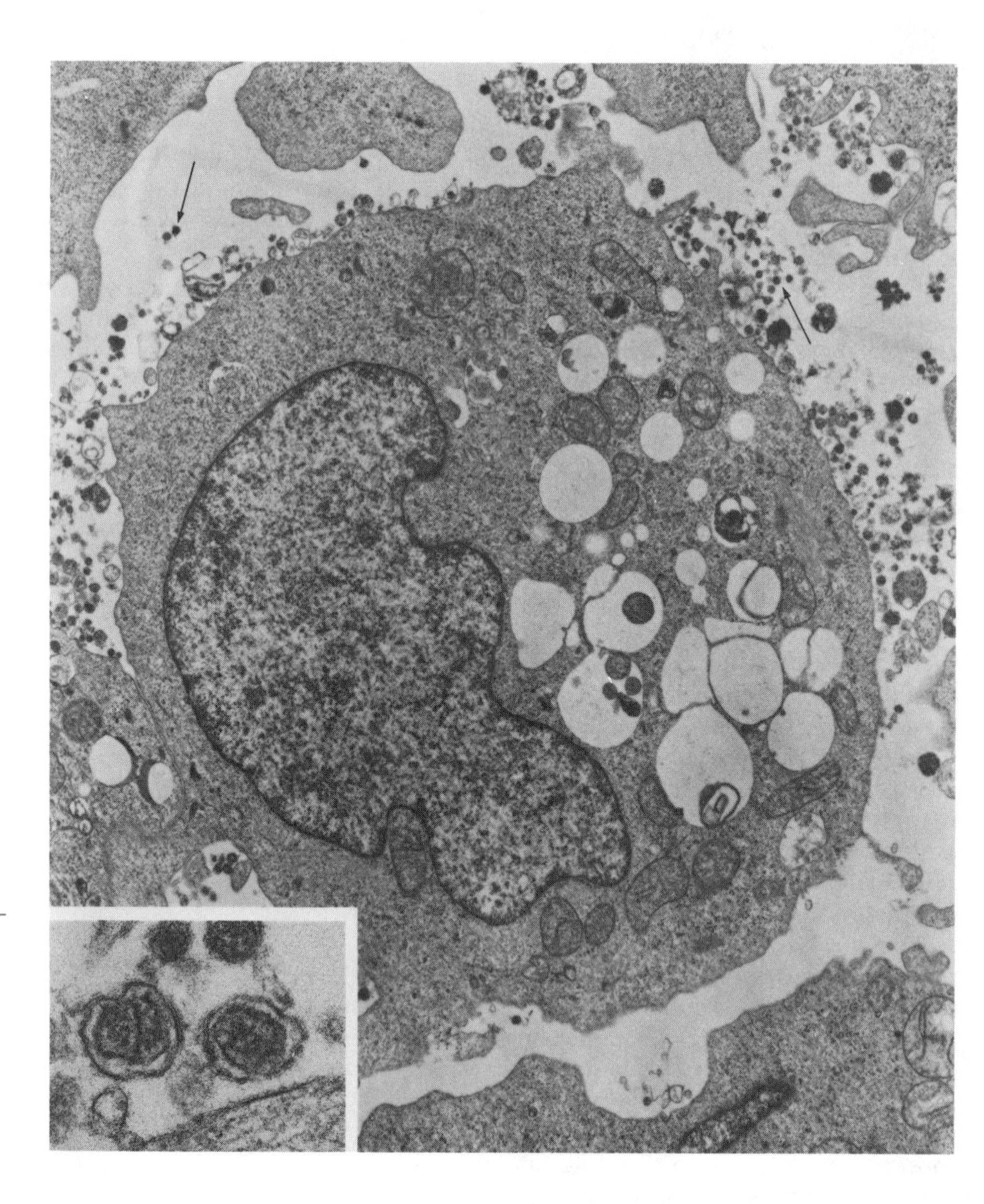

FIGURE 13-10 Electron micrograph of a human leukemia cell. No virus particles are discernible within this section, but many HTLV-I particles are present immediately outside (arrows). The inset contains an enlargement of two HTLV-I particles. [Courtesy of S. Z. Salahuddin and Robert C. Gallo.]

mia, lymphoma, and sarcoma in birds and mammals. C-type viruses have occasionally been seen by electron microscopy in human leukemia and sarcoma cells, but none has been proven to cause these cancers in humans. A few of the many extensively studied examples of C-type viruses are **mouse leukemia virus, Rous sarcoma virus** (originally found in chickens but capable of transforming mammalian cells), and **feline leukemia viruses (FLV).** About 50 percent of domestic cats in the United States are infected with FLV, but only about 40 cats per 100,000 develop leukemia per year. Why so few of the infected cats develop leukemia is not known. Feline leukemia virus is produced in the respiratory epithelium of healthy cats and is transmitted through saliva. A highly effective vaccine has been developed that prevents cat leukemia caused by this virus. There is no evidence that FLV causes disease in humans. A virus similar to FLV causes leukemia and lymphoma in cows, but it is not known whether a cat virus can infect a cow.

Human T-cell lymphotropic viruses. These viruses are so named because of their strong propensity to infect the class of human lymphocytes called T cells (see Chapter 15 for a discussion of T cells and other cells of the immune system). One of these viruses—**Human T-Cell Lymphotropic Virus I** or **HTLV-I** (Figure 13-10)—causes adult T-cell leukemia in humans, a disease that is rare in the United States and Europe, but more common in southern Japan and other parts of the world.

Another member of this group, HTLV-III, causes **Acquired Immune Deficiency Syndrome** or **AIDS.** In 1986 the virus was renamed **Human Immuno-Deficiency Virus** or **HIV** because it infects other cells in addition to T cells. T cells have surface receptors for HIV, and it is taken in by receptor-mediated endocytosis (Chapter 5). In T cells the virus reproduces abundantly, killing the host cell. Viruses released from a killed host cell infect other T cells, eventually severely depleting the infected individual of T cells, which are a crucial part of the immune system of defense against infection by bacteria, fungi, protozoa, and viruses. For example, individuals infected with HIV frequently die of subsequent infection by a protozoan called *Pneumocystis carinii,* which causes fatal pneumonia. Human immuno-deficiency virus also frequently invades the brain, causing death through neurological injury. HIV can infect another class of lymphocytes called **B cells** transforming them into cancer cells, giving rise to a fatal lymphoma. Further, HIV also contributes to the development of **Kaposi's sarcoma,** formerly a very rare cancer.

Human immuno-deficiency virus exhibits **latency;** that is, for an unknown reason it may infect an individual without causing any disease. Such individuals may carry the virus for several years without apparent ill effect, during which time they may spread it to other individuals. The phenomenon of latency of viruses in general is poorly understood. At first it was thought that most people who had been infected with HIV would fail to develop disease. By monitoring infected, disease-free individuals for several years, it has become apparent that most of such infected individuals eventually develop fatal disease.

Human immuno-deficiency virus undergoes frequent mutation, a property called **hypermutability.** By 1986 more than 80 different strains of HIV had been identified in the United States. Hypermutability results in frequent changes in the protein coat of the virus, the part of the virus that is recognized by the immune system. Frequent changes in the protein coat prevent the immune system from mounting an effective attack on the virus. Cold viruses (rhinoviruses) behave similarly, preventing the development of long-term immunity to colds. Hypermutability of the gene encoding the protein coat has made it impossible to develop vaccines against colds. For the same reason development of a vaccine against HIV is extremely difficult.

Some strains of HIV can apparently be transmitted only via blood, such as through blood transfusions. The virus is present in semen and is transmitted by homosexual contact. Some strains are also transmitted by heterosexual contact. HIV originated in Africa, probably in the 1950s, possibly by transmission from monkeys to humans. Adaptation of the virus to humans probably involved some mutational changes. The virus became an exploding medical problem in Africa in the late 1970s and in the United States in the early 1980s.

VIRUS INFECTION

When a virus successfully infects a cell, it multiplies, producing thousands of new virus particles in a single cell in a matter of hours (Figure 13-11(a)). This leads to bursting of the cell (cell lysis) and release of the new virus particles, which may in turn infect other cells. This behavior of a virus is called **productive** or **lytic infection.** Lytic infection is the mechanism by which many nononcogenic viruses (polio, measles, smallpox, cold, flu, etc.) reproduce in large numbers. The killing of cells in the process gives rise to the symptoms of a particular viral disease in an organism, depending on which tissues are infected by the virus.

Lytic infection is the only kind of infection that most animal viruses undergo. Oncogenic viruses, on

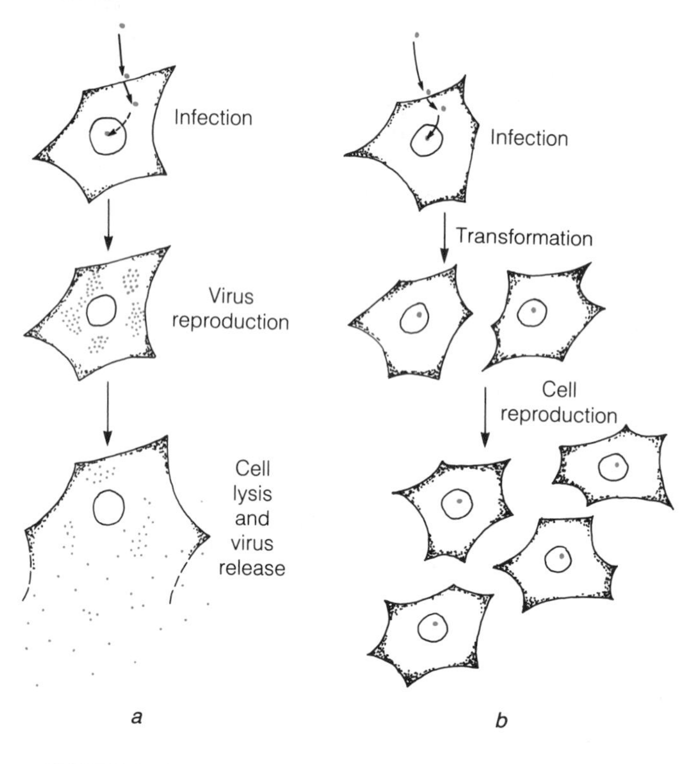

FIGURE 13-11
Two behaviors of viruses in cells. (a) Nononcogenic and some oncogenic viruses infect a cell and reproduce in the cytoplasm or nucleus, depending on the type of virus. The new virus particles are released by lysis of the cell. (b) Infection by an oncogenic virus may lead to stimulation of cell reproduction and transformation of the cell to a cancer cell. Some or all the viral genome remains in the nucleus and replicates in each cell cycle.

the other hand, take a different route within a cell; the virus does not reproduce extensively and does not cause cell lysis. Instead of killing the cell, the virus permanently alters the cell's growth characteristics, that is, the cell is transformed by the virus, and all descendants of the cell are also transformed (Figure 13-11(b)).

Whether or not a virus can successfully infect a cell and whether infection leads to a productive infection or transforms the cell depends on poorly understood virus-cell interrelations that are determined by the genetic properties both of the virus and the cell. When a virus undergoes a productive infection in a particular kind of cell, the cell is said to be **permissive** for that virus. For example, monkey cells are permissive for SV40 virus but mouse cells are nonpermissive for SV40. A cell that is **nonpermissive** for a particular virus may become transformed by that virus. Thus, permissive cells are generally killed by a virus as the virus achieves productive infection, and nonpermissive cells may remain unaffected or may be transformed. The general observation that permissive cells do not undergo transformation by a particular oncogenic virus is not a hard and fast rule; rather, permissiveness and nonpermissiveness refer to a usual pattern of virus-cell interaction. For example, mouse cells are permissive for polyoma virus, that is, *usually* they undergo lytic infection, but mouse cells may also sometimes be transformed by the polyoma virus.

Transformation by an Oncogenic DNA Virus

Much has been learned about the molecular events surrounding transformation of cultured cells by oncogenic viruses. Unfortunately, our understanding of the process is still incomplete, and we still do not know exactly how an oncogenic virus transforms a normal cell into a cancer cell.

After entry of an oncogenic virus into a transformable cell, RNA molecules that can hybridize with DNA prepared from pure virus begin to appear. This proves that genes of the viral chromosome are being transcribed. Transcription of viral genes is known to be a requirement for transformation, for example, inhibition of this transcription with a virus-specific inhibitor called **interferon** (Chapter 15) prevents the virus from transforming the cell.

A key effect of the viral gene activity is stimulation of the replication of cellular DNA. The effect may not be noticeable in cells that are already actively proliferating, but is striking when oncogenic viruses infect cells arrested in G1 by density-dependent inhibition. Within hours after infection the arrested cells begin DNA replication and subsequently divide. This stimulation by the virus is important because the *fixation* of the trans-

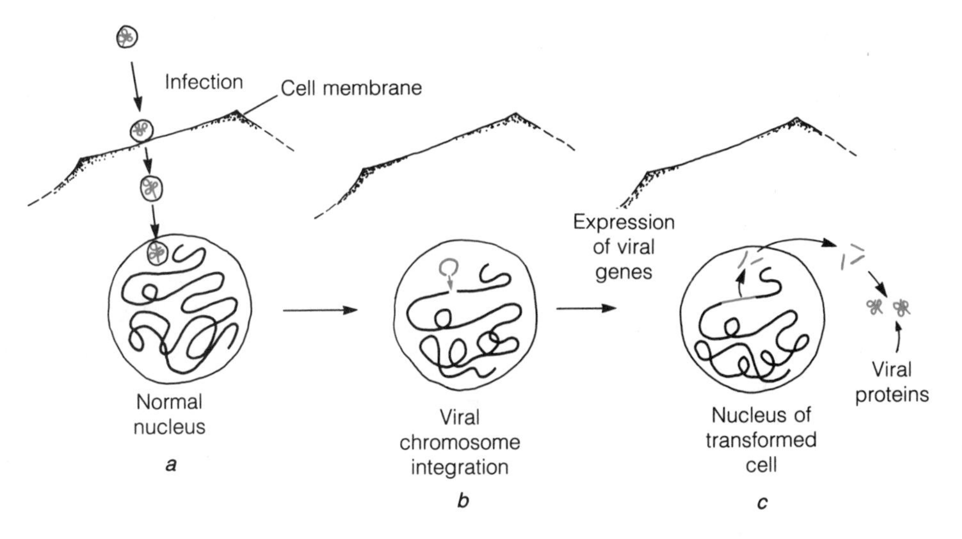

FIGURE 13-12 Cell transformation by a DNA oncogenic virus. (a) The virus infects a cell and enters the nucleus. (b) The protein coat of the virus particle is removed, releasing the circular viral chromosome. The viral chromosome breaks and its ends are joined to the ends at a break in one of the cell's chromosomes. (c) Genes of the integrated viral chromosomes are transcribed, and the mRNA is translated into proteins.

formed state of the cell can occur only during the S period of the cell cycle.

Once the transformed state has been induced (as soon as 24 hours after virus infection), the change is permanent, and all of the descendants of the original transformed cell are of the transformed type. Yet, no infectious virus particles can be recovered from a culture of transformed cells. (In this respect DNA and RNA oncogenic viruses differ because virus particles are usually found in and released from cells transformed by an RNA virus.) However, two kinds of experiments with cells transformed by SV40 or polyoma virus show that a complete viral genome is still present in the transformed cell.

In the first experiment two cells are fused together to form a hybrid cell. When a transformed cell (e.g., a hamster cell transformed with SV40) is fused to a permissive cell (e.g., a cell from the green monkey), virus particles begin to be produced in the nucleus contributed by the transformed cell. Subsequently, the binucleated cell bursts, releasing large numbers of infectious virus particles (in this case, SV40 particles). Since complete particles are formed, the complete viral genome must have still been present in the transformed cell.

In the second type of experiment DNA is extracted from transformed cells and added to a culture of permissive cells, a procedure called **DNA transfection.** The permissive cells take up the DNA and then begin to produce virus particles identical to the virus originally used to induce transformation, proving the presence of the viral genome in the transformed cell.

During transformation the viral genome becomes integrated into one of the chromosomes of the cell (Figure 13-12). Integration is random and can occur in any chromosome, and many viral genomes may become integrated. Integration occurs by breakage of the DNA molecule in a cellular chromosome, followed by joining of the two ends of the viral DNA to the two ends in the break of the cellular DNA. Since viral DNA molecules are closed circles, they also must break for integration to occur.

Viral transformation is an inefficient event. Even when 10^4 to 10^5 infectious virus particles *per cell* are added to a culture, only a few percent of the cells become transformed. The reason for this inefficiency is not known, but it is possible that the integration of the viral genome into a cell chromosome is the limiting step. This is suggested by the observation that conditions in which breakage of cell chromosomes is enhanced also favor a higher rate of viral transformation. Presumably, chromosome breakage provides increased opportunity for insertion of a viral chromosome. For example, cultured human cells are ordinarily transformed by SV40 with a very low frequency. Irradiation of human cells with a sublethal amount of X rays causes breakage of chromosomes and at the same time increases manyfold the susceptibility to SV40 transformation. Also, individuals with certain genetically based disorders such as Fanconi's anemia, Down's syndrome (mongolism, caused by the presence of an extra chromosome number 21), and Klinefelter's syndrome (males with two X chromosomes and one Y chromosome) have a higher frequency of chromosome breaks. Such individuals are not only more prone to develop cancer, but their cells grown in culture are also more susceptible to transformation by SV40.

Once the viral DNA is integrated, it replicates as any other part of the cellular chromosomes and is thus inherited from cell to cell, maintaining the transformed state of a cell from one cell generation to the next.

The genes carried in the chromosome of a virus code for proteins that function in the production of more virus particles. The double-stranded DNA of SV40 codes for two proteins that form the coat of the particle and another protein, the core protein, which is complexed with DNA to form the core of the virus particle. Oncogenic viruses contain in addition one or two genes that are responsible for transformation of their host cells. The SV40 virus contains two genes, designated *t* and *T*, that are essential for transformation. Adenoviruses carry a gene called *E1A* that is responsible for transformation. Indeed, cell transformation can be achieved by integration into a host cell chromosome of only the fragment of the viral chromosome that carries the transformation gene or genes but has none of the genes that function in viral reproduction.

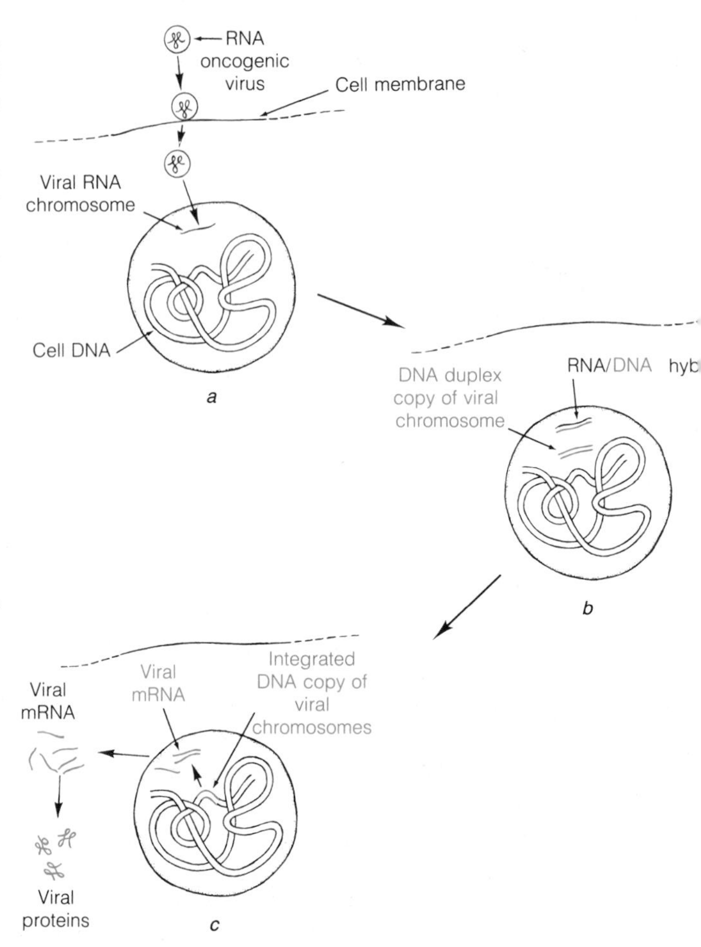

FIGURE 13-13
Transformation by a retrovirus. (a) Uncoating of the viral RNA chromosome. (b) Formation of a RNA/DNA hybrid followed by synthesis of a DNA duplex copy of the viral chromosome. (c) Integration of the DNA copy of the viral chromosome into a chromosome of the cell. Expression of the viral genome results in transformation of the cell.

Transformation by RNA Viruses

In general, RNA oncogenic viruses transform cells by the same principles as DNA oncogenic viruses. However, for transformation to occur, the single-stranded RNA chromosome must first be copied into a double-stranded DNA molecule (Figure 13-13). The key to this process is the enzyme reverse transcriptase (described in Chapter 11; see Figure 11-23), the gene for which is carried by all oncogenic RNA viruses. After infection of a susceptible cell, reverse transcriptase catalyzes the polymerization of a DNA chain using the single-stranded viral RNA as a guiding template. This creates a hybrid double helix, one strand of which is the original viral RNA molecule and the other strand of which is the newly synthesized DNA chain. The RNA is digested

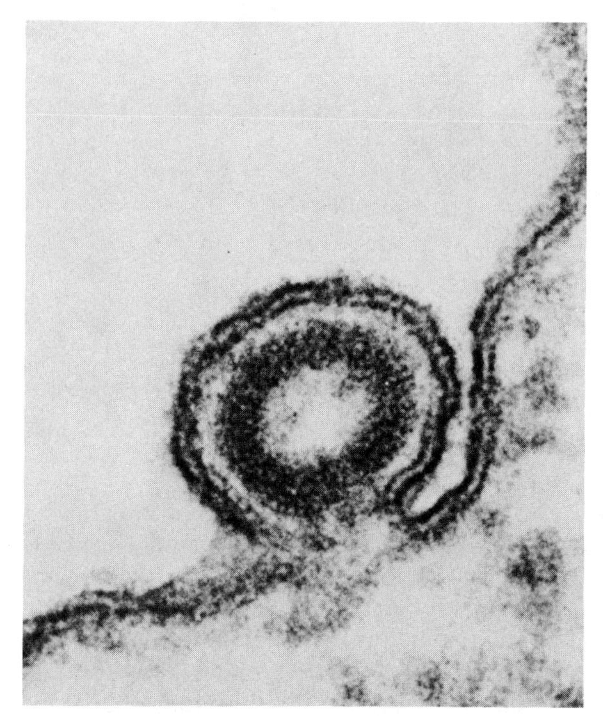

a

FIGURE 13-14
(a) Electron micrograph of a retrovirus (RNA oncogenic virus) particle budding out of a cell. (b) A drawing of the structures in the micrograph. This particular virus causes leukemia in animals. [(a) Courtesy of Etienne de Harven]

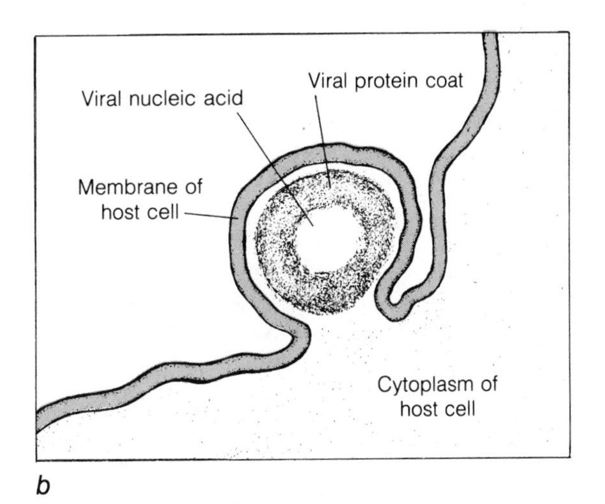

b

from the double helix and a DNA/DNA duplex is formed. The overall effect is to transfer the genetic information of the single-stranded viral RNA into a DNA duplex, the precise reverse of transcription. The DNA duplex is then integrated into a cellular chromosome (Figure 13-13). The reverse transcription of RNA into a DNA version as a part of transformation is the basis of the name retrovirus (retro = backward). The integrated DNA may be repeatedly transcribed in its entirety to yield single-stranded RNA molecules, which may become packaged into virus particles, which are released singly from the transformed cell without lysing the cell. Oncogenic RNA virus particles acquire an additional coat derived from the plasma membrane as they leave the cell by a budding process (Figure 13-14).

Oncogenes

As in the case of DNA oncogenic viruses, the integrated DNA version of viral RNA contains at least one gene whose activity converts the cell to the transformed state.

The chromosome of most retroviruses carries only a few genes. Many conform to the pattern shown for Rous sarcoma virus (RSV) in Figure 13-15. The ***gag*** gene codes an internal protein that complexes with the RNA to form the core of the virus. The ***env*** gene encodes the coat protein subunits. The ***pol*** gene codes for reverse transcriptase, and the ***onc*** gene encodes the protein responsible for cell transformation. The *onc* gene in RSV is given the more specific name ***src,*** for sarcoma.

The *src* gene was the first oncogene in a retrovirus to be identified and studied intensively. Its cancer-causing property was proven with a temperature-sensitive mutation. Cultured chicken cells infected with a virus containing a mutation in the *src* gene remained normal at 39°C (protein encoded by the gene is not functional) but immediately became transformed when cultured at 33°C (protein encoded by the gene is functional). The mutated virus reproduced normally in permissive cells

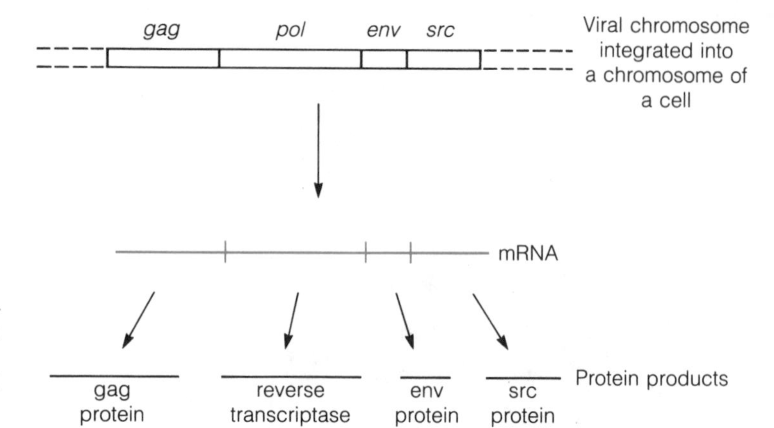

FIGURE 13-15
The chromosome of Rous sarcoma virus in its DNA duplex form integrated into a cellular chromosome. The DNA is transcribed into a single mRNA molecule that is translated into four proteins.

at both temperatures, which showed that the *src* gene product was unnecessary for virus reproduction.

Reverse transcriptase has been used to convert the RNA of a nonmutated Rous sarcoma virus into a double-stranded DNA copy (cDNA) in a cell-free system. A fragment of the cDNA containing the *src* gene has been prepared and shown to be sufficient by itself to transform a cell.

More than 25 different oncogenes have been identified in retroviruses that infect birds and mammals. Some of these, including the *src* gene, encode a **protein kinase.** The kinases encoded by these oncogenes are found in the cytoplasm bound to the inner side of the plasma membrane. These kinases bring about transformation of a cell by phosphorylating particular proteins, but little is known about these proteins or how their phosphorylation results in such dramatic changes in the cell. Other oncogenes encode proteins that are localized in the nucleus, but no enzyme activity has been identified for these as yet. Still others encode growth factors. Other kinds of oncogenes encode proteins that disrupt regulation of cell behavior by affecting some other part of the mechanisms that regulate cell behavior.

A normal cell can also be transformed to a cancer cell with carcinogenic chemicals or by radiation. Since most of these agents are proven mutagens, they presumably transform cells by mutating one or another gene involved in regulating cell behavior. At least some of the genes that are altered by chemical carcinogens and radiation in a transformed cell are closely related to the oncogenes of viruses, as shown by the following kind of experiment (Figure 13-16). Total DNA was isolated from human nonvirally induced cancer cells. One or more altered genes contained in that DNA were presumed to be responsible for the cancerous state of the cells. The presumption was tested by a **transfection assay** in which DNA purified from cancer cells was added to a confluent monolayer of mouse 3T3 cells in culture under conditions that promoted the uptake of the DNA by the cells. Mouse 3T3 cells are normal in that they show density-dependent inhibition, anchorage dependence, a high serum requirement, and ability to arrest in G1 and to enter G0. They have only two transformed characteristics—immortality and aneuploidy. A few of the millions of 3T3 cells in the density-arrested, confluent monolayer became fully transformed by uptake of the human cancer DNA. These cells were evident because they were released from density-dependent inhibition (arrest in G1) and resumed proliferation, forming foci of piled up growing cells.

In the transfection experiment with DNA from a human cancer it was proven that the transformed 3T3 cells had taken up human DNA because they now contained human Alu sequences. The Alu sequence is repeated in over 900,000 copies dispersed throughout human DNA, but the sequence is absent in mouse DNA (see Chapter 11). The transformed 3T3 cells had taken up other human DNA sequences in addition to the one causing transformation, making it impossible to identify which particular sequence was responsible for transformation (and the cancer). Therefore, DNA was isolated from the transformed 3T3 cells and added to another confluent culture of normal 3T3 cells. Because the added DNA was mostly mouse DNA with a very limited number of human DNA sequences, it could be reasonably expected that transformed cells now appearing would contain only that human DNA sequence or sequences necessary for transformation.

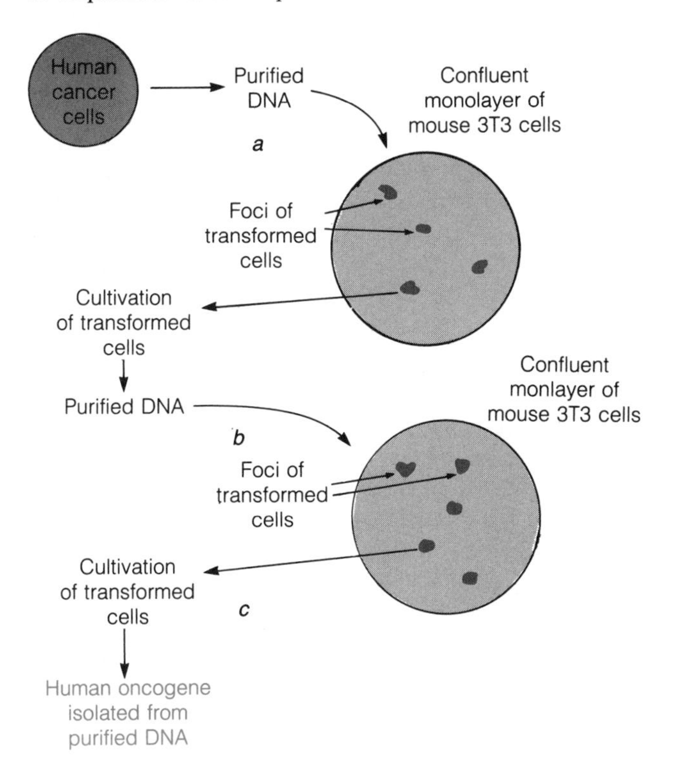

FIGURE 13-16
Identification of a human oncogene by DNA transfection. (a) DNA purified from human cancer tissue added to mouse 3T3 cells arrested by density-dependent inhibition results in transformation of a few cells, which form clones (foci). (b) DNA purified from transformed mouse cells is added to a second confluent culture of mouse cells. (c) The human DNA (oncogene) responsible for transformation is isolated from DNA purified from the transformed mouse cells.

Such proved to be the case. Only one human sequence was needed to transform mouse 3T3 cells, and moreover this sequence contained a copy of the sequence of the oncogene known as ***ras*** present in a virus that causes sarcoma in mice (murine sarcoma virus). Hybridization experiments have subsequently shown that all *normal* human and other mammalian cells contain a gene corresponding to the *ras* gene. However, the *ras* gene in the human cancer cells used in the above experiment has a mutation in one codon, changing a single amino acid in the protein encoded by the gene. Thus, mutation in the normal *ras* gene altered the function of the protein it encodes so as to convert a normal cell to a cancer cell.

Subsequently, over 20 other genes have been found in normal mammalian cells that correspond to oncogenes present in oncogenic viruses. Some of these genes have been found in a wide variety of unicellular and multicellular eukaryotes, such as, yeast and *Drosophila*. Often, small differences between the cellular versions and the viral versions are responsible for the oncogenic nature of the viral version. The genes in normal cells are called **protooncogenes.** These genes are believed to be important in regulation of cell reproduction and differentiation. The right kind of mutation in a protooncogene converts it to an oncogene. Such cellular oncogenes, found in cancer cells are called **c-oncogenes** or **c-oncs** to distinguish them from oncogenes, called **v-oncs,** carried by oncogenic viruses.

Many kinds of human cancers have been found to contain c-oncogenes derived by alteration of a protooncogene. In addition, using the DNA transfection assay described above in the original identification of the cellular *ras* oncogene (*c-ras*), several c-oncs have been found in human cancers that have no known viral counterpart.

In the transfection assay used to identify the *ras* gene the test cell was the 3T3. These cells are already transformed in two characteristics: they are immortal and aneuploid. A repeat of the transfection experiment with cancer cell DNA using normal test cells (diploid and lacking immortality) freshly explanted from mouse tissue failed to yield transformed cells. However, when two different oncogenes were introduced into normal cells, full transformation was achieved. Presumably each oncogene is responsible for a different part of transformation.

Protooncogenes are converted to cellular oncogenes in at least two ways: the first is mutation, as already cited. The mutation must occur at specific sites to alter the function of the gene product in the proper way. In the case of the *ras* protooncogene the mutation must change the amino acid either in the 12th or the 61st position of the polypeptide product. These two particular amino acid positions are important because they change the function of the protein encoded by the *ras* gene in a specific way.

The second kind of change in conversion of a protooncogene to an oncogene is one that results in overexpression of the gene without change in the encoded protein product. This can happen by any of the following: a mutation in the promoter that increases its effectiveness; movement of the protooncogene to a new location, where it comes under the influence of a strong promoter; or an amplification in the number of copies of the gene. Movement of genes occurs by breakage and rejoining of chromosomes in which two chromosomes exchange segments, a process called chromosome translocation, which can be induced by radiation and some kinds of chemical mutagens. For example, the protooncogene *abl* (present as the v-onc in a mouse leukemia virus) has been moved from its normal location in chromosome nine to chromosome 22 in almost all cases of chronic myelogenous leukemia in humans. In its new position it is expressed at a much higher rate. The protooncogene *myc* (present as a v-onc in avian myelocytomatosis virus) is amplified in some human cancers.

Viral oncogenes in retroviruses were almost certainly derived from mRNA copies of protooncogenes in the distant past. In a rare molecular accident an mRNA molecule of a protooncogene probably became incorporated into the RNA chromosome of a nononcogenic virus. Part of the support for this idea is the presence of introns in cellular protooncogenes and the absence of all such introns in the corresponding v-oncs of viruses.

Several of the protooncogenes present in mammals have been found in a wide variety of eukaryotes, including *Drosophila*, protozoa, and yeast cells. This indicates that these protooncogenes are ancient genes and probably encode proteins that are fundamental to regulation of cell reproduction in all eukaryotes. The gene in yeast known as *CDC-28*, which encodes a protein that functions in the Start switch in G1, has a high degree of homology of nucleotide sequence with the *src* gene of mammals. This homology indicates that the *src* gene evolved from an ancient gene important in regulating initiation of DNA replication.

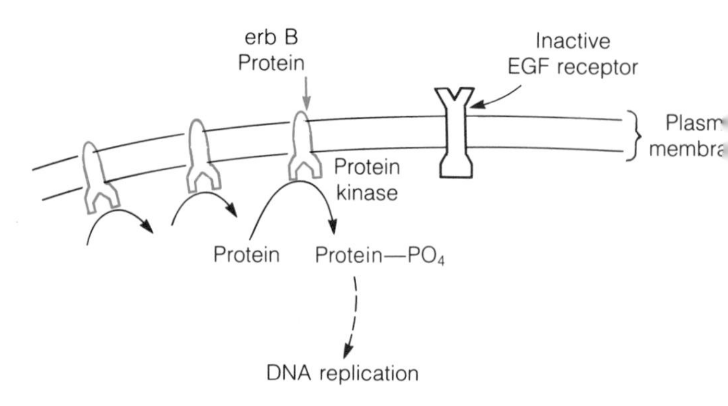

FIGURE 13-17
The erb B protein encoded by an oncogene in avian erythroblastosis virus is homologous to the protein kinase domain of the EGF receptor. The constitutive protein kinase activity of erb B protein results in cell transformation.

Growth Factors and Oncogenes

How conversion of protooncogenes to oncogenes disrupts normal cell behavior is not known, but clearly at least some protooncogenes encode proteins that are part of the mechanism by which growth factors regulate cell reproduction. For example, the receptor in the plasma membrane for epidermal growth factor (EGF receptor) is encoded by a gene that shows extensive homology with *v-erb B*, one of two oncogenes in avian erythroblastosis virus (Figure 13-17). The product of *v-erb B* lacks the EGF binding domain but does have the domain that acts as a protein kinase. Without the EGF binding domain to act as a modulator, the protein kinase domain is constantly expressed. This is equivalent to constant stimulation of a normal cell by EGF (although no EGF is present) and consequent loss of regulation of cell reproduction by EGF. The oncogene *sis*, present in simian sarcoma virus, is homologous to the normal cellular gene encoding platelet-derived growth factor (PDGF). Transformation with simian sarcoma virus results in a constant high level of synthesis of PDGF, which obliterates regulation of cell reproduction by extracellular PDGF.

Thus, recent research has tied together viral oncogenes, cellular protooncogenes, and mechanisms of regulation of animal cell reproduction by growth factors. Discoveries in each of these areas directly impinges on understanding of the other two, producing a cooperative effect that will lead to a molecular understanding of regulation of cell reproduction and the loss of regulated behavior in a cancer cell.

PROBLEMS

1. What are the two main reasons for the greatly increased occurrence of cancer in recent decades?
2. Why does cancer mainly occur in older people?
3. List three main properties of cancer cells that underlie the disease they cause.
4. What are the three principal methods used to eliminate cancer cells? What limits the usefulness of each of these methods?
5. What are the three classes of agents that are known to cause cancer?
6. What is the principal cause of
   1. Skin cancer?
   2. Lung cancer?
   3. Adult T-cell leukemia?
7. List two advantages of Ames-type tests.
8. In addition to lung cancer what other kinds of cancer are caused by tobacco?
9. What is aflatoxin and what kind of cancer is it believed to cause?
10. Two types of evidence indicate that dietary fat is a major cause of cancer. What are they?
11. How do cancer cells differ from normal cells in their growth characteristics in culture?
12. How are viral oncogenes believed to have originated?
13. Name three kinds of viruses that cause cancer in humans. What cancers do they cause?
14. Why is it so difficult to develop a vaccine against HIV?
15. Why is reverse transcriptase essential for transformation of a cell by an RNA oncogenic virus?
16. How was the first human cellular oncogene discovered?
17. What are two ways in which a protooncogene can be converted to an oncogene?

CHAPTER 14

Cell Motility

See Figure 14-3.

All organisms depend on movements of molecules from place to place. For example, ATP made in one region is often used elsewhere in the cell, and nutrients must be captured from the environment. Small organisms have no problem in randomizing their internal molecules since diffusion is a fast process over short distances like the few micrometers that span a bacterium. However, large cells need to mix their components, and they have developed special internal machinery to get molecules and ions from one place to another. This machinery is also used in big cells and multicellular organisms to position and move organelles and change cell shape. In addition, many organisms have evolved life-styles that depend on their ability to move around. Such mobility can be useful in feeding, in avoiding becoming food, and in sexual reproduction. Thus, natural selection has favored the evolution of motile systems.

Cells have evolved a variety of ways of moving. In some cases, movement is only intracellular, consisting of the active displacement or propulsion of materials and structures inside the cell. A common example of this is rapid **cytoplasmic streaming,** seen prominently in some algae and plant cells, in which it is called **cyclosis.** Another example is the movement of chromosomes during mitosis and meiosis. Other forms of intracellular movement are the fast propulsion of vesicles in the axon of a nerve cell and the rapid shuttling of pigmented granules in some kinds of pigment-containing cells.

In addition to strictly intracellular movement, many cells are capable of various forms of motility in which the cell changes its shape or moves in relation to its environment. An example is the gliding movement of certain species of bacteria and diatoms. The pinching in two of a cell during cytokinesis is another type of movement. Some cells, such as amoebae, move by cytoplasmic streaming (**amoeboid movement**). Some cells, like bacteria, sperm, some protozoa, and algae, move by means of flagella. Still others, including some protozoa and cells in many kinds of larvae, are propelled by cilia. Many kinds of bacteria are propelled by flagella, but the structure and operation of bacterial flagella are very different from the structure and operation of eukaryotic flagella. Certain animal tissue cells possess cilia (e.g., the ciliated epithelium of the respiratory tract), but since these ciliated cells are held stationary in the tissue, the motion of their cilia causes the external medium to flow across their surface. Cellular movement is developed to its highest degree in muscle cells. Since muscle cells remain in relatively fixed positions, their contraction produces displacement of other

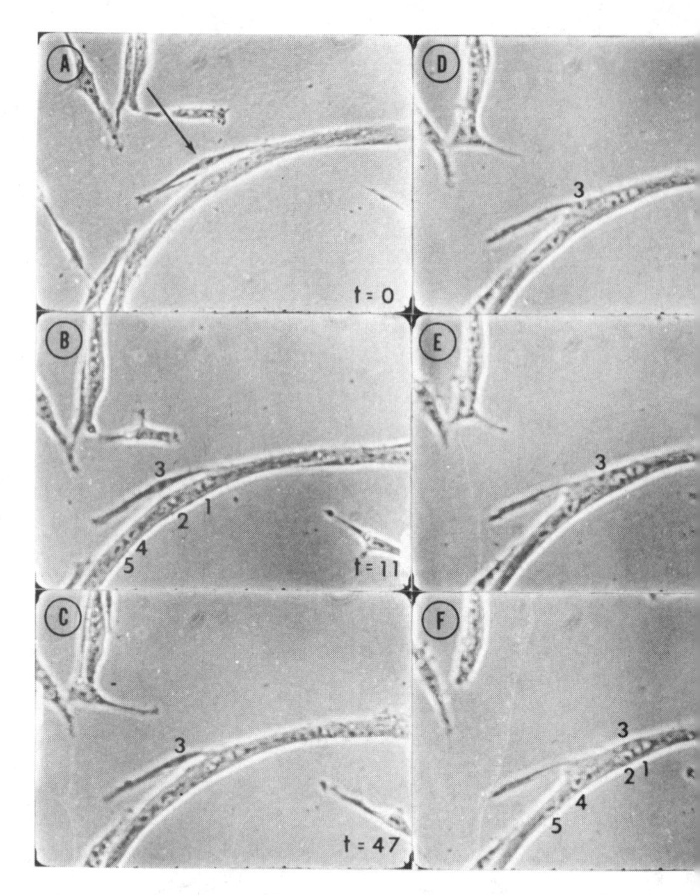

FIGURE 14-1
Fusion of myoblasts into multinucleated muscle cells. In this series of light micrographs a myoblast (arrow) at zero time in A has, in B at 11 minutes, fused at one end to a forming multinucleated muscle cell. The locations of four nuclei in the muscle cell and the nucleus in the myoblast are marked by numbers 1, 2, 4, and 5. At successively later intervals (C and D) the myoblast nucleus (3) becomes integrated into the muscle cell. In E and F the myoblast nucleus has moved well into the multinucleated muscle cell; fusion of the myoblast cytoplasm with the muscle cell is still incomplete in F. [Courtesy of Irwin R. Konigsberg. From *Develop. Biol.*, 63:11–26 (1978).]

parts of the organism, for example, the movement of bony parts (**skeletal muscle**), propulsion of blood (**cardiac muscle**), and the propulsive movements, **peristalsis,** of the alimentary tract (**smooth muscle**).

Movement of all cells is achieved by a mechanochemical process in which chemical energy is used to do mechanical work moving a cell or cell parts. In prokaryotes two kinds of movement that probably involve different mechanochemical processes are known: (1) gliding movements in some species of bacteria, and (2) flagellar movement.

In eukaryotes, three mechanochemical processes are known. Each involves the interaction of two kinds of proteins that produces force by hydrolyzing ATP to ADP and P_i. The three known processes in eukaryotes, which account for most kinds of cell movement, are the following:

1. Interaction of the protein actin with myosin. The actin-myosin system is the basis of all muscle contraction in invertebrates and vertebrates. It is also the mechanism for generating cyclosis in plant and algal cells, and cytokinesis in animal cells and most protozoa; it is probably responsible for cytoplasmic streaming in amoeboid movement.
2. Interaction of the proteins tubulin and **dynein.** The tubulin-dynein system is the basis of all ciliary and flagellar movement in eukaryotic cells.
3. Interaction of the proteins tubulin and **kinesin.** The tubulin-kinesin system is the basis for the fast transport of vesicles and probably other particulate structures, for example, transport of vesicles in nerve axons and other kinds of cells.

MOVEMENT BASED ON ACTIN-MYOSIN INTERACTION

Of all mechanisms of movement the interaction of actin and myosin in the contraction of muscle is the best understood. It is useful to begin discussion of cell movement with muscle contraction because knowledge of how it works can serve as a basis for understanding other forms of movement.

Structure and Function of Muscle Cells

Muscle cells contain large amounts of a few kinds of proteins that are highly ordered to form contractile structures. In skeletal and cardiac muscle, contractile structures are ordered in a way that gives rise to characteristic cross striations. The ordering of contractile elements in smooth or **involuntary muscle** does not produce cross striations. The molecular mechanisms of contraction of smooth muscle cells is fundamentally similar to that in striated muscle. However, the organization of the contractile machinery in smooth muscle is less well understood than for striated muscle, so only the striated muscle cell is discussed here.

Structure of a striated muscle cell. In the muscles of vertebrates the individual cells of skeletal muscle are multinucleated cylinders. They vary in diameter from 10 to 100 μm and vary in length from a few millimeters to several centimeters. These large cells are formed during muscle development by the fusion of many reproducing, uninuclear cells called **myoblasts.** Shortly before fusion, myoblasts stop reproducing and enter G0, where they remain permanently. Several thousand myoblasts may fuse to form a single, mature multinucleated muscle cell (Figure 14-1). Many thousands of multinucleated muscle cells are, in turn, packed together in bundles to form a muscle. At the two ends of the muscle, the muscle cells are attached to the collagenous fibers of tendons, and the tendons are usually attached to bones.

Myoblasts and the multinucleated cells produced by fusion of myoblasts synthesize enormous quantities of the few kinds of proteins that make up the contractile machinery. These proteins become organized in fibrous structures that give rise to the cross striations seen in a mature muscle cell. The many sets of cross striations in a single muscle cell reflect the presence of repeating contractile units called **sarcomeres** (Figure 14-2). An individual sarcomere occupies about 2.5 μm or 0.0l percent of the length of a cell, and hence one muscle cell may contain many thousands of sarcomeres. Each sarcomere contains a set of several kinds of cross striations that are visible by light microscopy but more effectively studied by electron microscopy (Figure

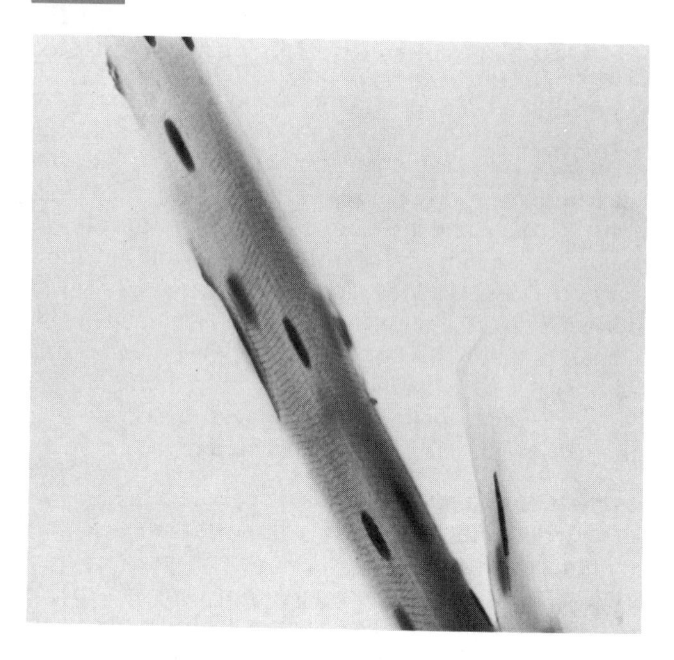

FIGURE 14-2
A light micrograph of a portion of a single skeletal muscle cell. Successive sarcomeres are marked by faint cross striations. The multiple nuclei are the dark, elongated oval structures near the surface of the cell. [Courtesy of Nadia Malouf. From *Exptl. Cell Res.*, 122:233–250 (1979).]

14-3). The cross striations consist of various bands and lines and are the result of a highly ordered arrangement of the proteins that make up the contractile machinery.

At each end of a sarcomere is a cross striation called the **Z line,** which forms the boundary shared by two successive sarcomeres. About 1.6 μm of the central portion of a sarcomere is occupied by a more densely staining structure called the **A band.** Between the end of an A band and the Z line is a less dense region called the **I band.** A single I band occupies portions of adjacent sarcomeres and is transected by a Z line (Figure 14-3). In short, the muscle cell consists of alternating A bands and I bands organized into sarcomeres, and each sarcomere consists of a central A band flanked on each end by one-half of an I band. The A band is marked through its center by a less dense region called the **H zone,** which contains a central line called the **M line.** The significance of these structural arrangements will become clear a little later in the discussion of the arrangement of the molecules that bring about contraction.

The structural basis of cross striations is found in the many **myofibrils** that run lengthwise in a muscle cell (Figure 14-4). The myofibrils are constructed from contractile proteins. They are all parallel and in register, so their cross striations are lined up with one another. A myofibril consists of lighter and darker staining regions; the lighter are the I bands and the darker staining regions are the A bands in a sarcomere. The myofibrils are surrounded by cytoplasm, which in muscle

FIGURE 14-3
Electron micrograph of part of a skeletal muscle cell of a fish. The cell is filled with fibrous bundles called myofibrils. Each myofibril is marked by a repeating pattern of cross striations. One repeat is called a sarcomere. The membranous vesicles between myofibrils are a specialized form of the endoplasmic reticulum called the sarcoplasmic reticulum. [Courtesy of Don W. Fawcett. From D. W. Fawcett (1981) *The Cell*, 2nd Ed. W. B. Saunders and Co.]

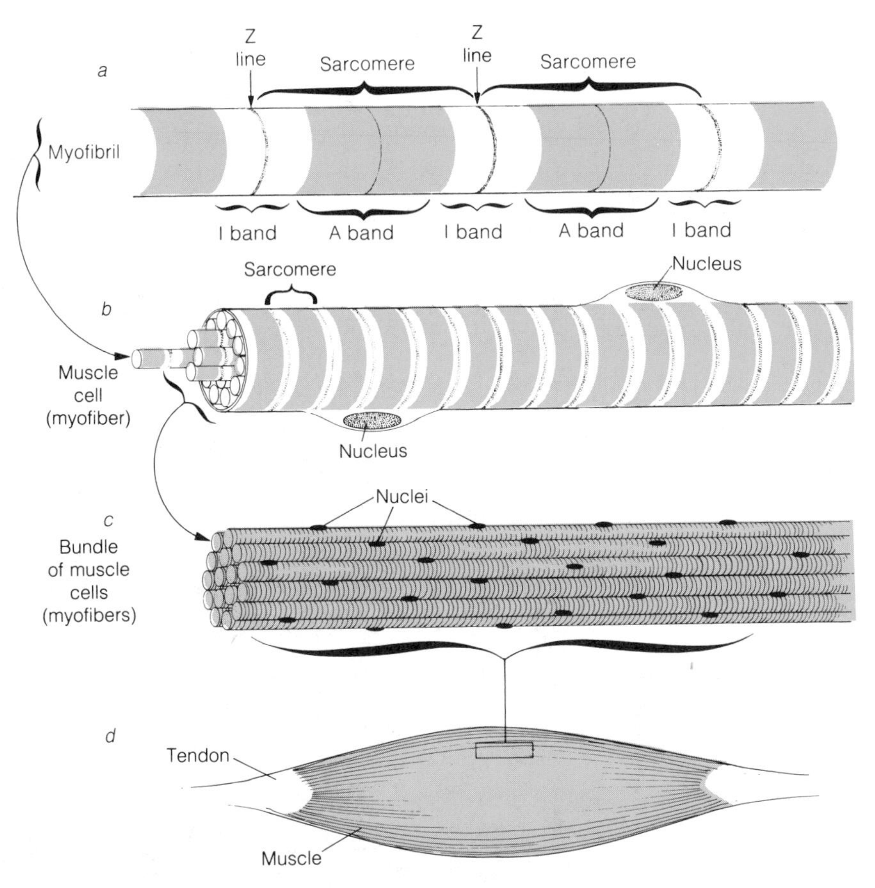

FIGURE 14-4
Structure of skeletal muscle. (a) A myofibril is a long cylinder consisting of tandemly arranged sarcomeres. A sarcomere consists of a darkly staining central region, the A band and two adjacent lightly staining regions, the I bands. Very darkly staining Z lines mark the borders of successive sarcomeres. (b) Many myofibrils are packed inside one muscle cell (myofiber). The myofibrils are in parallel register and impart the sarcomere banding pattern to the muscle cell. (c) A bundle of muscle cells. Muscle cells are multinucleated. (d) Bundles of muscle cells are in turn arranged into larger bundles within a muscle.

cells is called **sarcoplasm.** (Muscle cells are also called myofibers, but the term is avoided here because it can be confused with the term *myofibril.*) Much of the narrow space between the many myofibrils in a muscle cell is occupied by a system of membranes comparable to smooth endoplasmic reticulum, but which has a particular role in muscle contraction (see below); it is called the **sarcoplasmic reticulum.** The many nuclei occupy the outer region (Figure 14-4(b, c)) of the cylindrically shaped cell, positioned just inside the plasma membrane, which in muscle cells is called the **sarcolemma.** Organelles such as the Golgi complex, rough endoplasmic reticulum, free ribosomes and polysomes are scanty and lie concentrated around nuclei. Glycogen particles scattered through the cytoplasm provide a store of carbohydrate for glycolysis and oxidative phosphorylation. Muscle cells are richly supplied with glycolytic enzymes and possess numerous mitochondria scattered among and closely adjacent to the myofibrils, which allows for efficient transfer to myofibrils of the large amount of ATP needed to power contraction.

Each of the many myofibrils in a muscle cell is made up of a number of filaments called **myofilaments.** There are both thick and thin myofilaments.

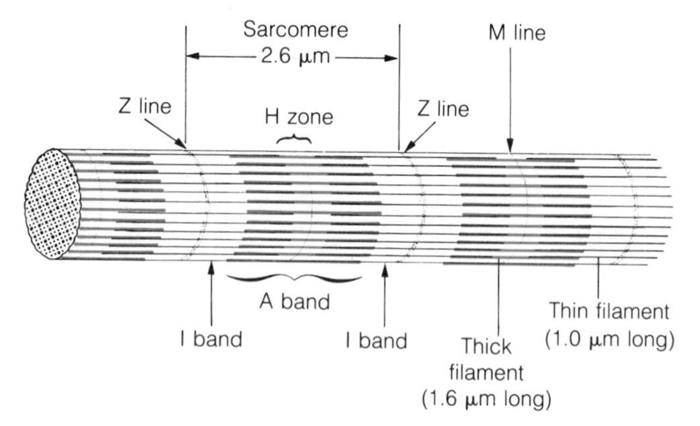

FIGURE 14-5
A portion of a myofibril. The Z lines mark the ends of sarcomeres. An A band consists of thick filaments. An I band consists of thin filaments only. The thin filaments extend part way into the A band leaving a central region, the H zone, free of thin filaments. A faintly staining M line is present in the center of the H zone.

Thick myofilaments in mammals are 1.6 μm long and are arranged in a single bundle with all of their ends in precise register (Figure 14-5). The bundle of thick myofilaments in a myofibril is what gives rise to the dark staining A band. The thin myofilaments are about 1.0 μm long and are arranged in two bundles, one extending into the sarcomere from each end, where they are attached to the Z lines. The thin myofilaments traverse one-half of an I band and enter the A band. In the A band the thin myofilaments interdigitate with the thick filaments and extend up to the edge of the **H zone** near the middle of the sarcomere (Figure 14-5). The two borders of the H zone are defined by the terminations of the two bundles of thin myofilaments that extend in from the two Z lines of the sarcomere. Thus, in a cross-sectional view of the sarcomere through an I band only thin myofilaments are visible (Figure 14-6(a)). Through part of an A band, both thick and thin myofilaments are seen (Figure 14-6(b)), but in the H zone only thick myofilaments are visible.

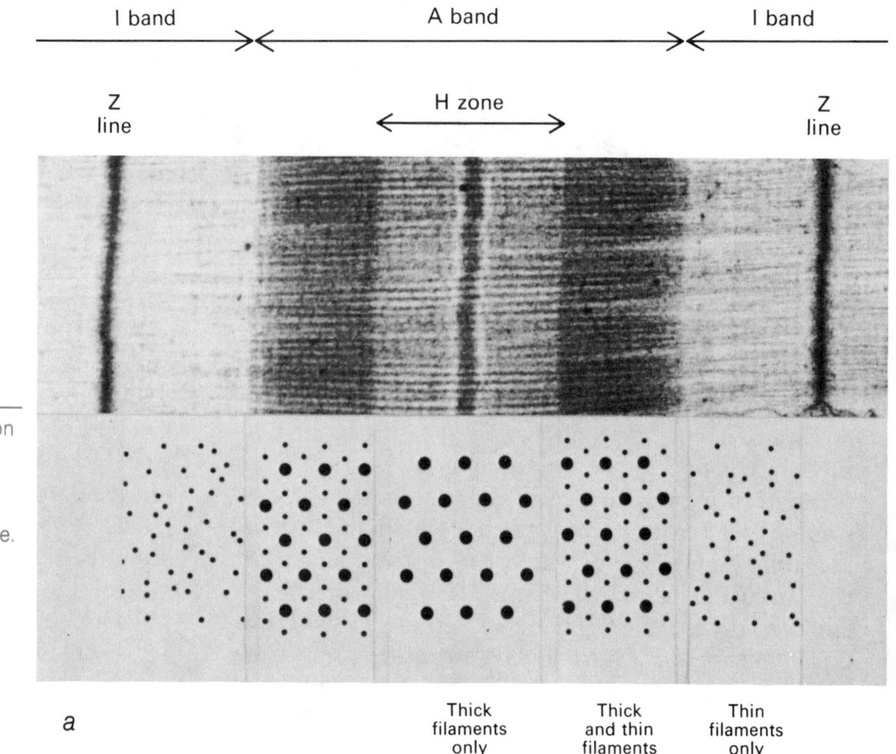

FIGURE 14-6
(a) Electron micrograph of a longitudinal section of a sarcomere. The diagrams below the micrograph show in cross section the arrangement of thick myosin filaments and thin actin filaments in various parts of the sarcomere. (b) This electron micrograph of a cross section through an A band corresponds to the center diagram in part (a). It shows six thin actin filaments arranged around each thick myosin filament. [(a) Courtesy of H. E. Huxley. From *Proc. Roy. Soc. London B.*, 178:131(1971). (b) Courtesy of Hans Ris.]

The thick and thin myofilaments interdigitate in the A band in a highly ordered way in vertebrate striated muscle. Each thick myofilament is surrounded by six thin myofilaments (Figure 14-6(b)).

Structural changes in a sarcomere during contraction. During maximum contraction of a skeletal muscle, each cell shortens about 40 percent. This shortening is accomplished by a uniform reduction in length of all the sarcomeres in the cell (from 2.6 μm to 1.7 μm). Shortening of a sarcomere brings the two Z lines closer together, without changing the length of the central A band (composed of thick myofilaments), but almost all of the H zone and the two I bands disappear (Figure 14-7). The disappearance of the H zone and I bands occurs because during contraction the thin filaments move farther into the A band. In full contraction the ends of the thin myofilaments extend in from the two Z lines and almost meet.

Electron microscopic observations on thin and thick myofilaments in relaxed and contracted muscle have revealed that muscle contraction occurs by the sliding of the thin myofilaments along the thick myofilaments. Thin and thick myofilaments interact with one another in a way that causes the thin myofilaments to be drawn into or to slide into the A band. The farther the thin myofilaments slide inward along the thick myofilaments, the greater the contraction.

Structure of thick myofilaments. Analysis of the proteins that make up the thin and thick myofilaments in a myofibril has enhanced understanding of the mechanisms by which the myofilaments are caused to slide along one another. When striated muscle cells are soaked in a physiological saline solution containing detergent to solubilize plasma membranes, the water-soluble proteins of the sarcoplasm are removed. The proteins that make up the myofibrils are not soluble in this solution; therefore, the sarcomeric structure of cells remains intact. If the muscle is now soaked in a cold alkaline solution of KCl (0.6 M), 50 percent of the protein is extracted, and the thick myofilaments disappear. Analysis of the extracted protein shows that thick myofilaments are composed almost entirely of a single type of protein, namely, myosin.

Myosin is a very large protein (molecular weight, 500,000) consisting of six polypeptide chains, two large ones and four smaller ones. The two large polypeptides are identical and have molecular weights of 200,000. They each contain about 1800 amino acids, which places them among the longer polypeptides known. A large portion of these chains is coiled into α-helices (Chapter 2). In myosin the α-helical portions of the two polypeptides are wound around each other, forming a long linear molecule or tail (Figure 14-8). The remaining part of each chain is folded into a globular configuration that forms a head portion of this large polypeptide. The four small polypeptides (molecular weight from 16,000 to 21,000) are bound in pairs to the two globular portions of the large polypeptides, thus enlarging the globular head of a myosin molecule.

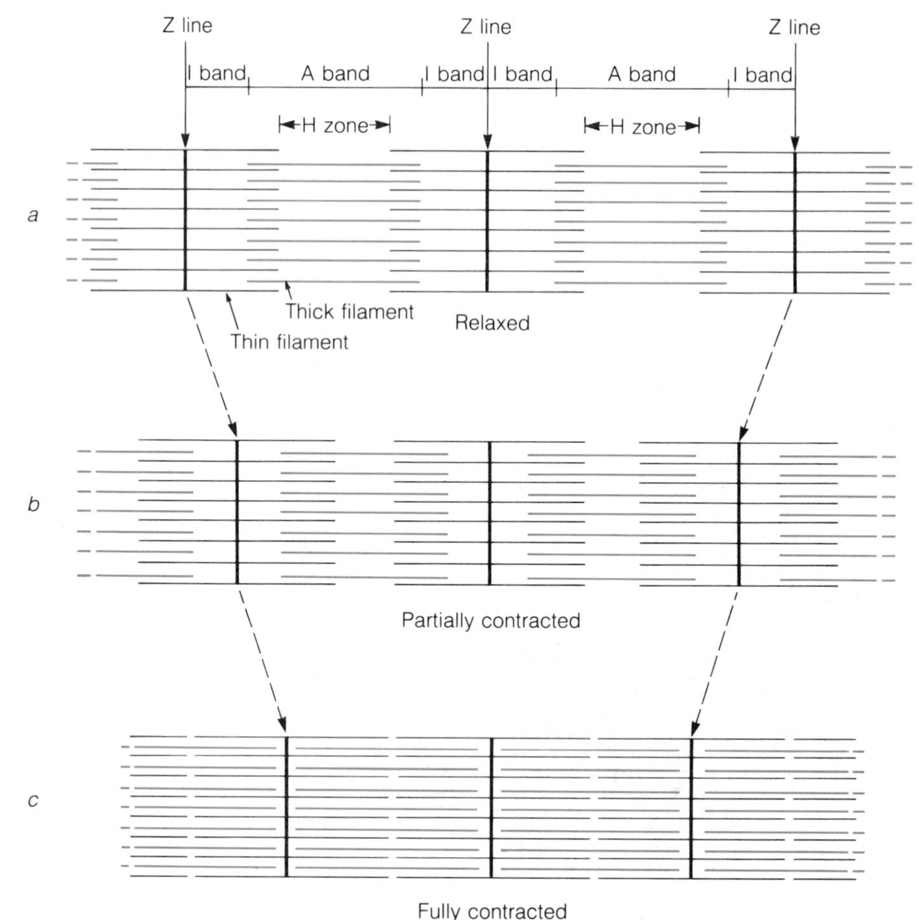

FIGURE 14-7 Contraction of muscle. (a) Relaxed myofibril. The I bands are at their longest and the Z lines are at maximum separation. The H zone in the A band is long. (b) Partially contracted myofibril. The thin microfilaments have been drawn further into the A band, resulting in shortening of I band and the H zone. The Z lines are now closer together. (c) Fully contracted myofibril. The I band has almost completely disappeared as a result of thin filaments being drawn into the A bands. The ends of thin filaments almost meet in the center of the sarcomere, and the H zone is barely visible.

About 500 myosin protein molecules are packed together in a highly ordered pattern to form a single thick myofilament in an A band. All the tails are arranged in parallel (Figure 14-9) and are held together by side-to-side associations due to interactions between amino acid side chains. The length of one myosin molecule is only a small fraction of the length of a thick myofilament, but the molecules are polymerized in a staggered pattern such that the tails form the continuous core of the thick myofilament and the globular

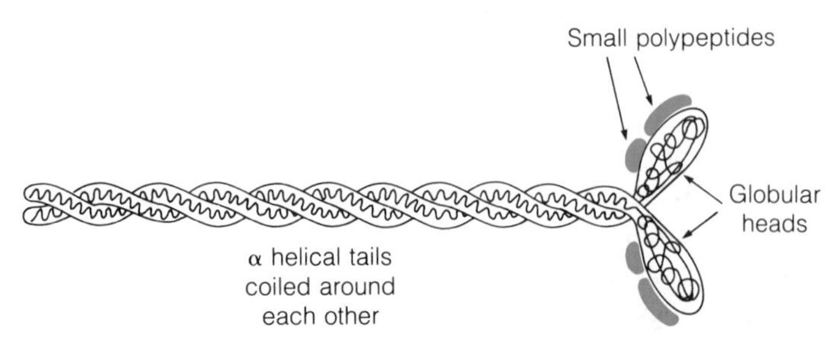

FIGURE 14-8 The arrangement of six polypeptides in a myosin molecule. The two large polypeptides have globular heads and long α-helical tails that coil around one another. Four smaller polypeptides increase the size of the globular heads.

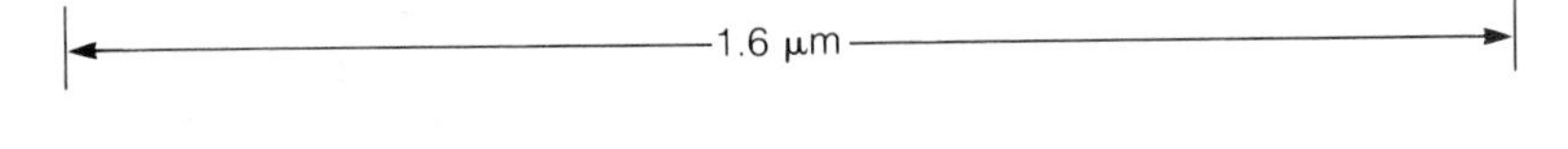

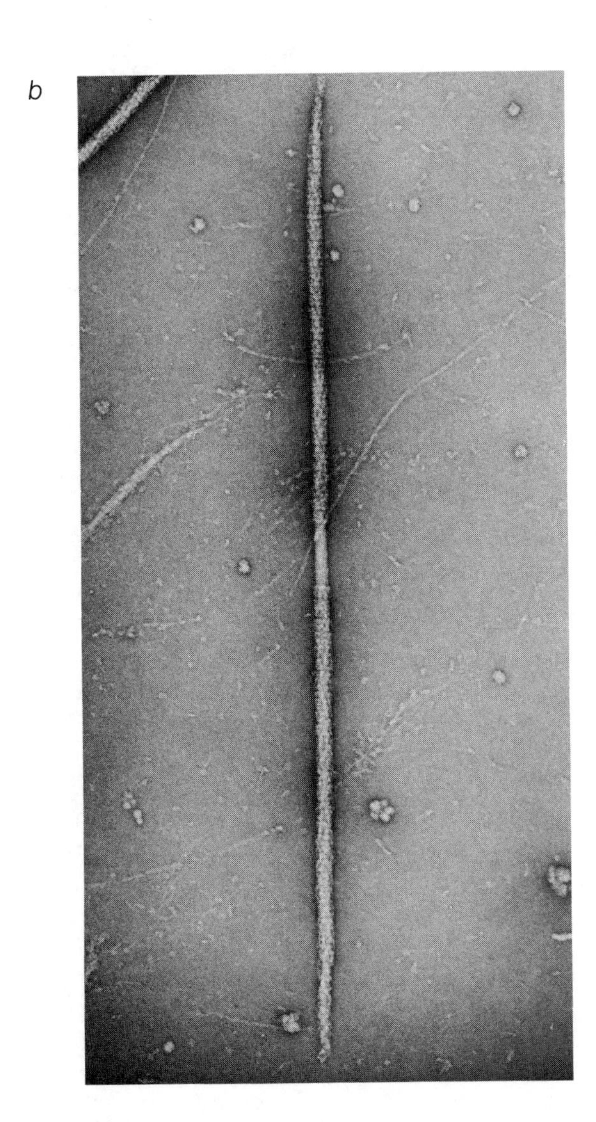

FIGURE 14-9
Structure of a thick filament. (a) Schematic diagram of myosin molecules aggregated to form a thick filament, with globular heads projecting outward. (b) Electron micrograph of a myosin thick filament isolated from scallop muscle. The clear zone in the center is in the H zone in a sarcomere. [(b) Courtesy of Roger Craig. From *J. Mol. Biol.*, 165:303–320 (1983).]

heads project from the side of the core at regular intervals (Figure 14-9). The molecules in one half of the thick myofilament are oriented in one direction, while those in the other half of the myofilament are turned around; the myosin tails always point toward the middle of the thick myofilament. The central section of a thick myofilament is composed only of tails of myosin molecules and looks smooth because it lacks the projecting heads. This is the clear zone in Figure 14-9. Looking down on a thick myofilament from the end, the heads are arranged helically and project from the core in six directions (Figure 14-10). This results in six lines of heads that are in positions to interact with the six thin myofilaments that surround each thick myofilament in a myofibril. These interactions between heads and thin myofilaments are, as we shall see, the crux of the sliding filament mechanism described later. Furthermore, the head of the myosin molecule, in addition to its interaction with thin myofilaments, possesses

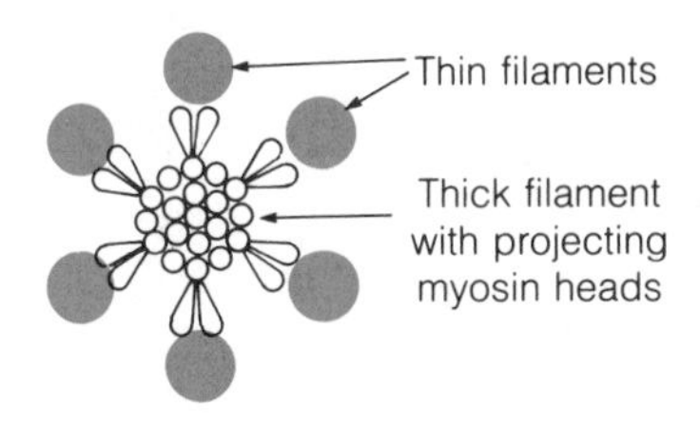

FIGURE 14-10
Schematic diagram of a thick filament seen end on. Pairs of heads of myosin molecules project radially toward six surrounding thin filaments.

an active enzymatic site that splits ATP into ADP + P_i. This splitting of ATP provides energy to fuel muscle contraction. The energetics of contraction are discussed later.

Myosin makes up at least 95 percent of the mass of thick filaments. Other proteins are present in thick myofilaments in trace amounts. One of the trace proteins, named the **C-protein,** is stretched out in parallel with the myosin tails, winding around the core of the thick myofilament. This arrangement suggests that C-protein serves to hold the myosin tails together, stabilizing the core of the thick myofilament.

Another protein present in small amounts is bound to the very center of the thick myofilament, where only myosin tails are present, and gives rise to the **M line** in the middle of the sarcomere (Figure 14-5). The function of this protein is not known, but it may have a role in connecting the thick myofilaments with one another and maintaining their precise order.

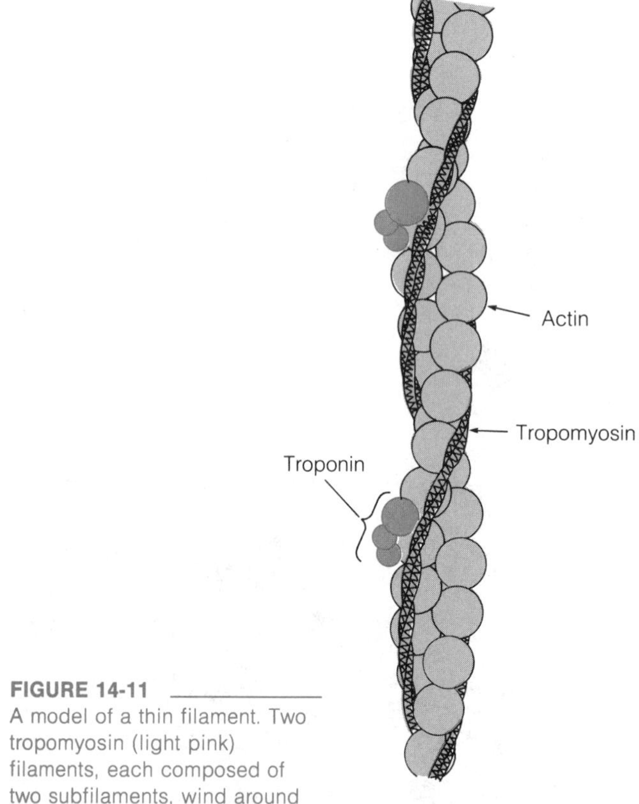

FIGURE 14-11
A model of a thin filament. Two tropomyosin (light pink) filaments, each composed of two subfilaments, wind around the actin chain. They block the binding of the globular heads of myosin molecules (in thick filaments) to the actin molecules. Troponin consists of three polypeptides and binds both to actin and to tropomyosin.

Structure of thin myofilaments. When muscle from which myosin has been extracted (with alkaline 0.6 M KCl) is subsequently soaked in cold, slightly alkaline water, the major proteins of the thin myofilaments dissolve. The principal protein in this extract is actin, which was described in the discussion of the cytoskeleton in Chapter 6. When actin is solubilized from the thin myofilament, it is recovered in individual globular molecules called G-actin. When the inorganic ion concentration of a solution of pure G-actin is raised to the level present in intact muscle, the globular molecules join together to form long chains (Figure 6-11) called F-actin. This is the form in which actin occurs in the thin myofilaments. In the thin myofilament, two chains of F-actin are wound around each other in helical fashion as described in Chapter 6. The double helix has a diameter of about 9.5 nm, which is the diameter of a thin filament.

Actin is the major component of thin myofilaments, but two other proteins, **tropomyosin** and **troponin,** are also present. The polypeptides of tropomyosin have an α-helical secondary structure, and are stretched out as long, thin molecules attached lengthwise to the F-actin. Purified tropomyosin molecules in solution stick also to each other in end-to-end fashion. This end-to-end association gives rise to long filaments that lie in the two grooves of the F-actin double helix (Figure 14-11). Each tropomyosin molecule is about 40 nm long and spans seven actin molecules.

The third protein of thin filaments, troponin, is a globular molecule consisting of three polypeptides. One molecule of troponin is bound to each molecule of tropomyosin. As described below, tropomyosin and troponin regulate contraction by regulating the interaction of thin and thick filaments, that is, the interaction of F-actin with myosin.

The thin myofilaments are all anchored at one end by strong attachment to a Z line. The Z line consists of a protein called α-actinin, which forms a multimolecular plate across the sarcomere (the Z line) that appears to bind the ends of F-actin chains (thin filaments). The α-actinin forms the connections between the thin myofilaments of one sarcomere and those in the next. It therefore plays an essential mechanical role in linking together the myofibril. The polarity of F-actin chains was noted previously in Chapter 6; F-actin chains are elongated by addition of G-actin monomers; the addition occurring far more readily at one end than the other. The polar nature of F-actin can also be demonstrated directly by the binding of modified myosin. Myosin molecules are digested briefly with the protease papain, which cuts them at the base of the double head, releasing the tail and the two halves of the head. When the purified heads are added to actin microfilaments, they bind in a highly regular pattern in which the myosin heads project from the microfilament at an angle (Figure 14-12). When observed by electron microscopy a microfilament decorated with myosin heads gives the impression of a chevron or a line of arrowheads, with the barbs of the arrowheads, formed by the myosin heads, all facing in the same direction. With this method all the microfilaments in the I band on one side of a Z line have the same orientation; the arrow tips point away from the Z line and the barbs are directed toward the Z line. During the formation of microfilaments, G-actin monomers are added more rapidly to the end corresponding to the direction of the barbs, that is, the end at the Z line.

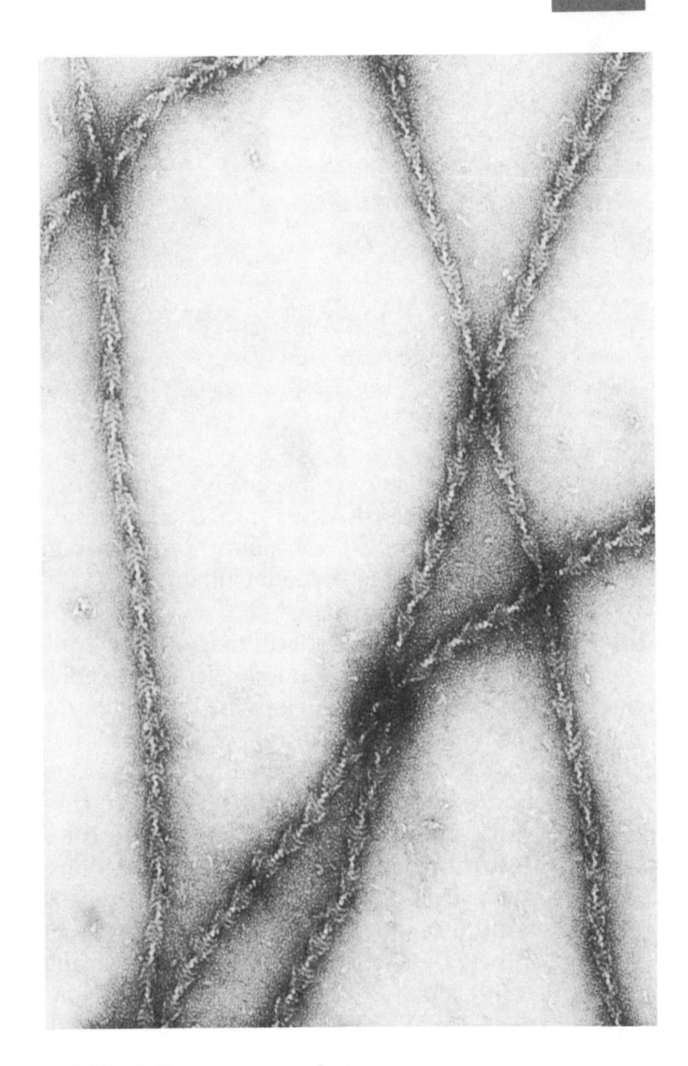

FIGURE 14-12
Electron micrograph of actin microfilaments decorated with heads of myosin molecules. All the heads attached to one microfilament tilt back in the same direction, giving the impression of arrowheads and showing that the microfilament has polarity. [Courtesy of Roger Craig.]

The interaction of actin with myosin in muscle contraction. Electron microscopic observations have indicated that muscle cells contract by an interaction of thick and thin filaments that causes the thin filaments to be drawn farther into the A band. By observing the interactions of the various purified muscle proteins, we have achieved considerable understanding of the molecular basis of the filament sliding.

When pure actin and pure myosin are mixed in solution, they form filamentous complexes called **actomyosin.** Actomyosin filaments can be made to aggregate into microscopically visible threads. One way to accomplish this is to layer actomyosin on the surface of water. The film of actomyosin can be skimmed from the water surface to make a fiber or thread, much as a film of scum from boiled milk can be collected into a threadlike mass. When such artifically prepared actomyosin threads are immersed in a solution of ATP, they contract. With an appropriate arrangement, actomyo-

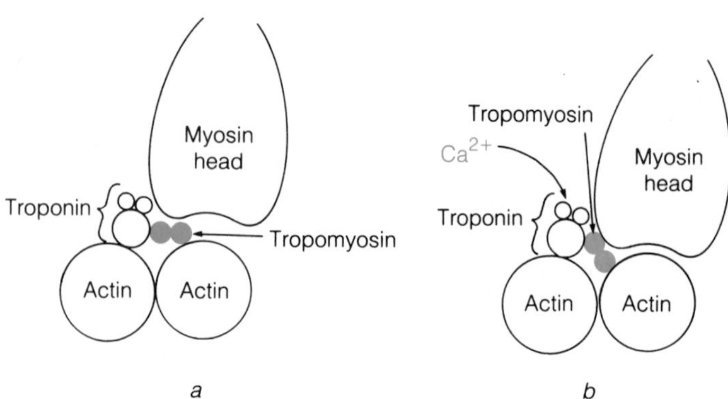

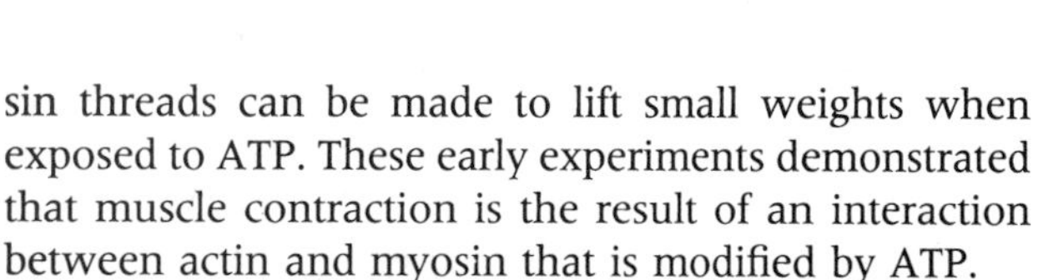
FIGURE 14-13
Schematic diagram of the interaction of myosin and actin. (a) In the absence of Ca^{2+} tropomyosin (red) prevents the binding of myosin heads to actin. (b) When the Ca^{2+} concentration rises, Ca^{2+} binds to a subunit of troponin, which causes the tropomyosin to shift slightly into the groove of the actin thin filament. The shift in position of tropomyosin allows the myosin heads to bind to actin. Lowering of the Ca^{2+} concentration results in reversal of these events.

sin threads can be made to lift small weights when exposed to ATP. These early experiments demonstrated that muscle contraction is the result of an interaction between actin and myosin that is modified by ATP.

Roles of Ca^{2+}, tropomyosin, and troponin in contraction. Ca^{2+} ions play a central role in regulating muscle contraction. When Ca^{2+} ions are injected into a muscle cell, it contracts. Also, direct evidence for intracellular release of Ca^{2+} ions associated with contraction has come from injection of **aequorin,** a protein obtained from a luminescent jellyfish. Aequorin emits photons (**bioluminescence**) when it binds Ca^{2+} ions. Aequorin has been injected into muscle cells and the muscle then stimulated to contract. A very brief burst of light is seen being emitted for a fraction of a second *before* the contraction. By reconstitution experiments researchers have been able to learn about the roles of Ca^{2+} ions, tropomyosin, and troponin in regulating muscle contraction. Actomyosin threads prepared from pure actin and pure myosin contract whether or not Ca^{2+} ions are present in the solution. When tropomyosin and troponin are added to actomyosin, ATP induces contraction only when Ca^{2+} ions are present. Thus, tropomyosin and troponin prevent contraction, but Ca^{2+} ions reverse this inhibition. It is now known that in the absence of Ca^{2+} ions, tropomyosin, which is arranged along the grooves of the F-actin double helix, blocks the interaction of F-actin with myosin and therefore blocks contraction. Troponin, which is bound to the tropomyosin, has a Ca^{2+} ion binding site. When added to the system, Ca^{2+} ions become bound to troponin. The binding of Ca^{2+} ions causes a change in the shape of the troponin molecule. However, the troponin molecule is bound to tropomyosin, and the Ca^{2+} ion-induced change in troponin shape causes the tropomyosin to move deeper into the grooves of F-actin (Figure 14-13). This shift in the location of tropomyosin uncovers sites on the F-actin chains to which myosin can now bind. The binding of actin and myosin allows contraction to take place.

The interaction between myosin and actin occurs between the globular head portion of myosin and actin molecules. By electron microscopy this interaction is observed as cross bridges between thick and thin filaments (Figure 14-14). The cross bridges, which correspond to the heads of myosin molecules, connect the thick and thin filaments only when the muscle is contracting. In the relaxed state the cross bridges are released because the interaction of the heads of myosin molecules with the F-actin of the thin filaments is blocked by tropomyosin. Indeed, not only is the interaction blocked in relaxed muscle, but the myosin heads also appear to be retracted closer to the thick filament.

Molecular mechanism of muscle contraction. The studies of intact cells by electron microscopy, physiological analyses of living muscle cells, and biochemical studies on isolated and recombined components have all contributed to a rather complete picture of muscle contraction.

Myosin is organized into thick filaments in the A band, with the heads of molecules projecting at precise spacings from the axis of the thick filament. The myosin heads bind ATP. Normally, the concentration of ATP (the energy charge) in the sarcoplasm is high, and in relaxed muscle the binding sites on myosin are occupied by ATP, which is cleaved into ADP and P_i by the ATPase of myosin heads as soon as they are bound. With ADP and P_i bound to them the myosin heads are now poised to interact with actin and bring about contraction, but the interaction is blocked by tropomyosin. The interaction is brought about by Ca^{2+} ions.

In relaxed muscle the concentration of Ca^{2+} ions is insufficient to bring about Ca^{2+} ion binding to the troponin of thin filaments. Under these circumstances, the tropomyosin of thin filaments is located slightly out of the groove of the F-actin chains in a position that blocks the interaction of myosin heads with F-actin. The arrival of a nerve impulse (Chapter 15) at the muscle cell causes an extremely rapid flux of Ca^{2+} ions into the sarcoplasm, and the myofibrils are bathed with these ions (control of the Ca^{2+} ion concentration in the sarcoplasm is discussed below). With the sudden increase in Ca^{2+} ion concentration in the sarcoplasm, the Ca^{2+} ion-binding sites of troponin become saturated. The configurational change in troponin molecules induced by Ca^{2+} ion-binding draws the tropomyosin molecules deeper into the grooves of the F-actin of the thin filaments, thereby unmasking the sites on F-actin to which myosin heads can bind. Myosin heads, each of which is already in an activated state by virtue of having bound ADP + P_i to it, bind to F-actin and form cross bridges between thick and thin filaments. The binding of myosin to actin induces the release of ADP and P_i from the myosin, causing the myosin heads to change their structure and revert to the nonactivated state.

The conversion of activated myosin to nonactivated myosin is associated with a change in molecular configuration, and this change somehow exerts a force

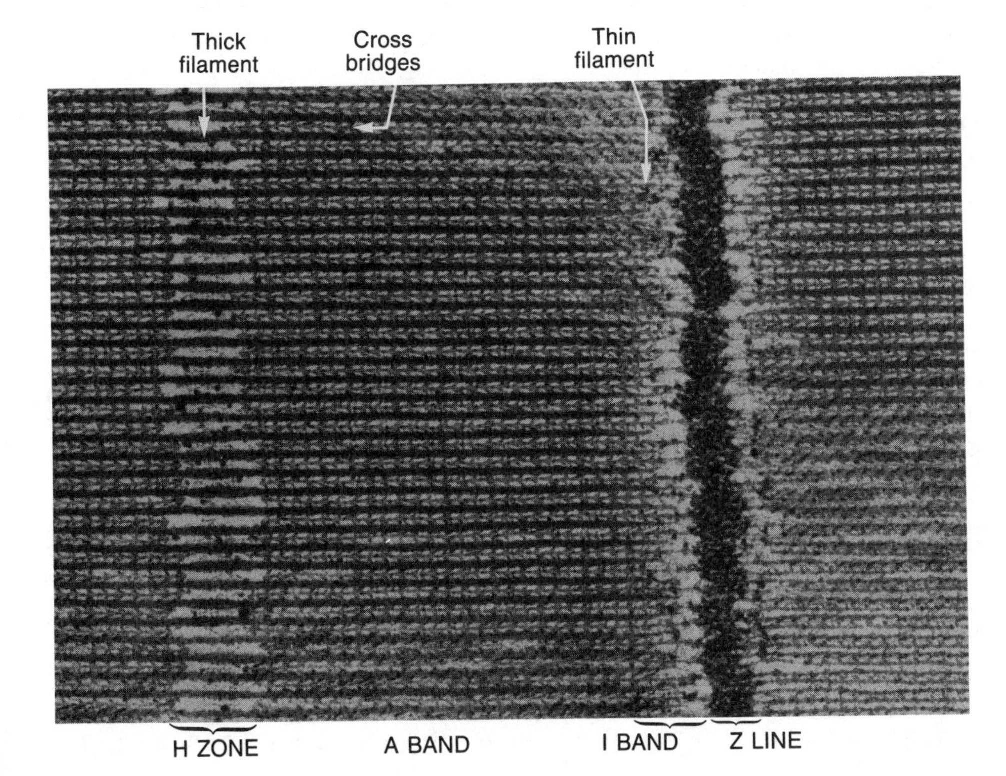

FIGURE 14-14 Electron micrograph of parts of two adjacent sarcomeres in a flight muscle of an insect (*Lethocerus*). The muscle is maximally contracted and I bands are almost completely obliterated. Thick filaments of myosin can be identified in the H zone and traced into the large A bands. Thin filaments of actin can be traced from the Z line through a narrow I band and into an A band. In the A band cross bridges formed of myosin heads project from thick filaments across to the thin filaments every 39 nm. The myosin heads project at an angle, giving rise to the chevron appearance of cross bridges with chevrons pointing toward the H zone. [Courtesy of Michael K. Reedy.]

on the thin filaments, pulling them a short distance further into the A band (about 10 nm per cross bridge). ATP can then bind once more to the empty site on the myosin molecule, and this causes myosin to fall off the actin, releasing the cross bridge, reexciting the myosin, and reextending it to a configuration in which it can cleave the ATP to ADP and P_i. The myosin is ready once again to bind to the thin filament and give it another pull. The action of multiple, asynchronous bridge cycles achieves a smooth, extensive sliding of the thin filaments into the A band, causing the sarcomere to shorten and thereby producing muscle contraction. This somewhat complicated chain of events, extending from the time of arrival of a nerve impulse at the muscle cell to the actual contraction of the muscle, takes place in a small fraction of a second. Researchers are still trying to determine how myosin accomplishes its pulling action on the thin filaments.

As long as nerve impulses continue to arrive at a muscle cell, the concentration of Ca^{2+} ions in the sarcoplasm remains high, and interaction of thick and thin filaments can continue. Thus, as rapidly as myosin heads revert to a nonactivated state, they are reactivated by binding of ATP molecules and again interact with F-actin to exert another pull. Repeated cycles of myosin interaction with actin, occurring in a small fraction of a second, results in a tendency for every sarcomere in a stimulated muscle cell to shorten. How much a sarcomere shortens depends on the load against which it is working. If contraction occurs freely, a sarcomere can shorten from 2.5 μm to 1.7 μm at maximum contraction. In a muscle cell with 20,000 sarcomeres (five cm long) this is equal to a shortening by 1.6 cm. Contraction continues until the nerve impulses cease or until the supply of ATP is used up. In contracting muscle, ATP is used up faster than it can be produced by glycolysis and mitochondrial respiration; thus, contraction cannot be maintained indefinitely.

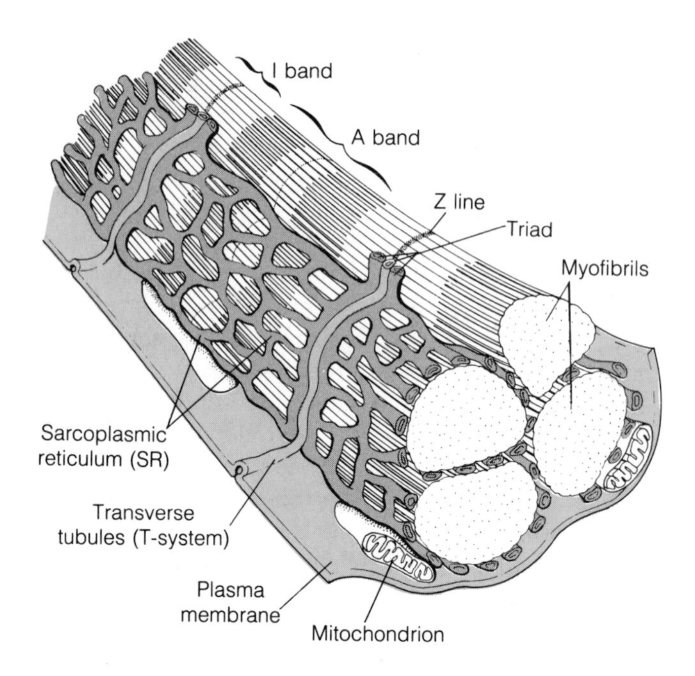

FIGURE 14-15
A drawing of part of a skeletal muscle cell showing the T system. Invaginations of the sarcolemma (plasma membrane) form a system of transverse tubules that relay a contraction signal from a neuron to the sarcoplasmic reticulum in the interior of a cell.

Control of muscle contraction. Muscle contraction is turned on and off by changes in the Ca^{2+} ion concentration in the sarcoplasm. Therefore, controlling the Ca^{2+} ion concentration regulates muscle contraction. A muscle contracts when a nerve impulse arrives at its sarcolemma (plasma membrane), but how does the nerve impulse cause the Ca^{2+} ion concentration to rise almost instantaneously throughout the entire sarcoplasm? Electron microscopic studies show that two membrane systems in the muscle cell, the **T system** and the **sarcoplasmic reticulum,** are the important elements.

The T system. The T system consists of tubular invaginations of the sarcolemma that penetrate deeply into the cell (Figure 14-15). These invaginations occur at regular intervals along the length of a muscle cell, usually at every Z line at multiple points around the circumference of the tubular-shaped muscle cell. When a nerve impulse reaches a muscle cell, the impulse is transferred to the sarcolemma at the point of contact between nerve cell and muscle cell, the **myoneural junction.** From the myoneural junction the impulse is propagated over the entire sarcolemma, including the invaginations of the sarcolemma that form the transverse tubule of the T system. In this way a nerve impulse transferred to a muscle cell is almost instantaneously propagated not only over the cell surface but along paths to the internal regions of the cell as well.

The T system ensures that an impulse arriving at one place on a muscle cell will produce virtually simultaneous responses in all the sarcomeres of all the myofibrils, causing an immediate, concerted contraction of the cell as a whole. The impulse is propagated over the sarcolemma at the same rate as a nerve impulse is conducted by a nerve axon, namely, at a rate measured in meters per second. Even though propagation down the T system is slightly slower, the result is far more rapid than would be possible if propagation depended on diffusion of molecules from the sarcolemma into the cell; diffusion would propagate a signal a few micrometers per second, a million times slower than the T system.

The sarcoplasmic reticulum. The sarcoplasmic reticulum resembles the endoplasmic reticulum in its structural organization. It consists of extensively branched membranous sacs that enclose spaces comparable to the cisternae of the endoplasmic reticulum (Figure 14-15). Each sarcomere contains a physically separate section of sarcoplasmic reticulum that branches to surround each myofibril at the two ends of the sarcomere. From these major portions of the sarcoplasmic reticulum thinner branches extend along all the myofibrils. At the two ends of the sarcomere the sarcoplasmic reticulum is somewhat expanded forming what are called **terminal cisternae.**

The sarcoplasmic reticulum and the T system are in close juxtaposition, usually at the Z line. The portion of

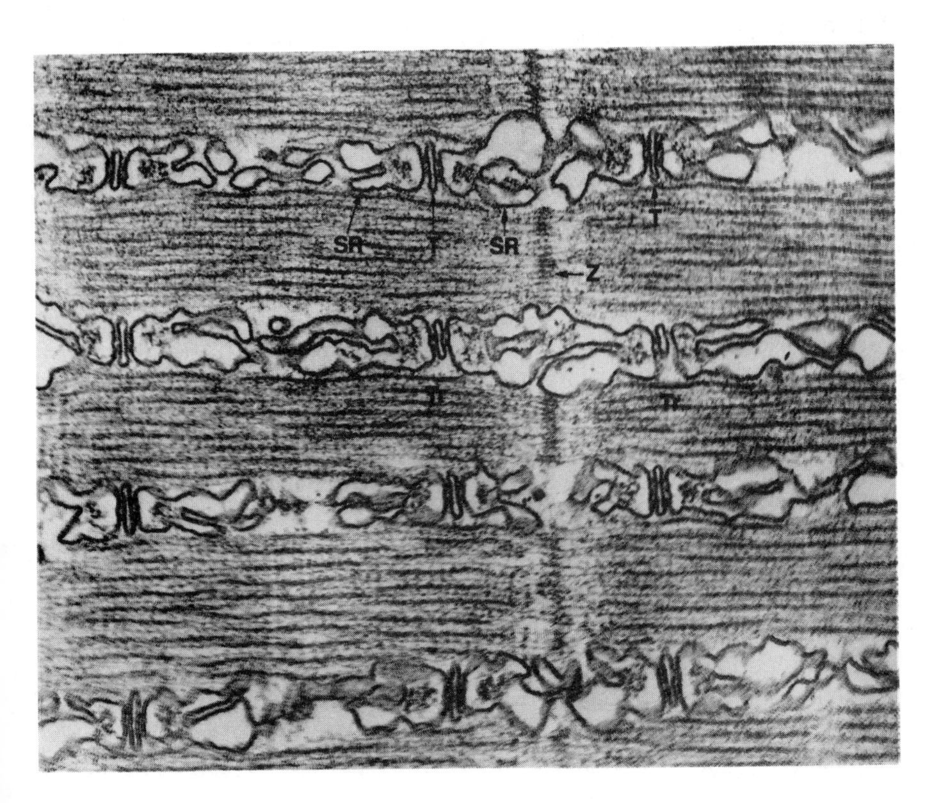

FIGURE 14-16
Electron micrograph of part of a muscle cell of the toadfish swimbladder. The muscle is fully contracted and therefore the I bands are almost obliterated. The myofibrils are separated by sarcoplasmic reticulum (SR). To each side of the Z line invaginations of the plasma membrane (sarcolemma) in the form of flattened tubes (T) are seen in cross section. These are intimately associated with the sarcoplasmic reticulum to each side, forming a triad (Tr). [Courtesy of Clara F. Armstrong.]

the sarcoplasmic reticulum at each end of the sarcomere is in close contact with the transverse tubules of the T system (Figure 14-15). Therefore, the transverse tubules lie between the terminal expansions of the sarcoplasmic reticulum of successive sarcomeres. In a longitudinal section through a muscle cell, the end portions of sarcoplasmic reticulum and the transverse tubule often appear as three vesicles in a line, a configuration known as a **triad** (Figure 14-16).

The function of the sarcoplasmic reticulum is to store Ca^{2+} ions. When an impulse arriving from a nerve cell is propagated along the transverse tubule, that impulse is transmitted in turn to membranes of the sarcoplasmic reticulum, causing a great increase in Ca^{2+} ion permeability of these membranes. As a result, Ca^{2+} ions stored in the cisternae are rapidly released into the sarcoplasm, where they trigger muscle contraction. When the nerve impulses cease, the membranes of the sarcoplasmic reticulum quickly regain their impermeability to Ca^{2+} ions. The membranes of the sarcoplasmic reticulum also contain an active transport system (Chapter 5) that carries Ca^{2+} ions from the sarcoplasm into the cisternae. Therefore, the cessation of nerve impulses is rapidly followed by reaccumulation of Ca^{2+} ions in the cisternae of the sarcoplasmic reticulum, and the muscle cell relaxes.

One ATP molecule is hydrolyzed by the Ca^{2+} ion transport system for each two Ca^{2+} ions pumped into the cisternae. Since muscle relaxation depends on the active removal of Ca^{2+} ions from the sarcoplasm, it is apparent that relaxation as well as contraction requires a supply of ATP.

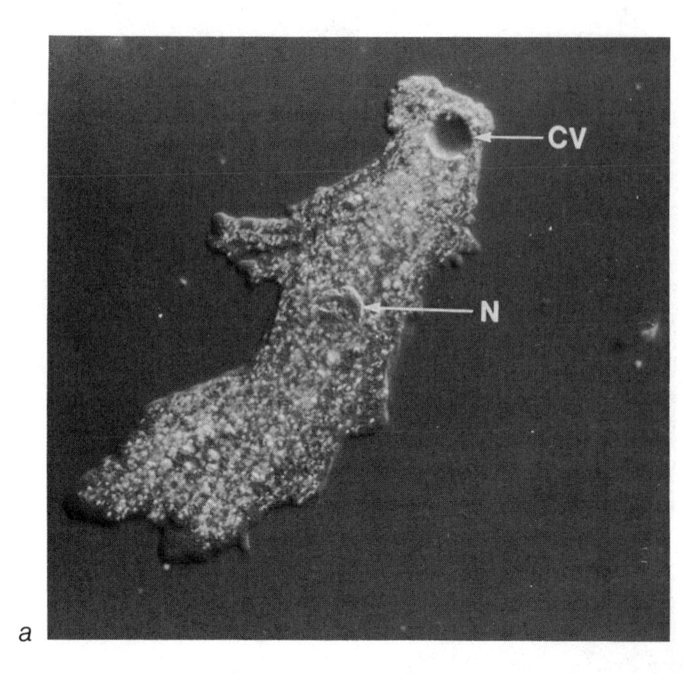

FIGURE 14-17

(a) Light micrograph of an actively moving amoeba (*Amoeba proteus*). N = nucleus. CV = contractile vacuole. (b) Series of light micrographs taken over four minutes showing movement of a small amoeba and function of the contractile vacuole. (1) The contractile vacuole (CV) has just emptied and has just begun to re-form. The two white dots (arrows) serve as reference points for movement in the subsequent pictures. N = nucleus. Nu = nucleolus. (2) The contractile vacuole is partially filled. (3) A full contractile vacuole. (4) The contractile vacuole has collapsed and the cell has moved 30 μm. [(b) Courtesy of Tim Spurck and Jeremy Pickett-Heaps, unpublished time-lapse micrographs.]

Reconstituted contractile system and formation of structure. The main purpose of experiments in which the various muscle proteins are extracted is to identify and analyze the role of each protein through reconstitutions like those described above. However, these reconstitution experiments also provide an example of the spontaneous association of macromolecules into functional structure. Purified myosin in solution spontaneously aggregates into thick filaments under appropriate ionic conditions. These reconstituted thick filaments have the same dimensions as and closely resemble the appearance of native thick filaments in the electron microscope. Thus, not only do myosin molecules spontaneously aggregate, they do so in the correct numbers to form a thick filament of correct size. When myosin, actin, tropomyosin, and troponin are mixed in solution, they aggregate into fibers that are capable of Ca^{2+} ion-dependent contraction.

The spontaneous aggregation of the component proteins presumably also underlies the formation of structure within the muscle cell. This example, as remarkable as it is, still leaves us short of understanding

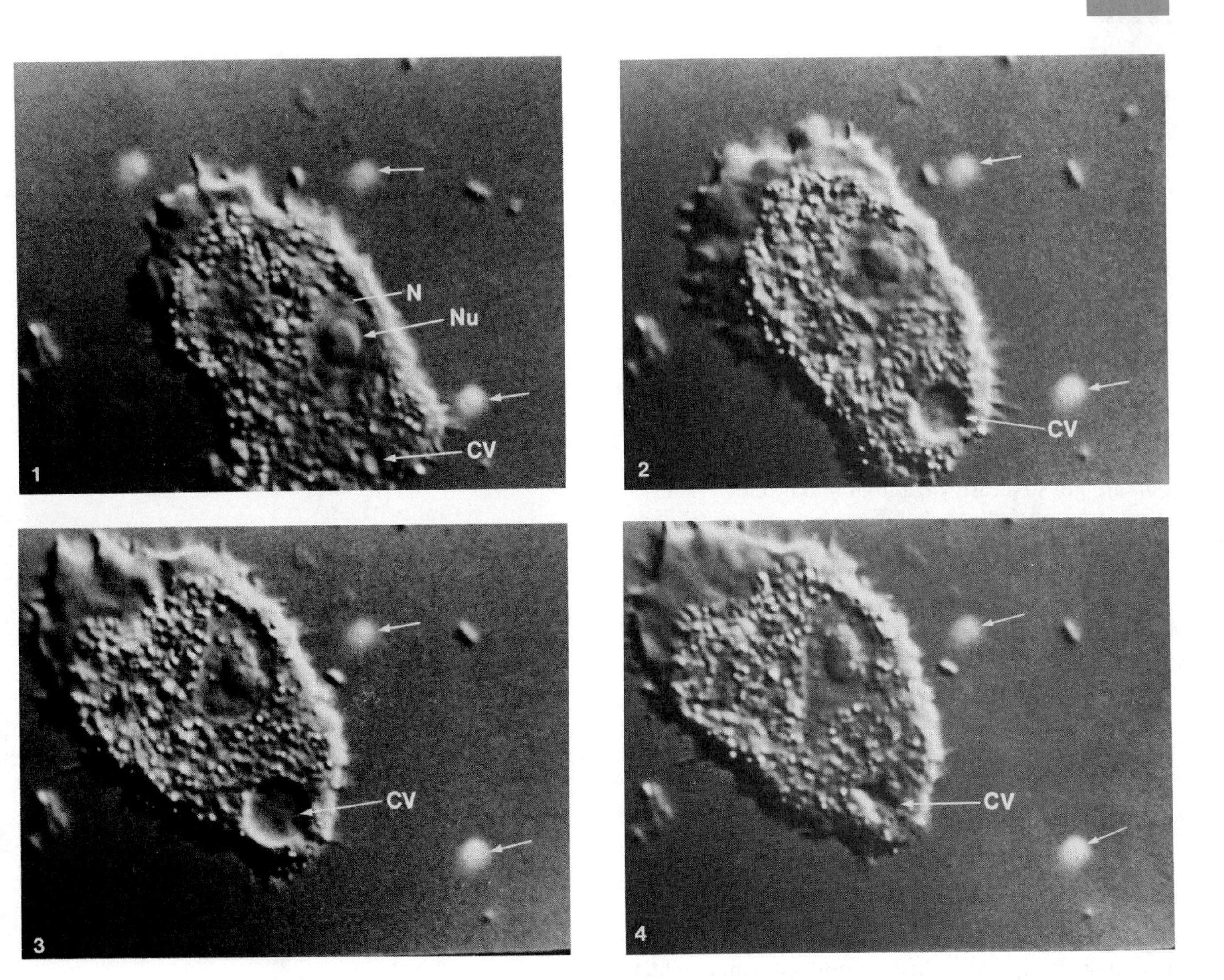

fully how a complicated structure like a myofibril or sarcomere is formed during differentiation of a muscle cell.

Amoeboid Movement

Some kinds of cells move by a flowing or streaming of their cytoplasm. This form of motility is particularly striking and rapid in the many species of amoeba, and hence was long ago named **amoeboid movement** (Figure 14-17). In intact invertebrate and vertebrate animals amoeboid movement is accomplished by various kinds of cells, for example, macrophages, fibroblasts, and white blood cells. During embryogenesis many kinds of cells move in a directed fashion by amoeboid locomotion to form various tissues. It is likely that virtually all cell types in animals are capable of amoeboid-like movement, although normally they remain stationary, at least in part because they are packed in tissues and restrained by their neighbors. The potential for amoeboid movement of tissue cells can be readily demonstrated by removing them from an animal and maintaining them in culture. All animal cells in culture, even those that fail to grow and reproduce, undergo some degree of amoeboid movement.

A major reason for the rapid progress in understanding contraction of skeletal muscle cells is that the contractile machinery is organized into well defined cytoplasmic structures, the thin and thick myofilaments, and the like. The structures serve as a guide for analysis of function. Smooth muscle cells also contain large quantities of much the same contractile machinery as do striated muscle, but the proteins for contraction are not so clearly organized into cytological structures. Consequently, it has been far more difficult to analyze the molecular basis of contraction of smooth muscle cells.

Similarly, amoeboid movement has been difficult to study because it is not based on a well defined cytological structure. However, the detailed information we now possess about the molecules involved in muscle contraction has, during the last several years, formed the basis for a molecular explanation of amoeboid movement. Amoebae are capable of a surprisingly rapid traverse across a surface. Larger species can move several hundred μm per minute. The cytoplasm of an amoeba is organized into an outer **ectoplasm** that is a gel (Figure 14-18). It contains networks of actin microfilaments that are likely responsible for the gelled state. The gelled ectoplasm surrounds the fluid portion or **sol** part of the cytoplasm, called the **endoplasm,** in which the nucleus, mitochondria, food vacuoles, and other cytoplasmic organelles tumble along during amoeboid streaming. The ectoplasm forms a rigid tube through which the endoplasm flows. As the endoplasm flows in the tube, it is converted into a gelled ectoplasm at the forward tip of the amoeba, thereby continuously extending the ectoplasmic tube. At the back end of the amoeba, gelled ectoplasm is converted into sol that then flows forward in the endoplasm.

Where is the motive force for amoeboid movement generated? Looking at a moving amoeba immediately

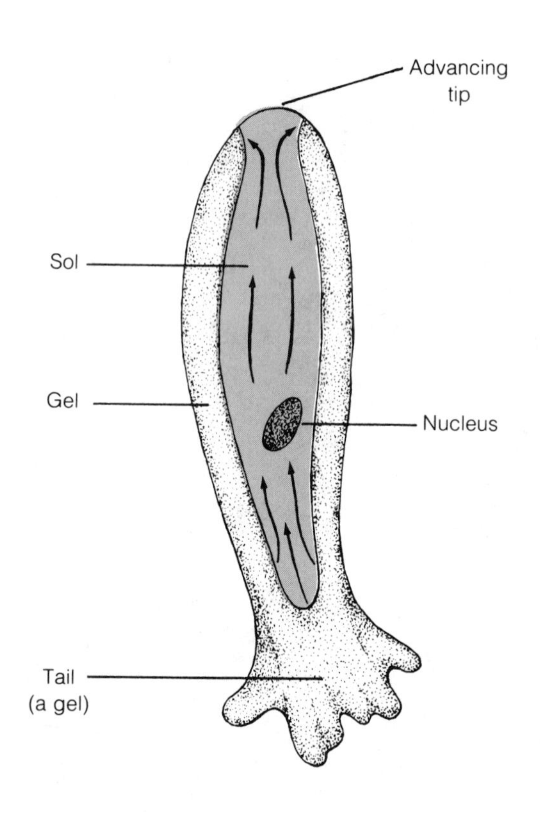

FIGURE 14-18
Diagram of the gel-sol structure of an amoeba. Cytoplasm in the sol state flows forward enclosed in a tube of cytoplasm in a gel state. At the advancing tip cytoplasm changes from sol to gel, and in the tail region cytoplasm is converted from gel to sol. The nucleus tumbles along with the stream of cytoplasm, but remains at the same relative position a little behind the middle of the cell.

gives the impression that the motive force for the forward streaming of the endoplasm is produced by a steady contraction of the ectoplasm near the back end of the cell. This is equivalent to an amoeba squeezing itself forward. The contractile mechanism that propels the endoplasm forward is believed to consist of actin and myosin. The generally accepted model is that the actin network in the ectoplasm toward the rear of the cell interacts with myosin to produce a contraction of the ectoplasmic tube that forces the endoplasm forward. This is accompanied by continuous dissolution of the actin networks and conversion of ectoplasm into endoplasm. At the advancing end of the cell, actin and myosin are recycled into networks that later interact in the ectoplasm in the back portion of amoeba. What regulates the interconversion of gel and sol at the two ends of the cell and the contraction of the actomyosin complexes is not known.

In its simplest form an amoeba moves as a single tube as just described. This kind of movement can be produced by placing an amoeba in a beam of light. An amoeba will move away from light (**negative phototaxis**) in a concerted fashion. In its more usual movement an amoeba continuously extends and retracts pseudopods that are branches of the main ectoplasmic tube. Pseudopods are transient structures that seem to form at more or less random locations on the cell, but they also form in phagocytic response to food organisms. Usually amoebae do not move in a single direction but follow a random walk pattern.

Amoeboid motion in animal cells is much slower than in amoebae, and the mechanism, although based on actin and myosin, is somewhat different. As a cell such as a fibroblast moves, sheetlike projections called **lamellipodia** are formed at its leading edge (Figure 14-19). The phenomenon is called **ruffling.** As the cell moves forward, the lamellipodia are displaced backward from the leading edge of the cell and disappear a short distance behind the leading edge. Lamellipodia are gel structures rich in actin networks. Their formation and disappearance reflect gel-sol interconversion similar to that seen in amoebae. It is believed that the actin fiber networks in lamellipodia also contain myosin, and the interaction of myosin with actin networks creates a contractile force that pulls the cell forward in a manner that is still poorly understood.

Cytoplasmic Streaming in Plant and Algal Cells

Plant and algal cells and other eukaryotic cells with rigid cell walls are incapable of amoeboid movement because they are physically restrained. However, the cytoplasm in these kinds of cells generally does move, a process called **cyclosis.** In some plants cyclosis is rapid, carrying cytoplasmic particles at rates up to 100 μm per second. Cytoplasmic streaming has never been observed in prokaryotes.

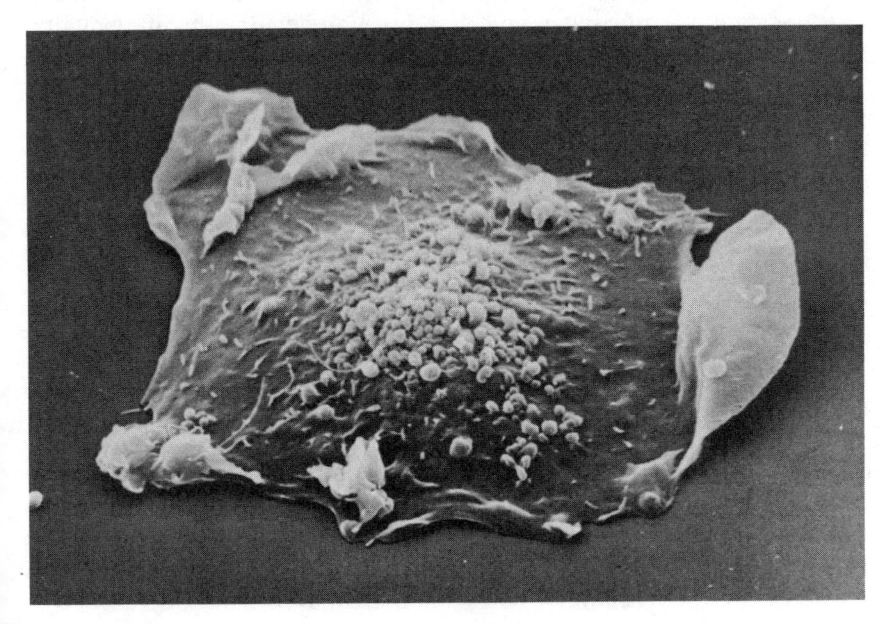

FIGURE 14-19
Scanning electron micrograph of a cultured mammalian cell with one large lamellipodium (ruffle) and several groups of smaller lamellipodia along the cell's edge. The cell was moving either to the right or to the upper left when fixed and may have been in the process of changing its direction. The center of the cell contains many spherical projections (blebs) of cytoplasm. The significance of these is not known. [Courtesy of Keith R. Porter.]

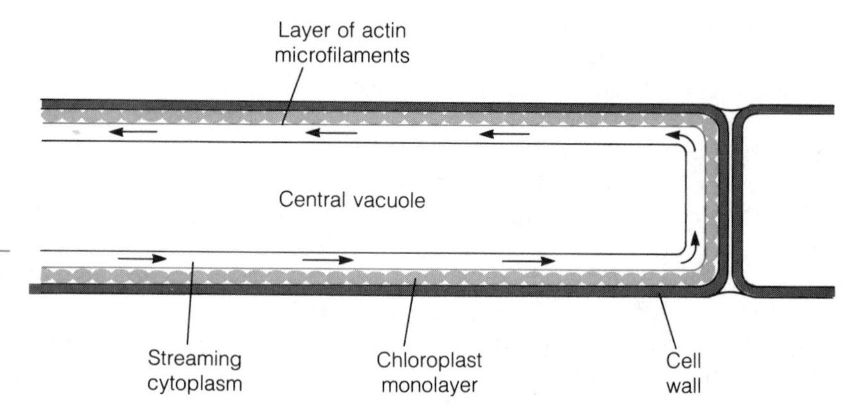

FIGURE 14-20
Drawing of a cell of the alga *Nitella*. Chloroplasts are arranged in a monolayer embedded in a cytoplasmic gel. A layer of actin filaments forms an interface between the chloroplasts and the streaming cytoplasm.

Giant, multinucleated algal cells such as *Nitella* and *Chara* are especially favorable for the study of cytoplasmic streaming. The cytoplasm immediately adjacent to the inner side of the plasma membrane is organized into a thin cortex or ectoplasmic gel (Figure 14-20). The cortex contains large numbers of chloroplasts immobilized as a single layer. Most of the internal part of the cell is occupied by a giant central vacuole (Chapter 1). Between the central vacuole and the cortex is a fluid endoplasmic region that undergoes continuous, unidirectional streaming at rates up to 100 μm per second. Mitochondria, nuclei, vesicles of various sizes, and other cytoplasmic organelles, are carried along in the stream. Streaming follows a fixed curved path along the length of the cylindrical cell, across the end of the cell, and back along the opposite side of the cell in the opposite direction, then around the other end, forming an endless moving belt. Between the oppositely moving streams is a narrow region of stationary, fluid endoplasm.

The route of streaming is determined by a layer of bundles of actin microfilaments bound to the inner surface of the cortex. These bundles are stationary and are arranged in parallel with the path of streaming. Each bundle contains 50 to 100 microfilaments. All the actin microfilaments are arranged with the same polarity (as in thin myofilaments of striated muscle), and this polarity determines the direction of streaming. The arrowheads produced by binding of myosin fragments point upstream. When the layer of microfilament bundles is experimentally disrupted in a localized region, for example, by microbeam irradiation, streaming stops or becomes passive in that region.

The force behind cytoplasmic streaming in *Nitella* is produced by interaction of myosin in the fluid endoplasm with the stationary bundles of actin microfilaments on the inner surface of the cortex. Myosin in the endoplasm is present on the surfaces of minute cytoplasmic vesicles of various sizes. The interaction of myosin with the actin microfilaments results in propulsion of the vesicles, apparently by a walking motion along microfilaments in a direction governed by the polarity of the microfilaments. When plastic beads (0.7 μm in diameter) coated with myosin purified from rabbit muscle are applied to the inner surface of the ectoplasm exposed by breaking open a cell, they move along the surface of the ectoplasm in a direction predicted from cytoplasmic streaming in the intact cell. Movement requires the presence of ATP. The propulsion of vesicles along the inner ectoplasmic surface is believed to cause the streaming motion of the entire endoplasm.

MOVEMENT BY CILIA AND FLAGELLA

Muscle contraction, amoeboid movement, cytoplasmic streaming, and cytokinesis are all achieved by interaction of myosin with actin. The movement of cilia and flagella is based on an entirely different pair of macromolecules, tubulin and dynein. However, the

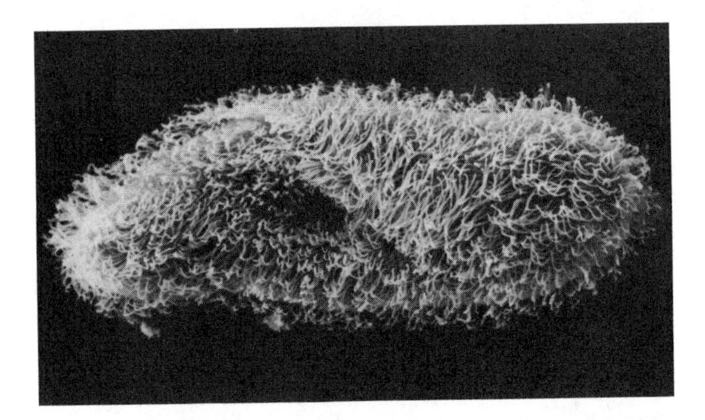

FIGURE 14-21
Scanning electron micrograph of the ciliated protozoan *Paramecium*. The hundreds of cilia present on the surface allow this cell to propel itself rapidly. [From G. Shih and R. Kessel 1982. *Living Images*. Jones and Bartlett.]

principle behind the movement is essentially the same; one kind of molecule, myosin or dynein, moves along a filament composed of the other macromolecule, that is, movement of myosin along an actin microfilament and movement of dynein along a microtubule.

Cilia and eukaryotic flagella have basically the same structure (see below), and these two organelles represent two versions of a single mechanism for motility. Cilia are present in large numbers on cell surfaces (Figure 14-21), whereas cells that move by means of flagella usually possess a single flagellum or a pair of flagella (Figure 14-22). Flagella are generally much

a

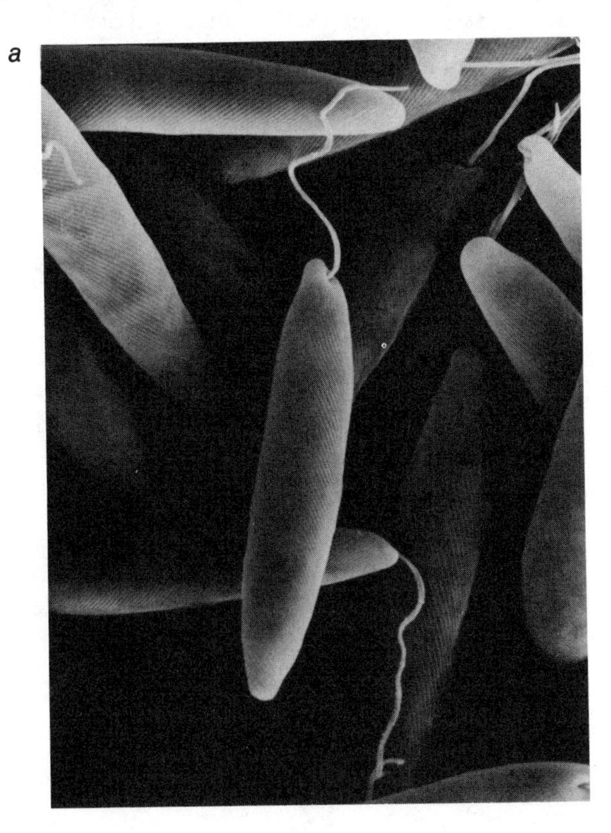

b

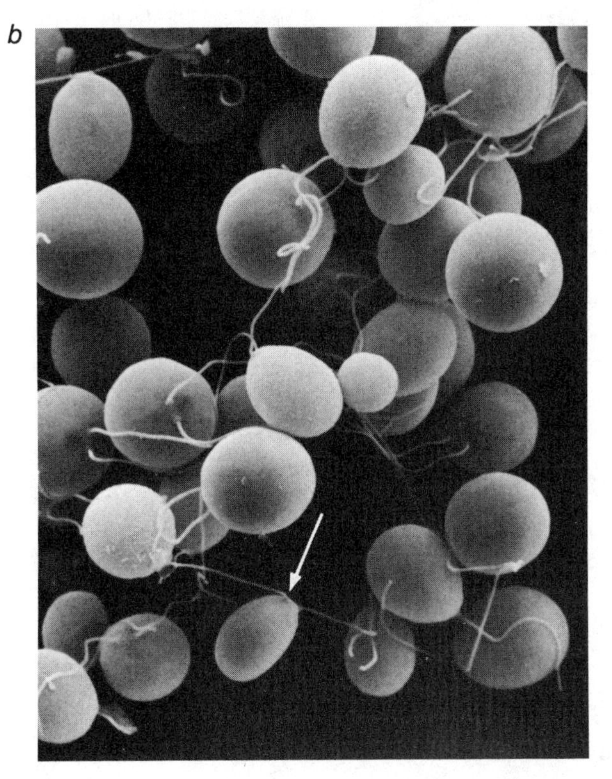

FIGURE 14-22
Scanning electron micrographs of flagella. (a) The single-celled alga *Euglena* has a single flagellum. (b) The single-celled alga *Chlamydomonas* possesses two flagella. The arrow indicates a clear example in this group of cells. [From G. Shih and R. Kessel. 1982. *Living Images*. Jones and Bartlett.]

FIGURE 14-23
Scanning electron micrograph of the surface of the hamster trachea. Large numbers of cilia project from each epithelial cell. Individual epithelial cells are not distinguishable in this micrograph. [Courtesy of Ellen R. Dirksen.]

longer than cilia. Cilia range from 0.2 μm to ten μm in length, and flagella range from ten μm to several mm. Both have a diameter of 0.2 μm. The flagella of eukaryotes are fundamentally different in both structure and mode of operation from the flagella of prokaryotes. Bacterial flagella are discussed in the final section of this chapter.

Cilia occur in a wide range of cell types among the protozoans, invertebrates, and vertebrates. They occur more rarely among algae, plant cells, and fungi. One of the major groups of protozoa, the **Ciliata,** consisting of many thousands of species, is partially defined and named by the possession of ciliary motility. Familiar examples are *Paramecium* and *Tetrahymena* (Figure 14-21). The cilia of most ciliated protozoa provide the means of cell motility but, of equal importance, are used to create currents to carry food organisms such as algae, bacteria, and other protozoa into an oral cavity where they are ingested by phagocytosis. Ciliated epithelial cells are common in animals, where in most cases they serve to propel fluid across the surface of epithelial tissues. For example, portions of the respiratory tract of mammals are lined with a ciliated epithelium (Figure 14-23) whose constant ciliary activity clears mucus, dust, bacteria, and so forth, from the respiratory passages.

Flagella are also widely distributed among eukaryotes. A major group of protozoa, the **Flagellata,** is defined by the presence of flagella. Many kinds of unicellular algae, for example, *Euglena* and *Chlamydomonas* (Figure 14-22), possess flagella. The sperm cells of most invertebrate and vertebrate animals and of a few plants are also propelled by flagella.

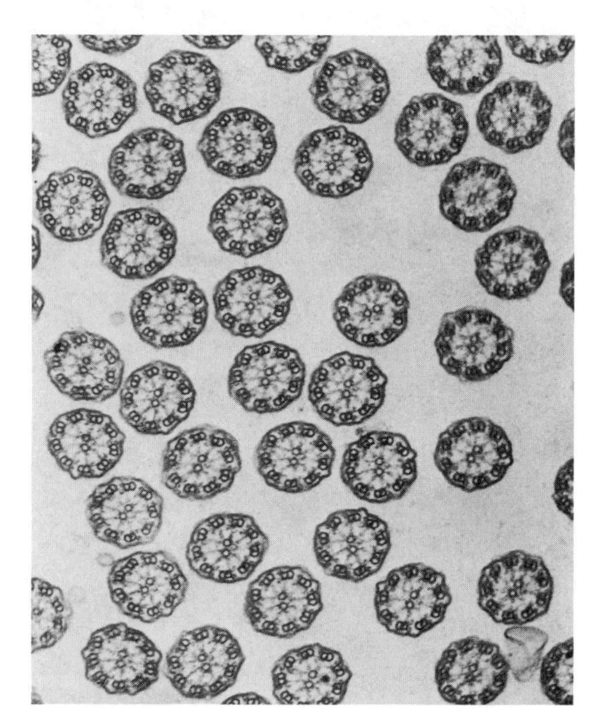

FIGURE 14-24
Electron micrograph of a group of cilia of an epithelial cell cut in cross section. Each cilium contains nine microtubule doublets surrounding a central pair of microtubules, forming a 9 + 2 pattern. [Courtesy of Fred D. Warner.]

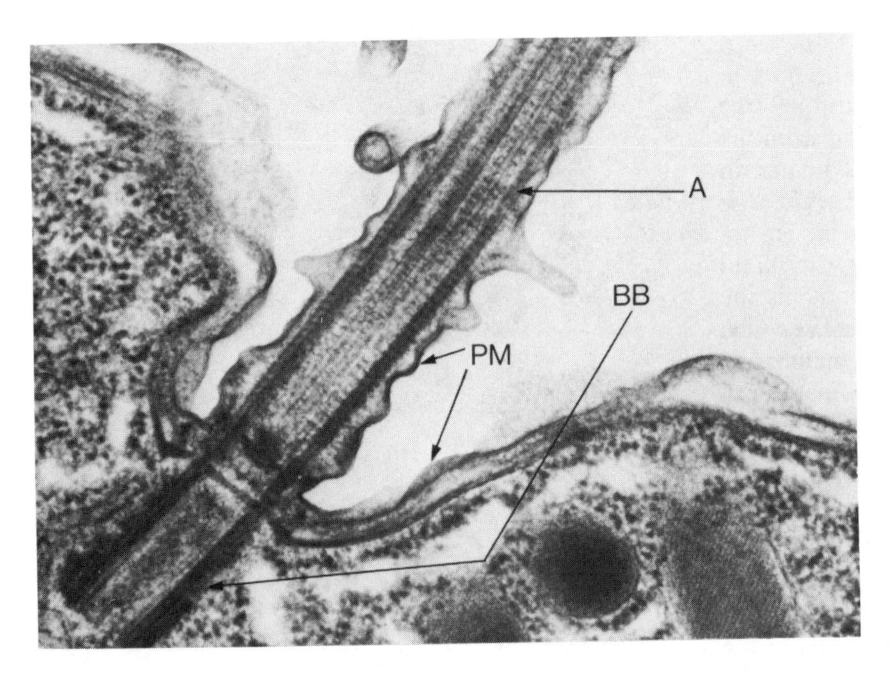

FIGURE 14-25
Electron micrograph of a cilium of *Tetrahymena*. The plasma membrane (PM) covers the axoneme (A) of the cilium. BB = basal body. [Courtesy of Ivan L. Cameron.]

Structure of Cilia and Flagella

Cilia and flagella are sufficiently alike that one description of structure can serve for both organelles. Cilia are remarkably constant in structural composition in eukaryotic species; the cilia of *Paramecium* are essentially identical to the cilia in the feeding apparatus of a clam and in the ciliated epithelium of the oviduct in a mammal. Microtubules form the core of cilia and flagella and run the length of these organelles. Nine pairs of microtubules are arranged in cylindrical fashion around two central microtubules. This 9+2 pattern is readily seen by electron microscopy in a cross section through the cilium (Figure 14-24). The set of microtubules, called collectively the **axoneme** (*axo* = axis; *neme* = threads), is enveloped in a cylindrical outpocketing of the plasma membrane (Figure 14-25). Hence, a cilium is a fingerlike, microtubule-containing projection of the cytoplasm.

At the base of a cilium is a structure called the **basal body** or **kinetosome**. The basal body has the same structure as a centriole, and these two organelles are considered to be one and the same. Indeed, the centriole in a developing sperm cell moves to the cell surface, where its component microtubules act as initiation sites for polymerization of the longer microtubules that form the flagellum, that is, the centriole becomes the basal body. The centriole is normally associated with the mitotic spindle organizing center (centrosome) in animal cells (Chapter 12). The basal body of a cilium or flagellum, like a centriole, is made up of 9 triplets of microtubules arranged in a circle to form a cylinder about 0.15 μm in diameter and about 0.5 μm long. In contrast to a cilium, there are no microtubules in the center of a centriole. The basal body is believed to serve as a starting or nucleating organelle in the formation of a cilium or flagellum. In the mature sperm cells of some insects the basal body is absent, suggesting that once a flagellum has been formed, the basal body is not needed for flagellar function.

Molecular Mechanism of Ciliary and Flagellar Movement

A variety of experiments has provided an explanation for the beating of cilia and flagella, although important aspects of the mechanism remain to be clarified. In an early experiment cilia were observed to continue to beat and independently move through the medium for a short time after being severed from the surface of a ciliated protozoan. Beating ceased when an isolated cilium had used up its energy reserves. The experiment proved that ciliary movement is not achieved by a motor at the base that propels the cilium back and forth; rather, the cilium itself contains the mechanism for movement. Treatment of isolated cilia with a mild detergent removes the plasma membrane, leaving the naked **axoneme** with the 9+2 pattern of microtubules. When the proper inorganic ions and ATP are added to the solution, the naked axoneme beats. Thus, it is clear that the plasma membrane around the cilium is also not necessary for movement, and second, the source of energy for movement is ATP, as in systems of motility based on a myosin-actin interaction.

A similar experiment can be done with flagella removed from sperm. As long as ATP is supplied in the medium, the flagella continue their wavelike beating. In ciliated cells such as *Tetrahymena,* mitochondria are concentrated just under the cell surface near the bases of cilia, a location favorable for provision of ATP to the cilia. In the sperm of most animals and plants the first part of the axoneme is surrounded by a single large mitochondrion or a group of mitochondria (Chapter 15), presumably to supply ATP for flagellar movement.

The ATPase that hydrolyzes ATP to ADP + P_i to provide energy for ciliary beating can be extracted from axonemes and still leave the axonemes intact. This ATPase has been named **dynein,** and its location in the axoneme was discovered in the following way. In normal axonemes a pair of projections or arms occurs at regular intervals on one side of each doublet of microtubules in the axoneme (Figure 14-26). In axonemes from which dynein has been extracted the arms are missing. If dynein is allowed to reattach to the axonemes, the arms and ATPase are simultaneously regained by the axonemes.

In a rare, genetically inherited disease described in human males the dynein arms are abnormally short or totally absent. A mutation is present in a gene carried by the X chromosome that causes synthesis of abnormal dynein or that causes a failure of dynein to attach normally to the microtubules. The cilia and flagella in these individuals are incapable of movement. As a result, the spermatozoa, whose flagella are normal in all structural aspects except for the lack of dynein, are nonmotile, and the individuals are infertile. In addition, the cilia of the epithelial cells lining the respiratory tract, paranasal sinus, and the Eustachian (auditory) tube are nonmotile. Therefore, these persons do not possess the normal mechanism for clearing dust, and debris, including viruses and bacteria, from the affected area, and they chronically suffer from bronchitis, sinusitis, common colds, and ear infections.

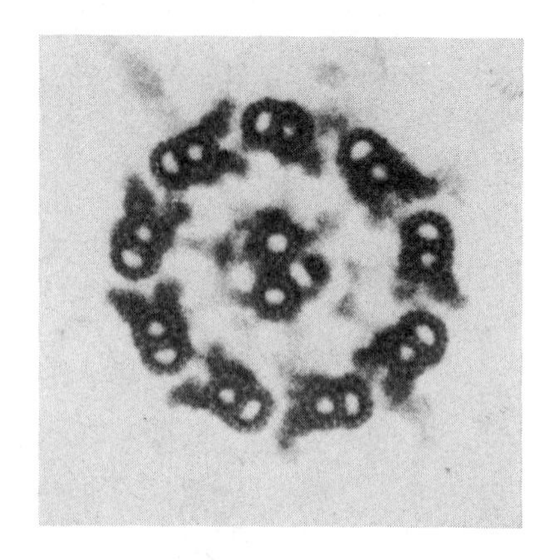

FIGURE 14-26
Electron micrograph of a cilium. A pair of dynein arms projects from each microtubule doublet towards the next doublet. Radial spokes connect each doublet to the central pair of microtubules. Not all the spokes are clearly visible in this case. [Courtesy of Bessie Huang.]

Microtubules in general are constructed of 13 protofilaments made of tubulin (Chapter 6). However, because the pair of microtubules in each of the nine doublets of the axoneme is partially fused in a lengthwise direction (Figure 14-27), the number of protofilaments is different. One of the two microtubules, called the **A subfiber,** contains a full set of 13 protofilaments. The **B subfiber** has only 10 or 11 protofilaments. At the site where the two or three protofilaments are missing, the B subfiber fuses with the A subfiber. Dynein molecules form pairs of arms on the A subfiber. The arms form transient bridges to the B subfiber in the next doublet during beating of the cilium.

Each of the nine doublet microtubules is connected by radial spokes to the central pair of microtubules (Figure 14-27). Spokes are present at regular intervals along the length of the axoneme, and curved rods project from the central pair of microtubules at regular intervals.

Over 250 proteins have been identified by electrophoresis from purified axonemes that have been dissociated with a strong ionic detergent. Two of these pro-

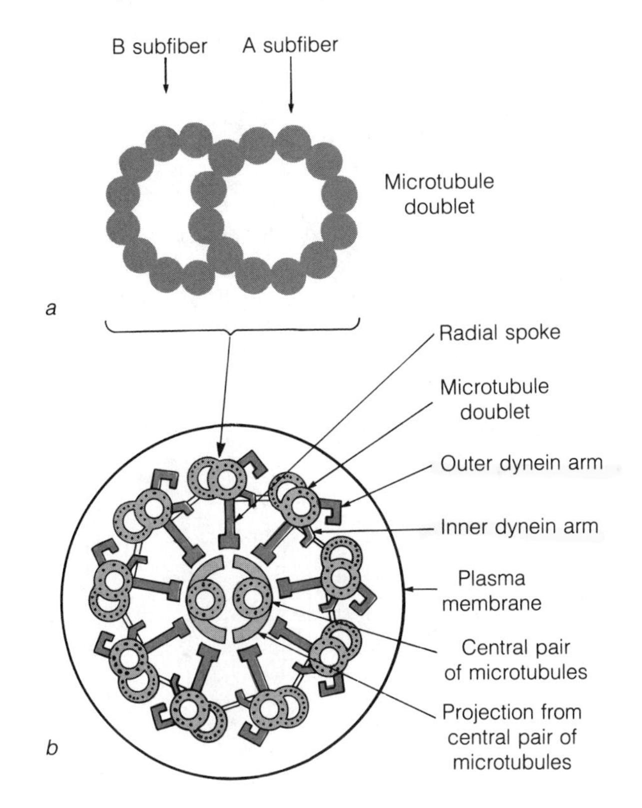

FIGURE 14-27
(a) A microtubule doublet consisting of an A subfiber with 13 protofilaments of tubulin and a B subfiber with 11 protofilaments. (b) A cilium contains nine microtubule doublets and a central pair of microtubules. The dynein arms provide for sliding of one doublet along another during the beating of the cilium. The functions of radial spokes and the projections from the central pair of microtubules have not been as well defined.

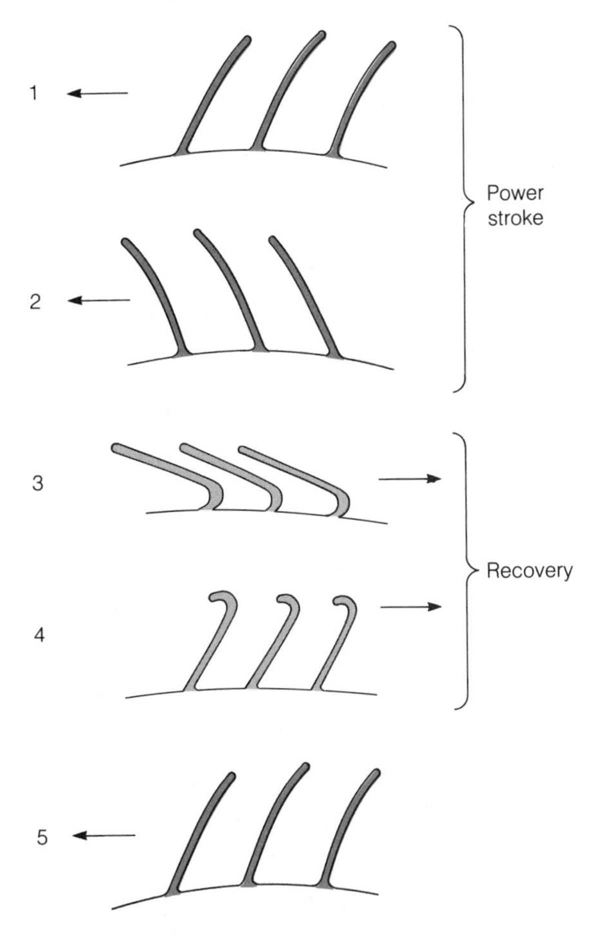

FIGURE 14-28
Cycle of ciliary stroke shown for three cilia beating in synchrony. (1) Beginning of the power stroke. The cilia are straight and stiff. (2) Half completion of the power stroke. (3) Start of recovery stroke. The cilia are flexible, and by bending reduce frictional drag. (4) Near completion of the recovery stroke. (5) Start of the next power stroke.

teins, α- and β-tubulin, account for about 70 percent of the total protein in an axoneme. The two dynein arms are each composed of multiple polypeptides. The outer arm contains about nine different polypeptides, two of which are ATPases that split ATP to ADP + P_i during generation of a ciliary beat. The inner arm is structurally similar, but seems to be made from completely different polypeptides. Each radial spoke contains

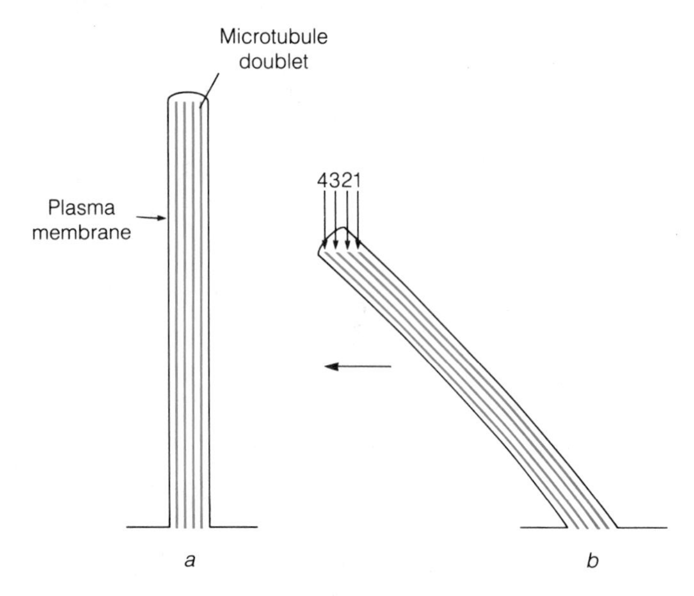

FIGURE 14-29
Sliding of one microtubule doublet along another causes a cilium to bend. (a) The cilium is perpendicular to the cell surface; all microtubule doublets extend the full length of the cilium. (b) The cilium is bent by a slight active sliding of doublet number 1 baseward along doublet number 2, number 2 baseward along number 3, and number 3 baseward along number 4. The doublets all remain the same length. However, they end at different points relative to the tip of the cilium because of the slight displacement of one doublet against another.

about 17 different polypeptides. The majority of axonemal polypeptides are unaccounted for. A few may be microtubule-associated proteins (MAPs, Chapter 6), and many are probably present in the matrix spaces between the recognizable structures. Except for the ATPases of dynein arms and for α- and β-tubulin of microtubules, nothing is known about the function of the more than 250 polypeptides of the axoneme.

The power stroke of a cilium consists of a bending motion (Figure 14-28). Bending is achieved by lengthwise sliding of microtubule doublets along each other (Figure 14-29). Sliding has been observed in the following experiment. Cilia were detached from *Tetrahymena*, the plasma membrane surrounding them was removed with nonionic detergent, and the radial spokes were broken by brief treatment with the enzyme

FIGURE 14-30
Electron micrograph of an axoneme isolated from a cilium of *Tetrahymena*. Addition of ATP has caused the microtubule doublets to slide along each other, producing an elongated axoneme. [Courtesy of Fred D. Warner.]

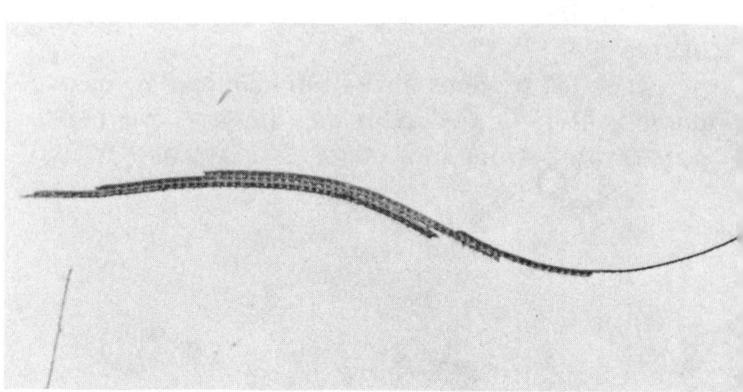

trypsin. Addition of ATP did not cause the axonemes to beat, but instead doublets slid along each other, causing the axoneme to elongate several fold (Figure 14-30). This elongation happens as follows.

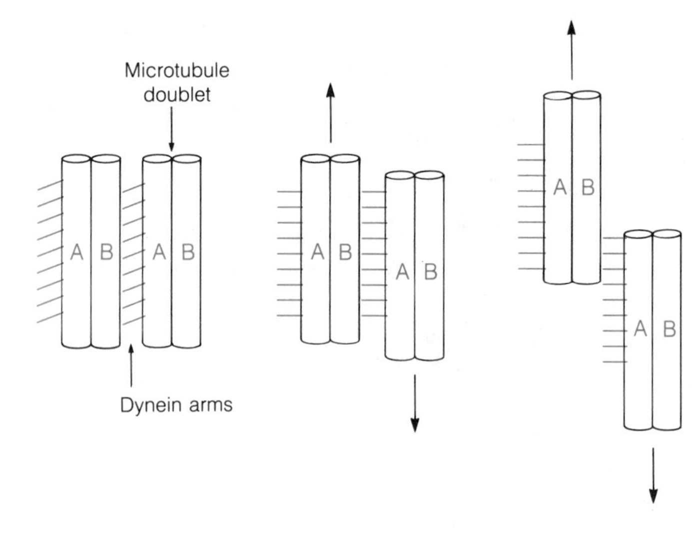

FIGURE 14-31
Diagram of the response of microtubule doublets in an isolated axoneme. One doublet continues to slide along another, causing the axoneme to elongate several fold (Figure 14-30). Eventually the doublets slide completely off one another.

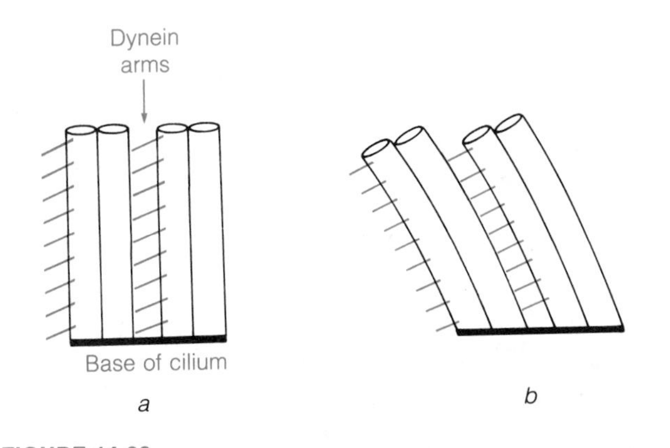

FIGURE 14-32
Sliding of one microtubule along another in an intact cilium causes the cilium to bend, creating a power stroke.

The dynein arms are permanently attached to the A microtubule in each doublet (Figure 14-31). At rest the arms are tilted downward toward the base of the cilium, as they form bridges to the B microtubule of the next doublet. The addition of ATP causes the dynein arms to move to a 90° angle relative to the A microtubule, which causes the two doublets to slide relative to each other. At the end of the stroke dynein detaches from the B microtubule, bends downward, reattaches to the B microtubule and repeats the stroke. In isolated axonemes repetition of the cycle eventually results in sliding of one doublet off the end of the other. In a cell the microtubule doublets are anchored at the base of the cilium (Figure 14-32). Hence, slight sliding of one doublet along another causes them to bend (Figure 14-32(b)).

Sliding mediated by dynein arms is the mechanism for generation of forces within axonemes, but how are these forces organized to produce the ciliary beat? The nine pairs of doublets are numbered 1 through 9 by their orientation relative to the central pair of microtubules. To define this orientation a line is drawn through the centers of the central microtubules (Figure 14-33). A second line drawn between the central microtubules at a 90° angle to the first line bisects one of the outer doublets. This outer doublet is number 1; the dynein arms of doublet 1 point toward doublet 2, the arms of 2 point toward doublet 3, and so on, around to doublet 9. All the dynein arms on all the doublets generate force in the same direction, and therefore every doublet slides in the same way. If all the dynein arms on all of the doublets were to operate at once, the resultant forces would almost cancel, resulting only in a twist of each doublet. This may be difficult to visualize, but note that doublet 1 would have two equal and almost opposite forces applied to it: doublet 2 would force it toward the base of the cilium and doublet 9 would force it toward the tip.

This dilemma appears to be solved by dividing the doublets into two functional groups. Doublets 1 through 4 are thought to act together during the power stroke in forward swimming, causing a force that tends to elongate the cilium on one side. The elongation is accommodated by bending of the cilium. Once the power stroke is done, doublets 5 through 9 slide against each other to straighten the cilium. In some way the operation of the two groups of doublets must be different since the cilium is rigid in the power stroke but bends in the recovery stroke (Figure 14-28). How the two groups of doublets are controlled to achieve the proper temporal pattern of sliding is not known. The two groups of doublets must operate in the opposite order when the direction of the power stroke is reversed in backward swimming of the cell.

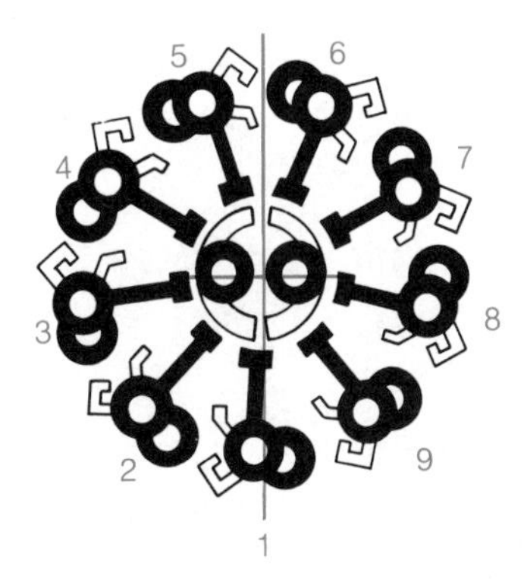

FIGURE 14-33
Numbering of microtubule doublets. Doublet number 1 is identified by a line that is perpendicular to a line that passes through the centers of the central pair of microtubules. Numbering 2 to 9 is in the direction of the dynein arms.

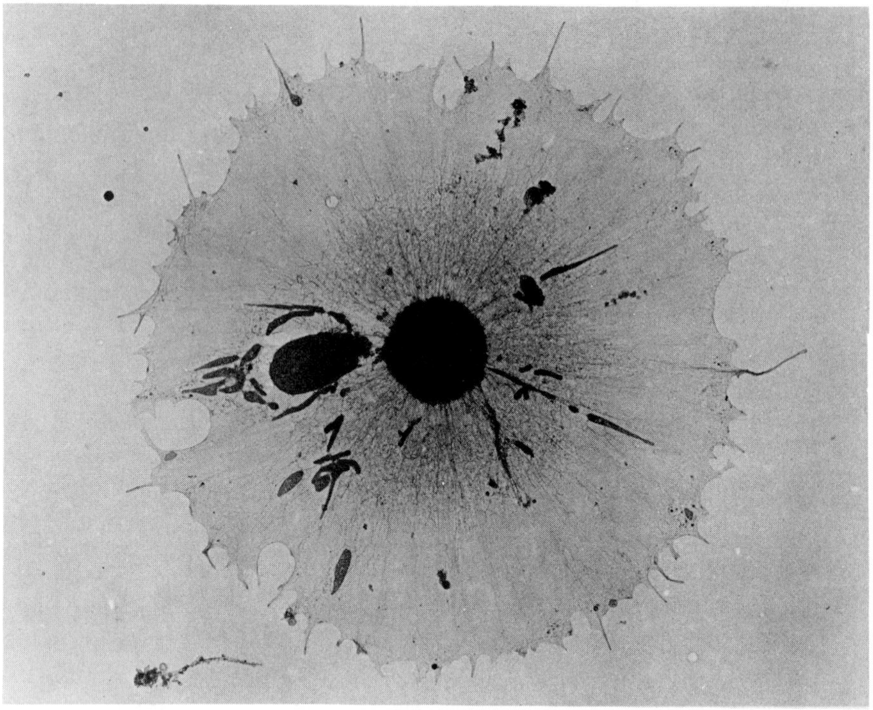

FIGURE 14-34
High voltage electron micrograph of a whole pigment cell (erythrophore) isolated from a squirrel fish. The pigment granules are retracted into a dense knot in the center of the cell. Compare this micrograph with the one in Figure 14-35. The nucleus is the dark oval body to one side of the pigment knot. [Courtesy of Mark McNiven. Reproduced from *The Journal of Cell Biology*, 99:152s–158s (1984) by copyright permission of the Rockefeller University Press.]

Finally, to accomplish cell movement of a motile cell or movement of fluid across the surface of a stationary cell, the cilia must all beat in approximate synchrony, perhaps by signals transmitted by the plasma membrane or by the complex network of fibers that interconnect all the cilia at their basal ends just inside the cell surface.

MOTILITY BASED ON THE KINESIN-MICROTUBULE INTERACTION

Vesicles, pigment granules, and other particulate structures in the cytoplasm often move rapidly from one part of the cell to another by a mechanism that does not involve cytoplasmic streaming. A dramatic example occurs in the rapid change in color of some species of fish. Fish coloration is due to large specialized cells called **chromatophores** that are present in the scales. In response to stimulation some fishes can switch from a pale appearance to bright or dark coloration in a few seconds. When the pigment granules in the chromatophores are all tightly gathered into a knot in the center of the cell, the fish has very little coloration (Figure 14-34). On stimulation, the granules rapidly disperse throughout the cell along a radially arranged network of microtubules (Figure 14-35), and the cell and the fish become colored. The movement of pigment granules is rapid, directional (along microtubules), and reversible.

Another example of rapid intracellular movement is the migration of vesicles along axons of nerve cells. Materials needed to maintain the structure and function of the axon of a nerve cell are synthesized in the body of the cell and must be transported down the axon. Substances that are released from the end of the axon in transmission of a nerve impulse to another nerve cell at a synapse (Chapter 15) or at a muscle-nerve junction are carried the full length of the axon. Some axons are more than two meters long.

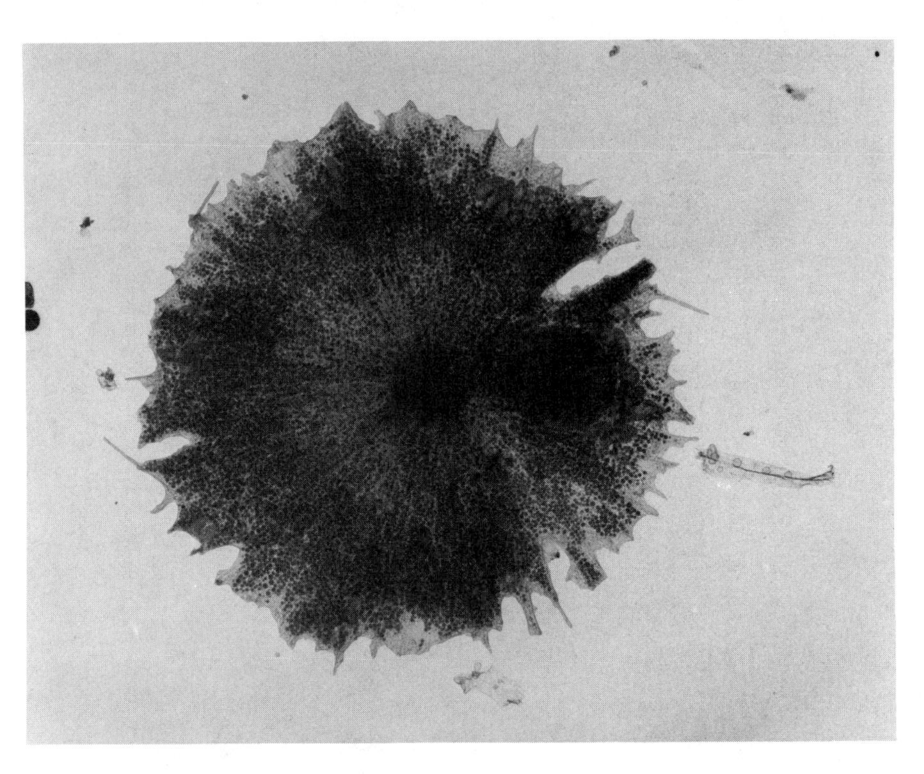

FIGURE 14-35 High voltage electron micrograph of a pigment cell (erythrophore) isolated from a squirrel fish. The pigment granules, which are red, are dispersed throughout the cytoplasm, giving the cell (and the fish) a bright red coloration. [Courtesy of Mark McNiven. Reproduced from *The Journal of Cell Biology*, 99:152s–158s (1984) by copyright permission of the Rockefeller University Press.]

Fast Axonal Transport

The squid contains an axon of particularly large diameter, which is favorable for observing axonal transport. Small vesicles move away from the cell body (**anterograde movement**) and back toward the cell body (**retrograde movement**) along microtubules. Anterograde movement delivers materials down the axon, and retrograde movement probably represents return of emptied vesicles for recycling. Vesicle movement is rapid—several μm per second—and is called **fast axonal transport.** Fast directional transport of vesicles along microtubules occurs in many kinds of cells other than axons, for example, fibroblasts. Organelles, including particles other than small vesicles, are transported. The mechanism of transport is likely the same in these various situations, and the name **fast organellar transport** is more appropriate than the term *fast axonal transport.*

Kinesin-Microtubule Interaction in Fast Organellar Transport

Current understanding of the molecular mechanism of fast organellar transport has come largely from study of the giant axon of the squid. The cytoplasm obtained from axons (axoplasm) has been separated into soluble supernatant and organelle-enriched fractions by centrifugation. Addition of the organelles to purified microtubules does not result in organelle movement, but when the soluble fraction from squid axons is added along with ATP, the organelles adhered to and migrated along individual microtubules. A translocator protein has been purified from axoplasm and from bovine brain and given the name kinesin (from the Greek word *kinein,* to move). Latex beads (0.15 μm diameter) coated with purified kinesin are observed to move along microtubules in the presence of ATP. Kinesin, like myosin and dynein, is composed of a

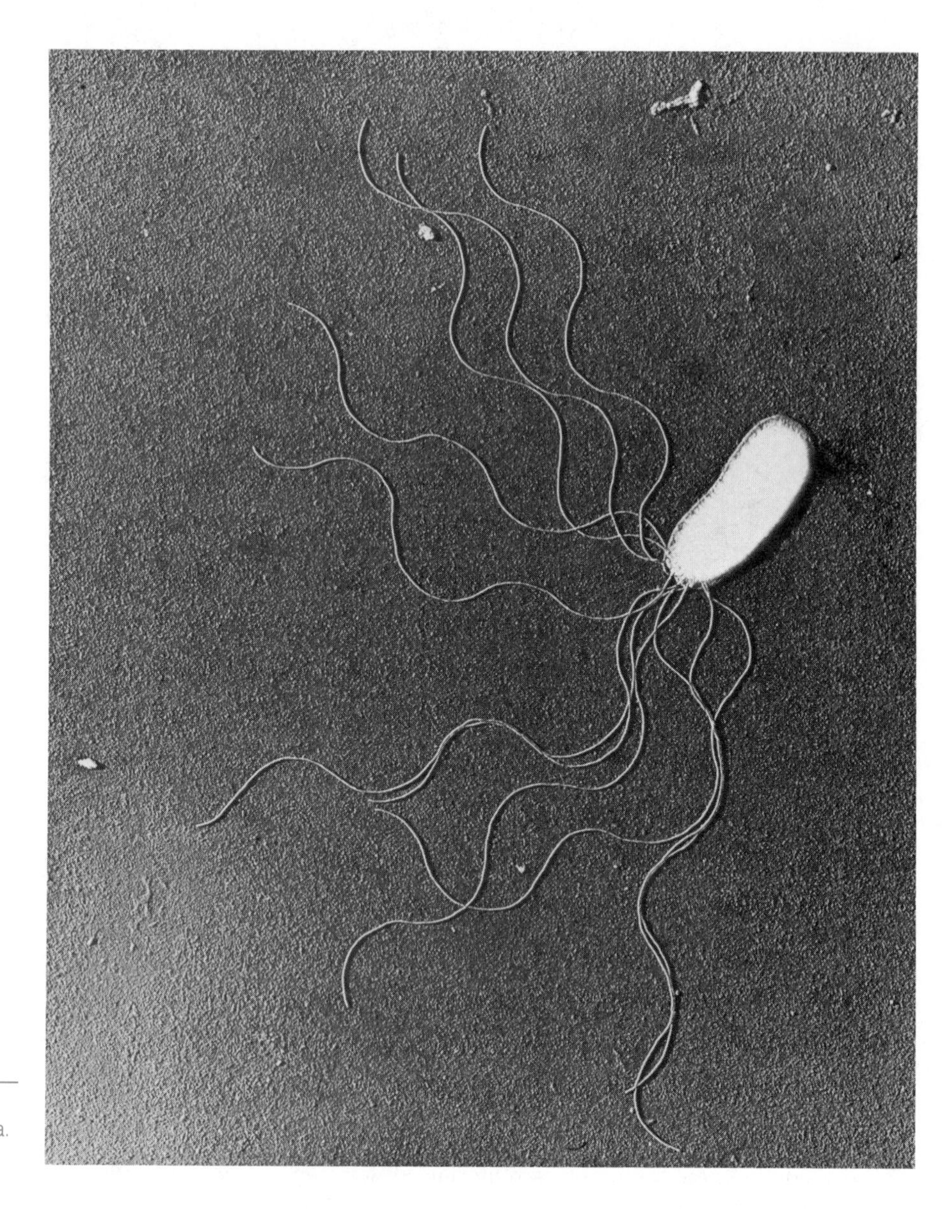

FIGURE 14-36
Electron micrograph of a bacterial cell *Pseudomonas marginalis* with 11 flagella. [Courtesy of Arthur Kelman.]

high molecular weight complex of polypeptides. Like dynein, it binds to microtubules (myosin binds to actin microfilaments), but it differs from dynein in several major ways and represents a different (a third) mechanochemical mechanism of movement.

The movement generated by kinesin along microtubules is unidirectional, as in the case of myosin-coated beads along actin microfilament bundles. Unidirectional movement probably reflects the polar nature of microtubules (Chapter 6). However, in reconstitution experiments in which the soluble fraction of axoplasm is mixed with microtubules, vesicles have been observed to move in opposite directions along a single microtubule. Since purified kinesin produces only unidirectional movement, this suggests that yet another translocator protein may be present that generates movement in the opposite direction.

In axons and in the cytoplasm of other cells, larger

vesicles and organelles (>0.25 μm in diameter) undergo **saltatory,** or jumping motions; vesicles and organelles move rapidly, stop transiently, and then move again, giving the impression of a jumping action. Smaller vesicles, for example less than 0.1 μm in diameter, move continuously without jumping. In isolated axoplasm the large vesicles and organelles move continuously and uninterruptedly along microtubules in the same manner and speed as the smaller ones. It may be that the saltatory movement is caused by cytoskeletal structures that transiently get in the way of larger vesicles and organelles but do not impede smaller ones. Alternatively, saltatory movement may be the result of periodic dissociation of particles and vesicles from microtubules. Dissociation is consistent with the observation that the rate of translocation of a vesicle during a single jump is the same as the rate of continuous nonsaltatory movement of smaller vesicles.

Pigment granules may move by means of kinesin or a kinesin-like protein, since movement is along microtubules and is rapid. Kinesin has also been identified in the mitotic spindle and may be responsible for the chromosome movement in anaphase.

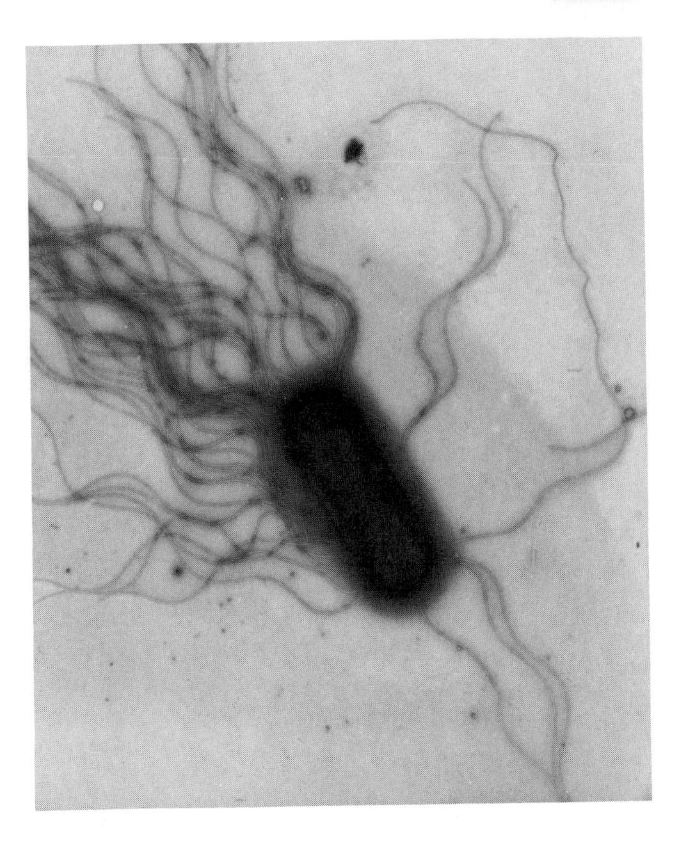

FIGURE 14-37
Electron micrograph of a bacterium with many flagella (*Proteus mirabilis*). [Courtesy of Fred D. Williams.]

FLAGELLAR PROPULSION IN BACTERIA

Many species of bacteria move by means of flagella. Others move by gliding across a surface, but little is known about the mechanism of gliding, and it will not be discussed. Some bacteria have a single flagellum, and others have several (Figure 14-36); and yet other bacterial species have dozens of flagella that project over the entire surface of the cell (Figure 14-37).

In *E. coli* at least 35 genes are involved in the assembly, structure, and function of a flagellum. One or more genes involved in flagellar formation in *E. coli* is subject to catabolite repression. In the presence of a plentiful supply of glucose, the synthesis of flagellar components is repressed. This underscores the importance of flagellar motility in cell nutrition. Recall that *E. coli* is a heterotroph that is able to synthesize all of its organic components from glucose. In the presence of glucose the need for motility is obviated, and flagella are not formed as cell reproduction proceeds.

Bacterial flagella are fundamentally different from eukaryotic cell flagella in both structure and function. A bacterial flagellum is about 20 nm in diameter, or about 1/20 the diameter of a eukaryotic cell flagellum, and is usually 10 to 20 μm long. Because of its small diameter, special optical methods must be used to see the beating flagellum of a living cell. The flagellum is not enclosed in an outpocketing of the plasma membrane, but instead projects from the cell as a naked protein filament. It is anchored in the plasma membrane by means of a base complex (Figure 14-38).

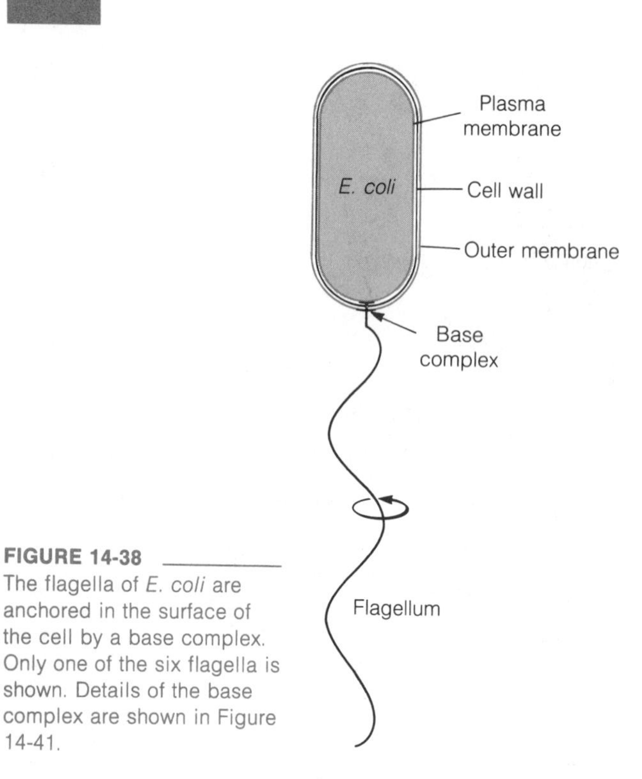

FIGURE 14-38
The flagella of *E. coli* are anchored in the surface of the cell by a base complex. Only one of the six flagella is shown. Details of the base complex are shown in Figure 14-41.

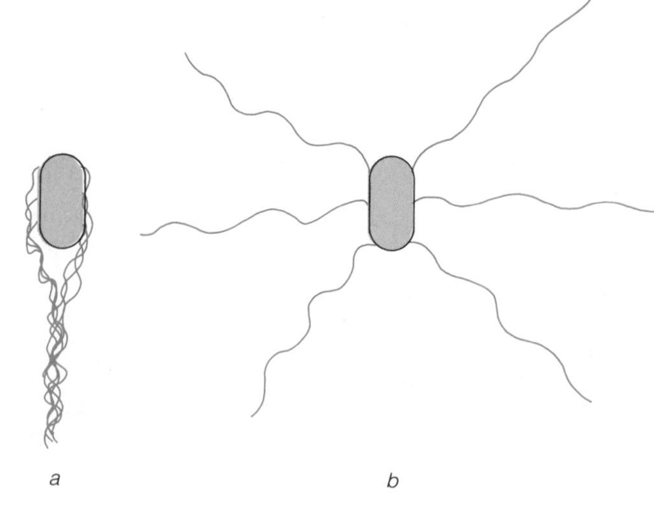

FIGURE 14-39
(a) Cooperative action of flagella during forward swimming. (b) Reversal in the direction of flagellar rotation results in separation of flagella and causes the bacterium to tumble.

When multiple flagella are present, they form a bundle and work cooperatively to propel the cell forward in a smooth swimming motion (Figure 14-39(a)). Reversal in the direction of rotation of the flagella results in separation of the flagella from the bundle, causing a tumbling motion of the bacterium (Figure 14-39(b)).

The bacterial flagellum is constructed of a protein known as **flagellin** with a molecular weight of about 53,000. The flagellum is made of five long chains of flagellin molecules wound around one another in a tight helical arrangement (a **pentahelix**), creating a tubular structure (Figure 14-40). Pure flagella can be prepared by severing them from bacteria. When purified flagella are gently heated or exposed to low pH, they dissolve, yielding a solution of flagellin. If the dissociating conditions are reversed, the flagellin molecules reaggregate into flagella. For this to happen, the solution of flagellin must be seeded by adding short fragments of flagella, which can be obtained by breaking up intact flagella with sound waves. The flagellin

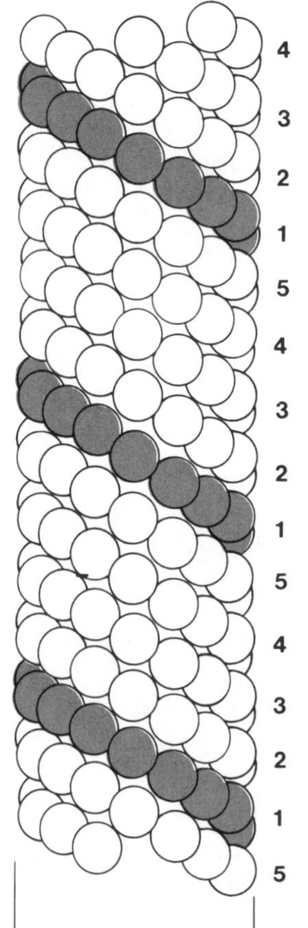

FIGURE 14-40
Five chains of flagellin molecules coil around a central axis to form a bacterial flagellum.

molecules always assemble at only one end of the flagellar fragment. The arrangement of flagellin molecules in a reconstituted flagellum is identical to the arrangement in a natural flagellum.

Formation of a flagellum in a living bacterium occurs by addition of flagellin molecules to the tip of the flagellum. The individual flagellin molecules pass down a channel in the center of the pentahelix that forms the flagellum and are assembled onto the distal ends of the five chains. The cilia and flagella of eukaryotes also grow by addition of protein molecules that pass down the organelle and assemble at the tip.

The bacterial flagellum is rigidly held in an inflexible helical form that is much like a corkscrew (Figure 14-36). Mutant bacteria have been obtained in which the flagellum is a straight, rigid rod. These mutants are immotile since the pushing or pulling action of the flagellum depends on a helical shape. The mutation is undoubtedly in the gene for the flagellin protein and causes a shift in the interaction of flagellin molecules with one another, so they form a straight structure rather than a helical one.

Propulsion of a bacterium occurs when the helical flagellum is rotated at the point of insertion into the cell. The force creating rotation is applied in the cell to the base plate (in the plasma membrane) to which the flagellum is attached (Figure 14-41). This is a totally different arrangement from that in the flagella and cilia of eukaryotes, where force is generated along the entire length of the structure. Rotation of the rigid, helically twisted bacterial flagellum exerts a pushing force against the surrounding fluid, which drives the cell forward like the propeller of a ship. Pushing occurs when the flagellum is rotated in the counterclockwise direction. Rotation in the clockwise direction results in a pulling force by the flagellum, although this is not efficient in producing movement. Clockwise rotation results in a tumbling motion of the cell. Normally a bacterium swims forward (counterclockwise rotation of the flagellum) in a smooth line for some seconds, reverses the direction of flagellar rotation for a fraction of a second, which causes a tumbling motion, and then swims forward in a new, randomly determined direction.

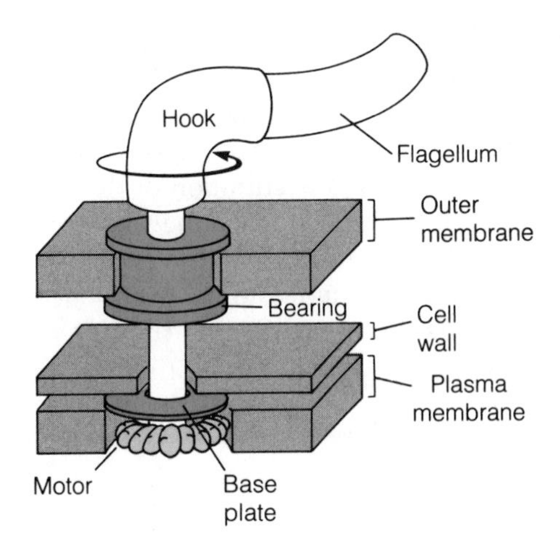

FIGURE 14-41
Schematic diagram of the base complex of the bacterial flagellum. The flagellum is rotated as a result of force generated by the motor in the plasma membrane.

The flagellum can rotate at speeds of 40 revolutions per second, which drives the cell forward at a rate of 50 μm per second. Energy for flagellar rotation is not derived from ATP, but rather comes from the proton gradient across the plasma membrane, generated by oxidative phosphorylation (Chapter 4). Remember that in the bacterial cell the machinery for oxidative phosphorylation is in the plasma membrane, which is also the location of the flagellar motor.

The overall migration of motile bacteria is not necessarily random in direction. For example, photosynthetic bacteria move in the direction of greater light intensity, a behavior known as **positive phototaxis.** At least some motile species have receptors in the plasma membrane for nutrients such as amino acids and sugars that allow them to move toward higher concentrations of these nutrients, a process called **positive chemotaxis.** The bacteria are not able to sense concentration gradients directly but move toward higher concentrations of nutrients by a trial-and-error method. When environmental conditions are constant, the bacterium swims in a straight path for a time, stops, tumbles, and swims another straight course in some other direction. When a bacterium encounters a gradient of an amino acid, the binding of the amino acid to receptors in the plasma membrane causes the cell to swim a straight course for a longer period before tumbling. If the course of swimming has taken the bacterium to a still higher concentration of the amino acid, its receptors signal it to swim on the straight course for longer than usual once again. However, if the bacterium has chanced to move to a lower concentration, fewer receptors are occupied, and the cell swims forward for a shorter course. Monitoring the concentration of the nutrient over time and swimming a longer straight path when the concentration of attractant is increasing results in net movement of the bacterium up a concentration gradient.

By a similar process of monitoring concentration over time, bacteria are able to migrate away from unfavorable conditions such as toxic concentrations of acetate, ethyl alcohol, and low or high pH. The process of migrating away from unfavorable conditions is called **negative chemotaxis.**

PROBLEMS

1. Three mechanochemical processes for producing motility are known in eukaryotes. What are they? Give a cellular example of each one.
2. How is contraction produced in a skeletal muscle cell?
3. What structure is responsible for the presence of the A band in muscle? The I band?
4. What is a myofibril? A sarcomere?
5. How are the heads of myosin molecules arranged in relation to actin thin filaments in the A band?
6. What other two proteins besides actin are present in thin filaments, and what are their functions?
7. How is it known that actin thin filaments in muscle have polarity?
8. What structure in a skeletal muscle cell makes it possible for all parts of all myofibrils to contract simultaneously in response to arrival of a nerve impulse?
9. What role does Ca^{2+} have in muscle contraction?
10. Describe the motility mechanism in amoebae.
11. Describe an experiment showing that cytoplasmic streaming in algal cells is based on an interaction between myosin and actin.
12. How do the microtubule doublets in a cilium or flagellum interact to produce motility?
13. Why is it assumed that the doublets in a cilium form two functional groups?
14. What two forms of intracellular motility contribute to the camouflage mechanism in fish?
15. Describe how a bacterium moves toward higher concentrations of nutrients.

CHAPTER 15

Function and Structure of Some Specialized Cells

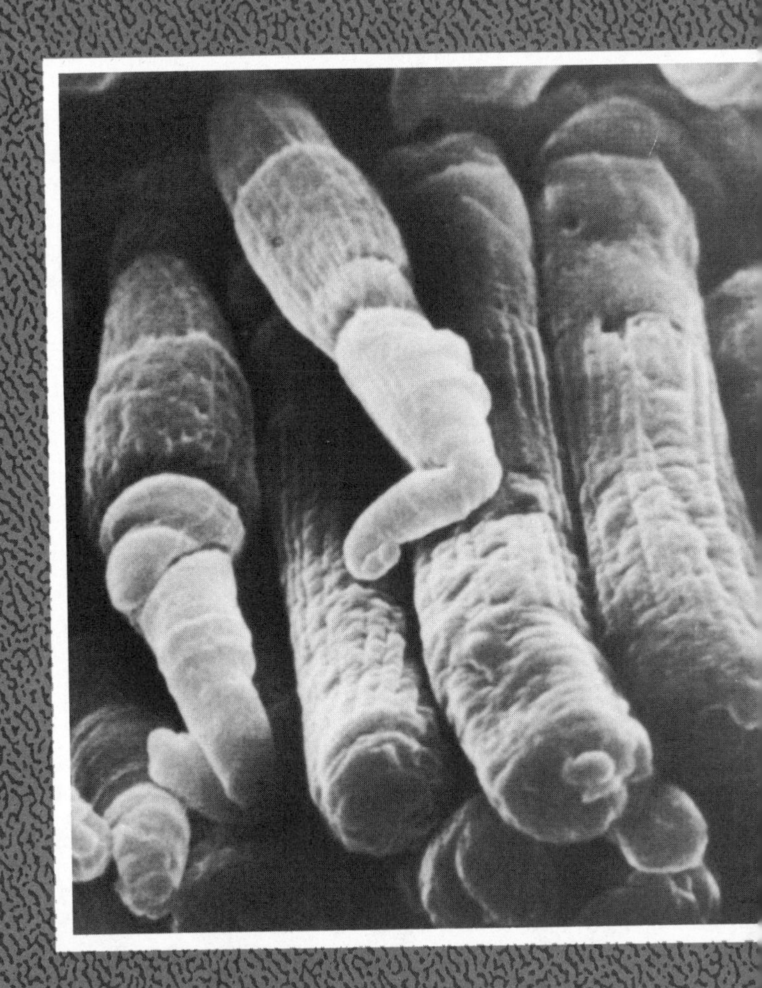

See Figure 15-24.

The cells of multicellular animals and plants are specialized for the performance of one or another particular function. Specialization, or differentiation of cells, consists of the synthesis of particular structures, such as the contractile fibers of muscle or the dendrites and axons of nerve cells, and synthesis of a particular biochemical machinery, (e.g., the machinery for synthesis and release of hormones by endocrine cells), the production of digestive enzymes by cells of the pancreas, the synthesis of hemoglobin in red blood cells, and so forth. Some 200 distinct cell types have been identified so far in mammals.

Cell specialization is achieved by the regulation of gene expression (Chapter 10). Clearly many of the same genes are expressed in nearly all cell types, such as the genes that code for RNA polymerases, glycolytic enzymes, and tRNA synthetases. The expression of other genes is restricted to particular cells. This process of cell differentiation is most often thought of in the context of embryological development. However, differentiation occurs continuously for newly produced cells in most adult organisms. A healthy adult human produces about ten million new cells every second, and these differentiate in one or another direction; for example, 2.5 million new cells complete differentiation into red blood cells every second, replacing red blood cells that wear out.

This chapter deals with a few major kinds of specialization in function and in structure that are found in various types of differentiated cells. The types included illustrate the specializations for sexual reproduction (egg and sperm cells), nutrient absorption (intestinal epithelial cells), impulse conduction (neurons), light reception (retinal rod cells), and immune response (lymphocytes). Specialization of function and structure for contraction by muscle cells was discussed in Chapter 14, and secretion of zymogen by pancreas cells was discussed in Chapter 5.

THE OOCYTE

Egg cells, or **ova**, are usually thought of as undifferentiated cells, yet they have special properties that enable them to develop into complicated multicellular organisms. First, oocytes, the precursors of ova, are specialized to undergo meiosis, a process that also occurs in **spermatocytes** but in none of the other cell types in a multicellular organism. We do not know what causes oocytes and spermatocytes to undergo meiosis but this is clearly a case of functional differentiation. Oocytes are also specialized for fusion with sperm cells, that is, to undergo fertilization. A particularly obvious specialization of oocytes is the storage of large amounts of reserve cytoplasmic materials such as lipids, glycogen, and proteins, much of which are in the form of yolk platelets. A less obvious part of the storage process is the large mass of ribosomes that is built up in the cytoplasm during oogenesis. The mRNAs that code for proteins needed during the initial phase of embryonic development are synthesized during oogenesis and stored in an inactive form in the oocyte (Chapter 10). In amphibian eggs, tRNAs are synthesized and stored in the cytoplasm during oogenesis. Mitochondria are also produced in large numbers in amphibian eggs and stored for later use.

The size of the ovum provides a measure of the magnitude of storage functions. Ova of mammals (Figure 15-1) tend to be small (e.g., about 100 μm in diameter), although still far larger than somatic cells. Oocytes of insects and fish are usually considerably larger, and the oocytes of amphibians, reptiles, and birds are enormous cells by any standard.

The materials built up during oocyte construction are used to carry the egg through at least early embryonic development. However, the egg cell is not a homogeneous sphere with materials evenly distributed throughout the cytoplasm. In the ova of most invertebrates and vertebrates materials are distributed differentially with respect to a polar axis. This is obvious at a gross level in amphibian eggs. The cytoplasm contains a gradient of pigment granules, with the high concentration defining, by convention, the top of the cell and the region of almost no pigment granules defining the bottom (Figure 15-2). The distribution of yolk platelets follows an opposite gradient, with a heavy concentration at the bottom and very few at the top.

Differential organization of cytoplasm is also evident in a profound functional sense. The first cleavage

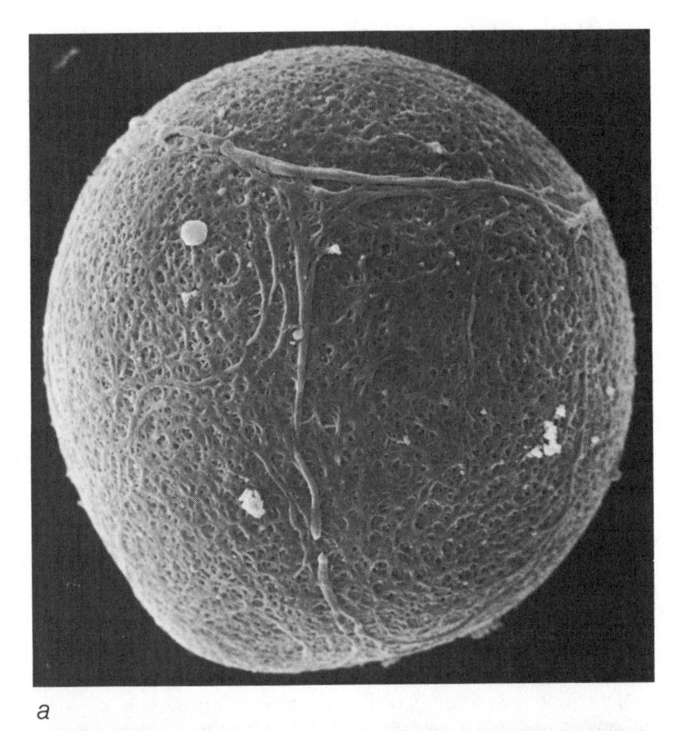

a

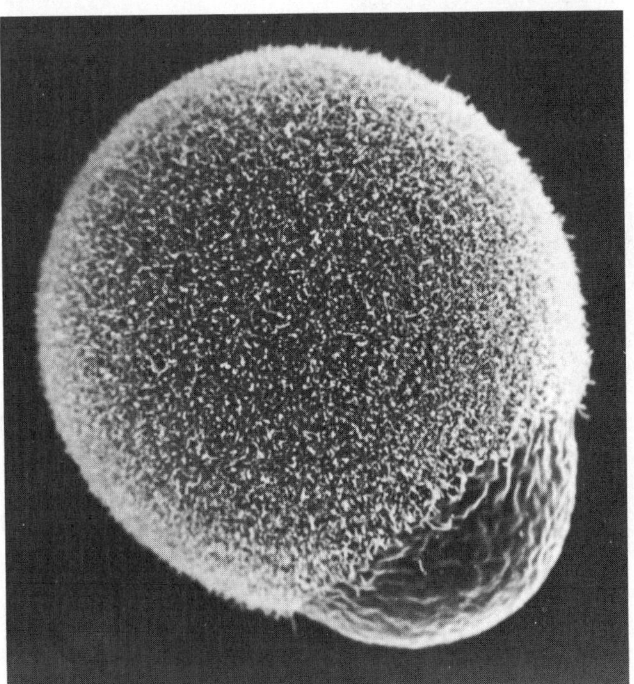

b

FIGURE 15-1
(a) Scanning electron micrograph of a mouse ovum with its outer protective covering (zona pellucida) still intact. (b) Mouse ovum with the zona pellucida removed, revealing the surface of the plasma membrane. The ovum is covered by fine microvilli except for one area. The smooth, slightly bulging area overlies the position of the nucleus, which is arrested in the second meiotic metaphase. [Courtesy of David M. Phillips. From *J. Ultrastructure Res.*, 72:1–12 (1980).]

division after fertilization is vertical, dividing the cell through the **polar axis** (Figure 15-3). In some species, for example the sea urchin, the two cells can be separated with a fine glass needle, and both will develop into completely normal but half-sized embryos. The second division is also vertical. The third division of the egg occurs equatorially, dividing the egg into four top cells and four bottom cells. The four top cells can be experimentally separated from the four bottom, but in this case the two halves of the egg develop into abnormal embryos. The embryo derived from the top four

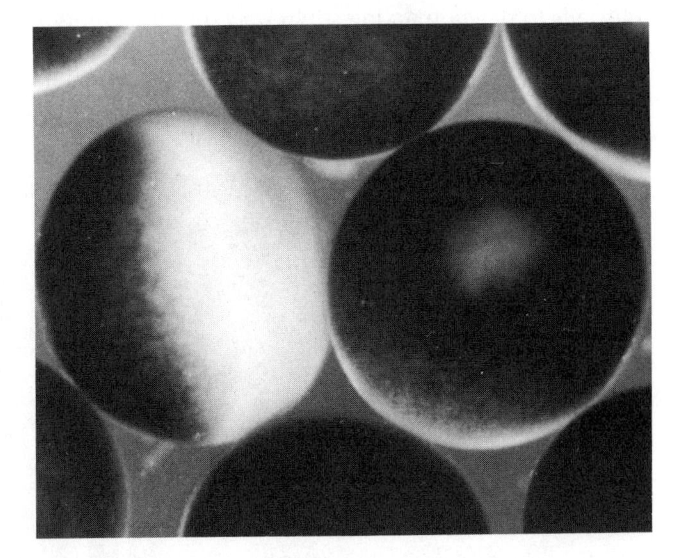

FIGURE 15-2
Light micrograph of two oocytes of *Xenopus*. The oocyte to the right is seen from the top. The oocyte to the left is tilted on its side. Black pigment granules are concentrated in the top and yolk granules are more concentrated in the bottom of the egg. [Courtesy of Michael W. Klymkowsky.]

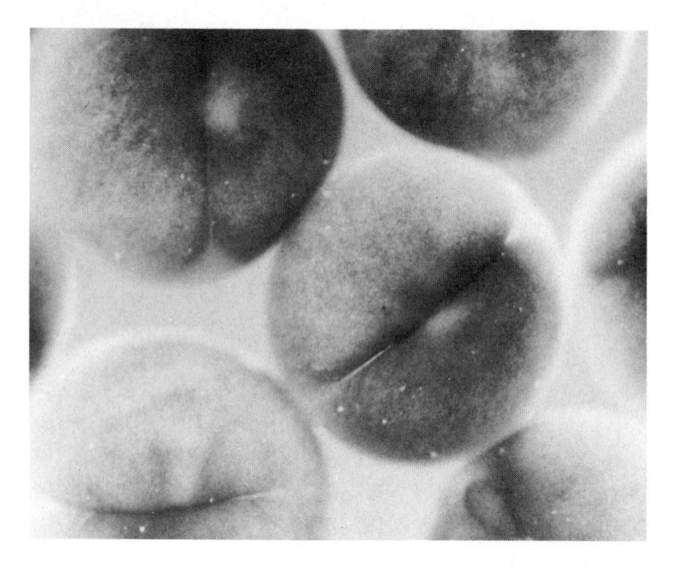

FIGURE 15-3
Light micrograph of *Xenopus* eggs viewed from the top in the first division after fertilization. The cleavage plane is vertical. [Courtesy of Michael W. Klymkowsky.]

cells fails to develop an intestinal tract and the embryo from the bottom half does not develop brain cells. From these and other experiments it is clear that the egg contains a **polar gradient** such that cells that are formed from the cytoplasm of the top half differentiate into certain types—for example, neurons—and cells that derive from cytoplasm in the lower half differentiate into other types, such as germ cells. Thus we see that even before fertilization a particular region of cytoplasm has properties that will influence what forms of differentiated cells will ultimately arise from it.

Striking proof that different regions of egg cytoplasm can have different properties has been obtained by transplanting cytoplasm from one region of an egg to another. *Drosophila* eggs are suitable for such experiments because the eggs are large enough (500 μm long and 150 μm across) to permit cytoplasm to be removed

FIGURE 15-4
Scanning electron micrographs of *Drosophila* eggs. (a) An unfertilized egg. The egg is covered by a tough membrane (chorion). The two projections at one end are used for intake of O_2 and release of CO_2 by the egg. (b) A fertilized egg with the chorion removed and containing several hundred nuclei. Separate cells have not yet formed; the location of nuclei just under the egg surface is indicated by the smooth oval patches. (c) An egg with several thousand nuclei. Individual cells have still not formed. (d) An egg in which plasma membranes have appeared and formed separate cells. Pole cells, which give rise to germ cells, are visible at one end. [Courtesy of Anthony P. Mahowald and F.R. Turner. From *Develop. Biol.*, 50:95–108 (1976).]

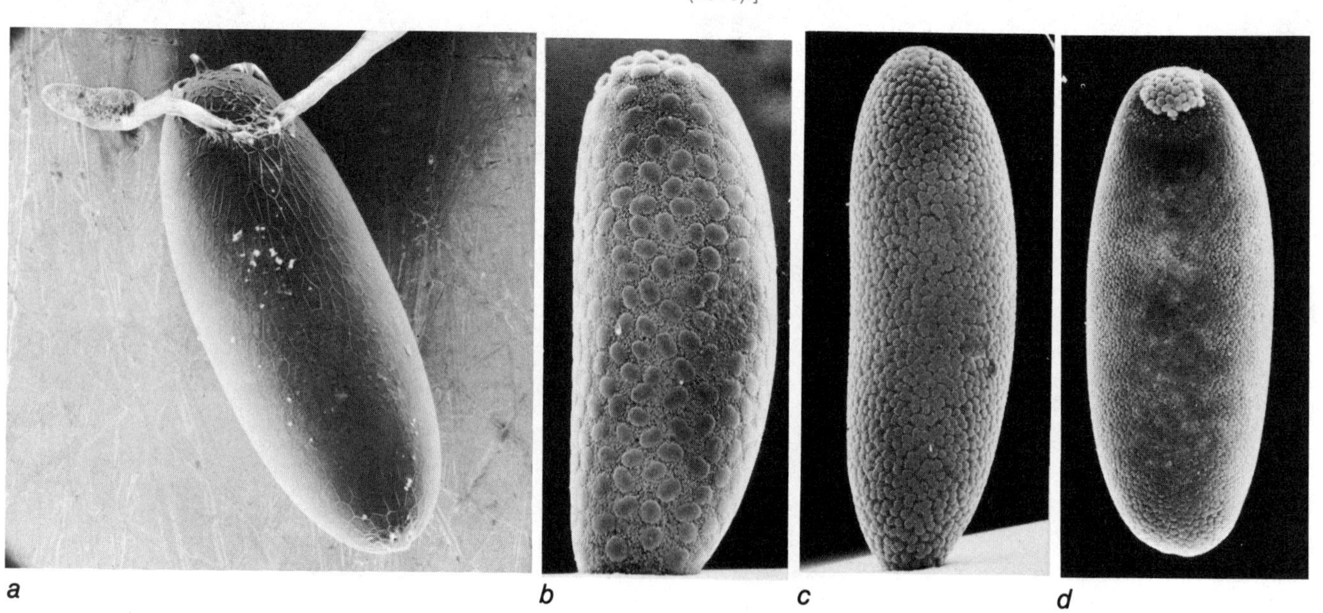

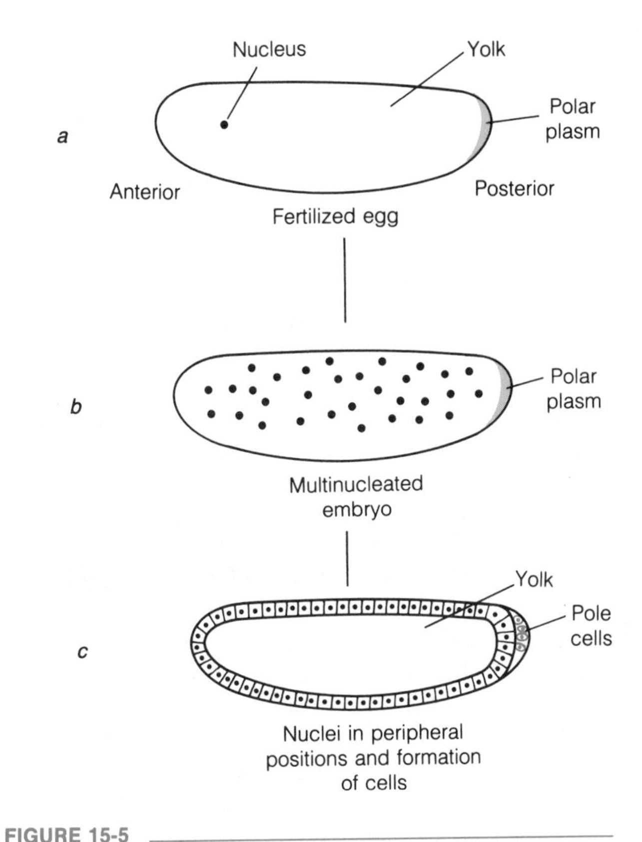

FIGURE 15-5
(a) Fertilized egg of *Drosophila*. (b) Development in *Drosophila* and other insects is somewhat unusual. The nucleus undergoes many mitoses without cell division. (c) The daughter nuclei migrate to the cytoplasm near the cell surface, where plasma membranes subsequently develop to form individual cells. Cells formed in the polar plasm become germ cells. The internal mass of the embryo is filled with yolk.

from a highly localized region with a micropipette and then injected into another region. Second, it is known that during embryogenesis the germ cells form from a particular region of the egg. The *Drosophila* egg has the shape of an elongated oval (Figure 15-4) and has an **anterior-posterior axis.** In development, cells at the anterior end ultimately differentiate into the head of the fly and its associated structures, cells in the midregion develop into thoracic structures, and cells in the posterior region give rise to abdominal structures. The gonads develop in the abdominal region, and the germ cells *always* arise from the cells that are formed in the posterior end of the egg (Figure 15-5). However, when cytoplasm is removed with a micropipette from the posterior pole (polar plasm) of an unfertilized egg and injected into the middle region of a newly fertilized egg, some of the cells that subsequently form in the midregion of the fertilized egg differentiate into germ cells. This shows that the cytoplasm in the posterior pole of the unfertilized egg, where germ cells normally arise, contains a factor that causes germ cell differentiation. Clearly, the cytoplasmic factor, whatever its nature, directs the nucleus to express the particular pattern of genes necessary for germ cell differentiation. This factor is called the **germ cell determinant.** Particular determinants for other specific differentiated cell types are not generally present in an egg. Rather, the cytoplasm in different regions of the egg has a general influence on what classes of cells can eventually differentiate from that portion of the cytoplasm. Only later in development do the determinants for specific cell types appear.

A clear example of regional differentiation of cytoplasm is found in eggs of the mollusc *Dentalium*. At the first cleavage after fertilization a large pocket of cytoplasm forms on one side of the egg and remains attached by a neck to one of the two daughter cells (**blastomeres**) (Figure 15-6). The pocket of cytoplasm, called a **polar lobe,** is acquired by one of the four blastomeres produced at the second cleavage of the egg (Figure 15-6). The cell with the attached polar lobe normally gives rise to the foot, eye, and shell gland of the mollusc. If the polar lobe is severed, these differentiated tissues do not appear or are grossly abnormal. Thus, the cytoplasm of the polar lobe influences what kinds of specific determinants may eventually appear in descendants of the fourth blastomere. In a variety of organisms one sees many other cases of cytoplasmic influence over gene expression during embryogenesis. To understand this regulation of gene expression we must first try to determine the molecular basis of such influence. Equally important, we must figure out how regional properties of the cytoplasm are established during construction of oocytes in the ovary.

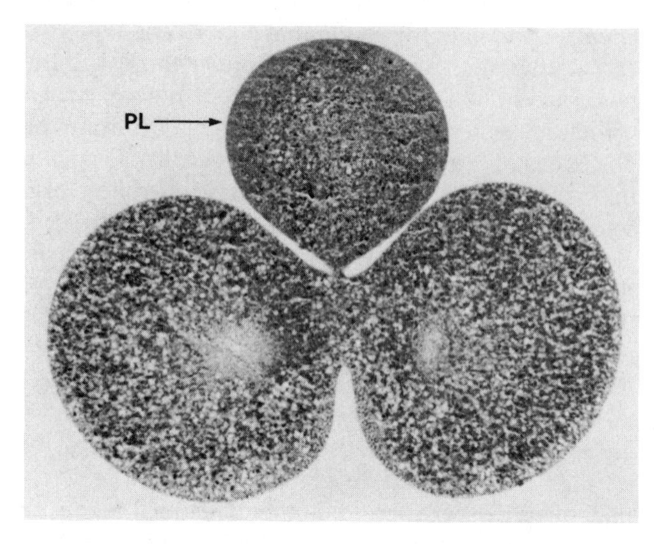

FIGURE 15-6
Light micrograph of a two-cell embryo of the mollusc *Dentalium*. A large polar lobe (PL) of cytoplasm forms during cell division and will subsequently be incorporated into one of the cells at the four-cell stage of the embryo. [Courtesy of René Dohmen. From *Determinants of Spatial Organization* (1979) S. Subtelny and I. R. Konigsberg, eds. Academic Press, New York.]

An extraordinary specialization that occurs in the oocytes of some species, notably frogs and salamanders, is the formation of many copies of the genes coding for the two large rRNAs (28S RNA and 18S RNA). Such selective replication of genes, called gene amplification, was discussed in Chapter 10.

SPERM CELLS

Like oocytes, **sperm cells** are specialized for fertilization. During their differentiation, precursors of both cell types undergo meiosis, become haploid, and develop specializations for fertilization and formation of a zygote. However, one major feature of the differentiation of sperm cells is reversed with respect to that of oocytes: formation of an oocyte is characterized by considerable synthesis and storage of materials, while production of a sperm cell consists of a stripping down of the precursor cell to a small size and a minimum of components. A sperm cell is a nucleus propelled by a mitochondrion-powered flagellum and specialized to penetrate the surface of an ovum.

In the formation of a sperm cell **spermatogonial cells** in the testis become spermatocytes and undergo meiosis to form four haploid cells known as **sper-**

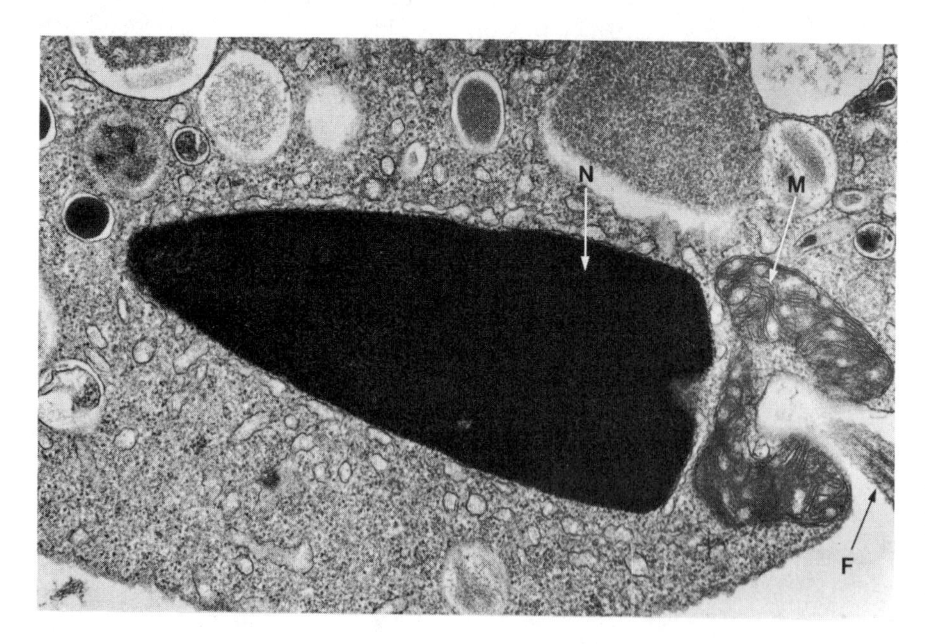

FIGURE 15-7
Electron micrograph of a sea urchin sperm that has just fertilized an ovum. The sperm nucleus (N) is surrounded by cytoplasm of the ovum. The chromatin is tightly condensed to form a compact nucleus. A mitochondrion (M) of the sperm and part of the flagellum (F) are visible. [Courtesy of Daniel S. Friend.]

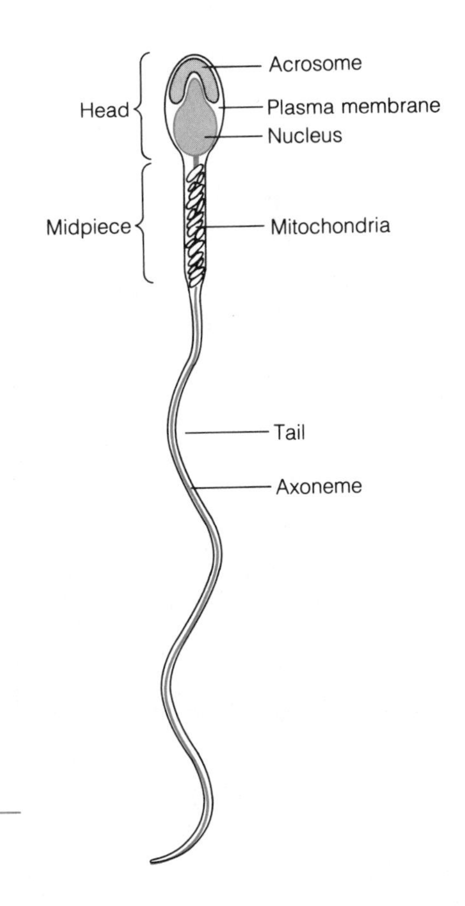

FIGURE 15-8
Drawing of a human sperm.

matids. In the differentiation of spermatids to sperm almost all of the cytoplasm is sloughed off. In this process the Golgi complex, endoplasmic reticulum, and ribosomes disappear from the cell. Prior to the loss of the cytoplasm all RNA transcription stops, and the chromatin gradually becomes condensed to form a compact nucleus (Figure 15-7). In some species this involves replacement of the histones by even more basic proteins called **protamines.** In the protamines of some species more than 50 percent of the amino acid residues are arginine, a strongly basic amino acid. The purpose of this replacement is still unclear. During the first cell divisions after fertilization the protamines disappear and are replaced by histones.

In addition to a compact nucleus, a mature sperm cell contains three other structures: the **flagellum,** the **middle piece,** and the **acrosome** (Figure 15-8). The flagellum consists of the usual 9+2 pattern (Chapter 14) of microtubules surrounded by the plasma membrane. There are several exceptions to the 9+2 pattern, all of which involve the two central microtubules. In the mayfly, the fish *Lycondontis,* and the annelid worm *Myzostomum* the two central microtubules are completely missing, yielding a 9+0 pattern. A few animals have a 9+1 pattern, and the caddis fly has a 9+7 pattern. These variations help shed light on the function of the central microtubules. The 9+0 pattern of some species indicates that the two central microtubules are not essential for movement of the flagellum. At the root of the flagellum is a pair of centrioles. In one of these the microtubules continue into the microtubules of the flagellum (Figure 15-9). The other is positioned at a right angle to the first centriole and probably plays no direct role in formation or function of the flagellum.

The midpiece of a sperm cell consists of one or

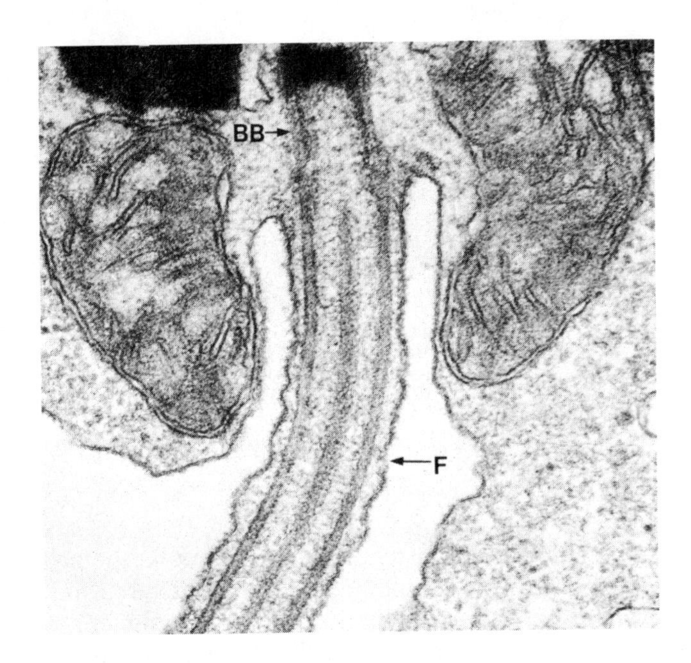

FIGURE 15-9
Electron micrograph of a sea urchin sperm that has entered an ovum. The basal body (BB), cut longitudinally, continues into the microtubules of the flagellum (F). The basal body lacks central microtubules. [Courtesy of Daniel S. Friend.]

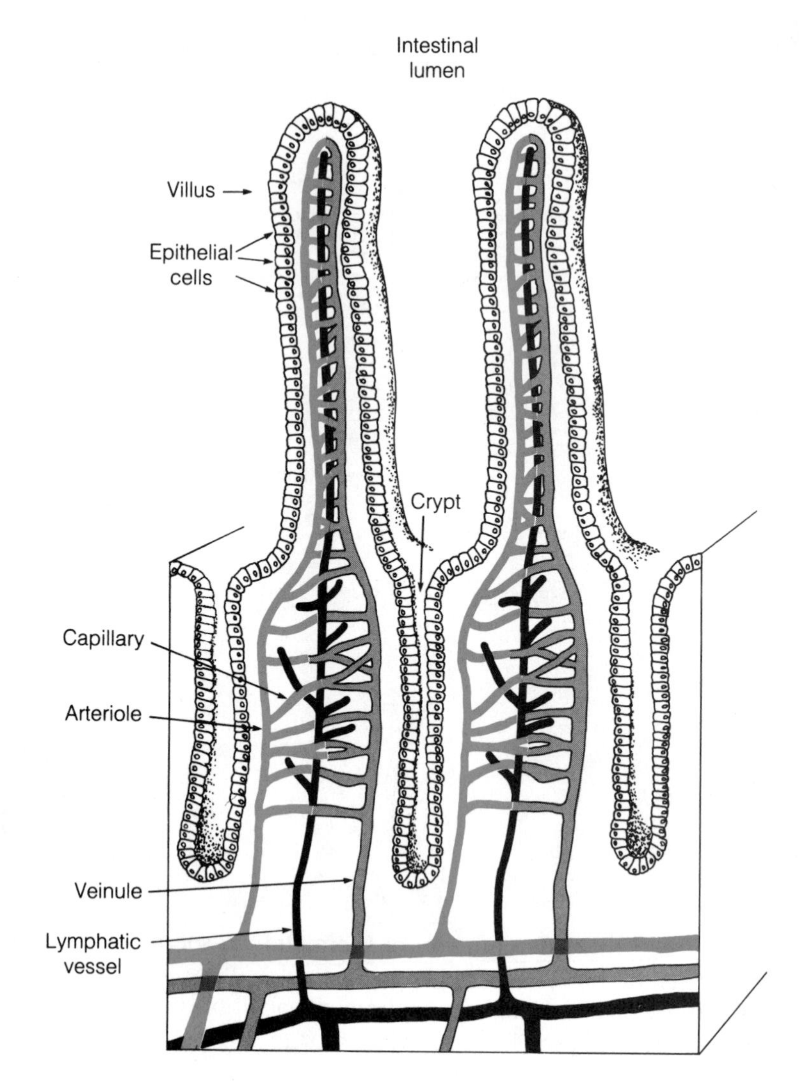

FIGURE 15-10
The lining of small intestine is made up of hundreds of thousands of villi that are covered with a layer of epithelial cells. Nutrients are absorbed from the intestinal lumen by the epithelial cells and passed to the blood vessels, or, in the case of some lipids, to lymphatic vessels. Epithelial cells in the lower part of crypts reproduce and move out onto villi by migration of the entire epithelial sheet. Cells are lost from the sheet into the intestinal lumen at the tips of villi.

more mitochondria wrapped helically around the first portion of the flagellum (Figure 15-8). The mitochondria furnish the ATP necessary for flagellar motion.

The acrosome is an organelle that is unique to the sperm cell. It develops from a collection of vesicles produced by the Golgi complex and comes to lie at the tip of the sperm head (Figure 15-8). Lysosomal enzymes are present in the acrosomes of some species, suggesting that the acrosome may be a highly modified lysosome. The acrosome contains enzymes important for penetration of the sperm head through the surface coats of the egg. One of these enzymes, hyaluronidase, digests hyaluronic acid, a major component of one of the extracellular coats of the egg. Another enzyme, zona lysin, locally digests the **zona pellucida,** a heavy membranous structure that surrounds the egg cell.

In some invertebrates the acrosome contains a large amount of G-actin. Contact of the sperm with the outer coat of the egg (eggs of these animals possess a coat of jellylike material many μm thick) causes the almost instantaneous polymerization of G-actin into a long actin filament that penetrates the egg coat ahead of the sperm. Contraction of the actin filament is believed to occur, drawing the sperm through the extracellular material to the surface of the egg.

EPITHELIAL CELL OF THE SMALL INTESTINE

Epithelial cells form continuous sheets in animal tissues, where they serve as coverings or linings. They perform different functions in different tissues. For example, the outer layer of skin (epidermis) is an epithelial sheet composed of many layers of cells (**stratified epithelium**) (Chapter 6). It provides a barrier to the loss of tissue fluids. Its outer portion consists of layers of flat cells filled with intermediate filaments of keratin. The cells are tightly bound to each other, making a tough layer that provides mechanical protection for underlying tissues.

The epithelium that lines the small intestine is a sheet one cell-layer thick. It carries out a number of functions, particularly the absorption of water, ions, and nutrients from the intestinal lumen and their transfer to the blood. The absorptive function is greatly increased by a very large number of projections of the lumenal surface of the intestine called **villi** (*singular,* villus) (Figure 15-10); each villus is covered with a continuous layer of epithelial cells. At the base of the villi are invaginations called **crypts.** Stem cells in the epithelial layer at the bottom of the crypts reproduce. As part of a steady shifting or sliding of the epithelial sheet, new cells are pushed out of the crypts. Once a cell moves out of the lower part of a crypt, it stops reproducing and completes its differentiation. At the top of the crypts mature epithelial cells, differentiated for absorption of water, ions, and nutrients move onto the villi. The constant flow of cells outward from the crypts onto the villi is balanced by a steady sloughing of cells from the tips of the villi. Sloughed cells are lysed and digested, and the resulting materials absorbed by the epithelium.

The entire transit from the base of a crypt to the tip of a villus, including differentiation along the way, occurs in just a few days. The epithelium of the intestine is one of the fastest renewing cell populations in the body. An adult human sheds several hundred grams of epithelial cells (10^{10} to 10^{11} cells) from villi into the intestinal lumen every 24 hours.

An important structural feature of an intestinal epithelial cell is the thousands of microvilli that extend from the surface of the cell (Figure 5-37) (Chapter 5). These are cylindrical projections of the cytoplasm about one μm long and less than 0.1 μm in diameter. There are hundreds of thousands of microvilli per mm^2 of lumenal surface. The microvilli greatly increase the absorptive surface area of the intestinal lumen. In the small intestine of a human the absorptive surface is about 180 m^2.

The epithelial cells are joined with each other by three kinds of junctions. At the base of microvilli adjacent cells are connected by **tight junctions** (Figure 15-11), which are continuous around the circumference of every cell. At tight junctions the plasma membranes tightly adhere to one another such that there is no intercellular space. These junctions prevent the diffusion of molecules or ions across the epithelial sheet in between adjacent cells. All movement of substances must take place through the cells themselves, a circumstance that allows for regulation of movement of particular substances by active transport (see below). At tight junctions, closely packed rows of transmembrane proteins in one cell stick to corresponding rows in the plasma membrane of an adjacent cell. The rows of transmembrane proteins are arranged in a matching network in the two adjacent cells, making a complete seal.

A short distance below the tight junctions, adjacent cells are held together by adhering junctions called **belt desmosomes** (Figures 6-24, 15-11). These are continuous around the cell circumference, joining each cell to all of its immediate neighbors. Belt desmosomes provide great mechanical strength to the epithelial sheet.

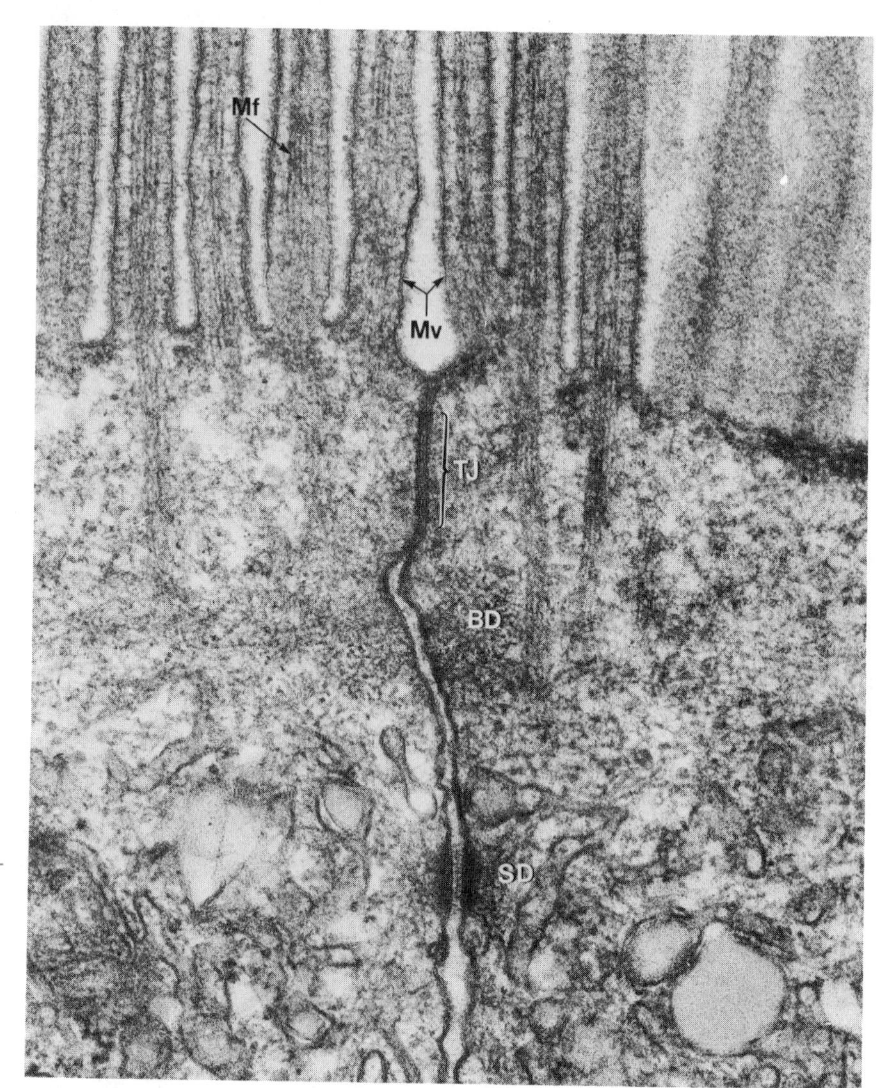

FIGURE 15-11

Electron micrograph of two adjacent epithelial cells in the small intestine of a tadpole. Microvilli (Mv) containing actin microfilaments (Mf) extend into the lumen of the intestine. The plasma membranes of two cells are joined by a tight junction (TJ), a belt desmosome (BD), and a spot desmosome (SD). [Courtesy of L. Andrew Staehelin.]

In belt desmosomes the plasma membranes of adjacent cells are separated by a narrow intercellular space filled with a filamentous material that presumably holds the two membranes together. Immediately inside the plasma membrane at belt desmosomes is a bundle of actin microfilaments that is thought to give mechanical strength to the junction.

In addition to belt desmosomes, the epithelial cells are held together by **spot desmosomes** (Figures 6-26, 15-11, 15-12). These are rivetlike structures in which adjacent plasma membranes are separated by a narrow intercellular space that is filled with a filamentous network. On the cytoplasmic surface of the plasma membrane at a spot desmosome is a disc of dense material to which intermediate filaments of keratin are attached. The keratin filaments form a network inside the cell joining all the spot desmosomes. The network in one cell is also joined, through the spot desmosomes, to the network in every adjacent cell. This arrangement is thought to provide another measure of physical strength to the epithelial sheet.

Each epithelial cell is in intimate contact with all of its neighbors through tight junctions, belt desmosomes, and spot desmosomes. These connections are formed after cell division in the deeper part of the crypts has ceased and are subsequently broken at the tips of villi when cells are sloughed into the lumen. Sloughing must include not only breakage of existing connections but establishment of new ones to seal the sheet where a cell has left.

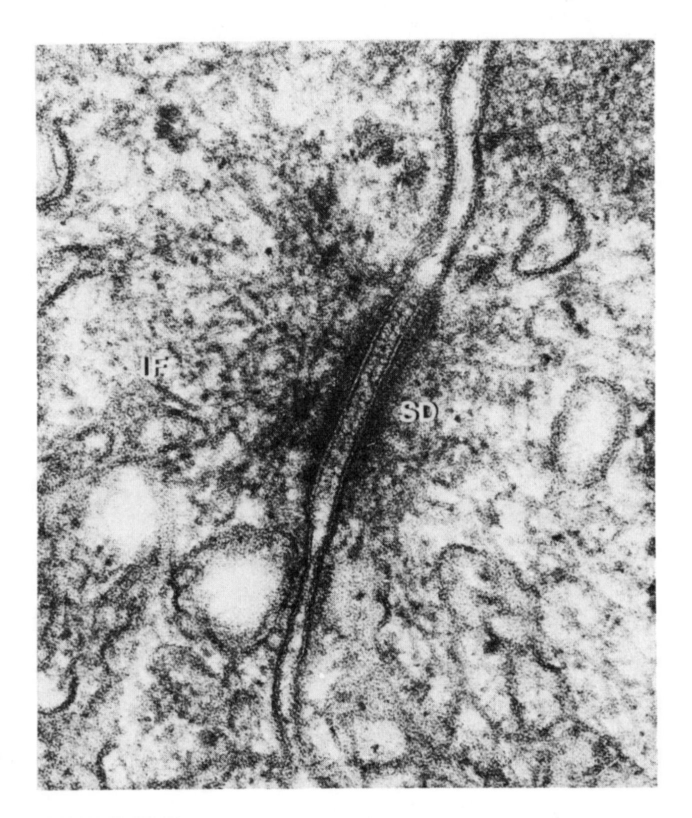

FIGURE 15-12
Electron micrograph of a spot desmosome (SD) at a higher magnification than in Figure 15-11 to show intermediate filaments (IF) of keratin. [Courtesy of L. Andrew Staehelin.]

The apical surface (surface facing the lumen) of the intestinal epithelial cell differs from the surface in the basolateral region of the cell. Microvilli of the plasma membrane of the apical surface possess enzymes involved in digestion and uptake of substances (e.g., disaccharidases), sugar- and amino acid-transport systems (Chapter 5), and alkaline phosphatase. These components are absent from the basolateral plasma membrane. The basolateral plasma membrane contains Na^+-K^+ ATPase, adenylate kinase, and hormone receptors; these enzymes are absent from the apical plasma membrane. The tight junctions between epithelial cells are thought to act as a barrier to diffusion of these various proteins between the apical and basolateral parts of the plasma membrane, maintaining the differences between the two parts of the plasma membrane. How the cell delivers particular proteins to a particular part of the plasma membrane is discussed in Chapter 8.

Apical-basolateral differences in the plasma membrane provide for unidirectional transport across the cell. For example, glucose, which is transported into the intestinal epithelial cell from the intestinal lumen (Figure 15-13), is exported from the cell by facilitated-diffusion protein carriers in the basolateral plasma membrane into the intercellular space, from which it enters the blood. Facilitated-diffusion protein carriers are absent from the apical plasma membrane and, in addition, glucose is prevented by the tight junctions from leaking from intercellular spaces back into the intestinal lumen.

NERVE CELLS

Nerve cells, or **neurons,** are specialized for the transmission of electrochemical signals, called **impulses,** over long distances. Through interactions with one another and with other cell types, such as muscle cells, neurons coordinate many of the activities of the different organs in multicellular animals. There are many kinds of neurons, most of which fall into one of

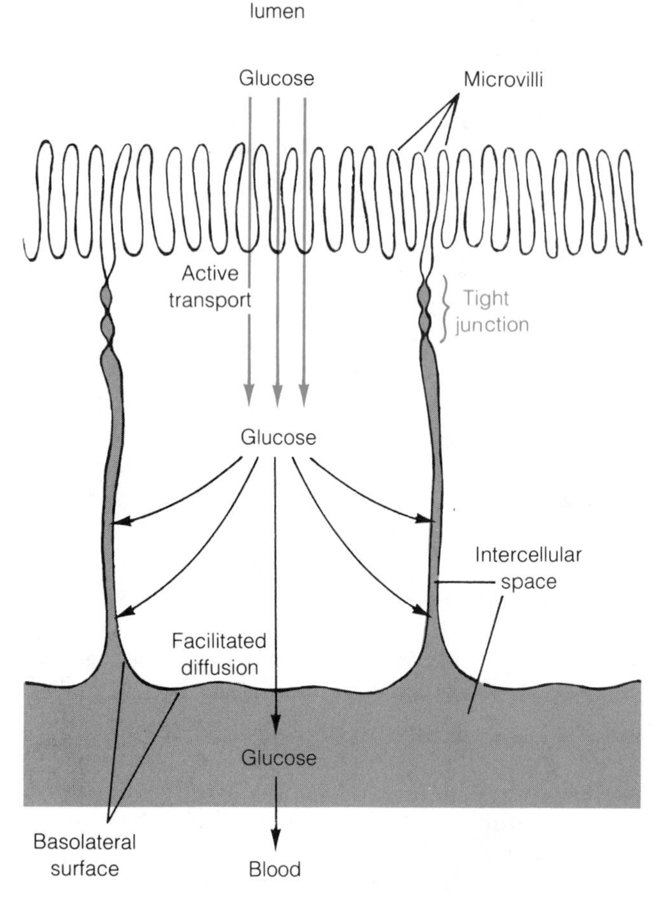

FIGURE 15-13
An intestinal epithelial cell. Glucose is actively transported from the intestinal lumen into the cell by carriers in the plasma membranes of the microvilli. At the basolateral surface of the cell glucose is moved into the intercellular space by facilitated diffusion, and from there into blood vessels. Tight junctions between epithelial cells prevent leakage of glucose back into the intestinal lumen.

three major classes: **sensory neurons,** which generate signals through sensory receptors; **effector neurons,** which carry impulses from the brain to various muscles; and **interconnecting neurons,** which receive impulses from one neuron and transmit them to another neuron. A single neuron may make functional connections with tens or hundreds of other neurons. Interconnecting neurons in particular may have many functional interactions with other neurons. There are about 1.4×10^{10} neurons in the human brain, and their functional interconnections create the incredibly complex neural networks that underlie brain functions.

Although different neurons have distinct roles, they all share the same basic structural and functional characteristics. A typical neuron has three principal parts—a cell body, multiple **dendrites** (afferent cyto-

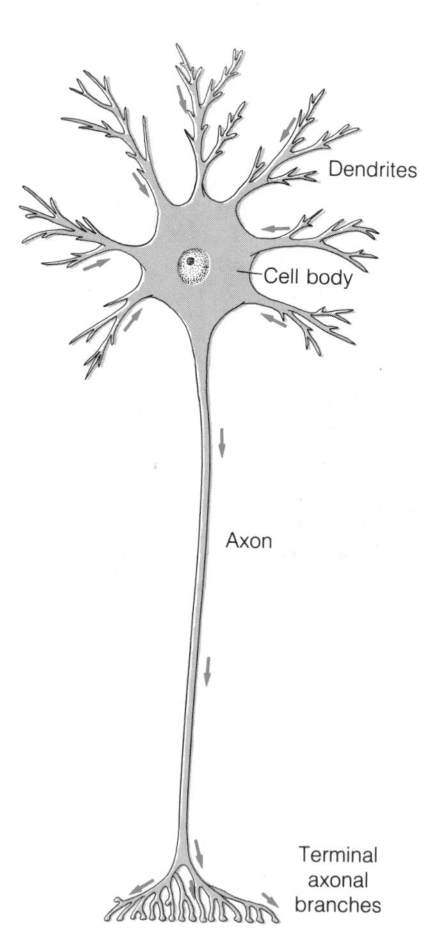

FIGURE 15-14
Diagram of a neuron. The red arrows show the direction of transmission of signals. Axons vary in length from a millimeter to several meters.

plasmic extensions) by which stimuli are received, and an **axon** (an efferent cytoplasmic extension) by which impulses are carried to other cells (Figure 15-14). For example, in a **motor neuron,** which is a neuron that innervates muscle, the cell body is located in the spinal cord. Through its dendrites the motor neuron receives impulses from interconnecting neurons and transmits impulses out to a muscle by means of its axon. Finally impulses are transmitted from one cell to another by means of special functional connections known as **synapses.**

Structure of a Nerve Cell

The cell body, which is also called the **perikaryon,** contains the nucleus and the usual cytoplasmic organelles. The nucleolus is large, reflecting a high rate of ribosome production. The cytoplasm is rich in ribosomes, many of which are attached to the endoplasmic reticulum to form an extensive rough endoplasmic reticulum. The rate of protein synthesis is also high, in part because the cell body supplies proteins to the axon, which lacks machinery for protein synthesis.

Axons range in length from a millimeter or two to a meter or more, depending on the type of neuron and the animal species. Axons are very long compared to the dimensions of most animal cells, which are typically 10 to 15 μm in diameter. These impulse-conducting cytoplasmic extensions do not contain rough endoplasmic reticulum or Golgi complexes, but do have microtubules, intermediate filaments, and vesicles. The microtubules and intermediate filaments serve as a cytoskeleton that maintains the narrow, highly elongated form of the axon. Proteins or other macromolecules are not synthesized in axonal cytoplasm. Instead, macromolecules needed for maintenance and function of the axon are provided by the cell body; these are carried down the axon by transport along microtubules.

If an axon is severed at some point, the severed part (distal part) degenerates since it can no longer receive the materials from the cell body that are required for maintenance. However, for the class of nerve cells called **peripheral neurons,** the proximal severed stump of the axon can regrow, usually following the path left by degeneration of the severed part of the neuron. Ability of axons to regrow has considerable practical importance. Nerves are bundles of axons carrying impulses to and from the spinal cord and other brain regions. Nerves are sometimes severed, for example in connection with fracturing an arm or leg. In this situation the cell bodies in the spinal cord (or spinal root ganglia in the case of sensory neurons) synthesize materials required for regrowth of the axons that make up the nerve. Typically, axons grow about one millimeter per 24 hours. In relation to the dimensions of cells this is rapid and represents a great amount of synthesis for a cell. In relation to gross anatomical structures the rate seems slow; nerve regeneration may take months to complete. Regeneration does not occur in central portions of the nervous system, such as after severing the spinal cord, in part because the growth of axons is blocked by the formation of scar tissue, which contains large amounts of collagen and other extracellular, tough fibrous material. Also, growth factors that stimulate regeneration of axons may be lacking in the spinal cord of adult animals.

Neurons are incapable of cell division (they are locked in G1 or G0; Chapter 12). Therefore, when the cell body of a neuron is destroyed, it cannot be replaced. Destruction of the cell body is followed by rapid disintegration of its dendrites and axon. During the adult life of a human, neurons are continually dying, which contributes to the process of aging of an organism.

Conduction of a Nerve Impulse

A nerve impulse is the propagation of an **electrochemical change** along the plasma membrane. The electrochemical change depends on the Na^+-K^+ pump that is present in the plasma membrane (Chapter 5). The Na^+-K^+ pump actively transports Na^+ ions outward across the plasma membrane and transports K^+ ions inward. As a result, the concentration of Na^+ ions is lower inside the cell than outside, and the concentration of K^+ ions is the reverse—higher inside than outside. Because of these concentration differences Na^+ ions tend to diffuse through the plasma membrane back into the cell, and K^+ ions tend to diffuse out. The rate of diffusion of each ion across the plasma membrane increases with steepness of the concentration differences created by the pump. Ignoring for the moment the effect of charges on the Na^+ and K^+ ions, a point is reached at which the rate of back diffusion of an ion equals the rate at which it is being pumped. This is the steady-state point, that is, there is no net movement of an ion across the plasma membrane. In the steady state, two opposite chemical concentration differences are present, one for Na^+ ions and one for K^+ ions.

If the rate at which Na^+ ions were pumped out were equal to the rate at which K^+ ions were pumped in, and if the intrinsic ability of the ions to penetrate the plasma membrane were the same, the concentration of $Na^+ + K^+$ ions just outside the cell would always be exactly equal to the concentration of $Na^+ + K^+$ ions inside. The concentration differences for Na^+ ions and for K^+ ions would be equal in steepness. Also, the total concentration of positive charges ($Na^+ + K^+$) just outside the plasma membrane would be the same as the total concentration of positive charges inside. Therefore, there would be no electrical difference, that is, no electrical gradient, only two chemical concentration differences.

Two phenomena upset the expected equivalence of the two chemical concentration differences, and this, in turn, results in creation of an **electrical gradient** across the plasma membrane. First, for every three Na^+ ions pumped out of the cell, two K^+ ions are pumped inward. This 3:2 ratio is a feature of the Na^+-K^+ pump (Chapter 5). For each cycle of the pump one ATP is used to transport three Na^+ ions out and two K^+ ions in. As a result, at equilibrium the concentration difference of Na^+ is steeper than the concentration difference of K^+ (Figure 15-15). This, in turn, means that the concentration of positive charges ($Na^+ + K^+$ ions) inside the cell is less than the concentration just outside the plasma membrane, that is, the outside is electrically more positive than the inside, or stated reciprocally, the inside is more negative than the outside. Thus, in addition to the unequal concentration differences for Na^+ and K^+ ions, an electrical gradient is established. Second, the rate of penetration of Na^+ through the plasma membrane is intrinsically much less than the rate for K^+ ions. Therefore, back diffusion of Na^+ ions occurs less readily than back diffusion of K^+ ions. This adds to the effect created by the Na^+-K^+ pump; the steepness of the Na^+ ion difference across the membrane at equilibrium relative to the steepness of the K^+ ion difference is increased. Thus, the Na^+-K^+ pump and the unequal rates of penetration of Na^+ and K^+ ions through the plasma membrane set up unequal concentration differences for Na^+ and K^+ ions, which result in an electrical difference (**potential**). The electrical potential, in turn, influences the chemical differences. The higher concentration of positive charges outside the plasma membrane tends to reduce the outward flow of K^+ ions and increase the inward flow of Na^+ ions. The concentration differences in Na^+ and K^+ ions inside and outside the cell and the magnitude of the electrical potential reached at steady state are therefore determined by an interplay of the ion and electrical differences set up by the Na^+-K^+ pump and the differential rates of penetration of Na^+ and K^+ ions through the plasma membrane.

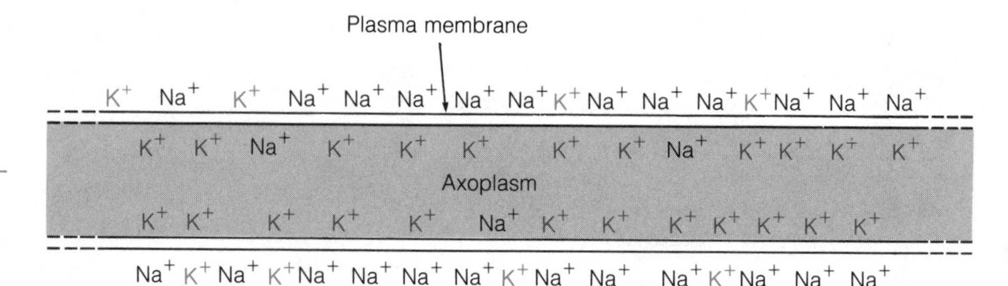

FIGURE 15-15
Differences in concentration of Na^+ and K^+ across the plasma membrane of a nerve axon created by the Na^+-K^+ pump.

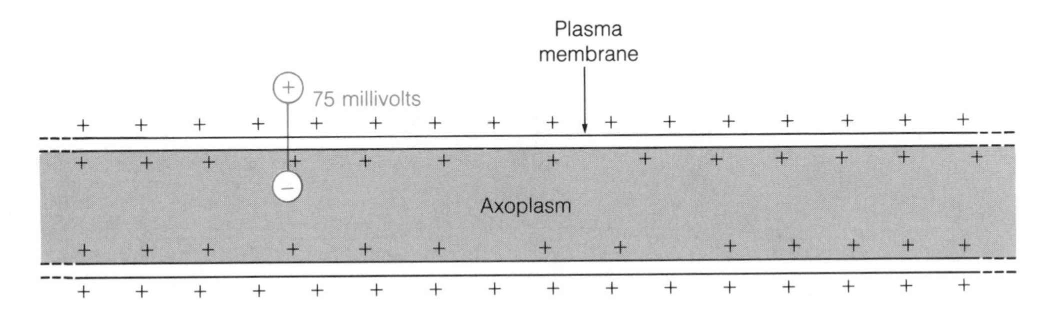

FIGURE 15-16
In the electrochemical gradient across the axonal plasma membrane the concentration of positive charges is higher just outside the plasma membrane than just inside. As a result the inside is electrically negative relative to the outside. The electrical potential across the membrane is about 75 millivolts.

The chemical and electrical gradients together comprise an **electrochemical gradient.** The electrical difference in the electrochemical gradient is called the **resting electrical potential** of the plasma membrane and is usually about 70 to 80 millivolts (Figure 15-16). Long-term maintenance of the resting electrical potential requires the continuous expenditure of ATP to operate the Na^+-K^+ pump.

Impulse conduction is made possible by the resting potential across the plasma membrane and occurs in the following way. The permeability of the plasma membrane to Na^+ ions can be caused to increase locally by application of an electrical or chemical stimulus, either of which cause Na^+ ion channels in the plasma membrane to be opened. The increase in Na^+ ion flow occurs initially only at the point where the stimulus is applied. Normally, the stimulation occurs on a dendrite. As a result of the increase in Na^+ ion flow, an inrush of Na^+ ions takes place. Since, under conditions of a resting potential, the Na^+ ion concentration difference across the membrane is greater than the K^+ ion concentration difference, the influx of Na^+ ions not only cancels the negativity inside the cell (**depolarizes** the membrane) but creates a net positive charge inside the cell relative to the outside (Figure 15-17). The increase in Na^+ ion flow is extremely transient, lasting less than one millisecond. Immediately after inrush of Na^+ ions, the Na^+ ion channels close and the rate of inflow of Na^+ ions returns to its original level, and the permeability of the plasma membrane to K^+ ions increases greatly. This permits a rapid efflux of K^+ ions, which causes the inside of the cell to become negative once again. The negativity created by K^+ ion efflux may transiently exceed the negativity (−75mv) present in a resting potential. Within a millisecond or two, permeability of the membrane to K^+ ions returns to its original level. The entire sequence of events from stimulation to

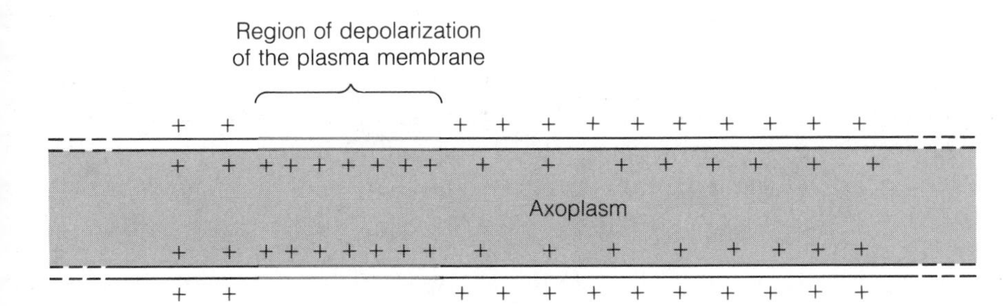

FIGURE 15-17
Local depolarization of the plasma membrane occurs by opening Na^+ channels and the inrush of Na^+ ions.

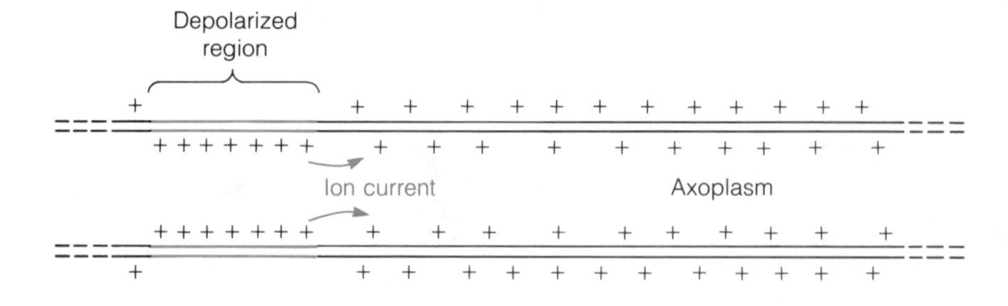

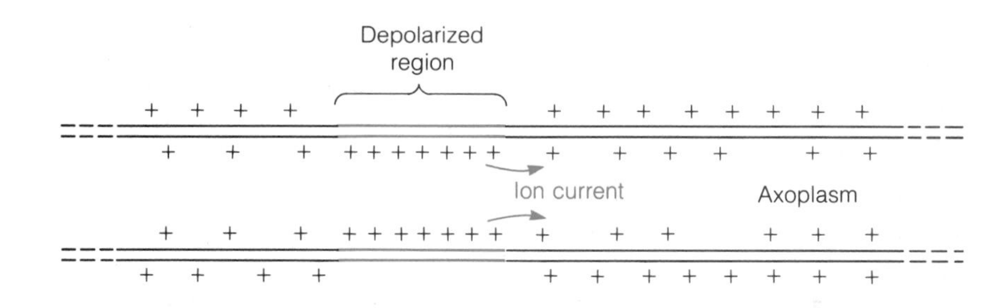

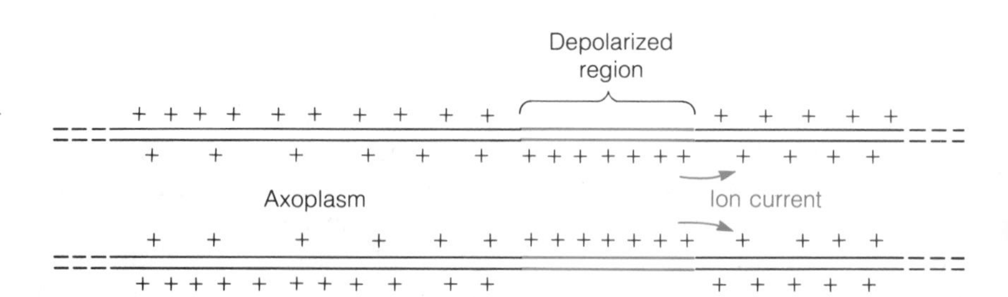

FIGURE 15-18
Propagation of a depolarized region along an axon. Depolarization of a short region of the plasma membrane sets up an ion current to the immediately adjacent membrane; the ion current causes depolarization of the adjacent membrane.

return to resting potential occupies only a few milliseconds. Some nerve cell poisons, or neurotoxins, such as **tetrodotoxin** of the puffer fish and the toxin in the venom of scorpions, work by blocking Na^+ ion channels, preventing depolarization of the plasma membrane in response to a stimulus.

The electrochemical changes just described occur in the highly localized region of the membrane where a stimulus is applied. When the inside of the cell in the region of the stimulus becomes transiently positive (Na^+ ion influx) with respect to outside of the cell, an electrical potential is necessarily created between the point of stimulus and adjacent cellular regions (Figure 15-18). The result is an ion current inside the cell that excites permeability changes to Na^+ and then to K^+ ions in the immediately adjacent regions of membrane, with the accompanying electrochemical changes and depolarization of the plasma membrane. This process of excitation continues, producing a propagated wave of localized electrochemical change (called the **action potential**) that spreads along the plasma membrane of axons at a rate of one to 100 meters per second; the larger the diameter of the axon, the more rapid the rate of travel of the impulse.

The conduction of impulses is not limited to nerve cells but occurs in some other cells with an electro-

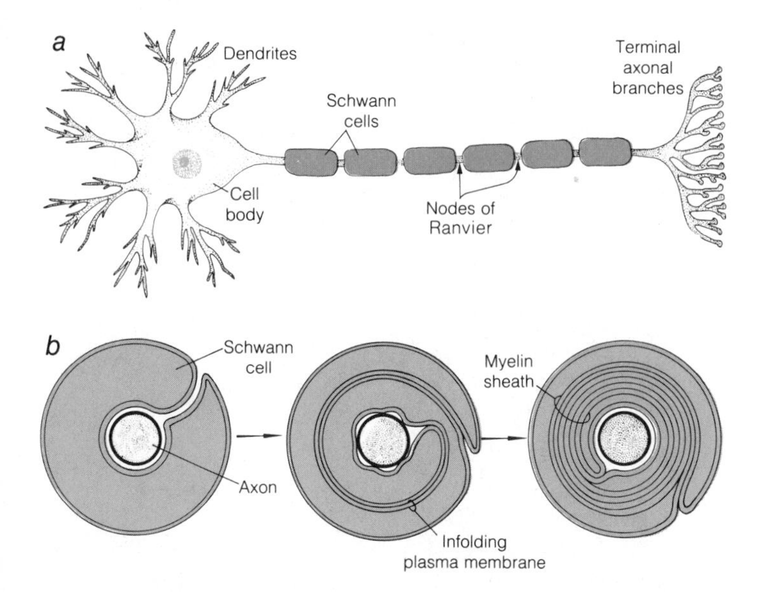

FIGURE 15-19
A myelinated axon. (a) Schwann cells insulate successive sections of the axon from extracellular fluids, causing an action potential to jump inside the axon from one Node of Ranvier to the next. (b) Formation of the myelin sheath seen in cross sections through an axon. The multiple layers of membrane in the myelin sheath are formed by infoldings of the plasma membrane. (c) Electron micrograph of a myelinated process from a cat cochlear neuron. An infolding (I) of the plasma membrane gives rise to the myelin membranes (MM). Cross sections of microtubules (Mt) are visible in the nerve process. [(c) Courtesy of Enrico Mugnaini. From D. W. Fawcett (1981) *The Cell*, 2nd Ed. W. B. Saunders and Co.]

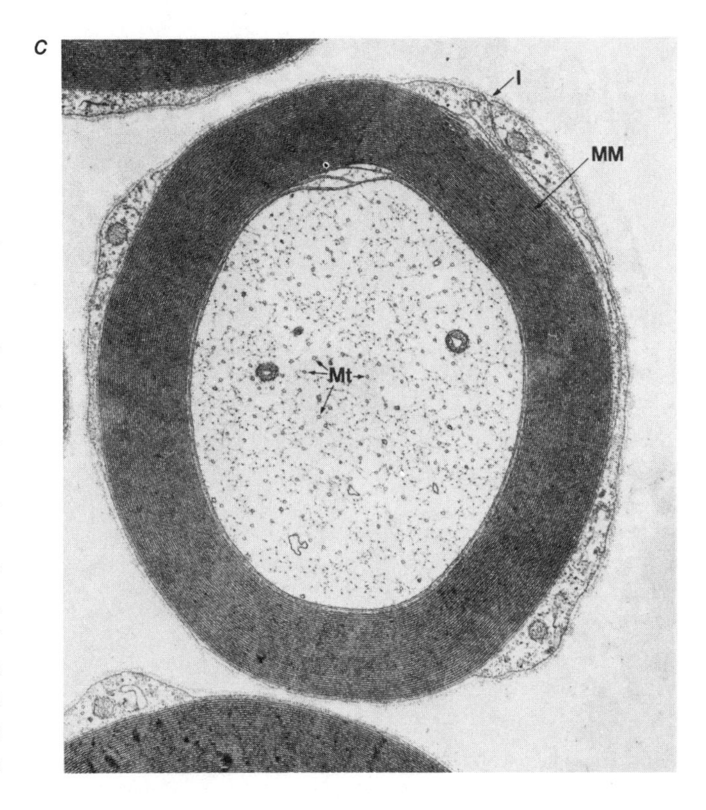

chemical gradient across their plasma membrane. Nerve cells are specialized for conduction by the possession of axons. An example of conduction by a non-nerve cell is the propagation of impulses by the plasma membrane and its invaginations (transverse tubules of the T system) in skeletal muscle cells.

Among invertebrates the rate of impulse conduction is increased by making a thicker axon. Among the thickest axons (0.5 mm in diameter) are those that control the swimming reflex in the squid. Because of their large size these axons have been used extensively in the study of conduction of action potentials and more recently for the study of fast axonal transport (Chapter 14).

In vertebrates the conduction of impulses by many nerves, for example, those that control the contraction of skeletal muscle cells, is sped up manyfold, not by larger diameter of axons, but by insulating axons with a **myelin sheath.** The myelin sheath of some neurons is formed from the plasma membranes of specialized cells called **Schwann cells** (Figure 15-19(a)). The plasma membrane of a single Schwann cell is wrapped around the axon many times, creating a multilayered sheath about one mm long (Figure 15-19 (b, c)). A succession of Schwann cells provides a succession of myelinated segments. Between segments is a gap of about one μm where the axon remains uncovered. The gaps are known as **Nodes of Ranvier** (Figure 15-19(a)). Instead of traveling continuously along a myelinated axon, an action potential jumps one mm from one node to the next—a phenomenon known as **saltatory conduction.** This occurs because the ion current set up by depolarization of the plasma membrane at one node is able to depolarize the plasma membrane exposed in the next node. In jumping, the action potential travels manyfold faster than it could if the plasma membrane were continuously depolarized. This results in much faster conduction (up to 100 meters per second) without the necessity of increasing the diameter of the axon.

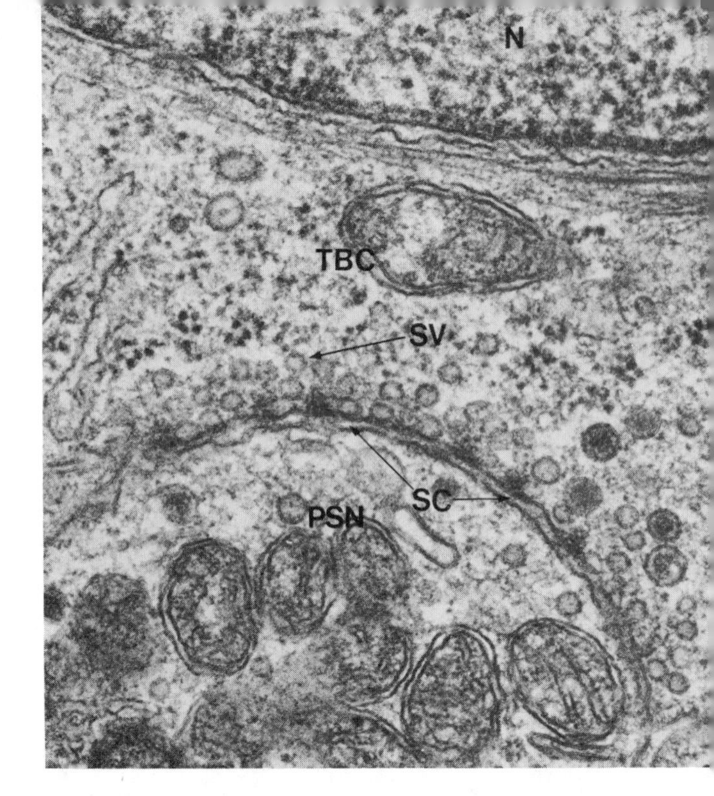

FIGURE 15-21
Electron micrograph of a synapse between a sensory cell of a taste bud and a neuron. Stimulation of the taste bud cell (TBC) results in fusion of synaptic vesicles (SV) with the plasma membrane and release of neurotransmitter molecules into the synaptic cleft (SC). The neurotransmitter molecules bind to receptors in the postsynaptic neuron (PSN), causing its excitation. N = nucleus of taste bud cell. [Courtesy of John C. Kinnamon.]

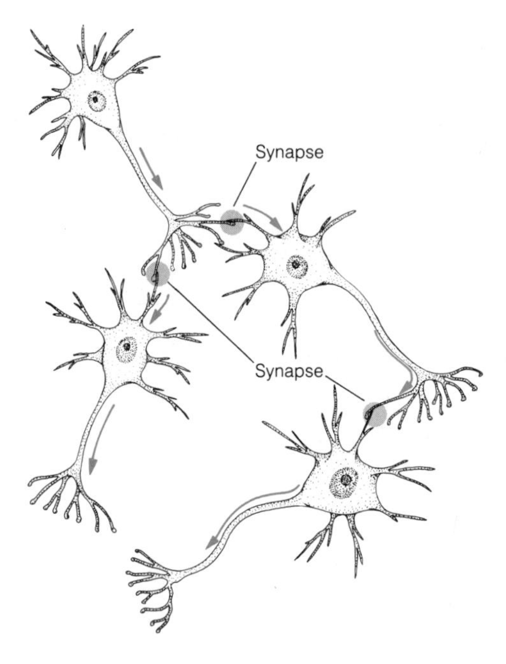

FIGURE 15-20
Transmission of a signal from one neuron to other neurons by synapses. Extremely complicated signal networks are created in three dimensions by synaptic connections.

Nerve Synapses

Impulses are passed from one cell to another through a specialized connection, the **synapse.** For example, an impulse traveling down an axon is transmitted by means of a synapse to the dendritic processes of another neuron (Figure 15-20). From the dendritic processes of the second neuron the impulse is propagated by the plasma membrane down its axon to yet another synapse. Synapses made by sensory neurons and by interconnecting neurons are always of the neuron-to-neuron type. However, the axons of effector neurons transmit impulses to other target cells, usually striated muscles and the smooth muscle cells surrounding glands.

At a synapse the signal is passed from cell-to-cell by chemical means. The membrane of the transmitting cell (**presynaptic cell**) is separated by a gap, called the **synaptic cleft,** from the membrane of the receiving cell (**postsynaptic cell**) (Figure 15-21). The synaptic cleft is less than 0.1 μm wide.

Transmission of a signal across a synapse is mediated by neurotransmitter molecules (Figure 15-22). The end of an axon branches; the tip of each branch has a bulb-shaped structure that engages in a synapse. Concentrated in the bulb are numerous small vesicles (**synaptic vesicles**) containing neurotransmitter molecules. Many neurons use **acetylcholine** or **norepinephrine** as chemical transmitters. The arrival of a nerve impulse at the end of an axon causes the membranes of many of these vesicles to fuse with the plasma membrane facing the synaptic cleft, resulting in the release of neurotransmitter molecules into the gap. Released neurotransmitter molecules diffuse quickly across the synaptic cleft and become bound to receptor sites for the neurotransmitter on the postsynaptic membrane of the cell receiving the impulse. Binding of neurotransmitter molecules to the receptor triggers depolarization of the postsynaptic membrane, setting up a nerve impulse that is then propagated by the plasma membrane of the postsynaptic cell. Transmission of a signal across a synapse takes less than one millionth of a second.

Neurotransmitter molecules in the synaptic cleft are rapidly broken down. Acetylcholine, for example, is cleaved into acetate and choline by the enzyme **acetylcholinesterase** at the surface of the plasma membrane of the postsynaptic cell. Rapid destruction of acetylcholine prevents continuous stimulation of the postsynaptic cell. The acetate and choline derived by acetylcholine breakdown are taken up by the presynaptic axon and are recycled in the synthesis of more acetylcholine. As a part of exocytosis in the release of acetylcholine, membranes of synaptic vesicles are added to the plasma membrane of the axon facing the synaptic cleft (Figure 15-22). Membrane is in turn retrieved from the plasma membrane by pinocytosis. The membranes taken in by pinocytosis are utilized for the production of new synaptic vesicles.

Nerve impulses accomplish the stimulation (contraction) of striated muscle by means of **neuromuscular junctions** that are much the same as the nerve-nerve synapsis just described. The transmitter molecule in neuromuscular junctions is acetylcholine. Transmission of nerve impulses across neuromuscular junctions is blocked by certain poisons, such as **curare.** Curare binds to receptors for acetylcholine, excluding binding

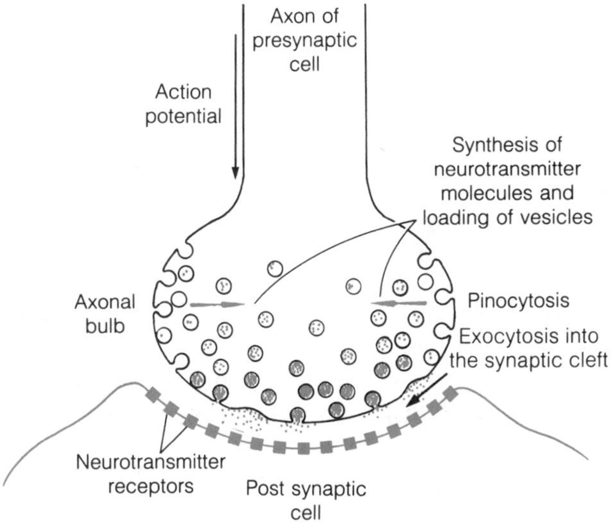

FIGURE 15-22
Diagram of a synapse. Arrival of an action potential at the terminal bulb of an axon results in fusion of synaptic vesicles with the plasma membrane, releasing neurotransmitter molecules into the synaptic cleft. Binding of neurotransmitter molecules to receptors in the membrane of the postsynaptic cell evokes depolarization and propagation of the signal. Membrane added to the cell surface by exocytosis of neurotransmitter molecules into the synaptic cleft is balanced by retrieval of membrane into vesicles by pinocytosis in regions of the axonal bulb outside the cleft.

a

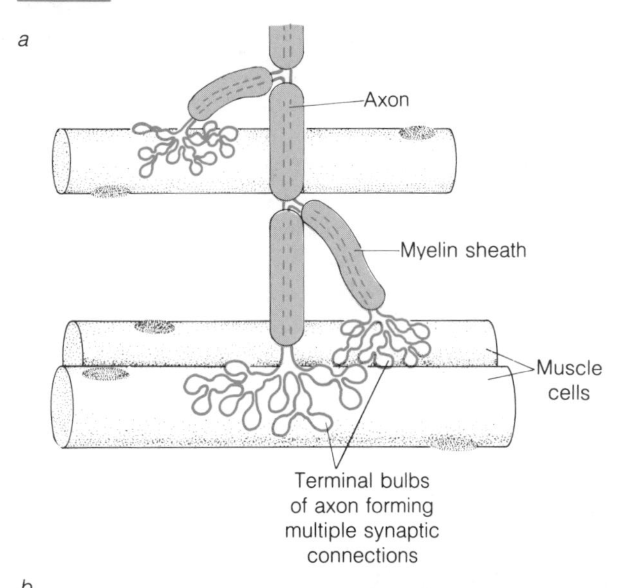

b

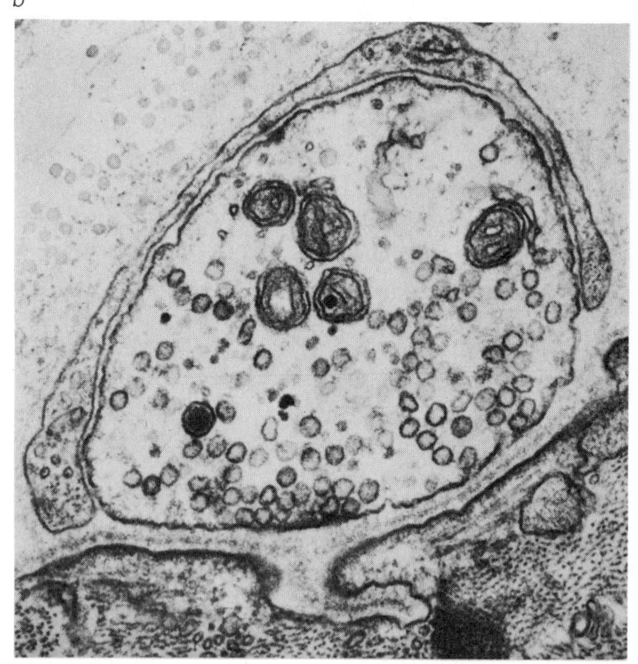

FIGURE 15-23
Neuromuscular junctions. (a) A diagram of a single axon, branching and innervating many muscle cells. Many bulbs are present at the terminus of an axonal branch as it contacts a muscle cell. (b) Electron micrograph of the synapse of a terminal bulb of a nerve axon with a frog skeletal muscle cell. Many synaptic vesicles are present and concentrated in the region facing the synapse. A thin portion of a Schwann cell covers the top side of the axon terminal bulb. [(b) Courtesy of Uel J. McMahan II.]

by acetylcholine, which prevents depolarization of the muscle cell plasma membrane. The result is paralysis of the muscle.

Nerve axons to muscle are normally highly branched with branches ending in synapses with different muscle cells (Figure 15-23). Thus a single neuron may bring about the synchronous contraction of many muscle cells. Recall that an impulse arriving at a skeletal muscle is carried deep into the cell by the T system of membrane invaginations.

SENSORY CELLS

Animals contain a variety of cells that are specialized to respond to various stimuli and to initiate a nerve impulse. Some sensory cells are stimulated by pressure (**tactile cells** of the skin); others respond to chemicals (**taste** and **smell sensory cells**). We consider here a sensory cell, the **retinal rod cell,** that is specialized to respond to light.

The Retinal Rod Cell

The process of vision is based on sensitivity and response of specialized cells, called **photoreceptor cells,** to light. Photoreceptor cells, such as those of the retina, contain a large amount of a light-absorbing receptor molecule that undergoes a chemical change as the result of absorbing light. This chemical change causes, in turn, a polarization change across the plasma membrane, creating an electrical signal. This signal is transmitted via a synapse to other neurons in the retina, and these neurons relay the signal to neurons making up the optic nerve to the central part of the brain.

There are two kinds of **photoreceptor cells** in the retinas of most vertebrates, **cone cells** and **rod cells** (Figure 15-24). Cone cells are stimulated by bright light and are responsible for color vision. Rod cells are responsible for black and white images and are more

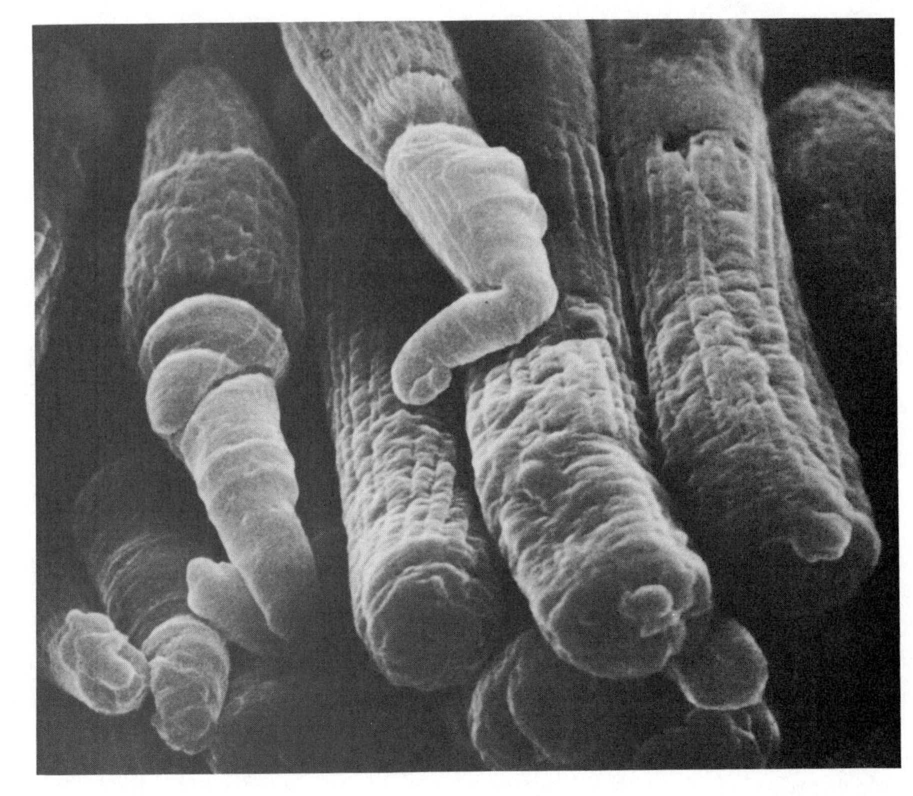

FIGURE 15-24
Scanning electron micrograph of rod and cone cells of the retina in a salamander eye. Rod cells are cylindrical and cone cells are tapered. In the retina the ends of these cells are embedded in the pigmented epithelial layer as shown in Figure 15-25. [Courtesy of Edwin R. Lewis.]

sensitive to dim light than are cone cells. Hence, in moonlight or dim light in general, colors are not well perceived and objects are seen in black, white, and gray. The human retina contains about 10^9 rod cells and about 3×10^6 cone cells. Rod cells are concentrated in the more peripheral regions of the retina. Therefore, in dim light objects are seen more clearly by peripheral vision.

A rod cell consists of two parts, a **cell body** containing a nucleus and the usual complement of cytoplasmic components and an **outer segment** specialized for reception of light (Figure 15-25). The two parts are connected by a **neck** that contains a centriole and a short ciliumlike structure. The function of the ciliumlike structure is not known, but it is generally believed that the outer segment, with its light-reception elements, is a modified cilium that forms from the neck. The outer segment is cylindrically shaped and packed with several hundred membrane discs arranged in a stack. The discs are continually formed by infoldings of the plasma membrane at the junction of the neck with the outer segment. The infoldings are severed from the plasma membrane, becoming in effect flattened vesicles. As new discs form, the existing discs move toward the tip of the outer segment. At the tip the discs are shed from the cell and are phagocytosed by the pigmented epithelial cells in which the tips are embedded (Figure 15-25). The transit time for a disc from formation to shedding varies from one to a few weeks depending on the species. In effect, the entire outer segment is replaced every one to several weeks.

The Cone Cell

The main structural difference between rod cells and cone cells is the persistence of continuity of the stacked discs with the plasma membrane in cone cells. There are three kinds of cone cells; these contain light-absorbing molecules that are most sensitive to blue, green, and red light, respectively. Recognition of all

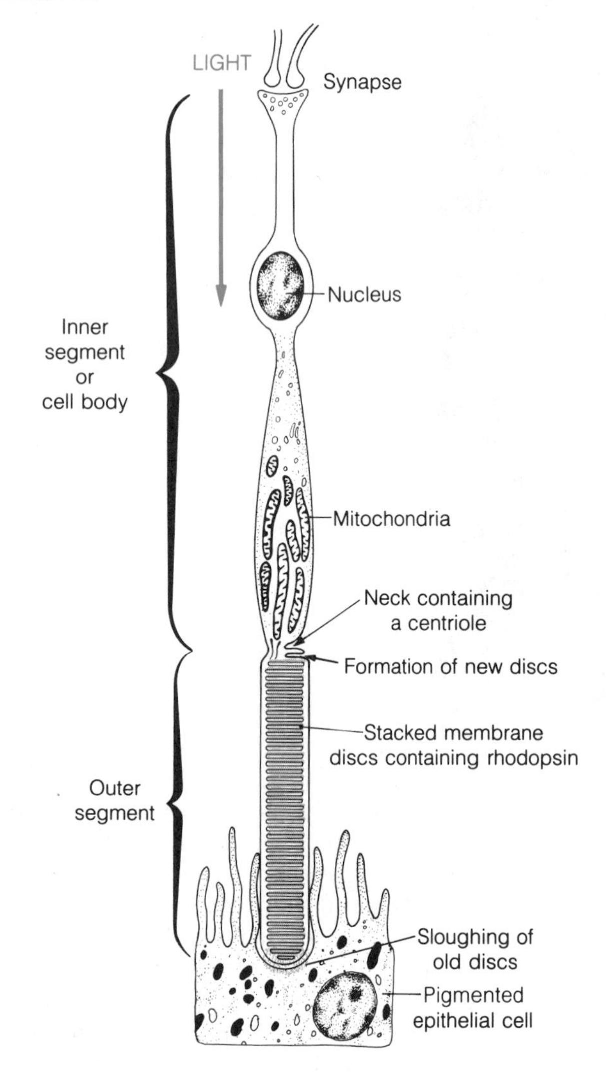

FIGURE 15-25
Drawing of a rod cell of the retina.

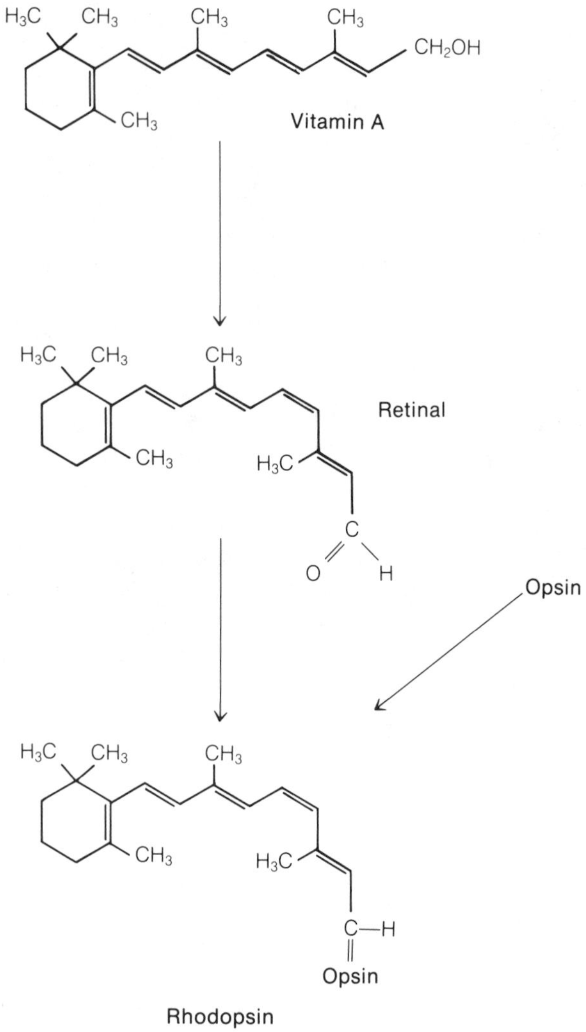

FIGURE 15-26
Vitamin A is modified to form retinal. Retinal is covalently joined to the protein opsin to form rhodopsin.

colors is achieved by stimulation of these three kinds of cells to various degrees in various combinations. Color blindness is the result of a recessive mutation in one of the genes (carried in the X chromosome) concerned with synthesis of light-absorbing molecules. Mutation results in the change or absence of one of the light-absorbing photoreceptor molecule types. About one percent of men are totally red-blind, and two percent are totally green-blind.

Function of the Rod Cell

Embedded in the membranes that form the discs in rod cells is a large amount of the protein **rhodopsin,** the light-receptor molecule of the rod cell. Rhodopsin is an integral membrane protein and is insoluble in water. It consists of a polypeptide portion called **opsin** (Greek, *opsis,* meaning vision) to which is bound a derivative of vitamin A called **retinal** (Figure 15-26). An individual receiving an inadequate amount of vitamin A in the diet is unable to synthesize sufficient rhodopsin as the new membrane discs are formed. Such a person suffers from **night blindness,** which is a loss of vision in dim light. About 80 percent of the protein in the membrane discs is rhodopsin, and a single rod cell contains about 40 million rhodopsin molecules. The planes of the rhodopsin-packed discs are oriented at right angles to the direction of light entering the eye, providing an extremely efficient arrangement for absorption of light.

Absorption of a single photon of light is sufficient to excite a rod cell. When light is absorbed by a rhodopsin molecule, the shape of the protein changes. How this initiates a nerve impulse is not well understood but experiments have suggested the following model (Figure 15-27).

The Na^+-K^+ pump in the plasma membrane of the cell body extrudes Na^+ ions from the cell. However, in the absence of light the plasma membrane of the outer segment of the rod cell is permeable to Na^+ ions, that is, Na^+ ion channels are open. Since the concentration of Na^+ ions is high outside the cell, there is an inward flux. Na^+ ions that enter the outer rod segment diffuse to the cell body and are continually pumped out, creating a circuit of Na^+ ion movement. When rhodopsin absorbs light, the configurational shift in the molecule

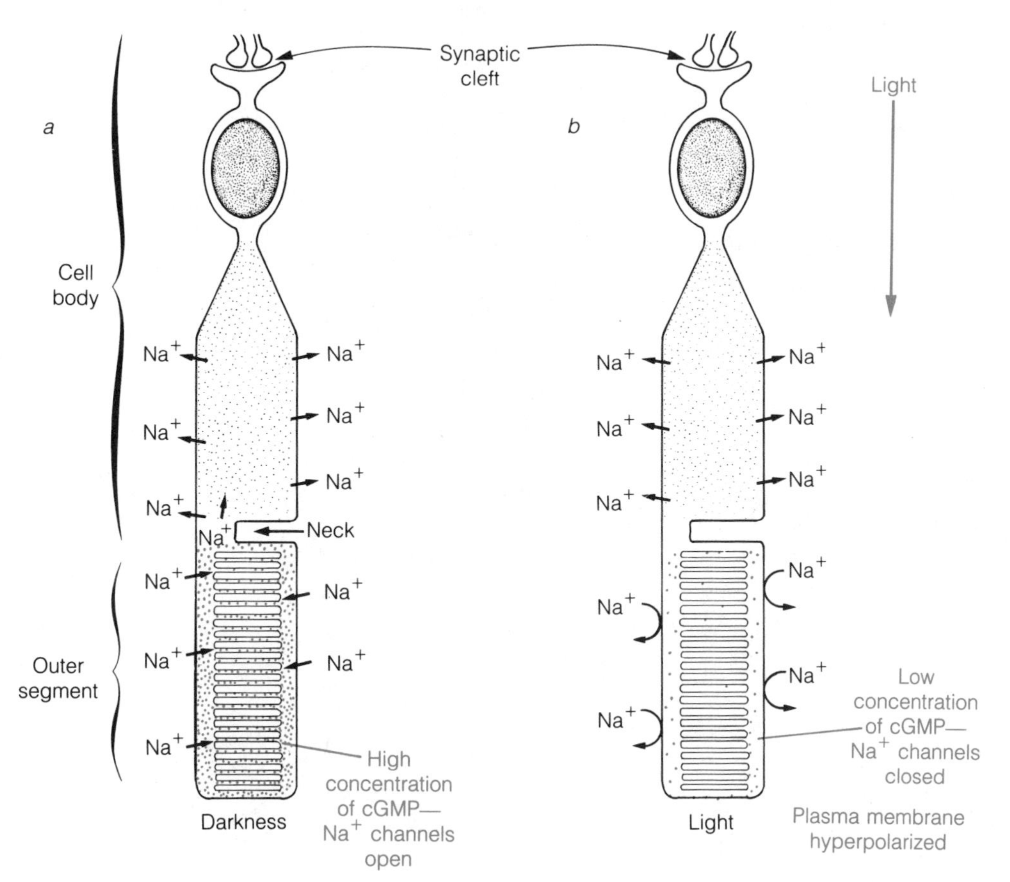

FIGURE 15-27
Schematic diagram of excitation of a rod cell by light. (a) In darkness the concentration of cGMP remains high, which causes Na^+ ion channels to remain open. (b) In light, cGMP is destroyed and the Na^+ ion channels are closed, resulting in hyperpolarization of the plasma membrane (high sodium outside the cell, low sodium inside). Hyperpolarization of the membrane results in discharge of neurotransmitter molecules into the synaptic cleft.

leads to blockage of the Na^+ ion channels in the plasma membrane. A hypothesis has been proposed to explain how blockage is achieved. Experiments suggest that cyclic GMP keeps the Na^+ ion channels open in the dark and that light leads to hydrolysis of cyclic GMP, which allows the Na^+ ion channels to close. (In cyclic GMP, like cyclic AMP, the phosphate group is covalently joined to both the 3′ and the 5′ carbon atoms of the ribose molecule.)

As a result of closing of the Na^+ ion channels, the inside of the plasma membrane becomes electrically more negative as Na^+ ions are continuously pumped out by the Na^+-K^+ pump in the plasma membrane in the region of the cell body, creating a greater electrical potential across the membrane; this is known as **hyperpolarization** of the membrane. The hyperpolarization is transmitted to the base of the cell body where, through a synaptic connection, the signal is transferred to an interneuron that passes it to the brain.

Eventually, the activated rhodopsin is returned to its original configuration in the membrane of the disc, and the Na^+ ion channels are again opened.

Some unicellular organisms show **positive phototaxis** (swim toward light). In at least one of these, the flagellated alga *Chlamydomonas,* sensitivity to light is based on the presence of rhodopsin molecules of the type found in the eyes of vertebrates. In some species of photosynthetic bacteria, rhodopsin is the light-absorbing pigment (Chapter 4).

LYMPHOCYTES

Lymphocytes are cells that respond immunologically to protect an animal from infection by viruses, bacteria, fungi, and protozoa. They respond not only to infectious agents but also to foreign objects and sub-

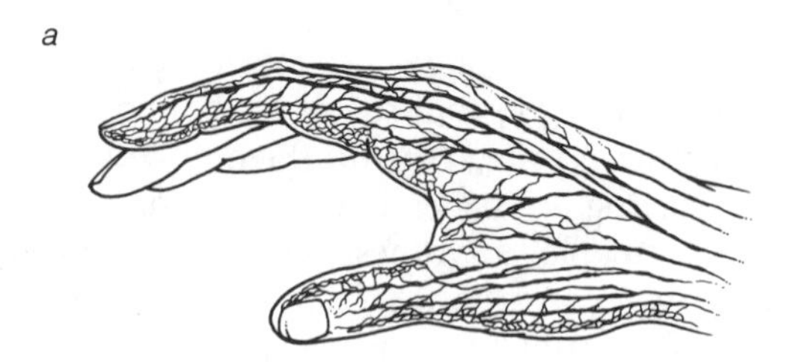

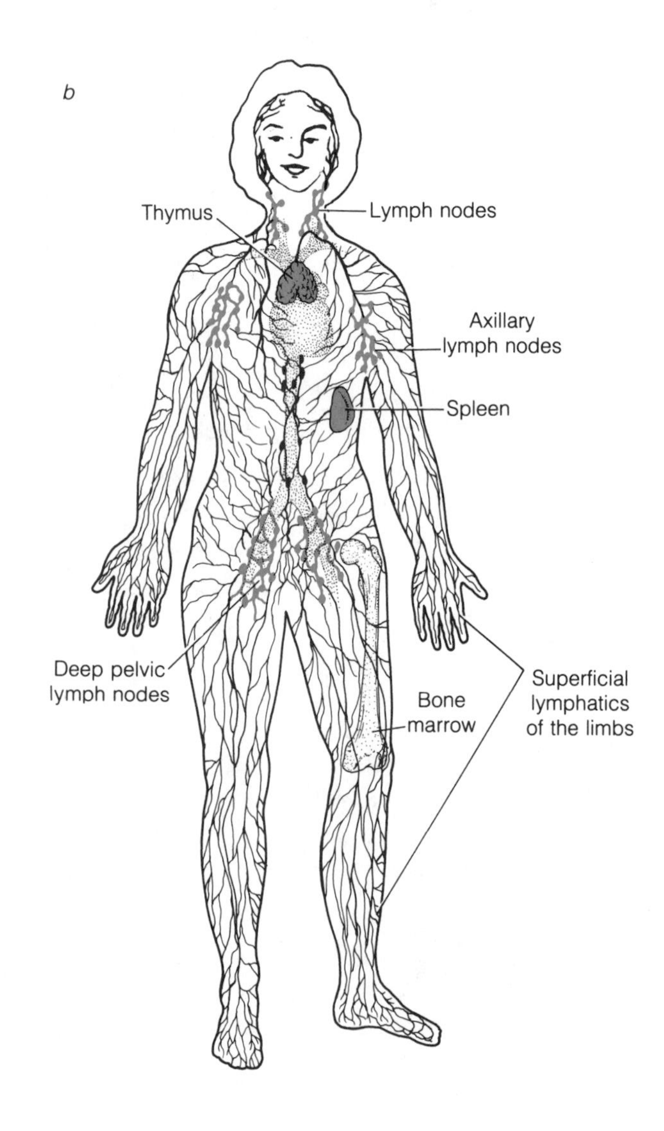

FIGURE 15-28
Schematic drawing of the lymphatic system. (a) Lymphatic vessels of the hand. (b) The lymphatic system includes the thymus, the spleen, lymph nodes, appendix, and lymphatic vessels. Lymphatic capillaries are not shown. Many more lymph nodes are present than shown here.

stances generally, particularly to foreign proteins and carbohydrates. Thus, lymphocytes are responsible for rejection and destruction of tissues or organs transplanted from one animal to another. They also provide the main defense against cancer cells, although they sometimes fail in this task.

The Immune System

The immune system of vertebrates consists of lymphocytes present in blood, tissue fluids, and the **lymphatic system.** The lymphatic system, whose main cellular components are lymphocytes, is made up of the thymus gland, the spleen, the appendix, lymphatic vessels, and hundreds of lymph nodes located at many sites in the body. Altogether in these tissues in an adult human there are about 2×10^{12} lymphocytes, accounting for one percent of total body weight. Lymphocytes continually circulate through the body. Those in the blood squeeze through the walls of tiny blood vessels (capillaries) and enter tissue spaces. From tissue spaces lymphocytes enter capillary vessels of the lymphatic system along with tissue fluid (Figure 15-28). This mixture of cells and fluid flows into successively larger lymphatic vessels that enter lymph nodes, percolate through the nodes, and leave in a continuation of the lymphatic vessel. Eventually, all lymphatic vessels coalesce into two major vessels that join with the venous system just before it reaches the heart. Lymph nodes contain mostly lymphocytes, some of which may join the lymphatic fluid as it flows through. Also, as blood circulates through the thymus gland, the spleen, and the appendix, these tissues contribute lymphocytes to the blood. Thus, lymphocytes derived from various parts of the lymphatic system reach almost every part of the body. The only regions excluded are the tissue and fluid spaces of the brain and spinal column.

Types of Lymphocytes

The three main types of lymphocytes are **B cells, T cells,** and **natural killer cells (NK cells).** Each is specialized to perform different functions. The body contains approximately equal numbers of B and T cells; both are present along with a much smaller number of NK cells throughout most of the lymphatic system of an adult.

Lymphocytes arise from **pluripotent hemopoietic stem cells** in the bone marrow during embryological development. These stem cells give rise not only to B, T, and NK lymphocytes but also to other types of white blood cells (e.g., neutrophils), to red blood cells, to **macrophages,** and to **megakaryocytes** (Figure 15-29). Macrophages are phagocytic cells present in most tissues and have the function of ingesting debris from various sources. For example, macrophages of lungs ingest microscopic particulates deposited in lungs from inhaled air. Megakaryocytes are giant cells present in the bone marrow. They are specialized for producing platelets, which are anucleate cytoplasmic fragments that carry factors essential for blood clotting. Platelets also contain a growth factor (platelet-derived growth factor) that stimulates reproduction of certain kinds of cells (Chapter 12).

In the production of all these cell types, the pluripotent hemopoietic stem cells give rise by cell reproduction to subpopulations of stem cells that have more restricted potentials for differentiation, for example, a subpopulation of stem cells that can only differentiate into red blood cells.

Origin and Function of B Lymphocytes

B lymphocytes have been extensively studied in birds. They are called B lymphocytes because during embryological development they form from hemopoietic stem cells that migrate from the bone marrow to an organ called the **b**ursa of Fabricius in birds. From the bursa of Fabricius they later migrate to lymph nodes and the spleen (Figure 15-30). The bursa of Fabricius is absent in mammals, where the cells that give rise to B lymphocytes migrate directly from the bone marrow to lymph nodes and the spleen.

Function of B Lymphocytes

The function of B lymphocytes is to produce and release antibodies, also called **immunoglobulins (Ig).**

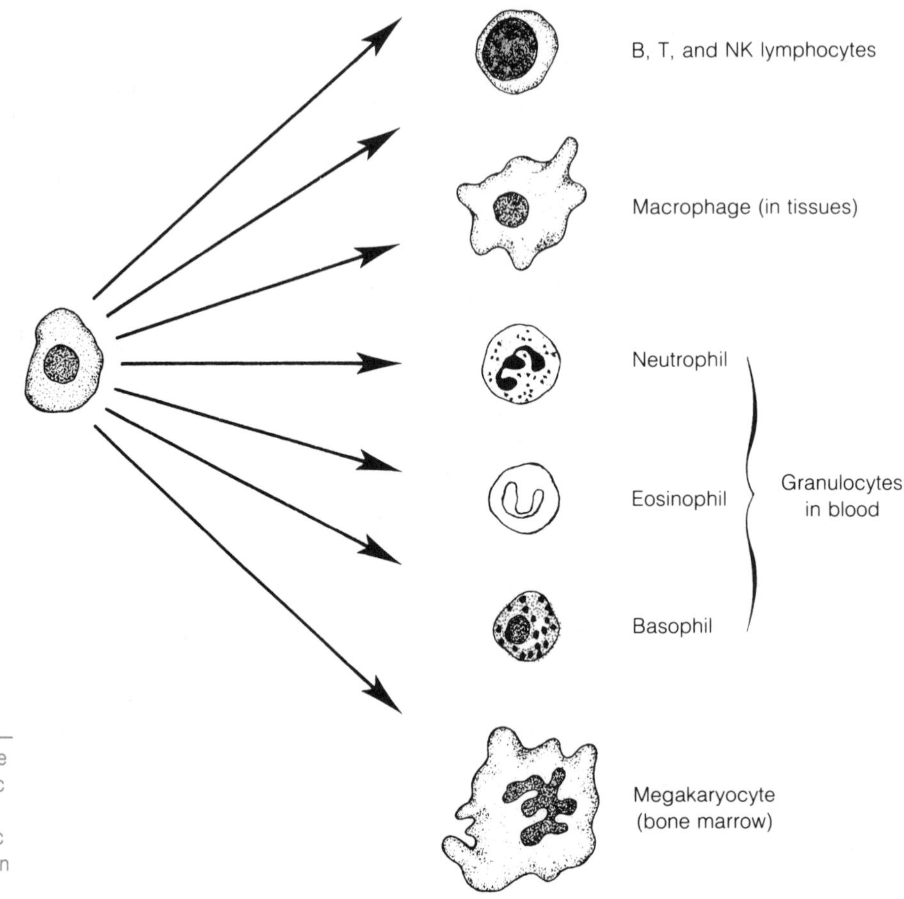

FIGURE 15-29
A pluripotent hemopoietic stem cell in the bone marrow gives rise during embryonic development to subpopulations of cells that in turn give rise to many hemopoietic cell types. Through a microscope one can identify various cell types by their size, shape, and staining properties.

Antibodies make up 20 percent of all the proteins by weight in blood plasma, the fluid part of blood. A human can produce 10^7 or more different kinds of antibody molecules, but an individual B lymphocyte synthesizes only a single kind of antibody. Therefore, among the 10^{12} B lymphocytes in an adult human are 10^7 or more kinds of B lymphocytes, each distinguished by the ability to produce a different kind of antibody. Each kind of antibody is capable of binding to a different antigen molecule.

Structure of an antibody. Antibody molecules all share the same structural pattern (Figures 2-19, 10-26 and Chapter 11). Each is composed of two polypeptides of about 220 amino acids each called **light (L) chains,** and two polypeptides of about 440 amino acids each, called **heavy (H) chains.** A light chain is composed of a variable region of approximately 110 amino acids in which the amino acid sequence varies from one kind of antibody to the next and a constant region of about 110 amino acids (Figure 15-31) in which the amino acid

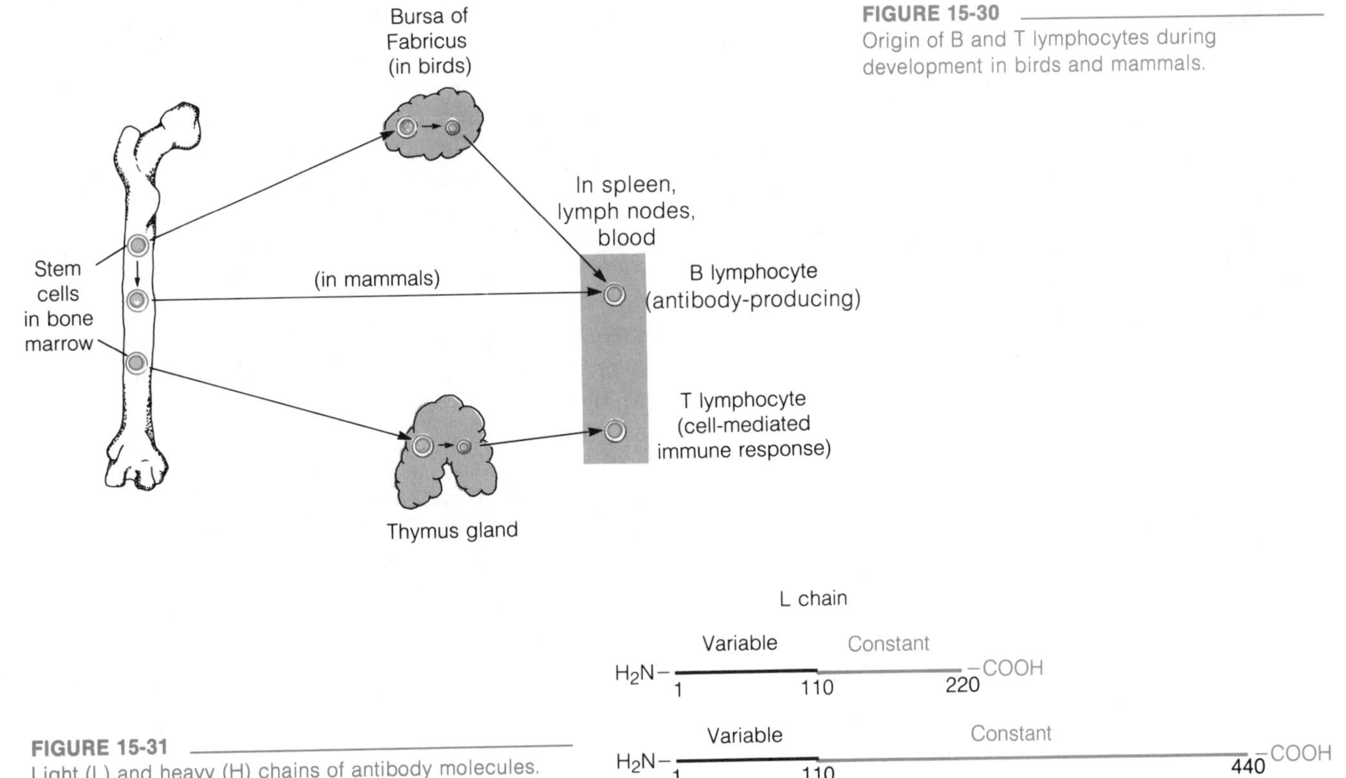

FIGURE 15-30
Origin of B and T lymphocytes during development in birds and mammals.

FIGURE 15-31
Light (L) and heavy (H) chains of antibody molecules. The numbers refer to amino acids counted from the NH_2 terminus.

sequence is the same in different antibodies. Heavy chains are composed of a variable region of about 110 amino acids and a constant region of about 330 amino acids. Two identical light chains and two identical heavy chains are joined by disulfide bonds (between the side chains of cysteine residues in the chains) to make one antibody molecule (Figure 10-26). In this arrangement the variable region of one light chain combines with the variable region of one heavy chain to form an antigen-binding site. Thus an antibody has two identical antigen-binding sites. The variable regions distinguish one kind of antibody from another and give the antibody its ability to bind a particular antigen.

Classes of antibodies. There are five classes of antibodies, **IgA, IgD, IgE, IgG** and **IgM,** defined by the presence of five different classes of H chains, α, δ, ε, γ, and μ. IgA antibodies have **α heavy chains,** IgD antibodies have **δ heavy chains,** and so on.

The five classes of heavy chains are defined by different constant regions. Within a single antibody class—for example, IgG antibodies—all the different kinds of antibodies have H chains with identical constant regions. The variable regions differ from one kind of antibody to the next within a class, as described earlier.

In addition, there are two classes of L chains, kappa **κ** and lambda **λ**, defined by different constant regions. Kappa and λL chains can combine with any of the five kinds of H chains to make an antibody. However, a given B lymphocyte produces only one kind of L chain and one kind of H chain. Thus, the L chains in an antibody molecule are always both κ or both λ and

TABLE 15-1
Classes of Antibody

	IgA	IgD	IgE	IgG	IgM
H-chain	α	δ	ε	γ	μ
L-chain	κ or λ	κ or λ	κ or λ	κ or λ	κ or λ
Percent of total antibodies in blood	15	<1	<1	80	5

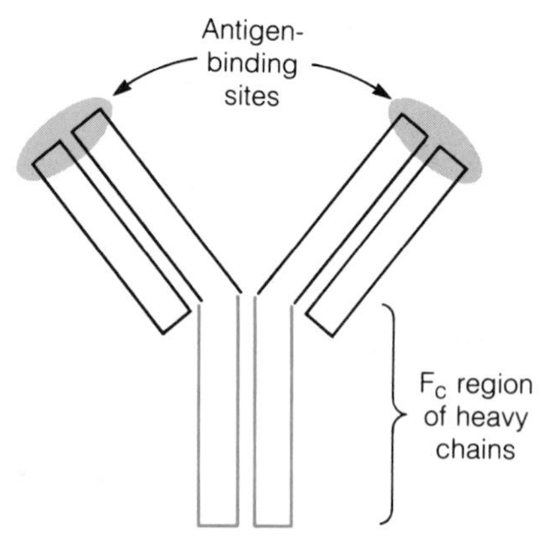

FIGURE 15-32
Diagram of an antibody molecule showing the Fc region and the two antigen-binding sites.

never a mixture, and the H chains are always both α chains, or δ chains, and so on, and never a mixture.

IgG antibodies account for about 80 percent of all antibodies in the blood. The IgA class accounts for approximately 15 percent. The remaining five percent is divided among IgD, IgE, and IgM. A summary of the H and L chain compositions of antibodies belonging to different classes and the abundance of antibodies in different classes is given in Table 15-1.

Functions of the different classes of antibodies. The tail-like part of an antibody molecule formed by the constant regions of H chains (Figure 15-32) is called the Fc region. In a given class of antibodies, for example IgG, all the antibodies have identical Fc regions (made of constant regions of γH chains in the case of IgG, αH chains for IgA, and so forth). The Fc

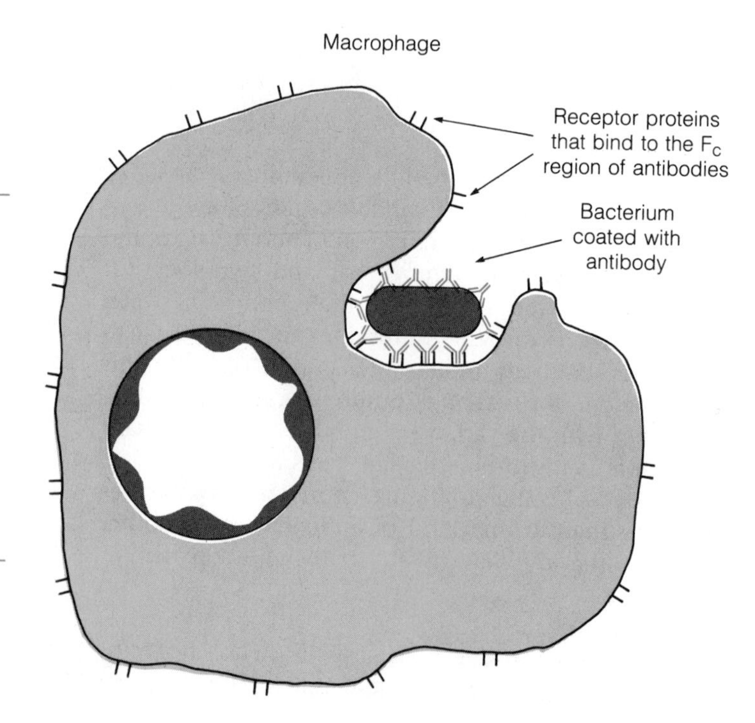

FIGURE 15-33
Schematic diagram of a bacterium coated with IgG antibody undergoing phagocytic engulfment by a macrophage. Receptor proteins at the surface of the macrophage bind to Fc regions of antibodies. Receptors move laterally in the membrane and aggregate as a result of binding.

regions of different classes of H chains are different, imparting different functions to the five classes of antibodies.

1. *IgG*. When IgG antibodies bind to antigens, for example, on the surface of a bacterium (Figure 15-33), the Fc regions projecting from the bound antibody are recognized by specific receptors on the surface of phagocytic cells (macrophages and neutrophils). The phagocytic cells then ingest and destroy the antigen-bearing object. The Fc regions of IgG bound to a microorganism can also bind to one of the proteins of the **complement system.** The complement system consists of nine different proteins present in blood plasma. Binding of the first component to the Fc region of an IgG antibody bound to a microorganism induces a cascade of interactions through the nine components. Through these interactions a complex of several of the protein components of the complement system is formed that binds to and causes lysis of the microorganism.
2. *IgA*. IgA antibodies occur in secretions such as saliva, milk, tears, and mucus of the respiratory tract. The Fc region of IgA antibodies is recognized by a secretory protein that is present on the surface of epithelial cells and acts as a receptor. The IgA antibodies in the blood are bound to the receptor, taken into the epithelial cells by receptor-mediated endocytosis (Chapter 5), and then released with the secretory product of the epithelial cell (saliva, milk, etc.).
3. *IgE*. The Fc region of IgE antibodies binds to the surface of white blood cells called **basophils** and to the surface of **mast cells.** When antigens subsequently bind to the IgE antibodies, the antibodies cause the basophil or mast cell to release amines (e.g., histamine). The amines, in turn, cause dilation of blood vessels, causing increased blood flow in the region where the antigen is present. This sequence of events is part of allergic reactions that underlie hay fever, hives, and asthma.
4. *IgM*. IgM antibodies are produced early in a primary response to an antigen. (Primary vs secondary responses are discussed in a later section on memory cells). Gradually the production of IgM antibodies is replaced by production of IgG antibodies.
5. *IgD*. IgD antibodies are bound by their Fc regions to some resting (nonantibody-producing) B lymphocytes. The function of the IgD class of antibodies is poorly understood.

Origin of different antibody-producing B lymphocytes. There are at least 10^7 different kinds of B lymphocytes, each capable of producing a particular antibody specific for a different potential antigen. The genome does not contain 10^7 genes encoding 10^7 different antibodies. Rather, a few hundred gene segments are joined together in various combinations to form genes encoding the wide spectrum of different antibodies, as explained in Chapter 11. Joining together of segments follows the rule that the portion of a gene encoding a variable region is always formed from a V segment and a J segment for L chains and from a V, D, and a J segment for H chains. These composite segments, together with a C DNA segment, yield an antibody gene. Which particular V, D, J, and C segments are brought together is apparently random; that is, the generation of antibody genes during B cell differentiation is random. This mechanism of combining a few hundred DNA segments generates more than 10^7 different antibody genes from a small amount of DNA.

A B lymphocyte produces only one kind of antibody. The DNA in only one of the two chromosomal homologues must undergo rejoining of segments to produce a gene that is expressed as an antibody protein. Expression of the corresponding allelic segments of DNA in the other homologue does not occur, a phenomenon known as **allelic exclusion;** allelic exclusion does not usually occur for other kinds of genes in autosomal chromosomes. (However, in female cells allelic exclusion occurs on a massive scale with the repression of almost all genes in one of the two X chromosomes.)

Once formed, every B lymphocyte produces a small amount of antibody that is anchored into the plasma membrane by means of a short tail of hydrophobic amino acid and with its antigen-binding site exposed at the cell surface. Such B lymphocytes do not secrete antibody, are therefore in a quiescent state, and are called **virgin** lymphocytes. Many B lymphocytes never encounter an antigen that could bind to the antibody in plasma membrane and eventually die without having been activated. If an antigen binds to the surface antibodies, the virgin cell is activated to proliferate and to produce large quantities of antibodies that lack hydrophobic, membrane-anchoring tails and that are released into the surrounding fluid (blood, tissue fluid, or lymph fluid). Virgin lymphocytes are small cells with little cytoplasm (Figure 15-34). Activation is accompanied by synthesis of a large amount of rough endoplasmic reticulum, which functions in the synthesis and secretion of antibody molecules. The result is an outpouring of antibody by an expanding population of B lymphocytes. Activation of a virgin cell is called the **primary response** to an antigen.

During their differentiation, virgin B lymphocytes reproduce, yielding clones of virgin cells; each clone is capable of producing a particular antibody. If the appropriate antigen is encountered, for example, as a result of a bacterial infection, members of the clone that bind that antigen are activated. This is known as the **clonal selection doctrine** of B lymphocyte response. Thus, an organism possesses a large repertoire of virgin clones of B lymphocytes, but only some of these clones are selected for activation during the life of the organism.

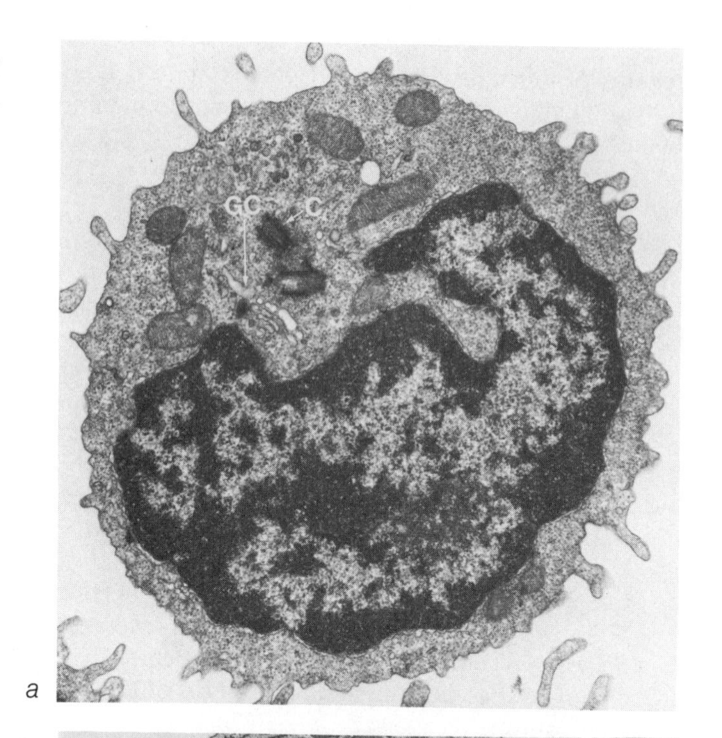

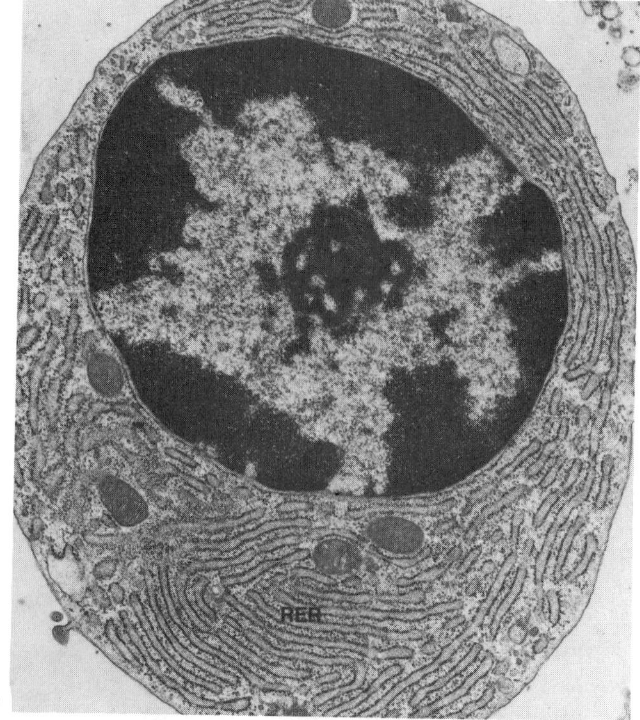

FIGURE 15-34

(a) Electron micrograph of an inactive lymphocyte. Centriole (C), Golgi complex (GC), and mitochondria are readily discerned. Note the abundant heterochromatin in the nucleus. Compare to (b) with respect to the RER. (b) Electron micrograph of an activated lymphocyte (plasma cell) from guinea pig bone marrow. This cell is synthesizing immunoglobulin molecules at a high rate, which accounts for its extensive RER. [(a) Courtesy of Dorothea Zucker-Franklin, from D. Zucker-Franklin et al. 1981. *Atlas of Blood Cells.* Edi Ermes, Milano, p. 351. (b) Courtesy of Don W. Fawcett. From D. W. Fawcett (1981) *The Cell*, 2nd Ed. W. B. Saunders and Co.

Memory cells. When a foreign antigen is cleared from an organism, the activated B lymphocytes, which are also called **plasma cells,** die. If a second infection by the same agent (same antigen) takes place, a **secondary response** to the foreign agent occurs that is more rapid and stronger than the primary response. The secondary response is the result of activation of **memory cells** formed during the primary response. In the primary response not all the virgin cells in a clone are stimulated to make antibody molecules. Some are stimulated only to proliferate for a while, forming a larger population of cells that can function in a secondary response. This accounts, in part at least, for the stronger nature of the secondary response. Other poorly understood changes occur in formation of memory cells that enable them to respond more quickly to antigens than do virgin cells.

A primary response may not be rapid enough to prevent development of a disease. However, an effective primary response ultimately eliminates the infecting agent. A secondary response is usually so strong and so rapid that no disease symptoms develop. The ability to mount secondary responses is the basis of permanent immunity to viruses and bacteria.

Recognition of self vs nonself. Every organism contains thousands of antigens that could potentially activate B lymphocytes to produce antibodies against the organism's own, normal molecular components. This does not normally happen because B lymphocytes do not recognize an organism's own antigens. By a mechanism that is not understood B lymphocytes (and T lymphocytes, see later) learn to distinguish between self and nonself. This is known as **acquired immunological tolerance.** Acquisition of tolerance occurs early in development, around the time of birth in mammals. Any antigen present in a fetus during this interval of development is normally accepted as self during the entire subsequent life of the organism. For example, if cells from an adult mouse are transplanted into a newborn mouse the foreign adult cells become accepted as self by the recipient. If the cells of an adult mouse are transplanted into another mouse at a later stage of development, they are recognized as nonself and are immunologically rejected.

The major reason for failure in transplantation of organs such as a kidney, a heart, or bone marrow from one human to another is immunological rejection. To prevent rejection recipients of organs are given an immunosuppressive drug that may, in some cases, prevent rejection. An individual with a suppressed immune system is much more prone to a successful infection by bacteria, viruses, protozoa, and fungi.

Tolerance to a particular self antigen sometimes fails, and B and/or T lymphocytes react against cells with that antigen. This is called an **autoimmune reaction,** and it is the basis of **autoimmune diseases.** Autoimmune diseases can arise at any age. **Myasthenia gravis** is an autoimmune disease in which B lymphocytes produce antibodies against acetylcholine receptors on skeletal muscle cells. This prevents response of the skeletal muscle cells to nerve impulses, leading gradually to paralysis and often ending in death. Other autoimmune diseases are rheumatoid arthritis, glomerular nephritis, and lupus erythematosis.

Monoclonal antibodies. Because of their high specificity for recognition of antigens, antibodies have become very important in many areas of research in cell biology. With an appropriate antibody, it is possible to detect the presence and location of components in cells, even those present in very small amounts. For example, antibodies labeled with fluorescent dyes have been extremely useful in the study of the cytoskeleton (Chapter 6), detecting the distribution of actin, myosin, intermediate filaments, and other components in muscle cells and a variety of nonmuscle cells. Antibodies can be injected into living cells to trace dynamic processes, such as microtubule assembly and disassembly, or to block activity of particular proteins. Antibodies can also be coupled to an inert support matrix like methylcellulose to make an affinity column that can be used to isolate and purify specific molecules.

Specific antisera (antisera containing antibody molecules to a particular type of antigen) are obtained by injecting a preparation of some molecule of interest into an animal, such as a rabbit, and then collecting

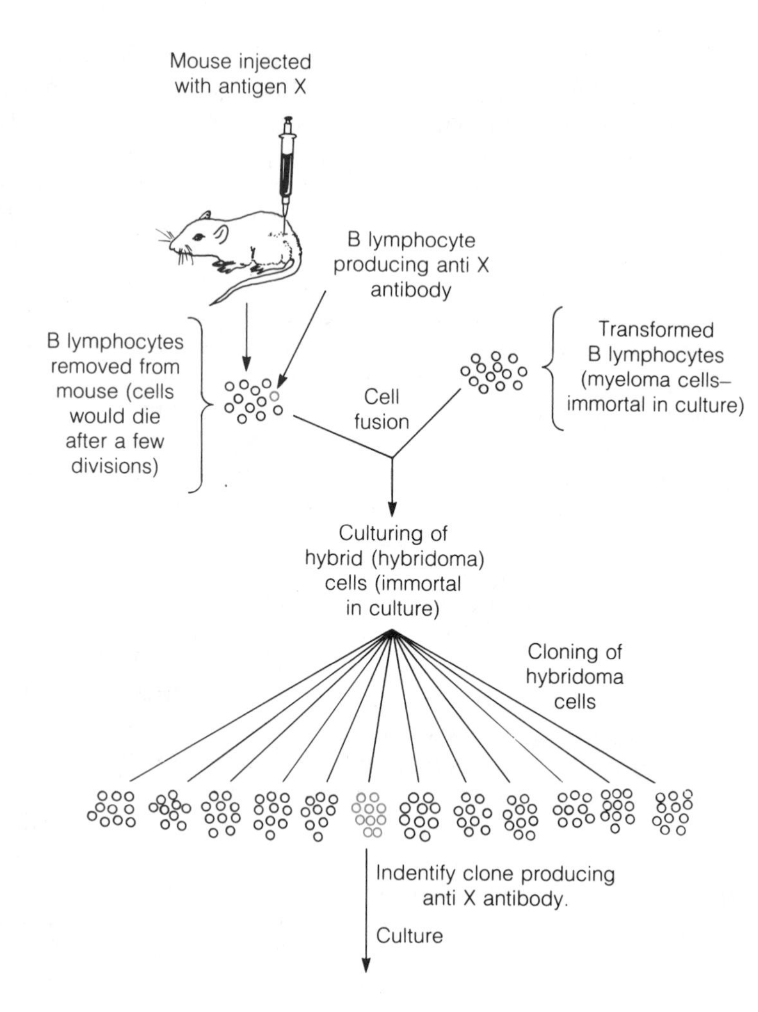

FIGURE 15-35

The monoclonal antibody or hybridoma technique. Normal B lymphocytes die after a few days in culture. Normal B lymphocytes fused with transformed B lymphocytes (myeloma cells) form hybrid cells (hybridomas) that are immortal in culture. Myeloma cells do not produce functional antibodies. Cloning of the hybridomas and selection of the clone producing the desired antibody provides an unlimited supply of a single type of antibody.

antibody-containing serum. This serum contains a wide variety of antibodies, the number of which depends on the antigens the animal has previously been exposed to during its life, plus a rich amount of antibody to the injected antigen. In fact, usually, several different antibodies are present that recognize different antigenic sites on a single kind of injected molecule. Most molecules possess multiple surface sites to which antibodies can be produced.

The heterogeneity of antibodies in antisera can often complicate the interpretation of experiments. An improved technique, the **monoclonal antibody technique,** has been developed that provides a pure preparation of a single antibody type against a single antigenic site on a particular molecule (Figure 15-35). The technique consists of culturing a clone of B lymphocytes. These cells normally reproduce only a few times and survive only a few days in culture, so, as they are obtained from an animal, usually from the spleen, they are not a suitable source of antibody. However, a B lymphocyte from an immunized mouse can be fused with a B lymphocyte that has previously been transformed into a cancer cell in an animal. B lymphocyte cancers, called **myelomas,** occur in humans and other animals and are relatively easy to induce in mice. The B lymphocytes of a myeloma are immortal and proliferate indefinitely in culture. Fusion of a normal B lymphocyte with a myeloma lymphocyte produces a hybrid cell that also is immortal. Such hybrid lymphocytes, called **hybridomas,** not only proliferate indefinitely, but also they continuously produce whatever antibody the normal B lymphocyte had been programmed to produce in the animal. Many clones of cells can be propagated from the original collection of hybrid cells. Each clone produces a single kind of antibody, reactive against a single antigenic site on a molecule; such antibodies are called **monoclonal antibodies.** Therefore, the clones provide an essentially inexhaustible supply of a particular antibody uncontaminated by other antibodies that could make experiments more difficult or impossible.

Origin of T Lymphocytes

T lymphocytes originally derive from the same pluripotent hemopoietic stem cell of the bone marrow as do all other kinds of cells of the blood and lymphatic system (Figure 15-30). T lymphocytes are so named because they migrate to the thymus gland, and that is where they develop T-lymphocyte characteristics. From the thymus gland they then move to other lymphatic tissues. This occurs during postnatal development, and by the end of adolescence the thymus gland becomes expendable and gradually shrinks to a nonessential remnant that persists for the life of the individual.

Function of T Lymphocytes

B lymphocytes are responsible for the **humoral responses** of the immune system because they secrete antibodies into body fluids. T lymphocytes are responsible for **cell-mediated responses** of the immune system because they destroy foreign or virus-infected cells by direct contact. T lymphocytes do not secrete antibodies but they do have antibody-like molecules (receptors) on their cell surface that recognize antigens and are the basis for specific immune responses.

Three different kinds of T lymphocytes have been identified: **cytotoxic T cells, helper T cells,** and **suppressor T cells.**

1. *Cytotoxic T lymphocytes* kill foreign cells or virus-infected cells. By killing cells infected with viruses they prevent viral replication from occurring and thereby help cure the host animal of virus infections. A single T lymphocyte can kill many virus-infected cells. Killing requires direct contact of the cytotoxic T lymphocyte with the target cell, but little is known about the molecular mechanism of killing. T lymphocytes, like B lymphocytes, *learn* to distinguish self from nonself antigens during development of the immune system. Cytotoxic T cells are important in the rejection of foreign tissues and hence represent an obstacle in organ transplantation. They have also been thought to provide defense against cancer, recognizing cancer cells as soon as they arise and destroying them. Some researchers believe that cancer cells frequently arise in the body but are recognized and killed by lymphocytes. This is known as the **immune surveillance theory** of cancer protection. According to this theory, the development of a cancer cell into a disease represents, by definition, a failure of cytotoxic T cells.

The immune surveillance theory was severely weakened by the development of a strain of mice that, because of mutation, totally lacks a thymus gland and never produces T cells. These mice, called **nude mice** because the mutation also results in failure of hair formation, are used in cancer research because they do not reject foreign tissues. For example, human cancers can be grown in nude mice, whereas normal mice reject human cancer cells. According to the immune surveillance theory nude mice should develop cancers at a high frequency because they have no cytotoxic T cells to protect them from any cancer cell that may arise. However, nude mice develop no more cancers than normal mice. Either the immune surveillance theory is incorrect or T cells do not provide the primary immunological defense against cancer cells. It may be that natural killer cells (discussed later) are the principal defenders against cancer.

2. *Helper T cells* provide an essential role in the response of B lymphocytes, cytotoxic T cells, and suppressor T cells to antigens. How they provide such help is poorly understood. However, it is clear that without helper T cells, B lymphocytes cannot respond to antigens and produce antibodies.

Helper T cells also secrete hormone-like substances called **lymphokines** that affect the activity of other cells. One such lymphokine, known as **macrophage migration inhibition factor (MIF),** inhibits the migration of macrophages, causing them to remain and therefore accumulate in regions where helper T cells have been activated by antigens. MIF also activates phagocytosis of invading microorganisms by macrophages.

T cells also secrete a lymphokine called **interleukin 2,** which binds to the surface of any T cell that has already been activated and causes it to proliferate. With interleukin 2, T cells can be propagated in culture indefinitely, providing clones of cytotoxic T cells, helper T cells, or suppressor T cells. This new technique makes it possible to study the function of T cells in culture, which will greatly facilitate research on them.

3. *Suppressor T cells,* when activated by an antigen, suppress the response of B and T lymphocytes to antigens. Suppressor T cells work only if they are assisted by an activated helper T cell. However, activated helper T cells are, in turn, suppressed by activated suppressor T cells.

The various interactions among helper T cells, suppressor T cells, cytotoxic T cells, and B lymphocytes form a complex network that regulates the activities of B and T lymphocytes. Many essential details of how the network functions remain to be discovered.

Natural Killer Cells

Natural killer cells (NK cells) are the most recent group of lymphocytes to be identified (in the mid-1970s). They arise from the pleuripotent hemopoietic stem cells of the bone marrow, but little is known about how they develop and populate the blood and lymphatic system. They represent a small subclass of lymphocytes that have in the cell surface antibody-like receptors resembling the receptors of T cells. However, they have structural and functional characteristics that clearly distinguish them from T cells.

Natural killer cells are able to attach to and to lyse cancer cells and virus-infected cells. There is some limited evidence that NK cells, and not cytotoxic T cells, may carry out immune surveillance against newly arising cancer cells. For example, they accumulate at the site of a cancer. Athymic mice (nude mice), which lack T cells, do not develop more cancers than normal mice, perhaps because they do possess NK cells.

PROBLEMS

1. What are some of the specialization characteristics of egg cells?
2. What is the function of the acrosome in a sperm cell?
3. Where are epithelial cells normally lost from the epithelial sheet of the small intestine, and where are new cells produced?
4. How is the lumenal surface of an epithelial cell specialized for uptake of nutrients?
5. What functions are provided by tight junctions, belt desmosomes, and spot desmosomes of intestinal epithelial cells?
6. Describe how intestinal epithelial cells transfer glucose from the intestinal lumen to the blood.
7. What is the role of the Na^+-K^+ pump in generating an electrical potential across the plasma membrane?
8. What changes occur in the plasma membrane of a nerve cell during conduction of an impulse?
9. How is the myelin sheath produced? What is its function?
10. How do nerve cells transmit signals to other cells?
11. What roles do exocytosis and endocytosis have in the function of a synapse?
12. Why are objects seen in black, white, and gray by the eye in dim light?
13. Why does a deficiency of vitamin A lead to night blindness?
14. Why are defective cone cells more common in human males than in human females?
15. What are the three main types of lymphocytes, and what does each type do?
16. Make a sketch of an antibody molecule showing variable and constant regions of light and heavy chains, disulfide bonds, and antigen-binding sites.
17. What kinds of diseases result from failure of tolerance to self antigens?
18. How are hybridoma cells that make monoclonal antibodies produced?
19. What is the immune surveillance theory?

CHAPTER 16

Origin and Evolution of Cells

See Figure 16-8.

Much of the discussion in the first fifteen chapters has dealt with an analysis of the molecular basis of cell function and structure. Many of the cellular properties that seemed a few decades ago to be so mysterious as to be beyond comprehension are now clearly understood in straightforward chemical terms. A prime example is the structure, replication, and decoding of genes. As knowledge of the chemical basis of cell function and structure has increased, so has our curiosity about the evolutionary route by which contemporary cells have come to be what they are. The path of cell evolution has been toward greater and greater efficiency and diversity of cellular operations, and for the most part this has been achieved by the evolution of continually greater complexity in cell function and structure. As a result, even the simplest contemporary cells we know about, the mycoplasmal bacteria, are still chemically very complex organisms.

Some reflections of the evolutionary path are observable in contemporary cells. For example, it is likely that the first eukaryotic cell evolved a long time ago from a progenitor cell that was probably not greatly different in structure from some contemporary prokaryotes. The progenitor cell, called a **progenote,** gave rise to both prokaryotes and eukaryotes. The progenote has long since disappeared, and the many functional and structural differences that now separate contemporary prokaryotes and eukaryotes attest to a long evolutionary series of intermediate cell types, particularly in the eukaryotic line of descent; little trace of the intermediates has remained.

The attempt to understand cell evolution ultimately leads us to the question of the origin of the first cell. Although we know roughly *when* in the Earth's history the first cell probably arose (more than 3.5×10^9 years ago), we do not know and probably never will be able to learn in specific terms *how* it arose. Still it is possible that we may gain a general understanding of the origin of the cell.

Not many years ago the question of how life originated seemed beyond serious scientific inquiry. The recent great increase in knowledge of the biochemistry and the molecular nature of cell function and structure now permits formulation of specific schemes of how the first cell might have come into existence. To some extent it is now possible to devise hypotheses about the origin of the cell that can be tested in the laboratory. Particularly important in all of this thinking is the concept of an **organic soup**, a complex mixture of organic molecules that formed and accumulated during the first one billion years after the Earth was formed, and within which the cell subsequently originated. Laboratory experiments designed to test the organic soup concept are discussed later. We will first consider briefly some early ideas about the origin of life.

SPONTANEOUS GENERATION OF LIFE

The ancient Greek scholars formulated the idea of **spontaneous generation** of life in which insects and other small animals were thought to arise spontaneously from mud or decaying organic matter. This idea persisted throughout the Middle Ages with descriptions of successful demonstrations of spontaneous generation of worms, flies, eels, frogs, and other organisms from mud and decaying materials. For example, mice were believed to arise spontaneously from cheese wrapped in rags and kept in a dark place. Frogs were said to form from decaying vegetation in ponds. The appearance of maggots in rotten meat seemed a particularly clear case of spontaneous generation.

The overthrow of the doctrine of spontaneous generation began in the late 1600s with the experiments of Francesco Redi, an Italian physician. In his most famous experiment he showed that maggots did not appear in rotting meat when the meat was protected from flies. On meat exposed to the open air, maggots develop from eggs laid by flies. Redi's experiments discredited the theory of the spontaneous generation of animals. However, microorganisms were discovered by Antonie van Leeuwenhoek, the father of microscopy, in 1677, about the same time as Redi's experiments, and the controversy about spontaneous generation shifted from animals to microorganisms.

The debate reached a peak in the 1700s and was led by an Englishman, John Needham, and an Italian,

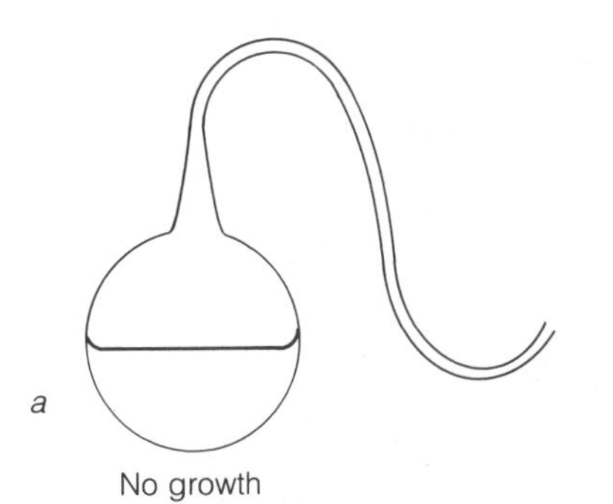

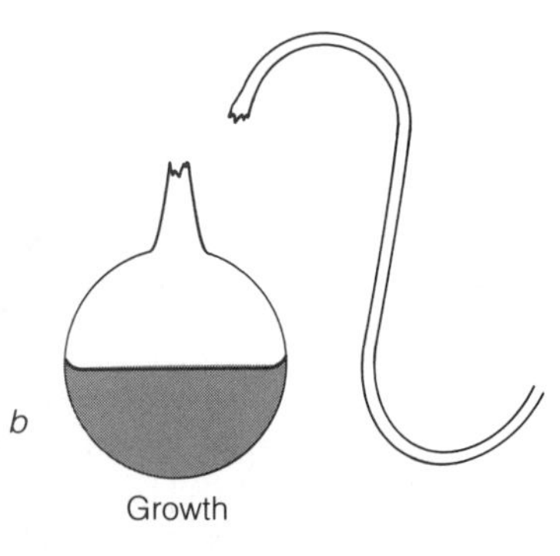

FIGURE 16-1
Pasteur experiment about the source of microorganisms that grew in boiled nutrient broth. (a) No growth in broth boiled in a flask with an S-shaped neck. (b) Growth in a flask with a broken neck.

Lazzaro Spallanzani. These two men did similar experiments but obtained different results. A solution of organic matter, for example mutton gravy, was boiled in a glass vessel, and the vessel was sealed. Needham sealed his vessels with corks and consistently observed the growth of microorganisms several days after boiling the solution. He concluded that the microorganisms had originated spontaneously out of the organic matter. Spallanzani boiled the organic solutions longer and sealed them more carefully. In some experiments he sealed vessels containing organic solutions by melting the neck of the glass vessel and *then* boiled the contents. The pressure in the closed vessels was increased by the heating, and therefore the boiling point of the enclosed organic solution was raised. This is equivalent to the modern method of sterilization called **autoclaving,** in which solutions are heated to 120°C using steam to generate a pressure of 15 to 20 lbs per in^2.

Spallanzani consistently observed that his organic solutions remained free of microorganisms. He concluded that the microorganisms that appeared in solutions heated and left open came from the air. However, others repeated Spallanzani's experiments and observed the appearance of microorganisms. We must conclude that these experimenters were less careful than Spallanzani and had not achieved sterilization of their solutions. The outcome was, however, that Spallanzani's experiments did not put an end to the controversy about spontaneous generation of microorganisms.

We now know that simple boiling of a solution (Needham's experiment) is insufficient to kill all forms of microorganisms. Many kinds of bacteria and fungi develop into spores when conditions are unfavorable for growth. In the form of a spore the cell is dehydrated, metabolically inert, and enclosed by a thick, tough cell wall. When placed in an environment that is favorable for growth, spores germinate into a metabolically active form and resume growth and reproduction. Spores can survive for many years in a dried state. They are not killed by brief treatment with boiling water but can be destroyed by prolonged boiling or by autoclaving. Bacterial and fungal spores are everywhere in our environment and are a common component of airborne dust. They are a major nuisance to those who work with animal and plant cell cultures, since a single contaminating spore germinates and grows rapidly in the nutrient media required for cell culture. A further complication is that autoclaving breaks down some of the components in nutrient media required for growth of animal cells. To avoid the deleterious effects of heat, sterilization is now often achieved by passing media through an extremely fine filter that removes any microorganisms or their spores.

The argument about spontaneous generation was settled to the satisfaction of most scientists only by the experiments of Louis Pasteur in the 1860s. One of his experiments consisted of sterilizing nutrient broth by prolonged boiling in a flask with an S-shaped neck (Figure 16-1). The flask was open to the air, but airborne spores were trapped in the S-shaped neck, and the broth remained sterile. When the neck of the flask was subsequently broken off, allowing airborne dust to enter, microorganisms began to grow in the broth (Figure 16-1). Pasteur also showed that microorganisms (presumably in spore form) could be collected by passing air through a filter made of cotton.

The disproof of spontaneous generation leaves us with an enigma. If cells can arise only from preexisting cells, where did the first cell on Earth come from? The experiments by Spallanzani and Pasteur showed that cells could not arise spontaneously from organic matter, at least under the conditions of their experiments. Modern experiments on this problem are designed to test the hypothesis that under certain conditions, particularly those that might have existed on Earth before the appearance of life, a simple primitive cell might have formed spontaneously. Thus, biologists accept that cells do not now arise spontaneously from organic material, but most believe that given the right conditions and sufficient time (many millions of years), spontaneous generation of a cell did happen at least once, and all contemporary cells have descended from that first cell.

In considering the current thinking and experimentation on the origin of life it is necessary to order real and postulated events onto a time scale beginning with the formation of the Earth.

FIGURE 16-2
A fossil trilobite. Trilobites were among the first arthropods, which includes crustaceans, insects, mites, and spiders. Trilobites existed for several hundred million years but died out about 250 million years ago. [Courtesy of Ryan Prescott.]

AGE OF THE EARTH

Scientists have estimated the age of the Earth by measuring the amount of decay of certain radioactive isotopes that has taken place in rocks. For example, radioactive potassium, ^{40}K decays into argon (^{40}A) and calcium (^{40}Ca).

$$^{40}K \nearrow {}^{40}A \quad \searrow {}^{40}Ca$$

About 90 percent of ^{40}K decays to ^{40}A and ten percent to ^{40}Ca. It is safe to assume that argon would not have been present when the rocks formed because this element occurs as a gas and would have been excluded from rocks. Any argon now present in rocks must therefore have derived from the decay of ^{40}K after their formation. The half-life of ^{40}K is 1.26×10^9 years. Knowing this, one can compute the age of the particular rock by measuring how much of the ^{40}K has decayed to ^{40}A, that is, by measuring the ratio of ^{40}A to ^{40}K.

In addition to the potassium-argon method of dating, the age of rocks can be estimated from the ratio of rubidium to strontium or by the uranium-thorium-lead method. The three dating methods all give about the same age for the oldest rocks, 4.6×10^9 years. This is generally accepted as the age of the Earth.

Formation of the Earth is believed to have occurred by aggregation of particles ranging in size from dust to asteroids over a period of millions of years. The process still continues with the capture from space of dust and meteors by the Earth's gravity. Contrary to earlier views, it is now generally believed that the surface of Earth was not molten when the Earth was formed, which means that the surface temperature remained below 900°C. This is important because it means that simple organic materials, which are known to be present outside the Earth, might have been included during formation of the Earth and survived to become a part of the primitive Earth.

THE SCALE OF BIOLOGICAL TIME

Until the late 1960s the oldest known fossils had been found in sedimentary rocks about 600 million years old. These represented highly evolved invertebrate animals such as **trilobites** (Figure 16-2), whose hard shells readily gave rise to fossils. The fossil record indicates that small invertebrates and plants were the only multicellular organisms in existence from about 600 to 500 million years ago. This interval of 100 million years is called the **Cambrian period** (Figure 16-3). The Cambrian period was followed by a succession of periods that are defined by the fossils of progressively more highly evolved plants and animals, extending to the **Quaternary period,** in which we now live.

The period before the appearance of the first fossilized invertebrates, extending from the origin of the Earth 4.6 billion years ago up to the beginning of the Cambrian period 600 million years ago, is called the

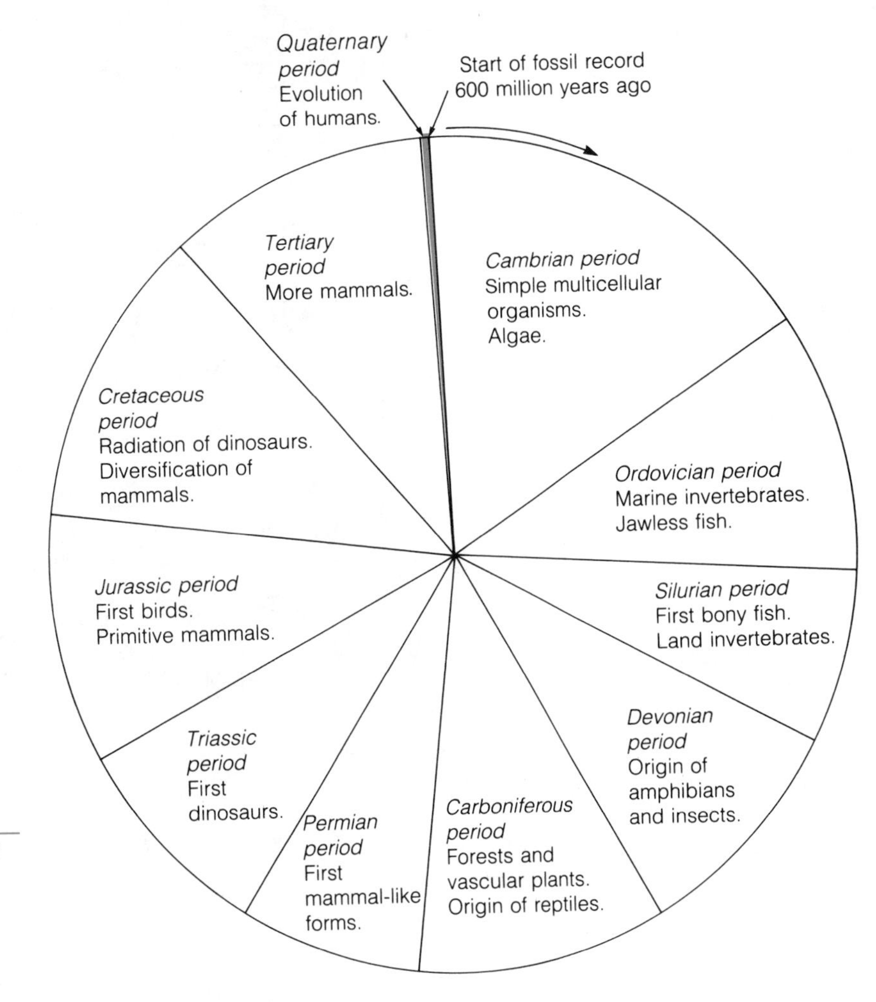

FIGURE 16-3
Major periods and events in the progression of life.

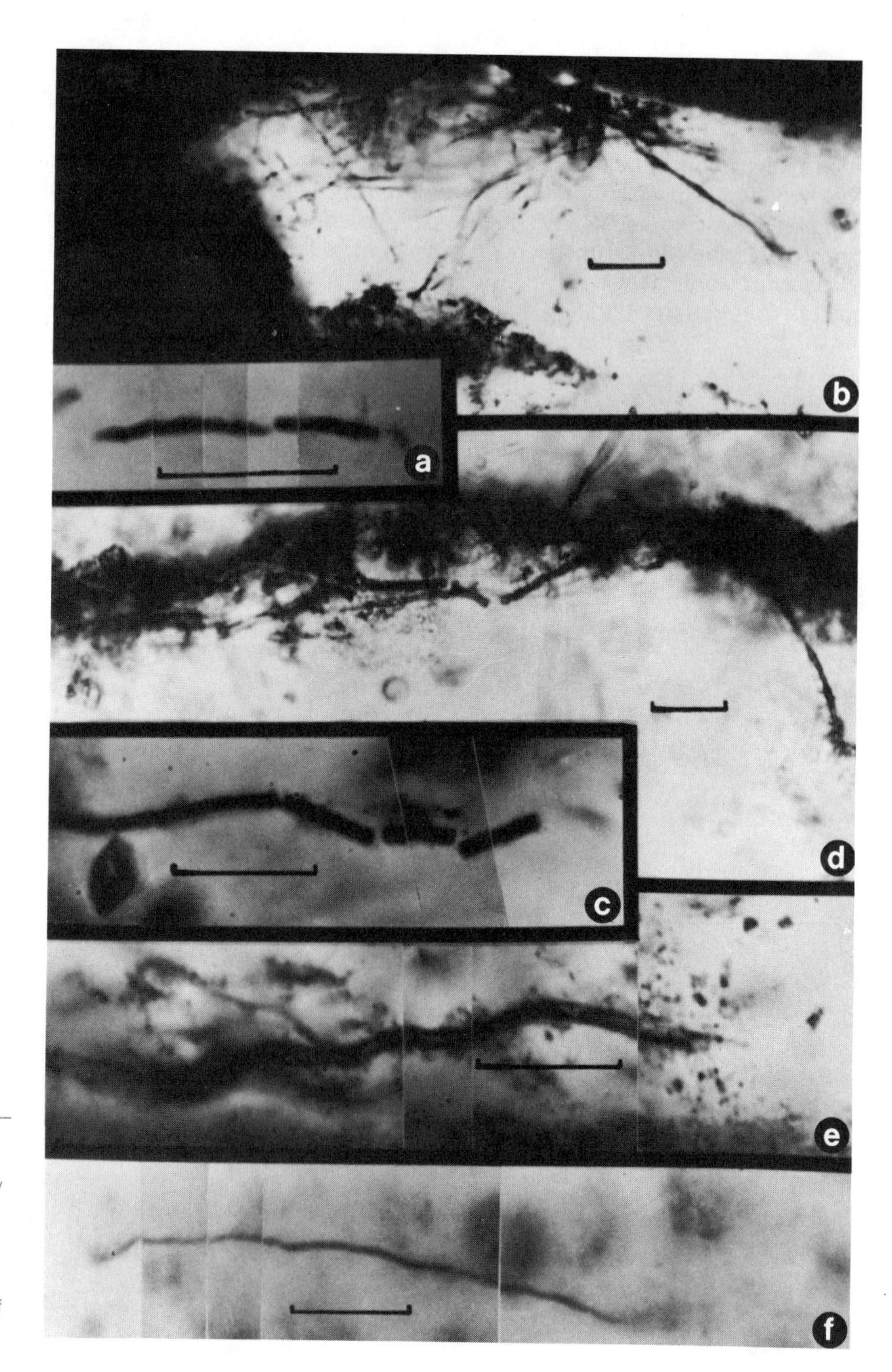

FIGURE 16-4

Electron micrographs of filamentous microfossils, believed to be bacteria, found in 3.5×10^6 year old sedimentary rocks in South Africa. (a) Cylindrical microfossil. (b) Threadlike and tubular filaments. (c, d, and e) Tubular filaments. (f) Threadlike filament. The bar in each is ten μm long. [Courtesy of Maud M. Walsh and Donald R. Lowe.]

Precambrian period. Six hundred million years is an underestimate of the time of appearance of the first multicellular organisms. Multicellular organisms that were structurally more simple, lacking hard shells, and leaving no apparent fossil record undoubtedly preceded the appearance of such highly evolved forms as hard-shelled invertebrates. Recently, worm tracks have been discovered in rocks at least 100 million years older than fossils from hard-shelled animals, showing that soft-bodied multicellular organisms were present at least 100 million years before the start of the Cambrian period, or 700 million years B.P. (before present). We do not know when the first multicellular animals evolved, but it must have been longer than 700 million years ago. Also, however long ago the first multicellular animal evolved, it must have been preceded by an interval in which all organisms were unicellular prokaryotes and eukaryotes. Furthermore, this interval was most likely preceded by a period in which only progenotes existed. All of this was largely supposition until the late 1960s, when the electron microscope was used in a successful search for microfossils of unicellular organisms in ancient rocks.

MICROFOSSILS OF UNICELLULAR ORGANISMS

Many descriptions of microfossils in rocks formed far back in Precambrian times have been published in scientific journals in the last 20 years. The oldest microfossils discovered so far are filamentous and spherical structures that strongly resemble bacteria, including blue-green bacteria. These microfossils occur abundantly in flintlike rocks called chert in South Africa and in rocks of Western Australia, both of which are 3.5×10^9 years old (Figure 16-4). Cherts formed in Precambrian times also typically contain 0.5 to 1.0 percent organic material, including substances that could be breakdown products of chlorophyll and proteins. These findings support the interpretation that microstructures such as those in Figure 16-4 are indeed fossils of unicellular organisms.

On the basis of these studies it is now generally believed that life existed on Earth at least 3.5×10^9 years B.P., and that photosynthetic organisms (blue-green bacteria) had already evolved by that time. For reasons discussed later, it is generally believed that the first cells were highly heterotrophic in their nutrition, obtaining energy and nutrients from an abundance of organic molecules in the environment. Evolution of photosynthetic prokaryotic organisms followed sometime later. Thus, we may conclude that the first cells, which were probably prokaryotic-like, arose sometime before 3.5×10^9 years ago, less than 10^9 years after formation of the Earth.

ORIGIN OF EUKARYOTIC CELLS

The fossil record provides an estimate of when eukaryotes originated. Some of the microfossils found in Bitter Springs chert of Australia contain structures resembling cytoplasmic organelles and nuclei and are believed to represent unicellular eukaryotes. This chert is 900 million years old, and unicellular eukaryotes therefore seem to have been well established by that time. Fossils of possibly nucleated cells have also been discovered in Beck Spring dolomite in California, which is about 1.3×10^9 years old. The oldest microfossils that might be remnants of eukaryotic cells have been found in rocks 1.4 to 1.6×10^9 years old from the Ural Mountains in the U.S.S.R. These microfossils are similar in form to existing kinds of unicellular green algae. Shale from Montana about 1.4×10^9 years old has yielded microfossils of filamentous blue-green bacteria as well as much larger spheroidal, thick-walled structures that are believed to be cysts of eukaryotes. From these and similar discoveries it is generally believed that eukaryotes originated at least 1.4×10^9 years ago. Thus, the origin of preprokaryotes (progenotes) preceded the origin of eukaryotes by about 2.1×10^9 years.

Many differences separate contemporary prokaryotes and eukaryotes (Chapter 1). They both probably evolved from progenotes, and it is virtually certain that not all the differences arose simultaneously. Mod-

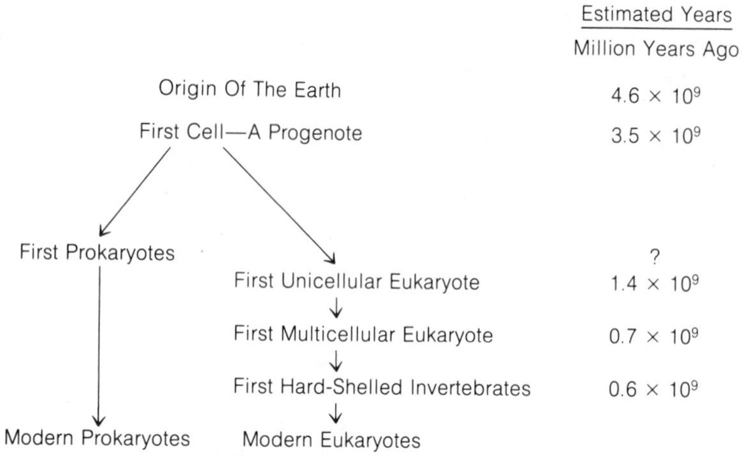

		Estimated Years Million Years Ago
Origin Of The Earth		4.6×10^9
First Cell—A Progenote		3.5×10^9
First Prokaryotes		?
	First Unicellular Eukaryote	1.4×10^9
	First Multicellular Eukaryote	0.7×10^9
	First Hard-Shelled Invertebrates	0.6×10^9
Modern Prokaryotes	Modern Eukaryotes	

FIGURE 16-5
Estimates of timing of appearance of major groups of organisms.

ern prokaryotes probably do not differ in structure nearly as much from the progenote ancestor as do modern eukaryotes. The origin of eukaryotes must have occurred with a single change, and this was followed by subsequent evolution of a succession of changes that now make eukaryotes very different from prokaryotes. We have little idea of which difference represents the first step in divergence of eukaryotes from the progenote. It might have been the acquisition of any one of a number of properties—new kinds of genes, multiple chromosomes, a larger content of DNA, a nuclear envelope, histones, a new principle of gene regulation, and so forth. The cell and molecular biology of only a small fraction of extant eukaryotes has been studied so far, but it is unlikely that any primitive eukaryotic cells have survived to give us insight into the origin of the eukaryotic cell line. Nevertheless, increased knowledge of the function and structure of contemporary eukaryotic cells may someday provide an indication of the origin and evolution of the eukaryotic cell.

Continued study of microfossils will no doubt result in more accurate estimates of the timing of such events as the origin of the first cells (progenotes), the evolution of the prokaryote and eukaryote lines, and evolution of the first simple multicellular organisms from unicellular eukaryotes. For the present the best estimate of the time scale of events may be summarized as shown in Figure 16-5.

Once the first cell had formed 3.5×10^9 or more years ago, we can understand, at least in principle, how the subsequent course of events was dictated by the evolution of an ever-increasing complexity in cell functions and structures. We know little about the particular steps, such as the evolution of regulatory genes, the evolution of photosynthesis, the evolution of the first eukaryotic cell, or the evolution of those genetic mechanisms that made possible the first multicellular organisms. Future research in molecular biology, genetics, and cell biology may yet give us a better idea of these processes of cellular evolution.

However, we are faced with a conceptually far more difficult problem than cellular evolution, and that is the matter of how the cell came into existence in the first place. It is impossible to formulate any reasonable scheme by which a cell might have formed *directly* from the *inorganic* materials present on the primitive Earth. The jump from inorganic chemicals to organic molecules capable of self-replication is simply too enormous. A solution to this conceptual dilemma was first proposed by the Russian biochemist, Alexander I. Oparin, and the British biologist J.B.S. Haldane beginning in the 1920s, and is now generally called the **organic soup concept.**

THE ORGANIC SOUP THEORY

The great contribution of Oparin and Haldane to the subject of the origin of life was based on the idea that in the period before life arose the atmosphere of Earth contained hydrogen (H_2), methane (CH_4), ammonia (NH_3), and water (H_2O), but no free oxygen (O_2). Thus, Oparin and Haldane proposed that the prelife atmosphere of Earth was highly reducing in a chemical sense. From a variety of evidence, geologists, cosmologists, and chemists now generally agree that primitive atmosphere was chemically reducing in nature. As an example of these lines of evidence, early Precambrian rocks contain ferrous iron, which is unstable in the presence of O_2. Therefore, the early Precambrian rocks must have been laid down in the absence of atmospheric O_2.

Oparin and Haldane both reasoned that a reducing atmosphere consisting of H_2, CH_4, NH_3, and H_2O would be favorable for the spontaneous formation of simple organic molecules, and these might then polymerize spontaneously into macromolecules. These macromolecules might then accumulate in the oceans and lakes of the time, giving rise to *organic soups*. It is doubtful that the accumulation would have been great because many kinds of organic molecules are unstable in aqueous solutions and are slowly and spontaneously hydrolyzed. However, one may suppose that organic molecules may have become concentrated by adsorption to solid surfaces (a common phenomenon) or through the rapid evaporation of lakes. It was Oparin's idea that the first cell arose not from inorganic substances but from a mass of prebiologically formed organic material.

Oparin's idea remained unknown in the West for a long time, probably because it was published in Russian. In 1938 Oparin's book entitled *The Origin of Life* was published in English, and immediately his theory attracted wide attention. Fifteen years later, in 1953, the Oparin-Haldane proposal about the spontaneous formation of organic molecules was tested directly.

With the apparatus shown in Figure 16-6, a gaseous mixture of H_2, CH_4, NH_3, and H_2O was exposed to electric spark discharges. Water was first added to the flask. The air was pumped out with a vacuum pump and the apparatus was then filled with a mixture of hydrogen, methane, and ammonia. The electric discharges, which were common in the primitive atmosphere (as lightning), provided energy for the synthesis of molecules from the four starting components. The water in the flask was boiled to cause circulation through the apparatus and remove any reaction products from the spark zone. Reaction products collected in the condensing water in the condenser and accumulated in the water phase. The experiment was run for one week, and the water then analyzed for any organic compounds that might have formed. The results of the analysis, which are shown in Table 16-1, are astonishing in at least two respects. A complicated mixture of small amounts of hundreds or even thousands of compounds is theoretically possible and might have reason-

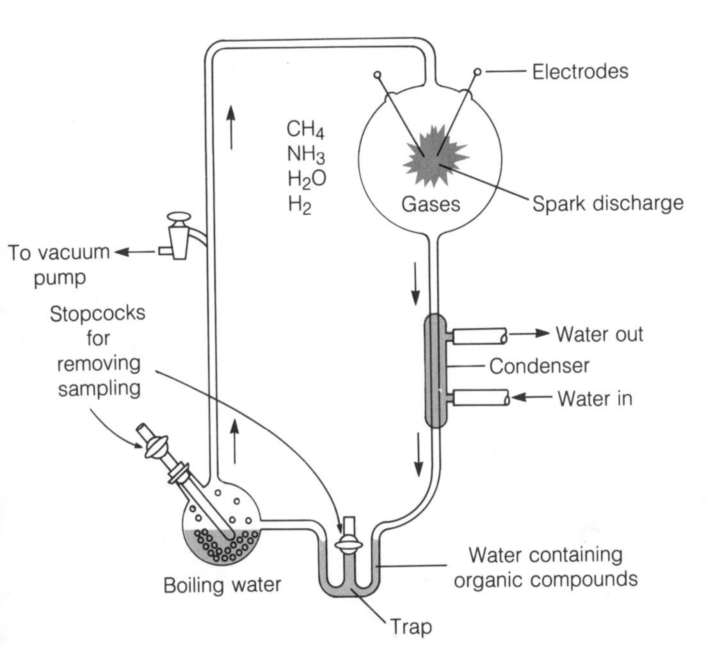

FIGURE 16-6
Apparatus for synthesis of amino acids and other organic compounds by spark discharges.

ably been expected. Instead a small number of compounds accounted for *most* of the reaction products. Second, the molecules produced included several of major biological importance, particularly the amino acids glycine, alanine, aspartic acid, and glutamic acid. Fifteen percent of the carbon added (as CH_4) to the apparatus was recovered in the compounds identified in Table 16-1. Additional carbon was converted into unidentified, tarlike, high-molecular-weight, organic polymers.

This experiment has been repeated with various modifications many times in other laboratories. As a result of these experiments almost all 20 amino acids, as well as purines, pyrimidines, ribose, nucleosides, and nucleotides have been produced abiologically under simulated early-Earth conditions.

In recent years radioastronomy has provided evidence by microwave spectroscopy that abiological synthesis of large quantities of biologically important molecules occurs commonly in the universe outside the Earth. These molecules include H_2, H_2O, NH_3, H_2S, CO, and HCN and the organic molecules cyanoacetylene (C_2HN), methanol (CH_3OH), ethanol (CH_3CH_2OH), formaldehyde (CH_2O), formic acid (HCOOH), formamide ($HCONH_2$), acetonitrile (CH_3CN), and acetaldehyde (CH_3CHO). This list is striking because it includes the very compounds that are the most important for the abiological synthesis of amino acids, purines, pyrimidines, and sugars. For example, formaldehyde, acetaldehyde, and hydrocyanic acid (HCN) react to form glutamic acid. Cyanoacetylene is a precursor of pyrimidines, and in particular can form a large amount of cytosine. Aldehydes, HCN, and NH_3 yield a variety of amino acids. Ribose, glucose, and other sugars are formed spontaneously in an alkaline solution of formaldehyde. Hydrocyanic acid is a precursor of glycine and purines.

A group of meteorites known as **carbonaceous chondrites** also contain organic molecules. The **Murchison meteorite,** which fell near Murchison, Australia in 1969, contains two percent carbon, much of which is present as a complex mixture of organic molecules. Among these are glycine, alanine, valine, proline, glutamic acid, and aspartic acid. Similar findings have been made with the **Murray meteorite,** which fell in the U.S. in 1950. Pyrimidines and possibly purines are also present in the Murchison meteorite.

TABLE 16-1
Some Organic Molecules Formed from CH_4, NH_3, H_2O, and H_2 as a Result of Spark Discharges

Glycine	Succinic acid
Glycolic acid	Aspartic acid
Sarcosine	Glutamic acid
Alanine	Iminodiacetic acid
Lactic acid	Iminoaceticpropionic acid
N-Methylalanine	Formic acid
α-Amino-n-butyric acid	Acetic acid
α-Aminoisobutyric acid	Propionic acid
α-Hydroxybutyric acid	Urea
β-Alanine	N-Methyl urea

The carbonaceous chondrites condensed from the same gaseous mass (the solar nebula) from which the sun and planets condensed. It is likely, therefore, that the organic molecules found in the meteorites were formed in the gaseous nebula. The Earth was also formed from dust and asteroids that condensed from the same gaseous mass, suggesting that organic molecules may have been present from the very beginning of the Earth's history. Of course, the presence of organic molecules in the solar nebula raises the possibility that organic molecules are present in other gaseous masses throughout the universe.

In sum, the synthesis of organic molecules under simulated conditions of prelife Earth, the detection in

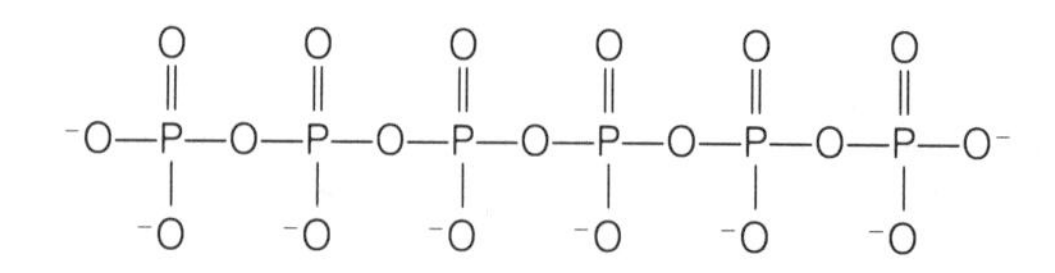

FIGURE 16-7
Polyphosphate molecules contain variable numbers of phosphate groups. Hydrolysis of polyphosphate molecules can be coupled to peptide bond formation between amino acids, a dehydration-condensation reaction.

other parts of the universe of substances that spontaneously react to form organic molecules, and the presence of organic molecules in meteorites all add plausibility to Oparin's idea that the origin of the first cell was preceded by the formation and accumulation of large amounts of organic molecules of major biological importance.

PREBIOLOGICAL FORMATION OF MACROMOLECULES

With prebiological synthesis of such monomers as amino acids, purines, pyrimidines, nucleotides, and sugars the origin of life no longer seems so inscrutable. However, to go from monomers to a cell, no matter how simple and primitive, is still a big leap. We now know, however, that under the appropriate conditions, a solution of amino acids can polymerize into large polypeptides, and nucleotides can polymerize into nucleic acid molecules. Thus, it is generally supposed that accumulation of monomers was followed by the prebiological formation of macromolecules and that the first cell arose by an aggregation of macromolecules.

What is required for abiological formation of macromolecules? Remember that the polymerization of amino acids into polypeptides and of nucleotides into polynucleotides occurs by dehydration-condensation of the monomers (Chapter 2). To form a peptide bond between two amino acids (condensation) a molecule of water must be removed (dehydration) (Figure 2-3). Similarly, water is removed in the formation of the 3′,5′-phosphodiester bond between two nucleotides.

These polymerizations do not occur readily in aqueous solutions because the presence of many water molecules opposes dehydration, driving the reaction toward depolymerization (hydration of the monomers) rather than polymerization. The problem has been solved in the cell by coupling the hydrolysis (bond breakage by addition of water) of ATP to the dehydration-condensation of amino acids or nucleotides into polymers. The dehydration is accomplished by forming an aminoacyl-tRNA at the expense of ATP in the case of peptide bond formation, and by splitting PP_i from nucleoside triphosphate in the case of nucleic acid synthesis (Chapter 2). The overall process, however, is a coupling of water removal from monomers to water addition (hydration) to ATP (to form ADP + P_i + H^+).

It is highly unlikely that ATP was present in a sufficient amount in the prebiological environment to drive polymerization reactions. However, **polyphosphates** (the simplest polyphosphate is pyrophosphate, Figure 16-7) can bring about polymerization of amino acids by a dehydration reaction in the same manner as does ATP, and polyphosphates could readily have formed under prelife conditions. Polyphosphates can form when orthophosphate (PO_4^{3-}) is warmed in the presence of urea and NH_4^+ (urea is a product of the organic soup experiment). Polyphosphates have been used experimentally to promote the synthesis of AMP from adenine, ribose, and phosphate, and to drive the formation of polynucleotides.

Other compounds might also have served as dehydration-condensation agents. These compounds include carbodiimide, cyanate, cyanogen, and cyanovinyl phosphate, all of which might readily have formed under prelife conditions. Monomers can also be caused to polymerize by adsorption on the surface of certain minerals such as clay and apatite compounds. Adsorption of amino acids to one common clay known as montmorillonite is followed by condensation of the amino acids into long polypeptide chains.

Another way to bring about polypeptide formation is to heat a mixture of dry amino acids to 130 to 180°C for a few hours. In the absence of water the necessary dehydration reaction is strongly favored. Large complex polypeptides formed by this method aggregate into microspheres when mixed with water (Figure 16-8). Protein-like molecules form when aminoacyl adenylates are mixed. These aminoacyl adenylates are the activated form of amino acids used in cellular protein synthesis. Aminoacyl adenylates readily form abiologically when AMP and amino acids are mixed in the presence of dehydration-condensation agents.

Not only do the abiologically formed protein-like polymers form stable microspheres but, remarkably, such polymers possess low levels of catalytic power, for

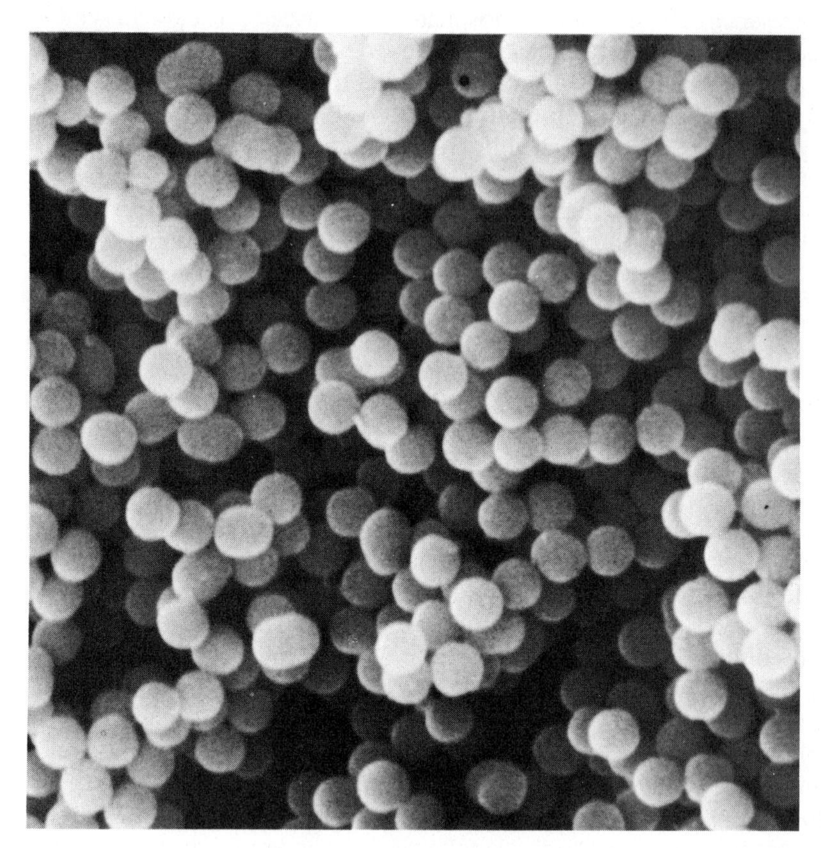

FIGURE 16-8
Scanning electron micrograph of proteinoid microspheres. Each is about 2 μm in diameter. [Courtesy of Sidney W. Fox and Steven Brooke Studios.]

example, in the decarboxylation of pyruvate to acetaldehyde and CO_2.

$$CH_3{-}\underset{\underset{O}{\|}}{C}{-}COOH \rightarrow CH_3{-}\underset{\underset{O}{\|}}{C}{-}H + CO_2$$

Thus, at least some of the protein-like polymers qualify as primitive enzymes.

FROM MACROMOLECULES TO CELLS

The strong probability that polypeptides and polynucleotides formed under prelife conditions further reduces the difficulties in conceptualizing how life originated. Yet we are still faced with formidable problems in trying to understand the origin of the cell.

Abiologically formed macromolecules may aggregate into cell-like structures, such as the proteinoid microspheres in Figure 16-8, and these may grow and subdivide. What additional properties must an aggregate of macromolecules possess to be called a cell? The answer to this is not so easy as it might seem initially. It seems essential that a genetic mechanism be present that allows the primitive cell to duplicate itself precisely. It is difficult to imagine how this could be achieved without a template-copying mechanism of the type present in nucleic acid replication. We may also add the requirement that the macromolecular components must have functions, however crude, that are useful for survival and reproduction. Enhancement by enzymatic catalysis of polymerization of monomers into more macromolecules would be one such function. These functions must be inherited through cell reproduction. The presence of a genetic inheritance carries with it the potential for chance improvement of macromolecular functions through mutation. Once genetically inherited functions have been acquired, the way is open for cellular evolution.

We might suppose that the first cell had to contain nucleic acids because only nucleic acids are known to replicate themselves with a reasonable amount of precision and therefore constitute a mechanism of genetic inheritance. As already stated, the formation of nucleotides under simulated prelife conditions has already been accomplished in the laboratory. In addition, these nucleotides can be caused to polymerize into short polynucleotides when a dehydration-condensation agent is added. With an appropriate polymerase obtained from cells, a single-stranded DNA or RNA chain acts as a template for the synthesis of a complementary chain from nucleoside triphosphates.

Essentially the same can be accomplished without a polymerizing enzyme, that is, under prelife conditions. For example, oligomeric chains of thymidylate residues can be generated under prelife conditions. In turn, oligomers composed of six thymidylate residues (hexamers) have been found to join together slowly to form longer polynucleotides of thymidylate residues in the presence of a dehydration-condensation agent. If polyadenylate is added to the mixture, the rate of joining of the thymidylic acid hexamers is greatly speeded up; apparently the polyadenylic acid acts as a template that binds the hexamers of thymidylate residues by base-pairing and thereby aligns the hexamers so that end-to-end joining takes place much more rapidly (Figure 16-9). In bringing together hexamers of thymidylate residues, and by promoting their end-to-end ligation, polyadenylate is actually functioning as a catalyst. Similarly, hexamers of polyadenylate residues may be coupled together into longer polynucleotides on a template of polyuridylate residues. Polycytidylate chains act as templates for polymerization of GTP into chains up to 40 nucleotides long.

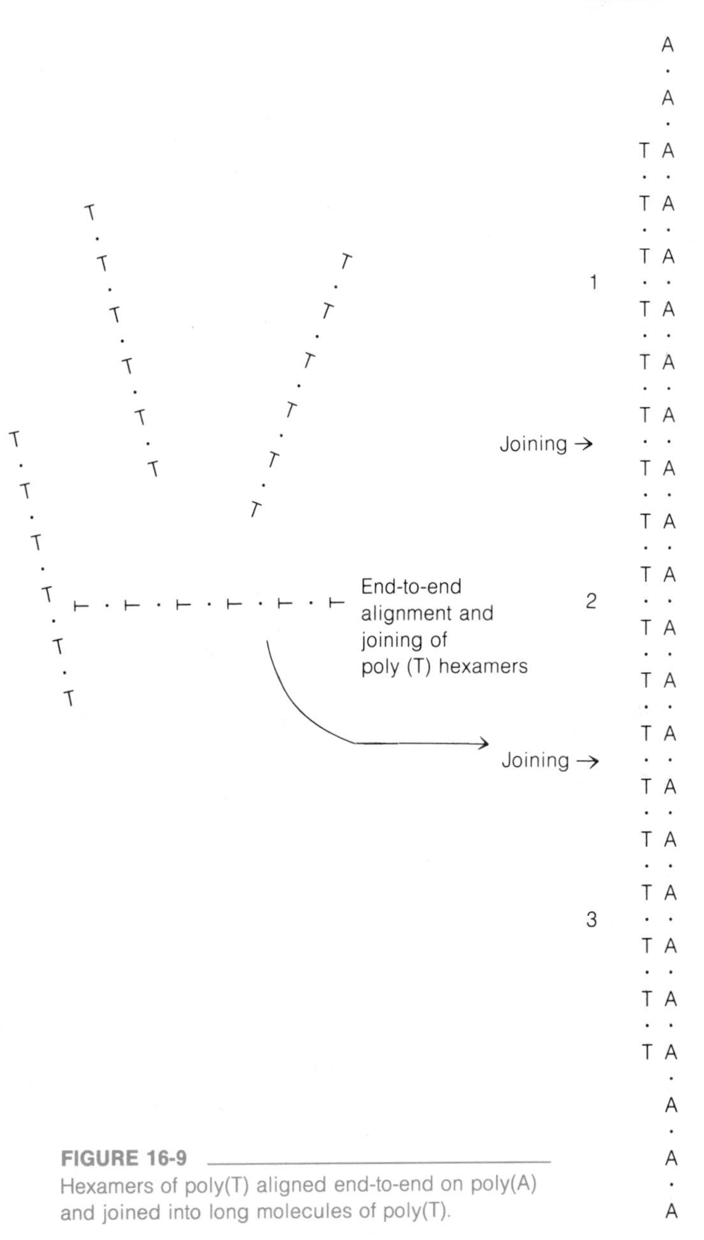

FIGURE 16-9
Hexamers of poly(T) aligned end-to-end on poly(A) and joined into long molecules of poly(T).

These experiments give evidence that polynucleotides might well be able to replicate themselves in the familiar template fashion without catalysis by protein enzymes. Therefore, it seems possible that cellular reproduction was from the beginning based on polynucleotide replication. We know of no way in which proteins can function as templates to guide their own reproduction. Not long ago, nucleic acids were thought to lack the kind of catalytic capabilities necessary to give the first cell its first functions. Thus it seemed that proteins (primitive enzymes) must have been present early in primitive cells. The supposition that proteins had to be present to provide enzymatic activity in primitive cells has changed dramatically as a result of recent studies on excision of introns from primary RNA transcripts in eukaryotes (Chapter 8). The precursor ribosomal RNA transcript in *Tetrahymena* undergoes catalytic self-splicing in which the RNA intron acts as an enzyme to remove itself from the transcript by cutting and splicing together of RNA ends. Moreover, the intron RNA can act as a ribonuclease to remove nucleotides from RNA, and astonishingly the intron RNA can also work as a polymerase, catalyzing the formation of polycytidylate chains on a poly(G) template that it itself contains. Segments of RNA such as the intron of precursor rRNA in *Tetrahymena* have been given the name **ribozymes.**

These discoveries favor strongly the idea that the first cells used RNA molecules both to hold genetic information and as enzymes to catalyze RNA metabolism. The subsequent switch from RNA to a DNA double helix as the genetic material is not a conceptually difficult problem. However, how self-catalyzed RNA replication evolved into the current genetic mechanisms in which information in nucleotide sequences is translated into amino acid sequences via tRNA adaptors is much harder to envisage.

THE FIRST CELL: AN EXTREME HETEROTROPH

We cannot reasonably suppose that the first cell was capable of more than a few enzymatic activities. Hence, it must have depended on abiologically formed organic molecules in its environment for growth and reproduction. With the evolution of anabolic capabilities, stringent dependence on heterotrophic existence presumably decreased; indeed, depletion of essential components (e.g., amino acids and nucleotides) in the environment necessitated the evolution of anabolic activities. Eventually, the exogenous supply of nutrients such as nucleotides and amino acids would begin to be depleted. Any primitive cell that acquired through chance the ability to catalyze the formation of one or another nucleotide or amino acid from some closely related molecule still available in the environment would have an advantage over its contemporaries. Hence early cellular evolution surely moved in the direction of **autotrophy.**

It seems unnecessary to postulate that in its earliest form the cell possessed an enveloping membrane. However, as soon as anabolic and energy-generating capabilities began to be acquired, a membrane became important for retention of valuable metabolic products. Hence, we may suppose that a plasma membrane of some sort appeared early in the evolution of metabolism.

EVOLUTION OF ENERGY-GENERATING MECHANISMS

The acquisition of anabolic capabilities carries with it a requirement for energy to drive synthesis of molecules. The primitive environment may have contained some energy-rich substances (e.g., polyphosphates), but these would probably have been quickly used up. Therefore, the ability of the cell to generate its own energy supply probably occurred early in cell evolution. For several reasons, it seems likely that the first energy-generating mechanism was a primitive form of glycolysis. All contemporary cells possess glycolysis, which indicates that it evolved before prokaryotic and eukaryotic cells diverged from a common ancestor. Glycolysis is also mechanistically simpler than photosynthesis and presumably preceded photosynthesis in evolution. Since the early atmosphere of Earth lacked oxygen, respiration was impossible. In fact, it is generally believed that the oxygen now present in the atmosphere is the result of photosynthesis. Hence, respiration probably evolved after photosynthesis. The order of evolution of energy-generating mechanisms is therefore believed to have proceeded from glycolysis to photosynthesis to respiration.

We can summarize the discussion about the origin and early evolution of the cell by elaborating on the scheme presented earlier (Figure 16-10).

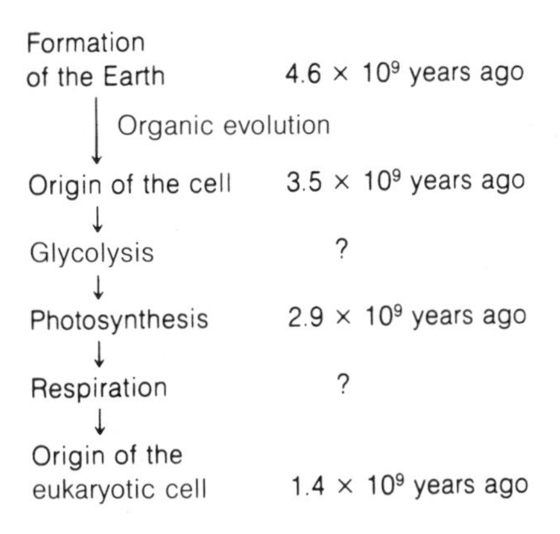

FIGURE 16-10
Estimate of occurrence of major cellular phenomena.

ORIGIN OF THE CHLOROPLAST AND THE MITOCHONDRION

The origin of these organelles represents one aspect of eukaryotic cell evolution about which we have gained some understanding. The mitochondrion is believed to have arisen through a symbiotic association between an early eukaryotic cell and a bacterium. According to this hypothesis a bacterium was engulfed by the cytoplasm of a eukaryotic cell, where it became a permanent resident, growing and reproducing with little if any detriment to its host. Such arrangements are not uncommon among contemporary fungi and protozoa. *Amoeba proteus,* for example, harbors a bacterium that grows and multiplies in the cytoplasm of the amoeba. A single amoeba contains thousands of these bacteria. The bacterium has become adapted to the environment provided by the amoeba cytoplasm to the extent that it can no longer grow and divide outside the amoeba. Whether the bacterium contributes anything to the well being of the amoeba (a symbiotic relationship) or whether it is strictly a parasite is not known.

The arrangement now present in *A. proteus* may resemble an early stage in mitochondrial evolution. The relationship between the early eukaryotic cell and its acquired bacterium evolved into a symbiotic one in which the bacterium became specialized for the respiratory synthesis of ATP, which it supplied to its host. The eukaryotic host, in turn, evolved in the direction of providing more and more of the proteins necessary for the structure and function of the symbiont, and this was accompanied by a corresponding loss of genes (DNA) in the symbiont. Assuming that the host cell was originally aerobic, it lost its own enzymatic machinery for respiration since all respiration is carried out in mitochondria in contemporary cells. The genetic and biochemical interrelationships between mitochondria and the rest of the cell are intricate and complex (Chapter 4). Chloroplasts are also likely to have evolved from blue-green bacteria acquired by an early eukaryotic cell.

According to the hypothesis about the symbiotic origin of the mitochondrion and the chloroplast, the DNA in these organelles represents all that remains of the original chromosome of an aerobic bacterium or blue-green bacterium. The strength of the symbiosis hypothesis lies in basic similarities between the organelles and free-living bacteria. First, the DNA of mitochondria and chloroplasts does not have histones associated with it, and neither does the DNA of bacteria or blue-green bacteria. The ribosomes of mitochondria and chloroplasts correspond in size to prokaryotic ribosomes rather than eukaryotic ribosomes. The rRNAs of organellar ribosomes, which are coded for by organellar DNA, are about the same size as prokaryotic rRNAs and hence are smaller than eukaryotic rRNAs. Moreover, organellar rRNAs have nucleotide sequences that are present in prokaryotic rRNAs but not in eukaryotic rRNAs. Finally, the drug chloramphenicol inhibits protein synthesis in bacteria and in mitochondria, but has no effect on protein synthesis in the cytoplasm of eukaryotic cells. Thus, mitochondrial and bacterial protein synthesis share a basic property that is not present in eukaryotic protein synthesis. There is a variety of other, corroborating observations involving similarities in other organellar and bacterial properties, but the above are the main ones.

PROBLEMS

1. When a rich nutrient medium like meat broth is boiled but left open to the air, the broth becomes spoiled by the growth of microorganisms. How do they arise?
2. What was the experiment that finally resolved the issue of spontaneous generation?
3. How old is the Earth? What is the earliest time that life is believed to have arisen on the planet?
4. What is the organic soup concept? Describe a laboratory test of the concept.
5. What is one kind of evidence that the early atmosphere of the Earth lacked O_2?
6. How has radioastronomy contributed to thinking about the origin of life?
7. What substance might have substituted for ATP during the origin and early evolution of life?
8. What kinds of macromolecules have been observed to form abiologically from mixtures of monomers?
9. Why is it argued that nucleic acids must have been present in the formation of the first cell?
10. How have the recent discoveries about excision of the intron in the rRNA of *Tetrahymena* influenced thinking about the origin of life?
11. Why is it believed that the first cells were extreme heterotrophs?
12. Cellular respiration almost certainly evolved after photosynthesis. Why is this thought to be so?

Answers to Problems
Glossary
Further Reading
Index

ANSWERS TO PROBLEMS

CHAPTER 1

1. i) Nucleoid. Contains the DNA and produces all the RNA for the bacterial cell.
ii) Ribosomes. RNA-protein particles on which proteins are synthesized.
iii) Cytosol. Contains enzymes for breakdown and synthesis of many kinds of molecules.
iv) Plasma membrane. A retentive barrier that regulates movement of molecules and ions in and out of the cell.
v) Cell wall. Provides mechanical protection.

2. i) All cells store information in genes made of DNA.
ii) The genetic code is, with minor exceptions, universal.
iii) All cell types decode DNA genes by an RNA system that translates genetic information into proteins.
iv) All cells use ribosomes to synthesize proteins.
v) Proteins govern cell function and structure.
vi) All cells use ATP as the currency of energy metabolism.
vii) All cells are enclosed by a plasma membrane composed of proteins and a double layer of lipid molecules.

3. The major differences between prokaryotic and eukaryotic cells are listed in Table 1-1, page 11.

4. The greater content of DNA in eukaryotes than in prokaryotes reflects the greater genetic complexity of eukaryotes.

5. Rich nutrient medium relieves the cell of the need to synthesize many molecules required for cell reproduction, thereby shortening the time taken to reproduce.

6. Photosynthetic bacteria generate much of the oxygen in the atmosphere.

7. Lysosomes contribute the enzymes needed to digest food organisms. The contractile vacuole serves in the excretion of water from an amoeba.

8. i) Nucleus. Carries the chromosomes and provides the RNA needed for cell function and structure.
ii) Nucleolus. An organelle within the nucleus where ribosomes are assembled.

iii) Plasma membrane. Serves as a retentive barrier and regulates traffic of molecules and ions in and out of the cell.
iv) Mitochondria. Convert chemical energy in food into a form (ATP) that is readily used by cells.
v) Ribosomes. RNA-protein particles on which proteins are synthesized.
vi) Lysosomes. Vesicles containing high concentrations of enzymes that break down many kinds of molecules, for example, in materials taken into a cell by phagocytosis.
vii) Golgi complex. A series of flattened, membranous vesicles that process, concentrate, and package proteins destined for other cell compartments or for export from the cell.
viii) Endoplasmic reticulum. A system of cytoplasmic, membranous sacs involved in synthesis of lipids and proteins.
ix) Microbodies or peroxisomes. Cytoplasmic vesicles containing enzymes that catalyze the oxidation of various molecules, especially fats and amino acids. Hydrogen peroxide is formed in microbodies and converted to water and oxygen by catalase in microbodies.
x) Microtubules. Tubular structures that form part of the cytoplasmic cytoskeleton and function in some kinds of cell motility.
xi) Microfilaments. Filaments that form part of the cytoplasmic cytoskeleton and function in some kinds of cell motility.
xii) Intermediate filaments. Filaments that form part of the cytoskeleton.

9. i) Cell wall. A structure containing cellulose and other molecules that protects cells from mechanical injury and various other forms of stress.
ii) Chloroplasts. Organelles in which photosynthesis is carried out.

10. A continuous source of energy is required for repair and maintenance of cell parts.

11. Nearly 10^{12} cells.

12. Heterotrophic cells fulfill their requirements for chemically reduced carbon in the form of organic molecules that are synthesized by autotrophic cells.

13. Cell fractionation enables the study of the function and structure of isolated, individual cell parts. Radioactive isotopes facilitate analysis of chemical events that make up cell functions and enable identification of the chemical nature of cell structures.

14. Flagella give *E. coli* the motility necessary to seek out a nutritionally more favorable environment.

15. A high concentration of ribosomes in a cell means that the cell has a high rate of protein synthesis, usually with secretion of proteins from the cell.

CHAPTER 2

1. The polar property of water, that is, its slight negative and positive charges, make it a good solvent for ions and many molecules.

2. Hydrolytic cleavage is so named because it proceeds with the covalent addition of water to the molecule undergoing cleavage. In a dehydration-condensation reaction the covalent joining of two molecules into one is accompanied by the removal of a water molecule (dehydration) from the two participating molecules.

3. A hydrogen bond is a weak bond that can form between a slightly positively charged hydrogen atom belonging to one molecule and a slightly negatively charged atom, such as oxygen or nitrogen, in another molecule. A gas.

4. Equilibrium is the state of a solution in which the rate of dissociation of water into hydrogen ions and hydroxyl ions is equal to the rate of reassociation of these ions into water.

5. For a 0.0001 M solution of HCl,

$$pH = \log_{10} (1/10^{-4}) = 4.0$$

For a 0.0001 M solution of NaOH: First, the molarity of H^+ ions must be calculated using the ion product of water K_w,

$$K_w = [H^+]\,[OH^-] = 10^{-14}$$

For a 0.0001 M solution of NaOH,

$$[H^+] = 10^{-14}/10^{-4} = 10^{-10}$$

$$pH = \log_{10}\,(1/10^{-10}) = 10$$

6. The dissociation of a weak organic acid like aspirin is so slight that it will cause a virtually imperceptible increase in $[H^+]$ concentration in a solution that already has a H^+ ion concentration of 0.001 M (pH 3).

7. Proteins, nucleic acids, lipids, and carbohydrates.

8. Tryptophan is poorly soluble in water because its large side group is polar in nature. Lysine is highly soluble in water because its side group (NH_3^+) is charged at neutral pH.

9.

```
H       H  O  H  H  O
 \      |  ‖  |  |  ‖
  N—C—C—N—C—C—OH
 /      |        |
H       R1       R2
```

10. α-helices and β-structure, which are stabilized by hydrogen bonds.

11. A polypeptide chain folds into a tertiary configuration that minimizes interaction of hydrophobic side groups with water and maximizes interaction of hydrophilic side groups with water. Hence, amino acids with hydrophobic side groups tend to be buried inside, and hydrophilic groups tend to be at the surface of a protein.

12. Addition of energy in the form of heat breaks the weak bonds, such as hydrogen bonds, which stabilize tertiary structure. In primary structure amino acids are held together by covalent bonds that are too strong to be affected by heating to 80 °C.

13. RNA contains ribose and uracil. DNA contains deoxyribose and thymine instead of uracil.

14. A nitrogen base, a ribose or deoxyribose sugar, and a phosphate group.

15. By a covalent bond between the oxygen atom attached to the carbon number 1 atom and the carbon number 4 atom (see Figure 2-29).

16. The carbons in the hydrocarbon chain of a saturated fatty acid are joined by single covalent bonds. In an unsaturated fatty acid at least one double bond is present between two carbon atoms (see Figure 2-32).

17. Hydrogen bonds, ionic or salt linkages, van der Waals forces, and hydrophobic interactions.

18. Aggregation of macromolecules into structures enables the macromolecules to work cooperatively to achieve a function, for example muscle contraction, that could not be achieved by macromolecules acting individually.

CHAPTER 3

1. Add energy in the form of heat. Lower the energy required by a substrate to achieve the activated state with a catalyst, such as an enzyme.

2. By synthesizing enzymes to catalyze particular reactions, a cell selects which reactions will be allowed to occur at significant rates.

3. Some enzymes provide binding sites that bring two molecules together into a particular orientation that allows them to react with one another.

4. The observation that higher and higher concentrations of substrate bring about less and less of an increase in the velocity of an enzyme-catalyzed reaction, suggesting that the enzyme molecules become saturated with substrate.

5. The Michaelis-Menten constant is an indicator of

the strength of binding between an enzyme and its substrate and of the catalytic power of the enzyme.

6. The concentration of fructose must be ten times higher than glucose.

7. Coenzymes are intermediate carriers of chemical groups or electrons that enter into specific enzyme-catalyzed reactions.

8. Eight vitamins have been shown to serve as building blocks in the formation of coenzymes.

9. An active site is formed by folding of a polypeptide chain in a specific pattern that brings together the side groups of a few amino acids into a particular topographical arrangement on the surface of the folded polypeptide. Sometimes two polypeptides may fold to form an active site cooperatively.

10. Substitution of a hydrophilic amino acid by a hydrophobic one will in general cause greater disruption of the correct folding of the polypeptide because folding patterns are favored that place hydrophilic amino acids at the surface of a protein and bury the hydrophobic amino acid inside. The greater the disruption in normal folding, the greater the effect on protein function.

11. The shape of its active site.

12. In the lock-and-key model the shape of the active site is perfectly complementary to the shape of the substrate. In the induced-fit-strain model the complementarity between active site and substrate is imperfect before binding of the substrate. Binding of the substrate induces a shift in the shape of the active site toward greater complementarity.

13. Binding sites for allosteric effectors and sites for binding to other macromolecules.

14. Increasing the concentration of the normal substrate increases the probability that the enzyme will collide with a substrate molecule to form an ES complex and thereby reduce the opportunity for binding of the competitive inhibitor to the enzyme.

15. A competitive inhibitor binds to the active site of an enzyme. A noncompetitive inhibitor binds to a site other than the active site.

16. A reversible inhibitor binds reversibly to the active site or some other location on the enzyme. An irreversible inhibitor binds so tightly to either location that it is essentially irreversible.

17. An allosteric site specifically binds an allosteric effector and in so doing shifts the tertiary configuration of the protein so as to alter slightly the shape of the active site.

18. Allosteric effectors that reduce enzyme activity are noncompetitive, reversible inhibitors.

19. Regulation of an enzyme pathway ensures an amount of product sufficient for cell needs without the wastefulness of overproduction.

20. The activities of some enzymes are regulated by covalent modification of the enzyme, for example by phosphorylation.

CHAPTER 4

1. When two reactions are coupled, some of the energy normally lost as heat in one reaction can be used to drive a second, energy-requiring reaction.

2. Photons from the sun provide the energy for the flow of energized electrons through a series of carriers by which ATP and glucose are produced in photosynthesis. The energy in glucose is in turn used by heterotrophs to generate an ATP-yielding electron flow.

3. By coupling of the energy-yielding hydrolysis of ATP to the energy-requiring reaction.

4. ATP is the immediate source of energy for almost all energy-requiring activities in a cell.

5. O_2 and glucose.

6. Electrons in the form of the reduced coenzyme NADPH and ATP produced by phase I of photosynthesis are used to synthesize glucose in phase II of photosynthesis.

7. Photosystem I can continue to transfer electrons to $NADP^+$ only if it also receives electrons from photosystem II.

8. H_2O.

9. Glucose.

10. To retain water, plants in hot, dry climates close their stomata. This reduces the intake of CO_2 needed for photosynthesis. C_4 plants solve the problem by specialization of mesophyll cells for capture of CO_2 for photosynthesis when the concentration of available CO_2 is extremely low.

11. Phase I reactions occur in the thylakoid membranes. Phase II reactions occur in the stromal compartment.

12. By glycolysis.

13. Pyruvate, ATP, and NADH + H^+.

14. Blockage of the electron transport chain prevents the oxidation of NADH to NAD^+. Without NAD^+ available to accept the electrons produced by the citric acid cycle, the cycle cannot proceed.

15. Energy charge is a measure of relative amounts of AMP, ADP, and ATP in a cell.

16. ATP, CO_2, and H_2O.

17. The reactions of the citric acid cycle occur in the matrix and those of oxidative phosphorylation in the inner membrane of a mitochondrion.

18. The flow of H^+ ions driven by a concentration difference on the two sides of the inner membrane of a mitochondrion provides the energy that allows the ATPase associated with the membrane to act as an ATP synthase. Without a flow of H^+ ions the enzyme hydrolyzes ATP.

19. To provide a passage for uptake of CO_2 and release of O_2 from the plant leaf. H_2O is lost from the leaf through stomata.

CHAPTER 5

1. Water taken into protozoa living in fresh water is excreted by the contractile vacuole.

2. The plasma membrane contains a bilayer of lipid molecules in the fluid state. Integral proteins are embedded in the lipid bilayer. Peripheral proteins are attached to the surface of the lipid bilayer and to exposed parts of integral proteins.

3. Substances diffuse from regions of higher concentration to regions of lower concentration. Cell swelling results from diffusion of water from the hypotonic medium (higher concentration of water) to the cell interior (lower concentration of water).

4. Integral membrane proteins contain stretches of hydrophobic amino acids that prevent solubilization in water. Detergent molecules bind by their hydrophobic regions to the hydrophobic regions of membrane proteins. The hydrophilic regions of detergent molecules, by their solubility in water, allow membrane proteins to be solubilized in water.

5. Freeze-fracture splits membranes between the two sheets of lipid molecules in the bilayer. Electron microscopic observations of the internal surfaces of lipid layers reveals the presence of particles that are protein molecules.

6. Some transmembrane proteins are receptors for hormones and growth factors at the surface of the cell. Binding of a hormone or growth factor creates a signal consisting of a change in protein tertiary configuration that is propagated to the portion of the protein on the cytoplasmic face of the plasma membrane, causing activation of an enzyme active site.

7. When two cells carrying different integral proteins

in their plasma membranes were fused, the integral proteins quickly intermixed in the membrane.

8. Size, shape, ionic charge, and lipid solubility.

9. They both show saturation kinetics, implying the presence of a carrier in the membrane. Active transport requires energy and can transport substances against a concentration difference. Facilitated diffusion has neither of these properties.

10. The higher concentration of Na^+ ions outside the cell created by the Na^+-K^+ ATPase results in inward facilitated diffusion of Na^+ ions. Diffusion is mediated by a carrier protein that binds both a Na^+ ion and an amino acid or sugar molecule at the cell surface. The higher concentration of Na^+ ions outside the cell drives the carrier to transport the Na^+ ion inward, accompanied by inward transport of the bound amino acid or sugar molecule.

11. Pinocytosis is the intake of fluid and dissolved materials by formation of vesicles at the cell surface. Phagocytosis is the intake of particulate material by formation of vesicles at the cell surface.

12. Radioactive amino acid was injected into guinea pigs. At increasing intervals after injection guinea pigs were killed and the intracellular location of radioactive amino acids incorporated in proteins in pancreas cells was determined by autoradiography. The labeled proteins could be followed in their movement from endoplasmic reticulum to the Golgi complex, then to zymogen granules and finally out of the cell into a duct.

13. LDL receptors are recycled to the plasma membrane (see Figure 5-26).

14. In autoradiography the photographic film records radioactive decay in the underlying cells.

CHAPTER 6

1. Microtubules, microfilaments, and intermediate filaments.

2. By staining cells with a fluorescent antibody against tubulin.

3. The centrosome serves as a nucleation site for microtubules.

4. i) Assembly of tubulin dimers into short protofilaments coiled into rings.
ii) Rings uncoil and the short protofilaments join side by side to make a sheet.
iii) The sheet roles up into a short microtubule with 13 protofilaments.
iv) The short microtubules serve as nucleation structures for addition of free dimers onto the ends.

5. In treadmilling, tubulin dimers are added to one end of a microtubule and subtracted from the other.

6. Binding of GTP to tubulin dimers promotes their addition to the ends of microtubules.

7. The mushroom poison phalloidin binds to microfilaments and prevents their disassembly. Amoeboid movement is blocked by phalloidin. The mold poison cytochalasin binds to the ends of microfilaments and prevents their elongation. Cytochalasin inhibits amoeboid movement.

8. Fibronectin and laminin help to anchor cells to surfaces and to each other and provide a molecular surface for cell migration.

9. Networks of intermediate filaments provide mechanical strength to sheets of epithelial cells.

10. Most microtubules and microfilaments undergo continuous assembly and disassembly in cells. Intermediate filaments, once assembled, do not disassemble, although they may change form, for example collapsing during mitosis.

CHAPTER 7

1. If 22 percent of bases are thymine, then 22 percent are adenine. The remaining 56 percent form GC base pairs, and hence 28 percent of the bases are guanines.

2. The pneumococcus transformation experiment is the most convincing. The use of protease to digest proteins minimizes the possibility that proteins act as the transformation material. In the blender experiment some protein enters the cells and could conceivably act as the transformation material. Stability of DNA is the weakest of the three studies because stability is a necessary but not sufficient property for the genetic material.

3. Forty percent of base pairs are AT, each of which are held together by two hydrogen bonds. Sixty percent of base pairs are GC, each of which is held together by three hydrogen bonds. Therefore,

$$0.4 \times 10^6 \times 2 + 0.6 \times 10^6 \times 3 = 2.6 \times 10^6 \text{ hydrogen bonds}$$

4. A phosphodiester bond connects the number 3 carbon in the deoxyribose of one deoxynucleotide with the number 5 carbon in deoxyribose of the next deoxynucleotide, which gives a deoxynucleotide chain polarity, described in shorthand as $3' \rightarrow 5'$ polarity. In a DNA duplex the two chains run in opposite directions, that is,

$$\begin{matrix} 3' \rightarrow 5' \\ 5' \leftarrow 3' \end{matrix}$$

5. Seventy-five percent of the DNA would contain one ^{15}N and one ^{14}N chain and have intermediate density. In twenty-five percent both chains would contain ^{14}N and be light.

6. The 3′OH end.

7. DNA polymerase cannot initiate polymerization of deoxynucleotides. RNA primers serve as starting points for DNA polymerase.

8. Topoisomerases relieve torsion generated by unwinding of the DNA duplex and allow intertwined daughter duplexes to separate at the end of replication.

9. Prophase. Chromosomes coil into condensed rods.
   Metaphase. Chromosomes are aligned on the metaphase plate in the center of the mitotic spindle.
   Anaphase. Chromatids separate and the daughter chromosomes move to opposite ends of the mitotic spindle.
   Telophase. Chromosomes uncoil and assume their interphase configurations.

10. The number of chromosomes, the chromosome lengths, the positions of the centromeres, and the chromosomal banding patterns.

11. Germ cells that have completed at least the first meiotic division.

12. In mitotic metaphase each duplicated chromosome lines up separately in the mitotic spindle. In the first meiotic metaphase, homologous chromosomes align with one another to form pairs in the meiotic spindle.

13. Meiosis reduces the chromosome number from diploid to haploid.

14. A pea plant heterozygous for yellow, smooth seeds would have the genotype *YySs*. A pea plant bearing green, wrinkled seeds would have to be homozygous for the two traits and have the genotype *yyss*. Since genes for the two traits are unlinked, a cross between these plants is described as,

		Green, wrinkled parent *yyss*	
	Gametes→ ↓	ys	Phenotypes of the F_1 generation ↓
Yellow, smooth parent *YySs*	YS	YySs	yellow, smooth
	Ys	Yyss	yellow, wrinkled
	yS	yySs	green, smooth
	ys	yyss	green, wrinkled

Phenotype ratios of the F_1 generation 1:1:1:1

15. A female fruit fly homozygous for vermilion eyes and miniature wings would have the genotype *rrnn*. A male fly with vermilion eyes and normal wings would be *rN*. A cross between them would produce the following offspring,

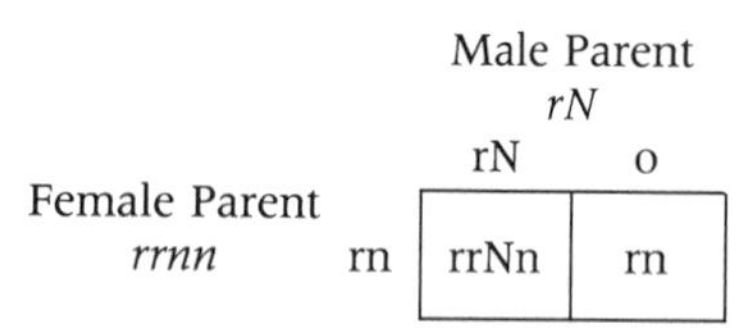

All F_1 females would have vermilion eyes and normal wings. All F_1 males would have vermilion eyes and miniature wings. Crossingover would have no effect on the F_1 phenotypes.

16.

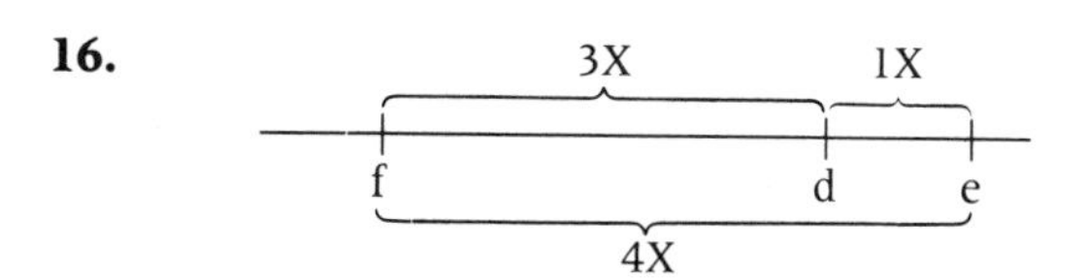

17. Mitochondria in nonphotosynthetic eukaryotes, and mitochondria and chloroplasts in photosynthetic eukaryotes.

CHAPTER 8

1. Incorporation of radioactive amino acids into proteins by enucleated cells demonstrated by autoradiography.

2. Incorporation of tritiated nucleosides into RNA was shown by autoradiography to occur in nucleated cells but not in enucleated cells. In nucleated cells radioactivity is initially found only in nuclei. With time radioactive RNA disappears in the nucleus and appears in the cytoplasm.

3. The prokaryotic promoter binds RNA polymerase.

4. RNA polymerase catalyzes addition of nucleotides to the end of an RNA chain in the $5' \rightarrow 3'$ direction along a $3' \rightarrow 5'$ DNA template. Since the chains in a DNA duplex have opposite polarities, RNA chains must be copied in opposite directions from the two DNA chains.

5. Transfer RNA and aminoacyl-tRNA synthetase.

6. Ribosomes have binding sites for mRNA and tRNA that ensure correct alignment of activated amino acids with the end of a growing polypeptide chain.

7. The 64 possible triplet codons were synthesized. Determining which aminoacyl-tRNA was caused to bind to ribosomes by each triplet allowed the assignment of amino acids to 61 of the 64 triplets.

8. *Wobble* refers to flexibility in the orientation of the first base in an anticodon such that it can hydrogen-bond with more than one kind of base in the third position of a complementary codon. For example a single tRNA, if it contains inosine in the first position of its anticodon, can recognize three different codons.

9. Missense mutation. Changing one or more nucleotides in a codon so that the codon now stands for a different amino acid.
Nonsense mutation. A change that alters a codon from specifying an amino acid to specifying termination; for example, changing the lysine codon AAA to UAA.
Frameshift mutation. Insertion or deletion of one or more base pairs in a gene so as to cause a shift in the reading frame of all triplet codons coming after the change.

10. Intragenic suppression. A second mutation in a gene restores, at least partially, the function of the gene product. For example, a frameshift mutation caused by deletion of a base pair may be corrected by insertion of a base pair near the deletion. A missense mutation may be suppressed by another missense mutation that restores correct folding of the polypeptide chain; that is, the first amino acid change is counteracted by a second amino acid change.
Intergenic suppression. Mutation in one gene may be counteracted by a mutation in a second gene. For example, a nonsense mutation in a gene may be at least partially negated by mutating a

tRNA gene so that it recognizes the termination codon of the nonsense mutation and prevents premature termination by adding an amino acid, allowing the chain to be completed.

11. The P site holds a tRNA molecule to which the growing polypeptide chain is attached. The A site binds incoming aminoacyl-tRNA molecules.

12. Eukaryotic mRNA molecules have a 5′ cap and a 3′ poly(A) tail. Prokaryotic mRNA molecules have neither.

13. Primary transcripts in eukaryotes generally contain introns and lack a 3′ poly(A) tail. Eukaryotic mRNA molecules lack introns and have a 3′ poly(A) tail.

14. Excision of introns must be precise to prevent disruption of the triplet codon reading frame in an mRNA molecule.

15. In the nucleolus.

16. Most mitochondrial proteins encoded by nuclear genes are guided into mitochondria by a leader sequence in the polypeptide chain of the protein. The leader sequence is thought to recognize a receptor protein on the surface of the mitochondrion.

17. Proteins destined for the plasma membrane are synthesized on the RER and transferred into the cisternae of the RER by means of a signal sequence, a signal recognition particle, and a docking protein. In the cisternae the proteins are glycosylated and carried by vesicles to the *cis* face of the Golgi complex. The sugar groups on the proteins are modified in the Golgi complex. The proteins are transferred to the plasma membrane as integral membrane proteins in the membranes of vesicles that bud from the *trans* face of the Golgi complex. Insertion of the proteins into the plasma membrane occurs by fusion of the vesicles with the plasma membrane.

CHAPTER 9

1. Different proteins are needed in the cell in different amounts, and the required amount of a protein may vary with cell activity. Regulation of synthesis of individual proteins permits variable requirements to be met with efficiency in use of resources.

2. By regulation of transcription and regulation of translation.

3. Reversibility of repressor binding must be reversible in order for transcription of an operon to be turned on.

4. The small amount of lactose permease present in the uninduced cell transports some lactose molecules into the cell. Some of the intracellular lactose is converted to allolactose. Allo-lactose binds to the lactose repressor proteins, preventing their binding to the operator of the *lac* operon. Unblocking the operator allows RNA polymerase to transcribe the operon.

5. If no β-galactosidase molecules were present in the uninduced cell, no lactose molecules could be converted to allo-lactose (the inducer) upon addition of lactose to the medium, and the *lac* operon could not be induced.

6. When glucose is present in the cell the concentration of cAMP remains low. When the concentration of cAMP is low, CRP cannot bind to the promoter of the *lac* operon. Without CRP protein bound to the promoter, the transcription of the *lac* operon remains very low.

7. Genes encoding catabolic enzymes are repressed when the catabolite, for example lactose, is *absent*. Genes encoding anabolic enzymes are repressed when the end product of the anabolic pathway, for example tryptophan, is *present*.

8. As soon as transcription starts at the 5′ end of the leader-attenuator, a ribosome binds to the partial transcript and begins translation. If the concentration of tryptophan is low, the ribosome stalls at the several tryptophan codons in the transcript of the leader-attenuator. This prevents segment 1 of the transcript from base-pairing with segment 2. Therefore, segment 2 can base-pair with segment 3. This prevents a termination signal being formed by base-pairing of segment 3 with segment 4, and the entire operon is transcribed. When the concentration of tryptophan is high, ribosome stalling does not occur, and the ribosome quickly reaches segment 2, preventing segment 2 from base-pairing with segment 3. Therefore, segment 3 can base-pair with segment 4, forming a termination signal for transcription. Transcription is aborted and the operon is not transcribed.

9. Ribosomal subunits and not rRNA molecules are the regulating elements in autoregulation of rRNA transcription.

10. Introduction of extra genes encoding r-proteins is followed by an increase in the synthesis of mRNA for r-proteins, but production of r-proteins does not increase, showing that increase in mRNA is accompanied by a lower rate of translation of mRNA molecules.

CHAPTER 10

1. Binding of a transcription factor to the TATA box directs RNA polymerase to the correct initiation site.

2. A TATA box, CAAT box, and GC box.

3. An enhancer is equally effective in either orientation of its sequence relative to the coding sequence it affects. An enhancer may be located upstream from the promoter or downstream from the 3′ end of a gene. Some genes lack enhancers, some may have several. These three properties are not shared with promoters.

4. See Figure 10-3.

5. The gene for coat color is carried in the X chromosome. Heterozygous females have a mixture of yellow and black patches of fur because inactivation of one or another of the two X chromosomes is random. Males have only one X chromosome and can only have all black or all yellow fur.

6. In experiments with chromatin, genes that have switched from an inactive to an active state become sensitive to digestion by DNase, particularly in 5′ regulatory regions.

7. In the absence of galactose the protein encoded by *GAL80* blocks the binding of the *GAL4* protein to the UASs of the four genes that encode galactose-metabolizing enzymes. The *GAL4* protein is a positively acting transcription factor whose binding to UASs is required for transcription of the galactose genes.

8. Histone genes are expressed primarily during the S phase of the cell cycle.

9. In cells treated with poisons that prevent assembly of tubulin into microtubules, free tubulin accumulates. Although transcription of tubulin genes continues, tubulin mRNA is rapidly destroyed, and tubulin synthesis is curtailed, indicating that free tubulin autoregulates synthesis of tubulin.

10. In the absence of light, expression of genes encoding many products required for the function and structure of chloroplasts is turned off. This avoids waste of resources in maintaining a structure that does not function in the absence of light.

11. Carrot plants can form from a single carrot cell derived from a carrot. Tadpoles develop from an ovum whose nucleus has been replaced by the nucleus from an erythrocyte.

12. Different kinds of cells, for example brain cells and liver cells, contain some mRNA molecules that are transcripts of different genes.

13. In *Xenopus* estrogen secreted by the ovaries binds to estrogen receptor proteins in liver cells and activates expression of the gene encoding the protein vitellogenin. Vitellogenin is released into the blood by liver cells and is taken up by growing oocytes in the ovaries.

14. Puffing of a band is accompanied by a great increase in RNA synthesis in the band, as demonstrated by autoradiography.

15. Fibronectin contains several functional domains. Alternative splicing of primary transcripts of the fibronectin gene yields different mRNA molecules encoding fibronectin molecules with different combinations of domains, thus providing fibronectin molecules with differences in functional properties.

16. In the construction of an immunoglobulin gene in a mammal, any one of 300 different V segments may be joined to any one of four different J segments. Therefore, the 300 V and four J segments can be joined in 1200 different combinations. Splicing of the V and J segments can occur in DNA at any site in a single triplet codon. Because the genetic code is degenerate, this alternative splicing generates an average of 2.5 different amino acids, increasing the possible number of V-J segments to 3000. L and H chains are assembled in this way, but an additional segment called D contributes to formation of the variable region of an H chain. More than 5000 different VDJ combinations are possible for H chains. Since an immunoglobulin molecule contains two identical H chains and two identical L chains, about 3000×5000, or 1.5×10^7 different immunoglobulin molecules are possible. The number is increased further by frequent mutations in V segments.

CHAPTER 11

1. Two $\times 10^6$ molecular weight units of DNA equals 3000 bp. Therefore, 4×10^9 molecular weight units equals 6×10^6 bp.

2. Negative supercoiling is produced by underwinding of the two chains of deoxynucleotides in DNA relative to each other. Positive supercoiling is produced by overwinding (see Figure 11-9).

3. DNA polymerase cannot catalyze initiation of a deoxynucleotide chain but can add deoxynucleotides to the 3′OH end of a ribonucleotide or deoxyribonucleotide chain. RNA polymerase can catalyze initiation of a ribonucleotide chain. Therefore, DNA replication must begin with formation of an RNA primer.

4. Since DNA polymerase can catalyze addition of deoxynucleotides in only the 3′OH direction on a DNA template, replication in the 5′ direction must be accomplished by repeated initiations with RNA primers as shown in Figure 11-11.

5. *E. coli* were allowed to initiate DNA replication in the presence of low specific activity ^{3}H-thymine and then switched to high specific activity ^{3}H-thymidine. The autoradiographic pattern of isolated DNA showed that DNA replication proceeds bidirectionally from a single origin.

6. Each replication fork replicates about half of the DNA molecule in *E. coli*, or about 2.35×10^6 bp. Therefore each of the two growing chains at each fork is extended by 2.35×10^6 deoxynucleotides in 40 minutes, or 980 deoxynucleotides per second.

7. i) All prokaryotes have a single chromosome. Almost all eukaryotes have more than one chromosome.
ii) The total amount of DNA in eukaryotic genomes is a few times to tens of thousands of times greater than in prokaryotic genomes.
iii) The chromosome in prokaryotes is always a covalently closed loop. DNA molecules in the nuclei of eukaryotes are linear.
iv) The DNA of most eukaryotic genomes contains nongene sequences that are repeated many times.
v) Eukaryotic genomes contain pseudogenes.

vi) The linear DNA molecules in eukaryotes terminate in special sequences called telomeres. Prokaryotic chromosomes have no ends.
vii) Eukaryotic DNA molecules contain centromeric sequences. Prokaryotic chromosomes do not.
viii) Eukaryotic chromosomes have many origins of replication. Prokaryotic chromosomes possess a single origin.
ix) Eukaryotic DNA forms a complex with histones, making nucleosomes. Prokaryotic DNA is complexed with a different kind of protein.
x) Many eukaryotic genes contain introns. Introns are rare in prokaryotic genomes.
xi) Much of the DNA in eukaryotes forms long spacers between genes. Little of the DNA in prokaryotes forms spacers, and spacers are short.

8. Satellite DNA consists of a deoxynucleotide sequence repeated many times. If the repeated sequence has an AT:GC ratio that is different from the main body of DNA, it forms a separate component called a satellite in a CsCl gradient.

9. See Figure 11-24, page 375.

10. Pseudogenes that lack promoters and introns are believed to have formed by reverse transcription of mRNA into DNA.

11. Total complexity of a genome is measured by determining the rate of reassociation of denatured DNA.

12. The S-value paradox is believed to stem largely from different amounts of DNA sequences present in spacers between genes in different species of organisms. Differences in intron sizes also contribute to the paradox.

13. Genes are separated by long untranscribed, noncoding DNA spacers. Spacers are believed to account for most of the DNA sequence complexity in most eukaryotes.

14. Two molecules each of four different histones aggregate to form a particle around which DNA is wound to form a nucleosome. The nucleosome represents the first level of DNA packing.

15. Satellite DNA was separated from the rest of the DNA in the genome. Radioactive transcripts of the isolated satellite DNA were made with RNA polymerase. Using autoradiography the radioactive transcripts were found to hybridize only to centromeric regions of mouse chromosomes.

16. In most eukaryotes mitochondrial DNA makes up less than one percent of the total cellular DNA.

CHAPTER 12

1. Approximately 3×10^{21} cells.

2. Tritiated thymidine was added to a proliferating population of cells for 10 minutes, and the cells were examined by autoradiography. None of the cells in mitosis were observed to be labeled, which means that DNA replication does not occur during mitosis.

3. Length of the S period = fraction of cells in DNA replication $\times$ generation time.
Length of the S period = 0.5×16 hours = 8 hours.

4. The rate of cell reproduction is regulated by blocking progression through the G1 period.

5. Fusion of a cell in metaphase with a cell in interphase caused the chromosomes in the interphase nucleus to undergo mitotic condensation, demonstrating a condensation factor (MPF) in the cytoplasm of a mitotic cell.

6. During mitosis transcription stops.

7. Microtubules.

8. Microtubules attach to mitotic chromosomes by attaching to the kinetochore.

9. Cytokinesis is achieved in plant cells by construction of a cell wall between the daughter cells and formation of new plasma membrane from vesicles of the phragmoplast. In animal cells cytokinesis occurs by a pinching of the cell in two by means of a contractile ring of actin microfilaments and associated myosin molecules.

10. The Golgi complex provides vesicles from which each daughter cell forms new plasma membrane in the plane between the daughter cells.

11. Thymidine is converted to TTP. A high concentration of TTP blocks synthesis of dCTP, without which DNA cannot replicate. Blocking DNA replication arrests cells at the G1/S border. When these cells are released from the block, they proceed through the cycle in synchrony.

12. Induction of a Ts mutation in a gene required for some event in the cell cycle causes cells to be blocked at the time of that event, thus identifying a cell-cycle gene.

13. Opposite mating types of yeast produce mating factors that arrest each other's cell cycle at Start in the G1 period.

14. Normal cells in culture exhibit,
i) anchorage dependence
ii) density-dependent inhibition of growth
iii) limited life span
iv) requirement for serum growth factors

15. Some transmembrane proteins serve as receptors for growth factor molecules. Binding of a growth factor molecule to the receptor site on the cell surface causes activation of an enzyme activity in the portion of the receptor molecule on the cytoplasmic side of the membrane. The enzyme activity initiates a chain of events that releases the cell from G0 and allows it to reproduce.

CHAPTER 13

1. An increase in the number of people over 65 years of age and increases in exposure to environmental and life-style carcinogens.

2. On the average, many years of exposure to carcinogens are required to evoke cancer. The disease is therefore often delayed to the 60s and beyond.

3. i) Defective regulation of cell reproduction.
ii) Immortality.
iii) Defective differentiation.

4. Surgery, radiation, and chemotherapy. Surgery and radiation usually cannot deal effectively with metastases. Chemotherapy is limited by toxic effects of the drugs on normal cells.

5. Radiation, chemicals, and viruses.

6. Ultraviolet rays of sunlight.
Tobacco smoke.
HLTV-I.

7. Inexpensive, quick, and sensitive.

8. Oral, throat, esophageal, kidney, bladder, and pancreatic cancers.

9. A carcinogen produced by a mold. Liver cancer.

10. Epidemiological studies of human populations worldwide show a strong correlation between dietary fat and several major kinds of cancer. Laboratory animals fed diets rich in fat have higher rates of several kinds of cancers.

11. Cancer cells lack density-dependent inhibition of reproduction, are anchorage-independent, have low or no requirements for serum growth factors, and in most cases cannot enter G0.

12. Viral oncogenes are believed to be derived from normal cellular genes.

13. HLTV-I (adult T-cell leukemia), hepatitis B virus (liver cancer), and Epstein-Barr virus (Burkitt's lymphoma).

14. One reason is that the virus mutates readily.

15. Reverse transcriptase is required to make a DNA copy of the RNA chromosome of the virus prior to its integration into a DNA molecule in a cellular chromosome.

16. By the transfection of DNA from human bladder cancer cells into mouse cells.

17. By mutation, by amplification, or by overexpression.

CHAPTER 14

1. i) Actin-myosin interaction in muscle.
ii) Dynein-microtubule interaction in cilia and flagella.
iii) Kinesin-microtubule interaction in fast axonal transport.

2. Crossbridges made by heads of myosin molecules making contact with actin thin filaments generate a sliding motion between thick and thin filaments.

3. Thick filaments composed of myosin.
Thin filaments composed of actin.

4. A myofibril is a contractile structure composed of a repeating series of thick and thin filaments. One repeat, consisting of two groups of thin filaments interdigitated with one central group of thick filaments, constitutes a sarcomere.

5. Myosin heads project radially from a thick filament in a hexagonal pattern to interact with six surrounding actin thin filaments.

6. Tropomyosin and troponin bound to thin filaments regulate contraction by regulating the interaction of thick and thin filaments.

7. Purified heads of myosin molecules bind to actin thin filaments in a highly regular chevron pattern with points of the chevrons all directed toward one end of the filament.

8. The sarcolemma T system.

9. A rise in Ca^{2+} ion concentration caused by arrival of a nerve impulse results in binding a Ca^{2+} ion to troponin. Ca^{2+} ion binding causes a configurational change in troponin, which then draws tropomyosin molecules deeper into the groove of the actin thin filament. This unmasks sites on thin filaments to which myosin heads can bind and bring about contraction.

10. A gelled cortical region of cytoplasm (the ectoplasm), containing actin microfilaments and myosin, is thought to contract and squeeze the central core of fluid endoplasm forward. At the forward end of the amoeba endoplasm becomes ectoplasm, and at the tail end, ectoplasm is continuously converted into endoplasm.

11. An algal cell can be broken open to expose a layer of bundles of actin microfilaments on the inner side of the cell cortex. Plastic beads coated with myosin applied to the layer of microfilaments move along the cortex in the direction predicted by cytoplasmic streaming previously observed in the intact cell.

12. Dynein arms extending from one microtubule doublet make contact with an adjacent doublet and produce a sliding movement of one doublet along the other.

13. If all the doublets worked in the same mode, the cilium would not perform a beating motion.

14. By increasing the time of straight swimming between tumbling episodes in response to environmental signaling, bacteria move up a concentration gradient of useful substances and away from toxic concentrations of other substances.

CHAPTER 15

1. Meiosis, ability to fuse with a sperm cell, storage of large amounts of reserve materials including lipids, glycogen, proteins, ribosomes, mRNA, and tRNA molecules.

2. The acrosome contains enzymes important for penetration of the sperm head through the surface coats of the egg.

3. Cells are shed from tips of villi. New cells are produced in the crypts.

4. Microvilli enormously increase the surface area.

5. Tight junctions prevent diffusion of molecules or ions between cells. Belt desmosomes provide great mechanical strength to the epithelial sheet. Spot desmosomes also provide mechanical strength to the epithelial sheet by interlinking cytoskeletons of adjacent cells.

6. Glucose is actively transported into an epithelial cell from the intestinal lumen and is exported into the tissue fluid at the basolateral plasma membrane by facilitated diffusion.

7. The pump exports three Na^+ ions for every two K^+ ions it imports. This pumping action, together with differences in permeability of the plasma membrane to Na^+ and K^+ ions, generates an electrical potential across the membrane.

8. The membrane is depolarized by an inrush of Na^+ ions lasting less than one millisecond and becomes transiently polarized in the opposite direction by the presence of a higher concentration of positive charges at the inside surface than at the outside. The inrush of Na^+ ions is immediately followed by a transient efflux of K^+ ions, which reestablishes the original polarity across the membrane. Depolarization at one point in the membrane stimulates depolarization of adjacent parts of the membrane, producing conduction of an impulse.

9. The myelin sheath is produced by enormous infolded extensions of the plasma membrane of a Schwann cell to form many membrane layers that wrap around a nerve axon. The sheath insulates the axon.

10. By release at a synapse of neurotransmitter molecules, which diffuse across the synaptic cleft and stimulate the plasma membrane of the adjacent cell.

11. Neurotransmitter molecules are released by exocytosis at a synapse. Membrane is retrieved from the plasma membrane into cytoplasmic vesicles by endocytosis.

12. Rod cells, which are responsible for black and white vision, are more sensitive to light than are cone cells (responsible for color vision) and rod cells therefore respond to dim light.

13. Vitamin A is a required precursor for synthesis of the visual pigment rhodopsin in rod cells.

14. Genes required for color vision are carried in the X chromosome. Hence, recessive mutations in such genes are manifested in males but masked in heterozygous females.

15. i) B lymphocytes produce and secrete antibodies. ii) T lymphocytes are responsible for cell-mediated responses of the immune system, destroying foreign cells and virus-infected cells by direct contact. iii) Natural killer cells attach to and lyse cancer cells and virus-infected cells.

16. See Figures 10-26, page 347, and 15-32, page 530.

17. Autoimmune diseases such as myasthenia gravis, rheumatoid arthritis, glomerular nephritis, and lupus erythematosis.

18. By fusion of a myeloma cell with a normal B lymphocyte.

19. Cancer cells may arise frequently in an individual, but they are recognized and killed by lymphocytes.

CHAPTER 16

1. From airborne spores and cysts.

2. The experiment of Pasteur in which boiled nutrient broth was protected from airborne contaminants by an S-shaped neck.

3. 4.6×10^9 years. About 3.5×10^9 years ago.

4. Biologically important molecules are believed to have formed in bodies of water before the origin of life. Boiling a mixture of H_2O, H_2, CH_4, and NH_3 and exposing the vapor to electrical spark dis-

charges results in formation of substantial amounts of biologically important molecules.

5. Rocks formed during early Precambrian times contain ferrous iron, which is unstable in the presence of oxygen. Therefore, oxygen cannot have been present when these rocks were formed.

6. Radioastronomy has shown that great quantities of biologically important molecules occur commonly in the universe outside the Earth.

7. Polyphosphates.

8. Polypeptides and oligonucleotide chains.

9. Only nucleic acids are known to have properties necessary for self-replication.

10. Excision of the intron in rRNA showed that RNA can act as an enzyme. This suggests the possibility that the first cells might have used RNA both for carrying genetic information and for catalyzing reactions, including replication of the RNA itself.

11. The first cells probably were heterotrophs because they lacked many anabolic capabilities.

12. All the oxygen in the atmosphere is believed to derive from photosynthesis. Respiration requires a source of oxygen.

GLOSSARY

A-band A band in a sarcomere of a skeletal or cardiac muscle cell, characterized by the presence of thick myofilaments.

A site One of two tRNA binding sites on a ribosome.

Acceptor stem The 3′OH end of a tRNA molecule, which accepts an amino acid in charging of a tRNA.

Acetyl-CoA Coenzyme A carrying the two-carbon fragment acetyl.

Acid A substance that increases the concentration of H^+ ions in a solution.

Acrosome An organelle in a sperm cell that develops from vesicles of the Golgi complex and comes to lie at the tip of the head of the sperm.

Actin The protein that makes up microfilaments.

Actin-binding protein A protein that cross-links microfilaments in orthogonal arrays.

Actinomycin D A mold poison that inhibits RNA polymerase.

Action potential A propagated wave of localized electrochemical change that spreads along the plasma membrane of an axon.

Active site A small area on the surface of an enzyme, often in the shape of a crevice, to which the substrate of the enzyme is bound and at which catalysis occurs.

Active transport Transport of molecules or ions across a membrane by carrier proteins in the membrane, requiring the expenditure of energy and capable of transport against a concentration difference.

Adenine A purine present in RNA, DNA, and ATP.

Adenosine A pyrimidine, consisting of adenine and ribose.

Adenosine monophosphate (AMP) A nucleotide consisting of adenine, ribose, and phosphate.

Adenosine triphosphate (ATP) A molecule used by cells for transfer of energy from energy-yielding to energy-requiring molecular events.

Adenoviruses A family of DNA-containing viruses, some of which are oncogenic.

Adenylate kinase An enzyme that catalyzes the synthesis of cAMP.

Adhesion plaque Points of attachment of animal cells to a substratum.

Adipocyte A cell specialized to store fat.

Aerobe An organism that can live only in the presence of oxygen.

Aflatoxin A carcinogenic substance produced by the mold *Aspergillus flavus*.

Allele Either of the two genes in homologous chromosomes encoding a particular product in a diploid cell.

Allo-lactose An alternate form of lactose, which is the immediate inducer of the *lac* operon.

Allosteric enzyme An enzyme whose activity is regulated by binding of an effector molecule to a site other than the active site of the enzyme.

α-actinin A microfilament-associated protein found at attachment sites of microfilaments to the cytoplasmic surface of the plasma membrane. Also a prominent constituent of the Z line in skeletal and cardiac muscle cells.

α-globin One of the two kinds of polypeptides in hemoglobin.

α-helix A helical form of secondary structure of a polypeptide chain, stabilized by intramolecular hydrogen bonds.

Ames test A test to determine mutagenicity of a chemical using bacteria.

Amino acid The monomer polymerized to make proteins.

Aminoacyl-tRNA A tRNA that has been charged with an amino acid.

Aminoacyl-tRNA synthetase An enzyme that catalyzes charging of a tRNA molecule with its proper amino acid.

Amino terminus The terminus of a polypeptide chain that has an amino group.

Amoeba proteus A large, freshwater amoeba.

Amoeboid movement Movement of a cell across a surface by flowing or streaming of the cytoplasm.

Amphipathic molecule A molecule with both polar and nonpolar regions and therefore with one part soluble in water and the other part insoluble in water.

Anabolism Synthesis of molecules from simpler precursors.

Anaerobe An organism capable of living in the absence of oxygen.

Anaphase The third part of mitosis, in which daughter chromosomes are separated into two groups at opposite ends of the mitotic spindle.

Anchorage dependence The requirement of normal cells in culture to be attached to a surface in order to survive and reproduce.

Aneuploidy An abnormal number of chromosomes in a cell.

Anion A molecule or ion with a negative charge.

Antibody A protein made by B lymphocytes with two identical surface sites that are complementary to a specific antigenic site on another molecule and can bind to the antigenic site.

Anticodon Three bases in the anticodon loop of a tRNA molecule that are complementary to a codon in mRNA.

Anticodon loop A loop of seven nucleotides in tRNA that contains a triplet anticodon.

Antigen A molecule capable of eliciting synthesis of a specific antibody by a B lymphocyte, and which can bind specifically to that antibody.

Archaebacteria One of the two main groups of bacteria.

Autolysis Self-digestion, mediated by lysosomal enzymes.

Autonomously replicating sequence (ARS) Autonomously replicating sequence in DNA that probably represents an origin of DNA replication.

Autoradiography A technique for detection of radioactivity in cells or cell parts with a photographic film.

Autoregulation of genes A form of regulation in which the product encoded by a gene regulates expression of that gene.

Autosome Any chromosome that is not a sex chromosome.

Autotrophs Cells that can build all their organic molecules from carbon dioxide only.

Auxotrophic mutation A mutation that creates a dependency on an exogenous source for a particular substance.

Axon A long cytoplasmic extension of a nerve cell.

Axonal transport Movement of vesicles up or down a nerve axon.

Axoneme The structural core of a cilium or eukaryotic flagellum, made of microtubule doublets.

5-Azacytosine An analog of cytosine in which the number-5 carbon of cytosine is replaced by a nitrogen atom.

B lymphocyte or cell An antibody-secreting lymphocyte.

Back mutation A mutation that reverses a previous mutation.

Bacteriophage A virus that infects bacterial cells.

Balbiani ring A particularly large puff in a polytene chromosome of the midge *Chironomus*.

Barr body A body of heterochromatin formed by one of the X chromosomes in a female mammalian somatic cell.

Basal body A cytoplasmic organelle, made of microtubules, at the base of eukaryotic flagella and cilia.

Base A substance that decreases the concentration of H^+ ions in a solution.

Belt desmosome A type of junction between epithelial cells characterized by a bundle of microfilaments on the cytoplasmic side of the plasma membrane in the area of the junction.

β-galactosidase An enzyme that catalyzes hydrolysis of lactose into galactose and glucose.

β-globin One of the two kinds of polypeptides in the protein hemoglobin.

β-structure A planar form of secondary structure in a polypeptide, stabilized by intramolecular hydrogen bonds.

Bidirectional replication Replication of a DNA double helix by two replication forks moving in opposite directions from a single initiation point.

Blastomere A cell in a cleavage-stage embryo.

Blender experiment One of the early experiments, with bacteriophage, demonstrating that DNA is the genetic material.

Brush border The lumenal surface of intestinal epithelium, whose many microvilli give the impression of a brush when observed by microscopy.

C_4 plant A plant capable of survival in hot dry climates because of a specialized mechanism for capturing CO_2.

CAAT box A short DNA sequence forming part of the promoter in many eukaryotic genes.

Calico cat A female cat in which one X chromosome specifies black fur and the other X chromosome specifies yellow fur.

Cancer A disease that starts with a single cell characterized by defective regulation of cell reproduction, defective differentiation, and immortality.

5′ Cap A cap on the 5′ end of mRNA molecules in eukaryotes, consisting of several modified deoxynucleotides.

Carbon fixation cycle A series of reactions in phase II of photosynthesis by which carbon dioxide is reduced to sugar molecules.

Carboxyl terminus The terminus of a polypeptide chain that has a carboxyl group.

Carcinogen Any agent that causes cancer.

Catabolism Degradation of molecules into simpler molecules.

Catabolite repression Repression of a gene(s) encoding catabolic enzymes for a substrate by a second catabolic substrate, e.g., repression of the *lac* operon by glucose.

Catalase An enzyme that catalyzes breakdown of hydrogen peroxide to water and oxygen.

Catalysis Increasing the rate of a reaction by lowering the activation energy.

Cation A molecule or ion with a positive charge.

Cell cycle The procession of events that makes up cell reproduction.

Cell-cycle genes Genes that encode products needed to complete particular steps in the cell cycle.

Cell senescence A property of normal animal cells characterized by a limited ability to reproduce in culture—usually 50 to 100 cell doublings.

Cellular respiration Breakdown of pyruvate to carbon dioxide and water inside mitochondria, yielding 36 ATP molecules.

Cellulose A polymer consisting of glucose molecules linked by glycosidic linkage.

Cell wall A tough layer of material outside the plasma membrane of a bacterium, a fungal cell, an algal cell, or a plant cell, which provides mechanical protection.

Central vacuole A large, storage vacuole in many kinds of plant cells.

Centriole A structure containing nine short microtubule triplets. Identical in structure to a basal body of a cilium or eukaryotic flagellum.

Centromere A short section of a eukaryotic chromosome where chromatids remain attached in a metaphase chromosome.

Centrosome A structure that serves as a nucleating center for the formation of microtubules.

Centrosome-to-kinetochore microtubules Microtubules that extend from a centrosome to the kinetochores of mitotic chromosomes.

Chemiosmotic coupling hypothesis A hypothesis that explains the synthesis of ATP in chloroplasts and mitochondria by the creation of a proton gradient across membranes.

Chemotaxis Movement of cells up or down a concentration gradient of a chemical.

Chemotherapy Treatment of disease with a chemical agent, used particularly in cancer treatment.

Chemotroph Cells that must derive energy by breaking down organic molecules.

Chiasma A crossover point between two chromatids in paired, homologous chromosomes, seen during the diplotene stage of the first meiotic division.

Chloramphenicol Another name for chloromycetin, an antibiotic that inhibits peptidyl transferase in bacteria.

Chloromycetin An antibiotic that inhibits peptidyl transferase in bacteria. Same as chloramphenicol.

Chlorophyll A green pigment in photosynthetic cells that absorbs photons.

Chloroplast A membranous organelle that contains the molecular machinery for photosynthesis in algae and plants.

Chloroplast DNA Molecules of DNA present in chloroplasts, which encodes many of the gene products in a chloroplast.

Cholesterol A sterol lipid found in the plasma membrane of eukaryotes.

Chromatid One of the two elements in a duplicated chromosome.

Chromatin A complex of DNA and proteins.

Chromatophore A skin cell containing pigment granules.

Chromomeres Knotlike structures composed of DNA and protein occurring at intervals along a chromosome.

Chromosome A DNA molecule with proteins bound to it.

Chromosome banding pattern A pattern of transverse bands in a mitotic chromosome produced by several staining techniques.

Chromosome mapping Mapping the position of genes in a chromosome.

Cilium An organelle with a core of microtubules, used for cell motility.

Cistron A segment of DNA encoding a polypeptide chain.

Citric acid cycle The first part of cellular respiration, in which a cycle of reactions breaks down two-carbon organic fragments to carbon dioxide, yielding energized electrons and GTP.

Clathrin A protein that forms a regular network on the cytoplasmic side of the plasma membrane at coated pits and around endocytic vesicles derived from coated pits.

Clone A population of cells derived by cell division from a single progenitor cell.

Coated pit A small region on the surface of a cell in which receptor proteins are concentrated.

Codon A sequence of three nucleotides in mRNA that specifies a particular amino acid during polypeptide synthesis.

Coenzyme An organic molecule that serves as an intermediate carrier of a chemical group or electrons transferred in an enzyme-catalyzed reaction.

Coenzyme A A coenzyme that transfers two-carbon fragments.

Colchicine A poison that binds to tubulin and prevents its assembly into microtubules.

Collagen The main extracellular, structural protein of connective tissue.

Competitive inhibition Inhibition of an enzyme by an inhibitor that binds reversibly to the active site of the enzyme.

Cone cell A sensory neuron of the retina responsible for color vision.

Constitutive gene A gene that is continuously and always transcribed.

Constitutive heterochromatin Heterochromatin that does not revert to euchromatin.

Contractile ring A microfilament bundle that forms a ring at the equator of a cell and, by contracting, pinches the cell in two during cytokinesis.

Contractile vacuole An organelle that excretes water from a cell.

Corepressor A molecule, such as an amino acid, that by binding to a repressor protein enables the repressor to bind to its target operator.

Cot A term (molar concentration of nucleotides in DNA × time) used in describing the reassociation of denatured DNA.

Coupled reactions Intimate joining of two reactions, for example joining an energy-yielding reaction with an energy-requiring reaction such that energy yielded by the one is used to drive the other.

Coupled transcription-translation Translation at the 5′ end of an mRNA molecule while the mRNA is still being transcribed from DNA.

Crista A sheetlike or tubular fold of the inner membrane of a mitochondrion.

Crossing over Exchange of segments between homologous chromosomes.

C segment or region Constant region of a polypeptide in an antibody.

C-type viruses Retroviruses, many of which are oncogenic.

C-value paradox A paradox based on wide variations in the amount of DNA in closely related species.

Cyanobacteria Blue-green bacteria, which carry out photosynthesis.

Cyclic AMP (cAMP) Adenosine monophosphate in which the phosphate group on the number-5 carbon of ribose is also bound covalently to the number-3 carbon, forming a ring.

Cyclic AMP-dependent kinase A protein kinase that is activated by cAMP.

Cyclic AMP-receptor protein A protein that, when complexed with cAMP, binds to a promoter and enhances transcription of an operon.

Cycloheximide A drug that blocks peptidyl transferase in eukaryotes, and which is used as a chemotherapeutic drug in treatment of cancer.

Cyclosis Streaming of cytoplasm in a cyclic pattern in plant and algal cells.

Cytidine A nucleoside consisting of cytosine and ribose.

Cytidine monophosphate (CMP) A nucleotide consisting of cytosine, ribose, and phosphate.

^{3}H-Cytidine A tritium-containing pyrimidine precursor of RNA.

Cytochalasin A mold poison that prevents elongation of microfilaments.

Cytochrome An iron-containing protein that forms part of the chain of electron carriers in oxidative phosphorylation and in photosynthesis.

Cytokinesis Division of the cytoplasm during cell division.

Cytoplast An enucleated animal cell.

Cytosine A pyrimidine present in DNA and RNA.

Cytoskeleton The structural framework of the cytoplasm, composed of microtubules, microfilaments, and intermediate filaments.

Cytosol The solution phase of the cytoplasm.

Decline phase The final period of a cell culture, in which cells die.

Degeneracy of the genetic code The circumstance whereby most of the 20 amino acids are prescribed by more than one triplet codon.

Dehydration-condensation reaction Joining of two molecules accompanied by the release of a water molecule.

Dendrite Short cytoplasmic extension of a nerve cell.

Density-dependent inhibition of cell reproduction The property of normal cells in culture to stop reproducing when they have formed a confluent monolayer.

Deoxyadenosine A deoxynucleoside consisting of adenine and deoxyribose.

Deoxyadenosine monophosphate (dAMP) A deoxynucleotide consisting of adenine, deoxyribose, and phosphate.

Deoxycytidine A deoxynucleoside consisting of cytosine and deoxyribose.

Deoxycytidine monophosphate (dCMP) A deoxynucleotide consisting of cytosine, deoxyribose, and phosphate.

Deoxyguanosine A deoxynucleoside consisting of guanine and deoxyribose.

Deoxyguanosine monophosphate (dGMP) A deoxynucleoside consisting of guanine, deoxyribose, and phosphate.

Deoxynucleoside A molecule consisting of a nitrogen base and deoxyribose.

Deoxynucleotide A molecule consisting of a nitrogen base, a deoxyribose molecule, and a phosphate group.

Deoxyribonucleic acid (DNA) The genetic material of cells.

Deoxyribose A 5-carbon sugar present in DNA.

Desmin filaments Intermediate filaments composed of the protein desmin, found primarily in muscle.

Differential centrifugation Centrifugation at successively higher speeds to sediment successively lighter cell parts in a cell homogenate.

Differentiation A specialization in molecular and usually structural makeup of a cell that specializes it for performance of a particular function.

Diffusion Movement of a substance from a region of high concentration to a region of low concentration.

Diploidy The presence of two of each of the chromosomes.

Diptera An order of insects, including flies and midges.

Disulfide bond A bond formed between two SH groups, for example, the SH groups in the side chains of cysteine molecules.

DNA (deoxyribonucleic acid) A polymer consisting of two chains of deoxynucleotides wound around each other to form a double helix.

DNA denaturation Separation of the chains in a DNA double helix by breaking the hydrogen bonds between complementary bases.

DNA gyrase An enzyme that catalyzes introduction of negative supercoiling into a DNA double helix.

DNA polymerase An enzyme that catalyzes addition of deoxynucleotides to the 3′OH end of a DNA chain.

DNA polymerase γ A DNA polymerase present in mitochondria.

DNase An enzyme that degrades DNA by cutting phosphodiester bonds.

DNase hypersensitivity Increased digestability of DNA in chromatin in a transcription-permissive state.

DNA spacers Stretches of noncoding DNA that separate genes in a chromosome.

DNA transfection Uptake by eukaryotic cells of DNA added to the medium.

Docking protein or SRP receptor An integral protein of the rough endoplasmic reticulum (RER) that binds a signal receptor protein to the membrane.

Double thymidine block A technique for synchronizing the cell cycle in a population of cells using a high concentration of thymidine to block DNA replication.

Drosophila A fruit fly.

Dynein A protein that is part of the force-generating mechanism in cilia and eukaryotic flagella.

Ecdysone An insect hormone that activates transcription of certain genes in development.

Effector neurons Nerve cells that carry signals from the brain to muscles.

Electrochemical gradient An interconnected difference in electrical charge and chemical concentration, for example, the electrochemical gradient across the plasma membrane in nerve cells.

Electrochemical proton gradient A gradient of protons across the inner membrane of a mitochondrion or the thylakoid membrane of a chloroplast.

Elongation phase The phase in translation during which the polypeptide is elongated by addition of amino acids.

Endocytosis Uptake of macromolecules bound to the outer surface of the plasma membrane by formation of vesicles from the plasma membrane. Endocytosis = pinocytosis and phagocytosis.

Endoplasmic reticulum A series of membranous vesicles, usually in the form of flattened sacs or tubules, with several metabolic functions.

Energy charge The relative proportions of ATP, ADP, and AMP in a cell.

$$\text{Energy charge} = \frac{[\text{ATP}] + \frac{1}{2}[\text{ADP}]}{[\text{ATP}] + [\text{ADP}] + [\text{AMP}]}$$

Enhancer sequence A sequence of 50 to 200 bp in DNA that functions in regulation of transcription of many eukaryotic genes.

Enzyme A catalyst, almost always a protein molecule but in a few cases an RNA molecule.

Epidermal growth factor (EGF) A protein, present in blood, that is required by some animal cell types in order to reproduce.

Epidermal growth factor receptor (EGF receptor) A transmembrane protein in some kinds of animal cells that works as a protein kinase in its cytoplasmic domain when activated by binding of EGF in its cell surface domain.

Epidermis An outer layer of cells of an organism, for example, the outer layer of skin cells.

Epstein-Barr virus (EBV) A herpes virus that causes infectious mononucleosis and is a causative agent of Burkitt's lymphoma.

Erythrocyte A red blood cell.

Erythromycin An antibiotic that binds to free 50S ribosomal subunits and prevents their attachment to 30S subunits.

Erythropoietin A protein, synthesized in kidney cells, that stimulates reproduction of red blood cell precursors in bone marrow.

E. coli A common species of bacteria found in the large intestine of mammals.

Etiolation Loss of chlorophyll and other parts of the photosynthetic machinery that occurs in plants and some algae kept in darkness.

Eubacteria One of the two main groups of bacteria. The eubacteria are most of the common bacteria.

Euchromatin Loosely packed chromatin; active in transcription.

Eukaryote Cells characterized by the presence of a nucleus defined by a nuclear envelope.

Exocytosis Export from a cell of bulk amounts of materials.

Exon A polypeptide-encoding segment of a gene.

Exon shuffling Recombination of exons in DNA to produce new genes, as in formation of immunoglobulin genes.

Exponential phase A period of time during which the cell number in a culture increases exponentially. Same as log phase.

Extracellular matrix Proteins and polysaccharides forming a matrix outside of cells.

Facilitated diffusion Transport of molecules or ions across a membrane by carrier proteins in the membrane, driven by a concentration difference for the substance on the two sides of the membrane.

Facultative anaerobe An organism that can live in the presence or absence of oxygen.

Facultative heterochromatin Heterochromatin that can revert to euchromatin and back to heterochromatin.

Fast axonal transport Rapid movement of vesicles in nerve axons.

Fatty acid A hydrocarbon chain with a carboxyl group at one end.

Feedback inhibition Inhibition of the first enzyme in an enzymatic pathway by the final product of the pathway in which the product binds away from the active site of the enzyme.

Ferritin An iron storage protein consisting of 24 identical polypeptides and about 2000 atoms of Fe^{+}.

Fibroblast A cell of connective tissue that synthesizes and releases collagen.

Fibronectin A glycoprotein of the extracellular matrix.

Fimbrin A microfilament cross-linking protein.

Flagellin A protein from which bacterial flagella are assembled.

Flagellum An organelle for propulsion of a cell.

Flavin adenine dinucleotide (FAD) A coenzyme that carries electrons from one chemical reaction to another.

Flow microfluorimetry (FMF) A method for determining the amount of DNA per cell in a population of cells.

Fluid mosaic structure Structure of a cell membrane in which lipid molecules are arranged in a fluid bilayer, with protein molecules inserted into the bilayer.

Fluorescent antibody An antibody to which a fluorescent dye has been covalently joined.

Fluorodeoxyuridine An analog of thymidine that is phosphorylated in cells to fluorodeoxyuridine monophosphate, a potent inhibitor of thymidylate synthase. Used as a chemotherapeutic cancer drug.

Fluorouracil A precursor of fluorodeoxyuridine.

Frameshift mutation Insertion or deletion of one or more base pairs in DNA such that the triplet codon reading frame is shifted.

Free energy Energy released in a chemical reaction.

Freeze-fracture A technique to prepare cells for observation by electron microscopy. Cells are rapidly frozen and then fractured. The fracture surfaces are etched by evaporation of water. A thin layer of metal, e.g., platinum, is deposited onto the etched surface to make it visible in the electron microscope.

Fructose A 6-carbon sugar.

Fungi A major group of organisms with many thousands of species, including yeasts, molds, and mushrooms.

G1 period The interval of time between the end of mitosis and the initiation of DNA replication.

G2 period The interval of time between the end of DNA replication and the start of mitosis.

G0 state A state of quiescence of normal cells in culture that have ceased to reproduce.

Galactose A 6-carbon sugar.

***Gal* genes** Genes encoding proteins involved in catabolizing galactose; studied particularly in yeast.

Galactose A 6-carbon sugar.

Gamete An egg or sperm cell.

GC box A short DNA sequence forming part of the promoter of some eukaryotic genes.

Gelsolin A protein that caps the (+) end of microfilaments and can also sever microfilaments.

Gene A sequence of deoxynucleotides in DNA that prescribes a sequence of amino acids in a polypeptide.

Gene amplification Multiple copies of a gene in a genome.

Gene family Multiple copies of a particular gene in a single genome, forming a family of genes, for example, histone genes.

Gene linkage The condition in which genes encoding different products are present in the same chromosome.

Generation time The time taken for a cell to complete one full cell cycle.

Gene-sized molecule A small DNA molecule containing a single gene—characteristic of all the genes in the macronucleus of hypotrichous ciliates.

Genetic code A code in which a group of three deoxynucleotides in DNA specifies a particular amino acid in a polypeptide chain.

Genotype The genes that prescribe a particular phenotypic trait in an organism.

Glial filaments Intermediate filaments made of the protein glial fibrillary acidic protein, found only in glial cells.

Glucose A sugar with six carbon atoms.

Glycogen A polysaccharide consisting of glucose molecules covalently joined by glycosidic linkages.

Glycolipid A molecule consisting of fatty acids and a hydrophilic head with one or more sugar groups.

Glycolysis Breakdown of a glucose molecule into two pyruvate molecules, with the net generation of two ATP molecules.

Glyoxysomes Microbodies or peroxisomes in plant cells.

Golgi complex A series of flattened membranous sacs in which proteins are accumulated, modified, and packaged for delivery to other parts of the cell or for export from the cell.

Grana Thylakoids in the shape of discs that are tightly arranged in uniform stacks.

Growth factors Proteins present in serum and blood platelets that are required for reproduction of normal cells in culture.

Guanine A purine present in DNA and RNA.

Guanine monophosphate (GMP) A nucleotide consisting of guanine, ribose, and phosphate.

Guanosine A pyrimidine, consisting of guanine and ribose.

Halobacteria Bacteria that grow only in solutions with a high salt concentration.

Haploidy A set of chromosomes containing one of each of the chromosomes.

Heat-shock genes Genes whose transcriptions are greatly increased by heat shock or other stress.

Heat-shock proteins Proteins synthesized at a high rate in response to a heat shock or other kinds of stress.

HeLa cell A cultured cancer cell derived from a human cervical carcinoma and widely used in research.

Helicase An enzyme that catalyzes unwinding of a DNA double helix at a replication fork.

Hemoglobin The oxygen transport protein in red blood cells.

Hemophilia An inherited disease characterized by a defect in blot clotting.

Hepatitis B virus A DNA virus that can cause liver cancer in humans.

Hepatocyte The principal kind of cell in the liver.

Herpesviruses A family of DNA-containing viruses, some of which are oncogenic.

Heterochromatin Tightly packed chromatin containing transcriptionally inactive DNA.

Heterokaryon A cell containing two or more kinds of nuclei.

Heterotroph A cell that must fulfill its carbon requirement by taking in organic molecules.

Heterozygous The condition in which the two alleles of a gene prescribe different forms of a trait.

Hexose A sugar with six carbon atoms.

High-energy bond A bond characterized by release of an unusually large amount of energy when broken.

Histone Types of protein molecules that aggregate into particles, which pack DNA into nucleosomes.

HnRNA Heterogeneous nuclear RNA or primary transcripts.

Homologues One of the two chromosomes of a pair in a diploid cell.

Homozygous A condition in which the two alleles of a gene that prescribe a trait are identical in a diploid organism.

Hormone A chemical messenger produced by one kind of cell and affecting the activity of a second kind of cell.

Human T-cell lymphotrophic viruses (HTLV) A family of viruses including one (HTLV-I) that causes adult T-cell leukemia and another (HTLV-III renamed HIV) that causes AIDS.

Hybrid cell A proliferating cell produced by fusing together two different kinds of cells, for example, a mouse-human hybrid cell.

Hybridoma A hybrid cell formed by fusion of a myeloma cell with a B lymphocyte.

Hydrogen bond A weak bond formed between a slightly positively charged hydrogen atom belonging to one molecule and a slightly negatively charged atom, such as oxygen or nitrogen, in another molecule or in another part of the same molecule.

Hydrolytic cleavage Breakdown of a molecule by addition of water.

Hydrophilic Soluble in water.

Hydrophobic Insoluble in water.

Hydrophobic interactions The interaction between hydrophobic molecules or portions of molecules stabilized by reduction in contact of the molecules with water molecules.

Hypertonic solution A solution containing a high concentration of dissolved molecules and/or ions relative to another solution; e.g., serum is hypertonic to water.

Hypotonic solution A solution containing a lower concentration of dissolved molecules and/or ions relative to another solution; e.g., fresh water is hypotonic to sea water.

I band A type of band in a sarcomere of a skeletal or cardiac muscle cell, characterized by the presence of thin actin myofilaments.

IgG An immunoglobulin.

Immunoglobulin An antibody molecule.

Immunoglobulin gene A gene encoding an antibody polypeptide.

Independent assortment Refers to two kinds of genes encoding two different traits assorting independently if they are on different chromosomes in sexual reproduction.

Independent segregation The rule that genes on different chromosomes are inherited independently in sexual reproduction.

Inducer In bacteria a molecule that binds to a repressor protein, thereby preventing the repressor from binding to the operator of an operon and allowing the operon to be transcribed.

Inducible enzyme An enzyme whose transcription can be turned on by the presence of an inducer.

Initiation The first step in creation of a cancer cell.

Initiation complex A complex formed by a small ribosomal subunit, an initiator tRNA molecule, and an mRNA molecule.

Initiation factors Factors that catalyze or otherwise participate in formation of an initiation complex in polypeptide synthesis.

Inosine A base present in some tRNA anticodons.

Integral proteins Proteins that are inserted into the lipid bilayer of a membrane.

Interconnecting neurons Nerve cells that receive impulses from one neuron and transmit them to another neuron.

Interdigitating microtubules Microtubules of the mitotic spindle that originate from a centrosome and extend toward the opposite centrosome, interdigitating with like microtubules extending from the opposite centrosome.

Intergenic suppression Reversal of the effect of a mutation in one gene by a mutation in a second gene.

Intermediate filaments A filament type forming part of the cytoskeleton in many kinds of eukaryotic cells.

Interphase The period between two successive cell divisions.

Intragenic suppressor mutation A second mutation in a gene that reverses the effect of a prior mutation in the same gene but occurring at a different site in the gene.

Intron A sequence of base pairs in a DNA molecule that does not encode amino acids and interrupts the coding sequence of a gene.

Ionic bond A bond formed by the electrostatic attraction between oppositely charged atoms.

Irreversible inhibition Inhibition by a substance that binds to the active site irreversibly.

Isotonic solution A solution with the same total concentration of dissolved molecules and/or ions relative to another solution; e.g., serum and seawater are nearly isotonic.

J segment Joining region between constant and variable regions of a polypeptide in an antibody molecule.

Karyotype The metaphase chromosome complement of a cell characterized by chromosome number, size, shape, and banding patterns of the chromosomes.

Keratin A type of protein that makes up one kind of intermediate filament.

Keratin filaments Intermediate filaments made of proteins called keratins or cytokeratins.

Kinase A generic term for enzymes that transfer phosphate groups, usually from ATP, to another molecule.

Kinesin A protein that interacts with microtubules to produce motility inside cells.

Kinetochore A plate-like structure on a chromatid to which microtubules of the mitotic spindle attach.

Labeled mitosis method A method for determining the length of the periods of the cell cycle using labeling of DNA with radioactivity.

***Lac* operon** The operon in *E. coli* encoding the enzymes β-galactosidase, lactose permease, and lactose acetylase.

Lactose A disaccharide present in milk.

Lagging strand The DNA chain that is synthesized as discontinuous pieces during DNA replication.

Lag phase A period of time after inoculation of cells into fresh culture medium during which cells adjust to the medium and do not proliferate.

Lamellipodia Ruffles that characterize the forward part of animal cells engaged in amoeboid movement.

Laminin A protein of the extracellular matrix.

Lamins Three kinds of proteins (lamins A, B, and C) that form the nuclear lamina.

Lampbrush chromosome A chromosome in the diplotene stage of the first meiotic division characterized by paired lateral loops—seen especially in amphibian oocytes.

Leader-attenuator A sequence upstream from the 5′ end of the coding region of certain bacterial operons that is the basis for regulating transcription of the operon.

Leader polypeptide A polypeptide translated from the leader-attenuator segment of a transcript in bacteria.

Leader sequence Twenty to 30 amino acids, usually on the amino-terminal end of a polypeptide, that direct the polypeptide into a mitochondrion.

Leading strand The DNA chain that is synthesized as one continuous polymer in DNA replication.

Log phase A period of time during which the cell number in a culture increases logarithmically. Same as exponential phase.

Low-density lipoprotein A complex of about 1500 cholesterol molecules encapsulated by a lipid bilayer containing several copies of a single large protein.

Lymphatic system The structural basis of the immune system; includes the thymus gland, the spleen, appendix, lymphatic vessels and lymph nodes.

Lymphocyte A cell of the immune system.

Lysosome A vesicle with a single enclosing membrane in which hydrolytic enzymes are stored.

Lytic infection Reproduction of a virus in a cell followed by cell lysis.

Macronucleus The somatic nucleus in a ciliated protozoan.

Macrophage A scavenger cell of the immune system that ingests foreign particles by phagocytosis.

Masked mRNA mRNA produced during oocyte formation and stored for use during early development.

Matrix The space enclosed by the inner membrane of a mitochondrion.

Maturation promoting factor (MPF) A protein that brings about mitotic condensation of chromosomes.

Megakaryocyte Blood platelet-forming cell in bone marrow.

Meiosis Two successive cell divisions that produce four haploid nuclei from a single diploid nucleus—characteristic of germ cells.

Memory lymphocyte or cell Lymphocytes responsible for the secondary response to a foreign agent.

Mesophyll cell A photosynthetic cell in C_3 plants and a CO_2-capturing cell in C_4 plants.

Metabolism The total of anabolic and catabolic reactions in a cell.

Metaphase The second part of mitosis, in which the chromosomes are aligned in the center of the mitotic spindle.

Metaphase plate A plane through the center of the mitotic spindle on which the metaphase chromosomes are aligned.

Metastases New foci of cancer cell growth established by cells that have spread from a primary cancer.

Metastasis Spread of cancer cells through the blood or lymphatic vessels to other tissues, where they form new growths.

Methotrexate A poison that inhibits the enzyme dihydrofolate reductase, preventing synthesis of thymidine monophosphate and hence inhibiting DNA synthesis.

Methylation of DNA Addition of methyl groups to DNA, usually to the 5-carbon of cytosine residues.

5-Methylcytosine Cytosine with a methyl group attached to the number 5-carbon.

Methyltransferase An enzyme that catalyzes methylation of DNA.

Micelle An aggregate of lipid molecules packed together to maximize interaction of the hydrophilic part of the molecules with water and to minimize interaction of the hydrophobic part of the lipid molecules with water.

Michaelis-Menten constant (K_m) The concentration of substrate in moles per liter required to elicit half maximal activity of an enzyme.

Microbody A vesicle bounded by a single membrane containing catalase and enzymes involved in oxidation of organic molecules. Also called peroxisomes.

Microfilaments Cytoplasmic fibers made of actin. One of the three fibrous elements of the cytoskeleton and part of the contractile machinery of a cell.

Micronucleus The germline nucleus in a ciliated protozoan.

Microsomes Fragments of rough endoplasmic reticulum that form vesicles during cell homogenization.

Microspike A long, thin cytoplasmic extension, usually terminating in an adhesion plaque.

Microtubule-associated proteins (MAPs) Proteins that specifically bind to microtubules and that are involved in microtubule behavior, for example, assembly and stability of microtubules.

Microtubule doublet Two microtubules fused along their length and sharing several protofilaments—present in cilia and eukaryotic flagella.

Microtubules Tubular structures composed of tubulin proteins. Microtubules are one of the three fibrous elements of the cytoskeleton and form the core of cilia and flagella.

Microvillus A fingerlike projection of cytoplasm on the lumenal surface of an intestinal epithelial cell.

Mid-body A remnant of the mitotic spindle in the cytoplasmic bridge between daughter cells in the last stage of cytokinesis.

Missense mutation A mutation that changes one or more nucleotides in a codon so that it specifies a different amino acid.

Mitochondrial DNA A small DNA molecule in a mitochondrion, usually circular, that encodes several of the gene products present in a mitochondrion.

Mitochondrion A membranous organelle in eukaryotes that contains the molecular machinery for cellular respiration.

Mitosis Distribution of chromosomes to daughter cells during cell division.

Mitotic apparatus A structure composed of microtubules and associated proteins that functions in the distribution of chromosomes during cell division.

Molecular weight The sum of the weights of all the atoms in a molecule. Molecular weight is given in avograms; an avogram is equal to 1.6×10^{-24} grams.

Monoclonal antibody An antibody of a single type produced by a clone of hybridoma cells.

Monomer The building block of a macromolecule.

Motor neuron A nerve cell that innervates muscle cells.

mRNA Messenger RNA—an RNA copied from a gene that encodes a protein.

Mutation A change in the sequence of deoxynucleotides in DNA.

Mycoplasma Bacteria without cell walls.

Myelination Formation of a myelin sheath by Schwann cells.

Myoblast An incompletely differentiated, uninucleated muscle cell still capable of reproduction.

Myofibril A fibril composed of thin actin myofilaments, thick myosin myofilaments, and several other kinds of proteins arranged in a repeating pattern, serving as the contractile element of skeletal and cardiac muscle cells.

Myofilament Actin or myosin filaments in muscle cells.

Myoneural junction A synapse of a motor nerve cell with a muscle cell.

Myosin A protein component of contractile machinery in nonmuscle and muscle cells, composed of two heavy and four light polypeptide chains.

NADH Chemically reduced form of nicotinamide adenine dinucleotide (NAD^+).

Na^+-K^+ ATPase or pump An active transport mechanism in membranes that moves three Na^+ ions out of a cell and two K^+ ions into the cell.

Natural killer cell (NK cell) A lymphocyte that kills foreign cells by contact.

Neomycin An antibiotic that binds to the 30S ribosomal subunit of bacteria and blocks translation.

Neurofilaments Intermediate filaments found only in nerve cells.

Nicotinamide adenine dinucleotide (NAD^+) A coenzyme that carries electrons from one chemical reaction to another.

Nicotinamide adenine dinucleotide phosphate ($NADP^+$) A coenzyme that shuttles high-energy electrons in photosynthesis.

Node of Ranvier A gap in the myelin sheath of a nerve cell axon.

Noncompetitive inhibition Inhibition of an enzyme by binding of an inhibitor away from the active site of the enzyme.

Nonsense mutation A mutation that alters a codon from specifying an amino acid to specifying termination of translation.

Nuclear envelope A double membrane structure enclosing the nucleus.

Nuclear lamina A layer on the inner side of the nuclear envelope composed of three kinds of proteins.

Nuclear matrix A framework or scaffold of proteins in the nucleus.

Nuclear pore A circular opening in the nuclear envelope formed by fusion of the inner and outer membranes of the envelope.

Nuclear pore complex A series of granules arranged radially and surrounding a central granule in a nuclear pore.

Nucleic acid A polymer consisting of nucleotides.

Nucleic acid hybridization Joining of one single-stranded nucleic acid with another by means of complementary base-pairing to form a double helix.

Nucleoid A body formed of the DNA and associated proteins in a prokaryotic cell.

Nucleolar organizer A region on a chromosome, containing genes encoding rRNA, where a nucleolus forms.

Nucleolus A roughly spherical structure in the nucleus in which ribosomal subunits are assembled.

Nucleoside A molecule consisting of a nitrogen base and the sugar ribose.

Nucleosome A spherical structure formed of eight histone protein molecules with a stretch of DNA (143 bp) wrapped around it.

Nucleotide A molecule consisting of a nitrogen base, a ribose sugar, and a phosphate group.

Obligate anaerobe An organism that cannot live in the presence of oxygen.

Okazaki fragments Short polymer chains consisting of an RNA primer and a short DNA chain. Intermediates in DNA replication.

c-Oncogene An oncogene formed from a cellular protooncogene.

v-Oncogene A gene carried in a viral chromosome, capable of transforming a normal cell into a tumor cell.

Oncogenic virus A virus capable of transforming a normal cell to a tumor cell.

Operator A short sequence of deoxynucleotides, 5′ to the coding region of an operon, to which a repressor specifically binds and prevents transcription.

Operon A gene with its promoter and operator sequences.

Organic soup theory The idea that organic molecules were formed in seas during prebiological times and led to the origin of the first cell.

Origin of replication A sequence in prokaryotic DNA at which replication of the chromosome is initiated.

Osmosis Movement of water driven by osmotic pressure.

Osmotic pressure Pressure responsible for movement of water through a membrane separating solutions with different tonicities.

Ovum An egg cell.

Oxidative respiration The second part of cellular respiration, in which the flow of energized electrons is coupled to the synthesis of ATP.

P site One of two tRNA binding sites on a ribosome.

Papovavirus A type of DNA-containing oncogenic virus.

Passive transport Same as facilitated diffusion.

Peptide bond A covalent bond between the carbon atom in the carboxyl group of one amino acid and the N in the amino group of another amino acid.

Peptidyl transferase An enzyme that catalyzes formation of peptide bonds during polypeptide synthesis.

Perinuclear space The narrow space between the inner and outer membranes of the nuclear envelope.

Peripheral proteins Proteins attached to the surface of the plasma membrane.

Petite yeast Yeast cells defective in aerobic respiration.

pH The $\log_{10}$ of the reciprocal of the molar concentration of H^+ ions.

Phagocytosis Intake of particulate material by a cell by means of infolding of the plasma membrane, forming a vacuole or vesicle in the cytoplasm.

Phalloidin A poison produced by a mushroom *Amanita phalloides* that binds to microfilaments and inhibits their disassembly.

Phase I reactions A set of chemical reactions in the first part of photosynthesis, taking place in the internal membranes of a chloroplast. The products are ATP and energized electrons.

Phase II reactions A set of chemical reactions in the second part of photosynthesis, taking place in the stromal compartment of a chloroplast and fixing carbon dioxide as sugar.

Phenotype The form of a genetically encoded trait displayed by an organism, for example, flower color or eye color.

Phosphodiester bond A type of covalent bond, which holds successive nucleotides together in an RNA or DNA chain.

Phospholipid An amphipathic molecule consisting of two or three fatty acid chains, a glycerol molecule, a phosphate group, and usually a fourth component, such as choline or ethanolamine.

Photoreceptor cell Sensory neurons specialized for response to light, for example, rod and cone cells of the retina.

Photorespiration The wasteful breakdown of ribulose 1,5-bisphosphate by oxidation under conditions of low carbon dioxide availability in photosynthetic organisms.

Photosystem I A functional and structural unit of phase I reactions in photosynthesis, containing a light-harvesting antenna and a reaction center for transfer of photon energy to electrons.

Photosystem II A functional and structural unit of phase I reactions of photosynthesis, containing a light-harvesting antenna and reaction center for transfer of photon energy to electrons.

Phototaxis Movement toward or away from light.

Phototroph An organism requiring only light, water, inorganic ions, and carbon dioxide to live.

Phragmoplast A collection of vesicles derived from the Golgi complex that accumulates in the center of a dividing plant cell and forms plasma membranes to separate the two daughter cells.

Physarum A genus of slime molds.

Phytochrome A protein molecule containing a light-absorbing pigment group. Phytochrome occupies a pivotal position in activation of transcription by light of genes in plants and algae that encode products used in the photosynthesis machinery.

Pinocytosis Intake of fluid by a cell by an infolding of the plasma membrane to form a cytoplasmic vacuole.

Plasma cell An antibody-producing cell. An activated B lymphocyte.

Plasma membrane A bilayer of lipid molecules with protein molecules inserted into it, forming a sheetlike structure that encloses the cell.

Plasmid A very small DNA molecule encoding only a few genes, present in some prokaryotes.

Plasmodium A single, large compartment of cytoplasm containing multiple nuclei.

Platelet-derived growth factor (PDGF) A protein, carried by blood platelets, that is necessary for reproduction of myoblasts, glial cells, and several other cell types.

Platelet-derived growth factor receptor (PDGF receptor) A transmembrane protein in some kinds of animal cells, possessing protein kinase activity in its cytoplasmic domain when it binds PDGF in its cell surface domain.

Platelets Packets of cytoplasm in the blood, formed by megakaryocytes in bone marrow, that carry blood-clotting and cell growth factors.

Polar gradient A gradient in an egg cell reflected by different developmental potentials of cells derived from the different poles of the egg.

Polar molecule A molecule that has one or more regions with electrical charge.

Polyadenylic acid tail A series of adenylic acid residues that form the 3′ end of an mRNA molecule in eukaryotes.

Polycistronic mRNA An mRNA molecule encoding more than one polypeptide.

Polypeptide A polymer of amino acids joined by peptide linkages.

Polysaccharide A polymer composed of sugars.

Polysome A structure formed when two or more ribosomes are bound to and engaged in translation of a single mRNA molecule.

Polytene chromosome An interphase form of a chromosome that has replicated many times without separation of the replication products, producing a clearly distinguishable structure characterized by a series of bands and interbands.

Premature chromosome condensation Induction of mitotic condensation of interphase chromosomes by fusion of a mitotic cell with an interphase cell.

Pribnow box A sequence of six base pairs in DNA that forms part of the promoter in bacteria.

Primary structure The sequence of amino acids in a polypeptide chain.

Primary transcript A nuclear RNA molecule that has not yet been modified.

Primase An RNA polymerase that catalyzes synthesis of RNA primers that prime DNA replication.

Profilin A cellular protein that binds to G-actin and inhibits its assembly in microfilaments.

Progenote A hypothetical progenitor cell in evolution of prokaryotic and eukaryotic cells.

Progression Development of increased malignancy of a cancer cell.

Prokaryotes Structurally simple cells that lack a nucleus defined by a nuclear envelope. Bacteria.

Prolamellar body A lattice structure of tubules in an etiolated chloroplast.

Promoter In bacteria a sequence of deoxynucleotides to which RNA polymerase binds in front of a transcription unit. In eukaryotes one or more sequences of deoxynucleotides that guide RNA polymerase to the correct initiation site for transcription.

Promotion The second step in creation of a cancer cell.

Prophase The first part of mitosis, in which the chromosomes condense into their mitotic form.

Protein A polymer made up of amino acids joined together by peptide linkage and folded into a three-dimensional configuration.

Protein kinase An enzyme that transfers a phosphate group from ATP to a protein.

Protists Unicellular eukaryotes.

Protofilament One of the 13 filaments made of tubulin that forms the wall of a microtubule.

Protooncogene A normal cell gene that can be converted to an oncogene by mutation, by increase in copy number, or by increased transcription.

Protozoa A major group of organisms with tens of thousands of species, characterized by motility with cilia, flagella, or amoeboid streaming.

Pseudogene A nonfunctional copy of a gene.

Pseudopod A rounded projection of cytoplasm at the surface of an amoeboid cell.

Puffing Extreme loosening of a band in a polytene chromosome that accompanies activation of transcription of DNA in the band.

Purine A nitrogen base, present in a nucleotide.

Pyrimidine A nitrogen base, present in a nucleotide.

Pyrophosphate A molecule composed of two covalently joined phosphate groups.

Quaternary structure The joining of two or more polypeptide chains to form a protein.

r-Protein A protein that is part of a ribosome.

Radioactivity Radiation produced by disintegration of a radioactive element.

Reassociation of DNA Formation of double-helix DNA from single-stranded DNA.

Receptor-mediated endocytosis Pinocytic uptake of macromolecules that bind to specific receptors on the cell surface.

Regulatory gene A gene that encodes proteins that act in regulation of transcription of other genes; for example, repressor proteins in bacteria and transcription factors in eukaryotes.

Repetitive DNA Sequences in DNA that are repeated in a few copies to several million copies.

Replication fork The point of separation of the chains in a DNA double helix as it unwinds during DNA replication.

Replication unit A segment of DNA in eukaryotic chromosomes that replicates separately, usually bidirectionally, from a single initiation point.

Repressor A protein that binds to the operator of an operon and prevents transcription.

Residue A term to describe a building block of a polymer after it has become part of a polymer; e.g., an amino acid is a residue in a polypeptide.

Retrovirus An RNA virus whose RNA is reverse transcribed into DNA and the DNA integrated into a cellular DNA molecule.

Reverse mutation Reversal of a mutation by a second mutation.

Reverse transcriptase An enzyme that catalyzes the synthesis of a DNA chain on an RNA template.

Reverse transcription Synthesis of a DNA chain on an RNA template.

Reversible inhibition Inhibition of an enzyme in which the inhibitor binds reversibly to the active site or another site on an enzyme.

Rhodopsin The light-absorbing purple pigment present in halobacteria and the eyes of animals.

Ribonuclease An enzyme that catalyzes breakage of phosphodiester bonds in RNA.

Ribonucleoprotein particle A complex formed in the nucleus by a primary RNA transcript and several kinds of proteins.

Ribose A 5-carbon sugar.

Ribosomal RNA (rRNA) RNA molecules that are constituents of a ribosome.

Ribosome A cytoplasmic particle composed of RNA and protein molecules. Ribosomes form the core of the machinery for carrying out protein synthesis.

Ribulose bisphosphate carboxylase An abundant enzyme in chloroplasts that catalyzes the fixation of carbon dioxide in photosynthesis.

RNA Nucleic acid consisting of nucleotides joined by phosphodiester bonds.

5S RNA A small ribosomal RNA.

RNA polymerase An enzyme that catalyzes the joining of nucleotides into RNA on a DNA template.

RNA polymerase I An enzyme that catalyzes synthesis of rRNA in eukaryotes.

RNA polymerase II An enzyme that catalyzes synthesis of mRNA in eukaryotes.

RNA polymerase III An enzyme that catalyzes synthesis of tRNA, 5S RNA, and other small RNA molecules in eukaryotes.

RNA primer A short polymer of nucleotides the 3′ OH end of which serves as the starting point for formation of a deoxynucleotide chain catalyzed by DNA polymerase.

RNA processing Formation of an mRNA molecule from a primary transcript by capping, intron excision, and addition of a poly(A) tail.

RNA transcript An RNA molecule transcribed from DNA.

Rod cell A sensory neuron of the retina responsible for black and white vision.

Rough endoplasmic reticulum That portion of the endoplasmic reticulum to which ribosomes are attached to synthesize proteins destined for packaging by the Golgi complex.

rRNA precursor A primary transcript containing one sequence for each rRNA molecule (28S, 18S, and 5.8S in many eukaryotes).

snRNA Small nuclear RNA molecules involved in RNA splicing.

Ruffles Folds of cytoplasm formed at the leading edge of an animal cell engaged in amoeboid movement.

S period The interval of time in interphase occupied by DNA replication.

30S subunit The smaller of two subunits in a bacterial ribosome.

40S subunit The smaller of two subunits in a eukaryotic ribosome.

50S subunit The larger of two subunits in a bacterial ribosome.

60S subunit The larger of two subunits in a eukaryotic ribosome.

70S particle A complete bacterial ribosome.

S-value The total sequence complexity of the DNA sequences in the genome of an organism.

Saccharomyces cerevisiae Brewer's yeast.

Sarcolemma The plasma membrane of a muscle cell.

Sarcomere A unit of organization of thin actin myofilaments and thick myofilaments in skeletal and cardiac muscle cells.

Sarcoplasm The cytosol of a muscle cell.

Sarcoplasmic reticulum A modified form of smooth endoplasmic reticulum in muscle cells, important in regulation of muscle contraction.

Satellite DNA A DNA sequence that is repeated many times in the genome of an organism and that forms a separate peak in a cesium chloride gradient because it has a different average base composition than the bulk of the DNA.

Schwann cell A type of glial cell that forms a myelin sheath around the axon of a nerve cell.

Secondary lysosome A vesicle formed by fusion of a pinocytic or phagocytic vesicle and a lysosome.

Self-splicing Removal of an intron from a primary RNA transcript through catalytic action of the RNA acting on itself.

Semiconservative replication Replication of a DNA double helix, in which each chain of the double helix becomes a chain in a daughter double helix.

Semipermeable membrane A membrane through which water molecules can diffuse but which blocks diffusion of ions and molecules.

Sendai virus A virus with a membrane coat, used experimentally to fuse adjacent animal cells into a single cell.

Sensory neurons Nerve cells that generate signals through sensory receptors.

Sex-linked gene A gene located in an X or Y chromosome.

Shine-Dalgarno sequence A sequence of nucleotides in an mRNA molecule that binds the mRNA molecule to a small ribosomal subunit.

Sickle cell anemia An inherited disease characterized by a mutation in the gene encoding β-globin.

Signal recognition particle (SRP) A particle consisting of six polypeptides and a small RNA molecule that recognizes the signal sequence on the amino-terminal end of a polypeptide and helps direct the polypeptide into a cisterna of the rough endoplasmic reticulum.

Signal sequence An amino acid sequence of a polypeptide that targets the polypeptide to a specific location in the cell, for example, to the nucleus.

Slime mold An organism consisting of a large, single mass of cytoplasm containing many thousands of nuclei or consisting of thousands of amoebae.

Sodium dodecyl sulfate An ionic detergent used to solubilize proteins in water.

Somatic cells Cells of the body of a multicellular organism, excluding the germ cells.

Spermatocyte An immature sperm cell.

Spore An inert form of a bacterium or fungus enclosed in a thick protective wall.

Spot desmosome A circular-shaped junction between cells characterized by a group of intermediate filaments attached to the cytoplasmic side of the plasma membrane in the area of the desmosome.

***Src* gene** The first viral oncogene to be identified.

Starch A polysaccharide consisting of glucose molecules joined covalently by glycosidic linkages.

Start A switching mechanism in yeast cells that regulates entry into DNA replication in response to modulating influences in the environment.

Stationary phase A period during which cell number in a culture remains stationary following exponential growth.

Stem cell An undifferentiated tissue cell that reproduces to provide replacement cells that differentiate.

Stoma (stomata) A closable pore in the surface of a leaf formed by two guard cells. Gases are exchanged between the leaf cells and the outside environment through stomata.

Stop codon A triplet codon that signals the end of translation of mRNA.

Streptomycin An antibiotic that binds to the 30S ribosomal subunit of bacteria and blocks translation.

Stress fibers Bundles of actin microfilaments.

Stroma The open space inside a chloroplast.

Supercoiling of DNA Writhing of DNA produced by overwinding (positive supercoiling) or underwinding (negative supercoiling) of a double helix.

Suppressor gene A gene that, by undergoing mutation, acquires the ability to reverse the effect of a mutation in another gene.

Suppressor mutation A second mutation in a gene that has undergone a first mutation or mutation in an entirely separate gene that reverses the effect of the first mutation.

Surface coat Glycoproteins on the outer surface of the plasma membrane.

SV40 Simian virus 40. Originally isolated from green monkey kidney cells; capable of transforming some species of animal cells.

Synapse A specialized connection in which neurotransmitter molecules transmit a signal from one nerve cell to another or from a nerve cell to a muscle cell.

Synapsis Joining of two homologous chromosomes in a tight pair.

Synaptic cleft The narrow space in a synapse between two cells.

Synaptic vesicles Nerve cell vesicles containing neurotransmitter substance.

Synaptonemal complex A structure formed in the tight pairing of homologous chromosomes during prophase of the first meiotic division.

Synchronization Treatment of cell populations to synchronize their cell cycles.

Talin A microfilament-associated protein found at the ends of microfilament bundles where they attach to the cytoplasmic side of the plasma membrane.

Targeting of proteins Specific distribution of proteins to different cell structures and compartments or for release from the cell.

TATA box A DNA sequence forming part of the promoter of most eukaryotic genes.

Telomere A specific DNA sequence at the end of a linear DNA chromosome.

Telophase The fourth part of mitosis, in which daughter chromosomes decondense into the interphase form.

Terminal web A network of microfilaments in which microfilaments in microvilli terminate in the cytoplasm.

Termination signal A sequence in DNA that signals the termination of transcription.

Tertiary structure Folding of a polypeptide chain into a 3-dimensional configuration.

Tetracycline An antibiotic that blocks protein synthesis in bacteria and mitochondria.

Tetrad The four chromatids present in a pair of homologous chromosomes joined in the diplotene stage of the first meiotic prophase.

Tetrahydrofolate A coenzyme that works with thymidylate synthase in the synthesis of TMP.

Tetrahymena A ciliated protozoan.

TGF-β (Tumor growth factor β) A protein normally present in a mammal that inhibits reproduction of epithelial cells but stimulates reproduction of fibroblasts and tumor cells in culture.

Thalassemia An inherited disease characterized by insufficient production of hemoglobin.

Thick filaments Filaments made of an aggregate of myosin molecules, prominent in skeletal and cardiac muscle cells.

Thin filaments Actin microfilaments in skeletal and cardiac muscle cells.

3T3 cell A type of cultured mouse cell that has the properties of a normal cell, except it does not senesce.

Thylakoid Flattened membranous sacs in a chlorplast, in which phase I of photosynthesis is carried out.

Thymidine A deoxynucleoside consisting of thymine and deoxyribose.

^{3}H-Thymidine A pyrimidine precursor of DNA containing tritium.

Thymidine kinase An enzyme that catalyzes phosphorylation of thymidine to form TMP.

Thymidine monophosphate (TMP) A deoxynucleotide consisting of thymine, deoxyribose, and phosphate.

Thymine A pyrimidine present in DNA.

Tight junction Tight adherence of the plasma membranes of adjacent epithelial cells.

T lymphocyte or cell A cytotoxic cell that kills foreign or virus-infected cells by contact.

Topoisomerase An enzyme that catalyzes transient breakage of DNA chains during DNA replication and in separation of daughter DNA molecules.

Totipotency The ability of a single starting cell or nucleus to support all types of differentiation during embryonic development.

Transcribed spacer A region of DNA that is transcribed to form a spacer in an rRNA precursor molecule. The spacer is destroyed to release separate 28S, 18S, and 5.8S rRNA molecules.

Transcription The synthesis of an RNA molecule on a template of DNA.

Transcription bubble A region of the DNA double helix that is transiently opened into single strands during transcription—17 base pairs long in *E. coli* DNA.

Transcription factors Proteins that bind to DNA sequences upstream from genes and regulate transcription in eukaryotes.

Transcription unit A segment of DNA made up of a gene, transcription regulatory sequences, a transcription termination signal, and a poly(A) addition signal.

Transfer RNA (tRNA) Small RNA molecules that work as adaptors to translate triplet codons in mRNA into a sequence of amino acids during polypeptide synthesis.

Transformation of animal cells Conversion of normal animal cells in culture to cells resembling cancer cells in culture.

Transformation of bacteria Alteration of the genome of a bacterial cell by uptake of DNA from the medium.

Transit peptide A segment of a polypeptide that targets the polypeptide for delivery to a chloroplast.

Translation The synthesis of a polypeptide chain under the direction of a messenger RNA molecule.

Translation-translocation coupling Translocation of a polypeptide across a membrane of the rough endoplasmic reticulum as the polypeptide is synthesized.

Translocation Exchange of segments of chromosomes between nonhomologous chromosomes.

Transmembrane protein An integral membrane protein with hydrophilic regions at each surface of a membrane and a connecting hydrophobic region that spans the lipid bilayer.

Tritium A radioactive form of hydrogen.

Tropomyosin A protein that functions in the regulation of muscle contraction. Also found in nonmuscle cells.

Troponin A protein that functions in the regulation of muscle contraction. Also found in nonmuscle cells.

Tryptophan operon An operon with five genes that encode enzymes catalyzing tryptophan synthesis.

Ts mutation A conditional mutation in which a gene product functions normally at one temperature but not at another.

T system A system of tubular invaginations of the sarcolemma (plasma membrane) of a muscle cell that penetrates deeply into the cell and transmit signals that trigger contraction.

Tubulin A type of protein that makes up microtubules.

Unique sequence A sequence that occurs only once in a genome.

Universality of the genetic code The dictum that almost without exception a given triplet codon has the same amino acid meaning in all cells and viruses.

Untranscribed spacer DNA sequences that are not transcribed and that separate transcription units in DNA.

Upstream activation site (UAS) A sequence in yeast upstream from the 5′ end of a gene that binds a transcription factor and avtivates transcription of the gene.

Uracil A pyrimidine present in RNA.

Uridine A nuceloside consisting of uracil and ribose.

^{3}H-Uridine A pyrimidine precursor of RNA labeled with tritium.

Uridine monophosphate (UMP) A nucleotide consisting of uracil, ribose, and phosphate.

V_{max} Maximum velocity of an enzyme-catalyzed reaction.

V segment Variable region of a polypeptide in an antibody.

Van der Waals force Attractive force between two atoms.

Vimentin filaments Intermediate filaments made of the protein vimentin.

Vinculin A microfilament-associated protein found at the end of microfilament bundles at their attachment sites to the cytoplasmic surface of the plasma membrane.

Volvox A simple multicellular alga.

Wobble Flexibility in the orientation of the first base in the anticodon of tRNA such that it can hydrogen-bond with more than one kind of base in the third position of the complementary codon.

X chromosome A sex chromosome present in two copies in female cells of most organisms.

Xenopus A froglike amphibian.

Y chromosome A sex chromosome present in male cells of most organisms.

Z line A structure containing the protein α-actinin that forms a cross striation in skeletal muscle cells and defines the ends of a sarcomere.

Zygote A fertilized egg cell.

Zymogen Digestive enzymes, particularly in pancreas cells, in the form of enzymatically inactive precursors.

FURTHER READING

Almost all of the articles listed here are reviews. Lists of research reports will be published in the *Study Guide for Cells.* The reviews here span a wide range in the level of difficulty. Many references are given to articles published in *Scientific American* since 1974. These are among the most general and simplest to read. Reviews published in the *Annual Review of Biochemistry, Annual Review of Cell Biology,* and *Annual Review of Genetics* are more difficult because they are addressed to other scientists. They are extremely valuable to those students who can manage them. Additional reviews will be listed in the *Study Guide for Cells.*

Chapter 1

Brock, T. D., D. W. Smith, and M. T. Madigan. 1984. *Biology of Microorganisms.* 4th Ed. Prentice-Hall.

de Duve, C. 1983. "Microbodies in the living cell." *Scient. Amer.,* 248 (May), 74-84.

de Duve, C. 1975. "Exploring cells with a centrifuge." *Science,* 189, 186-194.

Fawcett, D. W. 1981. *The Cell.* 2nd Ed. W. B. Saunders and Co.

Gunning, B. E. S., and M. W. Steer. 1975. *Ultrastructure of Plant Cells.* Arnold.

Ledbetter, M. C., and K. R. Porter. 1970. *Introduction to the Fine Structure of Plant Cells.* Springer-Verlag.

Margulis, L., and K. V. Schwartz. 1982. *Five Kingdoms.* W. H. Freeman and Co.

Porter, K. R., and J. B. Tucker. 1981. "The ground substance of the living cell." *Scient. Amer.,* 244 (March), 56-67.

Porter, K. R., and M. A. Bonneville. 1973. *Fine Structure of Cells and Tissues.* 4th Ed. Lea & Febiger.

Razin, S. 1978. "The mycoplasmas." *Microbiol. Rev.,* 42, 414-470.

Scheeler, P. 1980. *Centrifugation in Biology and Medical Science.* Wiley.

Chapter 2

Baker, J. J. W., and G. E. Allen. 1974. *Matter, Energy and Life.* Addison-Wesley.

Blake, C. C. F., and L. N. Johnson. 1984. "Protein structure." *Trends in Biochem. Sci.,* 9, 147-151.

Chothia, C. 1984. "Principles that determine the structure of proteins." *Ann. Rev. Biochem.,* 53, 537-572.

Creighton, T. E. 1983. *Proteins: Structure and Molecular Properties.* W. H. Freeman and Co.

Doolittle, R. 1985. "Proteins." *Scient. Amer.,* 253 (October), 88-96.

Eisenberg, D., and W. Kauzmann. 1969. *The Structure and Properties of Water.* Oxford University Press.

Felsenfeld, G. 1985. "DNA." *Scient. Amer.,* 253 (October), 58-67.

Ginsburg, V., and P. Robbins (eds.) 1984. *Biology of Carbohydrates.* Wiley.

Kendrew, J. C. 1961. "The three-dimensional structure of a protein molecule." *Scient. Amer.*, 205 (December), 96-110.

Kim, P. S., and R. L. Baldwin. 1982. "Specific intermediates in the folding reactions of small proteins and the mechanism of protein folding." *Ann. Rev. Biochem.*, 51, 459-489.

Lehninger, A. L. 1982. *Principles of Biochemistry*. Worth.

Lesk, A. M. 1984. "Themes and contrasts in protein structures." *Trends in Biochem. Sci.*, 9, v-vii.

Rossman, M. G., and P. Argos. 1981. "Protein folding." *Ann. Rev. Biochem.*, 50, 497-532.

Sharon, N. 1980. "Carbohydrates." *Scient. Amer.*, 243 (November), 90-116.

Stryer, L. 1981. *Biochemistry*. 2nd Ed. W. H. Freeman and Co.

Chapter 3

Fersht, A. 1985. *Enzyme Structure and Mechanism*. 2nd Ed. W. H. Freeman and Co.

Lipscomb, W. N. 1983. "Structure and catalysis of enzymes." *Ann. Rev. Biochem.*, 52, 17-34.

Newsholme, E. A., and C. Start. 1973. *Regulation in Metabolism*. Wiley.

Roach, P. J. 1980. "Principles of the regulation of enzyme activity." In D. M. Prescott and L. Goldstein (eds.), *Cell Biology: A Comprehensive Treatise. Vol. 4.* Academic Press.

Chapter 4

Atkinson, D. E. 1977. *Cellular Energy Metabolism and Its Regulation*. Academic Press.

Bogorad, L. 1981. "Chloroplasts." *J. Cell Biol.*, 91, 256s-270s.

Bridger, W. A., and J. F. Henderson. 1983. *Cell ATP*. Wiley.

de Duve, C. 1983. "Microbodies in the living cell." *Scient. Amer.*, 248 (May), 74-84.

Dickerson, R. E. 1986. "Cytochrome c and the evolution of energy metabolism." *Scient. Amer.*, 242 (March), 136-153.

Ernster, L., and G. Schatz. 1981. "Mitochondria: A historical review." *J. Cell Biol.*, 91, 227s-255s.

Harold, F. M. 1986. *The Vital Force: A Study of Bioenergetics*. W. H. Freeman and Co.

Hinkle, P. C., and R. E. McCarty. 1978. "How cells make ATP." *Scient. Amer.*, 238 (March), 104-123.

Hoober, J. K. 1984. *Chloroplasts*. Plenum Press.

Miller, K. R. 1979. "The photosynthetic membrane." *Scient. Amer.*, 241 (October), 102-113.

Nicholls, D. G. 1982. *Bioenergetics: An Introduction to the Chemiosmotic Theory*. Academic Press.

Tolbert, N. E., and E. Essner. 1981. "Microbodies: peroxisomes and glyoxysomes." *J. Cell Biol.*, 91, 271s-283s.

Tzagoloff, A. 1982. *Mitochondria*. Plenum.

Whittaker, P. A., and S. M. Danks. 1979. *Mitochondria: Structure, Function, and Assembly*. Longman.

Youvan, D. C., and B. L. Mars. 1987. "Molecular mechanisms of photosynthesis." *Scient. Amer.*, 256 (June), 42-48.

Chapter 5

Bainton, D. 1981. "The discovery of lysosomes." *J. Cell Biol.*, 91, 66s-76s.

Bretscher, M. S. 1985. "The molecules of the cell membrane." *Scient. Amer.*, 253 (October) 100-109.

Brown, M., and J. L. Goldstein. 1984. "How LDL receptors influence cholesterol and atherosclerosis." *Scient. Amer.*, 251 (November), 58-66.

Dautry-Varsat, A., and H. F. Lodish. 1984. "How receptors bring proteins and particles into cells." *Scient. Amer.*, 250 (May), 52-58.

Eisenberg, D. 1984. "Three-dimensional structure of membrane and surface proteins." *Ann. Rev. Biochem.*, 53, 595-623.

Farquhar, M., and G. Palade. 1981. "The Golgi apparatus (complex) (1954-1981) from artifact to center stage. *J. Cell Biol.*, 91, 77s-103s.

Finean, J. B., R. Coleman, and R. H. Michell. 1974. *Membranes and their Cellular Functions*. Wiley.

Goldstein, J. L., M. S. Brown, R. G. W. Anderson, D. W. Russell, and W. J. Schneider. 1985. "Receptor-mediated endocytosis: Concepts emerging from the LDL receptor system." *Ann Rev. Cell Biol.*, 1, 1-39.

Lodish, H. F., and J. E. Rothman. 1979. "The assembly of cell membranes." *Scient. Amer.*, 240 (January), 48-63.

Pastan, I., and M. C. Willingham. 1983. "Receptor-mediated endocytosis: Coated pits, receptosomes and the Golgi." *Trends Biochem. Sci.*, 8, 250-254.

Robertson, J. D. 1981. "Membrane structure." *J. Cell Biol.*, 91, 189s-204s.

Singer, S. J. 1974. "The molecular organization of membranes. *Ann. Rev. Biochem.*, 43, 805-833.

Singer, S. J., and G. L. Nicolson. 1972. "The fluid mosaic model of the structure of cell membranes." *Science*, 175, 720-731.

Tanford, C. 1980. *The Hydrophobic Effect: Formation of Micelles and Biological Membranes*. 2nd Ed. Wiley.

Unwin, N., and R. Henderson. 1984. "The structure of proteins in biological membranes." *Scient. Amer.*, 250 (February), 78-94.

Weissmann, G. and R. Claiborne (eds.) 1975. *Cell Membranes: Biochemistry, Cell Biology and Pathology.* HP Publishing Co.

Wickner, W. 1980. "Assembly of proteins into membranes." *Science*, 210, 861-868.

Wileman, T., C. Harding, and P. Stahl. 1985. "Receptor-mediated endocytosis." *Biochem. J.*, 232, 1-14.

Chapter 6

Allen, R. D. 1987. "The microtubule as an intracellular engine." *Scient. Amer.*, 256 (February), 42-49.

Brinkley, B. R. 1985. "Microtubule organizing centers." *Ann. Rev. Cell Biol.*, 1, 145-172.

Cleveland, D. W., and K. F. Sullivan. 1985. "Molecular biology and genetics of tubulin." *Ann. Rev. Biochem.*, 54, 331-365.

Cold Spring Harbor Laboratory. 1982. *Organization of the Cytoplasm. Cold Spring Harb. Symp. Quant. Biol.*, Vol. 46.

Dustin, P. 1980. "Microtubules." *Scient. Amer.*, 243 (August), 66-76.

Franke, W. W. 1987. "Nuclear lamins and cytoplasmic intermediate filament proteins: A growing multigene family." *Cell*, 48, 3-4.

Fulton, A. B. 1980. "How crowded is the cytoplasm?" *Cell*, 30, 345-347.

Haimo, L. T., and J. L. Rosenbaum. 1981. "Cilia, flagella, and microtubules." *J. Cell Biol.*, 91, 125s-130s.

Hynes, R. 1986. "Fibronectins." *Scient. Amer.*, 254 (June), 42-51.

Kirschner, M., and T. Mitchison. 1986. "Beyond self-assembly: From microtubules to morphogenesis." *Cell*, 45, 329-342.

Lazarides, E. 1982. "Intermediate filaments: A chemically heterogeneous, developmentally regulated class of proteins." *Ann. Rev. Biochem.*, 51, 219-250.

Marchuk, D., S. McCrohon, and E. Fuchs. 1984. "Remarkable conservation of structure among intermediate filament genes." *Cell*, 39, 491-498.

McIntosh, J. R. 1983. "The centrosome as an organizer of the cytoskeleton." *Mod. Cell Biol.*, 2, 115-142.

Mooseker, M. S. 1985. "Organization, chemistry, and assembly of the cytoskeletal apparatus of the intestinal brush border." *Ann. Rev. Cell Biol.*, 1, 209-241.

Olmsted, J. B. 1986. "Microtubule-associated proteins." *Ann. Rev. Cell Biol.*, 2, 421-457.

Pollard, T. D., and J. A. Cooper. 1986. "Actin and actin-binding proteins. A critical evaluation of mechanisms and functions." *Ann. Rev. Biochem.*, 55, 987-1035.

Pollard, T. D., and S. W. Craig. 1982. "Mechanisms of actin polymerization." *Trends Biochem. Sci.*, 7, 55-58.

Porter, K. R., and J. B. Tucker. 1981. "The ground substance of the cell." *Scient. Amer.*, 244 (March), 56-67.

Satir, B. 1975. "The final steps in secretion." *Scient. Amer.*, 233 (October), 28-37.

Schliwa, M. 1986. *Cytoskeleton: An Introductory Survey.* Cell Biology Monographs, Vol. 13, Springer-Verlag.

Schulze, E., and M. Kirschner. 1986. "Microtubule dynamics in interphase cells." *J. Cell Biol.*, 102, 1010-1031.

Sloboda, R. D. 1980. "The role of microtubules in cell structure and cell division." *Am. Sci.*, 68, 290-297.

Solomon, F. 1980. "Organizing microtubules in the cytoplasm." *Cell*, 22, 331-332.

Steinert, P. M., A. C. Steven, and D. R. Roop. 1985. "The molecular biology of intermediate filaments." *Cell*, 42, 411-419.

Steinert, P. M., and D. A. D. Parry. 1985. "Intermediate filaments: Conformity and diversity of expression and structure." *Ann. Rev. Cell Biol.*, 1, 41-65.

Stossel, T. P., C. Chaponnier, R. M. Ezzell, J. H. Hartwig, P. A. Janmey, D. J. Kwiatkowski, S. E. Lind, D. B. Smith, F. S. Southwick, H. L. Yin, and K. S. Zaner. 1985. "Nonmuscle actin-binding proteins." *Ann. Rev. Cell Biol.*, 1, 353-402.

Chapter 7

Freifelder, D. 1978. *The DNA Molecule: Structure and Properties.* W. H. Freeman and Co.

Parker, G., W. A. Reynolds, and R. Reynolds. 1977. *Heredity.* 2nd Ed. Educational Methods.

Ruddle, F. H. 1982. "A new era in mammalian gene mapping: somatic cell genetics and recombinant DNA methodologies." *Nature*, 294, 115-119.

von Wettstein, D., S. W. Rasmussen, and P. B. Holm. 1984. "The synaptonemal complex in genetic segregation." *Ann. Rev. Genet.*, 18, 331-413.

Watson, J. D., N. H. Hopkins, J. W. Roberts, J. A. Steitz, and A. M. Weiner. 1987. *Molecular Biology of the Gene.* 4th Ed. Benjamin/Cummings.

Chapter 8

Breitbart, R. E., A. Andreadis, and B. Nadal-Ginard. 1987. "Alternative splicing: A ubiquitous mechanism for the generation of multiple protein isoforms from single genes." *Ann. Rev. Biochem.*, 56, 435-495.

Cech, T. R. 1986. "The generality of self-splicing RNA: Relationship to nuclear mRNA splicing." *Cell,* 44, 207-210.

Cech, T. R. 1986. "RNA as an enzyme." *Scient. Amer.*, 255 (November), 64-75.

Cech, T. R., and B. L. Bass. 1986. "Biological catalysis by RNA." *Ann. Rev. Biochem.*, 55, 599-629.

Cech, T. R. 1983. "RNA splicing: Three themes with variations." *Cell,* 34, 713-716.

Chambliss, G., G. R. Craven, J. Davies, K. Kavis, L. Kahan, and M. Nomura (eds.) 1980. *Ribosomes: Structure, Function, and Genetics.* University Park Press.

Colman, A., and C. Robinson. 1986. "Protein import into organelles: Hierarchical targeting signals." *Cell,* 46, 321-322.

Darnell, J. E., Jr. 1983. "The processing of RNA." *Scient. Amer.*, 249 (October), 90-100.

Dingwall, C., and R. A. Laskey. 1986. "Protein import into the cell nucleus." *Ann. Rev. Cell Biol.*, 2, 367-390.

Douglas, M. G., M. T. McCammon, and A. Vassarotti. 1986. "Targeting proteins into mitochondria." *Microbiol. Rev.*, 50, 166-178.

Dunphy, W. G., and J. E. Rothman. 1985. "Compartmental organization of the Golgi stack." *Cell,* 42, 13-21.

Farquhar, M. G. 1985. "Progress in unraveling pathways of Golgi traffic." *Ann. Rev. Cell Biol.*, 1, 447-488.

Freifelder, D. 1987. *Molecular Biology*. 2nd Ed. Jones and Bartlett.

Gerace, L. 1985. "Traffic control and structural proteins in the eukaryotic nucleus." *Nature,* 318, 508-509.

Gilbert, W., Marchionni, M., and McKnight, G. 1986. "On the antiquity of introns." *Cell,* 46, 151-154.

Green, M. R. 1986. "Pre-mRNA splicing." *Ann. Rev. Genet.*, 20, 671-708.

Hay, R., Böhni, P., and Gasser, S. 1984. "How mitochondria import proteins." *Biochim. Biophys. Acta,* 779, 65-87.

Heddle, J. A. 1982. *Mutagenicity.* Academic Press.

Hershey, J. W. B. 1980. "The translational machinery: components and mechanisms." In D. M. Prescott and L. Goldstein (eds.), *Cell Biology: A Comprehensive Treatise. Vol. 4.* Academic Press.

Hirschberg, C. B., and M. D. Snider. 1987. "Topography of glycosylation in the rough endoplasmic reticulum and Golgi apparatus. *Ann. Rev. Biochem.*, 56, 63-87.

Kelly, R. B. 1985. "Pathways of protein secretion in eukaryotes." *Science,* 230, 25-32.

Lake, J. A. 1981. "The ribosome." *Scient. Amer.*, 245 (August), 84-97.

Lewin, R. 1982. "On the origin of introns." *Science,* 217, 921-922.

Maniatis, T., E. F. Fritsch, J. Lauer, and R. M. Lawn. 1980. "The molecular genetics of human hemoglobins." *Ann. Rev. Genet.*, 14, 145-178.

Miller, O. L. 1981. "The nucleolus, chromosomes, and visualization of genetic activity." *J. Cell Biol.*, 91, 15s-27s.

Moldave, K. 1985. "Eukaryotic protein synthesis." *Ann Rev. Biochem.*, 54, 1109-1150.

Neupert, W., and G. Schatz. 1981. "How proteins are transported into mitochondria." *Trends Biochem. Sci.*, 6, 1-4.

Nevins, J. R. 1983. "The pathway of eukaryotic mRNA formation." *Ann. Rev. Biochem.*, 52, 441-466.

Padgett, R. A., P. J. Grabowski, M. M. Konarska, S. Seiler, and P. A. Sharp. 1986. "Splicing of messenger RNA precursors." *Ann. Rev. Biochem.*, 55, 1119-1150.

Perry, R. P. 1981. "RNA processing comes of age." *J. Cell Biol.*, 91, 28s-38s.

Pfeffer, S. R., and J. E. Rothman. 1987. "Biosynthetic protein transport and sorting by the endoplasmic reticulum and Golgi." *Ann. Rev. Biochem.*, 56, 829-852.

Rich, A., and S. H. Kim. 1978. "The three-dimensional structure of transfer RNA." *Scient. Amer.*, 238 (January), 52-62.

Rothman, J. E. 1987. "Protein sorting by selective retention in the endoplasmic reticulum and Golgi stack." *Cell,* 50, 521-522.

Rothman, J. E. 1985. "The compartmental organization of the Golgi apparatus." *Scient. Amer.*, 253 (September), 74-89.

Rothman, J. E. 1981. "The Golgi apparatus: Two organelles in tandem." *Science,* 213, 1212-1219.

Schatz, G. 1987. "Signals guiding proteins to their correct locations in mitochondria." *Eur. J. Biochem.*, 165, 1-6.

Schatz, G., and R. A. Butow. 1983. "How are proteins imported into mitochondria?" *Cell,* 32, 316-318.

Siekevitz, P., and P. C. Zamecnik. 1981. "Ribosomes and protein synthesis." *J. Cell Biol.*, 91, 53s-65s.

Vanin, E. F. 1985. "Processed pseudogenes: Characteristics and evolution." *Ann. Rev. Genet.*, 19, 253-272.

Walter, P., and V. R. Lingappa. 1986. "Mechanism of protein translocation across the endoplasmic reticulum membrane." *Ann. Rev. Cell Biol.*, 2, 499-516.

Walter, P., R. Gilmore, and G. Blobel. 1984. "Protein translocation across the endoplasmic reticulum." *Cell,* 38, 5-8.

Watson, J. D., N. H. Hopkins, J. W. Roberts, J. A. Steitz, and A. M. Weiner. 1987. *Molecular Biology of the Gene*. 4th Ed. Benjamin/Cummings.

Wickner, W. T., and H. F. Lodish. 1985. "Multiple mechanisms of protein insertion into and across membranes." *Science*, 230, 400-407.

Wittman, H. G. 1983. "Architecture of prokaryotic ribosomes." *Ann. Rev. Biochem.*, 52, 35-66.

Chapter 9

Freifelder, D. 1987. *Microbial Genetics*. Jones and Bartlett.

Freifelder, D. 1987. *Molecular Biology*. 2nd Ed. Jones and Bartlett.

Maniatis, T., and M. Ptashne. 1976. "A DNA operator-repressor system." *Scient. Amer.* 234 (January), 64-76.

Miller, J. H., and W. S. Reznikoff. 1978. *The Operon*. Cold Spring Harbor Laboratory.

Nomura, M. 1984. "The control of ribosome synthesis." *Scient. Amer.*, 250 (January), 102-114.

Nomura, M., R. Gourse, and G. Baughman. 1984. "Regulation of the synthesis of ribosomes and ribosomal components." *Ann. Rev. Biochem.*, 53, 75-118.

Watson, J. D., N. H. Hopkins, J. W. Roberts, J. A. Steitz, and A. M. Weiner. 1987. *Molecular Biology of the Gene*. 4th Ed. Benjamin/Cummings.

Yanofsky, C., and R. Kolter. 1982. "Attenuation in amino acid biosynthetic operons." *Ann. Rev. Genetics*, 16, 113-134.

Yanofsky, C. 1981. "Attenuation in the control of expression of bacterial operons." *Nature*, 289, 751-758.

Chapter 10

Brown, D. D. 1981. "Gene expression in eukaryotes." *Science*, 211, 667-674.

Caron, J. M., and M. W. Kirschner. 1986. "Autoregulation of tubulin synthesis." *BioEssays*, 5, 211-216.

Corden, J., B. Wasylyk, A. Buchwalder, D. Sassone-Corsi, D. Kedinger, and P. Chambon. 1980. "Promoter sequences of eukaryotic protein coding genes." *Science*, 209, 1406-1413.

Darnell, J. E., Jr. 1982. "Variety in the level of gene control in eukaryotic cells." *Nature*, 297, 365-371.

Doerfler, W. 1983. "DNA methylation and gene activity." *Ann. Rev. Biochem.*, 52, 93-124.

Felsenfeld, G., and J. McGhee. 1982. "Methylation and gene control." *Nature*, 296, 602-603.

Gluzman, Y., and T. Shenk. 1983. *Enhancers and Eukaryotic Gene Expression*. Cold Spring Harbor Laboratory.

Khoury, G., and P. Gruss. 1983. "Enhancer elements." *Cell*, 33, 313-314.

Lewin, B. 1980. *Gene Expression*. Vol. 2 *Eucaryotic Chromosomes*. 2nd Ed. Wiley.

Lyon, M. 1974. "X-chromosome inactivation and developmental patterns in mammals." *Biol. Rev.*, 47, 1-36.

Martin, G. R. 1982. "X-chromosome inactivation in mammals." *Cell*, 29, 721-724.

McKnight, S. L., and R. Kingsbury. 1982. "Transcriptional control signals of a eukaryotic protein-coding gene." *Science*, 217, 316-324.

Nomura, M., Gourse, R., and Baughman, G. 1984. "Regulation of the synthesis of ribosomes and ribosomal components." *Ann. Rev. Biochem.*, 53, 75-117.

O'Malley, B. W., and W. T. Schrader. 1976. "The receptors of steroid hormones." *Scient. Amer.*, 234 (February), 32-43.

Rogers, B. L., and G. F. Saunders. 1986. "Transcriptional enhancers play a major role in gene expression." *BioEssays*, 4, 62-65.

Stark, G. R., and G. M. Wahl. 1984. "Gene amplification." *Ann. Rev. Biochem.*, 53, 447-491.

Struhl, K. 1987. "Promoters, activator proteins, and the mechanism of transcriptional initiation in yeast." *Cell*, 49, 295-297.

Chapter 11

Blackburn, E. H. 1986. "Structure and formation of telomeres in holotrichous ciliates." In K. W. Jeon (ed.), *Int. Review of Cytology. Vol. 99. Molecular Approaches to the Study of Protozoan Cells*. Academic Press.

Blackburn, E. H., and J. W. Szostak. 1984. "The molecular structure of centromeres and telomeres." *Ann. Rev. Biochem.*, 53, 163-194.

Borst, P., L. A. Grivell, and G. S. P. Groot. 1984. "Organelle DNA." *Trends Biochem. Sci.*, 9, 128-130.

Bostock, C. J., and A. T. Sumner. 1978. *The Eukaryotic Chromosome*. North-Holland.

Earnshaw, W. C., and U. K. Laemmli. 1983. "Architecture of metaphase chromosomes and chromosome scaffolds." *J. Cell Biol.*, 96, 84-93.

Franke, W. W. 1974. "Structure, biochemistry and functions of the nuclear envelope." *Int. Rev. Cytol. Supp.*, 4, 72-236.

Gall, J. G. 1981. "Chromosome structure and the C-value paradox." *J. Cell Biol.*, 91, 3s-14s.

Gerace, L. 1986. "Nuclear lamina and organization of nuclear architecture." *Trends Biochem. Sci.*, 11, 443-446.

Gerace, L., and G. Blobel. 1981. "Nuclear lamina and the structural organization of the nuclear envelope." *Cold Spring Harb. Symp. Quant. Biol.*, 46, 967-978.

Ghosh, S. 1976. "The nucleolar structure." *Int. Rev. Cytol.*, 44, 1-28.

Grivell, L. A. 1983. "Mitochondrial DNA." *Scient. Amer.*, 248 (March), 78-89.

Hancock, R., and T. Boulikas. 1982. "Functional organization in the nucleus." *Int. Rev. Cytol.*, 79, 165-214.

Kornberg, A. 1980. *DNA Replication.* W. H. Freeman and Co.

Kornberg, R. D., and A. Klug. 1981. "The nucleosome." *Scient. Amer.*, 244 (February) 52-64.

Kraut, H., H. J. Lipps, and D. M. Prescott. 1986. "The genome of hypotrichous ciliates." In K. W. Jeon (ed.), *Int. Review of Cytology. Vol. 99. Molecular Approaches to the Study of Protozoan Cells.* Academic Press.

Maul, G. G. 1977. "The nuclear and the cytoplasmic pore complex: structure, dynamics, distribution, and evolution." In *International Review of Cytology, Suppl. 8.* Academic Press.

McGhee, J. D., and G. Felsenfeld. 1980. "Nucleosome structure." *Ann. Rev. Biochem.*, 49, 1115-1156.

Newport, J. W., and D. J. Forbes. 1987. "The nucleus: Structure, function, and dynamics." *Ann. Rev. Biochem.*, 56, 535-565.

Palmer, J. D. 1985. "Comparative organization of chloroplast genomes." *Ann. Rev. Genet.*, 19, 325-354.

Rieder, C. L. 1982. "The formation, structure, and composition of the mammalian kinetochore." *Int. Rev. Cytol.*, 79, 1-58.

Ris, H., and P. L. Witt. 1981. "Structure of the mammalian kinetochore." *Chromosoma*, 82, 153-170.

Sanger, F. 1981. "Determination of nucleotide sequences in DNA." *Science*, 214, 1205-1210.

Stark, G. R., and G. M. Wahl. 1984. "Gene amplification." *Ann. Rev. Biochem.*, 53, 447-491.

Sumner, A. T. 1982. "The nature and mechanisms of chromosome banding." *Cancer Genet. Cytogenet.*, 6, 59-87.

Chapter 12

Campbell, J. L. 1986. "Eukaryotic DNA replication." *Ann. Rev. Biochem.*, 55, 733-771.

Carpenter, G. 1987. "Receptors for epidermal growth factor and other polypeptide mitogens." *Ann. Rev. Biochem.*, 56, 881-914.

Carpenter, G., and J. G. Zendegui. 1986. "Epidermal growth factor, its receptor, and related proteins." *Exptl. Cell Res.*, 164, 1-10.

Carpenter, G., and S. Cohen. 1984. "Peptide growth factors." *Trends Biochem. Sci.*, 9, 169-171.

DePamphilis, M. L., and P. M. Wassarman. 1980. "Replication of eukaryotic chromosomes: A close-up of the replication fork." *Ann. Rev. Biochem.*, 49, 627-666.

Hayflick, L. 1980. "The cell biology of human aging." *Scient. Amer.*, 242 (January), 58-65.

Huberman, J. A. 1987. "Eukaryotic DNA replication: A complex picture partially clarified." *Cell*, 48, 7-8.

Ingraham, J. L., O. Maaløe, and F. C. Neidhardt. 1983. *Growth of the Bacterial Cell.* Sinauer.

Inoue, S. 1981. "Cell division and the mitotic spindle." *J. Cell Biol.*, 91, 132s-147s.

James, R., and R. A. Bradshaw. 1984. "Polypeptide growth factors." *Ann. Rev. Biochem.*, 53, 259-292.

John, P. C. L. (ed.) 1981. *The Cell Cycle.* Cambridge University Press.

Kearsey, S. 1986. "Replication origins in yeast chromosomes." *BioEssays*, 4, 157-161.

Kornberg, A. 1980. *DNA Replication.* 2nd Ed.; Supplement, 1982. W. H. Freeman and Co.

Marcus, M., Fainsod, A., and Diamond, G. 1985. "The genetic analysis of mammalian cell-cycle mutants." *Ann. Rev. Genet.*, 19, 389-421.

Mazia, D. 1987. "The chromosome cycle and the centrosome cycle in the mitotic cycle." In K. W. Jeon (ed.), *Int. Review of Cytology. Vol. 100. Overviews: Thirty-Five Years of Cell Biology.* Academic Press.

Mazia, D. 1974. "The cell cycle." *Scient. Amer.*, 230 (January), 54-64.

McIntosh, J. R. 1984. "Mechanisms of mitosis." *Trends Biochem. Sci.*, 9, 195-198.

Mitchison, J. M. 1971. *The Biology of the Cell Cycle.* Cambridge University Press.

Nossal, N. G. 1983. "Prokaryotic DNA replication systems." *Ann. Rev. Biochem.*, 53, 581-615.

Pickett-Heaps, J. 1986. "Mitotic mechanisms: an alternative view." *Trends in Biochem. Sci.*, 11, 504-507.

Pickett-Heaps. J. D., D. H. Tippit, and K. R. Porter. 1982. "Rethinking mitosis." *Cell*, 29, 729-744.

Prescott, D. M. 1987. "Cell reproduction." In K. W. Jeon (ed.), *Int. Review of Cytology. Vol. 100. Overviews: Thirty-Five Years of Cell Biology.* Academic Press.

Prescott, D. M. 1976. *Reproduction of Eukaryotic Cells.* Academic Press.

Pringle, J. R., and L. H. Hartwell. 1981. "The *Saccharomyces cerevisiae* cell cycle." In J. N. Strathern, E. W. Jones, and J. R. Broach (eds.) *The Molecular Biology of the Yeast Saccharomyces Life Cycle and Inheritance.* Cold Spring Harbor Laboratory.

Zyskind, J. W., and D. W. Smith. 1986. "The bacterial origin of replication, *oriC.*" *Cell,* 46, 489-490.

Chapter 13

Bishop, J. M. 1985. "Viral oncogenes." *Cell,* 42, 23-38.

Bishop, J. M. 1983. "Cellular oncogenes and retroviruses." *Ann. Rev. Biochem.,* 52, 301-354.

Bishop, J. M. 1982. "Oncogenes," *Scient. Amer.,* 246 (March), 80-92.

Croce, C. M., and G. Klein. 1985. "Chromosome translocation and human cancer." *Scient. Amer.,* 252 (March) 54-60.

Devoret, R. 1979. "Bacterial tests for potential carcinogens." *Scient. Amer.,* 241 (August), 40-49.

Deuel, T. F. 1987. "Polypeptide growth factors: Roles in normal and abnormal cell growth." *Ann. Rev. Cell Biol.,* 3, 443-492.

Diet, Nutrition, and Cancer. 1982. Committee on Diet, Nutrition and Cancer, Assembly of Life Sciences, National Research Council. National Academy Press.

Evaluation of the Carcinogenic Risk of Chemicals to Humans, Tobacco Smoking. 1986. IARC Monographs, Vol. 38. International Agency for Research on Cancer, World Health Organization.

Gallo, R. 1987. "The AIDS virus." *Scient. Amer.,* 256 (January), 46-56.

Gallo, R. C. 1986. "The first human retrovirus." *Scient. Amer.,* 255 (December), 88-98.

Gross, L. 1983. *Oncogenic Viruses.* 3rd Ed. Pergamon.

Heldin, C.-H., and B. Westermark. 1984. "Growth factors: Mechanisms of action and relation to oncogenes." *Cell,* 37, 9-20

Hunter, T. 1984. "The proteins of oncogenes." *Scient. Amer.,* 251 (August), 70-79.

Jove, R., and Hanafusa, H. 1987. "Cell transformation by the viral *src* oncogene." *Ann. Rev. Cell Biol.,* 3, 31-56.

Knudson, A. G., Jr. 1986. "Genetics of human cancer." *Ann. Rev. Genet.,* 20, 231-251.

Land, H., L. F. Parada, and R. A. Weinberg. 1983. "Cellular oncogenes and multistep carcinogenesis." *Science,* 222, 771-778.

Nicolson, G. L. 1979. "Cancer metastasis." *Scient. Amer.,* 240 (March), 66-76.

Prescott, D. M., and A. S. Flexer. 1986. *Cancer, The Misguided Cell.* 2nd Ed. Sinauer Associates.

Roberts, L. 1984. *Cancer Today: Origins, Prevention, and Treatment.* National Academy Press.

Smoking and Health, a report of the Surgeon General. 1979. U.S. Department of Health, Education, and Welfare. DHEW Publication No (PHS) 79-50066.

Sporn, M. B., and A. B. Roberts. 1985. "Autocrine growth factors and cancer." *Nature,* 313, 745-747.

Tooze, J. (ed.) 1980. *Molecular Biology of Tumor Viruses.* 2nd Ed., part 2, *DNA Tumor Viruses.* Cold Spring Harbor Laboratory.

Varmus, H. 1987. "Reverse transcription." *Scient. Amer.,* 257 (September), 56-64.

Varmus, H. E. 1984. "The molecular genetics of cellular oncogenes." *Ann. Rev. Genet.,* 18, 553-612.

Vessey, M. P., and M. Gray (eds.) 1985. *Cancer, Risks and Prevention.* Oxford University Press.

Weinberg, R. A. 1983. "The molecular basis of cancer." *Scient. Amer.,* 249 (November), 126-142.

Chapter 14

Adler, J. 1976. "The sensing of chemicals by bacteria." *Scient. Amer.,* 234 (April), 40-47.

Albrecht-Buehler, G. 1978. "The tracks of moving cells." *Scient. Amer.,* 238 (April), 68-76.

Andrews, A. T. 1981. *Electrophoresis.* Oxford University Press.

Berg, H. 1975. "How bacteria swim." *Scient. Amer.,* 233 (August), 36-44.

Cohen, C. 1975. "The protein switch in muscle contraction." *Scient. Amer.,* 233 (November), 36-45.

Franzini-Armstrong, C., and L. D. Peachey. 1981. "Striated muscles—contractile and control mechanisms." *J. Cell Biol.,* 91, 166s-186s.

Gibbons, I. R. 1981. "Cilia and flagella of eukaryotes." *J. Cell Biol.,* 91, 107s-124s.

Haimo, L. T., and J. L. Rosenbaum. 1981. "Cilia, flagella and microtubules." *J. Cell Biol.,* 91, 125s-130s.

Harrington, W. F., and M. E. Rodgers. 1984. "Myosin." *Ann. Rev. Biochem.,* 53, 35-73.

Hynes, R. 1985. "Molecular biology of fibronectin." *Ann. Rev. Cell Biol.,* 1, 67-91.

Lazarides, E., and J. P. Revel. 1979. "The molecular basis of cell movement." *Scient. Amer.,* 240 (May), 100-113.

Lester, H. A. 1977. "The response to acetylcholine." *Scient. Amer.*, 236 (February), 106-116.

Luck, D. J. L. 1984. "Genetic and biochemical dissection of the eucaryotic flagellum." *J. Cell Biol.*, 98, 789-794.

Macnab, R. M. 1984. "The bacterial flagellar motor." *Trends Biochem. Sci.*, 9, 185-188.

Pollard, T. D. 1987. "The myosin crossbridge problem." *Cell*, 48, 909-910.

Pollard, T. D. 1981. "Cytoplasmic contractile proteins." *J. Cell Biol.*, 91, 156s-165s.

Rickwood, D., and B. D. Hames (eds.) 1982. *Gel Electrophoresis of Nucleic Acids.* IRL Press Ltd.

Satir, P. 1974. "How cilia move." *Scient. Amer.*, 231 (October), 44-52.

Schroer, T. A., and R. B. Kelly. 1985. "In vitro translocation of organelles along microtubules." *Cell*, 40, 729-730.

Squire, J. 1981. *The Structural Basis of Muscle Contraction.* Plenum.

Taylor, D. L., and D. S. Condeelis. 1979. "Cytoplasmic structure and contractility in amoeboid cells." *Int. Rev. Cytol.*, 56, 57-144.

Warner, F. D., and D. R. Mitchell. 1980. "Dynein, the mechanochemical coupling adenosine triphosphatase of microtubule-based sliding filament mechanisms. *Int. Rev. Cytol.*, 66, 1-43.

Warrick, H. M., and J. A. Spudich. 1987. "Myosin structure and function in cell motility." *Ann. Rev. Cell Biol.*, 3, 379-421.

Chapter 15

Ada, G. L., and Sir G. Nossal. 1987. "The clonal-selection theory." *Scient. Amer.*, 257 (August), 62-69.

Capra, J. D., and A. B. Edmundson. 1977. "The antibody combining site." *Scient. Amer.*, 236 (January), 50-59.

Eisen, H. N. 1981. *Immunology.* 3rd Ed. Harper & Row.

Epel, D. 1977. "The program of fertilization." *Scient. Amer.*, 237 (November), 128-138.

Golub, E. 1981. *The Cellular Basis of the Immune Response: An Approach to Immunobiology.* 2nd Ed. Sinauer.

Leder, P. 1982. "The genetics of antibody diversity." *Scient. Amer.*, 246 (May), 102-115.

McConnell, I., A. Munro, and H. Waldman. 1981. *The Immune System: A Course on the Molecular and Cellular Basis of Immunity.* 2nd Ed. Blackwell.

Milstein, C. 1980. "Monoclonal antibodies." *Scient. Amer.*, 243 (October), 66-74.

Moog, F. 1981. "The lining of the small intestine." *Scient. Amer.*, 245 (November), 154-176.

Mooseker, M. S. 1985. "Organization, chemistry, and assembly of the cytoskeletal apparatus of the intestinal brush border." *Ann. Rev. Cell Biol.*, 1, 209-242.

Morell, P., and W. T. Norton. 1980. "Myelin." *Scient. Amer.*, 242 (May), 88-118.

Paul, W. E. 1984. *Fundamental Immunology.* Raven.

Rushton, W. A. H. 1975. "Visual pigments and color blindness." *Scient. Amer.*, 232 (March), 64-74.

Schnapf, J. L., and D. A. Baylor. 1987. "How photoreceptor cells respond to light." *Scient. Amer.*, 256 (April), 40-47.

Schwartz, J. H. 1980. "The transport of substances in nerve cells." *Scient. Amer.*, 242 (April), 152-171.

Staehelin, L. A., and B. E. Hull. 1978. "Junctions between living cells." *Scient. Amer.*, 238 (May), 140-152.

Stevens, C. F. 1979. "The neuron." *Scient. Amer.*, 241 (September), 54-65.

Tonegawa, S. 1985. "The molecules of the immune system." *Scient. Amer.*, 253 (October), 122-131.

Yelton, D. E., and M. D. Scharff. 1981. "Monoclonal antibodies: a powerful new tool in biology and medicine." *Ann. Rev. Biochem.*, 50, 657-680.

Chapter 16

Cech, T. R., and B. L. Bass. 1986. "Biological catalysis by RNA." *Ann. Rev. Biochem.*, 55, 599-629.

Darnell, J. E., and W. F. Doolittle. 1986. "Speculations on the early course of evolution." *Proc. Natl. Acad. Sci.*, 83, 1271-1275.

Dickerson, R. E. 1978. "Chemical evolution and the origin of life." *Scient. Amer.*, 239 (September), 70-86.

Groves, D. I., J. S. R. Dunlop, and R. Buick. 1981. "An early habitat of life." *Scient. Amer.*, 245 (October), 64-73.

Kimball, A. P., and J. Oro. (eds.) 1971. *Prebiotic and Biochemical Evolution.* North-Holland/American Elsevier.

Lewin, R. 1986. "RNA catalysis gives fresh perspective on the origin of life." *Science*, 231, 545-546.

Margulis, L. 1982. *Early Life.* (Science Books International) Jones and Bartlett.

Oparin, A. I. 1953. *Origin of Life.* Dover Publications.

Schopf, J. W. 1978. "The evolution of the earliest cells." *Scient. Amer.*, 239 (September), 110-138.

Vidal, G. 1984. "The oldest eukaryotic cells." *Scient. Amer.*, 250 (February), 48-57.

INDEX